ERRATA

The title for Chapter 18, beginning on page 95, should read: "Effects of Dipyrimadole and Adenine/Ribose on ATP Concentration and Adenosine Production in Cardiac Myocytes."

The title of Chapter 149, beginning on page 781, should read: "Energy Metabolism in the Failing Human Heart with Dilated or Hypertrophic Cardiomyopathy."

Purine and Pyrimidine Metabolism in Man IX
Griesmacher, Chiba, Müller, eds.

0-306-45778-4
Plenum Press, New York, 1998

ERRATA

The title for Chapter 18, beginning on page 95, should read: "Effects of Dipyrimadole and Adenine/Ribose on ATP Concentration and Adenosine Production in Cardiac Myocytes."

The title of Chapter 149, beginning on page 781, should read: "Energy Metabolism in the Failing Human Heart with Dilated or Hypertrophic Cardiomyopathy."

Purine and Pyrimidine Metabolism in Man IX
Griesmacher, Chiba, Müller, eds.

0-306-45778-4
Plenum Press, New York, 1998

PURINE AND PYRIMIDINE METABOLISM IN MAN IX

ADVANCES IN EXPERIMENTAL MEDICINE AND BIOLOGY

A Continuation Order Plan is available for this series. A continuation order will bring delivery of each new volume
immediately upon publication. Volumes are billed only upon actual shipment. For further information please contact
the publisher.

PURINE AND PYRIMIDINE METABOLISM IN MAN IX

Edited by

Andrea Griesmacher

Emperor Franz Joseph Hospital
Vienna, Austria

Peter Chiba

University of Vienna
Vienna, Austria

and

Mathias M. Müller

Emperor Franz Joseph Hospital
Vienna, Austria

SPRINGER SCIENCE+BUSINESS MEDIA, LLC

Library of Congress Cataloging in Publication Data

Purine and pyrimidine metabolism in man IX / edited by Andrea Griesmacher, Peter Chiba, and Mathias M. Müller.
p.   cm. — (Advances in experimental medicine and biology; v. 431)
"Articles presented at the joint IXth International and 6th European Symposium on Purine and Pyrimidine Metabolism in Man, held in Gmunden, Austria, June 1 through 7, 1997"—Pref.
Includes bibliographical references and index.
ISBN 978-1-4613-7456-5   ISBN 978-1-4615-5381-6 (eBook)
DOI 10.1007/978-1-4615-5381-6
    1. Purines—Metabolism—Disorders—Congresses. 2. Pyrimidines—Metabolism—Disorders—Congresses. 3. Purines—Metabolism—Congresses. 4. Pyrimidines—Metabolism—Congresses. I. Griesmacher, Andrea. II. Chiba, Peter. III. Müller, Mathias M. IV. European Symposium on Purine and Pyrimidine Metabolism in Man (6th: 1997: Gmunden, Austria) V. International Symposium on Purine and Pyrimidine Metabolism in Man (9th: 1997: Gmunden, Austria) VI. Series.
    [DNLM: 1. Purines—metabolism congresses. 2. Pyrimidine—metabolism congresses. QU 58 P9856   1998]
RC632.P87P88 1998
616.3'9—dc21
DNLM/DLC
for Library of Congress   97-52565
                                                                      CIP

Proceedings of the joint Ninth International/Sixth European Symposium on Purine and Pyrimidine Metabolism in Man, held June 1–7, 1997, in Gmunden, Austria

ISBN 978-1-4613-7456-5

© 1998 Springer Science+Business Media New York
Originally published by Plenum Press, New York in 1998

IN MEMORY OF ANDRE DE VRIES

These proceedings are dedicated to the late Prof. Andre de Vries, who passed away last September at the age of 85. Andre De Vries was born in 1911 in Leeuwarden, Holland and received his M.D. from the University of Amsterdam in 1938. In 1940 he immigrated to Israel. He spent the first two years as an agricultural worker in a communal settlement prior to starting his medical career as an assistant in Internal Medicine at Hadassa Hospital, Jerusalem. From 1946 to 1948 he was a research fellow in the Department of Medicine, at Harvard. Upon his return to Hadassa, he advanced to the position of Associate Professor (1953) and Director of the Department of Internal Medicine (1954). In 1954 he moved to Beilinson Medical Center in Petah-Tikva, where he was appointed Director of a Department of Internal Medicine and of the Rogoff-Wellcome Medical Research Institute.

He was appointed Professor of Clinical Medicine at the Hebrew University of Jerusalem (1960), and Professor of Medicine and Experimental Biology at Tel Aviv University (1961). Prof. de Vries played a major role in the establishment of the Faculty of Medicine, leading it toward scientific eminence. He was elected Dean of Life Sciences (1962–1964), the first Dean of Medicine (1964–1969) and Rector (1969–1972). From 1972 to 1976 he was Director of the Beilinson Medical Center, from where he retired in 1976, to continue working in various medical responsibilities for an additional period of 16 years.

During his impressive career, Andre de Vries made important contributions in scientific and medical fields. In preparing the celebration of his 85th birthday, he chose to emphasize his achievements in the study of blood coagulation (Factors VII and XII), undertaken in Harvard and Jerusalem; in the study of the interaction of synthetic amino acids with blood cells (in cooperation with The Weizman Institute of Science, Rehovot); in the study of uric acid lithiasis and in his investigation of the biochemistry, immunology, pharmacology and physiology of the venoms of Vipera Palestinae and Echis Colorata (The Rogoff-Wellcome Medical Research Institute) and in the study of several new clinical syndromes. These achievements in science and medicine were published in several hundred papers, many of which, in leading journals.

In the field of purine metabolism he won respect for his studies concerning the epidemiology, etiology, prevention and treatment of uric acid lithiasis (1958–1968). These studies led to the publication of a book "Uric Acid Lithiasis" (Elsevier, 1963), and of numerous chapters on this subject in books. In addition he participated in the demonstration of the first case of purine overproduction, due to PRPP synthetase superactivity (feedback resistance mutation; 1972) and in the study of several families with rare inborn errors of purine metabolism, such as partial deficiency of HGPRT, Lesch-Nyhan syndrome, xanthinuria and hereditary renal hypouricemia.

In 1973, the first in the series of The International Symposia on Purine Metabolism in Man (the 9th of which is the subject of these proceedings) was organized by him and Oded Sperling, aided by an international team, including J.B. Wyngaarden, J.E. Seegmiller, and R.W.E. Watts. In 1976 he cooperated with R.W.E. Watts in organizing The Ciba Symposium on Purine and Pyrimidine Metabolism held in London. In 1977, he and Oded Sperling, aided by an international team, including J.E. Seegmiller, S. Segal, A. Szeinberg, R.W.E. Watts and J.B. Wyngaarden, organized The First International Symposium on Inborn Errors of Metabolism in Man in Tel Aviv. The 7th in this series was held last May in Vienna.

Andre de Vries was a giant in medicine and clinical sciences. For his important contributions to the field of purine metabolism in disease and for his initiative in the establishment of The Symposium on Purine and Pyrimidine Metabolism in Man, on behalf of his friends, colleagues and pupils throughout the world, the International Organizing Committee and the Local Planning Committee, dedicate this symposium and proceedings to his memory.

PREFACE

This volume comprizes articles presented at the joint IXth International and 6th European Symposium on Purine and Pyrimidine Metabolism in Man held in Gmunden, Austria, June 1 through 7, 1997. Since the first of this series of meetings was held in Israel in 1973, conventions were organized every three years in different parts of the world including the USA, Japan, and Europe.

The different aspects of purine and pyrimidine metabolism bring together researchers working in molecular genetics, biochemical pharmacology, biochemistry, developmental biology, immunology, epidemiology and the clinics. Oriented research in the field has been seminal for the development of potent anticancer and antiviral drugs. As the number of genes which are cloned, grows, the understanding of metabolism is increasingly enlarged and might provide leads to further improve therapeutic concepts and to better understand mechanisms responsible for the development of resistance against these drugs. In certain diseases purine and pyrimidine analogs represent not only the drugs of choice but in fact are the sole therapeutic alternative at present.

The field has also taken an early lead in attempting to correct inborn errors of purine and pyrimidine metabolism by gene therapy.

The organization of this meeting involved a large number of people who dedicated their time in an effort to make this symposium a success. We thank the Abstract Review Committee, the International Advisory Board and in particular the Symposium Secretariat for doing a wonderful job.

We look forward to the year 2000 which might bring the meeting back to Israel where scientists convened in 1973.

The Editors

CONTENTS

I. GOUT: CLINICS AND TREATMENT

II. ADENOSINE WORKSHOP

III. INBORN ERRORS OF PURINE AND PYRIMIDINE METABOLISM: WHAT WE CAN LEARN FROM THEM

IV. REGULATION AND ENZYMES OF PURINE
AND PYRIMIDINE METABOLISM

V. NMR IN STUDY OF PURINES AND ENERGY METABOLISM

VI. MUTATIONS: IMPACT ON FUNCTION

VIII. PURINES AND SIGNAL TRANSDUCTION

IX. HOT AREAS IN PP RESEARCH

X. IMMUNOMODULATIONS BY PURINES AND PYRIMIDINES

XI. PURINES AND PYRIMIDINES IN CELL DIFFERENTIATION (AND MALIGNANCIES)

XII. THE IMPACT OF PURINES AND PYRIMIDINES ON THERAPEUTIC STRATEGIES

XIII. MISCELLANEOUS TOPICS

PURINE AND PYRIMIDINE
METABOLISM IN MAN IX

GOUT

New Questions for an Ancient Disease

Juan G. Puig,[1] Felícitas A. Mateos,[2] Teresa Sancho,[1] Rosa J. Torres,[2]
Antonio Buño,[2] Eugenio de Miguel,[3] and Antonio Gil[1]

[1]Division of Internal Medicine
[2]Division of Clinical Biochemistry
[3]Division of Rheumatology
"La Paz" Hospital
Universidad Autónoma
Madrid, Spain

1. INTRODUCTION

Advances in the second part of this century have provided relatively satisfactory basic understanding, clinical classification, and guidelines for the management of hyperuricemia and gout. However, gout remains a significant health problem in many populations and several clinical problems have emerged to challenge clinicians, such as the atypical features and increasing prevalence of gout in elderly men and women, and the frequently aggressive course of cyclosporine induced gout.

Each gout patient that the clinicians meets nowadays still presents unanswered fundamental questions such as[1]: Why do only a minority of hyperuricemic subjects develop gout? What is the dominant molecular abnormality in most patients with primary gout who overproduce uric acid? What are the key factors in tophus formation? Which molecular mechanisms are most important in regulating renal urate handling in patients with primary gout? What are the key prophylactic and therapeutic effect of colchicine? Which factors initiate and spontaneously terminate most gouty attacks? These and other relevant questions may point the way to investigate the significant gap in our full understanding of hyperuricemia and gout.

From a clinical point of view we would like to raise two areas of interest related to uric acid metabolism and gout that have experienced some advances in recent years. These areas are: The syndrome of hereditary nephropathy associated with hyperuricemia or gout, and the X syndrome, hyperuricemia being one of the hallmarks of this recently described syndrome.

Purine and Pyrimidine Metabolism in Man IX,
edited by Griesmacher *et al.* Plenum Press, New York, 1998.

2. THE SYNDROME OF HEREDITARY NEPHROPATHY ASSOCIATED WITH HYPERURICEMIA OR GOUT

Several kindreds have been described in recent years in which either asymptomatic hyperuricemia or gout, usually appearing at an early age and often in both sexes, is associated with progressive renal failure[2-6] (Figure 1). The question as to whether hyperuricemia is the cause or the result of renal insufficiency has been a matter of debate and is not fully resolved. In fact, on the grounds that chronic hyperuricemia may be deleterious to the kidney, the term "familial juvenile gouty nephropathy" has been proposed for this disorder[7]. However, in our opinion, three lines of evidence argue against this view of pathogenetic events. First, the concept of gouty nephropathy is questionable since long-term follow-up studies have shown that serious renal damage, even after many years of sustained hyperuricemia is uncommon in the absence of additional risk factors, such as diabetes mellitus, arterial hypertension, chronic renal disease (glomerulonephritis, chronic pyelonephritis, amyloidosis), or atherosclerosis.[8] In addition, treatment with uric acid lowering drugs does not appear to hasten or inhibit the development of renal failure.[9] Second, urate crystal deposits, a distinctive feature of gouty nephropathy, were not evidenced in three kidney open biopsies obtained from our patients, and have been reported only in one patient with the disease[3]. Moreover, an increased kidney uric acid content in affected patients might be expected if the kidney damage were due to urate crystal deposition, but this was not found in the three kidney open biopsies analyzed from our patients.[6] Third, from longitudinal observations in extensive series of gout patients[9] it would be incorrect to conclude that there are no gout patients who have developed serious concomitant renal disease leading to renal failure. Berger and Yu reported 5 young patients (mean age 33 years), all but one with a family history of gout, who developed rapidly progressive renal disease in the absence of hypertension and despite adequate control of hyperuricemia.[10] Our follow-up studies in these families indicate that despite alopurinol and antihypertensive therapy, that provided an adequate control of hyperuricemia and arterial hypertension, renal deterioration proceeds[6] (Figure 1, A and B). In fact, the two elder siblings of our first family had had renal transplantation in the last two years. Is there an alternative pathogenic mechanism that could explain the clinical characteristics of this familial nephropathy characterized by an unusually reduced uric acid excretion? Our renal hemodynamic studies showed that all affected subjects did show a markedly reduced renal plasma flow and increased renal vascular resistances[6]. Moreover, the decrease in renal plasma flow and the increase in renal vascular resistances were significantly related to serum urate levels.[6] In fact, even obligate carriers of this familial nephropathy, with normal serum urate concentrations and inulin clearance, evidenced a diminished renal plasma flow and increased renal vascular resistances. Thus, hemodynamic studies in these patients supported the non-specific hystopathological findings of glomerular ischemic changes.[6] It appears, thus, that a severe disruption of renal hemodynamics may be of primary pathogenic relevance in this familial nephropathy, and we should look at the myriad of mediators that modulate the behaviour of vascular smooth-muscle cells resulting in renal vasoconstriction with consequent ischemic renal insufficiency, arterial hypertension, and purine underexcretion and hyperuricemia, accounting for the clinical expression of this disorder. This concept is supported by the observation that losartan (50 mg, single dose) an angiotensin II receptor antagonist markedly increased renal plasma flow and reduced renal vascular resistances in two patients with this familial nephropathy (Figure 1, C and D).

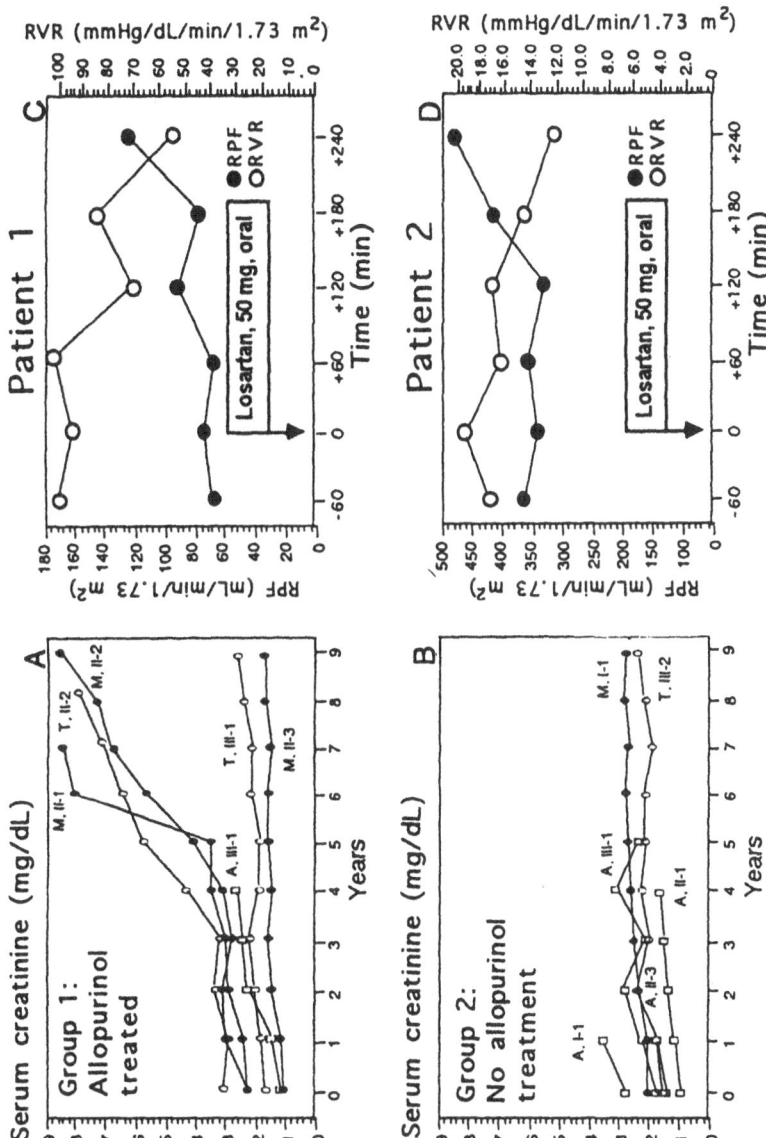

Figure 1. Serum creatinine over time in patients with Familial Nephropathy Associated with Hyperuricemia or Gout treated (A) and untreated (B) with allopurinol. Allopurinol therapy did not halt the progression of renal insufficiency. Losartan therapy markedly increased and reduced renal plasma flow (RPF) and renal vascular resistances (RVR), respectively, in 2 patients (C, D).

3. HYPERURICEMIA IN X SYNDROME

X syndrome comprises a cluster of metabolic abnormalities associated with insulin resistance, all of which have been shown to increase the risk of coronary heart disease. The main abnormalities include glucose intolerance, hyperinsulinemia, hypertrigly-ceridemia and diminished HDL-cholesterol levels, arterial hypertension, increased plasminogen activator inhibitor 1, and hyperuricemia. The well known associates of classic gout, such as obesity, glucose intolerance, hypertriglyceridemia, and hypertension, strikingly resemble those of insulin resistance. What is the link between hyperinsulinemia and hyperuricemia? To answer this question Facchini et al.[11] examined in normal volunteers whether serum urate concentrations varied as a function of insulin resistance and/or hyperinsulinemia. Their data showed a significant association between serum urate levels and both insulin resistance ($r = 0.69$; $p < 0.001$) and the plasma insulin response ($r = 0.61$; $p < 0.001$) to an oral glucose challenge (i.e. the more insulin resistant an individual the higher serum urate concentration). In addition, the resistance to insulin-mediated glucose uptake and the plasma insulin response were inversely correlated with the renal clearance of uric acid ($r = -0.49$; $p < 0.002$, and $r = -0.33$; $p < 0.05$, respectively) (i.e. the more insulin-resistant an individual the lower daily urinary uric acid excretion). In an epidemiological study carried out at the Olivetti factory in Naples, Italy, Capuccio et al.[12] demonstrated that serum urate level is independently associated with increased proximal tubular sodium reabsorption in men (i.e. the higher the serum urate level the greater the amount of sodium reabsorbed at nephron sites proximal to the distal tubule). It is well known that insulin infusion, while maintaining euglycemia, exerts a marked decline in sodium excretion which may reach 50% of preinfusion values.[13] This insulin-mediated antinatriuretic effect was observed in the absence of changes in the filtered load of glucose, glomerular filtration rate, renal plasma flow, and plasma aldosterone concentrations.[14] What is the possible mechanism by which insulin promotes an antinatriuretic effect in man? The plasma membranes of most if not all vertebrate cells contain a transport system that mediates the transmembrane exchange of sodium for hydrogen. This exchanger is crucial for the regulation of intracellular pH. Insulin slightly increases intracellular pH (pHi) and this effect has different actions, such as stimulation of glycolisis by the rate limiting enzyme phosphofructokinase which is

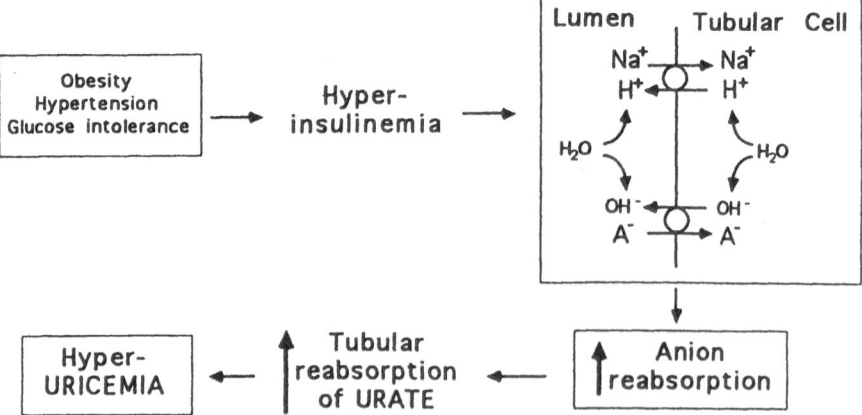

Figure 2. Proposed pathophysiological mechanism that links hyperinsulinemia and hyperuricemia. Hyperinsulinemia increases intracellular pH and stimulates the sodium-hydrogen ion exchanger, which in turn promotes anion reabsorption.

very sensitive to small changes in pHi. The insulin-mediated increase in pHi stimulates the sodium-hydrogen exchanger (i.e. activates proximal tubular sodium reabsorption and hydrogen secretion) thereby facilitating the active reabsorption of bicarbonate, chloride and organic anions, such as urate. Thus, stimulation of the sodium-hydrogen exchanger by hyperinsulinemia provides a putative mechanism by which insulin resistance and hyperuricemia could be linked (Figure 2).

ACKNOWLEDGMENTS

We are indebted to the Clinical Research Unit nursing staff and the dietetic staff for excellent patient care; to Mª Paz Canencia for valuable technical assistance; and to Erik Lundin for assistance in preparing the manuscript. Supported by grants from Caja de Madrid and Fondo de Investigación Sanitaria (FIS, 97/0458), Spain.

REFERENCES

1. Terkeltaub RA. Gout: questions that still need to be answered. Ann Rheum Dis 1995; 54:79–81.
2. Duncan H, Dixon AS. Gout, familial hyperuricemia, and renal disease. Q J Med 1960; 29:127–35.
3. Farebrother DA, Pincott JR, Simmonds HA, Warren DJ, Dillon MJ, Cameron JS. Uric acid crystal-induced nephropathy: Evidence for a specific renal lesion in a gouty family. J Pathol 1981; 135:159–68.
4. Leumann EP, Wegmann W. Familial nephropathy with hyperuricemia and gout. Nephron 1983; 34:51–7.
5. Calabrese G, Simmonds HA, Cameron JS, Davies PM. Precocious familial gout with reduced fractional urate clearance and normal purine enzymes. Q J Med 1990; 75:441–50.
6. Puig JG, Miranda MªE, Mateos FA, Picazo ML, Jiménez ML, Calvin TS, Gil AA. Hereditary nephropathy associated with hyperuricemia and gout. Arch Intern Med 1993; 153: 357–65.
7. Moro F, Ogg C, Simmonds HA, et al. Familial juvenile gouty nephropathy with renal urate hypoexcretion preceding renal damage. Clin Nephrol. 1991;35:263–69.
8. Beck LH. Requiem for gouty nephropathy. Kidney Int 1986; 30:280–7.
9. Yu T, Berger L. Impaired renal function in gout. Am J Med 1982; 72:95–100.
10. Berger L, Yu T. Renal functin in gout. IV. An analysis of 524 gouty subjects including long-term follow-up studies. Am J Med 1975; 59:605–13.
11. Facchini F, Chen Y-DI, Hollenbeck CB, Reaven GM. Relationship between resistance to insulin-mediated glucose uptake, urinary uric acid clearance, and plasma uric acid concentration. JAMA 1991; 266:3008–11.
12. Cappuccio FP, Strazzullo P, Farinaro E, Trevisan M. Uric acid metabolism and tubular sodium handling. Results from a population-based study. JAMA 1993; 270:354–9.
13. DeFronzo RA, Cooke CR, Andres R, Faloona GR, Davis PJ. The effect of insulin on renal handling of sodium, potassium, calcium, and phosphate in man. J Clin Invest 1975; 55:845–55.
14. DeFronzo RA. The effect of insulin on renal sodium metabolism. Diabetologia 1981; 21:165–71.

EFFICACY OF ALLOPURINOL IN AMELIORATING THE PROGRESSIVE RENAL DISEASE IN FAMILIAL JUVENILE HYPERURICAEMIC NEPHROPATHY (FJHN)

A Six-Year Update

M. B. Mc Bride,[1] H. A. Simmonds,[1] C. S. Ogg,[1] J. S. Cameron,[1] S. Rigden,[1] L. Rees,[2] W. Van't Hoff,[2] F. Moro,[3] and G. V. Raman[4]

[1]Purine Research, Renal and Paediatric Renal Units
UMDS, Guy's Hospital
[2]Paediatric Nephrology
Great Ormond Street Hospital, London
[3]Southampton General Hospital
[4]Renal Unit, St. Mary's Hospital
Portsmouth, Great Britain

INTRODUCTION

Familial juvenile hyperuricaemic nephropathy (FJHN) is a dominant disorder with high penetrance (Fig. 1). In the past it has been associated with progressive renal disease in young men, women and children, frequently spanning several generations and leading to premature death. The unusual presentation with gout in such young subjects, equally unexpected in women or in renal disease, has served to draw attention to kindreds with what was previously considered a 'familial renal disease' of unknown aetiology (Fig. 1). The biochemical hallmark of FJHN is now known to be hyperuricaemia due to a grossly reduced fractional uric acid clearance (1). Six years ago we reported the beneficial effect of allopurinol in 10 patients followed for up to 23 years (2). However, the efficacy of allopurinol found in our series has been disputed by others who have reported no such effect (3,4). We have now followed many more patients, including siblings recognised as having FJHN during biochemical screening. The results support our original contention (2), and suggest explanations for the discrepant findings reported by others (3,4).

Purine and Pyrimidine Metabolism in Man IX,
edited by Griesmacher *et al.* Plenum Press, New York, 1998.

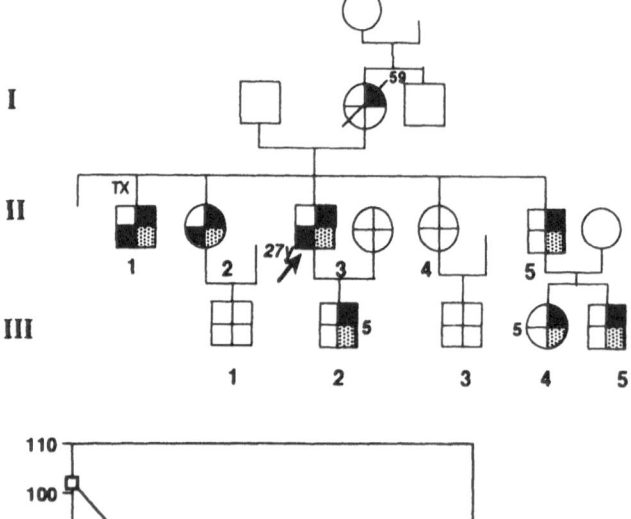

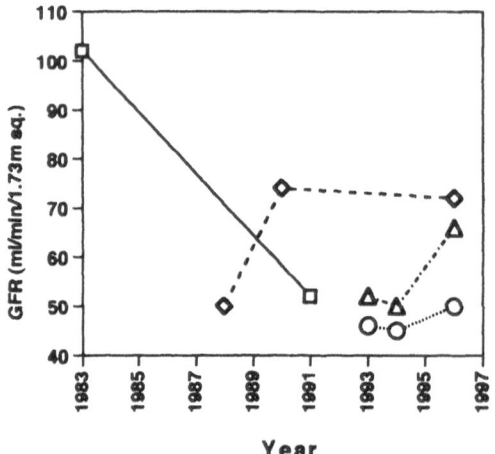

Figure 1. (top) Family tree of a representative kindred (Kindred 3) showing the dominant inheritance and high penetrance in a family originally considered to have a 'familial renal disease' until an attack of gout in a male aged 27 drew attention to the possibility of FJHN; (bottom) GFR versus time on allopurinol in this kindred. The fall in GFR in the propositus, II 3, a noncomplier, contrasts with the stability in the 3 of the 5 children found to be affected in III. All 5 were born since the family was first studied, enabling screening at an early age.

PATIENTS AND METHODS

This update of our earlier study of renal function and urate metabolism includes 21 additional evaluable subjects, making a total of 31 (17 f, 14 m) from 12 kindreds treated and followed up at regular intervals for periods ranging from 4 to 29 yrs. A representative family tree from Kindred 3 is given in Fig. 1, which shows the high penetrance, the fact that an attack of gout first drew attention to a kindred being treated for familial disease of unknown etiology and that biochemical screening confirmed that one of two seemingly healthy kindred members had also inherited FJHN. The clinical history in some of these 12 kindreds has been published (reviewed in 1). As anticipated in a dominant disorder biochemical screening identified FJHN in approximately 50% of seemingly healthy siblings.

The allopurinol dose has varied from 50 to 300 mg per day, depending on age and renal function. The twin of the index case in Kindred 2 was also found to have FJHN when her sister presented with gout at 9 yrs and has been followed up on allopurinol for 29 years. Both sisters showed a sharp increment in plasma creatinine following oestrogen therapy coupled with the development of hypertension (2). Methods used for assessment of uric acid and renal function status have been described [1,2].

Table 1. Effect of treatment with allopurinol in 31 patients from 12 kindreds with FJHN. The results show the mean plasma creatinine and uric acid concentrations in plasma and urine in subjects with FJHN treated with allopurinol for the periods indicated

Years (range)	Outcome	Creatinine/GFR* $\mu mol/l$ / $ml/min/1.73m^2$		Plasma Uric Acid $\mu mol/l$		Urine Uric Acid $mmol/l$	
		Start	Current	Start	Current	Start	Current
(5-8)	Normal RF (n =5)	82	80	382	306	1.56	1.32
(4-21)	GFR* 50-80 (n =13)	93	101	349	263	1.75	1.24
Nil	Non complier/ untreated (6yrs) (n=2)	90*	50*				
Increased plasma creatinine at start							
(5-29)	<200μmol (n= 5)	147	187	394	355	1.0	0.69
(4-11)	>200μmol (n=6) 4 dialysis/Tx 2 died	347	581§	333	337§		

§ = values at commencement of dialysis

RESULTS

Hyperuricaemia was found in all 31 subjects, consistent with the low FE_{ur} (Table 1). Twenty of the 31 had a plasma creatinine within the normal range at diagnosis. In 13 of these subjects the creatinine was in the upper limit of the normal range for age and sex. GFR when measured in the 13 patients was slightly reduced (50–80 ml/min/1.73m²). Renal function was severely impaired in 6 subjects at diagnosis.

Plasma uric acid in the latter subjects was disproportionately high considering the degree of renal dysfunction, consistent with the low FE_{ur} characteristic of FJHN, despite the severe renal disease where FE_{ur} normally increases with the fall in GFR (1).

Plasma uric acid in the 25 subjects who responded was reduced 30% by allopurinol to a value within the normal range for age and sex (Table 1). The absolute concentration of uric acid excreted in the urine on a mmol/l basis, mean initially 1 to 1.75 mmol/l, was reduced also by approximately 30%.

Five of the 20 subjects with a normal plasma creatinine at diagnosis have retained a normal GFR and 13 have shown relatively stable renal function. All were good compliers. The other two in this group with a normal GFR at start suffered a 50% reduction in GFR in 7 years. The first (K3, I 3, Fig. 1) was a poor complier, as reported earlier (10) and has been lost to further follow-up. The other patient, who was untreated by her doctor until a second referral 2 yrs ago, is now on allopurinol.

Five of the 11 subjects with an elevated plasma creatinine at start, but below 200 μmol/l, have retained low but relatively stable renal function for from 4 to 29 years. (Table 1).

A poor prognostic sign for allopurinol therapy (6 of the 11 subjects) was a value in excess of 200 μmol/l (3 index cases and 3 relatives). All 6 progressed to dialysis and transplantation between 3 to 11 yrs, two died.

DISCUSSION

The relative stability of renal function over long periods on allopurinol (4–29 years) in 23 of 31 subjects from the 12 kindreds studied here contrasts sharply with previous generations in the same kindreds who were treated only for a 'familial renal disease' (1). In those kindreds with a strong family history 13 subjects had died of 'familial renal disease' while of the 31 subjects in this study only 2 have died. Both were index cases with severe renal disease when FJHN was first recognised (1,2).

The noteworthy factor emerging from this study is that the 18 patients with a GFR >50 ml/min at diagnosis have continued to maintain stable renal function on allopurinol, five having a normal GFR. Of the 11 with an elevated creatinine at start allopurinol has ameliorated the progression of the renal disease in 5 subjects over periods from 5 to 29 years. In the remaining 6 with a plasma creatinine in excess of 200µmol/l at the start, diagnosis was too late for allopurinol to exert any beneficial effect. All 6 progressed to dialysis and transplantation over 4 to 11 years, two died.

The efficacy of allopurinol thus clearly depends on the renal function when treatment is first commenced (1). The importance of early diagnosis and instigation of allopurinol therapy is illustrated in one kindred K5 previously reported in detail (5), where the propositus had a plasma creatinine of 145 µmol/l at diagnosis aged 20, but treated only for renal disease progressed to dialysis in 10 years (1,2). An attack of gout in the latter stages drew attention to the possibility of FJHN. Screening identified FJHN in 2 seemingly healthy siblings and their 3 children (5). These 5 subjects have now been followed up for 9 years . One nephew with a creatinine in excess of 200µm/l when identified also progressed to dialysis in 4 years. By contrast the 3 others with a GFR >50ml/min have retained stable renal function on allopurinol over the past 9 years (Table 1).

Although early diagnosis is vital (and particularly important in the case of children) good compliance is equally essential. The patients in Kindred 3 (Fig. 1) illustrate the latter as well as the former points. The 50% drop in GFR in the father—a non-complier—contrasts with the relative stability of the GFR following allopurinol in the children (Fig. 1).

Other investigators report a similar experience with FJHN patients on allopurinol (personal communication). However, one group did not find any benefit from allopurinol (3,4). Deterioration in renal function in 12 subjects from 3 kindreds progressed similarly over periods from 1 to 6 years, whether the patients were treated with allopurinol or not. This led the investigators to conclude that this syndrome is an autosomal dominant nephropathy. Our results suggest that these discordant findings can be explained by the fact that one year (as in 40% of the patients reported) is not long enough to evaluate the effect of allopurinol. Second, 50% of the patients had plasma creatinines consistent with a low GFR initially, when allopurinol would be of little benefit.

At present the mechanism behind this putative beneficial effect of allopurinol in FJHN remains speculative. One possibility is that the demonstrated 30% lowering of the concentration of urinary uric acid in the tubular lumen following allopurinol (Table 1) and also in plasma urate would not only reduce the concentration in the renal vasculature, but also the absolute amount of the potentially nephrotoxic uric acid available for reabsorption.

In summary, this follow-up study confirms that allopurinol in 75% of subjects has ameliorated the rapid decline in renal function seen both previously and in this study in untreated members in kindreds with FJHN. However, the poor prognostic sign was a creatinine of >200 µmol/l at diagnosis. Our experience in these 12 kindreds shows that the efficacy of allopurinol depends equally on good compliance. Aggressive control of hypertension, where present, was also vital (1,2). These findings underscore the importance of early recognition, particularly in children. Further understanding of this disorder and resolution of the above debate must await the location of the gene for FJHN (1).

ACKNOWLEDGMENTS

We are greatly indebted to the Arthritis and Rheumatism Council for their support and to the families who have kindly participated in this long-term study.

REFERENCES

1. Cameron JS, Moro F, McBride MB, Simmonds HA. Inherited disorders of purine metabolism and transport. Chapter 16.5.3 in, *Oxford Textbook of Clinical Nephrology* 1997, pp 13–26.
2. Moro F, Simmonds HA, Mc Bride MB, Cameron JS, Williams DG, Ogg CS. Does allopurinol affect the progression in familial juvenile gouty nephropathy? Adv Exp Med Biol 1992;309A: 199–203.
3. Puig JG, Miranda ME, Mateos FA, Picazo ML, Jimenez ML, Calvin TS, Gil AA., Hereditary nephropathy associated with hyperuricaemia and gout Arch Intern Med 1993;153:357–365.
4. Miranda ME, Puig JG, Mateos FA, Herranz I, Martinez Ara J, Berrocal T, Prieto C, Picazo ML. The influence of allopurinol on renal deterioration in familial nephropathy associated with hyperuricaemia. Adv Exp Med Biol 1995; 370:61–64.
5. Moro F, Ogg C, Cameron JS, Simmonds HA, Duley JA, McBride MB, Davies PM. Familial juvenile gouty nephropathy with renal urate hypoexcretion preceding renal disease. Clin Nephrol 1991;35:263–269.

3

OPTIMAL RANGE OF SERUM URATE CONCENTRATIONS TO MINIMIZE RISK OF GOUTY ATTACKS DURING ANTI-HYPERURICEMIC TREATMENT

Hisashi Yamanaka,[1] Riko Togashi,[1] Masayuki Hakoda,[1] Chihiro Terai,[1] Sadao Kashiwazaki,[1] Takashi Dan,[2] and Naoyuki Kamatani[1]

[1]Institute of Rheumatology
Tokyo Women's Medical College
KS bldg. 9-12 Wakamatsu-cho
Shinjuku-ku, Tokyo, Japan
[2]Exploiting Laboratory
Chugai Pharmaceuticals
Gotenba, Shizuoka, Japan

1. SUMMARY

To find an optimal range of urate concentrations wherein the risk of attacks during the initial 6 months of treatment is minimized, data from 350 gouty patients treated with antihyperuricemic drugs were retrospectively analyzed. We determined the optimal range of urate concentrations to be 4.6–6.6 mg/dl. If urate concentrations were within this range, the risk ratio of an attack as opposed to outside of the range was 0.705 (95% confidence interval, 0.629–0.791). The increase (or decrease) in urate concentration in one month associated with minimal risk of gouty attacks was also determined. The lowest risk ratio of attack (0.451) occurred at a range of −0.1 to 0.6 mg/dl/month increase in urate concentrations (95% confidence interval, 0.310–0.655). In conclusion, we propose that urate concentrations during the initial 6 months of antihyperuricemic therapy should be maintained within a range of 4.6–6.6 mg/dl, and reduction in the urate concentrations during treatment should be as slow as possible.

2. INTRODUCTION

The occurrence of gouty attacks, especially during early phases of the treatment of hyperuricemia, is one of the major obstacles to patient care.[1-4] If gouty attacks are induced

Purine and Pyrimidine Metabolism in Man IX,
edited by Griesmacher *et al.* Plenum Press, New York, 1998.

13

by the treatment, patients understandably become anxious and annoyed, which probably accounts for the poor compliance of gouty patients during the early phases of antihyperuricemic treatments. Some earlier articles focused on methods to prevent these gouty attacks by the prophylactic administration of colchicine or non-steroidal anti-inflammatory drugs.[1-3] However, a precise analysis of the conditions that induced the attacks during the early phase of treatment would be important. Thus, we attempted to find the range of urate concentrations associated with the lowest incidence of gouty attacks during the initial 6 months of treatment, and also the rate at which the urate concentrations should be reduced. The present study represents an attempt to define the optimal range and the optimal rate of reduction, using a statistical approach in a retrospective study on 350 gouty patients.

3. MATERIALS AND METHODS

3.1. Patients

Data on 350 male patients who fulfilled the American Rheumatism Association preliminary criteria for the classification of acute gout[5] were analyzed. Nearly all subjects had both colchicine and non-steroidal anti-inflammatory drugs available to them during the treatment periods, for use when either aura or gouty attacks occurred. Thus, apparent gouty arthritis as well as aura followed by the ingestion of either colchicine or non-steroidal anti-inflammatory drugs were both enumerated as gouty attacks in this study.

3.2. Statistical Analysis

An optimal range of urate concentrations associated with a reduced occurrence of gouty attacks would be that wherein the incidence of attacks is significantly lower when urate concentration levels are within the range as opposed to outside of the range.

Let a certain range of urate concentrations be a–b mg/dl, and let a total number of values of urate concentrations in the sample be N. Among N values, some are within the range a–b while the others are outside of the range; i.e. < a or > b. Some values are associated with attacks but the others are not. Following the two criteria (+ or – attack, inside or outside of range), the N values are distributed as in a 2 × 2 contingency table: A, + attack, inside range; B, – attack, inside range; C, + attack, outside range; and D, – attack, outside range; N = A + B + C + D. The risk ratio and the odds ratio were calculated as follows.

$$\text{risk ratio (RR)} = (A/(A+B))/(C/(C+D))$$

$$\text{odds ratio (OR)} = (A/B)/(C/D) = AD/BC$$

RR values differ for different ranges, and there are two major variable parameters for the range: its extent, or spread from a to b, and the value of the lower limit, a. We calculated RR by changing both of the parameters and expressed the data by 3-dimensional graphs for the determination of the optimal RR or OR. 3-dimensional graphics were drawn using Mathematica (Wolfram Research, Champaign, IL). By examining the 3-dimensional graph, we broadly determined the areas where the surface of the graph was relatively smooth and, thus, the data were statistically reliable. In such an area (defined as a rectangle parallel to the plane that is perpendicular to the RR axis), we calculated the minimal RR as well as the 95% confidence interval.

4. RESULTS

4.1. Background of the Patients

The 350 patients with primary gout were separated into two groups, group A (n=132) and group B (n=218), respectively, by the presence or absence of gouty attack(s) during the initial 6-month period after the start of antihyperuricemic therapy.Backgrounds of the patients prior to the initiation of treatment were compared. Group A patients had a significantly lower incidence of diabetes mellitus as compared to group B patients (6.9:0.8%, $p < 0.05$, χ^2 test). There was no difference between groups in serum concentrations of urate (Student's t-test), suggesting that the concentration of urate before the antihyperuricemic treatment does not influence the frequency of gouty attacks.

4.2. Time of Gouty Attacks

In group A patients, attacks (including aura) occurred most commonly within the 1st (40%) and 2nd (27%) months. Although the incidence was much lower, attacks occurred at a steady pace during the 3rd through 6th months (6–11%).

4.3. Differences in Average Urate Concentrations during Treatment

Urate concentrations were determined every month in most patients. Average urate concentrations (mean ± SD) of group A patients ($6.15 ± 1.00$ mg/dl) was significantly lower ($p < 0.05$, Mann-Whitney's test) than that of group B patients ($6.42 ± 1.78$ mg/dl). There was no difference between groups with respect to both choice of antihyperuricemic drug and frequency of gouty attacks during treatment (χ^2 test).

4.4. RR Values

Although most patients visited our clinic regularly once a month, serum urate concentrations at the exact time of attack were not available. Thus, we statistically estimated that each urate concentration of group A patients as being associated with attacks. A three dimensional representation of RR values at varying range spreads and varying lower limiting values of ranges is shown in Fig. 1. An apparent large groove with straight banks at both sides is noted on the three domensional figure. The regions where the surface of RR are not smooth indicate relatively low reliability of the RR value from a statistical point of view. This is because the number of values within each range is small and a small change in the parameter causes a big change in RR. In contrast, areas within the groove and at range spreads larger than 1.0 mg/dl are generally smooth, indicating that the calculated RR values in these areas are relatively reliable.

Therefore, we attempted to determine the lowest RR value within the groove where the values are statistically reliable. Although the lowest value was at a range spread of 0.3 mg/dl, this is where surface of the graph in Fig. 1 was not smooth. To indicate this, at a range spread of lower than 0.8 mg/dl, 95% confidence intervals of RR values were rather wide and the upper limiting values of the confidence interval were close to 1.0. Except for these points, the lowest RR value was at a range spread of 2.1 mg/dl (Fig. 2A). Fig. 2B shows RR values at a fixed range spread of 2.1 mg/dl and varying lower limiting values of ranges. In this condition, the RR value within the groove was the lowest at a lower limiting value of 4.6 mg/dl . Therefore, when the range was set at 4.6–6.6 mg/dl, RR showed the local lowest value (0.705) in the proper area (95% confidence interval; 0.629–0.791).

Similarly, we prepared a 3-dimensional visualization (not shown) for OR values, and the lowest OR value 0.570 at statistically reliable 3-dimensional area was obtained at the range 4.6–6.6 mg/dl (95% confidence interval; 0.470–0.680).

4.5. Effect of Reduction in Serum Urate Concentration on Gouty Attacks

We examined the data to see if attacks tended to occur when changes in the serum urate concentration were within or outside of a determinable range. For all patients, changes in urate concentrations during a period of about 1 month were calculated when data for both months were available, and the relation between change in urate concentration and the occurrence of the attacks was examined. As before, RR was calculated for various lower limiting values of ranges and at various range spreads and represented in a 3-dimensional graphic (Fig. 1) and the lowest RR value within the groove with 95% confidence interval at each range spread was shown in Fig. 2A. In the statistically reliable area, RR value at the range spread of 0.8 mg/dl was considered the lowest. Fig. 2B shows RR values at a fixed range spread of 0.8 mg/dl and varying lower limiting value. Accordingly, the range −0.1 to 0.6 mg/dl was considered to give the lowest RR value of 0.451 (95% confidence interval, 0.310–0.655).

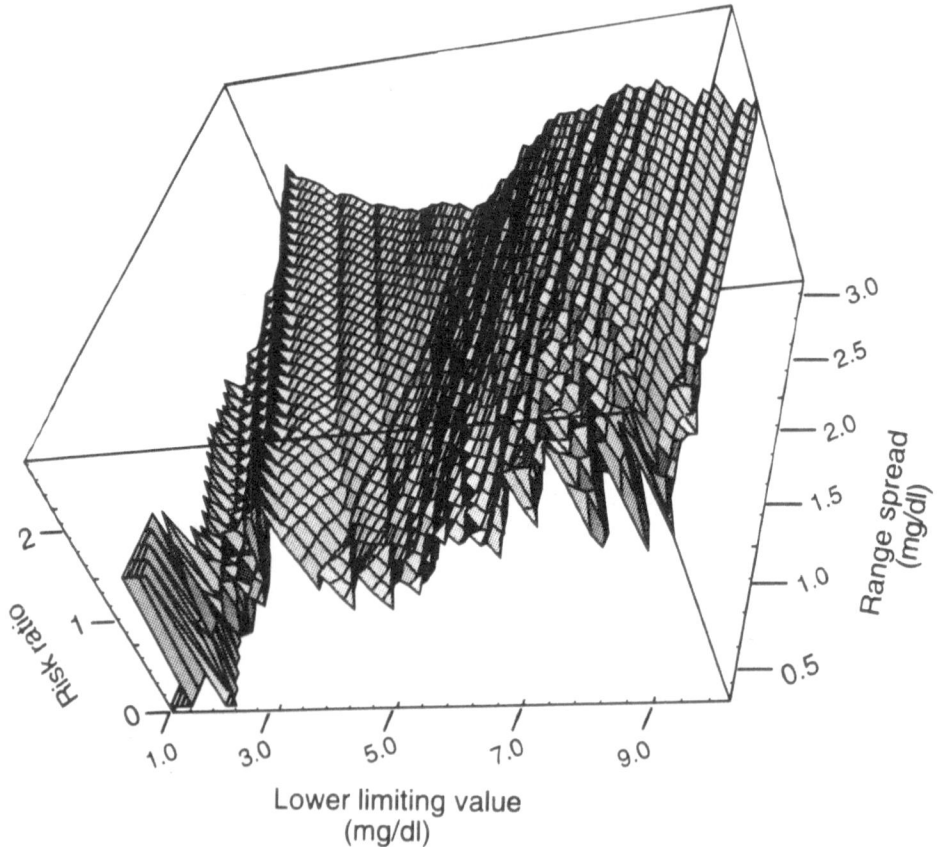

Figure 1. Three-dimensional graphics of RR values for gouty attack.

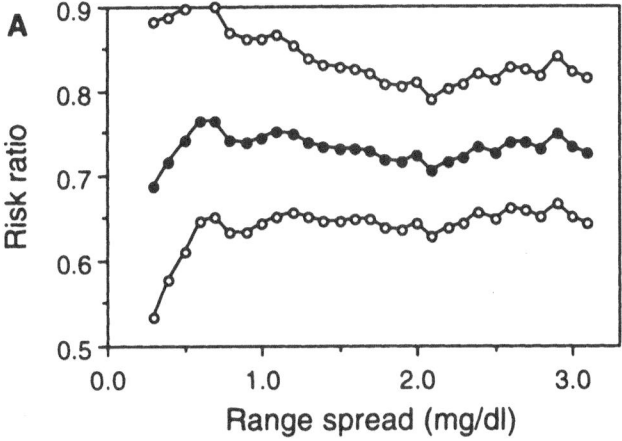

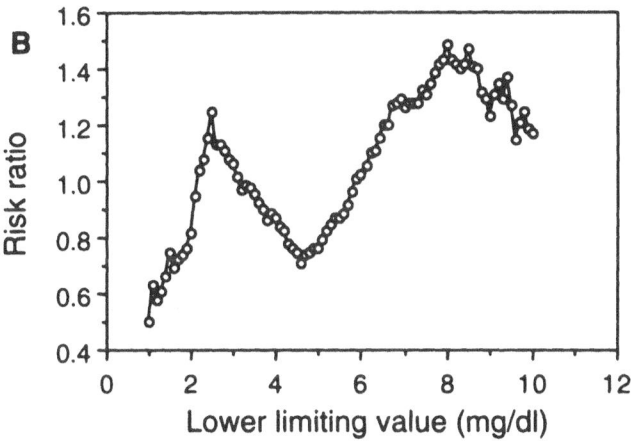

Figure 2. Two dimensional analysis of RR values for gouty attack.

5. DISCUSSION

Our results indicated that the range of the concentrations of serum urate during the antihyperuricemic treatment that minimize risks of gouty attacks was 4.6–6.6 mg/dl, when all the data within 6 months after the initiation of antihyperuricemic treatments for each patient were considered to be associated with the attacks. Although even lower RR values were obtained at very short ranges or very low and very high lower limiting values of the ranges, these data were either statistically unreliable or clinically impractical.

Interesting notion was that, concerning the association between change in urate concentrations and gouty attacks, the lowest RR was obtained when the range was set so that the urate value stayed at the same level or slightly increased (≤ 0.6 mg/dl). However, this is not surprising because one experiences gouty attacks most frequently when the urate concentrations decrease, and also this phenomenon could be well explained by the crystal shedding theory [4]. However, the optimal range for the change of the urate concentration thus determined does not meet the requirements of the antihyperuricemic treatments.

Therefore, although the RR of gouty attacks is not the lowest, a decrease in the urate concentration sufficient to achieve a urate concentration within the 4.6–6.6 mg/dl range by at least the 6th month should be recommend.

Of course, our standards are valid only for the initial 6 months of the treatment and should be considered solely from the aspect of the inhibition of the attacks. The optimal urate concentration for the prevention of such late-occurring disorders as renal insufficiency may be different from that for minimizing the occurrence of gouty attacks, and remains to be investigated in later studies.

REFERENCES

1. Wallace SL, Singer JZ. Therapy in gout. Rheum Dis Clin North Am 14, 441–457, 1988
2. Paulus HE, Schlosstein LH, Godfrey RC, Klinenberg JR, Bluestone R: Prophylactic colchicine therapy of intercritical gout. A placebo-controlled study of probenecid-treated patients. Arthritis Rheum 17, 609–614, 1974
3. Kot TV, Day RO, Brooks PM: Preventing acute gout when starting allopurinol therapy. Colchicine or NSAIDs? Med J Aust 159, 182–184, 1993
4. Terkeltaub RA: Pathogenesis and treatment of crystal induced inflammation. In, Arthritis and Allied Conditions. Twelfth edition. Edited by DJ McCarty, WJ Koopman. Philadelphia, Lea & Febiger, 1993
5. Wallace SL, Robinson H, Masi AT, Decker JL, McCarty DJ, Yu T-F: Preliminary criteria for the classification of the acute arthritis of primary gout. Arthritis Rheum 20, 895–900, 1977

4

HOW SHOULD WE TREAT TOPHACEOUS GOUT IN PATIENTS WITH ALLOPURINOL HYPERSENSITIVITY?

R. Grahame,[1] H. A. Simmonds,[2] M. B. McBride,[2] and F. P. Marsh[3]

[1]Department of Rheumatology
[2]Purine Research Laboratory
UMDS, Guy's Hospital
[3]Department of Nephrology
Royal London Hospital
London, United Kingdom

SUMMARY

We studied purine metabolism in gouty patients from three categories: primary gout, familial juvenile hyperuricaemic nephropathy (FJHN) and partial HPRT deficiency.

INTRODUCTION

Allopurinol has been the mainstay of the treatment of gout for the past 30 years. Its use has been of particular value in patients exhibiting tophaceous deposits or in the presence of renal failure. A gout patient's inability to tolerate allopurinol on account of hypersensitivity can create major therapeutic dilemmas for the physician bereft of suitable alternative therapy. Reactions may be severe, even life-threatening.

Desensitisation to allopurinol has been successfully applied, but is only suitable for milder reactions. Benzbromarone is a potent uricosuric agent which is effective in both gouty patients and healthy subjects (1,2). This study explores the possible use of benzbromarone in a series gout patients, who would normally be expected to respond satisfactorily to allopurinol, but who were precluded from doing so on grounds of intolerance. Patients with X-linked genetic gout with gross urate overproduction are also vulnerable to the effects of allopurinol because they are at risk from xanthine nephropathy.

Purine and Pyrimidine Metabolism in Man IX,
edited by Griesmacher *et al.* Plenum Press, New York, 1998.

PATIENTS STUDIED

Patient 1

Mr. R.S. aged 57. Tophaceous gout; normal renal function; allopurinol hypersensitivity.

In 1963 aged 27 he developed gouty arthritis every 2–3 months affecting the great toe, ankles, knee treated with Indomethacin. In 1992 the severity and frequency increased and the gout became polyarticular affecting his wrists, feet and elbows.

Probenecid resulted in a marked exacerbation. Allopurinol (tried on 2 occasions) produced an eruption on head and neck. He was suffering from severe gout, and felt very unwell. Allopurinol was abandoned. Between 1992 and 1994 he suffered frequent attacks of gout precipitated by wine, cheese, citrus fruits, mangoes and tomatoes. He was maintained on a low-purine diet and treated with naproxen. On examination he showed evidence of a tophaceous gouty deforming polyarthritis of his knees, ankles, feet, elbows.

The PUA was 448 μmol/l; C_{urate} 6.68 ml/min; C_{creat} 107.6 ml/min and $C_{urate}/C_{creat} \times 100$ 6.3%

In view of the history of allopurinol hypersensitivity he was commenced in March 1994 on benzbromarone 100 mg/day together with indomethacin, colchicine, dietary purine and alcohol restriction. Despite this he suffered a continuous run of migratory arthritis, becoming increasingly disabled and disillusioned. In August 1994 the dose of benzbromarone was raised to 200 mg/day. At this time the PUA was 194 μmol/l; C_{urate} 14.6 ml/min; C_{creat} 73.1 ml/min and $C_{urate}/C_{creat} \times 100$ 20%. Because of the profound uricosuric effect the dose of benzbromarone was subsequently reduced in November 1994 to 150 mg/day (Fig. 1). By November 1995 he was still suffering from moderately severe gouty attacks every 2 months. By June 1996 gout attacks ceased and he stopped colchicine. In August the benzbromarone was reduced to 100 mg/day, and it was noted that the tophi on his left heel and olecranon had disappeared. He remains well and on 29 January 1997 his GFR was 109 mls/min.

Patient 2

Mr. L.R. aged 70. Tophaceous gout, renal failure, allopurinol hypersensitivity.

He was diagnosed as suffering from hypertension in 1957 and in 1983 developed recurrent gout affecting his great toe, ankles, knees. In 1986 he was found to have non-insulin-dependent diabetes mellitus. In May 1986 allopurinol 100 mg/day was commenced. One month later he became jaundiced, oliguric, drowsy, and developed left ventricular failure and an extensive rash. His blood urea 23.2 mmol/l creatinine 415 mmol/l,

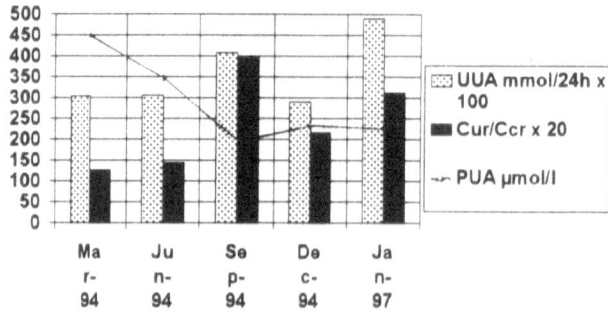

Figure 1. R.S. Tophaceous gout, normal renal function and allopurinol hypersensitivity response of PUA, UUA, C^{urate}/C^{creat} to benzbromarone.

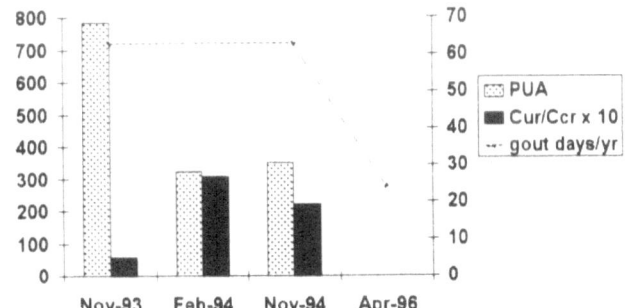

Figure 2. L.R. Tophaceous gout, renal failure and allopurinol hypersensitivity. Response of PUA, C_{urate}/C_{creat} and gout days per year to benzbromarone.

bilirubin 190 rising to 537 mmol/l, AST 2150 u/l, alkaline phosphatase 212 u/l, gamma GT 127 u/l. A liver biopsy revealed confluent hepatic necrosis with cholestasis. A diagnosis of hepato-renal failure, secondary to allopurinol hypersensitivity. He gradually recovered only after 10 days of intensive care.

He was advised not to take allopurinol again and relied on indomethacin treatment for attacks. From 1986 he was getting more than 6 gout attacks/year. Nevertheless there was a progressive increase in the size of his tophi, and he was troubled by chronic renal impairment and cardiac failure. When first seen in November 1993 there were obvious tophi on his fingers and toes and bilateral knee effusions. His PUA was 784 µmol/l; C_{urate} 1.35 ml/min; C_{creat} 23.9 ml/min and C_{urate}/C_{creat} ^{x100} 5.65%. In December 1993 benzbromarone 100 mg/day was commenced with colchicine. By February 1994 the PUA had fallen to 322 µmol/l, the C_{urate} risen fourfold to 7.63 ml/min. The C_{creat} was 24.9 ml/min and the C_{urate}/C_{creat} ^{x100} 30.6%. Over the period 1994–6 the size of the tophi diminished and the frequency of gouty attacks diminished (Fig. 2). He died of cardiac failure in late 1996.

Patient 3

L. van H. aged 14 developed recurrent acute arthropathy of toes and ankles due to Familial Juvenile Hyperuricaemic Nephropathy [FJHN] (3). By 22 years PUA was 529 mmol/l, renal function was impaired (s. creatinine 140 mmol/l. EDTA-GFR 46 ml/min/1.73m^2) and renal imaging was normal. Renal biopsy showed tubular atrophy, aggressive interstitial mononuclear infiltrate and a needle-shaped crystal in the collecting duct. She had a history of eythema multiforme after penicillin, and a generalised itchy rash after a single tablet of allopurinol. Treatment with benzbromarone was delayed when she became pregnant (successful Caesarian section for toxaemia at 35 weeks) after which

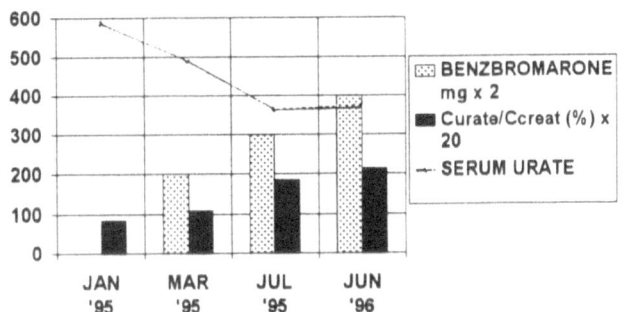

Figure 3. L. van H. Juvenile hyperuricaemic nephropathy. Response of PUA to benzbromarone.

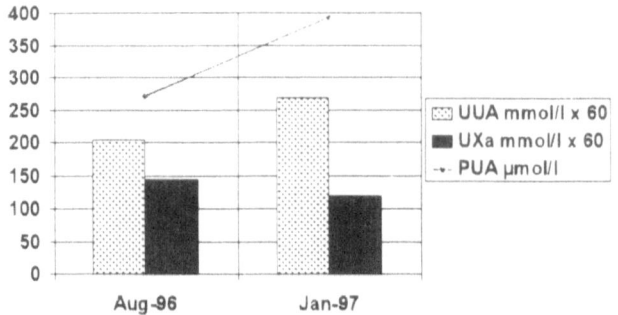

Figure 4. D.P. Early onset gout; partial HPRTase deficiency. Danger of effects of xanthinuria following allopurinol therapy in high dose are reduced by reducing the dose of allopurinol.

gout and hypertension became worse, EDTA-GFR was 40 ml/min/1.73m^2 and fractional urate excretion was reduced. Benzbromarone 100 mg/d, with colchicine was increased in stages to 200 mg/d and she became free of clinical gout after 12 months' therapy (Fig. 3).

Patient 4

D.P. presented at the age of 32 years with a history of early onset of gout in 1984 (aged 20) affecting the great toe, ankles, knees, right elbow, 2nd left PIP joint and right wrist. His PUA was found to be 660 μmol/l and he was treated with allopurinol in a dose of 600 mg/day. Between 1981 and 1986 his weight rose from 70 to 137 kg. From 1992 onwards he was troubled by renal stones and in October 1996 he underwent a pyelolithotomy operation. He was first seen by us in 1996 and reported that he had experienced 4 attacks of gout in 3 months. A maternal uncle also suffers from gout.

He denied taking a high purine diet or alcohol. On his current regime of allopurinol (dose of 600 mg/day) his PUA was 272 μmol/l and his UUA 3.47 mmol/24h. The urine xanthine 2.41 mmol/24h [twice its limit of solubility] and the C_{urate}/C_{creat} $^{\times 100}$ 5.9%. The level of erythrocyte HPRT enzyme was 3 nmol/mg Hb/h (normal 80–130). By contrast, the level of erythrocyte APRT enzyme was 24 nmol/mg Hb/h (normal 16–32). The allopurinol dose was accordingly reduced to 300 mg/d to reduce the risk of xanthine stone formation. On the revised regime the PUA rose 393 μmol/l and the UUA to 4.48 mmol/24h. The urinary xanthine fell to 1.99 mmol/24h. The C_{urate}/C_{creat} $^{\times 100}$ was virtually unchanged at 6.1% (Fig. 4).

DISCUSSION

Allopurinol has been the mainstay of the pharmacological treatment on hyperuricaemia and gout for the past 30 years (4). The incidence of serious adverse reactions is low. The Boston Collaborative Drug Surveillance Program revealed that with the exception of skin reactions 1.8% of 1835 allopurinol-treated patients experienced side-effects (5). Severe (even life-threatening) adverse reactions do occur, and this precludes the further use of the drug in affected patients. 23 fatalities thought to be associated with allopurinol were reported to the UK Medicines Control Agency between 1965 and 1997. In many instances an uricosuric agent such as probenecid or sulphynpyrazone will prove a satisfactory substitute. However, their use is unsuitable or ineffective in the presence of renal impairment or gross over production, and under these circumstances, no obvious alternative candidate drug may be forthcoming.

Benzbromarone is a uricosuric drug, which is absorbed from the gastrointestinal tract. It is metabolised in the liver to benzarone and bromobenzarone both of which are reported to have uricosuric properties. It is mainly excreted in the faeces, only small amounts being excreted in the urine (1,2).

In this study we have shown that it is an effective hypouricaemic agent in patients who are precluded from taking allopurinol, because they are hypersensitive to it. Importantly, unlike other uricosuric agents, benzbromarone appears to be effective and in patients in mild or severe renal impairment, including young subjects with FJHN. The single patient with HPRT presented here also confirms that Allopurinol should be prescribed with care for genetic overproducers of uric acid due to X-linked HPRT deficiency, whether partial as in our patient, or complete as in Lesch-Nyhan Syndrome. Such patients respond briskly with a rapid rise in urinary xanthine which parallels the drop in uric acid. This can precipitate either acute renal failure, or, as in our patient, xanthine lithiasis (3).

CONCLUSIONS

Successful lowering of plasma urate and resolution of tophi using benzbromarone in allopurinol hypersensitivity prompts us to recommend benzbromarone, even in renal failure for the treatment of either primary gout or FJHN. In HPRT deficiency the allopurinol dose must be adjusted to ensure a careful balance between the relatively insoluble xanthine and the more soluble uric acid.

REFERENCES

1. Fox IH, Sinclair DS. The Pharmacology of the hypouricemic effect of benzbromarone. Adv Exp Med Biol 1977; 76B: 328–323.
2. Zürcher R M, Bock H A, Thiel G. Excellent efficacy of benzbromarone in cyclosporine-A-treated renal transplant patients: a prospective study. Nephrol Dial Transplant 1994; 9: 548–551.
3. Cameron JS, Moro F, McBride M B, Simmonds H A. Inherited disorders of purine metabolism and transport. Chapter 16.5.3 in *Oxford Textbook of Clinical Nephrology*. 1997 pp 13–26.
4. Allopurinol in treatment of gout. Scott J T, Hall A P, Grahame R Brit Med Jl 1966; 2: 321–327.
5. McInnes et al. Acute adverse reactions to allopurinol in hospitalised patients. Ann Rheum Dis 1981; 40: 245–249.

5

THE SPECTRUM OF HGPRT DEFICIENCY

Clinical Experience Based on 20 Patients from 16 Spanish Families

Juan G. Puig,[1] Rosa J. Torres,[2] Felícitas A. Mateos,[2] Joaquín Arcas,[3] Antonio Buño,[2] and Ignacio Pascual-Castroviejo[3]

[1]Division of Internal Medicine
[2]Division of Clinical Biochemistry
[3]Division of Neuropediatrics
"La Paz" Hospital
Universidad Autónoma
Madrid, Spain

1. INTRODUCTION

Patients with HGPRT deficiency may be phenotypically classified into two differentiated entities: (a) Lesch-Nyhan syndrome, also named "complete HGPRT deficiency", characterized by spasticity, choreoathetosis, mental retardation, and self-mutilation behaviour,[1] and (b) Kelley-Seegmiller syndrome, also named "partial HGPRT deficiency."[2] Patients with this syndrome could present either mild to severe neurological manifestations including different degrees of spasticity and mental retardation, or no neurological symptoms at all. A differential characteristic between both syndromes is that patients with the Kelley-Seegmiller syndrome do not self-mutilate, although not all patients with the Lesch-Nyhan syndrome exhibit this dramatic manifestation.

The clinical expression of HGPRT deficiency may result in very different clinical phenotypes, with markedly different neurological manifestations.[3] Parents desperately demand whether their child is going to suffer the manifestations described by Lesch and Nyhan or will be affected by a not so severe disease. In other words, once the diagnosis of HGPRT deficiency is established, parents want to know how the enzyme defect will affect their child and how the disease is going to modify their lives.

To delineate the spectrum of HGPRT deficiency we reviewed our experience in 20 patients with this enzyme defect belonging to 16 different Spanish families.

Purine and Pyrimidine Metabolism in Man IX,
edited by Griesmacher *et al.* Plenum Press, New York, 1998.

2. MATERIALS AND METHODS

In the last 13 years (1984 to 1997) we have had the opportunity to study 20 patients with HGPRT defficency. All patients came to "La Paz" University Hospital, from different parts of the Spanish peninsula, Mallorca, and the Canary Islands. Patients were brought to our attention either because hyperuricemia or uric acid overproduction was incidentally discovered at another institution, or because HGPRT deficiency was already diagnosed and parents or patients demanded a thorough study of their families. Patients were studied at four levels: clinical, biochemical, enzymatic and at the molecular genetic level. Enzyme deficiency was diagnosed in all 20 patients by determining HGPRT and APRT activities in hemolysates.[4]

3. RESULTS

3.1. Clinical Characteristics

Table 1 shows the clinical characteristics of these 20 Spanish patients with HGPRT deficiency. Fourteen patients showed the typical characteristics of classic Lesch-Nyhan syndrome, including choreoathetosis, spasticity, self-mutilation and mental or psycomotor retardation. Six patients showed other clinical characteristics ranging from acute renal failure (patient 15), to hyperuricemia and crystaluria (patient 18), or mental retardation (two brothers, patients 19 and 20). From these clinical observations it was apparent that HGPRT defficiency imposed a markedly different dependence for daily activities. Besides the dicotomous classification of HGPRT defficiency into Lesch-Nyhan syndrome and Kelley-Seegmiller syndrome we attempted to classify the neurological disease as a function of the patient dependency for daily activities. Patients were classified into four groups: Group 1, normal development with no neurological symptoms (2 patients). HGPRT deficiency in these patients may be manifested by asymptomatic hyperuricemia, increased uric acid excretion, renal lithiasis or gout.

Group 2, mild neurological symptoms (3 patients). This group includes patients with mild neurological symptoms such as dystonic gait (patient 15) or some degree of spasticity which may be evident on performing any motor activity (patients 19 and 20). Group 2 patients may be differentiated from patients in Group 1 because the former are limited by their neurological symptoms, although they are independent for most activities. Group 3, comprises patients with neurological symptoms severe enough to preclude an independent life (1 patient). The intelligent patient reported by Catel and Smith[5] in 1959, later diagnosed as suffering HGPRT deficiency, could be included in this group. Group 4, classic Lesch-Nyhan syndrome (14 patients).

3.2. Biochemical Characteristics

Plasma and urinary concentrations of hypoxanthinje xanthine and uric acid were markedly elevated in all HGPRT deficient patients,[6] with no significant differences between polar phenotypes.[7] The most constant biochemical feature was an elevated urinary uric acid to creatinine ratio. However, these metabolites did not allow differentiation between HGPRT deficient groups.

Table 1. Clinical characteristics of 20 patients with HGPRT deficiency studied
at "La Paz" University Hospital, Madrid, Spain (1984–1997)

Pat. #	Family name	Initials	Age at diagnosis (yr, m)	Clinical characteristics	HGPRT**	APRT**
PHENOTYPE: Classic "Lesch-Nyhan syndrome"						
1	V1	ASV	31	Choreoathetosis, espasticity, self-mutilation, mental retard., tophy.	<0.01	80
2	V2	JJGS	6	Choreoathetosis, espasticity, mental retardation.	<0.01	45
3	V3	JGS	5	Choreoathetosis, espasticity, self-mutilation, mental retardation.	<0.01	54
4	LP	FMG	28	Choreoathetosis, espasticity, self-mutilation, mental retard., tophy.	0.14	80
5	B	JB	4	Choreoathetosis, espasticity, self-mutilation.	<0.01	45
6	A	JPA	3	Choreoathetosis, espasticity, self-mutilation, mental retardation.	<0.01	ND*
7	M	AM	6	Choreoathetosis, espasticity, self-mutilation, mental retardation.	<0.01	86
8	R	PR	19	Choreoathetosis, espasticity, self-mutilation, mental retardation.	<0.01	ND*
9	PP	APP	19m	Psicomotor retardation, crystaluria.	<0.01	68
10	AnD	JPR	11m	Psicomotor retardation, espasticity.	<0.01	52
11	H	JSSC	5m	Psicomotor retardation.	<0.01	60
12	C2	RMB	18m	Choreoathetosis, espasticity, mental retardation.	<0.01	40
13	C1	ABM	27	Choreoathetosis, espasticity, self-mutilation, mental retardation.	<0.01	61
14	Z2	JCJ	3	Choreoathetosis, espasticity, mental retardation.	<0.01	60
PHENOTYPE: "Partial HGPRT deficiency" (Kelley-Seegmiller syndrome)						
15	G	AGC	13	Acute renal failure, dystonia.	0.28	48
16	S	JGS	7m	Psicomotor retardation, crystaluria.	<0.01	58
17	F	ACM	30	Gout arthritis.	<0.01	ND*
18	Z	ASA	5m	Hyperuricemia, crystaluria.	9.38	45
19	SA1	AA	35	Mental retardation, hyperuricemia.	<0.01	41
20	SA2	JMA	33	Mental retardation, hyperuricemia.	<0.01	47

*ND, not determined.
**(nmol/h/mgHb)

3.3. Enzyme Studies

HGPRT activity in intact erithrocytes was determined in 12 patients (Figure 1). Detectable HGPRT activity (above 1% of normal HGPRT activity) was present in 4 patients: one classified into Group 1, and 3 belonging to Group 2. Overlapping values were evident among these 3 patients with different neurological involvement (Table 1, patients 15, 19 and 20).

3.4. Molecular Studies

Among the 16 Spanish families included in this series, and thanks to the cooperation of other research groups[8], a specific mutation was detected in 6 families (Table 2). HGPRT$_{Madrid}$ consisted of a substitution of glycine for valine in position 70 and corresponds to a patient (Table 1, number 16) with a severe form of the Kelley-Seegmiller syndrome (Group 3). In contrast, HGPRT$_{Yale}$, previously described,[9] corresponds to a

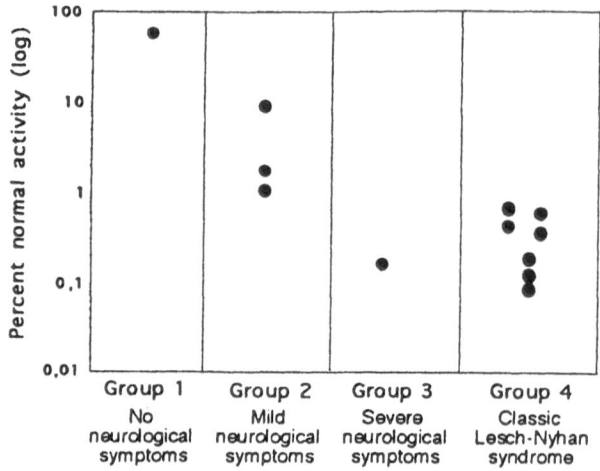

Figure 1. HGPRT activity in intact erithrocytes in 12 patients.

patient with the classic Lesch-Nyhan syndrome and therefore he would have been included in our Group 4. HGPRT$_{Yale}$ consists of a substitution of the same aminoacid glycine in position 70 for arginine.[9] It is possible that the more bulky side chain of arginine, as compared to valine, may disrupt protein folding thereby causing a causing the Lesch-Nyhan syndrome.[10]

4. CONCLUSIONS

1. Clinical classification of HGPRT deficient patients requires knowing whether the patient presents or not neurological symptoms.
2. The biochemical study of purine metabolism does not help to classify HGPRT deficient patients.
3. Enzyme studies in intact cells may differentiate amongst polar HGPRT deficient syndromes, but may not be useful for the central spectrum of the enzyme defect.
4. Identification of the gene mutation that causes HGPRT deficiency appears to be the most accurate procedure to establish whether an affected child will present the characteristics of Lesch-Nyhan syndrome or a less dramatic phenotype.

Table 2. HGPRT gene mutations in 6 Spanish families

Phenotype (abnormal protein size)	Type of mutation	Name	Amino acid
Lesch-Nyhan	Deletion	HGPRT$_{Murcia}$	119
Lesch-Nyhan	Deletion	HGPRT$_{Andorra}$	196
Lesch-Nyhan	Point	HGPRT$_{Almodovar}$	169
Lesch-Nyhan	Point	HGPRT$_{Sevilla}$	182
Kelley-Seegmiller	Point	HGPRT$_{Madrid}$	Gly70Val
Kelley-Seegmiller	Point	HGPRT$_{Salamanca}$	Met43Arg Asp44Asn

Gly, glycine; Val, valine; Met, methionine; Arg, arginine; Asp, aspartic acid; Asn, asparagine.

ACKNOWLEDGMENTS

We are indebted to the Clinical Research Unit nursing staff and the dietetic staff for excellent patient care; to Mª Paz Canencia for valuable technical assistance; and to Erik Lundin for assistance in preparing the manuscript. Supported by grants from the Wellcome Foundation, Glaxo-Wellcome, Caja de Madrid and Fondo de Investigación Sanitaria (FIS, 97/0458), Spain.

REFERENCES

1. Lesch M, Nyhan WL. A familial disorder of uric acid metabolism and central nervous system function. Am J Med 1964; 36:561–70.
2. Kelley WN, Greene ML, Rosembloom FM, Henderson JF, Seegmiller JE. Hypoxanthine-guanine phosphoribosyltransferase deficiency in gout. Ann Intern Med 1969; 70:155–206.
3. Emmerson RT, Thompson L. The spectrum of hypoxanthine-guanine phpshoribosyl-transferase deficiency. Quart J Med 1973; 42:423–40.
4. Rylance RC, Wallace RC, Nuki G. Hypoxanthine-guanine phosphoribosyltransferase assay using high performance liquid chromatography. Clin Chim Acta 1982; 127:159–65.
5. Catel W, Schmidt J. Uber familiäre gichtische diathese in verdendung mit zerebralen und renalem symptomen bei enem kleinkind. Dtsch Med Wschr 1959; 84:2145–7.
6. Puig JG, Mateos FA, Jiménez ML, Ramos TH. Renal excretion of hypoxanthine and xanthine in primary gout. Am J Med 1988; 85:533–7.
7. Mateos FA, Puig JG. Purine metabolism in Lesch-Nyhan syndrome versus Kelley-Seegmiller syndrome. J Inher Metab Dis 1994; 17:138–42.
8. Bouwens-Rombouts AGM, van den Boogaard, Puig JG, Mateos FA, Hennekam RCM, Tilanus MGJ. Identification of two new nucleotide mutations (HPRT Utrecht and HPRTMadrid) in exon 3 of the human hypoxanthine-guanine phosphoribosyltransferase (HPRT) gene. Hum Genet 1993; 91:451–4.
9. Fujimori S, Hidaka Y, Davidson BL, Palella TD, Kelley WN. Identification of a single nucleotide change in the hypoxanthine-guanine phosphoribosyltransferase gene (HPRT-Yale) responsible for Lesch-Nyhan syndrome. J Clin Invest 1989; 83:11–3.
10. Eads JC, Scapin G, Xu Y, Grubmeyer C, Sacchettini JC. The crystal structure of human hypoxanthine-guanine phosphoribosyltransferase with bound GMP. Cell 1994; 78:325–34.

6

PLASMA URIDINE AS WELL AS URIC ACID IS ELEVATED FOLLOWING FRUCTOSE LOADING

P. M. Davies,[1] H. A. Simmonds,[1] B. Singer,[2] T. G. Mant,[2] E. M. Allen,[2] A. B. Vassos,[3] and N. J. Hounslow[3]

[1]Purine Research Laboratory
UMDS Guy's Hospital
[2]Guys Drug Research Unit
London
[3]Parke-Davis Pharmaceutical Research
Eastleigh, Hampshire, Great Britain

INTRODUCTION

The hyperuricaemia following fructose loading was first demonstrated by Perheentupa and Raivio in 1967 (1). Subsequent studies (reviewed in 2) established that this elevation of plasma and urine uric acid concentrations in humans followed the rapid degradation of hepatic ATP used in the fructokinase reaction (Fig. 1). Experiments in rats given a parenteral fructose load showed that the hepatic concentration of ATP fell by 40% within 2 min, followed by an equivalent rise in ADP and AMP (3). These changes were accompanied by severe P_i depletion due to rephosphorylation of ADP in the mitochondria and were confirmed subsequently by NMR spectroscopy in humans (2). Further catabolism of AMP proceeded via AMP deaminase (AMPDA). AMPDA is normally activated by ATP, and strongly inhibited by GTP and Pi. The rapid fall in both the latter following fructose loading released these normal physiological controls, with the accumulation of IMP which was then further degraded via inosine by purine nucleoside phosphorylase (PNP) to hypoxanthine (Fig. 1). Rapid conversion of hypoxanthine to xanthine and uric acid followed because of the high activity of xanthine dehydrogenase in human liver.

Fructose loading has been used subsequently in a variety of clinical situations to evaluate the activity of enzymes involved in this catabolic cascade (2). The rationale behind its use in the present study was to evaluate the flux through PNP and hence the efficacy of a novel PNP inhibitor CI-1000 (4) in Phase 1 Clinical Trial. The results reported here were serendipitous findings during the evaluation of the effect of this combined regime on plasma purines and pyrimidines.

Purine and Pyrimidine Metabolism in Man IX,
edited by Griesmacher *et al.* Plenum Press, New York, 1998.

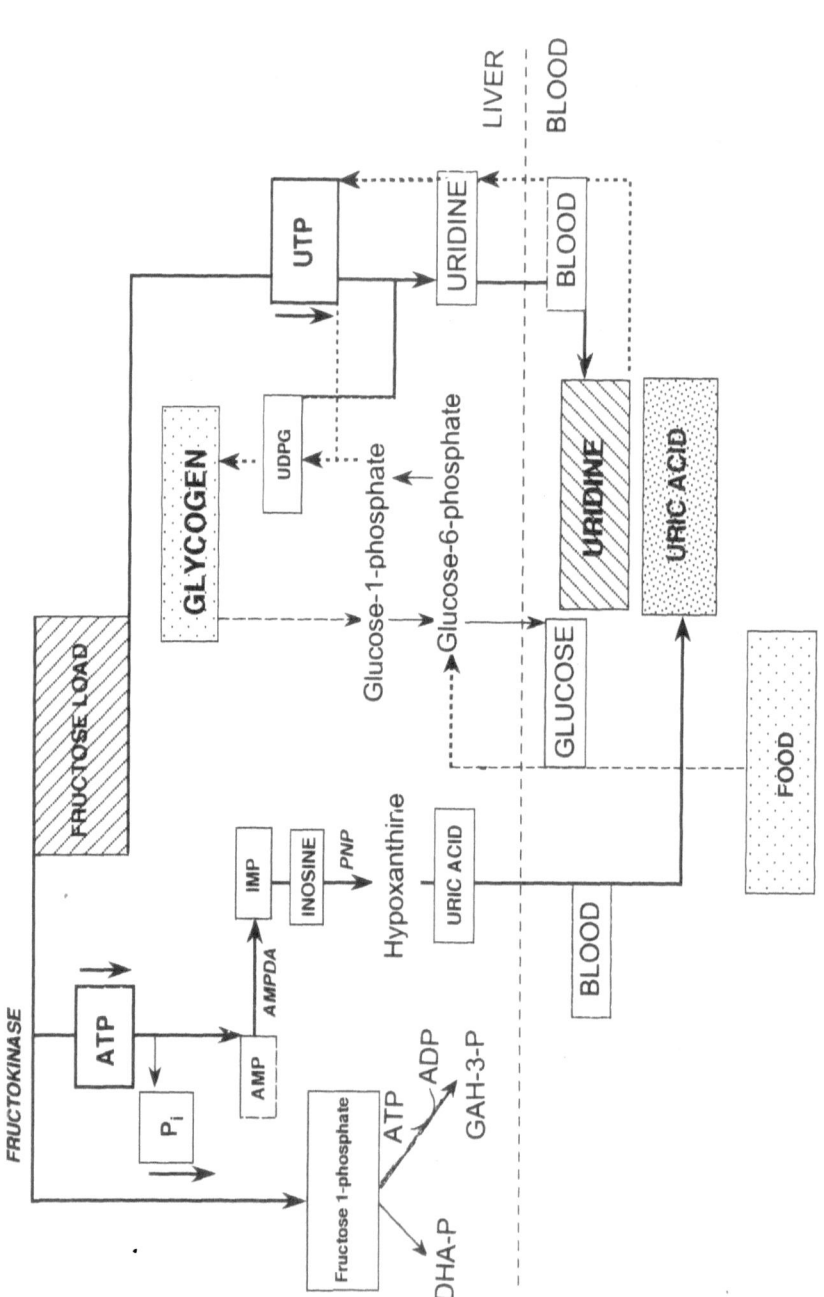

Figure 1. Effect of a fructose load, or food, on purine and pyrimidine metabolism in human liver. A fructose load induces severe ATP and Pi depletion, accompanied in turn by UTP and UDPG depletion. This is followed by further catabolism to the respective metabolic end-products, thereby elevating both plasma uric acid and uridine. Food by contrast stimulates glucose production, uptake and glycogen synthesis which utilises uridine for UDPG formation. Fasting will have the reverse effect, promoting the release of glucose to maintain constant blood levels. Catabolic pathways are indicated by unbroken lines, anabolic pathways by dashed lines. The relevant enzymes are shown in italics.

METHODS

Nineteen fasted healthy men received a 75-g oral fructose load in a protocol devised to evaluate the efficacy of the PNP inhibitor CI-1000 (4). The fructose load was administered 2 hours after a single dose of either placebo or CI-1000 to the 19 fasting healthy volunteers established on a low purine diet. Heparinised blood samples were taken before and at 2 hour intervals up to 6 hours post fructose, after which a meal was given and a further sample was taken 4 hours later. All samples were centrifuged and separated *immediately* (within one minute) at room temperature. Plasma nucleoside and base concentrations were measured in deproteinised extracts by HPLC with in-line photodiode-array detection as reported previously (5).

RESULTS

In this study nucleoside concentrations are reported following fructose challenge in subjects either at baseline or after receiving placebo in order to exclude any possible contribution of the CI-1000 to the findings noted. As anticipated plasma uric acid increased rapidly (Fig. 2) from a mean concentration of 311 ± 49.7 µmol/l over the first 2 hours following fructose to 367 ± 64.2 µmol/l (mean increment 18%, maximally up to 50%), with a gradual decline at 8 hours post dose to a mean of 353 ± 64.2 µmol/l (Fig. 2a). However, plasma uric acid increased again (21% above the basal value) 4 hours after the meal (377 ± 62.8 µmol/l).

The unexpected finding was the even more dramatic change in plasma uridine: 6.3 ± 1.1 µM pre-fructose which rose in the 2 hours following fructose by 43% to a mean peak at 2 hours of 9.0 ± 1.9 µM. Uridine likewise showed a gradual return to baseline at 6 hours

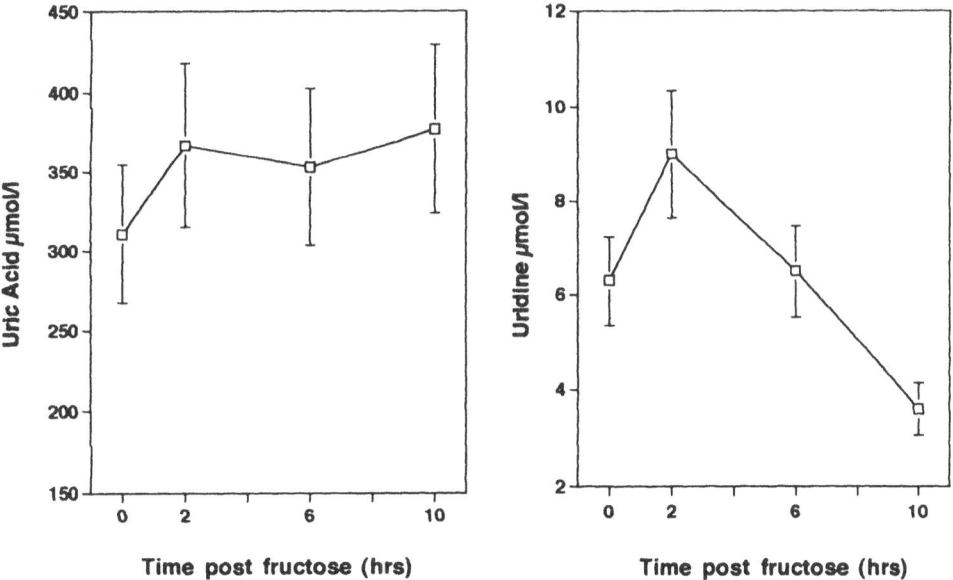

Figure 2. Mean plasma uric acid concentrations in heparinised blood (left) from 19 fasting healthy males taken prior to and at 2, 6 and 10 hours after an oral fructose load. A meal was given immediately after the 6 hour sampling. Mean plasma uridine concentrations (right) in the same samples. Uridine also rose after the fructose load, but in contrast to the slight additional increment in plasma uric acid after food, uridine concentrations fell dramatically in these previously fasted subjects.

(Fig. 2). However, in direct contrast to uric acid the mean uridine concentration decreased by 44% to 3.6 ± 0.8 µM at 4 hours after the meal. Uridine concentrations in the same subjects taken at the same time periods without a fructose load, did not vary significantly in the first two periods excluding diurnal fluctuation or operator/method error as the basis for the rise in plasma uridine after the fructose load. However, the fall in plasma uridine noted 4 hours after a meal was similar (data not shown).

DISCUSSION

Hyperuricaemia and hyperuricosuria are well documented consequences of oral or intravenous fructose administration in humans. As discussed above, this hyperuricaemia has been attributed to the rapid degradation of hepatic adenine nucleotides leading to ATP depletion (1–3). Subsequent depletion of P_i together with GTP, leads to a release of the normally controls on AMP deaminase, resulting in AMP degradation and catabolism via IMP to uric acid (Fig. 1). Differences in the magnitude of the changes observed have been reported, which may relate either to the different routes or the amounts of fructose administered.

In this study we observed a mean increment in plasma uric acid of 18% 2 hours following an oral 75 kg fructose load, accompanied by an even greater increment in plasma uridine (mean 44%) over the same time period. Although the focus in human studies has been on ATP and GTP degradation following fructose loading, early studies in rats (3) reported that concentrations of the pyrimidine nucleotides, UTP and UDPG, were also reduced following fructose infusion. The 44% increment in plasma uridine found in healthy humans in this study following an oral fructose load is consistent with the fructose-induced pyrimidine nucleotide depletion reported in the rat study. In that study marked depletion of liver UTP and UDPG, as well as ATP and P_i, with little glycogen synthesis, was reported. These earlier results were interpreted as indicating stimulation of pyrimidine as well as purine nucleotide catabolism, in consequence of P_i depletion. We believe this is the explanation for the increase in plasma uridine after fructose in this study.

The liver is considered to play a major role in maintaining plasma uridine concentrations (6). It has the ability to synthesise pyrimidines *de novo,* as well as possessing an active salvage pathway. Studies in rodents support a capacity for degrading uridine when the plasma concentration exceeds the physiological range (6). However, the relevance of such studies to the human situation is uncertain, given the high activity of uridine phosphorylase in rodent tissues (eg heart) compared with the undetectable levels in most tissues other than the liver in humans (7).

The fall in plasma uridine after the meal noted here in these previously fasted subjects can be attributed to the increased utilisation of pyrimidine ribonucleotides for glycogen synthesis. Clearly, under such circumstances, the liver needs a ready source of uridine to provide the UDPG necessary for conversion of glucose to glycogen (Fig. 1). The results in this study thus indicate that plasma uridine concentrations in healthy humans will vary depending on the diet of a given individual. Consequently, any study, particularly those evaluating the effect of pyrimidine analogues on plasma uridine concentrations, must include careful attention to diet and the time of blood sampling in relation to food intake.

REFERENCES

1. Perheentupa J, Raivio K. Fructose induced hyperuricaemia. Lancet 1967;ii:528–535.

2. Gitzelman R, Steinmann B, Van den Berghe G. Disorders of fructose metabolism. Chapter 23. In:Scriver CR, Beaudet AL, Sly WS, Valle D, eds. *The Metabolic and Molecular Basis of Inherited Disease.* New York:McGraw-Hill, 7th edition:1995;905–934.
3. Burch HB, Lowry OH, Meinhardt L, Max P, Chyu K. Effect of fructose, dihydroxyacetone, glycerol and glucose on metabolites and related compounds in liver and kidney. J Biol Chem 1970:245;2092–2102.
4. Gilbertson RB, Dong MiK. Blockade of nucleoside degradation in monkey whole blood in vitro by CL-1000, a purine nucleoside phosphorylase (PNP) inhibitor. Adv Exp Med Biol 1995;370:167–171.
5. Simmonds HA, Duley JA, Davies PM. Analysis of purines and pyrimidines in blood, urine and other physiological fluids. Chapter 25 In:Hommes F Ed, NY:Wiley-Liss. *Techniques in Diagnostic Human Biochemical Genetics: A laboratory Manual* 1991;p397–424.
6. Peters GJ, Codacci-Pisanelli G. Physiological and haematological sources of plasma uridine. Paths to Pyrimidines 1995;3:31–41.
7. Smolenski RT, de Jong JW, Janssen M, Lachno DR, Zydowo MM, Tavernier M, Huizer T, Yacoub MH. Formation and breakdown of uridine in ischaemic hearts of rats and humans. J Mol Cell Cardiol 1993;25:76–74.

7

INVERSE RELATIONSHIP BETWEEN SERUM TRIGLYCERIDE LEVELS AND FRACTIONAL URATE EXCRETION

C. P. Quaratino, M. Mellozzi, G. Tamburrano, and A. Giacomello

Dipartimento di Oncologia e Neuroscienze
Università di Chieti
Dipartimento di Endocrinologia
Università di Roma "La Sapienza," Italy

1. INTRODUCTION

Hyperuricaemia is known to be related to hyperlipidaemia. Evidence for an uricosuric effect of fenofibrate and for a lipid lowering effect of benzbromarone has been reported.[1] Furthermore, after dietary intervention, 15 hyperuricaemic-hyperlipidaemic subjects showed considerably decreased triglyceride and cholesterol concentrations and increased renal excretion of uric acid.[2] These observations suggest a relation between serum lipid levels and renal excretion of uric acid. This relation has been the objective of the present study. Some of the possible underlying mechanisms are discussed.

2. PATIENTS AND METHODS

The study population included 1941 outpatients: 1009 adult men, 298 premenopausal and 634 postmenopausal women. Morning blood and urine samples were taken after an overnight fast. The biochemical tests were performed by an autoanalyzer (Monarch, Instrumentation Laboratory) using standard methods. Only patients with a fasting blood glucose concentration less than or equal to 110 mg dl^{-1} were considered since serum urate levels are significantly related to development of glucose intolerance[3] and in diabetic subjects glycosuria can affect uric acid excretion.[4] Using these selection conditions, the number of patients decreased to 1420 subjects. In the three groups of patients statistical evaluation was performed using version 5 of the Systat software (Systat Inc., Evanston, IL). Normal probability plots were examined to determine the skewness of variables and

Purine and Pyrimidine Metabolism in Man IX,
edited by Griesmacher *et al.* Plenum Press, New York, 1998.

37

Table 1. Descriptive data of the examined variables in 707 adult (age ≥ 20 years) men, in 277 premenopausal (20 years ≤ age ≤ 45 years) (pre wom) and in 436 postmenopausal women (age ≥ 55 years) (post wom) with serum glucose level ≤ 110 mg/dl

	Geometric mean			Median			25th percentile			75th percentile		
	Men	Pre wom	Post wom	Men	Pre wom	Post wom	Men	Pre wom	Post wom	Men	Pre wom	Post wom
Age (years)	50.8	33.2	64.9	56	35.0	64.0	42.0	28.0	59.0	66.0	41.0	70.0
Cholesterol (mg/dl)	181	171	196	186	169	198	156	147	173	210	197	227
Creatinine (mg/dl)	1.14	0.98	1.04	1.10	1.00	1.00	1.00	0.90	0.90	1.20	1.10	1.10
Glucose (mg/dl)	94.3	89.8	93.0	95.0	91.0	94.0	89	84.0	88.0	102	97.0	101
Triglyceride (mg/dl)	132	87.5	124	125	82.0	117	90	55.0	89.0	186	125	167
Urate (mg/dl)	5.17	3.57	4.31	5.10	3.60	4.20	4.30	3.00	3.60	6.30	4.20	5.20
FE_{UA} (%)*	7.55	12.0	10.6	8.01	12.3	11.2	5.76	9.51	8.02	11.0	15.8	14.8

*FE_{UA} (%) = ([serum creatinine]/[serum urate]) × ([urinary urate]/[urinary creatinine]) × 100.

log transformation was used when necessary to approximate a normal distribution of data. Pearson correlation coefficients were derived to show the relation between variables. The following independent variables were entered into the fractional urate excretion (percentage ratio of urate and creatinine clearance) multiple regression model: age, serum cholesterol, glucose and triglyceride concentrations. For each variable, in the final model, the standardized beta coefficient and its significance are given. Preliminary data of uric acid renal excretion before and after treatment in three patients with insulinoma and with serum creatinine levels within the reference range are also reported.

3. RESULTS

Descriptive data obtained in men, premenopausal and postmenopausal women are presented in Table 1. The lower plasma urate levels in females appear to result, at least in part, from a higher fractional urate excretion as compared with their male counterpart. As shown in Table 2, in all groups serum triglyceride concentration was inversely correlated with fractional urate excretion as obtained by simple and multiple linear regression analysis. The magnitudes of Pearson correlation coefficients, however, were small. Although serum triglyceride levels had a significant negative independent association with fractional urate excretion, the small values of adjusted squared multiple R obtained (0.015 in men, 0.021 in premenopausal and 0.012 in postmenopausal women), denote that only a small proportion of variance in the dependent variable was accounted for by the predictors. In the subsets of diabetic (fasting serum glucose ≥ 140 mg dl^{-1}) patients (111 men and 104 postmenopausal women), multiple linear regression analysis showed that serum glucose was the only significant (p ≤ 0.01) predictor of fractional urate excretion (β coefficients were 0.248 in men and 0.275 in postmenopausal women). Only 9 outpatients had serum glucose levels ≥ 140 mg dl^{-1} in premenopausal women.

Sex, age, body mass index, serum insulin, glucose, triglyceride, urate, creatinine levels, creatinine and urate clearances, fractional urate excretion (%) before and after treatment in three patients with insulinoma are presented in Table 3. Decreased insulin levels were associated with an increase in fractional urate excretion in the three patients. In the two women (Patients 1 and 2) after treatment there was also an increase in urate clearance, while in patient 3 a slight decrease of the latter value was observed.

Table 2. Association between fractional urate excretion and age, serum glucose (SGlu), serum cholesterol (SChol), serum triglyceride (STriglyc) in outpatients as determined by simple and multiple linear regression analysis

		Age	SGlu	SChol	STriglyc
Simple linear regression	Men	0.08*	−0.02	0.02	−0.10***
	Women (Premenopausal)	−0.06	−0.10	0.07	−0.09
	Women (Postmenopausal)	−0.07	0.01	−0.01	−0.13***
Multiple linear regression	Men	0.07*	−0.02	0.07*	−0.13***
	Women (Premenopausal)	−0.04	−0.10	0.14*	−0.16*
	Women (Postmenopausal)	−0.06	0.01	0.03	−0.13***

Results are standardized β coefficient and significance.
All variables were normalized by logarithmic transformation except age.
Significance level: *P < 0.05; ***P ≤ 0.01

4. DISCUSSION

In this large cross-sectional study, a negative correlation between serum triglyceride concentration and fractional urate excretion has been demonstrated accounting for only a small proportion of the latter variable variance. Acute elevation of serum triglyceride concentrations have been shown not to modify uric acid synthesis or excretion.[5,6] However, these data are not conclusive since infused triglycerides have a different composition and metabolic origin than endogenous triglycerides. In addition prolonged triglyceride elevation might exhibit effects not elicited by this acute experimental model. Clustering of dyslipidaemia, hyperuricaemia and hypertension and the association of this clustering with insulin resistance, high insulin levels and central fat distribution has been repeatedly demonstrated.[7–9] An inverse correlation between insulin sensitivity and serum urate concentration and between insulin resistance and clearance of uric acid has been shown in healthy volunteers.[5,10] On the basis of this findings it has been suggested that changes in plasma insulin concentration could modulate urate renal handling.[10] This possibility was evaluated by a preliminary study of renal urate excretion in 3 patients with insulinoma before and after treatment. Decreased insulin levels were associated with an increase of fractional urate excretion in all subjects and also with an increase of urate clearance in two of the three patients, confirming a role of insulin in modulating the handling of uric acid by the kidney.

Table 3. Values of some variables obtained in 3 patients with insulinoma before ([b]) and after ([a]) treatment

PatN°	Sex	Age (yrs)	BMI (Kg/m²)	Insulin* (mIU/l)	Glucose* (mg/dl)	Triglyceride (mg/dl)	Urate (mg/dl)	Creatinine (mg/dl)	Clearance (ml/min) Creatinine	Clearance (ml/min) Urate	FE_{AU} (%)
1	F	65	26.5	12.00[b]	40.0[b]	173.0[b]	3.80[b]	0.8[b]	69.7[b]	3.7[b]	5.31[b]
				3.45[a]	100.0[a]	186.0[a]	3.30[a]	0.8[a]	47.8[a]	4.04[a]	8.45[a]
2	F	75	32.0	39.0[b]	40.0[b]	195.0[b]	6.0[b]	1.0[b]	87.5[b]	5.3[b]	6.1[b]
				13.6[a]	90.0[a]	196.0[a]	6.2[a]	0.9[a]	55.6[a]	5.5[a]	9.9[a]
3	M	27	26.4	30.0[b]	45.0[b]	113.0[b]	6.7[b]	1.1[b]	130.2[b]	5.8[b]	4.5[b]
				20.0[a]	120.0[a]	150.0[a]	6.7[a]	1.2[a]	81.3[a]	4.77[a]	5.9[a]

*Mean of at least 5 samples.

REFERENCES

1. Gresser U, Gathof B S, Gross M. Benzbromarone and fenofibrate are lipid lowering and uricosuric: a possible key to metabolic syndrome. Adv Exp Med Biol 1995; 370: 87–90.
2. Tinahones FJ, C-Soriguer FJ, Collantes E, Perez-Lindon G, Sanchez Guijo P, Lillo JA. Decreased triglyceride levels with low calorie diet and increased renal excretion of uric acid in hyperuricaemic-hyperlipidaemic patients. Ann Rheum Dis 1995; 54: 609–10.
3. Brand FN, McGee DL, Kannel WB, Stokes III J, Castelli WP. Hyperuricemia as a risk factor of coronary heart disease: the Framingham Study. Am J Epidemiol 1985; 121:11–18.
4. Herman JB, Medalie JH, Goldbourt U. Diabetes, prediabetes and uricemia. Diabetologia 1976;12: 47–52.
5. Vuorinen-Markkola H, Yki-Järvinen H. Hyperuricemia and insulin resistance. J Clin Endocrinol Metab 1994;78: 25–29.
6. Fox IH, John D, DeBruyne S, Dwosh I, Marliss EB. Hyperuricemia and hypertriglyceridemia: metabolic basis for the association. Metabolism 1985; 34:741–6.
7. Wyngaarden JB, Kelly WN. Gout and hyperuricemia. New York:Grune & Stratton,1976; 21–37.
8. Schmidt MI, Duncan BB, Watson RL, et al. A Metabolic Syndrome in Whites and Africans-Americans. Diabetes Care 1996; 19: 414–418.
9. Giacomello A, Di Sciascio N, Quaratino CP. Relation between serum triglyceride level, serum urate concentration, and fractional urate excretion. Metabolism 1997; in press.
10. Facchini F, Chen YD, Hollenbeck CB, Reaven GM. Relationship between resistance to insulin-mediated glucose uptake, urinary uric acid clearance, and plasma uric acid concentration. JAMA 1991; 266: 3008–3011.

8

MICROANALYSIS OF PATHOLOGICAL CRYSTALS AND URINARY CALCULI

Kiyoko Kaneko,[1] Shin Fujimori,[2] Naoyuki Kamatani,[3] Hisashi Yamanaka,[3] Noriko Yamaoka,[2] and Ieo Akaoka[2]

[1]Central Laboratory of Analytical Biochemistry
[2]Department of Internal Medicine
 Teikyo University School of Medicine
[3]Institute of Rheumatology
 Tokyo Women's Medical College
 Tokyo, Japan

INTRODUCTION

Pathological crystals are known to cause crystal-induced arthritis (1). The identification of crystals is necessary on the diagnosis of the disease (2). The identification of crystals is usually carried out with a polarizing microscope. However, it is difficult to determine the composition of crystals with the microscope. X-ray analysis is recommended on the determination of crystals.

Urinary calculi are often found in patients with gout or hypouriceamia (3). Infrared (IR) spectroscopy is a routine analytical method in studying urinary stones (4). It is described that best suited methods for the analysis of calculi are X-ray diffractometry and IR spectroscopy (5).

In this study, micro area analysis with X-ray diffractometer was carried out to determine fine structures of such small materials as pathological crystals and urinary calculi. Infrared analysis was also done on those small materials and was compared with micro area X-ray analysis.

MATERIALS AND METHODS

Materials

Crystalline materials from patients with arthritis, gout or hyperuricemia, were examined. White subcutaneous nodes were extirpated from patients with arthritis nodosa, which proved to be calcium deposit with the X-ray radiographic inspection. Urinary calculi were.

Purine and Pyrimidine Metabolism in Man IX,
edited by Griesmacher *et al.* Plenum Press, New York, 1998.

excreted in the urine from patients with gout or hyperuricemia. Synovial fluid with crystals were collected by a puncture needle from a patients with gout. After synovial fluid was incubated at 37°C for 60 min. with papain following 30 min. incubation with hyaluronidase, suspension was centrifuged for 1 min. at 11,000 g. Pellets containing crystals were dried and were examined.

Analysis

A micro area X-ray diffractometer (JEOL JDX-8030, DX-MAP2, Tokyo, Japan) with microscope was used in all experiments. Each specimens were analyzed with micro beam X-ray on several spots of those surface and cross sections. The X-ray diffraction pattern is recorded with an X-ray goniometer and is represented as an intensity as a function of twice the diffraction angle (2θ) curve. Analytical conditions were as follows, target: Cu; filter: Ni; voltage: 40 kV; current: 40 mA; diameter of the collimator: 100 μm. The diffraction pattern was compared with the data registered in JCPDS (Joint Committee on Powder Diffraction Standards) database. An infrared spectrophotometer (JASCO A-302, Tokyo, Japan) was used in IR analysis.

RESULTS

Subcutaneous Nodes

Subcutaneous nodes were washed several times with distilled water and were dried over 2 days. In Figure 1, the X-ray diffraction pattern of a white subcutaneous node (calcium deposit) is shown. When several spots of the calcium deposit were analyzed on the surface or the cross section, the obtained diffraction patterns were similar in every area analyzed. X-ray diffraction patterns agreed well with hydroxyapatite registered in JCPDS database, which was shown as (A) in Figure 1. IR spectra also supported those results. Furthermore, IR spectra additionally indicated carbonate groups in the deposits.

Monosodium Urate Crystals in Synovial Fluid

Synovial fluid from joints with inflammation contains many biological compounds, such as leukocytes, proteins, and proteoglycans. When synovial fluid from a patients with gout was examined by a polarizing microscope, a number of needle-shaped crystals and leukocytes were observed. Crystals were estimated to be monosodium urate (MSU) from their negative birefringence. Crystalline pellets were isolated according to Materials and Methods, and X-ray diffraction was carried out (Figure 2). The crystals were analyzed precisely with X-ray diffractometer even with much amount of interfering substances. The diffraction pattern was well in agreement with MSU which was shown as (C) in Figure 2. IR analysis could not determine the crystals in synovial fluid, because of a large amount of organic materials.

Urinary Calculi

A urinary calculus, which was analyzed with IR spectrophotometry first, was shown to contain mainly (more than 95%) calcium oxalate. But IR spectrophotometry could not distinguish clearly with two calcium oxalate salts; calcium oxalate monohy-

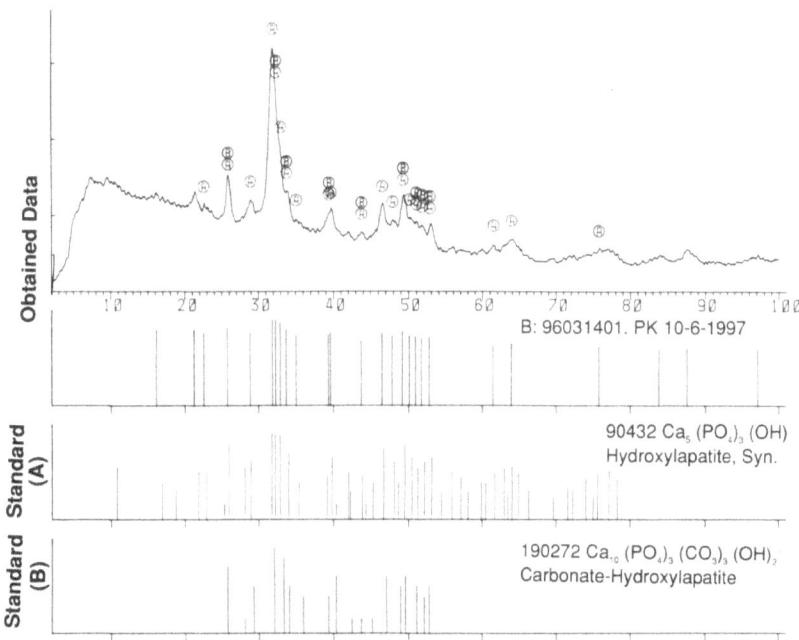

Figure 1. X-ray diffraction pattern of a subcutaneous node.

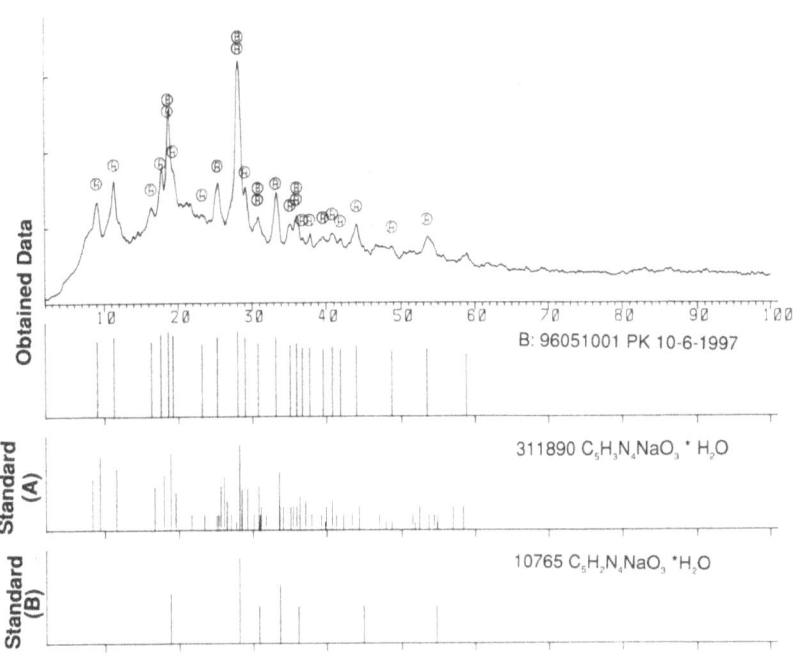

Figure 2. X-ray diffracion pattern of crystals in synovial fluid.

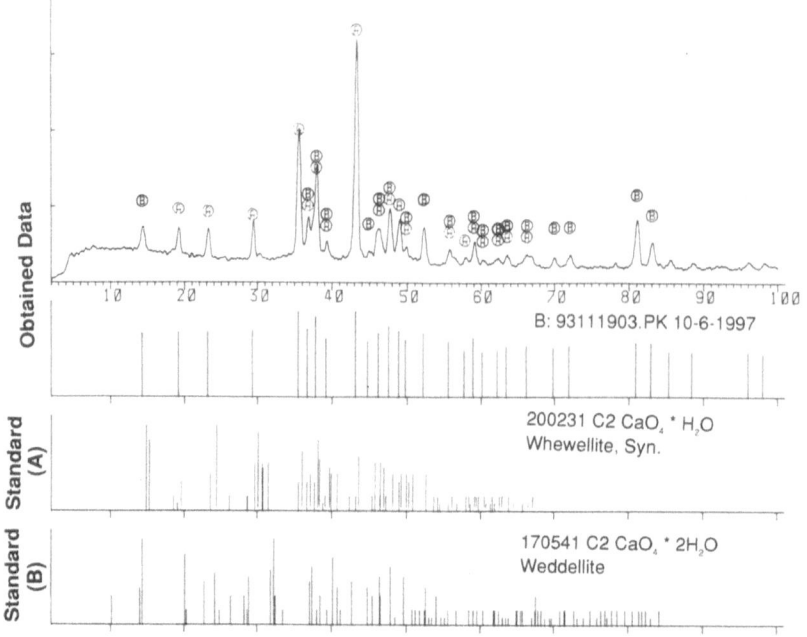

Figure 3. X-ray diffraction pattern of a urinary calculus.

drate (whewellite) and calcium oxalate dihydrate (weddellite). In Figure 3, one of X-ray diffraction patterns of the spots analyzed on the calculus is shown. It was demonstrated that both whewellite and weddellite were mixed in the calculus.

DISCUSSION

X-ray diffractometry is very reliable if the constituents are well crystallized. It ca detect crystalline minerals in low concentrations (6). IR spectroscopy can detect both cry talline and amorphous minerals Both methods can be used for semi-quantitative stor analysis and can analyze less than one mg of stone samples. X-ray diffractometry and I spectroscopy are also the best methods to identify "unexpected" stone constituents (5 Recently, analysis of urinary stones by means of Raman spectra has been reported (7).

In this study, a unique micro area X-ray diffractometer was employed. With the u of this instrument, X-ray diffraction on microscopic spots of small crystalline materia could be operated. To analyze these crystalline materials finely is considered to be helpt on studying the process of crystallization in the body and on investigating the preventi of recurrence of calculi or arthritis.

In conclusion, analysis with a micro area X-ray diffractometer could distinguish p cise structures of crystalline materials in the body.

REFERENCES

1. D.J.McCarty, N.N.Kohn and T.S.Faires, The significance of calcium phosphate crystals in the syn(fluid of arthritis patient; the pseudogout syndrome. Ann Int Med 56: 711–737 (1962)

2. J.B.Wyngaaden, W.N.Kelley, Diagnosis and differential diagnosis of gout. New York, Grune & Stratton, 270–283 (1976)
3. A.B.Gutman, T.F.Yu, A.Weinberger, et al, Uric acid nephrolithiasis. Am J Med 45: 756–779 (1968)
4. M.Volmer, A.Bolck, B.G.Wolthers, A.J.de Ruiter, D.A.Doornbos, and W.van der Slik, Partial least-squares regression for routine analysis of urinary calculus composition with Fourier transform infrared analysis, Clin Chem 39:948–954 (1993)
5. D.A.G.Vergauwe, R.M.H.Verbeeck and W.Oosterlinck, Analysis of urinary calculi, Acta Urologaca Belgica 62: 5–13 (1994)
6. L.Keller, X-ray powder diffraction patterns of calcium phosphates analyzed by the Rietveld method, J Biomed Mater Res 29: 1403–1413 (1995)
7. E.Takasaki, Carbonate in struvite stone detected in Raman spectra compated with infrared spectra and X-ray diffraction. Int J Urol 3: 27–30 (1996)

9

ZONAL DISTRIBUTION OF ALLOPURINOL-OXIDIZING ENZYMES IN RAT LIVER

Yuji Moriwaki, Tetsuya Yamamoto, Jun-ichi Yamakita, Sumio Takahashi, Zenta Tsutsumi, and Kazuya Higashino

Third Department of Internal Medicine
Hyogo College of Medicine Mukogawa-cho 1-1
Nishinomiya, Hyogo 663, Japan

SUMMARY

We describe an enzymatic histochemical localization of two allopurinol-oxidizing enzymes, xanthine oxidase and aldehyde oxidase in rat hepatic tissues. This method is based on the tetrazolium salt procedures by use of a tissue protectant, polyvinyl alcohol, with tetra-nitro BT as the final electron acceptor.[1] The present study demonstrated that both oxidases are present in the cytoplasm of hepatic cells. However, the distribution of the enzymes was uneven, being seen mainly in the pericentral rather than the periportal area. When allopurinol was used as a substrate, the specific staining by xanthine oxidase was more prominent than that of aldehyde oxidase. The results suggested that xanthine oxidase is more effective in oxidizing allopurinol than aldehyde oxidase.

INTRODUCTION

Xanthine oxidase and aldehyde oxidase have been reportedly oxidize allopurinol,[2] a xanthine oxidase inhibitor. Although the tissue localization of xanthine oxidase is extensively investigated, that of aldehyde oxidase has been little investigated. Information on the localization of aldehyde oxidase and its relative contribution to the oxidation of allopurinol is important in the understanding of pharmacological subgroup of xanthine oxidase deficiency. Therefore, we investigated the enzymatic histochemical localization of allopurinol-oxidizing enzymes, xanthine and aldehyde oxidases, and their relative contribution to the oxidation of allopurinol using rat liver.

Purine and Pyrimidine Metabolism in Man IX,
edited by Griesmacher *et al.* Plenum Press, New York, 1998.

MATERIALS AND METHODS

Male Wistar rat weighing 250 g was used in the experiment. The animal was killed with an overdose of diethyl ether anesthesia. The liver was removed immediately, blotted dry, cut into small pieces, mounted in OCT compound and frozen by immersion in a dry/ice ethanol mixture. The block was sectioned at 6 μm in a motor-driven Cryostat in which the ambient temperature was kept at −15°C. The sections were picked up onto clean glass slides and maintained in a refrigerator at 4°C until reaction.

The incubation medium contained as follows: polyvinyl alcohol (Sigma Chem Co.) was dissolved at 10–15% in phosphate buffer 100 mM, pH 8.0 by heating with continuous stirring. Just prior to sectioning the tissue block, 0.5 mM hypoxanthine, 0.5 mM *p*-dimethylaminocinnamaldehyde (DMAC) or 1.5 mM allopurinol was added respectively to the medium as a substrate, and 1.1 mM tetra-nitro BT and 0.45 mM phenazine methosulfate as final and intermediate electron acceptors.

The incubation medium was dropped onto the sections and reaction was performed at 37°C for 60 to 120 min. The reaction was stopped by immersing the slides in distilled water heated to 60°C for 10 sec; slides were fixed in 4% paraformaldehyde in PBS for 5 min, then rinsed in distilled water for 1 min and mounted with glycerol jelly. The control reactions were performed by incubating the sections without a substrate or by adding respective inhibitors; TEI-6720 for xanthine oxidase or benzamidine hydrochloride for aldehyde oxidase.

RESULTS AND DISCUSSION

The present study demonstrated that the distribution of xanthine and aldehyde oxidases in the liver was uneven, more prominent in the pericentral area than in the periportal area (Fig. 1), which was consistent with immunohistochemical localization study of the enzymes.[3] The specificity of the reactions was verified by control reactions that omitted

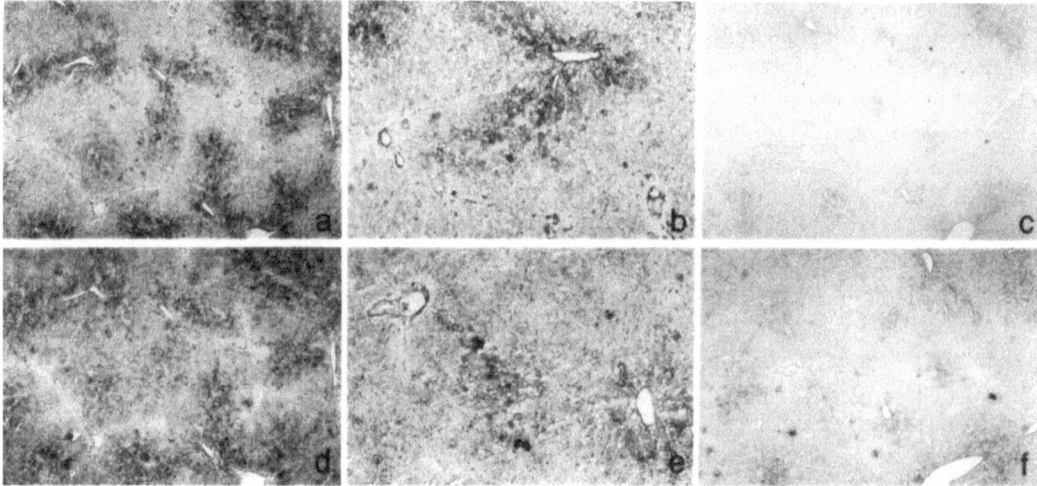

Figure 1. Moderate activities of xanthine and aldehyde oxidase were observed. Both xanthine oxidase (a,b) and aldehyde oxidase (d,e) activities were more prominent in the pericentral than in the periportal zone. Control reactions with TEI-6720 (c) or benzamidine hydrochloride (f). a,c,d,f: ×10; b,e: ×25 (original magnification).

substrates, hypoxanthine, DMAC or allopurinol and by the addition of TEI-6720, a specific inhibitor of xanthine oxidase[4] but not of aldehyde oxidase or benzamidine hydrochloride, a specific inhibitor of aldehyde oxidase[2] but not of xanthine oxidase. No activities of the enzymes were observed in the absence of the substrates. The addition to the incubation medium containing DMAC of 30 mM benzamidine markedly suppressed the reaction for aldehyde oxidase activity. The addition to the incubation medium containing hypoxanthine of 0.5 mM TEI-6720 markedly suppressed the reaction for xanthine oxidase activity. However, the addition of 30 mM benzamidine hydrochloride or 0.5 mM TEI-6720 to the incubation medium containing hypoxanthine or DMAC had almost no effect on xanthine or aldehyde oxidase activity (data not shown).

Xanthine oxidase deficiency is divided into two pharmacological subgroups.[5] One subgroup does not oxidize allopurinol to oxypurinol.[6] This subgroup is thought to have a combined deficiency of the allopurinol-oxidizing enzyme. The other subgroup is able to oxidize allopurinol, despite a lack of xanthine oxidase.[7] These findings indicate the existence of allopurinol-oxidizing enzyme. Aldehyde oxidase is reportedly oxidizes allopurinol.[2] However, as for allopurinol-oxidizing activity in the present enzyme-histochemical analysis, the distribution of allopurinol-oxidizing enzymes was rather even, and the addition to the incubation medium containing allopurinol of TEI-6720 markedly suppressed the deposit of reaction products, while that of benzamidine did not (Fig. 2). This result was in agreement with the previous in vivo study using guinea pig that xanthine oxidase plays a major role in the oxidation of allopurinol.[8]

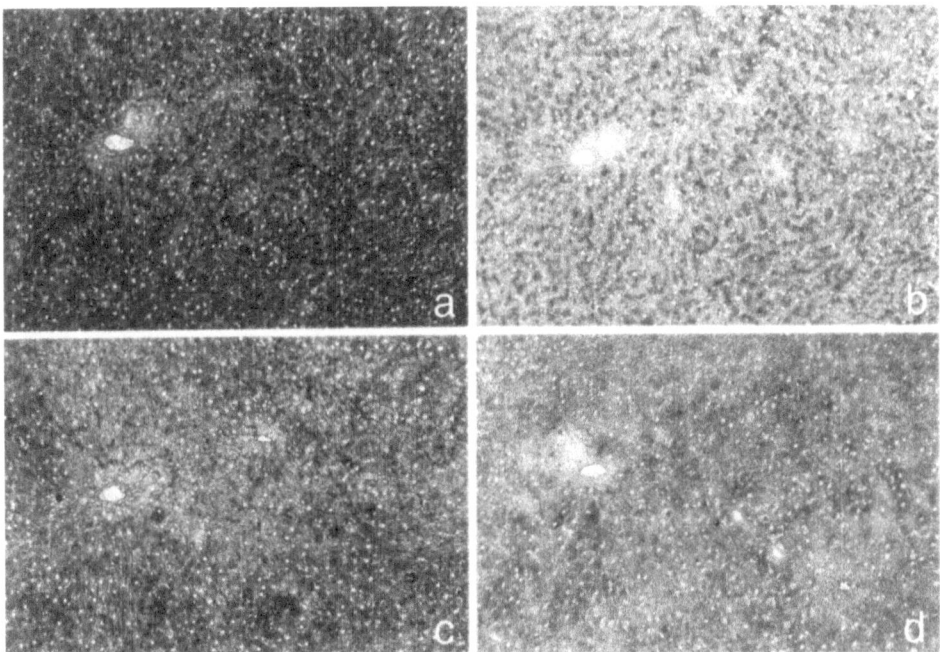

Figure 2. Diffuse reaction product of allopurinol-oxidizing enzymes was observed (a). TEI-6720 markedly suppressed the reaction (d). In contrast, benzamidine hydrochloride almost did not suppress the reaction (c). Control reaction (b) a,b,c,d: ×10 (original magnification).

REFERENCES

1. Moriwaki Y. Yamamoto T, Yamakita J, Takahashi S, Higashino K. Enzymatic histochemical localization of aldehyde oxidase in rat liver by the tetrazolium method. Acta Histochem Cytochem 30:113–115, 1997.
2. Moriwaki Y, Yamamoto T, Nasako Y, Takahashi S, Suda M, Hiroishi K, Hada T, Higashino K. *In vitro* conversion of pyrazinamide and allopurinol by rat liver aldehyde oxidase. Biochem Pharmacol 46:975–981, 1993.
3. Moriwaki Y, Yamamoto T, Yamaguchi K, Takahashi S, Higashino K. Immunohistochemical localization of aldehyde and xanthine oxidase in rat tissues using polyclonal antibodies. Histochem Cell Biol 105:71–79, 1996.
4. Komoriya K, Osada Y, Hasegawa M, Horiuchi H, Kondo S, Couch RC, Griffin TB. Hypouricemic effect of allopurinol and the novel xanthine oxidase inhibitor TEI-6720 in chimpanzees. Eur J Pharmacol 250,455–460, 1993.
5. Kojima T, Nishina T, Kitamura M, Hosoya T, Nishioka K. Biochemical studies on the purine metabolism of four cases with hereditary xanthinuria. Clin Chim Acta 137:189–198, 1984.
6. Yamamoto T, Kario K, Suda M, Moriwaki Y, Takahashi S, Higashino K. A case of xanthinuria: a study on the metabolism of pyrazinamide and allopurinol. Jap J Med 30:430–434, 1991.
7. Carpenter TO, Lebowitz RL. Nelson D, Bauer S. Hereditary xanthinuria presenting in infancy with nephrolithiasis. J Pediatr 109:307–309, 1986.
8. Yamamoto T, Moriwaki Y, Suda M, Nasako Y, Takahashi S, Hiroishi K, Nakano T, Hada T, Higashino K. Effect of BOF-4272 on the oxidation of allopurinol and pyrazinamide *in vivo*. Biochem Pharmacol 46:2277–2284, 1993.

10

SEVERE DEBILITATING POLYARTICULAR GOUT AND TERMINAL RENAL FAILURE IN AN ALLOPURINOL 'NON-RESPONDER'

S. Reiter,[1] R. Engelleiter,[2] H. Proske,[2] A. Müller,[2] F. J. van der Woude,[2] J. A. Duley,[3] and H. A. Simmonds[3]

[1]III. Medizinische Klinik
[2]V. Medizinische Klinik
Klinikum der Stadt Mannheim
[3]Purine Research Laboratory
Guy's Hospital, London

INTRODUCTION

Due to the availability of effective uricostatic (allopurinol), uricosuric (benzbromarone) and uricolytic (urate-oxidase) agents chronic polyarticular and tophaceous gout has become a very rare disease. The subject of this report came to our attention in 1993 and 1996 with severe debilitating gout and terminal renal failure despite continuous treatment with allopurinol for more than 20 years. Therefore he seemed to be an allopurinol non-responder.

CLINICAL HISTORY

This patient (1941–1996) experienced his first attack of gouty arthritis (podagra) in 1964 aged 23. He used to drink large amounts of beer (about 3 liters daily) until 1987. Allopurinol treatment was started in 1971, after a second attack of podagra, and continued in varying dosages (300–900 mg/d) until 1996 without effect on serum urate concentrations: the latter ranged from 9.4 to 15.3 mg/dl on ambulatory controls and on admissions to hospitals (Fig. 1). Subsequently gouty arthritis recurred up to once a week, and became polyarticular. Inflammation and pain were treated with colchicine, nonsteroidal anti-inflammatory drugs, phenacetine (3–6 g/d for 10 years) and opiates. Due to chronic inflammation, destruction and ankylosis of joints he was unable to walk from 1983 and confined to bed from 1985.

Purine and Pyrimidine Metabolism in Man IX,
edited by Griesmacher *et al.* Plenum Press, New York, 1998.

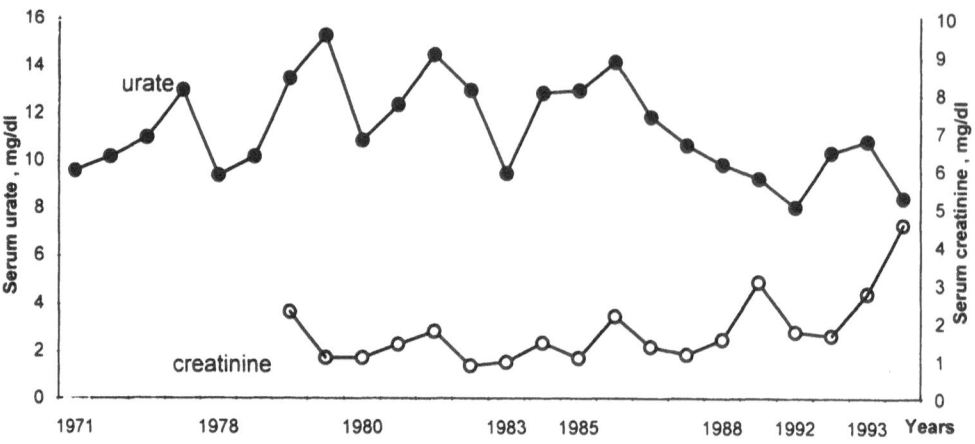

Figure 1. Serum urate and creatinine concentrations on ambulatory controls and on admissions to hospitals 1971–93. Allopurinol 300–900 mg/d.

Tophi on elbow and Achilles tendon were first noticed in 1984. Phlegmon of the big toes due to ulcerated tophi necessitated amputation in 1989 and 1992. Discharge of urate concrements from finger joints persisted until 1996.

Uric acid crystals (brick-dust) were found in the urine in 1980; repeated hematuria occurred in 1982, renal colic in 1983. Kidney stones (75–100% uric acid) passed spontaneously between 1986 and 1993. Serum creatinine levels increased slowly until 1993, then rapidly (Fig. 1). Creatinine clearances were 86ml/min (4/79), 31 ml/min (6/83), 27 ml/min (4/87) and 9 ml/min (9/93). Chronic hemodialysis was started in 10/93 and carried out with a maximum of 9hrs of dialysis per week (3 × 3 hrs).

Hypertension was documented in 1987 and was temporarily treated in 1989. Severe hypertension developed 8/93 together with renal insufficiency. Despite regular hemodialysis and combination therapy with 4 antihypertensive agents the blood pressure could not be normalized and he died of a stroke in 7/96.

METHODS

Because of preceding blood transfusions uric acid overproduction due to hypoxanthine phosphoribosyltransferase deficiency (3,5) or phosphoribosylpyrophosphate synthetase superactivity (1) had to be excluded in fibroblast culture instead of erythrocyte lysate. Oxipurinol was determined in deproteinised heparin plasma by HPLC (2). Serum urate and creatinine were measured by autoanalyzer.

Treatment of Hyperuricemia

1. Monotherapy with Allopurinol. During the first decade of allopurinol treatment the very high serum urate concentrations on ambulatory controls and on admissions to hospitals (Fig.1) could be lowered considerably during hospitalisation (e.g. 4/97 serum urate dropped from 15.3 mg/dl on admission to 11.4 mg/dl 24 hrs later without antihyperuricemic therapy, 5/81 serum urate decreased from 14.5 to 6.4 mg/dl within 6 days on 900 mg allopurinol/d, 10/84 from 13.0 to 6.9 mg/dl within 14 days on 300 mg allopurinol/d, 4/87 from 11.9 to 6.0 mg/dl within 14 days on 900 mg allopurinol/d).

From these data it might be suspected that allopurinol compliance was poor and/or that the consumption of beer (3l/day; purine content equivalent to 80–100 mg uric acid/l) had contributed much to the hyperuricemia. With the standard dose of 300 mg allopurinol/d it was not possible to lower serum urate concentration below 6.5 mg/dl, the limit of solubility in bradytrophic tissues. With 900 mg/d serum urate decreased to 6.0 mg/dl, but this dosage was not tolerated because of gastrointestinal symptoms.

With the end of beer consumption in 1987 and decreasing renal function monotherapy with 300 mg allopurinol/d influenced serum urate concentrations very little (e.g. 8–9/89 serum urate was 9.9 mg/dl on admission and 10.2 mg/dl after 7 days on 300 mg allopurinol/d). Serum creatinine ranging between 4.0 and 2.0 mg/dl at this time, allopurinol dosage was too high with respect to renal function (3) thus leading to very high serum oxipurinol concentrations: on his first admission to the V. Medizinische Klinik Mannheim 9/93 the oxipurinol concentration was 279 µM, the recommended upper level being 100 µM (3).

2. Combined Therapy with Allopurinol and Benzbromarone. The obvious ineffectiveness of allopurinol monotherapy prompted several clinicians to try combined therapy with benzbromarone (e.g. 4/79 serum urate was reduced to 4.0 mg/dl with 900 mg allopurinol plus 100 mg benzbromarone/d; 2–3/82 (Fig. 2) and 9/82 two trials with 300 mg allopurinol plus 100 mg benzbromarone/d brought serum urate down to 3.7–4.5 mg/dl).

Benzbromarone was withdrawn after each hospitalisation, possibly due to frequent gouty attacks with effective lowering of serum urate and crystal-induced renal colics or hematuria with increased excretion of uric acid.

Benzbromarone was reintroduced into therapy in 1987 but was far less effective than in 1979–1982, most likely due to decreased renal function (creatinine clearance 27 ml/min) (4).

3. Combined Therapy with Allopurinol and Hemodialysis. Hemodialysis started in 9/93 due to fluid and potassium retention with severe hypertension and recurring pulmonary edema. 200 mg allopurinol were given daily from 10/93 until 8/95 in order to increase the net elimination of urate by hemodialysis and inhibition of urate formation. With this regimen the pre-dialysis oxipurinol plasma concentration amounted to 148 µM (10/93) and within 10 months serum urate was below 6.5 mg/dl (Fig. 3). Nevertheless new tophi

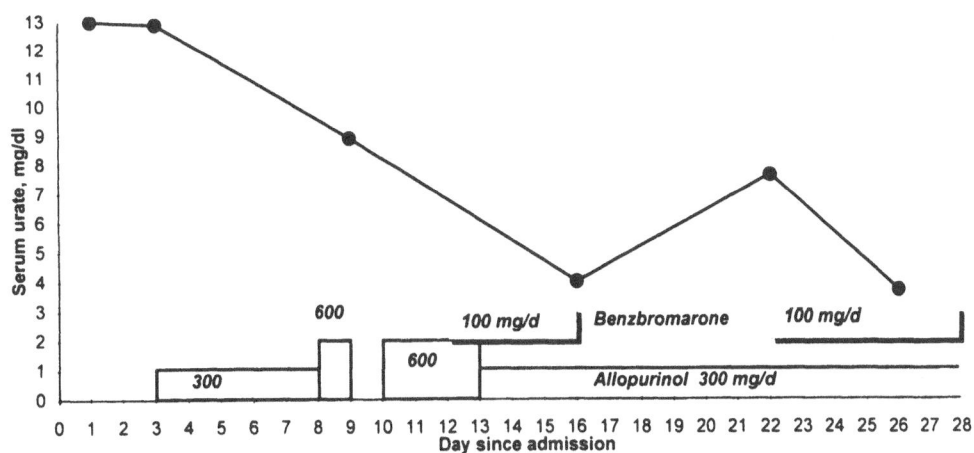

Figure 2. Serum urate concentrations on allupurinol monotherapy (300–600 mg/d) and combined therapy with allopurinol (300 mg/d) and benzbromarone (100 mg/d); hospital of Lampertheim.

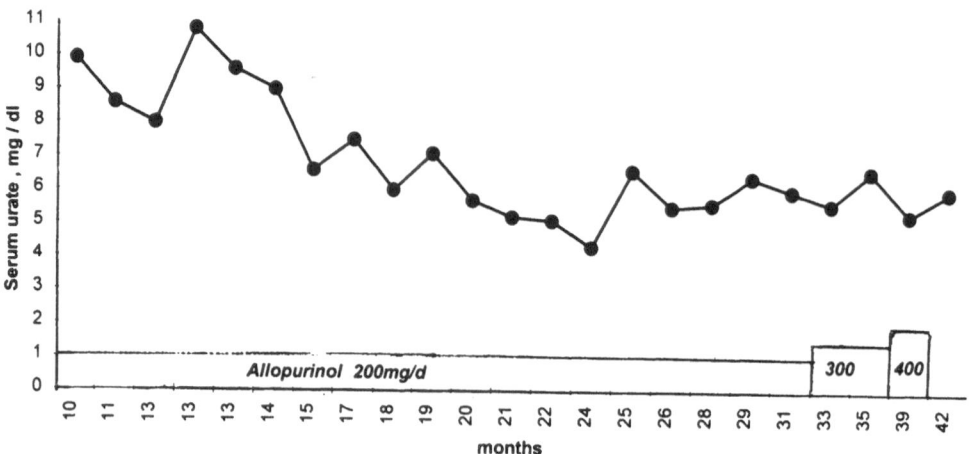

Figure 3. Serum urate concentrations (third day after dialysis) on combined therapy with allopurinol (200–400 mg/d) and ambulatory hemodialysis (3 × 3 h/w) 10/93–6/96.

appeared in 1995. Therefore allopurinol was increased to 300, then 400 mg/d without effect on serum urate concentrations (Fig. 3). We tried to optimize allopurinol treatment by giving 600 mg after hemodialysis in order to replace the eliminated amount of oxipurinol quickly, which resulted in pre-dialysis oxipurinol concentrations of about 100 μM, but in no better control of serum urate (ranging between 4.5 and 6.9 mg/dl).

The efficacy of urate elimination by hemodialysis was determined in 1996 by measuring the difference of urate concentration between influx and efflux of the hemodialyzer in intervals of 30 min. In a 4 hrs session with a pre-dialysis serum urate concentration of 8.1 mg/dl and a final post-dialysis serum urate concentration of 2.9 mg/dl a total of 2000 mg of urate were eliminated.

DISCUSSION

The devastating gout in this patient despite continuous treatment with allopurinol for more than 20 years and hemodialysis for 2.5 years seemed to suggest non- or impaired response to allopurinol (5,6). Unfortunately we were not able to prove or disprove this possibility because the patient only came to our attention with tophaceous gout in terminal renal insufficiency. Normally in tophaceous gout response to allopurinol/oxipurinol can be demonstrated by an increased excretion of the uric acid precursors hypoxanthine and xanthine (2). This was not feasible in our patient due to terminal renal failure, when such response is masked.

Non- or impaired response to allopurinol could be due to reduced absorption of allopurinol and/or oxidation to the main metabolite oxipurinol. However the high plasma concentrations of oxipurinol in this patient indicate, that absorption and metabolism of allopurinol must have been normal (2).

With increasing renal insufficiency from 1989 no effect of allopurinol on serum urate concentration could be observed, yet earlier medical reports dating back to the period before appearance of renal insufficiency and tophi (1979–84) show a clear reduction but no normalisation of serum urate concentrations on prolonged administration of

300 mg allopurinol/d (6/83, 10/84). The effect of beer withdrawal during hospitalisation might have contributed considerably to the decrease of serum urate concentrations.

The only determination of uric acid excretion on allopurinol monotherapy in this patient dates back to 4/87 (three years after the appearance of tophi): he excreted 500 mg uric acid/24hrs on 900 mg allopurinol/d, clearly more than healthy subjects on a similar dose of 500 mg/m^2 who excreted 154 ± 40 mg/24hrs on a purine free, and 283 ± 54 mg/24h on a purine rich diet (7). It is unclear whether this difference was due to an impaired response to allopurinol or to dissolution of urate from tophi.

Since xanthine and hypoxanthine were not measured, an underlying endogenous overproduction of uric acid due to hypoxanthine phosphoribosyltransferase deficiency or phosphoribosylpyrophosphate synthetase superactivity has been excluded in fibroblast culture. The only calculable renal clearance of uric acid was slightly reduced to 5.7 ml/min in 4/87 (normal range 7–9 ml/min).

From these data it might be concluded that the severe gout of this patient was due to an exogenous overload with purines (beer), a slightly reduced renal clearance of uric acid and an impaired response to allopurinol requiring high doses to reduce serum urate concentrations and urinary excretion of uric acid. The response to allopurinol was further impaired by decreasing renal function due to phenacetine nephropathy. The brisk response to benzbromarone suggests that the sad fate of this patient could have been avoided by early combination therapy with allopurinol plus benzbromarone adjusting serum urate to 5 mg/dl. Such a high excretion of uric acid due to the exogenous purine load, and apparently impaired response to allopurinol coupled with the uricosuric agent would require continuous neutralisation of urine pH to prevent urolithiasis.

ACKNOWLEDGMENTS

We are most grateful to Mrs. Herbst, Hautklinik, Klinikum der Stadt Mannheim for preparing the fibroblast cultures, and to Dr. Boppert, Marienkrankenhaus Lampertheim, Prof. Dr. Glückert, Rheumaklinik Wiesbaden, Dr. Leiling, Bürstadt, and Dr. Decker, Mannheim, for giving insight into their medical reports concerning our patient.

REFERENCES

1. Torres RJ, Mateos FA, Puig JG, Becker MA: A simplified method for the determination of phosphoribosylpyrophosphate synthetase activity in hemolysates. Clin Chim Acta 224, 55–63 (1994).
2. Reiter S, Simmonds HA, Zöllner N, Braun SL, Knedel M: Demonstration of a combined deficiency of xanthine oxidase and aldehyde oxidase in xanthinuric patients not forming oxipurinol. Clin Chim Acta 187, 221–234 (1990).
3. Simmonds HA, Cameron JS, Morris GS, Davies PM: Allopurinol in renal failure and the tumour lysis syndrome. Clin Chim Acta 160, 189–195 (1986).
4. Zürcher RM, Bock HA, Thiel G: Excellent uricosuric efficacy of benzbromarone in cyclosporin-A-treated renal transplant patients: a prospective study. Nephrol Dial Transplant 9, 548–551 (1994).
5. Simmonds HA, Gibson T, Huston GJ, Webster DR, Rodgers AV, Munro J: Gout resistant to allopurinol: poor compliance or non-response. Adv Exp Med Biol 165A, 171–174 (1984).
6. Löffler W, Bennhold I, Simmonds HA, Reiter S, Zöllner N: Impaired response to allopurinol (IRA) with high plasma oxipurinol. Int J Purine Pyrimidine Res 2, Suppl 1, 60 (1991).
7. Löffler W, Reiter S, Gröbner W, Zöllner N: Dose dependency of the uric acid lowering effects of allopurinol in the presence and absence of dietary purines. Adv Exp Med Biol 195A, 441–446 (1986).

11

DECREASED SERUM CONCENTRATIONS OF 1,25(OH)2-VITAMIN D3 IN PATIENTS WITH GOUT

Sumio Takahashi, Tetsuya Yamamoto, Yuji Moriwaki, Zenta Tsutsumi, Jun-ichi Yamakita, and Kazuya Higashino

Third Department of Internal Medicine
Hyogo College of Medicine
Mukogawa-cho 1-1
Nishinomiya, Hyogo 663, Japan

SUMMARY

We measured the serum concentrations of 1,25(OH)2-vitamin D3, 25(OH)-vitamin D3, parathyroid hormone (PTH) in 82 male patients with primary gout whose serum uric acid was significantly higher than that of 41 normal control male subjects (8.8 ± 0.2 vs 5.6 ± 0.2 mg/dL, $p < 0.001$). The serum 1,25(OH)2-vitamin D3 concentration was significantly lower in the patients with gout compared with the control subjects (39.6 ± 1.4 vs 44.8 ± 1.7 pg/mL, $p < 0.05$), while no differences were observed between the two groups in either the serum concentration of 25(OH)-vitamin D3 or PTH. The administration of uric acid lowering agent to the patients for 1 year caused a significant increase in their serum 1,25(OH)2-vitamin D3 concentration which was associated with a significant decrease in their serum uric acid concentration. In contrast, the serum concentrations of 25(OH)-vitamin D3 and PTH were not affected by these drugs. These results suggest that uric acid per se may directly decrease the serum concentration of 1,25(OH)2-vitamin D3 in patients with gout by inhibiting 1-hydroxylase activity.

INTRODUCTION

Uric acid has been reportedly suppresses 1,25(OH)2-vitamin D3 synthesis in patients with renal failure and hyperuricemia,[1] and their serum active vitamin D levels were inversely related to their serum uric acid,[2] suggesting a possible relationship between vitamin D and uric acid. However, the effect of uric acid on vitamin D metabolism in patients with gout, who have hyperuricemia, is unclear. Therefore, we determined the serum concentrations of 1,25(OH)2-vitamin D3 together with 25(OH)-vitamin D3, and PTH in

Purine and Pyrimidine Metabolism in Man IX,
edited by Griesmacher *et al.* Plenum Press, New York, 1998.

patients with gout, and also examined the effect of a decrease in the serum concentration of uric acid on the serum concentration of 1,25(OH)2-vitamin D3 in patients with gout.

SUBJECTS AND SAMPLE COLLECTION

The study included 82 male patients with primary gout aged 46.7 ± 1.3 years and 41 healthy male subjects aged 45.4 ± 1.8 years. The study was performed during the same season to minimize any seasonal changes in vitamin D levels.[3,4] During the study, all subjects were ambulatory and on their usual diet, and no instructions on their dietary or fluid intake were given. Blood samples were drawn after an overnight fast, and the serum was separated for the determination of 1,25(OH)2-vitamin D3, 25(OH)-vitamin D3, PTH, uric acid concentrations. In addition, 24-h urine samples were collected for the determination of the urinary excretion of uric acid and creatinine. Allopurinol (100 to 200 mg/day) was administered to 23 patients who excrete uric acid more than 800 mg/day in urine. The serum concentrations of uric acid, 1,25(OH)2-vitamin D3, 25(OH)-vitamin D3 and PTH were measured before and after therapy.

ANALYTICAL TECHNIQUES

The serum and urinary uric acid concentrations were measured by the uricase method. The 1,25(OH)2-vitamin D3 concentration was measured by a radioreceptor assay. The 25(OH)-vitamin D3 concentration was determined by a competitive protein assay. The PTH concentration was determined by a radioimmunoassay.

STATISTICAL METHODS

Data are expressed as the mean $\pm$ SE. Observed differences were tested by Student's t test for significance. Comparisons of the serum uric acid and 1,25(OH)2-vitamin D3 levels before and after uric acid lowering therapy were assessed by the two-tailed paired t test. A p value < 0.05 was considered to be significant.

RESULTS

The serum concentration of 1,25(OH)2-vitamin D3 was significantly lower in the patients than in the control subjects (39.6 ± 1.4 vs 44.8 ± 1.7 pg/mL, $p < 0.05$). However, the 25(OH)-vitamin D3 and PTH serum values were not different between the two groups (24.7 ± 0.7 vs 24.3 ± 1.0 ng/mL and 355.9 ± 14.3 vs 335.6 ± 15.8 pg/mL, respectively). Despite a difference in the serum 1,25(OH)2-vitamin D3 concentration, the serum concentrations of calcium and phosphate were not different between the two groups (Ca: 9.0 ± 0.0 vs 8.9 ± 0.0 mg/dL, P: 3.0 ± 0.1 vs 2.9 ± 0.1 mg/dL) (Table 1). The administration of allopurinol caused a significant increase in the serum concentration of 1,25(OH)2-vitamin D3 (from 34.7 ± 1.4 to 39.5 ± 2.0 pg/mL, $p < 0.05$) which was associated with a significant reduction in the serum uric acid levels (from 8.7 ± 1.0 to 6.5 ± 1.2 mg/dLl, $p < 0.001$) (Fig. 1). However, allopurinol did not change the serum concentrations of 25(OH)-vitamin D3 and PTH(data not shown).

Table 1. Clinical features and laboratory data of the subjects

	Primary gout (n=82)	Control (n=41)
Age (years)	46.7 ± 1.3	46.4 ± 1.8
Body mass index (kg/m^2)	24.9 ± 0.4	23.7 ± 0.4
Alcohol intake (g/day)	27.1 ± 2.8	20.4 ± 2.9
S-UA (mg/dl)	8.8 ± 0.2**	5.6 ± 0.2
S-Cr (mg/dl)	0.96 ± 0.01	0.94 ± 0.01
S-Ca (mg/dl)	9.0 ± 0.0	8.9 ± 0.0
S-Pi (mg/dl)	3.0 ± 0.1	2.9 ± 0.1
Cua (ml/min.)	5.59 ± 0.16**	9.25 ± 0.63
Ccr (ml/min.)	110.7 ± 2.3	111.1 ± 4.5
PTH (pg/ml)	355.9 ± 14.3	335.6 ± 15.8
25(OH)D$_3$ (ng/ml)	24.7 ± 0.7	24.3 ± 1.0
1,25(OH)$_2$D$_3$ (pg/ml)	39.6 ± 1.4*	44.8 ± 1.7

Values are expressed as mean ± SE.
*p < 0.05, **p < 0.001.

DISCUSSION

The present study demonstrated that the serum 1,25(OH)2-vitamin D3 level was significantly lower in the patients with gout than control, but that the serum 25(OH)-vitamin D3 and PTH levels were not different between the two groups. Furthermore, we found a significant rise in the 1,25(OH)2-vitamin D3 level associated with a reduction in the serum uric acid concentration after allopurinol administration. In addition, it was demonstrated that allopurinol did not affect the clearance or production rate of 1,25(OH)2-vitamin D3 in rat.[1] Therefore, it is conceivable that uric acid itself decreases the serum concentration of 1,25(OH)2-vitamin D3 in patients with gout.

The most important clinical manifestations of vitamin D deficiency are related to the development of renal osteodystrophy, rickets in children, and secondary hyperparathyroidism. However, to our knowledge there have been no studies suggesting a high incidence of

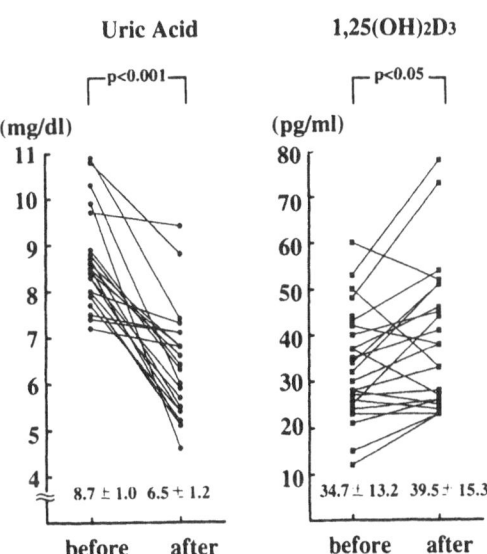

Figure 1. Changes in the serum uric acid and 1,25(OH)2-vitamin D3 levels in gout patients before and after allopurinol therapy. The serum 1,25(OH)2-vitamin D3 levels were significantly increased and were associated with a decrease in the uric acid concentration (mean ± SE).

osteoporosis in patients with gout. Furthermore, the serum 1,25(OH)2-vitamin D3 concentration in patients with gout was not markedly decreased compared to that in those with renal insufficiency in whom osteodystrophy is frequently observed due to a derangement of vitamin D metabolism. Therefore, the decrease in the serum 1,25(OH)2-vitamin D3 concentration in patients with gout may not be clinically important in this regard.

However, a substantial amount of information has accumulated that vitamin D metabolism is related to diabetes mellitus, impaired glucose tolerance,[3] obesity,[5,6] and hypertension,[7,8] all of which are frequently observed in patients with gout. Therefore, further studies are required to clarify the association, if any, between 1,25(OH)2-vitamin D3 levels and these conditions in patients with gout.

REFERENCES

1. Vanholder R, Patel S, Hsu CH: Effect of uric acid on plasma levels of 1,25(OH)2D3 in renal failure. J Am Soc Nephrol 4:1035–1038, 1993.
2. Lind L, Hanni A, Lithell H, Hvarfner A, Sorenson OH, Ljunghall S: Vitamin D is related to blood pressure and other cardiovascular risk factors in middle-aged men. Am J Hypertens 8:894–901, 1995.
3. Scragg R, Holdaway I, Singh V, Metcalf P, Baker J, Dryson E: Serum 25-hydroxyvitamin D3 levels decreased in impaired glucose tolerance and diabetes mellitus. Diabetes Res Clin Pract 27:181–188, 1995.
4. Stamp TCB, Round JM: Seasonal changes in human plasma levels of 25-hydroxyvitamin D. Nature 247:563–565, 1974.
5. Kerstetter J, Caballero B, O'Brien K, Wurtrman R, Allen L: Mineral homeostasis in obesity: effect of euglycemic hyperinsulinemia. Metabolism 40:707–713, 1991.
6. Liel Y, Ulmer E, Shary J, Hollis BW, Bell NH: Low circulating vitamin D in obesity. Calcif Tissue Int 43:199–201, 1988.
7. Scragg R, Holdaway I, Jackson R, Lim T: Plasma 25-hydroxyvitamin D3 and its relation to physical activity and other heart disease risk factors in the general population. Ann Epidemiol 2:697–703, 1992.
8. Kokot F, Pietrek J, Srokowska S, et al: 25-hydroxyvitamin D in patients with essential hypertension. Clin Nephrol 16:188–192, 1981.

12

LIPOPROTEINS IN PATIENTS WITH ISOLATED HYPERURICEMIA[*]

F. J. Tinahones,[1,†] F. Vazquez,[1] F. J. Soriguer,[1] and E. Collantes[2]

[1]Section of Endocrinology
Regional hospital of Málaga
Málaga, Spain
[2]Section of Rheumatology
Reina Sofía Regional hospital
Córdoba, Spain

SUMMARY

Background

There have been described abnormalities in the lipoprotein profile of hyperuricemic patients, it has not been clarified wether these abnormalities are due to the hyperuricemia or to the dyslipema often associated to these patients. Our aim is to study the apolipoprotein profile in hyperuricemic patients without dyslipemia compared to a control population.

Patients and Methods

30 hyperuricemic patients and 26 healthy controls.Measurements were of blood uric acid, total cholesterol, total triglycerides, creatinine, HDL-C, and VLDL cholesterol, trigliceride, Apo B, Apo CII and Apo CIII (1 and 2). Uric acid clearance and fractionated excretion were measured in 24 h. urine samples.

Results

No significant differences were found between hyperuricemic and control patients in cholesterol, triglycerides and apo B in VLDL, or LDL and HDL cholesterol. The levels of apo B, Apo AI levels and apo CIII/apo CII were similar in the hyperuricaemic and controls.

* This work was supported by grant from Junta de Andalucia (96/212).
† Correspondence to: F.J. Tinahones, c/ Ayala nº 28 4ª 6º A, Málaga 29002, Spain.

Purine and Pyrimidine Metabolism in Man IX,
edited by Griesmacher *et al.* Plenum Press, New York, 1998.

Conclusions

There are two types of hyperuricaemic patients, one group associated to hyper-lipideamia and would be included in the X Syndrome. The other group not associated to other metabolic abnormalities. Is important to distinguish between these two groups to define the prognosis of a given patient because the greater cardiovascular risk linked hyperuircaemic patients could be related to the association to others cardiovascular risks factors.

INTRODUCTION

Gout and hyperuricemia are frequently associated to several metabolic abnormalities, among them hyperlipemia and diabetes (1,2). From the moment that the association between hyperuricemia and hyperlipemia was described there has been a lot of discussion aimed to define the type of hyperlipemia thas is associated to hyperuricemia. Some studies have linked hyperuricemia to hypercholesterolemia (3), but most authors have found a relationship between hyperuricemia and hypertriglyceridemia, as an isolated hyper-triglyceridemia (5,6) or together with hypercholesterolemia (4).

Several hypothesis have been proposed to explain this association, one of them suggested the presence of exogenous concomitant factors like diet and obesity (7,8), nevertheless some studies have found the association between hyperuricemia and hyperlipemia with abscense of these exogenous factors (9).

In patients with hyperuricemia lipoprotein and apoprotein abnormalities have been described. Decreased levels of HDL and increased levels of VLDL have been repeatedly reported (10–12). About the levels of LDL the results are not clear (13). Few studies have focused on the apoprotein disturbances of the lipoproteins in hyperuricemic patients. Ulreich *et al.* (10) report an increase in Apo B as well as in Apo AII values. Macfarlane *et al.* (14) find a significant increase in Apo CIII and a decrease in Apo CII and in the CII/CIII ratio.

In hyperuricemic hyperlipemic patients abnormalities in apoproteins as well as in lipoproteins have been found (15), but wether these changes are present in pure hyperuricemic patients is not known. The aim of the present study is to determine the apolipoprotein profile in hyperuricemic patients without hyperlipemia.

PATIENTS AND METHODS

Subjects and Study Design

30 males diagnosed of primary isolated hyperuricemia (HU) and 26 healthy males as controls.

Hyperuricemic patients were selected in a consecutive manner from patients attending to the Section of Reumathology of the Cordoba Reina Sofia Regional Hospital that met the following criteria: 1) Plasma uric acid above 7 mg/dl and plasma cholesterol and triglycerides below 220 md/dl, 2) primary isolated hyperuricemia, 3) a commitment to follow a low purine diet for a week, 4) patients that did not normalize the uric acid levels after the above mentioned diet, 5) no medical treatments, 6) no altered carbohidrate metabolism, 7) no alcohol consumption.

The control group were healthy volunteers that met the following criteria: 1) Plasma uric acid below 7 mg/dl and plasma cholesterol and triglycerides below 220 md/dl, 2) no previous disease or medical treatment, 3) a commitment to follow a low purine diet for a week before blood was collected, 4) no altered carbohidrate metabolism, 5) no alcohol consumption.

Laboratory Measurements

Blood determinations were drawn after 12 hour fasting: uric acid (by uricase method), total cholesterol (by enzimatic method), total triglycerides (by enzimatic method), creatinine, HDL-cholesterol (by phosphotungstic acid precipitation method), total apoprotein AI and B (by nephelometry).

VLDL Separation

The VLDL fractions were isolated by sequential ultracentrifugation in a (Bekcman XL-90) ultracentrifuge equipped with a 90Ti rotor after 18 hours at 40,000 rpm.

Apoprotein C Separation

VLDL were delipidated by the Scanu's technique (16). Precipitated apoproteins were reconstituted by using a freshly prepared solution of 0.01 M Tris, 0.25% Nodinet (NP-40), 8 M urea, and 64.8 mM dithithreitol (DTT) solution. Apo C were separated by isoelectric focusing (IEF LKB Bromma) in acrylamide, methylenebisacrylamide, urea, ampholyte pH 4–6.5 and glycerol gels. The apoproteins bands were identified by their isoelectric point and were quantified in a densitometer at 633 nm. The apo CIII2/CII and apo CIII1+ 2/CII ratios were calculated.

Urine Analysis

Fractional excretion, excretion and clearance of uric acid were measured from 24 hour urine samples after a week following a low purine diet.

Statistical Analysis

To evaluate differences between the two groups Student's t test for non paired data or the Mann-Whitney test were used. The Shapyro test was used to determine the normality of the parameters.

RESULTS

Age (controls: 46.9 ± 15.6 and HU: 46.2 ± 12.1) and BMI (controls: 27.4 ± 4.2 and HU: 28.7 ± 4.1) did not differ significantly between the two groups controls and HU patients.

The mean levels of plasma uric acid in the HU group was 7.7 ± 0.8 mg/dl and in the control group 5.1 ± 1.1 mg/dl (p < 0.001). The parameters that measured the renal excretion of uric acid showed significant differences between the controls and the hyperuricemic patients, hyperuricemic patients had a decreased renal clearance and fractional excretion of uric acid with a p value < 0.001 for both parameters (Table 1).

Table 1. Uric acid plasma levels and renal excretion of uric acid

	Hyperuricemic	Control
Plasma uric acid (mg/dl)	7.7 ± 0.85	5.1 ± 1.1***
Uric acid renal clearance (ml/min)	7.00 ± 2.49	12.2 ± 4.46***
Fractional excretion of uric acid (%)	6.15 ± 1.76	10.4 ± 3.30***

Mean values ± standard deviation
*** p < 0.001

Mean values of cholesterol in the HU and control groups exhibited very similar levels and the same happened for the mean levels of triglycerides. Plasma apoprotein AI levels were slightly increased in the control group (149 ± 29.7 mg/dl) compared to the HU group (139 ± 23.2 mg/dl) but these differences were not significant. Total apoprotein B did not show significant differences either (Table 2).

The LDL-cholesterol mean values in both groups were similar (114 ± 21.2 mg/dl HU and 117 ± 22.9 mg/dl controls). HDL-cholesterol was lower in the HU group (42.9 ± 8.8 mg/dl vs 50 ± 15 mg/dl for the control group) but the great variability in the control group HDL-cholesterol results made these differences not significant. The VLDL were measured the HU patients did not show any significant compared to the controls (Table 2).

Measuring the apoprotein C of the VLDL the CIII/CII ratio did not show significant differences (controls CIII/CII ratio 2.39 ± 1.6 and HU 1.87 ± 0.7) (Table 2).

The cardiovascular risk ratios, cholesterol/ HDL-cholesterol and ApoB/ApoAI, did not show any significant differences. In both groups the ApoB/ApoAI ratio was lower than 1 (HU 0.78 ± 0.2 and controls 0.65 ± 0.2) and the total cholesterol/ HDL-cholesterol ratios were between 3.5 and 4.5 in the two groups (HU 4.3 ± 0.9 and controls 3.83 ± 1.1) (Table 2).

DISCUSSION

The HDL-cholesterol levels in the hyperuricemic patients group did not show any significant differences compared to the controls. Some authors have described a decrease in the HDL-cholesterol levels in patients with gout (10–12), but most studies did not select

Table 2. Mean values ± standard deviation of the hyperuricemic and control groups

	Hyperuricemic	Control
Cholesterol (mg/dl)	178 ± 19.4	180 ± 19.3
Triglycerides (mg/dl)	118 ± 35.3	95.3 ± 41.5
HDL-cholesterol(mg/dl)	42.9 ± 8.85	50.0 ± 15.0
LDL-cholesterol (mg/dl)	114 ± 21.2	117 ± 22.9
ApoAI (mg/dl)	139 ± 23.2	749 ± 29.7
ApoB (mg/dl)	105 ± 20.2	93.8 ± 23.8
VLDL-triglycerides (mg/dl)	63.9 ± 26.8	50.4 ± 46.2
VLDL-cholesterol (mg/dl)	18.5 ± 9.04	13.3 ± 7.43
VLDL-ApoB (mg/dl)	13.3 ± 7.97	9.79 ± 7.44
ApoCIII2+CIII1/ApoCII	5.10 ± 3.39	4.14 ± 1.25
ApoCIII2/CII	2.39 ± 1.65	1.87 ± 0.71
AoB/ApoAI	0.78 ± 0.21	0.65 ± 0.23
Cholesterol/HDL-cholesterol	4.30 ± 0.94	3.89 ± 1.16

patients with isolated hyperuricemia as this study does. Ulreich *et al.* (10) selected a population with gout with mean total cholesterol levels of 233 mg/dl, they evaluated a population that were hypercholesterolemic in most cases and found significant differences in the total cholesterol/HDL-cholesterol ratio when compared to a control group that was normocholesterolemic. When we examine the study by Jacobelli et al. (11) we also find that they select a gouty population with total cholesterol levels significantly increased (p < 0.01). Matsubara *et al.* (12) also described a decrease in the two most important HDL-cholesterol subfractions (HDL2 and HDL3), but they did not select their population either and they compared patients with gout who had elevated total triglycerides levels, when it is well known that hypertriglyceridemia is associated to low levels of HDL-cholesterol (17). This idea of decreased levels of HDL in gouty patients is not shared by all authors. Darlington *et al.* (13), in a group of patients with gout, did not find significant differences in HDL-cholesterol. Other studies using logistic regression methods have shown that the decrease in HDL-cholesterol in hyperuricemic patients could be explained by the hypertriglyceridemia itself (18).

When we look at the LDL our results are similar to other authors, not finding differences in the LDL-cholesterol levels of hyperuricemic patients when compared to controls (10–12) nevertheless increments (19,20) and decrements (13) of this lipoprotein have been described. All the these studies performed in gouty or hyperuricemic patients have not ruled out the cases associated to hyperlipemia, the discrepancies in their findings could be explained becouse of the great variability of hyperlipemia fenotipes present in hyperuricemic patients (21).

When we compare the VLDL composition we do not find any lipoproteic alteration in the hyperuricemic patients when compared to controls, other authors have reported increments in VLDL in gouty patients (13,19) although they did not select pure hyperuricemic patients.

There are no differences in the apoprotein levels in the hyperuricemic patients when compared to controls. The decrease in the ApoCII/CIII ratio reported in hyperuricemic patients (14) has been linked only to hyperuricemic hypertriglyceridemic patients in other studies (22).

There were no significant differences in the cardiovascular risks ratios: the Cholesterol/HDL-cholesterol ratio nor in the apoB/apoAI ratio. The ApoB/apoAI ratio has been used by several authors who report that the levels of apoproteins B and A have a predictive value for coronary heart disease equal or even higher than total cholesterol and that their predictive value is higher for population older than 50 years (23). Apopotein AI can be considered as a protective factor for coronary heart disease and on the other hand apoprotein B can be considered as a predictive marker for a future coronary event (24). Avogaro (25) considers the apoB/apoAI ratio more discriminative than isolated apoB and apoAI levels taken separatedly. Thus we can agree that our two populations had a similar cardiovascular risk derived from their similar lipid distribution.

We found differences between hyperuricemic patients and controls in the parameters of renal excretion of urates, these parameters were significantly lower in the hyperuricemic patients. It is known that 90% of primary hyperuricemias are caused by renal underexcretion of urates (26).

Hyperuricemia has been included as one of the components of the metabolic syndrome (27). Hiperuricemic patients showed hyperinsulinism when compared to normouricemic controls but the patients studied also had higher levels of triglycerides and cholesterol (28). A negative correlation has been described between insulin sensitivity and uric acid levels but in this study this correlation was present for the triglycerides levels too (29).

In conclusion, patients with isolated hyperuricemia have a lipoproteic profile similar to healthy normometabolic controls. Therefore it is possible that there are two different types of hyperuricemia, one associated to dyslipemia that could be included in the metabolic syndrome complex and another hyperuricemia that would not be associated to any other metabolic disturbance. To differenciate both types is important in order to define future risks becouse a higher cardivascular risk has been attributed to hyperuricemic patients (30), this higher risk could be modulated by its association to other cardiovascular risk factors (31).

REFERENCES

1. Berkowitz D. Gout hyperlipidemia and diabetes interrelationship. JAMA 1966;197:117–120.
2. Wiedmann E, Rose HG, Schwartz E. Plasma lipoproteins, glucose tolerance and insulin response in primary gout. Am J Med 1972;53:299–307.
3. Becker JH. Gout and serum cholesterol. Wis Med J 1960; 59:735.
4. Mielants H, Veges EM, Weerdy AD. Gout and its relation to lipid metabolism .I serum uric acid, lipid and lipoprotein levels in gout. Ann Rheum Dis 1973; 32: 501–505.
5. Emmerson BT, Knowles BR. Triglyceride concentrations in primary gout and gout of chronic lead nephropathy. Metabolism 1971; 20:721–729.
6. Collantes E, Añón J, Tinahones FJ, Sánchez P. Perfil lipídico asociado a hiperuricemia asintomática. Estudio protocolizado de 77 individuos en nuestro medio. Rev Esp Reumatol 1989; 16:129–131.
7. Faller J, Fox IH. Ethanol induced hyperuricemia: evidence for increased urate production by activation of adenine nucleotide turnover. N Engl J Med 1982; 307:1598–1603.
8. Gibson T, Grahame R. Gout and hyperlipidaemia. Ann Rheum Dis 1974; 33:298–303.
9. Collantes E, Pineda M, Añon J, Sanchez-Guijo P. Hyperuricemia-hyperlipemia association in the abscence of obesity and alcohol abuse. Clin Rheumatol 1990; 9:28–31.
10. Ulreich A, Korner GM, Pfeiffer KP. Serum lipids and lipoproteins in patients with primary gout. Rheumatol Int 1985; 5:73–77.
11. Jacobelli S, Arteaga A, Bidegain F. Cholesterol distribution among lipoprotein fractions in patients with gout and normal controls. J Rheumatol 1986; 13:774–777.
12. Matsubara K, Matsuzawa Y, Jiao S, Takama T, Masaharu K, Tami S. Relationship between hypertriglyceridemia and uric acid production in primary gout. Metabolism 1989; 38: 689–701.
13. Darlington L, Slak J, Scott JT. Family study of lipid and purine levels in gout patients. Ann Rheum Dis 1982; 41:253–256.
14. MacFarlane DG, Midwinter CA, Dieppe PA, Bolton CH, Hartg M. Demostration of an abnormality of C apoprotein of very low density lipoprotein in patients wiyh gout. Ann Rheum Dis 1985; 44:390–394.
15. Tinahones FJ, Collantes E, Gonzalez A, C-Soriguer F, Añón J, Sánchez P. Increased VLDL levels and disminished renal excretion of uric acid in hyperuricemic-hypertrigliceridemic patients.Br J Rheumatol 1995; 34: 920–924.
16. Scanu AM, Edelstein C. Solubility in aqueous solution or ethanol of the small molecular weight peptides of serum very low density and high density lipoproteins relevant to the recovery problem during delipidation of serum lipoprotein. Anal Biochem 1971; 44:576–588.
17. Deckel RJ. Plasma triglyceride determine structure composition in low and high density lipoproteins. Arteriosclerosis 1984; 4:225–231.
18. Takahashi S, Yamamoto T, Moriwaki Y, Tsutsumi Z, Higashino K. Impaired lipoprotein metabolism in patients with primary gout-influence of alcohol intake and body weight. Br J Reumatol 1994; 33:731–734.
19. Bouvenot G, Bartolin R, Sciara M, Delboy C, Arnaud C. Serum uric acid and serum lipids. Statistical correlations report of 1000 cases. Sem Hop Paris 1980; 56:263–264.
20. Kullich W, Ulreich A, Klein G. Hyperlipoproteinemia in primary gout and asymptomatic hyperuricemia. Wien Med Wochenschr 1988; 138:221–225.
21. Collantes E, Tinahones FJ, Cisnal A, Añón J, Sánchez-Guijo P. Variability of lipid phenotypes in hyperuricemic-hyperlipidemic patients. Clin Rheum 1994;13:244–247.
22. Collantes E, Tinahones FJ, Pineda M, Soriguer F, Sánchez P. Apoprotein C in hyperuricemic patients. Br J Rheum 1992;31(Suppl):21.
23. Thompson G. Apoproteins: determinants of lipoproteins metabolism and indices of coronary rick. Br Heart J 1984; 51:585–588.

24. Sveger T, Fex G. Apolipoproteins AI and B levels in adolescente. a trial to define subjects at risk for coronary heart disease. Acta Pediatr Scand 1983; 72:499–504.

25. Avogaro P. Lipid and lipoproteins of lipoproteins in humans atherosclerosis. En: Halpen MJ, editores. Lipid metabolism and its pathology. New York: Pleum Press, 1985:17–23.

26. Willian N, Kelley MD. Gout and Disorders of Purine Metabolism. En: Willian N, Kelley MD, Edward DH, Sham R, Clement BS, editores.Textbook of Rheumatology. Filadelfia: Sanders, 1989.

27. Moller DE, Flier JS. Mechanisms of disease: insulin resistance-mechanisms, syndromes and implications. N Engl J Med 1991; 325:938–948.

28. Zavaroti I, Mazza S, Fantuzzi M, Dall'Aglio E, Bonora E, Delsignore R, Passeri M, Reaven GM. Changes in insulin and lipid metabolism in males with asymptomatic hyperuricaemia. J Int Med 1993; 234:25–30.

29. Vuorinen-Markkola H, Yki-Järvinen H. Hyperuricemia and insulin resistance. J Clin Endocrinol Metab 1994; 78:25–29.

30. Fessel J. High uric acid as an indicator of cardiovascualr disease independence from obesity. Am J Med 1980; 68:401–404.

31. Brand. Hyperuricemia as a risk factor of coronary heart disease: The Framinghans study. Ann J Epidemiol 1985; 121:11–18.

13

ATHEROGENIC RISK FACTORS IN PATIENTS WITH GOUT

Zenta Tsutsumi, Tetsuya Yamamoto, Sumio Takahashi, Yuji Moriwaki,
Jun-ichi Yamakita, Yumiko Nasako, and Kazuya Higashino

Third Department of Internal Medicine
Hyogo College of Medicine
Mukogawa-cho 1-1
Nishinomiya, Hyogo 663, Japan

INTRODUCTION

In recent years, atherosclerotic diseases such as ischemic heart disease have become an important cause of death in Japanese patients with gout.[1] However, since it remained undetermined whether or not hyperuricemia per se is an independent risk factor for atherosclerosis, we investigated the possible risk factors for atherosclerosis including serum lipids in patients with gout.

SUBJECTS AND METHODS

Age-matched male patients with primary gout and healthy male adults with normal lipid profiles were included in the study (Table 1). After informed consent was obtained, any medication affecting serum lipid profiles was withheld at least two months before the study. Subjects who had diseases known to affect lipid metabolism such as hypothyroidism, renal failure, and hepatic diseases were excluded from the study. After an overnight fast, blood was drawn for the determination of serum concentrations of lipids, apolipoproteins, and other known risk factors for atherosclerosis. Visceral fat area was measured by a computed tomography at the level of umbilicus.[2] LPL and HTGL activities were measured as described previously.[3] LDL was isolated by sequential ultracentrifugation.[4] LDL particle size was determined using nondenaturing polyacrylamide gradient gel electrophoresis.[5]

Purine and Pyrimidine Metabolism in Man IX,
edited by Griesmacher *et al.* Plenum Press, New York, 1998.

Table 1. Clinical features of the subjects

	Gout (n=147)	Control (n=64)	p value
Age (yrs)	48.8 ± 12.8	47.0 ± 11.5	NS
S-UA (mg/dL)	8.8 ± 1.4	6.9 ± 2.3	p < 0.001
Alcohol intake (g/day)	31.2 ± 29.7	21.1 ± 19.7	p < 0.05

Values are expressed as mean ± SD.
NS: not significant.

RESULTS

Lipid Profiles

Serum triglyceride, apolipoprotein B, C-II, C-III and E were significantly higher in patients with gout than in control subjects (p < 0.01, p < 0.05, p < 0.05, p < 0.05 and p < 0.05, respectively), while HDL-cholesterol level was significantly lower in patients with gout than in controls (p < 0.05). Serum concentration of total cholesterol was not different between the two groups. Plasma concentration of Lp(a) was significantly higher in patients with gout than in control subjects (median: 19.8 mg/dL vs 13.0 mg/dL; p < 0.05). The median LDL particle diameter in patients with gout was not different from that in control subjects (26.5 nm vs 27.0 nm). However, the incidence of small-sized LDL particles (less than 27.3 nm) was more frequent in patients with gout as compared with control subjects (p < 0.05) (data not shown). The activities of postheparin plasma lipoprotein lipase (LPL) and hepatic triglyceride lipase (HTGL) were significantly lower in patients with gout (LPL: 5.43 ± 2.04 U vs 7.93 ± 2.97 U, p < 0.01; HTGL: 13.42 ± 6.57 U vs 17.85 ± 9.86 U, p < 0.05). Serum concentration of lipid peroxide was significantly lower in patients with gout than controls (Table 2).

Table 2. Lipid profiles of the subjects

	Gout (n=147)	Control (n=64)	p value
Triglyceride (mg/dL)	190 ± 122	136 ± 72	p < 0.01
HDL-cholesterol (mg/dL)	45 ± 11	51 ± 11	p < 0.05
Total cholesterol (mg/dL)	219 ± 40	213 ± 36	NS
apo-A1 (mg/dL)	133 ± 26	132 ± 20	NS
apo-AII (mg/dL)	39.0 ± 8.6	37.6 ± 6.5	NS
apo-B (mg/dL)	110 ± 25	98 ± 17	p < 0.01
apo-CII (mg/dL)	5.5 ± 2.2	4.6 ± 2.1	p < 0.05
apo-CIII (mg/dL)	15.0 ± 7.6	12.1 ± 5.4	p < 0.05
apo-E (mg/dL)	7.1 ± 2.3	6.1 ± 2.3	p < 0.05
LPL (U)	5.43 ± 2.04	7.93 ± 2.97	p < 0.01
HTGL (U)	13.42 ± 6.57	17.85 ± 9.86	p < 0.05
Lp(a) (mg/dL)	19.8	13.0	p < 0.01*
LDL-size (nm)	26.5	27.0	NS*
Lipid peroxide (mg/dL)	4.05 ± 0.88	4.61 ± 1.44	p < 0.01

Values are expressed as mean ± SD.
Lp(a) level and LDL size were expressed as median values.
NS: not significant; apo: apolipoprotein
*: Mann-Whitney U-test

Table 3. Other atherogenic risk factors

	Gout (n=147)	Control (n=64)	p value
vWF (%)	103 ± 45	88 ± 31	p < 0.05
PAI-1 (pg/mL)	49.7 ± 34.7	39.6 ± 18.7	NS
Homocysteine (μ mol/L)	8.90 ± 3.56	8.70 ± 4.02	NS
VFA (cm^2)	138.6 ± 52.9	113.2 ± 40.3	p < 0.001
BMI (kg/m^2)	24.9 ± 2.9	24.1 ± 3.1	NS
SBP (mmHg)	135.2 ± 16.5	127.3 ± 16.0	p < 0.01
DBP (mmHg)	79.4 ± 9.9	75.7 ± 8.5	p < 0.01
IRI (μ U/L)	79.4 ± 9.9	75.7 ± 8.5	NS
Endothelin (U/dL)	1.94 ± 0.65	1.81 ± 0.59	NS

Values are expressed as mean ± SD.
NS: not significant
BMI: body mass index; VFA: visceral fat area
SBP: systolic blood pressure; DBP: diastolic blood pressure

Other Atherogenic Factors

Table 3 indicated the heretofore known risk factors for atherosclerosis. The plasma concentration of von Willebrand factor was significantly higher in patients with gout than in controls. However, the plasma concentrations of endothelin, PAI-1 and homocysteine were not different between the two groups. Visceral fat area in patients with gout was significantly larger than in controls, although body mass index was not different. Blood pressure and plasma insulin level were significantly higher in patients with gout than in controls, and both hypertension and hyperinsulinemia were closely related with visceral fat area (data not shown).

DISCUSSION

Recently, hyperuricemia has been recognized as one of the elements of multiple risk factor clustering syndromes like syndrome X and visceral fat obesity.[6] On the other hand, the present study demonstrated that multiple risk factors for atherosclerosis were clustered in patients with gout. Among them, most well documented is hypertriglyceridemia, since hypertriglyceridemia was frequently observed in patients with gout.[7] Some risk factors for coronary heart disease, such as low HDL-cholesterol, small LDL size and visceral fat obesity which is also associated with blood pressure and insulin level, are closely related to hypertriglyceridemia. These findings suggest that triglyceride level may be an important factor for atherosclerosis in patients with gout, although the cause and effect relationship between hypertriglyceridemia and visceral fat obesity is unclear. Although the mechanism of hypertriglyceridemia in gout is still controversial, decreased LPL and HTGL activities may in part contribute to an increase in serum triglyceride level in patients with gout.[3] In addition to hypertriglyceridemia, increased levels of Lp(a)[7] in serum and von Willebrand factor in plasma may contribute to the atherogenecity in gout. However, homocysteine and lipid peroxide does not seem to contribute to atherosclerosis in patients with gout. No matter what are the causes of atherosclerosis, it is important to normalize these factors, besides hyperuricemia to prevent and/or treat atherosclerotic coronary disease. Thus, inclusive treatment against gout will be required for this purpose.

REFERENCES

1. Nishioka K, Mikanagi K: A retrospective study on the cause of death in Japanese patients with gout. The Ryumachi 21:29–33, 1981.
2. Tokunaga K, Matsuzawa Y, Ishikawa K, Tarui S: A novel technique for the determination of body fat by computed tomography. Int J Obesity 7:437–445, 1983.
3. Tsutsumi Z, Takahashi S, Yamamoto T, Moriwaki Y,Nasako Y, Hiroishi K, Hada T, Higashino K: Study on lipoprotein lipase and hepatic triglyceride lipase activities in patients with gout. Adv Exp Med Biol 370;83–86, 1995.
4. Hatch FT, Lees RS: Practical methods for plasma lipoprotein analysis. Adv Lipid Res 6:1–33, 1968.
5. Nichols AV, Krauss RM, Musliner TA: Nondenaturing polyacrylamide gel electrophoresis. In: Segrest JP, Albers JJ (eds) : Methods in Enzymology: Plasma lipoproteins 128:417–431, 1986.
6. Zimmet PZ: Kelley West lecture 1991 Challenges in diabetes epidemiology: From West to the Rest. Diabetes Care 15:232–252, 1992.
7. Wyngaarden JB, Kelley WN; Gout, in Stanbury JB, Wyngaarden JB, Fredrickson DS, Goldstein JL, Brown MS (eds): The metabolic basis of inherited disease, 5th ed. New York, McGraw-Hill, chap 50, p1043, 1983.

14

CARDIOVASCULAR ACTIONS OF ADENOSINE

Granulocyte and Blood Platelet Adhesion in the Reperfused Myocardium

B. F. Becker, S. Zahler, C. Kupatt, C. Seligmann, B. Heindl, and C. Kowalski

Department of Physiology
University of Munich
Pettenkoferstr. 12, 80336 Munich, Germany

Adenosine is a pluripotent bioactive metabolite, long known to be formed and released from ischemic and reperfused myocardium owing to enhanced catabolism of adenine nucleotides under these conditions.[2,6,16] However, beyond its familiar dilatory actions on coronary and peripheral vessels and its bradyarrhythmic potential in the heart, adenosine also exerts potent effects on the intracoronary adhesion of leukocytes and blood platelets and may be involved in the phenomenon of ischemic preconditioning.[3-8,10,23,33] Table 1 summarizes these actions according to the type of membrane receptor presumably involved, 4 kinds—designated A1, A2a, A2b and A3—having been pharmacologically characterized and cloned to date.[23]

CORONARY EFFECTS

It is of some surprise that the mechanism of vascular dilatation by adenosine is still controversially discussed, even though this was first observed about 70 years ago. However, general consensus exists in as far as this is an A2-receptor mediated effect. Indeed, we have never observed coronary dilatation using specific A1-agonists such as cyclopentyladenosine in the guinea pig heart.[6] There is some good evidence that intravascular adenosine at physiological concentrations, i.e., submicromolar levels, acts via the endothelial cells: owing to the high rate of uptake and metabolism by the endothelium, adenosine practically cannot permeate through to the smooth muscle.[5] With respect to intracellular signalling, we have previously shown this to be independent of cAMP in the guinea pig coronary endothelium, but perhaps to indirectly involve ATP-dependent K^+-channels and cellular hyperpolarization.[4,20,25] Interstitially formed adenosine, on the other hand, could act directly on the vascular smooth muscle, most likely making use of the adenylylcyclase

Purine and Pyrimidine Metabolism in Man IX,
edited by Griesmacher *et al.* Plenum Press, New York, 1998.

73

Table 1. Receptor-dependent cardiovascular actions of adenosine

A_1- and/or A_3-receptor mediated	A_2-receptor mediated
Bradycardia	Coronary dilatation
AV - block	Coronary steal
PMN adhesion	PMN inhibition
Platelet inhibition	Platelet inhibition
Anginal pain	
Inhibition of catecholaminergic effects	
Ischemic preconditioning	
Activation of ATP-dependent K^+-channels	

pathway.[23] Irrespective of the site and mode of action, dilatation of the coronary system by adenosine bears the great risk of inducing coronary steal, and so should not be regarded uncritically as a beneficial effect in each and every heart. Especially hearts with reperfused and critically stenosed coronary beds will suffer from maldistribution of coronary flow due to A2-mediated dilatation of the small resistance vessels.

REPERFUSION DAMAGE

Leukocytes, especially polymorphonuclear granulocytes (PMN), and blood platelets contribute to post-ischemic myocardial reperfusion damage, which bears markings of an inflammatory response.[9,12–14] However, in order to cause deleterious effects, platelets and PMN must interact with the tissue (primarily the endothelium) and possibly with one another, both processes being strongly susceptible to substances released from ischemic tissue as well as to shear stress.[8,22,32,33] Adenosine exerts an A2-receptor mediated platelet-stabilizing action and an anti-inflammatory effect on PMN themselves.[1,8,10,11] However, Cronstein et al. also reported an A1-receptor mediated chemotactic and proadhesive action of adenosine on isolated PMN.[11] Since A1-receptors exhibit a higher sensitivity towards adenosine than the A2-receptors, the net effect of adenosine on PMN will depend on the momentary nucleoside concentration and the relative preponderance of adenosine receptor subtypes.[6,7] The situation is further confounded by the fact that not only platelets and PMN, but also the coronary endothelial cells express A1- and A2-receptors.[27] Moreover, shear-stress dependency of adenosine effects on platelet adhesion to endothelial cells is only just beginning to be characterized.[8]

Adhesion of PMN

Intracoronary retention of PMN was elevated in postischemic isolated hearts of guinea pigs and correlated inversely with the coronary venous effluent adenosine concentration.[3,8,33] Elevated adhesion caused a pronounced fall in postischemic recovery of pump function and increased vascular permeability, largely due to spontaneous activation of the sequestered PMN and formation of oxygen radicals and hypochlorous acid.[19,29] Prevention of postischemic retention of PMN by any means (coinfusion of A1-antagonist DPCPX, of PAF-antagonist WEB 2086, or of adenosine deaminase) alleviated the loss in function, whereas enhancing adhesion (coinfusion of the A2-receptor antagonist DMPX) augmented reperfusion damage.[28] In the intact guinea pig coronary bed, the leukocytes becoming adherent upon application into postischemic hearts were located almost exclusively in small

(10–50 μm diameter) postcapillary venules.[17] This is the type of vessel known from other organs to contain the highest density of immunoglobulin-type membrane adhesion molecules (ICAMs). Surprisingly, there was no quantitative difference in adenosine-induced adhesion whether PMN of human or guinea pig origin were employed.[8,33]

Adenosine at submicromolar concentrations enhanced intracoronary PMN retention up to two-fold in guinea pig hearts by preferentially stimulating endothelial A1-receptors.[33] Higher (micromolar) levels of adenosine, in contrast, inhibited PMN adhesion via an A2-receptor dependent mechanism, apparently also mainly via the endothelium.[3,8,33] The proadhesive effect of adenosine was mimicked by the A1-agonist cyclopentyladenosine and antagonised by the A1-receptor antagonist dipropylcyclopentylxanthine.[33] With respect to the signalling cascade involved, adenosine-induced PMN adhesion was inhibited by the compound WEB 2086, a receptor antagonist of platelet activating factor (PAF).[3,8,28] PAF generated on the endothelial cells is known to activate the ß2-integrin (CD11/CD18) adhesion molecules of rolling PMN by interacting with a membrane receptor.[21] Moreover, CD11/CD18 integrins are the natural partners of ICAMs. As recently published, expression of CD11b integrin was strongly enhanced on human PMN passed through an acutely reperfused coronary bed, while preincubating human PMN with an anti-human CD18 antibody prevented postischemic adhesion and PMN-induced reperfusion damage.[34] Thus, adenosine may act as a pro-inflammatory mediator in the scanario of PMN-mediated reperfusion damage.

Adhesion of Platelets

In the presence of stimulants such as thrombin (elevated by coronary stasis, thrombotic occlusion, thrombolysis, mechanical endothelial damage, etc.), the endothelium undergoes prothrombotic, inflammatory changes.[15,26] It was possible to validate this in our model, adhesion of washed homologous thrombocytes in the guinea pig coronary system rising from just a few percent of the applied platelets to 30–40% in the presence of thrombin (0.3 U/ml) under low-flow conditions.[8] Human platelets showed somewhat lesser affinity.[18] However, platelets did not become adherent during early reperfusion (1–2 min), even in the presence of thrombin.[30] In the case of guinea pig platelets, this inhibition of adhesion under higher shear stress conditions was found to be due solely to the still elevated adenosine levels at that particular time: Theophylline (an adenosine A1- and A2-receptor antagonist) restored the capacity of thrombin to increase adhesion (32% platelet retention); the selective A1- and A2-antagonists (DPCPX and DMPX, resp.) were both partially effective, in an additive manner.[31] On the other hand, inhibition of NO-synthase (nitro-L-arginine) or cyclooxygenase (indomethacin) remained without effect.[8] Also, the adenosine uptake blocker dipyridamole prevented adhesion. Thus, adenosine prevents platelet adhesion to the coronary endothelium both via A1- and A2-receptor mediated mechanisms under conditions of higher shear stress.

An important issue concerns the effect of adenosine on postischemic adhesion of PMN and platelets when both blood constituents are simultaneously present. This is relevant because PMN and blood platelets can interact not only with the vascular endothelial cells and homotypically, but also heterotypically (e.g., directly via P-selectin on platelets and sialyl Lewis^x adhesion molecules on PMN, and indirectly via fibrinogen bridging between GPIIb/IIIa and CD11/CD18).[8,26] In vivo, platelets and PMN can release and bind components of the plasma coagulation system (e.g., factors I, V and X) and activate the endothelial cells (formation of tissue factor, thrombin and activated factors VII). Thus, formation of conglomerates between PMN and platelets in reperfused vessels localises coagulation at these sites, setting the stage for coronary reocclusion.[15,26] On the basis of our first experiments it seems clear that the proadhesive effect of adenosine on PMN and the subsequent spontaneous activation of these cells in the reperfused hearts suffices to induce platelet co-adhesion, even under reflow conditions.[8] Pertinently, retention of both blood constituents infringed postischemic heart function, already at cell numbers where each type alone was without an effect.[8]

PRECONDITIONING AND CARDIOPROTECTION

The phenomenon of ischemic preconditioning can be mimicked by infusion of adenosine and A1- and A2-agonists *in vivo* and has been prevented by use of A1- /A3-receptor antagonists in *ex vivo*, isolated heart models in some (but not all) instances.[23] However, Neely et al. recently reported the simulation of preconditioning by use of an A1-antagonist.[24] It now seems possible to reconcile these conflicting observations in an *in vivo* setting if adenosine effects on PMN and platelet adhesion are taken into consideration: Systemic application of adenosine or an A1-agonist before ischemia will lead to sequestration of PMN also in vascular beds other than the coronary system, leading to acute leukopenia, whereas an A1-antagonist or an A2-agonist will directly prevent postischemic recruitment of PMN to the heart. Thus, both interventions will mitigate PMN/platelet mediated reperfusion injury and lead to the same general benefit. This does not preclude additional effects of adenosine via A1- and A2-receptors which may contribute to preconditioning (e.g., activation of ATP-dependent K^+-channels) and serves to highlight the complexity of adenosine actions *in vivo*. However, in view of the possible negative side effects of application of A1- or A2-agonists (see Table 1), use of an A1-receptor antagonist would seem preferable for instigating cardioprotection.

ACKNOWLEDGMENT

This work was supported in part by the Friedrich-Baur-Foundation of the Medical Faculty, University of Munich.

REFERENCES

1. Agarwal KC (1993) Modulation of vasopressin actions on human platelets by plasma adenosine and theophylline: gender differences. J Cardiovasc Pharmacol 21: 1012–1018
2. Bardenheuer HJ, Fabry A, Höfling B, Peter K (1994) Adenosine: A sensitive marker of myocardial ischemia in man. Cardiovasc Res 28: 656–662
3. Becker BF, Kupatt C, Raschke P, Zahler S (1996) Microcirculatory disturbances due to activated leukocytes in ischemic and reperfusion injury. In: Kamada T, Shiga T, McCuskey RS (eds) Tissue perfusion and organ function: Ischemia/reperfusion injury. Elsevier Science Publishers, Amsterdam, pp 97–107
4. Becker BF, Leipert B, Gerlach E (1991) Alterations of microvascular flow responses in the heart by oxygen-derived radicals, HOCl and the K^+-channel blocker glibenclamide. In: Niimi H, Hori M, Naritomi H (eds) Microcirculatory disorders in the heart and brain. Harwood Academic Publishers, Chur London Paris New York Melbourne, pp 53–64
5. Becker BF, Leipert B, Schwartz L, Gerlach E (1991) The metabolic barrier of the coronary endothelium as a determinant of flow responses. In: Inoue M, Hori M, Imai S, Berne RM (eds) Regulation of coronary blood flow. Springer-Verlag, Tokyo, pp 206–216
6. Becker BF, Zahler S, Richardt G (1995) Endothelial adenosine receptors modulate leukocyte adhesion: relevance for myocardial reperfusion damage. In: Belardinelli L, Pelleg A (eds) Adenosine and adenine nu-

cleotides: from molecular biology to integrative physiology. Kluwer Acad Publ, Boston Dordrecht London, pp 379–387

7. Becker BF, Zahler S, Kupatt C, Münch G, Richardt G (1996) Importance of intracoronary adenosine concentration for modulation of leukocyte-endothelial interaction. Pflügers Arch 431(Suppl 6): R128

8. Becker BF, Zahler S, Seligmann C, Kupatt C, Habazettl H (1996) Interaction of adenosine with leukocytes and blood platelets. Z Kardiol 85 (Suppl 6): 161–170

9. Brady AJB, Williams FM, Williams TJ (1992) Inflammatory injury in myocardial ischemia. Clin Sci 83: 511–518

10. Cronstein BN, Bouma MG, Becker BF (1996) Purinergic mechanisms in inflammation. Drug Devel Res 39:426–435

11. Cronstein BN, Daguma L, Nichols D, Hutchison AJ, Williams M (1990) The adenosine/neutrophil paradox resolved: Human neutrophils possess both A1 and A2 receptors that promote chemotaxis and inhibit O_2^- generation, respectively. J Clin Invest 85: 1150–1157

12. De Servi S, Mazzone A, Ricevuti G, Fioravanti A, Bramucci E, Angoli L, Stefano G, Specchia G (1990) Granulocyte activation after coronary angioplasty in humans. Circulation 82: 140–146

13. Entman ML, Smith CW (1994) Postreperfusion inflammation: a model for reaction to injury in cardiovascular disease. Cardiovasc Res 28: 1301–1311

14. Gawaz M, Neumann F-J, Ott I, Schiessler A, Schömig A (1996) Platelet function in acute myocardial infarction treated with direct angioplasty. Circulation 93: 229–237

15. Gerlach E, Becker BF (1993) Interaktion von Blut und Gefäßwand: Hämostaseologische Aspekte. Z Kardiol 82 (Suppl 5): 13–21

16. Gerlach E, Deuticke B, Dreisbach RH (1963) Der Nukleotid-Abbau im Herzmuskel bei Sauerstoffmangel und seine mögliche Bedeutung für die Coronardurchblutung. Naturwissenschaften 50: 228–229

17. Habazettl H, Kupatt C, Zahler S, Becker BF, Gerlach E, Messmer K (1996) Nitric oxide modulates post-ischemic venular leukocyte adhesion in isolated guinea pig hearts. In: Messmer K, Kübler WM (eds) Proceedings 6th World Congress for Microcirculation. Monduzzi Editore, Bologna, pp 729–733

18. Heindl B, Becker BF, Zahler S, Möbert J, Conzen P, Gerlach E (1997) Volatile anesthetics reduce adhesion of platelets in the coronary system. Europ J Physiol (Pflügers Arch) 433(Suppl 6): R118

19. Kupatt C, Zahler S, Seligmann C, Massoudy P, Becker BF, Gerlach E (1996) Nitric oxide mitigates leukocyte adhesion and vascular permeability changes after myocardial ischemia. J Mol Cell Cardiol 28: 643–654

20. Leipert B, Becker BF, Gerlach E (1992) Different endothelial mechanisms involved in coronary responses to known vasodilators. Am J Physiol 262 (Heart Circ Physiol 31): H1676-H1683

21. Lorant DE, Patel KD, McIntyre TM, McEver RP, Prescott SM, Zimmerman GA (1991) Coexpression of GMP-140 and PAF by endothelium stimulated by histamine or thrombin: a juxtacrine system for adhesion and activation of neutrophils. J Cell Biol 115: 223–234

22. Lucchesi BR (1994) Complement, neutrophils and free radicals: mediators of reperfusion injury. Arzneim-Forsch/Drug Res 44(I): 420–432

23. Mubagwa K, Mullane K, Flameng W (1996) Role of adenosine in the heart and circulation. Cardiovasc Res 32: 797–813

24. Neely CF, DiPierro FV, Kong M, Greelish JP, Gardner TJ (1996) A1 adenosine receptor antagonists block ischemia-reperfusion injury of the heart. Circ 94(suppl II): II-376-II-380

25. Newman WH, Becker BF, Heier M, Nees S, Gerlach E (1988) Endothelium-mediated coronary dilatation by adenosine does not depend on endothelial adenylate cyclase activation: studies in isolated guinea pig hearts. Pflügers Arch 413: 1–7

26. Palabrica T, Lobb R, Furie BC, Aronovitz M, Benjamin C, Hsu Y-M, Sajer SA, Furie B (1992) Leukocyte accumulation promoting fibrin deposition is mediated in vivo by P-selectin on adherent platelets. Nature 359: 848–851

27. Parkinson FE, Clanachan AS (1991) Adenosine receptors and nucleoside transport sites in cardiac cells. Br J Pharmacol 104: 399–405

28. Raschke P, Becker BF (1995) Adenosine and PAF-dependent mechanisms lead to myocardial reperfusion injury by polymorphonuclear neutrophils after brief ischaemia. Cadiovasc Res 29: 569–576

29. Raschke P, Becker BF, Leipert B, Schwartz LM, Zahler S, Gerlach E (1993) Postischemic dysfunction of the heart induced by small numbers of neutrophils via formation of hypochlorous acid. Basic Res Cardiol 88: 321–339

30. Seligmann C, Becker BF, Kupatt C, Beblo S, Gerlach E (1995) Endogenous adenosine prevents platelet adhesion in the coronary system during early reperfusion, but not during low-flow ischemia. Pflügers Arch 429(Suppl 6): R109

31. Seligmann C, Becker BF, Kupatt C, Zahler S, Gerlach E (1995) Adenosine released in early reperfusion prevents platelet retention in the coronary system by a combined A1- and A2-receptor mediated mechanism. Europ Heart J 16 (Abstr Suppl): 361

32. Sixma JJ, de Groot PG (1994) Regulation of platelet adhesion to the vessel wall. Ann N Y Acad Sci 714: 190–199

33. Zahler S, Becker BF, Raschke P, Gerlach E (1994) Stimulation of endothelial adenosine A_1-receptors enhances adhesion of neutrophils in the intact guinea pig coronary system. Cardiovasc Res 28:1366–1372

34. Zahler S, Kupatt C, Seligmann C, Kowalski C, Becker BF, Gerlach E (1997) Leukocytes retained in postischemically perfused isolated guinea pig hearts do not cause capillary plugging. Pflügers Arch 433: 713–720

15

MAST CELL ADENOSINE RECEPTOR CHARACTERISTICS AND SIGNALING

Diana L. Marquardt

University of California
San Diego School of Medicine
Department of Medicine
San Diego, California 92093

1. IMPORTANCE OF ADENOSINE IN ALLERGIC INFLAMMATION

Adenosine appears to be an extremely important contributor to asthma and allergic inflammation. Mast cells, cells thought to be central to immediate hypersensitivity reactions, release markedly more histamine and other granule-associated mediators when activated in the presence of micromolar concentrations of adenosine.[1,2] The release of newly-generated mediators such as the leukotrienes and prostaglandins is potentiated modestly by adenosine as well. Hypoxic lung and stimulated mast cells[3] release adenosine, providing a ready local source of this molecule during allergic pulmonary inflammation.

In vivo studies have provided a fascinating confirmation of the pro-inflammatory effects of adenosine on the lung. When asthmatic subjects, but not normal controls, inhale adenosine or AMP, they develop an immediate bronchoconstrictor response as evidenced by a rapid fall in forced expiratory volumes.[4] This response can be largely abrogated by pre-treatment with anti-histamines,[5] suggesting that the mast cell may be the primary site for this effect of adenosine. In similar studies, atopic patients with rhinitis routinely develop rhinorrhea, sneezing, and nasal pruritus within minutes after inhaling adenosine or AMP.[6] Given the strong evidence that adenosine enhances allergic reactions both in vivo and in vitro, interest in determining the cellular site of action of adenosine, its receptor characteristics, and its intracellular signaling pathways has become intense.

2. SITE OF ACTION OF ADENOSINE IN ASTHMA

Because mast cells release adenosine and respond strongly to adenosine, the mast cell is a likely target for adenosine's effects in allergic inflammation. The facts that mast cells

Purine and Pyrimidine Metabolism in Man IX,
edited by Griesmacher *et al.* Plenum Press, New York, 1998.

and basophils are the primary sources of histamine in the body and anti-histamines block much of the pro-asthmatic effects of adenosine support this contention. However, a number of other cellular or tissue sites of action of adenosine are likely to be present and may serve to modulate the in vivo responses. Neutrophils possess cell surface adenosine receptors and exhibit a number of cellular responses to adenosine. Airway smooth muscle may constrict upon adenosine stimulation in some animal models of asthma. Thus, although the mast cell almost certainly plays a central role in adenosine-induced bronchospasm, it is not yet clear whether it is absolutely required for this reaction to take place. Studies using mast cell deficient animals may help to answer these questions in the near future.

3. SIGNAL TRANSDUCTION VIA MAST CELL ADENOSINE RECEPTORS

Although adenosine potentiates mast cell mediator release in cells from a variety of tissue sources and species, adenosine alone does not induce mediator release from normal mast cells. In most cases another secretagogue (such as anti-IgE, specific antigen, or a calcium ionophore) is required for mast cell degranulation to take place. However, adenosine alone does induce some direct mast cell biochemical responses. Within seconds after the addition of 1 μM adenosine to mast cells, a rapid increase in intracellular free calcium concentrations is observed.[7] Similar doses of adenosine induce the production of inositol trisphosphate[8] and the activation of adenylate cyclase.[9] In our laboratory we have demonstrated the activation and tyrosine phosphorylation of lyn kinase within a minute of adenosine administration and in the absence of other stimuli.[10] A cultured human mast cell line that fails to release histamine does produce interleukin-8 upon adenosine exposure, purportedly through A_{2b} receptor stimulation.[11] Given this myriad of intracellular responses to adenosine, it is not surprising to find that mast cells possess more than one adenosine receptor subtype.

4. MAST CELL ADENOSINE RECEPTOR SUBTYPES

There remains considerable controversy regarding the relative importance of specific mast cell adenosine receptor subtypes. Early studies revealed that adenosine induced a rapid rise in cyclic AMP, consistent with the action at an A_2 receptor subtype. However, blocking the cyclic AMP-dependent protein kinase activity failed to alter the ability of adenosine to augment mast cell mediator release, suggesting that this pathway was not necessary for adenosine's effects on degranulation.[12] Studies of relatively adenosine receptor subtype-specific agonists and antagonists (although few truly good ones exist) have been inconclusive for a number of reasons. More recent reports have shown that rat basophilic leukemia cells possess A_3 receptor subtypes[13] and mouse bone marrow-derived mast cells possess A_{2a} and A_{2b}[14] (and maybe A_3) receptor subtypes. This is based on pharmacologic data, Northern blotting studies with species non-identical probes, and reverse transcriptase polymerase chain reaction technology for the RBL cells and on Northern blotting with species identical probes and mast cell cDNA library screening for the mouse cells. Other interesting observations along these lines include a report of A_{2a} and A_{2b} receptors on human lung mast cells as assessed by in situ hybridization studies,[15] and the suggestion that dog mastocytoma cells respond to adenosine, but probably not through an A_3 receptor.[16] Some relatively A_3-specific agonists appear to promote RBL cell mediator release.

Specific A_{2a} agonists are ineffective in facilitating mouse bone marrow-derived mast cell mediator release. To further complicate the picture is the recent publication of experimental results in a rabbit model of asthma indicating that the A_1 receptor subtype may be important in adenosine-induced bronchoconstriction.[17] In a similar model, an A_3-specific agonist failed to induce bronchoconstriction.[18] Clearly this is an area ripe for further investigative efforts.

One reason that the adenosine receptor subtype situation in mast cells has been difficult to clarify is that the different subtypes may respond in ways that result in receptor interactions and crosstalk. For example, stimulation of mast cell A_{2a} receptors may cause adenylate cyclase activation at the same time stimulation of A_{2b} receptors (or perhaps A_3 receptors, or both) may cause increased intracellular free calcium and IP_3 generation. The effects of adenosine on the A_{2a} receptors could serve to dampen the A_{2b}-mediated effects, or perhaps both may be necessary for the ultimate enhancement of mediator release. In order to begin to dissect these issues, our laboratory is embarking on a study of mast cells derived from mice lacking specific adenosine receptor subtypes. By evaluating the ability of mast cells derived from bone marrow from A_{2a} –/–, A_{2b} –/–, and A_3 –/– mice, we hope to determine the relative contributions of each receptor to the various biochemical pathways affected by adenosine and to mediator release enhancement itself.

5. ANIMAL MODELS OF EOSINOPHILIC AIRWAYS INFLAMMATION

Returning to the question of the site of action of adenosine in asthmatic airways, a number of interesting lines of investigation have been pursued. Some studies have shown that adenosine induces a contractile response in animal airways, but others have suggested the possibility of adenosine-induced bronchial relaxation. These studies have the disadvantage of being based on the use of isolated large airways and tracheal rings—clearly not the primary sites of asthmatic pathophysiology. Bronchial hyperreactivity consistent with asthma may be induced in animals (guinea pigs, rats, and rabbits, among others) by various immunization protocols to allow the study of in vivo airway responsiveness. All of these animal models of asthma appear to be characterized by a bronchospastic response to inhaled adenosine.[19] Because adenosine does not directly induce mast cell mediator release but *does* induce the release of histamine into bronchoalveolar lavage fluid from asthmatic patients, one could postulate that local allergic inflammation in some way primes mast cells in asthmatic subjects to release mediators upon exposure to exogenous adenosine. To study this premise, we have induced eosinophilic airway inflammation by sensitizing mice with ovalbumin, initially by intraperitoneal injection and later with inhaled ovalbumin. Adenosine alone induces histamine release from lung tissue from mice challenged repeatedly with inhaled antigen but not from control saline-challenged mice.[20] To more definitively determine that local inflammation (characterized by increased chemokine and cytokine concentrations and a relative hypercellularity) somehow "primes" the mast cells, we are in the process of inducing either peritoneal and/or pulmonary inflammation using repeated ovalbumin challenges to the appropriate site to compare the adenosine responsiveness of mast cells obtained from each location. Our initial experiments suggest that pulmonary inflammation does not cause mouse peritoneal mast cells to develop direct mast cell secretory responses to adenosine, but whether the converse is true remains to be determined. In summary, adenosine potentiates the inflammatory response initiated with mast cell activation and is an important potential target for drug development for the treatment of asthma.

REFERENCES

1. Marquardt DL, Parker CW, Sullivan TJ. Potentiation of mast cell mediator release by adenosine. J Immunol 120:871–8, 1978.
2. Peters SP, Schulman ES, Schleimer RP, et al. Dispersed human lung mast cells. Pharmacologic aspects and comparison with human lung tissue fragments. Am Rev Respir Dis 126:1034–9, 1982.
3. Marquardt DL, Gruber HE, Wasserman SI. Adenosine release,from stimulated mast cells. Proc Natl Acad Sci USA 81:6192–6, 1984.
4. Cushley MJ, Tattersfield AE, Holgate ST. Inhaled adenosine and guanosine in normal and asthmatic subjects. Brit J Clin Pharmacol 15:161–9, 1983.
5. Rafferty P, Beasley R, Southgate P, Holgate ST. The role of histamine in allergen and adenosine-induced bronchoconstriction. Int Archs Allergy Appl Immunol 82:292–9, 1987.
6. Crimi N, Polosa R, Raccuglia D, et al. Nasal responses to adenosine 5'-monophosphate (AMP) in atopic rhinitic subjects: inhibition by cetirizine. J Allergy Clin Immunol 91:258 (abst), 1993.
7. Collado-Escobar D, Ali H, Beaven MA. On the mechanism of action of dexamethasone in a rat mast cell line (RBL-2H3). J Immunol 144:3449–54, 1990.
8. Ali H, Cunha-Melo JR, Saul WF, Beaven MA. The activation of phospholipase C via adenosine receptors provides synergistic signals for secretion in antigen-stimulated RBL-2H3 cells: evidence for a novel adenosine receptor. J Biol Chem 265:745–50, 1990.
9. Marquardt DL, Walker LL, Wasserman SI. Adenosine receptors on mouse bone marrow-derived mast cells: Functional significance and regulation by aminophylline. J Immunol 133:932–7, 1984.
10. Walker LL, Alongi JL, Marquardt DL. Activation of Lyn kinase by adenosine in mouse mast cells. (in review).
11. Feoktistov I, Biaggioni. Adenosine A_{2b} receptors evoke interleukin-8 secretion in human mast cells: an enprophylline-sensitive mechanism with implications for asthma. J Clin Invest 96:1979–83, 1995.
12. Marquardt DL, Walker LL. Inhibition of protein kinase A fails to alter mast cell adenosine responsiveness. Agents and Actions 43:7–12, 1994.
13. Ramkumar V, Stiles GL, Beaven MA, Ali H. The A_3 adenosine receptor is the unique adenosine receptor which facilitates release of allergic mediators in mast cells. J Biol Chem 268:16887–92, 1993.
14. Marquardt DL, Walker LL, Heinemann S. Cloning of two adenosine receptor subtypes from mouse bone marrow-derived mast cells. J Immunol 152:4508–15, 1994.
15. Jacobson M (personal communication).
16. Auchampach JA, Caughey GH, Linden J. Molecular cloning and pharmacological characterization of the canine A_3 adenosine receptor from BR mastocytoma cells. FASEB J A122, 1995.
17. Nyce JW, Metzger WJ. DNA antisense therapy for asthma in an animal model. Nature 385:721–5, 1997.
18. el-Hashim A, D'Agostino B, Matera MG, Page C. Characterization of adenosine receptors involved in adenosine-induced bronchoconstriction in allergic rabbits. Brit J Pharmacol 119:1262–8, 1996.
19. Thorne JR, Broadley KJ. Adenosine-induced bronchoconstriction in conscious hyperresponsive and sensitized guinea pigs. Am J Respir Crit Care Med 149:392–9, 1994.
20. Hoffman HM, Marquardt DL. The effect of adenosine on histamine release from allergen-sensitized mouse lung tissue. J Allergy Clin Immunol (abst.) 1997.

16

PURINE AND PYRIMIDINE NUCLEOSIDE CONTENT OF THE NEURONAL EXTRACELLULAR SPACE IN RAT

An *in Vivo* Microdialysis Study

Á. Dobolyi, A. Reichart, T. Szikra, and G. Juhász

Department of Comparative Physiology
Eötvös Loránd University
H-1088 Budapest, Múzeum krt. 4/a, Hungary

1. INTRODUCTION

Neurological side-effects of chemotherapeutic agents[1] known to act on purine and pyrimidine metabolism enzymes raised a possible role for nucleosides and deoxynucleosides in brain cell survival as it is demonstrated in recent investigations.[2–5] Some new data also demonstrated the involvement of nucleotides and nucleosides in various brain functions. Apart from the well-known transmitter function of ATP[6] and neuromodulator role of adenosine[7] other nucleotides and nucleosides have also been suggested to be neuroactive. Pyrimidine nucleotides have specific pyrimidinoceptors even in cells of brain origin.[8] Nucleosides could have their own modulatory actions as indicated by sleep modifying effect of uridine.[9] The *in vivo* measurement of nucleosides and related substances in behaving animals could be the following major contribution to understanding the mechanisms of neurological side-effects of chemotherapeutic agents and the functional roles of nucleosides in the brain. A recently developed method, the *in vivo* microdialysis technique was used in the present study. Dialysis measures the composition of local extracellular space under various experimental conditions in freely moving animals.[10] The main applications of microdialysis technique is measuring synaptic neurotransmitter overflow,[10] investigation of transport functions, and pharmacokinetics of various drugs,[11] and also gaining information about intracellular metabolic processes.[12] Microdialysis has already been applied in human patients[13] with subarachnoidal haemorrhage, head trauma, Parkinson's disease, brain tumors and epilepsy.

Present study is the first report on the extracellular concentrations of purine and pyrimidine bases, nucleosides and deoxynucleosides *in vivo*. Since there are some indica-

Purine and Pyrimidine Metabolism in Man IX,
edited by Griesmacher *et al.* Plenum Press, New York, 1998.

tions that nucleosides and nucleotides might have uneven distribution in the cerebrospinal fluid and interstitial space of the brain,[14] microdialysis and cerebrospinal fluid samples were compared. Dialysis experiments were also performed to address the possible neuromodulatory role of nucleosides. Interstitial levels of nucleosides and related compounds were correlated with experimentally induced changes of local neuronal activity.

2. METHODS

2.1. HPLC Technique

Pharmacia LKB SMART system with small-bore columns and UV detector (254 nm) was used. Eluent A was 0.02 M formiate buffer, pH=4.45. Eluent B was 0.02 M formiate buffer containing 40% acetonitrile, pH=4.45. The gradient profile was as follows: 0% B eluent at 0 minutes, 0% at 2.5 minutes, 10% at 20 minutes, 30% at 25 minutes, 100% at 38 minutes. The flow rate was 300 μl/min. The column and detector temperature was 10°C.

Three independent procedures were applied for identification and purity analysis of chromatographic peaks: 1./ retention times of chromatographic peaks of biological samples were compared to those of external standards. 2./ standard mixture samples and dialysates were analysed at 254 and 280 nm wavelengths. 3./ the peaks were collected in separate fractions and concentrations were measured by electrospray tandem mass spectrometry and compared to HPLC based concentration measurements.

2.2. Sample Collection and Preparation

Rats (350–400 g) were anaesthetised with 1% halothane-air mixture and then placed into a stereotaxic frame. Microdialysis probes were placed into the ventrobasal thalamic nuclei bilaterally (A:−1.9, L:2.5, V:−7 mm, according to the atlas of Pellegrino and Cushman[15]). The perfusion was done at 1 μl/min with artificial cerebrospinal fluid (144 mM NaCl, 3 mM KCl, 1 mM $MgCl_2$ and 2 mM $CaCl_2$ in amino acid free water, its pH was adjusted to 7.0 prior to experiments). Samples of 60 μl were collected in every 60 minutes from which 50 μl was used for the analysis As dialysis samples are relatively clean they were injected directly without a clean-up procedure.

Cerebrospinal fluid (CSF) (30 μl) was obtained from the cisterna magna of ketamine anaesthetised rats. The CSF was centrifuged at 1500 g for 10 minutes and 25 μl was used for HPLC analysis without further preparation.

3. RESULTS AND DISCUSSION

3.1. Interstitial Space

The optimised chromatographic method applied in present study allowed to measure many endogenous nucleoside compounds in less than twenty-five minutes in microdialysis samples (Figure 1 and Table 1). We first report the interstitial concentrations of many deoxynucleosides and pyrimidine nucleosides in the brain, namely uridine, 2′-deoxyuridine, uracil, cytidine, 2′-deoxycytidine, thymidine, 2′-deoxyadenosine and 2′-deoxyguanosine. Direct measurement of interstitial space composition is especially important in brain as it can not be estimated from blood concentrations of substances because of the blood–brain barrier.

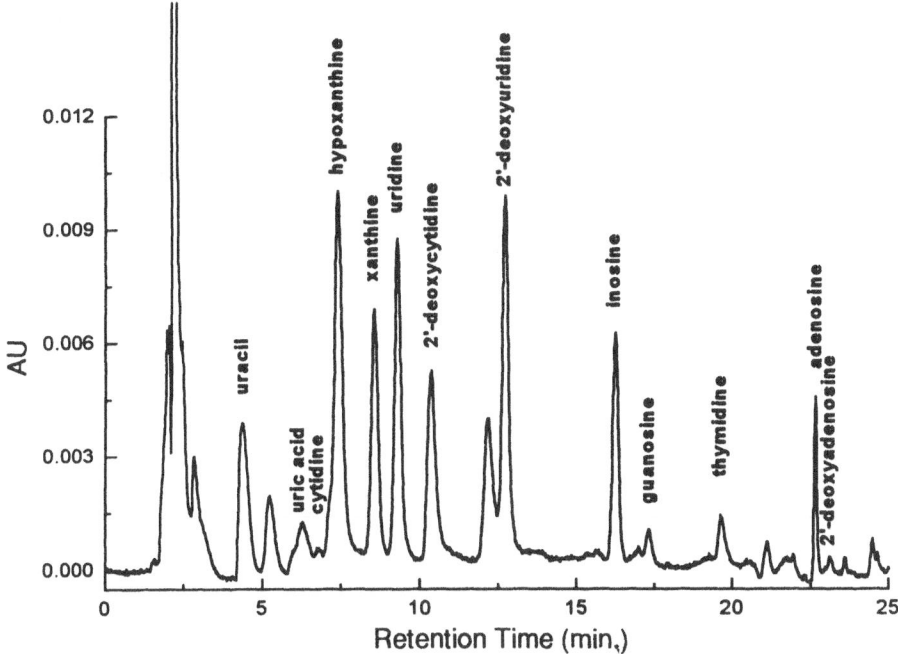

Figure 1. Elution profile of a 50 µl dialysate using the chromatographic method described in Methods.

Table 1. Detection limits (signal to noise ratio, S/N=3) were determined according to series of dilutions of standard mixtures. Interstitial concentrations of thalamus were determined from halothane anaesthetised rats using a relative recovery of 0.30.[18] Cerebrospinal fluid was obtained from the cisterna magna of ketamine anasethetised rats. Interstitial space and cerebrospinal fluid data represent the average of 13 and 7 animals, respectively

	Detection limit (µm)	Estimated interstitial concentrations (µm)	Cerebrospinal space concentrations (µm)
Purines			
Adenosine	0.002	0.95 ± 0.04	0.09 ± 0.01
Inosine	0.006	0.52 ± 0.03	0.14 ± 0.005
Hypoxanthine	0.016	4.48 ± 0.25	3.78 ± 0.27
Xanthine	0.022	1.17 ± 0.08	1.31 ± 0.08
Uric acid	0.046	1.15 ± 0.13	0.96 ± 0.06
Guanosine	0.004	0.17 ± 0.02	0.02 ± 0.004
2'-deoxyadenosine	0.004	0.17 ± 0.04	0.008 ± 0.002
2'-deoxyguanosine	0.004	0.01 ± 0.001	< 0.004
Pyrimidines			
Uracil	0.020	1.22 ± 0.22	2.66 ± 0.27
Uridine	0.010	0.76 ± 0.04	3.83 ± 0.40
Cytidine	0.026	0.11 ± 0.01	0.51 ± 0.05
2'-deoxycytidine	0.014	1.08 ± 0.06	7.35 ± 0.43
2'-deoxyuridine	0.016	1.06 ± 0.05	4.82±0.19
Thymidine	0.006	0.25 ± 0.02	0.76 ± 0.02

3.2. Comparison of Brain Interstitial Space and Cerebrospinal Fluid Composition

The concentrations of pyrimidines are about 5 times higher in cerebrospinal fluid than in interstitial space, the concentrations of purine nucleosides are about 5–10 times lower in the cerebrospinal fluid while purine bases have similar concentrations (Table 1). The marked difference between interstitial space and cerebrospinal fluid nucleoside compositions (Table 1) points out that cerebrospinal fluid analysis provides a limited if any information about the local extracellular concentrations of them. High concentrations of pyrimidine nucleosides and deoxynucleosides in cerebrospinal fluid may be a consequence of their significant active transport through choroid plexus.[14] Conversely purines which reach the brain cells predominantly through the blood–brain barrier,[14] have lower levels in the cerebrospinal fluid than in the interstitial space.

3.3. Effect of Neuronal Activity on Interstitial Nucleoside Composition

Tetrodotoxin, a blocker of action potential genesis caused a decrease in the levels of adenosine, its metabolites inosine and hypoxanthine and also guanosine and uridine (Figure 2). To depolarise cells, glutamate agonist (50 μM kainate) and high potassium concentration (120 mM) was applied. These concentrations in the perfusate result in about 10 μM and 25 mM extracellular concentrations, respectively, which is high enough to depolarise

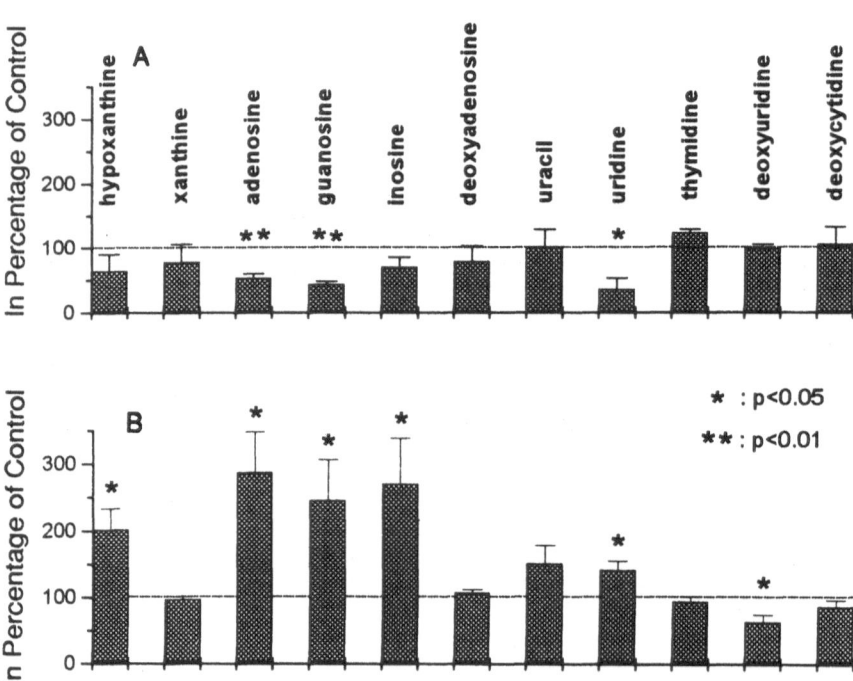

Figure 2. A. The effect of perfusing the action potential blocker tetrodotoxin (TTX) through the probe. **B.** The effect of perfusing 120 mM K⁺ through the probe. Data are expressed in the percentage of baseline levels measured from the same microdialysis probe before perfusing TTX or high K⁺. Values are averages ± SEM of data from 4–4 animals.

cells. The elevated level of adenosine and its metabolites can be explained by its known neuromodulator activity,[16] however an increased level of guanosine and uridine is also observed in our experiments (Fig. 2). Since extracellular nucleotides are metabolised to nucleosides in brain interstitial space,[17] our results might be a consequence of changes in nucleotide levels. Although metabolic effects or energy production could induce changes in nucleoside concentrations, we assume that guanosine and uridine or some of their precursors and metabolites have signalling role in the nervous system, however, further experiments are necessary to answer the question.

ACKNOWLEDGMENTS

This research was supported by OTKA T016552 Grant of Hungarian Scientific Research Fund for G. Juhász and Á. Dobolyi.

REFERENCES

1. Macdonald D.-R. (1991) Neurologic complications of chemotherapy. *Neurol. Clin.* **9(4)**, 955–967.
2. Wakade A.-R., Przywara D.-A., Palmer K.-C., Kulkarni J.-S. and Wakade T.-D. (1995) Deoxynucleoside induces neuronal apoptosis independent of neurotrophic factors. *J. Biol. Chem.* **270(30)**, 17986–17992.
3. Wallace T.-L. and Johnson E.-M. (1989) Cytosine arabinoside kills postmitotic neurons: evidence that deoxycytidine may have a role in neuronal survival that is independent of DNA synthesis. *J. Neurosci.* **9(1)**, 115–124.
4. Gysbers J.-W. and Rathbone M.-P. (1996) GTP and guanosine synergistically enhance NGF-induced neurite outgrowth from PC12 cells. *Int-J-Dev-Neurosci.* **14(1)**, 19–34.
5. Christjanson L.-J, Middlemiss P.-J. and Rathbone M.-P. (1993) Stimulation of astrocyte proliferation by purine and pyrimidine nucleotides and nucleosides. *Glia* **7(2)**, 176–182.
6. Zimmermann H. (1994) Signalling via ATP in the nervous system. *Trends Neurosci.* **17**, 420–426.
7. Williams M. (1990) Purine nucleosides and nucleotides as central nervous system modulators. Adenosine as the prototypic paracrine neuroactive substance. *Ann. N. Y. Acad. Sci.* **603**, 93–107.
8. Nicholas R.-A., Watt W.-C., Lazarowski E.-R., Li Q. and Harden K. (1996) Uridine nucleotide selectivity of three phospholipase C-activating P2 receptors: identification of a UDP-selective, a UTP-selective, and an ATP- and UTP-specific receptor. *Mol. Pharmacol.* **50(2)**, 224–229.
9. Inoue S., Honda K. and Komoda Y. (1995) Sleep as neuronal detoxification and restitution. *Behav. Brain Res.* **69(1–2)**, 91–96.
10. Benveniste H. and Huttemeier P.-C. (1990) Microdialysis: theory application. *Prog. Neurobiol.* **35**, 195–215.
11. Bonate P. L. (1995) Animal models for studying transport across the blood-brain barrier. *J. Neurosci. Methods* **56(1)**, 1–15.
12. Kekesi A.-K., Dobolyi A., Salfay O., Nyitrai G. and Juhasz G. (1997) Slow-wave sleep is accompanied by release of certain amino acids in the thalamus of cats. *Neuroreport* **8**, 1183–1186.
13. Kanthan R., Shuaib A., Goplen G., and Miyashita H. (1995) A new method of in-vivo microdialysis of the human brain. *J. Neurosci. Methods* **60(1–2)**, 151–155.
14. Spector R. (1989) Micronutrient homeostasis in mammalian brain and cerebrospinal fluid. *J. Neurochem.* **53**, 1667–1674.
15. Pellegrino L.-J., and Cushman A.-J., eds (1967) *A stereotactic atlas of the rat brain*. Appleton Century Crofts, New York.
16. Pazzagli M., Pedata F. and Pepeu G. (1993) Effect of K+ depolarization, tetrodotoxin, and NMDA receptor inhibition on extracellular adenosine levels in rat striatum. *Eur. J. Pharmacol.* **234(1)**, 61–65.
17. Zimmermann H. (1996) Biochemistry, localization and functional roles of ecto-nucleotidases in the nervous system. *Prog. Neurobiol.* **49(6)**, 589–618.
18. Juhasz G., Tarcali J., Pungor K. and Pungor E. (1989) Electrochemical calibration of *in vivo* brain dialysis samplers. *J. Neurosci. Meth.* **103**, 131–134.

17

EFFECT OF FMLP STIMULATION ON [³H]-NECA BINDING TO ADENOSINE RECEPTORS IN NEUTROPHILS MEMBRANES

C. Martini,[1] L. Trincavelli,[1] M. Fiorini,[1] M. Nardi,[1] L. Bazzichi,[2] and
A. Lucacchini[1]

[1]Istituto Policattedra di Discipline Biologiche
[2]Istituto Patologia Speciale Medica
University of Pisa
Pisa, Italy

1. INTRODUCTION

At physiological concentrations, adenosine can modulate a variety of biological activities by engaging specific surface receptors, termed A_1 and A_2, with different affinity for adenosine and its analogues.[1] Engagement of A_2 adenosine receptors induces an increase in cAMP level in several cells types, in contrast stimulation of A_1 receptors causes an opposite effect.[2] In neutrophils, adenosine and its analogues inhibit O_2^- generation, phagocytosis and adherence by occupancy of specific A_2 adenosine receptors, while occupancy of A_1 adenosine receptors enhance chemotaxis, phagocytosis and adherence.[3-7] In general, activation of leukocytes adenosine receptors reduces immune and inflammatory responses.[8] Therefore, it may be suggested that release of adenosine is one mechanism by which normal cells protect themselves from activated neutrophils. In fact, while stimulated neutrophils normally control bacterial infection, they may also contribute to the pathology of several inflammatory diseases including some rheumatoid diseases and emphysema.

Formylmethionylleucylphenylalanine peptide (FMLP) is one of the most potent bacterially derived neutrophils chemotaxis factors. FMLP stimulates many additional neutrophils function including superoxide anion (O_2^-) generation and adherence to endothelium.[9]

Previous investigation demonstrated that adenosine inhibits FMLP stimulated O_2^- generation, neutrophil aggregation, adherence and degranulation through an increase of intracellular cAMP levels via A_2 adenosine receptors.

Nevertheless Cronstein et al. found no change in intracellular cAMP upon adenosine or 5'-N-ethylcarboxamidoadenosine (NECA) additions during the first 2 minutes after FMLP stimulation.[10]

Purine and Pyrimidine Metabolism in Man IX,
edited by Griesmacher *et al.* Plenum Press, New York, 1998.

FMLP acts through two different G protein: Ga and Gp.[11] Ga leads to activation of PLA_2 and arachidonate release, Gp leads an increases of phosphatidylinositol turnover and intracellular Ca^{++} levels; however, neither of these signals appears to be a site at which adenosine controls FMLP-stimulated neutrophils function. In fact adenosine inhibits O_2^- generation by neutrophils both in presence or absence of extracellular calcium.

A possible mechanism by which adenosine could inhibit the FMLP stimulation by uncoupling FMLP receptors from signal transduction mechanism. Once occupied by ligand, FMLP receptors assume a high affinity configuration for FMLP and associate with cytoskeleton.[12] Subsequently these bound receptors are no longer capable of stimulating neutrophils function.

Occupancy of adenosine receptors promotes more rapid and complete association between FMLP receptors and the cytoskeleton. Thus adenosine A_2 receptor occupancy promotes more rapid desensitization of chemoattractant receptors by enhancing both the rate and the extent to which bound chemoattractant receptor associate with the cytoskeleton.

In conclusion adenosine inhibits FMLP stimulation by uncoupling G proteins from formyl peptide receptor. This interpretation is further supported by the observation that 5′ N-ethylcarboxamidoadenosine (NECA), an adenosine analogue, and adenosine inhibit G protein activity in neutrophils function.[13] On the other hand it could be possible that FMLP receptor occupancy may modulate the adenosine binding to its binding sites; so the aim of our studies was to investigate the FMLP modulation of [³H]-NECA binding to A_2 receptors in neutrophil membranes.

2. MATERIALS AND METHODS

[³H]-NECA (specific activity 25.7 Ci/mmol) were from New England Nuclear; NECA, N^6-(R) phenylisopropyladenosine (R-PIA), N^6-cyclopentyladenosine (CPA); 2-p-2-carboxyethylphenylamino- 5′ N-ethylcarboxamidoadenosine (CGS 21680); Theophylline were from Sigma. All other chemicals were from standard supplies.

Human neutrophils were isolated from heparinized whole blood of healthy donors essentially as describe by Boyum[14] using Lymphoprep (Nycomed Pharma, Oslo Norway). These cells, which are know to be essentially neutrophils (95 ± 2%) were washed twice with PBS at pH 7.4 containing 0.5% BSA and 0.13% EDTA.

Aliquots (50%) were diluted in Hanks' salt solution at pH 7.4 incubated at 4°C for 90 minutes; then they were stimulated with FMLP (10 μM/10^6 cells) and incubated at 37°C for 5 minutes.

Finally, neutrophils were centrifuged at 250 x g for 15 minutes at 25°C in a swing-out rotor.

For membrane preparations, cells were washed twice with 10 mM Tris-HCl buffer pH 7.4 containing 10 mM $MgCl_2$ in the presence of protease inhibitors (200 μg/ml bacitracine, 160 μ/ml benzamidine, 20 μg/ml trypsin inhibitor) (T1), centrifuged at 48,000 × g for 15 minutes at 4°C.

The pellet was diluted in 20 vol. of T1 buffer, treated with adenosine deaminase (2 U.I./ml) for 60 minutes at 37°C to remove endogenous adenosine, homogenized and washed twice with 50 mM Tris-HCl buffer pH 7.4 containing 10 mM $MgCl_2$ (T2).

Routine binding assays were performed as previously described.[15] Briefly incubation was performed on ice in a total volume of 250 μl containing 20 nM [³H]-NECA and 200–400 μg neutrophil membranes proteins in T2 buffer. Non specific binding was defined in the presence of 100 μM NECA or 100 μM R-PIA. The assay include 50 nM

CPA, a selective A_1 adenosine receptor ligand. After 90 min., the binding reactions were terminated either by vacuum filtration through Whatmann GF/C glass fiber filters, accompanied by three 4 ml washes with ice cold T2 buffer or, by centrifugation at 2900 × g for 15 min. at 4°C.

For saturation studies, membranes were incubated in T2 buffer, with 6–9 different concentrations of [³H]-NECA ranging from 10 nM to 500 nM.

Competition studies were carried out by incubation membranes in T2 buffer with 20 nM [³H]-NECA and up to 10–12 concentrations of NECA that ranged from 50 nM to2 mM. The maximum number of number of binding sites (Bmax) and affinity constant (Kd) were calculated by Scatchard analysis using EBDA, LIGAND program written by G.A. McPherson.[16]

Displacement studies were performed using adenosine analogues as agonist (CGS, R-PIA, CPA etc.) and xantine derivatives as antagonist (theophylline), in the absence and in the presence of 500 μM GTPγS.

3. RESULTS

[³H]-NECA binding was displaced by 100 μM R-PIA or 100 μM NECA to discriminate $A2_a$ adenosine receptor and A_2-like binding sites. The specific binding obtained in the presence of 100 μM NECA amounting to approximately 85% of total binding, only 20% [³H]-NECA bound to human neutrophil membranes could be displaced by N^6-substituted adenosine analogue such as R-PIA, that allow us to discriminate between A_2 adenosine receptor and A_2-like.

[³H]-NECA specific binding, obtained in the presence of 100 μM NECA, to human neutrophil membranes was saturable. Scatchard analysis were carried out both by saturation and competition experiments. The competition studies allowed to use a wider range of ligand concentration than those used in the saturation studies and therefore to investigate the low affinity binding site or the presence of more than one binding site.

Scatchard analysis of data, using a non linear curve-fitting program, revealed that a one-site model produce a significantly better fit than two site model ($p < 0.05$) both in control and in FMLP stimulated membranes.

Control neutrophil membranes showed the presence of a low-affinity [³H]-NECA binding site with a Kd value of 476 ± 34 nM and a Bmax value of 3696 ± fmol/mg (Fig. 1).

The data obtained showed also a modulatory effect of FMLP, in fact a Kd and Bmax of 322 ± 29 nM and 1550 ± 140 fmol/mg respectively were evidenced (Fig. 1).

The pharmacological profile of [³H]-NECA binding sites in our conditions were carried out using 2 Cl ADO, R-PIA, NECA and theophylline (Table 1).

Table 1. Competition equilibrium studies of ligand binding to adenosine binding sites

Ligand	Ki (μM)		
	Placenta	Platelets	Neutrophils
NECA	0.3	0.6	0.5–0.6
Theophylline	200	282	291
2-Cl-Adenosine	2	0.8	0.954
R-PIA	1000	1000	1000
GTPγ S	> 1000	> 1000	> 1000

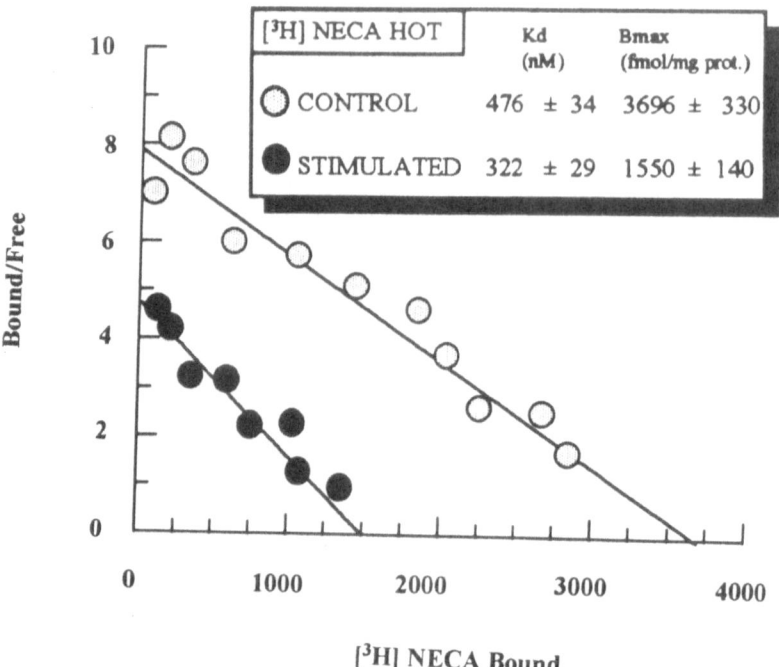

Figure 1. Representative Scatchard analysis of specific [³H]-NECA bound of control and stimulated neutrophil membranes.

The data obtained showed a different profile respect both A_1 than A_2 receptor subtypes and assimilable to an A_2-like activity.

In fact these values are comparable to the affinity values for the same ligands on adenotin receptors in platelets and placenta membrane.[17]

GTPγS modulation of [³H]-NECA binding site was induced as shown in the table, suggesting an exclusion of G protein coupling typical of A_2 adenosine receptor.

These results suggest that a low affinity [³H]-NECA binding site assimilable to adenotin protein coexist in neutrophil membranes with A_2 adenosine receptor. Moreover these sites appear to be mobulated by chemoattractant peptide FMLP.

4. DISCUSSION

Neutrophils play an important role on inflammation process, their chief function is phagocytosis and destruction of microorganism. However they also mediate many others events in acute, subacute and cronic inflammation.[18]

Adenosine and its analogues NECA, inhibit FMLP induced O_2^- generation, neutrophil aggregation, adherence and degranulation; this action is not mediate by cAMP increase.[1]

On the basis of several previous studies by different authors, Cronstein suppose a possible mechanism by which adenosine receptor occupancy might inhibit FMLP action.[13] He suggest that FMLP receptors assume a high affinity configuration for FMLP and associate with the cytoskeleton, so these bound receptors are no longer capable of stimulating neutrophil function.

A_2 adenosine receptor occupancy might uncouple FMLP receptor from signal transduction mechanism enhancing both the rate and the extent to which bound chemoattractant receptors associate with cytoskeleton.

On the other hand it could be possible that FMLP receptor occupancy may modulate the adenosine binding to its binding sites; so the aim of our studies was to investigate the FMLP modulation of [³H]-NECA binding to A_2 receptors in neutrophil membranes.

Nevertheless [³H]-NECA is able to label several binding sites such as A_1, A_2 and A_2-like ones.

To determine the A_2 binding component, we used CPA, an adenosine analogue as agonist, to exclude A_1 receptors component.

The data obtained showed that [³H]-NECA, in our experimental conditions, bound a binding sites that differ from typical A_1 and A_2 adenosine receptor subtypes suggesting an adenotin binding protein.

To evaluate the contribution of the A_2-like component to [³H]-NECA binding, the assay was carried out in the presence of 100 μM of R-PIA, that allow us to discriminate between A_2 adenosine receptor and A_2-like.[17] Under these conditions [³H]-NECA specific binding yielded a single straight line in both control and FMLP stimulated membrane preparation indicating the presence of a single class of binding sites. FMLP stimulation induced a decrease of the dissociation constant (Kd) from 476 to 322 nM and of maximum density (Bmax) from 36966 to 1550 fmol/mg.

These results suggested a peptide modulation on [³H]-NECA binding sites.

With the aim to evaluate the pharmacological profile of these low-affinity sites labeled by [³H]-NECA, specificity studies on unstimulated membranes were carried out using adenosine analogues as agonist and xantine derivatives as antagonist. 2-Cl-ADO, R-PIA, NECA and theophylline displaced [³H]-NECA binding with low affinities showing a pharmacological profile different from that observed for typical adenosine receptors and assimilable to an A_2-like activity. These values are comparable to the values for the same ligands on adenotin binding sites in platelet and placenta.[17]

Adenotin and other similar low-affinity adenosine binding proteins have binding properties that resemble the A_2 adenosine receptor.

These low affinity adenosine binding proteins are distinguished by submicromolar affinity for NECA and high binding density ranging from 4 to 26 pmol/mg protein.[19–20] This contrast with A_1 and A_2 adenosine receptors, which have nanomolar affinity for NECA and less than 1 pmol/mg protein density.[21–22] The lack of binding by N^6-substituted adenosine analogues and C^8-substituted xanthine derivatives distinguishes adenotin binding sites from A_1 and A_2 adenosine receptors, which bind these analogues.

Our data showed that in neutrophil membranes, such as in platelet membranes coexist two different adenosine binding sites: a high affinity sites, corresponding to about 10% of the total [³H]-NECA binding sites, and a low affinity binding sites, which represent the majority of [³H]-NECA binding sites, identifiable with an adenotine like protein.

To further confirm the hypothesis that [³H]-NECA binding sites modulated by FMLP are a low affinity adenosine binding site, we evaluated the modulation of the binding by the non hydrolyzable guanine-nucleotide GTP γS.

The data obtained showed that [³H]-NECA binding properties not change with the addiction of guanine nucleotide. Therefore, adenotine proteins cannot account for signal trasduction to a guanine nucleotide regulatory protein. Thus, since A_2 adenosine receptor seems to be linked only to Gs protein producing an increase of cAMP levels, and cAMP is not necessary to determine uncoupling between FMLP receptor and its G-protein, our results allow us to suppose that this uncoupling could be due by adenotin activity. In this way FMLP induced modulation on adenotin binding sites could represent a feed-back mechanism of control.

REFERENCES

1. Daval J.L., Nehlig A. and Nicolas F.1991. Physiological and pharmacological properties of adenosine: therapeutic implications. Life Science. 49:1435.
2. Van Calker D.M., Muller M. and Hamprecht B. 1979. Adenosine regulates via two different types of receptors, the accumulation of cyclic AMP in cultured brain cells. J. Neurochem. 33:999.
3. Robets P.A., Newby A.C., Hallet M.B. and Campbell A.K. 1985 Inhibition by adenosine of reactive oxygen metabolite production by human polymorphonuclear leucocytes. Biochem. J. 277: 669.
4. Cronstein B.N., Levin R.I., Belanoff J., Weissmann G. and Hirschhorn R. 1986. Adenosine : an endogenous inhibitor of neutrophil-mediated injury to endothelial cells. J. Clin. Invest. 78:760.
5. Cronstein B.N., Kramer S.B., Weissmann G. and Hirschhorn R. 1983. Adenosine: a physiological modulator of superoxide anion generation by human neutrophils. J. Exp. Med. 158:1160.
6. Marone G., Thomas L. and Lichtenstein L. 1980. The role of agonist that activate adenylate cyclase in the control of cAMP metabolism and enzyme release by human polymorphonuclear leukocytes. J. Immunol. 125:2277.
7. Cronstein B.N., Rosenstein E.D., Kramer S.B. Weissmann G.and Hirschhorn R. 1985. Adenosine: a physiologic modulator of superoxide anion generation by human neutrophils. Adenosine acts via an A2 receptor on human neutrophils. J. Immunol. 135:1366.
8. Cronstein B.N. 1994. Adenosine, an endogenous anti-inflammatory agent. J. Appl. Physiol. 76(1):5.
9. Verghese M.W., Smith C.D., Snyderman R. 1985. Potential role for a guanine nucleotide regulatory protein in chemoattractant receptor mediated polyphosphoinositide metabolism, Ca^{++} mobilization and cellular responses by leukocytes. Biochem. Biophys. Res. Commun. 127: 450.
10. Cronstein B.N., Kramer S.B., Rosenstein E.D., Korchak H.M., Weissmann G.and Hirschhorn R. 1988. Occupancy of adenosine receptors raises cyclic AMP alone and in synergy with occupancy of chemoattractant receptors and inhibits membrane depolarization. Biochem. J. 252:709.
11. Cockcroft S. and Stutchfield J. 1989. The receptors for ATP and fMetLeuPhe are independently coupled to phospholipases C and A_2 via G-protein(s).Biochem. J. 263:715.
12. Cronstein B.N. and Haines K.A. 1992. Stimulus-response uncoupling in the neutrophil. Biochem. J. 281:631.
13. Cronstein B.N., Hains K.A., Kolasinski S.L. and Reibman J.1992. Occupancy of G alpha s-linked receptors uncouples chemoattractant receptors from their stimulus-trasduction mechanisms in the neutrophil. Blood 80:1052.
14. Boyum A. 1968. Isolation of mononuclear cells and granulocytes from human blood. Isolation of mononuclear cells by cenrifugation and sedimentation at 1 x g. Scand. J. Clin. Lab. Invest. 21:77.
15. Martini C., Di Sacco S., Tacchi P., Bazzichi L., Soletti A., Bondi F., Ciompi M.L., and Lucacchini A. 1991. A_2 adenosine receptors in neutrophils from healthy volunteers and patients with rheumatic disease. Purine and Pyrimidine Metabolism in Man VI part A, Plenum Press, New York, 459.
16. McPherson G.A. Kinetic, EBDA, Ligand Lowry. A collection or radioligand binding analysis programs. Elsevier. Cambridge
17. Zolnierowicz S., Work C., Hutchison and Fox I.H. 1990. Partial separation of platelet and placental adenosine receptors from adenosine A_2-like binding protein. Mol. Pharmacol. 37:554.
18. Sandborg R.R and Smolen J.E. 1988. Biology of disease : early biochemical events in leukocyte activation. Laboratory Investigation 59(3):300.
19. Hutchison K.A. and Fox I.H. 1989. Purification and characterization of the adenosine A_2-like binding site from human placental membranes. J. Biol. Chem. 264:19898.
20. Hutchison K.A., Nevins B., Perini F.and Fox I.H.1990. Soluble and membrane associated human low-affinity adenosine binding protein (adenotin): properties and homology with mammalian and avian stress proteins. Biochemistry 29 :5138.
21. Gavish M., Goodman R.R. and Snyder S.H.1982. Solubilized adenosine receptors in the brain : regulation by guanine nucleotides. Science 215:1633.
22. Stiles G. 1985. The A_1 adenosine receptor. Solubilization and characterization of a guanine nucleotide-sensitive form of the receptor. J. Biol. Chem. 260:6728.

18

EFFECTS OF NUCLEOSIDE TRANSPORT INHIBITORS AND ADENINE/RIBOSE ON ATP CONCENTRATION AND ADENOSINE PRODUCTION IN CARDIAC MYOCYTES

Kameljit K. Kalsi, Ryszard T. Smolenski, and Magdi H. Yacoub

Heart Science Centre
National Heart and Lung Institute
Harefield Hospital, Harefield, United Kingdom

INTRODUCTION

Adenosine is an important catabolite of adenine nucleotides with potent coronary vasodilatory, antiplatelet, antileukocyte and antiarrhythmic properties.[1-4] The clinical relevance of these actions of adenosine make it an important therapeutic and diagnostic tool.

To produce adenosine in a localised manner, current investigations include studies on increasing endogenous adenosine concentration by preventing its re-uptake or by inhibiting the enzymes responsible for its degradation. Nucleoside transport inhibitors increase adenosine concentration by preventing its re-uptake.[5] Cardioprotective effects have been demonstrated with dipyridamole,[6] mioflazine,[7] NBTI,[8] and dilazep.[9] However little attention has been given to the possible deleterious consequences of increased adenosine efflux from the cell and the possible depletion of the nucleotide pool. The effect of the combined application of these compounds with precursors of nucleotides such as adenine and ribose are also not known. The aim of the present study was therefore to evaluate the consequences of nucleoside transport inhibition on ATP concentration and adenosine production in cardiomyocytes and to study metabolic effects of the combined application with adenine and ribose.

METHODS

Cardiomyocytes were isolated using collagenase perfusion technique from the hearts of Sprague-Dawley rats weighing 250–300 g as described previously.[10] The final preparation contained 70–85% of rod shaped cells. After isolation and purification cells were

Purine and Pyrimidine Metabolism in Man IX,
edited by Griesmacher *et al.* Plenum Press, New York, 1998.

finally suspended in buffer containing 120 mM NaCl, 2.6 mM KCl, 1.2 mM $MgSO_4$, 1.2 mM KH_2PO_4, 1 mM $CaCl_2$, 10 mM HEPES (pH 7.4), 11 mM glucose, 2 mM pyruvate and 2% bovine serum albumin. Five minutes prior to start of experiment, erythro-9(2-hydroxy-3-nonyl) adenine (EHNA) was added at 5 µM concentration.

Dipyridamole (DIPY) and nitrobenzylthioinosine (NBTI) at a final concentration of 10 µM of DIPY and 1µM of NBTI with and without the addition of 100 µM adenine and 2.5 µM ribose. The incubation tubes were set up as follows: **A.** EHNA alone (CONTROL), **B.** with EHNA plus adenine and ribose, **C.** with EHNA and DIPY, **D.** with EHNA and NBTI, **E.** with EHNA, DIPY plus adenine and ribose, **F.** with EHNA, NBTI plus adenine and ribose.

Incubation was carried out at 37°C in a shaking water bath for 480 min and initial samples were taken at the start. Incubations were terminated with the addition of $HClO_4$ and neutralized supernatants were analysed by HPLC. The reversed-phase method used for determination of ATP and adenosine concentration in the myocytes, has been described in detail previously.[11,12] The myocyte pellet was solubilized in 0.5 M NaOH for protein concentration using the method of Bradford.[13]

RESULTS

(Fig. 1) shows concentration of ATP after 480 minutes. There is a marked decrease in the ATP content when the cells are incubated with Dipyridamole. With the addition of adenine /ribose and dipyridamole there was a 12% increase in ATP content after 8 hours, while with NBTI plus adenine and ribose showed a 30% increase in ATP content.

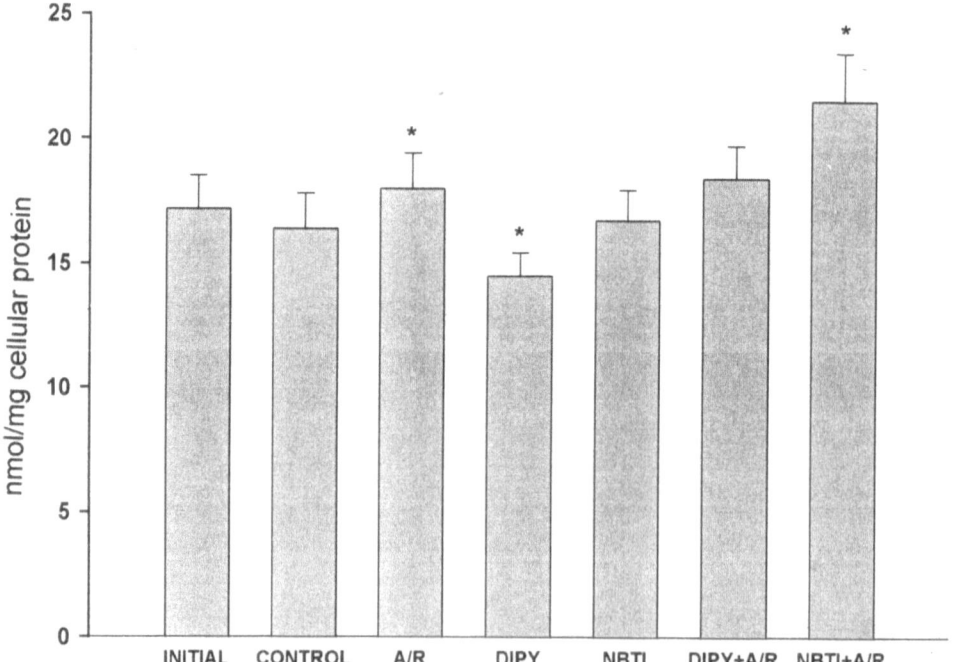

Figure 1. ATP concentration in isolated rat cardiomyocytes incubated for 480 min in HEPES-buffered physiological medium containing EHNA with additional presence of: A/R = Adenine/Ribose; DIPY = Dipyridamole; NBTI = Nitrobenzylthioinosine. Values represent the means ± SEM, n = 9–16. *p < 0.05 v initial.

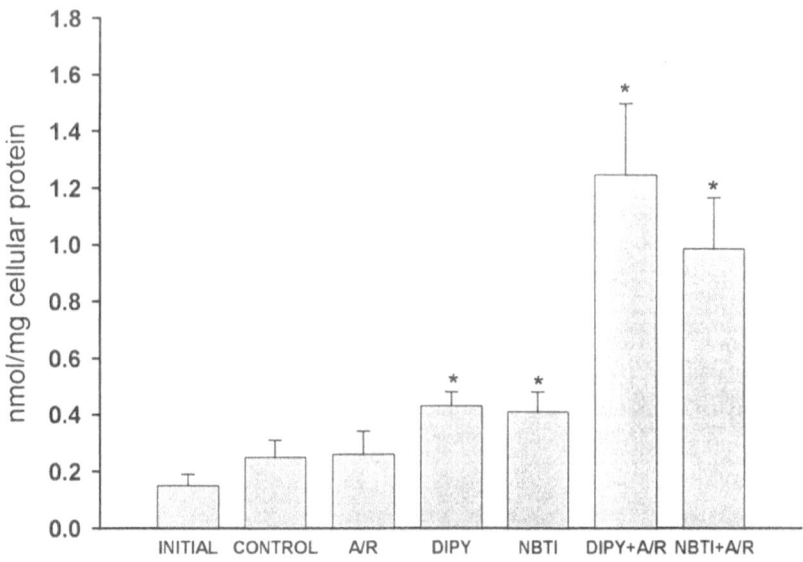

Figure 2. Adenosine concentration in isolated rat cardiomyocytes incubated for 480 min in HEPES-buffered physiological medium containing EHNA with additional presence of: A/R = Adenine/Ribose; DIPY = Dipyridamole; NBTI = Nitrobenzylthioinosine. Values represent the means ± SEM, n = 9–16. *p < 0.05 v initial.

Adenosine formation in the presence of EHNA alone or in the adenine and ribose group alone was not stimulated (Fig. 2). There was an increase in adenosine concentration in the nucleoside transport inhibitors groups but the greatest change seen was in the groups where adenine and ribose were present in addition to dipyridamole and NBTI.

DISCUSSION

It has been suggested that NTI could further increase adenosine release by inhibiting the uptake of adenosine to a larger extent than its release.[14,15] This in turn would lead to a deprivation in the cells of the nucleotide precursors. It is therefore important to establish what are the effects of long term exposure to NTI in single cells on both ATP levels and adenosine concentration. We have shown that the nucleotide pool decreased significantly in the cells treated with DIPY after 8 hours exposure. However, this does not seem to be the general effect of NTI since it was not observed with NBTI. DIPY exerts many other non-specific effects not related to nucleoside transport inhibition, such as inhibition of phosphodiesterases, increase in prostacyclin formation, interference with the transport of glucose, choline and phosphate[16] and these mechanisms could lead to further detrimental effects.

Supply of precursors alternative to the pathway of adenosine rephosphorylation of adenosine—ribose and adenine, which are involved in the pathway of adenine nucleotide synthesis—may provide several benefits in cells treated with NTI and adenosine deaminase inhibitors. These could prevent alterations of the nucleotide pool and further enhancement of adenosine production due to the increased nucleotide precursor pool. We have shown that the presence of adenine and ribose markedly increases ATP pool in the cardiac cells but no increase in adenosine concentration was observed. However, combined application of adenine/ribose with NTI increased not only ATP level but also adenosine concentration. This increase in adenosine concentration was markedly greater in cells treated with DIPY while ATP increase was smaller than with cells treated with NBTI. This could be again related to a number of known unspecific effects of DIPY.

Nucleoside transport inhibition offers major promise in the regulation of localised adenosine production. We have demonstrated that the metabolic consequences of the application of some of the drugs in this class, may induce deleterious effects in cardiac myocytes, such as a decrease in ATP concentration, but it is possible to reverse this by providing the precursors of nucleotides—adenine and ribose.

ACKNOWLEDGMENTS

We gratefully acknowledge the support of the British Heart Foundation (Grant No. PG-96/194) and the Harefield Heart Transplant Trust for this work.

REFERENCES

1. R.M. Berne, The role of adenosine in the regulation of coronary blood flow, *Circ. Res.* 47:807(1980).
2. R.J. Bache and D.J. Dymek, Local and regional regulation of coronary vascular tone, *Prog. Cardiovasc. Dis.* 24:191(1981).
3. K. Mubagwa, K. Mullane, and W. Flameng, Role of adenosine in the heart and circulation. *Cardiovascular Research* 32:797(1996).
4. A.C. Rankin, R. Brooks, J.N. Ruskin, and B.A. McGovern, Adenosine and the treatment of supraventricular tachycardia, *Am. J Med.* 92:655(1992).
5. H. Van Belle, Nucleoside transport inhibition: a therapeutic approach to cardioprotection via adenosine? *Cardiovasc. Res.* 27:68(1993).
6. Knabb R.M., J.M. Gidday, S.W. Ely, R. Rubio, and R.M. Berne, Effects of dipyridamole on myocardial adenosine and active hyperemia, *Am. J. Physiol.* 247:H804(1984).
7. A. Buchwald, B.R. Ito, and W. Schaper, Influence of Mioflazine on canine coronary blood flow and on adenine nucleotide and nucleoside content under normal and ischemic conditions, *J. Cardiovasc. Pharmacol.* 10:213(1987).
8. A. Abd Elfattah and A.S. Wechsler, Separtation between ischemic and reperfusion injury by site specific entrapment of endogenous adenosine and inosine using NBMPR and EHNA. *J. Card Surg* 9 [suppl]: 387(1994).
9. J.B. Gupta, S.D. Seth, U. Singh, and D.S. Vyas, Effects of dilazep in evolving myocardial infarction in dogs: a biochemical study. *Int. J. Cardiol.* 23:165(1989).
10. R.T. Smolenski, J. Schrader, H. de Groot, and A. Deussen, Oxygen partial pressure and free intracellular adenosine of isolated cardiomyocytes, *Am. J. Physiol.* 260:C708(1991).
11. R.T. Smolenski, D.R. Lachno, S.J.M. Ledingham, and M.H. Yacoub, Determination of sixteen nucleosides, nucleotides and bases using high performance liquid chromatography and its application to the study of purine metabolism in hearts for transplantation. *Journal of chromatography* 527:414(1990).
12. R.T. Smolenski and M.H. Yacoub, Liquid chromatographic evaluation of purine production in the donor human heart during transplantation, *Biomed. Chromatogr.* 7:189(1993).
13. M.M. Bradford, A rapid and sensitive method for the quantitation of microgram quantities of protein utilizing the principle of protein-dye binding, *Anal. Biochem.* 72:248(1976).
14. S. Afonso and G.S. O'Brien, Mechanism of enhancement of adenosine action by dypridamole and lidoflazine in dogs. *Int. Pharmocodyn. Ther.* 194:181(1971).
15. N. Kolassa, K. Pfleger, and W. Rummel, Specificity of adenosine uptake into the heart and inhibition by dipyridamole. *Eur. J Pharmacol.* 9:265(1970).
16. H. Van Belle, Myocardial protection using nucleoside transport inhibitors, in: "Purines and myocardial protection.", A.A. Abd-Elfattah et al., Kluwer Academic, Boston/Dordrecht/London, pp. 183–196 (1996).

19

PURIFICATION OF A PROTEIN INVOLVED IN ADP FORMATION

R. Leoncini, D. Vannoni, M. C. Di Pietro, I. Ceccarelli, F. Carlucci,
R. Guerranti, and E. Marinello

Istituto di Biochimica e di Enzimologia
Università degli Studi di Siena
Siena, Italy

1. INTRODUCTION

An attempt was made to purify a new enzyme from rat liver. The enzyme catalyzes the formation of ADP from AMP without consumption of ATP, and is widely distributed in different rat organs. The enzyme produces ADP and adenosine from two molecules of AMP and we have called it AMP-AMP phosphotransferase (AMP-AMP PT). *De novo* synthesis of ADP starting from lower energy compounds has never been previously reported. Here we report the procedure for partial purification of the enzyme using rat liver as biological source.

2. MATERIALS AND METHODS

2.1. Purification

fasted male Wistar rats were killed and their liver rapidly excised and homogenized at 25% in 50 mM TRIS HCl buffer (pH 6.5) containing 0.1 mM $MgCl_2$ and 0.1 mM DTT (buffer A). The homogenate was ultracentrifuged for 1h at 240,000 × g. The supernatant was precipitated with ammonium sulfate between 50–75% saturation. The pellet, obtained by centrifugation at 7,000 × g for 20 min, was resuspended in buffer A and run on Sephacryl S-200HR gel eluting with 10mM Hepes (pH 6.5) plus $MgCl_2$ and DTT (buffer B). The active fractions were pooled and loaded on a DEAE Sepharose FF column equilibrated with buffer B and eluted by a step-wise gradient of NaCl from 0.01 to 0.25 M. Active fractions were concentrated in vacuo, resuspended in a minimal amount of buffer A and then run on Sephadex G75SF eluting with buffer B. Active fractions were loaded on a PBE94 column equilibrated with the same buffer and developed with a step-wise gradient

Purine and Pyrimidine Metabolism in Man IX,
edited by Griesmacher *et al.* Plenum Press, New York, 1998.

99

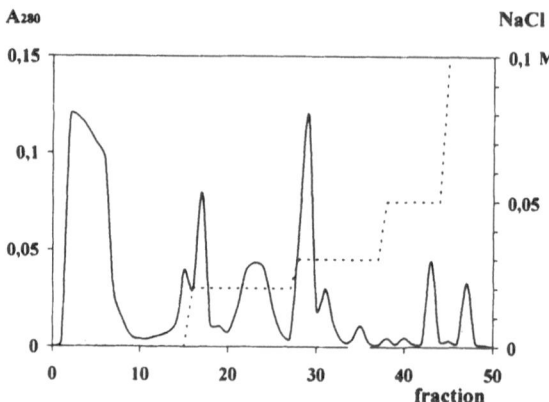

Figure 1. Elution pattern in PBE94 chromatography.

of NaCl from 0.01 to 0.25 M. The elution pattern is reported in Figure 1. A typical procedure is reported in Table 1.

2.2. Assay Mixture

The activity was followed in assay mixtures containing 4.0 mM AMP, 50 mM TRIS
HCl (pH 6.5), 1.0 mM $MgCl_2$, and protein to 0.1 mg. The mixtures were incubated for 60
min at 37°C. Substrate and products were quantified by HPLC.

2.3. HPLC Analysis

HPLC analysis and determination of nucleosides and nucleotides were carried out
using a Perkin Elmer 1020LC plus system, consisting of a Mod. 250 biocompatible binary
pump, a Mod. 235C DA detector, monitoring at 254 nm and a Mod. 1020 Nelson LC-plus
controller. A ready-to-use prepacked column (Partisil Sax Rac-II, 10 cm × 4.6 mm, Whatman) washed with 5.0 mM ammonium phosphate (pH 2.9) completed the analytical system. Elution was a linear gradient from 0 to 100% in 10 min with 0.5 M ammonium
phosphate buffer (pH 4.8). The flow rate was 1.5 ml/min. The volume injected ranged
from 10 to 50 µl. The retention times for adenosine, AMP, IMP, ADP, ATP were 1.0, 2.9,
4.5, 6.1 and 9.4 min respectively. The lowest detection limit for these metabolites was 10
µmoles/l (Figure 2).

Table 1. Purification steps of AMP-AMP phosphotransferase

Step	Total protein (mg)	Specific activity (nmoles/h/mg protein)	Total activity (units)	Yield (%)	Purification (fold)
Homogenate	3132.5	5.2	34036.4	100.00	1
P_{75}	1071.1	17.3	18581.8	54.61	3
Sephacryl S-200 HR	50.6	297.9	15093.6	44.36	44
DEAE Sepharose FF	16.7	425.9	7112.5	20.91	62
Sephadex G75 SF					
(Fraction A)	3.2	815.7	2610.2	7.67	117
PBE 94					
(Fraction B)	2.4	950.0	2320.5	6.8	183

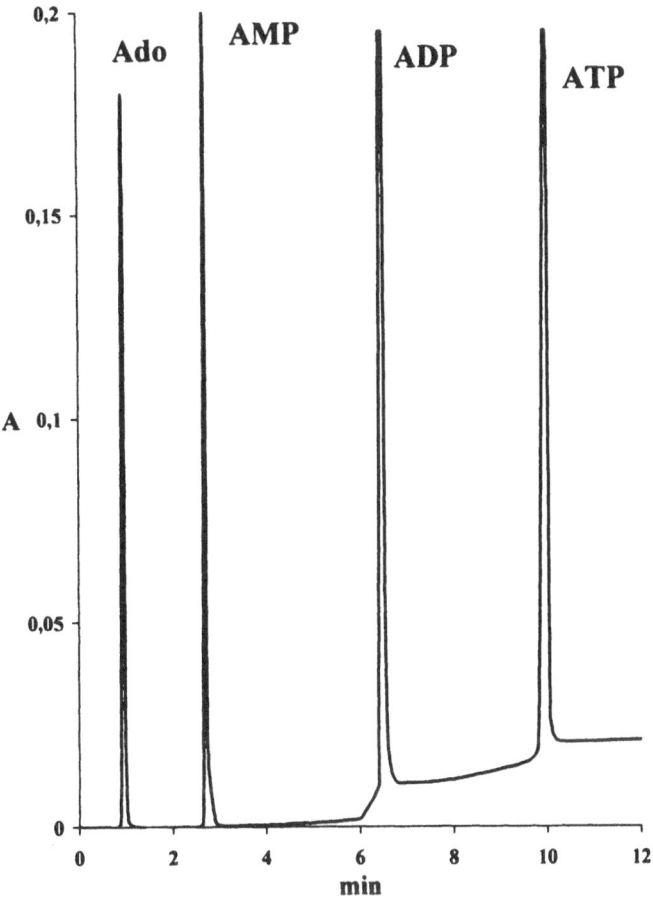

Figure 2. Typical HPLC separation of standards.

2.4. SDS-PAGE

A 7.5% SDS-PAGE was carried out under non-reducing conditions according to Laemmli.[1]

2.5. Protein Assay and Enzyme Activities

Protein content was measured according to Bradford[2] using BSA as standard. AMP-AMP PT specific activity was expressed in nmoles product/h/mg protein.

3. RESULTS

SDS PAGE analysis of the final preparation showed negligible protein contaminant. We tested for many enzymes involved in AMP metabolism, such as: adenosine deaminase, AMP deaminase, 5'-nucleotidase, adenosine kinase and ADPase, but none of them were detected.

4. CONCLUSION

Future research will include the obtaining of the homogenity of the new protein and the studying of its kinetic properties and structural analysis.

REFERENCES

1. Laemmli, U.K. (1970) Cleavage of structural proteins during the assembly of the head of bacteriophage T4. Nature *227*, 680–685
2. Bradford, M. M. (1976) A rapid and sensitive method for the quantitation of microgram quantities of protein utilizing the principle of protein-dye binding. Anal Biochem. *72*, 248–254.

20

ADP FORMATION IN RAT LIVER

D. Vannoni, R. Leoncini, M. C. DiPietro, M. D'Ercole, A. Tabucchi, F. Rosi, and E. Marinello

Istituto di Biochimica e di Enzimologia
Università di Siena, Italy

1. INTRODUCTION

The enzymes that phosphorylate nucleosides fall into two groups on the basis of the phosphate donors and the reaction mechanism. One group, the nucleoside kinases, utilizes nucleoside triphosphates as donors in direct displacement reactions; the other group, the nucleosides phosphotransferases, catalyzes the phosphorylation reaction with nucleoside monophosphates as donors, converting the acceptor nucleoside to its monophosphate and the nucleoside monophosphate to its unphosphorylated nucleoside.

In this report we describe a novel phosphotransferase which uses adenosine monophosphate (AMP) as both donor and acceptor of the phosphate group to produce ADP and adenosine. The energy balance of the reaction is discussed in relation to the coupled reaction with adenosine deaminase (ADA) which contaminated the enzyme preparation or was added to the incubation mixture. Some properties of the enzyme and the reaction are reported.

2. MATERIALS AND METHODS

2.1. Incubation Mixture

The incubation mixture contained 4.0 mM AMP, 50 mM TRIS HCl (pH 6.5), 1 mM $MgCl_2$. We used rat liver supernatants (obtained ultracentrifuging 25% homogenates in 20 mM TRIS HCl, pH 6.5) or other pure enzyme fractions, in concentrations ranging from 0.02 to 0.1 mg as enzyme source.

2.2. HPLC Analysis

Substrates and reaction products (AMP, ADP, ATP) were quantified by HPLC using a Partisil SAX RACII (10 cm × 4.6 mm, from Whatman). A Supelcosil LC_{18} (25 cm × 4.6 mm, Supelchem) was used for analysis of nucleosides and bases (adenosine, inosine,

Purine and Pyrimidine Metabolism in Man IX,
edited by Griesmacher *et al.* Plenum Press, New York, 1998.

hypoxantine). A HPLC Perkin Elmer 1020LC plus system, consisting of a Mod. 250 bio-compatible binary pump, a Mod. 235C DA detector, monitoring at 254 nm and a Mod. 1020 Nelson LC-plus controller was used. For separation of nucleotides, the column was eluted with a linear gradient of 0.005 M ammonium phosphate buffer (pH 2.9) and 0.5 M ammonium phosphate buffer (pH 4.8) from 0 to 100% in 10 min. The flow rate was 1.5 ml/min. For nucleosides and bases separation elution was performed with 10.0 mM ammonium phosphate buffer (pH 5.5) and methanol (10:1) at a flow rate of 1.0 ml/min. The volume injected ranged from 10 to 50 μl.

2.3. Enzyme Activity

Activity was expressed as nmoles of ADP formed in 1 hour of incubation at 37°C. Specific acrivity was per mg protein.

2.4. Chemicals

The reagents used were the purest commercially available. AMP purity was tested by several methods (HPLC, capillary electrophoresis, mass spectrometry). No compounds which could interfere with the reaction were found.

3. RESULTS

Male Wistar rat livers were the biological source of our enzyme for the experiments related to enzyme properties and purification. Other organs from different animals (calf, swine, lamb, kid) were tested. Table 1 compares enzyme activities in liver and spleen of various species in comparison to rat.

3.1. Reaction Mechanism

We hypotheized that a new reaction leading to ADP occurred and we called the enzyme involved AMP-AMP phosphotransferase. We assumed that two moles of AMP interacted transfering a phosphorus to form ADP and adenosine. ADA, present in the impure

Table 1. AMP-AMP phosphotranserase activity in liver and spleen of different animal species[a]

Organ	Animal	Protein mg/ml	Specific acitivity	Total activity (units)	Units/g tissue
Liver	Rat	15.33	23.25	12888	2.59
	Calf	12.93	14.23	2208	4.74
	Swine	18.33	5.89	1296	1.96
	Lamb	12.86	20.53	2112	10.26
	Kid	12.73	27.34	2784	13.67
Spleen	Rat	13.20	1.97	91	2.19
	Calf	9.20	1.30	36	0.11
	Swine	9.60	2.29	66	0.19
	Lamb	11.06	1.99	176	0.99
	Kid	14.86	2.02	240	1.01

[a]Specific activity was expressed as units/mg protein; 1 unit was 1 nmole of ADP formed in 1 hour in the incubation mixture. Assay mixture composition is reported in "Materials and Methods".

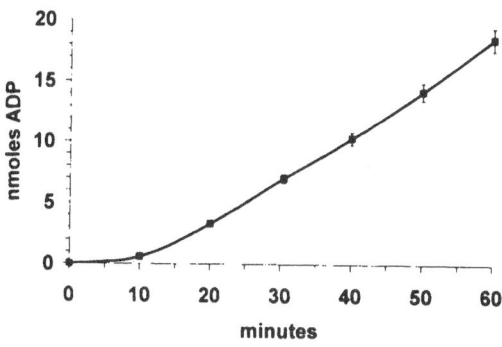

Figure 1. Progress curve of ADP.

enzyme preparation, immediatly converted adenosine to inosine which was detected in the reaction products. In purest enzyme fractions without ADA, the activity of the enzyme was remarkably reduced (by more than 90%). If standard ADA was added to the incubation mixture, activity was restored.

3.2. Enzyme Properties

We studied some characteristics of the phosphotransferase reaction.

1. Optimal pH was found to be 6.5. The enzyme was completely inactive at pH > 7.5 and < 5;
2. Progress curve (Figure 1) shows linearity from 10 to 70 minutes of incubation with an initial lag time;
3. Several AMP and Mg^{++} concentrations were tested. In both cases the activity followed a sygmoidal pattern. An apparent K_m of 0.85 mM was calculated for AMP; Mg^{++} ions were essential for the reaction: maximum activity was detected between 0.1 to 1.4 mM.

4. DISCUSSION

A new reaction forming a high energy level compound (ADP) starting from a low energy level compound (AMP) is reported. It is not yet clear whether the enzyme involved is already known. Regarding the energy balance of the reaction, we propose a functional association between AMP-AMP phosphotransferase and ADA, which contaminated the enzyme preparation or was added to the incubation mixture .The hydrolysis of adenosine, formed in the phosphotransferase reaction, to inosine and ammonia ($\Delta G°$ ranging from −4 to −10 Kcal/mol)[1] could fill the energy difference necessary to drive the reaction. This association could explain the loses of phosphotransferase activity observed during purification when ADA was removed and the initial lag time of the reaction: the activity increased only when enough adenosine had been accumulated as substrate for ADA activity.

Kid liver was the richest source of the enzyme among the species examined.

REFERENCE

1. Ma, P. F., and Fisher, J. R. (1966) Comp. Biochem. Physiol. 19, 799–807

21

A₁ ADENOSINE RECEPTORS IN HUMAN NEUTROPHILS

Electron Microscope Localization

C. Martini,[1] L. Trincavelli,[1] M. Fiorini,[1] F. Salvetti,[1] U. Montali,[2] A. Falleni,[3] V. Gremigni,[3] and A. Lucacchini[1]

[1]Istituto Policattedra di Discipline Biologiche
[2]Dipartimento di Scienze dell'Uomo e dell'Ambiente
[3]Dipartimento di Biomedicina Sperimentale
Infettiva e Pubblica
Università di Pisa
Pisa, Italy

1. INTRODUCTION

Adenosine is an ubiquitous nucleoside that mediates several important physiological effects, trough the binding of three pharmacologically distinct receptor subtypes, named A_1, A_2 and A_3.[1] Adenosine receptor subtypes has been distinguished on the basis of the effect of adenosine on the cellular content of cAMP: occupancy of A_2 adenosine receptors induces an increase in cAMP level in several cells types, by contrast stimulation of A_1 receptors causes the opposite effect. These effects are mediated by a coupling of these receptor subtypes with Gs and Gi transduction-proteins, respectively.[2]

In neutrophils adenosine and its analogues inhibit O_2^- generation, phagocytosis and adherence by occupancy of specific adenosine A_2 receptors, while occupancy of A_1 adenosine receptors enhances chemotaxis, phagocytosis and adherence.[3]

The susceptibility of adenosine A_1 receptors to pertussin toxin indicated that A_1 receptors were linked to Gi signal transduction proteins.[4] This observation was confirmed by cloning the A_1 receptors and by their characterization as a member of the seven transmembrane-spanning family of G protein linked receptors.[5] Recent findings showed that in neutrophils the effect of adenosine on chemotaxis was mediated by A_1 receptors which are coupled to Gi-proteins.

We have demonstrated using potent adenosine analogues that the direct labeling of A_1 adenosine receptors and their ultrastructural localization on the neutrophil surface by

Purine and Pyrimidine Metabolism in Man IX,
edited by Griesmacher *et al.* Plenum Press, New York, 1998.

107

electron microscopy are important tools to understand the effects of adenosine modulation on the cell function.[6]

The difficulty in detecting A_1 adenosine receptors by the radioligand binding technique, due to the low receptors concentration, stimulated us to develop a covalently linked ^{6}N-cyclohexyladenosine (CHA)ox-biotin that was easily recognizable by the conjugation with colloidal gold. This probe was biologically active and could easily bound to adenosine receptors on or within cells. In this investigation we used this probe to visualize A_1 adenosine receptors on human neutrophil surface and to study the traffic of ligand receptor complex inside the cell.

2. MATERIALS AND METHODS

2.1. Neutrophils Preparation

Human neutrophils were isolated from heparinized whole blood of healthy donors essentially as described by Boyum,[7] using a mixture of 9.6% (w/v) sodium metrizoate and 5.6% polysaccharide (Lymphoprep, Nycomed Pharma, Oslo Norway).

Bloods were collected into tubes containing anticoagulant (heparin), and were diluted by addiction of an equal volume of plasma substitute (Emagel, Behring) for 40 min to obtain a preliminary erythrocytes sedimentation.

The supernatant was carefully layered over Lympoprep (5:3) and was centrifuged at 800 × g for 20 min at room temperature in swing-out rotor.

After centrifugation neutrophils were sedimented to the bottom of the tube. The contaminating erythrocytes were lysed by 1% ammonium oxalate and 0.01% sodium azide solution.

These cells, essentially neutrophils (95 ± 2%), were washed twice with PBS at pH 7.4 containing 0.5% BSA and 0.13% EDTA.

Finally, neutrophils were centrifuged at 250 × g for 15 min at 25°C in a swing-out rotor.

2.2. Synthesis of the Probe

Oxidized CHA preparation was performed essentially as described by Gilham:[8] 1 mM CHA and 1 mM metaperiodate cold solution, in water pH 6.0, were mixed together and the reaction was allowed to proceed for 60 min at 0°C in the dark. The reaction was monitored by thin-layer cromatography on silica with methanol/chloroform 1:9 as developing solvent. Iodate was removed by filtration through Sep-Pack Cartridge, the oxidized CHA was eluted with acetonitrile and the solvent was evaporated to dryness.

Biotin XX hydrazide (1.0 mg) previously dissolved in DMF was diluted at 1 mM final concentration with 80 mM AcOH/AcO $^-$Na$^+$ buffer pH 5.5. To this solution, 1 mM CHAox dissolved in the same buffer was quickly added, and the reaction was carried out overnight with shaking at 4°C. The reaction mixture was analyzed at room temperature by HPLC and the biotin conjugate was isolated.

2.3. Electron Microscope Localization

Neutrophils (11 × 10^6 cells) were incubated with CHAox-biotin (5 μM) either in the absence (total binding) or in the presence (non specific binding) of 160 μM R-phenilisopropyladenosine (R-PIA) in T_1 buffer (10 mM Hepes-TRIS, 145 mM NaCl, 5 mM KCl, 5 mM $MgCl_2$, 1 mM $CaCl_2$, 10 mM Glucose pH 7.4) for 60 min at 0°C or for 5–60 min at 37°C. After incubation, the samples were centrifuged for 1 min in an Eppendorf centrifuge and rinsed three times with T_1 buffer.

Neutrophil pellets were cut in small pieces (1 mm³) and fixed with 0.5% glutaralde-hyde in 0.1 M cacodylate buffer (pH 7.2) for 60 min at 4°C and postfixed with 1% os-mium tetroxide in the same buffer for 60 min at room temperature. Following dehydration in a graded series of ethanols, specimens were infiltred with Unicryl resin and polymer-ized by UV light for two days at 4°C. Ultrathin sections (60–80 nm thick) were cut with a Reichert-Jung Ultratome E and placed on formvar-carbon coated nickel grids (200 mesh) then their sections were incubated face down on a drop of PBS containing 0.5% bovine serum albumin and 0.1% gelatin for 30 min. at room temperature, incubated on diluted (1:50) streptavidin-colloidal gold (10 nm) and then finally in 2% glutaraldehyde in PBS for 15 min. Sections were stained with uranyl acetate and lead citrate and examined in a Jeol 100 SX Electron Microscope.

3. RESULTS AND DISCUSSION

In the present work we demonstrated the presence of A_1 adenosine receptors using a covalently linked probe CHAox-biotin. The coupling of the biotin portion of the biotiny-lated CHA to streptavidin-colloidal gold conjugate provided us a tool for cytochemical localization of A_1 adenosine receptors via electron microscopy. Therefore this macro-molecular probe has been successfully used to visualize A_1 adenosine receptors on the sur-face of human neutrophils and to study the ligand-receptor complex internalization.

For this purpose CHA, an A_1 specific ligand, was covalently bound to biotin XX hydrazide (B-2600) by condensation of the hydrazidic group with aldheidic functions resulting from periodate oxidation of the cis-vicinal hydroxyl groups in the ribose moiety of CHA. So, the coupling of the biotin portion of the biotinilated CHAox to streptavidin-colloidal gold conjugate provided a macromolecular probe easily identifiable as an elec-tron-dense particle.

Neutrophils from untreated specimens were roundish (9–10 μm in diameter) and had a polilobed nucleus with heterocromatin characteristically massed on the inner nuclear membrane (Fig. 1). Ultrastructurally, the CHAox-biotin-streptavidin gold complex was viewed as electron-dense gold particles (10 nm) localized on the neutrophil surface and/or within membrane-bound cytoplasmic components.

At 0 °C the gold particles appeared distributed on the neutrophil surface as single particles (Fig. 2) or small groups of particles (Fig. 3). Following incubation at 37°C with CHAox-biotin, internalization of the complex occurred initially after 5 min and was more evident after 60 min.

Pinocytotic smooth-surfaced vesicles, commonly containing one to several gold particles, were visible under the plasma membrane (Figs. 4–5) as well as at different depth in the cytoplasm. After 60 min also some granules (0.6–0.8 μm in diameter) with a content of medium electrondensity appeared labelled with gold particles (Fig. 6).

Neutrophils incubated with R-PIA showed that the surface binding was scarce or absent suggesting the existence of a specific A_1 receptor-mediated system involved in the internalization of the macromolecular probe.

This method allowed us to confirm the presence of A_1 receptors on neutrophils as previously demonstrated using binding methods[6] and a specific monoclonal antibody.[9]

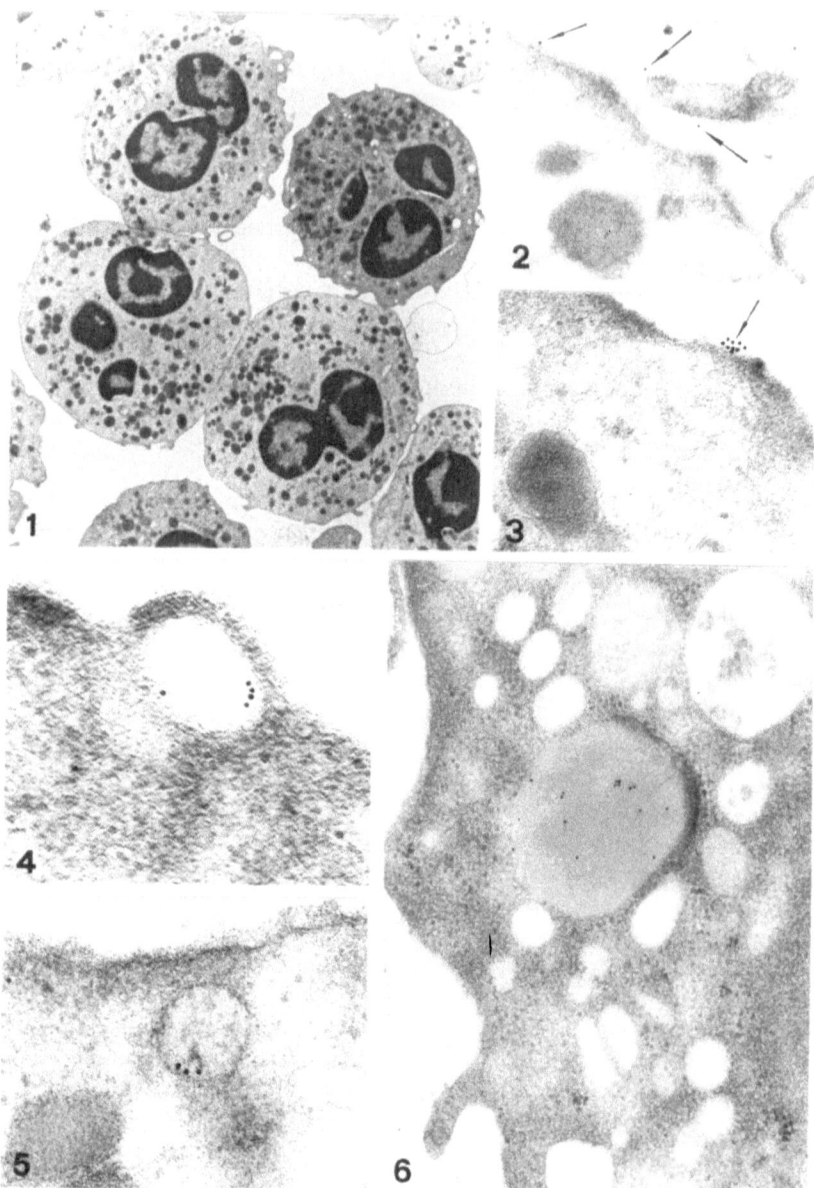

Figure 1-6. A group of untreated neutrophils × 4,400. Fig 2–6. CHAox-biotin-streptavidin-gold treated neutrophils. Fig. 2. Single gold particles (arrows) on the plasma membrane (0°C incubation in CHAox-biotin for 15 min) × 45,000. Fig. 3. Cluster of gold particles (arrow) on the plasma membrane (0°C incubation in CHAox-biotin for 60 min) × 80,000. Fig. 4. A newly formed uncoated vesicle with several gold particles (37°C incubation in CHAox-biotin for 15 min) × 120,000. Fig. 5. Some gold particles inside a smooth-surfaced vesicle (37°C incubation in CHAox-biotin for 60 min) × 100,000. Fig.6. A neutrophil granule labelled with gold particles (37°C incubation in CHAox-biotin for 60 min) × 43,000.

Moreover the specific localization of gold particle on neutrophil granules is in agreement with the role of A_1 receptors in the control of phagocytosis and chemotaxis through a decrease of cAMP level.[3]

REFERENCES

1. Fredholm B.B., Abbracchio M.P., Burnstock G., Daly J.W., Harden T.K., Jacobson K.A., Leff P. and Wiliams M. 1994. Nomenclature and classification of purinoceptors. Pharmacol. Rev. 46: 143

2. van Calker D., Muller M. and Hamprecht. B. 1979. Adenosine regulates, via two different type of receptors the accumulation of cyclic AMP in cultured brain cells. J. Neurochem. 33: 999.

3. Cronstein B. N., Duguma L., Nicholls D., Hutchison A. and Williams M. 1990. The adenosine/neutrophil paradox resolved. human neutrophil posses both A₁ and A₂ receptors which promote chemotaxis and inhibit O₂⁻ generation, respectively. J. Clin. Invest. 85: 1150.

4. Ramkumar V. and Stiles G.L. 1988. Reciprocal modulation of agonist and antagonist binding to A₁ adenosine receptors by guanine nucleotides is mediated via a pertussis toxin-sensitive G protein. J. Pharmacol. Exp. Ther. 246:1194.

5. Tucker A.L. and Linden J. 1993. Cloned receptors and cardiovascular responses to adenosine. Cardiovascular Reserch 27: 62.

6. Martini C., Montali U., Giusti L., Fiorini M., Falleni A., Gremigni V. and Lucacchini A. 1995. A₁ adenosine rceptors in human neutrophils: electron microscope localization using a colloidal CHA-gold-albumin probe.in Purine and Pyrimidine Metabolism in Man VIII. Edited by Sahota A. and Taylor M., Plenum Press, New York. 401.

7. Boyum A. 1968. Isolation of mononuclear cells and granulocytes from human blood. Isolation of mononuclear cells by centrifugation and sedimentation at 1 x g. Scand. J. Clin. Lab. Invest. 21: 77.

8. Gilham P. T. 1971. The covalent binding of nucleotides, polinucleotides and nucleic acids t cellulose. Methods. Enzimol. 21: 191.

9. Salmon J. E., Brownlie C., Brogle N., Edberg J. C., Chen B. X. and Erlanger B.F. 1993. Human mononuclear phagocytes express adenosine A₁ receptors: a novel mechanism for differential regulation of Fc-gamma receptor function. J. Immunol. 151: 2775.

22

PURIFICATION AND SOME MOLECULAR PROPERTIES OF PIGEON HEART AMP-SELECTIVE 5′-NUCLEOTIDASE

Andrzej C. Skladanowski,[1,2] Andrew C. Newby,[1,2] and
Wieslaw Makarewicz[1,2]

[1]Department of Biochemistry and Faculty of Biotechnology
Medical University of Gdansk, ul.
Debinki 1, 80-211 Gdansk, Poland
[2]Bristol Heart Institute
University of Bristol
Marlborough St.
Bristol BS2 8HW, United Kingdom

1. INTRODUCTION

Based on substrate specificity, kinetic characteristics and compartmentation in a cell, vertebrate 5′-nucleotidases (EC 3.1.3.5) have been classified into different groups.[1] Cytoplasmic isozymes are allosterically responding positively to ATP (c-N-II) or ADP (c-N-I), *ecto*-forms derived from the cell membrane (e-N) or cell solute (e-N$_s$) are regulated negatively by excess of ADP. c-N-I exhibiting strong preference towards 5′-AMP and located exclusively in cytosol of various hearts, is believed to be responsible for the formation of adenosine—one of the most important physiological vasodilators and cardioprotective substances.[2] The enzyme has been first time detected and later on purified from various hearts. Its native molecular mass was estimated as 150 kDa.[3]

The content of titratable His residues[4] and ability to form a phosphoprotein intermediate was reported only for c-N-II.[5] This has not been investigated in c-N-I. Neither the mechanism of the reaction is known.

We purified AMP-selective cytoplasmic 5′-nucleotidase (c-N-I) from pigeon heart using novel procedure involving preparative isoelectrofocusing and treated the enzyme with diethylpyrocarbonate (DEPC) in various conditions with aim to investigate essential residues in active centre. We also estimated nucleoside exchange rate catalysed by c-N-I.

Purine and Pyrimidine Metabolism in Man IX,
edited by Griesmacher *et al.* Plenum Press, New York, 1998.

2. METHODS

2.1. Preparation of AMP-selective 5'-nucleotidase (c-N-I)

Ventricles prepared from pigeon hearts were homogenized, centrifuged at 100,000 g; proteins were salted out from supernatant fluid and chromatographed on a phosphocellulose column as described in details elsewhere[3]. The further steps of procedure were modified and consisted of desalting by dialysis and affinity chromatography on AMP-Sepharose 4B (Pharmacia Fine Biochemicals, Uppsala, Sweden) with elution of the active enzyme with a combined gradient 0–0.3 M NaCl with 0–4 mM ADP. The pooled enzyme fractions had a specific activity equal to 20–30 U/mg protein. One unit of activity is defined as the amount of enzyme dephosphorylating 1 μmole of AMP during 1 min at 37°C.

The enzyme preparation used was proven to be almost free of AMP deaminase (< 10 mU/mg of protein) and completely devoid of adenosine deaminase and phosphodiesterase. The absence of secreted *ecto*-5'-nucleotidase (c-N_S) in the preparations was checked by using Concanavalin A (250 μg/ml).

2.2. Preparative Isoelectric Focusing of the c-N-I

7.0 ml of the pooled fractions from AMP-Sepharose 4B column containing 185 μg of soluble protein and 21.6 units of AMP-selective 5'-nucleotidase was mixed with 2.75 ml of Biolyte[R] (Bio-Rad Laboratories, Richmond, CA) ampholytes (pH range 3–10; 40% w/v) and 45.25 ml of ddH$_2$O. The sample contained in final volume of 55 ml: 2% Biolyte[R] 3–10, 6.4 mM Hepes (Sigma, St. Louis, MO), 10 mM NaCl, 0.125 mM ADP and stabilizing agents: 0.012 mM dithiotreitol, 0.12 mM EDTA, 3% glycerol, 25 μM phenylmethanesulphonyl fluoride. Twenty fractions were collected and their pH determined. Aliquots from each fraction were analyzed for protein and assayed for an AMP-specific 5'-nucleotidase activity. The assay conditions were as above but the incubation time was extended up to 30 min. Fraction of the highest specific activity was dialysed against 1 liter of 1 mM NaCl, liophyllized and analyzed by SDS-PAGE.

2.3. Enzyme Assays

The activity of AMP-selective 5'-nucleotidase during purification was assayed by a modification of the method described for purified rat liver soluble 5'-nucleotidase.[6] Incubations were conducted at 37°C from 2–10 min in a total volume 0.1 ml containing 100 mM Hepes/NaOH, pH 7.0, 10 mM MgCl$_2$, 30 mM NaCl, 1 mM ADP and 10 mM AMP (plus 2–3 kBq of [2-^{3}H]AMP) and protein concentrations in the range 0.1–10 μg/ml. Reactions were terminated by boiling for 3 min and the concentration of radioactive adenosine were determined.

Nucleoside exchange was determined in 0.1 ml containing 100 mM Hepes/NaOH, pH 7.0, 10 mM MgCl$_2$, 30 mM NaCl, 1 mM ADP with 5–15 mM adenosine (contained 6 kBq of [2-^{3}H]adenosine per assay) and 1 mM AMP. After the mixture had been boiled and centrifuged, 20 μl samples of supernatant were subjected to silica-gel t.l.c. AMP was located by u.v. fluorescence and its radioactivity determined. P$_i$ in the supernatant was determined colorimetrically.[7]

One unit of activity was defined as the amount needed to convert 1 μmol of substrate per min at 37°C. Protein was measured as described by Bradford,[8] with bovine serum albumine as standard.

2.4. Inactivation of c-N-I by DEPC

The enzyme was incubated with various concentrations of DEPC and other compounds as indicated in descriptions of experiments for 0–30 min at 23°C. The reaction was terminated with 10 mM imidazole-HCl (pH 7.4). In the reactivation experiments, after 15 min treatment with 0.2 or 1 mM DEPC, the enzyme was incubated for 1 h at 37°C with various concentrations of hydroxylamine hydrochloride and the remaining activity then determined.

3. RESULTS AND DISCUSSION

3.1. Isoelectric Point of c-N-I

AMP-selective 5'-nucleotidase (c-N-I) was prepared initially with use of phosphocellulose chromatography and gel filtration as a last step[3]. The further improvement of the preparation homogeneity was achieved by using AMP-Sepharose 4B column and preparative isoelectrofocusing. A single Rotofor run was sufficient to resolve the c-N-I 3–4 fractions which were substantially free of the bulk of contaminating proteins (Fig. 1). Protein recovery, hovever was low and only ca. 4 µg of pure enzyme was obtained in one run. Isoelectric point of the enzyme was equal 6.0 in the presented experiment (other data: 5.8; 6.1).

3.2. Essential Residues of c-N-I

The c-N-I was rapidly inactivated by diethylpyrocarbonate (DEPC, Fig. 2A). Inactivation of the enzyme was initially a first-order process (not shown). DEPC can modify Lys, His, Cys, Ser and Tyr residues in proteins. Blocking of histidine and serine is rapidly reversible by hydroxylamine, modified tyrosine is more resistant whilst modification of lysine and cystine is irreversible by NH_2OH.[9] The inhibition of cytosolic AMP-selective enzyme by DEPC was only partially reversed by NH_2OH (Fig. 2B) what means that except essential His residue(s), participation of Lys, Cys or Tyr in addition is likely. Inactivation of c-N-I with 0.2 mM DEPC was reduced in the presence of ADP, the main allosteric activator, but not when adenosine or AMP was present in the incubation mixture (not shown).

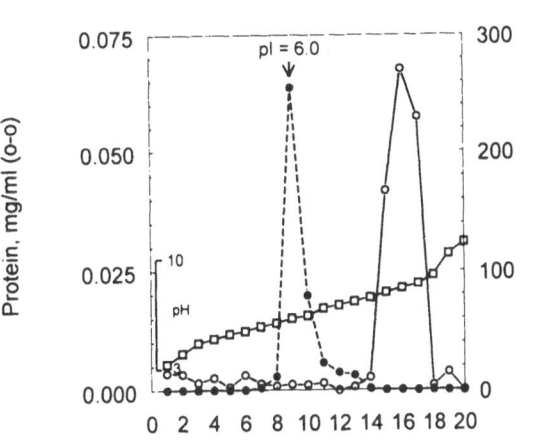

Figure 1. Isoelectric focusing of AMP-selective 5'-nucleotidase (c-N-I). Each of the 20 fractions collected from the Rotofor cell was analysed for protein, activity and pH.

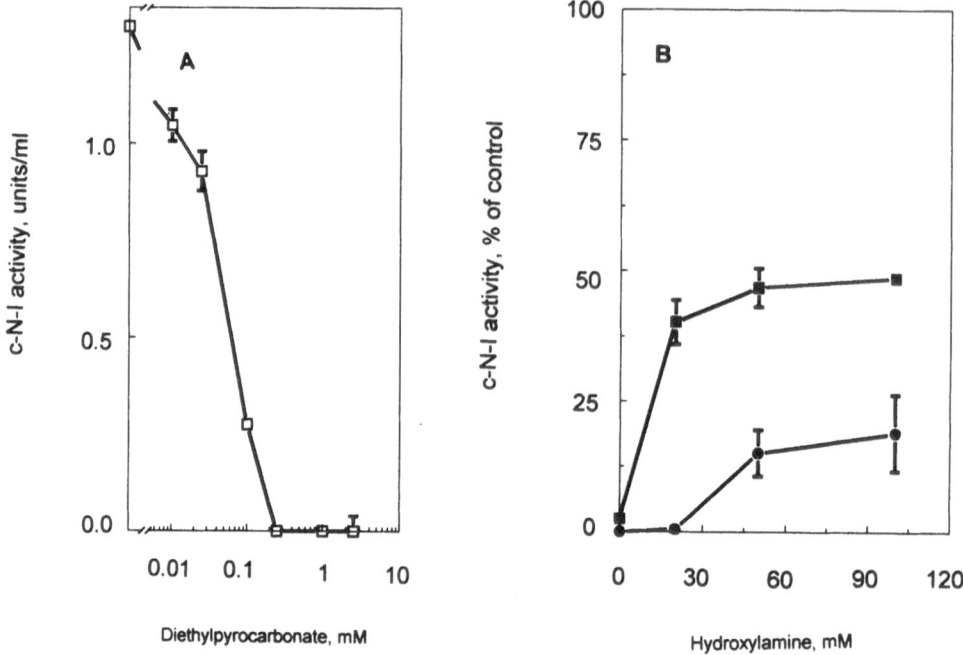

Figure 2. Inactivation by diethylpyrocarbonate (A) and reactivation by hydroxylamine (B) of AMP-selective cytosolic 5′-nucleotidase. 0.2 mM (squares) or 1 mM (circles) DEPC-treated enzyme through 15 min at 23°C was incubated with various concentrations of NH_2OH for 1 h at 37°C and then activity determined. Points represent means ± SE (n = 3).

3.3. Nucleoside Exchange Catalysed by c-N-I

Only slight inhibition of phosphate liberation from AMP was observed when 15 mM adenosine was added into incubation mixture of c-N-I. Simultaneously, a very small but stable incorporation of [2–³H]adenosine label into AMP fraction reaching 0.4% of total

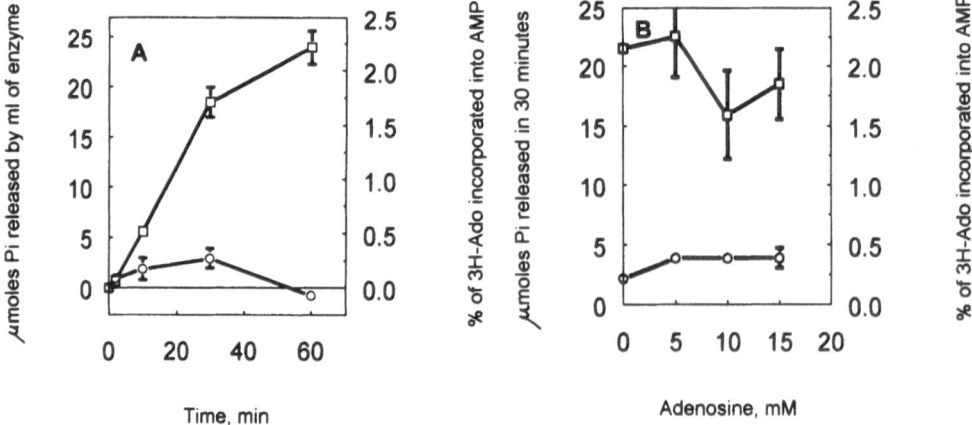

Figure 3. Dephosphorylation of AMP (squares) and nucleoside exchange (circles) catalysed by AMP-selective 5′-nucleotidase. Incubations were followed either with 15 mM adenosine in different times (A) or during 30 min with various concentrations of adenosine (B). Points represents mean ± SE (n = 3).

radioactivity could be found (Fig. 3). This dropped to zero after 60 min of incubation when reaction slowed down and the amount of enzyme-phosphate-adenosine complex diminished. We might say that the rate of exchange of [2-^{3}H]adenosine with cold adenosine in an intermediate is very low. Then the liberation of free adenosine and phosphate from AMP thereafter into medium is the only process catalyzed by c-N-I. In spite of this, c-N-II is able to exchange nucleoside in nucleoside monophosphates efficiently.[10,11] We might assume that in this respect two cytosolic isozymes of 5'-nucleotidase reveal a quite different mechanism of reaction.

ACKNOWLEDGMENTS

This study was supported by the Polish Committee for Scientific Research with grant 4 P05A 034 12 and by grants from the British Heart Foundation.

REFERENCES

1. Zimmermann, H. 5'-Nucleotidase: molecular structure and functional aspects. Biochem. J. **285**, 345–365 (1993)
2. Skladanowski, A.C. and Newby, A.C. 5'-Nucleotidases involved in adenosine formation. in Role of Adenosine and Adenine Nucleotides in the Biological Systems (Imai, S. and Nakazawa, M., eds.) pp. 289–299, Elsevier, Amsterdam, 1991
3. Skladanowski, A.C. and Newby, A.C. Partial purification and properties of an AMP-specific soluble 5'-nucleotidase from pigeon heart . Biochem. J. **268**, 117–122 (1990)
4. Worku, Y., Luzio, J.P. and Newby, A.C. Identification of histydyl and cysteinyl residues essential for catalysis by 5'-nucleotidase. FEBS Letters **167**, 235–240 (1984)
5. Baiocchi, C., Pesi, R., Camici, M., Itoh, R. And Tozzi, M.G. Mechanism of the reaction catalysed by cytosolic 5'-nucleotidase/phosphotransferase: formation of a phosphorylated intermediate. Biochem. J. **317**, 797–801 (1996)
6. Newby, A.C. The pigeon heart 5'-nucleotidase responsible for ischaemia-induced adenosine formation. Biochem. J. **253**, 123–130 (1988)
7. Itaya, K. and Ui, M. A new miromethod for the colorimetric determination of inorganic phosphate. Clin. Chim. Acta **14**, 361–366 (1966)
8. Bradford, M. A rapid and sensitive method for the quantitation of microgram quantities of protein utilizing the principle of protein-dye binding. Anal. Biochem. **72**, 248–254 (1976)
9. Miles, E.W. Methods in Enzymology **48**, 431–432 (1977)
10. Worku, Y. and Newby, A.C. Nucleoside exchange catalysed by the cytoplasmic 5'-nucleotidase Biochem. J. **205**, 503–510 (1982)
11. Banditelli, S., Baiocchi, C., Pesi, R., Allegrini, S., Turriani, M., Ipata, P.L., Camici, M., Tozzi, M.G. The phosphotransferase activity of cytosolic 5'-nucleotidase; a purine analog phosphorylating enzyme. Int. J. Biochem. Cell. Biol. **28**, 711–720 (1996)

23

SUCCINYLPURINES DO NOT MODIFY GLUTAMATE OR ADENOSINE EFFECTS IN THE CNS

T. W. Stone,[1] J. A. Duley,[2] H. A. Simmonds,[2] M. F. Vincent,[1] and
G. van den Berghe[3]

[1]West Medical Building
Division of Neuroscience and Biomedical Systems
University of Glasgow
Glasgow, Scotland
[2]Clinical Science Laboratories
Guys Hospital, London SE1
[3]Laboratory of Physiological Chemistry
UCL-ICP, Brussels, Belgium

1. INTRODUCTION

Children suffering from deficiency of adenylosuccinase (adenylosuccinate lyase, EC 4.3.2.2: ASase) are normal at birth, but profound psychomotor retardation, the principal symptom of the defect, becomes evident in the first two years (Jaeken & van den Berghe, 1984). The clinical picture is markedly heterogeneous, often manifesting epilepsy and autistic features, sometimes muscle wasting and growth failure (Jaeken et al., 1988). There is a recognised accumulation of the two dephosphorylated substrates of the enzyme: succinyl-aminoimidazole carboxamide riboside (SAICA riboside) and succinyladenosine (S-Ado). Dephosphorylation of SAICA ribotide and adenylsuccinate is performed by cytosolic 5'-nucleotidases. Neither SAICA riboside nor S-Ado are normally detectable in the blood or tissue samples of humans, but in patients with ASase deficiency, both compounds are present in the urine, blood and CSF. In the CSF, concentrations are frequently around 100–200 µM (Jaeken et al., 1993).

A study of brain glucose utilisation using fluorodeoxyglucose and Positron Emission Tomography (PET) in three ASase deficient patients revealed a consistent decrease of uptake in all areas of grey matter except the cerebellum, with the greatest abnormality occurring in anterior cerebral cortex (de Volder et al., 1988). While this supports the concept of a deterioration of neuronal function it does not reveal much about mechanism: these uptake changes are likely to be secondary to some other cause of neuronal dysfunction.

Purine and Pyrimidine Metabolism in Man IX,
edited by Griesmacher *et al.* Plenum Press, New York, 1998.

Adenosine itself, however, is well known to be a potent modulator of neuronal activity in the brain, decreasing transmitter release and inducing hyperpolarisation of neurones (Stone, 1989, 1991; Stone & Simmonds, 1991). We have, therefore, now tried to determine whether the succinylpurines could also interfere with synaptic transmission in the brain either directly or by modifying the effects of adenosine.

2. METHODS

Rats were killed by an overdose of urethane and the brain removed into artificial CSF at 4°C. The hippocampi were dissected out and cut into slices 500 μm thick using a McIlwain chopper. The slices were maintained in an atmosphere of water-saturated 5% CO_2 in O_2 at room temperature until use. Individual slices were then transferred to a 0.5 ml recording chamber and superfused with artificial CSF at 30°C. Recordings were made from the stratum pyramidale using glass microelectrodes, with stimulation in the stratum radiatum at 0.1 Hz, pulses of 30 μs duration and 50–300 μA amplitude.

As the compounds of interest were in limited supply, the perfusion was stopped for 10 minute periods to allow addition of the purines directly into the bath fluid. Controls for

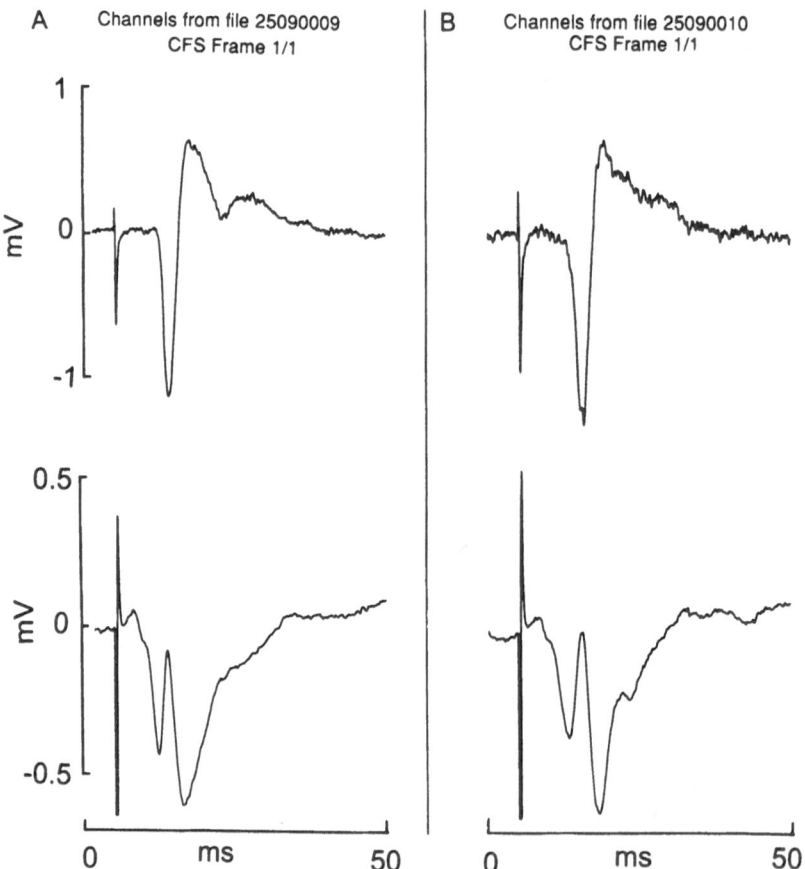

Figure 1. Recordings of the evoked population spike and population epsp in a hippocampal slice A: in normal medium and B: in the presence of SAICA riboside, 200 μM.

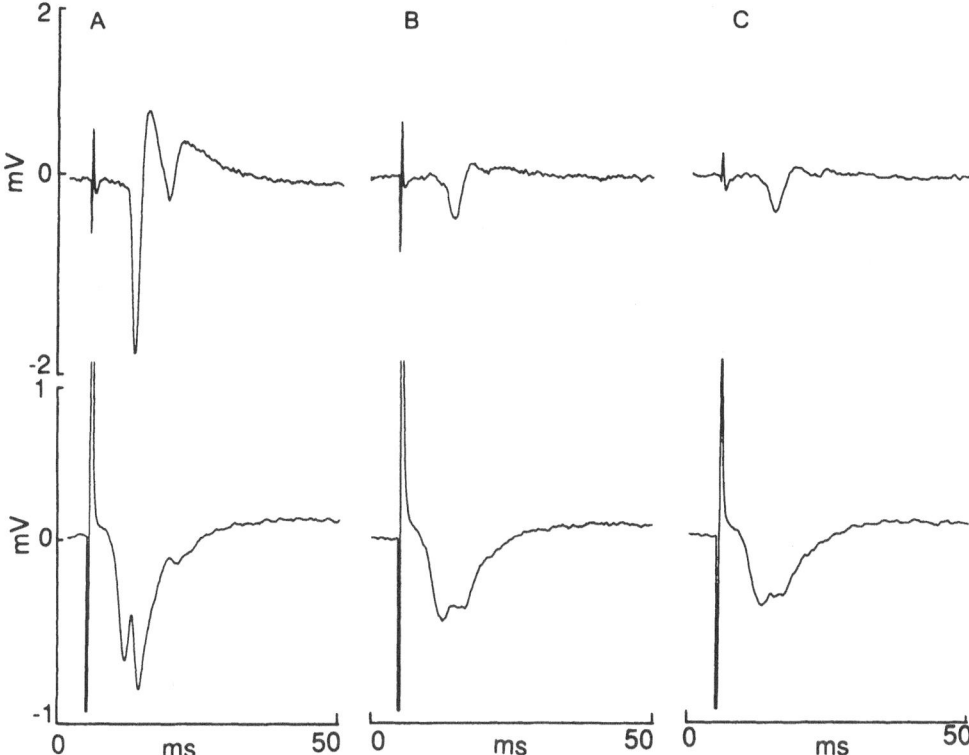

Figure 2. Recordings of the evoked population spike and population epsp in a hippocampal slice A: in normal medium, B: in the presence of adenosine, 50 μM and C: in the presence of adenosine plus SAICA riboside, 200 μM.

the effects of such a stoppage were performed, but no change in potential size was noted over this period. Succinylpurines were prepared as given in Van den Bergh et al., 1991.

3. RESULTS

Stimulation in the stratum radiatum evoked normal population synaptic potentials and population spikes in stratum pyramidale (Fig. 1A). Inclusion of SAICA riboside or succinyladenosine at concentrations between 2 and 200 μM did not modify the evoked responses (Fig. 1B).

Adenosine itself depressed evoked potential size (Fig. 2A,B) but neither SAICA riboside nor succinyladenosine at concentrations of up to 200 μM were able to modify this effect (Fig. 2C).

The application of glutamate, N-methyl-D-aspartate (NMDA) or kainate to single cells by microiontophoresis caused excitation with an elevation of firing rate, but neither succinyladenosine nor SAICA riboside altered these responses (Fig. 3).

4. DISCUSSION

Since adenosine is a potent inhibitor of neuronal function, depressing the release of several excitatory transmitters and causing direct hyperpolarisation of neurones, it was

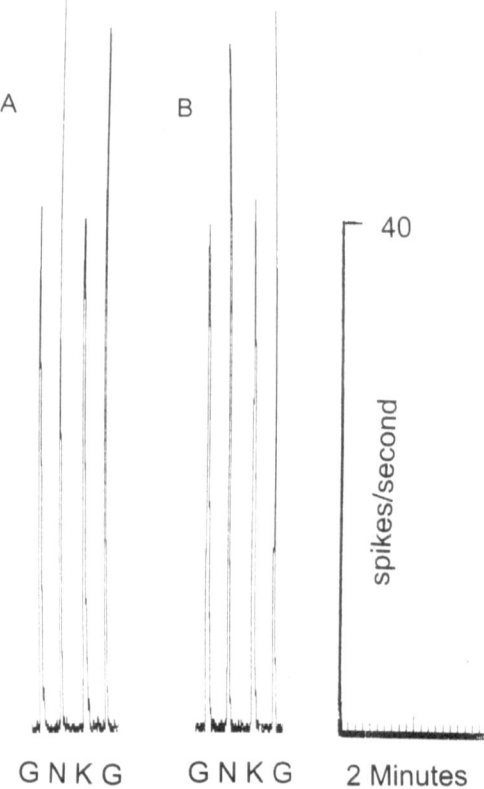

Figure 3. Ratemeter records of the firing rate of a hippocampal neurone during the microiontophoresis of glutamate (G: 60 nA), NMDA (N: 35 nA) and kainate (K: 25nA), A: in control conditions and B: in the presence of succinyladenosine, 200 μM.

possible that the symptoms seen in ASAse deficient patients could reflect an interference of succinylpurines with neurotransmission. The present results suggest that this hypothesis is not correct. At concentrations of up to 200 μM, similar to those found in the CSF of patients exhibiting ASAse deficiency, no change was seen of synaptic transmission at the synapses between the Schaffer collateral and commissural axons and the CA1 pyramidal neurones. This pathway involves a glutamatergic synapse, and any action of the succinylpurines to interfere with the release or postsynaptic actions of glutamate should be apparent in this system. The lack of effect on this synaptic pathway was supported by the demonstration that the succinylpurines has no effect on the excitation of pyramidal neurones induced by the direct administration of glutamate or the glutamate analogues kainic acid and NMDA by microiontophoresis (Stone, 1985).

An alternative explanation of the toxic effects of SAICA riboside was considered that it might interfere with the effects of adenosine itself. Since adenosine is an inhibitory and neuroprotective compound, interference with these actions could lead to a serious compromise of neuronal function. However, we have been unable to demonstrate any interaction, either potentiation or antagonism of adenosine in the hippocampal slice.

The possibilities remain that SAICA riboside may modify neurotransmission at non-glutamatergic sites, that it may have actions in other areas of the brain outwith the hippocampus, that there may be effects in humans which are not reproduced in the rat, or that the effects are primarily to modify the synthesis and release of adenosine from neurones or glial cells.

ACKNOWLEDGMENT

This work was supported by EC Biomed grant BMH1-CT94-1384.

REFERENCES

De Volder AG, Jaeken J, van den Berghe G, Bol A, Michel C, Cogneau M & Goffinet AM, 1988, Pediatr Res, 24:238–42.

Jaeken J & van den Berghe G. Lancet, 1984, 2:1058–61.

Jaeken J, Wadman SK, Duran M, van Sprang FJ, Beemer FA, Holl RA, Theunissen PM, De Cock P, van den Bergh F, Vincent MF & van den Berghe G, 1988, Europ. J. Pediatr, 1988, 148:126–31.

Jaeken J, Casaer P, De Cock P & van den Berghe G, 1993, In Molecular Genetics, Biochemistry and Clinical Aspects of Inherited Disorders of Purine and Pyrimidine Metabolism. Gresser, U (ed)., Springer, London, pp.140–143. ISBN 3-540-56774-7.

Stone TW, 1985, Microiontophoresis and Pressure Ejection. Wiley, Chichester.

Stone TW, 1991, Adenosine in the Nervous System. Academic Press, London.

Stone TW & Simmonds HA, 1991, Purines: Basic and Clinical Aspects. Kluwer Academic Press, Dordrecht. ISBN 0-7923-8925-5.

Van den Bergh F, Vincent MF, Jaeken J & van den Berghe G., 1991, Analyt. Biochem. 193, 287–291.

24

CLINICAL AND BIOCHEMICAL ASPECTS OF DIHYDROPYRIMIDINASE DEFICIENCY

Albert H. Van Gennip,[1] Ronney A. De Abreu,[2] Peter Vreken,[1] and André B. P. Van Kuilenburg[1]

[1]Academic Medical Center, University of Amsterdam
Department of Clinical Chemistry and Division
Emma Children's Hospital
P.O. Box 22700
1100 DE Amsterdam, The Netherlands
[2]University Hospital Nijmegen
Department of Pediatrics and Neurology
P.O. Box 9101
6500 HB Nijmegen, The Netherlands

1. INTRODUCTION

Dihydropyrimidinase (DHP, EC 3.5.2.2) catalyzes the second step in the degradation of uracil and thymine. The first step is catalyzed by dihydropyrimidine dehydrogenase (DPD, EC 1.3.1.2), the third step by β-ureidopropionase (UP, EC 3.5.1.6) and the fourth step is catalyzed by three transaminases (R)-(–)-β-aminoisobutyrate pyruvate aminotransferase (BAIBPAT, EC 2.6.1.40), β-alanine-pyruvate aminotransferase (BAPAT, EC 2.6.1.18) and β-alanine-α-ketoglutarate aminotransferase (BAKAT, EC 2.6.1.19). The first three steps of the catabolism of uracil and thymine are controlled by enzymes shared by both pathways and result in the production of the neurotransmitter acid β-alanine from uracil and the non-functional (R)-β-aminoisobutyrate from thymine. The thymine analogue 5-fluorouracil is degraded by the same pathway to fluoro-β-alanine. In contrast to DPD deficiency of which 50 cases have been reported[1,2] only six cases have been described with DHP deficiency and none with UP deficiency,[3–9] although secondary UP deficiency has been reported in patients with propionic acidemia.[10] The reason for this difference may be a lesser frequency of DHP compared to DPD deficient individuals, but another possibility may be that patients with DHP deficiency are overlooked. Therefore, we will focus on the clinical presentation and biochemical detection of patients with DHP deficiency.

Purine and Pyrimidine Metabolism in Man IX,
edited by Griesmacher *et al.* Plenum Press, New York, 1998.

2. CLINICAL ASPECTS

Symptomatology of DHP deficient individuals seems to be as variable as in DPD deficiency. Epileptic or convulsive attacks were mentioned in three[3,5,8] mental retardation, motor retardation and microcephaly in two of these individuals.[5,8] Dysmorphic features and growth retardation were seen in one case.[8] One patient suffered from intractable diarrhoea due to congenital microvillous atrophy, but had no other symptoms.[9] Two individuals are healthy; they were detected by mass-screening in Japan.[6,7] The clinical picture varies from severe to completely healthy. Inheritance is autosomal recessive.

3. BIOCHEMICAL ASPECTS

3.1. Metabolism

In DHP deficient patients dihydrothymine and dihydrouracil accumulate and because of the reversibility of the first step of pyrimidine degradation also thymine and uracil accumulate. These metabolites will appear in elevated concentrations in the body fluids. In contrast, the concentrations of the N-carbamyl-β-amino acids are low or absent. The reduced concentration of the neurotransmitter β-alanine may be of relevance with respect to the cerebral dysfunction and for this reason treatment with β-alanine supplementation is under investigation. Exposure of the nervous system to high concentrations of dihydropyrimidines and pyrimidine bases may be a contributing factor. Although not yet reported, increased sensitivity to fluorouracil toxicity can also be expected in individuals with (partial) DHP deficiency.

3.2. Detection and Diagnosis

The preferential material for the screening of DHP deficiency and the other pyrimidine degradation defects is urine, as all waste products accumulate in this body fluid. If urine is not available blood and cerebrospinal fluid can be used for screening, but the accumulation of abnormal metabolites in these body fluids is less prominent. DHP deficiency can be detected by procedures which are widely used for the screening of inborn errors of metabolism such as GC-MS analysis of urinary organic acid extracts[5] and amino acid analysis of urine before and after acid hydrolysis.[11] Quantification requires sophisticated methods such as HPLC with detection at various wavelengths in off- or on-line prepared fractions,[12] isolation of dihydropyrimidines followed by acid hydrolysis and amino acid analysis of the resulting β-amino acids[11] or proton NNR spectroscopy.[13] Concentrations of the relevant metabolites in urines from the individuals with DHP deficiency are shown in Table 1. As can be seen dihydrothymine and dihydrouracil are strongly elevated, thymine and uracil are moderately elevated in all cases. This pattern is highly characteristic for DHP deficiency, but the diagnosis can be missed if the urine is contaminated by bacteria e.g. due to urinary tract infection.[11] Except for uracil, which was found to be below the detection limit, increased concentrations for these metabolites were reported in CSF from two cases:[9,13] thymine 10 and 3, dihydrouracil 79 and 117, dihydrothymine 46 and 179 μmol/L, respectively.

3.3. Molecular Aspects

The diagnosis of DHP can only be confirmed by analysis of the enzyme activity in liver, because the enzyme is not expressed in other more accessible tissues. The enzyme

Table 1. Excretion values of relevant metabolites in the six individuals
with DHP deficiency and in controls

Metabolite	P1	P2[1]	P3	P4[2]	P5[3]	P6	C
Uracil	49	± 100	144	47	13	41	7–33
Thymine	12	± 100	230	13	2	38	0–3
HU	622	± 500	490	626	152	517	16–110
DHT	406	± 500	760	451	68	378	2–28

C = controls; n = 153 for uracil and thymine; n=6 for DHU and DHT.
Values taken from: [1] Henderson MJ et al. (1993); [2] Ohba S et al. (1995); [3] Sumi S et al. (1996).

assay is performed with radioactive dihydrouracil as the substrate.[8] The reaction is started with the liver homogenate and after incubation the radiolabelled substrate and products are separated by HPLC with on-line radioactivity detection. Radioactive carbondioxide is trapped with sodium hydroxide and counted by liquid scintillation technique. We had the opportunity to measure the DHP activity in liver of two patients with dihydropyrimidinuria. As can be seen in Figure 1A, the activity of DHP was undetectable low (< 0.3 nmol/h per mg protein) in these patients. For comparison DPD activity was also measured and was found to be slightly below 8 control values in one, above the control values in the other patient (Figure 1B). A plausible explanation may be that the control range is too narrow, because of the small number of controls.

In patients with DHP deficiency, gene defects have not yet been reported. However, the cDNA sequence for human DHP has been submitted recently[14] and therefore the elucidation of the molecular defects underlying DHP deficiency can be expected fairly soon.

4. FINAL REMARKS

As in many inborn errors of metabolism, the symptomatology in DHP deficient individuals is very non-specific. However, the biochemical detection of DPD deficiency is not too difficult and can be done by methods regularly used for the screening of inborn errors of metabolism. A systematic collection of clinical and biochemical data including the

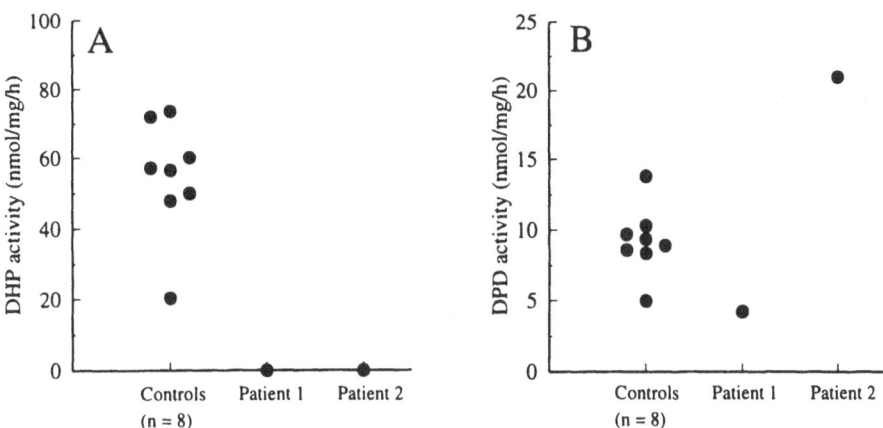

Figure 1. Activity of DHP in 8 control liver samples and in the liver biopsy specimens of 2 patients with dihydropyrimidinuria (A). DPD activity was measured for comparison (B).

determination of the responsible gene defect are needed in addition to basic studies in order to elucidate the pathophysiological mechanism in this defect.

REFERENCES

1. Van Gennip AH, Abeling NGGM, Vreken P, Van Kuilenburg ABP (1997) Genetic metabolic disease of pyrimidine metabolism: implications for diagnosis and treatment. International Pediatrics 12: 28–33.
2. Van Gennip AH, Abeling NGGM, Vreken P and Van Kuilenburg ABP (1997) Inborn errors of pyrimidine degradation: clinical, biochemical and molecular aspects. J. Inher Metab Dis 20: in press.
3. Duran M, Rovers P, De Bree PK, Schreuder CH, Beukenhorst H, Dorland L (1990) Dihydropyrimidinuria. Lancet 336: 817–818.
4. Van Gennip AH, Abeling NGGM, Stroomer AEM, Van Lenthe H, Bakker HD (1994) Clinical and biochemical findings in six patients with pyrimidine degradation defects. J Inher Metab Dis 17: 130–132.
5. Henderson MJ, Ward K, Simmonds HA, Duley JA and Davies PM (1993) Dihydropyrimidinase deficiency presenting in infancy with severe developmental delay. J Inher Metab Dis 16: 574–576.
6. Ohba S, Kidouchi K, Sumi S, Imaeda M, Takeda N, Joshizumi H, Tatematsu A, Kodama K, Yamanaka K, Kobayashi M and Wada Y (1995) Dihydropyrimidinuria: the first case in Japan. Adv Exp Med biol 370: 383–386.
7. Sumi S, Kidouchi K, Hayashi K, Ohba S and Wada Y (1996) Dihydropyrimidinuria without clinical symptoms. J Inher Metab Dis 19: 701–701.
8. Van Gennip AH, De Abreu RA, Van Lenthe H, Bakkeren J, Rotteveel J, Van Kuilenburg ABP (1997) Dihydropyrimidinase deficiency: confirmation of the enzyme defect in dihydropyrimidinuria. J Inher Metab Dis 20: in press. Abstract in: Van Gennip et al (1996) J Inher Metab Dis 19: suppl.1, P18.
9. Assmann B, Hoffmann G, Wagner L, Seyberth H, Berger R (1996) Microvillous atrophy and elevated urinary excretion of pyrimidines. J Inher Metab Dis 19: suppl. 1, P20.
10. Van Gennip AH, Van Lenthe H, Abeling NGGM, Scholten EG, Van Kuilenburg ABP (1997) Inhibition of β-ureidopropionase by propionate may contribute to the neurologic complications in patients with propionic acidemia. J Inher Metab Dis 20: in press. Abstract in: Van Gennip et al (1996) J Inher Metab Dis 19: suppl. 1, P86.
11. Van Gennip AH (1993) Application of simple chromatographic methods for the diagnosis of defects in pyrimidine degradation. Clin Chem 39: 380–385.
12. Van Gennip AH (1987) Screening for inborn errors of purine and pyrimidine metabolism by bidimensional TLC and HPLC. In Zweig G, Sherma J, Krstulovic AM, eds. Handbook of Chromatography. Boca Raton, FL: CRC Press, Vol. I, part A, 221–245.
13. Wevers RA, Engelke U, Rotteveel JJ, Heerschap A, De Jong JGN, Abeling NGGM, Van Gennip AH and De Abreu RA (1996) 'H-NMR spectroscopy of body fluids in patients with inborn errors of purine and pyrimidine metabolism. J Inher Metab Dis 19: Suppl 1, 06.
14. Hamajima N, Matsuda K, Sakata S, Tamaki N, Sasaki M, Nonaka M (1996) A novel gene family defined by human dihydropyrimidinase and three related proteins with differential tissue distribution. Gene 180: 157/163.

25

THE GENETIC BASIS OF MYOADENYLATE
DEAMINASE DEFICIENCY IS HETEROGENEOUS

E. Rötzer,[1] W. Mortier,[2] H. Reichmann,[3] and M. Gross[1]

[1]Medizinische Poliklinik Universität München
[2]Universitätskinderklinik Bochum
[3]Neurologische Klinik und Poliklinik Universität Dresden

1. INTRODUCTION

Myoadenylate deaminase (MAD) is the muscle specific isoform of adenosine monophosphate deaminase (AMPD). AMPD (EC 3.5.4.6.) is an aminohydrolase that catalyzes the deamination of AMP to inosine monophosphate (IMP) with liberation of ammonia from position 6 of the adenine ring. As part of the purine nucleotide cycle,[1,2] AMPD is involved in the regulation of the relative concentrations of intracellular purine nucleotides, the stabilization of the adenylate energy charge, the replenishment of citric acid cycle intermediates by the formation of fumarate, the deamination of amino acids (aspartate) and the regulation of the activity of the glycolytic enzymes phosphofructokinase and phosphorylase b.[3–5] The fact that AMPD is an ubiquitous enzyme found in eukaryotes and in all tissues of vertebrates[6–8] underlines the important role of this enzyme in energy metabolism. In man there are at least three isozymes. The isozyme with the highest activity is myoadenylate deaminase found in skeletal muscle.[6,9]

In 1978 MAD deficiency was first described in five patients with exercise-related muscle symptoms.[10] Nowadays, MAD deficiency is recognized as the most common muscle enzyme defect. There is a great variability of the clinical symptoms and asymptomatic MAD deficient subjects have been detected.

The inborn type of MAD deficiency is following an autosomal recessive trait according to the localization of the AMPD1 gene on chromosome 1 (1p13-p21)[11] and is caused by a single mutant allele harboring two mutations: C34-T (Gln → Stop) in exon 2 and C143-T (Pro48 → Leu) in exon 3. The nonsense mutation (C34-T) at the last nucleotide of exon 2 creates a stop-codon and leads to an early stop of protein synthesis producing a strongly truncated and catalytically inactive peptide. The second missense mutation in exon 3 would result in an amino acid exchange of proline48 to leucine if the translation would not stop earlier.

Purine and Pyrimidine Metabolism in Man IX,
edited by Griesmacher *et al.* Plenum Press, New York, 1998.

So far all patients with primary MAD deficiency were homozygous for the nonsense mutation in exon 2 and the missense mutation in exon 3.[12] We studied the genetic basis of this defect in a 24-year-old man with exercise induced myalgia and minimal residual MAD activity who was heterozygous for both mutations. The only pathological finding in the muscle biopsy was the absence of MAD activity indicating an inherited form of MAD deficiency.

2. METHODS

Genomic DNA was extracted from blood leukocytes as described elsewhere.[13] All 16 exons of the AMPD1 gene and the flanking intron regions were sequenced after amplification and purification. The primers for each exon were positioned in the flanking intron regions and designed with the Husar computer program. For sequencing, asymmetric PCR was performed with reduced amounts of the lower primer. The PCR mixture consisted of 10 mM Tris-HCl (pH 8.3), 50 mM KCl, 1.5 mM $MgCl_2$, 0.2 mg/ml gelatine, 0.2 mM of each dNTP, 50 pmoles upper and 15 pmoles lower primer, 250 ng of template DNA and 2.5 U Taq DNA Polymerase (Ampli Taq, Perkin Elmer Cetus) in a final volume of 100 μl. PCR amplification was carried out for 35 cycles at primer annealing temperatures ranging from 50°C to 60°C (94°C 1 min, annealing 1 min, 72°C 1min). After purification with glass milk (Bios 101 Geneclean II Kit), sequencing was performed according to the Sequenase Version 2.0 protocol (United State Biochemical Co.) with ^{35}S.

Messenger RNA was extracted from the muscle biopsy with the Micro-Fast Track mRNA Isolation Kit (Invitrogen Corp.) and cDNA synthesis was performed at 37°C for 2 hours using an Oligo-T 20mer primer and AMV reverse transcriptase (Sigma R-9376) in 50 mM Tris-HCl, 10 mM $MgCl_2$, 40 mM KCl, 1 mM EDTA, 1 mM DTT, 0.5 mM each dNTP, 2 μg mRNA, 500 ng primer, 40 U RNAse inhibitor (Boehringer Mannheim) and 200 U reverse transcriptase.

To answer the question if the mutations in exon 2 and exon 5 are located on the same or on different alleles, allele specific PCRs were performed. For this purpose, two lower primers were designed with the 3' nucleotide corresponding either to the wild type or mutant sequence in exon 5. The upper primer introduced a mismatch which creates a MaeIII restriction site in the allele with wild type sequence at position 34.

Lower primer for the wild type allele in exon 5:

3' CT CCA AGG GAT TTT GGG GAA 5'

Lower primer for mutant allele in exon 5:

3' AT CCA AGG GAT TTT GGG GAA 5'

Upper MaeIII restriction primer

5' CCT CTG TTC AAA CTC CCA GCT GAA GAG TAA 3'

PCR amplification was carried out for 35 cycles at 68°C annealing temperature with 50 pmoles of each primer. Reaction conditions were as described above. After precipitation the amplified allele was incubated with 6 U MaeIII (Boehringer Mannheim) in 20 mM Tris-HCl pH 8.2, 275 mM NaCl, 6 mM $MgCl_2$, 7 mM mercaptoethanol at 55°C for 1h.

3. RESULTS

After confirming the patient's heterozygosity for the known mutations C34-T and C143-T, we looked for a new mutation in the AMPD1 gene. Sequencing of the genomic DNA of this patient revealed a new mutation in exon 5. To exclude artifacts induced by PCR, the mutation was confirmed in the cDNA of this patient. The patient was heterozygous for a new G468-T mutation (Fig. 1). This missense mutation results in an exchange of the conserved and neutral amino acid glutamine at position 156 by the positively charged, basic amino acid histidine. Computer analysis of the amino acid secondary structure revealed no significant difference between mutant and wild type amino acid sequence. In both forms a helix was present at this position.

The finding that the patient was heterozygous for the two known mutations and for the new mutation raised the question of their allelic constellation. As any mutation downstream of the nonsense mutation in exon 2 would be ineffective, we presumed that the new mutation should be found on the other allele. Using the technique of allele specific PCR, we could demonstrate that that the new mutation G468-T in exon 5 and the C34-T mutation in exon 2 are indeed on different alleles (Figure 2).

Among 50 control DNA samples of healthy subjects the G468-T mutation was not present. This finding supports the hypothesis that this mutation is not a common polymorphism. Experiments with expression of the mutant enzyme are in progress to investigate in more detail the effects of this mutation on protein function.

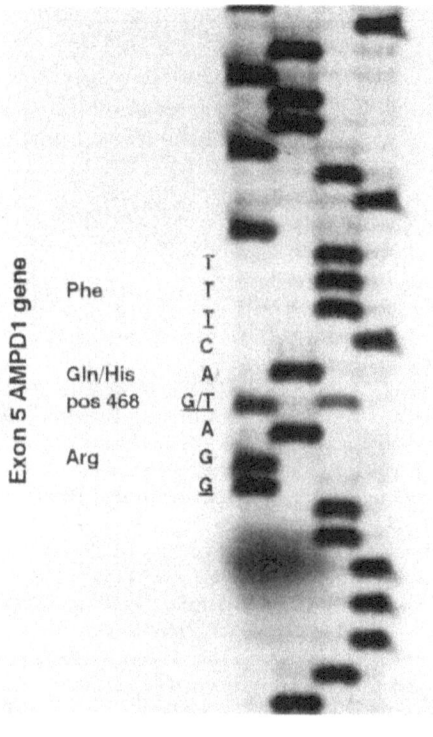

Figure 1. Sequencing gel showing part of the exon 5 sequence of the AMPD1 gene demonstrating heterozygosity for G468-T.

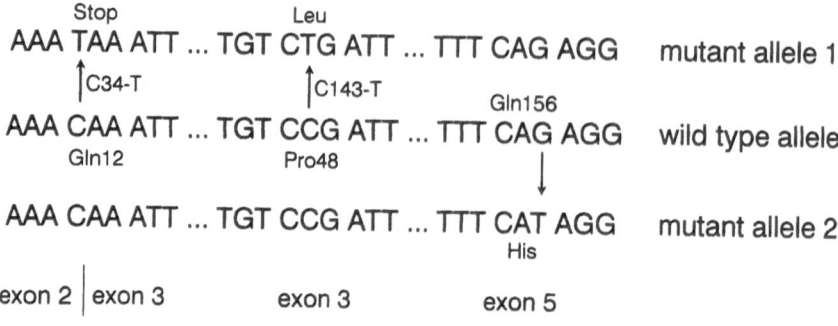

Figure 2. Mutations found in the patient with myoadenylate deaminase deficiency.

4. CONCLUSION

This is the first patient with primary MAD deficiency who is not homozygous for the C34-T mutation in exon 2. The G468-T mutation in exon 5 is assumed to cause MAD deficiency in combination with the nonsense mutation C34-T in exon 2. The fact that the exchanged amino acid glutamine at position 156 has been conserved during evolution might indicate that this amino acid is part of a functional site of the enzyme.

The detection of the new G468-T mutation shows that the genetic basis of primary MAD deficiency is heterogeneous. This mutation has to be considered when genetically studying patients presumed to have MAD deficiency.

ACKNOWLEDGMENTS

The research was supported by Deutsche Forschungsgemeinschaft Grant Gr1012/2-1. Parts of these results were obtained by E. Rötzer as part of her thesis at the medical faculty of the Ludwig-Maximilians-Universität München, Germany.

REFERENCES

1. Lowenstein J, Tornheim K (1971) Ammonia production in muscle: The purine nucleotide cycle. Science 171:397–400
2. Lowenstein JM (1972) Ammonia production in muscle and other tissues: the purine nucleotide cycle. Physiol Rev 52:382–414
3. Sabina RL, Swain JL, Olanow CW, Bradley WG, Fishbein WN, DiMauro S, Holmes EW (1984) Myoadenylate deaminase deficiency. Functional and metabolic abnormalities associated with disruption of the purine nucleotide cycle. J Clin Invest 73:720–730
4. Sabina RL, Swain JL, Patten BM, Ashizawa T, O'Brien WE, Holmes EW (1980) Disruption of the purine nucleotide cycle. A potential explanation for muscle dysfunction in myoadenylate deaminase deficiency. J Clin Invest 66:1419–1423
5. Flanagan WF, Holmes EW, Sabina RL, Swain JL (1986) Importance of purine nucleotide cycle to energy production in skeletal muscle. Am J Physiol 251:C795–802
6. Ogasawara N, Goto H, Yamada Y, Watanabe T, Asano T (1982) AMP deaminase isozymes in human tissues. Biochim Biophys Acta 714:298–306
7. Fishbein WN, Sabina RL, Ogasawara N, Holmes EW. Immunological evidence for three isoforms of AMP deaminase (AMPD) in mature skeletal muscle. Biochim. Biophys. Acta 1993;1163:97–104
8. Marquetant R, Desai NM, Sabina RL, Holmes EW (1987) Evidence for sequential expression of multiple AMP deaminase isoforms during skeletal muscle development. Proc Natl Acad Sci USA 84:2345–2349

9. Lowenstein JM (1990) The purine nucleotide cycle revised. Int J Sports Med 11:S37-S46
10. Fishbein WN, Armbrustmacher VW, Griffin JL (1978a) Myoadenylate deaminase deficiency: A new disease of muscle. Science 200:545–548
11. Sabina RL, Morisaki T, Clarke P, Eddy R, Shows TB, Morton CC, Holmes EW (1990) Characterization of the human and rat myoadenylate deaminase genes. J Biol Chem 265:9423–9433
12. Morisaki T, Gross M, Morisaki H, Pongratz D, Zollner N, Holmes EW. Molecular basis of AMP deaminase deficiency in skeletal muscle. Proc. Natl. Acad. Sci. USA 1992 89 : 6457–6461
13. Maniatis T, Fritsch EF, Sambrook J. Molecular cloning. Cold Spring Harbor NY, Cold Spring Harbor Laboratory Press, 2nd edition.

26

THE ORIGIN OF *APRT*J*

The Most Common Disease-Related Mutation of APRT Gene among Japanese Goes Back to a Prehistoric Era

Naoyuki Kamatani,[1] Chihiro Terai,[1] Seong Yoon Kim,[2] Ching-Lang Chen,[3] Hisashi Yamanaka,[1] Masayuki Hakoda,[1] Shin Totokawa,[1] and Sadao Kashiwazaki[1]

[1]Institute of Rheumatology
Tokyo Women's Medical College, KS Building
9-12 Wakamatsu-cho, Shinjuku-ku, Tokyo 162, Japan
[2]Rheumatism Center
Hanyang University Hospital
Seoul, Korea
[3]Special Clinic of Gout
Taipei Municipal Ho-Ping Hospital
Taipei, Taiwan

1. INTRODUCTION

Adenine phosphoribosyltransferase (APRT) converts adenine into AMP in the presence of 5-phosphoribosyl-1-pyrophosphate. When this enzyme is genetically deficient, 2,8-dihydroxyadenine (DHA) urolithiasis may occur.[1] It is usually considered that an individual develops DHA urolithiasis only when the subject is homozygous (both of the APRT alleles are defective) although one report described the development of the lithiasis in a heterozygote.[2]

Japanese are especially concerned with this genetic enzyme deficiency since the incidence of patients with APRT deficiency is apparently very high in this population as compared to other ethnic groups.[3] At least 150 individuals with homozygous APRT deficiency have been found among the Japanese.

The analysis of the mutant genes at the molecular level has disclosed that most of the disease-related mutations of the APRT deficiency among the Japanese are explained by only three gene changes.[3] The most frequent mutant allele had a ATG (Met) to ACG (Thr) nucleotide substitution at codon 136 of the APRT gene and was designated as *APRT*J* allele.[4] This mutant allele was special in that all the DHA lithiasis individuals

Purine and Pyrimidine Metabolism in Man IX,
edited by Griesmacher *et al.* Plenum Press, New York, 1998.

135

having this allele were only partially deficient in APRT activities (type II).[1] Those individuals exhibited enzyme activities because *APRT*J* codes for a mutant protein having an activity in vitro but not in vivo.

In the present investigation, we asked the origin of *APRT*J* mutation by examining the geographic distribution of the mutant allele in Japan and searching for the mutant allele among both Koreans and Taiwanese.[5]

2. MATERIALS AND METHODS

2.1. Diagnosis of Homozygotes

The diagnosis of all the homozygous individuals included in this study was performed in the Institute of Rheumatology, Tokyo Women's Medical College. The methods for the diagnosis were described previously.[6] Thus, heparinized blood was drawn from the individuals and mononuclear cells were separated. Then, the cells were cultured in the medium containing 6-methylpurine. If the individual was homozygously deficient in APRT, the cells were resistant to the adenine analog. However, if the individual was heterozygous or without any defective alleles, the cells incorporated 6-methylpurine into the cellular nucleotide pool and were killed.[6] In all the homozygous individuals included in this study, the above cellular study was performed to make a conclusive diagnosis.

2.2. Detection of Mutations

Blood was drawn into EDTA-containing tubes, and DNA was extracted from the samples using a DNA extraction kit (Isoquick, Microprobe, Bothell, Wash.). Part of genomic APRT gene was amplified using primers 5'-GCTGGAGATTCAGAAAGACG-3' (sense primer) and 5'-CAGCTTCTCCCTGCCCTTAA-3' (antisense primer) and the amplified DNA was subjected to SSCP analysis as described previously.[7] Then the gel was stained with silver. *APRT*J* allele was clearly separated from the most common allele.[5]

3. RESULTS

3.1. Geographical Distribution of Individuals with Homozygous APRT Deficiency

In 110 homozygous individuals whose diagnosis was done in our laboratory, the geographical distribution was studied. Fig. 1 shows the results. Each individual was shown with a genotype; i.e. when the subject had a genotype of *APRT*J/APRT*J*, it was shown by a closed square, while when the subject had a genotype of *APRT*Q0/APRT*Q0*, it was shown by an open square. When the subject was a compound heterozygote with a genotype of *APRT*J/APRT*Q0*, then it was shown by a half-closed square.

Among the 110 homozygous individuals, 64 were with a genotype of J/J while 17 and 29 individuals had genotypes of J/Q0 and Q0/Q0, respectively. As described previously, a homozygote exhibits partial APRT deficiency when the genotype of the subject is J/J or J/Q0. Therefore, approximately 74% (81/110) of the homozygotes were partially deficient (type II). The homozygotes lived in various regions of Japan, generally paralleling population densities. Importantly, *APRT*J* allele was found in all the 4 major islands of Japan as well as in Okinawa. Since the proportion of the homozygotes detected may

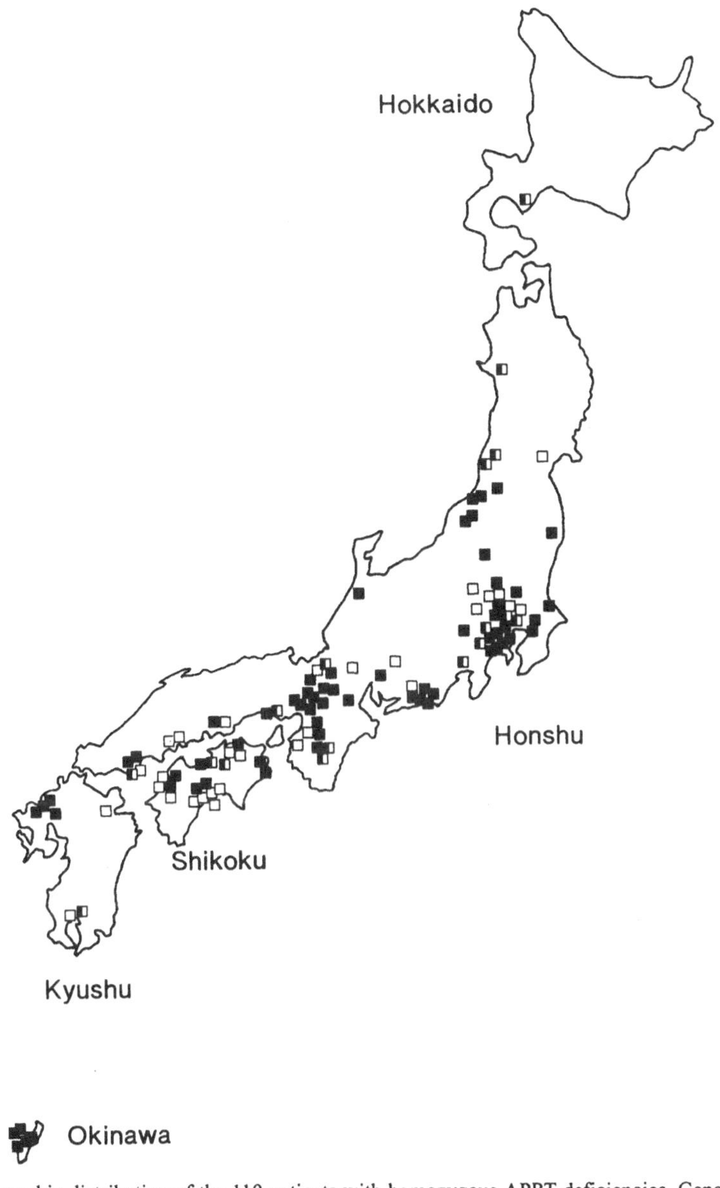

Hokkaido

Honshu

Shikoku

Kyushu

 Okinawa

Figure 1. Geographic distribution of the 110 patients with homozygous APRT deficiencies. Genotypes were J/J (closed square), Q0/Q0 (open square) or J/Q0 (half closed square). From Ref. 5.

vary between different regions of Japan, the proportion of the *APRT*J* alleles to the entire APRT deficient alleles may be more useful to examine whether non-uniform distribution of the *APRT*J* alleles is present. As shown in Fig. 1, the proportion of *APRT*J* alleles does not seem to be greatly different in different regions of Japan.

3.2. Analysis of DNA Samples from Various Regions

Using the PCR-SSCP method, we analyzed blood samples from Japanese, Koreans and Taiwanese. Among 955 Japanese blood samples, 7 (0.73%) were heterozygous (1/J in

Table 1. Numbers and percentages of different genotypes
in different populations (From Ref. 5)

Genotype*	Japanese	Koreans	Taiwanese
1/1	941	356	231
1/J	7 (0.73%)	2 (0.53%)	0
1/2	6 (0.63%)	0	1 (0.43%)
2/2	0	0	1 (0.43%)
1/3	1 (0.1%)	0	0
1/4	0	18 (4.8%)	0
1/5	0	1 (0.26%)	0
1/6	0	2 (0.53%)	0
Total	955	378	233

*1-6 represent different alleles.

Table 1) for the *APRT*J* mutation. Of the 378 Korean samples, 2 (0.53%) were heterozygous for the *APRT*J* mutation, while no *APRT*J* alleles were found in the 233 Taiwanese samples. Some aberrant bands different from those of *APRT*J* and the most common allele (*APRT*1*) were found (Table 1). Those bands presumably reflect polymorphisms rather than disease-related mutations.

4. DISCUSSION

The present investigation disclosed that the *APRT*J* alleles are present in all the 4 major islands of Japan as well as in Okinawa, and the distribution of the mutant alleles are generally uniform. In addition, *APRT*J* alleles were found in Koreans. By the analysis of the linkage disequilibrium concerning the polymorphic sites within and near the APRT gene, we have shown that all the *APRT*J* alleles were derived from a single ancestral gene.[8] Therefore, all the *APRT*J* alleles found in both Japan and Korea are likely to have a single origin. The presence of a hot spot in the APRT gene causing the same *APRT*J* mutation is excluded because none of the disease-related APRT alleles in non-Japanese populations were *APRT*J*.

The documented history of Japan begins about 1400 years ago. After that time, not many people have immigrated into Japan. Therefore, the presence of the *APRT*J* alleles in Korea and Japan suggests that the origin of the mutation goes back earlier than 1400 years ago. Nearly even distribution of the *APRT*J* alleles among the Japanese also suggests that the origin of the mutation is very old. Thus, previous genetic data have suggested that the people in Okinawa differ genetically from those living in the central part of Japan.[9,10] Before about 1400 years ago, a significant number of immigrants came from the Korean Peninsula (less likely from China) during the Yayoi era (third century BC to third century AD) to Japan where native people (Neolithic Jomon people) with a different genetic background had lived.[9,10] People in Ōkinawa are considered to be more closely related to the Jomon people than other Japanese.[9,10] Since people in Okinawa have *APRT*J* alleles at a rather high incidence, Jomon people are likely to have already possessed the alleles before the Yayoi era. Therefore, the origin of the *APRT*J* mutation probably goes back earlier than 2300 years ago when the Yayoi era began. Since both Jomon people and the immigrants from the Korean Peninsula had possessed the *APRT*J* alleles, the distribution of the mutational alleles among the Japanese has become relatively even. If the Jomon people had not possessed the *APRT*J* alleles, the people in Okinawa would have a lower incidence of the *APRT*J* alleles. Conversely, if the immigrants from the Korean Peninsula had not possessed the mutational alleles, the people in the central area of Japan would have a lower incidence of the *APRT*J* alleles. The latter assumption would not explain the presence of the *APRT*J* alleles among Koreans either.

The incidence of the *APRT*J* alleles among the Japanese was estimated to be 0.73% in this study. The incidence of the heterozygotes of APRT deficiency should be even higher because there are other mutant genes than *APRT*J* among the Japanese. Considering that about 68% of all the disease-related APRT alleles were *APRT*J*, the incidence of heterozygotes is calculated to be 1.1% (0.73/0.68). This figure is generally in accord with the previous data.[11,12]

Among the Caucasoid populations, disease-related mutations are known to be shared by different ethnic groups. The present study indicates that a disease-related mutation is also shared by different ethnic groups in Asia.

REFERENCES

1. Simmonds HA, Sahota AS, Van Acker KJ (1995) Adenine phosphoribosyltransferase deficiency and 2,8-dihydroxyadenine lithiasis. In: Scriver CR, Beaudet AL, Sly WS, Valle D (eds) Metabolic and molecular basis of inherited disease, 7th ed. McGraw-Hill, New York. pp 1799–1837
2. Sahota A, Chen J, Behzadian MA, Ravindra R, Takeuchi H, Stambrook PJ, Tischfield JA (1991) 2,8-Dihydroxyadenine lithiasis in a Japanese patient heterozygous at the adenine phosphoribosyltransferase locus. Am J Hum Genet 48: 983–989
3. Kamatani N, Hakoda M, Otsuka S, Yoshikawa H, Kashiwazaki S (1992) Only three mutations account for almost all defective alleles causing adenine phosphoribosyltransferase deficiency in Japanese patients. J Clin Invest 90: 131–136
4. Hidaka Y, Tarle SA, Fujimori S, Kamatani N, Kelley WN, Palella TD (1988) Human adenine phosphoribosyltransferase deficiency: Demonstration of a single mutant allele common to the Japanese. J Clin Invest 81: 945–950
5. Kamatani N, Terai C, Kim SY, Chen C-L, Yamanaka H, Hakoda M, Totokawa S, kashiwazaki S (1988) The origin of the most common mutation of adenine phosphoribosyltransferase among Japanese goes back to a prehistoric era. Hum Genet 98: 596–600
6. Kamatani N, Takeuchi F, Nishida Y, Yamanaka H, Nishioka K, Tatara K, Fujimori S, Kaneko K, Akaoka I, Tofuku Y (1985) Severe impairment in adenine metabolism with a partial deficiency of adenine phosphoribosyltransferase. Metabolism 34: 164–168
7. Terai C, Hakoda M, Yamanaka H, Kamatani N, Okai M, Takahashi F, Kashiwazaki S (1995) Adenine phosphoribosyltransferase deficiency identified by urinary sediment analysis: cellular and molecular confirmation. Clin Genet 48: 246–250
8. Kamatani N, Kuroshima S, Hakoda M, Palella TD, Hidaka Y (1990) Crossovers within a short DNA sequence indicate a long evolutionary history of *APRT*J* mutation. Hum Genet 85: 600–604
9. Horai S, Kondo R, Murayama K, Hayashi S, Koike H, Nakai N (1991) Phylogenetic affiliation of ancient and contemporary humans inferred from mitochondrial DNA. Philos Trans R Soc Lond Biol 333: 409–417
10. Hammer MF, Horai S (1995) Y chromosomal DNA variation and the peopling of Japan. Am J Hum Genet 56: 951–962
11. Kamatani N, Sonoda T, Nishioka K (1988) Distribution of the patients with 2,8-dihydroxyadenine urolithiasis and adenine phosphoribosyltransferase deficiency in Japan. J Urol 140:1470–1472
12. Hakoda M, Yamanaka H, Kamatani N, Kamatani N (1991) Diagnosis of heterozygous states for adenine phosphoribosyltransferase deficiency based on detection of in vivo somatic mutants in blood T cells: Application to screening of heterozygotes. Am J Hum Genet 48: 522–562

27

HGPRT AND APRT ACTIVITIES IN HEMOLYSATES DURING THE FIRST YEAR OF LIFE

A. Buño,[1] R. J. Torres,[1] D. Serfaty,[1] J. Tovar,[2] J. G. Puig,[3] and F. A. Mateos[1]

[1]Division of Clinical Biochemistry
[2]Division of Pediatric Surgery
[3]Division of Internal Medicine
Hospital "La Paz"
Universidad Autónoma
Madrid, Spain

1. INTRODUCTION

Early postnatal diagnosis of hypoxanthine guanine phosphoribosyl transferase (HGPRT) deficiency is frequently performed nowadays during the first year of life. The clinical expression of HGPRT deficiency is usually manifested by a variable psychomotor retardation between 6 and 12 months of age and may result in a variety of clinical syndromes, with markedly different neurological manifestations or even no neurological symptoms. At this age, or even earlier, parents or pediatricians may seek medical attention because they notice some developmental retardation in their child, sandy or coloured urine in the diapers, and/or because hyperuricemia and/or hyperuricosuria was discovered incidentally. Also babies belonging to a family with some male affected by this enzyme deficiency may seek early diagnosis (1).

It is known that in healthy newborns some erythrocytic enzymatic systems show a physiological immaturity, which reaches adult values throughout the first year of life (2). To our knowledge, there are a few studies in which HGPRT and adenine phosphorybosyl transferase (APRT) enzyme activities have been measured in blood of umbilical cord (3,4) and in children aged from 1 to 10 years (4). We performed this study to assess the physiological variations of HGPRT and APRT enzymatic activities during the first year of life in order to compare these normal values with the results obtained in babies who are suspected to have HGPRT deficiency.

Purine and Pyrimidine Metabolism in Man IX,
edited by Griesmacher *et al.* Plenum Press, New York, 1998.

2. SUBJECTS AND METHODS

2.1. Subjects

Hemolysates from venous blood samples of 39 healthy newborns were studied and, according to age, were divided into five groups: full-term newborns (n=15), 1–3 months (n=6), 4–6 months (n=7), 7–9 months (n=7) and 10–12 months (n=4).

2.2. Samples and Methods

Samples (1–2 ml) were taken from a peripheral vein of each subject without stasis and collected in tubes with litium heparin. After immediate cold centrifugation at 3,000 g, plasma was removed and the erythrocytes were washed three times with saline solution. The erythrocytes were stored at −20°C until the enzyme activities of HGPRT and APRT were assayed by an HPLC method using previously described techniques (5,6).

2.3. Statistical Analysis

Results were evaluated statistically with the SPSS program. None of the groups showed a normal Gaussian distribution. The ANOVA was used to test for significant differences among the groups.

3. RESULTS

Table 1 shows HGPRT (nmol/h/mg hb), APRT (nmol/h/mg hb) and HGPRT/APRT ratio values (mean ± S.D.) in all groups and their ranges (in parentheses). Normal adult controls were used as reference values in our laboratory.

HGPRT enzyme activity (Fig. 1) during the first year of life increased. Mean HGPRT activity was significantly different among groups. The lower values were found in newborns. There was also a significant correlation between HGPRT enzymatic activity and groups of age ($r = 0.47$; $p = 0.0026$).

In contrast, APRT enzyme activity decreased during the first year of life (Fig. 2). Mean APRT activity was significantly different among groups, showing a significant negative correlation with age ($r = −0.55$; $p = 0.003$).

According to these results, HGPRT/APRT ratio (Fig. 3) was significantly lower than normal adult values in all groups and reached adult normal values at the end of the first

Table 1. HGPRT (nmol/h/mg hb), APRT (nmol/h/mg hb) and HGPRT/APRT ratio values (mean ± S.D.) in all groups and their ranges (in parentheses)

Age (N)	Newborn (15)	1–3 months (6)	4–6 months (7)	7–9 months (7)	10–12 months (4)	Adult values (12)
HGPRT	64.6 ± 6.5	70.3 ± 10	71.5 ± 10	77.4 ± 4.7	72.2 ± 7.1	82.1 ± 6.4
	(61.0–68.2)	(59.2–81.6)	(62.9–72.9)	(73.1–81.8)	(60.9–83.5)	(78.1–86.2)
APRT	31.7 ± 3.8	30.2 ± 4.0	24.4 ± 4.0	28.1 ± 6.0	22.1 ± 3.1	25.8 ± 5.1
	(29.6–33.8)	(26.0–34.3)	(22.2–28.8)	(22.6–33.7)	(17.1–27.0)	(22.5–29.1)
HGPRT/APRT ratio	2.1 ± 0.3	2.4 ± 0.5	3.0 ± 1.0	2.8 ± 0.5	3.3 ± 0.1	3.25 ± 0.6
	(1.90–2.23)	(1.86–2.88)	(2.34–3.05)	(2.35–3.36)	(3.04–3.52)	(2.87–3.62)

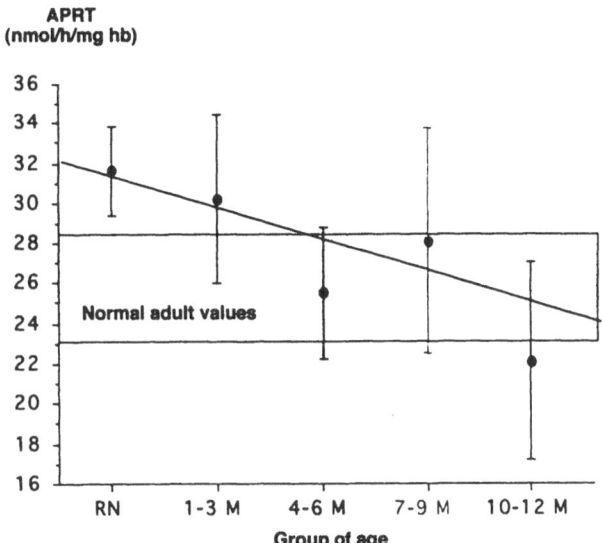

Figure 1. HGPRT activity in hemolysates of normal subjects during the first year of life according to age.

year of life. There was a significant correlation between HGPRT/APRT ratio and the age groups (r = 0.63; p = 0.,0001).

4. DISCUSSION

Recently, requirements for a diagnosis of HGPRT deficiency in a Clinical Biochemistry Laboratory have increased and are performed in younger patients than in the past. These patients usually are sent by doctors attending babies of families with some male affected by HGPRT deficiency or in whom hyperuricemia and/or hyperuricosuria is eventually discovered. The first situation is becoming less frequent, because female carriers

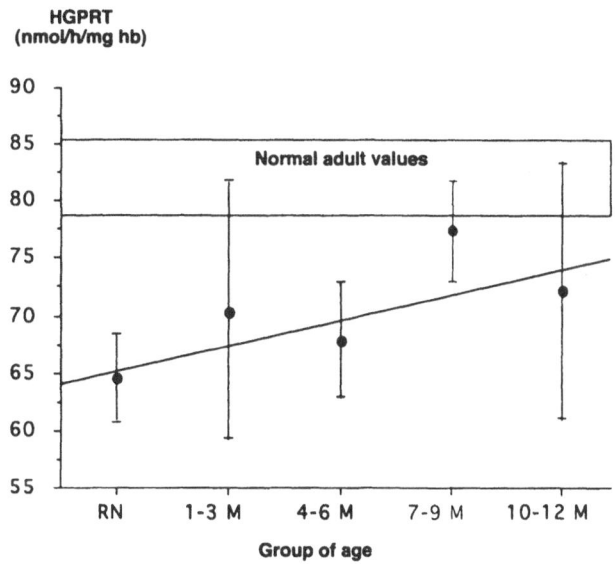

Figure 2. APRT activity in hemo lysates of normal subjects during th first year of life according to age.

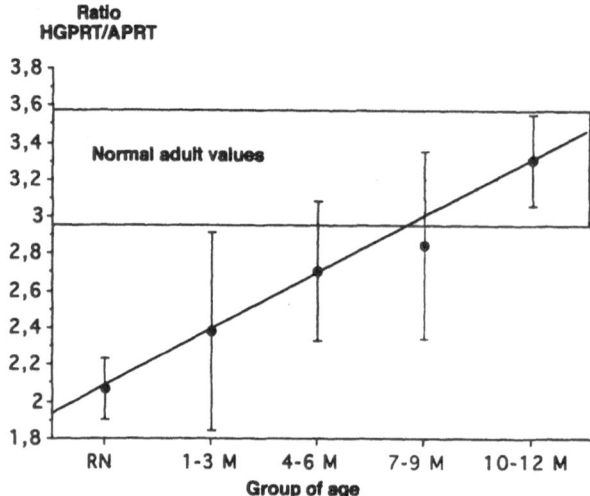

Figure 3. HGPRT/APRT ratio in normal subjects during the first year of life according to age.

usually know their possibility to transmit the enzyme defect to their offspring, and prenatal diagnosis is nowadays frequently performed. With respect to the second possibility, measurement of serum uric acid level is routinely performed in most of the multichanel analyzers and hyperuricemia may be an occasional finding (7). Hyperuricemia and/or hyperuricosuria, due to increased uric acid production, may occur in any disease with high cell turnover or tissue hypoxia and in some enzymatic defects (7). HGPRT enzymatic deficiency is responsible for uric acid overproduction resulting in high levels of blood and urine uric acid (8).

In HGPRT deficiency, the amount of residual enzyme activity in the hemolysates ranges from undetectable to 50% (9,10) and the clinical manifestations usually parallel the residual enzyme activity (9,11). However, patients with full (classic) Lesch-Nyhan syndrome have been reported with activities up to 7%, and an absence of HGPRT activity has been reported in gouty patients without the typical findings of the Lesch-Nyhan syndrome (9).

In newborns many enzymatic systems show an immaturity which may be present for a long time. For example, the phosphofructokinase enzymatic activity of the rate-controlling enzymes in red blood cells glycolysis and other non glycolytics enzymes (12), have been found to be lower in newborns and throughout the first year of life than normal adult values. Moreover, HGPRT enzyme activity measured in newborn and children aged from 1 to 10 years has been found lower than in adults and APRT activity was found to be higher in neonates but normal in children (3,4). Because most of our patients with HGPRT deficiency have been diagnosed during the first year of life we wish to determine the normal range of HGPRT enzymatic activity in this period.

The results of this study show that as compared to normal adult values, HGPRT activity in newborn hemolysates is reduced and these reduced values were evident throughout the first year of life when compared with adult values. This finding could indicate an immaturity of the enzyme during the first year of life but other possibilities may account for this finding. In contrast, APRT enzymatic activity was higher at birth than adult values and slowly decreased during the first year of life. In adult patients with HGPRT enzyme deficiency, APRT activity in hemolysates is increased (around twice the normal adult values) and this fact helps to provide an accurate diagnosis. The increased APRT activity seems to be related to an enhanced availability of intracellular PRPP, as documented in patients with Lesch-Nyhan syndrome (14) and in healthy neonates (3,4). The increased APRT activity could provide a larger quantity of adenine nucleotides in infants as happens in red blood cells of newborn infants (15).

As a result of decreased HGPRT and increased APRT activity, the HGPRT/APRT ratio was found to be reduced at birth and increased to adult normal values at the end of the first year of life. This ratio could also help to establish an accurate diagnosis of HGPRT deficiency. We conclude that it is necesary to know variations in physiological enzyme activities during the first year of life in order to perform a correct diagnosis of HGPRT enzymatic deficiency in this age group.

ACKNOWLEDGMENTS

We are indebted to the Clinical Research Unit nursing staff and the dietetic staff for excellent patient care; to Mª Paz Canencia for valuable technical assistance; to Erik Lundin for assistance in preparing the manuscript. Supported by grants from Caja de Madrid and Fondo de Investigaciones Sanitarias (FIS, 97/0458), Spain.

REFERENCES

1. Belinda JF Rossiter, C Thomas Caskey. Hypoxanthine-Guanine Phosphoribosyltransferase Deficiency: Lesch-Nyhan Syndrome and Gout, in Scriver CR, Beaudet AL, Sly WS, Valle D (eds): The Metabolic and Molecular Bases of Inherited Disease, 7th de. New York, McGraw-Hill, 1995, p1679–1706.
2. Oski AF, Komezane M. Metabolism of the erythrocytes of the newborn infant. Sem Hem 1975; 12:209–221.
3. Borden M, Nyhan WL, Bakay B. Increased activity of adenine phosphoribosyltransferase in erythrocytes of normal newborn infants. Pediatr Res 1974; 8:31–36.
4. Müller HM, Wafenbichler P. Activity of the salvage pathway in erythrocytes of newborn infants, children and adults. Adv Exp Med Biol 1977;76A:129–137.
5. Rilance HJ, Wallace RC, Nuhi G. Hypoxanthine-guanine phosphoribosyltransferase:assay using high performance liquid chromatography. Clin Chim Acta 1982; 127:159–165.
6. Rilance HJ, Wallace RC, Nuhi G. Adenine phosphoribosyl transferase:assay using high performance liquid chromatography. Clin Chim Acta 1985; 148:267–272.
7. Wilcox WD. Abnormal uric acid levels in children. J Pediatr 1996; 128:731–741.
8. Seegmiller JE, Rosenbloom RM, Kelley WN. An enzyme defect associated with sex linked human neurological disorder and excessive purine synthesis. Science 1967; 155:1682–1684.
9. de Bruyn CHMM. Hypoxanthine-guanine phosphoribosyl transferase deficiency. Hum Genet 1976; 31: 127–150.
10. Lorentz WB, Burton BK, Trillo A et al. Failure to thrive hyperuricemia and renal inssuficiency in early infancy secondary to partial hypoxanthine-guanine phosphoribosyltransferase deficiency. J Pediatr 1984; 104: 94–97.
11. Kelley WN, Wyngaarden JB. The Lesch-Nyhan syndrome. In: The Metabolic Basis Of Inherited Disease. JB Stanbury, JB Wingaarden, DS Fredrickson, eds. McGraw-Hill Booh Co. New York 4th edition. pp 1011 (1978).
12. Oski N. Trastornos del metabolismo de los eritrocitos. En: Problemas hematológicos en el Recién Nacido. FA Oski, Neiman JL eds. Editorial Panamericana. Buenos Aires 3ª edición. pp 111 (1984).
13. Rubin CS, Balin ME, Piomelli S et al. Elevated AMP pyrophosphorilase deficiency in congenital IMP pyrophosphorylase deficiency (Lesch-Nyhan disease). J Lab Clin Med 1969; 74: 732–741.
14. Seegmiller JE. Disease of Purine and Pirimidine Metabolism. In Metabolic Control and Disease. Boudy PK, Rosenberg LE eds. WB Sannders WB Co. Philadelphia 8th edition 1980: pp 777–937.
15. Gross RR, Schroeder EAR, Brounstein SA. Energy Metabolism in the erythrocytes of premature infants compared to a full term newborn infants and adults. Blood 1963; 21:755.

28

DELAYED HYPERSENSITIVITY TO 5-FLUOROURACIL ASSOCIATED WITH REDUCED DIHYDROPYRIMIDINE DEHYDROGENASE (DPD) ACTIVITY

John A. Duley[1] and Anthony B. W. Nethersell[2]

[1]Purine Research Laboratory
UMDS Guy's Hospital, London
[2]Department of Oncology
Royal Devon & Exeter Hospital and North Devon District Hospital
United Kingdom

INTRODUCTION

Fluorouracil (5FU) remains the primary chemotherapy agent for adenocarcinomas of the gastrointestinal tract. Toxicities are both dose and schedule dependent, and include myelosuppression, stomatitis and diarrhoea, and cutaneous manifestations. Side effects are usually easily manageable in normal clinical practice.

One practical problem with using the drug is the fact that the half-life of 5FU in humans is short. There is both an effective elimination of 5FU by the kidney, which also clears uracil very efficiently, and metabolic degradation is rapid, beginning with reduction by dihydropyrimidine dehydrogenase (DPD). The importance of DPD to the elimination of 5FU is demonstrated by inherited partial deficiency of this enzyme, which has been shown to explain many cases of 5FU sensitivity (1).

We present here the first reported case of 5FU sensitivity in the UK demonstrated to be associated with reduced DPD activity in peripheral blood lymphocytes (PBL). The case is interesting because the delayed hypersensitivity experienced by the patient to 5FU correlated with a level of DPD which was well within the range for heterozygosity for complete DPD deficiency.

PATIENT HISTORY

The patient, a 69-year-old man, was diagnosed with a metastatic adenocarcinoma of gastro-intestinal origin. Therapy was begun with 1.2 g 5FU by bolus injection. Following

Purine and Pyrimidine Metabolism in Man IX,
edited by Griesmacher *et al.* Plenum Press, New York, 1998.

147

this, the patient suffered only mild discomfort in his mouth and slight diarrhoea. His initial initial blood count showed haemoglobin 10.6, WBC 7.0, neutrophils 5.5 and platelets 242. A week after the first injection his haemoglobin was unchanged, with slight decreases in WBC 5.4, neutrophils 4.4 and platelets 178. As the patient was felt to be tolerating the drug well, a second injection was administered 8 days after the first. He also received a 2-unit packed cell transfusion, and had a mild febrile reaction following this.

Six days following the second injection, the patient presented with profound mucositis, a raised brick red rash all over the trunk and abdomen, exfoliation of his fingers with cracking of the epidermis, and a blood count showing profound bone marrow depression: haemoglobin 12.1, WBC 1.0, neutrophils 0.4 and platelets 29. He was immediately admitted and very nearly died of neutropenic sepsis. No neurological involvement was noted. The patient subsequently recovered, and interestingly a marker lymph node in his left neck, which was taken to be malignant, had completely regressed—an unexpected event after only 2 injections of 5FU.

METHODS

Whole EDTA blood from the patient was used for DPD analysis. PBL were collected from Ficoll-Hypaque and any contaminating erythrocytes were removed by lysis in ammonium chloride solution. Blood from a male control was taken at approximately the same time of day—to avoid reported problems with circadian changes in DPD activity (2)—and was treated identically.

DPD was assayed using a method modified from that of Bakkeren et al (3). In brief, PBL were lysed for 30 min on ice in 5 mM phosphate buffer pH 7.4 containing a non-ionic detergent (0.25% Triton X-100) then centrifuged briefly. Approximately 0.2 mg of PBL supernatant was added to reaction mixture of Tris-HCl buffer pH 7.4, containing magnesium chloride, 2-mercaptoethanol, freshly dissolved NADPH, and 6-[14C]-thymine, then incubated at 37°C. The reaction was stopped by addition of trichloroacetic acid, which was back-extracted with water-saturated diethylether.

The resulting reaction extract was injected onto a reverse-phase (ODS-1) HPLC column (4) which separated the DPD product, dihydrothymine, from the substrate in 11 minutes. Product and substrate radioactivity were monitored by an in-line radiodetector. Lymphocyte protein was determined in the trichloroacetic acid pellets by the method of Lowry et al (5). HPLC analysis of a urine specimen from the patient was also performed to analyse pyrimidine excretion (4).

RESULTS

The DPD activity in the patient's PBL, compared with a sex-matched control, is shown in Table 1. Patient DPD in fresh PBL supernatant was approximately half that of the normal mean, and was below the observed normal range for freshly-prepared PBL. DPD is known to be very labile to freezing and a lower normal range must be used for frozen cells. Interestingly, urine analysis (Table 1) revealed that the patient's excretion of uracil factored by pseudouridine was 0.38: this was above the upper limit of our normal range for age, but far below the grossly raised uracil excretion found in complete DPD deficiency (6).

Table 1. Analysis of patient's blood enzyme and urine

Leukocyte dihydropyrimidine dehydrogenase (DPD)	
Patient:	2100 pmol/h/mg protein
Control:	5500
(Normal range = 3000–5000, mean = 4146)	
Urine studies	
Pseudouridine:	0.654 mmol/24 h
Uracil:	0.249
Creatinine:	8.74
Orotic acid:	not detected
Uracil/pseudouridine = 0.38 (range < 0.3)	

DISCUSSION

The association of 5FU sensitivity with DPD partial deficiency in the USA has been documented extensively by Diasio and coworkers, who also showed that DPD activity in liver, the principal site of 5FU metabolism, correlated with the enzyme's activity in PBL (see 1,7,8). Patients sensitive to 5FU studied by these workers have been found to have PBL DPD activities varying between 6–28% of the mean normal activity. In contrast, the patient's DPD activity reported here was approximately half of the normal, suggesting that he was a heterozygotic carrier for complete DPD deficiency, although it has not been possible to perform family studies to confirm this. However, his delayed reaction to 5FU may possibly be explained by the mild degree of DPD deficiency observed. Our patient thus resembles another from the USA, mentioned by Gonzalez and Fernandez-Salguero (9).

Complete DPD deficiency, associated with gross thymine-uraciluria, was originally reported in association with relatively severe neurological and physiological symptoms and, in some cases, dysmorphia (6,10,11). More recently, however, a broad neurological spectrum of the disorder has emerged, some patients having the hallmarks of thymine-uraciluria but with 'milder' neurological deficits ranging from behavioural presentations (e.g., attention deficit disorder) to 'clumsiness' and mild retardation or regression.

One thing is becoming apparent: DPD deficiency appears to be more common than previous estimates of heterozygosity which ranged between one in three hundred to one in a thousand (8,12). While urine screening for thymine-uraciluria is simple and is becoming routine within the UK, assay of PBL DPD prior to 5FU therapy remains relatively difficult, and thus expensive. The human gene has now been mapped (13), but convenient genetic screening as part of the work-up to chemotherapy will depend upon there being a reasonably small number of common alleles for the gene. Development of a simple test for heterozygosity is important not only to explain 5FU toxicity in a patient, but for other family members who would also be at risk if they ever required 5FU therapy.

REFERENCES

1. Diasio RB, Beavers TL, Carpenter JT (1988). Familial Deficiency of Dihydropyrimidine Dehydrogenase. J. Clin.Invest. 81:47–51.
2. Zhang R, Lu Z, Lu T, Soong S-J, Diasio RB (1993). Relationship between circadian-dependent toxicity of 5-fluorodeoxyuridine and circadian rhythms of pyrimidine enzymes. Cancer Research 53:2816–22.

3. Bakkeren JAJM, De Abreu RA, Sengers RCA, Gabreels FJM, Maas JM, Renier WO (1984). Elevated urine, blood and cerebrospinal fluid levels of uracil and thymine in a child with dihydrothymine dehydrogenase deficiency. Clin. Chim. Acta 140: 247–56.

4. Simmonds HA, Duley JA, Davies PM (1990). Analysis of purines and pyrimidines in blood, urine and other physiological fluids. In: Techniques in Diagnostic Human Biochemical Genetics, Wiley-Liss, NY, pp397–424.

5. Lowry OH, Rosebrough NJ, Farr A, Randall RJ (1951). Protein measurement with the Folin phenol reagent. J. Biol. Chem. 193: 265–75.

6. Henderson MJ, Jones S, Walker P, Duley JA, Simmonds HA (1995). Heterogeneity of symptomatology in two male siblings with thymine uraciluria. J. Inher. Metab. Dis. 18:85–6.

7. Lu Z, Zhang R, Diasio RB (1993). Dihydropyrimidine dehydrogenase activity in human peripheral blood mononuclear cells and liver. Cancer Res. 53:5433–8.

8. Harris BE, Carpenter JT, Diasio RB (1991). Severe 5-fluorouracil toxicity secondary to dihydropyrimidine dehydrogenase deficiency. A potentially more common pharmacogenetic syndrome. Cancer 68: 499–501.

9. Gonzalez FJ, Fernandez-Salguero P (1995). Diagnostic analysis, clinical importance and molecular basis of dihydropyrimidine dehydrogenase deficiency. TiPS 16:325–7.

10. Van Gennip AH, Abeling AH, Stroomer NGGM, Van Lenthe H, Bakker HD (1994). Clinical and biochemical findings in six patients with pyrimidine degradation defects. J.Inherited Metabolic Diseases 17:130–2.

11. Webster DR, Becroft DM, Suttle DP (1995). Hereditary orotic aciduria and other disorders of pyrimidine metabolism. In: The Metabolic and Molecular Bases of Inherited Disease, ed. CR Scriver, AL Beaudet, WS Sly, D Valle, chapter 55, McGraw-Hill, NY, pp 1799–1837.

12. Milano G, Etienne M-C (1994) Potential importance of dihydropyrimidine dehydrogenase in cancer chemotherapy. Pharmacogen. 4:301–6.

13. Yokota H, Fernandez-Salguero P, Furuya H, Lin K, McBride OW, Podschun B, Schnackerz KD, Gonzalez FJ (1994). cDNA cloning and chromosome mapping of human dihydropyrimidine dehydrogenase, an enzyme associated with 5-fluorouracil toxicity and congenital thymine uraciluria. J. Biol. Chem. 269:23192–6.

29

HPRT-MUTATIONS IN ITALIAN LESCH-NYHAN PATIENTS

B. S. Gathof,[1] M. Rocchigiani,[2] V. Micheli,[2] Z. Gaigl,[1] and U. Gresser[1]

[1]Purine Laboratory
Medizinische Poliklinik
University of Munich, Germany
[2]Dipartimento di Biologia Molecolare
Università degli Studi
Siena, Italy

INTRODUCTION

The Lesch-Nyhan syndrome is a severe X-chromosomal disease caused by hypoxanthine-guanine phosphoribosyltransferase (HPRT) deficiency (Lesch and Nyhan 1964; Seegmiller et al. 1967). HPRT-deficiency leads to increased levels of uric acid in blood and urine with associated clinical manifestation of renal uric acid stones and gouty arthritis, and to neurological disturbances including choreoathetosis, spasticity, compulsive self-mutilation and mental retardation of variable severity. In individuals with partial deficiency of HPRT (Kelley-Seegmiller syndrome) hyperuricemia, nephrolithiasis and gouty arthritis are observed, but they are mostly spared the neurological symptoms.

Up to now more than 150 patients with HPRT-deficiency have been examined on the molecular genetic level and a variety of different mutations has been described (reviewed in Sege-Petersen 1992; Gathof in press). To our knowledge only a patient with partial HPRT-deficiency from Italy had been examined on the molecular genetic level (Rocchigiani, Gathof et al. in press).

PATIENTS AND METHODS

In this study we examined five unrelated Italian Lesch-Nyhan patients (and their respective families). For the families Z. und B. the propositi and the parents were studied. In family Co. the propositus and his mother were examined. In family Ca. and M. the propositus only was studied.

The coding region of the patients was examined by direct sequencing of PCR amplified genomic DNA using biotinylated primers. PCR products were purified with glasspow-

Purine and Pyrimidine Metabolism in Man IX,
edited by Griesmacher et al. Plenum Press, New York, 1998.

der (Geneclean), and single-stranded DNA was obtained by treatment with magnetic beads coated with streptavidine (Dynabeads, Dynal). Direct sequencing (modified according to Gibbs et al. 1990) was performed (Sequenase 2, USB, Amersham) followed by denaturing gel electrophoresis. Restriction enzymes were used to confirm mutations.

RESULTS

In patient M. we found the point mutation G307T leading to a change in aminoacid 70 from glycine to tryptophan. The mother of the propositus is heterozygous for this mutation.

A deletion of two bases at nucleotides 295 to 300 was found in patient Zu., which leads to a stop codon after 71 aminoacids. The parents are homozygous for the normal sequence.

In another patient B. the transition C607T was observed. In consequence aminoacid 170 Arg is replaced by a stop codon. The mother of the patient is heterozygous for the mutation, the sister is homozygous for the normal sequence.

In patient Ca. only exon 1 could be amplified by PCR and sequenced (normal sequence). So we assume a large deletion affecting exon 2 to 9.

Patient Co. shows a complete deletion of exon 4. In the mother of this patient we could amplify exon 4, but it remains unclear if only one or both allels carry exon 4, as RNA studies have not yet been performed.

DISCUSSION

In this study several different mutations in the HPRT-gene were detected in Italian Lesch-Nyhan patients: In a region coding for amino acid 66 to 71 two mutations were detected. This region is known to be a hot-spot for mutations (Cariello and Skopek 1993). In patient M. a point mutation, G307T was detected, which had not been described before. The transversion G307T is placed in the longest monotonic nucleotide run in the HPRT coding region, GGGGGG, in basepairs 306 to 311. Nearly 9% of all single base substitution mutations are found in this sequence (Cariello and Skopek 1993). As possible factors for the high number of mutations the authors discuss less efficient repair mechanisms in the GGGGGG sequence (Chen et al. 1990) and an increase in electronegativity in central guanine bases (Mattes et a.l 1986).

The deletion of the two bases in patient Zu. is placed in a repetitive sequence (bases 296 to 300), which is flanked by a CT sequence. Repeated DNA sequences are thought to be involved in spontaneous deletion in E. coli, and studies of other mammalian genes have shown that spontaneous deletions are often flanked by 2–7 bp repeats (Nalbantoglu et al. 1986).

The point mutation C607T, a stop codon mutation, has been described before in seven families from different countries (Gibbs et al., PNAS 86: 1919–1923, 1989; Tarlé et al., Genomics 10: 499–501; 1991, Marcus et al., Hum Genet 89: 395–400; 1992, Burgemeister et a.,l Adv Exp Biol Med 370: 331–335, 1995; Gathof unpublished).

In patient Co. a deletion of exon 4 and in patient Ca. of exon 2 to 9 is postulated. Deletions of one or several exons were already reported in several patients by different groups (reviewed in Sculley et al. 1992 and Gathof in press), but in most cases the mutation causing the deletion (as in our cases) has not been detected.

CONCLUSION

In this study the HPRT-gene of five Italian Lesch-Nyhan patients was examined. A new point mutation Gly70Trp was detected, as well as two stop codon mutations. In two patients deletions of exons (exon 4 and exons 2 to 9 respectively) are postulated. The mutation Arg170Stop has been observed before in unrelated families in different countries. So in Italian Lesch-Nyhan patients HPRT deficiency is caused by several different mutations. This reflects the variety of mutations in the HPRT-gene of Lesch-Nyhan patients reported in other studies worldwide.

ACKNOWLEDGMENT

We thank the patients and their families for their support of this study, and Dr. Hayek and Prof. Marinello, Siena, and Dr. Zambrino, Pavia, who provided the blood samples.

REFERENCES

NF Cariello and TR Skopek (1993). Analysis of mutations occurring at the human hprt locus. J Mol Biol 231: 41–57.

RW Chen, VM Maher, JJ McCormick (1990). Effect of excision repair by diploid fibroblasts on the kinds and locations of mutations induced by (±)-7β,8α-dihydroxy-9α-,10α-epoxy-7,8,9,10-tetrahydrobenzopyrene in the HPRT gene of diploid human fibroblasts. Proc Nat Acad Sci, 89: 5413–5417.

RA Gibbs, PN Nguyen, A Edwards, AB Civitello , CT Caskey (1990). Multiplex DNA deletion detection and exon sequencing of the hypoxanthine phosphoribosyltransferase gene in Lesch-Nyhan families. Genomics 7: 235–244.

M Lesch and WL Nyhan (1964). A familial disorder of uric acid metabolism and central nervous system function. Am. J. Med. 36: 561.

WB Mattes, JA Hartley, KW Kohn (1986). DNA sequence selectivity of guanine-N7 alkylation by nitrogen mustards. Nucl Acids Res 14: 2971–2987.

J Nalbantoglu, D Hartley, G Phear, G Tear, M Meuth (1986). Spontaneous deletion formation at the aprt locus of hamster cells: the presence of short sequence homologies and dyad symmetries at deletion termini. EMBO J. 5: 1199–1204.

JE Seegmiller, FM Rosenbloom, WN Kelley (1967). Enzyme defect associated with a sex-liked human neurological disorder and excessive purine synthesis. Science 155: 1682.

K Sege-Petersen, J Chambers T Page, OW Jones, WL Nyhan (1992). Characterization of mutations in phenotypic variants of hypoxynthine phosphoribosyltransferase deficiency. Hum Mol Gen 6: 427–432.

30

ERYTHROCYTE CDP-CHOLINE ACCUMULATION IN HAEMOLYTIC ANAEMIA AND RENAL FAILURE (RF)

Arian Laurence,[1] John A. Duley,[1] H. Anne Simmonds,[1] Mark Layton,[2]
Stephen J. Rose,[3] and Susan J. Kelly[4]

[1]Purine Research Laboratory
UMDS Guy's Hospital, London
[2]Department of Haematology
Kings College Hospital, London
[3]Birmingham Heartlands Hospital
Birmingham
[4]Wycombe General Hospital
High Wycombe, United Kingdom

INTRODUCTION

The accumulation of CDP-choline in high concentrations in the erythrocytes of a patient with haemolytic anaemia was first reported by Paglia et al. (1). A defect in CDP-choline phosphotransferase was suggested (Fig. 1). The selective nucleotide accumulation is assumed to be related to a deficiency of CDP-choline phosphotransferase in erythroid precursors, but is not yet proven, since the enzyme is not present in mature erythrocytes.

The clinical presentation of CDP-choline phosphotransferase is virtually identical with that for UMP-hydrolase (UMPH1: pyrimidine-5'-nucleotidase) deficiency (2), from which it must be distinguished by measurement of erythrocyte UMPH1. The subjects in this report also came to attention after the finding of normal UMPH1 activity following an abnormal result in the preliminary screening test routinely used for UMPH1 deficiency. This test is based on the characteristic UV absorption profile at 280 nm and 254 nm in deproteinised extracts of UMPH1 deficient erythrocytes, compared with normal cells, indicating raised pyrimidine nucleotides.

We report studies in 4 children (2 being siblings in a consanguineous kindred) who presented with haemolytic anaemia with an unusual shift in the UV screening test for pyrimidine 5'-nucleotidase (UMPH1) deficiency, but normal UMPH1 activity. Two adults with renal failure (RF) on treatment with allopurinol were also studied.

Purine and Pyrimidine Metabolism in Man IX,
edited by Griesmacher *et al.* Plenum Press, New York, 1998.

ENZYMES OF GLYCEROLIPID BIOSYNTHESIS

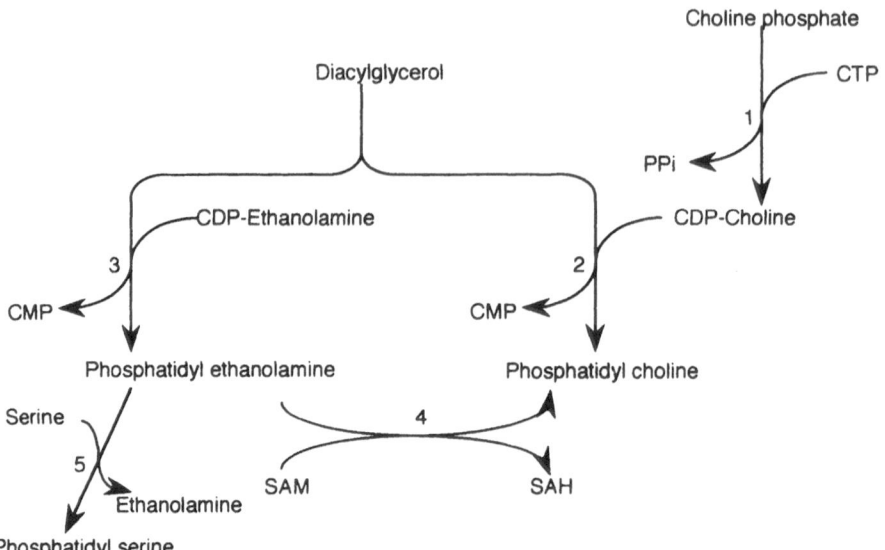

Figure 1. Summary of the pathways in which CDP-Choline and CDP-Ethanolamine are active intermediates. SAM S-adenosylmethionine. SAH S-adenosyl homocysteine. The enzymes listed are: 1. Choline phosphate cytidyl transferase; 2. Diacylglycerol cholinephosphotransferase; 3. Diacylglycerol ethanolaminephosphotransferase; 4. Phosphatidyl-ethanolamine phosphotransferase; 5. Phosphatidylethanolamine serinetransferase.

PATIENTS AND METHODS

1. NK. Child of a first cousin Asian marriage, presented at the age of 8 yrs with recurrent episodes of jaundice, fever and anaemia. On examination there was moderate hepatosplenomegaly. Investigations showed Hb 5.8 g/dL, MCV 67fL, WBC 9.2 × 10^9/L, platelets 232 × 10^9/L. Blood film showed polychromasia with 25% reticulocytes, and basophilic stippling. Serum bilirubin was 56 υM. Haemoglobin electrophoresis and assays for G6PD and UMPH1 were all normal. The patient recovered with blood transfusion but continues to require blood transfusion on an occasional basis

2. SA. Sister to NK noted subsequently to the diagnosis of her brother to have an asymptomatic anaemia at the age of 15 yrs (Hb 8.6 g/L, MCV 73.5, ferritin 45, normal haemoglobin electrophoresis). On examination there was no hepatosplenomegaly. Currently she has not required any blood products.

3. ZN. First child of a first cousin Asian marriage. Presented at birth with jaundice (bilirubin 157 υM) and a blood picture consistent with erythroblastosis fetalis (Hb 13.6 g/L, MCV 100fL, reticulocytes 3%, nucleated cell count 190 of which 95% were normoblasts). On examination there was mild hepatosplenomegaly (liver edge 1.5 cms below the costal margin, spleen tip just palpable). Haemoglobin electrophoresis and assays for G6PD, UMPH1, and pyruvate kinase were all normal. The child was treated with phototherapy and an exchange transfusion. Over the next year ZN remained chronically anaemic with Hb 7.0 g/L, WBC 11.7 × 10^9/L, platelets 636 × 10^9/L. The blood film showed poikilocytosis, polychromasia with red cell distortion and basophilic stippling and the occasional normoblast. A bone marrow aspirate was performed which showed a hypercellularity with marked expansion of the numbers of late normoblasts some of which showed paired nuclei and cytoplasmic basophilic stippling. Currently the child is maintained on supportive blood transfusions.

4. CW. First child of unrelated parents of European descent . Presented at the age of 22 mths with jaundice, vomiting and anaemia after a long running corysal illness. On examination there was moderate hepatosplenomegaly (liver edge 2cm below the right costal margin, splenic notch 4cm below the left). Investigations showed Hb 5.7 g/L, reticulocytes 20%, WBC 8.1×10^9/L, platelets 203×10^9/L. The blood film showed fragmented cells and basophilic stippling. Assays for G6PD, UMPH1 and pyruvate kinase were all normal. Haemoglobin electrophoresis showed a persistently high fetal haemoglobin of 20%. Haemoglobin mass spectrometry showed a β-chain variant Hb Southampton/Casper. Currently the patient has maintained his haemoglobin with folate supplements and the occasional blood transfusion.

5. TN. First presented at the age of 47 yrs with slowly declining renal function. He had an unremarkable family history and a past medical history of gout but no anaemia or jaundice. A renal biopsy showed changes consistent with IgA nephropathy. He was managed on allopurinol, frusemide, captopril and calcium carbonate and over the next three years developed end stage renal failure requiring haemodialysis. Investigations at this time showed Hb 9.3 g/L, WBC 5.8×10^9/L, platelets 189×10^9/L, serum creatinine 895 μmol/L. A year later he received a live-related renal transplant from his sister with no complications and is currently managed on prednisolone, cyclosporin A, azathioprine, enalapril, methyldopa and amilodipine. Six months post transplantation he has Hb 15.5, with serum creatinine of 149 μmol/L.

6. CF. First presented at the age of 70 yrs with end stage renal failure associated with gout and hypertension. He was started on peritoneal dialysis. His family history is unremarkable as is his full blood picture despite having a serum creatinine of 550 μmol/L.

HPLC. Red cell nucleotides were evaluated in erythrocyte lysates using HPLC with in-line diode-array detection as reported previously (3).
UMPH1 was also assayed by HPLC (4).

RESULTS

Several unusual features were noted in the nucleotide profiles in the four children: principally elevated concentrations of CDP-choline (mean 85 μmol/L, control 1.5) and CDP-ethanolamine (mean 35 μmol/L, control 2.7) (Fig. 2A).
The erythrocytes of the 2 patients with RF showed accumulation of the same unusual nucleotides (Fig. 2B), but concentrations were even higher (means of 150 and 48μmol/l for CDP-choline and CDP-ethanolamine respectively). These abnormal patterns of pyrimidine nucleotide accumulation were not observed in other renal patients. It is noteworthy that one of the 2 patients has recently undergone successful renal transplantation and that the erythrocyte nucleotide profile is now normal.
In contrast to UMPH1 deficiency, while CDP, UTP and UDP were also present in both the children and adults (normally absent) the concentrations were low (Fig. 2A,B).

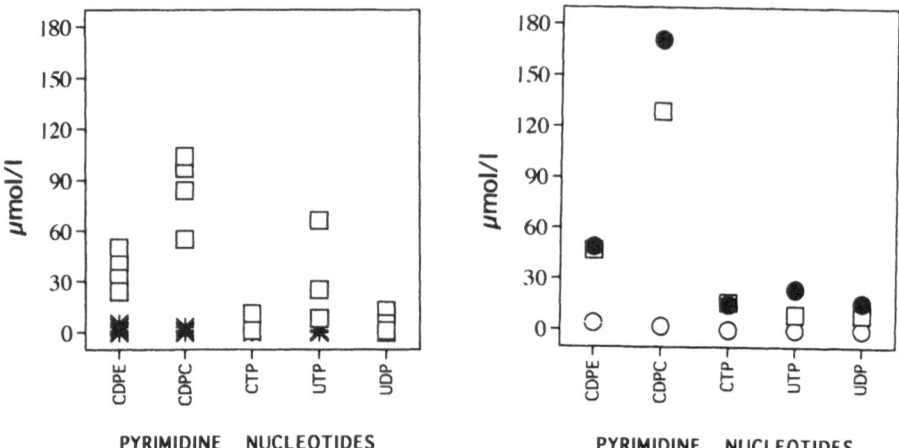

Figure 2. A. Concentrations of the 5 pyrimidine nucleotides accumulating in the erythrocytes of 4 children with haemolytic anaemia (open squares), compared with the values found in healthy children (cross). B. Concentrations of the same 5 pyrimidine nucleotides in the erythrocytes of two adult patients on renal support: One on haemodialysis (closed circles) and the same patient after transplantation (open circles), the other on CAPD (open squares).

No other abnormalities of purine or pyrimidine metabolism have been found in this defect. Purine production and excretion in the children was normal.

DISCUSSION

The results in the children and adults in this report are consistent with a putative deficiency of CDP-choline phosphotransferase. This diagnosis is presumed, not proven, from the finding by HPLC of of high CDP-choline and CDP-ethanolamine concentrations in erythrocytes of the six subjects here.

All values for UMPH1 were within the normal range (8–15 nmol/mg Hb/h), which contrasted with activities of less than 10% of normal which are characteristic of UMPH1 deficiency. Such a finding should alert the investigator to the possibility of this alternative explanation for the haemolytic anaemia. In the normal human erythrocyte, adenine nucleotides comprise over 90% of the total nucleotide pool and, apart from low levels of UDP-glucose, pyrimidine nucleotides are undetectable. In UMPH1 deficiency the concentration of pyrimidine nucleotides UTP and CTP may exceed those of the adenine nucleotides (2). Thus the other factor which distinguishes the two disorders is that UTP, CTP, UDP-glucose and UDP-N-acetylglucosamine, although also slightly elevated compared with controls, are much lower than the concentrations found in UMPH1 deficiency (1).

Two of the children were from a large consanguineous Asian kindred, one being identified only during the family study. One other child was also of Asian descent from a first-cousin marriage. These findings would be compatible with an autosomal recessive mode of inheritance for this putative deficiency of CDP-choline phosphotransferase. The fourth child reported here was of European descent, while the case of Paglia et al was of Irish/Scottish/Dutch ancestry, indicating a wide ethnic distribution. The molecular defect is unknown.

Secondary accumulation of CDP-choline has been reported in the erythrocytes of some patients with chronic renal failure, but the values were not as high as in the two adults in this report, who were also being treated with allopurinol for gout. The fact that CDP-choline accumulation is not a general finding in the erythrocytes of subjects with chronic renal failure suggested that the two patients here might be heterozygotes for the deficiency. In their case this is manifest through a combination of the uraemic state exacerbated by allopurinol therapy. This hypothesis is supported by the relatively normal erythrocyte nucleotide profile in one of the patients following successful renal transplantation.

What then is the metabolic basis for the haemolytic anaemia in these subjects? The last two steps in lecithin synthesis are mediated by CDP-choline synthetase (CTP: cholinephosphate cytidyltransferase; EC 2.7.7.15) and choline phosphotransferase (CDP-choline: 1,2-diacylglycerol choline phosphotransferase; EC 2.7.8.2). The first step results in the formation of CDP-choline and pyrophosphate (Fig. 1). Although potentially reversible, the presence of an active inorganic pyrophosphatase in human erythrocytes makes CDP-choline formation irreversible. The high concentrations of CDP-choline found in erythrocytes of the four children presenting with haemolytic anaemia, jaundice and evidence of splenomegaly suggests they are homozygotes for a deficiency of the latter enzyme (1).

Paglia hypothesised that the associated chronic haemolysis in the defect may be related to ineffective erythropoiesis due to intra-marrow death of erythroid precursors during maturation (1). An additional or alternative hypothesis would be unbalanced membrane lipid biosynthesis. However, the fact that CDP-ethanolamine concentrations were normal in his patient was considered to rule out the possibility that the same enzyme might mediate the formation of CDP-ethanolamine and CDP-choline. The elevated concentrations of CDP-ethanolamine in our patients would either support a dual role for this enzyme, or a secondary stimulation of CDP-ethanolamine synthesis consequent upon an accelerated stimulation of this section of the pathway.

The problem of treatment arises. Episodes of brisk haemolysis have required several successive transfusions, but subsequent growth in a case with neonatal onset has been normal. The adult studied by Paglia et al. (1) for 9 years eventually required splenectomy for splenomegaly and hypersplenism (personal communication).

In summary, both UMPH1 and CDP-choline phosphotransferase deficiency are associated with non-spherocytic haemolytic anaemia. Since the latter enzyme is not demonstrable in erythrocytes the basal defect remains unproven, but should be suspected in chronic haemolytic anaemia with normal erythrocyte UMPH1 activity.

REFERENCES

1. Paglia DE, Valentine WN, Nakatani M, Rauth B. Selective accumulation of cytosol CDP-choline as an isolated erythrocyte defect in chronic hemolysis Proc. Natl. Acad. Sci. 1983; 80:3081–5
2. Valentine WN, Fink K, Paglia DE, Harris SR, Adams WS. Hereditary haemolytic anaemia with human erythrocyte pyrimidine 5'-nucleotidase deficiency. J. Clin. Invest. 1974; 54: 866–9
3. Simmonds HA, Duley JA, Davies PM. Analysis of purines and pyrimidines in blood, urine and other physiological fluids. Chapter 25 In: Hommes F, ed, NY:Wiley-Liss. *Techniques in Diagnostic Human Biochemical Genetics: A Laboratory Manual* 1991; pp. 397–424
4. Duley JA, Simmonds HA. Superactive UMP hydrolase: cause or consequence of haemolytic anaemia? Adv Exp Med Biol 1992; 309B:315–319.

31

ERYTHROCYTE UMP SYNTHETASE ACTIVITY

An HPLC-Linked Non-Radiochemical Assay in Normal Subjects and in One Case of Oroticaciduria

V. Micheli,[1] G. Jacomelli,[1] E. Zammarchi,[2] and G. Pompucci[1]

[1]Dipartimento di Biologia Molecolare
Università di Siena
Policlinico Le Scotte, Siena, Italia
[2]Ospedale Pediatrico Meyer
Firenze, Italia

1. INTRODUCTION

Orotate phosphoribosyltransferase (OPRT) and OMP decarboxylase (ODC), the last two enzymes in the de novo synthesis of pyrimidine nucleotides, reside in two separate domains of a single polypeptide chain, forming the so called UMP synthetase (UMPs).[1] UMPs is responsible for the conversion of orotic acid to UMP, which in turn can be converted into the other pyrimidine nucleotides. One or both activities are known to be defective in human mutants with hereditary oroticaciduria, a rare disorder presenting with megaloblastic anemia, mental retardation and diverse additional features.[1] Diagnosis can be effected by measuring OPRT and ODC activities in the erythrocytes, which normally metabolise extracellular orotate through this salvage pathway, though lacking the de novo pyridine synthesis.

Radiochemical methods are commonly used for the assay of OPRT and ODC activities either measuring the release of $^{14}CO_2$ from labeled orotic acid or quantitating the production of labeled UMP or OMP by HPLC;[2] a spectrophotometric assay has also been developed.[3]

This paper presents an HPLC-linked, non-radiochemical method for the assay of OPRT and ODC activities, based on the direct measurement of the products OMP and UMP.

2. MATERIALS AND METHODS

Erythrocytes were isolated from whole blood anticoagulated with heparin. Controls were 14 children with no metabolic alterations, aged 4–10, and 13 adults, aged 22–66. The determinations were also performed on a 9 year old boy, presenting with oroticaciduria (26.5–130 mMol/Mol creatinine; normal values 0.5–3.3), mental retardation and convul-

Purine and Pyrimidine Metabolism in Man IX,
edited by Griesmacher *et al.* Plenum Press, New York, 1998.

sive seizures. Erythrocytes were washed twice with saline and lysed by freezing and thaw-ing twice. Lysates were shaken with activated charcoal (5% v/v) to discard most endo-genous nucleotides, nucleosides and bases and the suspension was centrifuged 10 min at 12,000 × g. The cell-free clear supernatant was used for enzyme activity assays. Hemoglo-bin (Hb) concentration was also measured for subsequent quantification.

2.1. OPRT Activity Assay

The incubation mixture contained, in a final volume of 200 µl, 0.28 mM orotate, 0.6 mM phosphoribosylpyrophosphate, 20 mM Tris-HCl pH 8, 3 mM $MgCl_2$, 15 mM potas-sium phosphate buffer pH 8, and lysate amount containing about 1–5 mg of hemoglobin. After 15–60 min of incubation at 37°C, perchloric acid extracts of incubation mixtures were obtained.

2.2. ODC Activity Assay

The incubation mixture contained, in a final volume of 200 µl, 0.1 mM OMP, 20 mM TRIS-HCl buffer pH 7.4, 50 mM potassium phosphate buffer, pH 7.4, and lysate amount containing about 1–5 mg of hemoglobin. After 15–60 min of incubation at 37°C, perchloric acid extracts of incubation mixtures were obtained.

Suitable blanks at zero time were performed for both enzyme assays.

2.3. HPLC Procedure

Separation of substrates and products was performed as previously described[4] by a System Gold Beckman Module 126 HPLC, supplied with a Scanning Detector module 167, using either a LC-18 Supelcosil column, or a Beckman ODS Ultrasphere column. Total run time was 15 min, achieving a good separation of substrates and products involved in both reactions (Figure 1).

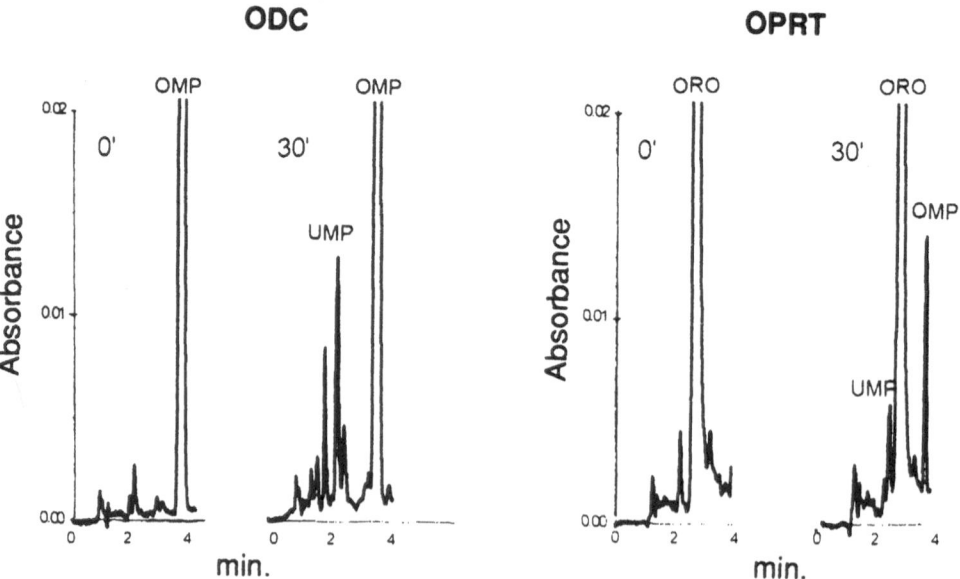

Figure 1. RP-HPLC trace after injection of OPRT (100 µl) and ODC (50 µl) assay mixtures at 0 time and after 30 min incubation.

3. RESULTS

The assay conditions for OPRT and ODC differed in the pH optimum (8 and 7.4, respectively) and in the presence of Mg^{++} ions which strongly activated OPRT but had no effect on ODC (data not shown). Products of both OPRT and ODC reactions, OMP and UMP, were always virtually absent in the zero time controls. Both nucleotides were found after incubation of the OPRT assay mixtures; UMP formation was prevented by adding 1.5 mM XMP, which is known to inhibit ODC (data not shown). In this case the amount of OMP produced equalled the amount OMP plus UMP obtained in the absence of the inhibitor. XMP was omitted in routine assay mixtures, and the amount of OMP plus UMP was assumed to reflect the activity of OPRT. UMP was the only nucleotide produced in ODC assay incubations. In both OPRT and ODC assays, considerable nucleoside formation (uridine and orotidine in OPRT, orotidine in ODC assay) took place, possibly due to the presence of active 5'nucleotidase in crude lysates. Nucleoside formation was prevented by addition of phosphate ions, which also enhanced OPRT and ODC activities in a concentration-dependent fashion (Fig. 2).

OPRT and ODC reactions were linear over the incubation time up to 60 min, hence 30 min was chosen for routine assays of both activities. Product formation increased linearly up to 100 μl of lysate in the assay, hence 75 μl was chosen for routine conditions (Fig. 3).

The activity values of both OPRT and ODC in healthy controls were in the range reported in the literature;[1] significant difference was found between children and adults for both enzymes ($P < 0.01$). Both activities were markedly lower in the oroticaciduric child J.A. than in age-matched controls (OPRT 20%, ODC 16%) (Table 1).

4. CONCLUSIONS

The HPLC-linked methods presented are easy to perform, sensitive and reliable enough to be used for both research and diagnostic purposes, avoiding the use of radiolabels. We have found significantly higher activity of both OPRT and ODC in children com-

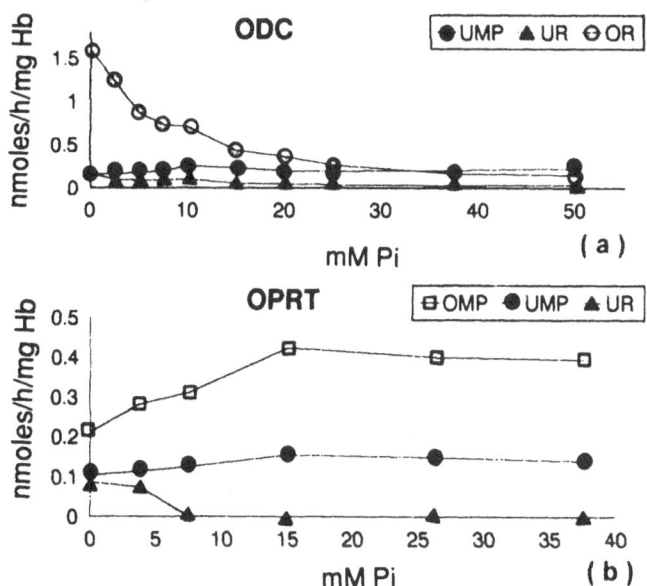

Figure 2. Dependence of the OPRT and ODC reactions on phosphate concentration.

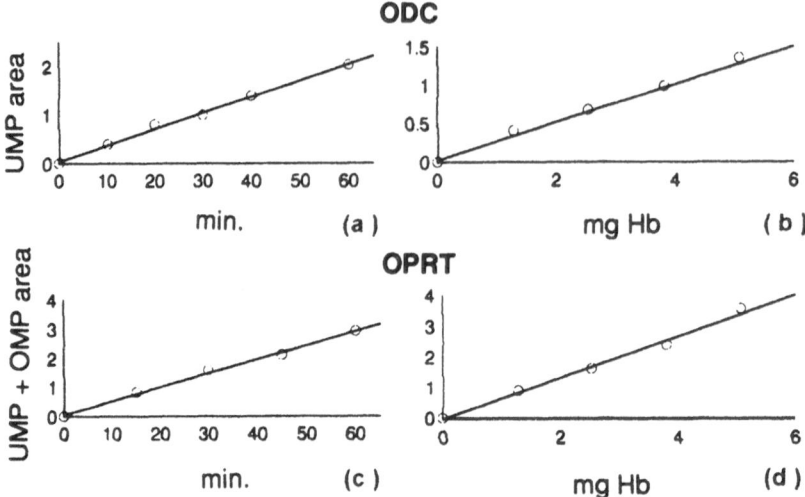

Figure 3. OPRT and ODC reactions: linearity with incubation time (a,c) and lysate amount (expressed as mg Hb) (b,d).

Table 1. OPRT and OCD activity in
the erythrocyte lysates of controls
and of the oroticoaciduric patient J.A.

	OPRT	ODC
Adults	0.206 ± 0.078	0.229 ± 0.106
Children	0.369 ± 0.152	0.323 ± 0.113
J.A.	0.072 ± 0.05	0.056 ± 0.02

Data expressed as nmoles.h^{-1}.mg^{-1}Hb (mean ± S.D.).

pared to adults. Comparable age-linked differences have been reported for other enzymes;[4] since remarkable activity decay during erythrocyte life span has been reported for both OPRT and ODC[5] such difference could be related to the age composition of circulating erythrocyte population. The activity of both OPRT and ODC is markedly lower in the propositus J.A. than in the age-matched controls. Data reported in the literature list different degrees of deficiency, the activity falling down to undetectable in some cases. Both activities ranged 20–16% of age-matched controls in our patient, thus suggesting a deficiency of both enzymes may account for the observed oroticoaciduria.

ACKNOWLEDGMENTS

This study was supported by the Italian M.U.R.S.T. (60% and 40% funds).

REFERENCES

1. Webster D.R, Becroft D.M.O., Parker Suttle D. Hereditary orotic aciduria and other disorders of pyrimidine metabolism. In "The metabolic and molecular bases of inherited disease", C.R. Scriver et al. Ed.,VII ed., McGraw-Hill, Inc., 1995.

2. Fairbanks L.D., Duley J.A., Shores A.J., Simmonds H.A. HPLC assay of uridine monophosphate synthetase (UMPS) in chorionic villus samples (CVS) and erythrocytes (RBC). Adv. Exp. Med. Biol. 309B:35 (1991).

3. Shostak K., Christopherson R.I., Jones M.E. Direct spectrophotometric assay for OPRT and ODC. Anal Biochem 191:365 (1990).

4. Micheli V., Rocchigiani M., Pompucci G. An HPLC-linked assay of phosphoribosylpyrophosphate synthetase activity in the erythrocytes of adults and children with neurological disorders. Clin Chim Acta 227:79 (1994).

5. Fox R.M., Wood M.H., O'Sullivan W.J. Studies on the coordinate activity and lability of OPRT and ODC in human erythrocytes, and the effect of allopurinol administration. J Clin Invest 50:1050 (1971).

32

NONRANDOM SELECTION OF HAIR FOLLICLES IN HPRT HETEROZYGOSITY

Theodore Page[1] and Robyn Broock[2]

[1]Department of Neurosciences
[2]Department of Pediatrics
University of California, San Diego
La Jolla, California 92093

INTRODUCTION

The gene for hypoxanthine phosphoribosyltransferase (HPRT) has been mapped to the X chromosome, and HPRT-deficiency syndromes are inherited in an X-linked recessive manner.[1] Two populations of cells can be demonstrated in the female heterozygous carriers of these syndromes. While great strides have been made in the detection of point mutations in cDNA, these methods are not yet sufficiently reliable for routine detection of heterozygotes.[2] At present, biochemical methods which detect HPRT$^+$ and HPRT$^-$ populations of cells are both technically simpler and more reliable. Cultured fibroblasts, lymphocytes, and hair roots have been used for this purpose.[1] Of these three, we have found the hair root method to be the least invasive, fastest, and most reliable.[3] In the past 15 years we have detected a total of 94 heterozygous carriers of HPRT-deficiency syndromes by this method.

Hair follicles are known to develop clonally from a single cell during embryogenesis, and thus the cells of each follicle contain almost exclusively maternal or paternal X chromosomes.[4] Theoretically, the distribution should be random, with each follicle having an equal probability of receiving the maternal or paternal chromosome. However, our data suggest that HPRT$^+$ follicles are slightly favored over HPRT$^-$ ones, and that distribution of the two types of follicles is far from perfectly random. This makes the detection of two populations of cells considerably less reliable, and indicates that larger sample is required than would be expected based on a perfectly random distribution.

MATERIALS AND METHODS

Detection of HPRT activity in individual hair follicles was done by thin-layer chromatography as previously described.[3] Both HPRT and adenine phosphoribosyltransferase

Purine and Pyrimidine Metabolism in Man IX,
edited by Griesmacher *et al.* Plenum Press, New York, 1998.

(APRT) activity were assayed in each hair follicle. Follicles which contained an amount of HPRT which was equal to or greater than the accompanying APRT activity were scored as HPRT⁺. Follicles which contained and amount of APRT activity equal to those follicles which were scored HPRT⁺ but no detectable HPRT activity were scored HPRT⁻. Follicles which contained none of either activity were deemed nonviable and were not included in the calculations. Rarely, follicles which contained a normal amount of APRT activity and small but detectable amount of HPRT were encountered; these follicles are thought to represent HPRT⁻ follicles which contain a small number of HPRT⁺ cells from neighboring follicles. These follicles also were not scored. In most cases, 30 follicles were collected, and varying number were subsequently found to be nonviable. If a given test scored less than 25 HPRT⁺ with 0 HPRT⁻, additional hair follicles were requested from the patient.

RESULTS AND DISCUSSION

The results of screening 94 different patients is shown in Figure 1. The upper panel shows the expected distribution based on the perfectly random "coin toss" model. In this model, the mean would be 50% with a standard deviation of 9.4%. The actual distribution of HPRT⁺/HPRT⁻ hairs from 94 heterozygotes is shown in the lower panel. The mean value was of this sample was calculated to be 60.80% positive with a standard deviation of 19.88%. Both of these numbers are distinctly different from those predicted by the perfectly random model.

The large actual standard deviation indicates that the probability of encountering a sample with high percentage of HPRT⁺ follicles is greater than expected, and thus the probability of finding no HPRT⁻ follicles in a given sample is also larger than expected. Since the finding of no HPRT⁻ follicles in a given sample would cause that patient to be incorrectly diagnosed as homozygous normal, the reliability of this assay is considerably less than that predicted by the perfectly random model. Taking into account the probility of encountering a sample with a high percentage of HPRT⁺ hair roots and the probability of finding no HPRT⁻ hairs in such a sample gives the actual reliability, which is showm in Table 1. For comparison, the reliability based on the perfectly random model is also shown. According to this calculation of the actual reliability, 95% certanty is obtained with 12 HPRT⁺ follicles; with 36 HPRT⁺ follicles the certainty increases to 99%.

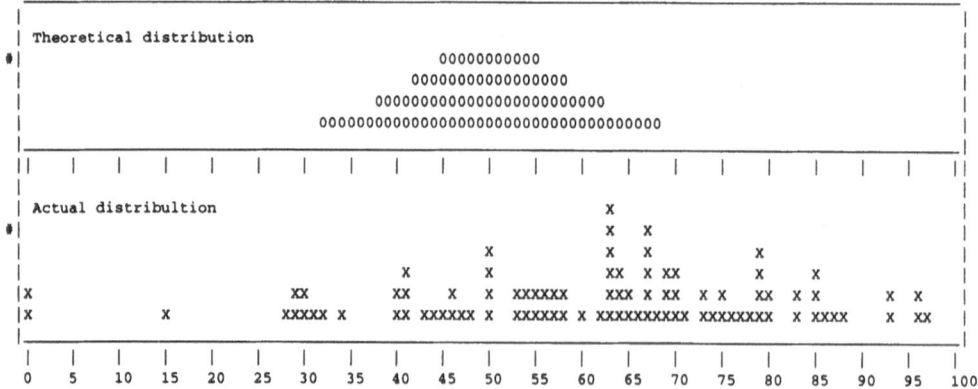

Figure 1. Distribution of HPRT⁺ and HPRT⁻ follicles.

Table 1. Actual and theoretical reliability of detection

P	X	Y	P	X	Y	P	X	Y
10	15	1024	27	62	1.3×10^8	44	137	1.8×10^{13}
11	17	2048	28	65	2.6×10^8	45	143	3.5×10^{13}
12	20	4096	29	69	5.4×10^8	46	149	7.0×10^{13}
13	22	8192	30	72	1.1×10^9	47	155	1.4×10^{14}
14	24	16384	31	76	2.1×10^9	48	162	2.8×10^{14}
15	27	32768	32	80	4.3×10^9	49	169	5.6×10^{14}
16	29	65536	33	84	8.6×10^9	50	176	1.1×10^{15}
17	32	131072	34	88	1.7×10^{10}	51	183	2.3×10^{15}
18	35	262144	35	92	3.4×10^{10}	52	190	4.5×10^{15}
19	37	534288	36	96	6.9×10^{10}	53	198	9.0×10^{15}
20	40	1.1×10^6	37	101	1.4×10^{11}	54	206	1.8×10^{16}
21	43	2.1×10^6	38	106	2.7×10^{11}	55	215	3.6×10^{16}
22	46	4.2×10^6	39	110	5.5×10^{11}	56	223	7.2×10^{16}
23	49	8.4×10^6	40	115	1.1×10^{15}	57	232	1.4×10^{17}
24	52	1.7×10^7	41	121	2.2×10^{12}	58	241	2.8×10^{17}
25	55	3.4×10^7	42	126	4.4×10^{12}	59	251	5.7×10^{17}
26	58	6.7×10^7	43	131	8.8×10^{12}	60	261	1.2×10^{18}

P = number of positive hair roots.
X = odds of heterozygosity with regard to nonrandomness.
Y = theoretical odds of heterozygosity with perfect randomness.

There are two separate issues with regard to the apparent lack of perfect randomness. One is an observed mean which is undeniably greater than 50%; the other is the larger-than-expected standard deviation. The mean value of 60.8% does not seriously comprimise the reliability of the assay, but suggests slight selection against HPRT⁻ hair follicles. Another possibility is selection against maternal X chromosomes, since HPRT⁻ chromosome is usually maternally inherited, but there is no evidence for this in humans.[5] The large standard deviation poses the more serious problem for assay reliability; it indicates that there is a higher-than-expected probability of encountering a subject with only a small percentage of HPRT⁻ follicles, failing to detect any of these HPRT⁻ follicles, and falsely assuming that the subject is homozygous normal. In the present sample, 5 of 94 samples (5%) had <10% HPRT⁻ follicles, and 13 of 94 (14%) had <20%. This "unfavorable Lyonization" may arise from nonrandom (i.e., a ratio much different from 1:1) inactivation of the X-chromosomes or a selective advantage or disadvantage conferred on cells bearing one of the X-chromosomes due to mutations in some other gene.

Thus, while the reliability of heterozygote detection by identification of the two X-chromosome-derived clones in hair follicles is considerably less reliable than predicted by a perfectly random distribution, it is still provides clinically useful information and will probably remain the mainstay of carrier detection until simple and reliable methods for the detection of point mutations are developed.

REFERENCES

1. WN Kelley, JB Wyngaarden: Clinical syndromes associated with hypoxanthine-guanine phosphoribosyl-transferase deficiency, in *The Metabolic Basis of Inherited Disease*, eds. JB Stansbury, JB Wyngaarden, DS Fredrickson, JL Goldstein, MS Brown, New York, McGraw-Hill, 1983, pp. 1115–1143.
2. M Ridanpaa, K Burvall, LH Zhang, K Husgafvel-Pursiainen, A Onfelt, Comparison of DGGE and CDGE in detection of single base changes in the hamster HPRT and human N-ras genes, *Mutation Res*, 334 (1995) 357–364.

3. T Page, B Bakay, WL Nyhan, An improved procedure for detection of hypoxanthine-guanine phosphoribo-
    syltransferase heterozygotes, *Clin Chem* 28 (1982) 1181–1184.
4. SM Gartler, RC Scott, JL Goldstein, B Campbell, Lesch-Nyhan syndrome: rapid detection of heterozygotes
    by use of hair follicles, *Science* 172 (1971) 572–574.
5. BR Migeon, X-chromosome inactivation: molecular mechanisms and genetic consequences, *Trends in
    Genetics*, 10 (1994) 230–235.

33

OPTIMISED MICELLAR ELECTROKINETIC CAPILLARY CHROMATOGRAPHY OF UV-ABSORBING COMPOUNDS IN URINE

Its Application to Studies on Purine Metabolism

Luwiza N. Alfazema,[2] Mark E. P. Hows,[1] Sian Howells,[2] and David Perrett[1]

[1]Department of Medicine
St Bartholomew's Hospital
London EC1A 7BE
[2]Department of Chemistry
University of Greenwich
London, SE18 6RZ

1. INTRODUCTION

Many UV components in urine have been separated and identified in normal human urine using chromatography. More than 150 UV absorbing components have been separated by column chromatographic methods including HPLC from human urine. However such methods require long analysis times up to a day, in the case of the highest resolution methods.[1–2] Capillary Electrophoresis (CE) is a simple and rapid separation technique offering both high resolution and sensitivity.[3–4]

The analysis of human urine by CE has been reported by several authors.[5–7] However most have concentrated on the separation of known drugs and their metabolites in urine whilst others have identified one or two UV-absorbing urinary endogenous components without generally optimising their separation. Some authors have studied pathological body fluids which have including urine.[8–11] Micellar electrokinetic capillary chromatography (MECC) is one of the derivatives of CE[12–13] which by incorporation of a surfactant, e.g. SDS, allows the separation of cationic, anionic and neutrals species in a single run.

Various instrumental and buffer parameters affect the separation·of compounds in CE especially in MECC. In order to separate the maximum number of peaks from any biological matrices and thereby gain the maximum amount of information, it is necessary to optimise these parameters. Trial and error optimisation can be tedious and time consuming. A chemometric approach,[14] like factorial design, is advisable as this minimises the

Purine and Pyrimidine Metabolism in Man IX,
edited by Griesmacher *et al.* Plenum Press, New York, 1998.

171

number of experimental analyses to be carried out. Factorial design determines important parameters and their interactions among the parameters that have a significant effect on the separation of the compounds.

The aim of this study, was to develop an optimised separation of UV urinary constituents by CE in human urine using a normal pool. The developed method was then applied for the analysis of pathological urine samples.

2. MATERIALS

Sodium dodecyl sulphate (SDS), β-cyclodextrin (β-CD), sodium tetraborate for the separation buffer and all standards were obtained from Sigma-Aldrich (Poole, UK) and were of analytical grade.

2.1. Instrumentation and Separation Conditions

All experiments were carried out on a CE 2000 Instrument (TSP, Stone, UK) using uncoated silica capillaries length 44 cm (effective length to the detector 37 cm) and I.D. 75 μm (Composite Metal Services, Hallow, UK).

On-column fast-scanning multi-wavelength UV-absorbance detection was employed. Spectral data of the migrating solutes was collected from 195 to 300 nm at 5 nm intervals (21 wavelengths) and analysed using SpectraSYSTEM Software. Hydrodynamic injection mode at 4 s was used. New capillary was conditioned for 30 min at 50°C with 0.5M NaOH. Initially prior to use the capillary was rinsed with 0.1M NaOH solution for 1.5 min and then 3.0 min with the running buffer. Before each run the capillary was filled with the run buffer for 2.0 min. At the end of each day the system was rinsed with water, the capillary washed and filled with distilled water over night.

Initially all electrophoresis was carried out at a temperature of 25°C, voltage was varied between 15–25 kV and separation time was limited to 15 minutes.

Factorial designs were generated using Stratgraphics (v.5).

2.2. Samples

Morning urine was collected from 6 healthy individuals chilled to +4°C and centrifuged to remove particulates. Equal volumes of the six samples were mixed thoroughly. The supernatant was divided into aliquots of 2.5 ml which were stored at −55°C until further analysis. Daily a new aliquot was thawed at room temperature and filtered through a 0.45 μm Whatman filter before injection.

Pathological samples were stored at −55°C and thawed at room temperature for an hour prior to use. They were analysed without any further treatment other than filtering through a Whatman 0.45 μm filter.

3. EXPERIMENTAL AND RESULTS

For maximum number of UV-absorbing urinary peaks the optimisation of the following parameters was studied, pH, voltage, β–cyclodextrin concentration in a 3^3 factorial design. Data was collected at all wavelenghts. Increasing pH gave a high number of peaks and sharper peaks and a better separation. Cyclodextrin had no significant effect but its inclusion improved overall separation of some peaks. It was found necessary to carefully optimise the separation of creatinine from urea at the capillary temperature (from 15–35°C) and injection time (1–5 sec) were optimised separately. This work has been described in detail elsewhwere.[17]

The final optimised conditions are given in the caption to Fig. 1a. They gave a separation of the normal pooled urine components with some 70 peaks using detection at 195 nm. The peaks were resolved in under twelve minutes. Some of the peaks have been tentatively identified using spectral analysis and spiking and are as labelled in the figure. Urea and creatinine were only resolved with MECC and not simple CZE.

3.1. Human Pathological Samples

The aim of the optimisation study was to produce a rapid screening system for purine and pyrimidine disorders. We have applied it already to a number of pathological samples. Purine Nucleoside Phosphorylase deficiency (PNP) is a purine disorder associated with neurological disorders and gross overproduction and excretion of inosine, guanosine, 2-deoxyinosine and 2′-deoxyguanosine in urine when detected at 254 nm.

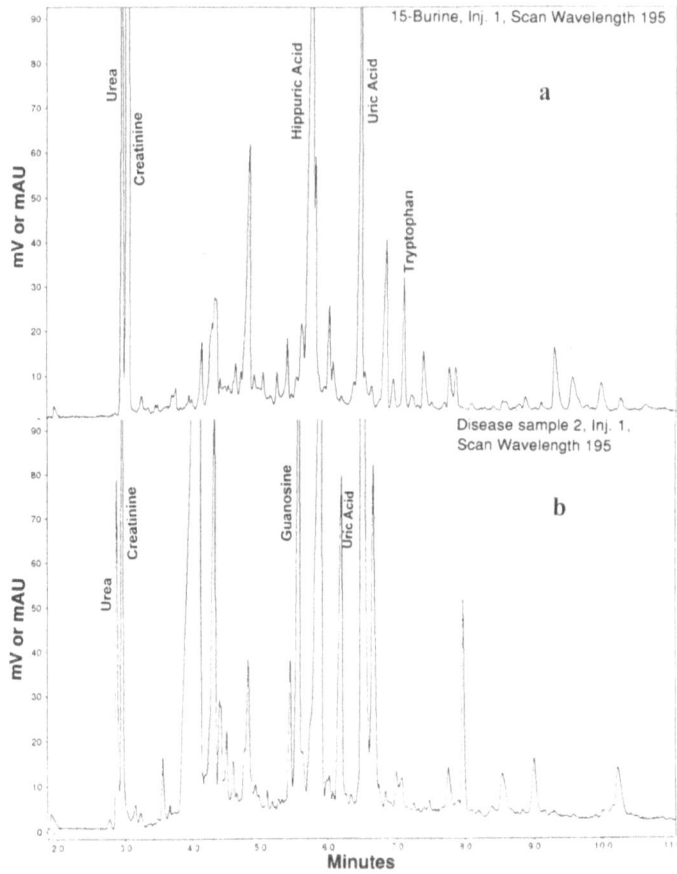

Figure 1. The optimised MECC separation of a normal pooled human urine (a) compared to a separation of pathological sample (b) from a patient with PNP deficiency. Conditions: 30 mM sodium tetraborate, pH 10, 75 mM SDS, 10 mM β-CD 20kV, 15°C, 4 s injection HD, detection at 195 nm. Peaks were identified using spectral analysis and spiking.

Fig. 1b shows the separation of a urine from a patient with PNP deficiency using the optimised MECC separation. Differences in the peak profiles were observed and some peaks were positively identified using spectral analysis and spiking as shown in the figure. Excessive peaks were seen in the pathological sample which were not observed in the normal pooled urine. These peaks could be some of the expected metabolites in this disorder. Studies are still going to identify all the major peaks some of which do not absorb at the standard detection wavelength (254 nm). One of the metabolites, guanosine was identified in Fig. 1b using spectral analysis.

These results demonstrated that differences in the urinary metabolites content due to certain disorders can be readily detected using MECC. This method was simpler, rapid and required minimal sample preparation.

3.2. Allantoin Determination in Urine from Species Other Than Man

Allantoin is a natural end-product of purine degradation in most mammals. Its formation involves a series of reaction involving hypoxanthine and xanthine. Another important end product of this reactions is uric acid, which is commonly found in man.

It is difficult to measure allantoin due to lack of the characteristic UV spectra of purines. Methods that have been used in the determination of allantoin include spectrophotometry alkalimetric titration and chromatography and HPLC with UV or electrochemical detection. Analysis of allantoin has been reviewed recently.[15] However, no work on determination of allantoin in biological fluids such as urine has been reported using CE. The ability of CE to operate at low UV wavelengths so should be ideal for the determination of allantoin and its precursors.

Fig. 2a shows separation of allantoin and other UV-endogenous compounds in dog urine compared to the same urine spiked with standard allantoin in Fig. 2b. Strong absorbance of allantoin was observed at 222 nm, which is in agreement with HPLC[16] whereas allantoic acid only absorbs at < 210 nm and therefore is not measurable by HPLC. Spiking and spectral analysis confirmed that the peak in Fig. 2a is allantoin as labelled. Allantoin was one of the major urinary components separated in dog urine along with urea and creatinine although there was still significant excretion of uric acid.

Allantoin in the dog urine was successfully separated using MECC conditions which were developed for the separation of normal human urinary components for maximum number of peaks.[17] The limit detection was 5 µg/ml. The MECC method described here has allowed rapid analysis of UV-urinary constituents in the dog urine with minimal sample treatment other than thawing, dilution and filtering. It is reliable and sensitive assay for understanding purine metabolism in other species other than man. This method has been applied to the determination of allantoin in urine from other mammal species, e.g. mice and horse.

4. CONCLUSION

The method described in this paper allowed the separation of a large number of UV-absorbing urinary components in under 12 minutes using MECC. This was much simpler and faster than gradient reversed phase HPLC and resolved some 20% more compounds under similar detection conditions. It was possible to detect cationic, anionic and neutral species in one single run.

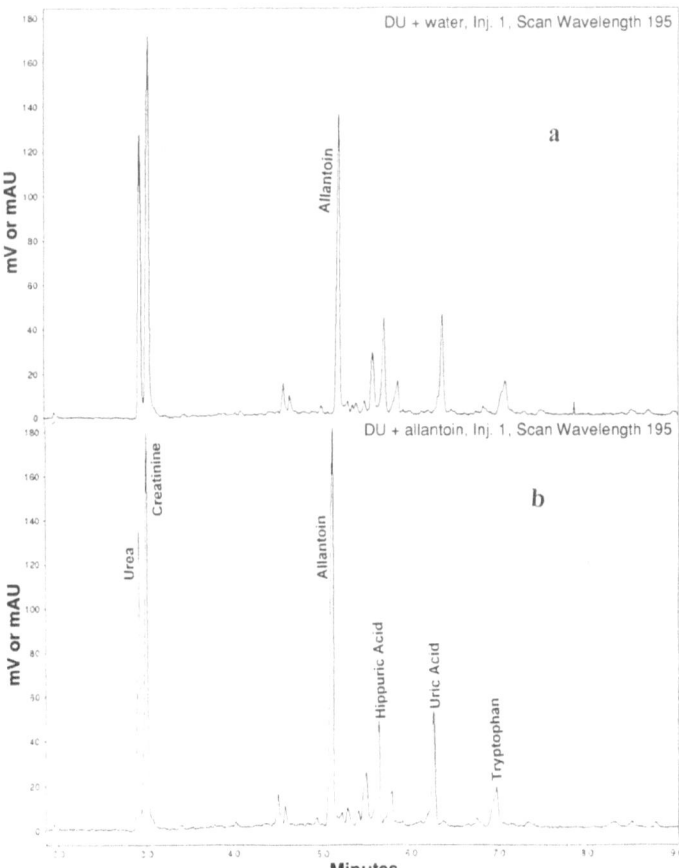

Figure 2. UV-urinary components separation from a dog urine (a) separated using conditions in Fig. 1 and compared to the same urine spiked with allantoin (b) Urea, creatinine and allantoin in Fig. 2a were positively identified using spectral analysis and spiking.

The present optimised conditions permits separation of some of major urinary metabolites and can be used as an analytical tool for determination of excess or new metabolites in pathological samples especially those resulting from defects in purine and pyrimidine metabolism.

In addition the method can be readily adapted to urine from other species. We have presented work here showing its use for the determination of allantoin in animal urine a compound normally difficult to measure by HPLC.

REFERENCES

1. Scott C.D., Atrill J.F and Anderson N.G. Automatic, high resolution analysis of urine for its UV-Absorbing constituents. *Proc. Soc. Exp. Biol.,* **1967**, 125, 181–184.
2. Scott C.D. Analysis of urine for its UV-absorbing constituents by High- Pressure Anion Exchange Chromatography. *Clin Chem* **1968**, 14(6), 521–528.
3. Guzman, N.A.; Hernandez, L. & Hoebel, B.G. Capillary electrophoresis: A new era in microseparations . *BioPharm.,* **1989**, 2, 22–37.

4. Xu Yan. Capillary Electrophoresis. *Anal. Chem.* **1995**, 67, 563R - 473R.
5. Miyake M, Shibukawa A and Nakagawa T. Simultaneous determination of Creatinine and Uric acid in Human Plasma and Urine by MEKC. *J. High Res. Chromatogr.* **1991**, 14, 181–185.
6. Lee.K-J, Lee.J.J & Moon D.C. Application of MEKC for monitoring of Hippuric and Methylhippuric acid in human urine. *Electrophoresis* **1994**, 15, 98–102.
7. Schafroth M, Thormann W & Allemann D. MECC of Benzodiazepines in human urine. *Electrophoresis* **1994**, 15, 72–78.
8. Jellum E, Thorsrud A K & Time E. Capillary electrophoresis for diagnosis and studies of human disease, particularly metabolic disorders. *J Chromatogr.,* **1991**, 559, 455–465.
9. Jellum E. Capillary electrophoresis for medical diagnosis. *J Cap Elec.,* **1994**, 1, 97–105
10. Jellum E, Dollekamp H & Blessum C. Capillary electrophoresis for clinical problem solving: Analysis of urinary diagnostic metabolites and serum proteins. *J Chromatogr B Biomed Appl.,* **1996**, 683, 55–65.
11. Guzman N.A, Berck C.M, Hernandez L & Advis J.P. CE as a diagnostic tool: determination of biological constituents present in urine of normal and pathological individuals. *J. Liq. Chromatogr.,* **1990**, 13(19), 3833–3848.
12. Vindevogel, J. & Sandra, P. Introduction to Micellar Electrokinetic Chromatography Huthig **1992**, pp 231.
13. Terabe, S. Electrokinetic chromatography: An interface between electrophoresis and chromatography. *Trends in Anal Chem.,* **1989**, 8, 129–134.
14. D.L. Massart, B.M.G. Vandeginste, S.N. Deming, Y.Michotte, L. Kaufman. Chemometrics: A textbook. Elservier **(1988)** 1st Edition. Chapters; 17 - 19 & 21.
15. Chen XB. Matuszewski W. Kowalczyk J. Determination of allantoin in biological, cosmetic and pharmaceutical samples. J Aoac Int. 1996, 79, 628–35
16. Balcells J, Guada JA, Peiro JM and Parker DS. Silmutaneous determination of allantoin and oxypurines in biological fluids by HPLC. J Chromatogr B Biomed Appl. 1992, 575, 153–157.
17. Alfazema L, Hows MEP, Sian Howells S & Perrett D. Micellar Electrokinetic Capillary Chromatography (MECC) of UV-Absorbing Constituents in Normal Urine: A Chemometric Optimisation of the Separation. (In press Electrophoresis submitted July 1997).

34

EFFECT OF D-RIBOSE ADMINISTRATION TO A PATIENT WITH INHERITED DEFICIT OF ADENYLOSUCCINASE

C. Salerno, M. Celli, R. Finocchiaro, P. D'Eufemia, P. Iannetti, C. Crifò, and O. Giardini

Institute of Paediatrics and
Department of Biochemical Sciences
University of Roma La Sapienza

1. INTRODUCTION

Adenylosuccinase (EC 4.3.2.2; ASase) deficiency is a newly discovered inborn error of metabolism that involves the purine de novo pathway and results in the accumulation of dephosphorylated substrate derivatives of the defective enzyme, namely succinylamino-imidazole carboxamide (SAICA) riboside and succinyladenosine (S-Ado).[1–3] Although substantial progress has been made regarding our knowledge of the inherited disease, including its characterisation at gene level,[4–5] much work remains to be done, particularly with respect to the mechanisms whereby the defect exerts its deleterious effects on brain function.

Accumulation of metabolites proximally from the enzyme defect, as well as deficiency of reaction products, has been considered in the pathogenesis of the symptoms. It has been suggested that, owing to their structural resemblance with adenosine, SAICA riboside and/or S-Ado could interfere with the cerebral adenosine receptors and thereby with the numerous physiological functions of adenosine.[6] Another hypothesis is that ASase deficiency could impair the energy metabolism because of the defective supply of aminoimidazole carboxamide (AICA) ribotide, AMP, and fumarate, which are necessary for activating the glycolytic pathway and the citric acid cycle.[7] Disruption of the purine nucleotide cycle could also account for the deterioration of cell bioenergetics.[8]

No therapies are unequivocally documented to be effective in treating ASase deficiency. Administration of high doses of allopurinol or sodium benzoate, which are known to inhibit purine synthesis,[9–10] did not decrease succinylnucleoside levels.[11] Oral supplements of adenine combined with allopurinol were ineffective, with the exception of promoting weight and height gain in two patients.[2] A slight improvement of the behaviour of just a patient was observed upon administration of AICA.[11]

Purine and Pyrimidine Metabolism in Man IX,
edited by Griesmacher *et al.* Plenum Press, New York, 1998.

2. EXPERIMENTAL PROCEDURES

ASase-deficient patient was described previously elsewhere.[3] At the time of study, she was 13 years old. The cardinal symptoms were a severe psychomotor retardation accompained by epilepsy. Stereotypies, unexplained screaming attacks and bouts of extreme physical agitation, especially involving arms and legs, were frequently observed. The gait was awkward. Episodes of enuresis and encopresis were rare. The speech was scantily intelligible; nevertheless, the patient was apparently able to write short sentences with sufficiently correct structure by using a computer-assisted program, but she required to be sustained and urged by the parents.

SAICA riboside and S-Ado were measured in plasma and urine by high-performance liquid chromatography.[1] Glucose, uric acid, and creatinine were measured by standard laboratory methods. Patient's studies were done after informed parental consent.

3. RESULTS

D-ribose (ICN Biomedicals, Costa Mesa, CA) was given orally in doses of 40 g/day (10 g four times daily) to the patient, who had been restricted to isocaloric purine-free diet. No appreciable toxic effects were observed over three months. Higher doses (up to 100 g/day) caused mild diarrhoea promptly reversed by reducing drug administration.

Oral load with the sugar led to a marked increase of urate excretion ([urate]/[creatinine] rose from 0.26 to 0.93), while urinary succinylnucleosides changed scarcely ([S-Ado]/[creatinine] increased from 0.088 to 0.094; by contrast, [SAICA riboside]/[creatinine] decreased from 0.062 to 0.052). As far as the concentration of serum metabolites is concerned, it was noteworthy the increase of urate (from 0.12 to 0.29 mM) and the decrease of glucose (from 5.4 to 4.8 mM), while no appreciable variations in the serum level of succinylpurines (2.5 µM SAICA riboside; 3.0 µM S-Ado) were observed. These changes in the biochemical parameters remained constant as long as D-ribose was being given.

D-Ribose administration improved appreciably the behaviour of the patient, who showed greater motor nimbleness and lower frequency of epileptic crisis, stereotypic body movements, and unexplained screaming attacks. According to her parents, she had greater wish of playing, expressing herself, and being independent.

4. DISCUSSION

The metabolic fate of D-ribose in man has been described in depth elsewhere.[12-13] It has been shown that D-ribose is largely phosphorylated to ribose 5-phosphate and then extensively metabolised, mainly to glucose by the sequence of reactions of the pentose phosphate shunt. The major pathway for stimulating purine de novo synthesis and salvage involves pyrophosphorylation of ribose 5-phosphate to form 5-phosphoribosyl-1-pyrophosphate. Another possibility is that ribose 5-phosphate is changed to ribose 1-phosphate and coupled with purine bases, to form nucleosides that may be phosphorylated to nucleotides.

Though converted to glucose, the supplemented sugar causes a lowering of glycaemia, presumably by inhibiting the enzyme phosphoglucomutase, thus preventing glycogen breakdown in liver.[14] D-Ribose was found to be insulin responsive, although its response is smaller than that observed with other pentoses.[12] On the other hand, if given orally, D-ribose leads to an increase in serum insulin concentration that strangely cannot be observed by administering D-ribose intravenously.[13]

The lack of symptomatic response to D-ribose-induced hypoglycaemia suggests that neural tissue could be able to utilise this sugar.[12] This hypothesis is in line with the observation that D-ribose can be metabolised to hexose phosphate by brain homogenates.[15] Nevertheless, it is doubtful that D-ribose itself obviates a hypoglycaemic response, since D-ribose is unable to cross the blood–brain barrier at least in the rat.[16]

Sustained urate overproduction upon administration of D-ribose to the patient with deficit of ASase indicates that this compound can promote purine de novo synthesis, very likely by activating the pathway in tissues where the structurally unstable enzyme is still present.[3] The slightly decreased excretion of SAICA riboside is of interest, because it does not parallel the changes in the metabolic flux, as expected.[17] Since cytoplasmic 5'-nucleotidase seems to be largely enough to dephosphorylate SAICA ribotide,[11] we can imagine that purine de novo synthesis is scantily operative in cells lacking in ASase activity, perhaps owing to the activation of the salvage pathway. The small increase in S-Ado excretion could be taken as an indirect evidence of the replenishment of the purine nucleotide pool in tissues suffering from the enzymatic defect. Another hypothesis involves a new synthesis of the enzymatic protein induced by supplementing the diet with sugars. Indeed, recovery of ASase activity with only a slight overshoot has been observed, at least in rat liver, after refeeding corn oil or glucose, while high protein diets depress the enzyme.[18] The recruitment of the defective enzyme could result also from the deinhibition of ASase activity due to the fall of intracellular phosphate, upon D-ribose loading, that reflects the phosphorylation of the pentose by the D-ribokinase.[12] A slow decrease in the plasma concentrations of succinylnucleosides has been already observed in response to intravenous fructose injection that is known to markedly reduce intracellular phosphate by means of similar mechanisms.[3]

ACKNOWLEDGMENTS

This research was done in the framework of the European Community Concerted Action contract no. BMH1-CT-1384.

REFERENCES

1. J. Jaeken and G. Van den Berghe. An infantile autistic syndrome characterized by the presence of succinylpurines in body fluids, Lancet 2:1058–1061 (1984)
2. J. Jaeken, S. K. Wadman, M. Duran, F. J. Van Sprang, F. A. Beemer, R. A. Holl, P. M. Theunissen, P. De Cock, F. Van den Bergh, M. F. Vincent, and G. Van den Berghe. Adenylosuccinase deficiency: an inborn error of purine nucleotide synthesis. Eur. J. Pediatr. 148:126–131 (1988)
3. C. Salerno, C. Crifò, and O. Giardini. Adenylosuccinase deficiency: a patient with impaired erythrocyte activity and anomalous response to intravenous fructose. J. Inher. Metab. Dis. 18:602–608 (1995)
4. R. J. Stone, J. Aimi, B. A. Barshop, J. Jaeken, G. Van den Berghe, H. Zalkin, and J. E. Dixon. A mutation in adenylosuccinate lyase associated with mental retardation and autistic features. Nature Genetics 1:59–63 (1992)
5. D. Verginelli, B. Luckow, C. Salerno, and M. Gross. Genetic basis of adenylosuccinase deficiency in an Italian patient. Adv. Exp. Med. Biol. (1997) this volume
6. M. F. Vincent and G. Van den Berghe. Influence of succinylpurines on the binding of adenosine to a particulate fraction of rat cerebral cortex. Adv. Exp. Med. Biol. 253B:441–445 (1989)
7. C. Salerno, S. Iotti, R. Lodi, C. Crifò, and B. Barbiroli. Muscle energy metabolism in human adenylosuccinase deficiency: an in vivo ^{31}P-NMR spectroscopy study. Adv. Exp. Med. Biol. (1997) this volume
8. C. Salerno, S. Iotti, R. Lodi, C. Crifò, and B. Barbiroli. Failure of muscle energy metabolism in a patient with adenylosuccinate lyase deficiency: an in vivo study by phosphorus NMR spectroscopy. Biochim. Biophys. Acta (1997) in press

9. G. H. Hitchings. Effects of allopurinol in relation to purine biosynthesis. Ann. Rheum. Dis. 25:601–610 (1966)
10. S. W. Brusilow, J. Tinker, and M. L. Batshaw. Amino acid acylation: a mechanism of nitrogen excretion in inborn errors of urea synthesis. Science 207:659–661 (1980)
11. G. Van den Berghe and J. Jaeken. Adenylosuccinase deficiency. Adv. Exp. Med. Biol. 195A:27–33 (1986)
12. S. Segal and J. Foley. The metabolism of D-ribose in man. J. Clin. Invest. 37:719–735 (1958)
13. M. Gross and N. Zöllner. Serum levels of glucose, insulin, and C-peptide during long-term D-ribose administration in man. Klin. Wochenschr. 69:31–36 (1991)
14. S. Segal, J. Foley, and J. B. Wyngaarden. Hypoglycemic effect of D-ribose in man. Proc. Soc. Exp. Biol. Med. 95:551–555 (1957)
15. H. Z. Sable. Pentose metabolism in extracts of yeast and mammalian tissues. Biochim. Biophys. Acta 8:687–691 (1952)
16. C. R. Park, R. L. Post, R. L. Kalman, J. H. Wright, L. H. Johnson, and H. E. Morgan. The transport of glucose and other sugars across cell membranes and the effect of insulin. Ciba Foundation Colloquia on Endocrinology 9:240–249 (1956)
17. J. S. Easterby. A generalized theory of the transition time for sequential enzyme reactions. Biochem. J. 199:155–161 (1981)
18. L. M. Brandt and J. M. Lowenstein. Effect of diet on adenylosuccinase activity in various organs of rat and chicken. J. Biol. Chem. 253:6872–6878 (1978)

35

ENZYME ACTIVITIES LEADING TO NAD SYNTHESIS IN THE ERYTHROCYTES OF HPRT DEFICIENT SUBJECTS

S. Sestini,[1] V. Micheli,[1] M. Rocchigiani,[1] G. Jacomelli,[1] F. Manzoni,[1] B. Gathof,[2] G. Hayek,[3] F. Cardona,[4] E. Zammarchi,[5] and G. Pompucci[1]

[1]Dipt. Biologia Molecolare
Università di Siena, Italia
[2]Ludwig Maximilians Universität
Munich, Deutschland
[3]Rip. Neuropsichiatria Infantile
USL 7, Siena, Italia
[4]Dip. Sci. Neurol. e Psic.
Università La Sapienza
Roma, Italia
[5]Osp. Pediatrico Mayer
Firenze, Italia

INTRODUCTION

The relationship between HPRT deficiency and the devastating neurologic dysfunction in Lesch-Nyhan syndrome is still unknown. Different metabolic alterations have been found, mainly in the erythrocytes, which could suggest involvement of different metabolic pathways; one of the described alterations is the high NAD level reported in patients with partial or total HPRT deficiency.[1] Whether NAD increase can contribute to the clinical features is not known, but due to the peculiar role played by this coenzyme, it is reasonable that its alteration may be reflected on various metabolic aspects. In order to understand the mechanisms causing NAD raised concentrations, we have looked for possible increase in the stability, which was demonstrated not to be altered in patient erythrocytes.[2] Possible alterations of NAD synthesis were thus hypothesized and "in vitro" incorporation of NAD precursors by intact erythrocytes was studied, demonstrating altered regulation in patients with partial or total deficiency.[3] In the present study we have assayed the enzyme activities of the so-called deamidated pathway, starting from nicotinic acid, i.e. nicotinate phosphoribosyl transferase (NAPRT), nicotinate mononucleotide adenylyltransferase (NAMN-AT) and NAD synthetase (NADs), in order to ascertain the involvement of the committed

Purine and Pyrimidine Metabolism in Man IX,
edited by Griesmacher *et al.* Plenum Press, New York, 1998.

181

enzymes leading to NAD synthesis. The possible effect of endogenous compounds and the heat resistance of the enzyme activities in crude lysates has also been examined in controls and patients.

PATIENTS AND METHODS

Patients were: Sa. aged 15, Si. aged 3, R. aged 9, M. aged 8, P. aged 36, all with L.N. syndrome; L. aged 3 and G. aged 30, both with partial HPRT deficiency. Age-matched controls were also examined.

Blood was drawn by venipuncture using heparine as anticoagulant; erythrocytes were separated from the plasma and the buffy coat, washed twice with PBS, then immediately used. For the enzyme assays, −80°C stored erythrocytes were sometimes used. Cell extracts were prepared on fresh erythrocytes by perchloric acid extraction and subsequent neutralization, then the clear supernatants were processed by HPLC. Lysates were prepared following different procedures for each enzyme assay, and the incubation mixtures performed according to the specific methods followed by HPLC quantitation. All these procedures have been described elsewhere.[4]

RESULTS

HPRT was virtually absent in all the L.N. patients, and was 17% and 9% of controls in the two partial HPRT deficient patients. The presence of high levels of NAD in all the HPRT⁻ patients (except G.) has been confirmed, its concentrations ranging 125–221 nmol/ml erythrocytes (controls 35–68). Increased activity of NAPRT and NADs, but no variations in NAMN-AT have been observed in all the patients tested. In the erythrocytes of the adult with partial HPRT deficiency (G.) no alteration of these enzyme activities has been highlighted (Table 1).

Possible activation of NAPRT and NADs by unknown effectors present in patient lysates has been investigated by incubating a mixture of equal amounts of lysates from one patient and one control. The activity measured on the mixture equalled the mean of the activities measured in the two lysates assayed separately (data not shown), thus demonstrating the absence of such activating compounds.

To check possible activating effect of NAD and PPribP (both presenting higher levels in L.N. patients in comparison with controls) each compound has been added to the

Table 1. Enzyme activity values in patients and in controls (nmol mg^{-1} Hb h^{-1}; *mean ± sd*)

Patient	NAPRT	NAMN-AT	NADs
Salvatore	3.00	0.41	0.64
Simone	2.14	0.29	0.49
Roberto	2.00	0.21	−
Michele	1.70	0.19	0.74
Paolo	2.28	−	0.50
Luca	2.50	0.29	0.73
Gino	1.19	0.22	0.11
Adults	*1.01 ± 0.33*	*0.27 ± 0.05*	*0.32 ± 0.06*
Children	*1.02 ± 0.26*	*0.37 ± 0.10*	*0.38 ± 0.11*

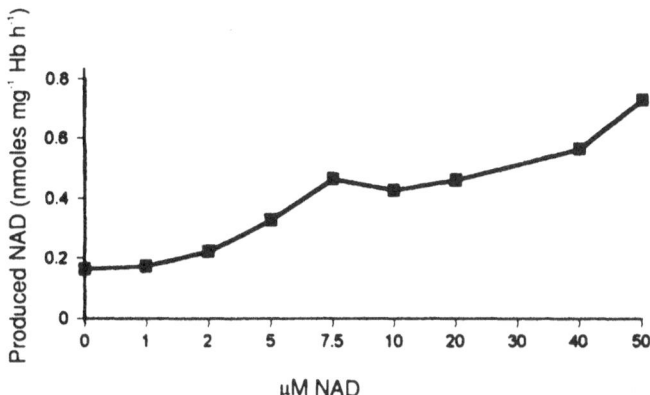

Figure 1. NADs activity in control lysates in the presence of NAD.

assay mixture at concentrations close to erythrocyte patient levels. NAPRT activity is not affected by NAD up to 0.3 mM. Surprisingly NAD seems to stimulate NADs activity in a control lysate, since NAD production (calculated as the difference between the final and the initial amount of NAD detected) increased with added NAD in a concentration-dependent fashion (up to 50 µM) (Fig. 1).

PPribP up to 88 µM has no influence on NADs activity. The effect of ATP, reported to inhibit NAPRT at physiological concentrations,[5] has been evaluated up to 2 mM; no difference has been found between patients and controls (data not shown).

The stability of NAPRT and NADs activity in lysates heated at 60°C up to 8 minutes has also been tested: no difference has been observed between patients (Si., Sa., L.) and controls for NAPRT, while NADs activity in one patient (L.) showed lower stability to heat than the control (Fig. 2).

PPribP (51 µM) slightly enhanced NAPRT stability to heating in control lysates (Fig. 3), while 0.3 mM NAD showed no effect.

DISCUSSION

Present study demonstrates that the activities of two enzymes of the deamidated pathway leading to NAD synthesis, i.e. NAPRT and NADs, are increased in the erythro-

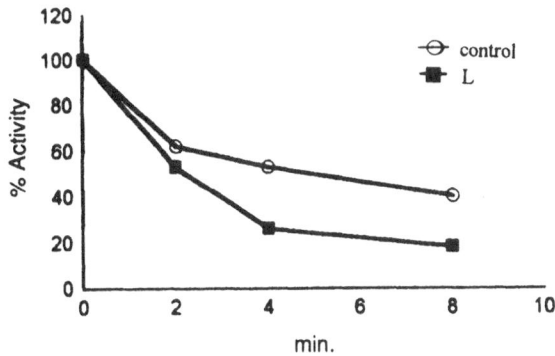

Figure 2. Heat (60°C) stability of NADs in a patient and a control.

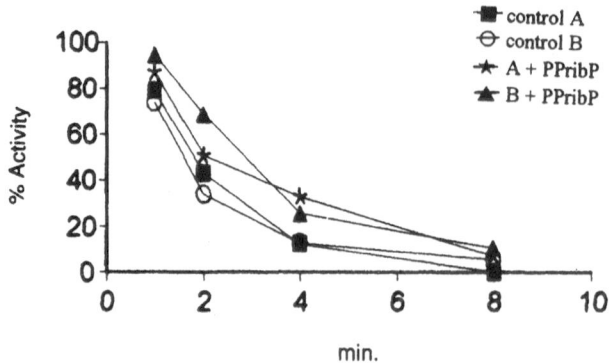

Figure 3. NAPRT heat (60°C) stability in the presence of 51 μM PPribP.

cytes of HPRT deficient patients; such increase may contribute to the higher NAD levels detected in these patients. Possible effects of endogenous compounds raising enzyme activities have been investigated, leading to rule out such hypothesis. Neither increased stability to heat, nor activation by high levels of PPribP seem to be responsible for the measured enzyme activity increase, though PPribP exerts a mild protection on NAPRT during heating treatment. The observed increase in NAPRT and NADs activities suggests that the raised NAD levels in the erythrocytes of HPRT deficient patients may be due to enhanced synthesis through the deamidated pathway. The final product NAD seems not to exert inhibition on this pathway, but rather to stimulate the last step of its own synthesis.

ACKNOWLEDGMENTS

This study was supported by the Italian M.U.R.S.T. (60% and 40% funds).

REFERENCES

1. H.A.Simmonds, L.D. Fairbanks, G.S. Morris, D.R. Webster, E.H. Harley. Altered erythrocyte nucleotide patterns are characteristic of inherited disorders of purine and pyrimidine metabolism. Clin. Chim. Acta, 171:197 (1988)
2. H.A. Simmonds, V.Micheli,P.M. Davies, M.B. Mc Bride. Erythrocyte nucleotide stability and plasma hypoxanthine levels: is 4°C really the best short-term storage temperature? Clin. Chim. Acta., 192:121 (1990).
3. V. Micheli, H.A. Simmonds and C. Ricci. Regulation of NAD synthesis in erythrocytes of patients with HGPRT deficiency and a patient with PRPP synthetase superactivity. Clin. Sci., 70: 239 (1990).
4. V. Micheli, S.Sestini. Methods for determining NAD synthesis in erythrocytes. Meth. Enzymol., 280: 211 (1997), in press.
5. V.Micheli, S.Sestini, M. Rocchigiani, M. Pescaglini, C. Ricci. Nucleotide synthesis in the human erythrocytes: correlation between purines and pyridines. Biomed. Biochim. Acta, 46:S268 (1987).

36

SUCCINYLPURINES INDUCE NEURONAL DAMAGE IN THE RAT BRAIN

T. W. Stone,[1] L. A. Roberts,[1] B. J. Morris,[1] P. A. Jones,[1] H. A. Ogilvy,[1]
W. M. H. Behan,[2] J. A. Duley,[3] H. A. Simmonds,[3] M. F. Vincent,[1] and
G. van den Berghe[4]

[1]West Medical Building
Division of Neuroscience and Biomedical Systems
University of Glasgow
Glasgow, Scotland
[2]Department of Pathology
University of Glasgow
[3]Clinical Science Laboratories
Guys Hospital, London SE1
[4]Laboratory of Physiological Chemistry
UCL-ICP, Brussels, Belgium

1. INTRODUCTION

Adenylosuccinase (adenylosuccinate lyase, EC 4.3.2.2: ASase) catalyses two steps in the purine metabolic pathways: a) the eighth step in the de novo pathway of purine synthesis: conversion of succinylaminoimidazole carboxamide ribotide (SAICAR) into aminoimidazole carboxamide ribotide (AICAR) and b) the second step in the formation of adenyl nucleotides from IMP: conversion of adenylosuccinate (S-AMP) into AMP. As a result of the enzyme deficiency, there is a recognised accumulation of the two dephosphorylated substrates of the enzyme: SAICA riboside and succinyladenosine (S-Ado). Dephosphorylation of SAICAR and S-AMP is performed by cytosolic 5'nucleotidases. Neither SAICA riboside nor S-Ado are normally detectable in the blood or tissue samples of humans, but in patients with ASase deficiency, both compounds are present in the urine, blood and CSF. In the latter, concentrations are usually around 100–200 μM (Jaeken et al., 1993).

Adenylosuccinase deficiency in humans was discovered by Jaeken & van den Berghe, (1984). Affected children are normal at birth, but profound psychomotor retardation, the principal symptom of the defect, becomes evident in the first two years. The defect is autosomal recessive, and its clinical picture is markedly heterogeneous, often manifesting epilepsy and autistic features, sometimes muscle wasting and growth failure (Jaeken et al., 1988). The enzyme deficiency is found in liver and kidney, in the muscle of the patients

Purine and Pyrimidine Metabolism in Man IX,
edited by Griesmacher *et al.* Plenum Press, New York, 1998.

with wasting and growth delay, in fibroblasts and lymphocytes, but not in granulocytes (Jaeken & van den Berghe, 1984, Jaeken et al., 1988; van den Berghe & Jaeken, 1986). This indicates genetic heterogeneity of the defect, and the existence of isozymes of ASase. Recently, the gene defect has been characterised in a first family (Stone et al., 1992). ASase deficiency may be well-suited in the future for gene therapy or enzyme replacement, similar to adenosine deaminase deficiency.

The underlying pathophysiological changes resulting from ASASe deficiency and leading to the psychomotor retardation are unknown. Several hypotheses have been put forward, including reduction of the nucleotides normally formed distally from the enzyme defect, and toxic effects of nucleotides or nucleosides accumulating proximally (Jaeken et al., 1988). However, assessments of nucleotides in ASase-deficient tissues (van den Berghe & Jaeken, 1986), and studies in intact patients' fibroblasts have failed to demonstrate nucleotide reduction (van den Berghe et al., 1993). Cellular concentrations of the major metabolic nucleotides ATP and GTP appear to be normal in affected patients. A study of brain glucose utilisation using fluorodeoxyglucose and Positron Emission Tomography (PET) in three ASase deficient patients revealed a consistent decrease of uptake in all areas of grey matter except the cerebellum, with the greatest abnormality occurring in anterior cerebral cortex (de Volder et al., 1988). While this supports the concept of a deterioration of neuronal function it does not, of course, reveal much about mechanism: these uptake changes are likely to be secondary to some other cause of neuronal dysfunction.

We have now tried to determine whether SAICA riboside is able to produce neurotoxic effects in the brain, by administering it directly into the brains of anaesthetised rats.

2. METHODS

Rats were anaesthetised with urethane (1.25 g/kg) if recovery was not required, or with ketamine/xylazine 2:1 for recovery experiments. Each animal was positioned in a stereotaxic frame, the scalp incised and reflected and a burr hole drilled over the hippocampus so as to allow insertion of the injection needle or probe to the coordinates AP –3.6 mm, L 3.0 mm and V 2.8 mm, measurements being taken with respect to the bregma suture. Rectal temperature was maintained at 36–37°C by a probe-controlled heating blanket.

Each of the succinylpurines or adenosine itself was applied initially by microinjection in a volume of 1 μl over 10 minutes. In later experiments the compounds were applied by reverse microdialysis over a period of 5 to 10 hours. The solution injected or dialysed contained 20 mmoles/l of the abnormal purines. The animals were either killed immediately after the experiment, or were allowed to recover for 7 days, when the brains were removed for histological analysis. The brains were stored in 10% w/v formalin until sectioned.

The brains were sectioned at 7 μm and alternate sections were stained with haematoxylin and eosin, and with cresyl violet. The amount of damage was determined in the CA1 and CA3 regions by estimating the number of damaged cells relative to the total number in a single field of view at 400× magnification.

In some cases the amount of SAICA riboside remaining in the brain at various times was measured by HPLC (Simmonds et al., 1990). Succinylpurines were prepared as given in Van den Bergh et al. (1991).

3. RESULTS

The initial, single injections of the purines did not induce any signs of neuronal damage or degeneration, perhaps because the compounds were not present for a suffi-

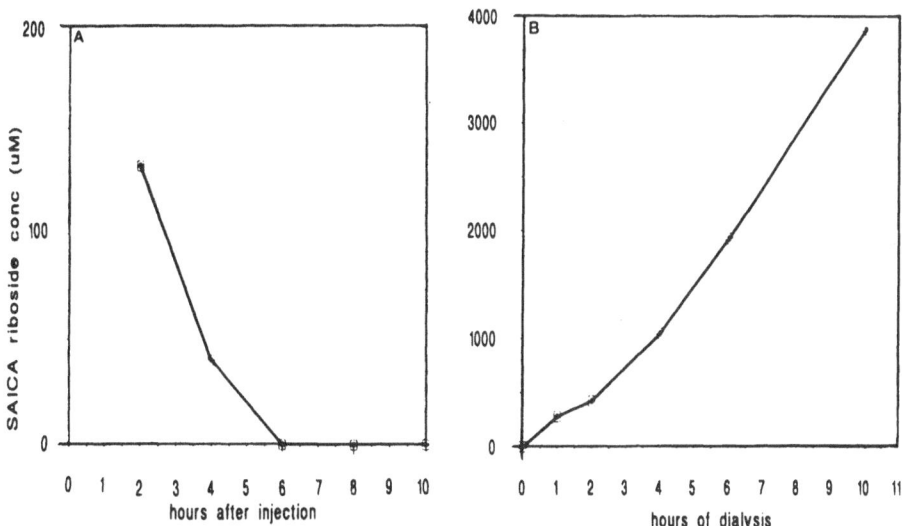

Figure 1. Graphs showing the amounts of SAICA riboside present in the rat hippocampus at various times after A: a single injection; B: reverse microdialysis. The SAICA riboside concentrations were assessed by HPLC.

ciently long period. This possibility was borne out by HPLC analysis of the hippocampus at various time points after the injection. As seen in Fig. 1A, the concentration of SAICA riboside in the brain fell to undetectable levels within 6 hours of an injection.

However, after reverse microdialysis, 6 of 9 animals receiving SAICA riboside showed significant degeneration of pyramidal neurones (Fig. 2). Neither succinyladenosine nor adenosine itself had any effect. As seen in Fig. 1B the levels of SAICA riboside in the hippocampus rose progressively with the duration of dialysis. This method of administration clearly allows high levels to be achieved and maintained, and presumably accounts for the greater success in inducing tissue damage.

In a further 6 animals, succinyladenosine was administered by reverse dialysis at the same time as the SAICA riboside. Damage was still seen in 5 of these, and no statistically significant difference was noted in the amount of damage seen compared with the damage seen with SAICA riboside alone.

In 4 animals, the glutamate receptor antagonist dizocilpine was injected intraperitoneally at a dose of 1 mg/kg simultaneously with the start of the SAICA riboside dialysis. All of these animals showed the same degree of damage as those with SAICA riboside alone.

4. DISCUSSION

The results suggest that SAICA riboside has a neurotoxic action on pyramidal neurones of the rat hippocampus. If a similar action occurs elsewhere, and in humans, such neuronal damage could contribute to the mental retardation, autism and epilepsy found in some patients with adenylosuccinase deficiency.

The mechanism of the damage remains obscure. It does not seem to be due to breakdown to adenosine, since this purine was not at all toxic and adenosine agonists are usually neuroprotective (Stone & Simmonds, 1991; MacGregor & Stone, 1993). Damage due

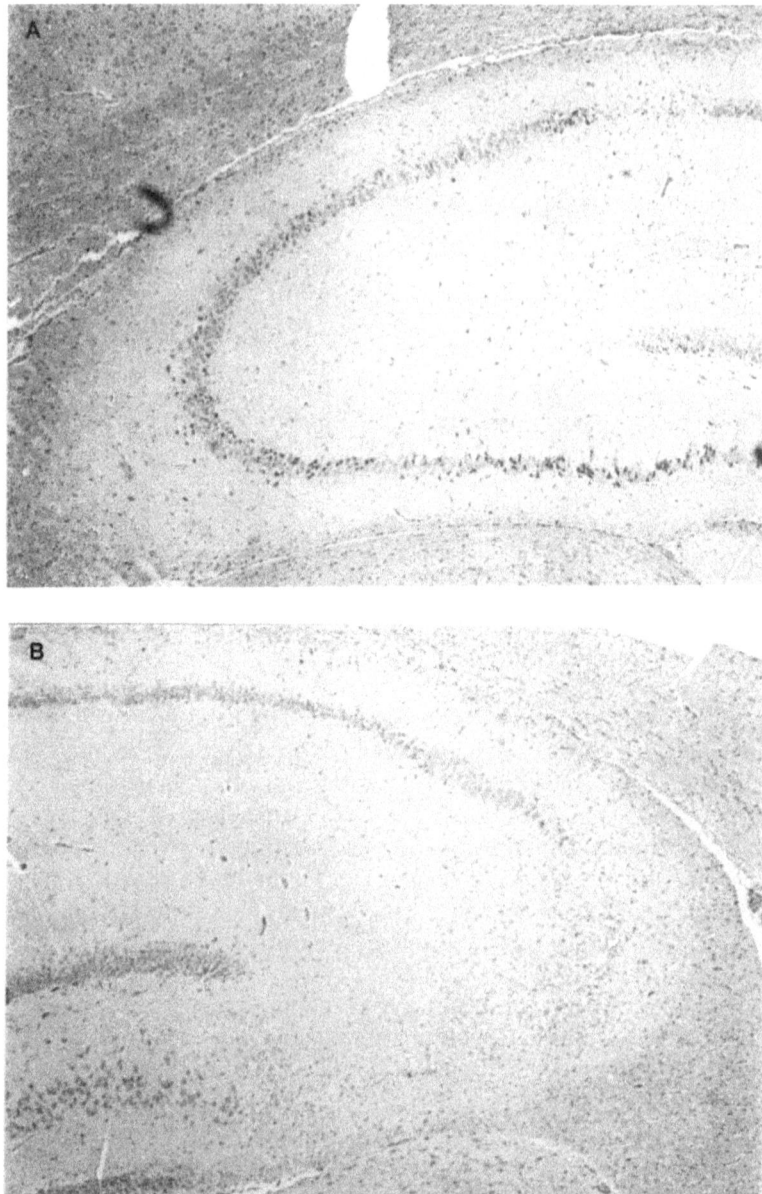

Figure 2. Photomicrographs of the rat hippocampus showing A: A control hippocampus showing the pyramidal cell layers intact; B: A specimen perfused with SAICA riboside by reverse microdialysis showing damage to the CA3 region.

to a wide variety of neurotoxic insults, including, hypoglycaemia, hypoxia, ischaemia and even the administration of glutamate-derived excitotoxins such as kainic acid has been shown to be due to the activation of receptors for the glutamate analogue N-methyl-D-aspartate (NMDA) receptors. However, this does not seem to be the mechanism of damage by SAICA riboside as dizocilpine, a non-competitive blocker of the ion channels associated with the NMDA receptor, afforded no protection.

It is now recognised that at least one variant form of ASase deficiency exists in which there is a greater loss of S-AMP metabolism than of SAICA riboside, resulting in a greater accumulation of S-Ado and a ratio [S-Ado: SAICA riboside] of around 4, rather than the ratio of around 1.5 in other patients (van den Bergh et al., 1993). This so-called 'Type II' deficiency results in much less mental retardation and autism. One interpretation of this finding is that the relatively high S-Ado may protect against neuronal loss or deterioration. However, we have found that succinyladenosine did not produce damage, and did not protect against the SAICA riboside damage. The significance of the increased levels of succinyladenosine, and the reason for the lower degree of impairment, therefore, remains unclear.

ACKNOWLEDGMENT

This work was supported by EC Biomed grant BMH1-CT94-1384.

REFERENCES

De Volder AG, Jaeken J, van den Berghe G, Bol A, Michel C, Cogneau M & Goffinet AM, 1988, Pediatr Res, 24:238–42.
Jaeken J & van den Berghe G. Lancet, 1984, 2:1058–61.
Jaeken J, Wadman SK, Duran M, van Sprang FJ, Beemer FA, Holl RA, Theunissen PM, De Cock P, van den Bergh F, Vincent MF & van den Berghe G, 1988, Europ. J. Pediatr, 1988, 148:126–31.
Jaeken J, Casaer P, De Cock P & van den Berghe G, 1993, In Molecular Genetics, Biochemistry and Clinical Aspects of Inherited Disorders of Purine and Pyrimidine Metabolism. Gresser, U (ed)., Springer, London, pp.140–143. ISBN 3-540-56774-7.
MacGregor DG & Stone TW, 1993, Brit. J. Pharmacol. 109, 316–321.
Simmonds HA, Duley JA & Davies PM, 1990, Analysis of purines and pyrimidines in blood, urine and other physiological fluids. In: Hommes FA (ed) Techniques in Diagnostic Human Biochemistry and Genetics. Wiley-Liss Publ., N.Y., pp.397–424.
Stone RL et al. Nature Genetics, 1992, 1:59–63.
Stone TW & Simmonds HA,1991, Purines: Basic and Clinical Aspects. Kluwer Academic Press, Dordrecht. ISBN 0-7923-8925-5.
Van den Berghe G & Jaeken J. Adv Exp Med Biol, 1986, 195A:27–33.
Van den Bergh F et al., 1993, J. Inher. Metab. Dis. 16, 425–434.
Van den Bergh F, Vincent MF, Jaeken J & van den Berghe G., 1991, Analyt. Biochem. 193, 287–291.

37

URINARY SCREENING FOR PYRIMIDINE METABOLISM DISORDERS

Reference Ranges for Dihydrouracil, Uracil, and Dihydrouracil/Uracil Ratio

Satoshi Sumi,[1] Kiyoshi Kidouchi,[2] Katsuo Hayashi,[3] Masayuki Imaeda,[4] Masami Asai,[4] and Yoshiro Wada[4]

[1]Department of Pediatrics
Nagoya City Child Welfare Center, Showa-ku
[2]Department of Pediatrics and
[3]Department of Medicine
Nagoya City Higashi General Hospital, Chikusa-ku
[4]Department of Pediatrics
Nagoya City University Medical School, Mizuho-ku
Nagoya 467, Japan

1. INTRODUCTION

Pyrimidine chemotherapy agent such as 5-FU are used widely but can occasionally cause serious adverse reactions in patients with pyrimidine metabolism disorders. The screening method which entailed measuring dihydropyrimidine dehydro- genase activity (DPD) has been reported. This method is acceptable with small groups, but difficulties arise when dealing with large populations owing to the complicated procedure for measuring enzyme activity. We have studied the urinary screening method using high-performance liquid chromatograpy, which is acceptable with large groups.[1] This method can diagnose not only dihydropyrimidine dehydrogenase deficiency but also dihydropyrimidinuria. Using this method, we have analyzed urinary dihydrouracil (DHU) and uracil concentrations in 167 healthy adults and 966 patients with malignancy, hypertension, cerebral infarction, etc. In order to establish the reference ranges of urinary pyrimidine, we used "log (concentration)". Additionally, we calculated dihydrouracil/uracil ratio, which seemed to be reflected of DPD activity.

Purine and Pyrimidine Metabolism in Man IX,
edited by Griesmacher *et al.* Plenum Press, New York, 1998.

2. SAMPLES AND METHODS

2.1. Samples

We collected single voided urine samples from 167 healthy adults on occasion of general medical check-ups, and 966 patients with various diseases, who treated at Nagoya City Higashi General Hospital. They were given informed concents to participate in this study. The patients were diagnosed, according to the general textbook of medicine. When a patient had many diseases, we selected one disease which caused the most serious damage for his health. In this study, we excluded pregnant women and patients treated with allopurinol. The urine samples were frozen and stored at –20°C. Immediately prior to analysis, each sample was passed throgh a 0.45 μm Centricut filter (Kurabo, Osaka, Japan) .

2.2. Methods

Urinary pyrimidine were analysed by high-performance liquid chromatography with column switching.[1]

Urinary creatinine was measured with an autoanalyzer (TBA80FR; Toshiba, Tokyo).

3. RESULTS AND DISCUSSION

Concentrations of dihydrouracil (DHU), uracil, and DHU/uracil ratio are shown in Table 1. The mean ± S.D. of DHU, uracil and DHU/uracil were 23.8 ± 35.5, 63.8 ± 66.0 and 0.586 ± 0.269. When the subjects less than 60 years and the subjects 60 years and older were compared, uracil concentration were higher in the subjects less than 60 years. When male and female were compared, dihydrouracil and uracil concentrations were higher in female. In comparison of diseases, uracil concentration was higher in the patients with malignancy.

Distributions of DHU, uracil, and DHU/uracil ratio are shown in Fig. 1. The distributions was not normal, showing asymmetrical. Therefore, we also examined the distributions using "log" (shown in Fig. 2). As the distribution using "log" showed normal, we speculate that it is better to use log(concentration) to make clinical reference ranges for urinary

Table 1. Urinary dihydrouracil (DHU), uracil, and DHU/uracil ratio

	Dihydrouracil (μ mol/g cre.)	Uracil (μ mol/g cre.)	DHU/uracil (ratio)
Total (n=1133)	23.8 ± 35.5	63.8 ± 66.0	0.586 ± 0.269
Age			
60yr < (n=507)	24.3 ± 21.4	71.4 ± 50.4	0.392 ± 0.042
60yr ≧ (n=626)	23.4 ± 21.4	57.7 ± 75.8	0.742 ± 0.359
Sex			
Male (n=666)	21.7 ± 41.7	54.8 ± 48.0	0.533 ± 0.119
Female (n=515)	27.6 ± 24.9	78.8 ± 87.2	0.650 ± 0.377
Diseases			
Healthy (n=167)	23.6 ± 78.5	66.6 ± 31.0	0.356 ± 0.596
Hypertension (n=216)	21.9 ± 14.8	53.5 ± 36.0	0.506 ± 0.396
Cerebral infarction (n=135)	21.7 ± 18.5	46.2 ± 36.6	0.805 ± 2.17
Malignancy (n=51)	25.0 ± 16.8	70.8 ± 117.1	0.521 ± 0.393
Liver dysfunction (n=28)	32.8 ± 41.5	51.2 ± 31.2	0.677 ± 0.636

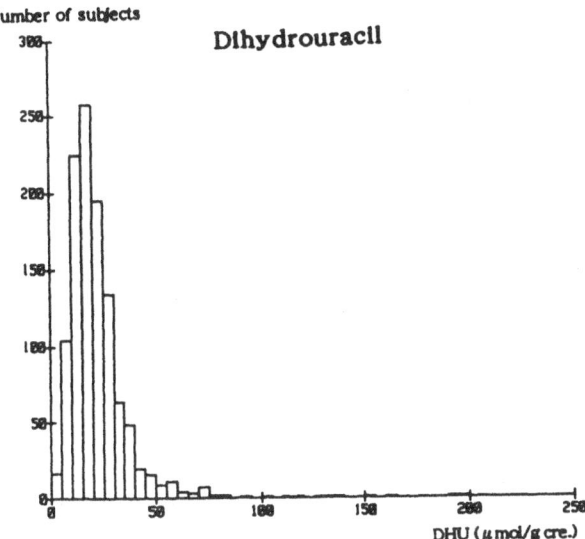

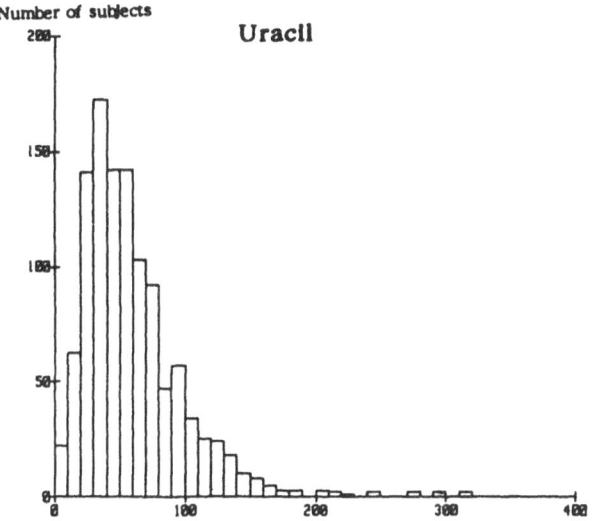

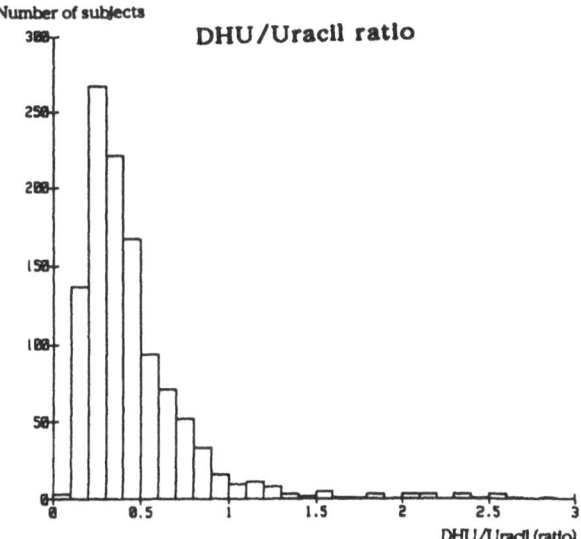

Figure 1. Distribution of urinary dihydro-uracil, uracil, and dihydrouracil/uracil ratio in the 1133 subjects.

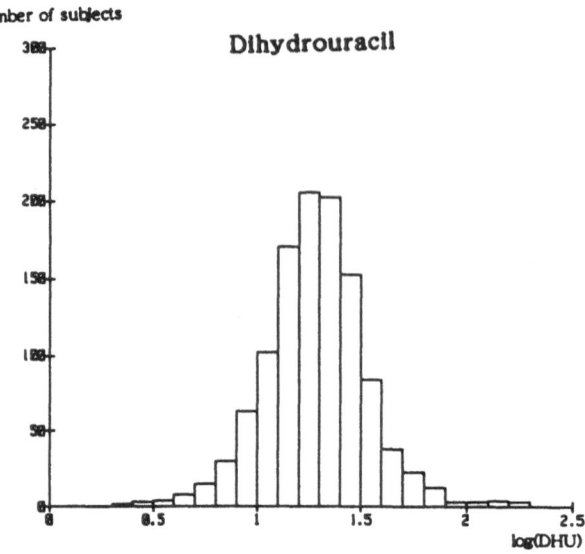

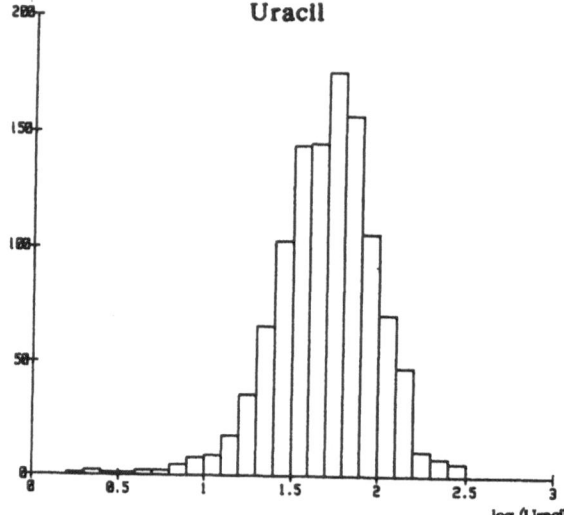

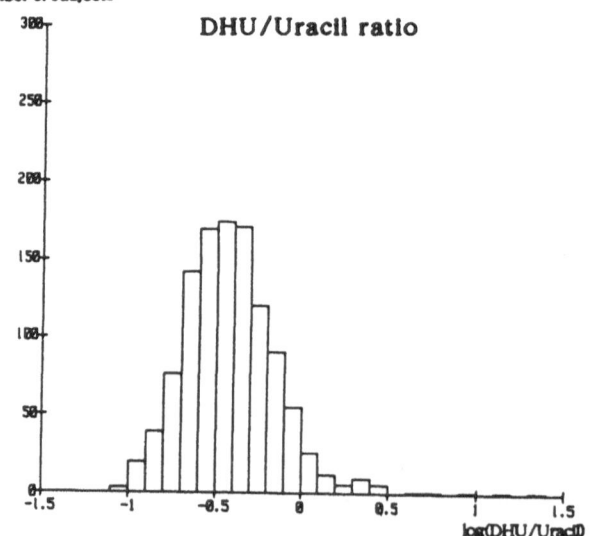

Figure 2. Distribution of log(dihyro-uracil), log(uracil), and log(dihydro-uracil/uracil ratio) in the 1133 subjects.

Table 2. Clinical reference ranges and reported cases

	Age	Dihydrouracil (μ mol/g cre.)	Uracil (μ mol/g cre.)	DHU/Uracil (ratio)
Normal range*	adults	10.7 ~ 34.5	25.1 ~ 99.8	0.21 ~ 0.74
Dihydropyrimidinuria				
Ohba et al. (1995)[2]	11 months	5535.8	421.3	13.3
Sumi et al. (1996)[3]	37 years	1171.9 ~ 1819.2	88.8 ~ 172.8	7.7 ~ 17.9
DPD deficiency (hetero?)				
Muro et al. (1997)[4]	48 years	8.0	94.9	0.084

*Range is between the log(mean − 1S.D.) and the log(mean + 1S.D.).

pyrimidine. We calculated log(mean − 1S.D.) and log(mean + S.D.), and re-converted them to concentrations which would be used as clinical reference ranges (shown in Table 2).

The urinary concentrations in the reported cases were shown in Table 2. The cases were analyzed by our group using the same HPLC method. The DHU concentrations in two dihydropyrimidinuria[2,3] cases were extremely higher than the normal range. Muro et al has reported quite recentry the first case with DPD deficiency in Japan.[4] The case showed a lethal 5-FU toxicity, with the liver DPD activity being about 1/3 of normal control. The DHU/uracil ratio in the case was decreased, although the uracil concentration was within normal range. We believe that urinary DHU/uracil ratio is an useful index for a predection of 5-FU toxicities.

REFERENCES

1. S.Sumi, K.Kidouchi, S.Ohba, et al. Automated screening system for purine and pyrimidine metabolism disorders using high-performance liquid chromatography. *J.Chromatogr.B*, 672, 233–239,1995.
2. S.Ohba, K.Kidouchi, S.Sumi et al. Dihydropyrimidinuria: the first case in Japan. *Adv.Exp.Med.Biol.*370, 383–386, 1995.
3. S.Sumi, K.Kidouchi, K.Hayashi et al. Dihydropyrimidinuria without clinical symptoms. *J. Inher. Metab. Dis.* 19, 901–702, 1996.
4. K.Muro, Boku N, Fujii H et al. Dihydropyrimidine dehydrogenase activity in Japanese patients treated with 5-FU based chemotherapy; pharmacokinetic and pharmacodynamic analysis. (abstract) *Am. Society Clin. Oncol.* Annual Meeting, 1997.

38

CARRIER STATE IN HGPRT DEFICIENCY

A Study in 14 Spanish Families

R. J. Torres,[1] A. Buño,[1] F. A. Mateos,[1] and J. G. Puig[2]

[1]Division of Clinical Biochemistry
[2]Division of Internal Medicine
"La Paz" Hospital, Madrid, Spain

1. INTRODUCTION

HGPRT deficiency is inherited in a chromosome X-linked recessive manner (1). Most of heterozygous females are asymtomatic and have normal serum urate concentrations (2). Moreover, HGPRT activity in hemolysate from HGPRT deficient carrier females is usually in the normal range (3). A number of methodologies have been designed for carrier testing in HGPRT deficient families, HGPRT activity in hair follicles, selective medium cultured skin fibroblasts and genetic analysis, being the most widely used. The purpose of this study was to review our experience with carrier diagnoses in 14 Spanish families with HGPRT deficiency by means of enzymatic and genetic tests.

2. DESIGN AND METHODS

2.1. Subjects

Since 1984 we have studied 14 HGPRT deficient families (Figure 1). 50 females and one female fetus were at risk of being carriers.

2.2. Methods

Carrier testing was performed by means of:

2.2.1. HGPRT Activity in Hair Follicles. HGPRT activity was determined in at least 90 hair follicles from each female according to the method described by Page et al. (4). APRT activity was also measured in these hair follicles as a cell viability control. A carrier

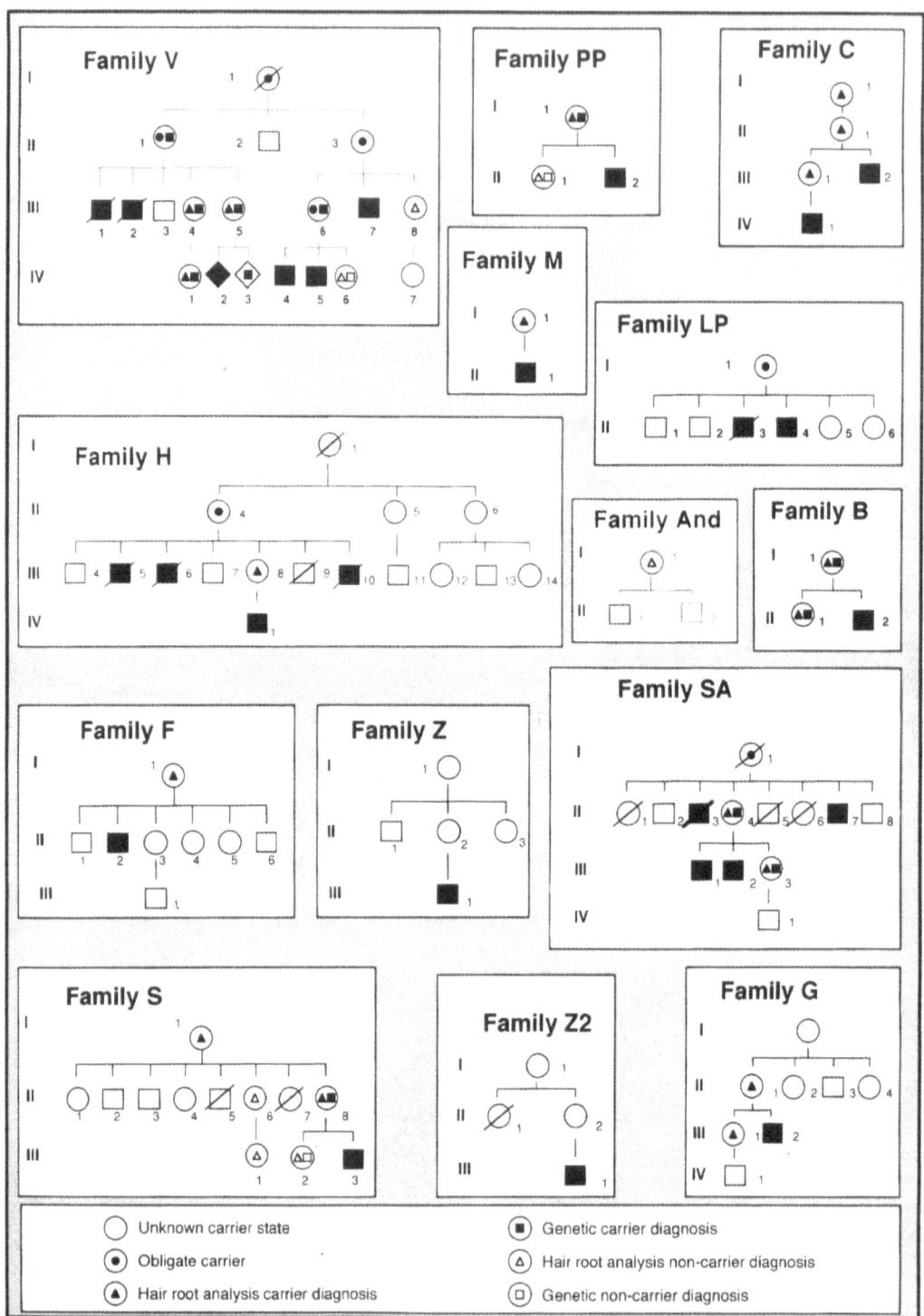

Figure 1. Carrier state diagnosis results in 14 HGPRT deficient Spanish families.

Table 1. Carrier state diagnosis result in females at risk among 14
HGPRT deficient Spanish families

Method	Carrier	Non-carrier	Total
Hair root analysis	18 (72%)	7 (28%)	25
Genetic testing	11 + fetus (80%)	3 (20%)	14 + fetus

diagnosis was established when at least one hair follicle showed APRT activity and did not show HGPRT activity.

2.2.2. Genetic Testing. In 5 of the 14 families the mutation responsible for the enzyme deficiency was assessed. This fact allowed genetic testing in females at risk by means of DNA sequencing (2 families) or restriction enzyme analysis with (2 families) or without (1 family) directed mutagenesis (see Torres RJ et al. in this issue).

3. RESULTS

Five females (10%) were obligate carriers, because they had two or more HGPRT deficient children and/or one affected child in a known HGPRT deficient family. Among the 50 females, the carrier state was assessed by hair root analysis in 25 and by genetic testing in 14 (Table 1). Among the former, 18 (72%) females were diagnosed as carriers and 7 (28%) as non-carriers. Fourteen females and a 9 weeks female fetus were tested by genetic methods. Among these, 11 females and the female fetus were diagnosed as heterozygous for the familial mutation responsible for the HGPRT deficiency (carriers, 80%), and 3 females were homozygous for the normal allele (non carriers, 20%) (Figure 1). Twelve (9 carriers and 3 carriers) females were tested by both hair root analysis and genetic testing. In all subjects the results were concordant. In only one case (7%) the propositus presented a "de novo" mutation. In two families we do not yet know whether the mutation could be classified as "de novo" or inherited, and in 11 families the mutation was inherited.

4. CONCLUSION

When possible, genetic tests are the faster and more reliable methods for carrier diagnosis. Hair root analysis of HGPRT and APRT activities is time consuming and the carrier state cannot be absolutely established; however it is reliable when genetic testing cannot be performed. These results show that the probability of being a carrier in HGPRT deficiency families is very high (about 80%). In our series, the incidence of "de novo" mutations responsible for HGPRT deficiency appears to be lower than expected, according to the Haldane hypothesis.

ACKNOWLEDGMENTS

We are indebted to the Clinical Research Unit nursing staff and the dietetic staff for excellent patient care; to Mª Paz Canencia for valuable technical assistance; to Erik Lundin for assistance in preparing the manuscript. Supported by grants from Caja de Madrid and Fondo de Investigaciones Sanitarias (FIS, 97/0458), Spain.

REFERENCES

1. Pai GS, Sprenke JA, Do TT, Mareni CE, Migeon BR. Localization of loci for hypoxanthine phosphoribo-syltransferase and glucose-6-phosphate dehydrogenase and biochemical evidence of non-random X-chromosome expression from human X-autosomal translocation. Proc Nat Acad Sci USA 1980; 77: 2810–2813.
2. Rossiter BJF, Caskey CT. Hipoxanthine-guanine phosphoribosyltransferase deficiency: Lesh-Nyhan syndrome and gout. In: Scriver CR, Beaudet AL, Sly WS, Valle D(eds). The Metabolic Basis of Inherited Diseases, 7th de. New York, Mc Graw-Hill, 1995; vol II: 1679–1706.
3. Migeon BR, Der Kaloustian VM, Nyhan WL, Young WJ, Childs B. X-linked hypoxanthine-guanine phosphoribosyltransferase deficiency: Heterozygote has two clonal populations. Science 1968; 160: 425–527.
4. Page T, Bakay B, Nyhan WL. An improved procedure for detection of hypoxanthine-guanine phosphoribosyl transferase heterozygotes. Clin Chem 1982; 28: 1181–1184.

39

GENETIC DIAGNOSIS OF HYPOXANTHINE-GUANINE PHOSPHORIBOSYLTRANSFERASE (HGPRT) CARRIER STATUS BY RESTRICTION ANALYSIS AND DIRECTED MUTAGENESIS

R. J. Torres,[1] A. Buño,[1] J. Molano,[1] F. A. Mateos,[1] and J. G. Puig[2]

[1]Division of Clinical Biochemistry
[2]Division of Internal Medicine
"La Paz" Hospital, Madrid, Spain

1. INTRODUCTION

Hypoxanthine-guanine phosphoribosyltransferase (HGPRT) deficiency is transmitted as an X-linked recessive trait. Female carriers are usually asymptomatic. X chromosome inactivation in heterozygous females hematopoietic system may not be random, and few HGPRT negative cells can be found in this tissue (1). The carrier diagnosis is usually performed by determining HGPRT activity in hair roots where HGPRT mosaicism is more evident (2). However, a non-carrier state diagnosis cannot be obtained with absolute certainty. Another limitation of this diagnostic method is the possibility of obtaining non-viable hair roots. Nowadays, genetic analysis must be favoured over biochemical analysis for carrier diagnosis, but not all laboratories have the necessary facilities for this purpose. The knowledge of the genetic mutation in three Spanish families with HGPRT deficiency enabled us to perform the genetic diagnosis of the carrier state in 10 females at risk, in one 9 weeks female fetus, and in a newborn of a carrier female, by using polymorphism restriction analysis and directed mutagenesis (3).

2. DESIGN AND METHODS

2.1. Design

Table 1 shows the mutations detected in the three Spanish families (4). Only family V mutation causes a change in the restriction pattern of HGPRT gene (Dra III site was sup-

Purine and Pyrimidine Metabolism in Man IX,
edited by Griesmacher *et al.* Plenum Press, New York, 1998.

Table 1. Genetic diagnosis of HGPRT deficient carrier state

Family	V	PP	SA
Mutation*	C 39837 to T	A 40031 to G	T 14881 to G G 14883 to A
PCR fragment	39770–40130 (361 bp)	40006–40130 (124 bp)	14857–14970 (114 bp)
Directed mutagenesis	No	Yes	Yes
Mutagenesis-induced change	None	New Dde I site	New Nla III site
Mutation-induced change	Dra III site suppressed and new Nla III site	Dde I site suppressed	Nla III site suppressed

*The first nucleotide in the genomic sequence is the first G of the EcoRI site 5′ to exon 1.

pressed and a new Nla III site was observed). This fact allows the distinction between the normal and the mutated allele by restriction endonuclease digestion. Directed mutagenesis was employed in the other two families in order to amplify a modified HGPRT gene fragment. Familial mutation causes a change in the restriction pattern of this modified HGPRT gene fragment which allows distinction between thenormal and the mutated allele.

2.2. Methods

DNA was obtained from venous blood or chorionic villus sample. An HGPRT gene fragment, including the mutation, was amplified by PCR (Table 1). Primers employed for PCR amplification are shown in Table 2. Base change implied in directed mutagenesis is marked. Amplified fragment was digested with the appropriate endonuclease and the fragment of the digestion separated by electrophoresis. The genetic diagnosis was performed because of the differences in the restriction pattern between the mutant and the normal allele.

3. RESULTS

Figures 1, 2 and 3 show the results obtained in these three families. All the diagnoses were established with great precision in a mean time of 24 to 48 hours. Among 10 females at risk of being carriers, 8 were diagnosed as carriers and 2 were non-carriers. The

Table 2. Primers used for PCR amplification

Family V	5′ CCT GTA GTC TCT CTG TAT GTT A
	3′ AAA GAA GCA ATT ACT TAC ATT C
Family PP	5′ CTA AAT GAT GAA TTA TGA TTC TTC* T
	3′ AAA GAA GCA ATT ACT TAC ATT C
Family SA	5′ TGT TTA TTC CTC ATG GAC TAA TC* A
	3′ TAG CAA GTA CTC AGA ACA

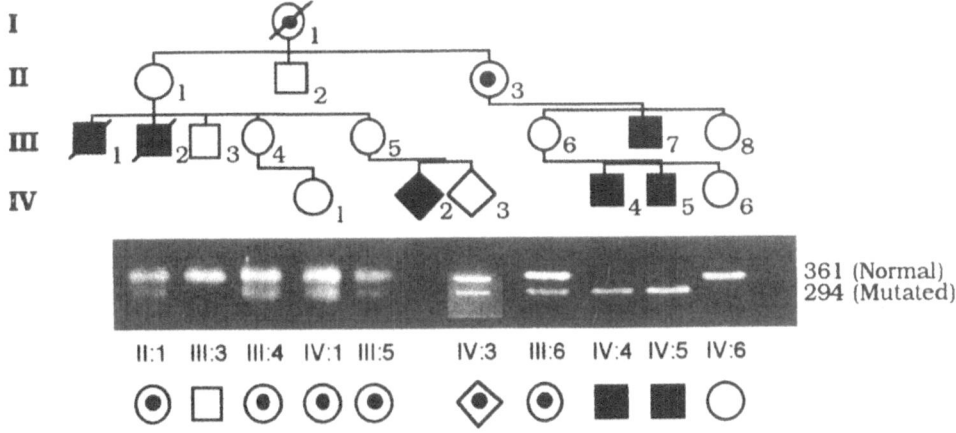

Figure 1. Electrophoresis from Nla III digestion of the 361 bp amplified fragment in family V. White squares represent normal males, black squares represent affected males. White circles represent a non- carrier female, and circles with a black point represent a carrier female.

female fetus was diagnosed as carrier, and the newborn of a known HGPRT deficient carrier female was healthy.

4. CONCLUSIONS

Restriction analysis has been previously employed to confirm or to perform carrier diagnosis when the familial mutation was known (5). This method is easy and quick, but often, HGPRT mutation does not cause a change in the genomic restriction pattern. In these cases, directed mutagenesis can help to provide a mutation-induced change in HGPRT genome restriction pattern. We conclude that directed mutagenesis and restriction analyses are valid tools in the genetic diagnosis of HGPRT deficiency carrier status, when the familial mutation is known.

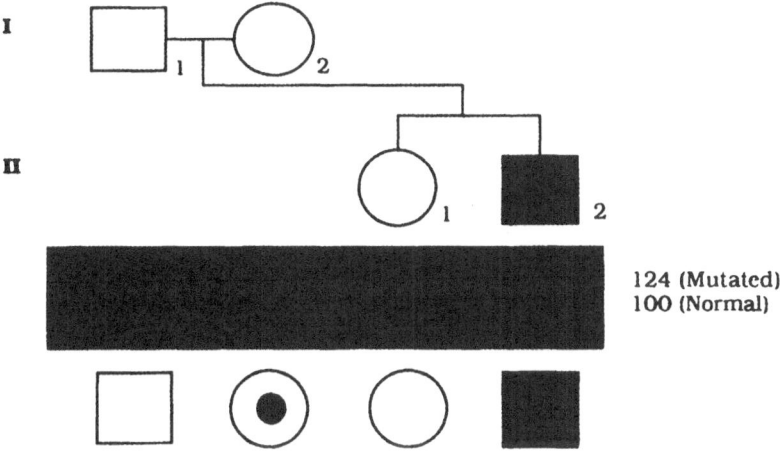

Figure 2. Electrophoresis from Dde I digestion of the 124 bp amplified fragment in family PP. Symbols represent the same as in Figure1.

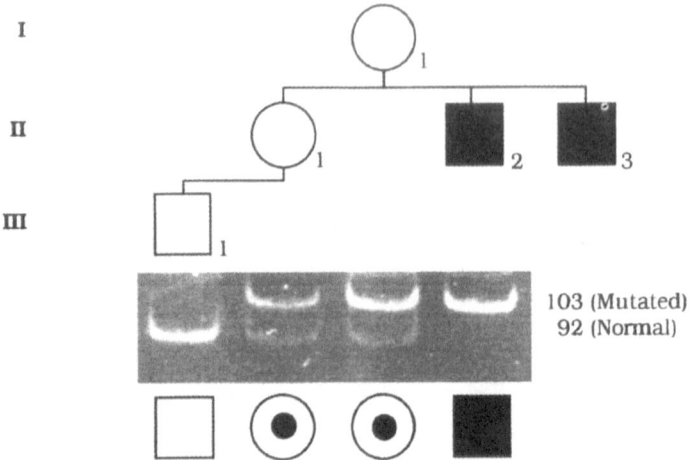

Figure 3. Electrophoresis from Nla III digestion of the 114 bp amplified fragment in family SA. Symbols repre-sentent the same as in Figure 1.

ACKNOWLEDGMENTS

We are indebted to the Clinical Research Unit nursing staff and the dietetic staff for excellent patient care; to Mª Paz Canencia for valuable technical assistance; to Erik Lundin for assistance in preparing the manuscript. Supported by grants from Caja de Madrid and Fondo de Investigaciones Sanitarias (FIS, 97/0458), Spain.

REFERENCES

1. Migeon BR, Der Kaloustian VM, Nyhan WL, Young WJ, Childs B. X-linked hypoxanthine-guanine phos-phoribosyltransferase deficiency: Heterozygote has two clonal populations. Science 1968; 160: 425–427.
2. Page T, Bakay B, Nyhan WL. An improved procedure for detection of hypoxanthine-guanine phosphoribo-syltranferase heterozygotes. Clin Chem 1982; 28: 1181–1184.
3. Haliassos A, Chomel JC, Tesson L, Baudis M, Kruh J, Kaplan JC, Kitzis A. Modification of enzymatically amplified DNA for the detection of point mutations. Nucleid Acids Res 1989; 17: 3606.
4. Sege-Peterson K, Chambers J, Page T, Jones OW, Nyhan WL. Characterization of mutations in phenotypic variants of hypoxanthine phosphoribosyltransferase deficiency. Hum Mol Genet 1992; 1: 427–432.
5. Marcus S, Steen AM, Andresson B, Lambert B, Kristoffersson U, Francke U. Mutation Analysis and prena-tal diagnosis in a Lesh-Nyhan family showing non-random X-inactivation interfering with carrier detection test. Hum Genet 1992; 89: 395–400.

40

MUSCLE PURINE NUCLEOTIDE CYCLE ENZYMES IN EXERCISE INTOLERANCE

Maria-Grazia Operti,[1,2] M. -Françoise Vincent,[1] Jean-Marie Brucher,[2] and Georges Van den Berghe[1]

[1]Laboratory of Physiological Chemistry
International Institute of Cellular and Molecular Pathology
Avenue Hippocrate 75-39
B-1200 Brussels, Belgium
[2]Laboratory of Neuropathology
University of Louvain Medical School

1. INTRODUCTION

Three enzymes of adenine nucleotide metabolism, adenylosuccinate synthetase (ASS), adenylosuccinate lyase (adenylosuccinase, ASL), and AMP deaminase (AMPDA), form the purine nucleotide cycle.[1,2] Deficiences of muscle AMPDA (often called myoadenylate deaminase in the clinical literature) are frequently diagnosed either as a primary genetic, or as a secundary defect in patients presenting with muscular symptoms.[3-5] In contrast, very few patient studies have been devoted to the two other enzymes of the purine nucleotide cycle. Only Kar and Pearson[6] reported normal activities of ASL in muscle from patients with Duchenne and other neuromuscular diseases. In the present study ASS, ASL, and AMPDA were assayed in a series of muscle biopsies taken from patients suffering from fatigue and cramps following moderate to vigorous exercise.

2. MUSCLE BIOPSIES

These had been taken for diagnostic purposes from 71 patients, aged 15–68 years. Specimens had been in part processed for histological and histochemical examination, in part immediately deep-frozen in liquid nitrogen and thereafter stored at –80°C. Eight biopsies were considered controls because they were provided by subjects who were ultimately found free of neuromuscular disease by combined clinical, electromyographic, and histologic criteria.

Purine and Pyrimidine Metabolism in Man IX,
edited by Griesmacher *et al.* Plenum Press, New York, 1998.

3. ENZYME ASSAYS

All procedures were carried out at 4°C. Frozen muscle (20–50 mg) was powdered in a micromortar and homogenized in 100 vol of bidistilled H_2O. After centrifugation for 5 min at 14,000 × g, activities of ASL and AMPDA were determined in the supernatant. Since ASS combines with debris in muscle homogenates prepared in water, because of binding with F-actin,[7] its activity was measured in the precipitate, taken up in 4–10 vol of bidistilled H_2O.

Enzyme activities were measured with radioactive substrates. [U-^{14}C]AMP (51Ci/mol) was purchased from Amersham International (Amersham, U.K.). [^{14}C]IMP was prepared from [^{14}C]AMP by incubation with purified muscle AMP deaminase (100 U/mg, Sigma, Saint Louis, MO, USA). [^{14}C]S-AMP was synthesized from [^{14}C]AMP as described by Van den Bergh et al.[8]

For measurement of ASS activity, 20 µl of the redissolved muscle precipitate was mixed with 25 mM Hepes buffer, pH 7.5, 0.2 mM [^{14}C]IMP, 5 mM aspartate, 7 mM GTP, 10 mM $MgCl_2$, and a GTP regenerating system consisting of 20 mM creatine phosphate and 10 mg/ml creatine kinase (Sigma, Saint Louis, MO, USA), in a total volume of 100 µl. After incubation at 37°C for 0, 15, 30, and 60 min, 10 µl of the reaction mixture were spotted on polyethyleneimine (PEI) thin-layer chromatography (TLC) plates on which 10 µl of 5 mM carrier solutions of IMP and S-AMP had been applied. After development in Bu/Me/H_2O (1/1/8, by vol), followed by 0.5 M $LiCl_2$, spots corresponding to IMP and S-AMP were localised under UV light. The spot corresponding to S-AMP was cut out, and its radioactivity counted by liquid scintillation.

For measurement of ASL activity, 40 µl of the supernate was mixed with 25 mM Hepes buffer, pH 7.5, 0.2 mM [^{14}C]S-AMP, 100 mM KCl, and 2.5 mM AOPCP in a total volume of 100 µl. After incubation for 0, 15, 30, and 60 min, 10 µl of the reaction mixture were spotted on PEI-TLC plates on which 10 µl of 5 mM carrier solutions of S-AMP and AMP had been applied. After development in 0.5 M $LiCl_2$, the spot corresponding to AMP was cut out, and its radioactivity counted.

For assay of AMPDA, 5 µl of the supernate was mixed with 25 mM Hepes buffer, pH 6.5, 100 mM KCl, and 0.2 mM [^{14}C]AMP, in a total volume of 100 µl. After incubation for 0, 5, 10, and 20 min, 10 µl of the incubation mixture were spotted on PEI-TLC on which 10 µl of 5 mM carrier solutions of AMP and IMP had been applied. After development in 1.4 M $LiCl_2$, the spot corresponding to IMP was cut out, and its radioactivity counted. It was verified that each assay was linear as a function of time and amount of protein. Protein was determined by the method of Lowry,[9] using bovine serum albumin as the standard.

4. RESULTS AND DISCUSSION

Mean activities of the three enzymes of the purine nucleotide cycle, obtained in the 8 control biopsies, are shown in Table 1. Deficient activites were defined as lower than the mean control activity minus 3 S.D. Activities of the three enzymes were normal in 52 out of the 63 pathological specimens (results not shown). Among the patients with normal activities, 46 had myogenic electromyography, associated with normal histology in 4, mitochondriopathy in 4, congenital myopathy in 5, and exercise intolerance with nonspecific histologic findings in 33. In 6 patients with neurogenic electromyographic and histologic findings, normal activities were also found.

Table 1. Activities of the purine nucleotide cycle enzymes

Diagnosis	ASS	ASL	AMPDA
Normal ± S.D.	2.1 ± 0.3	13.0 ± 0.2	132 ± 51
(range, n = 8)	(1.7–2.5)	(8.5–15.5)	(91–216)
Congenital myopathies			
Centronuclear	0.8*	6.7	48
With tubular aggregates (n = 2)	0.9*	13.4	285
	1.3	12.0	69
Central core	1.5	4.4*	64
Multiminicore (n = 2)	2.5	5.9*	214
	1.3	12.4	180
Nemaline (n = 3)	1.5	8.6	87
	1.9	10.5	125
	4.0	10.3	115
Nonspecific histologic alterations	0.5*	7.0	99
with exercise intolerance (n = 4)			
	0.9*	9.8	71
	1.0*	9.0	69
	1.0*	13.0	171
Other diagnoses			
Septicemia with hemiparesis	0.6*	4.5*	68
Progressive systemic sclerosis	0.8*	7.5	10
Deltoid atrophy	1.7	5.5*	46

Activities are expressed as nmol/min per mg of protein.
*Indicates activity below control mean minus 3 S.D.

Decreased activities of ASS and/or ASL were measured in 11 biopsies. As detailed in Table 1, these low activities were found in 4 subjects within a group of 9 patients with congenital myopathies, in 4 subjects out of a series of 37 with nonspecific histologic alterations, and in 3 patients with other diagnoses. A decreased activity of ASS was found in the biopsy of a patient with centronuclear myopathy, and in that of a patient with a myopathy with tubular aggregates. A decreased activity of ASL was measured in two biopsies revealing, respectively, central core myopathy and multiminicore myopathy. Nevertheless, a causal relationship between myopathy with tubular aggregates and multiminicore myopathy with ASS and ASL deficiency, respectively, was ruled out by the observation that for both enzymes, activities in the normal range were measured in other biopsies with the same histologic diagnosis. In 3 biopsies in which nemaline myopathy was diagnosed, activities of the three enzymes of the purine nucleotide cycle were within the normal range.

In muscle of the patients with congenital myopathies, histology had revealed excess of type 1 fibers, with type 2 fibers often decreased or absent. Several studies have evidenced that the activities of the purine nucleotide cycle enzymes are higher in fast-twitch, anaerobic, type 2 fibers than in the slow-twitch, oxidative, type 1 fibers.[10] Accordingly, we might have expected low activities of the three enzymes of the purine nucleotide cycle in these samples. However, our results show that there is no simple relationship between the loss of the latter fibers and the purine nucleotide cycle enzymes.

The activity of ASS was markedly below normal in 4 patients with nonspecific histologic alterations, myogenic electromyography, and exercise intolerance. Out of the three enzymes of the purine nucleotide cycle, ASS has the lowest activity in rat[10] and human (Table 1) muscle. Together with AMPDA, ASS thus probably exerts a rate-limiting role in the operation of the purine nucleotide cycle during contraction. Since a muscle-specific isozyme of ASS has been identified,[11] deficiencies of this rate-limiting isoform might potentially result in exercise intolerance. Although the present results do not allow as yet to corroborate this hypothesis, they suggest that searches are warranted for primary, genetic deficiencies of ASS as a cause of exercise intolerance.

Low activities of ASS were found in a sample from a patient with septicemia and hemiparesis, and from a subject with progressive systemic sclerosis. An association between progressive systemic sclerosis, skeletal muscle disease,[12] and muscle AMPDA deficiency,[13] has been documented. Although the frequency of AMPDA deficiency in muscle biopsies[14] and in the general population[4,5] is high, our finding of a decreased activity of muscle AMPDA in an additional patient with progressive systemic sclerosis, and its association with a decreased activity of ASS, warrant further investigations of the purine nucleotide cycle in patients with the latter disorder.

Decreased activity of ASL was also measured in the patient with septicemia and hemiparesis, and in a patient with deltoid atrophy. Besides the conversion of S-AMP into AMP within the purine nucleotide cycle, ASL also catalyzes the conversion of SAICAR (succinylaminoimidazole carboxamide ribotide) into AICAR along the *de novo* pathway of purine biosynthesis. A genetic deficiency of ASL in liver, kidney, and presumably brain, provokes variable psychomotor retardation, often accompanied by autistic features.[15,16] A number of the affected patients also suffer from growth retardation and muscular wasting, associated with decreased activity of ASL in muscle.[17] This observation indicates the existence of isozymes of ASL, and suggests that isolated genetic deficiencies of muscle ASL might also occur. The low activities of muscle ASL in two patients with acquired diseases indicate that, like the deficiencies of ASS and AMPDA, low activities of ASL can be secundary to various muscle disorders. Nevertheless, we believe that further investigations are also warranted to try to identify isolated, genetic deficiencies of a muscle isoform of ASL.

ACKNOWLEDGMENTS

We thank Ms T. Timmerman for expert technical assistance. This work was supported by grant 3.4557.93 of the Belgian Fund for Medical Scientific Research, by the Belgian Federal Service for Scientific, Technical and Cultural Affairs, and by the European Community Concerted Action n° BMH1-CT-1384. GvdB is Director of Research of the Belgian National Fund for Scientific Research.

REFERENCES

1. Lowenstein JM: Ammonia production in muscle and other tissues: the purine nucleotide cycle. *Physiol Rev* 52: 282–414 (1972).
2. Van den Berghe G, Bontemps F, Vincent MF, Van den Bergh F: The purine nucleotide cycle and its molecular defects. *Progr Neurobiol* 39: 547–561 (1992).
3. Fishbein WN: Myoadenylate deaminase deficiency: a new disease of muscle. *Science* 200: 545–548 (1978).
4. Gross M: Molecular biology of AMP deaminase deficiency. *Pharmacy World Sci* 16: 55–61 (1994).
5. Sabina RL, Holmes EW: Myoadenylate deaminase deficiency. In: Scriver CR, Beaudet AL, Sly WS, Valle D (eds) The metabolic and molecular bases of inherited disease. McGraw-Hill, New York, 7th ed, pp. 1769–1780 (1995).
6. Kar NC, Pearson CM: Adenylosuccinase in human muscular dystrophy. *Muscle Nerve* 4: 174–175 (1981).
7. Ogawa H, Shiraki H, Matsuda Y, Nakagawa H: Interaction of adenylosuccinate synthetase with F-actin. *Eur J Biochem* 85: 327–331 (1978).

8. Van den Bergh F, Vincent MF, Jaeken J, Van den Berghe G: Radiochemical assay of adenylosuccinase: demonstration of parallel loss of activity toward both adenylosuccinate and succinylaminoimidazole carboxamide ribotide in liver of patients with the enzyme defect. *Anal Biochem* 193: 287–291 (1991).

9. Lowry OH, Rosebrough NJ, Farr AL, Randall RJ: Protein measurements with the Folin phenol reagent. *J Biol Chem* 193: 265–275 (1951).

10. Reichmann H, DeVivo DC: Coordinate enzymatic activity of beta oxidation and purine nucleotide cycle in a diversity of muscle and other organs of rat. *Comp Biochem Physiol* 98 B: 327–331 (1991).

11. Stayton MM, Rudolph FB, Fromm HJ: Regulation, genetics and properties of adenylosuccinate synthetase: a review. *Curr Top Cell Regul* 22: 103–141 (1983).

12. Medsger TA: Progressive systemic sclerosis: skeletal muscle involvement. *Clin Rheum Dis* 5: 102–113 (1979).

13. Kar NC, Pearson CM: Muscle adenylate deaminase deficiency: report of six new cases. *Arch Neurol* 38: 279–281 (1981).

14. Mercelis M, Martin JJ, de Barsy T, Van den Berghe G: Myoadenylate deaminase deficiency: absence of correlation with exercise intolerance in 452 muscles biopsies. *J Neurol* 234: 385–389 (1987).

15. Jaeken J, Van den Berghe G: An infantile autistic syndrome characterised by the presence of succinylpurines in body fluids. *Lancet* 2: 1058–1061 (1984).

16. Stone RL, Aimi J, Barshop BA, Jaeken J, Van den Berghe G, Zalkin H, Dixon JE: A mutation in adenylosuccinate lyase associated with mental retardation and autistic features. *Nature Genetics* 1: 59–63 (1992).

17. Jaeken J, Wadman SK, Duran M, van Sprang FJ, Beemer FA, Holl RA, Theunissen PM, De Cock P, Van den Bergh F, Vincent MF, Van den Berghe G: Adenylosuccinase deficiency: an inborn error of purine nucleotide synthesis. *Eur J Pediatr* 148: 126–131 (1988).

41

PRENATAL DIAGNOSIS OF HPRT MUTANT GENES IN LESCH-NYHAN SYNDROME

Yasukazu Yamada,[1] Haruko Goto,[1] Kaoru Suzumori,[2] and
Nobuaki Ogasawara[1]

[1]Department of Genetics
Institute for Developmental Research
Aichi Human Service Center
Kasugai, Aichi 480-03, Japan
[2]Department of Obstetrics and Gynecology
Nagoya City University Medical School
Mizuho-ku, Nagoya 467, Japan

1. INTRODUCTION

Lesch-Nyhan syndrome is an X-linked recessive disorder characterized by hyperuricemia, physical and mental retardation, choreoathetosis, and compulsive self-mutilation.[1] This disease is associated with the complete absence of activity of an enzyme involved in purine metabolism, hypoxanthine guanine phosphoribosyltransferase (HPRT, EC 2.4.2.8).[2] Partial HPRT deficiency generally presents as gout and uric acid overproduction in early adulthood.[3] The marked genetic heterogeneity of HPRT deficiency is well known. As reviewed previously by Sculley et al.,[4] many different mutations at the HPRT gene locus (deletions, insertions, duplications, abnormal splicing, and point mutations at different sites of the coding region from exon 1 to 9) have been reported. We have identified Japanese and Korean HPRT mutants,[5-12] including two rare cases in female subjects,[5,9] by polymerase chain reaction of reverse-transcribed mRNA (RT-PCR) and the multiplex amplification technique of all nine HPRT exons from the genomic DNA coupled with direct sequencing. In this study, we report prenatal diagnoses of HPRT mutant genes which have been carried out in five Lesch-Nyhan families.

2. MATERIALS AND METHODS

2.1. Subjects

In three Japanese (family A, B, and C) and two Korean (family D and E) families with Lesch-Nyhan syndrome, the prenatal diagnoses of the mutant genes in seven fetuses

Purine and Pyrimidine Metabolism in Man IX,
edited by Griesmacher *et al.* Plenum Press, New York, 1998.

were carried out. The clinical aspects, the activities of purine salvage enzymes, and the HPRT gene mutations in respective families have been reported previously.[6,7,10,11]

2.2. Methods

All the methods for HPRT gene analysis, identification of the genomic mutation and the altered mRNA, were described previously.[6] DNA sequences were determined according to the simplified direct sequencing method as described previously,[6] and were recorded on a personal computer and analyzed with GENETYX version 9.0 software (SDC, Japan). The sex of fetus was diagnosed from analyses of SRY[13] and DYZ1.[14] The mutation of the HPRT gene in fetus was analyzed by PCR-RFLP method and direct sequencing using genomic DNA from chorionic villi samples (CVS) or amniotic fluid cells. The mutation was also confirmed by the existence of 6-thioguanine resistant cells, when amniotic fluid cells were used for the diagnosis.

3. RESULTS AND DISCUSSION

In a Japanese Lesch-Nyhan family A, a single nucleotide substitution of T to A was detected in exon 3, which resulted in a mis-sense mutation (L78Q), as described previously.[6] The prenatal diagnosis in this family was first performed by PCR-RFLP analysis utilizing the Pvu II (CAGCTG) restriction site created in the mutant gene (Figure 1A). The DNA fragment including exon 3 from the genomic DNA isolated form the CVS of the third conception was not digested by Pvu II, similar to that from normal subject. Direct

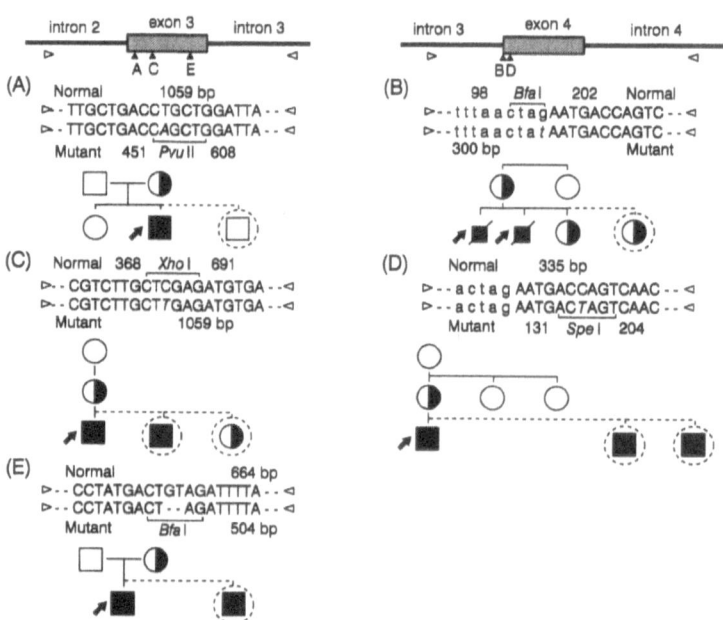

Figure 1. PCR-RFLP analysis for prenatal diagnosis to detect the mutant allele in Lesch-Nyhan families (A, B, C, D, and E). Fetuses are shown by dotted lines. The respective primer pairs (▷, ◁) were described previously.[6] Closed triangle (▲) pointed the position of the mutation and the arrows (➚) showed the proband.

sequencing of the fragment showed also a normal sequencing pattern. Therefore, the male fetus was diagnosed as inherited the normal HPRT gene, but had miscarriage.

The second Japanese family B had a point mutation of G to T at the 3'-end of intron 3 (318-1G→T)7, which generated RNA splicing error by the change of 3'-splice acceptor consensus of AG to AT. The *Bfa* I restriction site (CTAG) at 3'-end of intron 3 is lost in the mutant allele (CTAT). Family studies and prenatal diagnosis for the mutant were performed by PCR amplification of the exon 4 fragment and digestion by *Bfa* I (Figure 1B). By the PCR-RFLP and PCR-direct sequencing analyses of genomic DNA isolated from the CVS, a female fetus appeared to be carrier having both normal and mutant alleles, similar to the mother and the sister. The diagnosis was confirmed by the DNA analysis using the blood sample of umbilical cord when the baby was born.

The mutation in the third Japanese family C was a nonsense mutation R51X which occurred at a hot spot for germline mutations in the HPRT gene locus.[11] A single nucleotide substitution of C to T generating the nonsense mutation was detected in six unrelated Lesch-Nyhan families, not only Japanese[9,11,15,16] but also Caucasian.[17,18] In two Japanese male cases, HPRT$_{Fujimi}$[15] and HPRT$_{Kanagawa}$[16], and in the second female case,[9] the mothers of the probands were not heterozygotes. In only family C, the mother was heterozygous carrier, whereas the grandmother was not.[11] Thus, we carried out the prenatal diagnoses twice in family C by PCR-RFLP utilizing a Xho I restriction site (CTCGAG) which was lost in the mutant allele (CTTGAG). The first fetus was a male and inherited the mutant allele from the mother, and the second female fetus was diagnosed as a heterozygous carrier (Figure 1C). In the first case, analysis of genomic DNA from the fetal kidney obtained at autopsy after abortion demonstrated the same results.

Another nonsense point mutation (Q109X) of C to T at nucleotide 424 replacing Gln-109 (CAG) with a stop codon (TAG), as the same as RJK1930,[19] was found in a Korean family D.[10] The family study and prenatal gene diagnosis were performed utilizing *Spe* I (ACTAGT) restriction site created in the mutant gene (Figure 1D). The DNA fragments including exon 4 were amplified from genomic DNA from the family members and the CVS at the second and third conceptions, then the fragments were digested by *Spe* I. The grandmother and two aunts did not have the mutant gene. The mutation must be a *de novo* one occurring in the grandmother's or grandfather's germ cell similar to family C. Both two male fetuses inherited the mutant HPRT gene from the mother, the same as the patient.

In a Korean family E, a 2-bp deletion of GT at nucleotide position 289 and 290 on the exon 3, 289delGT, has been identified by the analysis of genomic DNA from the patient.[10] The detected mutation had been reported already in two different patients, RJK1332[20] and HPRT$_{Cheltenham}$.[17] By utilizing a *Bfa* I restriction site which was created in the mutant as an indicator, the family study and prenatal diagnosis were carried out (Figure 1E). The mother was predicted to be a heterozygous carrier. The amniotic cells of the mother's second conception were resistant for 6-thioguanine. Further, the genetic analysis of the cells revealed that the male fetus inherited the mutant HPRT gene from the mother.

Analyzing seven fetuses in five families, four male fetuses inherited the respective mutant genes, one male fetus was normal, and two female fetuses appeared to be carrier. HPRT gene mutation occurs at random and the marked genetic heterogeneity of the deficiency is well known. Therefore, the identification of the mutation in each family with Lesch-Nyhan syndrome is necessary in advance of the prenatal diagnosis. The appearance of new restriction sites and the loss of restriction sites by mutations were effectively used for the diagnosis.

ACKNOWLEDGMENTS

We thank Ms. Misao Iwanaga-Kato and Ms. Ikuko Iwamoto for their skillful technical assistance. This work was supported by a Gout Research Foundation grant, and an Intractable Disease grant from the Ministry of Health and Welfare of Japan.

REFERENCES

1. M. Lesch and W.L. Nyhan, A familial disorder of uric acid and central nervous system function., *Am. J. Med.* 36:561 (1964).
2. J.E. Seegmiller, F.M. Rosenbloom, and W.N. Kelley, Enzyme defect associated with a sex-linked human neurological disorder and excessive purine synthesis., *Science* 155:1682 (1967).
3. W.N. Kelley, F.M. Rosenbloom, J.F. Henderson and J.E. Seegmiller, A specific enzyme defect in gout associated with overproduction of uric acid., Proc. Natl. Acad. Sci. USA 57:1735 (1967).
4. D.G. Sculley, P.A. Dawson, B.T. Emmerson, and R.B. Gordon, A review of the molecular basis of hypoxanthine-guanine phosphoribosyltransferase (HPRT) deficiency, *Hum. Genet.* 90:195 (1992).
5. N. Ogasawara, J.T. Stout, H. Goto, S. Sonta, A. Matsumoto, and C.T. Caskey, Molecular analysis of a female Lesch-Nyhan patient., *J. Clin. Invest.* 84, 1024–1027 (1989).
6. Y. Yamada. H. Goto, K. Suzumori, R. Adachi, and N. Ogasawara, Molecular analysis of five independent Japanese mutant genes responsible for hypoxanthine guanine phosphoribosyltransferase (HPRT) deficiency., *Hum. Genet.* 90:379 (1992).
7. Y. Yamada, H. Goto, S. Tamura, and N. Ogasawara, Molecular genetic study of a Japanese family with Lesch-Nyhan syndrome: a point mutation at the consensus region of RNA splicing (HPRT$_{Keio}$)., *Jpn. J. Hum. Genet.* 38:413 (1993).
8. Y. Choi, J.W. Koo, I.S. Ha, Y. Yamada, H. Goto, and N. Ogasawara, Partial hypoxanthine-guanine phosphoribosyltransferase deficiency in two Korean siblings – a new mutation., *Pediatr. Nephrol.* 7:739 (1993).
9. Y. Yamada, H. Goto, T. Yukawa, H. Akazawa, and N. Ogasawara Molecular mechanisms of the second female Lesch-Nyhan patient., *Adv. Exp. Med. Biol.* 370:337 (1995).
10. Y. Yamada, Y. Choi, K.J. Kim, J.W. Koo, I.S. Ha, H. Goto, and N. Ogasawara, Hypoxanthine guanine phosphoribosyltransferase (HPRT) gene mutations in Korean families with Lesch-Nyhan syndrome., *Purine Pyrimidine Metab.* 19:13 (1995).
11. Y. Yamada, K. Suzumori, M. Tanemura, H. Goto, and N. Ogasawara, Molecular analysis of a Japanese family with Lesch-Nyhan syndrome: identification of mutation and prenatal diagnosis., *Clin. Genet.* 50:164 (1996).
12. Y. Yamada, H. Goto, M. Shiomi, T. Yamamoto, K. Higashino, and N. Ogasawara, A novel de novo mutation in HPRT gene responsible for Lesch-Nyhan syndrome (HPRT$_{Osaka}$)., *Jpn. J. Hum. Genet.* 41:427 (1996).
13. J. Gubbay, J. Collignon, P. Koopman, B. Capel, A. Economou, A. Munsterberg, N. Vivia, P. Goodfellow, and R. Lovell-Badge, A gene mapping to the sex-determining region of the mouse Y chromosome is a member of a novel family of embryonically expressed genes., *Nature* 346:245 (1990).
14. S. Nagafuchi, S. Seki, Y. Nakahori, T. Yamura, H. Numabe, and Y. Nakagome, PCR detection of structurally abnormal Y chromosomes., *Jpn. J. Hum. Genet.* 37:187 (1992).
15. S. Fujimori, N. Kamatani, Y. Nishida, N. Ogasawara and I. Akaoka, Hypoxanthine guanine phosphoribosyltransferase deficiency: nucleotide substitution causing Lesch-Nyhan syndrome identified for the first time among Japanese. *Hum. Genet.* 84:483 (1990).
16. S. Fujimori. T. Tagaya, N. Yamaoka, H. Saito, K. Suzuki, N. Kamatani, and I. Akaoka, Direct evidence for a hot spot of germline mutation at HPRT locus., *Adv. Exp. Med. Biol.* 370:679 (1995).
17. B.L. Davidson, S.A. Tarle, M. Van Antwerp, D.A. Gibbs, R.W.E. Watts, W.N. Kelley, and T.D. Palella, Identification of 17 independent mutations responsible for human hypoxanthine-guanine phosphoribosyltransferase (HPRT) deficiency., *Am. J. Hum. Genet.* 48:951 (1991).
18. S.A. Tarle, B.L. Davidson, V.C. Wu, F.J. Zidar, J.E. Seegmiller, W.N. Kelley, and T.D. Palella, Determination of the mutation responsible for the Lesch-Nyhan syndrome in 17 subjects., *Genomics* 10:499 (1991).
19. R.A. Gibbs, P.-N. Nguyen, and A. Edwards, Multiplex DNA deletion detection and exon sequencing of the hypoxanthine phosphoribosyltransferase gene in Lesch-Nyhan families., *Genomics* 7:235 (1990).
20. R.A. Gibbs, P.-N. Nguyen, L.J. McBride, S.M. Koepf and C.T. Caskey, Identification of mutations leading to the Lesch-Nyhan syndrome by automated direct DNA sequencing of in vitro amplified cDNA. Proc. Natl. Acad. Sci. USA 86:1919 (1989).

42

REGULATION OF HUMAN PRS ISOFORM EXPRESSION

Michael A. Becker, William Taylor, Patrick R. Smith, and Maqbool Ahmed

Department of Medicine
University of Chicago
Chicago, Illinois

PRPP is a substrate in the synthesis of virtually all nucleotides[1] and is also an important regulator of the *de novo* pathways of purine[2] and pyrimidine[3] nucleotide synthesis. Formation of PRPP is catalyzed in eukaryotes by a family of PRPP synthetase (PRS; EC2.7.6.1) isoforms.[4–6] The substrates in the reaction are MgATP and Rib-5-P, and Mg^{2+} and P_i are essential activators. Inhibitors of PRS activity include: purine, pyrimidine, and pyridine nucleotides; reaction products; and 2,3-diphosphoglycerate.[4,5,7] Human erythrocyte PRS is composed of a polypeptide subunit of 34.5 kD which undergoes concentration-dependent and effector-mediated reversible aggregation *in vitro* to active forms of 16 and 32 subunits.[4,8] The primary structure and kinetic and physical properties of purified human erythrocyte PRS are those of the purified recombinant PRS1 isoform.

Several lines of evidence indicate that synthesis of PRPP in mammalian cells is regulated in a complex manner that includes, but is not limited to, allosteric control of PRS enzymatic activity.[1,2,4,5,9–14] The existence of multiple PRS isoforms with differing kinetic properties,[4,5] apparent alterations in PRS activity in response to mitogens[12] and changes in intracellular divalent cation concentrations,[11] and tissue-specific differences in abundance of PRS transcripts[9] and isoforms suggest additional mechanisms of control of PRPP production, involving the selective expression of genes encoding PRS[9,10] and perhaps modifications of the structure or functional organization of the gene products by interaction with one another[8] or with specific PRS-associated modifier proteins.[13,14] For example, Professor Tatibana and colleagues have cloned and characterized the cDNAs for two PRS-associated proteins[13,14] that bind PRS isoforms and inhibit enzyme activity. Whether these interactions affect intracellular PRPP production is unknown, but they raise the possibility of substantial complexity in postranslational regulation of PRS isoform expression.

This presentation focusses on specific processes and potential mechanisms involved in control of PRS isoform expression in human cells. Some of this information comes from study of inherited PRS superactivity and is addressed after a review of the molecular genetics of human PRS isoforms.

Purine and Pyrimidine Metabolism in Man IX,
edited by Griesmacher *et al.* Plenum Press, New York, 1998.

Multiple isoforms of PRS were first identified by Professor Tatibana and colleagues[6,15] who cloned and sequenced 3 distinct PRS cDNAs (numbered 1, 2, and 3) encoding highly homologous polypeptides of identical length. Each human and rodent PRS cDNA is encoded by a separate *PRPS* gene: human *PRPS1* and *PRPS2* map to different regions of the X chromosome and are widely expressed;[16] *PRPS3* maps to human chromosome 7 and appears to be transcribed only in the testes.[6,9] X-linked human PRS1 and PRS2 cDNAs show 80% nucleotide sequence identity throughout their 954 bp translated regions but show no homology in the corresponding 5' and 3' untranslated regions. PRS1 and PRS2 cDNAs hybridize with transcripts of 2.3 and 2.7 kb, respectively.[16] Human organs, tissue, and cell lines contain both PRS1 and PRS2 transcripts and isoforms, but the relative abundances of these gene products vary with the cell source.[17] Studies of tissue-specific expression of *PRPS1* and *PRPS2* genes indicate that *PRPS1* may be a constitutively expressed gene[4,10,12] while *PRPS2* expression may be responsive to mitogenic stimulation and/or transformation.[12,17] Both X-linked *PRPS* genes exceed 30 kb and contain 7 exons with virtually identical exon-intron borders. 5'-Promoter regions of the genes are, however, structurally distinct.

Normal human PRS1 and PRS2 cDNAs have been expressed in *E. coli,* and purified recombinant PRS isoforms have been compared.[4] Despite 95% amino acid homology, recombinant PRS1 and PRS2 isoforms differ in several physical and kinetic properties, including: thermal stability and sensitivity to subunit disaggregation and inactivation in the absence of MgATP and Mg^{2+}; isoelectric points; pH optima; and kinetic constants for substrates, activators, and inhibitors.[4] The difference in isoelectric points of PRS1 and PRS2 has allowed the development of an isoelectric focusing (IEF)-immunoblotting method for separation and quantitation of the isoforms in tissue and cell samples.[17]

Inherited superactivity of PRS is an X chromosome-linked disorder associated with gout, purine nucleotide and uric acid overproduction and, in some families, neurodevelopmental impairment.[18,19] The metabolic aspects of PRS superactivity are consequences of increased availability of PRPP, leading to activation of purine nucleotide and uric acid biosynthesis,[2,18] but the kinetic defects underlying PRS superactivity are heterogeneous and include: regulatory defects in which allosteric control of PRS activity by purine nucleotides and Pi is altered;[19] and catalytic superactivity, characterized by increased maximal reaction velocity but otherwise normal kinetic constants.[17] In general, individuals in whom altered allosteric regulation comprises the sole or a major portion of the kinetic defect in PRS show the most severe biochemical[2] and clinical[18] phenotypes. The genetic heterogeneity suggested by differences in the kinetic abnormalities and phenotypic expressions of inherited PRS superactivity has been confirmed: point mutations in the translated region of *PRPS1* provide the genetic basis for altered allosteric control of PRS activity;[19] in contrast, PRS catalytic superactivity reflects altered regulation of the expression of the normal PRS1 isoform.[17]

RT-PCR analysis of patient and normal fibroblast and lymphoblast RNA identified single base substitutions in the PRS1 cDNAs derived from each of 6 unrelated male patients with superactive, purine nucleotide-resistant PRSs.[19] In each instance, the base change in PRS1 cDNA predicted a single amino acid substitution in PRS1, ranging from amino acid residue 51 to 192 of the 317 residue polypeptide. The functional significance of the mutation predicted for each PRS1, was established by demonstrating that each patient recombinant PRS1 showed the pattern and magnitude of aberrant allosteric responses to purine nucleotide inhibitors and Pi characteristic of PRS in cells from that patient.[19] Thus, the genetic basis of inherited PRS superactivity associated with altered allosteric regulation is point mutation in the PRS1-coding region of the *PRPS1* gene.

These variant PRS1s provide some insight into the molecular basis of regulation of PRS1 activity. The fact that single residue substitutions dispersed over a major portion of PRS1 result in diminished responsiveness to inhibition by ADP and GDP suggests that these compounds share a complex and probably extensive allosteric nucleotide inhibitory mechanism and that most or all of the mutations alter the transmission of allosteric effects to the active site of PRS1 rather than the primary structure of the nucleotide binding residues in the allosteric site. The MgATP binding properties of the recombinant mutant PRS1s are normal,[19] indicating that the active site is not directly altered. Overall, multiple residues and regions of the PRS1 polypeptide appear involved in determining isoform activity and sensitivity to GDP and ADP inhibition. Concurrent impairment of nucleotide inhibition and enhancement of Pi responsiveness in mutant human PRS1s[19] suggests that these effectors share a mechanism in which inhibited conformations favored by nucleotide binding are in equilibrium with more active conformations favored by Pi binding. In such a model, mutant PRS1s would likely be altered in residues involved in stabilizing the inhibited conformations, rather than directly in Pi or nucleotide binding.

The kinetics of ADP and GDP inhibition of purified recombinant normal and mutant PRS1s[19] also provide information regarding mechanisms by which inhibition is effected *in vivo*. ADP inhibition of normal PRS1 involves noncompetitive as well as competitive mechanisms, but only a competitive mechanism of ADP inhibition is demonstrable for recombinant mutant PRS1s. Despite a potent mechanism of competitive inhibition, however, mutant PRS1s clearly resist inhibition by ADP and the noncompetitive inhibitor GDP in the presence of saturating concentrations of ATP. These findings can be reconciled by the view that point mutations in PRS1 disrupt a major noncompetitive component of the allosteric nucleotide inhibitory mechanism without altering the ATP substrate binding region and that the competitive mechanism exerts little or no control on PRS1 activity under physiologic conditions in which ATP concentrations are saturating for the enzyme.

When RT-PCR analysis of PRS1and PRS2 transcripts was applied to 6 patients with X-linked catalytic superactivity of PRS, no alterations in the sequences of the translated regions of PRS1 and PRS2 cDNAs were identified,[17] showing that mutations in the translated region of X-linked *PRPS1* or *PRPS2* genes did not account for this class of PRS superactivity. These results were, however, compatible with the idea that inherited PRS catalytic superactivity involves increased expression of one or both X-linked PRS isoforms with normal primary sequence. Utilizing Northern blot analysis and quantitative IEF-immunoblotting, we compared levels of X chromosome-linked PRS transcripts and isoforms in cultured fibroblasts from normal and affected individuals in order to determine whether one or both PRS isoforms was involved in expression of catalytic superactivity and to limit the range of specific mechanisms likely to be involved in PRS overexpression.

In these studies,[17] we found that PRS catalytic superactivity is associated with 2- to 6-fold increased fibroblast concentrations of PRS1 isoform with physical and catalytic properties indistinguishable from those of normal PRS1. Total PRS activities in both normal and patient cells closely correlated with total PRS isoform contents (>80% of which is PRS1), so that superactivity is clearly associated with increased PRS1 content. In fact, PRS isoform specific activities (PRS activity per mg PRS isoform) in cell extracts are comparable in normal and patient cells and close to those of purified recombinant normal PRS isoforms.[4] PRS catalytic superactivity is thus not due to preferential inhibition of normal PRS, such as by specific PRS-associated proteins. In addition, relative steady state levels of PRS1 transcripts are increased 2- to 5-fold in patient cells, and PRS2 isoform and transcript levels are comparable in normal and patient cells. Together, these findings demonstrate that the enzymatic basis of PRS catalytic superactivity is selectively increased concentration of the normal PRS1 isoform. The accompanying increases in PRS1 transcript levels indicate that PRS1 overexpression is determined at least in major part by an altered pretranslational mechanism of control of *PRPS1* gene expression.

We have undertaken several approaches to identifying the aberrant pretranslational mechanism of *PRPS1* expression in PRS catalytic superactivity. Southern blots of restriction enzyme-digested genomic DNA isolated from normal male and patient fibroblasts were probed with PRS1 cDNAs and showed no differences in sizes or intensities of hybridizing bands, excluding *PRPS1* gene amplification as a likely basis for the increased *PRPS1* expression. Patient and normal fibroblast *PRPS1* genomic DNAs corresponding to the 5' (137 bp) and 3' (997 bp) untranslated portions of the PRS1 transcript were amplified and were identical in sequence.[17] In conjunction with the identity between normal and patient-derived PRS1 cDNA translated sequences, this finding excludes altered structure of the mature PRS1 mRNA as a determinant of differential *PRPS1* expression in PRS catalytic superactivity. In additional studies, relative PRS1 transcript levels were determined in lymphoblasts during incubation with the RNA polymerase II inhibitor actinomycin D (5 μg/ml). PRS1 mRNA stability in normal and patient cells is nearly identical: half-lives of the PRS1 transcript are 10.8 ± 1.2 and 11.1 ± 1.4 hours (mean $\pm$ SD in 4 experiments), respectively, in normal and patient cells. In contrast, rates of *PRPS1* transcription (by nuclear runoff analysis) were 2- to 4-fold greater relative to those of *GAPDH* and *PRPS2* transcription in patient lymphoblasts and fibroblasts than in corresponding normal cells (Figure 1). Taken together, these findings suggest that the altered pretranslational mechanism underlying inherited PRS catalytic superactivity involves, at least in part, increased rates of *PRPS1* gene transcription in the cells of affected individuals. Selectively increased *PRPS1* transcription rates should reflect altered structure and/or function of 5' *PRPS1* promoter region. However, we have found no differences in the 850 bp of DNA immediately 5' to the transcription initiation site of *PRPS1* when the promoter region sequences of 2 patient and 2 normal *PRPS1*s were compared, raising the possibility of altered regulation of structurally normal *PRPS1* promoters in PRS catalytic superactivity. Extended cloning and sequencing of the 5' DNA flanking *PRPS1* and measurements of *PRPS1* promoter activities by reporter gene analysis are in progress and should permit resolution of this question and perhaps identification of the molecular mechanism ultimately underlying increased PRS1 isoform expression in catalytic superactivity.

To summarize the results of studies of PRS isoform expression in human PRS superactivity, evidence has been provided for functionally significant metabolic consequences of defective allosteric regulation of PRS1 isoform activity.[18,19] In addition, altered control of levels of normal PRS1 result in purine nucleotide and uric acid overproduction despite

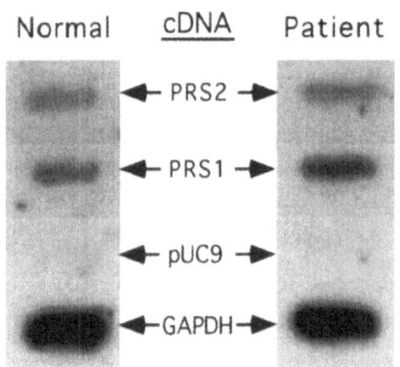

Figure 1. Slot blot analysis of rates of transcription of *PRPS1*, *PRPS2*, and *GAPDH* determined in normal and patient lymphoblasts by nuclear runoff. Arrows indicate sites of cDNA binding to nitrocellulose filters, with pUC9 representing the vector control. Radioactivities in the pUC9 slots were subtracted from those in the slots containing the corresponding cDNAs of interest. Rates of *PRPS* gene transcription expressed as percent of *GAPDH* transcription in the corresponding nuclear preparation are: normal lymphoblasts *PRPS1* 6.6, *PRPS2* 4.0; patient lymphoblasts *PRPS1* 14.9, *PRPS2* 4.9.

intact allosteric control,[17] and the fact that PRS catalytic superactivity is associated with increased rates of transcription of *PRPS1* identifies transcription as an important determinant of expression of this isoform.

With regard to the normal control of PRS isoform expression, evidence points to both tissue-differential and isoform-specific regulation of *PRPS1* and *PRPS2*.[4,9,10,17] We have confirmed the presence of PRS activity in all human and mouse organs, tissues, and cell lines studied and have examined the abundance of PRS1 and PRS2 transcripts and isoforms in many of the human tissues by Northern blot analysis and IEF-immunoblotting, respectively. Although there is wide tissue-specific variation in expression of *PRPS1* and *PRPS2*, with the latter more prominently displayed in organs (like testis and spleen) populated largely by rapidly dividing cells,[9,12] in all instances, PRS transcript levels are closely correlated with levels of the corresponding isoform.

For example, normal human fibroblasts and B lymphoblasts contain comparable levels of PRS1 transcript and isoform, but the corresponding PRS2 levels are substantially greater (4- to 6-fold) in lymphoblasts.[17] Moreover, PRS2 transcript and isoform levels in normal human peripheral blood lymphocytes are considerably lower than those of PRS1, but 72 h after addition of concanavalin A or after EB virus lysate-mediated establishment of B lymphoblast lines, comparable PRS1 and PRS2 transcript and isoform levels are demonstrable.[12] These findings support the concept of isoform-specific and tissue-differential expression of *PRPS1* and *PRPS2* gene products; and the coordinate relationship between transcript and isoform levels indicate that regulation is at least in part carried out at a pretranslational level. Whether the mechanisms involved in normal regulation operate at the structurally divergent 5′ promoter sequences of X-linked *PRPS* genes is uncertain. However, in preliminary studies of *PRPS1* and *PRPS2* promoter activities, measured in human and mouse cells transiently transfected with human *PRPS* promoter/reporter gene constructs, relative promoter activities for these genes correlate closely with the relative abundances of PRS transcripts and isoforms in the respective cell types, supporting the likelihood of transcriptional regulation.

Although some advances have thus been made in the study of the control of PRS isoform expression, much remains to be learned regarding: the molecular bases of the mechanisms involved; the existence and nature of additional levels of control of *PRPS1* and *PRPS2* expression; and the specific functions subserved by and the potential relationships between X-linked *PRPS* genes and their respective PRS gene products.

ACKNOWLEDGMENTS

This research was supported by USPHS Grant DK-28554 and by a grant from the Greater Chicago Chapter of the Arthritis Foundation. The authors acknowledge excellent manuscript preparation by Ms. Sandra Crane.

REFERENCES

1. Becker, M.A., Raivio, K.O., and Seegmiller, J.E., Synthesis of phosphoribosylpyrophosphate in mammalian cells. Adv. Enzymol. Relat. Areas. Mol. Biol. 49:281 (1979).
2. Becker, M.A., and Kim, M., Regulation of rates of purine synthesis *de novo* by purine nucleotides and phosphoribosylpyrophosphate. J. Biol. Chem. 262:14531 (1987).
3. Tatibana, M. and Shigesada, K., Two carbamyl phosphate synthetases of mammals: Specific roles in control of pyrimidine and urea biosynthesis. Adv. Enzyme Regul. 10:249 (1972).

4. Nosal, J.M., Switzer, R.L., and Becker, M.A., Overexpression, purification, and characterization of recombinant human 5-phosphoribosyl-1-pyrophosphate synthetase isozymes I and II. J. Biol. Chem. 268:10168 (1993).

5. Ishijima, S., Kita, K., Ahmad, I., Ishizuka, T., Taira, M., and Tatibana M., Expression of rat phosphoribosylpyrophosphate synthetase subunits I and II in *Escherichia coli.* Isolation and characterization of the recombinant isoforms. J. Biol. Chem. 266:15693 (1992).

6. Taira, M., Iizasa, T., Shimada, H., Kudoh, J., Shimizu, N., and Tatibana M., A human testis-specific mRNA for phosphoribosylpyrophosphate synthetase that initiates from a non-AUG codon. J. Biol. Chem. 265: 16491 (1990).

7. Fox, I.H., and Kelley, W.N., Human phosphoribosylpyrophosphate synthetase: Kinetic mechanism and end product inhibition. J. Biol. Chem. 247:2126 (1972).

8. Meyer, L. J., and Becker, M.A., Human erythrocyte phosphoribosylpyrophosphate synthetase. Dependence of activity on state of subunit association. J. Biol Chem. 252:3919 (1977).

9. Taira, M., Iizasa, T., Yamada, K., Shimada, H., and Tatibana, M., Tissue-differential expression of two distinct genes for phosphoribosylpyrophosphate synthetase and existence of the testis-specific transcript. Biochim. Biophys. Acta 1007:203 (1989).

10. Ishizuka, T., Iizasa, T., Taira, M., Ishijima, S., Sonoda, T., Shimada, H., Nagatake, N., and Tatibana, M., Promoter regions of the human X-linked housekeeping genes PRPS1 and PRPS2 encoding phosphoribosylpyrophosphate synthetase subunit I and II isoforms. Biochim. Biophys. Acta 1130:139 (1992).

11. Ishijima, S., Kita, K., and Tatibana, M., External Mg^{2+}-dependent early stimulation of nucleotide synthesis in Swiss 3T3 cells. Am. J. Physiol. 257 (Cell Physiol. 26): C1113 (1989).

12. Becker, M.A., Heidler, S.A., Nosal, J.M., Switzer, R.L., LeBeau, M.M., Shapiro, L.J., Palella, T.D., and Roessler, B.J., Human phosphoribosylpyrophosphate synthetase (PRS)2: An independently active, X chromosome-linked PRS isoform. Adv. Exp. Med. Biol. 309B:129 (1991).

13. Kita, K., Ishizuka, T., Ishijima, S., Sonoda, T., and Tatibana, M., A novel 39-kDa phosphoribosylpyrophosphate synthetase-associated protein of rat liver. J. Biol. Chem. 269:8334 (1994).

14. Sonoda, T., Ishizuka, T., Kita, K., Ishijima, S., and Tatibana, M., Cloning and sequencing of rat cDNA for the 41-kDa phosphoribosylpyrophosphate synthetase-associated protein has a high homology to the catalytic subunits and the 39-kDa associated protein. Biochim. Biophys. Acta 1350:6 (1997).

15. Taira, M., Ishijima, S., Kita, K., Yamada, K., Iizasa, T., and Tatibana, M., Nucleotide and deduced amino acid sequences of two distinct cDNAs for rat phosphoribosylpyrophosphate synthetase. J. Biol. Chem. 262:14867 (1987).

16. Becker, M.A., Heidler, S.A., Bell, G.I, Seino, S., LeBeau, M.M., Westbrook, C.A., Neuman, W., Shapiro, L.J., Mohandas, T.K., Roessler, B.J., and Palella, T.D., Cloning of cDNAs for human phosphoribosylpyrophosphate synthetases 1 and 2 and X chromosome localization of *PRPS1* and *PRPS2* genes. Genomics 8:555 (1990).

17. Becker, M.A., Taylor, W., Smith, P.R., and Ahmed, M., Overexpression of the normal phosphoribosylpyrophosphate synthetase isoform 1 underlies catalytic superactivity of human phosphoribosylpyrophosphate synthetase. J. Biol. Chem. 271:19894 (1996).

18. Becker, M.A., Puig, J.G., Mateos, F.A., Jimenez, M.L., Kim, M., and Simmonds, H.A., Inherited superactivity of phosphoribosylpyrophosphate synthetase: Association of uric acid overproduction and sensorineural deafness. Am. J. Med. 85:383 (1988).

19. Becker, M.A., Smith, P.R., Taylor, W., Mustafi, R., and Switzer, R.L., The genetic and functional basis of purine nucleotide feedback-resistant phosphoribosylpyrophosphate synthetase superactivity. J. Clin. Invest. 96:2133 (1995). Biophys. Acta 1207:126 (1994).

43

STRUCTURE AND FUNCTIONAL RELATIONSHIPS IN HUMAN *pur* H

G. Peter Beardsley,[1,2*] Elizabeth A. Rayl,[1] Karen Gunn,[2] Barbara A. Moroson,[1] Helen Seow,[2] Karen S. Anderson,[2] James Vergis,[3] Karen Fleming,[3] Steven Worland,[4] Brad Condon,[4] and Jay Davies[4]

[1]Department of Pediatrics
[2]Department of Pharmacology
[3]Department of Molecular Biophysics and Biochemistry
Yale University School of Medicine
New Haven, Connecticut 06520-8064
[4]Agouron Pharmaceuticals Inc.
La Jolla, California

1. THE *pur* H GENE AND ITS PRODUCT, Pur H

In both prokaryotic and eukaryotic cells the *pur* H gene encodes a bifunctional protein which carries the penultimate and the final activities of the *de novo* purine nucleotide biosynthetic pathway. These two steps are shown in Figure 1.

We wondered if the invariant bifunctionality of the *pur* H gene product throughout nature might have an important structural basis. This, along with the possibility that one of the activities might be an appropriate target for drug development, led us to more closely examine the structural and functional properties of this protein.

The cloning, sequencing, expression and purification of human *pur* H have previously been reported,[1] and will not be further discussed here.

2. CHROMOSOMAL LOCATION AND STRUCTURE OF THE HUMAN *pur* H GENE

FISH analysis using the cDNA from the cloning experiments above located the human *pur* H gene (h*pur* H) on the long arm of chromosome 2, between bands q34 and q35. Sequencing at the genomic level is as yet incomplete. One 278 base pair intron has thus

---

* Author to whom inquiries should be addressed.

Purine and Pyrimidine Metabolism in Man IX,
edited by Griesmacher *et al.* Plenum Press, New York, 1998.

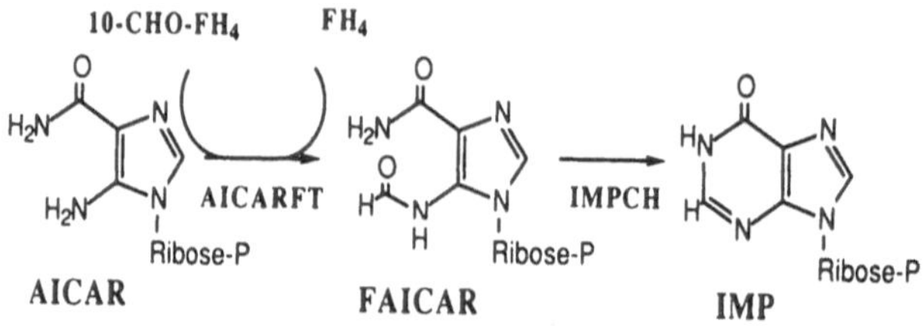

Figure 1. Reactions catalyzed by *pur* H.

far been detected near the 5' end of h*pur* H. The human genome project has re-named h*pur* H as ATIC (*A*ICAR *t*ransformylase/*I*MP *c*yclohydrolase) in order to avoid confusion with the nomenclature of some purine-rich transcription factor binding sites.

3. DOMAIN STRUCTURE OF hPurH

In our thinking about how the two activities of this protein might be arranged, we considered three possibilities:

3.1. Two independent domains, connected by a non-functional "linker" region. This is the simplest model, and would be consistent with what has been found in many other multifunctional proteins, including others in the *de novo* purine nucleotide biosynthesis pathway.

3.2. Two semi-independent domains connected by a functional element which could help "channel" the intermediate, FAICAR, from the AICARFT domain to the IMPCH domain. Such arrangements have been found in tryptophan synthetase and other well-investigated enzymes. A need for such channeling might underlie the invariant bifunctional nature of this protein in all organisms.

3.3. A single binding domain for the substrates of both reactions, *i.e.* AICAR and FAICAR, with two catalytic domains which could act sequentially. As the two reactions are at the end of a long synthetic pathway, the structural differences between the substrates for each are minimal and such an arrangement might be feasible. Many DNA polymerases, for example, have this kind of organization, with separate polymerase and exonuclease catalytic domains, and single DNA duplex binding regions.

3.4. Truncation Mutants. In order to distinguish between these and other possibilities, a series of N-terminal and C-terminal truncation mutants were constructed, expressed and purified. These are depicted in Figure 2.

They include two non-overlapping regions, MIN. IMPCH (amino acids 1–198) which has full IMPCH activity but lacks detectable AICARFT activity, and MIN.AICARFT (amino acids 199–591) which has full AICARFT activity but lacks IMPCH activity. These results strongly suggest that there are two domains which are structurally independent, *i.e.* model 3.3 can not be correct. The N-terminus of the AICARFT domain must lie near amino acid 199, as all shorter mutants we have made were inactive. The C-terminal limit of the IMPCH domain must lie at or before amino acid 198. Since MIN.IMPCH is the shortest N-terminal truncation mutant we have investigated however, we can not specify the extent of the IMPCH domain more precisely. The region from amino acid 1 through 198 may include, in addition to IMPCH activity, a "linker" region or the structural basis for a func-

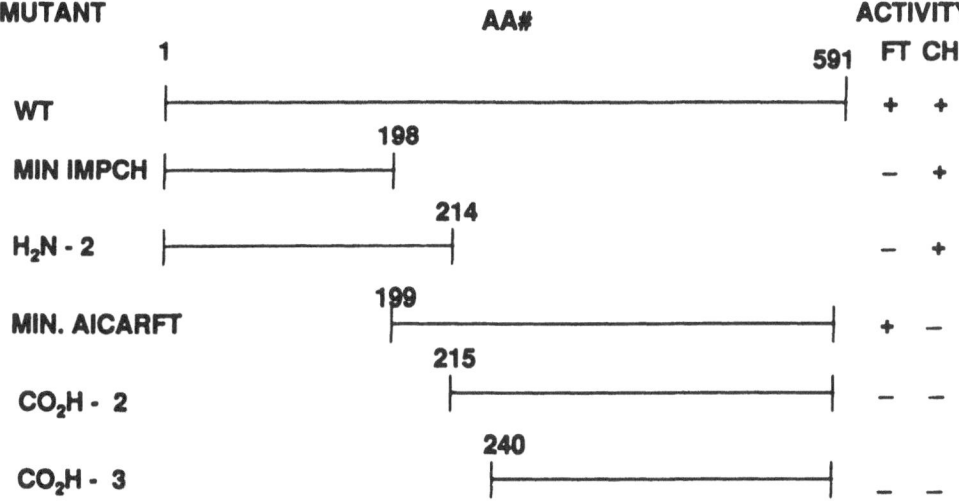

Figure 2. Human purH truncation mutants.

tional relationship between the two domains. Based on the above findings we hypothesized a rough model for human PurH which is shown in Figure 3.

4. REACTION MECHANISM

4.1. Steady State Kinetics

Investigation of the AICARFT steady state kinetics of the hPurH expressed in E. coli, showed kinetic constants which were basically identical to those of the enzyme

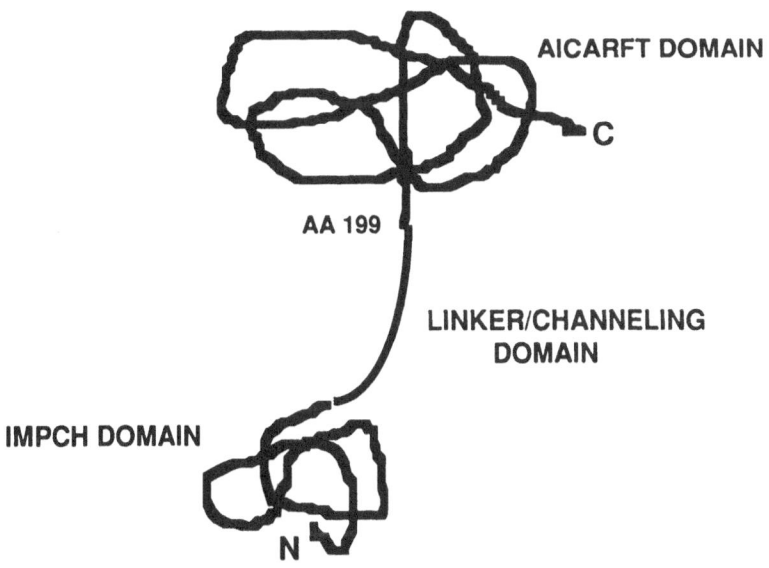

Figure 3. Hypothetical model of hPurH.

extracted from human cells.[1] These were also the same as values obtained using the MIN. AICARFT mutant. The kinetic constants for the IMPCH activity could not be compared, as the spectrophotometric assay employed lacks sufficient sensitivity to measure k_m values in the appropriate range, below 1uM.

4.2. Inhibition Kinetic Studies

Experiments of the Cleland type were performed using (6S)-DDATHF as a pure competitive inhibitor of the folate co-factor. The results of these experiments (not shown) demonstrated the reaction to be of the ordered, sequential type, with 10-formyl tetrahydrofolate binding first, followed by AICAR.

4.3. pH Profile

Studies of the reaction rate as a function of pH at constant ionic strength showed a broad maximum in the alkaline range at pH 8.2.

4.4. Transient Phase Kinetics

In order to understand the details of the PurH reaction mechanism(s), transient phase kinetic studies are necessary. A method for rapid quenching of the reaction was developed, as was a TLC system for separation of reaction substrates, products and intermediates. Results of a typical "burst" experiment in which substrates, in excess over enzymes, are exposed to each other for brief periods (0–1.2 sec.), quenched and subsequently analyzed are shown in Figure 4.

This plot of the product (IMP) concentration *vs.* time shows an initial steep slope over the first 0.1 sec, followed by a 40-fold more shallow slope over the next 1.0 sec. This

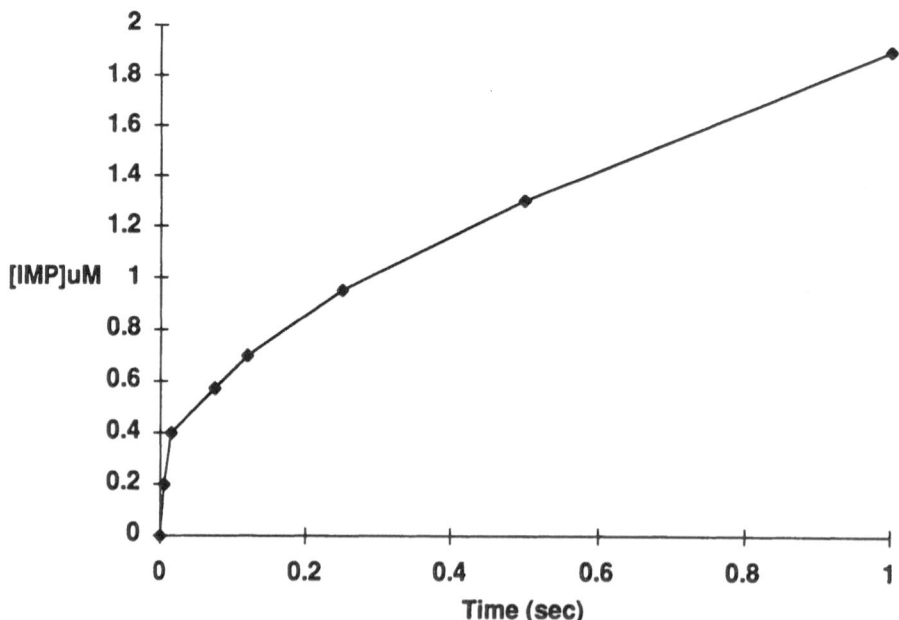

Figure 4. Pur H burst experiment.

pattern is that of a typical "burst", indicating that a step later than the chemistry step(s) in the reaction is rate limiting. Under the conditions utilized, only AICAR and IMP were found in the reaction mixture, and none of the intermediate, FAICAR, was detected.

5. CRYSTAL STRUCTURE OF Pur H

The crystal structure of Pur H is that of a homodimer. Each monomer includes a smaller, N-terminal globular portion, a 30 amino acid helical region, and a larger C-terminal globular region. The monomers are remarkably similar to the structure hypothesized in Figure 3.

Putative binding sites for AICAR and FAICAR can be located in the C-terminal and N-terminal globular domains of each monomer. A single binding site for 10-formyl tetrahydrofolate appears to be formed using elements from each monomer.

6. EQUILIBRIUM SEDIMENTATION STUDIES

Because we were concerned that the dimer of hPurH might represent only an artifact of the crystallization conditions, we carried out an independent study of its size in the solution phase using equilibrium sedimentation. The results of these studies showed that in the absence of either substrate, hPurH is in a monomer-dimer equilibrium with a K_D of 0.55 uM, strongly favoring the dimer.

7. CATALYTIC SITE RESIDUES

Examination of the crystal structure suggests a number of amino acid residues as candidates for catalytically active elements. Among these, we have thus far completed studies of two, Lysine 265 and Histidine 266. These mutants were constructed using PCR methodoloogy, expressed, and purified. The possibility that improperly folded products were produced was checked by comparison of their circular dichroism spectra with that of the wild type (WT). In each case the spectra were in essence, identical. The results of activity and binding studies are summarized in Table 1.

Both mutants had undetectable AICARFT activity, although each retained IMPCH activity. Furthermore, each also retained the capacity to bind to an AICAR-sepharose column. Although it is tempting to conclude that both K265 and H266 are involved in reaction catalysis, rather than binding of substrates, care must be taken in this regard as potential effects on binding of the reduced folate cofactor have not yet been investigated. Furthermore, the AICAR-sepharose binding observed may have occurred through inter-

Table 1. H purH mutants

| Protein | Activity | | Binding |
	AICARFT	IMPCH	AICAR-sepharose
WT	+	+	+
K265A	–	+	+
H266A	–	+	+

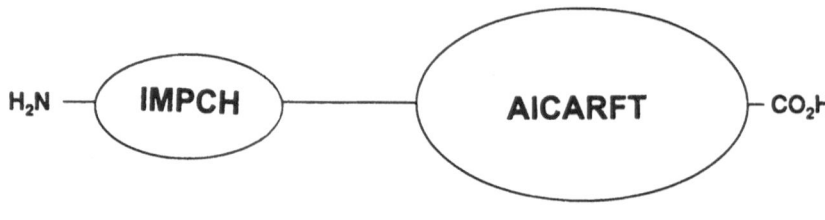

Figure 5.

actions with the FAICAR site in the N-terminal IMPCH domain. The structural differences between AICAR and FAICAR are relatively small, and it is certainly possible that AICAR could be accommodated in a region which normally binds FAICAR.

8. SUMMARY

1. The human *pur* H (ATIC) gene encoding a bifunctional protein, hPurH, which carries the penultimate and final enzymatic activities of the purine nucleotide synthesis pathway, AICARFT & IMPCH, has been cloned and sequenced. The gene product, hPurH has been overexpressed in E. coli, purified to homogeneity and crystallized.
2. The human *pur* H gene lies on chromosome 2, between band q34 and q35. There is at least one intron of 278 bp near the 5′ end.
3. Truncation mutant studies demonstrate two non-overlapping functional domains in the protein arranged as indicated in Figure 5. The existence of a linker or interaction region between the catalytic domains remains to be establiished.
4. Cleland-type kinetic inhibition experiments indicate that the AICARFT reaction is of the ordered, sequential type with the reduced folate cofactor binding first.
5. The reaction has a broad pH optimum in the alkaline range, with a maximum at about pH 8.2.
6. Preliminary transient phase kinetic studies show the presence of a "burst" indicating that a late step in the reaction sequence is rate limiting.
7. A PurH crystal structure is that of a dimer, with a putative single binding site for the reduced folate cofactor formed using elements from each of the monomer subunits. Probable binding sites for AICAR and FAICAR can be identified on each monomer.
8. Equilibrium sedimentation studies show hPurH apoprotein to be a monomer:dimer equilibrium mixture with a k_D of 0.55 uM.
9. The crystal structure has permitted identification of a number of candidate amino acid residues likely to be involved in catalysis and/or substrate binding. Among these, we have thus far completed studies on two, Lysine 265 and Histidine 266. These appear to be critically involved in the AICARFT reaction, although whether their role(s) are in catalysis or binding remains to be determined.

REFERENCE

1. Rayl, E, Moroson, BA, and Beardsley, GP. (1996) J. Biol. Chem. 271:2225–2233.

44

LOCALIZATION OF CARBAMOYL-PHOSPHATE SYNTHETASE II (CPS II) AND ASPARTATE CARBAMOYLTRANSFERASE (ACT) GENES IN *TRYPANOSOMA CRUZI* CHROMOSOMAL DNA

Takeshi Nara, Guanghan Gao, Junko Nakajima-Shimada, and Takashi Aoki

Department of Parasitology
Juntendo University School of Medicine
Hongo, Bunkyo-ku, Tokyo 113, Japan

INTRODUCTION

The de novo pathway for pyrimidine nucleotide biosynthesis represents one of the oldest metabolic pathways, and the six sequential steps leading to the production of UMP remain intact throughout evolution, although the organization of the enzymes involved deviate significantly in prokaryotes, protozoa, fungi, and animals. The first three enzymes in the pathway, CPS II, ACT, and dihydroorotase, occur as a multifunctional protein CAD in higher animals, while in the parasitic protozoan *Trypanosoma cruzi*, the mRNAs for the CPS II and ACT genes possess a spliced leader (SL) and a poly(A) sequence at their 5'- and 3'-terminals, respectively, indicating that these two enzymes are encoded independently (Aoki, 1994; Aoki et al., 1995). Since little is known about the location of closely related genes in *T. cruzi* genomic DNA, we have examined whether the CPS II and ACT genes are linked in the parasite chromosomal DNAs. We report here that two chromosomal DNAs of 1,000 kb and 800 kb in *T. cruzi* contain both of the genes simultaneously, and that the CPS II gene is located 14 kb upstream of the ACT gene. An additional copy of the ACT gene that may not be linked to the CPS II gene occurs in the 800-kb chromosomal DNA.

MATERIALS AND METHODS

Epimastigotes, the culture form of *T. cruzi* (Tulahuene strain), were used as a source of total DNA. Pulsed field gel electrophoresis (PFGE), Southern blot analysis, and construction of a parasitic genomic library were performed essentially as described (Sambrook et al., 1989). To collect genomic clones containing the CPS II or ACT gene, an

Purine and Pyrimidine Metabolism in Man IX,
edited by Griesmacher *et al.* Plenum Press, New York, 1998.

EMBL3 genomic library of *T. cruzi* DNA, which gave a titer of 1×10^6 pfu/ml with an average insert length of 10 kb, was screened using CPS II- or ACT-specific probes; the former probe was a digoxigenin-labeled cDNA that encodes the glutaminase domain of the parasite CPS II, and the latter probe corresponded to a cDNA that encodes full length *T. cruzi* ACT. Details of the experimental procedure will appear elsewhere.

RESULTS AND DISCUSSION

We first examined the localization of the CPS II and ACT genes in chromosomal DNAs from *T. cruzi* (Fig. 1A). PFGE demonstrated the separation of the parasite DNAs, the masses of which ranged from 580 to 2,200 kb. Upon Southern blot analysis, the CPS II- and ACT-specific probes hybridized to two chromosomal DNAs of 1,000 kb and 800 kb, indicating that both genes are located simultaneously on the two DNA molecules. The result also suggests that these DNAs are derived from homologous chromosomes with substantial size differences in *T. cruzi*.

In our preliminary experiments, the digestion of *T. cruzi* total DNA by various restriction enzymes, followed by gel electrophoresis, and hybridization with the CPS II- or ACT-specific probes resulted in a single band (data not shown), suggesting that both genes are present as single copy in the genome of the protozoan parasite. To further confirm the number of copies of these genes, the 1,000- and 800-kb DNAs were separated by PFGE, and the isolated DNAs were digested with Sal I and EcoR I, followed by gel electrophoresis, and then hybridized with the CPS II- or ACT-specific DNA probes (Fig. 1B). The

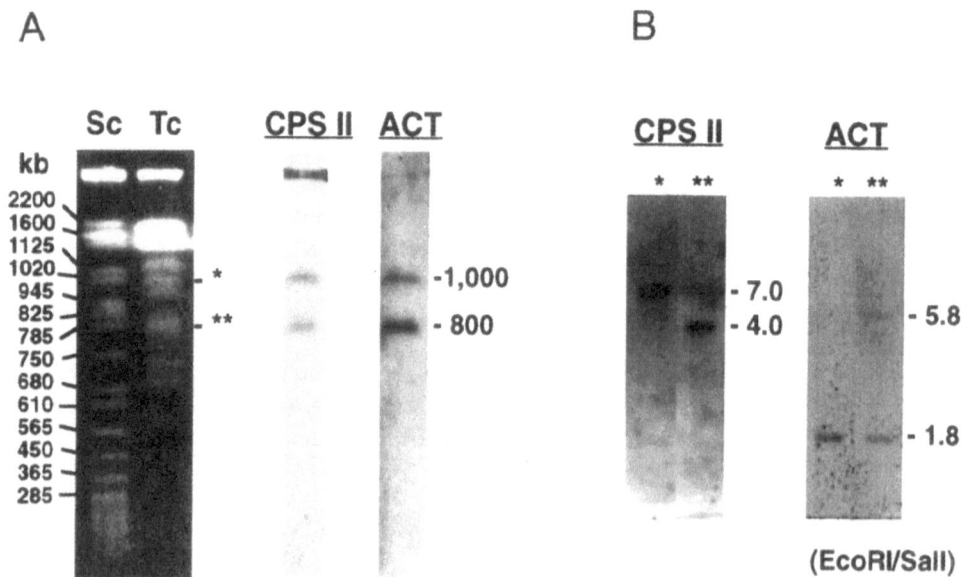

Figure 1. Pulsed field gel electrophoresis (PFGE) analysis of CPS II and ACT genes in *T. cruzi*. Panel A: Chromosomal DNAs from *T. cruzi* (Tc) were separated in a CHEFF apparatus (Bio-Rad) and hybridized with CPS II- or ACT-specific DNA probes. Sc: Chromosomal DNAs from *Saccharomyces cerevisiae* used as a molecular standard. Panel B: After separation of the *T. cruzi* chromosomal DNAs by PFGE, the isolated 1,000-kb (*) and 800-kb (**) DNAs were digested with Sal I and EcoR I, followed by gel electrophoresis, and hybridized with either the CPS II- or ACT-specific probe.

1,000- and 800-kb chromosomal DNAs contained one copy each of the CPS II gene, appearing as 7.0- and 4.0-kb restriction fragments, respectively. In contrast, the 1,000- and 800-kb DNAs harbored one and two copies of the ACT genes, recognized as 1.8-kb and 1.8-kb plus 5.8-kb restriction fragments, respectively.

We have attempted to isolate *T. cruzi* genomic DNA clones that carry both the CPS II and ACT genes, which would provide direct evidence for the close linkage of these genes. For this purpose, the parasite genomic library was screened with the CPS II-specific DNA probe. Of the 11 CPS II-positive clones obtained, however, none cross-hybridized with the ACT-specific DNA probe, indicating that these clones possessed no ACT gene. Since the library had an average insert length of 10 kb (see Materials and Methods), no ACT gene exists within the region approximately 10 kb upstream and downstream of the CPS II gene in these CPS II-positive clones.

Therefore, we rescreened the parasite genomic library with the ACT-specific DNA probe, yielding 23 positive clones from 80,000 plaques examined. These ACT gene-containing clones were classified into 3 groups by EcoR I/Sal I restriction mapping (refer to Fig. 2), A) 8 clones containing the 1.8-kb EcoR I fragment, B) 7 clones containing the 1.8-kb EcoR I fragment and having a different restriction pattern, and C) 8 clones containing the 5.8-kb EcoR I/Sal I fragment. As judged from Fig. 1B, the results imply that the ACT-positive clones in groups A and B are derived from either the 1,000- or 800-kb chromosomal DNA, and that the ACT-positive clones in group C are from the 800-kb chromosomal DNA. These ACT-positive clones again did not cross-hybridize with the CPS II-specific probe.

We then performed a hybridization analysis of the ACT-positive 23 clones using DNA probes corresponding to the extreme upstream and downstream regions of the CPS II-positive clones. A 3-kb probe upstream of the CPS II gene gave no positive reaction, whereas two probes (5 kb and 4 kb) downstream of the CPS II gene hybridized with several of the 23 ACT-positive clones. The results indicate that the CPS II gene exists at an upstream of the ACT gene, namely, both genes are linked in *T. cruzi* chromosomal DNA. Between these two genes, an intervening region should exist whose length is probably in the range of 10 to 20 kb, assuming the average insert length for both CPS II- and ACT-positive clones to be 10 kb.

Finally, an alignment, done to detect overlapping restriction fragments between the CPS II- and ACT-positive clones, revealed that the CPS II gene exists 14-kb upstream of the ACT gene in both the 1,000- and 800-kb *T. cruzi* chromosomal DNAs (Fig. 2). The

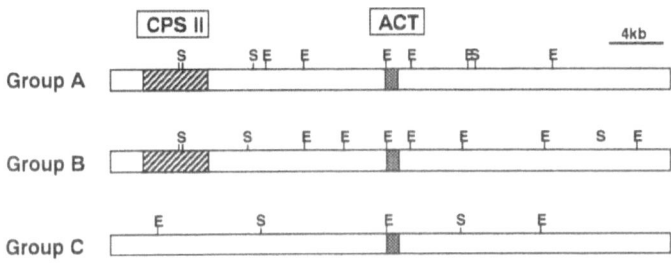

Figure 2. Restriction maps of CPS II and ACT genes from *T. cruzi*. The CPS II gene is illustrated by the shaded box, and the ACT gene by the dotted box. Groups A and B, consisting of 8 and 7 clones, respectively, in which the ACT gene is shown as a 1.8-kb EcoR I fragment, indicate the presence of the CPS II gene 14 kb upstream of the ACT gene in the 1,000-kb or 800-kb chromosomal DNA. Group C, consisting of 8 clones, shows an additional ACT gene, recognized as an EcoR I/Sal I fragment of 5.8 kb, that may not be linked to the CPS II gene and occurs in the 800-kb chromosomal DNA . E, EcoR I; S, Sal I. Scale bar represents 4 kb.

questions of whether these two genes are subjected to polycistronic transcription and whether the 14-kb intervening region contains an open reading frame (possibly for dihydroorotase) are now under investigation. The significance of an additional copy of the ACT gene not linked to the CPS II gene is of marked interest but remains unclear.

ACKNOWLEDGMENTS

This work was supported in part by a Grant-in-Aid for Scientific Research from the Ministry of Education, Science, and Culture of Japan (No. 07457068).

REFERENCES

Aoki, T. (1994): Initial steps of *de novo* pyrimidine nucleotide biosynthesis in parasites and mammalian tissues: Purificatiaon, regulation, adaptation, and evolution. Jap. J. Parasitol., 43: 1–10.
Aoki, T., Simogawara, R., Ochiai, K., Yamasaki, H., and Shimada, J. (1995): Molecular characterization of a carbamoyl-phosphate synthetase II (CPS II) gene from *Trypanosoma cruzi*. Adv. Exp. Med. Biol., 370: 513–516.
Sambrook, J., Fritsch, E.F., and Maniatis, T. (1989): Molecular cloning. A Laboratory Manual, 2nd ed., Cold Spring Harbor Laboratory, Cold Spring Harbor, N.Y.

45

EXPRESSION AND CHARACTERIZATION OF RECOMBINANT BOVINE CYTOSOLIC 5'-NUCLEOTIDASE IMP-GMP SPECIFIC

S. Allegrini,[1] R. Pesi,[2] M. G. Tozzi,[3] and S. Eriksson[1]

[1]Department of Veterinary Medical Chemistry
Swedish University of Agricultural Sciences
The Biomedical Center, Uppsala, Sweden
[2]Dip. Fisiologia e Biochimica
Università di Pisa, Pisa, Italy
[3]Dip. Scienze del Farmaco
Università di Sassari, Sassari, Italy

1. INTRODUCTION

Cytosolic 5'-nucleotidase specific for IMP-GMP and their respective deoxyderivatives (5'N) is an ubiquitous enzyme able to catalyse the hydrolysis of purine nucleoside monophosphates or the transfer of the phosphate to an acceptor nucleoside. The enzyme has been purified from several sources and its kinetic parameters and molecular characteristics have been studied.[1,2] All the cytosolic 5'-nucleotidases described, although sometimes differ remarkably in molecular mass, exhibit very similar properties, such as the activation by ATP and the inhibition by orthophosphate (Pi). A reaction mechanism proceeding via the formation of a covalent enzyme-phosphate intermediate was proposed by Worku and Newby.[3] The validity of this mechanism was recently demonstrated by the isolation of an enzyme-phosphate covalent complex. Even though there is indirect evidence for an involvement of histidine and cysteine residues in catalysis, nothing is known about the nature of the amino acid acceptor of the phosphate.[4]

Several hypothesis have been proposed on the physiological role of cytosolic 5'-nucleotidase in different cells and tissues, all of them implying its regulation of the intracellular IMP concentration, and thus directly or indirectly determining the destiny of the purine ring.[1]

Several authors reported that cytosolic 5'-nucleotidase is involved in the *in vivo* and *in vitro* activation of several purine prodrugs such as acyclovir, carbovir, 2'-3'-dideoxyinosine, and deoxycoformycin.[5,6,7] These results indicate that this activity may play a role either in the activation of prodrugs or in the drugs inactivation through hydrolysis.

Purine and Pyrimidine Metabolism in Man IX,
edited by Griesmacher *et al.* Plenum Press, New York, 1998.

231

The abundance of 5'N is low in most tissues and the purified enzyme is very unstable; therefore, a recombinant enzyme source is highly needed for further studies and particularly for the determination of the activity and regulatory sites of the enzyme. We have cloned the cDNA for bovine 5'N using the information from the human cDNA sequence.[8] The expression of the cloned bovine cDNA led to the synthesis of a protein whose properties were identical to these of the cytosolic 5'-nucleotidase phosphotransferase purified from calf thymus.[9]

2. METHODS

2.1. Expression and Characterisation of Recombinant 5'N

Using RT-PCR protocols a construct of cDNA flanked by a NheI and a EcoRI restriction sites respectively at the 5'-terminal and at the 3'-terminal was obtained. It was than ligated into the expression vector pET-28c (Novagen) and transfected into a maintaining bacterial stain (JM109). This vector provides in the expressed protein an His-tag, used for a single step purification, followed by a thrombin recognition site, fused to the N-terminal of the recombinant protein. The plasmid containing the 5'N's cDNA was then transformed into BL21(DE3) cells.[10] Expression of the recombinant protein was performed in LB medium containing 1 mM IPTG at 37°C for 3 h.

2.2. Purification of Recombinant 5'-Nucleotidase

Bacteria were harvested and lysated by freezing and thawing in the presence of lysozyme. From the cytosolic fraction recombinant 5'-nucleotidase was purified in a single step using Immobilised Metal Affinity Chromatography (IMAC) techniques. The affinity elution is achieved with 400 mM imidazole in 50 mM Tris-HCl, pH 7.4.

2.3. Enzyme Assays

Phosphotransferase activity was determined as previously described[11]. Hydrolysis of IMP was measured following the increase in absorbance of a solution containing IMP as substrate for the 5'N and both purine nucleoside phosphorylase and xanthine oxidase as helper system ($\Delta\varepsilon_{mM \, (IMP \rightarrow Uric \, acid)}$=11.26 mM^{-1}cm^{-1}).

3. RESULTS AND DISCUSSION

The isolation of the correct clone with the RT-PCR protocol presented some difficulties. After several attempt we were able to isolate a cDNA with the correct sequence. Most of the clones contained a 61 bp insertion at the 5'-terminal in the coding region. Performing the reverse transcriptase reaction at 52°C instead of 45°C as usual, we could isolate a single clone out of 14 containing the correct cDNA. The reasons of these results are not yet clear. The complete ORF isolated was 1683 bp long, with an ATG starting codon at the position 1→3 and a TGA stop codon at the position 1681→1683. The deduced amino acid sequence was 99.5% identical to that determined for the human enzyme.[8] Only two amino acids out of 560 were found to be different. This fragment was then inserted into the pET-28c expression vector and the contruct was transfected in two different E. coli strains commonly used for expression of recombinant proteins: BL21(DE3) and BL21(DE3)pLysS. We were able to detect phosphotransferase activity only in the extract obtained from the first strain, whereas the second one did not produce recombinant protein.

The crude extract obtained from BL21(DE3) transformed cells had a specific activity of 2.4 U/mg measured as phosphotransferase activity. Unfortunately most of the protein expressed was present as inclusion bodies. The native form of the recombinant protein purified in a single step from the crude extract, obtained by lysis of the harvested bacteria, showed a molecular mass of about 300 kDa. The monomer had a molecular mass of 61 kDa in SDS-PAGE gel (Figure 1) or 59 kDa after thrombin cleavage of the His–Tag fused at the N-terminal (data not shown), which is in agreement with the results obtained with the enzyme purified from calf thymus.[9] The molecular mass calculated from the amino acid sequence of the recombinant 5'N is 64,848 Da.

Recombinant enzyme cross-reacted with goose anti-pig lung cytosolic 5'N either before or after thrombin cleavage (data not shown).

The enzyme preparation obtained after the purification was stable at 4°C for several weeks in the presence of 2 mM dithiotreitol and either 50% glycerol or 20% ammonium sulphate, whereas the same samples kept at −20°C lost most of the activity after 24 hours.

The enzyme preparation used to determine the kinetic parameters had, after the purification, a specific activity of 83.3 μmol/min × mg protein measured as IMP hydrolase. This value resulted to be 11 times higher than the specific activity of the enzyme purified from calf thymus.[9]

The substrate specificity of the recombinant bovine 5'N is in complete agreement with that exhibited by the enzymes purified from both the calf thymus and the human colon carcinoma. Table 1 shows the phosphate donors whereas in Table 2 are listed the nucleoside acceptors. For all the enzyme preparations tested, the best substrates are the 5'-mononucleotides of 6-hydroxypurines and their respective deoxyderivatives, whereas the phosphate in the 3' position was not hydrolysed.

Furthermore, the cloned enzyme was activated by ATP and inhibited by phosphate at the same concentrations acting on the enzyme purified from different sources such man, calf, pig and chicken.[1,2]

The remarkable high similarity found in the cDNA sequence of 5'N from different species such as bovine, human,[8] chicken,[8] and the partially determined sequences of rat

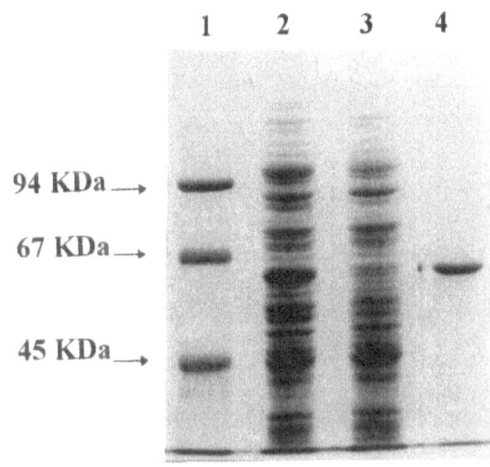

Figure 1. SDS-gel electrophoresis of recombinant bovine 5'N purified by ion metal affinity chromatography. Lane 1: molecular mass standards; lane 2: 10 μl of crude cell lysate; lane 3: 10 μl of the flow-through fraction; lane 4: 10 μl of the imidazole eluate.

94 KDa →
67 KDa →
45 KDa →

1 2 3 4

Table 1. Comparison between recombinant and purified 5′-nucleotidases substrate specificity of phosphate donors[a]

	Recombinant 5′N	Purified calf 5′N[9]	Purified human 5′N[11]
5′-dIMP	100	100	100
5′-IMP	95	85	78
5′-GMP	94	64	82
5′-dGMP	87	68	76
5′-UMP	35	31	26
5′-CMP	21	14	10
5′-AMP	19	9	11
5′-dAMP	7	4	3
3′-UMP	2	3	1.5
3′-IMP	1	5	0.3

[a]Values are expressed as % of maximal activity.

Table 2. Comparison between recombinant and purified 5′N substrate specificity of phosphate acceptors[a]

	Recombinant 5′N	Purified calf 5′N[9]	Purified human 5′N[11]
Inosine	100	100	100
Guanosine	28	28	16
dGuanosine	5.4	0	0
Adenosine	1.8	2	13
dAdenosine	1.2	1	1.7

[a]Values are expressed as % of maximal activity.

and mouse, gave reason of the substantial identity of the characteristics displayed by the enzymes purified from different sources. The wide distribution in different species and tissues of the 5′N,[1,2] together with the high level of similarity here reported, allow us to speculate that this cytosolic enzyme play a role of primary importance in the purine nucleosides and nucleotides metabolism.

ACKNOWLEDGMENTS

This work This work was supported from the following grants: Swedish National Board of Technical Development; Medivir AB, Huddinge; EU Commission BMH4-CT96-0479; Svenska Institutet; Blanceflor Boncompagni-Ludovisi, född Bilds Foundation, Sweden.

REFERENCES

1. Itoh, R. (1993) Comp. Biochem. Physiol. 105B, 13–1.
2. Zimmermann, H. (1992) Biochem. J. 285, 345–365.
3. Worku, Y. and Newby, A.C. (1982) Biochem. J. 205, 503–510.
4. Baiocchi, C., Pesi., R., Camici, M., Itoh, R. and Tozzi, M.G. (1996) Biochem. J. 317, 797–801.
5. Johnson, M.A. and Fridland, A. (1989) Mol. Pharmacol. 36, 291–295.
6. Keller, P.M., McKee, S.A. and Fyfe, J.A. (1985) J. Biol. Chem. 260, 8664–8667.
7. Banditelli, S., Baiocchi, C., Pesi, R., Allegrini, S., Turriani, M., Ipata, P.L. Camici, M. and Tozzi, M.G. (1996) Int. J. Biochem. Cell Biol. 28, 711–720.

8. Oka, J., Matsumoto, A., Hosokawa, Y. and Inoue, S. (1994) Biochim. Biophys. Res. Commun. 205 n°1, 917–922.
9. Pesi, R., Turriani, M., Allegrini, S., Scolozzi, C., Camici, M., Ipata, P.L. and Tozzi, M.G. (1994) Arch. Biochem. Biophys. 312 n°1, 75–80.
10. Studier, F.W., Rosenberg, A.H., Dunn, J.J. and Dubendorff, J.W. (1991) Methods Enzymol. 185, 60–89.
11. Tozzi, M.G., Camici, M., Pesi, R., Allegrini, S., Sgarrella, F. and Ipata, P.L. (1991) Arch. Biochem. Biophys. 291 n°2, 212–217.

46

MOLECULAR CHARACTERIZATION OF A CARBAMOYL-PHOSPHATE SYNTHETASE II (CPS II) GENE FROM *LEISHMANIA MEXICANA*

Guanghan Gao, Takeshi Nara, Junko Nakajima-Shimada, and Takashi Aoki

Department of Parasitology
Juntendo University School of Medicine
Hongo, Bunkyo-ku, Tokyo 113, Japan

INTRODUCTION

Carbamoyl-phosphate synthetase II (CPS II) is the first enzyme in de novo pyrimidine biosynthesis and catalyzes the formation of carbamoyl phosphate from L-glutamine, bicarbonate, and 2 moles of MgATP. In general, CPS II is composed of an N-terminal glutaminase domain and a C-terminal synthetase domain, the former producing ammonia from glutamine and the latter producing carbamoyl phosphate from ammonia, bicarbonate, and ATP. A short polypeptide linker exists between these two domains. The synthetase domain is subdivided into two apparently duplicated homologous amino acid sequences, CPS.A and CPS.B (Guy and Evans, 1996). CPS.A consists of subdomains, A1, A2, and A3, and CPS.B consists of B1, B2, and B3. A2 and B2 are assigned to the catalytic subdomains that retain the individual ATP-binding sites. The C-terminal subdomain, B3, is of allosteric importance, involving the allosteric ligand-binding sites for the inhibitor, UMP in *Escherichia coli* or UTP in animals, and activator, phosphoribosyl pyrophosphate (PRPP) or *N*-acetyl-L-glutamic acid (AGA). Previous work in this laboratory (Aoki et al., 1995) showed that in *Trypanosoma cruzi* the glutaminase and synthetase domains are covalently linked but separated from aspartate carbamoyltransferase (ACT). A trypanosomatid CPS II shows kinetic and regulatory properties different from those of the previously reported prokaryotic and eukaryotic CPSs II (Aoki, 1994). To gain insight into these biochemical and molecular peculiarities, we have determined and analyzed the nucleotide and deduced amino acid sequences of the CPS II gene from *Leishmania mexicana*, the pathogen that causes cutaneous leishmaniasis in Latin America.

Purine and Pyrimidine Metabolism in Man IX,
edited by Griesmacher *et al.* Plenum Press, New York, 1998.

MATERIALS AND METHODS

Culture forms of *Leishmania mexicana*, promastigotes, were propagated and harvested as described previously (Yamasaki et al., 1994). The preparation of parasite's genomic DNA and construction of a DNA library in EMBL3 (Stratagene) were essentially as described (Sambrook et al., 1989). From the two regions of highly conserved amino acid sequences in A1 and A2 in various CPSs II, a pair of oligonucleotide primers was synthesized and used for PCR, yielding an amplified DNA fragment 675 nucleotides long. This fragment was subcloned into pT7-Blue T-Vector (Novagen), labeled with a DIG-labeling kit (Boehringer Mannheim), and used as a probe for the screening of the DNA library.

RESULTS AND DISCUSSION

The *L. mexicana* CPS II gene is composed of 4560 nucleotides encoding 1520 amino acid residues. An alignment of the amino acid sequence with those from other sources, including *E. coli*, yeast, and mammals, highlights the following characteristic sequence motifs. The glutaminase domain consists of 358 amino acid residues and is divided into N-terminal (N-GAT) and C-terminal (C-GAT) halves, most probably corresponding to interaction and catalytic subdomains, respectively. The catalytic subdomain C-GAT is further divided into two halves, C-GAT 1 and C-GAT 2, showing markedly different pI values. For C-GAT 1 and C-GAT 2, the calculated pI values are 9.6 and 4.1 in hamster, 8.3 and 4.6 in *L. mexicana*, and 6.6 and 5.6 in *E. coli*; the parasite pI values are approximately between the hamster and *E. coli* values. In this context, it is of interest that the Km value of CPS II for L-glutamine is in the range of 0.01 to 0.02 mM in mammals, 0.1 to 0.2 mM in trypanosomatids (Aoki, 1994), and 0.4 mM in *E. coli*. In the *L. mexicana* CPS II, the glutaminase and synthetase domains are connected by a short hydrophilic polypeptide linker of 15 amino acid residues. These data are consistent with those reported for *Trypanosoma cruzi* CPS II (Aoki et al., 1995) and confirms our previous findings that a polypeptide in *T. cruzi* carries the glutaminase and synthetase domains, separated from both ACT and dihydroorotase, indicating a new type of CPS II that resembles neither the *E. coli* nor mammalian enzyme.

The *L. mexicana* synthetase domain, made up of a stretch of 1,147 deduced amino acid residues, is encoded by 3,441 nucleotides long. The amino acid sequence identities to *T. cruzi* CPS II, hamster *CAD*, yeast *CPA2* and *URA 2*, and *E. coli carB* are 50, 14, 27, 25, and 16%, indicating an intimate phylogenetic relationship between *Leishmania* and *Trypanosoma*. Similarly, the sequence identities for A2 and B2 are 69, 41, 45, 36, 23%, and 42, 29, 20, 33, 10%, respectively, suggesting that the *L. mexicana* synthetase domain is more closely related to the eukaryotic rather than the prokaryotic CPS II.

Figure 1 shows the deduced amino acid sequence of the *L. mexicana* allosteric subdomain, B3, aligned with those of various CPS II, highlighting some interesting characteristics. First, the C-termini are about 50 to 60 amino acid residues longer in *L. mexicana* and *T. cruzi*. The significance of these C-terminal extensions is unknown. Second, invariant residues (marked by asterisks) are also conserved in the two protozoan parasites, whereas an equivalent replacement occurs in the residue Asn (marked by +). Cervera et al. (1996) have shown that Lys 992 is critical for the binding of the inhibitor UMP in *E. coli*, and that the Lys residue is conserved in other prokaryotic CPSs II that utilize UMP as allosteric inhibitor. This residue is changed to Trp in mammalian CPS II that is inhibited by UTP. Interestingly, a Gln residue is present in this critical position in CPS II from *L. mexicana* and *T. cruzi*. These results are consistent with reports that in trypanosomatids,

```
HAM   1256   QFSFSRLAGADVVLGVEMTSTGEVAGFGESRCEAYLKAMLSTG-FKIPKK
DROS  1283   QFSFSRLRARSVQLGVEMTSTGEVACFGDNRYEAYLKAMMSTG-FQIPKN
CARB   893   VLPFNKFPGVDPLLGPEMRSTGEVMGVGRTFAEAFAKAQLGSNSTMKKHG
CPA2   916   QFSFTRLAGADPFLGPEMRSTGEVASFGRDLIESYWTAIQSTMNFHVPLP
TCRU  1259   VFSFNRLAGADPILGVEMASTGEIGVFGRDKKEVFLKAMLCQN-FRYPQR
LMA   1257   MFSFIPIAGADPILGVEMASTGEIGVFGRDKHEVFLKAMLCQN-FRIPKK
```

```
                                                  *
HAM   1314   NILLTIG-SYKNKSELLPTVRLLESLGYSLYASLGTADFYTEHGVKVTAV
DROS  1333   AVLLSIG-SFKHKMELLPSIRDLAKMGYKLXASMGTADFY-ANGVNVESV
CARB   943   RALLSVREGDKERVVDLAAKLLKQ--GFELDATHGTAIVLGEAGINPRLV
CPA2   966   PSGILFG-GDTSREYLGQVASIVATIGYRIYTTNETTKTYLQEHIK-EKN
TCRU  1308   GVFISCDVDAMAEDLCPTLSASDRFPVFTSK---QTSRVLADYGIPHTVL
LMA   1306   GVFFSIDVDSETEALCPYIQHLVRR-GLKVYGTANTAAVLHEYGIECEVL
```

```
              ↓                                   +
HAM   1363   DWHFEE-AVDGECPPQRSILDQL-AENH-FELVINLSMRGAG--GRR-LS
DROS  1380   QWTFDKTTPDDINGELRHLAEFL-ANKQ-FDLVINLPMSG--GVPRRV-S
CARB   991   NKVHEG-RPHIQDRIKNGEY---------TYIINTTS------GRR---
CPA2  1014   AKVSLIKFP-KNDKRKLRELFQEYD----IKAVFNLAS------KRA--E
TCRU  1355   TQRHEDSEPTFDT----AV-----AVKEKFDLVIQLRDKRQDFMLRRCTQ
LMA   1355   LQRSELPSGDACESNRPAVYDEEVAKKEKFDLVIQLRDKRRDFVLRRCTR
```

```
                  *
HAM   1407   SFVTKGYRTRRLAADFSVPLIIDIKCTKLFVEALGQIGPAPPLKVHVDC
DROS  1426   SFMTHGYRTRRLAVEYSIPLVTDVKCTKLLVESMRMNGGKPPMKTHTDC
CARB  1031   AIE-DSRVIRRSALQYKVHY--DTTLNGGFATAMALNADATEKVISVQEM
CPA2  1051   STDDVDYIMRRNAIDFAIPLFNEPQTALLFAKCLKAKIAEKIKILESHDV
TCRU  1396   ENATADYWIRRLAVDYNHSLLTEPNVVRMFCETLDVPVKEIEIEPFRLYV
LMA   1405   ETAPPDYWVRRLAVDYNIPLLTEPSLVKMFCEFMDLPASSIEVEPFRHYV
```

```
CARB  1078   HAGIK
CPA2  1101   IVPPEVRSWDEFIGFKAY
TCRU  1346   PRVYNKMENDNYTMLHRHKVGLCITSTNDSKVLAISLREEKIALTCFHAC
LMA   1455   PKIYHKVENNNCAMLRCHKVGLMITDNNGSKVLALRLSQEGLNITCFHGY
```

```
TCRU  1396   LGGIKNNSEEIAEQFRSIGSTSRAHRPPH
LMA   1505   LGGSDIGQFEQAFQRP
```

Figure 1. Alignment of the allosteric subdomain, B3, of various CPSs II. The Thr and Arg, marked by asterisks (*), are invariant residues that are also conserved in *L. mexicana* and *T. cruzi*. The (+) marks the conserved Asn that is replaced by a Gln residue in trypanosomatid parasites. The arrow points the position of critical residues, Lys 992 for the binding of UMP in *E. coli*, Trp for the binding of UTP in mammals, and Gln for the binding of UDP in trypanosomatids. The abbreviations are: HAM, hamster CPS II (*CAD*); DROS, *Drosophila melanogaster* CPS II (*CAD*) (according to Simmer et al., 1990); CARB, *Escherichia coli carB*; CPA 2, arginine-specific CPS II from *Saccharomyces cerevisiae*; TCRU, *T. cruzi* CPS II; LMA, *L. mexicana* CPS II.

UDP inhibits the CPS II activity most strongly (Aoki and Oya, 1987), and that the uncharged polar residue Gln, with intermediate electrostatic properties between the basic (charged) residue Lys and the nonpolar residue Trp, may be critically involved in binding the ligand UDP. Some of these conclusions, however, require testing on the expressed gene products.

ACKNOWLEDGMENTS

A part of this work was supported by a Grant-in-Aid for Scientific Research (07457068) from the Ministry of Education, Science, and Culture of Japan.

REFERENCES

Aoki, T. (1994): Initial steps of *de novo* pyrimidine nucleotide biosynthesis in parasites and mammalian tissues: Purification, regulation, adaptation, and evolution. Jap. J. Parasitol., 43: 1–10.

Aoki, T., and Oya, H. (1987): Regulatory properties of carbamoyl-phosphate synthetase II from the parasitic protozoan *Crithidia fasciculata* . Comp. Biochem. Physiol., 87B: 655–658.

Aoki, T., Shimogawara, R., Ochiai, K., Yamasaki, H., and Shimada, J. (1995): Molecular characterization of a carbamoyl-phosphate synthetase II (CPS II) gene from *Trypanosoma cruzi*. Adv. Exp. Med. Biol., 370: 513–516.

Cervera, J., Bendala, E., Britton, H.G.,. Bueso, J., Nassif, Z., Lusty, C.J., and Rubio, V. (1996): Photoaffinity labeling with UMP of lysine 992 of carbamyl phosphate synthetase from *Escherichia coli* allows identification of the binding site for the pyrimidine inhibitor. Biochemistry, 35: 7274–7255.

Guy, H.I., and Evans, D.R. (1996): Function of the major synthetase subdomains of carbamyl-phosphate synthetase. J. Biol. Chem., 271: 13762–13769.

Sambrook, J., Fritsch, E.F., and Maniatis, T. (1989): Molecular Cloning. A Laboratory Manual, Second Ed., Cold Spring Harbor Laboratory, NY.

Simmer, J.P., Kelly, R.E., Rinker, Jr., A.G., Scully, J.L., and Evans, D.R. (1990): Mammalian carbamyl phosphate synthetase (CPS): cDNA sequence and evolution of the CPS domain of Syrian hamster multifunctional protein CAD. J. Biol. Chem., 165: 10395–10402.

Yamasaki, H., Agatsuma, T., Pavon, B., Moran, M., Fukuya, M., and Aoki, T. (1994): *Leishmania*-major like parasite, a pathogenic agent of cutaneous leishmaniasis in Paraguay. Am. J. Trop. Med. Hug., 51: 749–757.

47

SUBSTRATE CHANNELLING BY HUMAN IMP SYNTHASE

Eve Szabados, Paul K. Wilson, and Richard I. Christopherson

Department of Biochemistry
University of Sydney
Sydney NSW 2006, Australia

1. INTRODUCTION

IMP synthase is a bifunctional enzyme containing the activities, 5-aminoimidazole-4-carboxamide ribotide (AICAR) transformylase (EC 2.1.2.3) and IMP cyclohydrolase (EC 3.5.4.10) and catalyses the ninth and tenth reactions of the pathway for *de novo* biosynthesis of purine nucleotides.

$$
\begin{array}{lll}
\text{AICAR} & \longrightarrow \quad \text{FAICAR} & \longrightarrow \quad \text{IMP} \\
+ \ N^{10}\text{-formyl} & + & + \ H_2O \\
\text{tetrahydrofolate} & \text{tetrahydrofolate} &
\end{array}
\qquad (1)
$$

We have purified AICAR transformylase-IMP cyclohydrolase (ACT-ICH) 780-fold from human CCRF-CEM leukaemia cells to homogeneity (Szabados *et al.*, 1994). ICH is subject to competitive inhibition by a series of purine nucleoside-5′-monophosphate analogues of which 2-mercaptoinosine-5′-monophosphate (MIMP) is the most potent with a K_i value of 94 nM (Szabados *et al.*, 1994). ICH is inactivated by N-ethylmaleimide and phenylglyoxal providing evidence for essential cysteine and arginine residues, respectively, at the active site. We have proposed a catalytic mechanism for ICH which involves ionised cysteine and arginine residues. Human ACT-ICH has been cloned, sequenced, expressed in *Escherichia coli* and purified (Rayl *et al.*, 1996). ACT-ICH has a subunit molecular weight of 64,425 Daltons and is found as a dimer.

The covalent association of the ACT and ICH activities on bifunctional IMP synthase should provide some selective advantage to the cell. Possible advantages include coordinate expression of ACT and ICH under all conditions of growth, coordinate regulation of the two activities by interaction of a purine metabolite at a single regulatory site, and substrate channelling of FAICAR produced by ACT to an ICH site on the same dimeric molecule of IMP synthase. Such channelling would minimise cellular concentrations of FAICAR in the

Purine and Pyrimidine Metabolism in Man IX,
edited by Griesmacher *et al.* Plenum Press, New York, 1998.

cytosol and enable rapid changes in flux through the purine pathway in response to changing conditions of growth. We have developed a radioassay for ACT (AICAR → FAICAR + IMP), ICH (FAICAR → IMP) and the overall reaction (AICAR → FAICAR → IMP, Szabados and Christopherson, 1994). This sensitive assay has enabled investigation of the spatial relationship of the ACT and ICH sites on IMP synthase.

2. EXPERIMENTAL PROCEDURES

2.1. Materials

AICAR, IMP, N^5-formyl tetrahydrofolate and N-ethylmaleimide were from the Sigma Chemical Co. (St. Louis, MO, USA). FAICAR was prepared from AICAR as the barium salt (Flaks *et al.*, 1957). N^{10}-formyl tetrahydrofolate was synthesised from N^5-formyl tetrahydrofolate by a modification of the method of Rabinowitz (1963). $[^3H]$FAICAR (15,000 Ci/mol) was prepared by tritium exchange of unlabelled FAICAR by Moravek Biochemicals Inc, (Brea, CA, USA). $[^3H]$AICAR was prepared by phosphorylation of ^3H-labelled 5-aminoimidazole-4-carboxamide riboside using a phosphotransferase prepared from *Serratia marcescens* with p-nitrophenyl phosphate as the phosphate donor (Fyfe *et al.*, 1978).

2.2. Assay of ACT and ICH and Modification of Cysteine Residues

Reaction mixtures contained 50 mM K.Hepes (pH 7.4), 5% (v/v) glycerol, 15 mM KCl, 500 µM N^{10}-formyl tetrahydrofolate and various concentrations of $[^3H]$AICAR or $[^3H]$FAICAR for assay of ACT and the overall reaction (AICAR → FAICAR → IMP) or ICH, respectively (Szabados and Christopherson, 1994). Inhibition of ACT and ICH by 2-mercaptoinosine-5′-monophosphate (MIMP) was determined under the same conditions in the presence of 15 mM KCl and 500 µM N^{10}-formyl tetrahydrofolate with 8 different inhibitor concentrations Reaction conditions for modification of cysteine residues with N-ethylmaleimide were as reviewed by Lundblad and Noyes (1988). Preincubation mixtures contained 5% (v/v) glycerol, 240 mM potassium phosphate (pH 7.0), 5 mM N-ethylmaleimide and purified IMP synthase. When substrate protection was studied, 50 µM AICAR with 500 µM N^{10}-formyl tetrahydrofolate, or 10 µM $[^3H]$FAICAR (5,000 Ci/mol) was included in the preincubation mixture. After 0.5, 3.5, 6.5 and 9.5 min, samples (5 µl) were removed for assay of residual AICAR transformylase and IMP cyclohydrolase activities.

2.3. Test for Channelling of FAICAR

Incubation mixtures (40 µl) contained all the components for the overall reaction, including 50 µM $[^3H]$AICAR (5,000 Ci/mol) and various concentrations of unlabelled FAICAR (0–5000 µM). Reactions were initiated by addition of pure IMP synthase (120 pg) and 7 µl samples were spotted onto poly(ethyleneimine)-cellulose chromatograms at 3, 6, 9, 12 and 15 min to separate and quantify the $[^3H]$IMP produced.

3. RESULTS AND DISCUSSION

3.1. Kinetic Evidence for a High Local Concentration of FAICAR

Progress curves were determined for the conversion of AICAR → FAICAR → IMP over a 20-min period using $[^3H]$AICAR as the precursor. During the incubation, FAICAR reached a steady-state concentration of 0.118 μM and the observed maximal rate of IMP formation was 0.218 μM/min. The predicted rate of IMP formation by ICH was calculated to be 0.0166 μM/min using the Michaelis-Menten equation, the steady-state FAICAR concentration of 0.118 μM and a K_m value for FAICAR determined under these conditions of 64 μM. The observed rate of 0.218 μM/min is 13.1-fold higher than the calculated rate, consistent with a higher local concentration of FAICAR at the ICH site. Further calculation showed that the observed rate of 0.218 μM/min would be sustained by a local concentration of FAICAR at the ICH site of 1.59 μM rather than the average concentration of FAICAR in the bulk solvent of 0.118 μM.

3.2. Inhibition by MIMP and N-Ethylmaleimide

To obtain information about the spatial relationship between the ACT and ICH sites on IMP synthase, MIMP was tested as an inhibitor of both activities. Szabados *et al.* (1994) have shown that MIMP is a potent inhibitor of ICH and if the same interaction of MIMP with the synthase resulted in inhibition of both activities with a similar dissociation constant (K_i value), then a common active site could be proposed. Inhibition constants obtained for the two activities under the same conditions were for ACT, $K_i = 3.1 \pm 0.7$ μM, and for ICH, $K_i = 0.12 \pm 0.05$ μM. We conclude that MIMP interacts with the two catalytic sites differently and that the sites are likely to be distinct.

A further test of the spatial relationship of the ACT and ICH sites was made using chemical modification of IMP synthase by N-ethylmaleimide in the absence and presence of FAICAR. Incubation of pure IMP synthase in the presence of N-ethylmaleimide resulted in time-dependent inactivation of ICH activity as reported by Szabados *et al.* (1994) but ACT was not inactivated. FAICAR protected ICH against this inactivation and it is therefore concluded that there is an essential cysteine residue at the active site of ICH but not ACT. These experiments are also consistent with distinct ACT and ICH active sites on bifunctional IMP synthase.

3.3. Substrate Channelling

A preliminary experiment to test for possible substrate channelling of FAICAR between the ACT and ICH sites was performed as described in Experimental Procedures. IMP synthase was incubated with 500 μM N^{10}-formyl tetrahydrofolate, 50 μM [^{3}H]AICAR and various concentrations of unlabelled FAICAR (0–5000 μM), and the rate of formation of [^{3}H]IMP was measured.

$$[^3\text{H}]\text{AICAR} \longrightarrow [^3\text{H}]\text{FAICAR} \longrightarrow [^3\text{H}]\text{IMP}$$
$$+$$
$$\text{FAICAR} \qquad\qquad (2)$$

The rate of formation of ^{3}H-labelled IMP at lower concentrations of exogenous FAICAR (0–200 μM) was relatively constant, but the [^{3}H]IMP formed then decreased to zero by 5000 μM exogenous unlabelled FAICAR (Figure 1). ICH has a V_{max} which is 36-fold higher than ACT and we conclude that the initial maintenance of [^{3}H]IMP synthesis at low levels of exogenous FAICAR could be attributed to an increasing saturation of ICH with both the endogenous and exogenous FAICAR. Under these conditions, the K_m of ICH for FAICAR is 64 μM. At higher concentrations of FAICAR, there is a progressive reduction in the rate at which endogenous [^{3}H]FAICAR is converted through to [^{3}H]IMP (Figure 1).

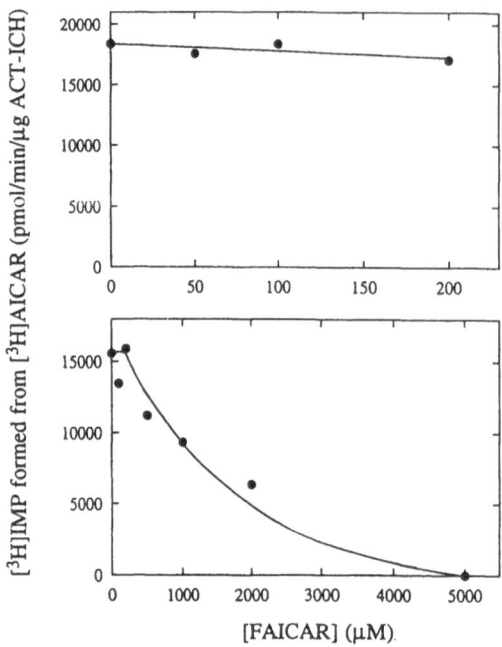

Figure 1. Rate of conversion of [³H]AICAR to [³H]IMP in the presence of increasing concentrations of unlabelled exogenous FAICAR. Details of the experiment appear in Experimental Procedures.

Thus, we have kinetic evidence for a high local concentration of endogenous FAICAR at the ICH site but substrate channelling, if it occurs, is not absolute. Inhibition experiments with MIMP and N-ethylmaleimide indicate that the ACT and ICH sites are distinct on bifunctional IMP synthase but they may be sufficiently close for there to be a higher local concentration of FAICAR at the ICH site than that prevailing in the bulk solvent.

ACKNOWLEDGMENT

This research was supported by project grant 950124 from the National Health and Medical Research Council of Australia.

REFERENCES

Flaks, J.G., Erwin, M.J., and Buchanan, J.M., 1957, Biosynthesis of the purines. XVIII. 5-Amino-1-ribosyl-4-imidazolecarboxamide 5'-phosphate transformylase and inosinicase. *J. Biol. Chem.*, 224: 603.

Fyfe, J.A., Keller, P.M., Furman, P.A., Miller, R.L., and Elion, G.B., 1978, Thymidine kinase from *Herpes simplex* virus phosphorylates the new antiviral compound, 9-(2-hydroxyethoxymethyl)guanine. *J. Biol. Chem.*, 253: 8721.

Lundblad, R.L. and Noyes, C.M., 1988, Chemical reagents for protein modification. CRC Press Inc., Florida.

Rabinowitz, J.C., 1963, Preparation and properties of 5,10-methenyltetrahydrofolic acid and 10-formyltetrahydrofolic acid. *Methods Enzymol.*, 6: 814.

Rayl, E.A., Moroson, B.A., and Beardsley, G.P., 1996, The human purH gene product, 5-aminoimidazole-4-carboxamide ribonucleotide formyltransferase/IMP cyclohydrolase, *J. Biol. Chem.*, 271:2225.

Szabados, E., and Christopherson, R.I., 1994, Radioassay of bifunctional 5- aminoimidazole-4-carboxamide ribotide transformylase-IMP cyclohydrolase by thin-layer chromatography. *Anal. Biochem.*, 221: 401.

Szabados, E., Hindmarsh, E.J., Phillips, L., Duggleby, R.G., and Christopherson, R.I., 1994, 5-Aminoimidazole-4-carboxamide ribotide transformylase-IMP cyclohydrolase from human CCRF-CEM leukemia cells: purification, pH dependence, and inhibitors. *Biochemistry*, 33: 14237.

48

INHIBITION OF ADENYLOSUCCINATE LYASE BY 2′, 3′-ACYCLIC SUBSTRATE ANALOGS

C. Crifò, A. Lomonte, and C. Salerno

Department of Biochemical Sciences and
Clinical Biochemistry Laboratory
University of Roma La Sapienza

1. INTRODUCTION

Adenylosuccinate lyase (EC 4.3.2.2; ASase) catalyzes the trans-elimination of fumarate from adenylosuccinate (S-AMP) to yield AMP. The other substrate in purine biosynthesis cleaved by ASase is 5-amino-4-imidazole-N-succinocarboxamide ribotide (SAICAR), which also has a succinyl group attached and closely resembles S-AMP.[1] Moreover, the enzyme plays a pivotal role in the conversion of dideoxypurine nucleosides into dideoxynucleotide triphosphates with known anti-HIV (human immunodeficiency virus) activity.[2–3] Increased ASase activities have been found in kidney tumors and in the liver following the administration of carcinogens.[4–5]

The catalytic reaction of ASase is postulated to involve the attack of an enzymic general base on the β-H and the elimination of the amino group facilitated by the protonation of the ring nitrogen by an amino acid residue of the enzyme functioning as a general acid.[6] Kinetic studies show that the catalysis follows an ordered uni-bi mechanism with fumarate leaving the enzyme before AMP.[7] The 6-amino group and the 5′-monophosphate group are important in substrate binding, as seen from the poor inhibition of IMP, thio-IMP, ATP, and inosine. That the succinyl portion of S-AMP plays a role in the binding is indicated by the fact that AMP has lower affinity for ASase than S-AMP.[8] The two carboxyl groups of S-AMP are likely neutralized in the active site by positively groups of the protein, since esterification of the carboxyl groups of the thio analogue facilitates the base-catalyzed elimination. This suggestion is also in line with the observation that the substitution of the β-carboxyl group of S-AMP with a phosphonate group gives rise to an analogue with enzyme affinity at least one order of magnitude greater than that of the natural compound.[9]

In this report, we show that the catalysis by ASase is not affected by opening the ribose ring of the substrate. The addition of aliphatic chains on the ring does not markedly affect the binding of the substrate analog, though preventing its cleavage.

Purine and Pyrimidine Metabolism in Man IX,
edited by Griesmacher *et al.* Plenum Press, New York, 1998.

2. EXPERIMENTAL PROCEDURES

ASase was prepared from human red blood cells according to published procedures with minor modifications.[10] Enzyme activities were assayed at 38°C in 10 mM K-phosphate buffered solutions, pH 7.0, containing 1 mM EDTA and 1 mM dithiothreitol by measuring the absorbance at 254 nm of the reaction products purified by high-performance zone electrophoresis in a fused-silica capillary.[11]

Substrate analogs of ASase were obtained by oxidizing S-AMP and AMP by sodium periodate (Fig. 1).[12] PMR spectra (200 MHz Gemini Varian spectrometer) of 2',3'-acyclic dialdehyde derivatives showed a high-field shift of some of the signals referred to the ribose, while the peaks reasonably assigned to C1'-proton appeared to be substantially unmodified. No significant changes in the signals of the purine ring and of the succinyl moiety (if present) were noticed. 2',3'-Acyclic dialcohol derivatives were obtained by treating the periodate-oxidized compounds with sodium borohydride, whereas 2',3'-acyclic (N,N-ethyl) diamine derivatives were synthetized by reducing the Schiff-bases formed with ethylamine. Succinyladenosine 2',3'-acyclic dialdehyde was obtained by incubating the 2',3'-acyclic dialdehyde nucleotide with E. coli alkaline phosphatase. The nucleotide analogs were purified by Dowex AG1x4 chromatography and appeared well resolved and homogeneous by high-performance zone electrophoresis. By contrast, the electropherogram of succinyladenosine 2',3'-acyclic dialdehyde showed three poorly-resolved major peaks.

3. RESULTS AND DISCUSSION

The susceptibility of S-AMP and of S-AMP analogs to undergoing ASase catalysis was tested analysing the incubation mixtures by a high-performance zone electrophoresis.

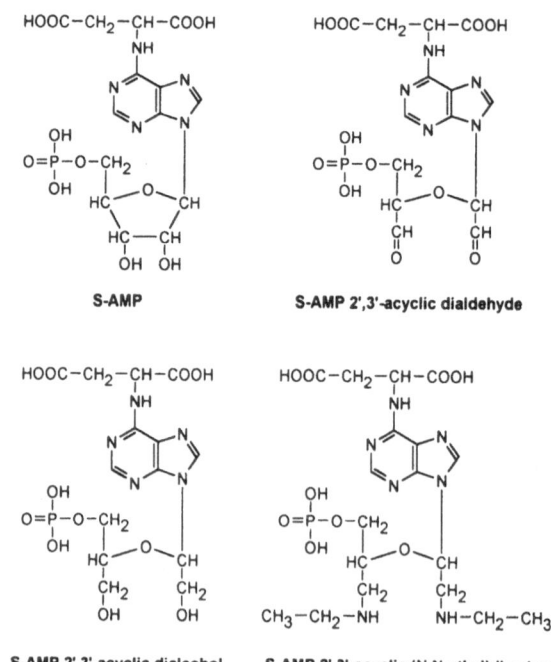

Figure 1. Structural formulae of S-AMP and acyclic S-AMP derivatives.

Table 1. Kinetic constants of the reactions catalyzed by
adenylosuccinate lyase. Vmax values were obtained
using the same enzyme preparation within 2 hours

	Vmax, %	Km, µM
S-AMP	100	22.9
S-AMP 2',3'-acyclic dialdehyde	79	21.3
S-AMP 2',3'-acyclic dialcohol	118	121.7

While the electropherograms of S-AMP 2',3'-acyclic (N,N-ethyl) diamine and of succinyladenosine 2',3'-acyclic dialdehyde did not change apppreciably in the presence of the
enzyme, new peaks with retention times equal to those of AMP, AMP 2',3'-acyclic dialdehyde, and AMP 2',3'-acyclic dialcohol were obtained by incubating ASase with micromolar
solutions of S-AMP, S-AMP 2',3'-acyclic dialdehyde, and S-AMP 2',3'-acyclic dialcohol,
respectively. The peak areas of the products accounted for the decrease in the peak areas of
the substrates in good agreement with the stoichiometry expected for the trans-elimination
of fumarate from the succinylnucleotides.[1] The initial rates of product formation were linearly related to the enzyme concentration and hyperbolic functions of succinylnucleotide
concentrations. As shown in Table 1 that summarizes the kinetic constants of the reactions,
only small changes in the catalytic properties of the enzyme were observed by comparing
the affinity constants and the turnovers with the three substrates.

Inhibition studies were performed by analysing the initial velocity pattern of S-AMP
cleavage in the presence of different fixed concentrations of acyclic derivatives. While
succinyladenosine 2',3'-acyclic dialdehyde did not inhibit the catalysis up to millimolar
concentrations, all the acyclic derivatives of S-AMP behaved as competitive inhibitors of
S-AMP cleavage. The inhibition constants of S-AMP 2',3'-acyclic dialdehyde, S-AMP
2',3'-acyclic dialcohol, and S-AMP 2',3'-acyclic (N,N-ethyl) diamine were equal to 4.7
µM, 16 µM, and 1.0 mM, respectively.

In conclusion, the present study indicates that S-AMP 2',3'-acyclic dialdehyde and
S-AMP 2',3'-acyclic dialcohol are alternate substrate inhibitors of ASase as efficient as the
natural substrate; S-AMP 2',3'-acyclic (N,N-ethyl) diamine is a weaker competitive
inhibitor that cannot be subjected to catalysis by ASase.

ACKNOWLEDGMENTS

This research was done in the framework of the European Community Concerted
Action contract no. BMH1-CT-1384.

REFERENCES

1. M. R. Redinbo, S. M. Eide, R. L. Stone, J. E. Dixon, and T. O. Yeates, Crystallization and preliminary
   structural analysis of Bacillus subtilis adenylosuccinate lyase, an enzyme implicated in infantile autism.
   Protein Science 5:786–788 (1996)
2. V. Nair and T. B. Sells. Interpretation of the roles of adenylosuccinate lyase and of AMP deaminase in the
   anti-HIV activity of 2',3'-dideoxyadenine and 2',3'-dideoxyinosine. Biochim. Biophys. Acta 1119:201–204
   (1992)
3. C. Salerno and C. Crifò. In vivo inactivation of adenylosuccinate lyase: pharmacological implications.
   Adv. Biosciences 92:101–108 (1994)

4. R. C. Jackson, H. P. Morris, and G. Weber. Enzymes of the purine ribonucleotide cycle in rat hepatomas and kidney tumors. Cancer Res. 37:3057–3065 (1977)

5. L. D. Smith, R. L. Emerson, and L. K. Dixon. Effect of hepatocarcinogens on the adenine purine nucleotide cycle during the initiation phase of carcinogenesis. Cancer Res. 39:2132–2138 (1979)

6. R. L. Stone, H. Zalkin, and J. E. Dixon. Expression, purification, and kinetic characterization of recombinant human adenylosuccinate lyase. J. Biol. Chem. 268:19710–19716 (1993)

7. W. A. Bridger and L. H. Cohen. The kinetics of adenylosuccinate lyase. J. Biol. Chem. 243:644–650 (1968)

8. A. Hampton. Studies of action of adenylosuccinase with 6-thio analogues of adenylosuccinic acid. J. Biol. Chem. 237:529–535 (1962)

9. L. M. Brand and J. M. Lowenstein. Inhibition of adenylosuccinase by adenylophosphonopropionate and related compounds. Biochemistry 17:1365–1370 (1978)

10. L. B. Barnes and S. H. Bishop. Adenylosuccinate lyase from human erythrocytes. Int. J. Biochem. 6:497–503 (1975)

11. C. Salerno and C. Crifò. Microassy of adenylosuccinase by capillary electrophoresis. Anal. Biochem. 226:377–379 (1995)

12. J. X. Khym. The reaction of methylamine with periodate-oxidized adenosine 5'-phosphate. Biochemistry 2:344–350 (1963)

49

REGULATION OF DEOXYCYTIDINE KINASE BY DEOXYCYTIDINE AND DEOXYCYTIDINE 5' TRIPHOSPHATE IN WHOLE LEUKEMIA AND TUMOR CELLS

V. Heinemann,[1] L. Schulz,[1] R. D. Issels,[1,2] and W. Wilmanns[1,2]

[1]Medizinische Klinik III
University Hospital Grosshadern
University of Munich
Marchionini-Strasse 15
81377 Munich, Germany
[2]GSF-Institute for Clinical Hematology
81377 Munich, Germany

INTRODUCTION

The clinical importance of the dCTP salvage pathway is given by the fact that deoxy-cytidine (dCyd) analogs like cytosine arabinoside (ara-C) are activated via the salvage path-way. The non-toxic prodrug ara-C gains its cytotoxic potential by phosphorylation to the 5'-triphosphate ara-CTP. Not only cellular accumulation, but also retention of ara-CTP are known determinants of response to antileukemic treatment. Deoxycytidine kinase is the rate-limiting enzyme for intracellular ara-CTP formation. Experiments performed with purified dCK (dCK) demonstrated that dCTP, the final product of phosphorylation, acts as a potent feed-back inhibitor of dCyd kinase.[1,2] In fact, depletion of intracellular dCTP as induced by inhibition of the de novo pathway increases of ara-C phosphorylation.[3] Therefore it was concluded that the release of dCTP-mediated feed-back inhibition was responsible for en-hanced dCK activity. The present study investigates the role of dCTP for regulation of dCK activity in the cell lines CCRF-CEM (T-cell line), Raji (B-cell line), HL-60 (promyelocytic), and CHO (Chinese hamster ovary). Compared to T-cells, B-cells are characterised by a fast turnover of dNTP and a short retention of ara-CTP. CHO cells were included because of their deficiency in deoxycytidylate (dCMP) deaminase activity. Cellular accumulation of ara-CTP was evaluated as a measure of salvage pathway activity. Inhibition of ribonu-cleotide reductase as induced by hydroxyurea may cause an activation of the salvage path-

Purine and Pyrimidine Metabolism in Man IX,
edited by Griesmacher *et al.* Plenum Press, New York, 1998.

way to maintain adequate dCTP levels for DNA synthesis. Cells with a fast dCTP turnover and a greater dependence on the de novo pathway should respond with a greater enhancement of the salvage pathway than cells with long retention of deoxynucleotides and a correspondingly lower dependence on the de novo pathway.[4]

MATERIALS AND METHODS

[³H]dCyd (specific activity, 24.8 Ci/mmol) and [³H]ara-C (specific activity, 29 Ci/mmol) were products of Amersham (Buckinghamshire, UK). Ara-C, deoxycytidine, hydroxyurea 3-deazauridine, and thymidine were purchased from Sigma Chemical Co., Inc. (St. Louis Miss).

Cell Lines. The human T-lymphoblast cell line CCRF-CEM, the erythroblastic K562 cells, and the promyelocytic HL-60 cells were obtained from the American Type Culture Collection (Rockville, MD). Chinese hamster Ovary (CHO) cells were provided by Dr. P. Saunders (M.D. Anderson Hospital and Tumor Institute).

Nucleotide Extraction and Analysis. Nucleotides were extracted from cells with 0.4 N $HClO_4$ as previously described.[5] Nucleoside triphosphates (NTP) and deoxynucleotide triphosphates (dNTP) in the neutralized acid-soluble extract were analyzed by HPLC as reported elsewhere.[5]

Determination of the Intracellular Deoxycytidine Concentration. A μ-Bondapak C_{18} column (Waters GmbH) was used for the separation of intracellular nucleosides. An initial buffer composition of 100% buffer A (0.5 M ammonium acetate, pH 6.8) was maintained over 10 min followed by a linear gradient reaching 40% buffer B (50% methanol) after 30 min. In all experiments analysing intracellular nucleoside concentratios, incubations were terminated by use of ice-cold PBS containing 10 μM dipyridamole.

RESULTS AND DISCUSSION

Modulation of Intracellular ara-CTP Accumulation by Inhibition of Ribonucleotide Reductase

This experiment was performed to activate the dCTP salvage pathway by inhibition of the de novo pathway. Cells were incubated for two hours with 5 mM HU, subsequently ara-C (10 μM) was added for further two hours, after which cellular dNTP and ara-CTP concentrations were analysed. In control experiments, accumulation of ara-CTP was greatest in CEM cells (412 ± 43 μM) followed by Raji-cells (116 ± 6 μM), HL-60 cells (114 ± 12 μM) and CHO cells (79 ± 20 μM). After a 2-hr preincubation with HU, intracellular ara-CTP formation remained essentially unchanged in CEM cells, while it increased in Raji cells (3.7-fold), HL-60 cells (3.5-fold), and CHO cells 8.1-fold). This analysis demonstrated that inhibition of ribonucleotide reductase by HU does not have a uniform effect in all cell lines. Specifically, in the T-cell line CCRF-CEM the dCTP salvage pathway was not modulated by inhibition of the de novo pathway. The following experiments were designed to clarify the mechanisms responsible for the apparent differences of dCTP salvage regulation.

Intracellular Half-Life of dCTP and ara-CTP

To provide an indicator for the cellular turnover of deoxynucleotides the intracellular half-life of dCTP was defined for each cell line. Cells were therefore exposed to [³H]dCyd for 2 hrs, and after washout, the decay of intracellular [³H]dCTP was determined. As expected, the intracellular half-life was greatest in the T-lymphoblastic CEM cells (75 min), while the respective half-lifes of dCTP elimination were considerably shorter in Raji cells (14 min), and HL-60 cells (19 min). When the intracellular half-life of ara-CTP was analysed, CEM cells again showed the longest half-life (290 min), followed by HL-60 cells (90 min), Raji-cells (26 min) and CHO cells (17 min). The cellular half-lifes of dCTP and ara-CTP were well correlated with an r-value of 0.987 indicating that the parameters functional for dCTP turnover also determined the turnover of ara-CTP. As expected, there was an inverse relationship between the cellular retention of ara-CTP and its capacity to accumulate ara-CTP. Since, in contrast to the report by Abe et al.,[6] HU did not significantly affect the velocity of ara-CTP elimination in Raji-, HL-60, or CHO cells it was concluded that HU-mediated enhancement of ara-CTP accumulation is essentially caused by an increased phosphorylation rate and not by a delayed elimination of ara-CTP.

Effect of Ribonucleotide Reductase Inhibition on the Cellular dCTP Pool

Since dCTP supposedly serves as a feed-back regulator of dCK it was of interest to analyse the cellular dCTP concentrations and relate these to the cellular capacity of ara-CTP formation. In CEM, Raji, HL-60, and CHO cells cellular dCTP concentrations of 25 µM, 6.1 µM, 3.6 µM, and 229 µM were measured under control conditions. There was no correlation between the cellular pool size of dCTP on the one hand and the cellular accumulation of ara-CTP on the other hand (r = −0.355). Further, the effect of ribonucleotide reductase inhibition on the dCTP pool was assayed after exposure of cells for 2 hours to 5mM HU. Inhibition of ribonucleotide reductase reduced intracellular dCTP to the greatest extent in in CEM cells (20% of control), while in Raji, HL-60, and CHO cells dCTP was decreased to 83%, 61%, and 41% of control, respectively. This experiment demonstrated that, although dCTP depletion was greatest in CEM cells, it did not affect dCK-mediated ara-C phosphorylation. Interestingly, there was also no correlation between the HU-induced enhancement of ara-CTP formation and the percentual (r = 0.150) decrease of the dCTP pool. It may therefore be concluded that at least in the comparative analysis between different cell lines, the cellular dCTP concentration is an inadequate predictor of cellular ara-CTP accumulation. In fact, Shewach et al.[7] demonstrated in vitro that in the presence of UTP cellular concentrations of dCTP up to 20 µM were insufficient to inhibit dCK. In accordance with this observation, we suggest that the decrease of dCTP noted in many cell lines during RR-inhibition is an accompanying effect reflecting the inhihibition of the de novo pathway but not necessarily responsible for the enhanced synthesis of ara-CTP by the salvage pathway. Inhibition of RR must, therefore, modulate cellular ara-CTP accumulation by a different mechanism.

Effect of Ribonucleotide Reductase Inhibition on the Cellular Pool of Deoxycytidine

Since dCTP, the final product of the salvage pathway, did not appear to be a good predictor of intracellular ara-CTP formation, the substrate level of the salvage pathway was analysed with regard to the cellular dCyd concentration. Deoxycytidine is a known competitive inhibitor of dCyd analog phosphorylation by dCK.[1] Under control conditions, the cellular concentration of dCyd amounted to 0.6 μM in CEM cells, compared to Raji cells (7.4 μM) and CHO cells (12.5 μM). There was an inverse correlation between the size of the dCyd pool the cellular synthesis of ara-CTP (r = −0.942). When cells were exposed for 2 hours to 5 mM HU, the concentration of dCyd decreased to 17% in CEM cells, 1.2% in Raji cells, and 21.6% in CHO cells. Again, the percentual decrease of the dCyd pool did not correlate to the HU-mediated enhancement of ara-C phosphorylation. However, the molar reduction of the dCyd concentration correlated to the enhancement of ara-CTP formation with an r-value of 0.918. The data suggest that inhibition of ribonucleotide reductase may activate dCyd analog phosphorylation rather by a decrease of cellular dCyd than by a depletion of the dCTP pool.

Modulation of dCyd- and dCTP Pools in Relation to ara-CTP Formation in Raji Cells

This analysis was performed only in Raji cells (B-cells) to analyse the differential impact of dCyd and dCTP on ara-CTP formation when the dNTP metabolism was modulated by various agents. Hydroxyurea (HU), thymidine (TdR), 3-deazauridine (3-DAU), and 3,4,5,6-tetrahydrodeoxyuridine (dTHU) were chosen as modulators of dNTP metabolism. HU (5mM), an inhibitor of ribonucleotide reductase, decreased dCyd and dCTP pools to 1.2% and 90% of control, respectively and enhanced ara-CTP synthesis by 3.7-fold. TdR acts as an inhibitor of CDP reduction. A 2-hr preincubation of Raji cells with 100 μM TdR and a subsequent exposure to 10 μM ara-C increased the dCyd pool by 3.4-fold, depleted the cellular dCTP concentration to 78% of control and decreased ara-CTP synthesis to 57% of control. 3-Deazauridine is a known inhibitor of CTP synthetase and indirectly depletes dCTP via a CTP depletion. Preincubation with 5 μM 3-deazauridine dCTP and subsequent exposure to 10 μM ara-C reduced dCyd and dCTP to 74% and 22% of control respectively, while ara-CTP formation remained at 90% of control. dTHU acts as an inhibitor of dCMP deaminase. Exposure to 100 μM dTHU increased the dCyd pool by 2.2-fold and enlarged the dCTP pool by 2.3-fold. These results indicate that also in a single cell line the degree of dCTP depletion does not predict the activity of ara-CTP formation by the salvage pathway. Moreover, depending on the respective enzyme inhibition, dCyd and dCTP may be modulated differently.

In conclusion, the here presented data support the understanding that the cellular concentration of dCTP is an insufficient predictor of ara-CTP accumulation. Accordingly, feed-back inhibition of dCK by dCTP does not play a significant role in whole cells. By contrast, cellular concentrations of dCyd provided a better basis to explain differential ara-CTP accumulation not only in different cell lines but also in a single cell line exposed to different modulators of dNTP metabolism. Specifically, a positive correlation was observed between the amount of dCyd pool depletion and the activation of ara-C phosphorylation. By contrast, inhibition of dCMP deaminase by some modulators may induce an expansion of the dCyd pool which may result in a competitive inhibition of ara-C synthesis.

ACKNOWLEDGMENT

This work was supported by the Deutsche Forschungsgemeinschaft, Grant He 2029/3-1.

REFERENCES

1. Momparler R L, Fischer G A. Mammalian deoxynucleoside kinases. J Biol Chem 243: 4298–4304, 1968
2. Kim M-Y, Ives D H. Human deoxycytdine kinase: kinetic mechanism and end product regulation Biochemistry 28: 9043–9047, 1989
3. Plagemann P G W, Marz R, Wohlhueter R M. Transport and metabolism of deoxycytidine and 1-β-D-arabinofuranosylcytosine into cultured Novikoff rat hepatoma cells, relationship to phosphorylation, and regulation of triphosphate synthesis. Cancer Res 38: 978–989, 1978
4. Kubota M, Takimoto T, Tanizawa A, Akiyama Y, Mikawa H. Differential modulation of 1-β-D-arabinofuranosylcytosine metabolism by hydroxyurea in human leukemic cells. Biochem Pharmacol 37: 1745–1749, 1988
5. Heinemann V, Plunkett W. Modulation of deoxynucleotide metabolism by the deoxycytidylate deaminase inhibitor 3,4,5,6-tetrahydrodeoxyuridine. Biochemical Pharmacol 38: 4115–4121, 1989
6. Abe I, Saito S, Hori K, Suzuki M, Sato H. Role of dephosphorylation of 1-β-D-arabinofuranosylcytosine 5′-triphosphate in human lymphoblastic cell lines with reference to their drug sensitivity. Cancer Res 42: 2846–2851, 1982
7. Shewach D, Reynolds K K, Hertel L. Nucleotide specificity of human deoxycytidine kinase. Mol Pharmacol 42: 518–524, 1992

50

KINETIC PROPERTIES OF CTP SYNTHETASE FROM HL-60 CELLS

André B. P. Van Kuilenburg, Lida Elzinga, and Albert H. Van Gennip

Academic Medical Center
University of Amsterdam
Department Clinical Chemistry and Division Emma Children's Hospital
P.O. Box 22700
1100 DE Amsterdam, The Netherlands

1. INTRODUCTION

CTP synthetase (EC 6.3.4.2) catalyses the formation of CTP from UTP with the concomitant deamination of glutamine to glutamate:

$$UTP + ATP + glutamine \xrightarrow{\text{GTP, Mg}^{2+}} CTP + ADP + P_i + glutamate$$

CTP synthetase is generally regarded as the rate-limiting enzyme in the synthesis of cytosine nucleotides from both *de novo* and uridine-salvage pathways. Increased activity of CTP synthetase has been observed in a variety of malignant cells.[1] Furthermore, the increased activity of CTP synthetase is often associated with the state of progression of tumours and hence provides a biochemical marker for the clinical aggressiveness of tumours.[1] In human lymphoblastic leukemia cells, we showed that the synthesis of CTP occurs predominantly via CTP synthetase, whereas in proliferating normal human T lymphocytes the salvage of cytidine is preferred.[2] Moreover, the increased activity of CTP synthetase proved to be associated with the process of transformation.[3] Previously, it was shown that the enzyme from Ehrlich *ascites* tumour cells possessed different kinetic properties as compared to the bovine liver enzyme.[4,5] Furthermore, mutations eliminating the allosteric regulation of CTP synthetase by CTP were demonstrated to cause a form of multidrug resistance in cultured Chinese hamster ovary cells.[6] However, it remains to be established whether altered kinetic properties of CTP synthetase in transformed cells provide an explanation for the increased activity of this enzyme in transformed cells. So far, studies of human CTP synthetase have not been reported. Therefore, we investigated the kinetic properties of CTP synthetase from HL-60 cells.

Purine and Pyrimidine Metabolism in Man IX,
edited by Griesmacher *et al.* Plenum Press, New York, 1998.

2. MATERIALS AND METHODS

2.1. Culturing and Harvesting of HL-60 Cells

Cultures of HL-60 cells were maintained in humidified air with 5% CO_2 at 37°C in RPMI-1640 medium supplemented with 2 mM glutamine, 100 U/ml penicillin, 100 µg/ml streptomycin and 10% (v/v) heat-inactivated fetal calf serum. Cells were kept in exponential growth by diluting the cells to a density of 3×10^5 cells/ml every 2–3 days. HL-60 cells were collected by centrifugation (223 g for 10 min) and washed once with PBS and the cell pellet was frozen in liquid nitrogen and stored at −80°C until further analysis.

2.2. Preparation of Cell Homogenates

The HL-60 cells were solubilised and lysed at a concentration of 20×10^6 cells/ml in a suspension buffer containing 35 mM Tris-Mops (pH 7.9), 1 mM EGTA, 2.5 mM phenyl-methylsulphonyl fluoride and 10 mM dithiothreitol. The HL-60 cells were lysed by rapid freeze-thawing twice in liquid nitrogen. The samples were centrifuged in a microfuge at 11000 g for 15 min at 4°C. CTP synthetase was precipitated from the clear supernatants with 45% ammonium sulphate. The resulting precipitate was collected by centrifugation and the pellet was dissolved in the suspension buffer. Finally, the CTP synthetase preparation was desalted by gel filtration.

2.3. Standard Assay Method

The reaction mixture contained an aliquot of cell sample, 35 mM Tris-Mops (pH 7.9), 4 mM ATP, 1 mM UTP, 1 mM GTP, 10 mM L-glutamine, 20 mM $MgCl_2$, 10 mM KCl, 1 mM EGTA, 10 mM dithiothreitol, 2 mM PEP and pyruvate kinase (20 U/ml) in a total volume of 50 µl. Separation of nucleotides was performed isocratically (0.594 M NaH_2PO_4 (pH 4.55) at a flow-rate of 1 ml/min) by anion-exchange HPLC on a Partisphere SAX column (125 × 4.6 mm, 5 µm particle size).[7]

2.4. Kinetic Analysis

The data were fitted according to the Hill equation: $v/V_{max} = [S]^n/(k_{app} + [S]^n)$. Apparant K_m values were calculated at half maximal velocity as $[S]_{0.5}$ according to $[S]_{0.5} = {}^n\tilde{A}K_{app}$.

3. RESULTS

Surprisingly, a high positive cooperativity was observed for UTP in the presence of saturating concentrations of ATP and GTP (Fig. 1A). Analysis of the kinetic data by the Hill equation showed a Hill number of 3.6 (Fig. 2A) and a $[S]_{0.5}$ value of 280 µM. In contrast, no cooperativity was observed for ATP in the presence of saturating concentrations of UTP and GTP (Fig. 1B) and normal Michaelis-Menten kinetics were followed (n = 1.0, Fig. 2B). A K_m value of 1788 µM for ATP could be calculated. A slight positive cooperativity was observed for GTP in the presence of saturating concentrations of UTP and ATP (Fig. 1C). The Hill equation showed a Hill number of 1.4 (Fig. 2C) and a $[S]_{0.5}$ value of 154 µM.

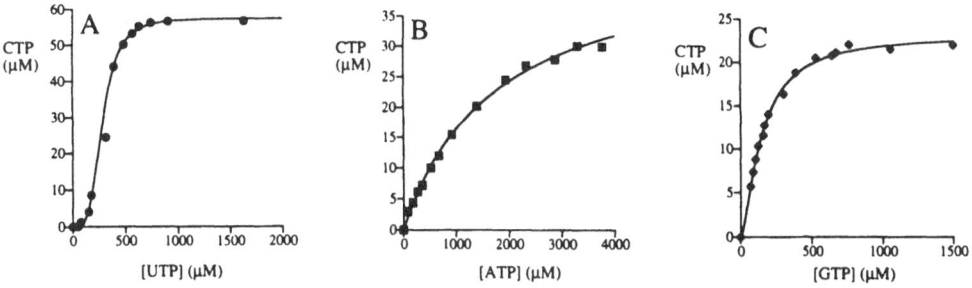

Figure 1. Kinetics of CTP synthetase from HL-60 cells. CTP synthetase activity was measured as a function of UTP (●), ATP (■) and GTP (◆). The reaction was allowed to proceed for 1 h at 37°C.

4. DISCUSSION

Studies of CTP synthetase from human tissues or tumour cells are usually hampered by the limited amount of human material available. Therefore, CTP synthetase was partially purified from HL-60 homogenates by ammonium sulphate precipitation and directly used in the assays to study the CTP synthetase kinetics. In this way, we observed a profound cooperative behavior of CTP synthetase from HL-60 cells for the substrate UTP. The Hill number of 3.6 indicates that at least 4 substrate binding sites are involved in the reaction catalyzed by CTP synthetase. This result suggests that under these conditions the active conformation of CTP synthetase from HL-60 cells is a tetramer. In contrast, it has been shown that CTP synthetase from Ehrlich ascites tumour cells lacked the ability to tetramerize in the presence of saturating amounts of UTP and ATP.[5]

Our results are not in line with those obtained for CTP synthetase from other mammalian species. For CTP synthetase from bovine liver normal Michaelis-Menten kinetics were observed for UTP at saturating or subsaturating concentrations of ATP.[8,9] Normal Michealis-Menten kinetics were also observed for the enzyme from Ehrlich ascites tumour cells[4] and murine leukemia L1210 cells.[10] However, at subsaturating concentrations of ATP a slight positive cooperativity was observed for UTP with a Hill number of 1.3 for the enzyme from Ehrlich ascites tumour cells.[4]

A characteristic common to these purified mammalian CTP synthetases is the inhibition of their activities by the product CTP.[5,9] CTP inhibited the activity of these enzymes by increasing the positive cooperativity of the enzyme for UTP. The regulation of the

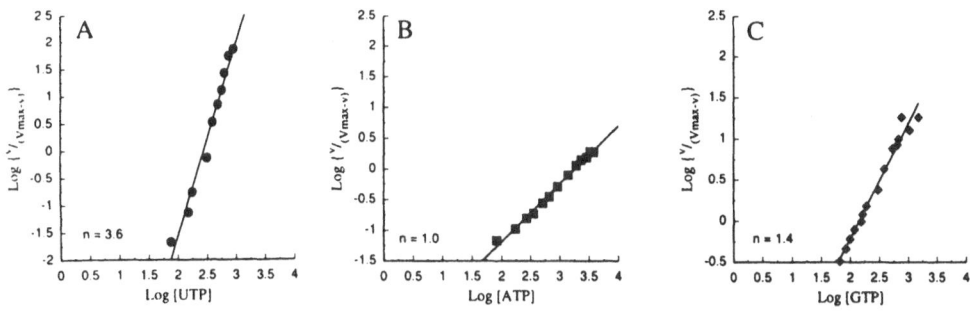

Figure 2. Hill plots of the reaction catalyzed by CTP synthetase. Panel A shows the Hill plot for UTP (●), panel B shows the Hill plot for ATP (■) and panel C shows the Hill plot for GTP (◆). The data were obtained from Figure 1.

activity of CTP synthetase by CTP plays in important role *in vivo*. Mammalian cell line cells possessing mutations that rendered CTP synthetase insensitive to the inhibition by CTP exhibited abnormal elevated intracellular levels of CTP and dCTP which were accompanied by resistance to various chemotherapeutics and an increased rate of spontaneous mutations.[6] Whether or not the profound positive cooperativity of CTP synthetase from HL-60 cells towards UTP, in the absence of added CTP, is shared by CTP synthetase from other human tissues as well remains to be investigated.

In the yeast *Saccharomyces cerevisiae*, two isoforms of CTP synthetase are present which are encoded by two different genes, URA7 and URA8. Although there is a high degree of homology between these two genes striking differences appeared to exist with respect to the kinetic properties of these enzymes.[11] Interestingly, the activity of CTP synthetase in *Saccharomyces cerevisiae* is regulated by phosphorylation which decreases the cooperativity in substrate binding.[12,13] Whether or not human CTP synthetase from tumours in particular and from normal tissues in general is also regulated by phosphorylation is currently under investigation.

REFERENCES

1. Ellims, P.H., Gan, T.E., Medley, G. (1983) Cytidine triphosphate synthetase activity in lymphoproliferative disorders. *Cancer Res.* **43**, 1432–1435.
2. Van den Berg, A.A., Van Lenthe, H., Busch, S., De Korte, D., Van Kuilenburg, A.B.P. and Van Gennip, A.H. (1994) The Roles of Uridine-Cytidine Kinase and CTP Synthetase in the Synthesis of CTP in Malignant Human T-Lymphocytic Cells. *Leukemia* **8**, 1375–1378.
3. Van den Berg, A.A., Van Lenthe, H., Busch, S., De Korte, D., Roos, D., Van Kuilenburg, A.B.P. and Van Gennip, A.H. (1993) Evidence ofr transformation-related increase in CTP synthetase activity *in situ* in human lymphoblastic leukemia. *Eur. J. Biochem.* **216**, 161–167.
4. Kizaki, H., Sakurada, T. and Weber, G. (1981) Purification and properties of CTP synthetase from Ehrlich ascites tumor cells. *Biochim. Biophys. Acta* **662**, 48–54.
5. Kizaki, H., Ohsaka, F. and Sakurada, T. (1985) CTP synthetase from Ehrlich ascites tumor cells. Subunit stoichiometry and regulation of activity. *Biochim. Biophys. Acta* **829**, 34–43.
6. Whelan, J., Phear, G., Yamauchi, M. and Meuth, M. (1993) Clustered base substitutions in CTP synthetase conferring drug resistance in Chinese hamster ovary cells. *Nat. Genet.* **3**, 317–321.
7. Van Kuilenburg, A.B.P., Elzinga, L., Verschuur, A.C., Van den Berg, A.A., Slingerland, R.J. and Van Gennip, A.H. (1997) Determination of CTP synthetase activity in crude cell homogenates by a fast and sensitive non-radiochemical assay using anion-exchange high-performance liquid chromatography. *J. Chromatogr. B*, in press.
8. Savage, C.R. and Weinfeld, H. (1970) Purification and Properties of Mammalian Liver Cytidine Triphosphate Synthetase. *J. Biol. Chem.* **245**, 2529–2535.
9. McPartland, R.P. and Weinfeld, H. (1979) Cooperative Effects of CTP on Calf Liver CTP Synthetase. *J. Biol. Chem.* **254**, 11394–11398.
10. Zhang, H., Cooney, D.A., Zhang, M.H., Ahluwalia, G., Ford Jr., H. and Johns, D.G. (1993) Resistance to Cyclopentenylcytosine in Murine Leukemia L1210 Cells. *Cancer Res.* **53**, 5714–5720.
11. Nadkarni, A.K., McDonough, V.M., Yang, W.-L., Stukey, J.E., Ozier-Kalogeropoulos, O. and Carman, G.M. (1995) Differential Biochemical Regulation of the URA7- and URA8-encoded CTP synthetases from *Saccharomyces cerevisiae*. *J. Biol. Chem.* **270**, 24982–24988.
12. Yang, W.-L., Bruno, M.E.C. and Carman, G.M. (1996) Regulation of Yeast CTP Synthetase Activity by Protein Kinase C. *J. Biol. Chem.* **271**, 11113–11119.
13. Yang, W.-L. and Carman, G.M. (1996) Phosphorylation and Regulation of CTP Synthetase from *Saccharomyces cerevisiae* by Protein Kinase A. *J. Biol. Chem.* **271**, 28777–28783.

51

CELLULOMONAS sp. PURINE NUCLEOSIDE PHOSPHORYLASE (PNP)

Comparison with Human and *E. coli* Enzymes

Beata Wielgus-Kutrowska,[1] Jan Tebbe,[2] Werner Schröder,[3] Marija Luic,[4]
David Shugar,[1] Wolfram Saenger,[2] Gertraud Koellner,[2*] and
Agnieszka Bzowska[1*]

[1]Department of Biophysics
Institute of Experimental Physics
University of Warsaw
Zwirki i Wigury 93, 02-089 Warsaw, Poland
[2]Institut für Kristallographie
Freie Universität Berlin
Takustr. 6, D-14195 Berlin, Germany
[3]Institut für Biochemie
Freie Universität Berlin
Thielallee 63, D-14195 Berlin, Germany
[4]Rudjer Boskovic Institute
Bijenicka 54, 10000 Zagreb, Croatia

1. BACKGROUND

The ubiquitous enzyme purine nucleoside phosphorylase (PNP, E.C. 2.4.2.1.) catalyzes the reversible phosphorolysis of naturally occurring purine nucleosides, and many analogues, as follows:

$$\beta\text{-purine nucleoside} + \text{orthophosphate} \Leftrightarrow \text{purine} + \alpha\text{-D-pentose-1-phosphate}$$

Inhibitors of PNP are potential immunosuppressive, antiviral and antiparasitic agents.[1,2,3] Properties of PNP from various sources are also of interest in light of reports that the product of the *E. coli* DeoD (PNP) gene mediates the killing of melanoma cells.[4]

---

* Corresponding authors.

Purine and Pyrimidine Metabolism in Man IX,
edited by Griesmacher *et al.* Plenum Press, New York, 1998.

Two main classes of PNPs have been characterized, so-called "low-molecular weight" ($M_r \sim 90$ kDa), trimeric, mainly mammalian enzymes, specific for 6-oxo-purines (e.g. human erythrocytes and calf spleen), and less specific "high-molecular weight" ($M_r \sim 110-140$ kDa), mainly hexameric, bacterial, phosphorylases (e.g. *E. coli, S. typhimurium*) that accept as substrates 6-oxo- and 6-aminopurines, as well as some unusual purine analogues like benzimidazole.[1,2,5-7] We herein describe some of the properties of the PNP from *Cellulomonas sp.* which differs from the foregoing in that some of its properties are common to both classes.

2. PURIFICATION

Commercially available enzyme, with specific activity ~8 U/mg, was purified on a BioCad station (PerSeptive Biosysytems) using a strong ion exchange poros 20HQ column, followed by a poros 20PI weak exchange column (both 4.6×100 mm). The enzyme was removed from the columns by an NaCl gradient, concentrated and desalted. The final specific activity of the purified enzyme was 107 U/mg, i.e. close to the specific activities of human and *E. coli* PNPs (95 U/mg and 104 U/mg, respectively).[8-10]

3. SOME PHYSICO-CHEMICAL PROPERTIES

The subunit molecular weight, determined by both mass spectrometry (MALDI method) and electophoresis under denaturating conditions, was found to be 29 kDa, i.e. very close to that of human PNP[1]. In solution the enzyme seems to be a tetramer, since the M_r obtained by gel filtration (FPLC, Superose 12 HR 10/30 column, Pharmacia) is 112 kDa. The subunit composition of *Cellulomonas* enzyme therefore differs from both human and *E. coli* PNPs that exist in trimeric and hexameric forms, respectively.[1,5]

Temperature stability (pH 7, no ligands present) of *Cellulomonas* PNP is similar to that of *E. coli* enzyme,[11] i.e. the enzyme is stable up to ~50°C and 50% decrease of activity is observed at 62°C, in contrast to mammalian PNPs[11] for which activity decreases rapidly above ~35°C and 50% decrease is observed at 40°C. When diluted and not stabilized by presence of ligands, the *Cellulomonas* enzyme is rather unstable in acidic media (pH < 6.5) but, in striking contrast to both mammalian and *E. coli* PNPs,[11] it is stable in alkaline media (pH 8.5–10) in the presence of orthophosphate.

4. SEQUENCING

Partial amino acid sequence (~23 kDa) of the *Cellulomonas* PNP was analyzed by automatic Edman-degradation on an Applied-Biosystem gas phase sequencer. Sequencing of peptides from different digests (with use of various proteases and cyanobromide) revealed no sequence homology of the 50 N-terminal amino acids with phosphorylases from other sources. In other parts of the sequence indication of some similarities to mammalian PNPs[12-14] were noted, in particular the Asn 243 of the purine binding site[15,16] seems to be conserved. For all peptides sequenced to date, closest homology was found to the putative PNP from *Mycobacterium leprae*.[17,18]

5. CRYSTALLIZATION AND X-RAY STUDIES

Crystals of the purified PNP were obtained from both organic solvents and high ionic strength solutions. Crystals grown from the latter appear to be more suitable for X-ray diffraction studies, and have the form of regular cubes with dimensions up to 0.6 mm. They diffract to at least 2.8 Å resolution with synchrotron radiation. The space group is cubic $P4_232$ with a = 158 Å and probably one tetramer in the asymmetric crystal unit. Complete data set was collected up to 3 Å under cryogenic conditions with R-merge = 0.11 (Daresbury Laboratory, station 9.5).

6. KINETIC PROPERTIES

6.1. Specificity

The kinetic properties of the *Cellulomonas* PNP, determined spectophotometrically by initial velocity method as previously described for PNPs form other sources,[19,20,6] are listed in Tables 1 and 2. From Table 1 it will be seen that the enzyme is specific for 6-oxopurines since adenosine is not phosphorolysed. Hence the base binding site appears at first sight to be similar to that for mammalian PNPs that have two H-bonding residues (Asn 243 and Glu 201) and one integral water molecule bridging N(1)–H and O^6 of the base,[15,16] while *E. coli* PNP has an open base binding site with only one H-bonding residue.[21] But inhibition studies revealed that adenosine and its structural analogue, formycin A, are both competitive inhibitors of *Cellulomonas* PNP with inhibition constants of 160 μM and 62 μM, respectively, in contrast to mammalian enzymes that are specific, in both binding and catalysis, for 6-oxopurines.[1,6]

6.2. Cooperativity

PNPs from different sources are known to show complex kinetic behaviour *vs* different substrates, manifested by nonlinear kinetic plots, e.g. v_o/S *vs* v_o or $1/v_o$ *vs* $1/S$ plots, indicating negative cooperativity or more than one binding site with different affinities.[22] The human enzyme shows this effect *vs* both nucleoside and orthophosphate[1], and the *E. coli* enzyme only *vs* phosphate.[5] The behaviour of *Cellulomonas* PNP is displayed in Figure 1 and Table 1. It will be seen that inosine phosphorolysis obeys simple Michaelis-Menten kinetics over a rather broad concentration range (0.01–1 mM), with $K_m^{app} \sim 43$ μM, very similar to that observed for the human enzyme (K_m^{app} = 28–46 μM) at low inos-

Table 1. Specificity and kinetic properties of PNP from *Cellulomonas sp*

Substrate	Cooperativity	Concentration range (mM)	Hill coefficient	K_m^{app} (μM)	V_{max}^{app} (U/mg)
Inosine[a]	not observed	0.010–1.0	1.00 ± 0.04	43	107
Guanosine[a]	not observed	0.005–0.4	0.98 ± 0.05	11	72
Adenosine[a]		not a substrate			
Orthophosphate (P$_i$)[b]	observed	0.010–0.1	0.55 ± 0.04	19	48
		1.0–25.0	1.21 ± 0.10	330	107

[a]In 50 mM phosphate buffer pH 7.
[b]With inosine at a fixed concentration of 1 mM and in 50 mM ammonium acetate buffer pH 7.

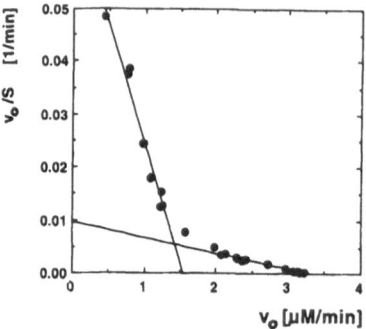 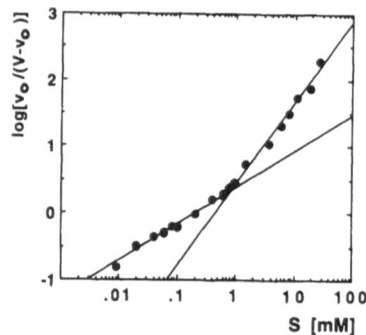

Figure 1. Kinetic properties of orthophosphate (P_i) with *Cellulomonas sp.* PNP (25°C, 50 mM ammonium acetate buffer pH 7, with fixed inosine concentration of 1 mM): v_o is initial velocity, S is the initial P_i concentration, V the maximal velocity of phosphorolysis observed experimentally. Left panel: Scatchard plot of v_o/S *vs* v_o. Although such a procedure is not correct for non-linear Scatchard plots,[23] straight lines were fitted to linear portions of the plot to obtain apparent kinetic constants comparable with those published for human PNP[1] (see also text). Right panel : Hill plot of $log[v_o/(V-v_o)]$ *vs* log S. Straight lines were fitted for low (< 0.1 mM) and high (> 1.0 mM) P_i concentrations to obtain Hill coefficients (see also Table 1).

ine concentrations.[1,6] By contrast, orthophosphate (P_i) shows complex kinetic behaviour with a non-liner Scatchard plot of v_o/S *vs* v_o (see Figure 1, left panel). Fitting straight lines to linear portions of the plot give apparent Michaelis constants of 19 µM and 330 µM, for low (< 0.1 mM) and high (> 1.0 mM) substrate concentrations, respectively (see Table 1). These values can be compared with 66 µM and 423 µM for the human enzyme,[1] also obtained by fitting straight lines for linear portions of the plot $1/v_o$ *vs* $1/S$. It is known, however, that such fitting is not correct for non-linear Scatchard plots[23] and therefore values obtained by such a method are apparent ones.[22]

The same data for orthophosphate as substrate of *Cellulomonas* PNP were analyzed by means of Hill plots (see Figure 1, right panel). The Hill plot is complex showing two regions that are more or less linear (correlation coefficients 0.978 and 0.985 for low and high concentrations of P_i, respectively). The corresponding Hill coefficients are 0.55 for low P_i concentrations and 1.21 for high concentrations. This points to a mixture of negative and positive cooperativity.[24] Detailed analysis of kinetic data for orthophosphate as substrate of *Cellulomonas* PNP in terms of number of binding sites, cooperativity between them and corresponding kinetic constants are under study.

6.3. Inhibition

Inhibition constants for some substrates and typical mammalian and bacterial PNP inhibitors are listed in Table 2. Guanine, a product of phosphorolysis, shows very potent inhibition (K_i = 3.6 µM), competitive *vs* inosine, a property of *Cellulomonas* PNP similar to mammalian enzymes that also bind purine bases strongly with µmolar inhibition constants,[1] in contrast to *E. coli* PNP, which displays noncompetitive weak inhibition by purine bases.[5]

Moreover *Cellulomonas* PNP is inhibited by typical mammalian PNP inhibitors, e.g. acyclovir, with an inhibition constant close to that for the human enzyme.[25] In addition typical *E. coli* PNP inhibitors like formycins are found to interact with *Cellulomonas* PNP binding site at µM concentrations. Inhibition constants, however, differ for formycins A and B by an order of magnitude (160 and 19 µM, respectively, see Table 2), while for *E.*

Table 2. Apparent inhibition constants
for inhibition of phosphorolysis by
Cellulomonas sp. PNP by some typical
inhibitors of mammalian and *E. coli* PNPs[a]

Inhibitor	K_i^{app} (μM)
Guanine	3.6
Adenosine	160
Acyclovir (ACV)	110
N(7)-acycloguanosine	97
N(9)-acycloguanosine	3.4
br^k-N(9)-acycloguanosine	4.1
Formycin A	62
Formycin B	19

[a]In all instances, inhibition was found to be competitive *vs* nucleoside substrate. Inhibition constants were determined with inosine as substrate in 50 mM phosphate buffer pH 7, at 25°C, from Dixon plots of $1/v_o$ *vs* inhibitor concentration. N(7)-acycloguanosine: Guo–N(7)–CH$_2$–CH$_2$–NH$_2$–CH–(CH$_2$OH)$_2$, with similar connotations for other acycloguanosines.

coli PNP both constants are about 5 μM (5.3 and 4.6 μM, respectively).[26] The human enzyme is not inhibited by formycin A, while formycin B is a weak inhibitor (K_i = 100 μM).[1]

In contrast to human PNP, the *Cellulomonas* enzyme has a much stronger affinity for N(9)-acycloguanosine relative to its N(7) counterpart (K_i 3.4 and 97 μM, respectively, see Table 2). With the human enzyme the opposite effect is observed (K_i is 14 μM and 5 μM for N(9) and N(7)-acycloguanosines).[27]

7. CONCLUSIONS

Cellulomonas PNP, with an interesting combination of properties of human and *E. coli* phosphorylases, differs significantly from the two latter in both physico-chemical and kinetic properties. Overall results, including ongoing X-ray diffraction studies, should contribute to a better understanding of the mode of action of this important class of enzymes, and to the design of more effective inhibitors for chemotherapy.

ACKNOWLEDGMENTS

Supported by the Polish Committee for Scientific Research (KBN grant 6 P04A 062 09), and, in part, by an International Research Scholar's award from the Howard Hughes Medical Institute (HHMI 79195-543401).

REFERENCES

1. J.D. Stoeckler (1984) in Developnemts in Cancer Chemotherapy (R.I. Glazer, ed.) 35–60, CRC Press, Boca Raton, FL.
2. J.A. Montgomery (1993) Med. Res. Rev. vol. 13, 209–228, John Wiley and Sons.

3. E.R. Giblett, A.J. Ammann, D.W. Wara and L.K. Diamond (1975) Lancet 1, 1010–1013.
4. B.W. Hughes, A.H. Wells, Z. Bebok, V.K. Gadi, R.I. Garver, Jr., W.B. Parker and E.J. Sorscher (1995) Cancer Res. 55, 3339–3345.
5. K.J. Jensen and P. Nygaard (1975) Eur. J. Biochem. 51, 253–265.
6. A. Bzowska, E. Kulikowska and D. Shugar (1990) Z. Naturforsch. 45c, 59–70.
7. N. Hori, M. Watanabe, Y. Yamazaki and Y. Mikami (1989) Agric. Biol. Chem 53, 3219–3224.
8. W.R.A. Osborne (1980) J. Biol. Chem. 255, 7089–7092.
9. W.J. Cook, S.E. Ealick, T.A. Krenitsky, J.D. Stoeckler, J.R. Helliwell and C.E. Bugg (1985) J. Biol. Chem. 260, 12968–12969.
10. B. Wielgus-Kutrowska, E. Kulikowska, J. Wierzchowski, A. Bzowska and D. Shugar (1997) Eur. J. Biochem. 243, 408–414.
11. T.A. Krenitsky, G.W. Koszalka and J.V. Tuttle (1981) Biochem. 20, 3615–3621.
12. A. Bzowska, M. Luic, W. Schroder, D. Shugar, W. Saenger and G. Koellner (1995) FEBS Letters 367, 214–218.
13. S.R. Williams, J.M. Goddard and D.W. Martin, Jr. (1984) Nucl. Acids. Res. 12, 5779–5787
14. J.P. Jenuth and F.F. Snyder (1991) Nucleic Acids Res. 19, 1708–1708.
15. G. Koellner, M. Luic, D. Shugar, W. Saenger, A. Bzowska (1997) J. Mol. Biol. 265, 202–216.
16. S.V.L. Narayana, C.E. Bugg and S.E. Ealick (1997) Acta Cryst. D53, 131–142.
17. D.R. Smith and K. Robinson (1994) EMBL Genbank, P46862.
18. K. Robinson, W. Gilbert and G.M. Church (1994) Nature Genetics 7, 205–214.
19. J.D. Stoeckler, R.P. Agarwal, K.C. Agarwal, R.E. Parks, Jr. (1978) Meth. Enzymol. 51, 531–538.
20. H.M. Kalckar (1947) J. Biol. Chem. 167, 429–443.
21. C. Li and S.E. Ealick (1996) Acta Cryst. A52, Suppl. C205
22. F.W. Dahlquist (1978) Meth. Enzymol. 48, 270–299.
23. K. Zierler (1989) Trends Biol. Sci. 14, 314–317.
24. A. Cornish-Bowden and D.E. Koshland, Jr. (1975) J. Mol. Biol. 95, 201–212.
25. J.V. Tuttle and T.A. Krenitsky (1984) J. Biol. Chem. 259, 4065–4069.
26. A. Bzowska, E. Kulikowska and D. Shugar (1992) Biochim. Biophys. Acta 1120, 239–247.
27. A. Bzowska, A. Ananiev, N. Ramzaeva, E. Alksnis, J.A. Maurins, E. Kulikowska and D. Shugar (1994) Biochem. Pharmac. 48, 937–947.

52

SUBDOMAIN STRUCTURE OF CARBAMYL PHOSPHATE SYNTHETASE

Modular Construction of the Enzyme

Hedeel I. Guy and David R. Evans

Department of Biochemistry and Molecular Biology
Wayne State University School of Medicine
Detroit, Michigan 48201

1. INTRODUCTION

The overall biosynthetic reaction, catalyzed by carbamyl phosphate synthetase (CPSase, EC 6.3.5.5) occurs in a series of four partial reactions which are catalyzed by different domains or subunits of the enzyme and must be precisely coordinated.[1]

$$glutamine \longrightarrow glutamate + NH3$$
$$bicarbonate + ATP \longrightarrow carboxyphosphate + ADP$$
$$carboxyphosphate + NH3 \longrightarrow carbamate + Pi$$
$$carbamate + ATP \longrightarrow carbamyl\ phosphate + ADP$$

The 40 kDa glutaminase domain or subunit (GLN) hydrolyzes glutamine and transfers ammonia to the 120 kDa synthetase domain or subunit (CPS) where carbamyl phosphate is formed.[2,3] The CPS domain, consists of two homologous halves,[4] designated CPS.A and CPS.B, which catalyze the two ATP dependent reactions,[5] the activation of bicarbonate and the phosphorylation of carbamate, respectively. The only known exception is mammalian CPSase I, the mitochondrial urea cycle enzyme,[1] which differs only in that it lacks a functional GLN domain and uses ammonia directly in the biosynthesic reaction.

In *E. coli* CPSase, the GLN and CPS domains are separate subunits,[2] while in the mammalian pyrimidine specific CPSase, the GLN and CPS domains are part of a 243 kDa multifunctional protein called CAD[6] which also carries the enzyme activities that catalyze the second and third steps in the *de novo* pathway, aspartate transcarbamylase (ATC) and dihydroorotase (DHO). Despite the diversity in structural organization among the CPSases from different organisms and pathways, the overall sequence similarity is strong and it is clear that these proteins share a common domain structure.

Purine and Pyrimidine Metabolism in Man IX,
edited by Griesmacher *et al.* Plenum Press, New York, 1998.

Sequence comparisons and controlled proteolysis suggest that the major structural domains are in turn comprised of subdomains.[7,8] The carboxyl half (G2) of the GLN domain contains the three conserved regions which are found in all amidotransferases, while the amino half (G1) of the domain is unique to carbamyl phosphate syntehtases. The CPS.A and CPS.B domains consist of three distinct regions, A1, A2, A3 and B1, B2, B3 respectively. Chemical modification[9–12] and site directed mutagenesis[5,13,14] indicate that nucleotide binding sites for the two ATP dependent partial reactions are located within A2 and B2. To confirm that these putative substructural elements represent a separately folded, autonomous subdomains and to determine their function, we have cloned and expressed segments of the cDNA or gene encoding these regions and characterized the resulting recombinant proteins.

2. RESULTS

2.1. Glutaminase Domain

The CAD GLN domain was expressed at high levels in *E. coli*,[3] but when purified was found to have barely detectable glutaminase activity (Table 1). However, when the mammalian GLN domain was mixed with a stoichiometric amount of the *E. coli* CPSase domain a fully functional hybrid complex was formed which catalyzed glutamine dependent CPSase activity at a rate comparable to that of the native enzyme. The putative GLN subdomains, the interaction subdomain (G1) and the catalytic subdomain (G2) were then separately cloned and expressed.[15] The interaction domain had no catalytic activity but formed a stable complex with the *E. coli* synthetase domain which could be isolated by gel filtration. The complex could catalyze the NH_3-dependent CPSase but lacking the G2 catalytic domain was unable to use glutamine as a source of nitrogen for carbamyl phosphate synthesis. The G2 subdomain was hyperactive (Table 1) but was unable to form a stable complex with the CPS subunit.

2.2. Synthetase Domain

Surprisingly the individual CPS.A and CPS.B domains of both the mammalian and bacterial proteins were found[16] to catalyze the overall synthesis of carbamyl phosphate from ammonia, ATP and bicarbonate. Moreover, if combined with the GLN domain, either covalently in the case of the mammalian enzyme or by formation of a non-covalent complex in the case of the *E. coli* enzyme, glutamine was hydrolyzed and the ammonia thus generated could be used to synthesize carbamyl phosphate. The kinetic parameters (not shown) were very similar to those measured for the native enzymes.

Table 1. Glutaminase activity of the mammalian GLN domain and subdomains

Protein	K_m^{gln} mM	k_{cat} s^{-1}
CAD (GLN-CPS)	95.3	0.14
GLN domain	1400	0.0165
GLN G1 subdomain	–	0
GLN G2 subdomain	110	5.73

Table 2. Synthetase activity of the *E. coli* CPS domains and subdomains

Protein	NH$_3$-CPSase		Gln-CPSase$_t$	
	K_m^{ATP} mM	k_{cat} s^{-1}	K_m^{ATP} mM	k_{cat} s^{-1}
GLN-CPSase	0.43	6.43	0.51	6.4
GLN-CPS.A1-A2	0.52	5.38	1.25	5.4
GLN CPS A2	0.36	75	nd	0

Deletion of A3 or B3 resulted in a fully functional molecule which could associate with the GLN subunit. The kinetic parameters of the A1-A2 (Table 2) and B1-B2 (not shown) were similar to that of the native enzyme. However, when the putative catalytic subdomain A2 and B2 were separately cloned and expressed the molecule could no longer form a functional complex with the GLN domain and thus could no longer catalyze glutamine dependent carbamyl phosphate synthesis. Moreover the ammonia dependent activity was 15-fold higher than that of the parent molecule. This hyperactivity establishes the role of A2 and B2 as the catalytic subdomains of CPS.A and CPS.B, respectively.

3. DISCUSSION

All of the domains and subdomains are expressed as stable, soluble proteins confirming that they are separately folded, autonomous structural entities. Moreover, each has a well-defined function. By employing this strategy to dissect the molecule, we may have retraced the course of evolution of the carbamyl phospahte synthetases in reverse. The core of the major domains, the GLN domain and the two synthetase domains, CPS.A and CPS.B, is a hyperactive catalytic domain, which lacks regulation and or the ability to form functional linkages with other domains or subunits. It is likely that the catalytic domains are descendants of an ancestral monofunctional glutaminase and kinase.

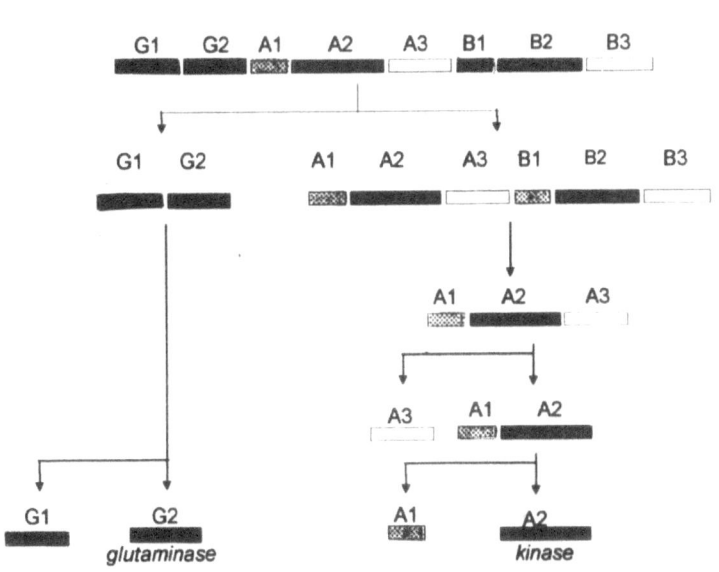

Figure 1. Dissection of carbamyl phosphate synthetase.

The domains and subdomains of CPSase have been systematically dissected by sub-cloning regions of the cDNA encoding them and expressing them in *E. coli*. Each of the two major functional domains of the GLN and duplicated CPS domains are comprised of a catalytic domain (black shading) and an interaction domain (gray shading). In addition, the CPS.A and CPS.B domains contain an additional subdomain (unshaded) which in CPS.B serves as a regulatory domain (Figure 1).

For coordinated catalysis and regulation, the intrinsically high activity of the catalytic domain must be attenuated. Thus, each of the GLN and CPS catalytic subdomains are fused to an interaction or attenuation domain which is required both for the physical association between the GLN and CPS domains and to suppress the activity of the catalytic domain. The consolidation of catalytic and interaction domains by gene translocation and fusion is likely to have occurred at some point early in the evolution of the protein. Carbamate kinase, an enzyme similar to CPSases in several respects, has been proposed[12] as an evolutionary precursor of the B1-B2 subdomains of carbamyl phosphate synthetase. Moreover, two recently characterized archebacterial CPSases[17,18] are similar in size and sequence to carbamate kinases. In the case of the CPS domain, a third subdomain was incorporated into the complex. The function of this domain in CPS.A is unknown, but in CPS.B it binds allosteric effectors and is the major locus of regulation.

While *E. coli* CPSase forms dimers and other higher oligmers, the monomer is fully active and regulated. In contrast, we have found that GLN-CPS.B,[16] CPS.B and B2 are dimeric and while the monomer can catalyze each of the ATP dependent partial reactions, the dimer is probably required to catalyze the overall reaction. However, in the evolution of the wild type enzyme, there was a gene duplication and fusion which resulted in the formation of the full synthetase domain, CPS.A-CPS.B. Once this occurred, catalysis of the overall reaction no longer required dimer formation. Although the CPS.A and CPS.B domains are functionally equivalent[16] in the sense that when isolated they can catalyze both partial reactions at an equivalent rate, once incorporated into the native enzyme they assume a specialized function,[5] the activation of bicarbonate or phosphorylation of carbamate, probably as a result of interdomain interactions.

The process of consolidation of catalytic activities into a single polypeptide chain continued in higher eukaryotic organisms. The GLN and CPS domains became fused and subsequently the ATC and DHO domains were recruited into the complex The structural organization of these molecules represents a nice example of the assembly of complex molecules into a mosaic of domains each having an important function.

ACKNOWLEDGMENT

This work was supported by GM47399.

REFERENCES

1. Anderson PM. (1995) Molecular Mechanism in Carbamyl Phosphate Synthesis. In: Nitrogen Metabolism and Excretion (eds. PJ Walsh, PA Wright), pp 33–55. CRC Press
2. Trotta, P. P., Burt, M. E., Haschemeyer, R. H., and Meister, A. (1971) *Proc Natl Acad Sci U S A* **68**(10), 2599–603
3. Guy, H. I., and Evans, D. R. (1994) *J Biol Chem* **269**(10), 7702–8
4. Nyunoya, H., and Lusty, C. J. (1983) *Proc Natl Acad Sci U S A* **80**(15), 4629–33
5. Post, L. E., Post, D. J., and Raushel, F. M. (1990) *J Biol Chem* **265**(14), 7742–7

6. Coleman, P., Suttle, D., and Stark, G. (1977) *J. Biol. Chem.* **252**, 6379–6385
7. Guillou, F., Rubino, S. D., Markovitz, R. S., Kinney, D. M., and Lusty, C. J. (1989) *Proc Natl Acad Sci U S A* **86**(21), 8304–8
8. Kim, H., Kelly, R. E., and Evans, D. R. (1992) *J Biol Chem* **267**(10), 7177–84
9. Powers-Lee, S. G., and Corina, K. (1987) *J Biol Chem* **262**(19), 9052–6
10. Kim, H. S., Lee, L., and Evans, D. R. (1991) *Biochemistry* **30**(42), 10322–9
11. Potter, M. D., and Powers-Lee, S. G. (1993) *Arch Biochem Biophys* **306**(2), 377–82
12. Alonso, E., and Rubio, V. (1995) *Eur J Biochem* **229**(2), 377–384
13. Stapleton, M. A., Javid-Majd, F., Harmon, M. F., Hanks, B. A., Grahmann, J. L., Mullins, L. S., and Raushel, F. M. (1996) *Biochemistry* **35**, 14352–14361
14. Javid-Majd, F., Stapleton, M. A., Harmon, M. F., Hanks, B. A., Mullins, L. S., and Raushel, F. M. (1996) *Biochemistry* **35**, 14362–14369
15. Guy, H. I., and Evans, D. R. (1995) *J. Biol. Chem.* **270**, 2190–2197
16. Guy, H. I., and Evans, D. R. (1996) *J Biol Chem* **272**, 13762–13769
17. Purcarea, C., Simon, V., Prieur, D., and Herve, G. (1996) *Eur. J. Biochem.* **236**, 189–199
18. Legrain, C., Demarez, M., Glansdorff, N., and Pierard, A. (1995) *Microbiology* **141**, 1093- 1099

53

THE EFFECTS OF NUCLEOTIDE POOL ON PURINE PRODUCTION IN THE POSTISCHEMIC HEART

Ryszard T. Smolenski, Jay Jayakumar, Anne-Marie L. Seymour, and Magdi H. Yacoub

Heart Science Centre
NHLI at Harefield Hospital
Harefield, United Kingdom

INTRODUCTION

Purine catabolite production in the heart during reperfusion after ischemia is markedly reduced.[1,2] Evaluation of the mechanism of this phenomenon is important for understanding of the regulation of nucleotide catabolism and the most important catabolite adenosine (ADO) production in the heart. On the other hand, reduced ADO production in the postischemic heart could be important mechanism of reperfusion injury due to numerous cardioprotective effects of ADO such as antiadrenergic, antiplatelet and antileukocyte action.[3,4] In this study we evaluated the relation between purine catabolite release and high energy phosphate metabolite concentrations in the heart before, during and after repeated ischemic intervals of varying duration.

MATERIALS AND METHODS

An isovolumic preparation of the rat heart perfused at constant pressure was used. As shown in Fig. 1, hearts were subjected to: (A) 1 min ischemia after 40 min of perfusion followed by reperfusion, 10 min of global (37°C) ischemia after 50 min of perfusion followed by reperfusion and to 1 min ischemia after 85 min of perfusion followed by reperfusion and (B) similar protocol with infusion of 30 μM adenosine during the first 15 min of reperfusion after 10 min ischemia at 50 min. At the end of experiment, hearts were frozen with aluminium clamps pre-cooled in liquid nitrogen. A ^{31}P NMR spectra were acquired throughout the experiment with 1 min time resolution allowing to quantitate ATP, PCr, in-

Purine and Pyrimidine Metabolism in Man IX,
edited by Griesmacher *et al.* Plenum Press, New York, 1998.

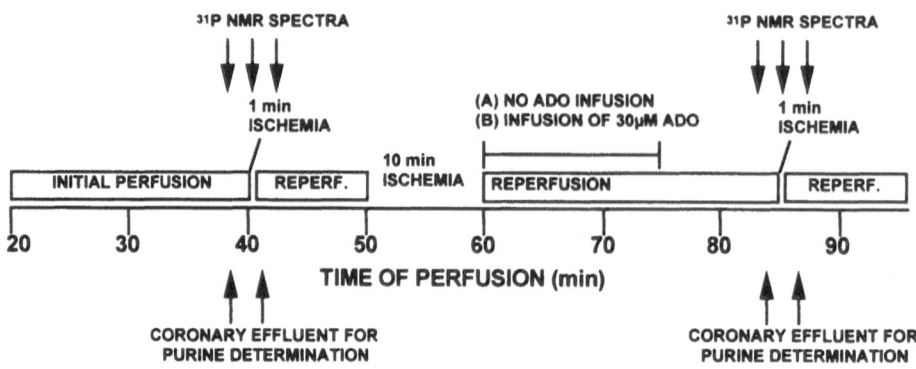

Figure 1. Experimental protocol.

tracellular inorganic phosphate and to estimate intracellular pH as described previously.[5,6] Coronary effluent for determination of purine catabolites was collected before, and immediately after 1 min ischemia. Hearts were freeze-clamped at the end of experiment to evaluate myocardial ATP concentration by HPLC.[7]

RESULTS

As may be seen in Table 1, phosphocreatine (PCr) decreased during 1 min ischemia at 40 min together with decrease of pH while ATP concentration remained constant in (A). These changes were similar during 1 min ischemia at 40 min in (B). During 1 min ischemia at 85 min in (A), PCr decreased from overshooting concentration to a level higher than during 1 min ischemia at 40 min, pH decreased less while ATP remained constant but at lower level than during 1 min ischemia at 40 min. In (B), ATP level before and during ischemia at 85 min was elevated as compared to B, while all other metabolic changes were similar to (A). Total purine catabolite release after 1 min ischemia at 40 min increased by 0.6–0.8 mM in both groups. This increase was reduced to 0.05 mM after ischemia at 85 min in (A) while in (B) release was 0.2 mM.

Table 1. Percentage changes in myocardial concentrations of phosphate metabolites and its relation to release of total purine catabolites (Pur = ADO + inosine + hypoxanthine + xanthine + uricacid) during 1 min ischemia (1'I) at 40 and at 85 min of perfusion[a]

	without ADO (Group A)					with ADO (Group B)				
	PCr	P,	ATP	pH	Pur µM	PCr	P,	ATP	pH	Pur µM
	initial = 100%			7.04	0.31	initial = 100%			7.05	0.36
1'I at 40'	64	163	90	6.92	0.81	52	159	92	6.90	0.92
prior 1'I at 85'	146	108	69	7.03	0.26	138	106	74	7.05	0.24
1'I at 85'	77	192	66	6.97	0.32	64	166	82	6.95	0.47
ATP at end	14.8 nmol/g dry wt					16.4 nmol/g dry wt				

[a]Values represent the mean, n=6 in A, n=5 in B. ATP concentration evaluated at the end of experiment by HPLC is also given.

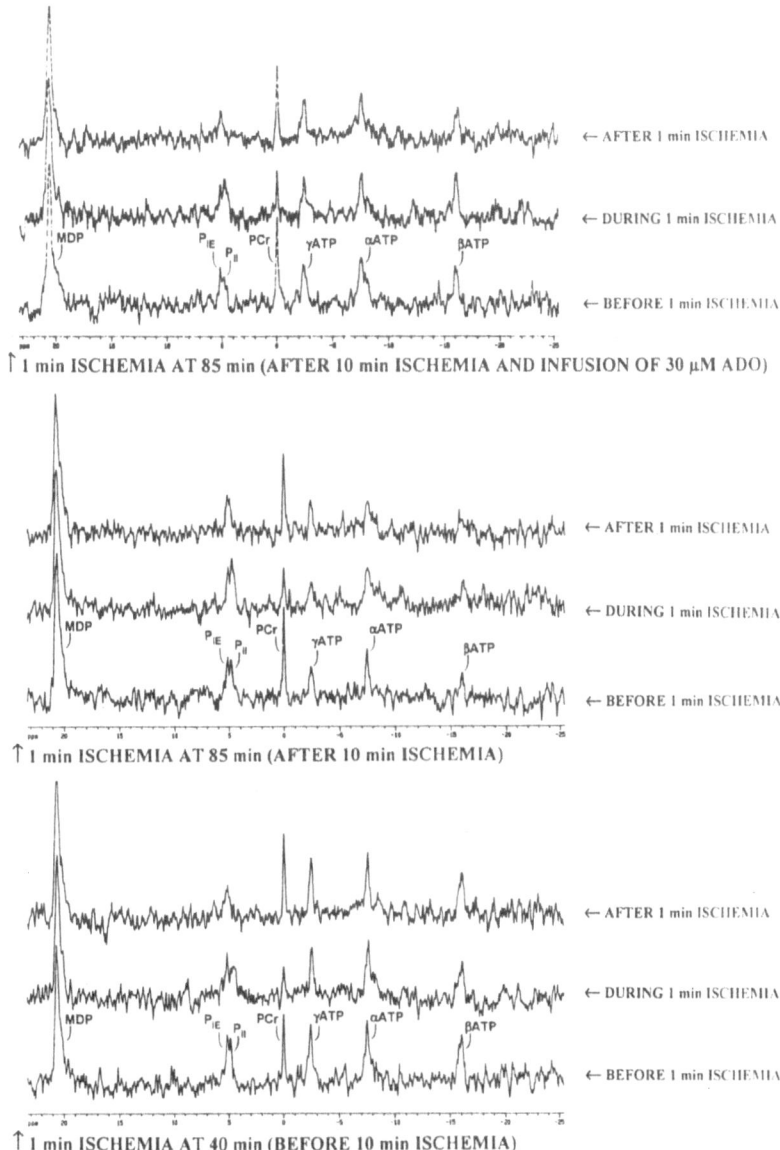

Figure 2. Typical ^{31}P NMR spectra acquired before during and after 1 min of ischemia at 40 or 85 min of experiment in controls (Group A) and hearts treated with adenosine (Group B).

DISCUSSION

Major finding of this study is the demonstration that the rate of nucleotide catabolite production could be affected not only by the energy status of the cell but also by nucleotide pool size in the heart. Although there is a general agreement that the role of phosphorylation potential in the regulation of the rate of purine production is critical,[8] importance of nucleotide pool size is questioned.[9] We have shown that in control group A, metabolic changes during 1 min ischemia at 85 min in comparison to that at 40 min were different by: higher preischemic PCr, higher PCr during 1 min ischemia and smaller decrease in pH, which could explain reduced purine production by energy status related mechanism. However, ATP was also lower before and during 1 min ischemia at 85 min. Infusion of ADO during the initial phase of reperfusion after 10 min ischemia in group B caused elevation of myocardial ATP without inducing any other major changes. Simultaneously, purine release after 1 min ischemia at 85 min was restored to about 30% of the release after 1 min ischemia at 40 min. Consequently, in addition to relation to the phosphorylation potential of the cell, purine catabolite production rate could be also affected by the size of adenine nucleotide pool.

ACKNOWLEDGMENTS

This study was supported by the British Heart Foundation (Grant No: PG 96194, and The Harefield Heart Transplant Trust.

REFERENCES

1. R.T. Smolenski, H.A. Simmonds, P.B. Garlick, G.E. Venn, and D.J. Chambers, Depressed adenosine and total purine catabolite production in the postischemic rat heart, *Cardioscience* 4:235(1993).
2. K.A. Reimer, C.E. Murry, I. Yamasawa, and R.B. Jennings, Four brief periods of myocardial ischemia cause no cumulative ATP loss or necrosis, *Am. J. Physiol.* 251:H1306(1986).
3. G.R. Gunther and M.B. Herring, Inhibition of neutrophil superoxide production by adenosine released from vascular endothelial cells, *Ann. Vasc. Surg.* 5:325(1991).
4. J. Schrader, G. Baumann, and E. Gerlach, Adenosine as inhibitor of myocardial effects of catecholamines, *Pflugers Arch.* 372:29(1977).
5. R.T. Smolenski, M.H. Yacoub, and A.-M.L. Seymour, Reduced purine catabolite production in the postischemic rat heart: A 31P NMR assesment of cytosolic metabolites, *Magma* 2:417(1994).
6. R.T. Smolenski, J.R. Jayakumar, A.-M.L. Seymour, and M.H. Yacoub, Energy metabolism and mechanica recovery after cardioplegia in moderately hypertrophied hearts, *Mol. Cell. Biochem.* in press(1997).
7. R.T. Smolenski, D.R. Lachno, S.J.M. Ledingham, and M.H. Yacoub, Determination of sixteen nucleotides, nucleosides and bases using high-performance liquid chromatography and its application to the study of purine metabolism in hearts for transplantation, *J. Chromatogr.* 527:414(1990).
8. J.P. Headrick and R.J. Willis, Adenosine formation and energy metabolism: a 31P-NMR study in isolated rat heart, *Am. J. Physiol.* 258:H617(1990).
9. R. Zucchi, G. Yu, S. Ronca Testoni, M. Mariani, and G. Ronca, Energy metabolism in myocardial stunning, *J. Mol. Cell Cardiol.* 24:1237(1992).

54

MUSCLE ENERGY METABOLISM IN HUMAN ADENYLOSUCCINASE DEFICIENCY

An in Vivo ³¹P-NMR Spectroscopy Study

C. Salerno, S. Iotti, R. Lodi, C. Crifò, and B. Barbiroli

Department of Biochemical Sciences
University of Roma
Clinical Biochemistry Laboratory
University of Bologna

1. INTRODUCTION

Adenylosuccinate lyase (EC 4.3.2.2; ASase) catalyses two steps in the synthesis of purine nucleotides: the conversion of succinylaminoimidazole carboxamide (SAICA) ribotide into aminoimidazole carboxamide (AICA) ribotide along the de novo pathway and the formation of AMP from adenylosuccinate in the conversion of IMP into adenine nucleotides. Both reactions involve the cleavage of a succinyl group, yielding fumarate.[1]

ASase deficiency is the first enzyme deficiency reported in humans along the de novo pathway of purine synthesis and results in the accumulation in body fluids of two normally undetectable compounds, SAICA riboside and succinyladenosine, that are the dephosphorylation products of the two substrates of the enzyme. The patients display severe psychomotor retardation, convulsions, autistic features, often cerebral and cerebellar hypotrophy and sometimes muscle wasting and growth failure.[2–4]

The pathogenesis of the symptoms is still debated, since deficiencies of metabolites which are normally formed distally from the enzyme defect, as well as accumulation of intermediates proximally thereof, could have deleterious effects.[3] In this report, we study the muscle energy metabolism in a patient with inherited ASase deficiency by means of phosphorus magnetic resonance spectroscopy. This is a non-invasive procedure that allows unlimited number of in vivo measurements at rest as well as during prolonged exercise and can be used to estimate ATP, phosphocreatine (PCr) and inorganic phosphate (P_i) more reliably than by chemical analyses.

Purine and Pyrimidine Metabolism in Man IX,
edited by Griesmacher *et al.* Plenum Press, New York, 1998.

2. EXPERIMENTAL PROCEDURES

ASase-deficient patient was described previously elsewhere.[4] Phosphorus magnetic resonance spectra of calf muscles were acquired by a GE 1.5 T Signa system by placing the surface coil (20.5 cm) directly on the skin.[5] The band width was 2 kHz. The stimulation-response sequence was repeated every 5 s. After optimising the magnetic field homogeneity, 60 transients were accumulated at rest in 5 min, while during recovery from exercise 2-FID data blocks (10 s) were recorded for 10 min. Intracellular pH was calculated from the chemical shift of P_i from PCr. Intracellular ATP concentration was calculated by comparison with an external standard solution measured immediately after each exam, assuming 8.2 mM the concentration of muscle ATP in healthy controls.[6] Informed consent was obtained in all cases.

3. RESULTS

Fig. 1 reports typical spectra of resting calf muscle from matched healthy volunteer and patient with ASase deficiency. If compared to the normal values (reported as means ± SD), the patient showed a markedly reduced intracellular ATP concentration (5.5 mM; controls: 8.2 ± 0.36 mM) as well as a very low PCr/P_i molar ratio (2.9; controls: 7.2 ± 1.03), due to low PCr and high P_i content in resting muscle. A relatively high intracellular pH (7.13; controls: 7.05 ± 0.02) was also observed in the patient during the rest.

Due to the severe psychomotor retardation, it was not possible to exercise the patient properly inside the magnet. Therefore, after data acquisition at rest the patient was induced to perform ischaemic work with her calf muscle by forcing plantar flexions outside the magnet and then quickly repositioned inside. At the end of exercise [PCr] fell to 66% of resting level, while intracellular pH dropped to a minimum value of 6.97 during recovery. PCr recovery behaved as a monoexponential process with a time constant significantly higher than those obtained with volunteers who reached the same percentile

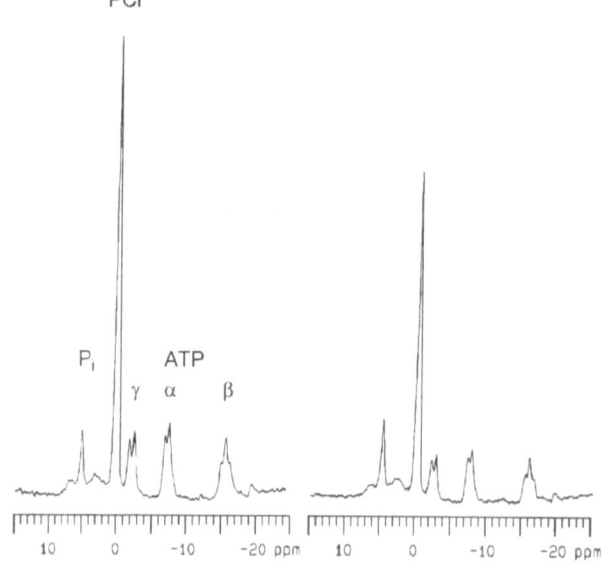

Figure 1. Phosphorus magnetic resonance spectra of resting calf muscle from age-, sex- and leg size-matched healthy control (left) and patient with ASase deficiency (right). Each spectrum consists of 60 scans over 5 min. The spectra are scaled to reflect the corresponding muscle ATP content.

depletion of PCr and the same cytosolic pH (38 s; controls: 14 ± 4 s). At the end of recovery, patient's muscle ATP content was lower than at rest (4.4 mM instead of 5.5 mM), while controls did not show any ATP change.

4. DISCUSSION

This study provides evidence of an impairment of muscle energy metabolism in a subject with deficit of ASase. Indeed, the patient shows a markedly reduced muscle energy reserve at rest, as judged by the decreased PCr/P_i ratio.[7] Moreover, unlike normal subjects, she depletes intracellular ATP after a relatively mild exercise.[8] Finally, the rate of post-exercise recovery of intracellular PCr is very slow, if compared to controls, thus suggesting a reduced ability of mitochondria to respond to metabolic needs.[6]

The lack of ASase could be responsible for the deterioration of muscle bioenergetics because of the defective supply of oxidable substrates either directly to the citric acid cycle from aminoacids or as pyruvate from the glycolytic pathway. The hypothesis is in line with previous observation that citrate, isocitrate, and succinate concentrations, at least in rat skeletal muscle, increase following exercise, and that about 70 percent of this expansion of the pool of citric acid cycle intermediates is derived from aspartate via the Asase-catalysed reaction.[9] On the other hand, it has been suggested that both AMP and AICA ribotide, the other products of the reaction catalysed by ASase, can induce lactate production in isolated soleus muscle preparations by stimulating glycogenolysis and glycolytic pathway through an allosteric activation of glycogen phosphorylase, phosphorylase kinase and/or protein kinase.[10]

The impairment of muscle energy charge could be also due to the disruption of the purine nucleotide cycle that is devoted to ammonia production from aspartate by coupling the reactions catalysed by adenylosuccinate sinthetase, ASase, and AMP deaminase. Several mechanisms have been proposed to explain the role of the purine nucleotide cycle in muscle function. One hypothesis is that the increase in flux through the purine nucleotide cycle during exercise prevents AMP accumulation following ATP catabolism, and this in turn displaces the adenylate kinase reaction toward ATP formation.[9] A second hypothesis is that the local production of NH_3 acts in conjunction with the decrease in ATP to stimulate the activity of phosphofructokinase and enhances the rate of glycolysis.[11] Finally, the accumulation of IMP may provide a mechanism for preserving the pool of purine nucleotides during exercise, and this reservoir of IMP may be used to restore adenylate pool rapidly during recovery.[12]

ACKNOWLEDGMENTS

This research was done in the framework of the European Community Concerted Action contract no. BMH1-CT-1384.

REFERENCES

1. M. R. Redinbo, S. M. Eide, R. L. Stone, J. E. Dixon, and T. O. Yeates, Crystallization and preliminary structural analysis of Bacillus subtilis adenylosuccinate lyase, an enzyme implicated in infantile autism. Protein Science 5:786–788 (1996)
2. J. Jaeken and G. Van den Berghe. An infantile autistic syndrome characterized by the presence of succinylpurines in body fluids, Lancet 2:1058–1061 (1984)

3. J. Jaeken, S. K. Wadman, M. Duran, F. J. Van Sprang, F. A. Beemer, R. A. Holl, P. M. Theunissen, P. De Cock, F. Van den Bergh, M. F. Vincent, and G. Van den Berghe. Adenylosuccinase deficiency: an inborn error of purine nucleotide synthesis. Eur. J. Pediatr. 148:126–131 (1988)

4. C. Salerno, C. Crifò, and O. Giardini. Adenylosuccinase deficiency: a patient with impaired erythrocyte activity and anomalous response to intravenous fructose. J. Inher. Metab. Dis. 18:602–608 (1995)

5. C. Salerno, S. Iotti, R. Lodi, C. Crifò, and B. Barbiroli. Failure of muscle energy metabolism in a patient with adenylosuccinate lyase deficiency: an in vivo study by phosphorus NMR spectroscopy. Biochim. Biophys. Acta (1997) in press

6. D. Arnold, P. M. Matthews, and G. K. Radda. Metabolic recovery after exercise and the assessment of mitochondrial function in human skeletal muscle in vivo by means of ^{31}P-NMR. Magn. Reson. Med. 1:307–315 (1984)

7. B. Chance, J. S. Leigh, J. Kent, K. K. McCully, S. Nioka, B. J. Clark, J. M. Maris, and T. Graham. Multiple controls of oxidative metabolism in living tissues as studied by phosphorus magnetic resonance. Proc. Natl. Acad. Sci. USA 83:9458–9462 (1986)

8. S. Iotti, R. Lodi, G. Gottardi, P. Zaniol, and B. Barbiroli. Inorganic phosphate is transported into mitochondria in the absence of ATP biosynthesis. An in vivo ^{31}P NMR study in the human skeletal muscle. Biochem. Biophys. Res. Commun. (1996) 225:191–194

9. J. J. Aragon and J. M. Lowenstein. The purine-nucleotide cycle: comparison of the levels of citric acid cycle intermediates with the operation of the purine nucleotide cycle in rat skeletal muscle during exercise and recovery from exercise. Eur. J. Biochem. 110:371–377 (1980)

10. M. E. Young, G. K. Radda, and B. Leighton. Activation of glycogen phosphorylase and glycogenolysis in rat skeletal muscle by AICAR—an activator of AMP-activated protein kinase. FEBS Letters 382:43–47 (1996)

11. K. Tornheim and J. M. Lowenstein. The purine nucleotide cycle: control of phosphofructokinase and glycolytic oscillation in muscle extracts. J. Biol. Chem. 250:6304–6308 (1975)

12. R. L. Sabina, J. L. Swain, B. M. Patten, T. Ashizawa, W. E. O'Brien, and E. W. Holmes. Disruption of the purine nucleotide cycle: a potential explanation for muscle dysfunction in myoadenylate deaminase deficiency. J. Clin. Invest. 66:1419–1423 (1980)

55

CARDIOPROTECTION BY ISCHEMIC PRECONDITIONING

Role of Adenosine and Glycogen

Robert de Jonge,[1] Silvia Bradamante,[2] and Jan Willem de Jong[1]

[1]Cardiochemical Laboratory
ThoraxCentre
Erasmus University Rotterdam, The Netherlands
[2]CNR-Centro Sintesi e Stereochimica Speciali Sistemi Organici
Università di Milano
Milan, Italy

1. ADENOSINE

Murry et al. reduced infarct size in dogs through repetitive brief occlusions, preceding prolonged ischemia, and coined this phenomenon "ischemic preconditioning."[13] ATP breakdown was attenuated during the prolonged period of ischemia, possibly due to a decrease in contractile function prior to this ischemic period. However, others observed no beneficial effects on ATP levels in the myocardium. Downey's group indicated the possible role of the ATP catabolite adenosine.[19] Subsequently, it was shown that A_1- and/or A_3-receptor activation or blockade by various agonists or antagonists, respectively, could augment or diminish the protective effect. Preconditioning through these receptors may involve a cascade of reations, including activation of cardiac protein-kinase C.

Apparently, adenosine acts both as an initiator and mediator of the protective effect.[19] In other words, it initiates protection but must also be present on the receptor during the prolonged ischemic period. Others, however, have stated that adenosine-mediated protection is most pronounced in the early reperfusion phase. Hypoxic preconditioning studies suggest that catabolite accumulation (read: adenosine formation) is not the effector of preconditioning, since there is a continuous washout. The role of adenosine in the preconditioning phenomenon has only been partly proven, and the exact relationship between A_1/A_3-receptor activation and protective effect still has to be elucidated. There are also indications that A_1-receptor activation is linked to the ATP-sensitive potassium channels.

Purine and Pyrimidine Metabolism in Man IX,
edited by Griesmacher *et al.* Plenum Press, New York, 1998.

Although the protective effect has been proven in various species, in man still controversial data exist, as discussed elsewhere.[5] Understanding the protective effect might give a powerful tool to improve myocardial cell viability and postischemic function during life-saving cardiac surgery by modulating adenosine levels through pharmacological means.

2. GLYCOLYSIS

The role of preconditioning on glycolysis/glycogenolysis during ischemia, as well as its relationship with ischemic injury, is controversial. Anaerobic glycolysis supports cell function by means of ATP production, but accumulated, glycolytic end products may determine the extent of ischemic damage; consequently, many studies have attempted to relate the beneficial effects of ischemic preconditioning to cardiac glycogen content or glycolytic rates.

2.1. Low-Flow Ischemia

Preconditioning before low-flow ischemia increases ischemic glycolytic flux and decreases cardiac injury.[12] These observations agree well with the results of studies showing that enhanced glycolysis and maintained glycolytic ATP production by various means lead to reduced damage and improved function upon reperfusion.[17,22]

One may pose the question: Why does anaerobic glycolytic flux protect the myocardium subject to hypoxia/low-flow ischemia? The answer seems to be that the ATP derived from anaerobic glycolysis is preferentially used for membrane-ion pumps and the preservation of cell integrity. During moderate ischemia, anaerobic glycolysis is stimulated (the Pasteur effect) whereas, during severe ischemia, this effect is inhibited by the accumulation of lactate and protons, and glycolysis ceases.[14] During low-flow ischemia, lactate and protons are continuously washed out of the ischemic heart, thus preventing their accumulation in tissue and the consequent inhibition of anaerobic glycolysis. Therefore, if the accumulation of the potentially toxic products of glycolysis is prevented by sustaining a moderate flow to the heart, glycolysis can be maintained and glycolytic ATP may preserve cell-membrane integrity and reduce Ca^{2+}-overload during reperfusion.

We believe that the maintenance of glycolysis during low-flow ischemia prevents myocardial injury (see ref. 14). So far, only one study[12] has reported that ischemic preconditioning increases glycolytic flux and reduces ischemic cell damage. We observed no difference in anaerobic glycolysis (lactate release) during low-flow ischemia between preconditioned and control hearts.[6] However, in preconditioned hearts glycogenolysis was reduced suggesting increased reliance on exogenous glucose. More research is necessary to clarify the important relationship between preconditioning and glycolysis during low-flow ischemia.

2.2. Stop-Flow Ischemia

Unlike low-flow ischemia, ischemic preconditioning causes a decrease in glycolytic flux during stop-flow global and regional ischemia in both isolated and in-vivo models.[13,16,23] Other strategies aiming at decreased glycolytic flux during regional stop-flow ischemia also induce less ischemic injury.[10]

Several groups, including Wolfe et al.,[23] assume that the beneficial effects of preconditioning are related to the reduced lactate accumulation and proton production that are the result of lower glycolytic rates—despite reduced ATP production. However, Vander Heide et al.[20] dissociated glycogen depletion and the reduction in lactate accumulation and anaerobic glycolytic flux from ischemic damage. Our own NMR studies also demonstrate cardioprotection by preconditioning in isolated rat hearts without any effects on pH (ref. 4), leading us to believe that acidosis is not a key factor in either functional recovery or ischemic injury (see also ref. 16).

Ischemic preconditioning of the heart reduces its pre-ischemic glycogen content. Wolfe et al.[23] correlated the decrease in infarct size caused by preconditioning with glycogen depletion before sustained ischemia, and with the attenuation of intracellular acidosis during ischemia. A longer period of preconditioning ischemia leads to glycogen depletion and a reduction in infarct size, that are both described by an exponential declining curve[2]. However, the hypothesis that diminished glycogen stores per se limit glycolysis, catabolite accumulation and cell damage is not supported by other results: Myocardial glycogen loading induced by fasting improves functional recovery of ischemic rat hearts,[18] and, in a retrospective study, the pre-ischemic glycogen content of rat hearts correlated with the time of day but not post-ischemic functional recovery.[1] Our ^{13}C-NMR studies indicate that ischemic preconditioning reduces glycogenolysis early during ischemia[3]. The question remains as to whether reduced glycolytic flux and accumulated glycolytic products mediate, or are secondary to preconditioning, but we do not believe that either lactate or proton accumulation is important in this respect.

3. GLYCOLYSIS AND ADENOSINE

Since adenosine has been implicated in ischemic preconditioning, a number of authors have studied its role on glycolytic flux. Various groups[11] have shown that, by binding to the A_1-receptor, adenosine increases glycolytic flux and induces cardioprotection in isolated rabbit and rat hearts, respectively subjected to ischemia and hypoxia, whereas others have found that it reduces glycolysis and improves functional recovery after ischemia in isolated working rat hearts perfused with glucose and palmitate.[8,9] These divergent results could be due to the models studied (low-flow vs. no-flow ischemia), the use of different substrates (glucose vs. glucose/fatty acids), differences in the preischemic metabolic status of the heart, and/or differences between species. We examined the role of adenosine A_1 receptor activation in the mechanism of ischemic preconditioning in rat hearts, and the reduced glycogenolysis observed at the start of no-flow ischemia[3]. Preconditioning with the adenosine A_1 agonist, 2-chloro-N^6-cyclopentyladenosine (CCPA) protected rat hearts from no-flow ischemia and mimicked ischemic preconditioning in the attenuation of glycogenolysis early during ischemia.[7]

ACKNOWLEDGMENTS

We gratefully acknowledge the support of the Netherlands Heart Foundation, the European Community (Biomed II) and NATO (project numbers NHS 94.043, BMH4-CT95 0838 and CRG 920235, respectively).

REFERENCES

1. Asimakis GK. Myocardial glycogen depletion cannot explain the cardioprotective effects of ischemic pre-conditioning in the rat heart. J Mol Cell Cardiol 1996;28:563–570.
2. Barbosa V, Sievers RE, Zaugg CE, Wolfe CL. Preconditioning ischemia time determines the degree of glycogen depletion and infarct size reduction in rat hearts. Am Heart J 1996;131:224–330.

3. Bradamante S, Marchesani A, De Jonge R, De Jong JW. Glycogen mobilization - Krebs cycle activity in preconditioned (PC) hearts. J Mol Cell Cardiol 1996;28:A27 (Abstr).

4. Bradamante S, Piccinini F, Delu C, Janssen M, De Jong JW. NMR evaluation of changes in myocardial high energy metabolism produced by repeated short periods of ischemia. Biochim Biophys Acta 1995;1243:1–8.

5. De Jong JW, De Jonge R, Marchesani A, Janssen M, Bradamante S. Controversies in preconditioning. Cardiovasc Drugs Ther 1996;10:767–773.

6. De Jonge R, Bradamante S, De Jong JW. Ischemic preconditioning-induced protection against low-flow ischemia: Role of (pro)glycogen, FASEB J 1997;11:A432 (Abstr).

7. De Jonge R, Bradamante S, De Jong JW. A_1 adenosine receptor activation mimicks ischemic preconditioning in rat hearts. Clin Biochem, in press (Abstr).

8. Finegan BA, Lopaschuk GD, Coulson CS, Clanachan AS. Adenosine alters glucose use during ischemia and reperfusion in isolated rat hearts. Circulation 1993;87:900–908.

9. Fralix TA, Murphy E, London RE, Steenbergen C. Protective effects of adenosine in the perfused rat heart: changes in metabolism and intracellular ion homeostasis. Am J Physiol 1993;264:C986-C994.

10. Goto M, Tsuchida A, Liu Y, Cohen MV, Downey JM. Transient inhibition of glucose uptake mimics ischemic preconditioning by salvaging ischemic myocardium in the rabbit heart. J Mol Cell Cardiol 1995;27:1883–1894.

11. Janier MF, Vanoverschelde J-LJ, Bergmann SR. Adenosine protects ischemic and reperfused myocardium by receptor-mediated mechanisms. Am J Physiol 1993;264:H163-H170.

12. Janier MF, Vanoverschelde J-LJ, Bergmann SR. Ischemic preconditioning stimulates anaerobic glycolysis in the isolated rabbit heart. Am J Physiol 1994;267:H1353-H1360.

13. Murry CE, Richard VJ, Reimer KA, Jennings RB. Ischemic preconditioning slows energy metabolism and delays ultrastructural damage during a sustained ischemic episode. Circ Res 1990;66:913–931.

14. Opie LH. The mechanism of myocyte death in ischaemia. Eur Heart J 1993;14, Suppl G:31–33.

15. Reimer KA, Murry CE, Yamasawa I, Hill ML, Jennings RB. Four brief periods of myocardial ischemia cause no cumulative ATP loss or necrosis. Am J Physiol 1986;251:H1306-H1315.

16. Schaefer S, Carr LJ, Prussel E, Ramasamy R. Effects of glycogen depletion on ischemic injury in isolated rat hearts: insights into preconditioning. Am J Physiol 1995;268:H935-H944.

17. Schaefer S, Prussel E, Carr LJ. Requirement of glycolytic substrate for metabolic recovery during moderate low flow ischemia. J Mol Cell Cardiol 1995;27:2167–2176.

18. Schneider CA, Taegtmeyer H. Fasting in vivo delays myocardial cell damage after brief periods of ischemia in the isolated working rat heart. Circ Res 1991;68:1045–1050.

19. Thornton JD, Thornton CS, Downey JM. Effect of adenosine receptor blockade: preventing protective preconditioning depends on time of initiation. Am J Physiol 1993;265:H504-H508.

20. Vander Heide RS, Delyani JA, Jennings RB, Reimer KA, Steenbergen C. Reducing lactate accumulation does not attenuate lethal ischemic injury in isolated perfused rat hearts. Am J Physiol 1996;270:H38-H44.

21. Vander Heide RS, Hill ML, Reimer KA, Jennings RB. Effect of reversible ischemia on the activity of the mitochondrial ATPase: Relationship to ischemic preconditioning. J Mol Cell Cardiol 1996;28:103–112.

22. Vanoverschelde J-LJ, Janier MF, Bakke JE, Marshall DR, Bergmann SR. Rate of glycolysis during ischemia determines extent of ischemic injury and functional recovery after reperfusion. Am J Physiol 1994;267:H1785-H1794.

23. Wolfe CL, Sievers RE, Visseren FLJ, Donnelly TJ. Loss of myocardial protection after preconditioning correlates with the time course of glycogen recovery within the preconditioned segment. Circulation 1993;87:881–892.

56

HIGH-ENERGY PHOSPHATE CHANGES IN THE NORMAL AND HYPERTROPHIED HEART DURING CARDIOPLEGIC ARREST AND ISCHEMIA

R. T. Smolenski, J. Jayakumar, A.-M. L. Seymour, and M. H. Yacoub

Department of Cardiothoracic Surgery
National Heart and Lung Institute
Imperial College, Harefield Hospital
Harefield, Middlesex UB9 6JH, United Kingdom

INTRODUCTION

The profile of metabolic changes in the ischemic heart following repeated infusions of cardioplegic solution has not been well characterized in cardiac hypertrophy. In this study we evaluated the effect of mild cardiac hypertrophy on the changes in phosphocreatine (PCr) and ATP throughout a cardioplegic arrest/ischemia/reperfusion experimental protocol (see Fig. 1). In addition to metabolic status, mechanical recovery of systolic and diastolic function was evaluated in the same hearts.

MATERIALS AND METHODS

Animals and Aortic Banding Procedure

Male Wistar rats weighing 200–250 g were subjected to constriction of the abdominal aorta according to a modified procedure previously described.[1] Animals were maintained for six weeks after surgery before commencing the experiments. All hearts in the hypertrophied group were excluded if heart dry weight [mg]/body weight [g] ratio was below 0.6 or above 0.8.

Heart Perfusion

Rats were anaesthetized with pentobarbital (60 mg/kg body weight) and heparinized with 200 IU sodium heparin via the femoral vein. The hearts were rapidly excised, placed

Purine and Pyrimidine Metabolism in Man IX,
edited by Griesmacher *et al.* Plenum Press, New York, 1998.

283

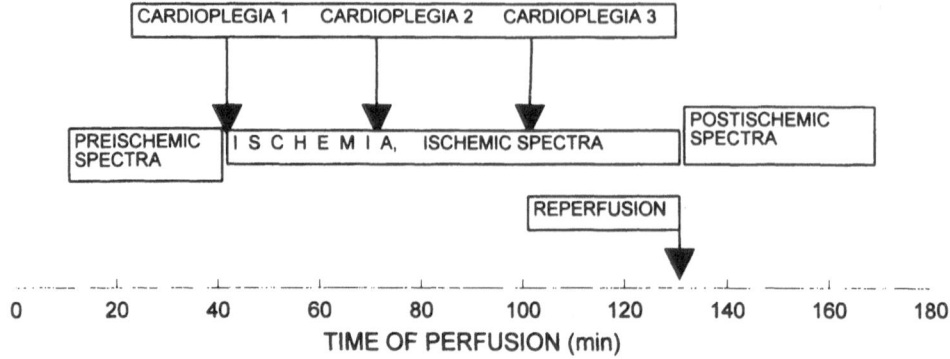

Figure 1. Experimental protocol.

in ice-cold perfusion buffer, immediately attached to a Langendorff perfusion system and perfused with Krebs-Henseleit buffer solution at a constant pressure of 85 mmHg at 37°C as described previously.[2]

Functional Assessment

Assessment of mechanical function was performed using a balloon catheter inserted into the left ventricle to determine systolic pressure and end-diastolic pressure–volume relations. The balloon was inflated with increasing volumes of water in 25 μl steps from 0 to 150 μl. Left ventricular systolic pressure (LVSP) and left ventricular end-diastolic pressure (LVEDP) were recorded at each loading of the balloon.[3]

^{31}P NMR and Metabolic Determination

The changes in ATP and PCr were followed using [31]P N.M.R. (Bruker AMX-400 wide bore vertical system, ^{31}P frequency 161.9 MHz), as described previously[2].

RESULTS

In controls there was an initial increase of 30% in PCr concentration. After the infusion, there was a gradual decrease in PCr. Subsequently, PCr was rapidly restored after each infusion of cardioplegic solution. However, depletion of PCr was more rapid after the second and the third infusion. The initial values of PCr in hypertrophied hearts were similar to the control group.

The profile of metabolic changes throughout the experiment was also generally similar to the control group. However, several important differences were observed. The increase in PCr during the first cardioplegic infusion was slightly smaller than in control hearts; the decrease of PCr concentration after cardioplegic infusion was faster and, especially, the recovery of PCr after ischemia was delayed in hypertrophied hearts. ATP concentration gradually decreased throughout cardioplegic arrest. The changes of phosphocreatine and ATP in control and hypertrophied groups are presented in Figure 2.

Before ischemia there was a tendency for higher values of LVSP in hypertrophied hearts. Similarly, the LVEDP curve tended to increase more steeply in this group. After

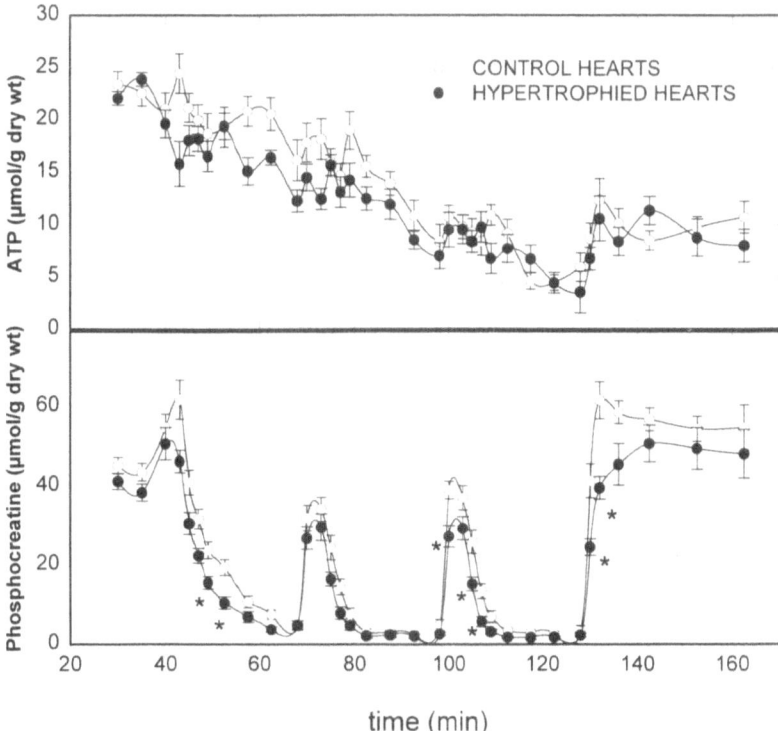

Figure 2. Changes in phosphocreatine and ATP in the normal and mildly hypertrophied rat heart subjected to cardioplegic arrest, ischemia at 25°C with repeated infusions of cardioplegic solution, and reperfusion. Values represent the mean ± SEM, *$p < 0.05$ vs. control.

ischemia and 40 min of reperfusion, there was a greater degree of functional deterioration in hypertrophied hearts. Although there was no change of systolic function represented by the LVSP curve, the diastolic function (LVEDP) in the hypertrophied hearts had deteriorated to a significantly greater extent than in controls.

DISCUSSION

This study demonstrates that even moderate hypertrophy, which does not induce metabolic changes under baseline conditions, could affect high energy phosphate metabolism and its response to cardioplegic arrest and ischemia. In the hypertrophied hearts, faster decrease in PCr during ischemia after cardioplegic infusion was shown, especially the delayed recovery of PCr concentration during reperfusion. This could play an important role in the impaired recovery of diastolic function in hypertrophied hearts.[4] Therefore attempts to increase high energy phosphate buffering capacity may be beneficial in hypertrophied hearts.

ACKNOWLEDGMENTS

This study was supported by the British Heart Foundation (Grant No. PG 96/194) and The Harefield Heart Transplant Trust.

REFERENCES

1. Scholkens BA, Linz W, Martorana PA. Experimental cardiovascular benefits of angiotensin-converting enzyme inhibitors: beyond blood pressure reduction. *J Cardiovasc Pharmacol.* 18 Suppl 2:S26-S30 (1991).
2. Smolenski RT, Yacoub MH, Seymour A-ML. Reduced purine catabolite production in the postischemic rat heart: A 31P NMR assesment of cytosolic metabolites. *Magma.* 2:417–420 (1994).
3. Weber KT, Janicki JS, Schroff S. Measurement of ventricular function in the experimental laboratory. In: Fozzard HA, Haber G, Jennings RB, Katz AM, Morgan ME, eds. *The heart and cardiovascular system.* New York: Raven Press. 856–886 (1986).
4. Zhang J, Merkle H, Hendrich K, et al. Bioenergetic abnormalities associated with severe left ventricular hypertrophy. *J Clin Invest* 92:993–1003 (1993).

57

ROLE OF GLUTAMATE-67 IN THE CATALYTIC MECHANISM OF HUMAN CYTIDINE DEAMINASE

Alessandra Cambi,[1] Silvia Vincenzetti,[1] Jan Neuhard,[2] and Alberto Vita[1]

[1]Dipartimento di Biologia M.C.A.
Università di Camerino, Italy
[2]Center for Enzyme Research
Institute of Molecular Biology
University of Copenhagen, Denmark

1. INTRODUCTION

Human cytidine deaminase (CDA, EC 3.5.4.5) is a tetramer of identical subunits (16KDa/subunit) each containing a single zinc atom[1] and catalyzes the deamination of cytidine (CR) and deoxycytidine (CdR) to their corresponding uracil compounds. Several cytosine-based drugs, such as cytosine arabinoside and 5-Aza-CdR,[2] which are anti-tumor agents used in the chemotherapy, are also deaminated by CDA. Thereby they loose their pharmacological properties. The deamination mechanism of CDA is similar to that of adenosine deaminase (ADA)[3,4] and consists of an enzyme-assisted direct water attack on the nucleoside C-4 and C-6 carbon atoms, respectively. In both enzymes the hydroxyl group thus formed interacts with an active-site zinc atom and a negatively charged carboxylate group. The amino acid sequences of ADA and CDA are unrelated, except for a short sequence within the active site: in Escherichia coli these sequences are **TVHAE** and **TVHAGE** for CDA and ADA, respectively[5] and the H residue is involved in zinc coordination in both deaminases. The E residue participates to the catalytic mechanism of CDA,[3] whereas in the crystal structure of ADA[4] the glutamic residue is shown to share a proton with N-1 of the purine ring, while an aspartic residue (D 295) plays the same role as the glutamate in CDA, suggesting the importance of the presence of a carboxyl group in the catalytic site of nucleotide deaminases. In fact, in both the human and the Bacillus subtilis CDA the E residue is conserved, whereas the H residue is replaced by a C residue. In the present paper we demonstrate the importance of the Glu-67 carboxylic group by studying the effect on enzyme activity of reacting wild-type CDA with the specific carboxyl reagent N-ethoxy-

Purine and Pyrimidine Metabolism in Man IX,
edited by Griesmacher *et al.* Plenum Press, New York, 1998.

287

carbonyl-2-ethoxy-1,2-dihydroquinoline (EEDQ), and by analyzing two mutant enzymes in which Glu-67 has been replaced with Asp and Gln (E67D and E67Q), respectively.

2. EXPERIMENTAL PROCEDURES

2.1. Recombinant DNA Procedures

Sub-cloning and sequencing of CDA cDNA constructs were performed as described by Vincenzetti et al. 1996.[1] Specific amino acid substitutions were obtained by site-directed mutagenesis of the wild-type CDA cDNA, previously cloned into pTrc99-A (Pharmacia) yielding pTrcHUMCDAwt, using a two-steps PCR procedure. In the first step, specific primers (DNA Technology, ApS, Aarhus, Denmark) containing the appropriate mutation, and the 5'-primer (P_{Nco}) caga**CCATGG**cccagaagcgtc[1] were used to produce specific "megaprimers" corresponding to the 5' half of the cDNA. In the second PCR step, the megaprimer was used together with the 3'-primer (P_{Bam}) cc**GGATCC**aggtggctgttac,[1] producing the whole CDA cDNA inserts with the desired mutations. Subsequently, the PCR fragment was digested with NcoI and BamHI and cloned into an appropriate vector. The primary structure of the inserts was confirmed both by restriction analysis and by DNA sequencing. When the pET expression system[6] was used, a 42-bp linker was cloned into *Nco*I site of pET-3d yielding pET-3d(His6). This linker fragment coded for a histidine hexapeptide followed by a tetrapeptide containing the cleavage site for protease FXa. The 446-bp NcoI-BamHI fragment containing the human CDA cDNA, from both wild type and mutants, was isolated from the vector pTrcHUMCDA[1] and inserted into pET-3d(His6) downstream of the His-tag, yielding pHUMCDA-H6 which was then transformed into the E. coli strain BL21(DE3) for expression of the His_6-CDA fusion proteins.

2.2. Expression and Purification of the His_6-CDA Fusion Proteins

Transformed E. coli BL21(DE3) strains were grown at 37°C in L broth supplemented with 200μg/ml ampicillin, and His_6-CDA expression was induced during late exponential growth (A_{600}=0.6) by the addition of 1mM IPTG. After 3 hrs vigorous shaking at 37°C, cells were harvested and His_6-CDA fusion proteins were purified by a metal chelate affinity chromatography on the iminodiacetic acid Sepharose 6B column (Sigma Chemical Co.) charged with nickel (Ni^{++}-IDA) as described by Van Dyke et al.[7] The final preparation was dialyzed against 50mM Tris-HCl pH 7.5, 5mM β-mercaptoethanol, 1mM EDTA, concentrated by ultra-filtration (Amicon Corporation) and analyzed by SDS-PAGE.[8]

2.3. HPLC Enzymatic Assay

To measure the slow conversion of CR to UR by the mutant enzymes a HPLC system was employed. Mixtures of 1 ml, containing 100 mM Tris-HCl pH 7.5, 100 mM KCl, mutant enzyme (3 μg) and 0.167 mM CR, were incubated at 37°C. At five or more timed intervals, aliquots of 100 μl were deproteinized and injected onto a reverse-phase LC-18 5-mm HPLC column (250 × 4.6 m-i.d.m, Supelco) equilibrated with 100 mM potassium hydrogen phosphate pH 6, 8mM tetrabutyl-ammonium hydrogen sulphate and eluted with the same buffer (1.3 ml/min.). CR and UR showed retention times of 3.4 and 4.8 min., respectively. Elution of substrate and product was monitored at 254 nm and the integrated peak areas of substrate and product were compared with standard samples of known concentration.

3. RESULTS AND DISCUSSION

3.1. EEDQ Inhibition Studies on Wild-Type CDA

The K_m and V_{max} values for CR were determined in presence of different concentration of EEDQ following preincubation of CDA and inhibitor for 10 min prior to addition of the substrate. EEDQ decreased the V_{max} value but did not affect the K_m for CR significantly, suggesting that binding of EEDQ to wild-type CDA did not modify the affinity of CDA for its substrate but inactivated the enzyme in an irreversible manner. Wild-type CDA was incubated with EEDQ at different fixed concentrations for different periods of time (EEDQ concentrations were in large excess over that of the enzyme). The time course of inactivation followed pseudo-first-order kinetics until 80% of inactivation was reached (Fig. 1a) and the half-time of inactivation ($T_{1/2}$) varied depending on the inhibitor concentration. A plot of log $1/T_{1/2}$ vs. log $[EEDQ]^{10}$ gave a straight line with a slope (n) corresponding to the number of EEDQ molecules reacting with each active site to form an inactive complex. As shown in Fig. 1b, n was very close to 1, suggesting that only 1 mol of EEDQ binds to 1 mol of wild-type CDA.

3.2. Mutations on Human CDA cDNA

Based on crystallographic studies of E. coli CDA,[3] which indicated involvement of the highly conserved glutamate residue (Glu 104) in the catalytic process, and the EEDQ inhibition experiments described above, site-directed mutagenesis was employed to change Glu67 of human CDA, corresponding to Glu104 of the E.coli enzyme, into either Asp (preserving the carboxylic group), or Gln (replacing the –COOH putatively involved in proton transfer during catalysis with an amide group).

3.3. Expression and Complementation Test of Wild-Type and Mutant CDA

The pyrimidine requiring E. coli strain SØ5201 (MC1061 *cdd::Tn10 pyr::*Kan) cannot grow with CdR as sole pyrimidine source due to the lack of CDA activity. SØ5201

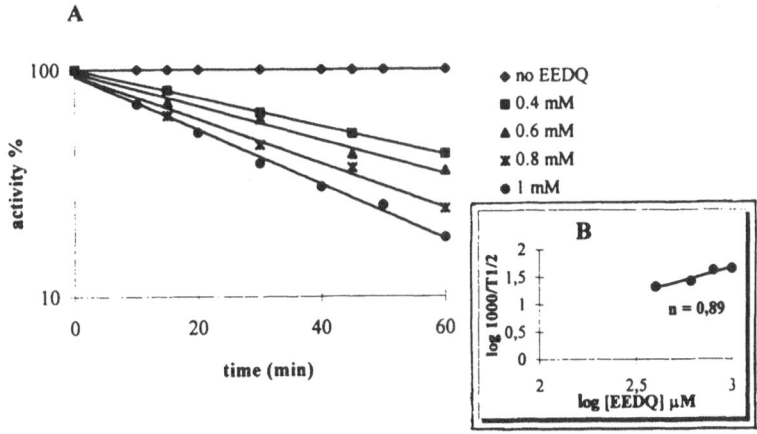

Figure 1. A) Kinetics of inactivation of CDA activity by EEDQ. CDA (0.5 µM) was incubated at 37°C in 50 mM Tris-HCl pH 7.5, 1 mM EDTA, 0.2 mM dithiothreitol with different concentration of EEDQ. At the indicated times, aliquot samples were diluted 50-fold in 0.1 M Tris-HCl pH 7.5 and immediately assayed for CDA activity.[1] B) Determination of the number, n, of reactive EEDQ per active site of CDA. $T_{1/2}$ is the half-time of inactivation (in min).

Table 1. Effect of Glu-67 replacement on CR deamination

	Wild-type	E67D	E67Q
K_m, CR (mM)	0.0092	0.012	0.042
V_{max}, CR (U^a/mg)	257	1.27	1.47×10^{-3}
V_{max}/K_m	27935	106	0.034

^{a}One enzyme unit is defined as the amount of enzyme which catalyzes the deamination of 1 mmol of CR per minute at 37°C.

harboring pTrcHUMCDAwt can grow with CdR as sole pyrimidine source in minimal medium even in the absence of IPTG, indicating that uninduced expression of CDA from the plasmid was adequate for complementation of CDA deficiency. When SØ5201 harboring each of the mutant recombinant plasmid was plated on glucose minimal medium,[11] supplemented with 40 µg CdR/ml and 0.6 mM IPTG, growth was only observed with SØ5201/pTrcHUMCDAE67D. SDS-PAGE of crude extracts of cells grown in L-broth, supplemented with 100 µg ampicillin/ml and 0.6 mM IPTG, indicated that the mutant enzymes were synthesized, but that E67Q CDA was expressed to a lesser extent than the wild-type and E67D enzymes. Any attempt to purify E67D and E67Q by CV6-affinity chromatography[1] lead to a low yield of the mutant enzymes. Therefore the E67D and E67Q cDNA was cloned into pET-3d(His6) yielding pHUMCDA-H6 plasmids which were subsequently transformed into E. coli BL21(DE3). Expression was induced by 1mM IPTG and the enzyme was purified by Ni^{++}-IDA Sepharose 6B affinity chromatography, as described in Materials and Methods.

3.4. Zinc Content and Kinetic Parameters of the Purified Enzymes

After prolonged dialysis against metal-free buffer to remove loosely bound metal ions, the zinc content of 20 µg of each purified mutant CDA was determined by Inductively Coupled Plasma-Optical Emission Spectrometry (ICP-OES), using a Jobin Yvon 24R model. The results revealed that the E67D and E67Q mutant enzymes, like wild-type CDA,[1] contained approximately 1 mol zinc per mol subunit, suggesting that the mutated residue was not involved in zinc coordination. The kinetic parameters for the E67D and E67Q enzymes were determined by HPLC assay and are shown in Table 1: K_m values are almost the same for the two mutants and the wild-type CDA, whereas V_{max} and V_{max}/K_m values were very different for the three enzymes. In fact the catalytic efficiency (V_{max}/K_m) was reduced approximately 260-fold for E67D and 800000-fold for E67Q. Since both E67D and E67Q contained normal amounts of zinc, the large difference between the CDA activities of the two Glu mutants and that of the wild-type is most likely due to lack of the appropriately localized carboxyl group of Glu-67 required for protonating both the leaving amino group and N-3 of the pyrimidine ring[3] during catalysis.

REFERENCES

1. Vincenzetti, S., Cambi, A., Neuhard, J., Garattini, E., and Vita, A. (1996) Prot. Expr. Purif. 8, 247–253.
2. Müller, W.E.G., and Zahn, R.K. (1979) Cancer Res. 39, 1102–1107.
3. Betts, L., Xiang, S., Short, S.A., Wolfenden, R., and Carter, C.W.Jr. (1994) J.Mol.Biol. 235, 635–656.
4. Wilson, D.K., Rudolph, F.B., and Quiocho, F.A. (1991) Science 252, 1278–1285.
5. Yang, C., Carlow, D., Wolfenden, R., and Short, S.A. (1992) Biochemistry 31, 4168–4174.
6. Studier, F.,W., Rosenberg, A.,H., Dunn, J.,J., and Dubendorff, J., W.(1990) Methods Enzymol. 185, 60–89.

7. Van Dyke, M., W., Sirito, M., and Sawadogo, M. (1992) Gene, 111, 99–104.
8. Laemmli, U.K. (1970) Nature (London), 277, 680–683.
9. Cacciamani, T., Vita, A, Cristalli, G., Vincenzetti, S., Natalini, P., Ruggieri, S., Amici, A., and Magni, G. (1991) Arch. Biochem. Biophys. 290, 285–292.
10. Levy, H.,M., Leber, P.,D., and Ryan, E.,M. (1963) J. Biol. Chem. 238, 3654–3659.
11. Clark, D.J., and Maaløe, O. (1967) J.Mol.Biol. 23, 99–112.

58

HETEROZYGOSITY FOR A POINT MUTATION IN AN INVARIANT SPLICE DONOR SITE OF DIHYDROPYRIMIDINE DEHYDROGENASE AND SEVERE 5-FLUOROURACIL RELATED TOXICITY

André B. P. Van Kuilenburg,[1] Peter Vreken,[1] Louk V. A. M. Beex,[2] Rutger Meinsma,[1] Henk Van Lenthe,[1] Ronney A. De Abreu,[3] and Albert H. Van Gennip[1]

[1]Academic Medical Center
University of Amsterdam
Department Clinical Chemistry and Division Emma Children's Hospital
PO Box 22700, 1100 DE Amsterdam, The Netherlands
[2]Department of Endocrine Diseases
[3]Department of Pediatrics
Academic Hospital Nijmegen
Nijmegen, The Netherlands

1. INTRODUCTION

Dihydropyrimidine dehydrogenase (DPD, EC 1.3.1.2) is the initial and rate-limiting enzyme in the catabolism of the pyrimidine bases and it catalyses the reduction of thymine and uracil to 5,6-dihydrothymine and 5,6-dihydrouracil, respectively. DPD is also responsible for the breakdown of the widely used antineoplastic agent 5-fluorouracil (5FU), thereby limiting the efficacy of the therapy. 5FU is one of the few drugs that shows some antitumour activity against various otherwise untreatable tumours including carcinomas of the gastrointestinal tract, breast, ovary and skin. Although the cytotoxic effects of 5FU are probably directly mediated by the anabolic pathways, the catabolic route plays a significant role since more than 80% of the administered 5FU is catabolised by DPD. The important role of DPD in the chemotherapy with 5FU has been shown in cancer patients with a complete or near-complete deficiency of this enzyme. These patients suffered from severe (neuro)toxicity including death, following 5FU chemotherapy.[1,2] It has been suggested that patients suffering from 5FU toxicities due to a low activity of DPD are genotypically het-

Purine and Pyrimidine Metabolism in Man IX,
edited by Griesmacher *et al.* Plenum Press, New York, 1998.

erozygous for a mutant allele of the gene encoding DPD.[3] Furthermore, the frequency of heterozygotes in the normal population has been estimated to be as high as 3%.[3] In this study we investigated the cDNA and a genomic region of the DPD gene of a cancer patient experiencing severe toxicity following 5FU treatment, for the presence of mutations.

2. MATERIALS AND METHODS

2.1. Isolation of Peripheral Blood Mononuclear Cells and Determination of the DPD Activity

Peripheral blood mononuclear cells were isolated from EDTA-anticoagulated blood by centrifugation over lymphopaque (spec. gravity 1.086 g/ml, 350 mOsm). Cells from the interface were collected and treated with ice-cold NH_4Cl to lyse the contaminating erythrocytes. The activity of DPD was determined as described before.[4]

2.2. PCR Analysis

RNA and DNA were isolated from cultured fibroblasts by the guanidinium thiocyanate method and standard procedures, respectively. cDNA synthesis and RT-PCR reactions for amplifying cDNA coding for DPD were carried out essentially as described before.[5,6] Four overlapping DPD cDNA fragments were amplified. Forward primers had an 5'-TGTAAAACGACGGCCAGT-3' extension whereas reverse primers had an 5'-CAGGAAACAGCTATGACC-3' extension at their 5'-ends. These sequences are complementary to the labeled M13 and M13 reversed primers used in the dye-primer sequence reaction. Amplification was carried out in 50 μl reaction mixtures containing 10 mM Tris-HCl (pH 8.3), 50 mM KCl, 3 mM $MgCl_2$, 1 μM of each primer and 0.2 mM dNTPs. After initial denaturation for 5 min at 95°C, 2 U of Taq polymerase was added and amplification carried out for 30 cycles (1 min 95°C, 1 min 65°C, 1 min 72°C). Amplification of the genomic region containing a DPD exon and its flanking sequences was carried out essentially as described before with the primers I_1 and I_2.[6] These primers also contained the above mentioned 5'-extensions for sequencing purposes. PCR products were separated on 1% agarose gels, visualized with ethidium bromide and purified using a Qiaquickspin PCR purification kit and used for direct sequencing. Sequence analysis was carried out on a Applied Biosystems model 377 automated DNA sequencer using dye-primer method for the DPD cDNA and genomic fragments.

3. RESULTS

3.1. Case Report

The patient is a 59-year-old female with breast cancer who received adjuvant chemotherapy in April 1990 with tamoxifen (20 mg) to be administered twice a day for 2 years, 3-amino-1-hydroprolylildene bis-phosphonate (150 mg) twice a day and 5FU (15 mg/kg) to be administered intravenously every week. The first two injections with 900 mg 5FU were tolerated well by the patient without complications. However, one day after the third intravenous injection with 5FU she developed a serious diarrhea without blood and suffered from nausea and vomiting. 9 days after admission of the third 5FU injection

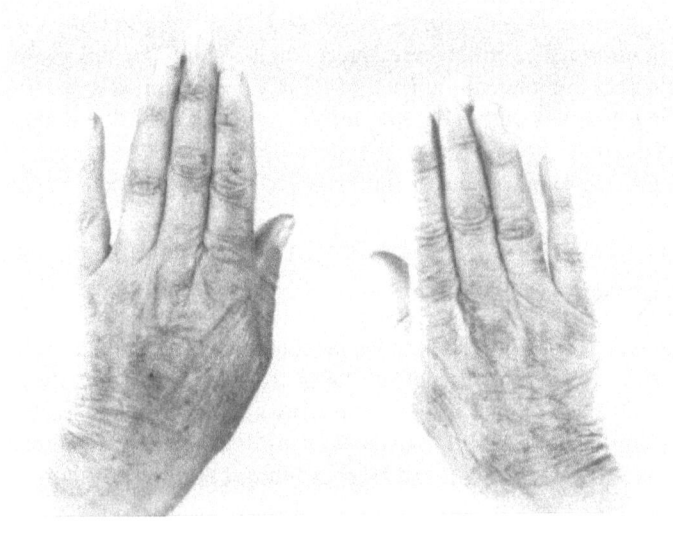

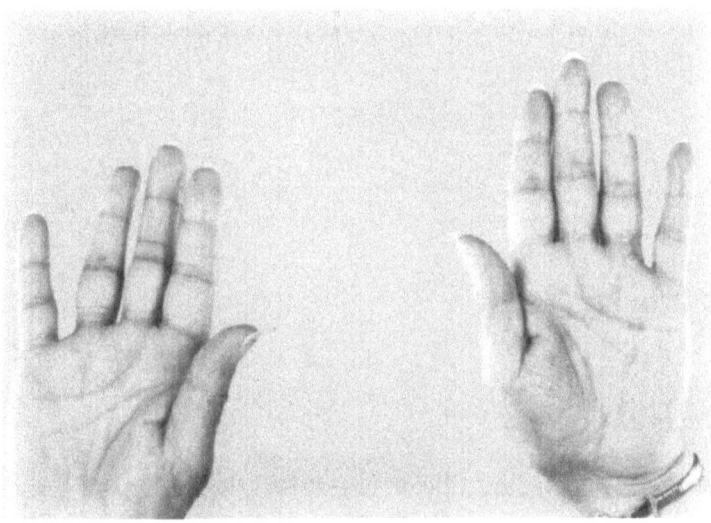

Figure 1. Pigmentation of the lines of the hands.

severe hypokalemia (2.1 mmol/l), hyponatremia (127 mmol/l) and leukopenia (1.6×10^9/l) with a normal number of thrombocytes (199×10^9/l) was observed. The next day, the number of leukocytes and thrombocytes decreased further to 0.9×10^9/l and 162×10^9/l, respectively. During this period the patient developed fever (39°C). The clinical picture and hematological parameters gradually improved and normalized during the subsequent days. A conspicuous pigmentation of the operation scar and lines of the hands was also observed (Figure 1) which disappeared during the subsequent weeks. Furthermore, ECG's showed extreme T wave inversions reflecting 5FU-associated cardiotoxicity. The observed cardiotoxicity reflects an increased coronary insufficiency in this patient who is familiar with coronary artery disease for which she underwent a by-pass operation in 1986.

3.2. Activity of DPD in Leukocytes

To investigate whether the severe fluorouracil toxicity in this patient might have been caused by a near-complete deficiency of DPD we determined the activity of DPD. In peripheral blood mononuclear cells the activity of DPD of the tumour patient (3.1 nmol/mg/h) was decreased compared to the mean activity of DPD of controls (9.6 nmol ± 3.7 nmol/mg/h) and was comparable to that observed for obligate heterozygotes.

3.3. RT-PCR Analysis of DPD mRNA in Cultured Fibroblasts and Genomic Sequence Analysis

Total RNA isolated from fibroblasts of the tumour patient and controls was subjected to RT-PCR and the coding sequence of the DPD cDNA was fully amplified in four fragments that span 737 bp (fragment A), 1010 bp (fragment B), 904 bp (fragment C) and 916 bp (fragment D) (Fig. 2, upper panel). Analysis of the PCR fragments by gel electrophoresis showed normal sized fragments A, B and D in all subjects (results not shown). However, the 904 bp fragment C was found together with a smaller sized fragment of 739 bp in the tumour patient whereas only the normal sized 904 bp fragment was detected in a control subject (Fig. 2 lower panel). Sequence analysis of the PCR fragments showed that the 739 bp fragment originated from the 904 bp fragment by a deletion of 165 bp which proved to be identical to the deletion described previously in two unrelated Dutch patients.[5,6]

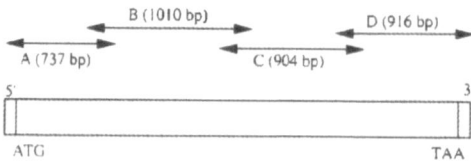

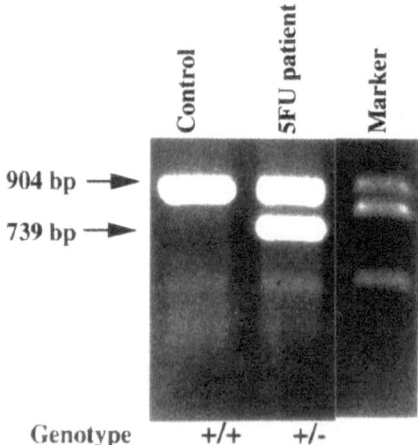

Figure 2. Amplification of DPD mRNA by RT-PCR and analysis by gel electrophoresis. The sequence of the cDNA coding for DPD was fully amplified in four fragments (upper panel). The lower panel shows the analysis of the C fragment from a control and the 5FU patient. The 904 bp and 739 bp bands correspond to the wild-type and the deleted cDNA fragments, respectively. The sizes of the three bands present in the marker lane are 947 bp, 831 bp and 564 bp, respectively. A plus indicates the presence of the wild-type allele and a minus indicates the presence of the mutant allele.

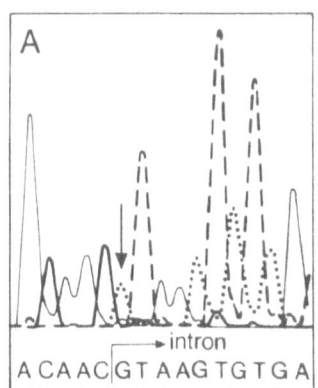
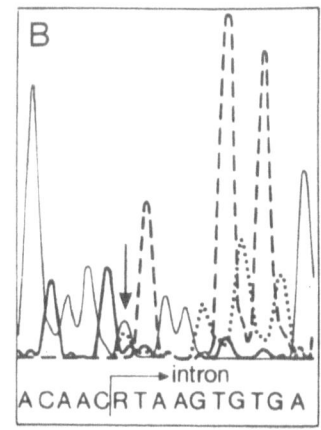

Figure 3. Sequence analysis of the intron-exon boundary. A genomic DNA fragment spanning the DPD exon and its flanking sequences was amplified and used for direct sequence analysis. A, control; B, 5FU patient. The sequence depicted below each panel shows the last five nucleotides of the exon (ACAAC) followed by the first 10 nucleotides of the intron.

In order to identify the genomic mutation leading to the observed exon skipping in our patients, two intron specific primers were used for the amplification of the complete exon and the upstream and downstream intron sequences.[6] Sequence analysis showed that the tumour patient proved to be heterozygous for a G → A point mutation in the invariant GT splice donor sequence in the intron downstream of the skipped exon (Fig. 3B) whereas a normal conserved GT splice donor site was observed in the control (Fig. 3A).

4. DISCUSSION

In this paper we provide unambiguous evidence at the molecular level for heterozygosity of a mutated allele of the gene encoding DPD in a tumour patient suffering from severe toxicity after administration of 5FU. Although hematologic (leukopenia) and gastrointestinal toxicities (nausea, vomiting, diarrhea) are common side effects of 5FU, the occurrence of hyperpigmentation and cardiotoxicity are rarer side effects.[7–9] Hyperpigmentation and cardiotoxicity are more often observed after protracted infusion of 5FU when compared to bolus injections as a result of the extended infusion of this drug.[7–9] Thus, the occurrence of these types of toxicities in our patient might be related to increased 5FU levels due to a decreased activity of DPD. Furthermore, our patient might be especially prone to the development of 5FU-associated cardiotoxicity since she is familiar with coronary artery disease.[9] Analysis of the DPD cDNA showed heterozygosity for a 165 bp deletion that results from exon skipping. Sequence analysis of the genomic region encompassing the skipped exon showed that the tumour patient was heterozygous for a G → A point mutation in the invariant GT splice donor sequence in the intron downstream of the skipped exon. So far, the G → A point mutation has also been found in 11 out of 19 patients suffering from a complete deficiency of DPD. Considering the frequent use of 5FU in the treatment of cancer patients, the severe 5FU-related toxicities in patients with a low activity of DPD and the high frequency of the G → A mutation in DPD deficient patients, analysis of the DPD activity and screening for the G → A mutation should be routinely carried out prior to the start of the treatment with 5FU.

REFERENCES

1. Diasio, R.B., Beavers, T.L. and Carpenter, J.T. (1988) Familial Deficiency of Dihydropyrimidine Dehydrogenase, Biochemical Basis for Familial Pyrimidinemia and Severe 5-Fluorouracil-induced Toxicity. *J. Clin. Invest.* **81**, 47–51.
2. Lu, Z., Zhang, R. and Diasio, R.B. (1993) Dihydropyrimidine Dehydrogenase Activity in Human Peripheral Blood Mononuclear Cells and Liver: Population Characteristics, Newly Identified Deficient Patients, and Clinical Implication in 5-Fluorouracil Chemotherapy. *Cancer Res.* **53**, 5433–5438.
3. Fernandez-Salguero, P., Gonzalez, F.J., Etienne, M.C., Milano, G. and Kimura, S. (1995) Correlation between catalytic activity and protein content for the polymorphically expressed dihydropyrimidine dehydrogenase in human lymphocytes. *Biochem. Pharmacol.* **50**, 1015–1020.
4. Van Kuilenburg, A.B.P., Van Lenthe, H. and Van Gennip, A.H. (1996) Identification and Tissue- Specific Expression of a NADH-Dependent Activity of Dihydropyrimidine Dehydrogenase in Man. *Anticancer Res.* **16**, 389–394.
5. Meinsma, R., Fernandez-Salguero, P., Van Kuilenburg, A.B.P., Van Gennip, A.H. and Gonzalez, F.J. (1995) Human Polymorphism in Drug Metabolism: Mutation in the Dihydropyrimidine Dehydrogenase Gene Results in Exon Skipping and Thymine Uraciluria. *DNA Cell Biol.* **14**, 1–6.
6. Vreken, P., Van Kuilenburg, A.B.P., Meinsma, R., Smit, G.P.A., Bakker, H.D., De Abreu, R.A. and Van Gennip AH. (1996) A point mutation in an invariant splice donor site leads to exon skipping in two unrelated Dutch patients with dihydropyrimidine dehydrogenase deficiency. *J. Inher. Metab. Dis.* **19**, 645–654.
7. Perlin, E. and Ahlgren, J.D. (1991) Pigmentary Effects from the Protracted Infusion of 5-Fluorouracil. *Int. J. Dermatol.* **30**, 43–44.
8. Freeman, N.J. and Costanza, M.E. (1988) 5-Fluorouracil-Associated Cardiotoxicity. *Cancer* **61**, 36–45.
9. Keefe, D.L., Roistacher, N. and Pierri, M.K. (1993) Clinical Cardiotoxicity of 5-Fluorouracil. *J. Clin. Pharmacol.* **33**, 1060–1070.

59

SITE DIRECTED MUTAGENESIS OF THE *SACCHAROMYCES CEREVISIAE* APT1 GENE

A Functional and Enzymatic Analysis

Timothy R. Crother and Milton W. Taylor

Department of Biology
Indiana University
Bloomington, Indiana 47405

INTRODUCTION

Adenine phosphoribosyltransferase (APRT) catalyzes the formation of 5'-adenosine monophosphate (AMP) from adenine and 5'-phosphoribosyl-pyrophosphate (PRPP). Exogenous adenine can thus be incorporated into the nucleotide pools. This enzyme appears to be ubiquitous to all eubacteria, archea, and eukaryotes examined to date. Although the crystal structure of APRT has not been determined, much is known about it biochemically. APRT functions as a dimer, requires Mg^{++}, and follows second order kinetics (also known as Ping-Pong kinetics).[1] APRT has been cloned and sequenced from many organisms. The amino acid alignment shows approximately 60–70% similarity. The PRPP binding site has been identified using crystallographic data from related enzymes and by mutational analysis.[2,3]

We have previously reported the cloning of APRT from *Saccharomyces cerevisiae*, its purification, and characterization.[4,5]

Utilizing *in vitro* mutagenesis, we have initiated a project to determine the role of individual amino acids in APRT. Using sequence alignment to determine conserved residues, site directed mutagenesis was performed to create single and multiple amino acid substitutions. The mutant proteins were tagged with 6 histidines at the amino terminus, overexpressed in *E. coli* and purified on a Ni^{++} column. The apparent Km adenine and kcat was found for several mutants. Our data suggests that the residues E106 and Y107 may be important for catalysis. In addition to this, residue G108 may play a role in adenine binding.

METHODS AND RESULTS

APT1 was previously cloned into the 6xhis tagged expression vector pQE-A1. (Qiagen).[6] Sequence alignment of various APRT's was done by hand using the program

Purine and Pyrimidine Metabolism in Man IX,
edited by Griesmacher *et al.* Plenum Press, New York, 1998.

299

	5	15	25	35	45
E.coli	...MTATAQQ	LEYLKNSIKS	IQDYPKPGIL	FRDVTSLLED	PKAYALSIDL
H.influenzae	......MTTQ	LDLIKSSIKS	IPNYPKEGII	FRDITTLLEV	PAAFKATIDL
S.cere-Apt2	...MSISESY	AKEIKTAFRQ	FTDFPIEGEQ	FEDFLPIIGN	PTLFQKLVHT
S.cere-Apt1	....MSIASY	AQELKLALHQ	YPNFPSEGIL	FEDFLPIFRN	FGLFQKLIDA
A.thaliana	MATEDVQDPR	IAKIASSIRV	IPDFPKPGIM	FQDITTLLLD	TEAFKDTII-A
D.melanogaster	MSPSISAEDK	LDYVKSKIGE	YPNFPKEGIL	FRDIFGALTD	PKACVYLRDL
Mouse	.....MSEPE	LKLVARRIRV	FPDFPIPGVL	FRDISPLLKD	PDSFRASIRL
Human	.....MADSE	LQLVEQRIRS	FPDFPTPGVV	FRDISPVLKD	PASFRAAIGL

	55	65	75	85	95
E.coli	L-VERYKNA-	G-ITKVV-GT	EARGFIFGAP	VALGLGVGFV	PVRKPGKLPR
H.influenzae	I-VEQYRDK-	G-ITKVL-GT	ESRGFIFGAP	VALALGLPFE	LVRKPKKLPR
S.cere-Apt2	FKTHLEEKFA	KEKIDFIAGI	EARGLLFGPS	LALALGVGFV	PIRRVGKLPG
S.cere-Apt1	FKLHLEEAFP	EVKIDYIVGL	ESRGFLFGPT	LALALGVGFV	PVRKAGKLPG
A.thaliana	LFVDRYKDK-	G-ISVVA-GV	EARGFIFGPP	IALAIGAKFV	PMRKPKKLPG
D.melanogaster	L-VDHIRESA	PEAEIIV-GL	DSRGFLFNLL	IATELGLGCA	PIRKKGKLAG
Mouse	LASHLKSTHS	GKIDYIA-GL	DSRGFLFGPS	LAQELGVGCV	LIRKQGKLPG
Human	LARHLKATHG	GRIDYIA-GL	DSRGFLFGPS	LAQELGLGCV	LIRKRGKLPG

	105 * * * * 115	125	135	145
E.coli	ETISETYDIE YGTDQLEIHV	DAIKPGDKVL	VVDDILATGG	TIEATVKLIR
H.influenzae	ETISQSYQLF YGQDTLEMHV	DAISEGDNVL	IIDDLLATGG	TVEATVKLVQ
S.cere-Apt2	ECASITFTKL DHEEIFEMQV	EAIPFDSNVI	VVDDVLATGG	TAYAAGDLIR
S.cere-Apt1	ECFKATYEKE YGSDLFEIQK	NAIPAGSNVI	IVDDIIATGG	SAAAAGELVE
A.thaliana	KVISEEYSIE YGTDTIEMHV	GAVEPGERAI	IIDDIIATGG	TLAAAIRLLE
D.melanogaster	EVVSVEYKLE YGSDTFELQK	SAIKPGQKVV	VVDDLLATGG	SLVAATELIR
Mouse	PTVSASYSLE YGKAELEIQK	DALEPGQRVV	IVDDLLATGG	TMFAACDLLH
Human	PTLWASYSLE YGKAELEIQK	DALEPGQRVV	VVDDLLATGG	TMNAACELLG

	155	165	175	185	195
E.coli	RLGGEVADAA	FIINLFDLGG	EQRLEKQGIT	SYSLVPFPGH	....
H.influenzae	RLGGAVKHAA	FVINLPELGG	EKRLNNLGVD	CYTLVNFEGH	....
S.cere-Apt2	QVGAHILEYD	FVLVLDSLHG	EEKLSAPIFS	ILHS......	....
S.cere-Apt1	QLEANLLEYN	FVMELDFLKG	RSKLNAPVFT	LLNAQKEALK	K.........
A.thaliana	RVGVKIVECA	CVIELPELKG	KEKLGETSLF	VLVKSAA...	....
D.melanogaster	KVGGVVVESL	VVMELVGLEG	RKRLDGKVHS	LIKY......	....
Mouse	QLRAEVVECV	SLVELTSLKG	RERLGPIPFF	SLLQYD....	....
Human	RLQAEVLECV	SLVELTSLKG	REKLAPVPFF	SLLQYE....	....

Figure 1. Amino acid sequence alignment of various APRT's. Highlighted residues indicate 70% or greater indentity. Mutagenized residues are denoted by *. PRPP binding site is underlined.

Seq App.[7] A region around amino acids 103–108 was identified as being conserved in all known functional APRT's (Fig. 1). This region, however, is not conserved in APT2, an inactive APRT in S. cerevisiea. Mutagenesis was performed on pQE-A1 to substitute the conserved amino acids in this region with those of APT2. Using the Quickchange mutagenesis kit from Stratagene, nucleotide substitutions were made. Upon confirmation by sequencing, the mutant gene (pQE-A1-1) was overexpressed and purified from E. coli using a Ni++ column. Protein amounts and purity were assessed by Bradford assay (Biorad) and coomassie stained PAGE (Fig. 2). Enzymatic assays were performed on the mutant protein using the method of Hershey and Taylor.[8] APT1-1 was found to contain no APRT activity. Since we have found that the APT2 protein does not form homodimers, we decided to examine whether APT1-1 is inactive due to its inabilty to dimerize.[9] It was found, using a yeast two-hybrid system, that APT1-1 was still able to dimerize (data not shown).

Figure 2. 6XHis tagged recombinant APRT proteins were purified on a Ni column. Purifcation was carried out according to the Qiagen protocol. Eluted products were dialyzed against 50 mM Tris pH 7.4, 5 mM MgCl, and 20 mM KCl. Purity of recombinant proteins was assesed by PAGE and coomasie staining.

Individual mutants were made at amino acid positions 106, 107 and 108 (Fig. 1) utilizing APT2 as a template. Upon over expression and purification, these mutants were found to be enzymatically active, although at a much lower specific activity (Table 1). The Km apparent for adenine and kcat was calculated for the mutants. Enzymatic assays were performed as above, except varying concentrations of adenine were used against supersaturating amounts of PRPP to determine the Km apparent adenine. Fig. 3 shows a Lineweaver-Burk plot of the data. Two of the mutants, 106E-L and 107Y-D were found to have similar Km app binding constants as wild type. There was a profound change in the apparent adenine binding in mutant 108G-H. The kcat/Km ratio was calculated and all three mutants were found to have much smaller numbers than wild type. Table 2 summarizes the kinetic results.

DISCUSSION

Using sequence conservation of APRT protein as a guide for determining important amino acids, mutant APRT's were made. The amino acid substitution His-for Gly at posi-

Table 1. The specific activity of recombinant APRT proteins in umoles ^{3}H- adenosine monophosphate/mg/min

	Specific activity (μmoles/mg/min)
APT1	6.18
106E-L	2.00E-05
107Y-D	0.012
108G-H	0.147

Table 2. Summary of kinetic data from wildtype and mutant recombinant APRT proteins

	Km app (μM)	Vmax app (nmoles/min)	kcat app (min^{-1})	kcat/Km
APT1	40	7.30E-05	1.14	0.02
106E-L	18	3.40E-07	1.06E-06	5.90E-08
107Y-D	56	4.00E-06	2.00E-03	3.50E-05
108G-H	198	3.00E-05	0.079	3.90E-04

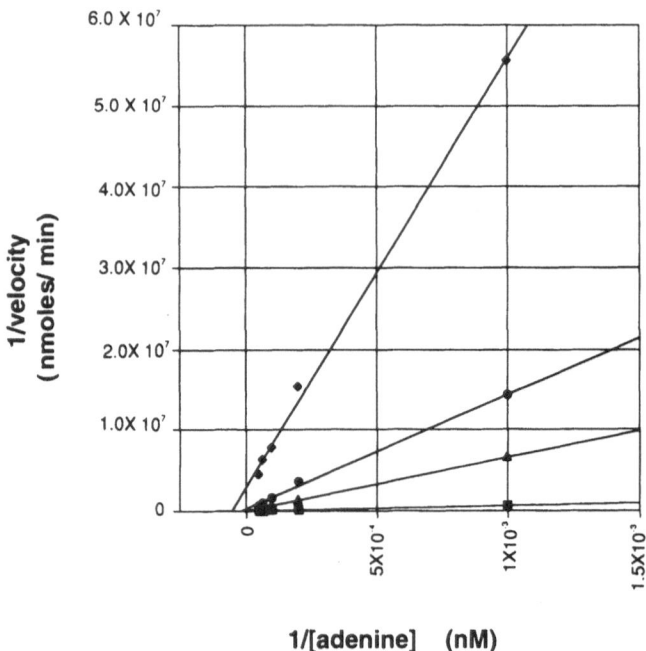

Figure 3. Double reciprical plot of velocity vs adenine concentration. Reactions were carried out at 100 uM PRPP and 1–20 uM adenine.

tion 108 alters the apparent binding of adenine. This finding is in good agreement with previous models and findings.[10–12] Busetta et al. 1988, proposed the adenine binding site to be between the ends of two beta sheets in APRT. Position Glycine 108 in *S. cerevisiae* is located in this region. A Gly to Glu mutation at this position in mouse has also been shown to affect adenine binding.[12]

The other two mutants, 106E-L and 107Y-D seem to have normal adenine binding, even though they are in the same region. However, in these two mutants, the ratio of kcat/Km is drastically altered. This might imply that these amino acids are involved in catalysis. However, this could also mean that the structure of the protein has been altered so that the efficiency of the reaction has decreased, or that the Mg^{++} binding site has been affected. No catalytic mutants are known for this enzyme. Further study needs to be done on these mutants, as well as the generation of new mutants to determine the extent of the adenine binding site and the catalytic core of the enzyme.

REFERENCES

1. Hori, M., and Henderson, J. F., Kinetic Studies of Adenine Phosphoribosyltransferase, J. Biol. Chem., 241: 3404 (1966).
2. Eads, J. C., Scapin, G., Yiming, X., Grubmeyer, C., and Sacchetini, J. C., The Crystal Structure of Human Hypoxanthine-Guanine Phosphoribosyltransferase with Bound GMP, Cell, 78: 325 (1994).

3. Scapin, G., Grubmeyer, C., and Sacchetini, J. C., Crystal Structure of Orotate Phosphoribosyltransferase, Biochemistry, 33(6): 1287 (1994).

4. Alfonzo, J. D., Sahota, A., Deeley, M. C., Ranjekar, P., and Taylor, M. W., Cloning and Characterization of the Adenine Phosphoribosyltransferase Gene (APT1) from *Saccharomyces cerevisiae,* Gene, 161:82 (1995).

5. Alfonzo, J. D., Sahota, A., and Taylor, M. W., Purification and Characterization of Adenine Phosphoribosyltransferase from *Saccharomyces cerevisiae*, Biochem. Biophy. Acta, In Press.

6. Alfonzo, J. D., Crother, T. R., and Taylor, M. W., In Prep.

7. Gilbert. D., SeqApp 1.9: A Biosequence Editor and Analysis Application, Indiana University, (1996).

8. Hershey, H. V., and Taylor, M., W, Purification od Adenine Phosphoribosyltransferase by Affinity Chromatography, Prep. Biochem., 8: 453 (1978).

9. Crother T. R., PhD. Thesis, In prep.

10. Busetta, B., The Use of Folding Patterns in the Search of Protein Structural Similarities; A Three-Dimensional Model of Phosphoribosyltransferases, Biochem. Biophy. Acta, 957: 21 (1988).

11. De Boer, J. G., and Glickman, B. W., Mutational Analysis of the Structure and Function of the Adenine Phosphoribosyltransferase Enzyme of Chinese Hamster, J. Mol. Biol., 221: 163 (1991).

12. Khattar, N.H., and Turker, M.S., A Role for Certain Mouse *APRT* Sequences in Resistance to Toxic Adenine Analogs, Som Cell Mol Gen, 23:51 (1997).

60

STUDIES ON CYSTEINE RESIDUES INVOLVED IN THE ACTIVE SITE OF HUMAN CYTIDINE DEAMINASE

Silvia Vincenzetti,[1] Alessandra Cambi,[1] Jan Neuhard,[2] and Alberto Vita[1]

[1]Dipartimento di Biologia M.C.A.
Università di Camerino, Italy
[2]Center for Enzyme Research
Institute of Molecular Biology
University of Copenhagen, Denmark

1. INTRODUCTION

The three-dimensional structure of homodimeric E. coli cytidine deaminase (CDA)[1] showed a zinc atom coordinated to histidine (H102) and two cysteine residues (C129 and C132) in the active site. The role of zinc in catalysis is to bind and activate a water molecule for attack on the pyrimidine ring with consequent elimination of ammonia. The human CDA cDNA has been cloned and the protein expressed in E. coli is a tetramer composed of identical 16.2 KDa subunits each containing a firmly bound zinc atom in the active site.[2] Comparison of the deduced amino acid sequence of the human recombinant CDA with that of E. coli enzyme[1] identified the putative zinc coordinating residues as C99, C102 and C65, the latter replacing H102 of the E. coli enzyme. To confirm the proposed role of the three indicated cysteines in catalysis of human CDA, we have studied the effect of specific thiolic reagents, and we have constructed and characterized three mutant proteins by site directed mutagenesis in which each of the cysteines, C99, C102 and C65 were replaced by an alanine residue.

2. EXPERIMENTAL PROCEDURES

2.1. Recombinant DNA Procedures

The sub-cloning and sequencing of CDA cDNA constructs were performed as described previously.[2] To facilitate production of the wild-type protein, the CDA cDNA was cloned into pTrc-99A yielding plasmid pTrcHUMCDA.[2] Specific amino acid substitu-

Purine and Pyrimidine Metabolism in Man IX,
edited by Griesmacher *et al.* Plenum Press, New York, 1998.

tions were obtained via site-directed mutagenesis of the wild-type CDA cDNA, by a two-steps PCR procedure. At first specific "megaprimers" were produced by using the 5' primer (P_{Nco}) cagaCCATGGcccagaagcgtc or the 3' primer (P_{Bam}) ccGGATCCaggtggctgttac together with specific primers containing the appropriate mutation: ctctccagctggggcctgc for C99A mutant, ggggccgccaggcaagtc for C102A mutant, gttcagcagcgatgcccag for C65A mutant. Each "megaprimer" was then used in a second PCR step together with either (P_{Nco}) or (P_{Bam}) primers producing the whole CDA cDNA insert with the desired mutations. The primary structure of the different inserts was confirmed by specific restriction analysis and DNA sequencing.

For purification of the C99A, C102A, C65A mutant proteins the corresponding cDNAs were inserted into the His-tag vector pET-3d (His6)[3] and the resulting pHUM-CDA-H6 plasmid transformed into E. coli strain BL21 (DE3). Expression was induced by adding 1mM IPTG to exponentially growing cultures. Crude cellular extracts[3] were loaded on Ni^{2+} IDA-Sepharose 6B columns and the enzyme was recovered from the column by elution with an imidazole gradient.

2.2. Enzyme Assay and Zinc Content

CDA activity was determined by a direct spectrophotometric assay at 37°C.[2] The reaction mixture contained 0.167 mM CR, 100 mM Tris-HCl pH 7.5 in a final volume of 1 ml. The reaction was started by adding 0.02–0.04 enzyme units.

One enzyme unit is defined as the amount of enzyme that catalyzes the deamination of 1 μmol of cytidine per minute at 37°C.[2]

20 μg of wild type or mutant His_6-CDA purified under metal-free conditions, were analyzed for zinc content by ICP-OES, using a Jobin Yvon 24R model. All measurements were performed checking the absence of exogenous metal contamination as described by Auld.[4]

2.3. CDA Wild Type -SH Groups Titration

The number of exposed sulphydryl groups in the native enzyme and in the enzyme denatured by 8 M urea and reduced by sodium borohydryde, was determined by titration with DTNB.[5] The native CDA was also titrated with the strongly dissociating sulphydryl reagent p-hydroxymercuriphenylsulfonic acid (PMPS).[6] The release of zinc that occurs during this titration was followed spectrophotometrically at 500 nm by performing the PMPS titration in the presence of 0.1mM of the metal indicator dye 4-(2-pyridylazo) resorcinol (PAR).

3. RESULTS AND DISCUSSION

3.1. Complementation Test, Expression and Purification of the Mutants Proteins

The pyrimidine requiring E. coli strain SØ5201(MC 1061 cdd::Tn 10 pyr:kan) cannot grow with CdR as sole pyrimidine source due to the lack of CDA activity. However, SØ5201 harboring pTrcHUMCDAwt can grow with CdR as sole pyrimidine source in minimal medium even in absence of IPTG, indicating that uninduced expression of CDA from plasmid is adequate for complementation of the CDA deficiency of SØ5201. In contrast, plasmids encoding the C99A, C65A and C102A mutant CDAs did not complement the CDA mutation of SØ5201, and crude extracts prepared from cultures of these mutants did not show any CDA activity when assayed either with or without the addition of zinc. In order to purify the inactive mutant enzymes the corresponding cDNAs were inserted into the His-tag vector pET-3d (His6) and the His-tagged proteins were expressed and purified as described in Experimental Procedures. The enzyme, recovered from the column after an imidazole gradient elution, was homogeneous as judged by SDS-PAGE.

3.2. Zinc Content of the Mutant Enzymes

The zinc content of purified CDA wild-type and mutant enzymes was determined by ICP-OES after prolonged dialysis against metal-free buffer to remove loosely bound metal ions. The results indicated that the wt enzyme contained 1 mol of zinc per mol of subunit while the C99A and C102A enzymes contained only about 0.2 mol of zinc per mol of subunit. The C65A mutant protein was found to contain 1 mol of zinc per subunit although it was found to be catalytically inactive both in crude extract and when purified to homogeneity.

3.3. Effect of Sulphydryl Reagents on Wild-Type CDA

According to the nucleotide sequence, human CDA contains nine cysteine residues. Titration of the wild-type enzyme with DTNB[5] showed the presence of six –SH groups. Zinc analysis of the titrated enzyme after extensive dialyses against metal free buffer revealed the presence of 1 mol of zinc per mol of subunit. DTNB titration of the enzyme following denaturation with 8 M urea yielded still six sulphydryl groups per subunit. The remaining three cysteine residues, probably buried in the catalytic site of the enzyme, were made accessible to DTNB following denaturation in the presence of a strong reducing agent such as sodium borohydryde. By using the strong dissociating sulphydryl reagent PMPS[6], nine –SH groups per subunit were titrated. The enzyme activity decrease during PMPS titration as shown in Fig. 1. PMPS also promotes the release of zinc coordinated to cysteines residues. By performing the PMPS titration in the presence of the high affinity metal indicator dye (PAR), the release of zinc was followed spectrophotometrically at 500 nm. Fig. 2 shows the results of a PMPS titration of wild -type CDA conducted in the presence of PAR.

In the E. coli CDA, zinc is coordinated by two cysteines, C99 and C102, of the highly conserved region CysXXCys together with an histidine residue (H102).[1] In human

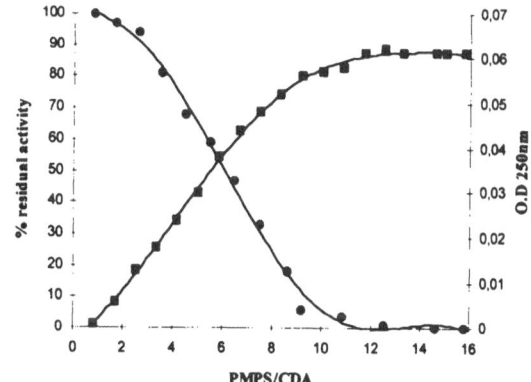

Figure 1. (■) titration of human CDA at 250 nm with PMPS. CDA (3 nmol in 1 ml of 50 mM Hepes/KOH pH 7.0) was teatred with PMPS (0.25 mM solution in the same buffer) to give the indicated molar ratios of PMPS to enzyme. (●) CDA activity during PMPS titration.

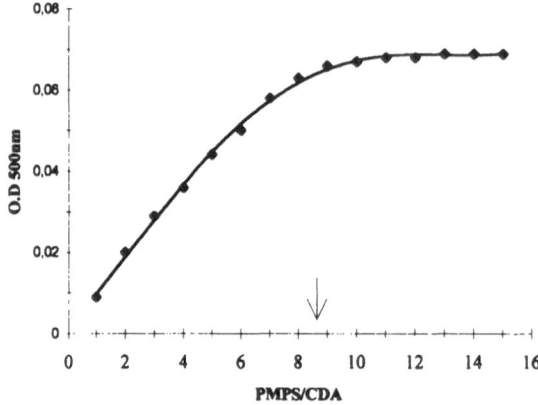

Figure 2. Titration of CDA with PMPS in presence of PAR. The absorption developing at 500 nm is due to the formation of a zinc-dye complex.

CDA the residues C99 and C102 may play the same role since mutation changing these cysteines into Ala residues lead to inactive mutant proteins with very low zinc contents.The properties of C65A mutant enzyme suggest that, C65, corresponding to H102 of the E. coli enzyme, is required for the enzyme activity but not for the maintenance of zinc in the active site. It is plausible that the main forces holding zinc in the catalytic site are the coordinations to C99 and C102, whereas C65 is important in guiding the zinc ion to the right position and orientation within the active site.

REFERENCES

1. Betts, L., Xiang, S., Short, S.A., Wolfenden, R. and Carter, Jr C.W., (1994) J.Mol.Biol. 235, 635–656.
2. Vincenzetti, S., Cambi, A., Neuhard, J., Garattini, E.and Vita, A. (1996) Prot. Expr. Purif. 8, 247–253.
3. Cambi A., Vincenzetti, S., Neuhard, J.,and Vita, A., this book.
4. Auld, D.S. (1988) Methods in Enzymology 158, 71–79.
5. Habeeb, A.F.S.A. (1973) Anal. Biochem. 56, 60–65.
6. Giedroc, D.P., Keating, K.M., Williams, K.R., Konigsberg, W.H. and Coleman, J.E. (1986) Proc. Natl. Acad. Sci. USA 83, 8452–8456.

61

RELATIONSHIP OF THE TWO APRT GENE PRODUCTS FROM *SACCHAROMYCES CEREVISIAE*

Timothy R. Crother and Milton W. Taylor

Department of Biology
Indiana University
Bloomington, Indiana 47405

INTRODUCTION

Adenine phosphoribosyltransferase (APRT) is a key component of the nucleotide salvage pathway converting free adenine plus 5'-phospho-ribosylpyrophosphate (PRPP) to 5'-adenosine monophosphate (AMP) plus pyrophosphate. This enzyme is ubiquitous in all organisms examined. In some parasitic organisms, which lack a *de novo* purine synthesis, APRT is the only means by which AMP is formed.[1] APRT deficiency in humans and mice leads to the accumulation of highly insoluble 2,8-dihydroxy adenine.[2,3] This can lead to kidney stones and possible kidney failure.[2,3]

This laboratory has previously cloned and characterized APRT (APT1) from *S. cerevisiae*.[4] At the same time, a second APRT sequence was accidentally cloned from *S. cerevisiae* (APT2) by Yurev and Cordon, 1993.[5] This clone does not compliment an APRT strain. In addition to this, overexpressed APT2 had no detectible APRT activity. Amino acid sequence alignment showed approximately 55% identity and 70% similarity between APT1 and APT2 (Fig. 1). It was speculated that APT2 protein might be able to heterodimerize with APT1 protein and perhaps modify its activity.[6] Alternately, it might represent a duplication event that has lost activity with time.

We have made use of the yeast two-hybrid system to determine if APT2 can interact with APT1 and form a heterodimer. In addition to this, northern and western analysis were performed on APT2. Using the two-hybrid system, we have shown that APT2 and APT1 can interact with each other. APT2's transcription levels are similar to APT1, but no protein is detectable by western analysis.

To address the question of whether APT2 represents a duplication event of APT1, a phylogenetic tree analysis was performed. It indeed shows that APT2's closest relative is APT1. Flanking DNA sequences were also examined to determine the extent of the dupli-

Purine and Pyrimidine Metabolism in Man IX,
edited by Griesmacher *et al.* Plenum Press, New York, 1998.

309

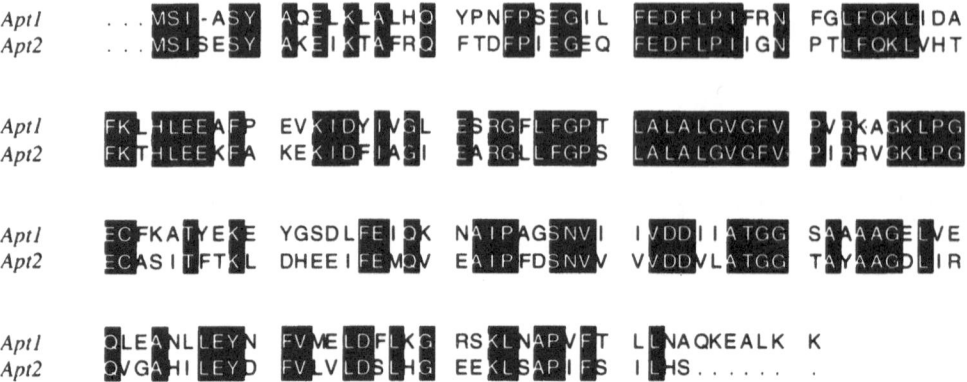

Figure 1. Amino acid sequence alignment of APT1 and APT2.

cation event. Finally, to determine when the duplication event took place, southern analysis was utilized on the closely related yeast, *Pichia pastoris*. Our data indicates that there may be two forms of APRT in *P. pastoris*.

METHODS AND RESULTS

The yeast two-hybrid system was used to determine if the APT2 gene product could interact with the APT1 protein. The APT1 and APT2 coding regions were cloned into the two-hybrid constructs pBTM116 and pGAD424. PBTM116 contains the LexA DNA binding domain. APT1 and APT2 were ligated in frame using EcoR1 and Sal1 sites. PGAD424 contains the gal activation domain. APT1 and APT2 were cloned in frame using the same sites as before. If the two proteins interact, the Gal activation domain will be in close proximity with the Lex A DNA binding domain. The Lex A DNA binding domain will then bind to its binding site and drive a chromosomal lacZ gene. Positive interactions will therefore lead to a blue colony in an Xgal assay. Using this system, APT1 protein was found to interact with APT2 protein. APT1 was also found to interact with itself, as was expected, since APRT function as a homodimer. However, APT2 does not form dimers with itself. Table 1 summarizes these results. There was no substantial difference in the interactions between APT1, itself, and with APT2 (Fig. 2).

In order to discern if APT2 was a duplication event, a phylogenetic tree analysis was performed using various APRT's with APT1 and APT2. The program PAUP was used to generate a tree using a heuristic search.[7] The results of this analysis clearly show that APT2 is most closely related to APT1 (Fig. 3). The bootstrap value for APT1 treeing with APT2 was 100%. Using sequences from Genbank, the extent of the duplication event was investigated. It appears that at least 5 other genes are present at both APT chromosomal regions (Fig. 4).[8] Due to the unavailability of flanking sequences, further analysis of this type could not be performed. Southern analysis was used to discern if this duplication event was solely limited to *Saccharomyces cerevisiae*. The closely related yeast, *Pichia pastoris*, was examined for the presence of a homolog of APT2. Genomic DNA was digested with EcoR I. The southern analysis was performed under low stringency conditions using APT1 and APT2 as probes. Fig. 5 shows the results. Using an APT1 probe, a band is seen between 4.3 and 6.4 kB. However, when an APT2 probe is used, this same band is seen, in addition to another band between 6.5 and 9.4 kB.

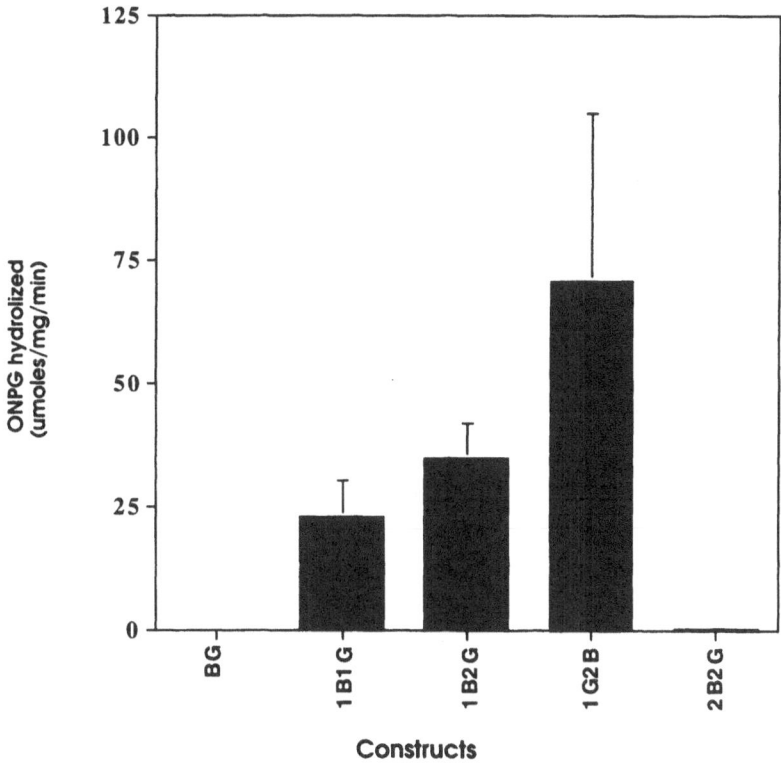

B=DNA BINDING DOMAIN, G=GAL ACTIVIATION DOMAIN

Figure 2. The relative strengths of the APT1 APT2 interactions in the Yeast two-hybrid system measured in ONPG hydrolyzed (umoles/mg/min).

Table 1. PBTM116 contains the LexA DNA binding domain. PGAD424 contains the Gal activation domain. Yeast strain Y835 was electroporated with 50 ng of each construct and plated on Leu-,Trp-, minimal plates. Colonies were lifted with nitrocellulose filters and lysed with liquid nitrogen. The colonies were then scored for lacZ activity with Xgal. Blue colonies indicate lacZ activity

Constructs	Lac Z activity
pBTM116 + pGAD424	−
pBTM116-A1 + pGAD424	−
pBTM116 + pGAD424-A1	−
pBTM116-A1 + pGAD424-A1	+
pBTM116-A2 + pGAD424-A2	−
pBTM116-A1 + pGAD424-A2	+
pBTM116-A2 + pGAD424-A1	+
pBTM116 + pGAD24-A2	−
pBTM116-A2 + pGAD24	−

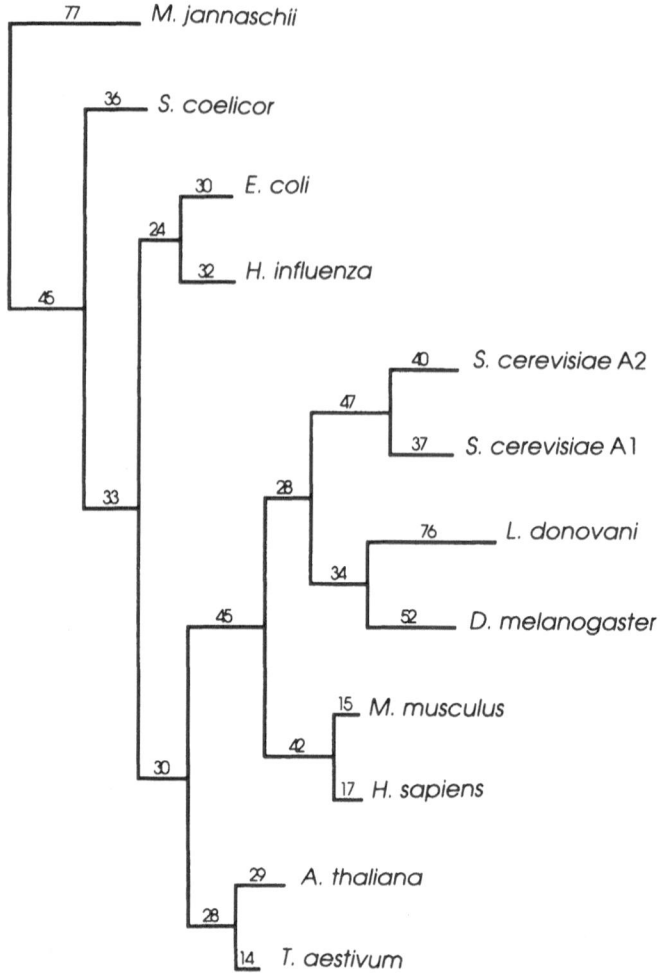

Figure 3. Phylogenetic tree analysis of various APRT amino acid sequences using Paup v3.11. Numbers indicate changes between each protein.

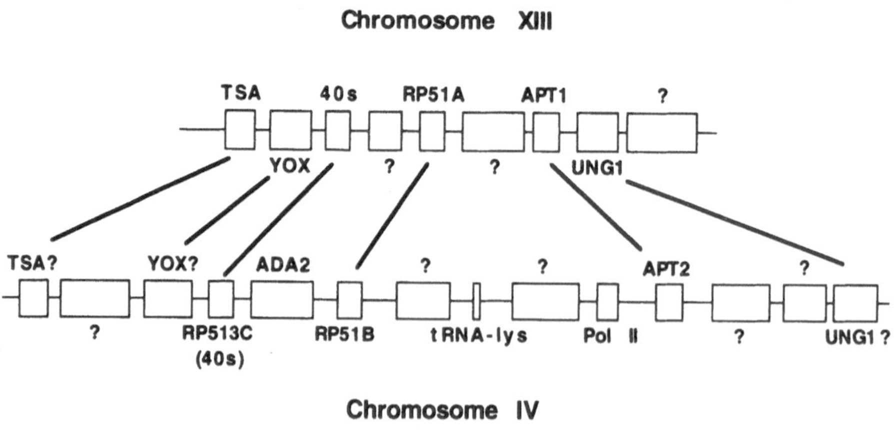

Chromosome XIII

Figure 4. Comparison of *S. cerevisiae* chromosome XIII and chromosome IV regions around APT1 and APT2 respectively. ? denotes ORF with no known function.

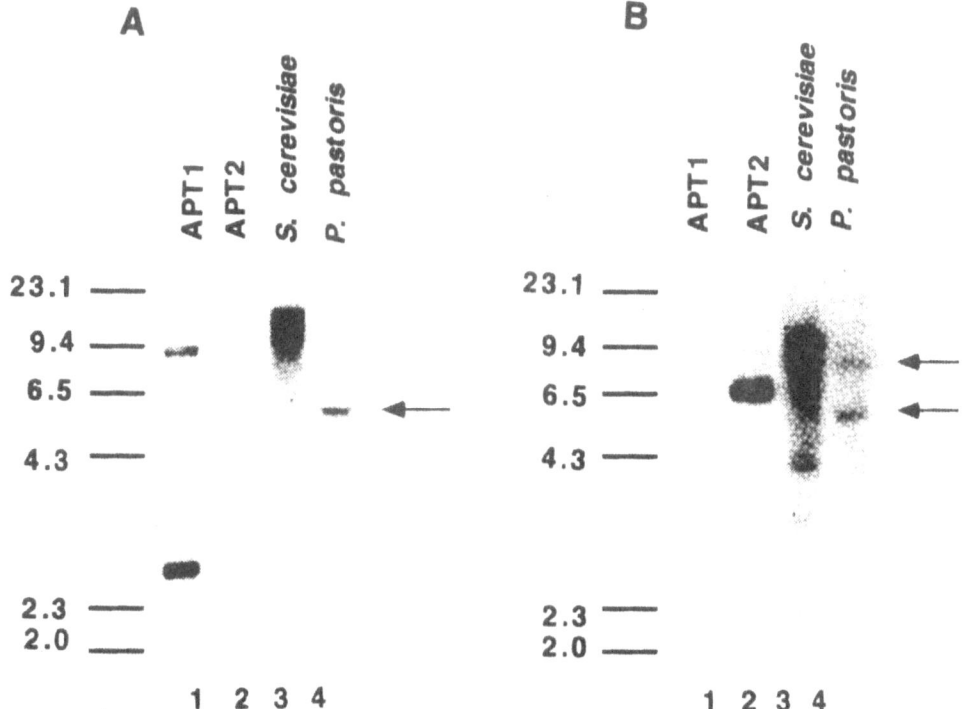

Figure 5. Southern analysis of APT1 and APT2 in *S. cerevisiae* and *P. pastoris*. Full length APT1 and APT2 were random primer labelled with ^{32}P. 10 ug of *S. cerevisiae* and 25 ug *P. pastoris* were digested with EcoRI and run in a .8% agarose gel (lanes 3 and 4 respectively). The DNA was transferred to a Zeta Probe membrane (BioRad) and hybridization was carried out at 420C. Lanes 1 and 2 are positive controls loaded with 10 pg plasmid DNA. Panel A was probed with APT1. Panel B was probed with APT2.

DISCUSSION

We have shown that the APT2 protein can interact with the APT1 protein in the yeast two-hybrid system. Whether this heterodimer is functional is not known. However, since native APT2 cannot be detected by western analysis, it may have no functional role in *S. cerevisiae* metabolism. Tree analysis indicates that APT2 is most likely a recent duplication event of the APT1 gene. APT2 is much more closely related to APT1 than to APRT's from different organisms. Sequences flanking APT2 are also present next to APT1. APT2 may be the result of a very large chromosomal duplication event. Arabidopsis thaliana is known to have two APRT's, both of which are active.[9] Southern analysis of the closely related yeast Pichia pastoris indicates that the APT2 duplication event may have occurred in a common yeast ancestor. Since APT2 seems to have no detectable activity when overexpressed, and is not detectable via western analysis, it is our conclusion that APT2 is a recent duplication event that is not being maintained by selection.

However, the presence of APT2 does provide excellent information for the study of functional APRT's. It's close sequence identity and similarity with other APRT's can add valuable information when identifying conserved sequences. Conserved sequences which are missing in APT2 can be identified, and from this information we can extrapolate to regions required for activity. In addition to this, by making calculated amino acid substitutions, an active APT2 might possibly be produced.

REFERENCES

1. Dovey, H. F., McKerrow, T. H., and Wang, C. C., Purine salvage in Schistosoma mansoni, Mol. Biochem. Parasitol, 11:157 (1984).
2. Simmonds, A.,and Muller, M., Purine metabolism in man-IV. Clinical and regulatory mechanisms, Plenum Press, New York, NY, 1982.
3. Engle, S.J., Chen, J., Stockelman M., Stambrook, P.J., and Tischfield, J.A., Targeted disruption of the mouse adenine phosphoribosyltransferase (APRT) gene and the production of APRT deficient mice, American Journal of Human Genetics, 55:171 (1994).
4. Alfonzo, J. D., Sahota, A., Deeley, M. C., Ranjekar, P., and Taylor, M. W., Cloning and Characterization of the Adenine Phosphoribosyltransferase Gene (APT1) from *Saccharomyces cerevisiae,* Gene, 161:82 (1995).
5. Yuryev, A., and Corden, J., A *Saccharomyces cerevisiae* gene encoding a potential adenine phosphoribosyltransferase, Yeast, 10:659 (1995).
6. Alfonzo, J. D., Sahota, A., and Taylor, M. W., Purification and Characterization of Adenine Phosphoribosyltransferase from *Saccharomyces cerevisiae*, Biochem. Biophy. Acta, In Press.
7. PAUP v3.11, Phylogentic Analysis Using Parsimony, Smithsonian Institute, (1993).
8. Entrez accession U33007, and Z46659.
9. Schnorr, K.M., Gaillard, C., Biget, E., Nygarrd, P., and Laloue, M., A second form of adenine phosphoribosyltransferase in Arabidopsis thaliana with relative specificity towards cytokinins, The Plant Journal, 9:891 (1996).

62

STRUCTURAL AND FUNCTIONAL ANALYSIS OF THE HUMAN TPMT GENE PROMOTER

M. Y. Fessing, E. Y. Krynetski, J. D. Schuetz, and W. E. Evans

St. Jude Children's Research Hospital
332 N. Lauderdale
Memphis, Tennessee 38101
and University of Tennessee
Memphis, Tennessee 38105

1. INTRODUCTION

Thiopurine S-methyltransferase (TPMT) catalyzes S-methylation of aromatic and heterocyclic sulfhydryl compounds, including the cytotoxic and immunosuppressive thiopurine medications mercaptopurine, azathioprine and thioguanine. Variations in TPMT activity are an important determinant of the therapeutic and toxicologic effect of these drugs.[1] Recently our lab and others have isolated and characterized the human gene encoding TPMT,[2] permitting us to study the transcriptional regulation of the gene, by structural and functional analysis of its promoter region.

2. MATERIALS AND METHODS

2.1. Plasmid Construction

The TPMT promoter region was cloned in direct orientation into promoterless luciferase reporter plasmid pGL-3Basic (Promega, USA). The resulting plasmid was used to generate 5' truncated constructions of TPMT promoter region by ExoIII based technique and 3' truncated constructions of the promoter by PCR, with subsequent subcloning.

2.2. Transient Transfection Experiments

HepG2 cells were transfected with the reporter plasmids by calcium phosphate precipitation and CCRF-CEM cells by electroporation. The cells were harvested and lucifirase activity was determined in the lysate by Lucifirase Assay System (Promega, USA)

Purine and Pyrimidine Metabolism in Man IX,
edited by Griesmacher *et al.* Plenum Press, New York, 1998.

315

according to the supplier's recommendations. The lucifirase activity was normalized per total protein concentration in the lysate. For each construction at least three independent experiments were performed and the data were collected in duplicate.

3. RESULTS AND DISCUSSION

3.1. TPMT Gene Expression in Different Human Tissues

Human multi-tissue Northern analysis with a part of TPMT cDNA containing the open reading frame as a probe (Fig. 1), established that this gene is transcribed in all the tissues tested and revealed the highest mRNA level in liver, skeletal muscle and kidney and the lowest level in brain and placenta. White blood cells, the main targets of thiopurine therapy, had intermediate levels of the TPMT mRNA.

3.2. TPMT 5′ Untranslated Region Is Transcriptionally Active

To assess promoter activity of the 5′-untranslated region of the TPMT gene, we cloned the gene fragment spanning −873 to +736 b.p. relative to previously reported transcription start site[2] into promoterless luciferase reporter vector pGL-3Basic. When the resulting reporter construct was transiently transfected into hepatocellular carcinoma HepG2 and T-cell leukemia CCRF-CEM cells, it exhibited about 10% of the reporter gene activity in comparison with the construction containing the luciferase gene under the control of the SV40 early gene promoter with enchancer.

3.3. Sequencing Analysis of the 5′UTR of the Human TPMT Gene

The sequencing analysis of 873 b.p. upstream from the previously reported transcription start site revealed that it is 71% GC rich and does not contain consensus sequences for TATA box or CCAAT element, similar to a number of TATA-less promoters often found in "housekeeping" genes (e.g. human DHFR). 14 putative binding sites for Sp1[3] and other putative cis-regulatory elements were found within the region.

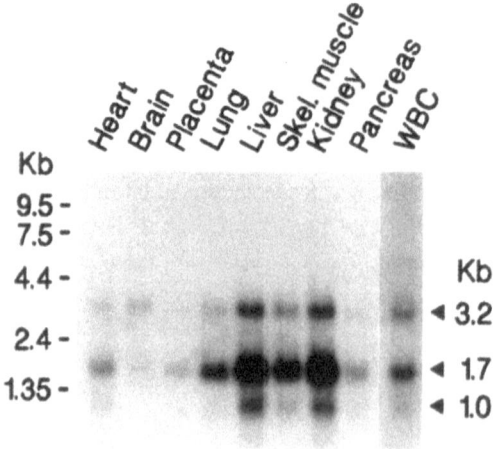

Figure 1. Northern blot analysis of polyA⁺ RNA isolated from various human tissues and hybridized with wild-type TPMT cDNA. Each lane contained approximately 2 μg of polyA⁺ RNA. Reproduced with publisher's permission from Krynetski et al., PNAS 1995, 92: 949–953.

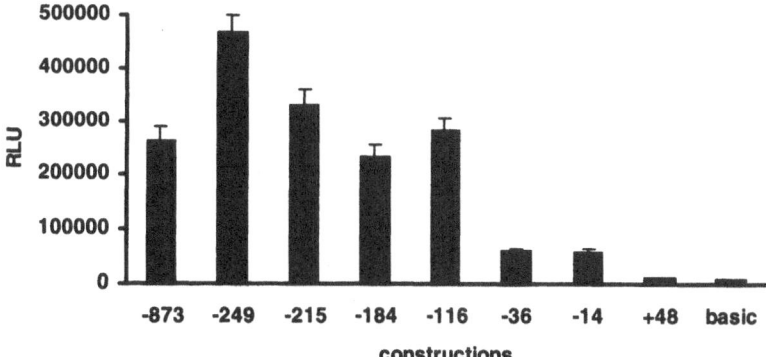

Figure 2. Relative activity of 5'-truncated TPMT promoter constructions. A set of 5' truncated TPMT promoter constructs coupled to the reporter lucifirase gene were transiently transfected into HepG2 cell line. Basic – parental promoterless plasmid pGL3-Basic.

3.4. Isolation of the Minimal Promoter Region Essential for Constitutive Expression of the TPMT Gene

To localize the sequence essential for the basal activity of the human TPMT promoter, we prepared a set of 5' truncated promoter constructions with the same 3' end fixed at the position +262 b.p. Next, we analyzed the activity of a reporter lucifirase gene under the control of the resulting TPMT promoter constructs in HepG2 and CCRF-CEM cell lines. These experiments revealed that progressive deletions down to position -116 relative to the transcription start site exhibit minimal effect on the full activity of TPMT promoter in HepG2 cells (Fig. 2), but further deletion of the fragment from −116 to −36 b.p. led to 5-fold reduction in promoter activity. This reduced activity remained in constructs with the deletion down to position −14 and was completely abolished in the construct +48. Similar results were obtained in CCRF-CEM cells (data not shown).

To determine the sequence after the transcription start site required for full TPMT promoter activity, we prepared a set of 3' truncated constructions of the promoter with the same 5' end fixed at the position −318 b.p. The transient expression of a reporter gene under control of the resulting constructs in HepG2 cells (Fig. 3) revealed that the sequence after the

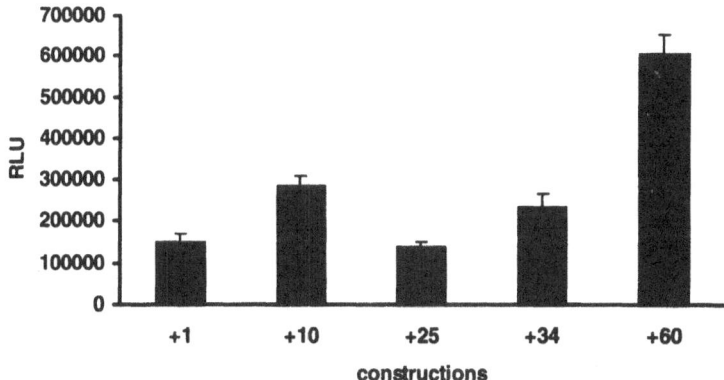

Figure 3. Relative activity of 3'-truncated TPMT promoter constructions. A set of 3'-truncated TPMT promoter constructions coupled to the reporter lucifirase gene was transiently transfected into HepG2 cell line.

transcription start site does not play a dramatic role in TPMT promoter activity. Deletion of the 3′ region up to the position +60 did not affect the promoter activity significantly. Further deletion of the region from +60 to +34 led to 2.5-fold reduction in the promoter activity, and the constructs +1, +10, +25 and +34 had a similar level of promoter activity.

This result allows us to suggest that the core promoter region of the TPMT gene that is essential for its full level basal expression is located downstream of −116 b.p. according to the previously reported transcription start site , designated as position +1. In this region the sequence downstream of position −14 supplies a low level of constitutive expression, and a region from −116 to −36 exhibits a strong stimulation of activity. The latter region contains multiple GC boxes, putative binding sites for the transcription factor Sp1.[3] These GC boxes are the most probable cis-elements determining the full level of basal transcription of the TPMT gene, a possibility currently under investigation in our lab. Interestingly, we didn't find any sequences similar to the known initiator consensus sequences[4] near the transcription start site. Moreover, the sequence downstream from this start site appeared to have minor effect on the basal promoter activity, in good agreement with the absence of a strong functional initiator in this region.

Further studies are necessary to characterize cis-elements critical for its basal and regulated activity, work that is currently ongoing in our lab.

REFERENCES

1. Krynetski E.Y.,Tai H.L.,Yates C.R., Fessing M.Y., Loennechen T.,Schuetz J.D., Relling M.V. and Evans W.E. Genetic polymorphism of thiopurine S-methyltransferase:clinical importance and molecular mechanisms. Pharmacogenetics 6 (1996) 279–290.
2. Szumlanski C., Otterness D., Her C., Lee D., Brandriff B., Kelsell D., Spurr N., Lennard L., Wieben E. and Weinshilboum R. Thiopurine methyltransferase pharmacogenetics: human gene cloning and characterization of a common polymorphism. DNA Cell Biol. 15 (1996) 17–30.
3. Kadonaga J.T., Jones K.A. and Tjian R. Promoter-specific activation of RNA polymerase II transcription by Sp1. TIBS 11 (1986) 20–23.
4. Smale S.T. Transcription initiation from TATA-less promoters within eukaryotic protein-coding genes. BBA 1351 (1997) 73–88.

63

A FIRST EVIDENCE OF AN ASYMPTOMATIC GERMLINE MISSENSE BASE SUBSTITUTION IN THE HYPOXANTHINE PHOSPHORIBOSYLTRANSFERASE (HPRT) GENE IN HUMANS

Shin Fujimori,[1] Noriko Yamaoka,[1] Ryozo Sakuma,[2] Masayuki Hakoda,[3] Hisashi Yamanaka,[3] Ieo Akaoka,[1] and Naoyuki Kamatani[3]

[1]Second Department of Internal Medicine
Teikyo University School of Medicine
[2]Department of Clinical Chemistry
Toranomon Hospital
[3]Institute of Rheumatology
Tokyo Women's Medical College

INTRODUCTION

Germline mutations at hypoxanthine phosphoribosyltransferase (HPRT) locus in humans cause HPRT deficiency. Since HPRT gene is on the X chromosome, this genetic disorder is inherited as an X-linked recessive fashion. In the sever form of the disease, designated Lesch-Nyhan syndrome, the affected males develop characteristic neuropsychological symptoms including self-mutilation in addition to hyperuricemia due to urate overproduction. In the milder form, designated partial HPRT deficiency, the patients develop gout, urolithiasis or hyperuricemia but not neurological symptoms. Germline mutations involved in HPRT deficiencies have been extensively analyzed,[1] and the accumulated data concerning the molecular abnormalities have been stored in a database.[2] Although this database contains numerous germline mutations at the HPRT locus in humans, none of them are asymptomatic mutations; i.e. they are associated with either Lesh-Nyhan syndrome or gout (partial deficiency). Here, we describe, for the first time, a missense base substitution in the germline HPRT gene that caused a reduction in the amount of enzyme per cell but was associated with no recognizable adverse conditions, thus providing the proof that some missense base substitutions in the HPRT gene are asymptomatic.

Purine and Pyrimidine Metabolism in Man IX,
edited by Griesmacher *et al.* Plenum Press, New York, 1998.

MATERIALS AND METHODS

HPRT activity in the hemolysates and B lymphoblast extracts was determined by non-radioactive and radioactive methods. cDNA for HPRT was amplified by the reverse transcription(RT)-PCR method previously described.[3] The amplified cDNA was cloned into pCR-Script cloning vector and sequenced by automated sequencer. The 580-bp fragment containing exon 3 of the HPRT gene was amplified from genomic DNA by using the following primer set: 5'-CCTTATGAAACATGAGGGCAAAGG-3', 5'-ACCTAC-TGTTGCCACTA-3'.

Western blot analysis of the HPRT protein was performed as described previously[4] with a modification. The antibody against human HPRT prepared in Dr. W. N. Kelley's laboratory at Michigan University was kindly provided by Dr. Hidaka of Ichihara Hospital of Teikyo University. Northern blot analysis was performed by the standardized method with ^{32}P labeled amplified HPRT cDNA as a probe.

RESULTS

During a screening of over 1000 normal volunteers for hemolysate HPRT activities, we found a male (S) with a relatively low enzyme activities. The HPRT activity in his hemolysate (0.92 μmol/min/g Hb) was about 46% of the normal activities (1.99 ± 0.31 μmol/min/g Hb). He was 40 year old and had suffered from no major diseases including gout, urolithiasis or neurological abnormalities. None of his relatives suffered from gout or hyperuricemia. His serum urate level was 5.5 mg/dl. The HPRT activity in his B lymphoblast extract (145 nmol/min/mg protein) was about 37% of the normal activities (392 ± 12 nmol/min/mg protein; data from 5 lymphoblast lines). Sequencing of HPRT cDNA from subject S showed A to G base substitution leading to the amino acid substitution from His to Arg in codon 60 (the initiation codon ATG was counted as 1) (Fig. 1). Since this base substitution should generate a new Tsp45 | site, the mutant sequence is likely to be distinguished from the normal sequence by using this restriction enzyme. The amplified

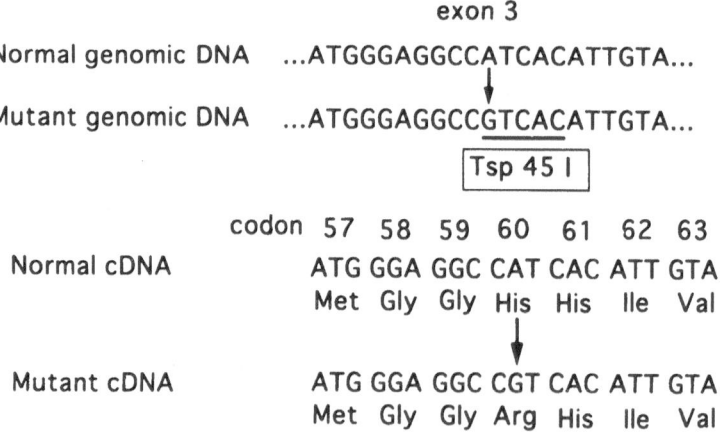

Figure 1. Normal and mutant DNA sequences as well as predicted amino acid sequences. Arrows indicate a base substitution observed in the genomic DNA of subject S as well as the expected from the cDNA sequence. Codon number 1 corresponds to the initiation codon ATG.

580-bp fragment containing exon 3 of the HPRT gene of subject S was cut into two fragments of 395-bp and 185-bp by Tsp45 |; this was not the case in normal subjects (data not presented). About half of the amplified DNA from the mother was cut by Tsp45 |, indicating that the mother was heterozygous for the mutation. A Western blot analysis revealed that the amount of HPRT protein in the B lymphoblasts from subject S was 25–50% of normal cells. Northern blot analysis showed essentially normal concentration of HPRT mRNA in the B lymphoblasts from subject S.

DISCUSSION

In a review article, Rossiter and Caskey[5] raised the possibility that alterations in the HPRT gene could occur with no accompanying clinical symptoms. However, such a mutation has not been reported as yet. In the present study, we demonstrate for the first time that such a gene alteration indeed exists. The missense base change reported here does not seem to be a polymorphism, as reported previously.[6] Thus, unlike the latter, the frequency of the mutant allele of this type is very low since only one individual among more than 1000 people had low HPRT activity. The prediction by Rossiter and Caskey[5] was reasonable because accumulated data on mutations in patients with HPRT deficiencies strongly suggested the presence of such asymptomatic mutations. Thus, while random base substitution in the coding region of the HPRT sequence should give 74%, 4.1%, and 22%, missense, nonsense, and synonymous mutations,[3] respectively, base substitutions observed in Lesh-Nyhan syndrome and partial HPRT deficiency deviate significantly from the distributions; synonymous substitutions are not observed and the proportion of nonsense substitutions to missense substitutions is higher than expected.[3] Further, the distribution of amino acid substitutions in the HPRT sequence is far from uniform, suggesting against that they deviate from random mutations.[7] The presence of asymptomatic missense base substitutions was expected, and their exclusion would yield biased data. Nevertheless, such asymptomatic germline HPRT mutations were hard to identify because the frequency of the mutation was expected to be extremely low. Of course, the mutations leading to Lesh-Nyhan syndrome and gout are also infrequent, but affected individuals can be identified by the specific symptoms. Our case was identified only because we performed a large scale screening and he was found to show low enzyme activity. Germline HPRT mutations that cause neither symptoms nor reductions in the enzyme activity are conceivable but would be extremely hard to identify. Although the HPRT mutation database describes human somatic HPRT mutations in addition to germline mutations, the missense mutation described in the present study would be undetectable as a somatic mutation since the present mutation does not render the cells resistant to 6-thioguanine. The mechanism for the low HPRT activity in S is also of interest. Data from the Western blot analysis indicate that the amount of HPRT protein per cell is reduced by the mutation. Since the amount of HPRT protein (25–50% of normal) and the enzyme activity (about 37% of normal) in the B lymphoblasts are broadly equivalent, the reduced amount of HPRT enzyme protein rather than a decrease in the enzyme activity per molecule is likely to be responsible for the low enzyme activity. Much less likely, the amino acid change observed in the mutant enzyme may have reduced the immunoreactivity to the antibody. The amino acid substitution from His to Arg is interpreted as a conservative amino acid change since both are typical basic amino acids. Therefore, it is not surprising that this amino acid change scarcely impairs the functions of the enzyme. In fact, the calculated Km for 5-phosphoribosyl-1-pyrophosphate or hypoxanthine for the mutant enzyme was not different from the normal enzyme (data not presented). It will be of interest to consider the amino acid substitution with regard to the recently established crystal structure of human HPRT.[8]

REFERENCES

1. Sculley DG, Dawson PA, Emmerson BT, Gordon RB (1992) A review of the molecular basis of hypoxanthine-guanine phosphoribosyltransferase (HPRT) deficiency. Hum Gent 90: 195–207
2. Cariello NF, Craft TR, Vrieling H, van Zeeland AA, Adams T, Stopek TR (1992) Human HPRT mutant database: software for data entry and retrieval. Environ Mol Mutagen 20: 81–83
3. Fujimori S, Kamatani N, Nishida Y, Ogasawara N, Akaoka I (1990) Hypoxanthine guanine phosphoribosyltransferase deficiency: Nucleotide substitution causing Lesh-Nyhan syndrome identified for the first time among Japanese. Hum Genet 84: 483–486
4. Wilson JM, Stout T, Palella TD, Davidson BL, Kelley WN, Caskey CT (1986) A molecular survey of hypoxanthine-guanine phosphoribosyltransferase deficiency in man. J Clin Invest 77: 188–195
5. Rossiter BJF, Caskey CT (1995) Hypoxanthine-guanine phosphoribosyltransferase deficiency: Lesh-Nyhan syndrome and gout. In: Scriver CR, Beaudet AL, Sly WS, Valle D (eds) The metabolic and molecular basis of inherited disease, 7th ed. McGraw-Hill, New York, pp 1679- 1706
6. Nussbaum RL, Crowder WE, Nyhan WL, Caskey CT (1983) A three allele restriction-fragment-length polymorphism at the hypoxanthine phosphoribosyltransferase locus in man. Pro Natl Acad Sci USA 80: 4035–4039
7. Cariello NF, Stopek, TR (1993) Analysis of mutations occurring at the human hprt locus. J Mol Biol 231: 41–57
8. Eads JC, Scapin G, Xu Y, Grubmeyer C, Sacchettini JC (1994) The crystal structure of human hypoxanthine-guanine phosphoribosyltransferase with bound GMP. Cell 73:325–334

64

GENETIC BASIS OF ADENYLOSUCCINASE DEFICIENCY IN AN ITALIAN PATIENT

D. Verginelli, B. Luckow, C. Salerno, and M. Gross

Purine Research Laboratory, Medical Polyclinic
University of München
Department of Biochemical Sciences
University of Roma La Sapienza

1. INTRODUCTION

Adenylosuccinase (EC 4.3.2.2; ASase) deficiency is an autosomal recessive disorder that results in the accumulation in body fluids of two normally undetectable compounds, succinylaminoimidazole carboxamide (SAICA) riboside and succinyladenosine (S-Ado). These succinylpurines are the products of dephosphorylation, by cytosolic 5'-nucleotidase, of the two substrates of the defective enzyme.[1-3] The inherited defect has been diagnosed in a small group of patients of different nationalities with psychomotor retardation and autistic features.

A single T-to-C transition in the first base of codon 413 was identified by Stone et al.[4] in human ASase gene. The mutation predicted a Ser-to-Pro alteration in the enzyme. All three of the mentally-retarded children of a Moroccan family were homozygous for this mutation, while the parents and the unaffected children were heterozygous. The alteration was absent in control subjects. The Ser[413]Pro substitution led to decreased enzyme stability, as evidenced by its greater heat lability and its increased sensitivity to guanidine HCl and urea. The kinetic constants and the pH profiles of the Asase-catalysed reactions were not affected by the mutation.[5]

In this report, we analyse the ASase cDNA from an Italian patient with reduced ASase activity in blood cells, who is the offspring of unrelated parents and is suffering from severe psychomotor retardation. This study allow to identify two new missense-transitions in ASase gene.

2. CASE REPORT

The Asase-deficient patient was described previously elsewhere.[3,6] The concentrations of SAICA riboside and S-Ado in the urine were 0.06 mol/mol creatinine and 0.075

Purine and Pyrimidine Metabolism in Man IX,
edited by Griesmacher *et al.* Plenum Press, New York, 1998.

Table 1. Primer sequences

Primer	Sequence 5′ ⟶ 3′	Direction	Position
AS1	TTTCGTTTAATTCTCTTCCA	Antisense	1480–1461
AS2	CATTTCTGAAGGATTGGTCG	Sense	1073–1092
AS3	AACAGGGGATACACCTCCTC	Antisense	1408–1389
AS4	CTGTGGTTAAGCAGGAAGGG	Sense	1237–1256
AS5	TGGCAATCCTGGCGGCTACC	Antisense	1201–1182
AS6	GGCTAGCTTGGGGGCATCAG	Sense	767–786
AS7	GGGTCATCAGGTGGCGGGCA	Antisense	942–923
AS8	GATGACCTGCGCTTCCGGGG	Sense	573–592
AS9	TCCTGCCTTTTCTGTCACCA	Antisense	695–676
AS10	CACGTGCACACATTTGGCCA	Sense	273–292
AS11	TGGGTAGACTGGCTCGTTCC	Antisense	462–443
AS12	GTTCGCCCGACAGCTACCGC	Sense	25–44
AS13	GCAGTGGCCAAATGTGTGCA	Antisense	296–277
AS14	GCCAAAGCTTGCCAGAGTGA	Sense	398–417
AS16	TAAGCGGAATCCCATGCGTT	Sense	884–903
AS17	CCAACATAGCAAGAAGTAGC	Antisense	349–330
AS18	CAAGATGGCAGCTGAGGAAG	Sense	224–243
AS19	GAAGAAGGATCCAGTAAATG	Antisense	1345–1326
AS20	AGCATCTGTCCAGTGGTTTG	Sense	965–984
ASE28	GAGATGTGCTTCGTGTTTAGCGA	Sense	75–97
ASE29	AACACAATTTTCATGCCTCAGCA	Antisense	1515–1493

mol/mol creatinine, respectively. ASase activity in erythrocytes and mixed peripheral blood lymphocytes was reduced to 30–40% of that in controls. The erythrocyte enzyme showed normal substrate affinity, but impaired thermal stability. The relatives of the patient were asymtomatic, did not excrete appreciable amounts of succynilpurines, but elicited abnormal heat denaturation curves of erythrocyte ASase with values between the patient and healthy subjects.[3]

3. RESULTS AND DISCUSSION

To determine the DNA sequence of ASase gene from the patient and her family, we isolated the cDNA encoding human ASase from cultured lymphoblasts. Thus, lymphoblast mRNA was purified and reversed transcribed with AMV Reverse Transcriptase.[7] The cDNA was amplified by using appropriate 20-mer primers based on the published sequence of human ASase (Table 1) to allow the synthesis of four fragments with nucleotide coordinates 25–942, 25–1201, 273–1201, and 1073–1480. These PCR products were subjected to a subsequent asymmetric amplification using limiting or internal sequence-specific primers, purified, and sequenced in both directions with the Sequenase kit (Version 2.0; US Biochemicals, Cleveland, OH) according to manufacturer's instructions.[8]

While PCR amplification of control samples gave rise to products with encoding sequences identical to the published sequences, the patient was found to be compound heterozygous for two mutations not yet described: a C_{300}-to-G transversion in the codon 75 corresponding to a Pro-to-Ala substitution and a G_{1266}-to-T transversion in the codon 397 corresponding to an Asp-to-Tyr substitution. The father and the healthy sister were heterozygous only for the first mutation, whereas the mother was heterozygous only for the second mutation. The point mutations in codons 75 and 397 were confirmed by recognising the new restriction endonuclease sites for MwoI and Tsp509I, respectively.

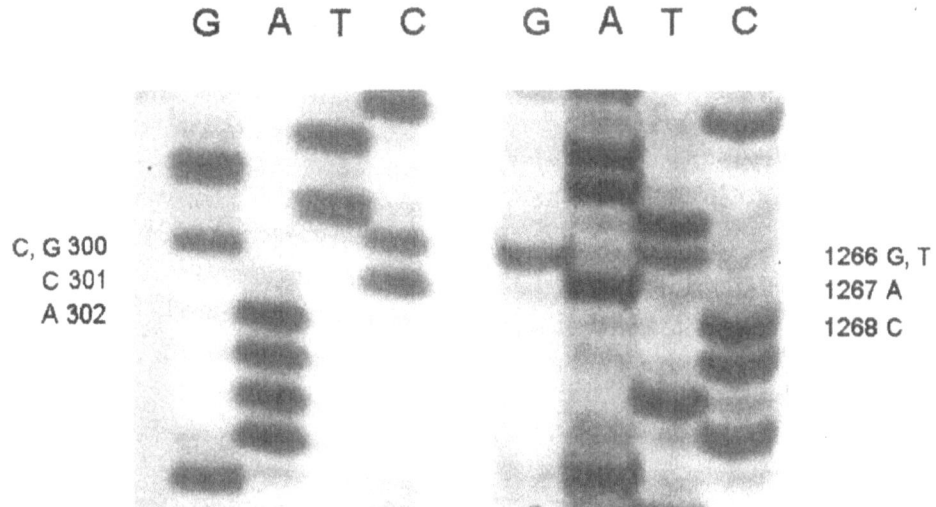

Figure 1. Partial nucleotide sequence of PCR-amplified ASase cDNA from the patient compound heterozygous for a C_{300}-to-G transversion in codon 75 (left) and for a G_{1266}-to-T transversion in codon 397 (right).

It is conceivable that these amino acid substitutions would significantly alter the flexibility of ASase structure,[9] thus accounting for the decreased thermal stability reported for the erythrocyte enzyme.[3]

ACKNOWLEDGMENTS

This research was done in the framework of the European Community Concerted Action contract no. BMH1-CT-1384.

REFERENCES

1. J. Jaeken and G. Van den Berghe. An infantile autistic syndrome characterized by the presence of succinylpurines in body fluids, Lancet 2:1058–1061 (1984)
2. J. Jaeken, S. K. Wadman, M. Duran, F. J. Van Sprang, F. A. Beemer, R. A. Holl, P. M. Theunissen, P. De Cock, F. Van den Bergh, M. F. Vincent, and G. Van den Berghe. Adenylosuccinase deficiency: an inborn error of purine nucleotide synthesis. Eur. J. Pediatr. 148:126–131 (1988)
3. C. Salerno, C. Crifò, and O. Giardini. Adenylosuccinase deficiency: a patient with impaired erythrocyte activity and anomalous response to intravenous fructose. J. Inher. Metab. Dis. 18:602–608 (1995)
4. R. L. Stone, J. Aimi, B. A. Barshop, J. Jaeken, G. Van den Berghe, H. Zalkin, and J. E. Dixon. A mutation in adenylosuccinate lyase associated with mental retardation and autistic features. Nature Genetics 1:59–63 (1992)
5. R. L. Stone, H. Zalkin, and J. E. Dixon. Expression, purification, and kinetic characterization of recombinant human adenylosuccinate lyase. J. Biol. Chem. 268:19710–19716 (1993)
6. C. Salerno, S. Iotti, R. Lodi, C. Crifò, and B. Barbiroli. Failure of muscle energy metabolism in a patient with adenylosuccinate lyase deficiency: an in vivo study by phosphorus NMR spectroscopy. Biochim. Biophys. Acta (1997) in press
7. J. Sambrook, E. F. Fritsch, and T. Maniatis. Molecular cloning: a laboratory manual. Cold Spring Harbor Laboratory Press, Cold Spring Harbor (1989)
8. E. Rhéaume, J. Simard, Y. Morel, F. Mebarki, M. Zachmann, M. G. Forest, M. I. New, and F. Labrie. Congenital adrenal hyperplasia due to point mutations in the type II 3β-hydroxysteroid dehydrogenase gene. Nature Genetics 1:239–245 (1992)
9. Geourjon and G. Deléage. SOPMA: significant improvements in protein secondary structure prediction by consensus prediction from multiple alignments. Comput. Applic. Biosci. 11:681–684 (1994)

65

MUTATIONS IN XANTHINE DEHYDROGENASE GENE IN SUBJECTS WITH HEREDITARY XANTHINURIA

K. Ichida,[1] N. Kamatani,[2] T. Nishino,[3] M. Saji,[1] H. Okabe,[1] and T. Hosoya[1]

[1]2nd Department of Internal Medicine
The Jikei University School of Medicine
Tokyo, Japan
[2]Institute of Rheumatology
Tokyo Women's Medical College
Tokyo, Japan
[3]Department of Biochemistry and Molecular Biology
Nippon Medical School
Tokyo, Japan

1. INTRODUCTION

Xanthine dehydrogenase is an enzyme that catalyzes the oxidation of hypoxanthine to xanthine and uric acid from xanthine during the final stage of purine metabolism. Xanthine dehydrogenase exists as a dimer and each subunit has a molecular weight of 145000. Human xanthine dehydrogenase cDNA consists of 4002 nucleotides (1) and the gene is mapped to chromosome 2p23 (2). Mapping of the functions on xanthine dehydrogenase was performed for 3 peptide domains generated by the protein cleavage (1). The N-terminal 20 kD domain includes a 2Fe/2S non-heme iron binding site while the adjacent 40 kD and the C-terminal 85 kD domains include flavin binding and molybdenum cofactor binding domains, respectively (1). Most of xanthine dehydrogenase usually exist as the dehydrogenase form and the enzyme is converted to the oxidase form, xanthine oxidase, by the proteolytic cleavage and the oxidation of cystein residues under certain conditions (3). Xanthine dehydrogenase has recently been attracting attention for its possible involvement in triggering tissue damage by producing free radicals. It is exhibited in many studies that xanthine oxidase injured the tissues on the conditions, such as post-ischemic reperfusion tissue injury, adult respiratory distress syndrome and lung injury resulting from influenza virus infection (4–8).

Classical xanthinuria, inherited xanthine dehydrogenase deficiency, is valuable because of the lack of xanthine dehydrogenase which may produce free radicals. It is rare

Purine and Pyrimidine Metabolism in Man IX,
edited by Griesmacher *et al.* Plenum Press, New York, 1998.

327

disease and transmitted as an autosomal recessive disease with no chromosomal abnormalities (9). Recently, classical xanthinuria was classified into the following two types (10): type 1, a deficiency of xanthine dehydrogenase alone and type 2, deficiencies of xanthine dehydrogenase and aldehyde oxidase. Classical xanthinuria type 1 and 2 were regarded as a single clinical entity because of similar clinical symptoms and blood chemical analyses. We recently reported the mutations in xanthine dehydrogenase of subjects with classical xanthinuria type 1 (11). From accumulation of the variation of mutation, the function of xanthine dehydrogenase can be analyzed in detail. In this paper, we exhibited the influence of the mutation in xanthine dehydrogenase gene reported (11) to the activity of the enzyme.

In subjects with xanthinuria type 1, it is sometimes assumed that there is an overactivity of aldehyde oxidase instead of xanthine dehydrogenase because changes of serum and urinary oxyprinol in subjects with xanthinuria type 1 who have only aldehyde oxidase are similar to those in healthy subjects who have the two enzyme activities in allopurinol loading test and the two enzymes have biochemical features in common. The amounts of the aldehyde oxidase mRNAs in duodenal mucosa of subjects with xanthinuria type 1 were also reported in this study.

2. MATERIALS AND METHODS

2.1. Materials

We examined 4 subjects with xanthinuria. The subject 1 and 2 are brothers. Though subject 1, 2, and 4 were with the type 1, the type of xanthinuria in subject 3 were not confirmed. The quantity of xanthine dehydrogenase mRNA in duodenal mucosa of subjects except 3 were similar to that in healthy subjects. The subject 1, 2, and 3 had a C to T base substitution at nucleotide 682 (nucleotide number in cDNA) and it should cause a CGA (Arg) to TGA (Ter) nonsense mutation at codon 228. The subject 4 had a deletion of C at nucleotide 2567. The subject 1, 2, and 4 were homozygous and the subject 3 was a compound heterozygote for this mutation.

2.2. Methods

2.2.1. Measurement of Xanthine Dehydrogenase. The activity of xanthine dehydrogenase was measured with high performance liquid chromatography according to the method described earlier (12).

2.2.2. Quantification of mRNA for Aldehyde Oxidase. Total RNAs were isolated from duodenal mucosae of subject 1, 4, and control subject using Isogen (Nippon Gene, Tokyo, Japan) according to the manufacturer's recommendations. Aldehyde oxidase mRNAs in the samples were quantified by competitive reverse transcriptase (RT)-PCR using a MIMIC construction kit (Clontech, Palo Alto, CA). Primers for competitive PCR were designed from sequence of aldehyde oxidase cDNA reported (13). The competitive template was constructed from 2 ng of Bam HI/Eco RI fragment of v-erbB using primers A1-mic (5'-GAGCTGCTCTTCTACGTGAACGGCCCAAGTTTCGTGAGCTGATTG) and A2-mic (5'-TGGCAATCCATTGATTCCTTGATCCTCTGTCAATGCAGTTTGTAG) in the first 16-cycle PCR at 94°C, 45 s; 60°C, 45 s; and 72°C, 90 s. After the RT reaction using RNA from the mucosae, the competitive PCR was performed. Thus, the solution containing cDNA of aldehyde oxidase synthesized by the RT reaction was mixed with various concentrations of the competitive template, and 18 cycles of PCR reactions were performed using primers A1 (5'-GAGCTGCTCTTCTACGTGAACGGCC) and A2 (5'-TGGCAATCCATTGATTCCTTGATCC).

3. RESULTS AND DISCUSSION

3.1. Influence of Mutation in Xanthine Dehydrogenase Gene to the Function

No activities of xanthine dehydrogenases in the duodenal mucosa of the four subjects were detected. Both the nonsense and the deletion mutations may cause the cells to synthesize the truncated peptides. However, even if they are synthesized, normal functions are not likely to be retained.

The base substitution from C to T at nucleotide position 682 causes a CGA to TGA nonsense substitution at codon 228. Flavin and molybdenum cofactor binding sites would be missing in the peptide (Figure 1). The truncated peptide would accordingly have no activities.

The deletion of C at nucleotide 2567 in the human xanthine dehydrogenase gene should cause a frameshift from codon 856 and a termination codon is encountered at codon 928 (Figure 1). Rosy locus of Drosophila melanogaster includes xanthine dehydrogenase gene. Mutations in rosy locus are available studying the structure and function of xanthine dehydrogenase. It was reported that even some point mutations in the putative molybdenum domain make the protein lose the activity (14). Moreover, it was recently reported that more segments in xanthine dehydrogenase protein contact molybdenum cofactor than we expected (15), so that many regions are influenced by mutations in the molybdenum domain. It is reasonable that these mutant xanthine dehydrogenase protein don't have activity, because the truncated peptide would lack molybdenum cofactor binding sites.

3.2. Quantitation of mRNA for Aldehyde Oxidase

While the primers A1 and A2 should amplify a 552 bp fragment of human aldehyde oxdase cDNA, the same primers are expected to amplify a 330 bp fragment from a competitor template. From the known concentrations of the competitor template that gave nearly equal fluorescent intensities of the amplified 552 and 330 bp fragments compensated by size, we could detect the amounts of aldehyde oxdase mRNA. In quantitative RT-PCR analyses on duodenal mucosa from Subjects 1 and 4 and the control subject, the amount of aldehyde oxidase mRNA from subjects 1 and 4 were $0.5 \times 10^{-3} \sim 1 \times 10^{-3}$ attomole competitor/tube and similar to it in normal subject. These data may indicate that aldehyde oxdase in

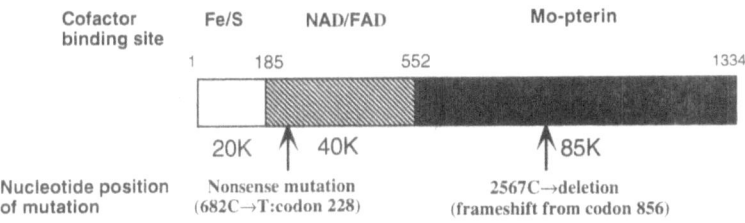

Figure 1. The mutation in human xanthine dehydrogenase gene.

subject with xanthinuria type 1 is not activated instead of xanthine dehydrogenase. But we could not ultimately confirm the results because we could not retry the competitive PCR for the lack of the samples and aldehyde oxdase activity is mainly found in liver, not duodenum mucosa.

4. CONCLUSION

From the mutations of xanthine dehydrogenase gene in xanthinuric subjects, we attempted to analyze the function of xanthine dehydrogenase protein. By comparing to biochemical findings and rosy mutant strains in Drosophila melanogaster that were reported, we exhibited the reason of no activitys of mutant xanthine dehydrogenase protein. Moreover, amount of aldehyde oxdase mRNA could be measured using competitive PCR.

REFERENCES

1. Ichida K., Amaya Y., Noda K., Minoshima S., Hosoya T., Sakai O., Shimizu N., and. Nishino T. Cloning of the cDNA encoding human xanthine dehydrogenase (oxidase): Structural analysis of the protein and chromosomal location of the gene. Gene 133: 279–284, 1993.
2. Minoshima S., Wang Y., Ichida K., Shimizu N., Nishino T., and Shimizu N. Mapping of the gene for human xanthine dehydrogenase (oxidase) (XDH) to the band p23 of chromosome 2. Cytogenet Cell Genet 68: 52–53, 1995.
3. Della Corte, E., Stirpe, F. The regulation of rat liver xanthine oxidase. Involvement of thiol groups in the conversion of the enzyme activity from dehydrogenase (type D) into oxidase (type O) and purification of the enzyme. Biochem J, 126: 739–745,1972.
4. MacGowan S.W., Regan M.C., Malone C., Sharkey O., Young L., Gorey T.F., and Wood A.E. Superoxide radical and xanthine oxidoreductase activity in the human heart during cardiac operations. Ann Thorac Surg 60: 1289–1293, 1995.
5. McCord J.M. Oxygen-derived free radicals in post-ischemic tissue injury. N Engl J Med 312: 159–163, 1985.
6. Grum C.M., Ragsdale R.A., Ketai L.H., and Simon R.H. Plasma xanthine oxidase activity in patients with ARDS. J Crit Care 2: 22–26, 1987.
7. Oda T., Akaike T., Hamamoto T., Suzuki F., Hirano T., and Maeda H. Oxygen radicals in influenza-induced pathogenesis and treatment with pyran polymer-conjugated SOD. Science 244: 974–976, 1989.
8. Akaike T., Ando M., Oda T., Doi T., Ijiri S., Araki S., and Maeda H. Dependence on O2-generation by xanthine oxidase of pathogenesis of influenza virus infection in mice. J Clin Invest 85: 739–745, 1990.
9. Simmonds H.A., Reiter S., and Nishino T. Hereditary xanthinuria. In The Metabolic Basis of Inherited Disease, 7th ed. C.R. Scriver, A.L. Beaudet, W.S. Sly, D. Valle, editors. McGraw Hill, New York. 1781–1797, 1995.
10. Reiter S., Simmonds H.A., Zollner N., Braun S.L., and Knedel M. Demonstration of a combined deficiency of xanthine oxidase and aldehyde oxidase in xanthinuric patients not forming oxypurinol. Clin Chim Acta 187: 221–234, 1990.
11. Ichida K., Amaya Y., Kamatani N., Nishino T., Hosoya T., and Sakai O. Identification of two mutations in human xanthine dehydrogenase gene responsible for classical type I xanthinuria. J Clin Invest 99: 2391–2397, 1997.
12. Kojima T., Nishina T., Kitamura M., Hosoya T., and Nishioka K. Biochemical studies on the purine metabolism of four cases with hereditary xanthinuria. Clin Chim Acta 137: 189–198, 1984.
13. Wright R.M., Vaitaitis G.M., Wilson C.M., Repine T.B., Terada L.S., and Repine J.E. cDNA cloning, characterization, and tissue-specific expression of human xanthine dehydrogenase/xanthine oxidase. Proc Natl Acad Sci. U.S.A. 90: 10690–10694, 1993.
14. Hughes R.K., Doyle W.A., Chovnick A.,Whittle J.R., Burke J.F., and Bray R.C. Use of rosy mutant strains of Drosophila melanogaster to probe the structure and function of xanthine dehydrogenase. Biochem J 285, 507–513, 1992.
15. Romao M.J., Archer M., Moura I., Moura J.J.G., LeGall J., Engh R., Schneider M., Hof P., and Huber R. Crystal structure of the xanthine oxidase-related aldehyde oxido-reductase from D. gigas. Science 270: 1170–1176, 1995.

66

ROLE OF CYSTEINE AND LYSINE RESIDUES IN HUMAN HYPOXANTHINE-GUANINE PHOSPHORIBOSYLTRANSFERASE

D. T. Keough,[1,2] C. L. Gee,[1] B. T. Emmerson,[2] and J. de Jersey[1]

[1]Department of Biochemistry and
[2]Department of Medicine
University of Queensland
Brisbane, Australia 4072

INTRODUCTION

Hypoxanthine-guanine phosphoribosyltransferase (HPRT) is a purine salvage enzyme which is widely distributed in human tissues. It catalyses the synthesis of the purine mononucleotides, IMP and GMP. In humans, partial enzyme deficiency is associated with gout and hyperuricemia while complete enzyme deficiency results in the Lesch-Nyhan syndrome.

Human erythrocyte HPRT is only available in small amounts. However, recombinant HPRT has been expressed in *E. coli* cells and purified to homogeneity in our laboratory using the expression system described by Free *et al.* (1990). This has allowed purification of mg quantities of the enzyme and site-directed mutagenesis of residues which may be involved in catalysis and/or substrate binding.

An intimation that a cysteine residue may be involved in either the structure and/or function of human HPRT is the finding of a naturally-occurring mutation (C205Y) in two Lesch-Nyhan patients, HPRT$_{Reading}$ (Davidson *et al.*, 1991) and RJK1727 (Gibbs *et al.*, 1989). This mutation has also been found in a 6-thioguanine resistant diploid human fibroblast cell line and a 6-thioguanine resistant human B-lymphoblast cell line (Lukash *et al.*, 1991; Bronstein *et al.*, 1991). There have also been a number of reports citing mutations of C65 (C65R, C65Y, C65W, C65S) (Harbach *et al.*, 1992, Bronstein *et al.*, 1991, Yang *et al.*, 1991, Recio *et al.*, 1990). The cell lines containing these mutant HPRTs were also resistant to 6-thioguanine, suggesting that these enzymes are inactive.

Reagents that react with thiol residues (such as mercuric ion and organomercurials) have been reported to inhibit the activity of human HPRT (Keough *et al.*, 1991) and *Tritrichomonas foetus* HPRT (Kanaani *et al.*, 1996). The crystal structures of both these

Purine and Pyrimidine Metabolism in Man IX,
edited by Griesmacher *et al.* Plenum Press, New York, 1998.

enzymes have now been determined (Eads *et al.*, 1994, Somoza *et al.*, 1996). Human HPRT was crystallized in the presence of the product of the reaction, GMP. However, these structures have not clarified the reason for inhibition by organomercurial compounds.

The crystal structure of human HPRT suggests roles for a number of lysine residues in binding and catalysis. The role of cys and lys residues in the structure and function of human HPRT has been investigated by chemical modification and site-directed mutagenesis.

MATERIALS AND METHODS

Enzyme Activity. Enzyme activity was determined spectrophotometrically by measuring the rate of conversion of guanine to GMP at 257.5 nm in 0.1 M Tris-HCl, 0.11 M $MgCl_2$, pH 8.5; 25°C (Keough *et al.*, 1987). The $\Delta\varepsilon$ is $5816.5 \ M^{-1} \ cm^{-1}$.

Expression of Active Wildtype HPRT. Human HPRT was expressed in the Sϕ606 cell line as described by Free *et al.* (1990) and was purified using GMP-Sepharose chromatography.

Preparation of Mutant Recombinant Enzymes. Site-directed mutagenesis was carried out by the method of splicing by overlap extension (Horton *et al.*, 1989) or by the Eckstein method (Olsen *et al.*, 1989). The mutant recombinant enzymes were purified to homogeneity using the same procedure used for wildtype HPRT.

Inhibition by TNBS. Studies of the inhibition of HPRT by TNBS were performed in 0.05 M Tricine buffer, pH 8.5.

RESULTS AND DISCUSSION

Disulfide Bond Formation in Wildtype Enzyme

There are four cys residues per subunit in human HPRT, at positions 22, 65, 105 and 205 in the primary sequence. In the crystal structure of human HPRT-GMP complex, none of the SH groups is close to the GMP binding site. Further, the measured distance between any two of these groups is too great to allow the formation of disulfide bonds without a significant structural change.

Purified recombinant and erythrocyte human HPRTs lose $\geq 90\%$ of their activity in the absence of the substrate 5-phospho-α-D-ribosyl-1-pyrophosphate (*P*-Rib-*PP*). This activity can be fully restored by the addition of a fresh solution of 10 mM dithiothreitol (DTT), indicating that the loss of activity is due to disulfide bond formation and is reversible.

Mass spectrometry shows inter-subunit disulfide bonds are present in the low specific activity samples (Fig. 1). The calculated molecular mass of the non-acetylated subunit is 24449 Da, with the expected mass of the covalently-linked dimer being 48898 Da, in close agreement with the predominant species observed. Dimer formation was confirmed by SDS-PAGE, which also provided evidence for the formation of intra-subunit disulfide bonds. Therefore, the structure in solution is sufficiently flexible to allow forma-

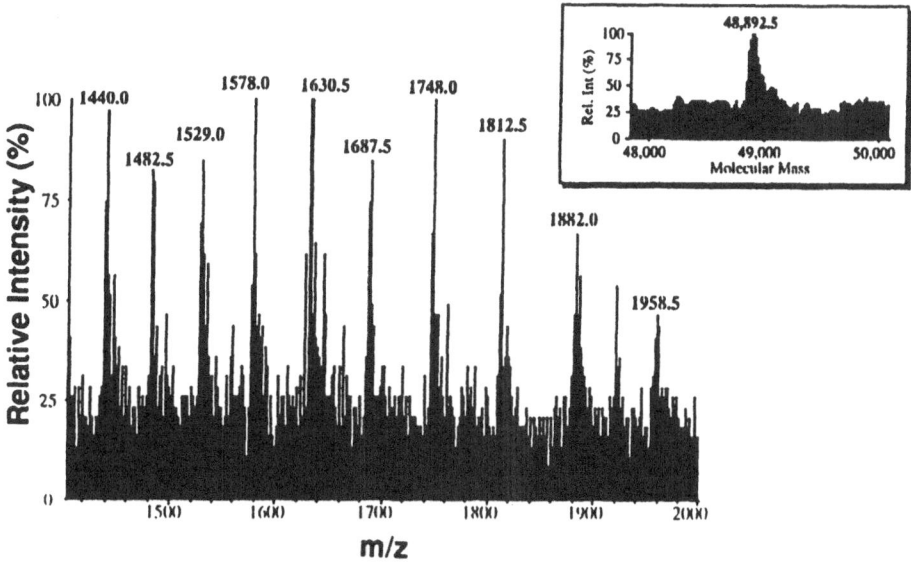

Figure 1. Mass spectrometry of human HPRT with a specific activity of 6 μmol min^{-1} A$_{280}$$^{-1}$. Mass spectrometry of high specific activity enzyme (70 μmol min^{-1} A$_{280}$$^{-1}$) showed the absence of inter-subunit disulfide bonds.

tion of disulfide bonds in the absence of the substrates. When *P*-Rib-*PP* is present, however, these bonds do not form.

Properties of Cysteine Mutants

Five cysteine mutants (C22A; C65A; C105A; C205A and C205Y) were prepared, expressed, purified to homogeneity and analysed for catalytic activity and effect of Hg^{2+}. Each was active, with a specific activity similar to that of the wildtype enzyme, indicating that none of the cysteines is necessary for catalysis. The finding that the C205Y mutant was fully active was surprising in view of the earlier identification of this mutation in Lesch-Nyhan patients and in 6-thioguanine resistant cells. It seems possible that this mutant enzyme is unstable when expressed in mammalian cells. Future experiments are planned to reveal which cysteine residues are involved in the disulfide bond formation which occurs in the wildtype enzyme.

Human wildtype HPRT was rapidly and completely inactivated by Hg^{2+} wher as the C65A mutant was only slightly inactivated. Therefore, Hg^{2+} reacts with C65 (possibly bridging C65 and another cysteine) to give completely inactive enzyme. This reaction does not occur in the presence of the substrates (during catalysis), suggesting that C65 is not accessible to Hg^{2+} when substrates are bound.

The Lysine Residues of Human HPRT

The lysine specific reagent TNBS causes partial inactivation of HPRT activity in the absence but not in the presence of *P*-Rib-*PP* (Fig. 2). The crystal structure of HPRT has revealed three lysine residues which may play a role in structure or function of HPRT. They are K68, K140 and K165. Tryptic hydrolysis and peptide mapping by reverse phase HPLC indicated that K140 reacts with TNBS only in the absence of *P*-Rib-*PP*.

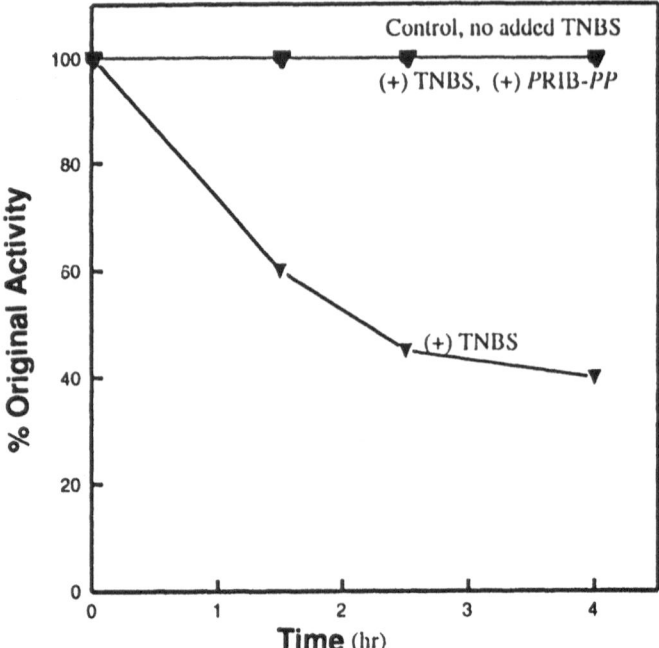

Figure 2. TNBS inhibition of human HPRT activity.

The K140A mutant was prepared, expressed and purified to homogeneity. This enzyme expresses at low levels (4 μmol min^{-1} ml^{-1} of cell culture compared with 30 μmol min^{-1} ml^{-1} of cell culture for wildtype enzyme). The specific activity was 70 μmol min^{-1} A$_{280}$$^{-1}$ and its K$_m$ for *P*-Rib-*PP* was 200 μM, similar to wildtype. Therefore, the K140 side chain is not involved in catalysis nor does it bind to *P*-Rib-*PP*. The binding of *P*-Rib-*PP* does however prevent the reaction of K140 with TNBS. In contrast to the partial inactivation of wild type enzyme by TNBS, the K140A mutant was rapidly and completely inactivated. A possible explanation for this unexpected result is that reaction of K140 with TNBS blocks reaction of a catalytically essential lysine residue with TNBS. In the absence of the K140 side chain, TNBS modifies this essential lysine resulting in the loss of activity. A likely candidate is K165 which, in the crystal structure of the enzyme-GMP complex, binds to the exocyclic oxygen of guanine.

CONCLUSIONS

None of the four cysteine residues is involved in catalysis. However, disulfide bonds form in the HPRT enzyme, both inter-subunit and intra-subunit, when *P*-Rib-*PP* is absent. When *P*-Rib-*PP* is bound, it presumably stabilises a conformation of the enzyme in which disulfide bonds cannot form. Disulfide bonding results in a conformationally constrained enzyme with limited (~10%) activity. Extensive inactivation of human HPRT by Hg^{2+} only occurs when C65 is present. The crystal structure suggests that this inactivation could be due to the presence of a bulky metal ion at the dimer interface. K140 reacts with TNBS leading to partial loss of enzyme activity. Replacement of K140 with ala produces a catalytically active protein which is completely inactivated by TNBS. This suggests that TNBS can now access another lysine residue in the structure, possibly K165.

ACKNOWLEDGMENTS

The authors would like to thank Alun Jones from the Centre for Drug Design and Development, University of Queensland, for the mass spectrometric analysis.

REFERENCES

1. Bronstein, S.M., Cochrane, J.E., Craft, T.R., Swenberg, J.A. and Skopek, T.R. (1991) *Cancer Research* 51: 5188–5197
2. Davidson, B.C., Tarle, S.A., Van Antwerp, M., Gills, R.A., Watts, R.W.E., Kelley, W.N. and Palella, T.D. (1991) *Am. J. Hum. Genet.* 48: 951–958
3. Eads, J.C., Scapin, G., Xu, Y., Grubmeyer, C. and Sacchettini, J.C. (1994) *Cell* 78: 325–334
4. Free, M.L., Gordon, R.B., Keough, D.T., Beacham, I.R., Emmerson, B.T. and de Jersey, J. (1990) *Biochim. Biophys. Acta* 1097: 205–211
5. Gibbs, R.A., Nguyen, P.-N., McBride, L.J., Koepf, S.M. and Caskey, C.T. (1989) *Proc. Natl. Acad. Sci. USA* 86: 1919–1923
6. Harbach, P. R., Filipunas, A. L., Wang, Y. and Aaron, C. S. (1992) *Environ. and Mol. Mutagen.* 20: 96–105
7. Horton, R.M., Hunt, H.D., Ho, S.N., Pullen, J.K. and Pease, L.R. (1989) *Gene* 77: 61–68
8. Kanaani, J., Somoza, J.R., Maltby, D. and Wang, C.C. (1996) *Eur. J. Biochem.* 239: 764–772
9. Keough, D.T., Emmerson, B.T. and de Jersey, J. (1991) *Biochim. Biophys. Acta* 1096: 95–100
10. Keough, D.T., McConachie, L.A., Gordon, R.B., de Jersey, J. and Emmerson, B.T. (1987) *Clin. Chim. Acta* 163: 301–308
11. Lukash, L.L., Bolt, J.J., Dolan, M.E., Maher, V.M. and McCormick, J.J. (1991) *Mutation Res.* 250: 397–409
12. Olsen, D. B., Sayers, J. R. and Eckstein, F. (1989) *Methods Enzymol.* 217: 189–217
13. Recio, L, Simpson, D., Cochrane, J., Liber, H. and Skopek, T. R. (1990) *Mutation Res.* 242: 195–208
14. Somoza, J.R., Chin., M.S., Focia., P.J., Wang, C.C. and Fletterick, R.J. (1996) *Biochemistry* 35: 7032–7040
15. Yang, J., Hu, M. and Wu, C. (1991) *J. Mol. Biol.* 221: 421–430

67

MOLECULAR ANALYSIS OF MOUSE *Ampd3* GENE ENCODING HEART-TYPE ISOFORM OF AMP DEAMINASE*

Takayuki Morisaki,[1] Kannika Sermsuvitayawong,[1] Xudong Wang,[1]
Akira Nagabukuro,[2] Yoichi Matsuda,[2] Nobuaki Ogasawara,[3] Ikuo Mineo,[4]
Hiroko Morisaki,[1] and Tsunehiro Mukai[1]

[1]Department of Bioscience
National Cardiovascular Center Research Institute
Suita, Osaka, Japan
[2]Nagoya University
Nagoya, Aichi, Japan
[3]Aichi Prefectural Colony
Kasugai, Aichi, Japan
[4]Otemae Hospital
Osaka, Osaka, Japan

1. INTRODUCTION

Purine metabolism is one mechanism to maintain cellular ATP concentration. The pathway to degrade adenosine nucleotide directly involves in stabilization of the energy charge during ATP consumption. AMP deaminase (E.C.3.5.4.6.; AMPD), which catalyzes the hydrolytic deamination of AMP and converts it to IMP, is thought to play an important role in purine metabolism in all eukaryotic cells. In higher eukaryotes, AMPD has multiple isoforms encoded by a family of AMPD multigenes.[1] There are three genes in human: *AMPD1*, *AMPD2*, *AMPD3*, whose products are predominant in muscle (M), liver (L) and erythrocyte (E), respectively.[2–4] In rodent, two genes, *Ampd1* and *Ampd2* that are equivalent to the human *AMPD1* and *AMPD2*, respectively, have been known,[2,5] but the gene corresponding to *AMPD3* has not been reported. From the analogy of human AMPD, this hypothetical gene has been inferred to code for heart(H)-type AMPD, but molecular studies with rodent heart AMPD has not been done. In this study, mouse cDNA and the gene encoding H-isoform of AMPD have been isolated and characterized for the first time.

---

* The sequence data have been submitted to EMBL/Genbank/DDBJ databases (D85596, D88984-D88994).

Purine and Pyrimidine Metabolism in Man IX,
edited by Griesmacher *et al.* Plenum Press, New York, 1998.

2. EXPERIMENTAL

2.1. cDNA Cloning of Mouse *Ampd3*

A fragment corresponding to human *AMPD3* was first amplified from mouse heart poly(A)+ RNA by reverse transcription-polymerase chain reaction (RT-PCR). The isolated 400 bp fragment was used for screening of mouse heart cDNA library, and positives were subcloned and subjected to sequencing analysis. With all isolated clones, 4078 bp of cDNA including poly(A)+ tail was identified. This 4 kb cDNA contains an open-reading frame of 2301 nt corresponding to a predicted peptide of 766 amino acids. Sequence comparison revealed that this cDNA was highly homologous to human *AMPD3* as 78.3% in the nucleotide level and 88.1% for the coding region and 94.1 % in the amino acid level of predicted peptides.[6]

2.2. RNA Analysis of *Ampd3* Transcripts in Mouse Tissues

Northern blot and RNase protection were performed to analyze the distribution of each *Ampd* transcripts in mouse tissues. Expression of *Ampd3* RNA was found in almost all tissues except liver. Northern analysis revealed a transcript of about 4 kb for *Ampd3*, confirming that the identified cDNA was indeed a full-length clone for *Ampd3* cDNA. In addition, there was relatively high expression of *Ampd3* transcripts in heart or kidney. In skeletal muscles, slow-twitch skeletal muscle showed higher level of *Ampd3* transcripts than did fast-twitch skeletal muscles.

2.3. Bacterial Expression of *Ampd3* cDNAs

Prokaryotic expression vectors were constructed to produce AMPD peptides. The transformed bacterial extracts were prepared and tested for AMPDÊactivity as well as immunoreactivity for the specific antibody[7] by immunoprecipitation. Bacterial lysates showed that both GST-fused mouse AMPD3 and GST-fused rat AMPD1 were catalytically active, while the mock-transformed bacteria showed no AMPD activity. Immunoprecipitation revealed that GST-fused mouse AMPD3 was immunoreactive to the specific antibody raised against rat H-isoform of AMPD, but not to that raised against rat M-isoform of AMPD.

2.4. Assignment of Chromosomal Locus of Mouse *Ampd*

Fluorescent in situ hybridization (FISH) were first performed to assign the chromosomal locus for *Ampd3*. FISH analysis revealed that the *Ampd3* gene was localized to mouse chromosome 7 band E2 and E3 and to rat chromosome 1q35–36, where conserved linkage homology to mouse chromosome 7 has been identified.[8] The interspecific backcross analysis revealed a fine linkage map shown in Figure 1. These results indicate that the identified region for mouse *Ampd3* is located in the syntenic region of human chromosome 11 where human *AMPD3* is located.[9]

2.5. Isolation of Mouse *Ampd3* Gene and Genomic Organization

Mouse genomic DNA libraries were screened with a radiolabeled probe of *Ampd3* cDNA fragment. Positives were isolated, and DNA fragments were subcloned and sequenced. The identified DNA for the *Ampd3* gene spans more than 40 kb in length. It

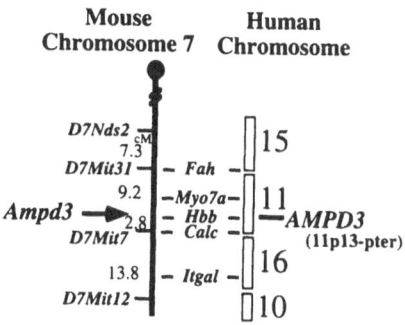

Figure 1. Fine linkage mapping of mouse *Ampd3*.

revealed that mouse *Ampd3* gene consists of 15 exons and that the exon-intron boundaries are quite conserved between human *AMPD3* and mouse *Ampd3*. However, the 5' untranslated region of mouse *Ampd3* showed no similarity to that of human *AMPD3*.

2.6. Isolation and Comparison of All Three Mouse *Ampd* cDNAs

Mouse *Ampd1* and *Ampd2* cDNAs were also amplified by RT-PCR from RNA prepared from skeletal muscle cells and embryonal cells using oligonucleotides designed from the sequences of human *AMPD1* and *AMPD2* cDNAs. Isolated cDNAs for mouse *Ampd1* and *Ampd2*, which are predicted to encode peptides of 746 amino acids and 798 amino acids, respectively, were quite homologous to those for human *AMPD1* and *AMPD2*. Northern analysis revealed that expression patterns of *Ampd1* and *Ampd2* RNA were found to be the same as those reported for the rodent M-isoform and L-isoform of AMPD.[7]

2.7. Comparison between Mouse AMPD3 Peptide and Human AMPD3 Peptide

Prokaryotic expressed peptides for mouse AMPD3 and human AMPD3 were compared to evaluate immunoreactivity and kinetical properties. Mouse AMPD3 and human AMPD3 showed cross-reactiveness to both anti-E antibody and anti-H antibody, but they showed different reactivities upon higher dilution of antibodies. In addition, mouse AMPD3 showed higher affinity to substrate AMP than human AMPD3.

3. DISCUSSION

Molecular studies on H-isoform of AMPD are important because AMPD is thought to play an important role in the tissue requiring high-energy including heart as well as skeletal muscle. Rodent heart has been reported to have relatively high AMPD activity, while human heart has been reported to have little AMPD activity which is composed of a mixture of isoforms.

3.1. Mouse *Ampd3* Encodes H-Isoform of AMPD

Our studies showed that a cDNA clone, *Ampd3*, homologous to human *AMPD3*, isolated from mouse heart encodes the H-isoform of AMPD upon bacterial expression and tests in immunoprecipitation. RNA analysis demonstrated that the tissue distribution of mouse *Ampd3* transcripts was compatible with the distribution of H-isoform of AMPD reported before, while the expression pattern of mouse *Ampd1* and *Ampd2* was the same as those of M-isoform and L-isoform, respectively. In addition, the assigned region for *Ampd3* on mouse chromosome 7 has a conserved linkage homology with that reported for *AMPD3* on human chromosome 11p13-ter. Therefore, we conclude that the *Ampd3* encoding H-isoform of AMPD is a homologue of human *AMPD3* encoding E-isoform of AMPD.

3.2. Mouse *Ampd3* Has Some Different Characters from Human *AMPD3*

Although mouse *Ampd3* is thought to be a homologue of human *AMPD3*, several characters of *Ampd3* are different from those of human *AMPD3*. First, different expression pattern between mouse *Ampd3* and human *AMPD3* were observed. Second, sequences of non-coding regions of mouse *Ampd3* and human *AMPD3* were found to be quite different. In addition, there were several different characters of the proteins expressed in bacteria for mouse AMPD3 and human AMPD3.

4. CONCLUSION

Mouse cDNA and gene for the H (heart) type-isoform of AMPD were isolated and characterized. RNA analysis and immunoprecipitation have confirmed that the isolated *Ampd3* cDNA encodes the mouse H-isoform of AMPD. The mouse *Ampd3* gene is a homologue of human *AMPD3*, although the products of these genes differ to some extent.

REFERENCES

1. Sabina RL. Swain JL, Holmes EW: Myoadenylate deaminase deficiency. In Metabolic and Molecular Basis of Inherited Diseases. (C. Scriver, L. Beaudet, W.S. Sly, D. Valle, Eds.) Vol. 2. pp1769–1780, McGraw-Hill, New York, NY, 1995.
2. Sabina RL. Morisaki T, Clarke P, Eddy R, Show TB, Morton CC, Holmes EW: Characterization of the human and rat myoadenylate deaminase genes. J.Biol.Chem. 265:9423–9433, 1990.
3. Bausch-Jurken MT, Mahnke-Zizelman DK, Morisaki T, Sabina RL: Molecular cloning of AMP deaminase isoform L. Sequence and bacterial expression of human *AMPD2* cDNA. J.Biol.Chem. 267:22407–22413, 1992.
4. Mahnke-Zizelman DK, Sabina RL: Cloning of human AMP deaminase isoform E cDNAs. Evidence for a third AMPD gene exhibiting alternatively spliced 5'-exons. J.Biol.Chem. 267:20866–20877, 1992.
5. Morisaki T, Sabina RL, Holmes EW: Adenylate deaminase. A multigene family in humans and rats. J.Biol.Chem. 265:11482–11486, 1990.
6. Wang X, Morisaki H, Sermsuvitayawong K, Mineo I, Toyama K, Ogasawara N, Mukai T, Morisaki T: Cloning and expression of cDNA encoding heart-type isoform of AMP deaminase. Gene 188: 285–290, 1997.
7. Ogasawara N, Goto H, Yamada Y, Watanabe T: Distribution of AMP-isozymes in rat tissues. Eur.J.Biochem. 87:297–304, 1978.
8. Sermsuvitayawong K, Wang X, Nagabukuro A, Matsuda Y, Morisaki H, Toyama K, Mukai T, Morisaki T: Genomic organization of *Ampd3*, heart-type AMPD gene, located in mouse chromosome 7. Mammal.Genom (in press).
9. Mahnke-Zizelman DK, Eddy R, Show TB, Sabina R: Characterization of the human AMPD3 gene reveals that 5' exon usage is subject to transcriptional control by three tandem promoters and alternative splicing. Biochem.Biophys.Acta 1306:75–92, 1996.

68

DIHYDROPYRIMIDINE DEHYDROGENASE DEFICIENCY

Identification of Two Novel Mutations and Expression of Missense Mutations in *E. coli*

Peter Vreken,[*] André B. P. van Kuilenburg, Rutger Meinsma, and Albert H. van Gennip

Academic Medical Center
University of Amsterdam
Departments of Pediatrics and Clinical Chemistry F0-224
P. O. Box 22700
1100 DE Amsterdam, The Netherlands

1. INTRODUCTION

Dihydropyrimidine dehydrogenase (DPD) deficiency (McKusick 274270) is a, clinically heterogeneous, autosomal recessive disease. DPD (EC 1.3.1.2) catalyzes the first and rate-limiting step in the catabolism of uracil, thymine and the analogue 5-fluorouracil. Patients with a nearly complete enzyme defect show convulsive disorders in about 50% of cases whereas patients experiencing acute 5-fluorouracil toxicity usually show DPD enzymatic activities in the heterozygous range.[1-9]

So far, four mutations in the DPD gene have been described. The first mutation is a G→A point mutation in an invariant GT splice donor sequence leading to skipping of an 165 bp exon.[10,11] This mutation has been identified in 17/26 alleles from patients with complete DPD deficiency. In addition, heterozygosity for this mutation was shown in patients experiencing acute 5-fluorouracil toxicity.[9] Three other mutations, a frameshift (ΔC1897) and two missense (C29R and R886H) mutations were identified in two patients from Turkey.[12]

We now report a four base deletion (delTCAT$_{296-299}$) in a patient with DPD deficiency. Interestingly, the brother of this patient proved to be homozygous for this mutation as well, whereas he had no clinical phenotype.[7,13] In addition we identified a novel mis-

---

[*] FAX: +31 206962596, E-mail: p.vreken@amc.uva.nl.

Purine and Pyrimidine Metabolism in Man IX,
edited by Griesmacher *et al.* Plenum Press, New York, 1998.

sense mutation (R235W) in a Turkish DPD deficient patient. Both the latter and previously identified missense mutations were introduced into a wild-type DPD cDNA and subcloned in a pSE420 expression vector. Expression vectors containing either wild-type DPD cDNA or mutated alleles were tested for DPD activity in order to establish whether C29H, R885H and R235W are responsible for the DPD deficient phenotype. The expression studies show that C29R and R235W lead to a mutant DPD protein without significant residual enzymatic activity, whereas the R886H mutation leads to a DPD protein with about 25% residual activity.

2. MATERIALS AND METHODS

2.1. PCR and Sequence Analysis

RNA and DNA were isolated from cultured fibroblasts by the guanidinium thiocyanate method and standard procedures respectively. cDNA synthesis and RT-PCR reactions for amplifying DPD cDNA were carried out essentially as described previously.[11,13,14] Four overlapping DPD cDNA fragments were amplified using sense primers AF, BF, CF and DF corresponding to nt −21–1, 606–629, 1473–1496 and 2254–2276 respectively and antisense primers AR, BR, CR and DR complementary to nt 694–716, 1590–1616, 2353–2376 and 3145–3169 respectively (*A*TG=1). Forward primers contained an 5′-TGTAAAACGACG-GCCAGT-3′ extension whereas reverse primers contained an 5′-CAGGAAACAGCTAT-GACC-3′ extension at their 5′ ends. The latter sequences are complementary to fluorescent labeled primers used in the dye terminator sequence reaction. Amplification was carried out in 50 μl reaction mixtures containing 10 mM Tris/HCl pH 8.3, 50 mM KCl, 3 mM MgCl$_2$, 1 μM of each primer, 0.2 mM dNTPs. After initial denaturation for 5 minutes at 95 °C, 2 U of Taq polymerase (AmpliTaq, Perkin Elmer (San Jose, CA)) was added and amplification carried out for 30 cycles (1 min 95°C, 1 min 65°C, 1 min 72°C). Amplification of the genomic region containing a DPD exon and its flanking sequences was carried as described previously.[11] PCR products were purified using a Qiaquickspin PCR purification kit (Qiagen, Hilden, Germany) and used for direct sequencing. Sequence analysis was carried out on a Applied Biosystems model 377 automated DNA sequencer using the dye-primer method for the DPD cDNA and genomic fragments (dye-primer cycle-sequence-ready reaction kit, Applied Biosystems (San Jose, CA)).

2.2. DPD Activity Assay

DPD activity in fibroblasts and crude E. coli lysates was measured as described previously.[14]

2.3. Expression of Missense Mutations in E.coli

Expression plasmids were constructed in the vector pSE420 (Invitrogen Corp., San Diego, CA) and designated pSE420ΔSH, pSE420ΔSH-DPD, pSE420ΔSH-DPD-C29R, pSE420DSH-DPD-R886H, pSE420DSH-DPD-C29R/R886H and pSE420DSH-DPD-R235W indicating the mutation(s) present in each construct. Plasmids were introduced into E. coli strain BL21 (F-,ompT, hsdS$_b$(r$_b$-m$_b$-), gal, dcm). A 20 ml LB broth culture, supplemented with 100 μg/ml ampicillin, 100 μM uracil, 100 μM of each FAD and FMN, and 10 μM each of Na$_2$S and Fe(NH$_4$)$_2$(SO$_4$) was inoculated with 200 μL of an overnight preculture grown in LB broth. Cells were grown to OD_{550nm} = 0.5–0.6 at 30°C and induction was performed by the addition of 1 mM IPTG. Cells were sedimented at time points between 1 and 3 hours, washed with isolation buffer (35 mM potassium phosphate (pH=7.3), 10 mM EDTA, 1mM DTT, 0.1 mM PMSF, 2 µM leupeptin and 2 µM pepstatin) and resuspended in isolation buffer. The cell suspension was frozen at −20°C for at least 16 hours, thawed on ice and lysed by sonication. The crude lysate was centrifuged at 20,000 × g for 15 minutes. The resulting supernatant was used for the DPD activity assay, Laemli gel electroforesis or protein quantitation (bicinchoninic acid (BCA) procedure, Pierce, Rockford, IL).

3. RESULTS

3.1. Clinical and Biochemical Characterization

Detailed clinical descriptions of patients A, B, C and their family members have been published elsewhere.[7,12,13] These patients were shown to be completely DPD deficient as judged from DPD enzymatic studies in cultivated fibroblasts, whereas the father of patient A was also completely DPD deficient The mother of patient A showed intermediate DPD activity in accordance with her obligate carrier status (Table 1). Patient C showed epileptic manifestations during childhood but is now, at the age of 29 years, clinically normal. The parents of patients B were not available for study.

3.2. cDNA sequence analysis

Total RNA isolated from fibroblasts of the patients and the parents of patient A was subjected to RT-PCR. The mother of patient A was shown to be heterozygous for a novel single base substitution C703T leading to the amino acid substitution R235W (Figure 1). Patient A was previously shown to be heterozygous for the single base deletion ΔC1897.[12] This allele exhibits a reduced stability, as judged from the very weak mRNA signal obtained from the father of this patient who is homozygous for this mutation.[12] In accordance with this observation, sequence analysis in patient A showed an approximately 3:1 ratio for T:C at position 703, implicating that the steady-state concentration of the R235W allele in this patient is significantly higher than that of the ΔC1897 allele (Figure 1).

Patient B is homozygous for both the T85C and G2658A mutation, leading to amino acid substitutions C29R and R886H.[12] Sequence analysis of the amplified fragments in fam-

Table 1. Genotype and DPD enzymatic activity

Patient	DPD activity*	Genotype	Result**
A	0.0015	ΔC1897/C703T	fr/R235W
Father A	nd	ΔC1897/ΔC1897	fr/fr
Mother A	0.47	C703T/wt	R235W/none
B	nd	T85C,G2658A/T85C,G2658A	C29R,R886H
C	nd	delTCAT$_{296-299}$/delTCAT$_{296-299}$	fr/fr
Father C	0.12	delTCAT$_{296-299}$/wt	fr/none
Mother C	nd	delTCAT$_{296-299}$/delTCAT$_{296-299}$	fr/fr
Brother C	nd	delTCAT$_{296-299}$/delTCAT$_{296-299}$	fr/fr

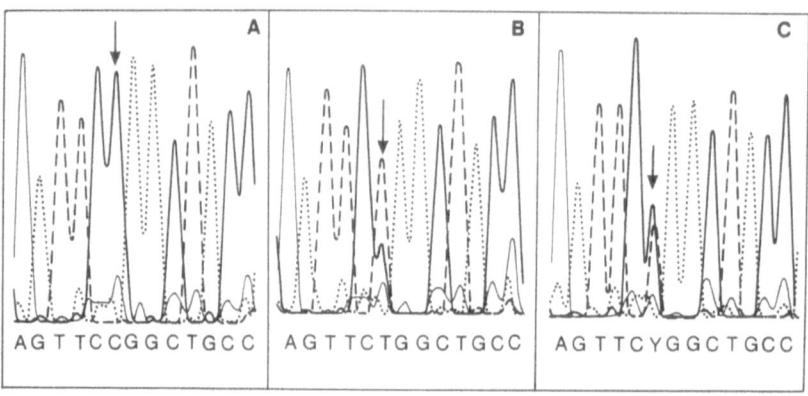

Figure 1. Identification of the C703T (R235W) mutation. Direct sequence analysis was performed on amplified cDNA fragments in a control (A), patient A (B) and the mother of patient A (C). The C703T mutation is indicated with an arrow. Note that the C/T ratio at position 703 is 1:3 in panel B and 1:1 in panel C. For explanation see text.

ily C revealed that the propositus, her mother and her brother were homozygous for a four base deletion (delTCAT$_{296-299}$).[13] The father was shown to be heterozygous for this deletion.

3.3. Expression of Missense Mutations

In order to establish whether the missense mutations C29R, R886H and R235W are responsible for the DPD deficient phenotype, these mutations were introduced in the pSE420ΔSH-DPD expression vector (see Materials and Methods). Expression of pSE420ΔSH-DPD-C29R and pSE420ΔSH-DPD-C29R/R886H yielded no residual activity, whereas pSE420ΔSH-DPD-R886H and pSE420ΔSH-DPD-R235W showed about 25% and 1% residual activity, respectively (Table 2). All mutant proteins were expressed properly as judged by western blotting and hybridisation with a polyclonal antibody against pig DPD (data not shown). Therefore we conclude that C29R and R235W are causative mutations. R886H alone is not sufficient to explain the DPD deficient phenotype in patient B but reduces DPD catalytic activity about 75%.

4. DISCUSSION

Mutation analysis in three DPD deficient patients and their family members revealed the presence of two frameshift mutations (ΔC1897 and delTCAT$_{296-299}$) and three amino

Table 2. Enzymatic activity of expressed mutant DPD

Construct	DPD enzymatic activity* (mean ± 1SD, n=4)
pSE420ΔSH	< 0.01
pSE420ΔSH-DPD	16.7 ± 1.5
pSE420ΔSH-DPD-C29R	< 0.01
pSE420ΔSH-DPD-R886H	4.4 ± 0.4
pSE420ΔSH-DPD-C29R/R886H	< 0.01
pSE420ΔSH-DPD-R235W	0.2 ± 0.05

*DPD enzymatic activity measured in E. coli lysates is expressed as nmol product /hr/mg protein.

acid substitutions (C29R, R235W and R886H). The frameshift mutations lead to premature termination of translation before the uracil binding site, which is encoded by nucleotides 1960–1993.[15] Therefore it is clear that these mutant proteins cannot express any residual DPD activity in accordance with the absence of activity as observed in fibroblasts of the father of patient A, patient C and the mother and brother of patient C (Table 1). Analysis of the missense mutations identified in patient A (R235W) and patient B (C29R and R886H) in an E. coli expression system revealed that both C29R and R235W result in an inactive mutant DPD protein. The R886H mutation does decrease the catalytic activity of the resulting mutant DPD protein, but is unlikely to be responsible for the lack of residual activity in patient B, who is both homozygous for the C29R and R886H mutation. So far, screening of 18 unrelated DPD deficient individuals did not reveal the latter mutation in any patient. Screening for this mutation in the general population and in a group of patients experiencing acute 5-fluorouracil toxicity is currently in progress.

Alignment of the DPD sequences from pig, bovine and human revealed that the arginine residue at position 235 is conserved between the three species, but that the cysteine at position 29 and the arginine at position 886 are not.[15] The expression studies, however, clearly show that mutations at the non-conserved positions 29 and 886 do lead to a severe reduction of DPD catalytic activity.

The E. coli expression system presented here is a valuable tool for examining enzymatic DPD variants obtained through screening of DPD deficient individuals. In addition, this system can also be used for preparing large quantities of both wild-type and mutant DPD protein for physicochemical analysis.

REFERENCES

1. Van Gennip AH, Van-Bree-Blom EJ, Wadman SK, De Bree PK, Duran M, Beemer FA (1981) Liquid chromatography of urinary pyrimidines for the evaluation of primary and secondary abnormalities of pyrimidine metabolism. In Hawk GL, Champlin PB, Hutton RF, Mol C, eds. Biological/Biomedical Applications of Liquid Chromatography II. New York and Basel: Marcel Dekker, Chromatographic Science Series **18**: 285–296.
2. Van Gennip AH, Abeling NGGM, Elzinga-Zoetekouw L, Scholten LG, van Cruchten A, Bakker HD (1989) Comparative study of thymine and uracil metabolism in healthy persons and in a patient with dihydropyrimidine dehydrogenase deficiency. *Adv Exp Med Biol* **253A**: 111- 118.
3. Van Gennip AH, Abeling NGGM, Stroomer AEM, van Lenthe H, Bakker HD (1994) Clinical and biochemical findings in six patients with pyrimidine degradation defects. *J Inher Metab Dis* **17**: 130–132.
4. Van Gennip AH, Abeling NGGM, Vreken P, van Kuilenburg ABP (1997) Genetic metabolic disease of pyrimidine metabolism: implications for diagnosis and treatment. *Int. Pediatrics* **12**: 28–33.
5. Van Gennip AH, Abeling NGGM, Vreken P, van Kuilenburg ABP (1997) Inborn errors of pyrimidine degradation: Clinical, biochemical and molecular aspects. *J Inher Metab Dis* **20**: 203–213.
6. Van Gennip AH, Van Lenthe H, Abeling NGGM, Bakker HD, Van Kuilenburg ABP (1995) Combined deficiencies of NADPH- and NADH-dependent dihydropyrimidien dehydrogenases, a new finding in a family with thymine- uraciluria. *J Inher Metab Dis* **18**: 185–188.
7. Braakhekke JP, Renier WO, Gabreëls FJM, De Abreu RA, Bakkeren JAJM, Sengers RCA (1987) Dihydropyrimidine dehydrogenase deficiency; neurological aspects. *J Neurol Sciences* **78**: 71–77.
8. Harris BE, Carpenter JT, Diasio RB (1991) Severe fluorouracil toxicity secondary to dihydropyrimidine dehydrogenase deficiency. A potentially more common pharmacogenetic syndrome. *Cancer Res* **68**: 499–501.
9. Wei XX, Mcleod HL, Mcmurrough J, Gonzalez FJ, Fernandezsalguero P (1996) Molecular basis of the human dihydropyrimidine dehydrogenase deficiency and 5-fluorouracil toxicity. *J Clin Invest* **98**:610–615.
10. Meinsma R, Fernandez-Salguero P, van Kuilenburg ABP, van Gennip AH, Gonzalez FJ (1995) Human polymorphism in drug metabolism:Mutation in the dihydropyrimidine dehydrogenase gene results in exon skipping and thymine uraciluria. *DNA Cell Biol* **14**: 1–6.

11. Vreken P, van Kuilenburg ABP, Meinsma R, Smit GPA, Bakker HD, De Abreu RA, van Gennip AH (1996) A point mutation in an invariant splice donor site leads to exon skipping in two unrelated dutch patients with dihydropyrimidine dehydrogenase deficiency. *J Inher Metab Dis 19*: 645–654.
12. Vreken P, van Kuilenburg ABP, Meinsma R, van Gennip AH (1997) Identification of novel point mutations in the dihydropyrimidine dehydrogenase gene. *J Inher Metab Dis* **20**: in press.
13. Vreken P, van Kuilenburg ABP, Meinsma R, De Abreu RA, van Gennip AH (1997) Identification of a four base deletion (delTCAT296–299) leading to dihydropyrimidine dehydrogenase deficiency. *Human Genetics*: in press.
14. Van Kuilenburg ABP, van Lenthe H, van Gennip AH (1996) Identification of a NADH- dependent activity of dihydropyrimidine dehydrogenase in man. *Anticancer Res* **16**: 389–394.
15. Yokota H, Fernandez-Salguero P, Furuya H et al (1994) cDNA cloning and chromosome mapping of human dihydropyrimidine dehydrogenase, an enzyme associated with 5- fluorouracil toxicity and congenital thymine uraciluria. *J Biol Chem* **269**: 23192–23196.

69

GENE MUTATIONS RESPONSIBLE FOR HUMAN ERYTHROCYTE AMP DEAMINASE DEFICIENCY IN POLES

Yasukazu Yamada,[1] Wieslaw Makarewicz,[2] Haruko Goto,[1] Noriko Nomura,[1] Hiroshi Kitoh,[1] and Nobuaki Ogasawara[1]

[1]Institute for Developmental Research
Aichi Human Service Center
Kasugai, Aichi 480-03, Japan
[2]Faculty of Biotechnology
Medical University of Gdansk
80-11 Gdansk, Poland

1. INTRODUCTION

In human, there are three fundamental tissue-specific isozymes of AMP deaminase (EC 3.5.4.6),[1,2] and three different genes for the isozymes have been identified:[3–6] AMPD1 encoding isozyme M found in skeletal muscle, AMPD2 encoding L in liver, and AMPD3 encoding erythrocyte AMP deaminase (E1). The erythrocyte AMP deaminase deficiency that we initially identified,[7,8] is clinically asymptomatic. Erythrocyte ATP contents in the complete deficiency are about 150% of normal,[8] which can be explained by impairment of the degradation of AMP by way of AMP deaminase. The inheritance of the deficiency was autosomal recessive and the heterozygote frequency was estimated at about 1/30 in Japan, Korea and Taiwan.[7,8] The deficiency was also reported in Europe,[9] and the frequency in northern Poland was almost the same as that in east Asia. The nonsense mutation on AMPD1 responsible for human myoadenylate deaminase (muscle type AMP deaminase) deficiency was found in 12% of Caucasians and 19% of African-Americans, whereas none of the 106 Japanese subjects surveyed has the mutant allele.[10] Human erythrocyte AMP deaminase deficiency in Japanese is associated with 75% of the same mutation (R573C) and 25% of other heterogeneous mutations on AMPD3.[11,12] It is of interest to know whether the same mutation occurs in Poles, since their frequency of the enzyme deficiency was not significantly different from Japanese.

Purine and Pyrimidine Metabolism in Man IX,
edited by Griesmacher *et al.* Plenum Press, New York, 1998.

2. MATERIALS AND METHODS

2.1. Screening of AMP Deaminase Activity

Samples from healthy population of blood donors were prepared according to the methods described previously.[7,8] AMP deaminase activity was assayed by estimating the production of ammonia.[1] One unit of enzyme activity was defined as the amount of enzyme which catalyzed the formation of 1 μmole of ammonia/min.

2.2. Genomic Analysis

Each exon of *AMPD3* was PCR-amplified from genomic DNA extracted from blood samples, according to the methods described previously.[11,12] The mutations detected in Japanese[11,12] were screened by RFLP and direct sequencing. The nucleotide sequences of amplified fragments were analyzed to identify the mutation by direct sequencing. The protocol for direct sequencing was slightly modified from that of Sequenase Version 2.0 (United State Biochemical Co.), as described previously.[13] DNA sequences were recorded into personal computer and analyzed with a software of gene analysis, GENETYX version 9.0 (SDC, Japan).

3. RESULTS AND DISCUSSION

We screened blood samples from local healthy Polish population for the possible heterozygotes of erythrocyte AMP deaminase mutant gene. By screening of 227 samples, seven samples were found to have the activity roughly at the level of 50 % as compared to the average. The frequency of heterozygotes were similar to those from previous studies in Poles[9] and Asian.[7,8] The coding region of the *AMPD3* was analyzed by PCR-RFLP method and direct sequencing using genomic DNA from these seven Polish individuals.

We first screened a Japanese major mutation (R573C)[11] in exon 11 of AMPD3 by the analysis of genomic PCR amplification and digestion at the *Pst* I site created in the mutant gene, but the R573C was not detected in seven Polish samples. Direct sequencing of the fragments showed also normal sequencing patterns. Then, the others mutations found in the Japanese deficiency, 600delT, N310K, A320V, M324T, R331C, R402C, Q434X, W450R, and P585L reported previously,[12] and a recently identified Q712P, were screened by PCR-RFLP or the direct sequencing of the respective exons. Neither the major mutation R573C nor other mutations found in Japanese were detected in any alleles of seven Polish samples.

By the analyses of each exon of the *AMPD3* amplified from genomic DNA of the seven Polish samples, a new mutation was identified in three samples (Figure 1). Direct sequencing of the exon 6 from one of samples showed a heterozygous sequence pattern showing both G and T bands at the nucleotide position (nt.) 931. A single nucleotide substitution of G to T was confirmed by cloning study using a T-A cloning vector. The substitution resulted in a mis-sense mutation V311L, an amino acid change of valine to leucine at codon 311. The mutation can be screened by digestion with a restriction enzyme *Tai I* (ACGT), since its digestion site was lost in the mutant (ACTT). When the DNA fragments including exon 6 of *AMPD3* were PCR-amplified from genomic DNA samples using a primer pair (E7A and E7B) described previously[12] and digested by *Tai* I, the fragment from the mutant allele should result in a 371 bp, whereas that from normal was digested to 263 and 108 bp. Three samples had the mutant allele which was not digested by *Tai* I (Fig-

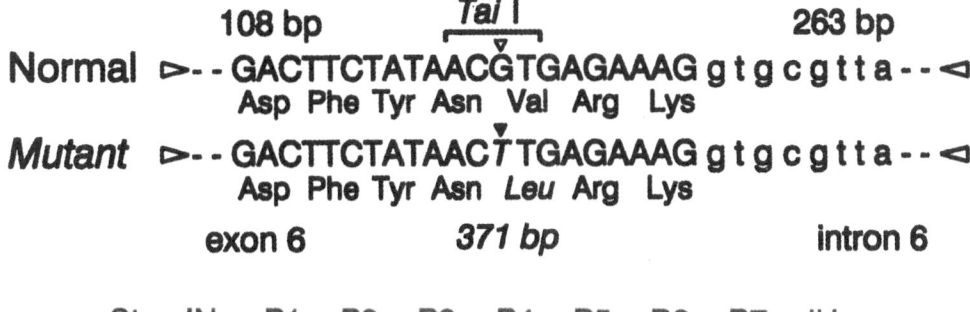

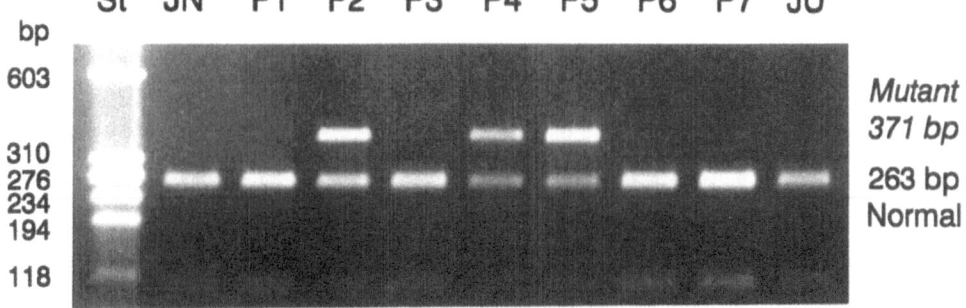

Figure 1. A mutation V311L on *AMPD3* for erythrocyte AMP deaminase deficiency in Poles (A), and its detection by PCR-RFLP using a restriction enzyme *Tai* I (B). The DNA fragments including exon 6 of *AMPD3* were PCR-amplified from genomic DNA samples using a primer pair (E7A and E7B) described previously12 and digested by *Tai* I. FX174/HaeIII digest was used as standard DNA size markers (ST). P1 to P7 are Polish samples, JN is a Japanese normal, and JU is a Japanese sample of unknown mutation.

ure 1). These results were confirmed by direct sequencing analyses. Direct sequencing analyses of each exon of *AMPD3* from other four samples are now in progress.

The decrease of hydrophilicity and secondary structural change (random coil to β-turn) in the mutant (V311L) enzyme protein are predicted using the analysis system by Chou and Fasman in the PROTEIN STRUCTURE program (GENETIYX). Expression study for the mis-sense mutation (V311L), described previously in studies of Japanese mutants11,12, is nesessary to confirm that the mutation causes the enzyme deficiency. The possibility that the substitution is neutral polymorphism remains. However, we consider that the V311L must affect AMP deaminase activity, since the mutation is located in the region well-conserved among three isozymes (*AMPD1*, 2, *and* 3), corresponding to exons 6 to 14 in *AMPD3*,[3-6] and some structural changes are predicted.

The Polish mutation V311L detected in one of alleles of three Polish samples has not been found in Japanese samples of unknown mutation. On the other hand, none of eleven Japanese mutants[11,12] was found in Poles. Thus the gene mutations of *AMPD3* in Poles are different from those in Japanese, although the frequency of the enzyme deficiency was similar.

ACKNOWLEDGMENTS

We thank Ms. Misao Iwanaga-Kato and Ms. Ikuko Iwamoto for their skillful technical assistance. This work was supported by a Gout Research Foundation grant, and an Intractable Disease grant from the Ministry of Health and Welfare of Japan.

REFERENCES

1. N. Ogasawara, H. Goto, Y. Yamada, T. Watanabe and T. Asano, AMP deaminase isozymes in human tissues., *Biochim. Biophys. Acta* 714:298 (1982).
2. N. Ogasawara, H. Goto, Y. Yamada, and T. Watanabe, Distribution of AMP deaminase isozymes in various human blood cells., *Int. J. Biochem.* 16:269 (1984).
3. R.L. Sabina, T. Morisaki, P. Clarke, R. Eddy, T.B. Shows, C.C. Morton, and E.W. Holmes, Characterization of the human and rat myoadenylate deaminase genes., *J. Biol. Chem.* 265:9423 (1990).
4. M.T. Bausch-Junken, D.K. Mahnke-Zizelman, T. Morisaki, and R.L. Sabina, Molecular cloning of AMP deaminase isoform L. Sequence and bacterial expression of human AMPD2 cDNA., *J. Biol. Chem.* 267:22407 (1992).
5. Y. Yamada, H. Goto, and N. Ogasawara, Cloning and nucleotide sequence of the cDNA encoding human erythrocyte-specific AMP deaminase., *Biochim. Biophys. Acta* 1171:125 (1992).
6. D.K. Mahnke-Zizelman, and R.L. Sabina, Cloning of human AMP deaminase isoform E cDNAs. Evidence for a third AMPD gene exhibiting alternatively spliced 5'-exons., *J. Biol. Chem.* 267: 20866 (1992).
7. N. Ogasawara, H. Goto, Y. Yamada, I. Nishigaki, T. Itoh, and I. Hasegawa, Complete deficiency of AMP deaminase in human erythrocytes., *Biochem. Biophys. Res. Commun.* 122:1344 (1984).
8. N. Ogasawara, H. Goto, Y. Yamada, I. Nishigaki, T. Itoh, I. Hasegawa, and K.S. Park, Deficiency of AMP deaminase in human erythrocytes., *Hum. Genet.* 75:15 (1987).
9. M.M. Zydowo, J. Purzycka-Preis, and N. Ogasawara, Deficiency of AMP deaminase in human erythrocytes., *Adv. Exp. Med. Biol.* 253A:31 (1989).
10. T. Morisaki, M. Gross, H. Morisaki, D. Pongratz, N. Zollner, and E.W. Holmes, Molecular basis of AMP deaminase deficiency in skeletal muscle., *Proc. Natl. Acad. Sci. USA* 89:6457 (1992).
11. Y. Yamada, H. Goto, and N. Ogasawara, A point mutation responsible for human erythrocyte AMP deaminase deficiency., *Hum. Mol. Genet.* 3:331 (1994).
12. Y. Yamada, H. Goto, T. Murase, and N. Ogasawara, Molecular basis for human erythrocyte AMP deaminase deficiency: screening for the major point mutation and identification of other mutation., *Hum. Mol. Genet.* 3:2243 (1994).
13. Y. Yamada, H. Goto, K. Suzumori, R. Adachi, and N. Ogasawara, Molecular analysis of five independent Japanese mutant genes responsible for hypoxanthine guanine phosphoribosyltransferase (HPRT) deficiency., *Hum. Genet.* 90:379 (1992).

70

RESPONSE OF PURINE METABOLISM TO HYPOXIA AND ISCHEMIA

Wiesław Makarewicz

Department of Biochemistry and Faculty of Biotechnology
Medical University of Gdańsk
ul. Dębinki 1, 80-211 Gdańsk, Poland

1. INTRODUCTION

There are three types of O_2 supply disturbances leading to tissue hypoxia: ischemia, hypoxemia and anemia.[1] Ischemia means critically reduced blood flow. Upon reduction of blood flow the oxygen concentration in venous blood is decreased leading to ischemic anoxia. During ischemia the washout of catabolic end-products is also decreased.

2. DEGRADATION PATHWAYS AND ENZYMES OF PURINE CATABOLISM IN DIFFERENT CELL TYPES

The pathways of adenylate catabolism are different in different cell types within the same organ.

Figure 1 represents the situation existing in heart where cardiomyocytes, endothelial cells and erythrocytes participate in the overall purine catabolism. In cardiomyocytes the ATP degradation leads mainly to adenosine and inosine which could be transported out of the cell. Endothelium contains high activity of purine nucleoside phosphorylase and degrades ATP until hypoxanthine. Endothelial cells characterize also an active extracellular ATP degradation pathway which can degrade the external adenylates to nucleosides. These cells are relatively resistant to hypoxia but vulnerable to oxidative stress during reperfusion. Physiological circumstances under which ATP is degraded to adenosine in an erythrocyte are uncommon. To the contrary erythrocyte uptakes adenosine and metabolize it either to ATP or hypoxanthine. In general one may conclude that hypoxia tolerance of different cells does not depend on a specific degradation pathway but rather on the ability to generate ATP anaerobically (i.e. in glycolysis) and the level of energy requirement for contraction and/or ion transport.

Purine and Pyrimidine Metabolism in Man IX,
edited by Griesmacher *et al.* Plenum Press, New York, 1998.

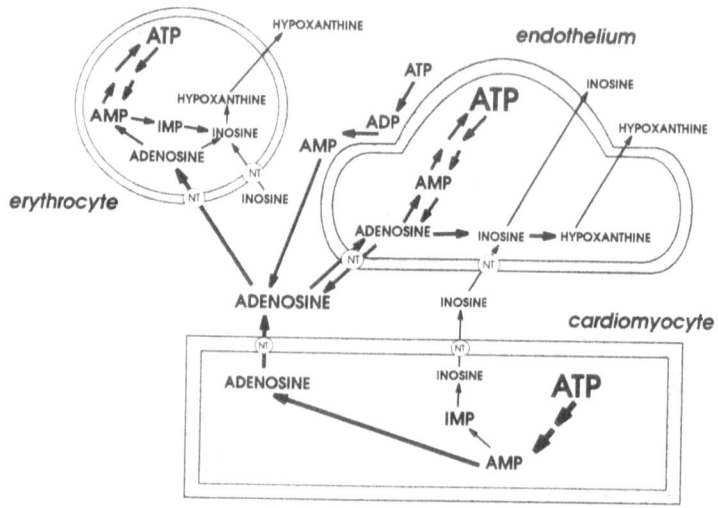

Figure 1. Main pathways of adenylate catabolism in the heart.

The degradation pathways may differ quite substantially in the same cell types in different species, for example rat vs human heart,[2] or in the same organ under different experimental conditions.

Data presented in Fig. 2. show the purine catabolite profile in perfused rat heart during normothermic and hypothermic ischemia. During the experiment comparable degree of ATP depletion (30%) at both temperatures investigated was observed. As compared to normothermic (37°C) conditions, hypothermia (4°C) results in much less production of adenosine (25 and 4.5% respectively) and increased inosine formation (52 and 80% respectively). The total amount of oxypurines formed decrease during hypothermic perfusion from 23 to 15.5%.

3. CELLULAR RESPONSE TO ENERGY DEPRIVATION

ATP depletion affects not only the metabolism of high energy phosphates but also the ionic balance accross the cell membrane and the stability of cellular proteins. In particular the energy deprivation results in:

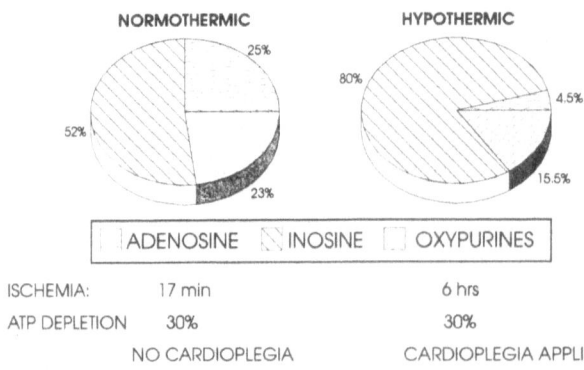

ISCHEMIA:	17 min	6 hrs
ATP DEPLETION	30%	30%
	NO CARDIOPLEGIA	CARDIOPLEGIA APPLIED

Figure 2. Adenylate catabolite profile in the rat heart effluent after normothermic and hypothermic ischemia (M. Zych et al., this volume).

- Dephosphorylation of high energy phosphates
- Accumulation of adenosine and hypoxanthine
- Decrease of TAN pool
- Glycogen depletion
- Acidosis
- Lactate accumulation
- NADH accumulation
- Ca^{+2} accumulation
- K^+/Na^+ imbalance
- Protein aggregation and cytoskeleton disruption
- Metabolic arrest

4. HOW COULD THE DEGRADATION OF HIGH ENERGY PHOSPHATES BE DIMINISHED OR REPLENISHMENT ENHANCED?

Several strategies are described in the literature along with some experimental data from our laboratory. It has been shown in human volunteers using two experimental designs:

1. rapid fructose infusion which causes acute degradation of hepatic ATP,
2. foream ischemic exercise which stimulates ATP consumption in skeletal muscle,

that infusion of β-hydroxybutyrate may alter the metabolic profile from ATP degradation toward ATP resynthesis in muscle and liver by providing an immediate source of fuel and reducing equivalents under specific metabolic conditions.[3] Similarly, in the isolated rabbit skeletal muscle, after 3.5 hrs ischemia and 24 hrs reperfusion, the infusion of 200 μmoles of PEP fortified with 6.6 μmoles of ATP resulted in increased viability of muscle cells and elevated ATP content.[4]

Preservation of high-energy phosphates during ischemia likely contributes also to the cerebroprotective potency of oxypurinol.[5] Oxypurinol pretreatment (40 mg/kg iv. 20 min before the onset of ischemia) of cerebral ischemic/reperfused rats resulted in:

- improvement functional recovery of the CNS,
- preservation of cerebral cortical ATP and ADP levels,
- elevated adenylate energy charge.

It has been shown in our laboratory in post-ischemic perfused rat heart that infusion of adenosine at the beginning of reperfusion augments the ATP content in the heart.[6]

Under some specific conditions (when the activities of adenosine deaminase and adenosine kinase were inhibited) cultured endothelial cells may synthetize ATP from exogenous adenine (Fig. 3). The experiments presented above indicate clearly that there are several possible ways to protect the cells from ATP depletion.

5. MOLECULAR MECHANISMS TRIGGERED BY HYPOXIA

5.1. Induction of Heat Shock Proteins (HSPs)

One of the main primary consequences of ATP-depletion is the protein aggregation and cytoskeleton disruption within a cell. The concept of this "proteotoxic effect" of

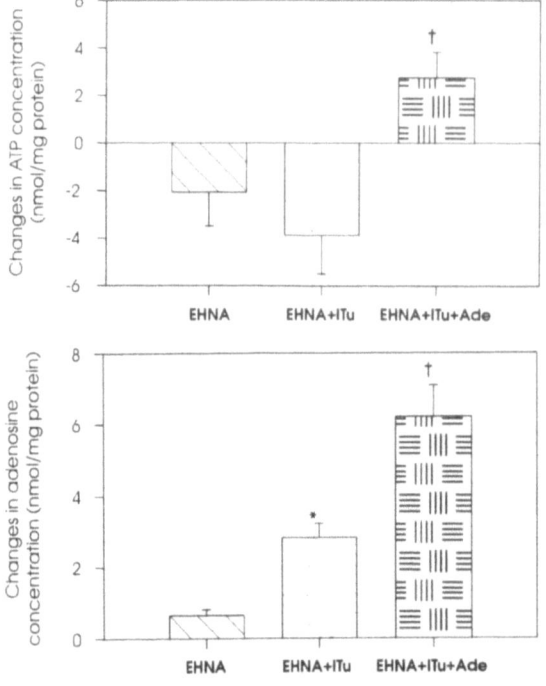

Figure 3. Stimulation of adenylate synthesis and adenosine production in rat capillary endothelial cells (R. T. Smoleński et al., this volume). EHNA (erythro-9-(2-hydroxy-3-nonyl)adenine — 5 μM; Itu (5-iodotubercidin) — 10 μM.; Ade (adenine) — 100 μM. Incubation time 180 min.

energy deprivation and supporting it experimental evidence was recently nicely collected and presented by A.E.Kabakov and V.L.Gabai.[7] The main message of their book is that high energy phosphate homeostasis *per se* is obligatory for the prevention of protein aggregation in a cell. According to these authors the probable sequence of events resulting in a rapid cell death through ATP deprivation involves protein aggregation in cytosol, collapse of cytoskeleton, destabilization of plasma membrane, loss of potassium, influx of sodium and water, swelling and blebbing. These changes could be reversible but if pronounced or sustained, can lead to the necrotic death of the cell. Transient energy deprivation is usually well tolerated by cells and several lines of evidence indicate that this adaptive response is governed by the induction of heat shock proteins [HSPs]. Among several families of HSP the most thoroughly investigated is the group classified as HSP70 or molecular chaperones[8]. These proteins perform the following functions:

- control of cellular protein folding under normal conditions,
- prevent the denaturation and aggregation of intracellular proteins during stressful exposures,
- accelerate the disaggregation, renaturation and/or degradation of damaged proteins within stressed cells.

There are several treatments producing "stress response" of the cell and induction of HSPs which are listed below:

- Hyperthermia
- Ischemia
- Hypoxia
- Low pH
- Oxidants

- Detergents
- Ethanol
- Heavy metals
- Amino acids analogs (receptor agonists)
- Restraint stress (elevated blood pressure)
- Exercise
- Surgical stress

Of a great importance is the observation that stress proteins induced by one type of stress can protect against a differing subsequent stress (cross-tolerance). The close correlation has been shown between the amount of heat-induced HSP72 and the infarct size after 35 minutes of ischemia and 2 hours of reperfusion in rat hearts.[9] It has been also reported that in rat embryonic heart-derived cell line H9c2, the rate of survival under conditions of simulated ischemia (hypoxia and glucose deprivation) is markedly enhanced in stable transfected cells overexpressing the human HSP70 gene and cells are rendered thermotolerant by previous, heat shock induced, HSP synthesis.[10]

5.2. Regulation of HSP Synthesis

Regulation of HSP synthesis may be exerted on transcriptional and/or translational level.

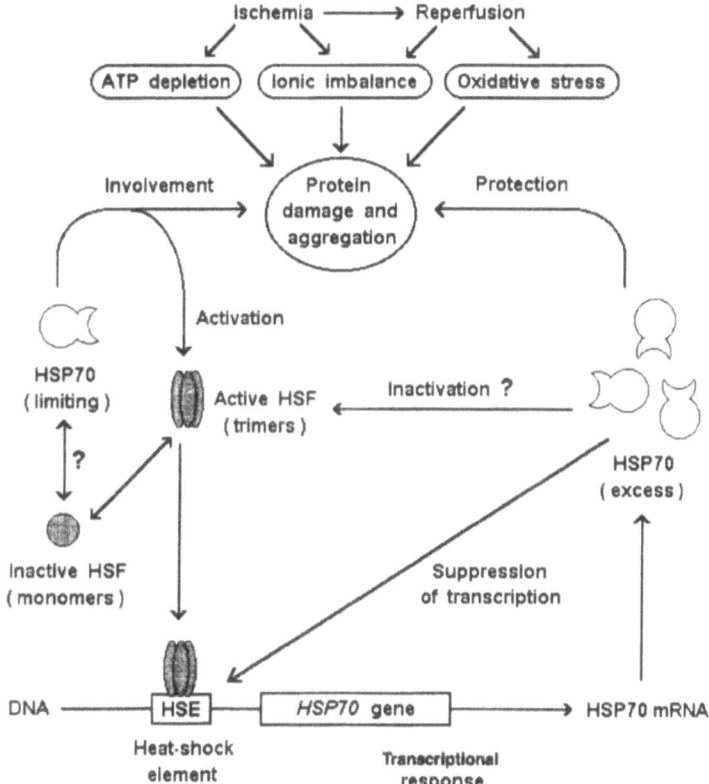

Figure 4. General scheme showing molecular mechanisms of cellular adaptation to energy deprivation [from ref.[7] with permission].

Transcriptional regulation involves activation of specific heat shock protein transcription factor (HSF) which recognizes and binds to the specific DNA sequence (heat shock element — HSE) to initiate transcription.[11] HSPs themselves negatively regulate heat shock gene transcription via an autoregulatory loop.[7] At a translational level, selective stabilization of HSP mRNA by thermal stress has been observed; the half-life of HSP70 mRNA increases 10-fold in heat-shocked HeLa cells.[12]

Ischemia and reperfusion results in ATP-depletion, ionic imbalance and oxidative stress. They all contribute to protein damage and aggregation within the cell. When HSP70 content decreases and becomes limiting, activation of HSF1 follows, and the transcription of HSP70 gene is induced resulting in accumulation of HSP70 protein in excess. This in turn inactivates HSF1 (trimer to monomer transition) and leads to suppression of transcription.

5.3. Induction of Other Proteins

Except HSPs hypoxia induces the synthesis of several other proteins including catalase,[13] superoxide dismutase,[14] and other transcription factors HIF-1[15] and AP-1.[16]

These transcription factors seems to be involved in a hypoxia signal transduction pathway and regulation of several hypoxia-responsive genes. This provides the cell with "an adaptive response" to various stimuli including hypoxia and ischemia characterized by cellular reprogramming and metabolic reorganization.[17]

5.4. Stabilization of mRNA Induced by Hypoxia

It has been shown in rat PC12 cells that hypoxia induces synthesis of 66 kDa protein which binds to the pyrimidine rich tract located between bases 1552–1578 in the 3′-untranslated region of mRNA for tyrosine hydroxylase (TH) and increases 3-fold its stability. The optimal binding site is represented by the motif: (U/C)(C/U)CCU which is present also in 3′-UTRs of TH mRNA from mouse, human, bovine and quail.[18,19]

6. PERSPECTIVES AND SIGNIFICANCE FOR MEDICINE

As HSP induction may be beneficial to the cell under various pathological conditions, the possibility of its pharmacological induction is very promising. On the other hand in cancer cells hypoxia activates genes which render these cells resistant to therapeutic treatment. Preventing this may offer new opportunities in effective cancer treatment. Finally the HSP transgene therapy may be possible in the future.

ACKNOWLEDGMENTS

I am grateful to Drs R. T. Smoleński, A. C. Składanowski and M. Zych for providing their experimental data and for criticism and helpful discussion. This work was supported by the State Committee for Scientific Research, within Medical University in Gdańsk grant W-144.

REFERENCES

1. Piper, J. Oxygen Supply and Energy Metabolism. In: R.Greger and U.Windhorst (Eds.), Comprehensive Human Physiology. Vol.2, Springer-Verlag, Berlin, Heidelberg, 1996, pp.2063–2069

2. Tavenier, M., Skĺadanowski, A.C., De Abreu, R.A. and De Jong, J.W. Kinetics of adenylate metabolism in human and rat myocardium. Biochim.Biophys.Acta. 1244, 351–356 (1995)
3. Lestan, B., Walden, K., Schmalz, S., Spychala, J. and Fox, I.F. β-Hydroxybutyrate decreases adenosine triphosphate degradation products in human subjects. J.Lab.Clin.Med. 124, 199–209 (1994)
4. Hickey, M.J., Knight, K.R., Hurley, J.V. and Lepore, D.A. Phosphoenolpyruvate/ adenosine triphosphate enhances post-ischemic survival of skeletal muscle. J.Reconstr.Surg. 11, 415–422 (1995)
5. Phillis, J.W., Perkins, L.M., Smith-Barbour, M. and O'Regan, M.H. Oxypurinol-enhanced postischemic recovery of the rat brain involves preservation of adenine nucleotides. J.Neurochem. 65, 2177–2184 (1995)
6. Smolenski, R.T., Simmonds, H.A., and Chambers, D.J. Exogenous adenosine, supplied transiently during reperfusion ameliorates depressed endogenous adenosine production in the post-ischemic rat heart. J.Moll.Cell.Cardiol. 29, 333–346 (1997)
7. Kabakov, A.E. and Gabai, V.L. Heat Shock Proteins and Cytoprotection: ATP-Deprived Mammalian Cells. Springer-Verlag, Heidelberg (Germany) and R.G.Landes Company, Austin; 1997, pp. 237.
8. Xu, Q. and Wick, G. The role of heat shock proteins in protection and patophysiology of the arterial wall. Mol.Med.Today 2, 372–379 (1996)
9. Hutter, M.M., Sievers, R.E., Barbosa, V. and Wolfe, C.L. Heat-shock protein induction in rat hearts. Circulation 89, 355–360 (1994)
10. Mestril, R., Chi, S-H.,Sayen, M.R., O'Reilly and Dillmann, W.H. Expression of inducible stress protein 70 in rat heart myogenic cells confers protection against simulated ischemia-induced injury. J.Clin.Investig. 93, 759–767 (1994)
11. Higashi ,T., Nakai, A., Uemura, Y., Kikuchi, H. and Nagata, K. Activation of heat shock factor 1 in rat brain during cerebral ischemia or after heat shock. Mol.Brain Res. 34, 262–270 (1995)
12. Theodorakis, N.G., Morimoto, R.I. Post-translational regulation of HSP70 expression in human cells: Effects of heat shock, inhibition of protein synthesis, and adenovirus infection on translation and mRNA stability. Mol.Cell.Biolog. 7, 4357–4368 (1987)
13. Das, D.K., Engelman, R.M. and Kimura, Y. Molecular adaptation of cellular defences following preconditioning of the heart by repeated ischemia. Cardiovasc.Res. 27, 578–584 (1993)
14. Uyama, O., Matsuyama, T., Michishita, H., Nakamura, H. and Sugita, M. Protective effect of human recombinant superoxide dismutase on transient ischemic injury of CA1 neurons in gerbils. Stroke 23, 75–81(1992)
15. Wang, G.L. and Semenza, G.L. Purification and characterization of hypoxia-inducible factor 1. J.Biol.Chem. 270, 1230–1237 (1995)
16. Rupec, R.A. and Baeuerle, P.A. The genomic response of tumor cells to hypoxia and reoxygenation. Differential activation of transcription factors AP-1 and NF-kB. Eur.J.Biochem. 234, 632–640 (1995)
17. Hochachka, P.W., Buck, L.T., Doll, C.J. and Land, S.C. Unifying theory of hypoxia tolerance: Molecular/metabolic defense and rescue mechanisms for surviving oxygen lack. Proc.Natl.Acad.Sci.USA 93, 9493–9498 (1996)
18. Czyzyk-Krzeska, M.F., Dominski, Z., Kole, R. and Millhorn, D.E. Hypoxia stimulates binding of a cytoplasmic protein to a pyrimidine-rich sequence in the 3'-untranslated region of rat tyrosine hydroxylase mRNA. J.Biol.Chem. 269, 9940–9945 (1994)
19. Czyzyk-Krzeska,M.F. and Beresh, J.E. Characterization of the hypoxia-inducible protein binding site within the pyrimidine-rich tract in the 3'-untranslated region of rat tyrosine hydroxylase mRNA. J.Biol.Chem. 271, 3293–3299 (1996)

71

ENDOTHELIAL PURINE METABOLISM AND REPERFUSION INJURY

G. Weigel,[1] A. Griesmacher,[2] and M. M. Müller[2]

[1]Department of Cardiothoracic Surgery
University of Vienna, Austria
[2]Ludwig Boltzmann Institute for Cardiosurgery Research
Vienna, Austria

The endothelium shows high metabolic activity and thus influences a great number of physiological processes. Eicosanoids, thrombomodulin, protein S, and von Willebrand factor are important for the regulation of blood coagulation. By releasing eicosanoids and nitric oxide the endothelium is involved in the regulation of the vascular tone. Growth factors, cytokines and adhesion molecules deriving from endothelial cells are strongly involved in the regulation of the immune response. Since another important endothelial function is the production of the extracellular matrix, which is responsible for the adhesion of the endothelial cells to the underlying strata, metabolic impairment of the cells can cause detachment. Loss of endothelial cells induces blood coagulation, alterations of vascular tone, uncontrolled proliferation of smooth muscle cells, and leukocytic infiltration.

Free radical injury to the endothelium is proposed as initial event for atherogenesis[1] which should be defined by identification of markers and functional changes.

Risk factors for atherosclerotic lesions are hyperlipidemia, tabacco smoking, hypertension, diabetes mellitus (increased nonenzymatic and auto-oxidative glycosylation) as well as chronic infections, which can be defined as states with increased free radical activity. Also the reperfusion injury is closely associated with the generation of reactive oxygen species (ROS) during reoxygenation following hypoxia (e.g. aortic cross clamping during cardiac surgery).[2,3] Potential sources of toxic reactive oxygen species include activation of monocytes/macrophages,[4,5] eicosanoid biosynthesis, mitochondrial electron transport systems and the xanthine/xanthine oxidase system. Oxidation of proteins, lipids and deoxyribonucleic acids occurs[6-8] and play a general role in ROS induced endothelial cell (EC) damage thereby promoting atherogenesis. Concentrations of H_2O_2 formed during reoxygenation are reported to be up to 1 mmol/l, possibly generated by stimulated neutrophils in the immediate vicinity of ECs.[9] Consequently, an increase in vascular permeability, expression of endothelial adhesion molecules followed by the activation of leukocytes, as well as enhanced eicosanoid release and platelet aggregation occurs. Inflammatory processes are amplified and the cellular metabolism altered.

Purine and Pyrimidine Metabolism in Man IX,
edited by Griesmacher *et al.* Plenum Press, New York, 1998.

Table 1. Influence of H_2O_2 on intracellular ATP and CP levels

H_2O_2 (mM)	ATP (nM/10^6 cells)	CP (nM/10^6 cells)
	Incubation time: 60 min	
0.00	11.5 ± 0.7	21.2 ± 1.0
0.01	$17.4 \pm 1.5^*$	$24.3 \pm 2.8^*$
0.10	12.7 ± 2.4	21.5 ± 2.4
1.00	$0.6 \pm 0.2^*$	$0.2 \pm 0.2^*$
	Incubation: 60 min + Equilibration: 60 min	
0.00	11.5 ± 0.7	21.2 ± 1.0
0.01	11.0 ± 0.5	21.0 ± 1.9
0.10	12.6 ± 0.8	$25.2 \pm 1.5^*$
1.0	$0.3 \pm 0.1^*$	$0.5 \pm 0.3^*$

Data are given as mean ± STD (n=12).
*Statistically significant vs. 0.00 mM H_2O_2 (p < 0.001).

Incubation experiments with endothelial cells show that 10 mM H_2O_2 over 60 min leads to a significant loss of cells. In the presence of 0.1 and 1 mM H_2O_2 the reduction of cells is very modest (by approx. 10%). H_2O_2 dose dependently decreases the incorporation of ^{3}H-thymidine, indicating inhibition of DNA-synthesis and cell proliferation. Addition of 1 mmol/l H_2O_2 to ECs leads to a dramatic decrease of energy rich phosphates (Table 1). Incubations with lower concentrations of H_2O_2 result in unchanged or even increased ATP and CP levels. H_2O_2 concentrations of between 1 and 10 µmol/l lead to strong elevations of ATP and CP. These elevated levels might be explained either by a metabolic disturbance based on decreased ATP consumption or by an elevated rate of ATP synthesis. In a second step HUVEC were exposed to increasing doses of hydrogen peroxide and thereafter allowed to regenerate for 60 min whereby it was demonstrated that the 1 and 10 µmol/l H_2O_2 induced ATP and CP elevations are reversible.

Investigating the time dependent effects of 0.1 mmol/l H_2O_2 as well as 0.1 mmol/l cumene hydroperoxide (CuOH) we could observe a remarkable transient increase in intracellular ATP. The increase in case of CuOH occurs earlier which might be probably due to its fat solubility. Elevated ATP levels indicate metabolic disturbances. It might be speculated that the significant increase in the ATP content is a result of either an increased ATP formation or a massive inhibition of ATP-requiring processes such as DNA synthesis and transmembrane transport mechanisms. In order to clarify, whether the observed ATP-elevations are caused by an increased ATP-formation, we investigated the uptake and salvage of extracellularly added radioactive bases and nucleosides. 0.01 mmol/l H_2O_2 leads to an approx. 30% increase of ^{14}C-AD uptake and salvage after 240 min, whereas 0.1 mmol/l H_2O_2 showed no influence on the incorporation of ^{14}C-AD (Fig. 1a). 0.01 mmol/l and 0.1 mmol/l H_2O_2 induce an increase in the ^{14}C-HX uptake and salvage (Fig. 1b). However, no statistically significant differences to the control can be found. Both H_2O_2 concentrations induce a strong and statistically significant increase in the ^{14}C-ADO uptake and salvage (Fig. 1c), whereby the observed time course is comparable to that observed for ^{14}C-HX. Additionally, we could demonstrate that the uptake and incorporation of ^{3}H-glycine into purine nucleotides via the de novo synthesis is not affected by H_2O_2 (data not shown).

Experiments with cells pretreated with H_2O_2 for 60 min and allowed to regenerate show no marked changes in the characteristic time-dependent uptake and incorporation processes for the precursors investigated (Fig. 2a-c). However, it should be mentioned that the amount of ^{14}C-AD incorporated during the first 30 min is reduced.

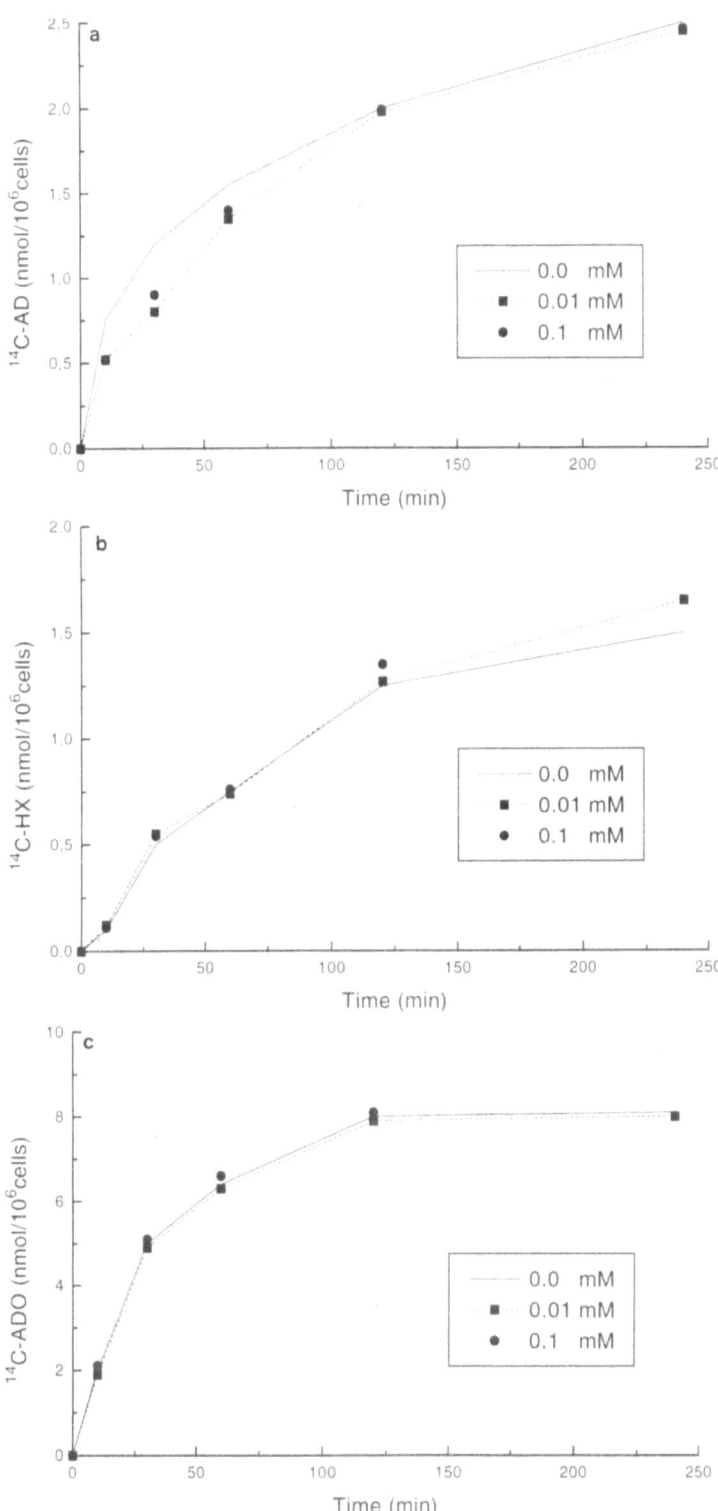

Figure 1. Uptake and salvage of ^{14}C-AD (a), ^{14}C-HX (b), and ^{14}C-ADO (c) during 240 min of incubation with H$_2$O$_2$. Data are given as mean ± STD (n=9). *=statistically significant compared to 0.0 mM H$_2$O$_2$ (p < 0.005).

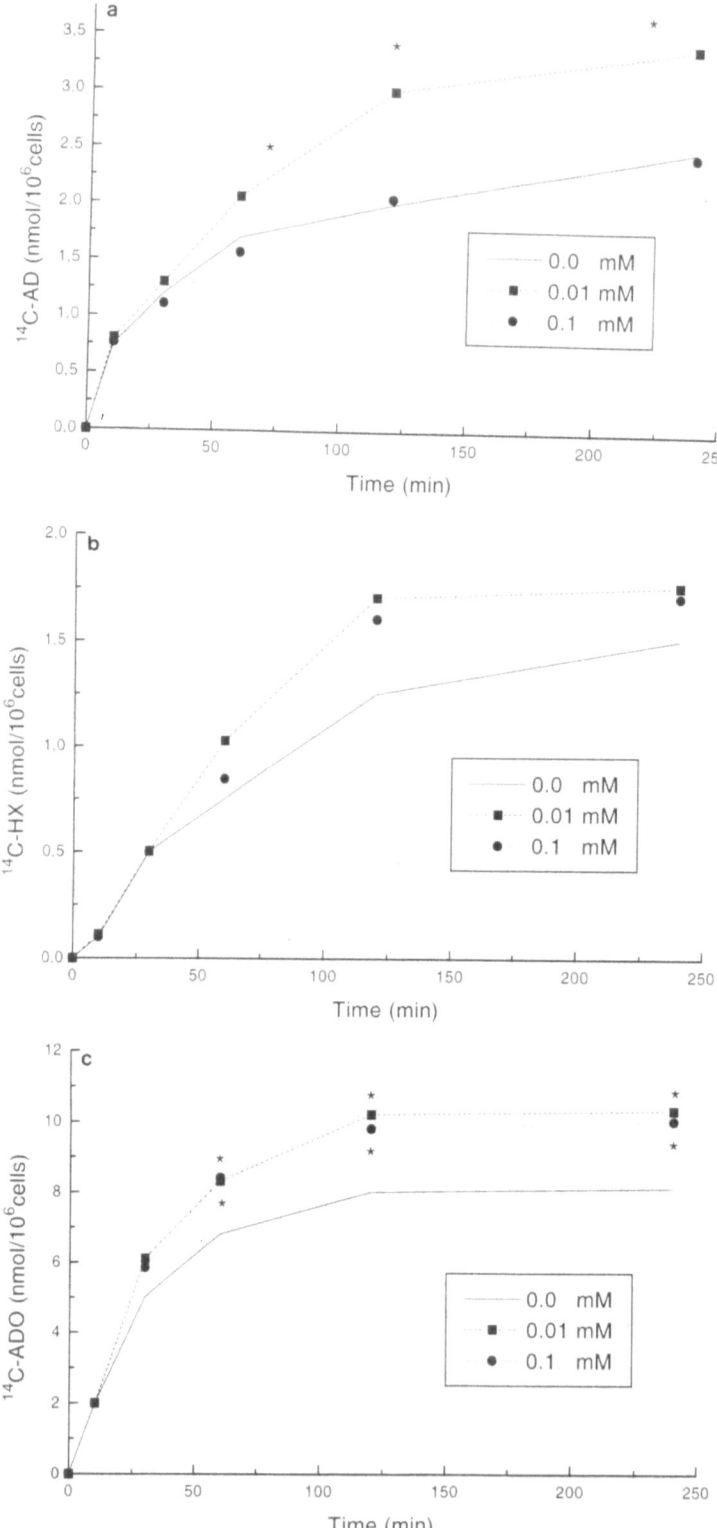

Figure 2. Uptake and salvage of ^{14}C-AD (a), ^{14}C-HX (b), and ^{14}C-ADO (c) during 240 min after 60 min of pre-treatment with H_2O_2. Data are given as mean ± STD (n=9). *= statistically significant compared to 0.0 mM H_2O_2 (p < 0.005).

Table 2. Influence of H_2O_2 on key enzyme's activities
of purine metabolism

Enzyme	Control	0.01 mM H_2O_2	0.1 mM H_2O_2
PRPP-S	20 ± 8	12 ± 4*	8 ± 4*
ADA	177 ± 22	159 ± 24	155 ± 22
PNP I (Ino → Hx)	228 ± 29	283 ± 44*	231 ± 30
PNP II (Hx → Ino)	1861 ± 159	2043 ± 326	1930 ± 258
AK	23 ± 2	13 ± 2*	10 ± 2*
APRT	31 ± 7	42 ± 8*	40 ± 9*
HGPRT	16 ± 4	14 ± 4	13 ± 6

Enzyme activities are given in nmol/10^6 cells/hour (mean ± STD, n=9).
*Statistically significant vs. control (p < 0.005).

For investigating the influence of oxygen free radicals on key enzymes of purine metabolism endothelial cells were incubated with hydrogen peroxide. The activities of 5-phosphoribosyl-1-pyrophosphate synthetase (PRPP-S) and adenosine kinase (AK) were reduced, whereas purine nucleoside phosphorylase (PNP) and adenine phospho-ribosyl-transferase (APRT) were found to be increased by H_2O_2 (Table 2). Hypoxanthine guanine phosphoribosyltransferase (HGPRT) and adenosine deaminase (ADA) activities remained unchanged.

Incubating the cells with 0.1 mmol/l H_2O_2 no significant effects on intracellular ATP, GTP or UTP levels after 60 min of incubation were detected but slight increases in the releases of HX, inosine (INO) and guanine (GUA) were measured.

Due to the enhanced nucleotide catabolism in presence of 1 mmol/l H_2O_2 an extracellular accumulation of nucleotides and bases is observed. The main accumulating purines are HX and INO, only small amounts of ADO are detected. These data indicate that the increase in the purine uptake and salvage observed during oxidative stress is not sufficient to neutralize the nucleotide depletion from HUVEC caused by high doses of H_2O_2. Since HX is found to be the main degradation product, it can be concluded that all enzymes, responsible for the conversion of adenine and guanine nucleotides to HX, are not inhibited by ROS. Moreover, it can be suggested that no xanthine oxidase activity exists in HUVEC since no xanthine or uric acid are detectable. Data from clinical studies show that during hypoxic conditions several purines are released to a greater extent into the blood stream. For this reason not only their uptake and salvage by ECs but also their influence on the absolute levels of high energy phosphates in endothelial cells seems to be of interest. Although at the end of the incubation experiments considerable amounts of purines in the supernatants were detected, indicating an excess of purine bases for the uptake and salvage processes, no toxicity was observed. AD is found to be the most incorporated base, whereas the incorporation of GUA and HX is very weak. AD is directly salvaged to its corresponding mononucleotide by the adenine phosphoribosyltransferase. It causes only a slight increase by approximately 8% in ATP. The total adenine nucleotide content is increased by 7% in presence of adenine. When investigating the effects of nucleosides on extracellular purines, it revealed that HUVEC predominately uptake and salvage ADO, whereas the uptake of INO is low. Incubation with ADO, which is converted to AMP by the enzyme adenosine kinase (AK), results in a pronounced increase in ATP (+18%) and a small rise in intracellular ADP. The total adenine nucleotide content is 17% higher compared to controls. HX and INO fail to influence ATP (and also GTP) levels. Only ADP was found to be slightly elevated in presence of hypoxanthine. Similar to AD, GUA increases GTP. The total guanine nucleotide content is elevated by 11% in presence of GUA. HX and INO also do not influence GTP levels.

In summary, we demonstrated that H_2O_2 (1–10 µM) leads to increased ATP levels in human endothelial cells. This effect can neither be explained by decreasing CP levels, since the changes observed for ATP were paralleled by CP changes nor by a contribution of the de novo synthesis, since the ^3H-glycine incorporation was not affected. Higher concentrations of H_2O_2 (≥ 10 µM) lead to a reduction of ^3H-thymidine incorporation in HUVEC pretreated with H_2O_2, indicating the inhibition of ATP requiring processes and to an increased uptake and salvage of purines (^{14}C-AD, ^{14}C-HX, ^{14}C-ADO). Furthermore, changes in enzyme activities are observed: AK and PRPP-S inhibition, APRT and PNP increase, as well as unchanged activities of ADA and HGPRT.

REFERENCES

1. R. Ross, and J.A. Glomset, The pathogenesis of atherosclerosis. New Engl. J. Med. 295:420 (1976).
2. M. Shlafer, P.F. Kane, and M.M. Kirsh, Superoxide dismutase plus catalase enhance the efficacy of hypothermic cardioplegia to protect the globally ischemic, reperfused heart. J. Thorac. Cardiovasc. Surg. 83:830 (1982).
3. L. Koyama, G.B. Bukley, G.M. Williams, and M.J. Im, The role of oxygen free radicals in mediating reperfusion injury of cold-preserved ischemic kidneys. Transplantation 40:590 (1985).
4. O. Palluy, C. Bonne, and G. Modat, Hypoxia/reoxygenation alters endothelial prostacyclin-synthesis protection by superoxide dismutase. Free Radic. Biol. Med. 11:269 (1991).
5. O. Palluy, L. Morliere, J.C. Gris, C. Bonne, and G. Modat, Hypoxia/reoxygenation stimulates endothelium to promote neutrophil adhesion. Free Radic. Biol. Med. 12:21 (1992).
6. C.G. Cochrane, R.G. Spragg, and S.D. Revak, Studies on the pathogenesis of the adult respiratory distress syndrome: evidence of oxidant activity in bronchoalveolar lavage fluid. J. Clin. Invest. 71:754 (1983).
7. E.W. Kellog, and L. Fridovich, Liposome oxidation and erythrocyte lysis by enzymatically generated superoxide and hydrogen peroxide. J. Biol. Chem. 752:6721 (1977).
8. K. Brawn, and L. Fridovich, DNA strand scission by enzymatically generated oxygen radicals. Arch. Biochem. Biophys. 206:414 (1981).
9. I.U. Schraufstatter, D.B. Hinshaw, P.A. Hyslop, R.G. Spragg, and C.G. Cochrane, Oxidant injury of cells: DNA strand breaks activate polyadenosine diphosphate-ribose polymerase and lead to depletion of nicotinamide adenine dinucleotide. J. Clin. Invest. 77:1312 (1986).
10. E.A. Jaffe, R.L. Nachman, C.G. Becker, and C.R. Minck, Culture of human endothelial cells derived from umbilical veins. J. Clin. Invest. 52:2745 (1973).

72

ADENOSINE-INDUCED PRECONDITIONING OF RAT NEURONAL CULTURES AGAINST ISCHEMIA-REPERFUSION INJURY

A. Reshef,[1] O. Sperling,[1,2,3,*] and E. Zoref-Shani[1]

[1]Department of Clinical Biochemistry
Sackler Faculty of Medicine
Tel Aviv University, Tel Aviv
[2]Departments of Clinical Biochemistry, and
[3]Felsenstein Medical Research Center
Rabin Medical Center (Beilinson Campus)
Petah-Tikva, Israel

1. INTRODUCTION

Murry et al.,[14] were the first to demonstrate the phenomenon of preconditioning of a tissue against ischemic injury. In dogs, a series of short coronary artery occlusions, followed by reperfusion, protected the heart from a subsequent lethal ischemia and reperfusion. This phenomenon was shown to be initiated by the binding of adenosine, the ischemia-induced degradation product of ATP, to the A_1 receptor on the cell outer membrane, activating a signal transduction pathway, resulting finally in improved ability to resist the ischemic damage.[13] The 'adenosine mechanism' in the heart tissue was found to be characterized by a relatively short 'time window of protection', appearing immediately but extending for only 60–120 min.[1,14] Very recently, however, a delayed 'second time window of protection' was reported in the heart model, appearing and lasting days after the induction of preconditioning.[9,12,19]

The brain tissue can be preconditioned against a lethal ischemic insult by several mechanisms,[17] including sublethal ischemia or hypoxia,[5,7] adenosine and synthetic agonists for adenosine receptors,[3,6,8] heat shock and other stresses[16] and growth factors.[2] The 'time window of protection' induced in the brain by adenosine or by sublethal ischemia, appears to be relatively long,[3,5,6,8,10,11] associated in some of the cases with synthesis of heat shock proteins (HSP), or other stress proteins.[5,10,11,15] The aim of the study was to characterize the adenosine-induced time window of protection in cultured neurons.[17]

Purine and Pyrimidine Metabolism in Man IX,
edited by Griesmacher *et al.* Plenum Press, New York, 1998.

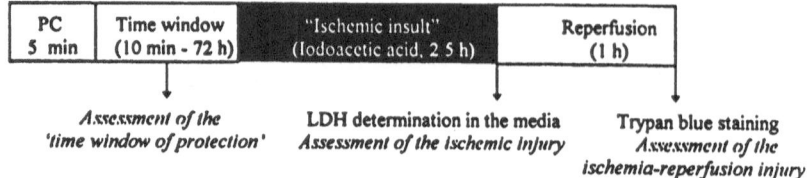

Figure 1. The experimental protocol (Reprinted from Brain Research, 741, Reshef et al.,[17] Preconditioning of primary rat neuronal cultures against ischemic injury: characterization of the 'time window of protection', pp. 252–257, 1996, with kind permission of Elsevier Science NL, Sara Burgerhartstraat 25, 1055 KV Amsterdam, The Netherlands).

2. EXPERIMENTAL PROTOCOL

In Figure 1, 9- to 13-days-old neuronal cultures,[18] were exposed first for 5 min to the preconditioning substances. The preconditioning medium, as well as the regular medium of the control cultures, were then discarded, and the cultures incubated in fresh medium for various time periods, as specified, to assess the duration of the 'time window of protection'. The cultures were then exposed to a sustained (150 min) chemical ischemic insult (100 μM iodoacetic acid). The medium was removed and analyzed for LDH activity, to assess the ischemic injury. The cultures were washed with fresh medium and incubated for 1 h (reperfusion), at the end of which the trypan blue exclusion test[4] was performed in order to assess the injury induced by the exposure to the combined ischemia-reperfusion insult.

3. RESULTS AND DISCUSSION

The ischemic insult resulted in 5.4-fold increase in LDH release, and the combined ischemia-reperfusion insult resulted in 8.5-fold increase in the proportion of cells stained by trypan blue. Preconditioning (Fig. 1) of the neuronal cultures with iodoacetic acid (100 μM), adenosine (1 mM) or R-PIA (100 μM), resulted in a significant protection against the injury. The degree of protection exhibited dependence on the mechanism of preconditioning, the type of insult and on the time window separating between the preconditioning period and the onset of insult (Fig. 2).

The phenomenon observed in our study complies with the original definition of the preconditioning phenomenon,[14] in that the induced protection was demonstrated following a time window, separating between the preconditioning period and the ischemic insult. The 'time window of protection' obtained by the various preconditioning processes, started in all cases at 10 min following preconditioning (the first time monitored) and lasted from 1 h to 48 h, depending on the preconditioning process and on the insult, whether ischemia alone or in combination with reperfusion. The 'time window of protection' induced in the present study by adenosine and R-PIA against ischemia-reperfusion injury (Fig. 2a) is markedly longer in comparison to the relatively short 'first window of protection', obtained by the classical 'adenosine mechanism' in the heart.[14] This could reflect the operation in the neurons of an entirely different 'adenosine mechanism', or neuronal-specific modification of this mechanism, manifested in extenuation of the 'time window of protection'. It could also reflect combination of the classical, short term adenosine window with additional window(s) of protection, characterized by delayed response. Such windows may be associated with HSP, as was already demonstrated in the brain[7,11,15]

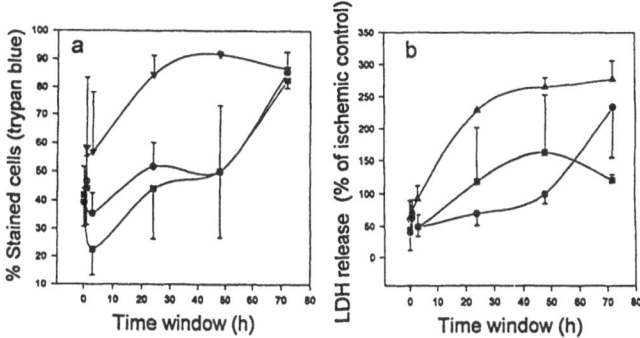

Figure 2. Preconditioning of primary neuronal cultures against ischemic and ischemia-reperfusion injury. (a) The injury caused to the cells by the exposure to the ischemia-reperfusion insult and (b) that by ischemia alone. ■ – R-PIA, 100 μM; ● – adenosine 1 mM; ∇ – iodoacetate, 100 μM. Each point represents the mean ± S.D. (n for each point 8–17). (Reprinted from Brain Research, 741, Reshef et al.,[17] Preconditioning of primary rat neuronal cultures against ischemic injury: characterization of the 'time window of protection', pp. 252–257, 1996, with kind permission of Elsevier Science NL, Sara Burgerhartstraat 25, 1055 KV Amsterdam, The Netherlands).

and in the heart, in association with the phenomenon of the 'second window of protection.'[19] Indeed, a combination of the two protective mechanisms was suggested recently for the heart.[19]

Apparently, the mechanism causing the damage at the ischemic period differs from that operating at the reperfusion period. Accordingly, also the mechanism of protection against each of the insults may be partly or totally different. Support for this assumption may be drawn from the difference between the time response curves obtained against the two insults (Fig. 2). The finding that the preconditioning process provided a better protection capacity against the ischemia-reperfusion insult may be taken to suggest that the preconditioning process provides protection especially against the mechanism causing the injury at the reperfusion period.

An unexpected phenomenon was that of increased sensitivity to the damage observed after the longer time periods studied (Fig. 2).

ACKNOWLEDGMENTS

This work partially fulfills the requirements of the Ph.D. thesis of A. R. at the Sackler Faculty of Medicine, Tel Aviv University.

REFERENCES

1. Armstrong, S., Downey, J.M. and Ganote, C.E., *Cardiovascul. Res.*, 28 (1994) 72–77.
2. Cheng, B., McMahon, D.G. and Mattson, M.P., *Brain Res.*, 607 (1993) 275–285.
3. Daval J.L., and Nicolas, F., *Neurosci. Lett.*, 175 (1994) 114–116.
4. Dessi, F., Chariaut-Marlangue, C., Khrestchantiski, M. and Ben-Ari, Y., J. Neurochem., 60 (1993) 1953–1955.
5. Gidday, J.M., Fitzgibbons, J.C., Shah, A.R. and Park, T.S., *Neurosc. Lett.*, 168 (1994) 221–224.
6. Goldberg, M.P., Monyer, H., Eeiss, J.H. and Choi, D.W., *Neurosc. Lett.*, 89 (1988) 323–327.
7. Kato, H., Kogure, K., Liu, Y., Araki, T. And tytoyama, Y., *Brain Res.*, 652 (1994) 71–75.
8. Krieglstein, J. and Rischke, R., *Euop. J. Pharmacol.* 205 (1991) 7–10.

9. Kuzuya, T., Hoshida, A., Yamashita, N. et al., *Circ. Res.* 72 (1993) 1293–1299.
10. Liu, Y., Kato, H., Kato, N. and Kogure, K., *Brain Res.*, 586 (1992) 121–124.
11. Liu, Y., Kato, H., Nakat, N. and Kogure, K., *Neuroscience,* 56 (1993) 921–927.
12. Marber, M.S., Latchman, D.S., Walker, J.M. and Yellon, D.M., *Circulation*, 88 (1993) 1264–1272.
13. Mullane, K., *Circulation,* 85 (1992) 845–847.
14. Murry C.E., Jennings, R.B. and Reimer, K.A., *Circulation*, 74 (1986) 1124–1136.
15. Nishi, S., Taki, W., Uemura, Y., Higashi, T., Kikuchi, H., Kudoh, H., Satoh, M. and Nagata,K., *Brain Res.,* 615 (1993)281–288.
16. Rordrof, G., Koroshetz, W.J. and Bonventre, J.V., *Neuron,* 7 (1991) 1043–1051.
17. Reshef, A., Sperling, O. and Zoref-Shani, E., Brain Res. 741 (1996) 252–257,
18. Yavin, Z. and Yavin, E., *Exp. Brain Res.,* 29 (1977) 137- 147.
19. Yellon, D.M. ang Baxter, G.F., *J. Mol. Cell. Cardiol.*, 27 (95)1023–1034.

73

MYOCARDIAL ISCHEMIC INJURY DURING CARDIO-PULMONARY BY-PASS

Evaluation of Purine Compounds by Capillary Electrophoresis

F. Carlucci,[1] A. Tabucchi,[1] B. Biagioli,[2] M. Maccherini,[2] G. Sani,[2] F. Simeone,[2] D. Perrett,[3] and E. Marinello[1]

[1]Institute of Biochemistry and Enzymology
University of Siena
[2]Insitute of Thoracic and Cardiovascular Surgery and Biomedical Technologies
University of Siena
[3]Department of Medicine
St. Bartholomew's Hospital Medical College, London

1. INTRODUCTION

Prolonged in vivo and ex vivo myocardial protection is one of the primary factors for the success of the procedure in the therapy of patients with end-stage cardiac disfunction. Advances in myocardial preservation for routine heart surgery and transplantation have been closely linked. Future advances in the understanding of myocardial energy metabolism and the limitation of oxygen-derived free radical induced injury will have wide application in both fields.

Changes in high energy-phosphates of the myocardium are closely associated with ischemic damage following aortic cross-clamping during cardiac surgery. Since this may affect the contractile function of heart and membrane ionic transport, it is important to monitor myocardial purine metabolism during the preservation period in order to evaluate the effects of cardioplegic solutions.

Capillary electrophoresis (CE) has become one of the most efficient separation techniques. The separation of nucleotides nucleosides and base can be performed by micellar electrokinetic chromatography (MEKC)[1–3] and by free solutions;[4] the second method was been used in our study. The simple and fast CE procedure is able to separate the complete spectrum of purine metabolites found in myocardial extracts. The small sample volumes used in CE make it possible to solve a number of biological and clinical problems, in which the amount of biological material is limited.

Purine and Pyrimidine Metabolism in Man IX,
edited by Griesmacher *et al.* Plenum Press, New York, 1998.

2. MATERIALS AND METHODS

2.1. Patients

We studied six patients (age 64–72 years mean 68) undergoing coronary artery by-pass surgery (CABG) for ischemic cardiopathy, four had a plurivasal and two monovasal ischemic cardiopathy.

2.2. Surgical Technique

Antiangina medication (nitrates and calcium-blocking agents) were continued until morning of operation. Anaesthesia was induced with fentamyl (5 μg/kg) combined with the diazepam (0.3 mg/kg) and maintained with fentanyl infusion (25 $μg·kg^{-1}·min^{-1}$). Muscle relaxation was achieved with pancuronium bromide (0.1 mg/kg). Ventilation with a mixture of oxygen and nitrous oxide (50:50) was adjusted to maintain normocapnia. During by-pass the haematocrit level was maintained between 20% and 25% and pump flows were maintained between 2.0 and 2.2 $l·min^{-1}·m^{-2}$. Mean arterial pressure was maintained between 60 and 70 mmHg with sodium nitroprusside (0.5–5 $μg·kg^{-1}·min^{-1}$), or norepinephrine (0.01–0.5 $g·kg^{-1}·min^{-1}$).

2.3. Cardioplegic Techniques

Cold blood cardioplegia was performed with a 4:1 ratio of oxygenated circuit blood and crystalloid solution. The physical and biochemical composition of our blood cardioplegic solution were according to the protocol of the UCLA Medical Centre[5] the only difference was that our solution did not contain L-Aspartate and L-Glutamate. The delivery was divided between antegrade and the retrograde through the aorta and through the coronary sinus. Diastolic arrest was achieved with an initial cold (4 to 8°C) blood high-potassium solution (20 to 25 mEq/l) at a flow rate of 200 to 300 ml/min for 2 minutes through the aorta and for 2 minutes through the coronary sinus. The coronary sinus perfusion pressure was maintained below 40 mm Hg. On completion of each distal-proximal anatomosis or at 20-minutes intervals, a flow rate of 150–200 ml/min potassium solution (8 to 10 mEq/l) was administred for 1 minute through the aorta and for 1 minute through the coronary sinus. Before cross-clamp release warm (37°C) blood low-potassium solution was infused at 150 ml/min and 50 mm Hg for 3 to 5 minutes and alternating between the aortic root and the coronary sinus (modified cardioplegic reperfusion according Buckberg).[6] Moderate systemic hypothermia (28 to 30°C) and topical hypothermia were used.

2.4. Myocardial Biopsies

Transmural left ventricular biopsy specimens were obtained with a Tru-cut biopsy needle. Specimens were immediately washed of blood in an ice-cold, isotonic buffer solution before freezing in liquid nitrogen (within 20 sec. of biopsy). Speciments were taken immediately before aortic cross-clamping, immediately before its removal and 30 minutes after reperfusion.

The samples were homogenized at 10% with 0.4 N perchloric acid in a 1 ml-Potter pestle. For practical handling of the extract, not less than 20 μl was used for biopsy specimens under 2 mg wet-weight. Extracts were then centrifuged (12,000 g × 10 min) in a cooled microfuge. The supernatants were neutralized with 5N potassium carbonate. Potassium perchlorate was removed with a subsequent centrifugation at 12,000 g for 3 minutes. Aliquots of the extracts were analyzed by CE.

2.5. CE Procedure

We used a Waters Quanta 4000 capillary electrophoresis apparatus, equipped with 60(53) cm × 75 μm i.d. uncoated fused silica capillary. The conditions were: 20 mM Na-borate buffer (pH 9.97), 22 kV voltage ($\cong 90$ μA current). The electropherogram was read at 254 nm.

Statistical analysis was performed by one-way analysis of variance.

3. RESULTS AND DISCUSSION

We were able to separate the complete spectrum of purine metabolites found in myocardial extracts from 0.8–10 mg biopsy speciments. Nucleotide contet was referred to wet weight since no appreciable protein precipitate was present after perchloric acid extraction.

Basal levels of purine compounds showed dispersed values presumibly related to the different myocardial alterations in the patients studied, however the pattern of purine metabolism was clear. Before aortic cross-clamp the energy charge was low (0.55) and IMP and inosine content high, showing an imbalance in purine metabolism. In these patients catabolism presumibly occurs via AMP deamination rather than through the dephosphorylation of AMP to adenosine. In prolonged myocardial hypoxia due to ischemic cardiopathy, low pH values and ATP levels, could inhibit the cytosolic isoforms of 5'-nucleotidase acting on AMP and on IMP.[7] After the ischemic period (50'–120') ATP and GTP levels decreased, and their dephosphorylation products, AMP and GMP respectively; IMP, nucleosides and bases accumulated. ADP and GDP levels and energy charge were essentially unchanged (Tab. 1); ATP/ADP and nucleotide/(nucleoside + base) ratios decreased. It is evident that the energy charge is a less sensible index of ATP cathabolism to respect to ATP/ADP ratio since AMP doesn't accumulate being converted to IMP or adenosine (Fig. 1).

An increase in adenosine (400%) during ischemia shows that purine catabolism proceeds through a different pathway in this situation.

After thirty minutes of reperfusion nucleotides, nucleosides and bases tended to return to t basal values. Energy charge did not increase but the nucleotide/(nucleoside + nucleobase) ratio, which represents the prevalence of synthesis over degradation showed a rebound.

Table 1. Values in pmoles/mg wet weight of tissue ± standard error. *p ≤ 0.05

	Ado	Hx	Ino	AMP	ATP	ADP	GMP	GTP	GDP	IMP
Basal										
mean	38.36	150.46	462.50	611.45	1269.63	1148.56	64.20	85.71	83.50	827.63
s.e.	21.32	31.47	102.84	159.21	566.53	272.18	14.17	22.21	22.53	247.96
Isc.										
mean	151.81*	341.81*	883.34*	977.14*	653.18*	1077.07	119.70*	54.48*	76.65	1200.38*
s.e.	44.99	98.25	145.19	417.30	179.08	271.92	18.65	3.97	16.63	313.33
Rep.										
mean	79.34*	171.85	620.93	708.27	1226.74	1184.56	54.94	105.42	73.35	1098.79*
s.e.	30.89	82.01	321.77	204.12	718.35	398.53	7.11	44.82	30.90	622.08

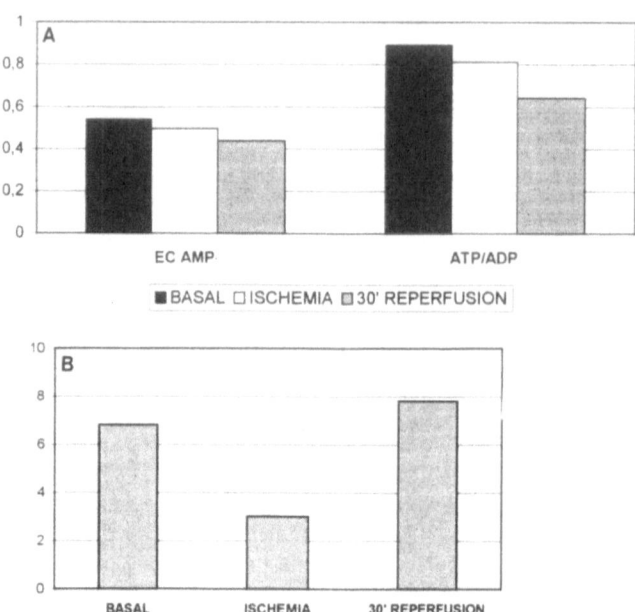

Figure 1. A) Energy charge values and ATP/ADP ratio. B) nucleotide/(nucleoside + base) ratios. *p ≤ 0.05.

This study raises the problem of myocardial protection during heart surgery, showing the importance of maintaining adequate purine compound levels during cardiac ischemia following aortic cross-clamping. Metabolic pathways of purines are thus a possible target for drugs in cardioplegic solutions.

In this kind of investigation capillary electrophoresis is an extremely adaptable technique for the evaluation of ischemic damage.

REFERENCES

1. M. Ng, T.F. Blaschke, A.A: Arias and R.N. Zare. Analysis of free intracellular nucleotides using high-performance capillary electrophoresis. *Anal Chem*, 64: 1682 (1992)
2. Huang M.X., Liu S.F., Murray B.K., Lee M.L. High resolution separation and quantitation of ribonucleotides using capillary electrophoresis. *Anal. Biochem.*, 207: 231 (1992).
3. Perrett D., Ross G. Capillary electrophoresis for the analysis of cellular nucleotides. *Purine and Pyrimidine Metabolism in Man VII Part B*, 1–5, Plenum Press, New York (1991).
4. T. Grune, D. Perrett. Rapid simultaneous measurement of nucleotides, nucleosides and bases in tissues by capillary electrophoresis. *Purina and Pyrimidine Metabolism in Man VIII*, 805–810, Plenum Press, New York (1995).
5. Buckberg G.D. Antegrade/retrograde blood cardioplegia to ensure cardioplegic distribution:operative techniques and objectives. J. Card. Surg. 4: 216–238, 1989.
6. Buckberg G.D. Strategies and logic of cardioplegic delivery to prevent, avoid and reserve ischemic and reperfusion damage. J. Thorac. Cardiovasc. Surg. 1987, 93: 127–39.
7. M.I. Bak &J.S. Ingwall. Acidosis during ischemia promotes adenosine triphosphate resyntesis in postischemic rat heart. In vivo regulation of 5'-nucleotidase. *J. Clin. Invest.* 93: 40–9 (1994).

74

NUCLEOTIDE METABOLISM IN THE HEART SUBJECTED TO HEAT STRESS

Ryszard T. Smolenski, Caroline Gray, Jay Jayakumar, Mohamed Amrani, and Magdi H. Yacoub

Heart Science Centre
National Heart and Lung Institute at Harefield Hospital
Harefield, United Kingdom and
Department of Biochemistry
Medical University of Gdansk, Poland

INTRODUCTION

Heat shock increases cardiac resistance to ischemia but the mechanism of this effect is unknown.[1] Whole body hyperthermia has been recognised to induce synthesis of a group of proteins, termed "heat shock proteins" (HSP). The predominant function of HSP is protection against protein denaturation and facilitation of desired protein folding at the time of ribosomal polypepide synthesis. Furthermore HSP are able to unfold the polipeptide chains, by associating with proteins being incorporated into mitochondria and placing them in a translocation-competent configuration. These roles are referred to as "chaperone" functions.[2] There is strong evidence to suggest that increased expression of heat shock protein HSP70 is directly involved in the cardioprotective effect of heat stress.[3,4] However it is not yet known which enzymes or structural proteins and which metabolic pathways are protected by HSP70. To investigate the possible role of alterations in nucleotide metabolism in this beneficial effect, we analysed the concentrations of adenine, guanine, pyrimidine and pyridine nucleotides and phosphocreatine concentration in the rat heart subjected to heat shock. The concentrations of nucleotide metabolites were evaluated under normoxic conditions and after 5 min of global normothermic ischemia

METHODS

Male Wistar rats weighing 250–350 g were used in this study. Rats were subjected to heat stress at 42°C for 15 min (HS) under pentobarbital anaesthesia, while control animals (C) did not undergo heat shock as described in detail previously.[5] In one series of experiments, hearts were collected after 24 hours, connected to perfusion apparatus, perfused for

Purine and Pyrimidine Metabolism in Man IX,
edited by Griesmacher *et al.* Plenum Press, New York, 1998.

373

5 min under normoxic conditions in Langendorff mode and than were freeze clamped (n=10 and n=7 in C and HS respectively). In a second series of experiments of both HS and C animals hearts were similarly connected to Langendorff perfusion apparatus, subjected to 15 min of normoxic perfusion and then subjected to 5 min of normothermic global ischemia (n=6 and n=6 in C and HS respectively). Nucleotide, nucleoside and base concentrations, as well as creatine metabolites were analysed using reversed-phase[6] or anion-exchange[7] HPLC methods described in detail previously.

RESULTS

GTP levels were significantly higher in the HS group when compared to C by about 20%. Likewise, UTP and CTP concentrations were higher. UTP was increased by about 35% while CTP was elevated by about 45% in HS as compared to C group. NADP levels were also higher in HS by about 50% when compared to C, while NAD levels was similar in both groups. No changes were observed in ATP, phoshocreatine or creatine concentrations. The ATP/ADP and phosphocreatine/creatine ratios were also found to be similar. However, important differences were observed after 5 min ischemia, ATP was significantly higher in HS group than in the C and purine catabolite, including adenosine accumulation was lower (not shown).

DISCUSSION

This study demonstrated profound alterations in nucleotide metabolism in heart of rats which were induced by heat stress. Under normoxic conditions, an increased levels of guanine and pyrimidine nucleotides were demonstrated while ATP and phosphocreatine concentrations were similar. NADP levels were also found to be elevated in hearts of rats subjected to heat stress. There are studies which demonstrated that heat shock enhances cardioprotection without changes in major high energy phosphate metabolite concentration.[8] The relation between the effect of heat shock and energy metabolism may include a several mechanisms. ATP binding seems to play an important role in the mechanism of action of HSP 70 and dissociation of the mt-HSP70.Mim44 complex, which is a crucial step of protein transport across the inner mitochondrial membrane.[9] An increased thermostability of lactate dehydrogenase LDH-A4 isoenzyme was demonstrated after exposure to heat shock protein.[10] HSP 90 was found to facilitate folding of citrate synthase at the time of ribosomal synthesis.[11] A recent study showed increased pyruvate utilisation in the heart subjected to heat stress followed by ischemia.[12] The latter finding is of special interest in view of the data showing importance of pyruvate dehydrogenase complex activity in the mechanical recovery of the heart subjected to ischemia.[13] Our study provided additional information allowing to speculate on the mechanism of cardioprotective effect of heat shock. It is possible that increased guanine and pyrimidine nucleotide levels in HS hearts could be a consequence of stimulation of RNA synthesis. Simultaneously, increased UTP concentration may facilitate glycogen metabolism or increased CTP concentration may facilitate phospholipid turnover. On the other hand GTP is required for normal operation of regulatory mechanisms involving G-proteins and enhanced GTP concentration may facilitate these processes. Our results demonstrated also higher NADP concentration. This may incdicate an increased antioxidant capacity of the heart after heat stress. An important observation is decreased rate of ATP depletion during ischemia in heat stressed hearts. It indicates better handling of energy resources during ischemia in hearts subjected to heat stress.

Table 1. Concentrations of adenine, guanine, cytosine, uracil and pyridine nucleotides in the heart of rats subjected to heat stress and in control animals. Values are means ± SEM. Significant differences are highlighted in bold. (µMol/g dry weight)

	Pcreatine	Creatine	GTP	GDP	GMP	ATP	ADP	ADPR	AMP	NADP	NAD	CTP	UTP
Control hearts under normoxia	30.70 ±1.64	41.20 ±1.21	1.17 ±0.03	0.15 ±0.04	0.04 ±0.01	21.70 ±0.47	5.13 ±0.25	0.34 ±0.05	0.84 ±0.15	0.31 ±0.03	4.29 ±0.05	0.34 ±0.02	0.61 ±0.05
5 min ischemia						11.90 ±0.27							
Heat stressed under normoxia	30.09 ±1.56	41.39 ±1.40	**1.43 ±0.06**	0.10 ±0.04	0.04 ±0.01	21.81 ±0.52	5.46 ±0.29	0.42 ±0.09	0.92 ±0.13	**0.46 ±0.07**	4.37 ±0.24	**0.50 ±0.07**	**0.84 ±0.11**
5 min ischemia						**14.11 ±0.49**							

ACKNOWLEDGMENTS

This study was supported by the British Heart Foundation (Grant No: PG 96194, the Polish State Committee for Scientific Research (St-42 and W-922) and The Harefield Heart Transplant Trust.

REFERENCES

1. R.W. Currie, Effects of ischemia and perfusion temperature on the synthesis of stress-induced (heat shock) proteins in isolated and perfused rat hearts, *J. Mol. Cell Cardiol.* 19:795(1987).
2. W.L. Kelley and C. Georgopoulos, Chaperones and protein folding, *Curr. Opin. Cell Biol.* 4:984(1992).
3. R.J. Heads, D.M. Yellon, and D.S. Latchman, Differential cytoprotection against heat stress or hypoxia following expression of specific stress protein genes in myogenic cells, *J. Mol. Cell Cardiol.* 27:1669(1995).
4. M.S. Marber, R. Mestril, S.H. Chi, M.R. Sayen, D.M. Yellon, and W.H. Dillmann, Overexpression of the rat inducible 70-kD heat stress protein in a transgenic mouse increases the resistance of the heart to ischemic injury, *J. Clin. Invest.* 95:1446(1995).
5. M. Amrani, J. Corbett, N.J. Allen, J. O'Shea, S.Y. Boateng, A.J. May, M.J. Dunn, and M.H. Yacoub, Induction of heat-shock proteins enhances myocardial and endothelial functional recovery after prolonged cardioplegic arrest, *Ann. Thorac. Surg.* 57:157(1994).
6. R.T. Smolenski, D.R. Lachno, S.J.M. Ledingham, and M.H. Yacoub, Determination of sixteen nucleotides, nucleosides and bases using high-performance liquid chromatography and its application to the study of purine metabolism in hearts for transplantation, *J. Chromatogr.* 527:414(1990).
7. E. Harmsen, P.P. De Tombe, and J.W. de Jong, Simultaneous determination of myocardial adenine nucleotides and creatine phosphate by high-performance liquid chromatography, *J. Chromatogr.* 230:131(1982).
8. R.W. Currie and M. Karmazyn, Improved post-ischemic ventricular recovery in the absence of changes in energy metabolism in working rat hearts following heat-shock, *J. Mol. Cell Cardiol.* 22:631(1990).
9. O. von Ahsen, W. Voos, H. Henninger, and N. Pfanner, The mitochondrial protein import machinery. Role of ATP in dissociation of the Hsp70.Mim44 complex, *J. Biol. Chem.* 270:29848(1995).
10. M.S. Zietara and E.F. Skorkowski, Thermostability of lactate dehydrogenase LDH-A4 isoenzyme: effect of heat shock protein DnaK on the enzyme activity, *Int. J. Biochem. Cell Biol.* 27:1169(1995).
11. U. Jakob, H. Lilie, I. Meyer, and J. Buchner, Transient interaction of Hsp90 with early unfolding intermediates of citrate synthase. Implications for heat shock in vivo, *J. Biol. Chem.* 270:7288(1995).
12. M.S. Marber, J.M. Walker, D.S. Latchman, and D.M. Yellon, Myocardial protection after whole body heat stress in the rabbit is dependent on metabolic substrate and is related to the amount of the inducible 70-kD heat stress protein, *J. Clin. Invest.* 93:1087(1994).
13. E.D. Lewandowski and L.T. White, Pyruvate dehydrogenase influences postischemic heart function, *Circulation* 91:2071(1995).

75

DIFFERENCES IN NUCLEOTIDE METABOLISM AND MECHANICAL RECOVERY AFTER CARDIOPLEGIC ARREST OF THE HEART AT DIFFERENT AGE

Mohamed Amrani, Ryszard T. Smolenski, Andrew T. Goodwin, Caroline C. Gray, Jay Jayakumar, Piotr Jagodzinski, and Magdi H. Yacoub

Department of Cardiothoracic Surgery
National Heart and Lung Institute
Imperial College
Harefield Hospital
Harefield, Middlesex UB9 6JH, United Kingdom

INTRODUCTION

Number of metabolic processes in the heart such as calcium handling or nucleotide metabolism, are known to be modified during ageing.[1-3] Although several studies have evaluated the effect of differences in nucleotide metabolism of on the recovery of the heart after normothermic ischemia, little is known about its relationship with cardiac function after cardioplegic arrest and hypothermic ischemia, mimicking preservation for cardiac transplantation. In this study, we evaluated the association between concentration of purine nucleotides and functional recovery of the rat heart at different ages, after cardioplegic arrest and hypothermic ischemia.

MATERI1ALS AND METHODS

Male Sprague-Dawley rats were used in all experiments. Six hearts were studied in each group. Animals were divided into the following 3 age groups: group A: 1 month; group B: 5 months; group C: 20 months. The animals were anaesthetised with halothane mixed with 95% O_2 plus 5% CO_2. The femoral vein was exposed and heparin (200 IU) was injected. One minute later, the heart was excised and immediately placed in cold (4ºC) Krebs Buffer. Approximately 30 seconds later, the aorta was cannulated and perfusion was started. The isolated working rat heart preparation which has been described in

Purine and Pyrimidine Metabolism in Man IX,
edited by Griesmacher *et al.* Plenum Press, New York, 1998.

377

detail elsewhere,[4] was used in this study. Oxygenated Krebs-Henseleit (KH) bicarbonate buffer (NaCl 118.5 mmol/l; NaHCO$_3$ 25 mmol/L; KCl 4.75 mmol/L; MgSO4 1.19 mmol/L; KH$_2$PO$_4$ 1.18 mmol/L; CaCl$_2$ 2.5 mmol/L), pH 7.4, containing glucose (11.1 mmol/L) and gassed with 95% O$_2$ and 5% CO$_2$ at 37°C, enters the heart via the cannulated left atrium at a constant pressure of 15 cm H$_2$O and passes into the left ventricle. From which it is spontaneously ejected via an arotic cannula against a hydrostatic pressure of 100 cm H$_2$O. Using this preparation, which is essentially that described by Langendorff, the heart will continue to beat, but does not perform external work. After initial Langendorff perfusion for 15 min, conditions were switched into working mode for 20 minutes and at the end of this phase pre-ischemic indices of coronary flow (CF), cardiac output (CO) and aortic pressure (AP) were assessed. After taking the readings, hearts were arrested by infusion of 10 ml of St Thomas' Hospital (St1) cardioplegic fluid and maintained under hypothermia (4°C) for 4 hours. Then, hearts were reperfused for 15 minutes in Langendorff mode followed by perfusion in a working mode for the next 20 minutes. At the end of this phase the post-ischemic function (CF, CO, AP) was evaluated. Hearts were freeze-clamped at the end of the experiment and nucleotides were analysed by HPLC according to procedure previously described in detail.[5]

RESULTS

The postischemic recovery (% of the pre-ischemic level) of cardiac power was $48.89 \pm 7.78\%$ for group A, which was higher ($p < 0.05$) than the functional recovery in group B ($24.11 \pm 3.50\%$) or group C ($21.41 \pm 4.69\%$). There was no difference in the

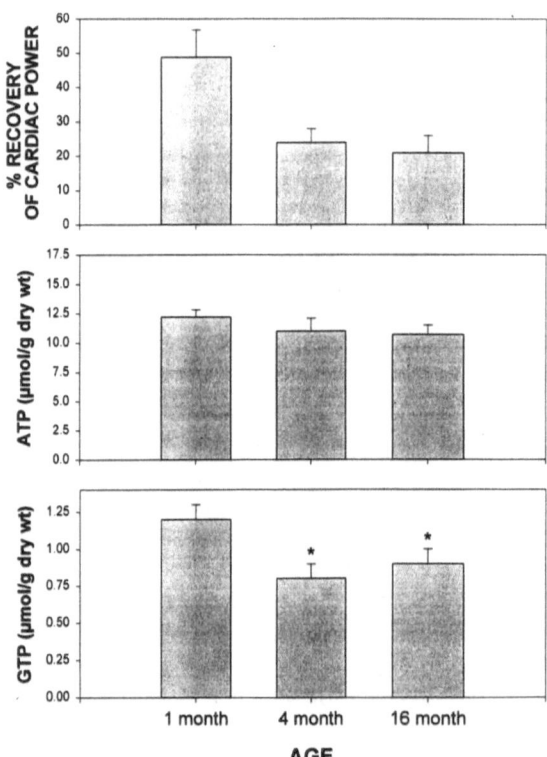

Figure 1. Recovery of cardiac power and concentrations of ATP and GTP in hearts of three age groups after cardioplegic arrest, hypothermic (4°C) ischemia for 4 hours and reperfusion for 35 min.

post-ischemic ATP concentration between groups (12.26 ± 0.65, 10.97 ± 1.11 and 10.69 ± 0.79 μmol/g dry weight) or in total adenine nucleotide pool (16.6 ± 0.6, 16.13 ± 0.6 and 15.7 ± 0.7 μmol/g dry weight) in A, B and C respectively. In contrast GTP concentration was higher ($p < 0.05$) in A (1.204 ± 0.08 mmol/g dry weight) than in B or C (0.778 ± 0.062 mmol/g dry weight and 0.903 ± 0.074 mmol/g dry weight respectively). On the other hand total NAD content (sum of NAD and ADPR - adenosine diphophoribose) was highest ($p < 0.05$) in the B group (4.10 ± 0.13 μmol/g dry weight) as compared to A and C (3.72 ± 0.05 and 3.82 ± 0.12 μmol/g dry weight respectively). Calculated total adenine nucleotide/total NAD ratio was highest ($p < 0.05$) in group A (4.65 ± 0.14, 3.94 ± 0.13 and 4.09 ± 0.08) in groups A, B and C respectively.

DISCUSSION

The present study has shown that following a prolonged hypothermic cardioplegic rrest, hearts from young rats displayed a better recovery of cardiac mechanical function than adult or elderly animals. GTP content but not ATP was better preserved in the youngest group. Cardiac myocytes are known to undergo profound physiological, ultrastructural and biochemical alterations during the ageing process. These include a decreased activity of the calcium pump, altered receptor system linked to adenylate cyclase, and a reduction of the fast isoform of myosin. The changes in nucleotide levels after hypothermic ischemia could be thus secondary to the other metabolic alterations. However, its significant role in the impaired functional recovery cannot be excluded. GTP is essental for normal operation of G protein of adenylate cyclase system and reduced concentration of GTP may impair regulatory mechanisms in the heart. The role of altered NAD content in hearts at different age merit further investigation.

ACKNOWLEDGMENTS

This study was supported by the British Heart Foundation (Grant No: PG 96194) and The Harefield Heart Transplant Trust.

REFERENCES

1. R.T. Smolenski, J. Swierczynski, M. Narkiewicz, and M. Zydowo, Purines, lactate and phosphate release from child and adult heart during cardioplegic arrest, *Clin. Chim. Acta* 192:155(1990).
2. J.W. de Jong, E. Keijzer, T. Huizer, and B. Schoutsen, Ischemic nucleotide breakdown increases during cardiac development due to drop in adenosine anabolism/catabolism ratio, *J. Mol. Cell. Cardiol.* 22:1065(1990).
3. M. Amrani, A.H. Chester, J. Jayakumar, and M.H. Yacoub, Aging reduces postischemic recovery of coronary endothelial function, *J. Thorac. Cardiovasc. Surg.* 111:238(1996).
4. M. Amrani, R. Shirvani, N.J. Allen, S. Ledingham, and M.H. Yacoub, Enhancement of low coronary reflow improves postischemic myocardial function, *J. Thorac. Cardiovasc. Surg.* 104:1375(1992).
5. R.T. Smolenski, D.R. Lachno, S.J.M. Ledingham, and M.H. Yacoub, Determination of sixteen nucleotides, nucleosides and bases using high-performance liquid chromatography and its application to the study of purine metabolism in hearts for transplantation, *J. Chromatogr.* 527:414(1990).

76

DIFFERENT MECHANISM OF MYOCARDIAL NUCLEOTIDE BREAKDOWN DURING NORMOTHERMIC AND HYPOTHERMIC ISCHEMIA

Marek Zych, Ryszard T. Smolenski, Wiesław Makarewicz, and Magdi H. Yacoub

Department of Biochemistry
Medical University of Gdansk, Poland and
Heart Science Centre
Harefield Hospital, Harefield, United Kingdom

INTRODUCTION

Mechanisms of nucleotide breakdown in the heart were studied extensively under normothermic conditions and the view that low temperature simply slow down to the same extent all processes involved is common. However, it is possible that low temperature may exert different effect on catabolic enzymes or membrane transport systems. This problem is of clinical relevance especially during heart transplantation when the temperature during ischemia lasting up to four hours is maintained at 4°C. The aim of this study was to evaluate the differences in the mechanism of nucleotide breakdown during normothermic unprotected ischemia and during hypothermic ischemia following cardioplegic arrest.

MATERIALS AND METHODS

Langendorff perfused rat hearts[1] were subjected to: (1) 17 min of global ischemia at 37°C followed by 30 min reperfusion and (2) cardioplegic arrest (St. Thomas' No. 1) followed by 6 hr of ischemia at 4°C and 30 min reperfusion. Coronary effluent was collected at multiple time points over 15 min of reperfusion. Hearts were freeze-clamped at the end of experiment. Nucleotides and catabolites were analysed by HPLC in the coronary effluent and heart extracts according to methods described previously.[2,3]

Purine and Pyrimidine Metabolism in Man IX,
edited by Griesmacher *et al.* Plenum Press, New York, 1998.

RESULTS

Both normothermic and hypothermic ischemia caused similar degree of nucleotide breakdown as measured at the end of experiment. ATP decreased from preischemic 25.1 ± 0.7 (µmol/g dry wt, $\pm$ SEM) to 19.3 ± 0.6 and 18.1 ± 0.9 in (1) and (2) respectively. Adenosine, accounted for 25% of all catabolites released in (1) and only for 5% in (2). Purine release during reperfusion rapidly decreased with time in (1) while this release was prolonged in (2).

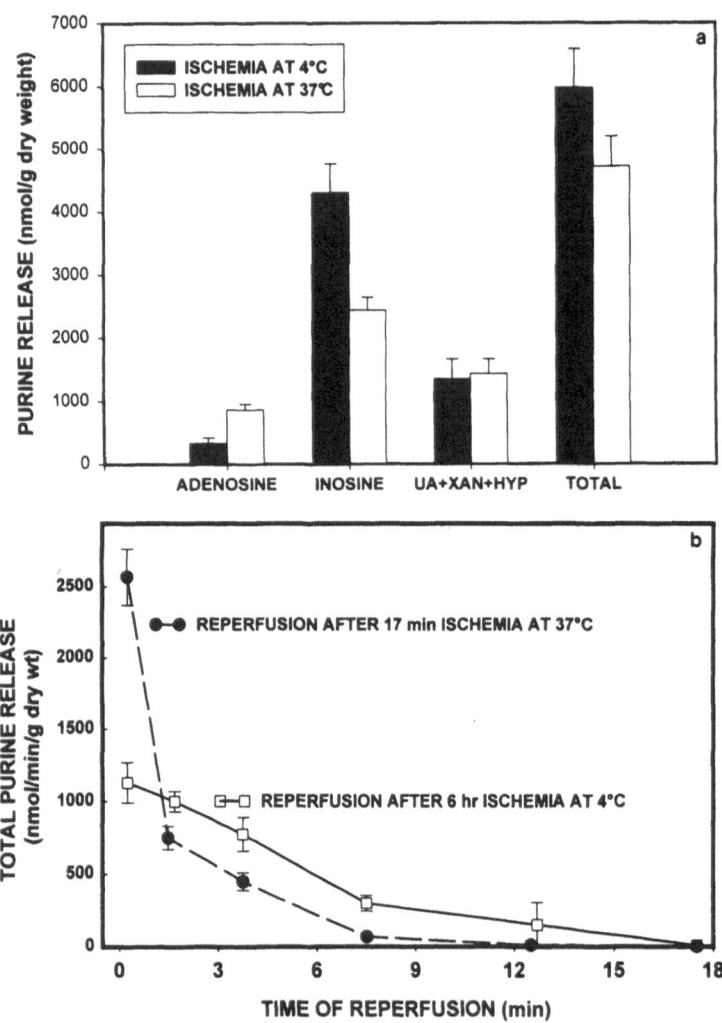

Figure 1. (A) Concentrations of adenosiene, inosine and sum of uric acid (UA), xanthine (XAN) and hypoxanthine (HYP) in the coronary effluent collected throughout 15 min of reperfusion and (B) concentration of total purine catabolites (adenosine + inosine + hypoxanthine + xanthine + uric acid) in the coronary effluent at different time of reperfusion of the rat heart after unprotected normothermic (37°C) ischemia or hypothermic (4°C) ischemia following cardioplegic arrest. Values are means $\pm$ SEM.

DISCUSSION

Cardioprotection during cardiac surgery remains still a significant clinical problem. Adequate understanding of the metabolic processes involved is essential to develop better cardioptotective strategies. We have shown in this study that nucleotide catabolism under hypothermic conditions is markedly different in many aspects from this process under normothermia. Different effects of temperature on enzyme activities and nucleoside transport is the most likely explanation for higher proportion of adenosine under normothermic conditions. Prolonged purine release after hypothermic ischemia may indicate continuation of nucleotide breakdown during reperfusion. The later phenomenon could be observed also under clinical conditions. A significant release of purine catabolites after heart transplantation was demonstrated even after 30 min of reperfusion.[4] It indicates that different approaches should be undertaken to protect nucleotide pool in the heart after normothermic and hypothermic ischemia.

ACKNOWLEDGMENTS

This study was supported by the British Heart Foundation (Grant No: PG 96194, the Polish State Committee for Scientific Research (W-922) and The Harefield Heart Transplant Trust.

REFERENCES

1. R.T. Smolenski, D.J. Chambers, and H.A. Simmonds, Exogenous adenosine supplied transiently during reperfusion ameliorates depressed endogenous adenosine production in the post-ischemic rat heart, *J. Mol. Cell. Cardiol.* 29:333(1997).
2. R.T. Smolenski, D.R. Lachno, S.J.M. Ledingham, and M.H. Yacoub, Determination of sixteen nucleotides, nucleosides and bases using high-performance liquid chromatography and its application to the study of purine metabolism in hearts for transplantation, *J. Chromatogr.* 527:414(1990).
3. E. Harmsen, P.P. De Tombe, and J.W. de Jong, Simultaneous determination of myocardial adenine nucleotides and creatine phosphate by high-performance liquid chromatography, *J. Chromatogr.* 230:131(1982).
4. R.T. Smolenski, A.-M.L. Seymour, and M.H. Yacoub, Dynamics of energy metabolism in the transplanted human heart during reperfusion, *J. Thorac. Cardiovasc. Surg.* 108:938(1994).

77

EFFECTS OF ADENINE/RIBOSE SUPPLY ON ADENOSINE PRODUCTION AND ATP CONCENTRATION IN ADENOSINE KINASE INHIBITED CARDIAC CELLS

Ryszard T. Smolenski,[1,2] Kameljit Kalsi,[1] Marek Zych,[2] Zdzislaw Kochan,[2] and Magdi H. Yacoub[1]

[1]Heart Science Centre
National Heart and Lung Institute at Harefield Hospital
Harefield, United Kingdom
[2]Department of Biochemistry
Medical University of Gdansk, Poland

INTRODUCTION

Adenosine is one of the most important metabolites of endogenous cardioprotective mechanisms. Coronary vasodilatation has long been recognized as the primary physiological activity of adenosine[1] but the range of its other cardioprotective properties extends to being antiadrenergic, antiplatelet and antileukocytic agent. In addition, adenosine plays an important role in angiogenesis and preconditioning.[2–6] Among factors which could stimulate adenosine production in the heart under normoxic conditions, combined inhibition of the adenosine kinase and adenosine deaminase activities was found to be the most efficient.[7–9] However, the exact metabolic effects and their duration in the cell have not been clearly defined. The objective of the present study was to develop the method to increase endogenous adenosine production in cardiomyocytes and endothelium without subjecting them to ischemia and avoiding any other deleterious metabolic effects. For this purpose we have evaluated the effect of simultaneous inhibition of adenosine kinase using 5′-iodotubercidin (ITu) and adenosine deaminase using erythro-9(2-hydroxy-3-nonyl)adenine (EHNA) with or without supply of adenine and ribose in isolated cardiomyocytes and cultured endothelial cells.

MATERIALS AND METHODS

Cardiomyocytes were isolated from rat hearts of Sprague-Dawley rats weighing 250–300 g using collagenase perfusion technique as described previously.[10,11] The final

Purine and Pyrimidine Metabolism in Man IX,
edited by Griesmacher *et al.* Plenum Press, New York, 1998.

preparation contained 70–85% of rod shaped cells. After isolation and purification, cells were finally suspended and incubated in a buffer containing 120 mmol/L NaCl, 2.6 mmol/L KCl, 1.2 mmol/L MgSO$_4$, 1.2 mmol/L KH$_2$PO$_4$, 1 mmol/L CaCl$_2$, 10 mmol/L HEPES (pH 7.4), 11 mmol/L glucose, 2 mmol/L pyruvate and 2% bovine serum albumin. Incubation was carried out for 15, 180 or 480 min under conditions indicated in the figure. Incubation was terminated by the addition of 25 μl of 10 mol/L HClO$_4$. Perchloric acid extracts were analyzed by HPLC as described below. Total cellular protein content was evaluated in myocyte suspension washed twice with albumin free incubation medium, solubilized in 0.5 mol/L NaOH using the method of Bradford. Endothelial cells were cultured after collagenase digestion of the heart. Endothelial cells after separation from cardiomyocytes by centrifugation were plated in plastic culture dishes and used after 2–3 passages. Cells were used for experiments after reaching confluence. Before experiments incubation medium was replaced by Hank's balanced salt solution (HBSS). Incubation was carried out for 15 or 180 min under conditions indicated in the figures or was immediately terminated in zero time samples. At the end of incubation, the medium was collected and cells extracted with 0.6 mol/L HClO$_4$. Acid extracts were then processed as described for cardiomyocytes and analyzed by HPLC. Protein concentration was evaluated as for

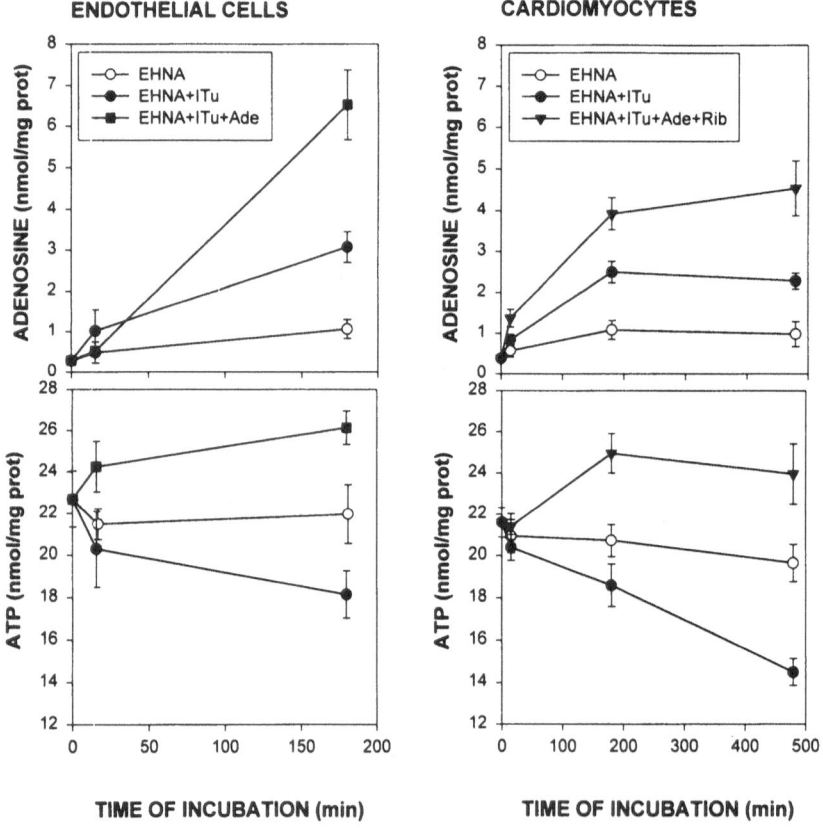

Figure 1. Concentration of adenosine and ATP in cultured rat heart endothelial cells incubated for 180 min in Hanks balanced salt solution or in isolated rat cardiomyocyte suspension incubated for 480 min in HEPES-buffered physiological medium. Incubations were carried out as follows: with 5 μM erythro-9(2-hydroxy-3-nonyl)adenine (EHNA) with EHNA and 10 μM 5'-iodotubercidin (EHNA + ITu), with EHNA, ITu and 100 μM adenine (EHNA + ITu + Ade), with EHNA, ITu, adenine and 2.5 mM ribose (EHNA + ITu + Ade + Rib). Values represent the means ± SEM, n = 6–12.

cardiomyocytes. The reversed-phase HPLC method used for determination of ATP, ADP, AMP, adenosine, inosine, hypoxanthine xanthine, uric acid, in the incubation medium and cell extracts has been described in detail previously.[12]

RESULTS

Adenosine production in endothelial cells with ITu and EHNA was increased to 2.8 ± 0.4 nmol/mg prot compared to 0.7 ± 0.2 nmol/mg prot with EHNA alone after 3 hr while ATP was decreased slightly (18.8 ± 1.6 vs. 20.6 ± 1.4 nmol/mg prot respectively) compared to initial 22.7 ± 1.3 nmol/mg prot. Adenosine production increased in the presence of adenine with ITu and EHNA after 3 hr (6.2 ± 0.9 nmol/mg prot) while ATP was elevated to 25.4 ± 1.1 nmol/mg prot. In the presence of ribose in addition to adenine, EHNA and ITu, changes in adenosine concentration and ATP were similar to incubation with adenine, EHNA and ITu.

Adenosine production in cardiomyocytes incubated with ITu and EHNA for 3 hr was enhanced compared to EHNA alone (2.2 ± 0.3 vs. 0.7 ± 0.2 nmol/mg prot respectively). ATP was decreased from initial 21.6 ± 0.7 to 18.6 ± 1.0 nmol/mg prot in the presence of ITu and EHNA, while ATP was maintained at 20.7 ± 0.8 nmol/mg prot with EHNA. With adenine, ITu and EHNA the changes were similar to those observed with ITu and EHNA (not shown). However, with ribose, adenine, ITu and EHNA, ATP increased to 24.5 ± 1.0 nmol/mg prot and adenosine production was elevated to 3.5 ± 0.4 nmol/mg prot. ATP further decreased after 8 hr of incubation with ITu while adenosine concentration did not increase. With ribose, adenine, ITu, and EHNA concentrations of ATP and adenosine were similar to concentrations after 3 hr.

DISCUSSION

We have demonstrated, that sustained increase in endogenous production of adenosine could be induced in the heart cells without disturbing energy equilibrium or altering nucleotide levels by combined application of the inhibitors of adenosine metabolism and precursors of purine base reutilization pathway. We have demonstrated that adenosine kinase and adenosine deaminase inhibition, a previously established approach to increase adenosine production under normoxic conditions, exerts only transient effect on adenosine production and causes depletion of ATP after prolonged exposure. The results of the present study demonstrated that concurrent administration of adenine (in endothelium) or adenine with ribose (in cardiomyocytes) totally prevented depletion of ATP after inhibition of adenosine kinase and further increased endogenous adenosine production. Our results highlighted also important differences between cardiomyocytes and endothelial cells in adenine reutilization, resulting most likely from differences in pentose phosphate pathway activity. Exogenous ribose administration was important to allow adenine incorporation into the nucleotide pool in cardiomyocytes, due to a very low endogenous production, while it was not essential in endothelial cells due to high endogenous supply.

ACKNOWLEDGMENTS

This study was supported by the British Heart Foundation (Grant No: PG 96194, the Polish State Committee for Scientific Research (St-42) and The Harefield Heart Transplant Trust.

REFERENCES

1. R.M. Berne, The role of adenosine in the regulation of coronary blood flow, *Circ. Res.* 47:807(1980).
2. G.R. Gunther and M.B. Herring, Inhibition of neutrophil superoxide production by adenosine released from vascular endothelial cells, *Ann. Vasc. Surg.* 5:325(1991).
3. D.C.B. Mills, D.E. MacFarlane, B.W.G. Lemmex, and R.J. Haslam, Receptors for nucleosides and nucleotides on blood platelets, in: "Regulatory Functions of Adenosine", R.M. Berne et al., Nijhof Publs, The Hague/Boston/London, pp. 277–289 (1983).
4. J. Schrader, G. Baumann, and E. Gerlach, Adenosine as inhibitor of myocardial effects of catecholamines, *Pflugers Arch.* 372:29(1977).
5. G.J. Grover, P.G. Sleph, and S. Dzwonczyk, Role of myocardial ATP-sensitive potassium channels in mediating preconditioning in the dog heart and their possible interaction with adenosine A1-receptors, *Circulation* 86:1310(1992).
6. M.B. Grisham, L.A. Hernandez, and D.N. Granger, Adenosine inhibits ischemia-reperfusion-induced leukocyte adherence and extravasation, *Am. J. Physiol.* 257:H1334(1989).
7. F. Bontemps, G. Van den Berghe, and G. Hers, Evidence for a substrate cycle between AMP and adenosine in isolated hepatocytes, *Proc. Natl. Acad. Sci. USA* 80:2829(1983).
8. D.R. Wagner, F. Bontemps, and G. Van den Berghe, Existence and role of substrate cycling between AMP and adenosine in isolated rabbit cardiomyocytes under control conditions and in ATP depletion, *Circulation* 90:1343(1994).
9. B.N. Cronstein, D. Naime, and G. Firestein, The antiinflammatory effects of an adenosine kinase inhibitor are mediated by adenosine, *Arthritis Rheum.* 38:1040(1995).
10. R.T. Smolenski, J. Schrader, H. de Groot, and A. Deussen, Oxygen partial pressure and free intracellular adenosine of isolated cardiomyocytes, *Am. J. Physiol.* 260:C708(1991).
11. R.T. Smolenski, A. Suitters, and M.H. Yacoub, Adenine nucleotide catabolism and adenosine formation in isolated human cardiomyocytes, *J. Mol. Cell Cardiol.* 24:91(1992).
12. R.T. Smolenski, D.R. Lachno, S.J.M. Ledingham, and M.H. Yacoub, Determination of sixteen nucleotides, nucleosides and bases using high-performance liquid chromatography and its application to the study of purine metabolism in hearts for transplantation, *J. Chromatogr.* 527:414(1990).

78

A NUCLEOTIDE AND NUCLEOSIDES MIXTURE SOLUTION IMPROVES RECOVERY OF ENERGY METABOLISM AFTER HEPATIC ISCHEMIA IN RATS

Makoto Usami, Mitsutoshi Ogino, Hiroshi Kasahara, Seiji Haji,
Akihiro Hirai, George Kotani, Yuko Kitamura, Yoshihiro Tagawa,
Hiroyoshi Sen, Atsunori Iso, Kazuya Sakata, Masahiro Yamamoto,
and Yoichi Saitoh

First Department of Surgery
Kobe University School of Medicine
7-5-2 Kusunoki-cho, Chuo-ku, Kobe, 650, Japan

INTRODUCTION

Hepatic ischemia/reperfusion is a common problem encountered under many clinical conditions, and it occurs in a variety of surgical procedures. Possible consequences include liver failure and/or multi-organ system failure, both of which have high rates of morbidity and mortality.[1,2] Efforts to minimize the extent of injury incurred during the ischemic period or to maximize reparative process following ischemia have importance for survival.[3] In vivo ^{31}P-nuclear magnetic resonance (NMR) spectroscopy has been used to follow the response of hepatic high-energy phosphates during hepatic ischemia/reperfusion.[4] It measures the continuous changes of hepatic adenosine triphosphate (ATP) levels and their metabolites, and is an effective tool to evaluate the serial data in the individual following drug treatment.

Drugs containing purine nucleotides have been used to improve hepatic energy metabolism during ischemia/reperfusion, but ATP itself has adverse haemodynamic effects.[5] A nucleoside-nucleotide mixture solution has been developed to supply purine and pyrimidine nucleosides in the organs. OG-VI is a solution composed of 30 mmol/l inosine, 30 mmol/l sodium 5'-guanylate, 30 mmol/l cytidine, 22.5 mmol/l uridine and 7.4 mmol/l thymidine.[6] We have previously reported the effects of OG-VI to facilitate DNA and RNA syntheses of hepatocytes and hepatic regeneration after partial hepatectomy of the normal liver, and to improve protein fractional synthetic rates in the cirrhotic liver.[7-9] Recently, a protective effect on stunned myocardium has been reported.[10-12] OG-VI has a positive

Purine and Pyrimidine Metabolism in Man IX,
edited by Griesmacher *et al.* Plenum Press, New York, 1998.

inotropic action, a marked cardiac protective effect, and an effect to improve the myocardial metabolic derangement induced by ischemia

The aim of this study is to evaluate the effect of intravenous administration of OG-VI on hepatic purine metabolism after ischemia.

MATERIALS AND METHODS

Male Wistar rats (each weighing 200–300 g) under general anesthesia using intraperitoneal secobarbital injection were laparotomized. A BEM 250/80 system (2.0 Tesla, Otsuka Electric Co., Tokushima, Japan) was used for in vivo NMR analysis. A single turn 15-mm diameter surface coil tuned to 34.5 MHz for ^{31}P was placed on the left lateral lobe of the liver. ^{31}P-NMR spectra were collected with a 30 microsec pulse width, 60 W pulse, 0.5 sec repetition time, and 600 aquisitions. Beta-ATP, phosphomonoester (PME), and Pi were evaluated as the percent changes of value before treatment. Intracellular pH was calculated using the chemical shift of Pi and alpha-ATP as the internal standard.

Vascular clamping of the left branch of portal triad for 30 or 60 minutes was carried out to avoid portal congestion2. Changes in ^{31}P-NMR spectra of the left lateral lobe during ischemia/reperfusion were evaluated. 4.16 ml/kg/hr of diluted OG-VI solution (a kind gift from Otsuka Pharmaceutical Factory, Tokushima, Japan, OG group), with a final nucleoside concentration of 2.4 mM, or physiologic saline (C group) was injected via cervical vein continuously from 30 min before vascular clamping until sacrifice. Biochemical analyses of adenine nucleotides of the freeze clamped liver by the Jaworek's enzymatic method and routine blood biochemical tests by the automated analyzer were carried out at the endpoint of reperfusion.

All data was expressed as mean ± standard deviation and Student's *t* test was used to compare two samples.

RESULTS

Representative ^{31}P NMR spectra in a rat that underwent 60-min ischemia and 90-min reperfusion are shown in Fig. 1. Six peaks of PME, Pi, phosphodiester, gamma-ATP, alpha-ATP and beta-ATP were identified. Decreased ATP peak, increased Pi and PME peak, right shift of Pi peak were observed during ischemia. However, these changes recovered during reperfusion.

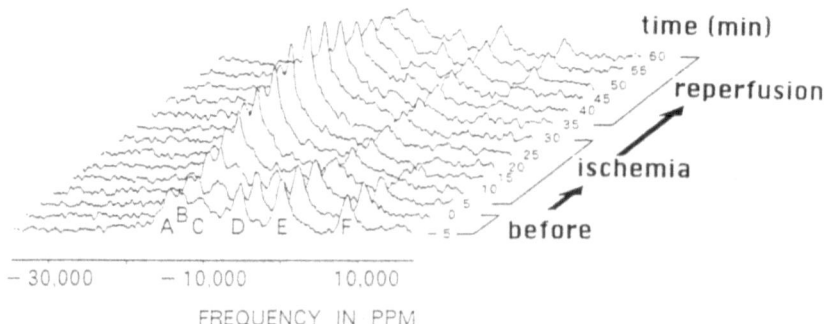

Figure 1. ^{31}P NMR spectra of the liver in the OG group during 60 min ischemia and 90 min reperfusion. A: phosphomonoester (PME), B: Pi, C: phosphodiester, D: gamma-ATP, E: alpha-ATP, F: beta-ATP peaks are identified.

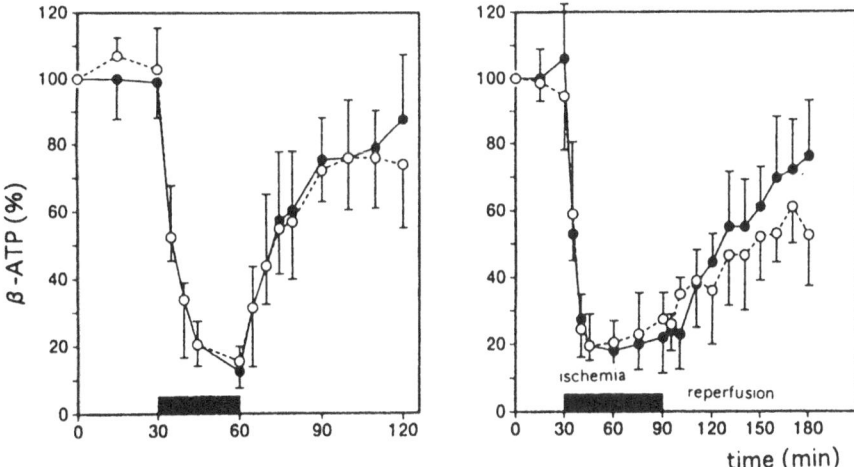

Figure 2. Changes in hepatic beta-ATP level during ischemia/reperfusion. Left: 30 min ischemia and 60 min reperfusion, right: 60 min ischemia and 90 min reperfusion n = 6, mean ± SD.

The ischemia caused beta-ATP depletion to approximately 20% of the preischemic state (Fig. 2). Recovery of the beta-ATP level was better in the OG group. Percent changes of beta-ATP in the OG group was 70.8 ± 18.1% versus 57.3 ± 23.3% in the C group after 60-min ischemia and 90-min reperfusion. Hepatic ATP level by the enzymatic measurement in the OG group at 1.88 ± 0.29 micromol/g liver was higher than the C group, 1.38 ± 0.42 micromol/g liver (p < 0.05). AMP level was lower in the OG group, 0.22 ± 0.06 micromol/g liver, than in the C group, 0.30 ± 0.08 micromol/g liver (p < 0.01).

Changes in PME + Pi peak level during reperfusion in both the 30-min ischemia group and the 60-min ischemia group showed facilitated recovery in the OG group (p < 0.05, Fig. 3). Chemical shift in Pi also showed facilitated recovery during reperfusion in the OG group after 60-min ischemia (p < 0.01–0.05, Fig. 4). Results of blood biochemical

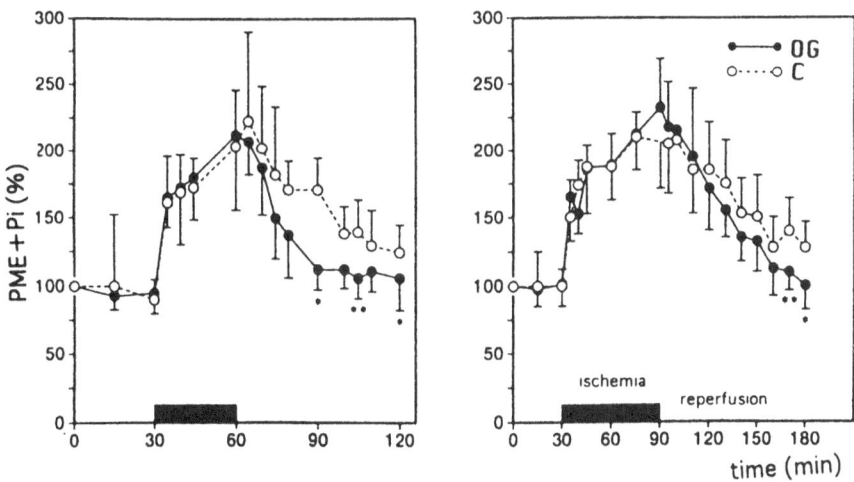

Figure 3. Changes in hepatic PME + Pi level during ischemia/reperfusion. Left: 30 min ischemia and 60 min reperfusion, right: 60 min ischemia and 90 min reperfusion, n = 6, mean ± SD, *: p<0.05 vs. C group.

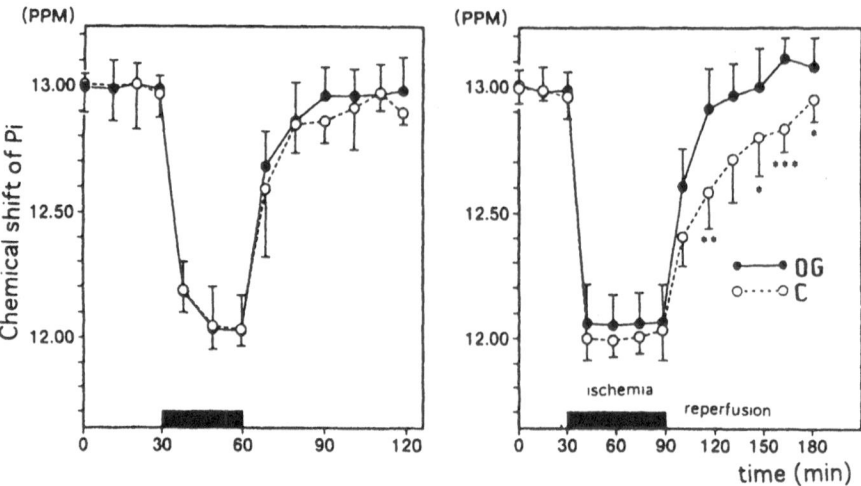

Figure 4. Changes in hepatic chemical shift of Pi level during ischemia/reperfusion. Left: 30 min ischemia and 60 min reperfusion, right: 60 min ischemia and 90 min reperfusion, mean ± SD, **: $p < 0.01$, *: $p < 0.05$ vs. C group.

testing showed the improvement of increased GOT, GPT, CPK, ALP, lactate and pyruvate levels had no statistical significance.

DISCUSSION

Hypoxia stops oxidative phosphorylation in the mitochondria, then decreases ATP level during ischemia. Subsequent dephosphorylation increases Pi level and the increased anaerobic glycolysis increases PME level. Signal intensity in ^{31}P NMR spectra reflects the tissue levels of each component during ischemia/reperfusion, however, the sum of PME + Pi was used to cover the difficulty in distinguishing the peaks in young rats with sensitive metabolic changes.

OG-VI increased ATP level 1.3-fold against the C group in both the beta-ATP peak and hepatic ATP contents after reperfusion. Also, facilitated recovery of PME + Pi accumulation and metabolic acidosis shown by the chemical shift of Pi was observed in the OG group. These effects are considered metabolic effects, because OG-VI did not increase hepatic blood flow measured by laser doppler flowmetry (data not shown). The increased amount of ATP roughly calculated from rat body weight using the estimate of 68% of the liver as the ischemic lobe weight and 40 g/kg body weight as the liver weight, was 13.6 micromol/kg body weight. However, this amount was two times the total dose of inosine, 7.45 micromol/kg. That is, the increased amount of ATP exceeds the administered amount of inosine, the substrate for salvage ATP synthesis. And, it is not realistic to speculate that all nucleotide and nucleoside materials in the OG-VI solution are metabolized to ATP.

From these results, it is suggested that OG-VI not only improves energy metabolism in the mechanism to supply substrates for the salvage purine synthesis pathway, but also facilitates the glycolytic process and metabolism in the TCA cycle, that is, the pharmacological effects to improve high energy phosphate metabolism during ischemia/reperfusion. This speculation is in accordance with those of short time cardiac ischemia. In myocardial ischemia, OG-VI caused a dose-dependent improvement in the myocardial contractile function during reperfusion after brief ischemia.[10-12] OG-VI decreased ischemia-induced derangement of carbohydrate metabolism and attenuated the ratio ([G6P] + [F6P])/[FDP], indicating that inhibition of glycolysis after ischemia occurs via the phosphofructokinase reaction.[12]

In conclusion, continuous infusion of OG-VI is effective in improving hepatic energy metabolism during reperfusion after ischemia.

REFERENCES

1. D.V. Feliciano, K.L. Mattox, G.L. Jordan, J.M. Burch, C.G. Bistondo, P.A. Cruse, Management of 1000 consecutive cases of hepatic trauma (1979–1984), *Ann Surg.* 204:438 (1986).
2. M. Usami, K. Furuchi, H. Shiroiwa, Y. Saitoh, Effect of repeated portal-triad cross-clamping during partial hepatectomy on hepatic regeneration in normal and cirrhotic rats, *J Surg Res .57*: 541 (1994).
3. M.G. Clemens, P.F. McDonagh, I.H. Chaudry, and A.E. Baue, Hepatic microcirculatory failure after ischemia and reperfusion: improvement with ATP-MgCl2 treatment, *Am J Physiol .* 2481:H804 (1985).
4. C.R. Malloy, C.C. Cunningham, and G.K. Radda, The metabolic state of the rat liver in vivo measured by ^{31}P NMR spectroscopy, *J Biochem Biophys Acta* 885:9276 (1986).
5. W. Ertel, G. Singh, MH. Morrison, A. Ayala, IH. Chaudry, Chemically induced hypotention increases PGE2 release and depresses macrophage antigen presentation, *Am J Physiol* 264:R655 (1993).
6. S. Ogoshi, M. Iwasa, T. Yonezawa, and T. Tamiya, Effect of nucleotide and nucleoside mixture on rats given total parenteral nutrition after 70% hepatectomy, *J Parenter Enteral Nutr.* 9:339 (1985).
7. H. Ohyanagi, S. Nishimatu, Y. Kanbara, M. Usami, and Y. Saitoh, Effect of nucleosides and a nucleotide on DNA and RNA syntheses by the salvage and de novo pathway in primary monolayer cultures of hepatocytes and hepatoma cells, *J Parenter Enteral Nutr.* 13:51 (1989).
8. M. Usami, K. Furuchi, M. Ogino, H. Kasahara, T. Kanamaru, Y. Saitoh, H.Yokoyama, and S. Kano, The effect of a nucleotide-nucleoside solution on hepatic regeneration after partial hepatectomy in rats *Nutr.* 12:797 (1996).
9. M. Usami, A. Iso, H. Kasahara, G. Kotani, S. Haji, T. Kanamaru, and Y. Saitoh, Effect of a parenteral nucleoside-nucleotide mixture on hepatic metabolism in partially hepatectomized cirrhotic rats, *Nutr.12:436* (1996).
10. K. Satoh, T. Nakai, H. Kohri, and K. Ichihara, Limitation of stunning in dog myocardium by nucleoside and nucleotide mixture, OG-VI. *Coron Artery Dis.* 4:1007 (1993).
11. S. Kano, T. Nakai, H. Kohri, and K. Ichihara, Effects of OG-VI, a nucleoside/nucleotide mixture, and its constituents on myocardial stunning in dogs. *Coron Artery Dis.* 6:811 (1995).
12. Y. Okazaki, S. Kano, S. Ogoshi, and K. Ichihara, Effects of OG-VI, a nucleoside-nucleotide mixture, on ischemic myocardial metabolism in dogs. *Coron Artery Dis.* 8:39 (1997).

79

CAN ADENOSINE DEAMINASE INHIBITORS BE CYTOPROTECTIVE AGENTS?

Jerzy Barankiewicz, Anne M. Danks, and Paul J. Marangos

Cypros Pharmaceutical Corporation
2714 Loker Ave. West
Carlsbad, California 92008

INTRODUCTION

Adenosine (Ado) is a well recognized nucleoside that possesses a number of beneficial properties including cardioprotective, cerebroprotective, anti-inflammatory and analgesic activities.[1]

Since elevation of Ado concentration can affect many important physiological parameters, the biochemical challenge to pharmacologically manipulate Ado concentrations has attracted many laboratories, both pharmaceutical and academic. Although a number of attempts have been made to pharmacologically elevate Ado concentrations, none has been clinically accepted, mainly due to insufficient elevation of Ado concentrations or unacceptable toxicity.

It has been shown in many different cells and tissues that inhibition of adenosine deaminase (ADA) results in high elevation of Ado.[2-6] ADA is predominantly a cytoplasmic enzyme, but can also aggregate with extracellular ADA-binding protein (ADA-BP) to form extracellular ADA. The extracellular ADA seems to be responsible for controlling Ado level near Ado receptors.[7] Although ADA inhibitors are well-known Ado elevating agents, they have not been considered for development as cytoprotective agents, mostly due to the toxic effects expressed by 2'-deoxycoformycin (DCF), the most potent and specific ADA inhibitor.[8] The depletion of T-cells observed with DCF and development of SCID (severe combined immunodeficieny disease) in children having a genetic lack of ADA activity[9] have led to consideration of ADA inhibitors as immunotoxic, and have led to their use as an antileukemic drugs.[10] Toxicity of DCF seems obvious, since it is a tight-binding ADA inhibitor which inhibits ADA to the extent that enzyme resynthesis is required to regenerate its activity.[11] However, in contrast to DCF, which is practically an irreversible ADA inhibitor, other ADA inhibitors: erythro-9-(2-hydroxy-3-nonyl)adenine (EHNA) and its analogs produce a reversible inhibition of ADA. These compounds are potent and specific ADA inhibitors, they potently elevate Ado and when removed from the

Purine and Pyrimidine Metabolism in Man IX,
edited by Griesmacher *et al.* Plenum Press, New York, 1998.

cell environment, ADA activity is recovered. They have also been shown to exert cardio- and cerebro-protective effects.

INHIBITION OF ADA BY EHNA AND ITS ANALOGS

A number of EHNA analogs are potent ADA inhibitors. They inhibit ADA activity in cell-free environments (enzyme isolated from calf spleen), as well as in intact cells (human astrocytoma cells-HAC) (Table 1).

Similar range of Ki has been reported by others for calf intestinal ADA (6.5 nM)[12] or for human erythrocyte ADA (1.6 nM).[13] The most potent EHNA analogs in both the cell-free and cell-based studies were 9'-phthalimido-EHNA and 9'-chloro-EHNA. The compounds studied so far seem to be specific ADA inhibitors, since 9'-chloro-EHNA and 9'-phthalimido-EHNA do not affect adenosine kinase activity or nucleoside transport.[14]

ELEVATION OF ADENOSINE CONCENTRATIONS BY EHNA AND ITS ANALOGS

EHNA and its analogs are potent agents which are able to elevate Ado concentrations in stress conditions. They stimulate Ado accumulation when degradation of ATP is accelerated in anaerobic conditions or by addition of 2-deoxyglucose (Table 1) In contrast, no elevation of Ado by EHNA and its analogs is observed under normoxic conditions. Elevation of Ado by these compounds is dose-dependent, and elevation of Ado release is well correlated with the degree of ADA inhibition. In addition, in ischemic tissues, the degree of Ado accumulation when ADA is inhibited by EHNA or its analogs can depend on the ratio of alternative AMP catabolism: via AMP deamination or AMP dephosphorylation and subsequent Ado deamination. It is also known that ADA is in the predominant pathway in adenine nucleotide degradation in stressed brain or in hypoxic heart.[15,16] In the cases where the predominant pathway involves Ado deamination, ADA inhibition will result in high elevation of Ado under stress conditions, whereas in the cells in which AMP deami-

Table 1. Inhibition of ADA and elevation of Ado concentrations by EHNA and its analogs

| | ADA inhibition | | Ado release from HAC[3] | |
| | Isolated type IX calf spleen ADA Ki(nM)[1] | Intact HAC IC$_{50}$ (μM)[2] | | |
Compound			Hypoxia[4]	dGlu[4]
EHNA	9.3	1.45 ± 0.1	327 ± 47	328 ± 25
9'-hydroxy-EHNA	8.0	12.72 ± 3.2	207 ± 33	289 ± 12
9'-chloro-EHNA	2.7	2.61 ± 0.5	470 ± 26	375 ± 25
9'-phthalimido-EHNA	0.95	0.75 ± 0.3	416 ± 61	559 ± 19
8',9'-didehydro-EHNA	7.4	4.08 ± 0.9	323 ± 79	300 ± 48

[1]ADA inhibition was determined spectrophotometrically and Ki was calculated from Lineweaver-Burk plots.
2IC$_{50}$ was determined in human astrocytoma cells (HAC) incubated with radiolabelled Ado in the presence of AK inhibitor: 10 μM 5'-iodotubercidin.
[3]Ado release was determined in cells containing radiolabelled ATP and anaerobically stressed by addition of 5.5 mM 2-deoxyglucose in the presence or absence (control) of EHNA or its analogs.
[4]% control.

nation is the predominant pathway, inhibition of ADA will result in only minor Ado increase, especially since AK activity will consume the small amounts of Ado formed.[17] Since a significant portion of Ado is deaminated by extracellular ADA in the brain and coronary arterial endothelial cells,[18,19] inhibition of extracellular ADA by EHNA and its analogs would also highly elevate Ado concentration in these tissues.

Therefore, the extent to which Ado level rises in tissues when ADA is inhibited depends upon a multitude of factors, such as energy state, degree of ischemia, and will vary among tissues.

REVERSIBILITY OF ADA INHIBITION AND IMMUNOTOXICITY

Although DCF elevates Ado concentration, it can not be considered for development as a cardioprotective or cerebroprotective compound due to practically irreversible ADA inhibition and toxic immunosuppresive effects. The mechanism of DCF toxicity includes: accumulation of 2'-deoxyAdo which inhibits S-adenosylhomocysteine hydrolase and, consequently transmethylation reactions; conversion of dAdo to dATP and as a result the inhibition of ribonucleotide reductase and DNA synthesis, especially in T-lymphocytes.[9]

In contrast, inhibition of ADA by EHNA or its analogs is reversible, and ADA activity is recovered after simply removing the inhibitors from the cell culture medium by washing. In a similar manner, EHNA or its analogs would be removed from tissues during blood perfusion. The ADA activity recovered after EHNA or its analogs inhibition seems to result in reversing the toxicity to T-lymphocytes (Table 2).

FUNCTIONAL ROLES OF EHNA AND ITS ANALOGS

The question: *Does ADA inhibition protect ischemic myocardium* was discussed previously by Gallagher and coworkers.[20] These authors concluded that ADA inhibition leads to a striking elevation of Ado concentration in myocardium that can be beneficial in ischemic injury in terms of functional recovery. However, no beneficial effects of elevation of Ado by ADA inhibitors was observed when ischemia was more severe, ending in infarction (irreversible injury). An explanation for this may be that nucleosides trapped in myocardium during ADA-inhibition can be re-utilized as nucleotides for re-establishment

Table 2. Reversibility of ADA inhibition and T-cell toxicity

Compound	ADA activity in intact HAC after removing inhibitor[1]*	Incorporation of thymidine into T-Cell DNA after removing inhibitor[2]*
DCF (1 μM)	4.7 ± 0.1	6.8 ± 5.2
EHNA (40 μM)	97.9 ± 2	128.6 ± 22
9'-chloro-EHNA (40 μM)	90.0 ± 1	133.7 ± 30
9'-phthalimido-EHNA (40 μM)	81.0 ± 1	66.0 ± 16

[1]HAC cells were incubated for 1 hour with or without (control) ADA inhibitors and then washed 4 times with fresh medium. ADA activity was measured after incubation with radiolabelled Ado in the presence of AK inhibitor: 10 μM 5'-iodotubercidin.
[2]After incubation of Jurkat cells with or without (control) ADA inhibitors for 2 hours, cells were washed 4-5 times and then incubated with 50 μM deoxyAdo for 24 hours and with radiolabelled thymidine for 18 hours.
*Denotes % of control.

Table 3. Cardiac protection by EHNA and its analogs in Langendorff ischemia-reperfusion injury in rat hearts (% recovery [20 minutes post-reperfusion])

	EHNA		9'-Chloro-EHNA		9'-Phthalimido-EHNA	
	10 μM	30 μM	10 μM	30 μM	10 μM	30 μM
LVDP	191.7	125.0	191.7	220.8	191.7	241.7
LVEDP	92.8	89.9	81.2	66.7	79.7	71.0
+dp/dt max	186.3	131.5	210.7	238.1	200.8	240.0
−dp/dt max	142.0	113.8	173.4	186.6	170.1	181.2

Rat hearts, after a stabilization period, were perfused for 10 minutes with EHNA or its analogs. Global ischemia was achieved by clamping the aortic cannula. After 35 minutes of global ischemia, hearts were reperfused and measurements of hemodynamic heart performance were taken at 5-minute intervals. n = 8, $p < 0.05$.

of energy status only in reversible ischemia; not in irreversible injury. Although both ADA inhibitors, DCF and EHNA, lead to striking elevation of Ado within ischemic areas, thereby fulfilling the criteria of site and event specificity, the concept of using them for cardio- or cerebroprotection has not been systematically studied and developed due to the immunotoxicity observed for the most potent and specific ADA inhibitor, DCF.[8,9]

We have developed the concept[14] that because the EHNA-based ADA inhibitors block ADA activity in a reversible manner, they should be less toxic. Such inhibitors should be specific ADA inhibitors which highly elevate Ado concentrations only in ischemic conditions, show cardio/cerebroprotection, and are less toxic due to their reversible action on ADA activity.

Considering that EHNA and its analogs elevate Ado concentration and are apparently lacking immunotoxicity, their protective effects on ischemic tissues were recently studied in two models: Langendorff isolated heart and hypoxic hippocampal slices.[14] In both models, EHNA derivatives showed protective effects (Table 3 and Table 4). One plausible explanation for the observed protective effects is their action via Ado elevation. Use of ADA inhibitors results not only in the elevation of Ado concentrations, but also in reduced loss of adenine bases from ischemic myocardium[6] where ADA reaction is a major route of ATP catabolism.[16] It has also been reported that when EHNA was administrated together with NBMPR (p-nitrobenzylthioinosine) the nucleoside trapping was correlated with improved functional recovery after sustained or repetitive ischemia.[21] EHNA also improved functional preservation of the hypothermically stored rat hearts.[22]

In addition, inhibition of ADA can have protective effects on ischemic tissues by preventing free radical-mediated injury.[23] It has been reported that inhibition of ADA by

Table 4. Protection of hippocampal slices during hypoxic injury by EHNA and 9'-phthalimido-EHNA

Compound	Concentration (μM)	Hippocampal slices*
EHNA	10	50.00 ± 5.00
	20	116.00 ± 19.00
9'-phthalimido-EHNA	10	57.10 ± 1.80
	20	93.00 ± 4.10

Rat hippocampal slices were incubated with EHNA or its analog for 30 minutes before hypoxia and through the first 15 minutes of return to oxygenated conditions. $p < 0.05$.

*(% recovery of population spike ± SEM)

EHNA reduced formation of hypoxanthine and xanthine (substrates for xanthine oxidase) and blocked free radical generation. This resulted in greater than 2-fold recovery of contractile function of post-ischemic heart.

Additional mechanistic questions remain regarding the cytoprotective action of adenosine. However, it is clear that specific and selective alteration of its degradation in metabolically stressed tissues should be of therapeutic benefit. The reversible ADA inhibitors, EHNA and its analogs, represent such a pharmacological tool. They may represent a class of reversible ADA inhibitors that have the potential to be less toxic than DCF. This, coupled with the cardioprotective and neuroprotective activities observed with 9'-chloro-EHNA and 9'-phthalimido-EHNA suggests that this series of compounds may have clinical utility in a range of ischemic disorders.

In conclusion, use of EHNA and its analogs as cardio or cerebroprotective agents may be a promising approach to cytoprotection of ischemic tissues, and additional investigation in animal models should facilitate the practical pharmacological development of these compounds.

REFERENCES

1. Adenosine and Adenine Nucleotides: From Molecular Biology to Integrative Physiology Eds. Belardinelli, L., Pelled, A., Kluwer Academic Publishers, Boston, Dordrecht, Lonodn, 1994.
2. Achtenberg P.W.. Harmsen E.E. and De Jong J. W., Cardiovasc. Res. 19, 593, 1985.
3. Barankiewicz, J. and Cohen A., Biochem. J. 219, 197, 1984.
4. Zoref-Shani, E., Kessler-Icekesoen, G., and Sperling, O., J. Mol. Cell. Cardiol. 20. 23,1988.
5. Van den Berghe, G., Bontemps, F. and Hers, H.G., Biochem. J. 188, 913, 1980
6. Sandhu, G.S., Burrier, A.C. and Janero, D.R., Am. J. Physiol. 265, H1249, 1993.
7. Schrader, W.P., West, C.A., J. Histochem. Cytochem. 33, 508, 1985.
8. Dostal, L.A., Brown, S., Bleck, J., Anderson, J.A., Teratology 44, 325,1991.
9. Gelfand E. W. and Cohen,A., Adv. Host Defense Mechanisms 2, 43,1983.
10. Spears A.S.D.,Moor, D., Cassileth, P.A., Harrington, D.P., Cummings, F.J., Neiman, R.S., Bennett, J.M., O'Connell, M.J., N. Engl. J. Med. 316, 825, 1987.
11. Padua, R.A., Geiger, J.D., Delaney, S. M., and Nagy, J.I., J. Neurochem. 58, 421, 1992.
12. Agarwal, R.P., Spector, T., Parks Jr, R.E., Biochem.Pharmac. 26, 359, 1977.
13. Lambe, C.U., Nelson, D.J., Biochem. Pharmac. 31, 535, 1982.
14. Barankiewicz, J., Danks, A.M., Abushanab, E., Makings, L., Wieman, T., Wallis, R.A., Pragnacharyulu, P.V.P., Fox, A., Marangos, P., Sumitted to: J. Pharm. Exp. Therapeutics, 1997.
15. Maire, J.C., Medilanski, J., Straub, R.W., J. Physiol. 357, 67, 1984.
16. Smolenski, R. T., Suitters, A., Yacoub, M.H., J. Mol. Cell. Cardiol. 24, 91, 1992.
17. Barankiewicz, J., Ronlov, G., Jimenez, R., Gruber, H.E., J. Biol. Chem. 265, 15738, 1990.
18. Franco, R., Cane;a, E.J., Bozal, J., Neurochem. Res. 11, 423, 1986.
19. Meghji, P., Middleton, K., Hassall, C.J.S., Phillips, M.I., Newby, A.C., Int. J. Biochem. 20,1335, 1988.
20. Gallagher K.P.,McClanahan, T.B., Martin, B.J., Saganek, L.J., Ignasiak, D.P., Mertz, T.E., Van Wylen, D.G.L., Vinten-Johansen, J., Adv. Exp. Med. Biol. 370, 291, 1995.
21. Abd-Elfattah, A.S., Ding, M., Wechsler, A.S., J. Thorac Cardiovasc. Surg. 110, 328, 1995.
22. Zhu, Q., Yang, X., Glaydon, M.A., Hicks, G.L., Wang, T., Transplantation 57, 35, 1994.
23. Xia, Y., Khatchikian, G. and Zweier, J. L., J. Biol. Chem. 271, 10096, 1996.

80

ROLE OF PURINE METABOLISM IN REGULATION OF SIGNAL TRANSDUCTION IN HUMAN CARCINOMA CELLS

George Weber,[1] Fei Shen,[1] and Wei Li[2]

[1]Laboratory for Experimental Oncology
[2]Walther Oncology Center
Indiana University School of Medicine
Indianapolis, Indiana

The purpose of this state-of-the-art lecture is to outline the evidence that there is an increased capacity for purine biosynthesis in cancer cells and that this heightened ability is linked with the up-regulation of signal transduction in human cancer cells.

INCREASED CAPACITY FOR IMP PRODUCTION

Extensive evidence provided in this Laboratory[1-4] indicates that in various cancer cells the enzymic activities and amounts in the *de novo* biosynthesis of IMP (amidophosphoribosyl-transferase, FGAR amidotransferase and adenylosuccinase) were increased. By contrast, the activities and amounts of the enzymes of IMP degradation, inosine phosphorylase and xanthine oxidase, were markedly reduced. There is evidence for the operation of reciprocal regulation of opposing rate-limiting enzymes in the elevation of amidophosphoribosyl-transferase and the decrease of xanthine oxidase activities and amounts. As a result of this imbalance in gene expression, there is an increased ability to produce IMP in cancer cells.

INCREASED CAPACITY FOR THE SYNTHESIS OF GTP AND dGTP

In the utilization of IMP for GTP biosynthesis, there is an increased activity and amount of IMP dehydrogenase (IMP DH), the rate-limiting enzyme of GTP production. There is also an increase in the activity of GMP synthase and nucleosidediphosphate kinase and in the activity of the salvage enzyme for guanine, guanine phosphoribosyltransferase. Ribonucleotide reductase activity is also markedly increased in cancer cells.[1] As a result of this reprogramming of gene expression there is an increased capacity to make

GTP and dGTP and a decreased ability to degrade the nucleotides. These alterations should confer selective advantages to cancer cells. Therefore, we proposed that IMP DH should be a sensitive target to chemotherapy.[2,5] This idea was tested in the clinical treatment of leukemia with good therapeutic results (see below).

STRATEGIC BIOLOGICAL FUNCTIONS OF GTP

Apart from its role in intermediary metabolism, in biosynthesis of RNA, proteins, biopterins, UTP and tubulin, GTP is an intricate part of the signal transduction mechanisms, production of cGMP and adenylates, G-protein action and the expression of the *ras* oncogene family. Since guanylates are indispensable in DNA biosynthesis a selective curtailing of GTP and dGTP pools is an important chemotherapeutic objective. The biological significance of guanylate biosynthesis is also apparent in that in liver cancer cells the concentration of GTP is the lowest as compared to that of ATP, CTP and UTP. The concentration of dGTP is the lowest among the deoxynucleosidetriphosphates. Thus, GTP and dGTP are rate-limiting pools in DNA biosynthesis in hepatoma cells.[1] In GTP biosynthesis in various normal and neoplastic tissues the activities of the salvage enzymes were markedly higher than those of the rate-limiting enzymes, amidophoribosyltransferase and IMP DH. During the transition of cancer cells from plateau and lag phases to logarithmic proliferation the ratio of guanylate to adenylate synthesis for IMP sharply changed[5], as indicated by a marked increase in the relative labelling of guanylates with a concurrent marked decrease in that of the adenylates. The striking redirection of the label to the preferential synthesis of guanylates indicates the significance of the guanylate biosynthetic pathway as a target for chemotherapy (Figure 1).

Treatment of hepatoma cells with various inhibitors of IMP DH, including tiazofurin or ribavirin to block *de novo* guanylate synthesis, accelerated the flux activity of the guanine salvage pathway. The rise of the salvage activity in response to drugs targeted against the *de novo* pathway highlights the important role salvage synthesis might play in circumventing the impact of inhibitors of *de novo* purine synthesis in cancer chemotherapy.

The significance of GTP and dGTP in cancer biochemistry and chemotherapy was highlighted by discovering in this laboratory that IMP DH activity was increased in various murine and human cancers and was particularly high in rapidly growing neoplastic cells such as leukemic cells.[1-4] The increased ability of cancer cells to produce GTP and dGTP is underlined by the elevated activities of the enzymes involved in their biosynthesis and by the decrease in the activities of the catabolic enzymes.

MOLECULAR BIOLOGY OF IMP DH

Because of the significance of IMP DH extensive work was carried out on the molecular biology of this rate-limiting enzyme.[6-8] In this laboratory two distinct cDNA clones were isolated, types I and II, encoding IMP DH. This was the first report suggesting the existence of two distinct types of human IMP DH molecular species. Subsequent work showed that the total amounts of type I and II enzymes increased in human leukemic cell lines K562 and HL-60 in parallel with the increase in IMP DH activity to 7.8- and 9.4-fold, respectively, above that of normal lymphocytes. The expression of type I isozyme was unchanged indicating that the increase in the total IMP DH amount in leukemic cells was due to specific up-regulation of the type II enzyme. Northern blot analysis showed also the specific and predominant expression of type II enzyme in the leukemic cells.

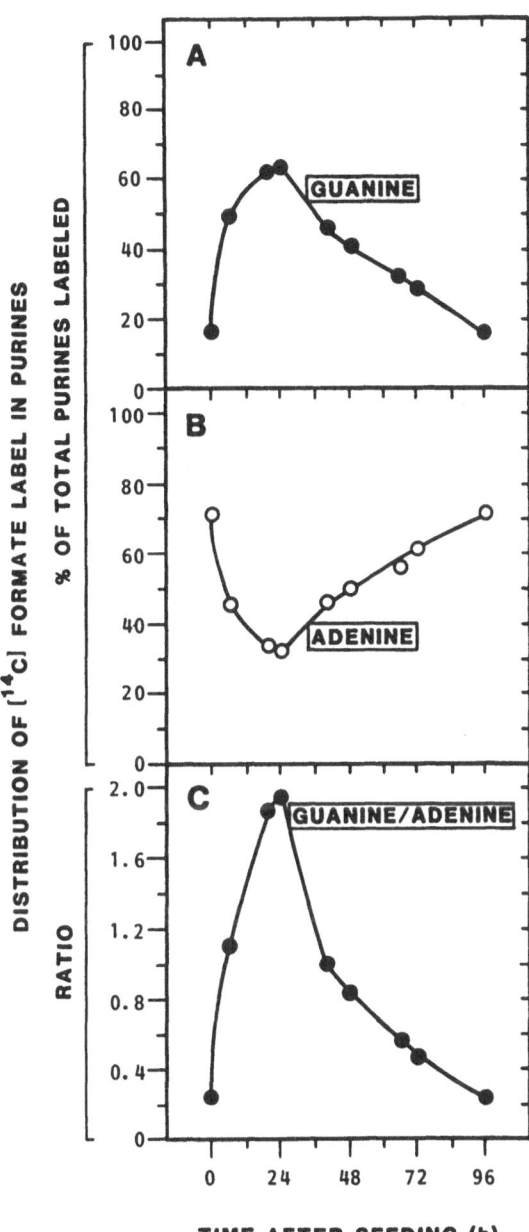

Figure 1. Relative utilization of IMP for adenylate and guanylate synthesis in proliferating hepatoma cells.

Northern blots revealed that type II IMP DH was more active transcriptionally (1.5- to 5.1-fold) in all the leukemic cells examined than in normal lymphocytes, whereas type I expression was not changed. The increased expression of type II mRNA in human leukemic cells was closely linked with the increase in total IMP DH activity (r = 0.92).[8] Treatment of leukemic blasts with the differentiating agent phorbolacetate resulted in a 90% decrease in the expression of type II mRNA with macrophage-like maturation; the expression of type I mRNA was relatively stable. Thus, the expression of type II IMP DH is stringently linked with the immature characteristics of leukemic cells and, therefore, it should be a sensitive target for antileukemic chemotherapy.[8]

The low level of type II mRNA in normal lymphocytes was up-regulated by phyto-hemagglutinin stimulation (3.2-fold) and Epstein-Barr viral transformation (5.7-fold) in quiescent HL-60 cells and type II mRNA expression was 2.8-fold elevated by serum stimulation. By contrast, the up-regulated level of type II IMP DH mRNA in HL-60 cells was reduced to less than 5% along with differentiation induced by retinoic acid, phorbol myristate acetate or DMSO independently of the end-stage phenotype. Type I IMP DH mRNA was constitutively expressed in the various states of proliferation and differentiation. These observations underline the proliferation-linked expression of type II IMP DH and indicate different regulatory mechanisms that govern expression of the genes for type I and type II enzymes.[8]

CLINICAL APPLICATION OF THE DISCOVERY OF BIOCHEMICAL IMBALANCE IN CANCER CELLS

The discovery of increased IMP DH activity in cancer cells prompted us to suggest that this enzyme should be a sensitive target to selective chemotherapy.[2,5] The activity of the guanine salvage enzyme was also increased in leukemic cells. In normal cells the activities of IMP DH and the salvage enzyme, GPRT, were 3.1 and 389 nmol/h/mg protein, respectively. In various human leukemic blast cells IMP DH activity increased 15- to 41-fold and GPRT activity 2.8- to 6.8-fold. After reading our suggestion that inhibitors of IMP DH were needed for clinical studies, Roland Robins prepared a series of nucleosides which included tiazofurin and ribavirin.

ROLE OF GTP IN TIAZOFURIN ACTION

We showed that tiazofurin killed rat hepatoma cells in culture with an $LD_{50} = 5$ μM. The tightness of linkage of IMP DH activity with guanylate metabolism was elucidated in rats carrying transplanted hepatomas injected i.p. with tiazofurin. In rats carrying subcutaneously transplanted hepatoma 3924A, tiazofurin (150 mg/kg i.p. daily for 5 days) caused an 85% inhibition of tumor growth as measured 14 days after tumor implantation. A single tiazofurin i.p. injection (200 mg/kg) caused a rapid reduction to 35% in IMP DH activity in the hepatoma and a concurrent 15-fold rise in the concentration of the substrate, IMP. The pools of GMP, GDP and GTP declined to 30% and dGTP concentration to 15% of the preinjection values. All parameters returned to normal by 24 to 48 h, but the dGTP pool remained depressed for 72 h.[9] That the GTP concentration explains, in a large extent, the chemotherapeutic action of tiazofurin is supported by *in vitro* studies in various cancer cells. The oncolytic action of tiazofurin, the induction of differentiation and apoptosis and the down-regulation of *ras* and *myc* oncogenes can be abrogated by maintaining GTP concentration by adding guanine[10,11] (see below). The relevance of these events to human disease was tested in patients treated with tiazofurin.

BASIS FOR SELECTIVITY OF TIAZOFURIN ACTION IN LEUKEMIC PATIENTS

Tiazofurin in sensitive cells is activated by a kinase to tiazofurin monophosphate. Then NAD pyrophosphorylase converts it into an NAD analogue which is the active metabolite, TAD (thiazole-4-carboxamide adenine dinucleotide). TAD may be degraded by phosphodiesterase. The sensitivity of murine tumors was shown to depend on the concentration of TAD achieved in tumor cells. Tiazofurin inhibits IMP DH at a concentration of 10^{-6}; however, TAD inhibits at a concentration of 10^{-7} M. Our studies indicated that TAD is a tight binding inhibitor attacking at the NAD/NADH site of the IMP DH molecule. TAD has orders of magnitude better affinity to the NAD site of IMP DH than the normal endogenous coenzyme, NAD.[5]

The selectivity of tiazofurin action is based, in part at least, on evidence that there was a high conversion of TAD from tiazofurin in leukemic cells whereas normal leukocytes synthesized only low concentrations of TAD. The clinical consequence was that in patients treated with tiazofurin the normal leukocytes were largely preserved and the patient's bone marrow was not compromised. Therefore, after tiazofurin treatment the patients could be released without any requirement for antibiotic treatment. This is in sharp contrast with the conventional treatment of leukemia.

LEUKEMIA: A PARTICULARLY SUITABLE DISEASE FOR TIAZOFURIN TREATMENT

In human leukemic cells IMP DH activity was increased 5- to 20-fold over that of normal leukocytes and TAD had higher affinity (0.1 μM) to the enzyme than the normal substrate NADH (150 μM). IMP DH activity was inhibited with the TAD concentrations available in the blast cells of the patients. The increased GPRT activity in the blast cells was competitively inhibited by hypoxanthine and hypoxanthine blood levels can be elevated by allopurinol treatment. The TAD accumulation was 20- to 30-fold higher in leukemic blast cells than in normal leukocytes. Incubation of human leukemic blast cells with tiazofurin decreased GTP. Tiazofurin was known to inhibit the growth of murine leukemias P-388 and L1210.

This reasoning for using tiazofurin had an important role in the decision of the National Cancer Institute to approve our clinical trial for leukemia treatment at Indiana University School of Medicine. The treatment with tiazofurin and allopurinol was limited to leukemic patients in blast crisis where no other treatment was available. Initially 30 patients were treated with this protocol using daily 1 h infusion of tiazofurin (2,000/mg/m^2 as starting dose) through a permanent line. Allopurinol was initially given as a pill of 300 mg in one dose; subsequently, to better maintain serum hypoxanthine levels 600 to 800 mg per day were given in 4 divided doses. The best clinical results were in chronic granulocytic leukemia in blast crisis where 75% responses were observed including complete returns to the chronic phase. In this difficult group of patients in end-stage leukemia these clinical results are the best ones ever reported. Daily 3 samples were taken from the patients which allowed the monitoring in the blast cells of the IMP DH activity and GTP concentration and the concentration of tiazofurin, TAD, hypoxanthine, and other parameters in the serum. In this biochemically monitored protocol the response to treatment, particularly in IMP DH activity and GTP concentration, was evident in hours whereas the hematological signs lag far behind. Without this biochemistry the hematological impact is not known for 3 to 7 days.

In patients with high sensitivity to tiazofurin there was a reduction in IMP DH activity and GTP concentration to 20% or lower within 24 h of beginning treatment along with down-regulation of *ras* and *myc* oncogene expression in the blast cells. In some patients rapid tumor lysis occurred and in one such patient 2 infusions resulted in remission with disappearance of blast cells from the periphery.

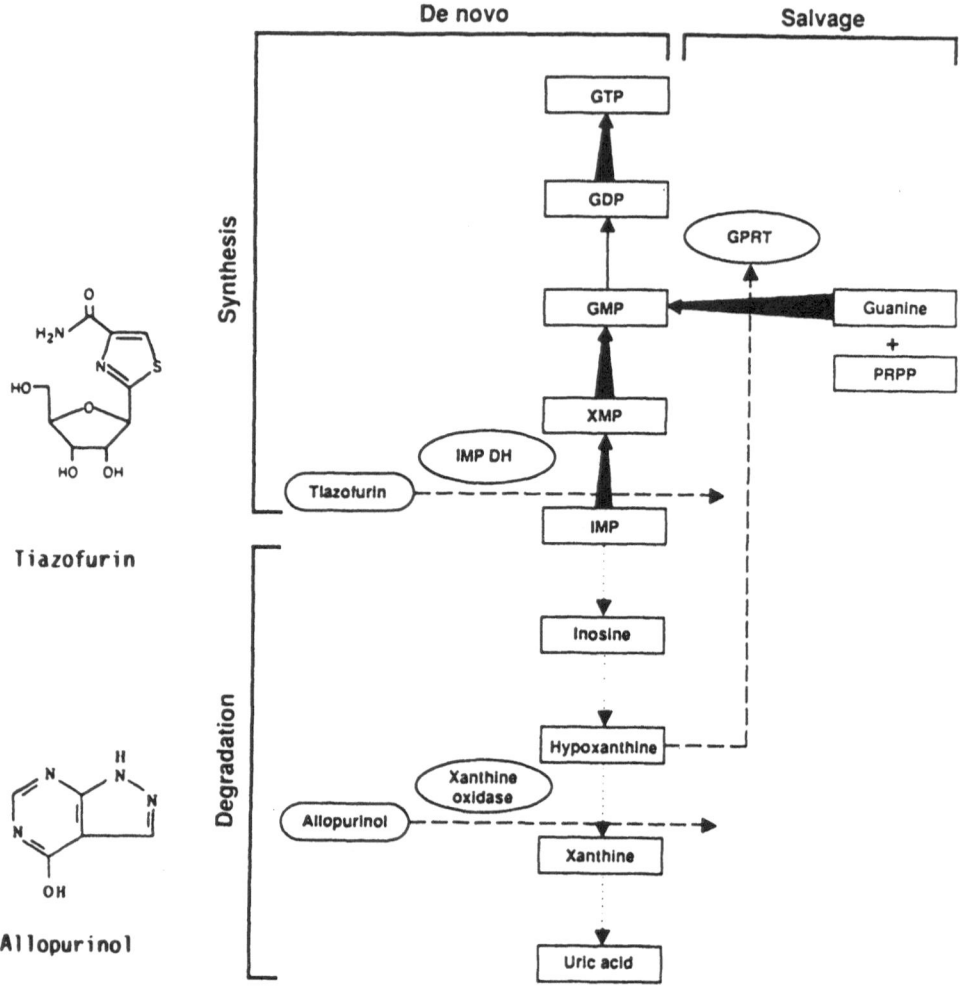

Figure 2. Targets of tiazofurin and allopurinol therapy in human leukemic blast cells.

TARGETS OF TIAZOFURIN TREATMENT

In animals and in the patient IMP DH is rapidly inhibited by tiazofurin treatment with a $t_{1/2}$ = about 30 min. There is evidence in tissue culture and in patients that as the GTP concentration drops there is an immediate attempt of the cells to produce more IMP DH molecules, as indicated by elevated mRNA production. However, in the continued presence of tiazofurin and its active metabolite, TAD, the total activity of IMP DH remains inhibited. The speed of decay of GTP concentration is determined by the utilization of this nucleotide in the various cells. That GTP reduction is the key to tiazofurin action is supported by the following observations. When tiazofurin was added to various cell lines GTP concentration decreased and induced differentiation and apoptosis occurred. However, if GTP concentration was restored or the reduction was prevented by adding exogenous guanine, the down-regulation of oncogenes, the induction of differentiation and apoptosis were abrogated even in presence of tiazofurin.[10,11] These data underline the role of GTP in the chemotherapeutic response and the significance of the guanine salvage pathway. Clinical results support this observation since if high serum hypoxanthine level (which inhibits guanosine salvage) cannot be maintained because of rapid excretion by the kidneys, the therapeutic results are not satisfactory. This supports the ideas of enzyme pattern targeted chemotherapy that for best oncolytic results both the *de novo* and the salvage pathways must be inhibited.[1]

Among various aspects of tiazofurin treatment the superiority of this management protocol is evidenced by the induced maturation of blast cells and the preservation of normal leukocyte population. Tiazofurin, through monitoring drug targets (IMP DH, GTP, *ras* oncogene), has contributed to a novel rational treatment of leukemia and provides a new paradigm for clinical treatment of cancer. These clinical and biochemical results were recently independently confirmed.[12] Tissue culture studies in this laboratory showed that additivity or synergism may be obtained when the following drugs were used in combination with tiazofurin: taxol, quercetin, retinoic acid, ribavirin, gemcitabine, dipyridamole, brefeldin and an inhibitor of *ras* protein expression.

ROLE OF PURINE METABOLISM IN SIGNAL TRANSDUCTION

In the conversion of PI to IP_3 the activities of PI and PIP kinases depend on ATP as a cofactor. PIP kinase can also utilize GTP. However, phospholipase C activity is entirely dependent on GTP. Thus, the concentrations of ATP and particularly GTP are critical for the processes of signal transduction. We showed that tiazofurin reduced the activities of PI and PIP kinases and the concentration of GTP. The mechanism of action relates to the inhibitory action of tiazofurin on macromolecular synthesis. By limiting macromolecular biosynthesis, tiazofurin caused a decrease in activities of enzymes with short half-lives. We showed in cycloheximide studies that PI and PIP kinases have very short half-lives ($t_{1/2}$ = 12 min in bone marrow cells). A single injection of tiazofurin (150 mg/kg, i.p.) caused a marked reduction in the activities of the PI and PIP kinases and in the concentration of IP_3 in rat bone marrow cells. Similar reduction of signal transduction activity was observed in rat liver and transplanted hepatoma.[3] The half-lives of these indicators were between 23 and 48 h.

The linkage of purine metabolism with signal transduction activity can be markedly influenced by agents that reduce GTP concentration, such as tiazofurin, or by compounds that inhibit PI and PIP kinases or phospholipase C directly.

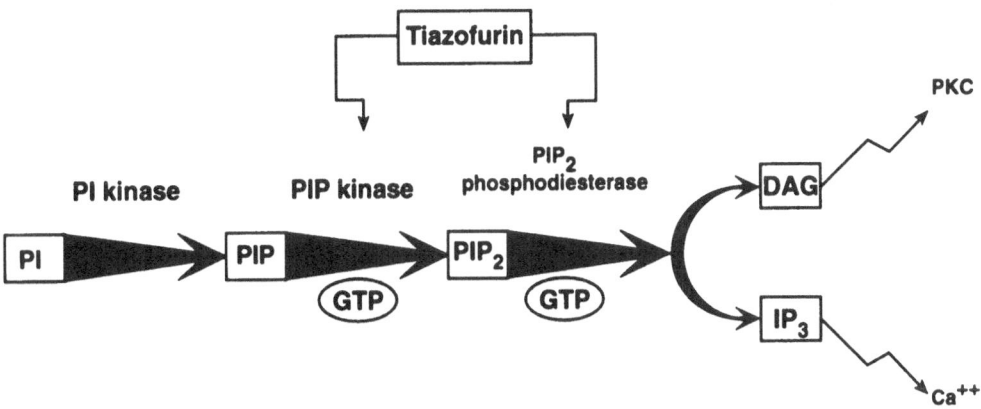

Figure 3. Targets of tiazofurin in signal transduction.

ACKNOWLEDGMENTS

This work was supported in part by a Milan Panič Professorship and a grant from the Women's Auxiliary to the Veterans of Foreign Wars to GW; American Cancer Soc. Inst. grant IRG-161-J to FS.

REFERENCES

1. G. Weber, Biochemical strategy of cancer cells and the design of chemotherapy: G. H. A. Clowes Memorial Lecture, *Cancer Res.* 43:3466–3492 (1983).
2. G. Weber, Enzymology of cancer cells, Parts 1 & 2, *New England J. Med.* 296:486–493 & 541–551 (1977).
3. G. Weber, N. Prajda, M. Abonyi, K.Y. Look, and G. Tricot, Tiazofurin: Molecular and clinical action, *Anticancer Res.* 16:3313–3322 (1996).
4. G. Weber, N. Prajda, and R.C. Jackson, Key enzymes of IMP metabolism: Transformation- and proliferation-linked alterations in gene expression, *Advan. Enzyme Regul.* 14:3–24 (1976).
5. Y. Natsumeda, T. Ikegami, K. Murayama, and G. Weber, *De novo* guanylate synthesis in the commitment to replication in hepatoma 3924A cells, *Cancer Res.* 48:507–511 (1988).
6. Y. Natsumeda, S. Ohno, H. Kawasaki, Y. Konno, G. Weber, and K. Suzuki, Two distinct cDNAs for human IMP dehydrogenase, *J. Biol. Chem.* 265:5292–5295 (1990).
7. M. Nagai, Y. Natsumeda, and G. Weber, Proliferation-linked regulation of type II IMP dehydrogenase gene in human normal lymphocytes and HL-60 leukemic cells, *Cancer Res.* 52:258–261 (1992).
8. M.S. Lui, M.A. Faderan, J.J. Liepnieks, Y. Natsumeda, E. Olah, H.N. Jayaram, and G. Weber, Modulation of IMP dehydrogenase activity and guanylate metabolism by tiazofurin (2-ß-D-ribofuranosylthiazole-4-carboxamide), *J. Biol. Chem.* 259:5078–5082 (1984).
9. E. Olah, Y. Natsumeda, T. Ikegami, Z. Kote, M. Horanyi, J. Szelenyi, E. Paulik, T. Kremmer, S.R. Hollan, J. Sugar, and G. Weber, Induction of erythroid differentiation and modulation of gene expression by tiazofurin in K562 leukemia cells, *Proc. Natl. Acad. Sci. U.S.A.* 85:6533–6537 (1988).
10. M. Vitale, L. Zamai, E. Falcieri, G. Zauli, P. Gobbi, S. Santi, C. Cinti, and G. Weber, IMP dehydrogenase inhibitor, tiazofurin, induces apoptosis in K562 human erythroleukemia cells, *Cytometry* 30:61–66 (1997).
11. G. Tricot, H. N. Jayaram, G. Weber, and R. Hoffman, Tiazofurin: biological effects and clinical uses, *Intl. J. Cell Cloning* 8:161–170 (1990).
12. D.G. Wright, M.S. Boosalis, L.J. Oshry, and K. Waraska, Tiazofurin effects on IMP dehydrogenase activity and expression in the leukemia cells of patients with CML blast crisis, *Anticancer Res.* 16:3349–3354 (1996).

81

ROLE OF ADENOSINE DEAMINASE AND dATP LEVELS IN THYMOCYTE APOPTOSIS

Amos Cohen, Weimin Zhu, and Patricia Benveniste

Division of Immunology and Cancer Research
The Hospital for Sick Children
Department of Immunology
University of Toronto
Toronto, Ontario, Canada

1. INTRODUCTION

While ADA is a ubiquitous enzyme in the purine degradation pathway. Surprisingly, inherited deficiency in ADA in human, results in a specific severe combined immunodeficiency decease, where the tissue damage is restricted to the lymphoid lineage.[1] Previous studies have demonstrated that during T lymphocyte development ADA activity peaks at immature cortical CD4[+]CD8[+] double positive (DP) thymocytes.[2,3]

Within the thymus the majority of thymocytes (>90%) die by apoptosis. DP positive cells in the thymic cortex undergo programmed cell death unless rescued by positive selection. These thymocytes express very low levels of the anti-apoptotic protein Bcl2 and thus easily activate their programed cell death. Low affinity engagement of the T cell antigen receptor (TCR) with the appropriate peptide, presented by MHC class I or II, results in re-expression of Bcl2, rescue of these T cells from apoptosis, migration into the thymic medulla and further development into mature CD8[+] or CD4[+] single positive T cells respectively (Figure 5).

Deficiency of ADA activity results in accumulation of its substrate deoxyadenosine which is then phosphorylated to yield dATP.[4] It can be demonstrated that the excessive accumulation of dATP in ADA deficiency results in imbalance of deoxynucleoside triphosphates (dNTP) pools that causes DNA breaks.[5] Immature thymocytes are especially efficient in accumulating dATP and express low levels of Bcl2 making them especially susceptible to ADA deficiency.[6,7] We have hypothesized that dATP accumulation in ADA deficiency causes immature thymocytes to undergo rapid apoptosis due to the accumulation of DNA breaks.[5] This hypothesis was supported by results obtained using an animal model of ADA deficiency with the ADA inhibitor deoxycoformycin and genetically altered mice either expressing thymic Bcl2 transgene or lacking the p53 gene.[8] In the pre-

Purine and Pyrimidine Metabolism in Man IX,
edited by Griesmacher *et al.* Plenum Press, New York, 1998.

sent report, we further analyzed the relationship between the expression of ADA and Bcl2 proteins in the course of normal thymic development.

2. MATERIALS AND METHODS

2.1. Animals and *in Vivo* Injections

C57B1/6 mice were used as controls. p53 knockout mice together with their wild type controls were purchased from Genpharm International (Mountain View, CA.). Mice overexpressing the bcl-2 transgene in thymocytes[6] were obtained from Hang Siah Teh (University of British Columbia Vancouver BC) and were bred and screened in the animal facility of the Hospital for Sick Children. C57B1/6 mice were purchased from Jackson Laboratories and used at 4–6 weeks of age. Deoxycoformycin (Pentostatin, a generous gift of Parke Davis, An Arbour, Michigan) was injected from day 1 to day 3 at 300 µl/day/mouse of a 2 mM solution intraperitoneally. Control mice received similar treatment using saline. Analysis was performed 24 hours after the last injection.

2.2. Cell Isolations and Cultures

CD4⁻ thymocytes were isolated by negative selection using avidin-activated magnetic beads (Advanced Magnetics, Cambridge, MA.) to deplete biotin conjugated anti-CD4 mAb stained total thymocytes (Pharmingen San Diego, CA.) as previously described.[8]

2.3. Flow Cytometry

Flow cytometry was performed with either a dual laser FACSCAN (Becton-Dickenson) or a dual laser FACSTAR PLUS (Becton Dickenson) as previously described.[8]

2.4. Flow Cytometric Analysis of Apoptosis

Total thymocytes or isolated thymic subpopulations were washed once in Phosphate Buffer solution (PBS) after culture in single cell suspension. Cells were then fixed at 2×10^6 cells/ml in 70% ethanol for 1–12 hours at 40°C. Pellets were then resuspended in 50 µl propidium iodide (0.1 mg/ml Sigma) with 0.6% NP40, followed by 500 µl RNAase (2 mg/ml Bovine Pancreas type II, Sigma).

Nuclei were then analyzed by flow cytometry and frequencies of subdiploid DNA were determined.

2.5. Deoxynucleoside Triphosphates Levels

Cellular nucleotides were extracted using Alamine Freon. The DNA polymerase assay was employed for dNTP analysis.[9]

3. RESULTS AND DISCUSSION

3.1. Effect of Inhibition of ADA on Apoptosis in the Course of Thymocyte Differentiation

The sensitivity of thymocytes to apoptosis, induced by inhibition of ADA activity, varies as a function of intrathymic T cell differentiation (Figure 1). Apoptosis induced by

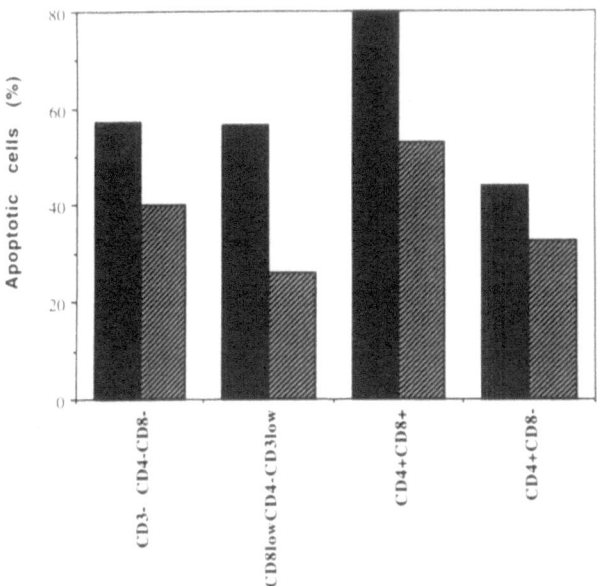

Figure 1. Effect of deoxyadenosine in the presence of deoxycoformycin on apoptosis in differentiating thymocytes. Subdiploid DNA frequencies of total thymocytes and/or isolated purified by flow cytometry cell sorting thymocyte subsets are shown in each panel. Cells were cultured for 14 hours in the presence (black bars) or absence (gray bars) of deoxycoformycin (20 μM) and deoxyadenosine (60 μM dADO) as indicated and stained with propidum iodide for subdiploid DNA quantification by flow cytometry. The effective concentration of deoxycoformycin and deoxyadenosine as well as the optimal time of treatment were determined in separate experiments. Results are from a single experiment performed with thymocytes obtained from five mice.

deoxyadenosine in the presence of the ADA inhibitor deoxcoformycin peaks at the $CD4^+CD8^+$ double positive stage of differentiation (80% apoptotic cells) and decreases after positive selection in single positive thymocytes (40% apoptotic cells). These changes in sensitivity to apoptosis follow the increase in Bcl2 expression during intrathymic T cell differentiation.[10]

3.2. Deoxyadenosine-Induced Apoptosis Depends on p53 Expression and Is Prevented by Bcl2 Overexpression

To examine whether thymocyte apoptosis, induced by ADA inhibition, is the result of DNA breaks, we have used $p53^{-/-}$ mice. In the absence of p53 these mice are unable to undergo apoptosis following DNA damage.[13] Indeed $p53^{-/-}$ mice were less sensitive to apoptosis induced by the combination of deoxyadenosine and deoxycoformycin (Figure 2). As expected, forced expression of the Bcl2 transgene in thymocytes from Bcl2 transgenic mice resulted in complete protection (<10%) from ADA-deficiency-induced apoptosis as compared to over 60% apoptotic cells in non-transgenic litter mates with low Bcl2 expression. These results support the hypothesis that in ADA deficient patients there is increased thymocyte apoptosis that is a result of increased levels of intracellular dATP and accumulation causing DNA breaks. The tissue and differentiation stage specificity of the disease is thus due to low levels of expression of Bcl2 during double positive stage of thymocyte differentiation. The low expression of Bcl2 at this stage is necessary to allow the process of T cell selection following activation of programmed cell death.

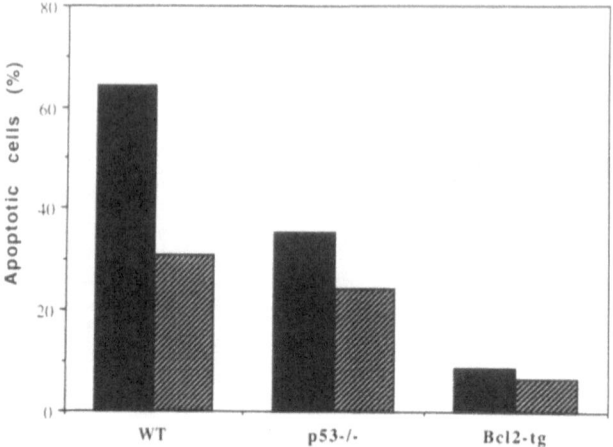

Figure 2. Induction of apoptosis following inhibition of ADA activity in p53$^{-/-}$, Bcl2 transgenic and wild type mice. Frequencies of subdiploid DNA derived from thymocyte cultures of p53 knock-out and wild type control mice. Thymocytes were cultured for 14 hours in single cell suspension cultures in the presence (black bars) or absence (gray bars) of indicated deoxycoformycin (20 µMDC) and deoxyadenosine (60 µM dADO). Recovered cells were then lysed and stained with propidium and analyzed by flow cytometry for the presence of subdiploid DNA. Results are from a single experiment performed using two wild type and two p53$^{-/-}$ mice. Three additional experiments gave similar results.

3.3. Thymocytes from Bcl2 Transgenic Mice Have Increased dATP Levels

Recently, dATP was found to participate in the activation of apoptosis.[11] This observation raise the possibility tha changes in the levels of intracellular dATP levels may be a requirement of the apoptotic process controlled by Bcl2 in thymocytes. To approach this question we analyzed dATP levels in thymocytes from Bcl2 transgenic mice and compared these levels to those of their non-transgenic litter mates (Figure 3). Surprisingly, it was found that thymocytes obtained from Bcl2 transgenic mice have markedly increased levels of dATP as compared to their non-transgenic litter mates. This increase in dATP levels is not

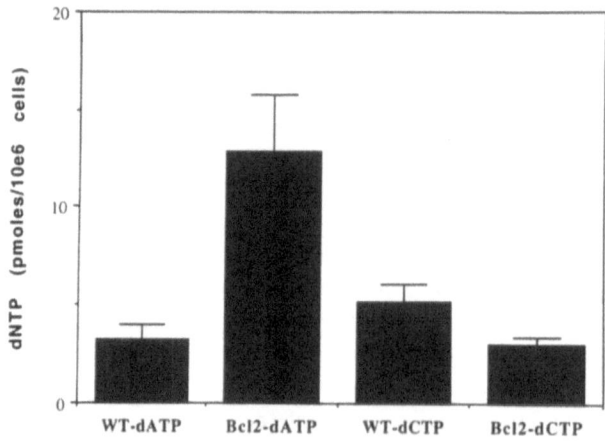

Figure 3. Intracellular dATP levels in thymocytes from Bcl2 transgenic mice and their non-transgenic litter mates. Deoxynucleoside triphosphate levels were determined by the DNA polymerase method.[9]

due to a general increase in dNTPs levels since dCTP levels were slightly lower in Bcl2 transgenic mice. These differences are not likely due by differences in thymocyte populations between the two mice strains since no significant differences in the distribution of thymocyte subpopulations exist between Bcl2 transgenic mice and their litter mates.[10]

3.4. Thymocytes from Bcl2 Transgenic Mice Express Lower Levels of ADA mRNA

To determine whether the increase in dATP levels in thymocytes from Bcl2 transgenic mice can be explained by changes in ADA expression, we have analyzed the levels ADA mRNA in thymocytes from Bcl2 transgenic mice and their non-transgenic litter mate using RNAse protection assay (Figure 4). The results of this analysis reveal that indeed ADA mRNA levels are markedly higher in Bcl2 transgenic mice as compared to their non transgenic litter mates. The levels of β-actin mRNA are comparable in these samples and thus provide a suitable control for equal amounts of RNA analyzed. The differences in ADA mRNA levels between the two mice strains is not likely to be due to differences in apoptosis since induction of apoptosis by dexamethason or γ radiation does not affect ADA mRNA levels in thymocytes from the transgenic mice or their non transgenic litter mates. On the other hand, phorbol ester causes a decrease in ADA mRNA as expected by inducing the terminal differentiation of thymocytes as previously reported.[4] It thus seems that forced expression of the Bcl2 transgene in thymocytes suppresses the expression of ADA mRNA either directly or indirectly. This result raises the possibility that the dramatic increase in ADA activity in immature thymocytes may be controlled by the simultaneous decrease of Bcl2 during normal intrathymic T cell differentiation.

3.5. Interplay of ADA and Bcl2 Expression in the Course of Thymocyte Differentiation

The results reported here, combined with previous data,[4] can be summarized in a scheme that represents the interaction of Bcl2 and ADA during normal intrathymic T cell

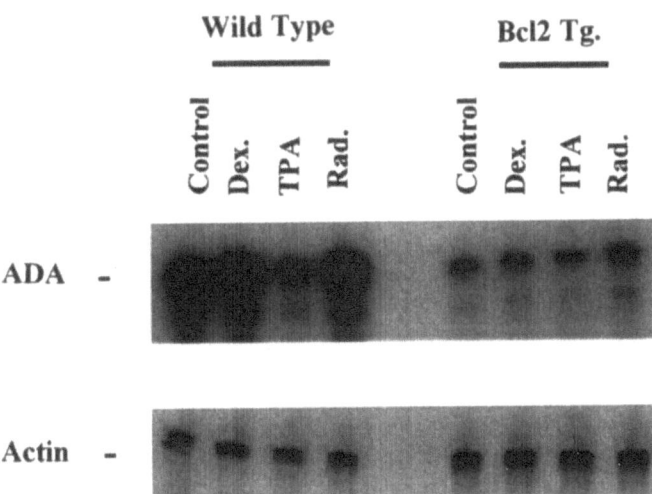

Figure 4. Levels of ADA mRNA in thymocytes from Bcl2 transgenic and wild type mice. mRNA levels in thymocytes were determined using RNAse protection assay. Actin mRNA levels were used to control for equal levels of total RNA in thymocyte extracts.

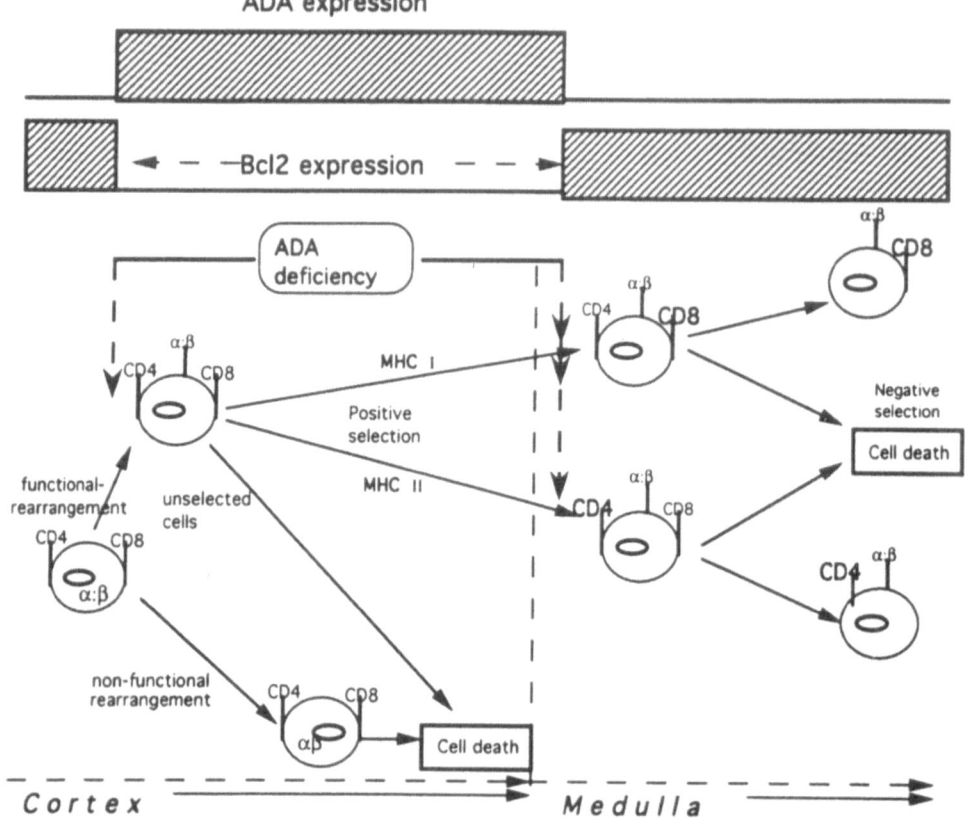

Figure 5. Schematic representation of the role of ADA and Bcl2 in the course of intrathymic T cell differentiation.

development, protection against apoptosis and ths consequences of ADA deficiency (Figure 5). It is clear that there is a reverse relationship between ADA and Bcl2 levels in the course of thymocyte differentiation. During the double positive stage of differentiation, cortical thymocytes express very low levels of Bcl2 combined with the highest ADA levels. During this stage of differentiation the majority of thymocytes undergo apoptosis and can be rescued by engagement of the T cell receptor. This positive selection process results in increased Bcl2 expression, migration into the medulla and terminal differentiation into single positive T cells. It is our hypothesis that cortical thymocytes that express low levels of Bcl2 have to be protected from non-specific apoptosis due to dATP accumulation. This protection is achieved by overexpression of ADA activity. The protection from apoptosis is lacking in ADA deficiency, hence, the increased thymocyte apoptosis and combined severe immunodeficiency observed in these patients.

REFERENCES

1. Giblett, E.R., Anderson, J.E., Cohen, B. & Meuwissen, H.J. (1972) Lancet 2, 1067.
2. Ma, D.D.F., et al. (1982) J. Immunol. 129, 1430–1435.

3. Martinez-Valdez, H. & Cohen, A. (1988) Proc. Natl. Acad. Sci. USA 85, 6900–6903.
4. Cohen, A., et al. (1975) Proc. Natl. Acad. USA 75, 472–476.
5. Cohen, A. & Thompson, E. (1986) Cancer Res 46, 1585–1588.
6. Sentman, C.L., Shutter, J.R., Hockenbery, D., Kanagawa, O. & Korsmeyer, S.J. (1991) Cell 67, 879–888.
7. Cohen, A., Barankiewicz, J., Lederman, H.M. & Gelfand, E.W. (1984) Cantor, H., L. Chess And E. Sercarz (Ed.). Ucla (University Of California Los Angeles) Symposia On Molecular And Cellular Biology New Series 18, 0–8451.
8. Benveniste, P. &Cohen, A. (1995) Proc. Natl. Acad. Sci. USA 92, 8373–8377.
9. Gao, W-Y, Johns, D.G. &Mitsuya, H. (1994) Anal. Biochem. 222, 116–122.
10. Linette, G.P., Grusby, M.J., Hedrick, S.M., Hansen, T.H., Glimcher, L.H. & Korsmeyer, S.J. (1994) Immunity, 1, 197–205.
11. Liu, X., Kim, C.N., Yang, J., Jemmerson, R. &Wang, J. (1996) Cell, 86, 147–157.

82

STRUCTURE AND FUNCTION OF P2Y$_2$ NUCLEOTIDE RECEPTORS IN CYSTIC FIBROSIS (CF) EPITHELIUM

G. A. Weisman,[1] R. C. Garrad,[1] L. J. Erb,[1] M. Otero,[3] F. A. Gonzalez,[3] and L. L. Clarke[2]

[1]Department of Biochemistry
[2]Department of Veterinary Biomedical Sciences and Dalton Research Center
University of Missouri-Columbia
Columbia, Missouri
[3]Department of Chemistry
University of Puerto Rico
Rio Piedras, Puerto Rico

1. INTRODUCTION

Cystic fibrosis (CF) is caused by mutations in the gene encoding the cystic fibrosis transmembrane-conductance regulator (CFTR), a cAMP-dependent chloride transporter expressed in epithelial tissue. A novel treatment being developed for CF utilizes inhaled nucleotides, particularly uridine 5'-triphosphate (UTP), to activate calcium-dependent chloride channels in airway epithelial cells. This therapy is based on earlier studies from the laboratory of Dr. Richard C. Boucher at the University of North Carolina-Chapel Hill, which have shown that luminal administration of UTP[1] and the Na$^+$ channel blocker amiloride,[2] together overcome the ion transport defects in CF airway epithelial cells by promoting increased Cl$^-$ secretion and decreased Na$^+$ absorption.[3] UTP stimulates Cl$^-$ secretion in epithelial cells by activating a P2Y$_2$ (formerly named P$_{2U}$) nucleotide receptor subtype[4,5] that is coupled to phospholipase C via G$_{q\alpha}$ protein resulting in inositol 1,4,5 trisphosphate (IP$_3$)-dependent increases in the cytoplasmic calcium concentration ([Ca^{2+}]$_i$) that serve to activate calcium-dependent chloride channels.[6] Evidence that activation of this non-CFTR Cl$^-$ secretory conductance may be beneficial in CF therapy was provided by studies with CFTR knockout mice in which the presence of a large, basal, non-CFTR Cl$^-$ conductance was associated with the lack of respiratory disease in this species.[4] Early clinical studies involving administration of aerosolized UTP and amiloride indicate that this therapy has a positive effect on mucociliary clearance in CF nasal epithelia.[7] However, little is known about the mechanisms of nucleotide regulation of epithelial ion transport. It is clear that fundamental infor-

Purine and Pyrimidine Metabolism in Man IX,
edited by Griesmacher *et al.* Plenum Press, New York, 1998.

mation concerning $P2Y_2$ receptor biology in epithelia will be useful in optimizing nucleo-tide therapy for CF. We have utilized the CFTR knockout mouse and its normal counterpart to isolate cultures of gallbladder epithelial cells that form tight junctions and represent a good model for studying the roles of $P2Y_2$ receptors in mediating signal transduction path-ways leading to increased Cl^- secretion. In addition, we have cloned the human airway epi-thelial cell $P2Y_2$ receptor cDNA and expressed the receptor in a heterologous cell system, human 1321N1 astrocytoma cells, that lack any endogenous nucleotide receptors.[8] In the 1321N1 cell transfectants, the human airway $P2Y_2$ receptor is distinguished pharmacologi-cally by its ability to be activated equipotently by the purine nucleotide ATP and the pyrimidine nucleotide, UTP. The investigation of the molecular and pharmacological prop-erties of the $P2Y_2$ receptor expressed in 1321N1 cells is being undertaken to develop meth-ods to maximize calcium-dependent chloride conductance in CF epithelial cells through stimulation of endogenous $P2Y_2$ receptors.

2. $P2Y_2$ RECEPTOR FUNCTION IN CF EPITHELIAL CELLS

P2 receptors are unique among G protein-coupled receptors (GPCRs) whose activa-tion may be therapeutic in CF. With the exception of a small complement of bradykinin receptors,[9] only nucleotide receptors are prominently localized to the apical membrane of airway epithelial cells[3] where they can signal the mobilization of intracellular calcium that activates calcium-dependent chloride channels.[4] We have investigated the effect of UTP on the activation of IP_3-dependent calcium mobilization coupled to $P2Y_2$ receptors in anion-secreting murine gallbladder epithelial (MGE) cell lines. These cell lines were formed by SV40 transformation of the relatively homogenous epithelial cells of the murine gallbladder and cell lines have been developed from the CFTR knockout mouse[10] (i.e., CFTR negative, MGEN cells) and from normal mice (i.e., CFTR positive, MGEP cells). The MGE cells have been continuously grown on permeable collagen supports to maintain their polarity. The cells form electrically-resistive tight junctions and have not changed their morphologi-cal or bioelectrical properties in over 70 passages. We have demonstrated a UTP-dependent transepithelial short-circuit current (Isc) in these cells that can be attributed to electrogenic anion secretion.[5,11–13] The inability of cAMP agonists to increase anion secretion was shown to be characteristic of the absence of functional CFTR in MGEN cells.[11]

The anion secretory response to UTP in the MGE cells has a similar profile to changes in intracellular Ca^{2+} mobilization and is characterized by a rapid increase in the Isc, followed by a decline to a steady-state level within minutes.[5,12,13] An EC_{50} of 2×10^{-6} M was obtained for $P2Y_2$ receptor activation by UTP of the peak Isc response in MGEP cells (Figure 1). The role of calcium signaling in this Isc response is supported by experi-ments showing that UTP stimulated IP_3 production and Ca^{2+} mobilization in MGEP cells (Figure 1).

In our studies, the kinetics and dose-dependence of these responses in MGEP and MGEN cells were comparable to the activation of a recombinant $P2Y_2$ receptor expressed in 1321N1 cells that is coupled to IP_3-mediated Ca^{2+} mobilization.[8,11,12,15] In addition to the Ca^{2+}-dependent Isc response in MGE cells, 10^{-4} M UTP also stimulated a small, 4,4'-di-isothiocyanatostilbene-2,2'-disulfonic acid (DIDS)-inhibitable, increase in Isc in the con-tinued presence of a maximally effective concentration of a Ca^{2+} ionophore, indicative of the co-activation of a Ca^{2+}-independent Cl^- conductance.[12] These results suggest that in addition to CFTR, MGE cells have both Ca^{2+}-dependent and Ca^{2+}-independent secretory pathways coupled to $P2Y_2$ receptor activation. Recently, we have found that UTP evokes a

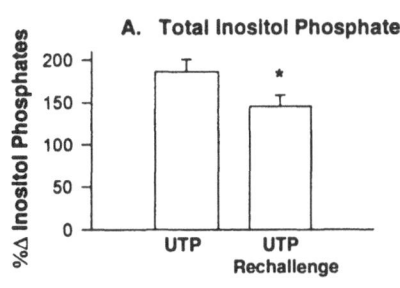

A. Total Inositol Phosphates

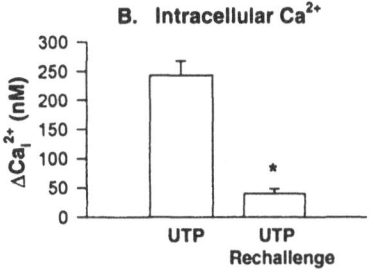

B. Intracellular Ca²⁺

Figure 1. Activation and desensitization of signal transduction pathways regulating anion secretion induced by UTP in MGEP cells. Cells were pretreated with 10^{-4} M UTP, washed and rechallenged with 10^{-4} M UTP after 10 min. **A.** Cells were prelabeled with [^{3}H]-inositol and increases in inositol phosphate after pretreatment or rechallenge with UTP were determined, as previously described,[8] and normalized to basal levels of inositol phosphate formation (n=5). **B.** Cells were loaded with fura-2 and changes in the intracellular calcium concentration induced by UTP were determined as previously described[8] (n=4). **C.** Cell cultures on permeable collagen matrices were mounted in modified Ussing chambers and bathed on both the mucosal and serosal sides with a physiologic Ringers solution. The data indicate the magnitude of the anion current (ΔIsc, i.e., short circuit current) induced by 10^{-4} M UTP, as previously described[5,14] (n = 3). Data are means ± S.E.M.

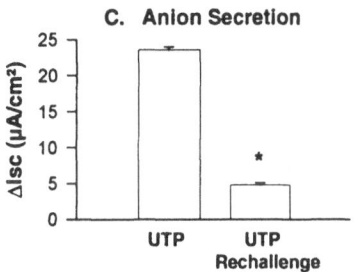

C. Anion Secretion

4–5-fold greater Isc response in MGEN than MGEP cells.[11] This apparent up-regulation of P2Y₂ receptor activity in MGEN cells may be due to a variety of factors including differences in P2Y₂ receptor mRNA and protein expression, inositol phosphate production, calcium mobilization and chloride secretory responses between MGEN and MGEP cells.

MGE cells also responded with anion secretion when treated on the luminal membrane with 2-methylthioATP (2-MeSATP; EC₅₀ ~ 10^{-5} M)[5] or α,β- methylene ATP (α,β-MeATP; EC₅₀ ~ 5 × 10^{-3} M) (unpublished results), agonists respectively at the G protein-coupled P2Y₁ receptor subtype[6,16] and some P2X receptor subtypes, members of the ligand-gated ion channel superfamily.[17–19] Previous studies have shown that the P2Y₁ and the P2X receptor subtypes are not activated by UTP whereas the P2Y₂ receptor is not effectively activated by 2-MeSATP or α,β-MeATP.[20] The co-expression of multiple subtypes of P2 receptors has also been reported for primary cultures of human airway epithelium.[14] To date, 11 subtypes of P2 nucleotide receptors have been cloned and distinguished structurally and functionally as either P2Y or P2X receptors.[21] We have identified P2Y₁, P2Y₂ and P2X₁ receptor mRNA in MGEP cells using oligonucleotide primers designed to selectively amplify, by reverse transcription-polymerase chain reaction (RT-PCR), first strand cDNA produced from isolated MGEP cell mRNA.[5,11] Recent studies also indicate that human airway epithelial cells express a uridine 5- diphosphate (UDP)-selective receptor in addition to the P2Y₂ receptor.[22] These findings suggest that CF pharmacotherapies could target more than one P₂ receptor subtype to potentiate Cl⁻ secretion in CF epithelia.

3. DESENSITIZATION OF P2Y$_2$ RECEPTORS

Consistent with findings from studies on agonist-dependent (homologous) and -independent (heterologous) regulation of other GPCRs,[23,24] our studies indicate that preincubation of MGE cells with UTP can desensitize the subsequent Cl$^-$ secretory response to a second UTP treatment. The Cl$^-$ secretory response in MGEP cells is partially desensitized by a 10 minute preincubation with 10^{-5} to 10^{-7} M UTP, whereas the receptor almost completely desensitizes in response to 10^{-3} M UTP.[5,11] P2Y$_2$ receptor-mediated responses (i.e., IP$_3$-dependent increases in the intracellular calcium concentration, ([Ca^{2+}]$_i$), in 1321N1 cell transfectants expressing the cloned P2Y$_2$ receptor also desensitize at UTP concentrations greater than 10^{-7} M[15] and, recently, we have found that UTP-induced increases in [Ca^{2+}]$_i$ and IP$_3$ desensitize in response to repeated UTP treatment in MGE cells (Figure 1). The desensitizing effect of UTP on Ca^{2+} mobilization in these cells was not due to depletion of intracellular Ca^{2+} stores because the increase in [Ca^{2+}]$_i$ in response to the intracellular Ca^{2+}-ATPase inhibitor, thapsigargin, was not blocked by prior exposure to UTP at concentrations up to 100 mM (unpublished results). Other results indicate that recovery of P2Y$_2$ receptor activity after desensitization is dependent upon the UTP concentration that induces desensitization and the time period between successive applications of UTP.[12,13] The chloride secretory response to UTP in MGE cells recovers rapidly (< 30 min) after desensitization by 10^{-6} M UTP as compared to 10^{-5} or 10^{-4} M UTP (> 60 min), whereas cells do not recover from desensitization induced by 10^{-3} M UTP even after 3 hours. The desensitization of the intracellular Ca^{2+} response to UTP in 1321N1 cell transfectants expressing the cloned P2Y$_2$ receptor occurred with a similar time course as UTP-induced Cl$^-$ secretion in MGEP cells, but the recovery of the Ca^{2+} mobilization response after desensitization was more rapid than in the epithelial cell line.[12,15]

Because of the strong implication of phosphorylation in the mechanism of desensitization of G protein-coupled receptors,[23,24] we performed studies examining the effects of protein kinase activation on P2Y$_2$ receptor desensitization in 1321N1 cell transfectants. Addition of the protein kinase C (PKC) activator, phorbol myristate acetate (PMA, 1 μM), was found to decrease by > 50% the Ca^{2+} response to a subsequent application of UTP.[25] Preincubation with both PMA and 10^{-6} M UTP virtually abolished the response to a second dose of UTP at the EC$_{50}$ concentration. These effects could be overcome by addition of the non-selective protein kinase inhibitor, staurosporine or the selective PKC inhibitor GF109203X obtained from Calbiochem (unpublished results). Other results indicate that modulation of PKC by these compounds also affects UTP-induced Cl$^-$ secretion in MGEP cells (unpublished results).

Sequence analysis of the cloned P2Y$_2$ receptor indicates the presence of three consensus phosphorylation sites for PKC in the intracellular carboxy-terminal domain of the receptor (Figure 2A), whereas there are no sites for cAMP-dependent protein kinase.

In total, there are 13 serine/threonine residues in the C-terminus of the receptor, which suggests the role of other G protein-coupled receptor kinases (GRK) in P2Y$_2$ receptor desensitization, a characteristic of other GRK-regulated receptors.[24] To determine the role of protein kinases in P2Y$_2$ receptor desensitization, we constructed mutant receptor cDNAs that encode a series of receptors with truncation and substitution mutants of amino acids in the C-terminal domain. The truncated receptors expressed in 1321N1 cells exhibited desensitization of the UTP-induced calcium response (Figure 2B). Other results indicated that replacement of the three conserved PKC sites in intracellular domains of the P2Y$_2$ receptor (Figure 2A) produced receptors that only partially desensitized.[25] Staurosporine and GF109203X prevented a significant percentage of this desensitization

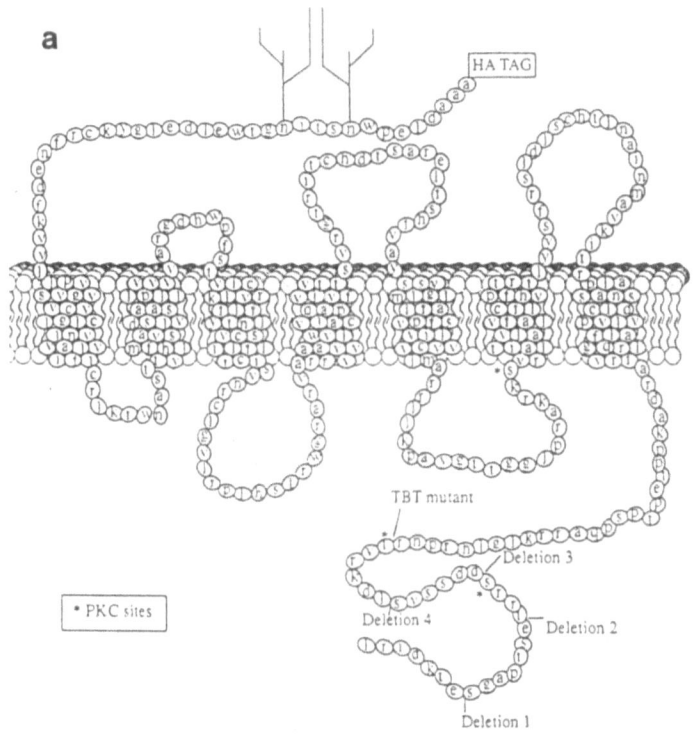

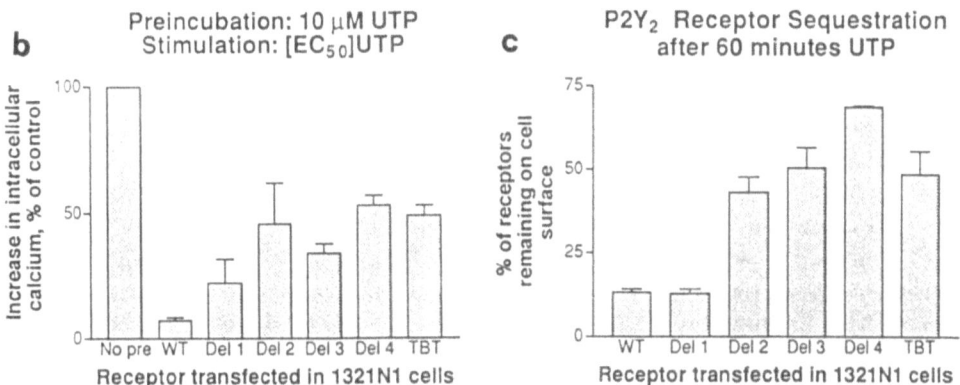

Figure 2. A. Predicted 2D structure of the murine P2Y$_2$ nucleotide receptor with deletion and PKC sites indicated. B and C. Hemagglutinin-tagged wild type murine P2Y$_2$ nucleotide receptors and indicated deletion mutants were expressed in human 1321N1 astrocytoma cells, as previously described[8]. B. Cells were loaded with fura-2 and the $[Ca^{2+}]_i$ was determined, as previously described,[8] after exposure to 10 μM UTP for 5 minutes, washing of the cells and re-exposure to the EC$_{50}$ concentration of UTP. Cells not exposed to UTP for 5 minutes were rechallenged with the EC$_{50}$ concentration of UTP and the subsequent response taken as 100%. C. Cells were exposed to 1 mM UTP for 60 minutes, then stained with mouse monoclonal anti-HA antibody (12CA5; Boehringer-Mannheim) followed by FITC-labeled goat anti-mouse IgG (Fc specific). Cells were then fixed and analyzed in a flow cytometer. Cells that were not exposed to UTP were taken as having 100% of the receptors available for staining. Data are the mean ± S.E.M. for three determinations.

response in the mutant receptors indicating that 1) other PKC phosphorylation sites remain in the P2Y$_2$ receptor, 2) sites remain for GRKs, or 3) PKC acts by mechanisms other than direct receptor phosphorylation to cause desensitization.

Another series of experiments utilized wild type and mutant P2Y$_2$ receptor cDNA that were modified at the N-terminus by the addition of a hemagglutinin epitope tag to which an antibody is commercially available. Expression of these epitope-tagged receptors in 1321N1 cells enabled us to monitor the sequestration of receptors from the cell surface using a secondary fluorescent antibody and flow cytometry. Figure 2C indicates that the desensitization resistant truncation mutants described above also showed decreased rates of sequestration as compared to the wild type epitope-tagged receptors.[26] Because resistance to desensitization would likely be beneficial in P2Y$_2$ receptor-based pharmacotherapies for CF, the full elucidation of these mechanisms is likely to identify intervention points for preventing desensitization *in vivo*. Furthermore, desensitization-resistant, epitope-tagged P2Y$_2$ receptor mutants could be expressed in CFTR-negative epithelial cells to directly determine the role of P2Y$_2$ receptor phosphorylation and sequestration on desensitization of Ca^{2+}-dependent Cl$^-$ secretion in an endogenous system.

4. INVESTIGATION OF THE LIGAND BINDING SITE OF THE P2Y$_2$ RECEPTOR

It is likely that a variety of other yet unidentified compounds instead of UTP will prove to be the optimum drug for activation of P2Y$_2$ receptors in CF epithelia. Knowledge of the amino acid determinants of the P2Y$_2$ receptor may prove useful in the design of high affinity agonists for the treatment of CF. Previous studies have compared primary sequences of P2 receptor subtypes to identify several conserved positively charged amino acids in the 6th and 7th transmembrane domains (TM6 and TM7) of the P2Y$_2$ receptor that were postulated to participate in the binding of the negatively charged phosphate groups of UTP and ATP.[27] Site-directed mutagenesis of P2Y$_2$ receptor cDNA to replace nucleotides encoding negatively charged amino acids with neutral residues at several positions produced P2Y$_2$ receptors with higher EC$_{50}$s than the wild type receptor when expressed in 1321N1 cells. Using this data, computer models of TM6 and TM7 of the P2Y$_2$ receptor based on the structurally similar protein bacteriorhodopsin were produced that indicated the feasibility of triphosphate binding to Arg^{262}, His^{265} and Arg^{292}. Other results indicated that a mutation of Lys^{289} to Arg in the P2Y$_2$ receptor, conserved the positive charge at that site but decreased the EC$_{50}$ for the nucleoside 5'-diphosphates, ADP and UDP, while increasing the EC$_{50}$ for ATP and UTP, as compared to the wild type receptor. Lys^{289} is predicted to lie at the cell surface region of TM7 of the receptor and may regulate the accessibility of agonist to the ligand binding site. Jacobson and colleagues have performed molecular modeling studies for the P2Y$_1$ receptor based on the structure of rhodopsin that have confirmed the importance of the positively charged amino acids in TM6 and TM7 for ligand binding and have suggested the possible participation of amino acids in the 2nd and 3rd transmembrane domains in nucleoside binding.[28]

5. CONCLUSION

The ability of P2Y$_2$ nucleotide receptor activation in CF epithelial cells to increase chloride conductance offers a means to directly overcome the primary defect in chloride transport in CF epithelia that is due to a mutation in the gene for CFTR, an apical membrane chloride transporter. Optimization of this nucleotide therapy will likely necessitate inclusion of other agents that could minimize P2Y$_2$ receptor desensitization, a common property of G protein-coupled receptors. Elucidation of the protein kinases that regulate short term desensitization as well as the pathways that mediate P2Y$_2$ receptor sequestration and down regulation are important goals towards the development of pharmacological agents for preventing desensitization. Furthermore, identification of the ligand binding determinants of the P2Y$_2$ receptor should facilitate the design of potent agonists for a receptor that can be activated equivalently by molecules as diverse as purine and pyrimidine nucleotides. The presence of multiple nucleotide receptor subtypes in epithelial cells suggests the possibility that several different pathways could be targeted for promoting CFTR-independent chloride secretion in CF epithelia. Finally, epithelial cells express significant ectoATPase activity on the apical cell membrane, and this enzyme can degrade both extracellular ATP and UTP to the nucleoside 5'-diphosphate (unpublished results). Inhibition of these enzymes could potentially prolong the effect of nucleotide agonists on activation of P2Y$_2$ receptors in CF epithelial cells, although increased stability of nucleotide agonists may exacerbate the desensitization phenomenon. Thus, our studies suggest that through the co-regulation of P2 receptors, protein kinases and ectoATPases in epithelial cells, an ideal pharmacotherapy for the treatment of CF may ultimately be achieved.

REFERENCES

1. Knowles, M.R., Clarke, L.L., Boucher, R.C. (1991) Activation by extracellular nucleotides of chloride secretion in the airway epithelia of patients with cystic fibrosis. New Eng. J. Med. 325:533–538.
2. Knowles, M.R., Church, N.L., Waltner, W.E., Yankaskas, J.R., Gilligan, P., King, M., Edwards, L.J., Helms, R.W., and Boucher, R.C. (1990) A pilot study of aerosolized amiloride for the treatment of lung disease in cystic fibrosis. N. Engl. J. Med. 322:1189–1194.
3. Clarke, L.L. and Boucher, R.C. (1993) Ion and water transport across airway epithelia. In: Pharmacology of the Respiratory Tract. Experimental and clinical research, (K.F. Chung and P.J. Barnes, eds.). pp. 505–550. Marcell Decker, New York.
4. Clarke, L.L., Grubb, B.R., Yankaskas, J.R., Cotton, C.U., McKenzie, A., Boucher, R.C. (1994) Relationship of a non-CFTR-mediated chloride conductance to organ-level disease in Cftr(-/-) mice. Proc. Natl. Acad. Sci. (USA) 91:479–483.
5. Harline, M.C., Turner, J.T., Garrad, R.C., Weisman, G.A., and Clarke, L.L. (1995) Desensitization of P$_{2U}$ purinoceptors regulating transepithelial Cl$^-$ secretion. Ped. Pulmonol. Suppl. 12:190.
6. Boarder, M.R., Weisman, G.A., Turner, J.T., and Wilkinson, G.F. (1995) G Protein-coupled P$_2$ Purinoceptors: From Molecular Biology to Cellular and Tissue Responses. Trends in Pharmacol. Sci. 16:133–139.
7. Knowles, M.R., Olivier, K.N., Hohneker, K.N., Bennett, W.D., and Boucher, R.C. (1995) Aerosolized uridine triphosphate (UTP) +/- amiloride for the treatment of cystic fibrosis lung disease. Ped. Pulmonol. Suppl. 12:S12.4.
8. Parr, C.E., Sullivan, D.M., Paradiso, A.M., Lazarowski, E.R., Burch, L.H., Olsen, J.C., Erb, L., Weisman, G.A., Boucher, R.C., and Turner, J.T. (1994) Cloning and expression of a human P$_{2U}$ nucleotide receptor, a target for cystic fibrosis pharmacotherapy. Proc. Natl. Acad. Sci. (USA) 91:3275–3279.
9. Clarke, L.L., Paradiso, A.M., Mason, S.J., Boucher, R.C. (1992) Effects of bradykinin on Na$^+$ and Cl$^-$ transport in human nasal epithelium. Am. J. Physiol. 262:C644-C655.
10. Snouwaert, J.N., Brigman, K.K., Latour, A.M., Malouf, N.N., Boucher, R.C., Smithies, O., and Koller, B.H. (1992) An animal model for cystic fibrosis made by gene targeting. Science 257:1083–1088.
11. Harline, M.C., Price, E.M., Glover, G.G., Garrad, R.C., Weisman, G.A., Turner, J.T., and Clarke, L.L. (1996) Increased Ca^{2+}-mediated Cl$^-$ conductance in a transformed epithelial cell line from the CFTR knockout mouse. Ped. Pulmonol. Suppl. 13:271.
12. Glover, G.G., Harline, M.C., Otero, M., Camden, J.M., Turner, J.T., Weisman, G.A., and Clarke, L.L. (1996) Mechanisms of P$_{2U}$ receptor desensitization in an epithelial cell line. Ped. Pulmonol. Suppl. 13:272.
13. Weisman, G.A., Turner, J.T., Erb, L., Garrad, R.C., Harline, M.C., Otero, M.A., Glover, G.G., Camden, J.M., Gonzalez, F.A., and Clarke, L.L. (1996) P$_{2U}$ nucleotide receptor structure and function: relevance to cystic fibrosis pharmacotherapy. Ped. Pulmonol. Suppl. 13:273.

14. Mason, S.J., Paradiso, A.M., and Boucher, R.C. (1991) Regulation of transepithelial ion transport and intracellular calcium by extracellular adenosine triphosphate in human normal and cystic fibrosis airway epithelium. Brit. J. Pharmacol. 103:1649–1656.

15. Otero, M, Camden, J.M., Erb, L., Garrad, R.C., Gonzalez, F.A., Clarke, L.L., Weisman, G.A., and Turner, J.T. (1995) Regulation of native and heterologously expressed P_{2U} purinoceptors. Ped. Pulmonol. Suppl. 12:198

16. Burnstock, G. and Kennedy, C. (1985) Is there a basis for distinguishing two types of P_2-purinoceptor? Gen. Pharmacol. 5:433–440.

17. Trezise, D.J., Bell, N.J., Kennedy, I., and Humphrey, P.P.A. (1994) Effects of divalent cations on the potency of ATP and related agonists in the rat isolated vagus nerve: Implications for P_2 purinoceptor classification. Br. J. Pharmacol. 113:463–470.

18. Brake, A.J., Wagenbach, M.J., and Julius, D. (1994) New structural motif for ligand-gated ion channels defined by an ionotropic ATP receptor. Nature (London) 371:5119–523.

19. Valera, S., Hussy, N., Evans, R.J., Adami, N., North, R.A., Surprenant, A., and Buell, G. (1994) A new class of ligand-gated ion channel defined by P_{2X} receptor for extracellular ATP. Nature (London) 371:516–519.

20. Fredholm, B.B., Abbracchio, M.P., Burnstock, G., Daly, J.W., Harden, T.K., Jacobson, K.A., Leff, P., and Williams, M. (1994) Nomenclature and classification of purinoceptors. Pharmacol. Rev. 46:143–156.

21. Weisman, G.A., Gonzalez, F.A., and Turner, J.T. (1997) Molecular Biology of P2Y Receptors. In: P2 Nucleotide Receptors (J.T. Turner, G.A. Weisman, and J.S. Fedan, eds.). pp. 231–237, Human Press, Totowa, NJ, USA (in press).

22. Lazarowski, E.R., Paradiso, A.M., Watt, W.C., Harden, T.K., and Boucher, R.C. (1997) UDP activates a mucosal-restricted receptor on human nasal epithelial cells that is distinct from the $P2Y_2$ receptor. Proc. Natl. Acad. Sci. (USA) 94:2599–2603.

23. Dohlman, H.G., Thorner, J., Caron, M.G., Lefkowitz, R.J. (1991) Model systems for the study of seven-transmembrane-segment receptors. Ann. Rev. Biochem. 60:653–688.

24. Inglese, J., Freedman, H.J., Koch, W.J., and Lefkowitz, R.J. (1993) Structure and mechanism of the G protein-coupled receptor kinases. J. Biol. Chem. 268:23735–23738.

25. Garrad, R.C., Otero, M.A., Camden, J., Clarke, L.L., Gonzalez, F.A., Turner, J.T., and Weisman, G.A. (1996) Directed mutagenesis of P_{2U} nucleotide receptor PKC phosphorylation sites decreases receptor desensitization. Ped. Pulmonol. Suppl. 13:274.

26. Garrad, R.C., Erb, L., Otero, M., Barnett, L., Gonzalez, F.A., Clarke, L.L., Turner, J.T., and Weisman, G.A. (1997) Deletions of the C-terminal domain of the $P2Y_2$ nucleotide receptor alter receptor desensitization and sequestration: implications in the therapy of cystic fibrosis. J. Biol. Chem. (Submitted).

27. Erb, L., Garrad, R.C., Wang, Y., Quinn, T., Turner, J.T., and Weisman, G.A. (1995) Site-directed mutagenesis of P_{2U} purinoceptors. Positively charged amino acids in transmembrane helices 6 and 7 affect agonist potency and specificity. J. Biol. Chem. 270:4185–4188.

28. Van Rhee, A.M., Fischer, B., Van Galen, P.J.M., and Jacobson, K.A. (1995) Modeling the P_{2Y} purinoceptor using rhodopsin as template. Drug Des. and Disc. 13:133–154.

83

DEOXYCYTIDINE KINASE CAN BE ALSO POTENTIATED BY THE G-PROTEIN ACTIVATOR NaF IN CELLS

M. Staub, Zs. Csapó, T. Spasokukotskaja, and M. Sasvári-Székely

Semmelweis Medical University
Department of Medical Chemistry
Molecular Biology and Pathobiochemistry
Budapest, Hungary

1. SUMMARY

Recently, it has been shown, that 2-Chloro-deoxyadenosine (1), a series of analogues, and other DNA synthesis inhibitors, increased the deoxycytidine kinase (dCK) enzyme activity in different cells, without influencing thymidine kinase isoenzymes (TK1, TK2), dCMP-deaminase and thymidylate synthase (TS) activities (2,3). The dCK activity was 2–4 times higher in analogue treated cells, than in controls, which can not be explained by metabolic pool imbalance induced by the drugs. New mRNA and protein synthesis of dCK could not be detected, thus post-translational modification has been suggested for potenciatiation the activity of the dCK (1). Because secondary modifications of enzymes usually involve the signalling processes in cells, the universal G-protein activator fluorine ions were tested. dCK activity of human lymphnode lymphocytes were increased 2-times, if cells were incubated in the presence of NaF for 1–2 hrs in cultures, while TK activity was not changed. The formation of dUTP from dCyd, was also enhanced by NaF, in paralel of dCK potentiation.

2. INTRODUCTION

Deoxycytidine kinase (dCK EC.2.7.1.) is one of the most important salvage enzyme, with a broad substrate specificity, phosphorylating extracellular deoxycytidine (dCyd), deoxyadenosine (dAdo) deoxyguanosine (dGuo) and most of their analogues used in anti-cancer and anti-viral therapy (4,11). Even deoxythymidine can be synthesised via dCK after deamination and methylation of dCyd, supply of all dNTPs also in

Purine and Pyrimidine Metabolism in Man IX,
edited by Griesmacher *et al.* Plenum Press, New York, 1998.

resting cells, where no de novo nucleotide biosynthesis, no thymidine kinase is available. This might be the reason, why dCK is constitutively expressed in tissues, without pronounced changes in activity during the cell cycle (4,5). The cytoplasmic TK1 has opposite properties, it is expressed only in S phase, phosphorylates only deoxythymidine as natural nucleoside, has only one nucleotide pool in the cells (6), while there are two dCTP pools in different cells (6,7) one for replication the other for repair of DNA (8).

The enhancement of dCK in cancer therapy would have been an extraordinary importance, because the response to the nucleoside analogues usually correlates with the level of dCK activity in the cells (4,10). The adjuvant effect of CdA pre-treatment in leukemic patients has already been observed on treatment with Citarabine (11,12). A 2–3 times increase in dCK activity has been observed after 1–2 hrs treatment by CdA in primary human tonsillar lymphocyte cultures (1–3), without changes in TK1, TK2 isoenzymes, TS, and dCMP-deaminase activities (2,3). Neither the mRNA nor the protein level of dCK were enhanced in treatments where the activity of dCK increased 2–3 times (1), thus the secondary modification of the enzyme was suggested. Different inhibitors and activators influencing the signalling pathways leading to the secondary modification(s) of the proteins, were tested concerning the potentiation of the dCK activity by CdA. It has benn found, that NaF itself can also enhance the dCK activity 2 times compered to the untreated cell enzyme. NaF is accepted as an "universal G protein activator" inhibiting the GTPase activity (13,14), thus the imbalance of the signalling pathway, the function of the small G proteins, seem to be involved in the activation of dCK, the universal deoxynucleoside kinase, under cell conditions when the DNA repair became important.

3. RESULTS AND DISCUSSION

Tonsillar lymphocytes were obtained from surgically removed 3–6 years old children tonsills as described eralier (7,9). Primary cell cultures (10^7/ml) were incubated in serum free Eagle's MEM medium at 37°C for 30, 60 and 120 minutes with, or without 5, 10, 15, and 20 mM NaF as indicated on Figures. The metabolism of 5-^{3}H-dCyd was measured into the ethanol (70%) soluble pool after TLC chromatography on Silica plates as described (7,9). The labeling of the dCyd nucleotides is shown in Figure 1, part *Nucleotides*, expressed as % of the total labeling in separated peaks. The labeled dCyd will be metabolised via mono-, di- and triphosphates into lipo-nucleotides (LN) (LN: dCDP-choline, dCDP-ethanolamine, dCDP-diacylglycerol), and a substantial part of dCyd will be deaminated by dCMP-deaminase into dUMP, further converted into TMP, which is no longer labeled. The third part of dCyd will be incorporated into DNA (not shown). As it can be seen, the labeling of LN, dCTP, dCDP and dCMP decreased above 10 mM of NaF treatment, while the the proportion of dUMP was proportionally increased by incresing NaF concertration. The incorporation of dCyd into DNA did not change at 5 and 10 mM NaF treatment, but it decreased about 50% at the higher concentrations (data not presented). The increase of the dUMP labeling might be explaned by the inhibition of methylation of dUMP by thymidylate synthase (TS), which might be a consequence of the destruction of the dCMP-deaminase-TS complex, demonstrated (15).

The enzyme activities of the two deoxynucleoside kinases, dCK and TK, were also measured in cell free extracts. Lymphocytes treated by NaF were wased twice by PBS and lysed in an extraction buffer as described (4). dCK and TK enzyme activities were assayed with 5–3H-dCyd or 3H-dThd as substrates (both 10 mM) in a reaction mixture containing 50 mM TRIS-HCl pH 7.6, 5 mM $MgCl_2$, 5 mM ATP, 2 mM DTT, 10 mM NaF and cell free

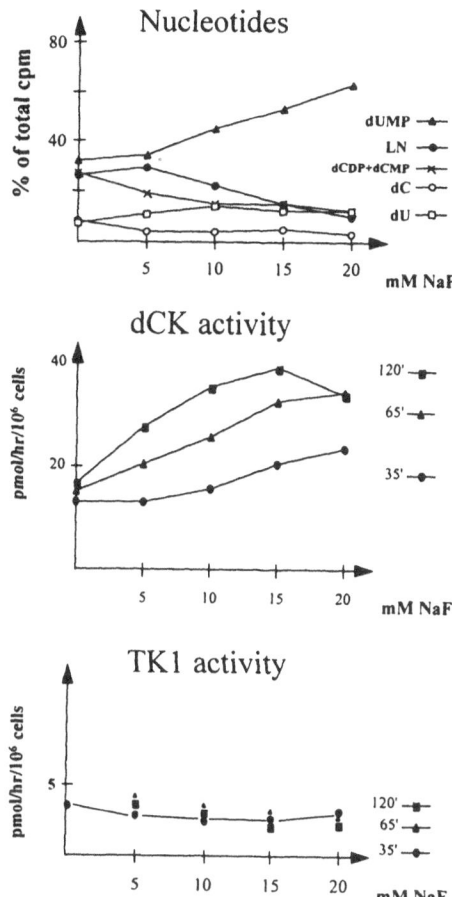

Figure 1. NaF increases dUMP formation and dCK activity in lymphocytes.

extracts incubated at 37°C for 30 min. Aliquots were spotted to DEAE-cellulose filters to bind the phosphorylated substrates, free nucleosides were eluted by ammonium phormate, filters were dried and labeling was counted. Kinase activities were expressed as pmol monophosphate/10^6 cells/30 min. The *dCK activity* has been enhanced by NaF at all concentrations used for cell treatment. The activity was doubled after 120 min treatment by 10–15 mM NaF. The *TK activity* did not increase during the same treatment, shown also in Fig. 1.

The enhancement of the phosphorylation of nucleoside analogues (ara-C) has been already shown after the pretreatment of pations by an other drug i.e. CdA (12). We have recently shown the potentiation of dCK enzyme activity by CdA treatment in human lymphocytes (1–3) and an even higher activity was gained in peripheral blood cells (2). The mRNA and the protein level did not change during the treatment, suggesting a secondary modification of the dCK protein. NaF is known as a potent phosphatase inhibitor, used very often as an general G protein stimulator in cell cultures (13,14). NaF inhibits the GTP-ase activity of corresponding G proteins, keeping in active, GTP binding form, the signaling small protein(s).

Concerning our results, dCK seems to bee the most important deoxynucleoside kinase in resting phase cells, or in cells with DNA demages to be repaired. The DNA demage might be induced by different nucleoside analogues, inhibitors of DNA synthesis or other influences. Deoxyadenosine and its analogues, their triphosphates are inhibitors of the DNA repair itself, might be pushing the corresponding salvage kinase (dCK) in a "more" active form, to produce deoxynucleotides for reparing DNA. In the regulations of the salvage and DNA repair signalling molecules semm to be participating. The similar effect of NaF, to potentiate also dCK activity after treatment, drew the attention to the GTP binding proteins involved in cell cycling and signal transduction pathways might be involved in the activation of dCK. The imbalance of nucleotide pools might be the one of first events in DNA damage too.

REFERENCES

1. Staub, M., Spasokukotskaja T., Sasvári-Székely M., and Keszler G, (1996) Drug Development Res. 37/3, 178 Special Issue, Purines'96, Milan.
2. Spasokukotskaja T., Sasvási-Székely m., Hullán L., Albertioni F., Eriksson S., and Staub M. in the same issue.
3. Sasvári-Székely M., Csapó Zs., Spasokukoctskaja T., Eriksson S., and Staub M. in the same issue.
4. Arnér A.S. and Eriksson S. (1995) Pharmacol Theor. 67, 155–186.
5. Hegstschlager M., Mudrak I., Wintersberger E., and Wawra E., (1994) Cell Growth and Differentiation 5, 1389–97.
6. Reichard P., (1989) Ann. Rev. Biochem. 57, 349–77.
7. Spasokukotskaja T., Sasvári-Székely M., Taljanidis J. and Staub M., (1992) FEBS Letters 297, 151–54.
8. Xie, K. Ch., and Plunkett W. (1996) Cancer Res. 56, 3030–37.
9. Spasokukotskaja T., Spyrou G., and Staub M. (1988) Biochem.Biophys.Res.Commun. 155, 923–927.
10. Björklund S., Skog S., Tribukait B. and Thelander L. (1990) Biochemistry 29, 5452–58
11. Xu Y.,Z., and Plunkett W. (1992) Biochemical Pharmacol. 44, 1819–27.
12. Gandhi V., and Plunkett W., (1988) Cancer. Res. 48, 329–334.
13. Zelles T., Chernaeva L., Baranyi M., Déri Z., Ádám-Vizi V. and Vizi E.S. (1995) J. Neurosci. Res. 42, 242–5177.
14. Miwa M., Kozawa O., Suzuki A., Watanabe Y., Shinoda J., and Oiso Y. (1995) Biochem-Cell Biol. 73, 191–99.
15. McGaughey K.M., Wheeler L.J., Moore J.T., Maley G.F., Maley F., and Mathews Ch.K. (1996) J. Biol. Chem. 271. 23037–42.

84

LOW GTP INDUCES PKC-β-DEPENDENT MYELOID CELL DIFFERENTIATION

Hiroshi Tsutani,[1] Kunihiro Inai,[1] Takahiro Yamauchi,[1] Toru Nakamura,[1] Eliezer Huberman,[2] and Takanori Ueda[1]

[1]First Department of Internal Medicine
Fukui Medical University, School of Medicine
Matsuoka, Fukui 910-11, Japan
[2]Center for Mechanistic Biology and Biotechnology
Argonne National Laboratory
Argonne, Illinois 60439

1. INTRODUCTION

Inosine-5'-monophosphate dehydrogenase (EC 1.1.1.205) catalyzes the formation of xanthosine-5'-monophosphate from inosine-5'-monophosphate (IMP). In the purine synthetic pathway, IMP dehydrogenase is the rate-limiting enzyme in the *de novo* synthesis of guanine nucleotides. Changes in IMP dehydrogenase activity have been implicated in the regulation of cell growth and differentiation (1). Inhibitor of this enzyme has been reported to inhibit cell growth and to induce cell differentiation in myeloid, monocytic, erythroid, or T-cell leukemia cell lines through a decline of intracellular GTP level (2–5). However, it has still not been clear how decreases in intracellular GTP levels induce differentiation. In parallel, protein kinase C (PKC) is a family of serine-threonine protein kinase that has been implicated as one of the prime regulatory enzyme that is activated in response to many external stimuli (6). Studies from a numerous number of laboratories have shown that PKC-mediated signal transduction pathway handles cell growth and differentiation. In this study, we investigate that reduction of intracellular GTP level promotes myeloid differentiation through PKC-β-mediated signal transduction pathway using a PKC-β deficient variant myeloid cell line.

2. MATERIALS AND METHODS

2.1. Cells

The human myeloid HL-60 leukemia cell line was originally obtained from R. C. Gallo (National Cancer Institute), and the differentiation-resistant HL-525 cell lines were

Purine and Pyrimidine Metabolism in Man IX,
edited by Griesmacher *et al.* Plenum Press, New York, 1998.

obtained by cloning HL-60 cells subcultured in the presence of increasing concentrations (up to 3 mM) of phorbol 12-myristate 13-acetate (PMA) (7). These HL-60 cell types were incubated in 100-mm Petri dishes with a RPMI-1640 (Nissui Pharmaceutical Co., Tokyo, Japan) supplemented with 15% fetal bovine serum , glutamine (2 µM) (GIBCO, Grand Island, NY), and kanamycin (100 U/ml) (Meijiseika Co., Tokyo, Japan) at 37°C in an atmosphere of 5% CO_2 in humidified air.

2.2. Differentiation Markers

Morphologic differentiation was evaluated using cytocentrifuged smears prepared using a Cytospin 3 (Shandon, Cheshire, UK). The cytocentrifuged smears were stained using the May-Gruenwald-Giemsa technique. Expression of cell surface antigen was studied using a direct immunofluorescence technique with anti-CD11b antibody (DAKO, Glostrup, Denmark) as a marker of myeloid differentiation. Immunofluorescence analysis was performed using an FACScan flow cytometer (Becton-Dickinson, San Jose, CA). To examine the ability of individual cells to generate superoxide, a slide test using nitro blue tetrazolium (NBT) was performed. Cells containing blue-black formazan deposits were detected by direct microscopic inspection of at least 200 cells for each experimental condition.

2.3. Intracellular GTP Level

To determine intracellular GTP level by high-performance liquid chromatography (HPLC), an acid-soluble fraction was extracted by adding trichloroacetic acid to cell pellets and then neutralized by addition of an equal volume of cold freon containing 0.5 M of tri-n-octylamine. HPLC was performed using a JASCO intelligent pump 880-PU and a JASCO intelligent UV/VIS detector 875-UV (Japan spectroscopic Co., Tokyo, Japan), and a TSK gel DEAE-2 SW column (250 × 4.6 mm, TOSOH, Tokyo, Japan). The elution was performed in 0.05 M Na2HPO4 (pH 6.9), and 20% CH3CN at a constant flow rate of 0.7 ml per min. Elution from the column was monitored at 269 nm. The intracellular GTP level was quantified from detected peak areas using external standards (8).

3. RESULTS

3.1. Cell Growth

Human myeloid leukemia cell line HL-60 and the PMA-resistant variant HL-525 cell were exposed to mycophenolic acid for 72 hours and determined the growth rate among 48 and 72 hours. The inhibitory effect was observed over a range of 0.1 µM to 100 µM of mycophenolic acid dose-dependently in a same fashion. In contrast, when these cell lines were treated with mycophenolic acid for 24 h and then re-suspended mycophenolic acid-free fresh medium, HL-525 cells were less inhibited to grow than HL-60 cells (Fig. 1).

3.2. Cell Differentiation

Following exposure to 2 µM of mycophenolic acid for 24 h and incubation in the drug-free medium for 48 h, HL-60 cells showed significant changes in the morphology of 60 to 80% of the cells. These cells developed a decreased nuclear cytoplasmic ratio, looser chromatin, and less prominent nucleoli. The expression of CD11b determined by flow

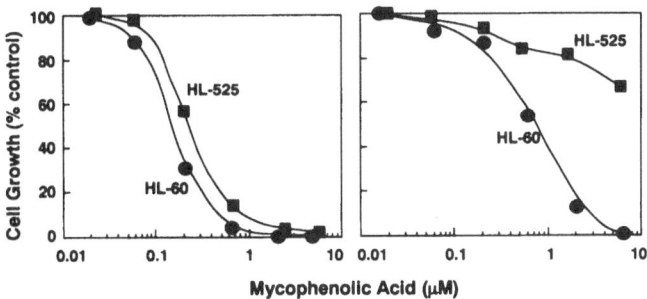

Figure 1. Cell growth of HL-60 cells and the PMA-resistant variant HL-525 cells in the presence of mycophenolic acid. Following exposure to mycophenolic acid for 72 h (left) or 24 h (right), the both cell lines were determined the growth rate among 48 and 72 hours. Data are the mean of triplicate cultures.

cytometric analysis and superoxide generation assessed by NBT slide test were also significantly increased ($p < 0.01$) (Fig. 2). In contrast, HL-525 cells did not show any change in appearance or any increases in number of NBT-positive cells and CD11b expressing cells.

3.3. Intracellular GTP Level

To investigate the relationship between induction of differentiation and guanylate biosynthesis, intracellular GTP levels were quantified by HPLC. Following treatment for 24 h with 2 μM of mycophenolic acid, intracellular GTP levels in both HL-60 and HL-525 cells decreased to approximately 20–30% of control levels.

4. DISCUSSION

In the present study, treatment with mycophenolic acid for 94 h inhibited both HL-60 and the PMA-resistant variant HL-525 cells to grow in a same dose-dependent fashion. However, when these cells were treated for 24 h and then incubated in drug-free medium, HL-525 cells revealed more resistant to arrest the cell growth than HL-60 cells. In addition, although HL-60 cells developed mature appearance as well as expression of CD11b antigens and the ability to oxidize NBT, HL-525 cells did not show any changes in the differentiation markers. These findings suggest that short-term treatment with mycophenolic acid is able to differentiate HL-60 cells but not HL-525 cells. In the two cell lines, 24 h-exposure to mycophenolic acid suppressed *de novo* guanylate synthesis less than 50% of control level. Thus, short-term reduction in intracellular GTP levels may affect some cellular mechanism

Figure 2. Cell differentiation of HL-60 cells and HL-525 cells exposed to mycophenolic acid. The two cell lines were incubated in the absense (filled bar) or in the presense (open bar) of 2 μM of mycophenolic acid for 24 h, then re-suspended mycophenolic acid-free fresh medium for 48 h. The expression of CD11b and superoxide generation were determined by flow cytometric analysis and NBT slide test respectively. Data are the mean of triplicate cultures.

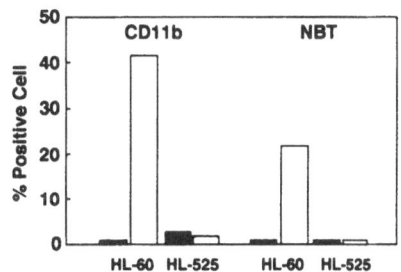

to switch on the induction of differentiation (9). Several studies have suggested a relationship between intracellular GTP levels and GTP-dependent signal transduction (10,11). However, no studies have shown PKC is linked to GTP-related cell differentiation. HL-525 cell is resistant to PMA-induced differentiation and is deficient in PKC-β gene expression (12). Thus, our results suggest that PKC-β-mediated signal transduction system is a candidate of the mechanism associated with low-GTP induced cell differentiation.

REFERENCES

1. Weber G., Cancer Res. 43:3466–92 (1983).
2. Collart F. R. and E. Huberman, Blood. 75:570–6 (1990).
3. Kiguchi K., Collart F. R., Henning-Chubb C., et al., Exp. Cell Res. 187:47–53 (1990).
4. Collart F. R., Chubb C. B. Mirkin B. L., et al., Cancer Res. 52:5826–8 (1992).
5. Tsutani H., Inai K., Imamura S., Adv. Exp. Med. Biol. 370:757–60 (1994).
6. Nishizuka Y., Nature 334:661–5 (1988).
7. Homma Y., Henning-Chubb C., Collart F. R., et al., Proc. Natl. Acad. Sci. USA 83:7316–9 (1986)
8. Yamauchi T., Ueda T., Nakamura T., Cancer Res. 56:1800–4 (1996).
9. Inai K., Tsutani H., Nakamura T., et al., Int. J. Hematol. (in press).
10. Olah E., Natsumeda Y., Ikegami T., et al., Proc. Natl. Acad. Sci. USA 85:6533–7 (1988).
11. Senda M., DeLustro B., Eugui E., et al., Transplantation 60:1143–8 (1995).
12. Tonneti D., Henning-Chubb C., Yamanishi D. T., et al., J. Biol. Chem. 269:23230–5 (1994).

85

EXOGENOUS S-ADENOSYL-METHIONINE METHYLATES PHOSPHOLIPIDS LOCATED ON THE OUTER CELL SURFACE OF ISOLATED RAT HEPATOCYTES

Françoise Bontemps and Georges Van den Berghe

Laboratory of Physiological Chemistry
International Institute of Cellular and Molecular Pathology
Avenue Hippocrate 75-39
B-1200 Brussels, Belgium

1. INTRODUCTION

S-adenosyl-methionine (AdoMet) is a naturally occurring molecule, produced endogenously from ATP and methionine (Met) by AdoMet synthetase. It acts as a methyl donor for a large variety of substrates, including hormones, proteins, nucleic acids, phospholipids and several small molecules, such as glycine or guanidinoacetate.[1] Transmethylation from AdoMet yields S-adenosylhomocysteine (AdoHcy), of which the homocysteine moiety serves as precursor of glutathione via the transsulfuration pathway.

Therapeutic administration of AdoMet has been shown to have favourable effects in liver disorders, including cholestasis, alcoholic and nonalcoholic cirrhosis, and drug-induced hepatotoxicity, as reviewed by Mato et al.[2] The mechanism of action of AdoMet has, however, remained elusive, mainly owing to controversies with respect to its capacity to enter intact liver cells. In the present work, the metabolism of AdoMet in isolated hepatocyte suspensions has been reinvestigated.

2. METHODS

Hepatocytes were isolated as described by Van den Berghe et al.[3] and incubated at the concentration of 100 ± 10 mg of cells/ml in Krebs-Ringer bicarbonate buffer, supplemented with 1% (w/v) albumin and 10 mM glucose, and gassed with O_2/CO_2 (95/5). They were preincubated 15 min before the addition, at 0.25 µCi/ml, of [methyl-^{14}C]AdoMet or [methyl-^{14}C]Met. For analytical determinations on the cell suspension as a whole, 0.5 ml

Purine and Pyrimidine Metabolism in Man IX,
edited by Griesmacher *et al.* Plenum Press, New York, 1998.

aliquots were transferred into 0.125 ml of ice-cold 1 N HClO$_4$. For the determination of metabolites belonging specifically to the cellular or to the extracellular medium, 0.5 ml of the cell suspension were centrifuged through a 0.5 ml-layer of silicone into 0.5 ml of ice-cold 1 N HClO$_4$. For the determination of intracellular AdoMet in hepatocytes incubated with exogenous AdoMet, cells (0.5 ml of suspension) were washed 3 times in 10 ml of ice-cold PBS before the addition of 300 µl of 1 N HClO$_4$ to the washed cell pellet.

AdoMet and AdoHcy were quantified by HPLC, as described by Stet et al.[4] Labelled AdoMet was determined with a Berthold LB 507A HPLC radioactivity monitor coupled to a HPLC apparatus. Radioactivity associated with AdoMet was eluted 10 min after other unidentified HClO$_4$-soluble [methyl-^{14}C]products. Radioactivity of the latter compounds was added up and collectively termed HClO$_4$-soluble products.

Radioactive acid-insoluble products were quantified after two washes with 1 ml of 0.2 N HClO$_4$ of the HClO$_4$-precipitated material, and digestion of the latter with 1 ml of soluene. Phospholipids were extracted by the method of Folch et al.[5] and identified by chromatography on silica-gel TLC plates and development in chloroform/methanol/water (55/25/4; v/v).

3. RESULTS AND DISCUSSION

3.1. Metabolism of Exogenous AdoMet in Hepatocyte Suspensions

As depicted in Fig. 1, [methyl-^{14}C]AdoMet, added at the initial concentration of 2 µmol/ml, was clearly utilized by the cell suspension. Calculation shows that it disappeared from the extracellular medium at an initial rate of about 0.5 nmol/min per g of cells. The major part of the radioactivity which had disappeared was recovered in acid-insoluble products with a minor part in acid-soluble products. Analysis of the acid-insoluble products showed that these corresponded nearly entirely to phospholipids and that 80% of these phospholipids were phosphatidylcholine (results not shown). It was furthermore verified that these [methyl-^{14}C]labelled phospholipids belonged to intact cells which pass through the silicone and not to cellular debris which remained above the silicone layer. Similar experiments, performed with 50 µM AdoMet, gave similar results (not shown).

Despite obvious metabolism of exogenous AdoMet by intact hepatocytes, nearly no labelled AdoMet was found inside the cells. Furthermore, the intracellular concentration

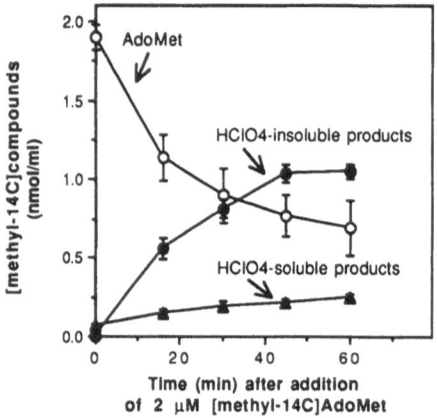

Figure 1. Metabolism of 2 µM [methyl-^{14}C]AdoMet in isolated hepatocyte suspensions. AdoMet was determined in the extracellular medium. Incorporation of ^{14}C-methyl into HClO$_4$ insoluble and soluble products was measured in the total suspension. Values shown are means ± S.E.M. of 3 separate experiments; they were calculated from the specific radioactivity of exogenous [methyl-^{14}C]AdoMet.

of AdoMet which was initially about 40 nmol/g of cells was not increased upon incubation of hepatocytes with increasing concentrations of exogenous AdoMet, at least up to 200 μM (results not shown). The concentration of AdoHcy was also not elevated. These results support the hypothesis that metabolism of exogenous AdoMet occurs without entry of AdoMet inside the hepatocytes.

3.2. Effect of Inhibition of AdoHcy Hydrolase on Intra- and Extracellular Accumulation of AdoHcy

Hydrolysis by AdoHcy hydrolase of AdoHcy, the demethylation product of AdoMet, can be strongly inhibited by adenosine dialdehyde.[6] When hepatocytes were incubated in the presence of this inhibitor, AdoHcy accumulated in their intracellular medium (Fig. 2A), indicating the occurrence of endogenous transmethylations. When 50 μM methionine (Met), which penetrates easily into hepatocytes, was added to the cell suspension, a clear additionnal increase of intracellular AdoHcy was observed. In contrast, when 50 or even 500 μM AdoMet was added to the suspension, no additionnal increase of AdoHcy was observed. On the other hand, there was no elevation of AdoHcy in the extracellular medium in the presence of adenosine dialdehyde alone or with 50 μM Met, but an increase of Ado-Hcy in the extracellular medium was evident after addition of AdoMet (Fig. 2B).

These results indicate that extracellular AdoMet, contrary to Met, is not a substrate for intracellular transmethylation in hepatocytes, and confirm that AdoMet does not cross the plasma membrane of these cells. These observations led to postulate, as already proposed by Van Phi and Söling,[7] that phospholipids methylated by exogenous AdoMet are located on the outer surface of the cell membrane.

3.3. Effect of SAH on Phospholipid Methylation from Exogenous AdoMet

If phospholipids methylated from exogenous AdoMet are located on the outer cell surface, a phospholipid methyltransferase (PLMT) activity should be present on the external

intracellular medium

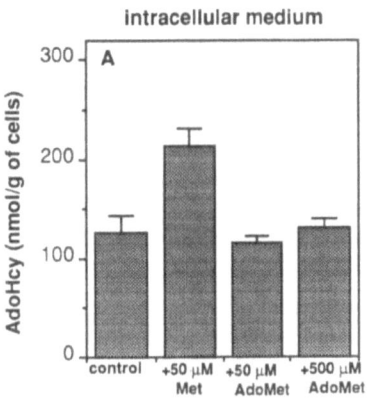

extracellular medium

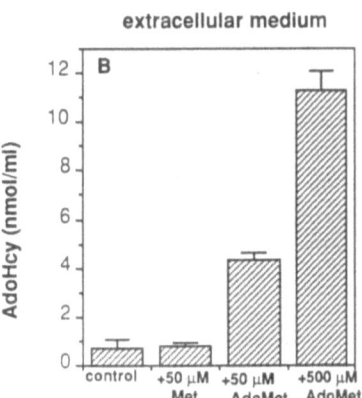

Figure 2. Effect of adenosine dialdehyde on the intra- (A) and extracellular (B) accumulation of AdoHcy. Hepatocytes were incubated for 20 min with 20 μM adenosine dialdehyde, without other addition, or with the substrates indicated. Values shown are means ± S.E.M. of 3 separate experiments. At time 0, the intracellular concentration of AdoHcy was 6.3 ± 0.5 nmol/g (n = 6).

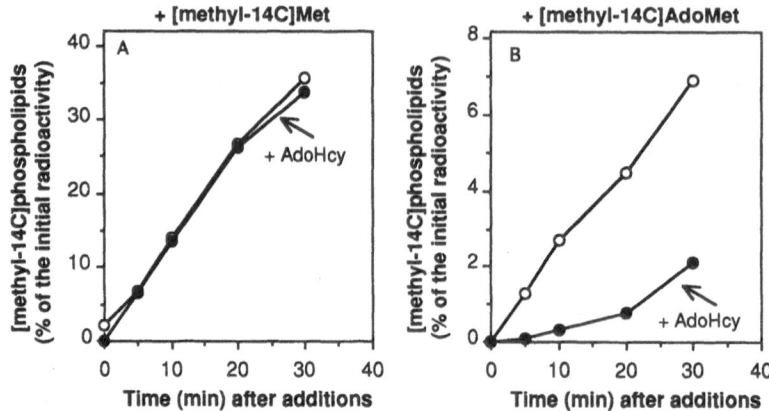

Figure 3. Effect of 50 μM extracellular AdoHcy on phospholipid methylation from 50 μM [methyl-^{14}C] Met (A) or [methyl-^{14}C] AdoMet (B). Phospholipids were extracted from the cellular medium obtained by centrifuging the cells through a silicone layer. Results are expressed as percentage of the initial radioactivity of extracellular AdoMet.

face of the plasma membrane of hepatocytes. PLMT activity is known to be inhibited by AdoHcy.[8] On the other hand, AdoHcy is known not to be taken up by intact cells, including isolated hepatocytes.[9] Accordingly, addition of AdoHcy to hepatocyte suspensions did not inhibit methylation of intracellular phospholipids when Met was the methyl donor.[10]

We have confirmed these findings (Fig. 3A), but have observed (Fig. 3B) that addition of AdoHcy had a strong inhibitory effect on phospholipid methylation when extracellular [methyl-^{14}C]AdoMet was the methyl donor. Since AdoHcy does not enter the cells, it should act on a PLMT which is accessible from the outside of the cell. The latter observation implies that the outer face of the liver plasma membrane contains a PLMT activity, probably different from phosphatidylethanolamine N-methyltransferase 1 (PEMT 1), the major PEMT activity in liver, found on the endoplasmic reticulum, and from PEMT 2, recovered in a mitochondria associated membrane fraction.[11]

4. CONCLUDING REMARKS

Taken together, our results indicate that exogenous AdoMet, despite the fact that it cannot enter the cells, is utilised by isolated hepatocytes to methylate phospholipids present on the outer face of the plasma membrane. Our observations suggest that the protective effects of AdoMet are, at least initially, mediated by a direct effect on membrane structure and function. How these membrane changes result in protection and improvement of hepatic function remains to be determined.

ACKNOWLEDGMENTS

We thank Mrs A. Delacauw for expert technical assistance. This work was supported by grant 3.4539.93 of the Fund for Medical Scientific Research (Belgium), by the Belgian Federal Service for Scientific, Technical and Cultural Affairs, and by Knoll Farmaceutici (Liscate, Italy). G. V. d. B. is Director of Research of the Belgian National Fund for Scientific Research.

REFERENCES

1. Cantoni, G. L. (1975) Biological methylation: selected aspects. Annu. Rev. Biochem. **44**, 435–451
2. Mato, J. M., Alvarez, L., Corrales, F. J. and Pajares, M. A. (1994) S-adenosylmethionine and the liver. In The liver: biology and pathobiology (Arias, I. M., Boyer, J. L., Fausto, N. , Jakoby, W. B., Schater, D. A. and Shafritz, D. A., eds), pp. 461–470, Raven Press, Ltd., New York
3. Van den Berghe, G., Bontemps, F. and Hers, H.G. (1980) Purine catabolism in isolated rat hepatocytes. Influence of coformycin. Biochem. J. **188**, 913–920
4. Stet, E. H., De Abreu, R. A., Bökkerink, J. P. M., Blom, H. J., Lambooy, L. H. J., Vogels-Mentink, T. M., de Graaf-Hess, A.C., van Raay-Selten, B. and Trijbels, F. J. M. (1994) Decrease in S-adenosylmethionine synthesis by 6-mercaptopurine and methylmercaptopurine ribonucleoside in Molt F4 human malignant lymphoblasts. Biochem. J. **304**, 163–168
5. Folch, J., Lees, M. and Sloane-Stanley, G. H (1957) A simple method for the isolation and purification of total lipids from animal tissues. J. Biol. Chem. **226**, 497–509
6. Bartel, R. L., and Borchardt, R. T. (1984). Effects of adenosine dialdehyde on S-adenosylhomocysteine hydrolase and S-adenosylmethionine-dependent transmethylations in mouse L929 cells. Mol. Pharmacol. **25**, 418–424
7. Van Phi, L. and Söling, H. D. (1982) Methyl group transfer from exogenous S-adenosylmethionine on to plasma-membrane phospholipids without cellular uptake in isolated hepatocytes. Biochem. J. **206**, 481–487
8. Schanche, J. S., Schanche, T. and Ueland, P. M. (1981) Inhibition of phospholipid methyltransferase(s) from rat liver plasma membranes by analogues of S-adenosylhomocysteine. Mol. Pharmacol. **20**, 631–636
9. Aarbake, J., and Ueland, P.M. (1981) Interaction of S-adenosylhomocysteine with isolated rat hepatocytes. Mol. Pharmacol. **19**, 463–469
10. Schanche, J. S., Schanche, T. and Ueland, P. M. (1982) Inhibition of phospholipid methylation in isolated rat hepatocytes by analogues of adenosine and S-adenosylhomocysteine. Biochim. Biophys. Acta **721**, 399–407
11. Cui, Z., Vance, J. E., Chen, M. H., Voelker, D. R. and Vance, D. E. (1993) Cloning and expression of a novel phosphatidylethanolamine N-methyltransferase. J. Biol. Chem. **268**, 16655–16663

86

IMP-DEHYDROGENASE (IMPDH), HYPOXANTHINE-GUANINE PHOSPHORIBOSYLTRANSFERASE (HGPRT), AND PHOSPHODIESTERASES (PDEs) EXPRESSION DURING MYCOPHENOLIC ACID (MPA)-INDUCED DIFFERENTIATION IN HUMAN NEUROBLASTOMA CELL LINES

C. P. Quaratino, E. Messina, A. Arduini, G. Angelini, G. Spoto, F. Gizzi, I. Ruffini, M. Odorisio, and A. Giacomello

Dipartimento di Oncologia e Neuroscienze and
   Dipartimento di Scienze Biomediche
Università di Chieti, Italy

The IMP dehydrogenase (IMPDH) inhibitor mycophenolic acid (MPA), at concentrations ranging from 5 to 50 nM, induces differentiation in various human neuroblastoma cell lines.[1] Purpose of the present study was to investigate guanylate synthesis and cyclic nucleotides metabolism during mycophenolic acid-induced differentiation in the human neuroblastoma cell line LA-N-5.

DESIGN AND METHODS

Exponentially growing cells were treated with MPA (ranging from 0 to 50 nM) from 0 to 6 days. Media were replaced every 3 days. Enzyme activities and nucleotide concentrations were measured after 3 and 6 days of incubation with MPA. Cellular enzyme activities were measured on the supernatant of sonicated cells. Hypoxanthine-guanine phosphoribosyltransferase (HGPRT) and IMPDH activities were measured after desalting using spectrophotometric methods.[2,3] Cyclic-AMP phosphodiesterase (cAMP PDE), cGMP phosphodiesterase (cGMP PDE), adenylate cyclase (ACase) and guanylate cyclase (GCase) activities were determined by HPLC procedures.[4] Nucleotide pools were quantitated by HPLC.[4,5] Cell cycle phases were measured by flow cytometry.

Purine and Pyrimidine Metabolism in Man IX,
edited by Griesmacher *et al.* Plenum Press, New York, 1998.

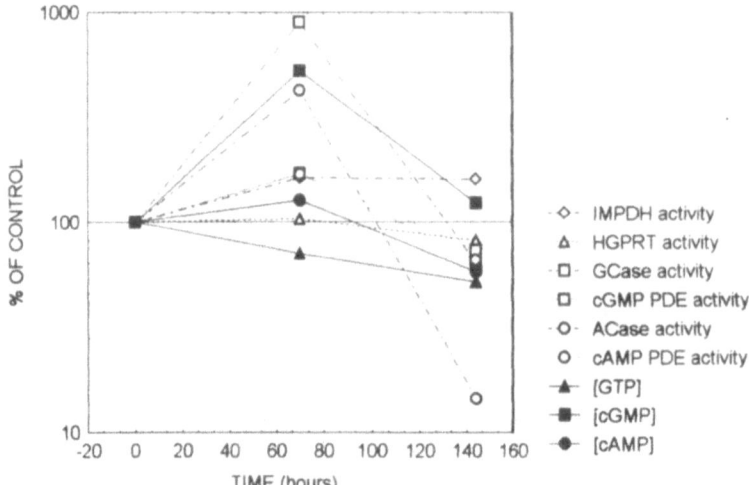

Figure 1. Enzyme activities and nucleotide levels during MPA-induced LA-N-5 differentiation. Results are expressed as percentage of control cells grown in the absence of MPA. Data shown are representative of two to four experiments.

RESULTS AND DISCUSSION

Obtained results are summarized in Figure 1, where enzyme activities and nucleotides levels are expressed as percentage of control cells grown in the absence of MPA. As the Figure shows, after three days, at the onset of observable differentiation, there was a slight decrease of GTP levels, an increase of cAMP concentration and a marked increase of cGMP levels. An increased amount of cellular IMPDH was observed, perhaps, because of compensation for the inhibitor-mediated decrease in enzyme activity. Similar results were obtained in T-lymphoid CEM-2 leukemia cells, studying the kinetics of IMPDH expression during MPA-induced differentiation.[6] After three days there was also an increase of PDEs and above all of cyclases activity. After six days of incubation with MPA, alterations of enzyme activities and nucleotide concentrations with respect to control cells were less striking with a decrease of all measured variables, but cGMP concentration and IMPDH activity. Cell cycle phases were comparable in treated and in control samples. Although it is known that agents which elevate intracellular cAMP level induce neuronal differentiation,[7] we put forward the hypothesis that an increase of cGMP concentration could also have a critical role in neuronal differentiation.

REFERENCES

1. Savini F, C Rucci C, Messina E, Quaratino CP, Scarpa S, Vasaturo F, Modesti A, and Giacomello A Differentiating and biochemical effects of a reduction of intracellular GTP levels induced by mycophenolic acid (MPA) in human neuroblastoma (NB) cell lines. Adv Exp Med Biol (This volume).
2. Giacomello A, Salerno C. A continuous spectrophotometric assay for hypoxanthine guanine phosphoribosyltransferase. Anal Biochem, 1977; 79: 263–267.
3. Atkinson MR, Morton RK, Murray AW. Inhibition of inosine 5'-phosphate dehydrogenase from Ehrlich ascites tumor cells by 6-thioinosine 5'- phosphate. Biochem J, 1963; 89: 167–172.
4. Spoto G, Whitehead E, Ferraro A, Di Terlizzi PM, Turano C, Riva F. A reverse-phase HPLC method for cAMP phosphodiesterase activity. Anal Biochem, 1991; 196:207–10.

5. Di Pierro D, Tavazzi B, Perno CF et al. An ion-pairing high performance liquid chromatographic method for the direct simultaneous determination of nucleotides, deoxynucleotides, nicotinic coenzymes, oxypurines, nucleosides, and bases in perchloric acid cell extracts. Anal Biochem, 1995; 231: 407–12.

6. Kiguchi K, Collart FR, Henning-Chubb C, Huberman E. Cell differentiation and altered IMP dehydrogenase expression induced in human T-lymphoblastoid leukemia cells by mycophenolic acid and tiazofurin. Exp Cell Res, 1990; 187: 47–53.

7. La Rosa FG, Kumar S, Prasad KN Increased expression of ubiquitin during adenosine 3',5' monophosphate-induced differentiation of neuroblastoma cells in culture. J Neurochem. 1996; 66: 1845–50.

87

DIFFERENTIATING AND BIOCHEMICAL EFFECTS OF A REDUCTION OF INTRACELLULAR GTP LEVELS INDUCED BY MYCOPHENOLIC ACID (MPA) IN HUMAN NEUROBLASTOMA (NB) CELL LINES

F. Savini, C. Rucci, E. Messina, C. P. Quaratino, S. Scarpa, F. Vasaturo, A. Modesti, and A. Giacomello

Dipartimento di Oncologia e Neuroscienze
Università di Chieti
Dipartimenti di Medicina Sperimentale
Università di Roma I e II, Italy

Purpose of the present study was to investigate in human neuroblastoma (NB) cell lines the effects of GTP depletion induced by mycophenolic acid (MPA) (the most specific IMP dehydrogenase inhibitor):

1. on biopterin biosynthesis, using cytotoxic drug concentrations in short-term experiments;
2. on cellular differentiation and extracellular matrix remodelling, using drug concentrations below those inducing apoptosis.[1]

DESIGN AND METHODS

1. Short-term biochemical experiments were performed with IMR-32 NB cells in the logarithmic phase of growth. Cells in RPMI 1640 medium supplemented with 10% dialysed FCS and 2mM L-glutamine were treated with various concentrations of MPA (from 10 to 1000 nM) and/or 10 μM guanosine for 20 hours at 37°C in an atmosphere of 5% CO_2. Analysis of nucleotides and pteridines was performed following published HPLC methods on perchloric acid extracts.[2,3]
2. Differentiation studies were performed at low MPA concentrations (0 to 50 nM) after 3 and 6 days of incubation in the presence or in absence of 50 μM guanosine on IMR-32, LA-N-5 and SK-N-SH NB cells. Cell differentiation was

Purine and Pyrimidine Metabolism in Man IX,
edited by Griesmacher *et al.* Plenum Press, New York, 1998.

assessed by morphological changes and by cytoskeletal protein expression. Immunoperoxidase or immunofluorescence analysis of NB cells was performed using the following primary MAbs: MAP-5 (microtubule associated protein), NF-200 (intermediate filaments), GAP-43 (growth associated protein 43), Vimentin (VM), NCAM (neural adhesion molecule), Fibronectin (FN), Laminin (LM). In some experiments expression of p53 was analyzed by Western blotting. The extracellular matrix glycoprotein laminin and fibronectin synthesis was evaluated by ^{35}S-methionine metabolical radiolabelling, immunoprecipitation and SDS-PAGE.

RESULTS

1. In IMR-32, MPA decreased both GTP and total biopterin levels in a parallel and dose-dependent manner up to about 20% of the control values. These effects were reversed by the simultaneous addition of guanosine suggesting that the decrease in the biopterin level was the result of the decrease in GTP concentration.
2. While in LA-N-5 and in SK-N-SH, treatment with MPA resulted in marked differentiation toward the neuronal phenotype, with extension of neurite like processes, in IMR-32 the morphological changes were less evident. In all NB cells up-regulation of neuronal differentiation markers was demonstrated: NF-200, GAP-43, NCAM, LM in SK-N-SH and in IMR-32; NF-200 and MAP in LA-N-5 [Figure 1 (A–H)]. Furthermore, the extracellular matrix was partially modified by MPA: SK-N-SH synthesised intracellular and secreted form of LM and FN and both glycoproteins were increased by MPA; IMR-32 synthesised only traces of intracellular and secreted LM and FN and only FN was increased by MPA.

   During MPA-induced differentiation, at least in LA-N-5, a decrease of intracellular GTP concentration (up to around 50% of the control cells), a slight induction of p53 with its shuttling into the nucleus, were also observed [Figure 1 (I–L)]. Guanosine, added simultaneously with MPA, reversed, at least partially, the observed effects.

DISCUSSION

1. A dopaminergic reduction in Lesch-Nyhan syndrome has been documented both in postmortem and in vivo studies.[4,5] Evidence for guanine nucleotide depletion in Lesch-Nyhan disease comes from the observation of a reduction of these nucleotides in erythrocytes[6] and of an abnormal plasma guanosine response following fructose infusion with no increase of plasma concentration of this nucleoside.[7] According to the findings obtained in the present study, guanine nucleotide depletion could lead to a depletion of biopterin perhaps owing to a

---

Figure 1. Morphological evaluation, differentiation markers and p53 expression assessed by immunocytochemical analysis of NB cell lines after 6 days of treatment with 50 nM MPA (B, D, F, H, L) or control medium (A, C, E, G, I). Immunofluorescent staining of SK-N-SH cells with anti GAP-43 (A, B) and of IMR-32 cells with anti NF-200 (C, D); immunoperoxidase staining of LA-N-5 cells with anti NF-200 (E, F), anti MAP-5 (G, H) and anti p53 (Ab6, Oncogene) (I, L).

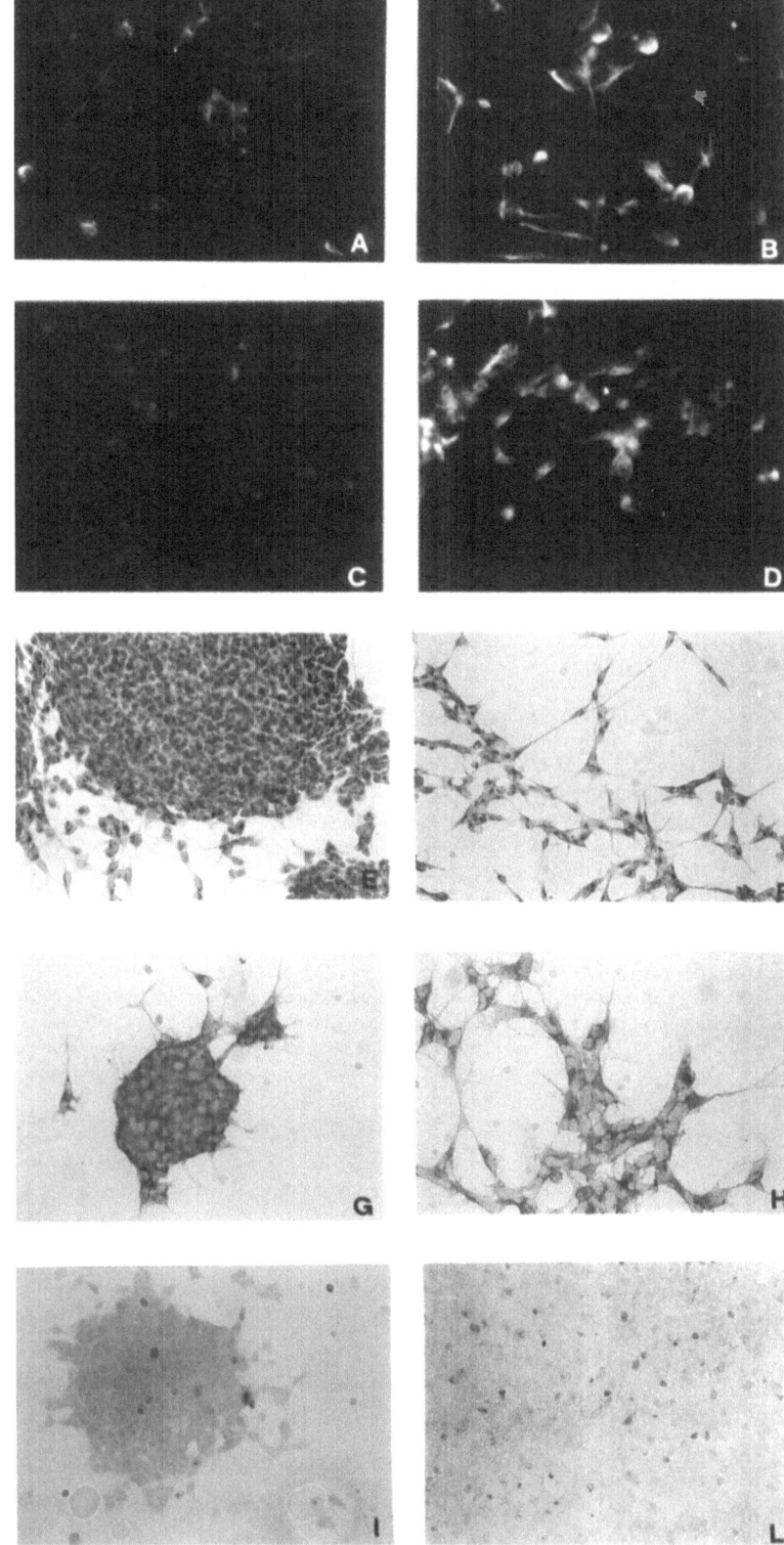

change in GTP- cyclohydrolase I activity, the rate-limiting enzyme of tetrahydrobiopterin biosynthesis. It is well established that tetrahydrobiopterin is an essential cofactor for three aromatic amino acid monooxygenases and in particular for tyrosine hydroxylase, the rate-limiting step in the biosynthesis of catecholamines in vivo. Thus, guanine nucleotide depletion could, at least in part, account for the dopaminergic dysfunction observed in Lesch-Nyhan disease.

2. In the present study it has been shown that MPA induces a differentiation of the human neuroblastoma cell lines LA-N-5, SK-N-SH and IMR-32, associated with a slight increase of p53 at least in LA-N-5. The up-regulation of LM in SK-N-SH was correlated with a neuronal differentiation, and the up-regulation of FN in SK-N-SH and IMR-32 suggests a modification of NB cell adhesion characteristics and perhaps of NB cell invasion behaviour.

ACKNOWLEDGMENTS

This study was supported in part by the "Associazione Malattie Rare Mauro Baschirotto", Via P. Lioy, 13 – Vicenza (Italy).

REFERENCES

1. Messina E, Savini F, Lisio R, Quaratino CP and Giacomello A. Neuroblastoma cell apoptosis induced by mycophenolic acid . Adv Exp Med Biol (This volume).
2. Di Pierro D, Tavazzi B, Perno CF et al. An ion-pairing high performance liquid chromatographic method for the direct simultaneous determination of nucleotides, deoxynucleotides, nicotinic coenzymes, oxypurines, nucleosides, and bases in perchloric acid cell extracts. Anal Biochem, 1995; 231: 407–12.2.
3. Fukushima K, and Nixon JC.Analysis of reduced forms of biopterin in biological tissues and fluids. Anal Biochem 1980; 102: 176–188.
4. Lloyd KG, Hornykiewicz O, Davidson L et al. Biochemical evidence of dysfunction of brain neurotransmitters in the Lesch-Nyhan syndrome. N Engl J Med 1981; 305:1106–11.
5. Ernst M, Zametkin AJ, Matochik JA et al. Presynaptic dopaminergic deficits in Lesch-Nyhan disease. N Engl J Med 1996; 334:1568–72.
6. Simmonds HA, Faibanks LD, Morris GS et al. Altered erythrocyte nucleotide patterns are characteristic of inherited disorders of purine or pyrimidine metabolism. Clin Chim Acta 1988; 171: 197–210.
7. Puig JG, Jiménez ML, Mateos FA, Fox IH. Adenine nucleotide turnover in hypoxanthine-guanine phosphoribosyl-transferase deficiency: evidence for an increased contribution of purine biosynthesis de novo. Metabolism 1989; 38: 410–418.

88

NEW MODELS FOR THE STUDY OF ADENOSINE DEAMINASE DEFICIENCY

Linda F. Thompson

Immunobiology and Cancer Program
Oklahoma Medical Research Foundation
Oklahoma City, Oklahoma 73104

Mutations in the structural gene for adenosine deaminase (ADA) which lead to a loss of enzymatic activity cause severe combined immunodeficiency (SCID) in humans. Although this disease was discovered approximately 25 years ago[1] and there have been many experiments to elucidate the biochemical and immunological mechanism by which ADA deficiency leads to SCID (reviewed in 2), it is fair to say that this issue is still unsettled. For the purpose of this discussion, my remarks will be limited to the consequences of ADA deficiency on T cells, since T cells are more severely and uniformly affected in ADA-deficient patients and T cells are the focus of the next two presentations.

It is generally agreed that T cells fail to develop in ADA-deficient patients because of the accumulation of ADA substrates or their metabolites. Both adenosine and deoxyadenosine are elevated in the serum and urine of ADA-deficient patients. Deoxyadenosine is generally thought to play a more significant role because of the high concentrations of dATP in the red blood cells of ADA-deficient patients and the fact that dATP concentrations generally correlate with the severity of the immunodeficiency.[2] However, there is more than one mechanism by which dATP can exert toxicity. First, dATP is a potent feedback inhibitor of ribonucleotide reductase,[3] the enzyme responsible for the *de novo* synthesis of deoxynucleotides needed for DNA synthesis. There is much evidence for this mechanism from *in vitro* studies by Ullman, Mitchell, and others during the late '70s and early '80s.[4,5] Second, Carson and colleagues showed that elevated dATP can cause DNA strand breaks in resting lymphocytes which can lead to NAD^+ depletion and apoptosis.[6] More recently, Cohen and colleagues showed that deoxyadenosine-mediated apoptosis of ADA-inhibited murine thymocytes required the presence of p53.[7] Not only can dATP be deleterious to T cells, but the deoxyadenosine from which it is derived can also be harmful. Hershfield and co-workers showed that deoxyadenosine is a potent suicide inactivator of the enzyme S-adenosylhomocysteine (SAH) hydrolase.[8] Elevations in SAH can lead to an inhibition of methylation reactions and growth inhibition of lymphoid cell lines. SAH hydrolase has been shown to be inhibited in the red blood cells of ADA-deficient patients.[9]

Purine and Pyrimidine Metabolism in Man IX,
edited by Griesmacher *et al.* Plenum Press, New York, 1998.

Although the role of adenosine in ADA deficiency is often overlooked, there are also multiple mechanisms by which elevated adenosine could potentially interfere with T cell development. First, although it is not as potent as deoxyadenosine, adenosine is also a suicide inactivator of SAH hydrolase.[8] Second, Kizaki[10] and McConkey,[11] as well as our own group,[12] showed that adenosine and certain adenosine analogues are capable of inducing apoptosis of thymocytes through adenosine receptor engagement. A2a adenosine receptors which are expressed in the thymus are coupled to G-proteins and can mediate increases in cAMP. Lalli and colleagues recently reported that cAMP elevating agents can block thymocyte development in fetal thymic organ culture.[13] Furthermore, Resta and colleagues demonstrated that treatment of mice with the ADA inhibitor dCF *in vivo* leads to elevations in intrathymic adenosine which should be sufficient to engage A2a adenosine receptors.[12]

Progress in evaluating the relative importance of the above mechanisms in the pathogenesis of SCID due to ADA deficiency has been limited by a lack of suitable experimental systems. Most of the work has been done with normal T cell lines or mature peripheral blood T cells treated with an ADA inhibitor (reviewed in 2). However, treatment with an ADA inhibitor alone does not inhibit the growth or function of T cells. It is necessary to add an ADA substrate, either adenosine or deoxyadenosine, to see an effect. This makes it impossible to determine the relative contributions of these two nucleosides to the pathogenesis of SCID from such *in vitro* studies. Mice have been treated with the ADA inhibitor 2'-deoxycoformycin in an attempt to create a model system for the disease.[14] However, ADA inhibition has multi-organ system consequences in mice and it is difficult to separate the consequences of ADA substrate accumulation from those of stress-induced glucocorticoids. Conventional ADA knock-out mice were also disappointing because they died at birth of liver failure.[15,16]

The next two presentations describe two new models which show great promise as systems for elucidating the mechanism by which a loss of ADA leads to a failure of T cell development. The first, described by Dr. Regina Resta of The Oklahoma Medical Research Foundation, utilizes murine fetal thymic organ culture treated with 2'-deoxycoformycin. In this system, it is not necessary to add an ADA inhibitor to see a block in thymocyte differentiation. The title of her presentation is: "Insights into Adenosine Deaminase Deficiency Provided by Murine Fetal Thymic Organ Culture with 2'-Deoxycoformycin." The second model system will be described by Dr. Rod Kellems of The Baylor College of Medicine, and utilizes ADA knock-out mice which have been targeted with a minigene directing ADA expression only to the placenta. This minigene rescues the mice from liver toxicity and allows the postnatal development of the immune system in an ADA-free environment. The title of Dr. Kellems' presentation is: "Making Mice More Like People: The Case of Adenosine Deaminase Deficiency."

ACKNOWLEDGMENTS

This work was supported by grants AI18220 and GM39699 from the National Institutes of Health.

REFERENCES

1. Giblett, E. R., J. E. Anderson, F. Cohen, B. Pollara, and H. J. Meuwissen. 1972. Adenosine-deaminase deficiency in two patients with severely impaired cellular immunity. *The Lancet* 2:1067.

2. Hershfield, M. S. and B. S. Mitchell. 1995. Immunodeficiency diseases caused by adenosine deaminase deficiency and purine nucleoside phosphorylase deficiency. In The metabolic and molecular basis of inherited disease. C. R. Scriver, A. L. Beaudet, W. S. Sly, and D. Valle, editors. McGraw-Hill, Inc. New York. 1725–1768.

3. Moore, E. C. and R. B. Hurlbert. 1966. Regulation of mammalian deoxyribonucleotide biosynthesis by nucleotides as activators and inhibitors. *J. Biol. Chem. 241*:4802.

4. Ullman, B., L. J. Gudas, A. Cohen, and D. W. Martin, Jr. 1978. Deoxyadenosine metabolism and cytotoxicity in cultured mouse T lymphoma cells: a model for immunodeficiency disease. *Cell 14*:365.

5. Mitchell, B. S., E. Mejias, P. E. Daddona, and W. N. Kelley. 1978. Purinogenic immunodeficiency diseases: Selective toxicity of deoxyribonucleosides for T cells. *Proc. Natl. Acad. Sci. 75*:5011.

6. Seto, S., C. J. Carrera, M. Kubota, D. B. Wasson, and D. A. Carson. 1985. Mechanism of deoxyadenosine and 2-chlorodeoxyadenosine toxicity to nondividing human lymphocytes. *J. Clin. Invest. 75*:377.

7. Benveniste, P. and A. Cohen. 1995. P53 expression is required for thymocyte apoptosis induced by adenosine deaminase deficiency. *Proc. Natl. Acad. Sci. USA 92*:8373.

8. Hershfield, M. S. 1979. Apparent suicide inactivation of human lymphoblast S-adenosylhomocysteine hydrolase by 2'-deoxyadenosine and adenine arabinoside. *J. Biol. Chem. 254*:22.

9. Hershfield, M. S., N. M. Kredich, D. R. Ownby, and R. Buckley. 1979. In vivo inactivation of erythrocyte S-adenosylhomocysteine hydrolase by 2'-deoxyadenosine in adenosine deaminase-deficient patients. *J. Clin. Invest. 63*:807.

10. Kizaki, H., K. Suzuki, T. Tadakuma, and Y. Ishimura. 1990. Adenosine receptor-mediated accumulation of cyclic AMP-induced T lymphocyte death through internucleosomal DNA cleavage. *J. Biol. Chem. 265*:5280.

11. McConkey, D. J., S. Orrenius, and M. Jondal. 1990. Agents that elevate cAMP stimulate DNA fragmentation in thymocytes. *J. Immunol 145*:1227.

12. Resta, R., S. W. Hooker, A. B. Laurent, S. M. J. Rahman, M. Franklin, T. B. Knudsen, N. L. Nadon, and L. F. Thompson. 1997. Insights into thymic purine metabolism and adenosine deaminase deficiency revealed by transgenic mice over expressing ecto-5'-nucleotidase (CD73). *J. Clin. Invest. 99*:676.

13. Lalli, E., P. Sassone-Corsi, and R. Ceredig. 1996. Block of T lymphocyte differentiation by activation of the cAMP-dependent signal transduction pathway. *EMBO J. 15*:528.

14. Ratech, H., R. Hirschhorn, and G. J. Thorbecke. 1985. Effects of deoxycoformycin in mice: III. A murine model reproducing multi-system pathology of human adenosine deaminase deficiency. *Am.J.Pathol. 119*:65.

15. Migchielsen, A. A. J., M. L. Breuer, M. A. van Roon, H. te Riele, C. Zurcher, F. Ossendorp, S. Toutain, M. S. Hershfield, A. Berns, and D. Valerio. 1995. Adenosine-deaminase-deficient mice die perinatally and exhibit liver-cell degeneration, atelectasis and small intestinal cell death. *Nature Genet. 10*:279–287.

16. Wakamiya, M., M. R. Blackburn, R. Jurecic, M. J. McArthur, R. S. Geske, J. J. Cartwright, K. Mitani, S. Vaishnav, J. W. Belmont, R. E. Kellems, M. J. Finegold, C. A. J. Montgomery, A. Bradley, and C. T. Caskey. 1995. Disruption of the adenosine deaminase gene causes hepatocellular impairment and perinatal lethality in mice. *Proc. Natl. Acad. Sci. USA 92*:3673.

89

INSIGHTS INTO ADENOSINE DEAMINASE DEFICIENCY PROVIDED BY MURINE FETAL THYMIC ORGAN CULTURE WITH 2'-DEOXYCOFORMYCIN

Regina Resta, Hong Jiang, Scott W. Hooker, Aletha B. Laurent, and Linda F. Thompson

Immunobiology and Cancer Program
Oklahoma Medical Research Foundation
Oklahoma City, Oklahoma 73104

INTRODUCTION

Adenosine deaminase (ADA) deficiency as a cause of severe combined immunodeficiency (SCID) was first described twenty-five years ago.[1] One striking feature of this disease is that although the enzyme is missing in every cell in the body, the immune system is most severely affected. There is an almost complete absence of T cells and a variable decrease in the number of B cells. Gene therapy trials for ADA deficiency are well underway,[2,3] and ADA knockout mice have been generated by two groups[4,5] as a model to study this disease. Yet even after twenty-five years, the selective lymphotoxicity of ADA deficiency is still not understood. Understanding the mechanism of lymphotoxicity will not only teach us about the pathophysiology of this uncommon disease, but may lend insights into the therapy of this disease and the therapy of other lymphoid developmental defects. It may also give us clues to the design of new therapeutic options for more common lymphoid diseases such as malignancies, and will certainly teach us more about normal lymphocyte development.

For these reasons, a variety of models for ADA deficiency have been established. As outlined earlier in this volume (Thompson, L.F.), each system has its shortcomings. Murine fetal thymic organ culture (FTOC) seemed like an ideal system to examine the mechanisms of T cell toxicity in this disorder. This is the only system that can support the complete program of T cell development *in vitro*.[6] In FTOC, a thymus is removed from a day 15 embryo and cultured *in vitro* for several days. T cells develop in the explanted fetal thymic rudiment grown at an air-liquid interface on cellulose filters placed on gelfoam sponges floating on media. T cell precursors maintain contact with stromal cells as *in vivo*.

Purine and Pyrimidine Metabolism in Man IX,
edited by Griesmacher *et al.* Plenum Press, New York, 1998.

The maturation of T cells from the primitive CD4⁻CD8⁻ (double negative) precursors to the mature CD4⁺CD8⁻ or CD4⁻CD8⁺ (single positive) T cells is recapitulated *in vitro* in FTOC. All the cell-cell and major histocompatibility antigen (MHC)-peptide-T cell receptor (TCR) interactions required for both positive and negative selection of thymocytes are maintained. However, the confounding effects of systemic toxicity from ADA deficiency or the ADA inhibitor 2′-deoxycoformycin (dCF) are not present. We report our initial results using FTOC with dCF as a model system for dissecting the biochemical and cellular mechanisms of the block imposed by ADA deficiency on T cell development

MATERIALS AND METHODS

Mice and FTOC

Fetal thymuses were removed from timed pregnant C57BL/6 mice on day 14 or 15 of gestation (day 0 = plug day). They were separated into individual lobes, randomized, and cultured at 37°C in 24 well culture plates on cellulose filters (Millipore) resting upon Gelfoam sponges (Upjohn) in 2 mls of RPMI 1640 + 10% fetal calf serum + 2mM glutamine + 5×10^{-5} M 2-mercaptoethanol + penicillin at 100 U/ml + streptomycin at 100μg/ml + amphotericin B at 2.5 μg/ml + non-essential amino acids ± 5 μM dCF (Parke Davis) ± 100 μM hydroxyurea (HU). Media was changed after 3 days and the lobes were harvested on day 5.

Antibodies and Immunofluorescent Staining

Single cell suspensions were made by pushing the lobes through nylon screens. Cells were counted and their phenotype determined by multicolor immunofluorescence. The following monoclonal antibodies (mAbs) were used: hamster mAb 2C-11 anti-CD3, FITC goat anti-hamster Ig, biotinylated hamster anti-δγ TCR + streptavidin Cy-chrome or FITC-avidin, FITC rat anti-CD4, PE rat anti-CD8, Cy-chrome rat anti-CD4, Cy-chrome rat anti-CD8, FITC rat anti-CD25, and PE rat anti-CD44. Staining was performed as previously described[7] using isotype matched rat myeloma proteins or irrelevant hamster mAbs as negative controls. Propidium iodide (PI) was added to all single or two color stains to eliminate dead cells from analysis. Data were collected on 10,000 cells for two color stains and 20,000 cells for three color stains using a Becton-Dickinson FACScan, and analyzed with Lysis software.

RESULTS AND DISCUSSION

Effects of dCF on Thymocyte Development in FTOC

5μM dCF consistently inhibited the production of thymocytes by 85% in FTOC initiated on day 15 of gestation (Figure 1). Thymocyte differentiation proceeds from the earliest CD4⁻CD8⁻ double negative stage, through a CD4⁻CD8$^{wk\,+}$ intermediate, to a double positive CD4⁺CD8⁺ stage, and finally to the mature single positive CD4⁺CD8⁻ and CD4⁻CD8⁺ cells. In dCF-treated FTOC, the majority of thymocytes were arrested in the CD4⁻CD8⁻ stage. The numbers of CD4⁺CD8⁺ and CD4⁺CD8⁻ cells were decreased by 97 and 99%, respectively. In contrast, the number of CD4⁻CD8⁺ cells was only slightly reduced. Even though the cell yield was dramatically reduced by dCF, the number of CD3^{hi} cells

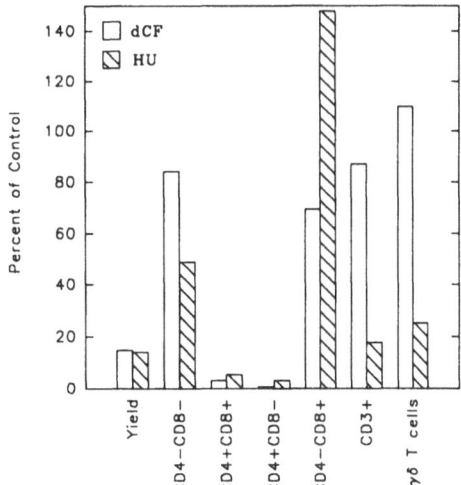

Figure 1. Yield and phenotype of thymocytes recovered from FTOC in the presence of 5 μM dCF (open bars) and 100 μM Hydroxyurea (hatched bars). Cultures were initiated on day 15 and harvested 5 days later. Results are expressed as a percentage of yield from FTOC in media alone cultured for the same period of time.

was 87% of control values and corresponded closely to the number of δγ T cells. The number of δγ cells was approximately 8–30-fold increased over the number of cells present at the time the cultures were initiated. Three color immunofluorescence (data not shown) revealed that most of the CD4⁻CD8⁺ cells were δγ cells. Similar results were obtained when FTOC was initiated at day 14 of gestation and harvested 6 days later.

In the double negative stage many events critical for T cell development occur; therefore, it has been intensively studied and subdivided on the basis of expression of two other T cell differentiation markers, CD44 and CD25. Thymocyte differentiation in the earliest CD4⁻CD8⁻ cells proceeds from CD44⁺CD25⁻ → CD44⁺CD25⁺ → CD44⁻CD25⁺ → CD44⁻CD25⁻ before the cells proliferate and enter the CD4⁺CD8⁺ double positive compartment. In order to precisely delineate the block in thymocyte differentiation caused by dCF, the CD4⁻CD8⁻ cells were characterized for expression of CD44 and CD25. We found that differentiation was blocked at the CD44⁻CD25⁺ stage (data not shown). In order to progress to the next stage (CD44⁻CD25⁻), a productively rearranged T cell receptor β–chain must pair with a pre-T cell receptor α-chain, allowing critical signalling through this pre-T cell receptor complex to occur. This signals the CD44⁻CD25⁺ cells to both proliferate and enter the next stage of differentiation. Of interest, this is the same stage where Lalli *et al* found that cAMP blocked differentiation in FTOC.[8] This raises the question of whether ADA inhibition inhibits thymocyte differentiation by causing an elevation of adenosine which in turn elevates cAMP, perhaps via adenosine receptors.

The Effect of the Ribonucleotide Reductase Inhibitor Hydroxyurea on FTOC

One of the profound metabolic abnormalities in ADA deficient patients is the massive accumulation of dATP in their erythrocytes, to levels as high as those of ATP. This results from the successive phosphorylation of deoxyadenosine → dAMP → dADP → dATP. One mechanism by which dATP accumulation has been suggested to cause a block in thymic development is by inhibition of the enzyme ribonucleotide reductase (RR). RR catalyzes the conversion of ribonucleoside diphosphates to the corresponding deoxyribonucleoside diphosphates. This is the only route for synthesizing deoxyribonucleotides *de novo.* dATP-mediated inhibition of RR is believed to play a major role in some cell culture models of ADA deficiency. For example, deoxyadenosine resistant clones of the a murine T cell lymphoma line S49 were noted to have mutations in RR which made the enzyme insensitive to feedback inhibition by dATP[9]. Thus, to determine whether the mechanism by which dCF inhibited thymocyte differentiation in FTOC involved RR inhibition, we compared the effects of the RR inhibitor hydroxyurea (HU) with those of dCF (Figure 1). HU at 100 μM inhibited the overall cell yield and the production of double positive and single positive CD4[+] thymocytes in FTOC to the same extent as dCF. However, the cells were blocked at a later stage with HU, the CD4[-]CD8[wk+] stage that occurs after the double negative stage. In addition, there was no sparing of δγ T cells with HU as there was with dCF (Figure 1). These results suggest that inhibition of RR may not be the primary mechanism by which T cell development is blocked in ADA-deficient patients.

In summary, dCF inhibits murine thymic differentiation in FTOC without the addition of exogenous substrates for ADA. This system has several advantages over earlier systems for studying the way in which a loss of ADA enzyme activity leads to a failure of T cell development. By using specific enzyme inhibitors to RR and S-adenosylhomocysteine hydrolase, as well as adenosine receptor agonists and antagonists, for example, we plan to delineate the metabolic mechanisms responsible for the block in thymic differentiation seen in ADA deficiency.

ACKNOWLEDGMENTS

This work was supported by NIH grants CA08906 (R.R.) and AI18220 (L.F.T) and GM39699 (L.F.T.).

REFERENCES

1. Hershfield, M. S. and B. S. Mitchell. 1995. Immunodeficiency diseases caused by adenosine deaminase deficiency and purine nucleoside phosphorylase deficiency. In *The metabolic and molecular basis of inherited disease.* Seventh ed. C.R. Scriver, A.L. Beaudet, W.S. Sly and D. Valle, eds. McGraw-Hill, Inc. New York, p. 1725.
2. Blaese, R. M., K. W. Culver, A. D. Miller, C. S. Carter, T. Fleisher, M. Clerici, G. Shearer, L. Chang, Y. Chiang, P. Tolstoshev, J. J. Greenblatt, S. A. Rosenberg, H. Klein, M. Berger, C. A. Mullen, W. J. Ramsey, L. Muul, R. A. Morgan, and W. F. Anderson. 1995. T lymphocyte-directed gene therapy for ADA- SCID: initial trial results after 4 years. *Science 270*:475.
3. Bordignon, C., L. D. Notarangelo, N. Nobili, G. Ferrari, G. Casorati, P. Panina, E. Mazzolari, D. Maggioni, C. Rossi, P. Servida, A. G. Ugazio, and F. Mavilio. 1995. Gene therapy in peripheral blood lymphocytes and bone marrow for ADA- immunodeficient patients. *Science 270*:470.
4. Migchielsen, A. A. J., M. L. Breuer, M. A. van Roon, H. te Riele, C. Zurcher, F. Ossendorp, S. Toutain, M. S. Hershfield, A. Berns, and D. Valerio. 1995. Adenosine-deaminase-deficient mice die perinatally and exhibit liver-cell degeneration, atelectasis and small intestinal cell death. *Nature Genet. 10*:279.
5. Wakamiya, M., M. R. Blackburn, R. Jurecic, M. J. McArthur, R. S. Geske, J. J. Cartwright, K. Mitani, S. Vaishnav, J. W. Belmont, R. E. Kellems, M. J. Finegold, C. A. J. Montgomery, A. Bradley, and C. T. Caskey. 1995. Disruption of the adenosine deaminase gene causes hepatocellular impairment and perinatal lethality in mice. *Proc. Natl. Acad. Sci. USA 92*:3673.
6. Jenkinson, E. J. and J. J. T. Owen. 1990. T cell differentiation in thymus organ cultures. *Semin. Immunol. 2*:51.
7. Fox, R. I., L. F. Thompson, and J. R. Huddlestone. 1981. T cells express T-lymphocyte associated antigens. *J. Immunol. 126*:2062.
8. Lalli, E., P. Sassone-Corsi, and R. Ceredig. 1996. Block of T lymphocyte differentiation by activation of the cAMP-dependent signal transduction pathway. *EMBO J. 15*:528.
9. Ullman, B., A. Cohen, and D. W. Martin, Jr. 1976. Characterization of a cell culture model for the study of adenosine deaminase- and purine nucleoside phosphorylase-deficient immunologic disease. *Cell 9*:205.

90

THYMIDINE KINASE EXPRESSION

A Marker for Malignant Cells

Markus Hengstschläger,[1] Michael Pfeilstöcker,[2] and Edgar Wawra[1]

[1]Institute of Molecular Biology
University of Vienna, Vienna Biocenter
Dr. Bohrgasse 9, A-1030 Vienna, Austria
[2]Ludwig Boltzmann Institute for Leukaemia Research and Haematology
Hanusch Hospital, Heinrich Collin-Str. 13
A-1140 Vienna, Austria

1. SUMMARY

The expression of thymidine kinase—an enzyme of the DNA precursor pathway—is strictly regulated during the normal cellular cycle, but is much higher and permanently expressed in malignant growing cells. Using a new cytofluorometric assay for thymidine kinase in single cells, we were able to discriminate between normal growing cells and virally transformed cells or lines derived from tumours. In material (blood and bone marrow) taken from leukaemia patients, we identified the leukaemic cells in a surplus of normal leucocytes. From cell cultures representing a tumour progression model, only the later, and malignant, stages showed enhanced fluorescence, whereas benign tumour cells looked normal.

2. WHAT CAUSES A CELL TO BECOME A TUMOUR CELL?

The replication cycle of eukaryotic cells may generally be considered as a series of transitions or checkpoints, where the cells decide what to do next: to synthesise DNA, to enter mitosis, simply continue to growth, to differentiate etc. One of these is the point where the cell makes the decision "to divide or not to divide". This is dependent on the metabolic state of the cell (is it able to perform the energy consuming steps of DNA synthesis and mitosis?) but in a multicellular organism it is also a function of signals from the surrounding (is it good for the organism to have two cells instead of one?). In case the cell decides to divide, it has to activate a number of genes—so-called "S-phase genes"—which

Purine and Pyrimidine Metabolism in Man IX,
edited by Griesmacher *et al.* Plenum Press, New York, 1998.

455

are necessary for DNA synthesis, but remain silent during any other period of the cell cycle and also in resting or differentiated cells. Some members of these S-phase gene-group obviously are the DNA polymerases and the enzymes of the DNA precursor metabolism which supply the ongoing synthesis with the necessary deoxynucleotide triphosphates. The activation of these (and probably of many other unidentified genes) is a matter of few hours, after that time the cell possibly has to pass another checkpoint to ensure that everything is ready for DNA synthesis—and then it starts.

Tumour cells behave differently. They have obviously lost the checkpoint where the decision to divide is made and they grow independently and—if malign—as fast as they can. So they should also have lost control over their S-phase genes! For such a cell, two possibilities arise: (i) either it has all its S-phase genes constantly switched off—not able to replicate, it will remain latent somewhere in the organism; (ii) or it has all these genes permanently active—is therefore able to divide and to form a tumour.

We already have shown that genes encoding thymidine kinase, dihydrofolate reductase, and both subunits of nucleosidediphosphate reductase, which are well regulated on mRNA level in normal cells, lose this regulation and are constantly enhanced in tumour virus transformed cells [1]. Moreover, we have shown that polyoma virus large T antigen (a protein necessary for polyoma virus-caused cellular transformation) is alone sufficient to change in otherwise normal cells the expression pattern of S-phase genes from the regulated phenotype to the unregulated one [2].

3. HOW DO ONCOGENES AND TUMOUR SUPPRESSORS FIT IN THIS PICTURE?

Recently, two families of regulatory proteins have been described: the cyclines, which fluctuate during the cellular cycle and are therefore able to act as cellular clocks, and the cyclin-dependent kinases (CDK), which—in concert with the cyclines—specifically phosphorylate other regulatory proteins. The current model invokes successive waves of different CDK activity, regulated in turn by different cyclins. One subgroup of these—the cyclines D—reach their maximum intracellular level few hours before S-phase and are therefore most probably involved in the onset of the replicative cycle.

The S phase regulated genes share common properties in respect to their promoters. DNA polymerase α, dihydrofolate reductase, thymidylate synthase and thymidine kinase are all regulated by the transcription factor E2F, which itself is regulated by the retinoblastoma gene product (pRb) (see Fig. 1). pRb binds and inactivates E2F, but phosphorylation of pRb sets E2F free [3]. This pRb is target for the tumour antigens of DNA tumour viruses, so that in virus transformed cells, pRb allows E2F to activate these genes. It was demonstrated that the thymidine kinase gene is switched on after viral infection via exactly this mechanism [4].

But pRb is cell cycle regulated by the cyclin D/CDK system, and this is known to be a target for two of the most prominent tumour suppressor gene products p16 and p53. p16 is an inhibitor of the cyclin D/cdk4 kinase, p53 is an activator of p21 which inhibits cyclin D/cdk2 and cyclin E/cdk2. Therefore, the described deregulation of S-phase genes is an indicator for many different tumour causing mechanisms: presence of viral coded tumour antigen, defect in pRb, and defect in either p53, p21 or p16. This explains why we found this effect in so many different neoplastic cell populations: in various tumours, in leukaemia samples, in cell lines derived from tumours or obtained after transformation by tumour virus.

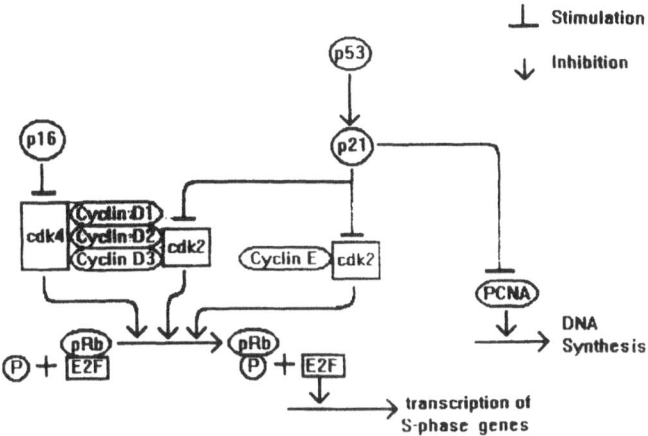

Figure 1. Simplified mechanisms for the regulation of S-phase genes (see text for abbrevations).

4. THYMIDINE KINASE IS A CONVENIENT MARKER FOR THE DEREGULATION OF S-PHASE GENES

Thymidine kinase (TK) catalyses the ATP-dependent phosphorylation of thymidine and deoxyuridine. The activity of TK is strictly regulated during the normal cell cycle, peaking at the onset of DNA synthesis but remaining extremely low in resting cells. We already have been shown that the change in TK gene expression during the cell cycle is accompanied by a similar change in TK enzyme activity. Obviously, activity of TK is dominated by the amount of its mRNA, and the difference between normal and malignant cells at the level of transcription is mirrored by the enzyme activity. Compared with other S phase regulated genes, TK appears to be the most convenient indicator for this effect: (i) the half live of the enzyme is short enough to reflect variations during the cell cycle, (ii) there is no interfering cell-cycle dependent regulation on protein level, and (iii) TK activity is easy to detect.

We used elutriation centrifugation for the separation of cell populations from different phases of the cellular cycle. In the fractions so obtained, we measured enzyme activities and mRNA levels and could detect the differences between normal and malignant cells described above [2]. However, it would be much desirable to measure this effect on a single cell basis. Besides the statistical advantage, it would also enable us to analyse cellular mixtures and identify few transformed cells in a surplus of normal ones. The idea for such an assay was to synthesise an artificial but fluorescent substrate for thymidine kinase, which is phosphorylated by the enzyme like thymidine. the accumulation of the phosphorylated products in the cells should therefore be visible by their fluorescence. Moreover, simultaneous measurement of this fluorescence together with that of a DNA sensitive stain should provide us with information for both thymidine kinase activity and phase of the cell cycle for each individual cell.

Finally, we succeeded to synthesise such a dye, and we found that the intracellular accumulation of its fluorescence was strictly correlated with the overall thymidine kinase activity of the cells [5]. This implies that from the series of necessary events (uptake into the cell, phosphorylation to the mono-, di-, and triphosphate), the step catalysed by thymidine kinase is the bottle neck of the whole pathway and therefore rate limiting. Using a cytofluorometer, this assay allows to discriminate between normal growing cells,

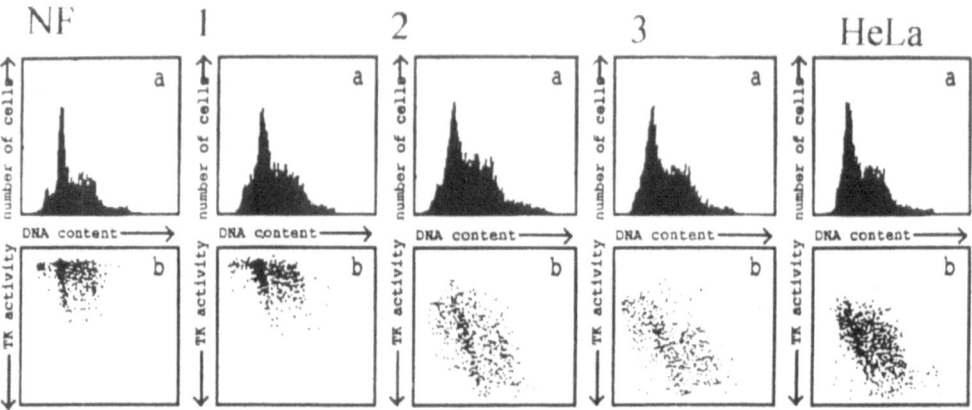

Figure 2. Examples for the simultaneous cytofluorometric determination of DNA and intracellular TK activity. Top: distribution of DNA content during the analysed population. Bottom: two dimensional presentations of DNA content (abscissa) versus thymidine kinase (TK) activity (ordinate, note that increasing enzyme activity goes down the axis!) NF = normal fibroblasts; 1,2,3 = three progressive states of the same Ewing tumour; HeLa = HeLa cervix carcinoma cells.

like diploid fibroblasts, and virally transformed cells or lines derived from tumours. But we also had identified leukaemic cells in patients bone marrow or peripheral blood [6].

In order to show how our results have to be interpreted, a few examples are shown in Figure 2. The DNA distribution presented on top of each diagram represents the distribution of cells in G1, S and G2 phase (the G2 peak on the right contains cells with exactly twice the amount of DNA than the G1 peak on the left, between the two peaks are the S phase cells with intermediate DNA content.). The two dimensional presentation below shows simultaneous measurement of two fluorescences, reflecting DNA amount and thymidine kinase (TK) activity for each cell. Normally growing cells have low TK activity in G1, this activity increases during early S and returns to about the original level in G2. (Resting cells have even lower TK activity.) Malignant growing cells always exhibit more

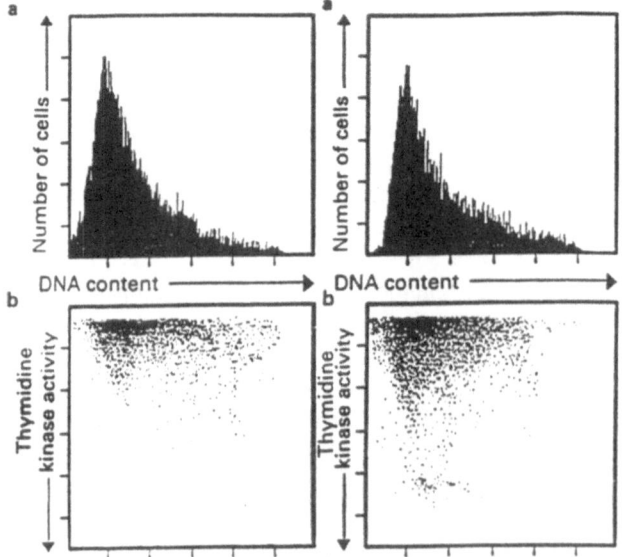

Figure 3. Visualisation of a low fraction of neopalstic cells. (See Fig. 2 for explanation of graphical presentation) Left: EBV transformed lyphocytes; right: mixture of KG-1 cells and EBV transformed lymphocytes in a ratio of 1: 10 000.

Table 1. Cytofluorometric measurement of thymidine kinase
in different forms of leukaemia

Diagnosis	Sex	Sample	Result
AML/FAB M0	M	BM	+
ALL/FAB L2	F	BC	++
ALL/FAB L3	M	BM	++
		PB	++
AML/FAB M2	M	BC	++
		PB	++
AML/FAB M2	M	BM	++
Plasma cell leukaemia	F	PB	++
AML/FAB M3	F	BM	+
		PB	++
AML/FAB M2/4	F	BC	++
		PB	++
Normal	M	stimulated lymphocytes	neg.
		unstimulated lymphocytes	neg.
Normal	M	unstimulated lymphocytes	neg.

++ means, that positive identification would be possible even in the absence of a negative control sample. BM: bone marrow, BC: buffy coat, PB: peripheral blood.

TK activity than S phases of normal cells do. These cells are therefore found further down the axis in the diagram.

The examples in Fig. 2 are normal fibroblasts (extreme left) and HeLa cells (a classical tumour cell line, on the extreme right). The other diagrams represent cell lines obtained from three different subsequent biopsies from a patient suffering on Ewing tumour. The first biopsy (1) was taken from the benign growing tumour, the other samples (2, 3) were taken from the same tumour during its rapid progression to malignancy. Cytogenetic analysis did not indicate any alteration in the areas of the gene locus for thymidine kinase nor for pRb. But the later two biopsies show a deletion of the region where the p16 gene is localised [7]. (Indeed, analysis on RNA levels confirmed that the lines 2 and 3 were p16 negative). Obviously, the defect in p16 caused the cells to become malignant. We further conformed this idea by genetically manipulating p16-negative cells, i.e. we transiently transformed the cells with the p16 gene, and were able to find the expected reversion in the thymidine kinase expression pattern when using our fluorescence assay [8].

When applied on artificial mixtures of logarithmically growing normal and transformed cells, this method enabled us to detect the malignant in a 10 000 fold excess of normal ones (Fig. 3) [9]. In order to test the practical application of this method, we run a series of samples taken from patients with different form of leukaemia [6]. In all these samples, we found a population of cells with high activity of thymidine kinase which was neither found in control lymphocytes nor in PHA stimulated lymphocytes (Table 1). The benefits of such a simple and general tumour marker are obvious: this opens the possibility to use this method widely for clinical diagnosis, but it may also serve as a tool in cell biology for the study of the mechanism of cell transformation.

REFERENCES

1. HENGSTSCHLÄGER, M., MUDRAK, I., WINTERSBERGER, E., and WAWRA, E.(1994) A Common Regulation of Genes Encoding Enzymes of the Deoxynucleotide Metabolism is Lost after Neoplastic Transformation Cell Growth and Differentiation 5, 1389–1394.

2. HENGSTSCHLÄGER, M., KNÖFLER, M., MÜLLNER, E.W., OGRIS, E., WINTERSBERGER, E., and WAWRA. E. (1994) Different Regulation of Thymidine Kinase during the Cell Cycle of Normal versus DNA Tumor Transformed Cells J. Biolog. Chem. 269, 13836–13842.
3. NEVINS, J.R. (1992) E2F: a link between the Rb tumor suppressor protein and viral oncoproteins. Science 258, 424–429.
4. OGRIS, E., ROTHENEDER, H., MUDRAK, I., PICHLER, A., and WINTERSBERGER, E. (1993) A binding site for transcription factor E2F is a target for transactivation of murine thymidine kinase by polyomavirus large t antigen and plays an important role in growth regulation of the gene. J.Virol. 67, 1765–1771.
5. HENGSTSCHLÄGER, M., and WAWRA, E. (1993) Cytofluorometric Assay for the Determination of Thymidine Uptake and Phosphorylation in Living Cells Cytometry 14, 39–45.
5. HENGSTSCHLÄGER, M., PFEILSTÖCKER, M., and WAWRA, E. (1996) Identification of leukaemic cells in bone marrow and blood samples by a new cytofluorometric assay Brit. J. Cancer 73, 1237–1240.
6. HENGSTSCHLÄGER, M., PUSCH, O., HENGSTSCHLÄGER-OTTONAD, E., AMBROS, P.F., BER-NASCHEK, G., and WAWRA, E. (1996) Loss of p16/MTS1 Tumor Suppressor Gene Causes E2F-Mediated Deregulation of Essential Enzymes of the DNA Precursor Metabolism DNA and Cell Biology 15, 41–51.
7. HENGSTSCHLÄGER, M., HENGSTSCHLÄGER-OTTONAD, E., PUSCH, O., and WAWRA, E. (1996) The role of p16 in E2F-dependent thymidine kinase regulation Oncogene 12, 1635–1643.
8. HENGSTSCHLÄGER, M., and WAWRA, E. (1993) Cytofluorometric Determination of Thymidine Kinase Activity in a Mixture of Normal and Neoplastic Cells Brit. J. Cancer 67, 1022–1025.

91

NEUROBLASTOMA CELL APOPTOSIS INDUCED BY MYCOPHENOLIC ACID

E. Messina, F. Savini, R. Lisio, C. P. Quaratino, and A. Giacomello

Dipartimento di Oncologia e Neuroscienze
Università 'G. D'Annunzio'
Chieti, Italy

IMP-dehydrogenase (IMPDH) inhibitors lead to cellular GTP depletion, G_0/G_1 arrest and often differentiation of malignant cells.[1] Guanine nucleotides depletion has been involved in the pathogenesis of the neurological dysfunction in Lesch-Nyhan disease.[2] Furthermore it has been shown that malignant brain tumors express higher levels of IMPDH than normal brain tissue.[3] In the present study the cytotoxicity of mycophenolic acid (MPA) (the most specific IMPDH inhibitor) has been examined in human neuroblastoma cell lines.

DESIGN AND METHODS

Experiments were carried out using exponentially growing human cell lines (LA-N-5, SK-N-AS and SH-EP) in RPMI medium supplemented with 10% heat-inactivated FCS and 2 mM glutamine in an atmosphere of 5% CO_2 at 37°C. Proliferation assays were performed using cell proliferation ELISA BrdU Labeling and Detection Kit (Boehringer Mannheim Italia SpA). Quantitative assessment of the cell cycle phases and of apoptotic cells was carried out by flow cytometry. Apoptosis was measured by flow cytometry using the In Situ Cell Death Detection Kit, Fluorescein (Boehringer Mannheim Italia SpA). Nucleotide pools were quantitated by HPLC.[4] Determinations of p53 were carried out by Western blotting.

RESULTS

MPA induced time- and dose-dependent inhibition of proliferation (Fig. 1), decrease in GTP pools, G0/G1 arrest and cell death by apoptosis (Fig. 2) in LA-N-5. The fraction of apoptotic cells was 45% (mean of three experiments) at 1μM MPA after 48 hours of treatment. Apoptosis was associated with up regulation of p53 (Fig. 3). The effects of MPA on

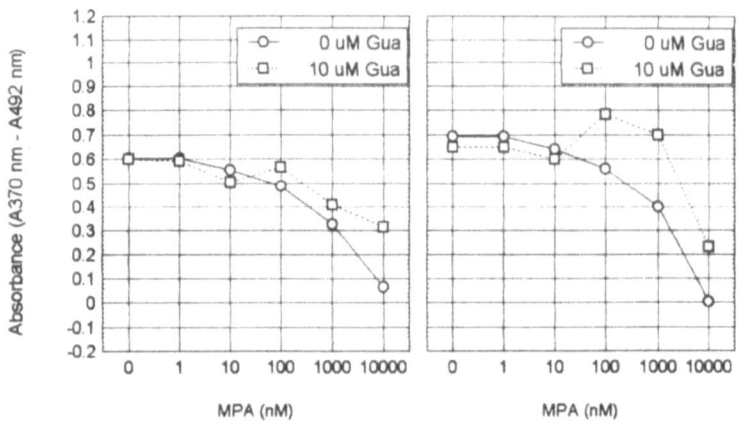

Figure 1. BrdU incorporation in LA-N-5 NB cell line after 16 and 40 hours of incubation with 0 to 10 μM MPA with or without the simultaneous addition of 10 μM guanine (Gua).

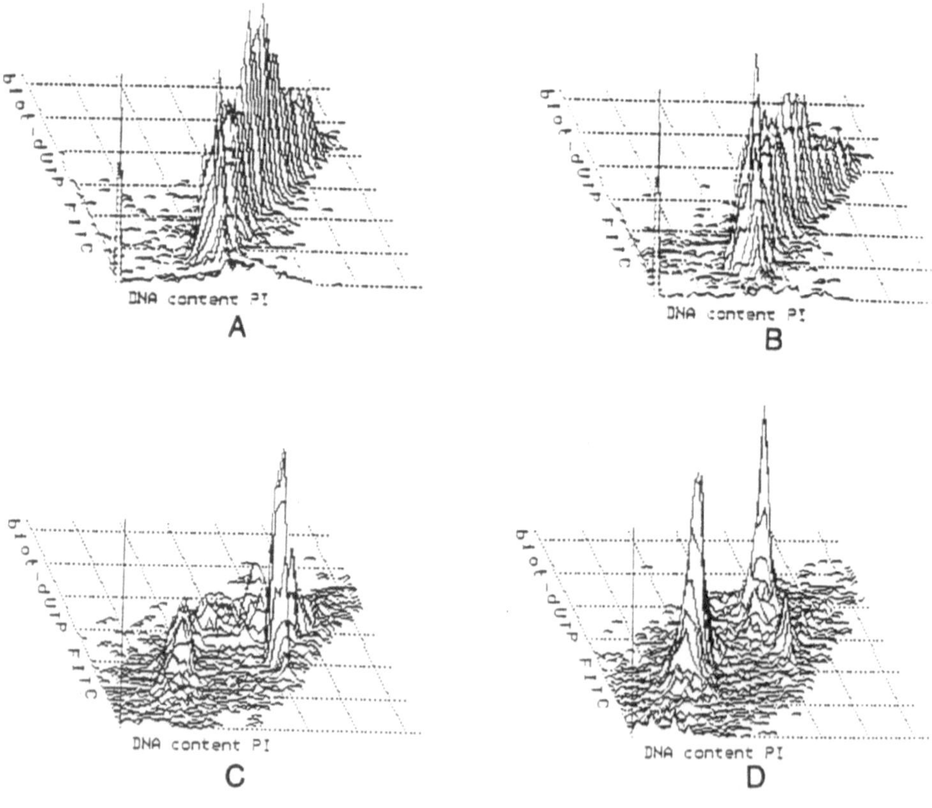

Figure 2. 3D-evaluation of cell cycle and apoptosis in LA-N-5 NB cell line treated for 48 hours with 1 μM MPA (D), 10 μM MPA (C), 100 μM guanosine (B), 100 μM guanosine and 1 μM MPA (A). Apoptosis was reversed by the simultaneous addition of guanosine (A). Apoptosis (%): 3.4 (A), 3.0 (B), 60.0 (C), 57.9 (D).

Figure 3. Espression of p53 protein. Western-blot analysis of protein estracts from LA-N-5 NB cell line treated for 48 hours with 10 µM MPA (lane 1), 1 µM MPA (lane 2), 0 µM MPA (lane 3), 50 µM guanosine (lane 4), 1µM MPA and 50 µM guanosine (lane 5).Thirty micrograms of total protein were loaded on each lane on a SDS/4–20% gradient polyacrylammide gel and p53 was detected with monoclonal antibody Ab-6 (Oncogene) by the enhanced chemiluminescence procedure.

BrdU incorporation was, at least partially, reversed by coincubation with 10 µM guanine (Fig. 1), while GTP depletion, up regulation of p53 and apoptosis were completely reversed by the simultaneous addition of 50–100 µM guanosine (Figs. 2, 3) to the culture medium.

Similar results were obtained using the human SK-N-AS and SH-EP neuroblastoma lines.

DISCUSSION

In human fibroblasts it has been shown that CTP, GTP or UTP depletion elicit a p53-dependent reversible early G1 arrest in the absence of detectable DNA damage.[5] Overexpression of p53 may force the cell to exit the cell cycle and, depending on the cell type lineage, progress toward apoptosis or cell differentiation.[6,7] At least in HL-60, p53 protein levels may determine whether a given cell should prefer one pathway over the other to exit the cell cycle.[7] There is mounting evidence that p53 is often latent in non-stimulated cells, becoming activated only upon exposure to a stress signal.[6] In some tumors, such as in undifferentiated neuroblastomas,[8] latency is maintained through retention of p53 in the cytoplasm, and activation entails its shuttling into the nucleus.[6] We have shown that such a mechanism is operating when neuroblastoma cells are treated with low, non apoptotic MPA concentrations leading to cell differentiation.[9] In the present study, it has been shown that in neuroblastoma cell lines MPA, at concentrations ranging from 10^{-7} to 10^{-5} M, induces an increase in p53 levels in a dose-dependent manner, a G_0/G_1 arrest and apoptosis. These effects are, at least partially, reversed by the simultaneous addition of guanine or guanosine, suggesting that their cause is GTP depletion. Our findings indicate a potential role of MPA or of his ester pro-drug (Mofetil) in the menagement of neuroblastoma patients.

REFERENCES

1. Franklin TJ, Edwards G and Hedge P. Inosine 5'-monophosphate dehydrogenase as a chemotherapeutic target. Adv Exp Med Biol 1994; 370: 155–60.
2. Rossiter BJF, Caskey CT. Hypoxanthine-guanine phosphoribosyltransferase deficiency: Lesch-Nyhan syndrome and gout. In: Scriver CR, Beaudet AL, Sly WS,Valle D (eds) The metabolic and molecular bases of inherited disease, 7th edn.,Vol II, 1995, McGraw-Hill, New-York, pp 1679–1706.
3. Pillwein K, Chiba P, Knoflach A, et al. Purine metabolism of human glioblastoma in vivo. Cancer Res 1990; 50: 1576–79.

4. Di Pierro D, Tavazzi B, Perno CF et al. An ion-pairing high performance liquid chromatographic method for the direct simultaneous determination of nucleotides, deoxynucleotides, nicotinic coenzymes, oxy-purines, nucleosides, and bases in perchloric acid cell extracts. Anal Biochem, 1995; 231: 407–12.

5. Linke SP, Clarkin KC, Di Leonardo A, et al. A reversible, p53-dependent G_0/G_1 cell cycle arrest induced by ribonucleotide depletion in the absence of detectable DNA damage. Genes & Development 1996; 10: 934–47.

6. Oren M, Prives C. p53: upstream, downstream and off stream. Review of the 8th p53 workshop Dundee, July 5–9, 1996. BBA 1996; 1288: R13-R19.

7. Ronen D, Schwartz D, Teitz Y et al. Induction of HL-60 cells to undergo apoptosis is determined by high levels of wild-type p53 protein whereas differentiation of the cells is mediated by lower p53 levels. Cell Growth & Differentiation 1996; 7: 21–30.

8. Moll UM, LaQuaglia M, Bénard J, Riou G. Wild-type p53 protein undergoes cytoplasmic sequestration in undifferentiated neuroblastomas but not in differentiated tumors. Proc Natl Acad Sci USA 1995; 92:4404–4411.

9. Savini F, Rucci C, Messina E, et al. Differentiating and biochemical effects of a reduction of intracellular GTP levels induced by mycophenolic acid (MPA) in human neuroblastoma (NB) cell lines. Adv Exp Med Biol (This Volume).

92

THE SINGLE DEOXYNUCLEOSIDE KINASE IN *DROSOPHILA MELANOGASTER*, *Dm*-dNK, IS MULTIFUNCTIONAL AND DIFFERS FROM THE MAMMALIAN DEOXYNUCLEOSIDE KINASES

Birgitte Munch-Petersen,[1] Jure Piskur,[2] and Leif Søndergaard[3]

[1]Department of Life Science and Chemistry
Roskilde University
P. O. Box 260, Roskilde, Denmark
[2]Department of Microbiology 301
Denmarks Technical University
Lyngby, Denmark
[3]Department Gen.
Copenhagen University
Copenhagen, Denmark

1. INTRODUCTION

In mammalian cells, four enzymes are responsible for the salvage of deoxyribonucleosides. The thymidine kinases TK1 and TK2, the deoxycytidine kinase dCK and the deoxyguanosine kinase, dGK. TK1 is expressed only in dividing cells[1] and phosphorylates thymidine and deoxyuridine. The other three kinases are constitutively expressed. TK2 is located in the mitochondria and is also an efficient deoxycytidine and deoxyuridine kinase.[2] dCK phosphorylates deoxycytidine, deoxyadenosine and deoxyguanosine,[3] and the newly isolated mithochondrial dGK can also phosphorylate deoxyadenosine.[4] The genes of the four human enzymes have been cloned,[5–8] and homology and enzymatic studies indicate that human TK1 together with mammalian TK1 from mouse and chicken are closely related to pox-viral TK's and TK from *E. coli*,[9] whereas human dCK, dGK and TK2 shares many properties with Herpetic TK's.[8]

To our knowledge, deoxynucleoside kinases from insects have not previously been characterized, and we know of no information about these enzymes in non-chordate animals. In the present study we have investigated the deoxynucleoside kinase pattern in the cultured *Drosophila* S-2 cells. Our results indicate that the deoxynucleoside kinase pattern differs from the mammamlian pattern.

Purine and Pyrimidine Metabolism in Man IX,
edited by Griesmacher *et al.* Plenum Press, New York, 1998.

2. METHODS

2.1. Cells

The *Drosophila* S-2 embryonic cells were grown in the Schneider medium (Sigma) supplemented with 10% fetal calf serum at 25°C. 8.6 × 10^{10} actively dividing S-2 embryo cells were pelleted by centrifugation, resuspended in hypotonic phosphate buffer (50 mM pH 7.6) containing 2 mM dithiotreitol, 20% glycerol, 5 mM benzamidine, 0.5 mM phenyl-methylsulfonyl fluoride and 50 mM ∈-aminocaproic acid, and disrupted by sonication. The cell debris was pelleted by centrifugation for 20 min at 15,000 G.

2.2. Enzyme Purification

The purification of kinase activity was performed according to previously described procedures:[2] After removal of nucleic acids with streptomycinsulfate and precipitation with ammoniumsulfate, the desalted extract was chromatographed on DEAE. The peak with thymidine/deoxycytidine kinase activity was applied on dTMP-Sepharose and the activity was eluted with 2 mM thymidine in the buffer. Thymidine was removed and the enzyme concentrated by phenyl-Sepharose chromatography, where the enzyme was eluted with 10 mM CHAPS in the buffer. The enzyme activity was kept at 4°C during the purification procedures and stored at −80°C.

2.3. Sub-Unit and Native Size

The sub-unit size was determined by SDS-PAGE electrophoresis by standard procedure. The size of the native enzyme was determined by gel-filtration on Superose 12 connected to a Gradifrac with a P-50 pump. The buffer contained 50 mM imidazol, 1 mM dithiothreitol, 5 mM $MgCl_2$ and 0.1 M KCl.

2.4. Enzyme Assays

Deoxyadenosine kinase (dAK), deoxyguanosine kinase (dGK), deoxycytidine kinase (dCK) and thymidine kinase (TK) activities were determined from initial velocity measurements by the DEAE-cellulose 81 filter paper method as described.[2] The standard assay conditions were: 50 mM Tris-HCl, pH 8.0 (22°C), 2.5 mM $MgCl_2$, 10 mM dithiothreitol, 3 mg/ml BSA, 2.5 mM ATP and 10 µM radiolabelled compound.

TMP kinase activity was examined by thin-layer chromatography on PEI-cellulose. The standard assay conditions were as described for the TK activity, except that ^{3}H-thymidine was replaced with ^{3}H-TMP. The reaction was terminated by addition of a mixture of unlabelled thymidine, dTMP, dTDP and dTTP, 20 mM of each, and 5 µl of this mixture was applied on PEI-cellulose F thin-layer plates, and chromatoghraphed with 0.5 M $LiCl_2$. The spots containing nucleosides and nucleotides were identified and counted. The radioactive substrates were from Amersham, UK and the radioactive analogs from Moravek. Biochemicals Inc. Brea.

3. RESULTS AND DISCUSSION

3.1. Purification of Deoxynucleoside Kinase Activity

When extracts from human cells are separated by DEAE chromatography, three kinases are obtained: TK1, TK2 with both TK and dCK activity and dCK with both dCK, dAK and dGK activity.[2] From the *Drosophila* cells only one peak of kinase activity was obtained, containing TK (100%), dCK (100%), dAK (15%) and dGK (4%) activity. This kinase was eluted at 50 mM KCl which is the same condition used for elution of human TK2. The four kinase activities could not be separated by TMP-Sepharose and phenyl-Sepharose chromatography, and the proportion between the activities associated with the pure enzyme was the same as found in the DEAE fractions. SDS-PAGE gel-electrophoresis showed a single band at 29 kDa. This indicates that only one deoxynucleoside kinase activity with the capacitiy to phosphorylate all four deoxynucleosides exists in *Drosophila melanogaster*. We have designated this kinase as *Dm*-dNK. In comparison with the three human kinases, the specific activity was high, 30 000 nmol/min/mg. This means 3, 50 or 207 fold higher activity than the specific acitivities of TK1, TK2 or dCK, respectively (Table 1).

3.2. Characterisation of *Dm*-dNK

The properties of *Dm*-dNK were compared with those of TK1, TK2 and dCK from human cells using the well described characteristic differences between the human enzymes, i.e. the native size, substrate kinetics, phosphate donor and nucleoside analog specificity. The results are outlined in Tables 1 and 2.

3.2.1. Molecular Weights. The sub-unit size was similar to the native size, and therefore *Dm*-dNK apparently does not oligomerize (Table 1). This behaviour is similar to TK2.

3.2.2. Phosphate Donor and Acceptor Specificity. Dm-dNK phosphorylated deoxycytidine and thymidine with the same efficiency. With both substrates, the specific activities were 30,000 nmol/min/mg, the K_m values were about 1 µM, and the kinetics was hyperbolic. This was in contrast to the non-hyperbolic behaviour of human TK1,[10] TK2[2] and dCK.[11] Furthermore, the K_m values of TK1 and TK2 were 15–17 fold higher.[2,10,11]

Thin-layer chromatography on PEI-cellulose showed that there was noTMP kinase activity associated with *Dm*-dNK.

A characteristic feature of TK2 is the ability to use CTP as phosphate donor with about 60% efficiency of ATP. With *Dm*-dNK, CTP had the same phosphotransfer capability as ATP, i.e. 100%.[2]

Table 1. Size and activities of deoxynucleoside kinases

	Human cells			Drosophila
	TK1	TK2 (TK/dCK)	dCK	*Dm*-dNK
Sub-uni (kda)	24	28	30	29
Native size (kDa)	55	33	60	30
+ATP (kDa)	110	33	60	30
Cellular activity (nmol/min/10^{10} cells)	500	10	400	35
Spec. act. (µmol/min/mg)	10	0.6	0.145	30
k_{cat} (sec^{-1})*	17	0.3	0.16	15

TK1 and TK2 are from phytohemagglutinin stimulated human lymphocytes,[2] dCK from human spleen[3] and lymphocytes.[3] k_{cat} is calculated for the tetramer form of TK1 and the dimer form of dCK. The data are averages from several experiments. The standard deviations are below 15%.

Table 2. Phosphate acceptor specificity

	Human cells			Drosophila
	TK1	TK2 TK/dCK	dCK	*Dm*-dNK
Thymidine	1.0	1.0	0	1.0
Deoxycytidine	0	0.6	1.0	1.0
Deoxyadenosine	0	0	0.26	0.15
Deoxyguanosine	0	0	0.11	0.04
Ara-C	0	0.02	0.35	0.5
Ara-T	0	0.6	0	0.15
AZT	0.4	0.05	0	0.003
ACV	0	0	0	0

The activites are measured with 10 µM of substrate and given as relative to the activity with 10 µM thymidine. Data for TK1, TK2 and dCK are from Munch-Petersen et al.[2], Eriksson et al.[14] and Bohman and Eriksson.[3]

3.2.3. Specificity Towards Anti-Viral and Anti-Cancer Nucleoside Analogues. TK1, TK2 and dCK phosphorylate a number of medically important nucleoside analogues with characteristic different specificity. As shown in Table 2, *Dm*-dNK showed a mixed specificity pattern. The activity with Ara-C (arabinosylcytosine) was 50% of that obtained with deoxycytidine. In this respect *Dm*-NK is like human dCK. The activity of *Dm*-dNK with Ara-T (arabinosylthymidine) was 10% and AZT (3'-azidothymidine) was only 0.3% of the activity with thymidine. This is less efficient than with TK2. ACV (acyclovir) which is a good substrate for Herpes virus TK[12], was not phosphorylated by *Dm*-dNK.

3.2.4. Substrate Competition. The ability of *Dm*-dNK to phosphorylate both thymidine and deoxythymidine as well as the sub-unit and native sizes indicated a closer resemblance with TK2 than with TK1 and dCK. A characteristic feature of TK2 is the preference of thymidine to deoxycytidine. Thus, K_i for deoxycytidine inhibition of the TK activity is 630 µM, whereas K_i for thymidine inhibition of the dCK activity is only 6 µM. When this competition pattern was examined for *Dm*-dNK, the K_i values for both inhibitions were 1 µM. This means that deoxycytidine and thymidine are equally preferred as substrates for *Dm*-dNK.

3.2.5. Feed-Back Inhibition. TTP was an efficient competitive inhibitor of the *Dm*-dNK with a K_i of 5 µM.

4. CONCLUSION

In *Drosophila melanogaster* S-2 cells only a single deoxynucleoside kinase, *Dm*-dNK, was found. This kinase was able to phosphorylate all four deoxynucleosides. According to the DEAE elution profile, the sub-unit and native size, and the capacity to use CTP as phosphate donor, *Dm*-dNK resembles TK2. On the other hand, the capacity to phosphorylate Ara-C, deoxyadenosine and deoxyguanosine indicated a resemblance to dCK, and the high turnover indicated resemblance to TK1. As *Dm*-dNK did not phosphorylate TMP and acyclovir, and was efficiently feed-back inhibited by TTP (K_i = 5 µM) there was no indication of a resemblance to the Herpes viral thymidine kinases.

The conclusion drawn from these enzymatic studies is that the multifunctional *Dm*-dNK is different from the known mammalian kinases and Herpes viral TK, and may not be a member of either of these families.

REFERENCES

1. M.G. Kauffman and T.J. Kelly, Cell cycle regulation of thymidine kinase: Residues near the carboxyl terminus are essential for the specific degradation of the enzyme at mitosis, Mol. Cell. Biol. 11, 2538–2546 (1991).
2. B. Munch-Petersen, L. Cloos, G. Tyrsted, and S. Eriksson, Diverging substrate specificity of pure human thymidine kinases 1 and 2 against antiviral dideoxynucleosides, J. Biol. Chem. 266, 9032–9038 (1991).
3. C. Bohman and S. Eriksson, Deoxycytidine kinase from human leukemic spleen: Preparation and characterization of the homogenous enzyme, Biochemistry 27, 4258–4265 (1988).
4. L.Y. Wang, A. Karlsson, E.S.J. Arnér, and S. Eriksson, Substrate specificity of mitochondrial 2'-deoxyguanosine kinase-efficient phosphorylation of 2-chlorodeoxyadenosine, J. Biol. Chem. 268, 22847–22852 (1993).
5. H.D. Bradshaw, Jr. and P.L. Deininger, Human thymidine kinase gene: Molecular cloning and nucleotide sequence of a cDNA expressible in mammalian cells, Mol. Cell. Biol. 4, 2316–2320 (1984).
6. E.G. Chottiner, D.S. Shewach, N.S. Datta, E. Ashcraft, D. Gribbin, D. ginsburg, I.H. Fox, and B.S. Mitchell, Cloning and expression of human deoxycytidine kinase cDNA, Proc. Natl. Acad. Sci. USA 88, 1531–1535 (1991).
7. L.Y. Wang, U. Hellman, and S. Eriksson, Cloning and expression of human mitochondrial deoxyguanosine kinase cDNA. FEBS Lett. 390, 39–43 (1996).
8. M. Johansson and A. Karlsson, Cloning of the cDNA and chromosome localization of the gene for human thymidine kinase 2, J. Biol. Chem. 272, 8454–8458 (1997).
9. M.E. Black and D.E. Hruby, Nucleotide sequence of the *Eschericia coli* thymidine kinase gene provides evidence for conservation of functional domains and quaternary struture, Molec. Microbiol. 5, 373–379 (1991).
10. B. Munch-Petersen, G. Tyrsted, and L. Cloos, Reversible ATP-dependent transition between two forms of human cytosolic thymidine kinase with different enzymatic properties, J. Biol. Chem. 268, 15621–15625 (1993).
11. B. Kierdaszuk, R. Rigler, and S. Eriksson, Binding of substrates to human deoxycytidine kinase studied with ligand-dependent quenching of enzyme intrinsic fluorescence, Biochemistry 32, 699–707 (1993).
12. G.B. Elion, Acyclovir - Discovery, Mechanism of Action, and Selectivity, J. Med. Virol. Suppl.1, 2–6 (1993).
13. E.S.J. Arnér, M. Flygar, C. Bohman, B. Wallström, and S. Eriksson, Deoxycytidine kinase is constitutively expressed in human lymphocytes: Consequences for compartmentation effects, unscheduled DNA synthesis, and viral replication in resting cells, Exp. Cell Res. 178, 335–342 (1988).
14. S. Eriksson, B. Kierdaszuk, B. Munch-Petersen, B. Öberg, and N.G. Johansson, Comparison of the substrate specificities of human thymidine kinase 1 and 2 and deoxycytidine kinase toward antiviral and cytostatic nucleoside analogs, Biochem. Biophys. Res. Commun. 176, 586–592 (1991).

93

EFFECT OF DEOXYNUCLEOSIDES ON THE REPAIR OF UV INDUCED DNA BREAKS

Birgitte Munch-Petersen

Department of Life Sciences and Chemistry
Roskilde University
DK 4000 Roskilde, Denmark

1. INTRODUCTION

The accuracy of replication and repair of DNA is dependent on a balanced supply of the four DNA precursors, the dNTP's.[1] In dividing cells, the dNTP pools are limited to a few minutes of DNA synthesis.[2] In quiescent cells, the pools are many fold lower—in the range of $0.1–0.5 \ 10^{-6}$ M.[3,4] The low levels may be ascribed to the fact that two important enzymes in the dNTP production, ribonucleotide reductase and thymidine kinase, are not expressed in quiescent cells.[5,6] When DNA is damaged by ultra-violet light and alkylating agents, the damages are repaired by nucleotide-excision repair with repair patch sizes of about 35–40 nucleotides.[7] Thus, a limited supply of dNTP's may be a hindrance for the DNA repair process.

In this study the effect of micromolar, non-carcinogenic deoxyribonucleosides on rejoining of UV-induced DNA strand-breaks is examined.

2. METHODS

2.1. Cells and Irradiation

Lymphocytes were isolated from blood samples taken from healthy donors, by Isopaque-Ficoll density gradient centrifugation. The cells were washed three times in PBS+5% FCS, suspended in PBS (5×10^6 /ml) and irradiated with two Philips UV lamps (TUV 6 W, 254 nm peak emission). In all experiments, the UV dose was 4 J/m^2 given with a dose rate of 0.2 J/m^2/s. The irradiated cells were incubated in duplicate at 37°C in RPMI+10% FCS (8×10^5 cells/ml) and the indicated concentration of deoxyribonucleoside. For each experimental set, an unexposed control sample was included.

Purine and Pyrimidine Metabolism in Man IX,
edited by Griesmacher *et al*. Plenum Press, New York, 1998.

471

2.2. Determination of DNA Single Strand-Breaks

The relative amount of DNA single strand-breaks was determined by the FADU (Fluorometric Analysis of DNA Unwinding) technique.[8,9] Briefly, after repair-incubation the cells were suspended in 1 ml 10 mM Na-phosphate buffer (pH 7.2) with 0.25 M mesoinositol and 1 mM $MgCl_2$ and divided in aliquots of 0.1 ml in T, B and P tubes (in triplicates). The B (blank) and P (sample) tubes were exposed to alkaline unwinding at 16–17°C for 60 min. After sonication for 2 s, ethidium bromide was added to a concentration of 4×10^{-7} M, and the fluorescence was read with excitating and emitting wavelengths of 520 and 590 nm, respectively. The percentage double-stranded DNA, D, was determined as $100 \times (P-B)/(T-B)$. The relative amount of DNA strand-breaks, ΔD was obtained as the difference between the D values of an exposed sample and an unexposed control sample from the same individual.

The percentage stimulation was calculated as the relative difference between the ΔD values for untreated cells and cells exposed to deoxyribonucleosides.

3. RESULTS AND DISCUSSION

3.1. The Effect of Deoxyribonucleosides on DNA Repair

The effects of the presence of the four deoxyribonucleosides during rejoining of UV-induced DNA strand-breaks was examined with healthy donors, and the results are shown in Fig. 1.

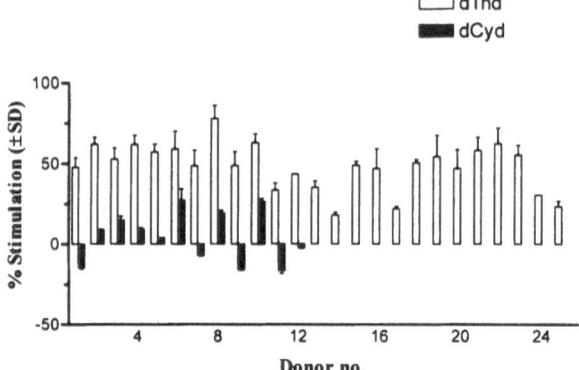

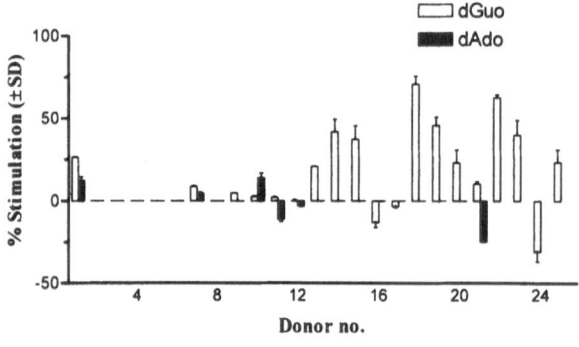

Figure 1. The effect of the presence of 2 μM deoxyribonucleoside on the rejoining of UV-induced DNA strand breaks.

3.1.1. The Effect of Thymidine. Thymidine (2 μM) stimulated the rejoining in each of the examined 25 donors. The stimulating effect was extremely significant (p < 0.0001, N=25). The average stimulation was 48% ranging from 18% (donor 8) to 78% (donor 14).

3.1.2. The Effect of Deoxyguanosine. The effect of deoxyguanosine (2 μM) was examined for 19 donors and was stimulatory with 16 donors and inhibitory with 3. The effect was extremely significant (p < 0.0004, N=19), the mean stimulation was 28% ranging from −31% to 71%.

3.1.3. The Effect of Deoxycytidine. Deoxycytidine (2 μM) exhibited stimulating effects with 7 donors and inhibitory with 5 donors. The overall effect was not significant (p=0.6, N=12). The mean stimulation was 4.5% ranging from −16% to 27%. Furthermore, the presence of 2 μM deoxycytidine decreased the stimulatory effect of 2 μM thymidine by about 1.7 fold (p < 0.05), as examined with 12 of the donors (results not shown).

3.1.4. The Effect of Deoxyadenosine. Deoxyadenosine did not show any significant effect (p=1, N=6). The mean effect was −1.3%, ranging from −25% to 14%.

3.2. Thymidine Prevents the Inhibitory Effect of Azidothymidine

The effect of thymidine on the inhibition of DNA strand-break rejoining by azidothymidine was examined for three donors. 100 μM azidothymidine inhibited rejoining about 50%. This effect was completely abolished by the presence of less than 10 μM thymidine.

4. CONCLUSION

The high stimulating effect of thymidine, and to a less degree of deoxyguanosine, may be related to the levels of the deoxyribonucleoside triphosphate pools in quiescent cells. The pools are known to be many fold lower in quiescent cells than in dividing cells,[4,10] and the dTTP pools has been shown to be 5–10 fold lower thant the other three pools.[4] Thus, among the four deoxyribonucleoside triphosphates, dTTP may be the most limiting factor in the DNA repair process.

As the stimulatory effects are exhibited by non-carcinogenic concentrations of deoxynucleosides[11] within physiological range,[12] it would be worthwhile examining the therapeutic use in prevention of sunlight induced damages to the skin.

REFERENCES

1. P. Reichard, Interactions between deoxyribonucleotide and DNA synthesis, Ann. Rev. Biochem. 57, 349–374 (1988).
2. K.L. Skoog, B.A. Nordenskjöld, and K.G. Bjursell, Deoxyribonucleoside-triphosphate pools and DNA synthesis in synchronized hamster cells, Eur. J. Biochem. 33, 428–432 (1973).
3. R.D. Snyder, Deoxyribonucleoside triphosphate pools in human diploid fibroblasts and their modulation by hydroxyurea and deoxynucleosides, Biochem. Pharmacol. 33, 1515–1518 (1984).
4. B. Munch-Petersen, G. Tyrsted, and B. Dupont, The deoxyribonucleoside 5′-triphosphate (dATP and dTTP) pools in phytohemagglutinin-stimulated and non-stimulated human lymphocytes, Exp. Cell Res. 79, 249–256 (1973).
5. S. Eriksson, A. Gräslund, S. Skog, L. Thelander, and B. Tribukait, Cell cycle-dependent regulation of mammalian ribonucleotide reductase. The S phase-correlated increase in subunit M2 is regulated by *de novo* protein synthesis, J. Biol. Chem. 259, 11695–11700 (1984).

6. B. Munch-Petersen, Differences in the kinetic properties of thymidine kinase isoenzymes in unstimulated and phytohemagglutinin-stimulated human lymphocytes, Mol. Cell. Biochem. 64, 173–185 (1984).

7. I.G. Walker and J.P.H. Th'ng, Excision-repair patch size in DNA from human KB cells treated with UV-light, or methyl methanesulfonate, Mutat. Res. 105, 277–285 (1982).

8. H.C. Birnboim and J.J. Jevcak, Fluorometric method for rapid determination of DNA strand breaks in human white blood cells produced by low doses of radiation, Cancer Res. 41, 1889–1892 (1981).

9. B. Munch-Petersen, Azidothymidine inhibits mitogen stimulated growth and DNA-repair in human peripheral lymphocytes, Biochem. Biophys. Res. Commun. 157, 1369–1375 (1988).

10. W.Y. Gao, D.G. Johns, and H. Mitsuya, Enzymatic assay for quantification of deoxynucleoside triphosphates in human cells exposed to antiretroviral 2', 3'-dideoxynucleosides, Anal. Biochem. 222, 116–122 (1994).

11. V. Bianchi and L. Celotti, Accuracy of UV-induced DNA repair in V79 cells with imbalance of deoxynucleotide pools, Mutat. Res. 146, 277–284 (1985).

12. L. Holden, A.V. Hoffbrand, and M.H.N. Tattersall, Thymidine concentrations in human sera: Variations in patients with leukemia and megaloblastic anaemia, Eur. J. Cancer 16, 115–121 (1980).

94

A NEW MODEL ORGANISM FOR STUDYING THE CATABOLISM OF PYRIMIDINES AND PURINES

Zoran Gojkovic,[1,2] Silvia Paracchini,[1] and Jure Piskur[1]

[1]Department of Microbiology
Technical University of Denmark
Building 301, DK-2800 Lyngby, Denmark
[2]Department of Genetics
University of Copenhagen, Denmark

1. INTRODUCTION

Pyrimidines and purines are constituents of nucleic acids, but they also play an important role in other aspects of metabolism. The sizes of their pools in the cell are determined by *de novo* biosynthetic, salvage and catabolic pathways. While the *de novo* biosynthetic and salvage pathways are relatively well characterized, so far very little is known about the degradation of pyrimidines and purines at the genetic level.

During catabolism pyrimidines are most commonly subjected to reduction rather than oxidation. Oxidative degradation of pyrimidines has been found only in some bacteria (Lara, 1952). The reductive pathway involves three enzymatic steps where uracil and thymine are converted to β-alanine and β-aminoisobutyric acid, respectively (Piskur *et al.*, 1993; Traut and Loechel 1984). The first step is catalysed by dihydrouracil dehydrogenase (DHPDase, EC 1.3.1.1/EC 1.3.1.2). Formation of dihydropyrimidines is followed by oxidative cleavage by dihydropyrimidinase (DHPase, EC 3.5.2.2) to yield β-ureidopropionic/β-ureidoisobutyric acid. In the third enzymatic step these two compounds are converted by β-ureidopropionase (UPase, EC 3.5.1.6) into β-amino acids. Thereafter β-alanine may be catabolized by β-alanine-aminotransferase (BAAase, EC 2.6.1.18) to malonate semialdehyde.

A number of enzymes participate in conversions of purine nucleotides by dephosphorylation, deamination, cleavage of glycosidic bonds and oxidation. However, the purine degradation pathway has been only partially characterized by genetic studies (Glatingy and Scazzocchio, 1995; Nygaard *et al.*, 1996; Wright *et al.*, 1993).

So far only a few studies have examined catabolism of pyrimidines and purines in yeast. While *Saccharomyces cerevisiae* is unable to utilize pyrimidines as a nitrogen

Purine and Pyrimidine Metabolism in Man IX,
edited by Griesmacher *et al.* Plenum Press, New York, 1998.

source (LaRue and Spencer, 1968), in this study we describe a yeast *Saccharomyces kluyveri*, closely related to *S. cerevisiae*, which can degrade pyrimidines and purines. A simple genetic system has been developed to study the regulation of both catabolic pathways in this yeast.

2. MATERIALS AND METHODS

2.1. Yeast Strains, Growth Conditions, and Media

The type strains of different yeast species used in this study are listed in Table 1. Strains used for mutagenesis were *S. kluyveri* GRY1175 (MATα, ura3⁻) and *S. kluyveri* GRY1183 (MATa, ura3⁻) (Weinstock and Strathern, 1993). YPD was the rich medium and SD was the minimal medium. N minimal medium contains: 10 g/l succinic acid, 6 g/l NaOH, 1.7 g/l yeast nitrogen base without amino acids or any other nitrogen source (Difco), 20 g/l glucose. To this medium a source of nitrogen was added as 1 g/l of uracil, dihydrouracil, β-ureidopropionic acid or β-alanine, or 0.5 g/l of adenine, hypoxanthine, xanthine, uric acid or urea. All yeast strains were grown at 25°C. The growth curves were determined in liquid media by following the optical density at 600 nm.

2.2. Mutagenesis and Mutant Selection

1 ml of *S. kluyveri* overnight culture ($1–2 \times 10^8$ cells/ml) was washed and the cells resuspended in 1 ml of sodium phosphate buffer (pH 7.0). 10 μl ethylmethane sulphonate

Table 1. Growth of various species of yeasts from the *Saccharomyces* genus on media containing pyrimidines or purines as the only nitrogen source (++ good growth, + moderate growth, +/– very weak growth, – no growth)

Strain	ura	DHU	UP	β-ala	ade	hyp	xan	uric acid	urea	SD
S. cerevisiae NRRL Y-12632	–	–	–	–	–	–	–	–	++	++
S. bayanus CBS 380	–	–	–	–	–	–	–	–	++	++
S. paradoxus NRRL Y-17217	–	–	–	–	–	–	–	–	++	++
S. servazzii NRRL Y-12661	–	–	–	–	–	–	–	–	–	++
S. castellii NRRL Y-12630	–	–	–	++	–	–	–	–	++	++
S. dairensis NRRL Y-12639	–	–	–	–	–	–	–	–	–	++
S. exiguus NRRL Y-12640	–	–	–	–	–	–	–	–	++	++
S. unisporus NRRL Y-1556	–	–	–	++	–	–	–	–	++	++
S. kluyveri NRRL Y-12651	++	++	++	++	++	++	++	++	++	++

Abbreviations used: ura - uracil, DHU - dihydrouracil, UP - β-ureidopropionic acid, β-ala - β-alanine, ade - adenine, hyp - hypoxanthine, xan - xanthine, SD - minimal medium containing NH_4^+.
CBS stands for Centraal-Bureau voor Schimmelcultures (Delft, The Netherlands), NRRL stands for Agricultural Research Service Culture Collection (Peoria, USA).

was added and the tube was incubated at 28°C for 30 min. Thereafter cells were transferred to 8 ml 5% sodium thiosulfate to quench the reaction, washed with water and recovered in 1 ml YPD for 1 hour. Subsequently the cultures were plated on YPD and replica-plated to various appropriate media to look for mutants.

3. RESULTS AND DISCUSSION

3.1. *Saccharomyces kluyveri*

Since it has been shown that *S. cerevisiae* cannot utilize pyrimidines and purines as sole sources of nitrogen, the first goal was to find a yeast species able to do so. Among nine members of the genus *Saccharomyces*, only *S. kluyveri* was able to utilize pyrimidines, as well as purines, as the only sources of nitrogen (Table 1).

The growth of *S. kluyveri* on dihydrouracil and ureidopropionic acid may indicate that *S. kluyveri* degrades pyrimidines in a manner similar to higher eukaryotes by employing the reductive pathway leading to β-alanine. Growth on hypoxanthine, xanthine and uric acid likewise suggests that purines are degraded via a pathway similar to that in higher organisms. Therefore, *S. kluyveri* could be a useful model organism in which to study the degradation of pyrimidines and purines. Possibly this will enable to deduce how the same systems function in higher eukaryotes.

3.2. Pyrimidine Degradation Mutants

The next step was to isolate mutants of *S. kluyveri* which were unable to utilize pyrimidines as sole sources of nitrogen. Two different haploid strains, GRY 1175 and GRY 1183, were used for this purpose. These strains are auxotrophs for uracil. Since they need an externally provided precursor of pyrimidine, such as uracil, this acted as a selection against mutants with impaired uptake of pyrimidines. Several mutants defective in the utilization of tested compounds were isolated. They belonged to several different classes with respect to inability to grow on uracil, DHU, UP or β-alanine (Table 2).

Growth of these mutants in liquid media was also examined. Growth curves of *S. kluyveri* GRY1183 and the mutant Y1004, *pyd1*, are shown in Figure1.

Among the presumed pyrimidine catabolic mutants, we characterized three independent loci, *pyd1*, *pyd2* and *pyd3* which could encode DHPDase, DHPase and UPase. Uracil carries two N atoms and the first one is released only in the third step of the putative reductive degradation pathway. The second N is released by transamination of β-alanine. Therefore, mutants affected in the first step of pyrimidine catabolism, *pyd1*, can grow on DHU and UP, whereas *pyd2* only grows on UP and *pyd3* cannot grow on uracil, DHU

Table 2. Growth of mutants defective in utilization of pyrimidines or UP (*pyd*), and β-alanine (*bac*) on various media.
+ and – symbols are used as in Table 1

Mutant	Mutatedlocus	ura	DHU	UP	β-ala
Y1004	*pyd1*	–	++	++	++
Y1020	*pyd2*	–	–	++	++
Y1022	*pyd3*	–	–	–	++
Y1024	*bac1*	++	++	++	–

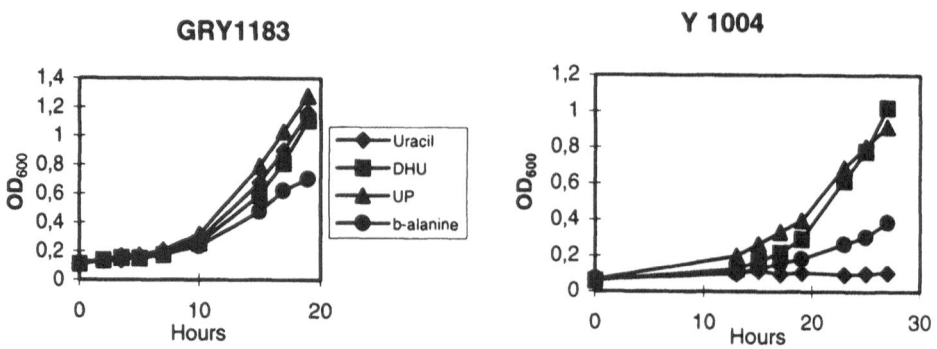

Figure 1. Growth curves for GRY1183 (PYD1⁺) and Y1004 (*pyd1⁻*) strains of *S. kluyveri* in various media. The mutant grows on DHU, UP and β-alanine, but not on uracil. Note that growth on β-alanine is slower than on the other media, possibly because only one nitrogen is obtained from this compound.

Table 3. Growth of purine degradation mutants (*pud*) on various media.
\+ and – symbols are used as in Table 1

Mutant	ade	hyp	xan	uric acid	urea	SD
Y1.5	–	–	+/–	++	++	++
Y2.13	++	++	+/–	+/–	++	++

or UP. All three classes of mutants can grow on β-alanine. These results strongly suggest that *S. kluyveri* uses the reductive pathway of pyrimidine degradation.

Another class of mutants, represented here by the mutant Y1024, has a defect in the *bac1* locus which presumably encodes BAAase. These mutants can utilize pyrimidines but growth on β-alanine is impaired. Mutants belonging to this class were slightly leaky.

3.3. Purine Degradation Mutants

Mutagenized cells of *S. kluyveri* were also screened for mutants impaired in the utilization of purines. So far two different classes of such mutants have been obtained (Table 3). One of these mutants, Y1.5, does not grow with adenine, hypoxanthine or xanthine as sole nitrogen source. The other class, represented by Y2.13, grows with adenine and hypoxanthine, but not with xanthine, uric acid or urea as sole nitrogen source. Both of these mutants can use pyrimidines as sole nitrogen source. Therefore they are expected to be impaired in the transport or catabolism of purines.

3.4. Conclusion

Several different mutants with presumed defect in the pyrimidine or purine catabolic pathway have been isolated. Currently efforts are being made to isolate the corresponding genes by complementation using a genomic library.

REFERENCES

Glatigny A. and Scazzocchio C., 1995, Cloning and molecular characterization of *hxA*, the gene coding for the xanthine dehydrogenase (purine hydroxylase I) of *Aspergillus nidulans. J. Biol. Chem.* 270: 3534–3550.

Lara F.J.S., 1952, On the decomposition of pyrimidines by bacteria. I. Studies by means of the technique of simultaneous adaptation *J. Bacteriol.* 64: 271–277.

LaRue T.A. and Spencer J.F.T., 1968, The utilization of purines and pyrimidines by yeasts. *Can.J.Microbiol.* 14:79–86.

Nygaard P., Duckert P., and Saxild H.H., 1996, Role of adenine deaminase in purine salvage and nitrogen metabolism and characterization of the *ade* gene in *Bacillus subtilis*. *J.Bacteriol.* 178:846–853.

Oestreicher N. and Scazzocchio C., 1993,Sequence, regulation and mutational analysis of the gene encoding urate oxidase in *Aspergillus nidulans*. *J.Biol.Chem.* 268:23382–23389.

Piskur J., Kolbak D., Søndergaard L. and Pedersen M.B., 1993, The dominant mutation *Suppressor of black* indicates that *de novo* pyrimidine biosynthesis is involved in the *Drosophila* tan pigmentation pathway. *Mol. Gen. Genet.* 241: 335–340.

Traut T.W. and Loechel S., 1984, Pyrimidine catabolism: individual characterization of the three sequential enzymes with a new assay. *Biochemistry* 23: 2533–2539.

Weinstock K.G. and Strathern J.N., 1993, Molecular genetics in S*accharomyces kluyveri*: the *HIS3* homolog and its use as selectable marker gene in *S. kluyveri* and *Saccharomyces cerevisiae*. *Yeast* 9:351–361.

Wright R.M., Vaitaitis G.M., Wilson C.M., Repine T.B., Terada L.S. and Repine J.E., 1993, cDNA cloning, characterization, and tissue-specific expression of human xanthine dehydrogenase/xanthine oxidase. *Proc. Natl. Acad. Sci. USA* 90: 10690–10694.

95

CAD OVEREXPRESSION IN MAMMALIAN CELLS

Yu Qiu* and Jeffrey N. Davidson

Department of Microbiology and Immunology
Albert B. Chandler Medical Center & Lucille P. Markey Cancer Center
University of Kentucky
Lexington, Kentucky 40536-0084

1. INTRODUCTION

Hamster CAD is a multifunctional protein which catalyzes the first three enzymatic steps of *de novo* pyrimidine biosynthesis.[1,2] The major catalytic domains of CAD are carbamyl phosphate synthetase (CPSase) (EC 6.3.5.5), dihydroorotase (DHOase) (EC 3.5.2.3), and aspartate transcarbamylase (ATCase) (EC 2.1.3.2). The first enzyme of the pathway, carbamyl phosphate synthetase, is regulated by UTP, PRPP and phosphorylation.[3,4,5,6] While the monomeric size of CAD is 243 KDa,[7] the native protein exists as an oligomer of identical subunits (mostly hexamers but also, trimers and larger oligomers).[1,7,8]

In order to study the function and structure of CAD, a CAD overexpression system would be invaluable. Expression of CAD in *E.coli* is not efficient and multiple-sized CAD bands are observed on SDS-PAGE. Moreover, expression of a mammalian protein in a bacterial system may provide an incomplete picture. Appropriate phosphorylation would be an example of this point. The mammalian expression system used in our lab previously was also not efficient. Usually, after transfection of a CAD deficient Chinese hamster ovary (CHO) cell line, complementation was observed for only 10% of the G418 resistant colonies. Furthermore, CAD expression in these transfectants was very low, usually about 20–35% of the wild-type level. Even for the highest expressing colony, CAD was not detectable on SDS-PAGE using coomasie blue staining. It should be noted that wild-type CHO cell crude extracts also do not yield a visible band.

Use of a bicistronic vector[9] as an efficient, high expression system for CAD in mammalian cells is described here. The bicistronic vector utilizes an IRES—the internal ribosome entry site of a picornavirus. An IRES allows translation of two proteins from a single

* Tel. (606) 323-5231; FAX (606) 257-8994; email:yqiu1@pop.uky.edu

Purine and Pyrimidine Metabolism in Man IX,
edited by Griesmacher *et al.* Plenum Press, New York, 1998.

messenger RNA in mammalian cells. Furthermore, when a polyhistidine tag is added to the N-terminus of CAD, CAD can be purified from cell extracts in a single step.

2. DESIGN AND METHODS

2.1. Constructs

2.1.1. pCIN•CAD. Hamster CAD cDNA (6.7 kb) was subcloned into a bicistronic vector downstream of the CMV promoter for the first cistron, and the neomycin resistant gene was located in the second cistron.

2.1.2. pCIN•His•CAD. In order to purify CAD more easily, a PCR product encoding a 6xhistidine tag and a spacer region (with or without a rTEV protease cleavage site) was engineered into pCIN•CAD at the N-terminus of hamster CAD. The final construct was called pCIN•His•CAD. The resulting constructs were verified by DNA sequencing. The additional N-terminal sequence of the 6xhis tag without protease cleavage site is MGHHHHHHDYDN*MAALV*..., while the construct encoding a protease cleavage site is MGHHHHHHDYDIPTTENLYFQGN*MAALV*... (normal CAD N-terminal sequence underlined).

2.2. Cell Growth

G9C cells[10] were used for transfection. G9C cells are a CHO cell line deficient in CAD and are unable to grow in medium lacking exogenous uridine. Maintaining G9C cells require 3 mM uridine with 10% fetal bovine serum in the medium, at 37°C, 5% CO_2.

K1 cells are the wild-type CHO cell line. These cells are used as a control for CAD expression. K1 cells were grown in medium containing 10% fetal bovine serum at 37°C, 5% CO_2. The LD_{50} of N-phosphonacetyl-L-aspartate (PALA) for K1 cell is 5 μM.

2.3. Transfection and Selection

pCIN•CAD and pCIN•His•CAD were transfected into G9C cells by calcium phosphate. After transfection, cells were selected by growing in medium plus 0.7 mg/ml G418 and 30 μM Uridine. The G418 resistant colonies of pCIN•CAD were further tested in medium without uridine. Additionally, a nearly irreversible inhibitor of aspartate transcarbamylase, PALA[11], was used to select for colonies overexpressing CAD. For the transfectants of pCIN•His•CAD, isolated G418 resistant colonies were grown in 96-well plates until confluence on medium containing uridine. Cells from each 96-well plate were harvested and screened by SDS-PAGE for expression of a 243 KDa band.

2.4. Enzyme Assays

2.4.1. ATCase Assay. The ATCase assay[10] follows the conversion of ^{14}C-aspartate to ^{14}C-carbamyl aspartate, in 50 mM Hepes buffer (pH 8.3), 37°C. Carbamyl phosphate concentration was 2.4 mM.

2.4.2. CPSase Assay. CPSase assay[10] measured the conversion of ^{14}C-ornithine to ^{14}C-citrulline in the presence of ornithine transcarbamylase (OTCase).

Table 1. Characterization of G9C cells transfected
with CAD bicistronic vector

	Growth			
Cell line	G418+Uridine	G418−Uridine	G418−Uridine 5µM PALA	ATCase relative activity (folds)
K1	−	−	−	1[a]
G9C	−	−	−	0.001[b]
#1	+	+	+	9[c]
#3	+	+	+	6[c]
#6	+	−	−	0.01[c]
#10	+	+	+	10[c]
#7	+	±	−	0.2[c]
#18	+	+	+	2[c]
#15	+	+	+	7[c]

[a]Extract prepared from cells growing in F12 medium with 10% Fetal Bovine Serum.
[b]Extract prepared from cell growing in F12 medium with 10% Fetal Bovine Serum plus Uridine (30 µM).
[c]Extract prepared from cells growing in F12 medium with 10% Fetal Bovine Serum plus G418 (0.7 mg/ml) and Uridine (30 µM).

2.4.3. DHOase Assay. DHOase activity was assayed with the reverse reaction by the conversion of ^{14}C-dihydroorotate to ^{14}C-carbamyl aspartate.

2.5. Purification

Crude extract was prepared from 2×10^7 transfectants of pCIN•His•CAD by sonication and centrifugation in 20 mM Tris-HCl, pH 7.9, 0.5 M NaCl, 5 mM imidazole and 25 mM MgCl$_2$. Crude extract was loaded on a 1 ml Ni^{2+} column equilibrated with the same buffer. The column then was washed in 20 mM Tris-HCl, pH 7.9, 0.5 M NaCl, with increasing concentrations of imidazole: 5 mM, 60 mM, and 100 mM. The 6xHis tagged CAD protein was eluted with 1 M imidazole in the buffer above.

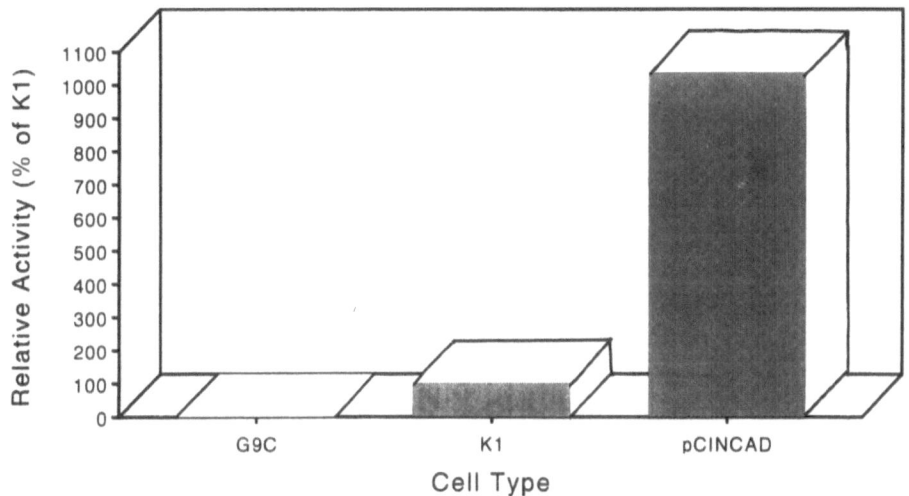

Figure 1. Relative ATCase activity of G9C, K1, and pCIN•CAD crude extracts. K1 represents 100% activity.

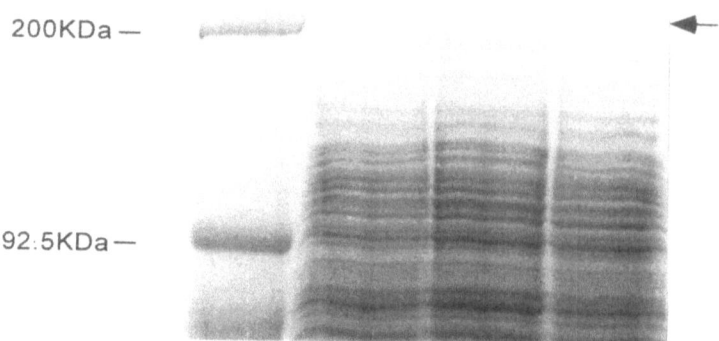

Figure 2. SDS-PAGE of crude extract stained by coomassie blue. Arrowhead marks position of CAD band. M is molecular weight standards.

3. RESULTS

After transfection, almost all isolated G418 resistant colonies grew in the uridine deficient medium. Many colonies were able to grow in 5 μM PALA which is the LD_{50} for the wild-type CHO cell (K1). Some colonies even grew in 1 mM PALA (Table 1). ATCase activity for most colonies was ≥ 100% of K1 cells. Some colonies expressed levels of CAD 10 times higher then K1 cells (Fig. 1). CAD from some colonies is visible on a SDS-PAGE gel stained with coomassie blue (Fig. 2). The CAD from such a transfectant can be successfully purified in one step by affinity purification (Fig. 3).

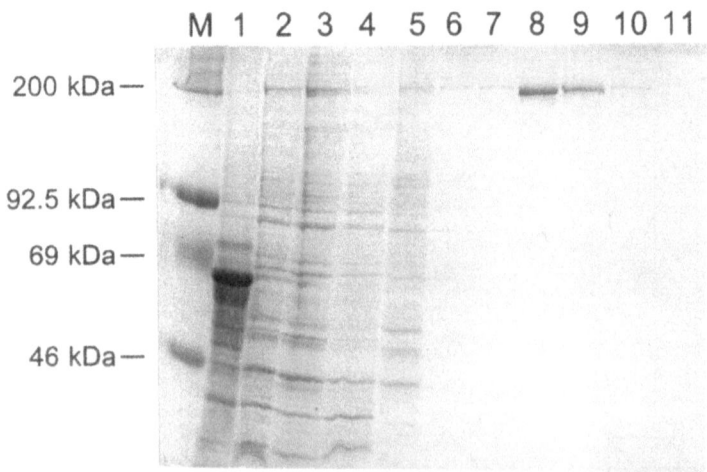

Figure 3. Coomassie blue stained SDS-PAGE showing intermediate steps in His•CAD purification. Lane 2, crude extract; lane 3, loading flow through; lane 4, 5 mM imidazole wash; lane 5, 60 mM imidazole wash; lane 6, 100 mM imidazole wash; lane 7 through 11, consecutive fractions eluted with 1M imidazole; lane 1, untransfected G9C cells extracted directly from tissue culture plate-prominent band below 69 kDa is from fetal calf serum.

From 2×10^7 cells, ~93 μg of CAD can be purified to near homogeneity (~95% purity). The purification only takes two hours from cell lysis to eluting purified fractions. The rapid procedure is helpful in retaining the enzymatic activities and in maximally reducing the opportunity for proteolysis. Preliminary enzymatic assays proved that the purified protein harboured activities for all of the three enzymes.

Preliminary results reported here showed that an efficient, high expression and rapid one-step affinity purification system has been developed. This will make the studying of hamster CAD easier and allow for studies using the homologous cell system.

ACKNOWLEDGMENTS

This work was supported by funding from the National Science Foundation (MCB-9418413) and the National Institutes of Health (GM 47644).

REFERENCES

1. Coleman, P. F., Suttle, D. P., and Stark, G. R. Purification from hamster cells of the multifunctional protein that initiates *de novo* synthesis of pyrimidine nucleotides. *J. Biol. Chem.* 252:6379–6385 (1977)
2. Davidson, J. N., Chen, K. C., Jamison, R., Musmanno, L. A. and Kern, C. B. The evolutionary history of the first three enzymes in pyrimidine biosynthesis. *Bioessays* 15:157–164 (1993)
3. Carrey, E. A. Double regulation by UTP of the mammalian carbamyl phosphate synthase II activity. *Biochem. Soc. Trans.* 17:761–762 (1989)
4. Carrey, E. A., Campbell, D. G. and Hardie, D. G. Phosphorylation and activation of hamster carbamyl phosphate synthetase II by cAMP-dependent protein kinase. A novel mechanism for regulation of pyrimidine nucleotide biosynthesis. *EMBO. J.* 4:3735–3742 (1985)
5. Shaw, S. M. and Carrey, E. A. Regulation of the mammalian carbamoyl-phosphate synthetase II by effectors and phosphorylation. Altered affinity for ATP and magnesium ions measured using the ammonia-dependent part reaction. *Eur. J. Biochem.* 207:957–965 (1992)
6. Evans, D. R. and Guy, H. Regulation of carbamyl phosphate synthetase. *Paths to pyrimidines* 5:1–10 (1997)
7. Davidson, J. N., Rumsby, P. C. and Tamaren, J. Organization of a multifunctional protein in pyrimidine biosynthesis: analyses of active, tryptic fragments. *J. Biol. Chem.* 256:5220–5225 (1981)
8. Lee, L., Kelly, R. E., Pastra-Landis, S. C. and Evans, D. R. Oligomeric structure of the multifunctional protein CAD that initiates pyrimidine biosynthesis in mammalian cells. *Proc. Natl. Acad. Sci. USA* 82:6802–6806 (1985)
9. Rees, S., Coote, J., Stables, J., Goodson, S., Harris, S. and Lee, M. G. Bicistronic vector for the creation of stable mammalian cell lines that predisposes all antibiotic-resistant cells to express recombinant protein. *Biotechniques* 20:102–110 (1996)
10. Patterson, D. and Carnright, D. V. Biochemical genetic analysis of pyrimidine biosynthesis in mammalian cells: I. Isolation of a mutant defective in the early steps of *de novo* pyrimidine synthesis. *Som. Cell. Genet.* 3:483–495 (1977)
11. Kempe, T. D., Swyryd, E. A., Bruist, M., and Stark, G. R. Stable mutants of mammalian cells that overproduce the first three enzymes of pyrimidine nucleotide biosynthesis. *Cell* 9:541–550 (1976)

96

AZATHIOPRINE TREATMENT AND THIOPURINE METABOLISM IN RHEUMATIC DISEASES

Introduction and First Results of Investigation

J. N. Stolk,[1] A. M. Th. Boerbooms,[1] R. A. De Abreu,[2] and
L. B. A. van de Putte[1]

[1]Department of Rheumatology
[2]Department of Pediatrics (Laboratory of Purine and Pyrimidine Metabolism)
University Hospital Nijmegen
P. O. Box 9101
6500 HB Nijmegen, The Netherlands

1. INTRODUCTION

1.1. Clinical Picture and the Immune System of Rheumatoid Arthritis and Systemic Lupus Erythematosus

Most rheumatic diseases are systemic autoimmune disorders in which not only the musculoskeletal system but other organ systems as well may be involved in the disease process. The primary cause of many of these diseases is still not clarified but many have a certain genetical susceptibility as appears from stronger associations with several HLA-types. Two well-known examples are Rheumatoid Arthritis (RA) and Systemic Lupus Erythematosus (SLE). In RA, the classical clinical picture is that of symmetric polyarthritis and erosive destruction of the joints by chronic inflammation. Extraarticular manifestations like e.g. scleritis, serositis or vasculitis, may accompany the joint symptoms. In SLE the joint involvement is often less symmetric and non-erosive and frequently there is arthralgia instead of arthritis. Symptoms of other organs e.g. skin, kidneys, neurologic system and gastrointestinal tract, diseased by inflammation, vasculitis or immune complex deposition, are more prominent in the clinical picture of SLE.(1)

Both diseases are characterized by dysregulation of the immune system as demonstrated by the presence of autoantibodies, e.g. rheumatoid factor in RA and anti-dsDNA antibodies in SLE, and changes in lymphocytes e.g. concerning their activation state, fre-

Purine and Pyrimidine Metabolism in Man IX,
edited by Griesmacher *et al.* Plenum Press, New York, 1998.

487

quency distribution of subsets and produced cytokine profiles. Perhaps one might say that there is an immune deficiency in RA and SLE, since both display an increased death rate due to infectious diseases (irrespective immunosuppressive therapy) when compared to healthy controls.(1)

1.2. Treatment

In the treatment of both diseases the use of immunosuppressive agents is important to decrease the chronic autoimmune mediated inflammatory activity. One of the drugs quite often used is azathioprine (AZA), a nitro-imidazole derivate of 6-mercaptopurine (6-MP) which in turn is an analogue of hypoxanthine. After intake, AZA is rapidly converted to 6-MP by thiolysis activity of glutathione. Subsequently, 6-MP is metabolised by enzymes of purine metabolism. Formation of thionucleotides is initiated by action of hypo-xanthine-guanine-phosphoribosyl-transferase (HGPRT) and catabolism is catalysed by xanthine oxidase (XO) (2). In general, it is believed that cytotoxicity is caused by 6-thioguanine nucleotides (3). Another specific thiopurine pathway involves S-methylation by thiopurine-methyltransferase (TPMT) and possibly leads to some inactivation of thiopurines (4). However, there are data indicating that methylated thiopurines are not inactive since they can inhibit purine de novo synthesis and additionally may play a role in thioguaninenucleotide independent cytotoxicity (5–7).

The outcome of AZA treatment is influenced by the enzyme activity of TPMT. Low or deficient TPMT activities are reported to result in higher levels of 6-thioguanine nucleotides and increased cytotoxicity (8). In addition, when AZA is used together with co-medication of allopurinol (an inhibitor of XO), it is ne-cessary to lower the dose of AZA to approximately 25% of the conventional dose to prevent its toxicity (9).

The goal of thiopurine (AZA, 6-MP, thioguanine) treatment is related to the disease. In childhood leukemia one aims for cytotoxicity but in autoimmune disorders like RA, SLE, autoimmune hepatitis and inflammatory bowel disease and also in organ transplant recipients the goal is immunosuppression and reduction of inflammation. The anti-inflammatory and immunosuppressive effects of AZA most likely results from several biologic activities: it inhibits prostaglandin E_2 production, lymphocyte responses, IL-2 effects and natural killer cell activity (10,11).

It is known from childhood leukemia that the concentration of intracellular 6-thioguanine nucleotides is inversely related to the TPMT activity which may be used to monitor treatment (8). However, there is an ongoing discussion about the optimal way to predict toxicity of thiopurine treatment.

2. RATIONALE FOR INVESTIGATING PURINE ENZYMES IN RHEUMATIC DISEASES

The main reason for our investigations on purine salvage enzyme and TPMT activities came from the observation that the outcome of AZA treatment in RA was very divergent, ranging from good tolerance to the drug and substantial reduction of disease activity to absence of any beneficial effect and development of severe adverse reactions. In addition, and like-wise the characteristics of immunodeficiency syndromes of which some are related to deficiency of a purine enzyme, we took the view that alterations in purine enzyme activities could have a relation with the altered immune status of patients with rheumatic diseases. They have not only an increased susceptibility to infections but also an increased mortality due to that cause. As far as we know there are no data about smaller changes in enzyme activities instead of deficiencies, and a possible relation with the outcome of AZA treatment or altered lymphocyte function.

There are two pilot-studies on purine enzyme activities in patients with RA (12,13). They show that RA patients have different activities of ADA, PNP and purine 5'NT. However, these results have to be interpreted with caution since in both studies the patients use several kinds of disease modifying antirheumatic drugs (DMARDs) and/or prednisone and there is no evaluation for possible confounders like disease duration, age and gender. We also reported that low activity of 5'NT may be a cause of AZA related myelosuppression (14).

So, on the basis of these results we decided to continue our investigations with optimized study designs and after improvement of some laboratory techniques (15).

3. AIMS OF OUR STUDIES

To detect possible differences in thiopurine metabolism between patients and healthy controls, that could be related to the varying results of AZA treatment and/or the altered immune status with increase susceptibility to infections. If such an alteration is present then we are interested to know whether this could have also a predictive value for the outcome of AZA treatment.

4. METHODS

To avoid influences of possible circadian rhythms, clinical assessments and blood sampling were done between 9 and 11 a.m.

4.1. Clinical Assessments

Parameters of disease activity were measured with standardized and validated methods and included patients assessment of general health, several joint scores and the morning stiffness.

4.2. Laboratory Assessments

Venous blood samples for determination of purine enzyme activities, IgM-Rheumatoid Factor (ELISA, positive if >10 IU/ml), HLA-types and routine blood tests (complete blood cell count, blood biochemistry profile consisting of kidney- and liver function tests and serum uric acid, ESR (Westergren, mm in 1 hour) and CRP (mg/ml))(patients only) were collected immediately after the clinical assessments.

To determine purine enzyme activities of 5'NT, PNP and HGPRT, blood was collected in 10 ml Vacutainer® tubes containing polystyrene granules (Becton and Dickinson, Rutherford, NJ). Mononuclear cells were obtained by Ficoll-Isopaque (density 1.077 g/ml, Nycomed) gradient centrifugation of defibrinated blood and in these mononuclear cells the enzyme activities were measured by a non-radiochemical HPLC method (15). Enzyme activities are expressed in $nmol/10^6$ cells/hour of incubation. To measure the TPMT activity in red cell lysate as described by Weinshilboum (16), peripheral blood was collected in 10 ml Monoject® tubes containing 150 USP Units of Lithium Heparin (Sherwood Medical,

Ballymoney, N. Ireland). TPMT activity is expressed in pmol/10^6 erythrocytes/hour of incubation. The enzyme assays have good reproducibility with mean coefficients of variation ± S.E.M. for HGPRT, 5'NT, PNP and TPMT of respectively $7.6 ± 0.7$, $5.3 ± 0.9$, $6.6 ± 0.8$ and $4.0 ± 0.5$. All enzyme assays were carried out in quadruplicate.

5. STUDIES AND RESULTS

5.1. Cross-Sectional Study in Patients with Early RA

Twenty-three patients with recent onset and active RA, diagnosis less than 1 year ago, not previously treated with DMARDs or prednisone and 28 healthy controls were studied. Clinical and laboratory assessments were done as mentioned above. The results show that purine enzyme activities did not differ between patients and controls. Enzyme activities had no significant relations with indices of disease activity or rheumatoid factor nor with the RA associated HLA-types. In agreement with a previous study, activity of 5'NT significantly decreased with age in patients and controls (one-way ANCOVA; $p=0.05$ and $p=0.009$ respectively), the estimates of the yearly decrease ranging from 0.21 to 0.37 nmol/10^6 cells/hour for patients and controls respectively. In addition, 5'NT activity was found to be significantly lower (about 27%) in males compared to females.

In conclusion: because of the absence of changes in the enzyme activities of 5'NT, PNP, HGPRT and TPMT, early in the disease and prior to the use of DMARDs/prednisone, neither the variability in the effects of AZA treatment in RA patients nor their elevated death rate due to infections, can be explained by pre-existing alterations in activities of these purine enzymes. Results of studies including purine 5'NT activity should be adjusted for age and gender (17).

5.2. Prospective Longitudinal Study in Patients with Established RA Starting with AZA Treatment

Patients with longstanding RA (n=33) enrolled a prospective longitudinal study for treatment with AZA. Before start and at month 1 and 6 of treatment clinical and laboratory assessments were done as mentioned before. Between these times patients were monitored every 2 weeks by blood sampling for hematologic indices, liver-enzymes and kidney function. Their previous DMARDs were withdrawn at least 2 weeks before starting AZA. A stable dose of prednisone (maximum daily dose 10 mg) was accepted as was the use of fixed doses of non-steroidal anti-inflammatory drugs or paracetamol.

During the first month, AZA was given in an increasing dose: the first week 1 tablet of 50 mg each day; the second week 50 mg b.i.d. and so on, till a target value of approximately 2 mg/kg/day. In case of a possible and persistent side effect, the AZA dose was reduced or stopped until disappearence. If after rechallenge, the side effect reoccurred AZA was withdrawn definitely.

After completing the study, AZA treated patients could be divided into two subgroups. One group had a complete follow-up during 6 months without persistent side effects. A second group developed side effects which were defined as severe because of the need to withdraw AZA, and hence there was an incomplete follow-up. Results were compared with those of cross-sectionally and partly previously studied groups of 24 early RA patients who did not use a DMARD yet and 42 healthy controls (17). All participants were Caucasian.

Table 1. Age adjusted purine enzyme activities by group

Month	RA AZA treatment					Early RA cs	Controls cs
	No side effects			Side effects			
	0	1	6	0	1		
HGPRT	9.1 ± 3.5	9.6 ± 4.0	7.9 ± 2.2	8.6 ± 3.4	7.3 ± 3.1	11.0 ± 3.5	9.9 ± 3.8
n	19	19	18	14	13	23	30
TPMT	22.4 ± 3.8	23.0 ± 3.9	24.8 ± 4.3	17.3 ± 5.9	17.1 ± 6.4	24.3 ± 4.8	22.1 ± 4.3
n	19	19	18	14	13	24	39

Means ± SD. 'CS': cross-sectionally studied. Adjustment to an overall mean age of 50.4 years.
HGPRT activity in nmol/10^6 cells/hour and TPMT activity in pmol/10^6 cells/hour.

Nineteen patients completed the study without side effects; 14 patients (42%) experienced severe side effects with need to withdraw AZA, in 11 within the first 6 weeks of treatment. The side effects were: 10 patients with gastro-intestinal intolerance; in 1 associated with hepatotoxicity, in 2 with a flu-like syndrome, in 2 with headache, in 1 with fever and in 1 with pancytopenia; of the other 4 patients, 1 developed vasculitis; 1 patient had recurrent airway-infections and sinusitis, 1 patient pancytopenia and 1 hepatotoxicity. We found no significant relationships between enzyme activities and indices of disease activity. In comparison with the other groups, the group of patients with side effects had a significantly lower TPMT level already before the start of AZA and at month 1 (Kruskal-Wallis, p=0.004; Mann-Whitney U test, p<0.01) and it also had a lower HGPRT activity before start (insignificant) with further decrease at month 1 (Table 1). None of the patients was TPMT deficient. The distribution of TPMT acitivity was clearly trimodal as shown before by other investigators (TPMT activity is 'undetectable', 'intermediate', or 'high'). Fischer's Exact test showed a highly significant (p=0.005) relation between 'intermediate' TPMT activity and the occurrence of side effects with a Relative Risk of 3.1 (95% CI: 1.6–6.2). AZA-treated patients without side effects, initially showed similar enzyme levels to those of the control groups.

In conclusion: in this study, rheumatoid patients quickly developing severe side effects, have lower TPMT and HGPRT activities already before the start of AZA. Consequently, a relative overload of the catabolic purine pathway (catalysed by XO) may take place. This enzyme has high activities in gastointestinal mucosa and liver. We hypothetise that on conditions of AZA treatment and lower baseline levels of TPMT and/or HGPRT, local accumulation of catabolic purine and thiopurine metabolites possibly plays a role in the occurrence of gastrointestinal or hepatotoxic side effects. However, the most important finding is the significant Relative Risk of 3.1 to develop severe side effects on AZA treatment if the pre-treatment inherited TPMT activity is in the 'intermediate' range. As clinical implication one should consider TPMT measurement prior to treatment to improve the proper use of AZA.

5.3. Cross-Sectional Study in Patients with Established SLE

Seventeen patients with established SLE (2 males, 15 females) entered the pilot-study. Most of them were under second line treatment: 15 used prednisone and 7 AZA. Results were compared with those of partly previously studied healthy controls (n=43) (17). Clinical and laboratory assessments were done as mentioned above. In addition in several patients (n≥10) and controls (n≥12) the enzyme activities were also measured separately in B- and T-cells obtained by immunofluorescent double-staining and FACS cell sorting procedures.

The most important finding is a significantly lower 5'NT activity peripheral blood mononuclear cells (PBMNC) and T-cells and a trend to a lower 5'NT activity in B-cells of SLE patients. Mean activities ± SD (nmol/10^6 cells/hour) in SLE and [controls] being respectively: PBMNC: 9.4 ± 6.4 [21.9 ± 8.3], T-test P < 0.000 ; T-cells: 15.3 ± 19.6 [22.2 ± 6.4], Mann-Whitney U test p=0.006 ; B-cells: 41.1 ± 32.6 [56.1 ± 25.4], not significant. Because 5'NT is known to be influenced by gender and since there were only 2 males, they were excluded from the analyses. Furthermore, SLE patients had a lower percentage of B-cells (T-test: p=0.01), a higher percentage of T-cells (T-test: p=0.001) and similar mononuclear cell counts. Enzyme activities did not correlate with indices of disease activity.

In conclusion: compared to healthy controls, SLE patients seem to have significantly lower mononuclear cell and T-lymphocyte activities of 5'NT. Further investigations have to point out the clinical relevance of these findings. However, it is conceivable that the low 5'NT activities may have a pathogenic role in the altered lymphocyte function in SLE, possibly by reduced breakdown of nucleotides. In addition, in case of AZA treatment, low 5'NT activity may cause intra-cellular increase of thionucleotides and possibly, through that, effective immunosuppression.

6. CONCLUSION

In RA we have found no alterations in purine enzyme activities of 5'NT, PNP and HGPRT, nor in early disease, neither in established disease. Like in other diseases, in RA the inherited TPMT activity plays an important role during AZA treatment because deficiency of this enzyme is related to severe cytotoxic adverse reactions. In addition, our study in established RA demonstrates that an intermediate activity also is associated with an increased risk to develop side effects. For a safer use of AZA one should consider TPMT measurement prior to treatment. Because the TPMT activity is inherited the question arises whether our findings also hold true for other non-malignant AZA treated conditions.

In SLE, the lower 5'NT activity we found, may be related to the efficacy of immunosuppression during thiopurine treatment and possibly plays a role in the altered immune status.

Needless to say that our results have to be confirmed in larger studies and in different designs.

REFERENCES

1. McCarthy DJ. Arthritis and allied conditions. A textbook of rheumatology. Twelfth edition. Lea & Febiger, Philadelphia, London.
2. Elion GB, Hitchings GH. Azathioprine. In: Handbook of experimental pharmacology, vol 38; Sartorelli AC, Johns DG (eds). Springer, New York, 1975;404–25.
3. Tidd DM, Paterson AR. A biochemical mechanism for the delayed cytotoxic reaction of 6-mercaptopurine. Cancer Res 1974;34:738–46.
4. Krynetski EY, Tai HL, Yates CR, et al. Genetic polymorphism of thiopurine s-methyltransferase: clinical importance and molecular mechanisms. Pharmacogenetics 1996;6:279–90.
5. Stet EH, De Abreu RA, Janssen YP, et al. A biochemical basis for synergism of 6-mercaptopurine and mycophenolic acid in Molt F4, a human malignant T-lymphoblastic cell line. Biochem Biophys Acta 1993;1180:277–82.
6. Bökkerink JP, Stet EH, De Abreu RA, et al. 6-Mercaptopurine: cytotoxicity and biochemical pharmacology in human malignant T-lymphoblasts. Biochem Pharmacol 1993;45:1455–63.

7. Janka-Schaub GE, Erb N, Harms D. Randomized comparison of 6-mercaptopurine vs 6-thioguanine in maintenance treatment of childhood acute lymphoblastic leukemia (ALL): differing metabolism and hematologic toxicity. Med Ped Oncol 1994;23:197 (abstract).

8. Lennard L. The clinical pharmacology of 6-mercaptopurine. Eur J Clin Pharmacol 1992;43:329–39.

9. Calabresi P, Chabner BA. Antineoplastic agents. In: Goodman Gilman A, Rall ThW, Nies AS, Taylor P (eds). The pharmacological basis of therapeutics. Oxford: Pergamon Press, 1990:1209.

10. Harth M. Mechanisms of action of disease modifying antirheumatic drugs. J Rheumatol 1992;(suppl 32)19: 100–3.

11. Alamartine E, Sabido O, Berthoux FC. The influence of prophylactic immunosuppressive regimens on natural killer and lymphokine activated killer cells in renal transplant recipients. Transplantation 1990;- 50:969–73.

12. Kerstens PJ, Stolk JN, Boerbooms AM, et al. Purine enzymes in rheumatoid arthritis: possible association with response to azathioprine. A pilot study. Ann Rheum Dis 1994;53:608–11.

13. Appelboom T, Mandelbaum I, Vertongen F. Purine enzyme levels in rheumatoid arthritis. J Rheumatol 1985;12:1075–8.

14. Kerstens PJ, Stolk JN, Hilbrands LB, et al.5'-Nucleotidase and azathioprine related bone marrow toxicity. Lancet 1993;342:1245–6.

15. Stolk JN, De Abreu RA, Boerbooms AM, et al. Purine enzyme activities in perpheral blood mononuclear cells: comparison of a new non-radiochemical high-performance liquid chromatography procedure and a radiochemical thin-layer chromatography procedure. J chromatogr B 1995;666:33–43.

16. Weinshilboum RM, Raymond FA, Pazmino PA. Human erythrocyte thiopurine methyltransferase: radiochemical microassay and biochemical properties. Clin Chim Acta 1978;85:323–33.

17. Stolk JN, Boerbooms AM, De Abreu RA, et al. Purine enzyme activities in recent onset rheumatoid arthritis: are there differences between patients and healthy controls? Ann Rheum Dis 1996;55:733–8.

97

IDENTIFICATION OF MULTIPLE FORMS OF THE CYTOSOLIC 5'-NUCLEOTIDASE/PHOSPHOTRANSFERASE IN RAT TISSUES

R. Pesi,[1] S. Allegrini,[2] S. Golfarini,[1] C. Baiocchi,[1] E. Moretti,[1] M. Camici,[1] S. Eriksson,[3] and M. G. Tozzi[2]

[1]Dipartimento di Fisiologia e Biochimica
Università di Pisa, via S. Maria
55 56100 Pisa, Italia
[2]Dipartimento di Scienze del Farmaco
Università di Sassari, via Muroni
23/A, 07100 Sassari, Italia
[3]Department of Veterinary Medical Chemistry
SLU, Uppsala, Sweden

1. INTRODUCTION

Cytosolic 5'-nucleotidase, an ubiquitous IMP preferring nucleotide phosphatase has been purified and characterised from different sources.[1] The enzyme acts through the formation of a phosphorylated enzyme intermediate, and catalyses both the hydrolysis of the mononucleotides and the transfer of the phosphate to a nucleoside acceptor, thus operating a monucleotide interconversion.[2] The enzyme is specific for the 6-hydroxypurine nucleotides and is able to phosphorylate inosine, deoxyinosine, guanosine and many purine nucleoside analogues including 2',3'-dideoxyinosine and deoxycoformycin.[3] The enzyme is activated by a number of nucleoside diphosphates and triphosphates such as ATP and ADP, and by 2,3-bisphosphoglycerate (BPG), furthermore it is inhibited by phosphate.[1,4] The sequence of human, avian and bovine cytosolic 5'-nucleotidase has been determined, showing a degree of homology higher than 90%, among the different species.[5,6] A similar homology has been described for proteins such as actin and histones, whose structure and function are of fundamental importance in the living cells. The high conservation of protein sequence, the widespread distribution of the enzyme and its complex regulation indicate that the activity of cytosolic 5'-nucleotidase must be of fundamental importance probably in the regulation of the intracellular IMP concentration. In the course of our work we became aware that our final enzyme preparation contained two polypeptides with different electrophoretic mobil-

Purine and Pyrimidine Metabolism in Man IX,
edited by Griesmacher *et al.* Plenum Press, New York, 1998.

ity, both cross-reacting with specific antibodies raised against cytosolic 5'-nucleotidase. Furthermore, both polypeptides were able to form the phosphorylated intermediate, indicating that they were active 5'-nucleotidases.[2] The two forms were separated by affinity chromatography on ADP-Agarose and their regulatory characteristics were studied. Recently we noticed that even though ADP and BPG, separately, act at a too high concentration to be considered physiological regulators, there is a synergistic effect among the two compounds. In fact, the presence of physiological ADP concentration significantly lowers the amount of BPG necessary to exert an activatory effect on the form with the lower apparent molecular weight, and viceversa.[7] The other form, on the contrary, was activated at a greater extent by ADP and BPG but at different concentrations with respect to the other form, and the synergistic effect was absent.[8] With the aim to collect more information on the nature, origin, and physiological function, if any, of these enzyme forms we studied the distribution of enzyme activity, mRNA and electrophoretic mobility in rat tissues.

2. EXPERIMENTAL

2.1. mRNA Distribution

A ^{32}P labelled 1.1 Kb fragment corresponding to the 5'-terminus of the calf thymus 5'-nucleotidase cDNA was used to probe a Rat Multi-Tissue Blot (Clontech) to determine 5'-nucleotidase mRNA amount in different tissues. The blot was prehybridised 30' at 68°C in ExpressHyb hybridisation solution. Hybridisation was carried out in the same solution containing the probe at 68°C for 50'. The blot was then rinsed three times for 40' at room temperature in 2x SSC 0.05% SDS, washed once 40' at 37°C and once 40' at 42°C in the same solution and autoradiographed.

2.2. Preparation of Crude Extracts

Tissues were taken from freshly sacrificed Wistar male rats (200–250 g), washed in a 0.9% NaCl solution and then homogenised with a Politron apparatus in three ml/g of 50 mM Tris HCl buffer pH 7.4, 1 mM dithiothreitol, 250 µg/l leupeptin and pepstatin (Standard buffer). Homogenates were centrifuged 60' at 10400 g and the 5'-nucleotidase specific activity was immediately measured.

2.3. Enzyme Assay

The phosphotransferase activity associated with cytosolic 5'-nucleotidase was measured as the rate of 8[^{14}C]IMP formation in the presence of 8[^{14}C]inosine according to Tozzi et al.[4] Protein was measured according to Bradford.[9]

2.4. Immunoblotting

About 33 µg of protein of each tissue crude extract were loaded on a freshly prepared Matrex Green-A column (4 ml) equilibrated with standard buffer. After extensive washing with standard buffer containing 0.6 M KCl, cytosolic 5'-nucleotidase was eluted with a linear gradient of KCl (0.6–2.1 M) in the same buffer. The procedure usually took no more than 2 h and resulted in a purification factor of about 40 and gave a yield of about 60%. Fractions containing enzyme activity were collected, dialysed, concentrated, and 10 µg protein of each enzyme preparation were used for electrophoresis on 12% polyacrylamide gel according to

Laemlli.[10] After electrophoresis the samples were blotted on a PVDF membrane, finally immunostaining with specific antibody was carried out as previously described.[2]

3. RESULTS

3.1. Northern Blot Analysis of Rat mRNA

Northern blot analysis of rat mRNA with bovine 5'-terminal cDNA showed a single band of 1.8 Kb in all tissues. Testis and kidney showed a higher content of mRNA and also same additional longer transcripts (Fig. 1).

3.2. Distribution of Cytosolic 5'-Nucleotidase in Rat Tissues

Table 1 shows the distribution of the phosphotransferase activity associated with the cytosolic 5'-nucleotidase. Since mammalian tissues possess very low cytosolic inosine and guanosine kinases, the phosphotransferase activity associated with cytosolic 5'-nucleotidase seems to be the main responsible for inosine phosphorylation, thus its measure allows a more correct evaluation of enzyme specific activity compared to the evaluation of the rate of phosphate or nucleoside formation from a mononucleotide substrate, which might be due to the action of many different phosphatases. The 5'-nucleotidase distribution is in agreement with the activity of 5'-nucleotidase observed earlier in mammalian and avian tissues by immunotitration.[1]

3.3. Electrophoresis and Immunoblotting of Rat Tissue Extracts

Our results after electrophoresis and immunoblotting showed that testis, brain, liver and spleen partially purified preparation contained two polypeptides identified with specific antibodies for cytosolic 5'-nucleotidase. The molecular weight of these forms was similar to the molecular weight of the enzyme forms present in thymus treated in the same way (Figs. 2, 3, 4). As well as in calf thymus, the form with the higher mobility is more

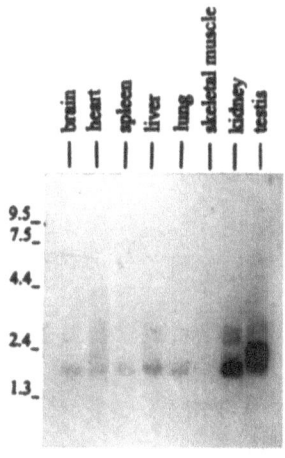
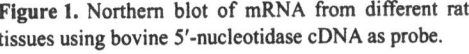
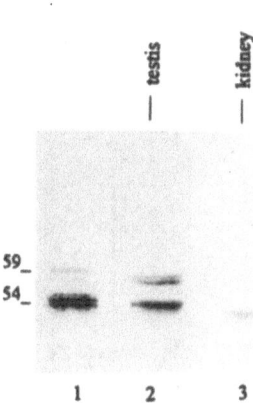

Figure 1. Northern blot of mRNA from different rat tissues using bovine 5'-nucleotidase cDNA as probe.

Figure 2. Immunoblotting of partially purified cytosolic 5'-nucleotidase. Line 1. Enzyme preparation from calf thymus. Lines 2 and 3. Enzyme preparation from rat tissues.

Table 1. Specific activity in rat tissue crude extracts

	Phosphotransferase specific activity (nmol/min × mg protein)
Testis	4.90
Liver	3.16
Spleen	2.35
Kidney	2.07
Brain	2.06
Lung	1.65
Pancreas	0.80
Skeletal muscle	0.34

represented in these organs. In kidney, the polypeptide with the lower mobility was unde-tectable. Finally, skeletal muscle which displayed the lower specific activity and no detectable mRNA, showed a single band at a molecular weight around 66000.

4. DISCUSSION

Our results show that cytosolic 5'-nucleotidase mRNA is present in all rat tissues tested although in different amount with the exception of skeletal muscle. The distribution of enzyme activity is in agreement with the distribution of the same enzyme in mammal-ian tissues determined by immunotitration, indicating that the determination of the phos-photransferase activity as described by Tozzi et al.[4] is a reliable, specific and relatively fast and easy way to determine cytosolic 5'-nucleotidase activity in crude extracts. Even though the mRNA distribution indicates a remarkable difference among tissues, which is

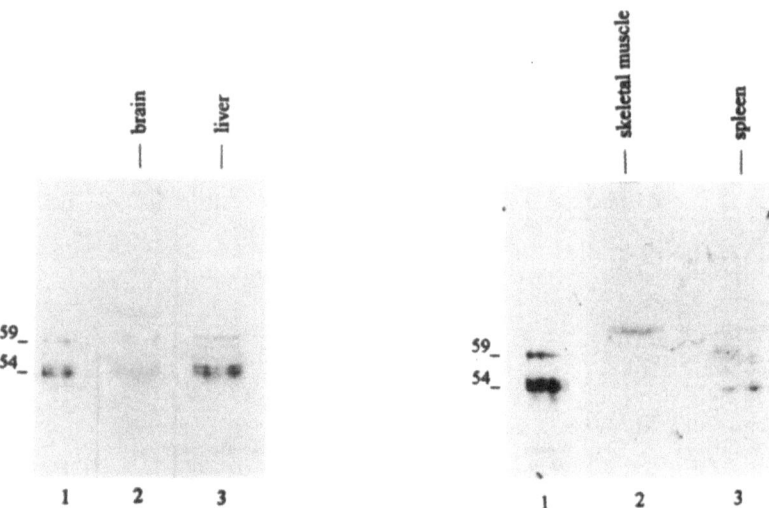

Figure 3. Immunoblotting of partially purified cytoso-lic 5'-nucleotidase. Line 1. Enzyme preparation from calf thymus. Lines 2 and 3. Enzyme preparation from rat tissues.

Figure 4. Immunoblotting of partially purified cytoso-lic 5'-nucleotidase. Line 1. Enzyme preparation from calf thymus. Lines 2 and 3. Enzyme preparation from rat tissues

not detectable in the distribution of enzyme activity, the results of both analysis indicate that testis contained the higher and skeletal muscle the lower enzyme concentration. Finally most of the tissues contained two polypetides showing electrophoretic behaviour similar to that reported for the two forms described in bovine thymus, both cross-reacting with specific antibody, indicating that, the presence of multiple forms of 5′-nucleotidase is a widespread phenomenon and that probably the faster moving form is related with the higher enzyme specific activity.

ACKNOWLEDGMENTS

This work was supported by a grant from the Italian MURST.

REFERENCES

1. Itoh, R. (1993) Comp. Biochem. Physiol. 105B, 13–19.
2. Baiocchi, C., Pesi., R., Camici, M., Itoh, R. and Tozzi, M.G. (1996) Biochem. J. 317, 797–801.
3. Banditelli, S., Baiocchi, C., Pesi, R., Allegrini, S., Turriani, M., Ipata, P.L. Camici, M. and Tozzi, M.G. (1996) Int. J. Biochem. Cell Biol. 28, 711–720.
4. Pesi, R., Turriani, M., Allegrini, S., Scolozzi, C., Camici, M., Ipata, P.L. and Tozzi, M.G. (1994) Arch. Biochem. Biophys. 312 n°1, 75–80.
5. Oka, J., Matsumoto, A., Hosokawa, Y. and Inoue, S. (1994) Biochim. Biophys. Res. Commun. 205 n°1, 917–922.
6. Allegrini S., Jhonson, R.B., Fiol, C.J., aand Eriksson, S. 5th Symposium ESSPPMM 1955, Vasto-Termoli.
7. Pesi, R., Baiocchi, C., Tozzi, M.G. and Camici, M. (1996) Biochim. Biophys. Acta 1294, 191–194.
8. Pesi, R., Baiocchi, C., Camici, M., and Tozzi, M.G. (1996) Ital. J. Bioc. 45, 166–167.
9. Bradford, M.M. (1979) Anal. Biochem. 72, 248–254.
10. Laemmli, U.K. (1970) Nature (London) 227, 680–685.

98

REGULATION OF DEOXYNUCLEOTIDE POOLS BY SUBSTRATE CYCLES

Vera Bianchi[*]

Department of Biology
University of Padova
Viale Colombo 3, 35121 Padova, Italy

1. SUBSTRATE CYCLES

In mammalian cells the pools of deoxyribonucleoside triphosphates (dNTPs) are finely regulated in relation to the needs of DNA synthesis.[1] When cells enter the S phase of the cell cycle dNTP pools expand and are continuosly replenished by *de novo* synthesis until DNA replication is completed. Throughout interphase dNTP pools are low, adequate for the limited requirement of precursors for DNA repair.

The regulation of dNTP pools during the cell cycle results from the interplay of anabolic and catabolic enzymes. The former carry out the *de novo* synthesis of dNTPs and the salvage of deoxynucleosides and are well characterized. Here, I will deal with a class of catabolic enzymes, the 5'-nucleotidases, which are still poorly known, especially with respect to their involvement in dNTP pool regulation.

The reaction catalyzed by 5'-nucleotidases, i.e. the dephosphoryl- ation of nucleoside-5'-monophosphates, reverses the effects of nucleoside kinases, the enzymes that phosphorylate nucleosides to nucleotides. Each of the two reactions is in itself irreversible, but the coexistence of the two kinds of enzymes gives rise to a cycle, a futile or *substrate* cycle (Figure 1). I will limit my discussion to substrate cycles between deoxynucleosides and their 5'-monophosphates.

The cycles influence the concentration of deoxynucleotides in the cell because the nucleotidase reaction produces a chemical species, the deoxynucleoside, which readily permeates the plasma membrane and is excreted when the concentration in the cell exceeds the extracellular concentration. When the concentration gradient is directed inward, the deoxynucleoside enters the cell and is trapped by phosphorylation by the kinase reaction. Thus a substrate cycle can have an overall anabolic or catabolic function depending on the relative efficiency with which the two opposite reactions occur in any given situation.

---

[*] Telephone: -39 49 827 6282; fax: -39 49 827 6280; email: vbianchi a civ.bio.unipd.it

Purine and Pyrimidine Metabolism in Man IX,
edited by Griesmacher *et al.* Plenum Press, New York, 1998.

A Substrate Cycle

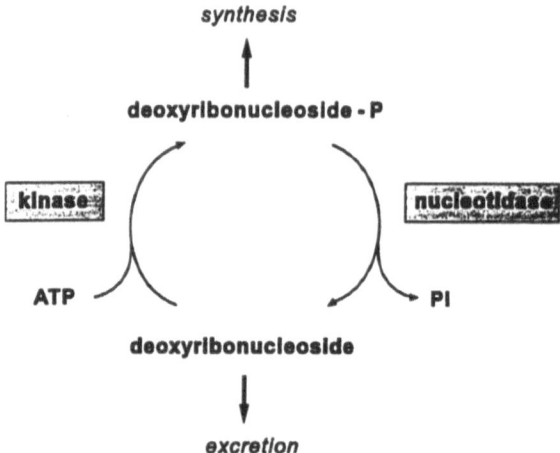

Figure 1. Schematic representation of a substrate cycle between a deoxynucleoside and its 5'-monophosphate.

Two main kinases are involved in the phosphorylation of deoxynucleosides, thymidine kinase which is S-phase restricted and phosphorylates thymidine and deoxyuridine, and deoxycytidine kinase which is constitutively expressed and phosphorylates deoxycytidine and the purine deoxynucleosides. Both enzymes are known in much detail and are still being studied for their role in the phosphorilation of physiological deoxynucleosides and the activation of therapeutic nucleoside analogs. Although the same reasons of interest apply to 5'-nucleotidases, these enzymes are almost neglected. Yet a body of experimental evidence, together with the features of some genetic syndromes caused by defects in nucleotide pool regulation,[2] clearly show that substrate cycles operate in cells and whole organisms. A better understanding of their mechanisms through a deeper knowledge of both classes of enzymes involved—kinases and nucleotidases—would be of advantage in the treatment of disease.

2. EVIDENCES FOR SUBSTRATE CYCLES

In cultured cells substrate cycles are easily detected by measuring their catabolic activity on the basis of the accumulation of deoxynucleosides in the medium. By this means we and others have shown that the cycles are continuosly active, and change their catabolic function depending on the availability of the nucleotidase substrate. The anabolic effect of the cycle can be measured *in vitro* by following the uptake of deoxynucleosides, but some problems are encounterd with this approach, as we will see below.

Table 1 summarizes experimental data on which the model of substrate cycles is built. The increased excretion of deoxynucleosides in cells lacking either thymidine[3] or deoxycytidine[4] kinases shows that deoxynucleosides are continuosly produced inside the cells and that kinase proficient cells rephosphorylate them before they diffuse out of the plasma membrane. Expansion of the dNTP pools, due to either overproduction caused by an enzyme mutation[5] or to reduced utilization of dNTP caused by inhibitors of DNA synthesis,[6,7] leads to enhanced nucleotidase activity and excretion of deoxynucleosides. When the pools are

Table 1. The concept of substrate cycles between deoxynucleosides and their 5'-monophosphates as components of the regulation of dNTP pools is based on several lines of evidence obtained with cultured mammalian cells

Evidence for substrate cycles in dNTP regulation
1. Mutations in deoxynucleoside kinases increase excretion of deoxynucleosides.
2. dNTP pool expansion by mutation increases excretion of deoxynucleosides.
3. Inhibitors of DNA synthesis favour excretion of deoxynucleosides.
4. Inhibitors of dNTP synthesis favour uptake of deoxynucleosides.

reduced in size due to inhibition of *de novo* synthesis, e.g. by hydroxyurea,[8] uptake and phosphorylation of deoxynucleosides by the kinases is favoured over the 5'- nucleotidase reaction and the cycles operate in the anabolic direction. The efficiency of anabolism varies with the deoxynucleoside substrate, because the kinases have different affinity for different substrates and other enzymatic activities compete with the kinases. The competition is particularly strong in the case of purine deoxynucleosides, which are easily degraded.

2.1. Substrate Cycles Are Differently Active in Different Cell Types

To measure the catabolic activity of substrate cycles we incubate cells with a radio-active ribonucleoside or purine base, thus labelling the corresponding dNTP pool by the *de novo* pathway, and quantitate the excretion of the deoxynucleoside by dividing the radioactivity recoverd in the medium as deoxynucleoside by the specific activity (s.a.) of the dNTP measured in the nucleotide pool fraction. In the case of pyrimidine cycles the deoxynucleosides are sufficiently stable and can be measured directly as they accumulate in the incubation medium. In the case of the deoxyadenosine/dAMP cycle, instead, additional precautions are needed. In fact, deoxyadenosine excretion can be measured experimentally only in the presence of inhibitors of adenosine deaminase, whose km for deoxyadenosine is much lower than that of the kinase. In addition, we use HPRT⁻ cells which do not recycle hypoxanthine. With such a protocol we could detect the operation of the deoxyadenosine/dAMP cycle in hamster fibroblasts and show that it is affected by the loss of deoxycytidine kinase activity,[4] responsible for deoxyadenosine phosphorylation, or by the addition of inhibitors of DNA replication which expand the dATP pool[7] (Table 2).

A similar experiment performed with human T lymphoblasts gave similar results in terms of enhancement of deoxyadenosine excretion caused by mutational loss of the relevant kinases.[4] Excretion increased 6–8 fold relative to the parental line in cells which had lost deoxycytidine kinase or adenosine kinase, which also phosphorylates deoxyadenosine, or both kinases. However the catabolic activity of the cycle was here at least 50 times lower in the kinase-proficient cells compared to the wild type fibroblasts in Table 2, and 25 times lower in the kinase-deficient cells. Clearly the balance between phosphorylation and de-phosphorylation is shifted towards the former in lymphoid cells, and this is believed to be the reason for the lympho-specific toxicity of adenosine deaminase deficiency.[9] In the absence of deamination of deoxyadenosine, the deoxynucleoside is recycled and dATP toxicity occurs. Our data indicate that in lymphoid cells the accumulated dATP derives from deoxyadenosine produced by other cell types in the body and circulated by the blood. In fact, lymphoblasts devoid of both kinases that phosphorylate deoxyadenosine still release very little deoxynucleoside compared to fibroblasts. Thus the nucleotidase(s) which participate in the deoxyadenosine/dAMP cycle are less active in T lymphoblasts than in fibroblasts. Which are such nucleotidases?

Table 2. Detection of the dAMP/deoxyadenosine
substrate cycle in V79 fibroblasts

Additions	AdR excretion (pmol/min/10^6 cells)	
	WT cells	dCK⁻ cells
None	ND	ND
EHNA	0.15	0.51
deoxycoformycin	0.20	–*
+ aphidicolin	2.1	–*
+ araC	1.0	–*

ND: not detectable; –*: not measured. dCK = deoxycytidine kinase,
araC= arabinosyl-cytosine.
Cells were incubated for 3 h with ^{3}H adenine and then chased for 2 h
in isotope-free medium. Both incubations were done with the indicated combinations of drugs. The average specific activities of dATP
between 60 and 120 min of the chase were used to calculate the rates
of excretion of deoxyadenosine.

3. 5'-NUCLEOTIDASES IN SUBSTRATE CYCLES

Three major classes of 5'-nucleotidases have been described in mammalian systems. The most studied enzyme is the ecto-nucleotidase, also known as CD73, because it was described as a lymphocyte surface antigen. It has a role in lymphocyte differentiation which is apparently independent of its catalytic activity.[10] The other two enzymes are cytosolic. The high km nucleotidase has been sequenced,[11] and its catalytic properties characterized, showing that it is also endowed with a phosphotransferase activity responsi- ble, i.al., for the phosphorylation of the anti HIV drug dideoxyinosine.[12] The third enzyme can be considered a deoxynucleotidase, as it is the only nucletidase known to prefer deoxy-5'-monophsphate substrates.[13] The catalytic properties are sufficiently different to allow to distinguish the three enzymatic activities in subcellular fractions assayed *in vitro*.

At present nothing is known about the nature of the nucleotidases involved in deoxynucleotide substrate cycles. We have started to approach this question by measuring the excretion of deoxycytidine and deoxyadenosine in Jurkat cells overproducing the ectonucleotidase, obtained from Linda Thompson's laboratory.[14]

Figure 2 shows that basically the same amount of deoxynucleosides were released by the control cells and the ectonucleotidase overproducers which expressed about 20 fold higher enzyme activity. Such observations suggest that the ectonucleotidase has no access to the dNTP pools, even during the intracellular sorting of the enzyme to the plasma mem- brane. Cytosolic nucleotidases are therefore more likely candidates for participation in substrate cycles. We are currently working on such enzymes.

4. DEOXYNUCLEOSIDE KINASES AND THE ANABOLIC FUNCTION OF SUBSTRATE CYCLES

To measure the anabolic function of substrate cycles one can measure the incorpora- tion of a deoxynucleoside into the end-products, the dNTP and DNA. The results are strongly affected by the nature of the deoxynucleoside one uses, and thus by the kinase catalyzing the first phosphorylation step. The reason resides in the different cell-cycle regulation of the two major deoxynucleoside kinases and in the fact that cultured cell populations are asynchronous. Our recent work illustrates this point.[15] Labelled thymidine is incorporated into the dTTP pool of cells expressing thymidine kinase, i.e. S phase-cells,

A) AdR excretion

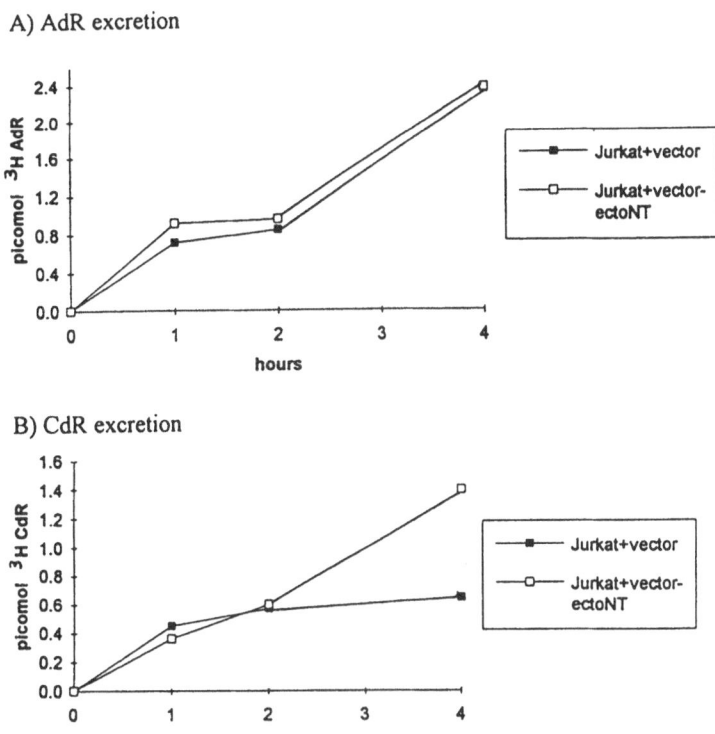

B) CdR excretion

Figure 2. The ecto-5'-nucleotidase is not involved in the dAMP/deoxyadenosine nor in the dCMP/deoxycytidine substrate cycles in Jurkat cells. Control Jurkat cells transfected with the empty vector, or Jurkat BNT 5.1 cells overexpressing the ectonucleotidase were incubated with 1 μM ^{3}H adenine (A) or 1 μM ^{3}H cytidine (B) and excretion of ^{3}H deoxyadenosine (AdR) or ^{3}H deoxycytidine (CdR) was measured by HPLC. Ectonucleotidase activity in the membrane fraction was 3.4 U in the controls and 62 U in BNT 5.1 cells. [1U = 1 nmol CMP dephosphorylated/min/mg protein].

where it is diluted by unlabelled dTTP made *de novo* and used for DNA replication. One can determine the specific activity of the dTTP pool and the rate of incorporation of radioactive dTMP into DNA by dividing the accumulation of isotope into DNA by the calculated dTTP s.a.. How accurate is this evaluation of DNA replication and of the anabolism of extracellular thymidine? We separated by centrifugal elutriation exponential populations of CEM lymphoblasts after incubation with tritiated thymidine, and measured the dTTP pool size and s.a. in G1 and S phase cells. As expected, efficient incorporation of thymidine occurred only in S phase cells, which contained a dTTP pool that was 3-fold larger and almost 10-fold more radioactive than the dTTP pool of G1 cells.[15] Therefore, when the pool is extracted from an unseparated population the dTTP pool of S phase cells is negligibly modified by the pool of non-S phase cells, and the rate of DNA replication can be estimated quite accurately also from the average value of dTTP s.a. in the unsynchronized population (Table 3), which is what is commonly done.

When instead exponentially growing cells are incubated with deoxycytidine, the deoxynucleoside is phosphorylated by deoxycytidine kinase in all phases of the cycle, although only S phase cells make dCTP *de novo* and replicate DNA. Our elutriation experiments[15] demonstrate that incorporation of radioactive deoxycytidine into the dCTP pool of G1 cells occurs with very little dilution by endogenous non labelled dCTP, while the dCTP of S phase cells has a 30-fold lower s.a., due to active *de novo* synthesis of

Table 3. Rate of DNA replication calculated from average or S-phase specific activity of precursor dNTP in CEM lymphoblasts incubated with ^{3}H thymidine or 5-^{3}H deoxycytidine

| Precursor | dNTP | dNTP s.a.(cpm/pmol) | | DNA radioactivity Δcpm/min | DNA synthesis (pmol/min) | |
		Average	S-phase		From average s.a.	From s-phase s.a.
^{3}H thymidine	dTTP	1150	1270	3200	2.8	2.5
5-^{3}H deoxycytidine	dCTP	6150	430	730	0.12	1.7

dCTP which competes with the kinase reaction. An additional outlet of radioactivity from the dCTP pool comes from the conversion of dCMP to dTMP. Due to the ability of non-S phase cells to phosphorylate deoxycytidine, and the large differences in s.a. of the dCTP pool in the different phases of the cycle, the average s.a. of dCTP in the unsynchronized population is very far from the actual s.a. of the dCTP used to make DNA in S phase. Only by considering the latter value one can obtain a correct evaluation of the rate of DNA replication (Table 3). These results solve the long-debated question of the ability of deoxycytidine kinase to provide dNTPs for DNA replication. In fact, both with deoxycytidine and deoxyadenosine labelling incorporation of radioactivity into DNA is poor compared to thymidine incorporation.[7,16] The different cell cycle regulation of the two kinases involved provides the mechanism for such observations.

5. CONCLUSIONS

The regulation of anabolic enzymes in deoxynucleotide metabolism occurs at different levels, from gene transcription to allosteric regulation of the proteins. Although these mechanisms are known in much detail, they still represent a fascinating field of investigation because of their connection with cell cycle progression and their implications for anticancer and antiviral therapy. The role of the 5'-nucleotidases in the regulation of dNTP pools is as yet not understood in detail. While the cloning of the cytosolic enzymes is proceeding, time is getting ripe to understand their physiological significance in cell metabolism and to manipulate their functions for therapeutic purposes.

REFERENCES

1. Reichard P. (1988) Annu. Rev. Biochem. 57, 349–374.
2. Martin DW & Gelfand EW (1981) Annu. Rev. Biochem. 50, 845–877.
3. Höglund L, Pontis E & Reichard P (1988) Cancer Res. 48, 3681–3687.
4. Bianchi V et al., (1994) J. Biol. Chem. 269, 16677–16683.
5. Bianchi V, Pontis E & Reichard P (1987) Mol. Cell. Biol. 7, 4218–4224.
6. Nicander B & Reichard P (1985) J. Biol. Chem. 260, 5376–5381.
7. Bianchi V, Pontis E & Reichard P (1992) Exp. Cell Res. 199, 120–128.
8. Bianchi V, Pontis E & Reichard P (1986) Proc. Natl. Acad.Sci.USA 83, 986–990.
9. Carrera CJ & Carson DA (1987) in The Molecular Basis of Blood Diseases (Stamatoyannopoulos G ed.) pp. 407–449, WB Saunders Co, Philadelphia, PA.
10. Gutesohn W et al., (1995) Cell. Immunol. 161, 213–217.
11. Oka J et al., (1994) Biochem. Biophys. Res. Comm. 205, 917–922.
12. Johnson MA & Fridland A (1989) Mol. Pharmacol. 36, 291–295.
13. Höglund L & Reichard P (1990) J. Biol. Chem. 265, 6589–6595.
14. Resta R et al., (1994) J. Immunol. 153, 1046–1053.
15. Bianchi V et al. (1997) J. Biol. Chem. 272, in press.
16. Nicander B & Reichard P (1983) Proc. Natl. Acad.Sci.USA 80, 1492–1495.

99

DIHYDROOROTATE DEHYDROGENASE

Profile of a Novel Target for Antiproliferative and Immunosuppressive Drugs

Monika Löffler,* Klaus Grein, Wolfgang Knecht, Astrid Klein, and Ute Bergjohann

Institute for Physiological Chemistry
School of Medicine
Philipps-University
Karl-von-Frisch-Str. 1
D-35033 Marburg, Germany

1. INTRODUCTION

This article offers a brief perspective on some of the research undertaken in this laboratory and presents original data to be of particular concern to the understanding of drugs with 'anti-pyrimidine' efficasy.

Whilst the structure and function of both cytosolic enzymes in the de novo pyrimidine biosynthesis—the multifunctional CAD and UMPsynthase—are quite well understood, the fourth enzyme, dihydroorotate dehydrogenases (DHOdehase; DHODH; EC 1.3.99.11) has evaded biochemical elucidation. With respect to the topochemistry and the proximal electron acceptor ubiquinone (Q) and the final electron acceptor oxygen, mammalian DHOdehase resembles succinate dehydrogenase known as complex II of the respiratory chain (Figure 1). Due to the translocation of dihydroorotate to and of orotate from the inner mitochondrial membrane, DHOdehase links functional mitochondria in the nucleic acid precursor metabolism.[1] In previous work, we could allocate this mechanism to be of high relevance for the well-known oxygen dependence of DNA and RNA synthesis and proliferation of eukaryotic cells.[2,3] The membrane association and the low abundance of DHOdehase in tissues may have been an impediment to purify the enzyme in sufficient amounts for characterization and cristallization studies.

* Requests to Dr. Monika Löffler, Institute for Physiological Chemistry, School of Medicine, Philipps-University, Karl-von-Frisch-Str. 1, D-35033 Marburg, Germany. Tel +6421-285022; Fax +06421-286957. E-mail: loeffler@mailer.uni-marburg.de.

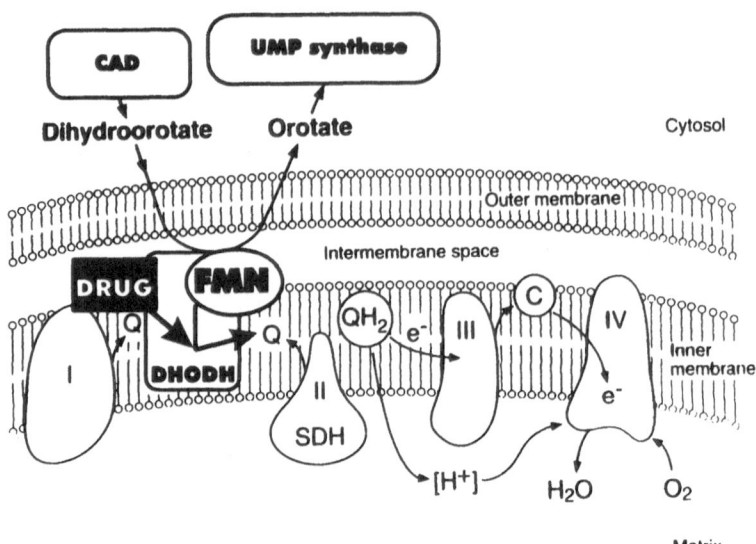

Figure 1. Scheme of cellular compartmentation of the *de novo* pyrimidine synthesis. The cytosol contains the CAD (multifunctnal protein with the first three enzymes) and the UMP synthase (multifunctional protein with the last two enzymes) and the inner mitochondrial membrane holds the flavoprotein DHODH (dihydroorotate dehydrogenase). FMN (flavinmononucleotide), Q (ubiquinone), QH2 (reduced ubiquinone), I, II, III, IV (enzyme complexes of the respiratory chain), C (cytochrome c). The action of an inhibitory drug is indicated by the arrow at the DHODH.

2. DIHYDROOROTATE DEHYDROGENASE IN SITU AND IN VITRO

Catalytic enzyme histochemistry with cryo-sections demonstrated a tissue-function-specific pattern of DHOdehase activity and revealed considerable disparity of activity with succinate dehydrogenase, which is a prominent marker enzyme of the inner mitochondrial membrane.[4] Until 1995, by multi-step purification protocols, the isolation of mammalian DHOdehase from fresh or frozen tissues—predominantly from liver—was the sole way to get the enzyme for biochemical in vitro studies.[5-8] Since cultured cells can be advantageous for protein extraction, and because DHOdehase activity was found to be quite high in proliferating cells,[4] efforts to collect the enzyme from cells were performed likewise but made large amounts of cell cultures necessary.[9] Recently, we communicated the overexpression of the human DHOdehase (42 kDa) to high level in the baculovirus expression vector system.[10] A considerable improvement of the purification protocol in comparison to the former long sequence of steps allowed to obtain a high yield of the enzyme (Table 1). A truncated (39 kDa) human DHOdehase with much lower specific activity was obtained by overexpression in Escherichia coli.[11]

The baculovirus expression vector system was also used for producing the rat DHOdehase (42.7 kDa).[12] The rodent enzyme received preferential treatment, since rodents, and predominantly rats, are preferred animal models for research in human disease and for investigating of drugs for therapeutic strategies. From the cDNA of rat liver DHOdehase cloned in our laboratory the first complete sequence of a mammalian enzyme was deduced.[13] We achieved a high level of biologically active enzyme and the purification protocol resulted in the highest specific activity noted so far for a rodent enzyme (Table 1).

Table 1. Survey on purification protocols of mammalian dihydroorotate dehydrogenase

Source	Number of steps in the protocol	Activity yield (%)	Specific activity (U/mg)	References
Bovine liver	8	3.4	100	[6]
Porcine liver	6	0.02	0.04	[9]
Rat liver	6	8.7	0.7	[7]
Human recombinant (E. coli)	4	0.7	1.4	[11]
Human recombinant (BEVS)*	3	58	55	[10]
Rat recombinant (BEVS)*	3	26	15	[12]

*(BEVS), baculovirus expression vector system

3. DIHYDROOROTATE DEHYDROGENASE AS TARGET FOR DRUGS

The enzymes of pyrimidine metabolism were determined long ago to be of great importance for the development of drugs to combat malignant cell proliferation. Inhibitors of de novo pyrimidine biosynthesis such as PALA, pyrazofurin, lapachol have had marginal success in the clinic as single agents in anti-cancer therapy.[14,15] Brequinar sodium, originally discovered as cytostatic agent with strong effect on DHOdehase, is currently evaluated for its effectiveness as immunosuppressive agents for preventing graft rejection.[16] In view of testing other cinchoninic acid derivatives, isoxazol derivatives and naphoquinone compounds, DHOdehase is of increasing interest lately as a target for drugs to reduce aberrant immunological reactions and to interfere in the multiplication of animal parasites, parasitic protozoa in malaria-infected people and pathogenes of opportunist diseases.[17-19] The mode and mechansims of interaction of such compounds with the cellular target are mostly not understood. Figure 2 shows the formulas of A771726, brequinar sodium, atovaquone and two lawsones, which were described to impair the oxidation of dihydroorotate with concommitant cell growth reduction.

Atovaquone has been known as chemotherapeutic agent against Plasmodium falsiparum and Pneumocystis carinii.[18] The isoxazol derivative leflunomide (with the in vivo active metabolite A771726) is currently in clinical phase III trials for the treatment of rheumatoid arthritis which is believed to have an underlying autoimmune pathology. Phase II studies assessed the safety and effectiveness of this new drug.[20] Its malononitrilamide derivatives have been found to block rejection after allograft and xenograft transplantation in animals.[21] However, it is not known so far, whether the enzyme DHOdehase is the main intracellular target of these compounds. As reported recently, the recombinant and purified human DHOdehase was a valuable tool to evaluate more precisely the kinetic features of the A771726–enzyme interaction.[10] This showed a characteristic uncompetitive type of inhibition with respect to the substrate dihydroorotate and a non-competitive inhibition with respect to ubiquinone. The IC_{50} value, which was determined to be 1 μM A771726 will be of importance in connection with results from bioavailability and pharmacokinetic studies raised in other laboratories. As for the rodent DHOdehase, here, experiments are in progress on the kinetics with various inhibitors. In comparison to the human enzyme, the recombinant rat DHOdehase exhibited a 50 fold higher affinity for A771726, whereas brequinar sodium was 30 fold more effective in damping the activity of the human enzyme activity than that of the rat enzyme. Likewise, species-specific characteristics were found with the

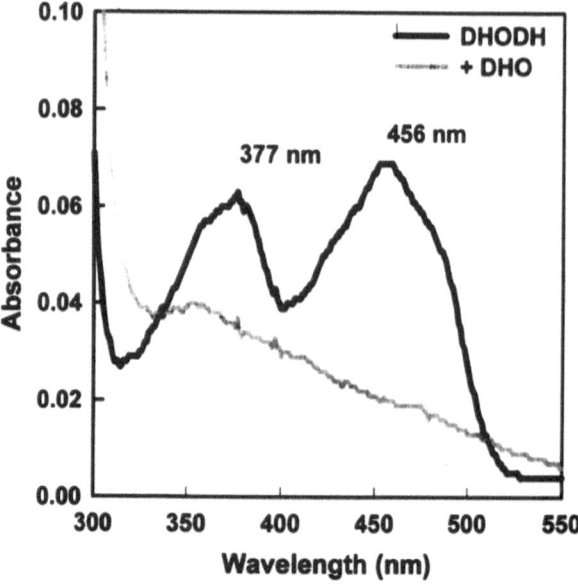

Figure 2. Formulas of drugs with proven anti-dihydroorotate dehydrogenase activity.

other compounds: the studies with atovaquone exhibited a 40 fold higher affinity of the rat versus the human enzyme, and dichloroallyl lawsone and lapachol, respectively, had a 10 fold and a 20 fold higher effect on the rat enzyme. Further validation of these observations will be necessary for establishing the pharmacological profile of these drugs. In addition, the approval will be of great relevance for structure-activity relation studies, since the sequence of the human enzyme shows 88% amino acid identity and 94% similarity with the sequence of the rat enzyme.

The availability of high amounts of catalytically active DHOdehase allowed the final identification of flavinmononucleotide (FMN) as the main redox center. Figure 3 shows the UV-vis spectrum of the native human enzyme with maxima at 377 nm and 456 nm. This typical absorbance of the flavin was decreased on addition of the substrate dihydroorotate which, intermediately, reduced the flavin. The flavin cofactor would be re-oxidized on transfer the electrons to the acceptor ubiquinone which was not present in the cuvette. In combination with others, this spectrometric technique will be very useful to get information

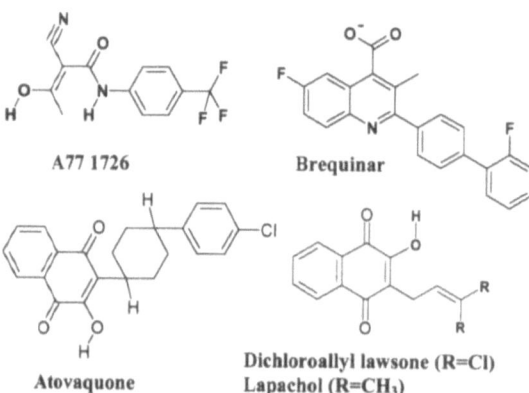

A77 1726

Brequinar

Atovaquone

Dichloroallyl lawsone (R=Cl)
Lapachol (R=CH₃)

Figure 3. Ultraviolet-visible absorption spectrum of recombinant human dihydroorotate dehydrogenase. The cuvette contained 150 µg of the enzyme in 25 mM potassium phosphate buffer, 400 mM potassium chloride, 0.1% Triton, pH 7.4 (DHODH spectrum). On addition of 200 µM dihydroorotate (spectrum + DHO) the reduction of the flavin redox-cofactor was accompanied by a decline in its characteristic absorbance.

on the stage of the interference of drugs with the redox processing along the DHOdehase enzyme molecule.

4. ANTI-DIHYDROOROTATE DEHYDROGENASE IMMUNOGLOBULINS

Since antibodies provide an excellent analytical tool for the purpose of protein identification throughout purification procedures and of enzyme localisation in tissues, it was the rationale of the present work to generate polyclonal immunoglobulins. These were obtained from an immunization protocol with a truncated recombinant human DHOdehase protein (32 kDa) in a chinchilla-bastard rabbit. For Western blots, the same protein at different concentrations was taken as standard for semiquantitative determination of the DHOdehase protein in whole cells or in mitochondria isolated from rat tissues. After electrophoretic separation and semi-dry blotting the detection of the antibodies bound to DHOdehase was by means of the Amersham Enhanced Chemical Luminescence detection kit (Figure 3). The content of DHOdehase protein in total mitochondrial protein was evaluated to be 2.3 ng/10 μg (heart), 4.9 ng/10 μg (kidney) and 10.5 ng/10 μg (liver). The ratio of 1 : 2.2 : 4.7 (heart / kidney / liver) was in good accordance with the previously described ratio of DHOdehase activity determined with mitochondria of rat tissues as 1 : 2 : 3.8 (heart / kidney / liver).[21]

Antibodies against DHOdehase were applied here for cytochemical demonstration of DHOdehase in the human lymphoma cell line U937.[23] Figure 4 shows a comparison of photographs of the cell smears. Cells incubated with the anti-DHOdehase IgG (Figure 4A) clearly revealed a positive signal after binding of the antibodies to the intracellular target protein allover the cytosol but not over the nuclei. An identification of mitochondria as organelles in the cytosol of cells was not possible by light microscopy. The results obtained from these and other immunochemical experiments[10,12] give evidence that the anti-DHOdehase immunoglobulins obtained so far met the requirements of high titre and specificity which are a prerequisite for successful application in protein chemistry and immunohistochemistry.

5. CONCLUSION

The present and other methods from our laboratory could provide useful analytical tools to detect and follow the effect of continuous drug exposure on the DHOdehase

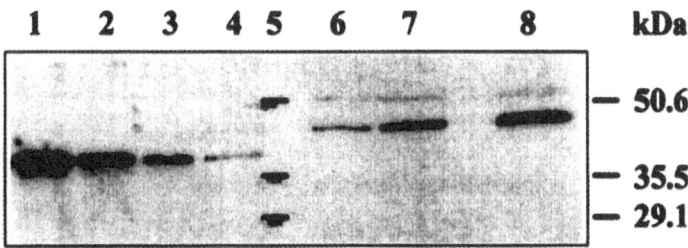

Figure 4. Detection of dihydroorotate dehydrogenase in mitochondria by Western blotting. Samples were separated by sodium dodecyl sulfate-polyacrylamide gel electrophoresis, electrotransferred to ImmubilonP, and analyzed by Western blotting using rabbit anti-DHOdehase IgG and peroxidase conjugated anti-rabbit IgG in combination with the ECL detection kit (Amersham). 10 μg mitochondrial protein of each tissue was taken for the analysis. Lane 1 - 4: 15 - 2 ng protein of truncated recombinant human DHOdehase as standards. Lane 6: heart mitochondria. Lane 7: kidney mitochondria; Lane 8: liver mitochondria.

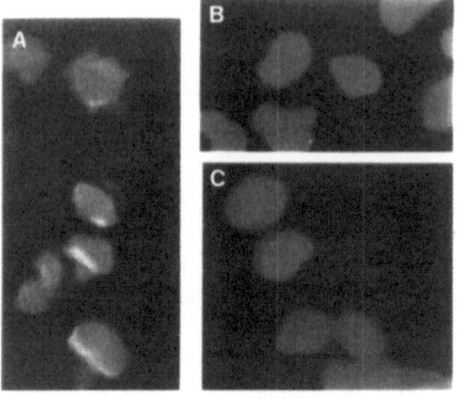

Figure 5. Immunocytochemical detection of dihydroorotate dehydrogenase. Smears of U937 cells were fixed in cold methanol and immediately frozen at −70°C. Blocking of thawed samples was performed with bovine serum albumine; incubation was with anti-DHOdehase IgG raised in Chinchilla bastard rabbits against truncated human recombinant DHOdehase protein; Cy3-anti-rabbit IgG was used as secondary antibody; the specific binding at the intracellular target was detected by fluorescence microscopy at > 580 nm. DAPI stained nuclei were visible at > 425 nm. (A), with anti-DHOdehase IgG. (B), without anti-DHOdehase IgG. (C), unrelated IgG preparation from rabbits (Sigma).

protein expression in cells and tissue. This may be of vital importance in view of putative development of drug resistance.

ACKNOWLEDGMENTS

A771726, the active metabolite of the pro-drug Leflunomide was a gift of Hoechst-Kalle Albert (Wiesbaden), Atovaquone of Glaxo-Wellcome (UK), Brequinar sodium of DuPont Merck Pharma (Bad Homburg), dichlorallyl lawsone of NHI (Bethesda, USA) and lapachol was obtained from Sigma (Munich).

Supported by Deutsche Forschungsgemeinschaft, Graduiertenkolleg Enzymchemie Marburg and by a research grant from Hoechst-Marion-Roussel Inc., Germany.

REFERENCES

1. Jones, M.E. (1980) Pyrimidine nucleotide biosynthesis in animals: genes, enzymes and regulation of UMP biosynthesis. Annu. Rev. Biochem. 49, 253–279.
2. Löffler, M. (1980) On the role of dihydroorotate dehydrogenase in growth cessation of Ehrlich ascites tumor cells cultured under oxygen deficiency. Eur. J. Biochem. 107, 207–215.
3. Amellem, O., Löffler, M. & Pettersen, E.O. (1994) Regulation of cell proliferation under extreme and moderate hypoxia: the role of pyrimidine (deoxy)nucleotides. Br. J. Cancer 70, 857–866.
4. Löffler, M., Becker, C., Wegerle, E. & Schuster, G. (1996) Catalytic enzyme histochemistry and biochemical analysis of dihydroorotate dehydrogenase / oxidase and succinate dehydrogenase in mammalian tissues, cells and mitochondria. Histochem. Cell Biol. 105, 119–128.
5. Forman, H.J. & Kennedy, J. (1977) Purification of the primary dihydroorotate dehydrogenase (oxidase) from rat liver mitochondria. Prep. Biochem. 7, 394–401.
6. Hines, V. Keys, L.D. & Johnstone M. (1986) Purification and properties of the bovine liver mitochondrial dihydroorotate dehydrogenase. J. Biol. Chem. 261, 11386–11392.
7. Lakaschus, G. & Löffler, M. (1992) Differential susceptibility of dihydroorotate dehydrogenase / oxidase to brequinar sodium (NSC 368390) in vitro. Biochem. Pharmacol. 43, 1025–1030.
8. Williamson, R.A., Yea, C.M., Robson, P., Curnock, A., Suresh, G., Hambleton, A.B., Moss, D., Thomson, T.A., Spinella-Jaegle, S., Morand, P. Courtin O., Sautéts, C., Westwood, R., Hercend, T., Kuo, E.A. & Ruuth, E. (1995) Dihydroorotate dehydrogenase, a high affinity binding protein for A77 1726 and mediator of a range of biological effects of the immunomodulatory compound. J. Biol. Chem. 270, 22467–22472.
9. Lakaschus, G., Krüger, H., Heese, D. Löffler, M. (1991) Dihydroorotate dehydrogenase may be a source of toxic oxygen species. Adv. Exp. Med. Biol.309, 361–364.
10. Knecht, W., Bergjohann, U., Gonski, S., Kirschbaum, B. & Löffler, M. (1996) Functional expression of a fragment of human dihydroorotate dehydrogenase by means of the baculovirus expression vector system, and kinetic investigation of the purified recombinant enzyme.Eur. J. Biochem. 240, 292–301.

11. Copeland, R.A., Davis, J.P., Dowling, R.L. Lombardo, D., Murphy, K.B. & Patterson, T.A. (1995) Recombinant human dihydroorotate dehydrogenase: expression, purification, and characterization of a catalytically functional truncated enzyme. Arch. Biochem. Biophys. 323, 79–86.

12. Knecht, W., Altekruse, D., Rotgeri, A. Gonski, S. Löffler, M. (1997) Dihydroorotate dehydrogenase: isolation of the recombinant enzyme from mitochondria of insect cells. Prot. Expr. Purif. 10, 89–99.

13. Rotgeri, A. & Löffler, M. (1995) Molecular cloning and sequence analysis of rat liver dihydroorotate dehydrogenase. Adv. Exp. Med. Biol. 370, 693–698.

14. Cysk, R.L., Malinowski, N., Marquez, V., Zaharevitz, D., August. E.M. Moyer, J.D. (1995) Cyclopentenyl uracil: an effective inhibitor of uridine salvage in vivo. Biochem. Pharmacol. 49, 203–207.

15. Benett, L.L., Smithers, D., Rose, L.M., Adamson, D.J. & Thomas, H.J. (1979) Inhibition of synthesis of pyrimidine nucleotides by 2-hydroxy-3-(3,3-dichloroallyl)-1,4-naphtoquinone. Cancer Res. 39, 4868–4874.

16. Cramer, D.V. (1996) Brequinar Sodium. Transplant. Proced. 28, 960–963.

17. Greene, S., Watanabe, K., Braatz-Trulson, J. & Lou, L. (1995) Inhibition of dihydroorotate dehydrogenase by the immunosuppressive agent leflunomide. Biochem. Pharmacol. 50, 861–867.

18. Harder, A. & Haberkorn, A. (1989) Possible mode of action of toltrazuril: studies on two Eimeria species and mammalian and Ascaris suum enzymes. Parasitol. Res. 76, 8–12.

19. Ittarat, I., Asawamahasaka, A., Bartlett, M.S., Smith, J.W. & Meshnick, S.R. (1995) Effects of Atovaquone and other inhibitors on Pneumocystis carinii dihydroorotate dehydrogenase. Antimicrob. Agents Chemother. 39, 325–328.

20. Bartlett, R.R., Dimitrijevic, M., Mattar, T., Zielinski, T., Germann T., Rüde, E., Thoenes, G.H., Küchle, C.C. A., Schorlemmer, H.U., Bremer, E., Finnegan, A. & Schleyerbach, R. (1991) Leflunomide (HWA 486), a novel immunomodulating compound for the treatment of autoimmune disorders and reactions to transplantation rejection. Agents and Actions 32, 10–21.

21. Schorlemmer, H.U., Bartlett, R.R. Kurrle, R. (1997) Analogues of Leflunomide's primary metabolite, the malononitrilamides, prevent the development of graft-versus-host disease.Transplant. Proc. 29, 1298–1301.

22. Löffler, M., Jöckel, J. Schuster, G., & Becker , C. (1997) Dihydroorotate dehydrogenase links mitochondria in the biosynthesis of pyrimidines. Mol. Cell. Biochem. in press.

23. Nilson, K. & Sundstrom, P. (1976) Int. Nat. J. Cancer 17, 565–577.

100

PURINE NUCLEOSIDE PHOSPHORYLASE DEFICIENT MICE EXHIBIT BOTH AN AGE DEPENDENT ATTRITION OF THYMOCYTES AND IMPAIRED THYMOCYTE DIFFERENTIATION

F. F. Snyder, J. P. Jenuth, E. R. Mably, R. K. Mangat, and A. Pinto-Rojas

Departments of Medical Genetics, Medical Biochemistry, and Pathology
Faculty of Medicine
University of Calgary
Calgary, Alberta T2N 4N1, Canada

INTRODUCTION

Purine nucleoside phosphorylase deficiency is characterized by T-cell immunodeficiency with a majority of patients also showing signs of neurological disorders ranging from spasticity to mental retardation (1,2). The accumulation of dGTP in T lymphocytes is thought to be of importance in inhibiting ribonucleotide reductase and thereby potentially affecting T cell differentiation or function (2). In addition the observations of reduced GTP levels in erythrocytes have suggested the possibility that reduced levels of GTP in the brain might correlate with neurologic deficits (3).

We have recovered three mutants at the PNP locus from the progeny of mutagen-treated male mice. The mutations have been extensively backcrossed with the C57BL/6J strain in order to establish an identical genetic background and to eliminate potential extraneous mutations. The three mutations represent differences in single amino acid substitutions and vary in the severity of their biochemical and immunological phenotype.

MATERIALS AND METHODS

The methods used in the recovery and characterization of the mutant strains have been previously communicated (4–7).

Purine and Pyrimidine Metabolism in Man IX,
edited by Griesmacher *et al.* Plenum Press, New York, 1998.

515

Table 1. Enzymatic and molecular characterization of three
mutations at the PNP locus

Strain	Relative erythrocytic PNP activity	Altered residue	Amino acid substitution
C57BL/6J	100		
B6-NPE	4.6	87	Met → Lys
B6-NPF	1.5	228	Ala → Thr
B6-NPG	1.0	16	Trp → Arg

RESULTS AND DISCUSSION

Three point mutations were recovered on the C57BL/6J mouse strain Np^b allele and have been designated B6-NPE, B6-NPF and B6-NPG in increasing order of severity (Table 1). Each of these represent a unique amino acid substitution, however, neither of the mutations have altered an active site residue of PNP (8). Erythrocytic PNP activity relative to that of C57BL/6J shows a range from 1 to 4.5% and indicates that neither of these mutations represents a complete deficiency of PNP.

Cell numbers per thymus were essentially normal at 2 months of age but began to decline for the two severe mutations between 2 and 3 months of age to 61 and 41% of control for B6-NPF and B6-NPG mice respectively. At 8 months all mutants showed a reduction in total thymocyte numbers: B6-NPE, 65%; B6-NPF, 59% and B6-NPG, 29% of control. Examination of thymocyte subpopulations at two intervals for the B6-NPG mutant showed a progressive increase in the CD4⁻CD8⁻ T cell precursor population between 5 to 10 months, to an 8-fold accumulation, and a corresponding decline in the intermediate CD4⁺CD8⁺ population to less than half its normal level (Table 2).

Thymuses of PNP deficient mice of various ages having the severe mutations, B6-NPF and BG-NPG, and normal controls were examined histopathologically by light microscopy. The PNP deficient thymuses show an irregular cortico-medullary junction, overall narrowing of the cortex and relative expansion of the medullary compartment which appears hypocellular. These changes were more prominent at 8 months of age than at earlier times. Histopathology of the thymus in human PNP deficiency has revealed a parallel but more severe reduction in size. Fatty tissue is present and the stromal component is comprised of epithelial cells (9).

T cell development in the thymus proceeds from CD4⁻CD8⁻ precursors through double-positive CD4⁺CD8⁺ intermediates into single-positive CD4⁺CD8⁻ and CD4⁻CD8⁺ mature T cells (10). The majority of thymocytes are normally the immature or pre-T cells which coexpress both CD4 and CD8 and reside in the cortex. The PNP deficient B6-NPF and B6-NPG mice, exhibit an accumulation of the precursor double negative thymocytes and reduction of double positive thymocytes. These findings may be due to increased attrition of double positive thymocytes during T cell receptor rearrangement and transition to mature T cells. Alternatively there may be restriction of the progression of double negative to double positive thymocytes.

The consequence of these changes in the periphery are a reduction in Thy-1⁺ spleen leucocytes to approximately half normal for both B6-NPF and B6-NPG mutations (Table 2). In addition there is a selective decrease in CD4⁻CD8⁺ to less than half with an apparently normal or enhanced proportion of CD4⁺CD8⁻ cells. Spleen leucocyte response to concanavalin A and IL-2 were also reduced for all three mutations (Table 3) in proportion to the severity of the enzyme deficiency.

Table 2. Thymocyte and spleen leucocyte profile for the B6-NPG mutant

	5–6 months		8–10 months	
	C57BL/6J	B6-NPG	C57BL/6J	B6-NPG
A. Thymocyte				
Thy-1$^+$	99.9 ± 0.2	93.3 ± 6.3	99.8 ± 0.3	93.7 ± 4.2
CD4$^-$CD8$^-$	3.2 ± 0.9	11.7 ± 6.2	2.9 ± 0.6	24.0 ± 9.8
CD4$^+$CD8$^+$	86.7 ± 0.7	63.8 ± 11.6	88.0 ± 2.8	42.6 ± 18.4
CD4$^+$CD8$^-$	6.3 ± 0.5	15.2 ± 5.0	6.3 ± 1.9	23.6 ± 9.2
CD8$^-$CD8$^+$	3.8 ± 0.2	9.3 ± 1.7	2.8 ± 1.0	9.8 ± 1.4
B. Spleen lymphocyte				
Thy-1$^+$	43.5 ± 1.6	20.6 ± 5.8	4.97 ± 11.5	25.0 ± 13.9
CD4$^+$	28.0 ± 3.0	22.9 ± 0.2	30.2 ± 6.1	37.0 ± 7.0
CD8$^+$	14.3 ± 2.1	8.5 ± 0.4	14.3 ± 3.8	5.8 ± 2.1

Table 3. Response of spleen leucocytes from PNP deficient mice to concanavalin A and interleukin-2

	Response as a percentage of C57BL/6J		
	N6-NPE	B6-NPF	B6-NPG
Con A	62.3	17.3	18.3
Con A + IL2, 6.5 U/ml	63.9	22.4	17.4
IL2, 60 U/ml	59.3	55.1	26.9

We have not found evidence for a decrease of GTP in erythrocytes, spleen leucocytes or thymocytes of PNP deficient mice (6). We have found 2.5–5 fold increases in thymocyte dGTP for the severe mutants (7). The change in dGTP is not marked but reflects a secondary deficiency of deoxyguanosine kinase which occurs in the PNP deficient mouse (6). The decrease in deoxyguanosine kinase is suggestive of a compensatory mechanism which may protect the cell from the potentially deleterious effects of markedly increased dGTP pools. Further studies are required to assess the metabolic perturbations and their consequence on thymocyte differentiation.

Our analysis of the three point mutations at the PNP locus in the mouse have provided evidence for an age dependent progressive attrition of total thymocyte numbers, characterized by an accumulation of double negative CD4$^-$CD8$^-$ T cell precursors. A corresponding reduction of spleen T cells numbers and response was found in proportion to the severity of the enzyme deficiency. The mutations in the mouse exhibit a T-cell deficient phenotype with significant parallels to human PNP deficiency thereby providing a valuable animal model for the human disorder.

ACKNOWLEDGMENTS

This work was supported by grant MT-6376 from the Medical Research Council of Canada.

REFERENCES

1. Markert, M.L. (1991) Immunodeficiency Reviews 3, 45–81.

2. Hershfield, M.S. and Mitchell, B.S. (1995) in The Metabolic and Molecular Bases of Inherited Disease, 7th Ed., Eds, Scriver, C.R., Beaudet, A.L., Sly, W.S., Valle, D., McGraw-Hill, pp.1725–1768.
3. Simmonds, H.A., Watson, A.R., Webster, D.R., Sahota, A. and Perret, D. (1982) Biochem. Pharmacol. 31, 941–946.
4. Mably, E.R., Fung. E. and Snyder, F.F. (1989) Genome 32, 1026–1032.
5. Jenuth, J.P., Mangat, R.K. and Snyder, F.F. (1993) Mamm. Genome 4, 598–603.
6. Snyder, F.F., Jenuth, J.P., Dilay, J.E., Fung, E., Lightfoot, T. and Mably, E.R. (1994) Biochim. Biophys. Acta 1227, 33–40.
7. Snyder, F.F., Jenuth, J.P., Mably, E.R. and Mangat, R.K. (1997) Proc. Natl. Acad. Sci. USA 94, 2522–2527.
8. Ealick, S.E., Rule, S.A., Carter, D.C., Greenhough, T.J., Babu, Y.S., Cook, W.J., Habash, J., Helliwell, J.R., Stoeckler, J.D., Parks, R.E., Jr., Chen, S.-F. and Bugg, C.E. (1990) J. Biol. Chem. 265, 1812–1820.
9. Ammann, A.J., Wara, D.W. and Allen, T. (1978) Clin. Immunol. Immunopath. 10, 262–269.
10. Kisielow, P. and von Boehmer, H. (1995) Adv. Immunol. 58, 87–209.

101

ACTIVATION OF DEOXYCYTIDINE KINASE DURING INHIBITION OF DNA SYNTHESIS IN HUMAN LYMPHOCYTES

Maria Sasvári-Székely,[1] Zsolt Csapó,[1] Tatjana Spasokoukotskaja,[1] Staffan Eriksson,[2] and Maria Staub[1]

[1]Semmelweis University of Medicine
Department of Medical Chemistry, Molecular Biology, and
    Pathobiochemistry
Budapest, Hungary
[2]Swedish University of Agricultural Sciences
Department of Veterinary Medical Chemistry
Biomedical Centre, Uppsala, Sweden

1. SUMMARY

Deoxycytidine kinase was shown to be activated during 2-chlorodeoxyadenosine (CdA) treatment of human lymphocytes, under the conditions when the DNA synthesis is inhibited. As the increase of dCK activity was shown in crude protein extracts, without an increase in the amount of dCK protein, shown by immunostaining after SDS-PAGE, a secondary modification of the protein structure was considered. NaF treatment of cells in the concentration range of 5–20 mM gave a similar activation of dCK, suggesting a possible role of phosphatases and/or a possibility of a G-protein related phenomenon. Using the same conditions, no effect of CdA or NaF was found on the thymidine kinase activity of cell extracts. Alternatively, activation of catabolic pathways could be considered, however, the increase in dCK activity was not influenced either by the removal of 5'-nucleotidases, or by the inhibition of deaminases.

2. INTRODUCTION

Cells acquire deoxyribonucleotides, the building blocks of DNA, via *de novo synthesis* or via *salvage*. These two separate pathways may lead to compartmentation of the produced dNTP pool, as it was shown for dCTP (1–3). De novo synthesis needs ribonucleotide reductase, however, in terminally differentiated cells resting in G_0, transcription

Purine and Pyrimidine Metabolism in Man IX,
edited by Griesmacher *et al.* Plenum Press, New York, 1998.

519

of ribonucleotide reductase is undetectable (4), resulting a lack of de novo synthesis in resting lymphocytes. Deoxycytidine kinase (dCK, EC 2.7.1.74.) is the main human deoxynucleoside *salvage* enzyme, capable to supply resting lymphocytes with all the four deoxynucleotides, needed for DNA synthesis: three out of the four are directly the substrates of dCK (deoxycytidine (dCyd), deoxyadenosine (dAdo) and deoxyguanosine (dGuo)), while deoxythymidine nucleotides can be formed from dAMP via deamination and methylation. Beside supplying lymphoid cells with precursors of DNA synthesis, dCK activates many chemotherapeutically important deoxynucleoside analogues, giving a special importance of this enzyme in pharmacologycal investigations (for a recent review see 5). Lymphoid tissues contain a high level of dCK, which explains why this type of tissue has been used to purify the protein (6,7). Beside its high expression in lymphoid tissues (6–10), dCK can be detected in some solid tumors (8). 2-Chlorodeoxyadenosine (CdA, Cladribine) is an anti-leukemic drug with a remarkable therapeutic effect, most pronounced in chronic lymphocytic leukemias and hairy cell leukemia. dCK is the main enzyme in lymphocytes, which activates CdA, however, the mitochondrial deoxyguanosine kinase is also efficiently phosphorylates CdA in other cell types (9). The level of dCK enzyme activity in crude extracts of peripheral blood mononuclear cells was shown to correlate with the response of leukemic patients to CdA treatment (9,10). On the other hand, CdA pretreatment of patients had an adjuvant effect for araC (citarabide) treatment, based on the in vivo activation of dCK (11).

3. METHODS

Human tonsillar lymphocytes were treated with 3 µM of 2-chloro-deoxyadenosine (CdA) for 2 hours. Then cells were washed and lysate with a throwing-freezing procedure, the lysate was centrifuged for 30 min at 10,000 rpm, and the supernatant was measured for dCK activity by a standard method (12) under optimal conditions at saturating substrate concentration (13). dCMP deaminase was measured in crude extracts by ^{3}H-dCMP as a substrate. Separation of ^{3}H-dCMP from ^{3}H-dUMP was made by thin layer chromatography, and the results were expressed as % of dUMP formed from ^{3}H-dCMP.

4. RESULTS

Tonsillar lymphocytes were treated with 3 µM for 2 hours, and an activation of dCK was found using either ^{3}H-deoxycytidine (dCyd), or ^{3}H-CdA, as substrate (Fig. 1).

Activation of dCK was more pronounced if measured by dCyd, than with CdA as substrate. This raised the possibility that the increase in the ^{3}H-dCyd phosphorylating capacity is rather due to the activation of thymidine kinase 2 (TK2) than to the increase of dCK activity, as dCyd is a substrate for both dCK and TK2, but CdA is not phosphorylated by TK2. To check this possibility, cells were incubated with 2-bromo-deoxyadenosine, and the ^{3}H-dCyd phosphorylating capacity was determined in the presence or absence of 1 mM thymidine (Fig. 2).

TK2 can be measured as a difference between the ^{3}H-dCyd phosphorylation capacity in the presence or absence of excess of non-labeled thymidine. As it can be seen on Fig. 2, no significant change was found in TK2 activity during 2-Br-Adenosine treatment. Similar results were found in case of CdA treatment, too (data not shown). Another possibility to explain the increase in the ^{3}H-dCyd phosphorylating capacity without a real increase in dCK activity could be the decreased activity of the catabolic enzymes, like dCMP deaminase.

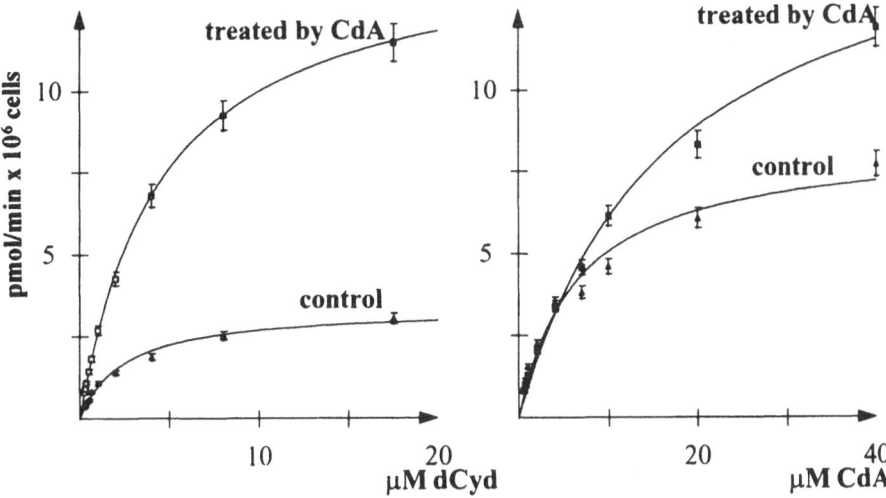

Figure 1. Substrate saturation curves of dCK in crude extracts of cells: A comparison of CdA treated and control lymphocytes. Human tonsillar lymphocytes were treated with 3 μM CdA for 2 hours, and dCK activity was determined in the crude protein extracts of the cells, using different concentrations of ³H-dCyd or ³H-CdA.

As it can be seen on Fig. 3, activity of dCMP deaminase, as measured in the crude extracts of CdA treated and untreated lymphocytes, was not changed during 2 hours of treatment. These data shows that the increased ³H-dCyd phosphorylating capacity of the extracts cannot be explained by the decreased activity of catabolic enzymes.

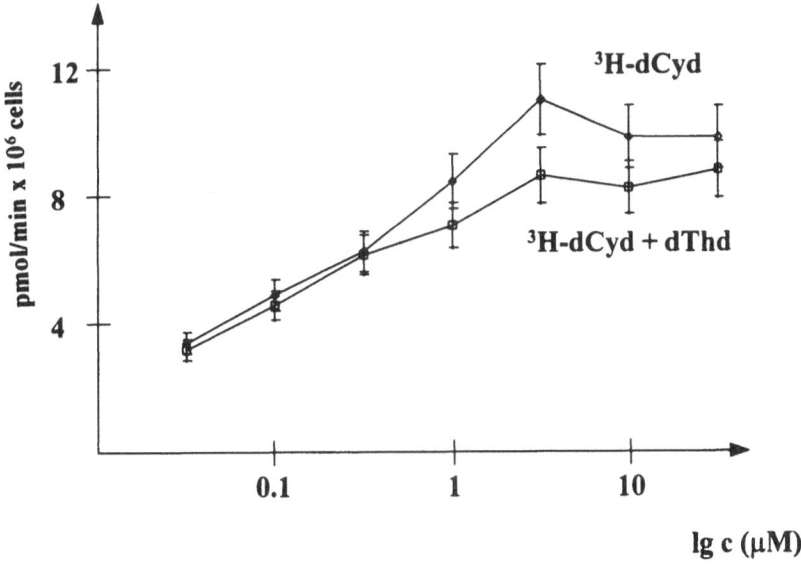

Figure 2. Increase of ³H-dCyd phosphorylating capacity in extracts of cells treated by 2-bromo-deoxyadenosine in the presence or absence of non-labeled thymidine. Human tonsillar lymphocytes were treated for 2 hours by different concentration of 2-bromo-deoxyadenosine, as indicated. Phosphorylation of ³H-dCyd was determined in the presence or absence of 1 mM thymidine using the crude extracts of treated and control cells.

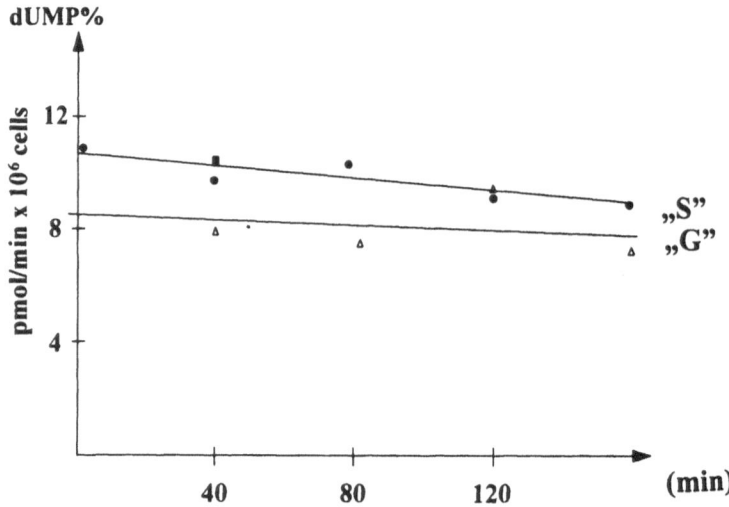

Figure 3. dCMP deaminase in crude extracts of lymphocytes treated with CdA. "S phase" and "G phase" enriched lymphocyte subpopulations were prepared from human tonsillar lymphocytes by centrifugation through an albumin gradient. Then cells were treated with 3 μM of CdA for different time periods, as indicated, and crude extracts of the cells were measured for dCMP deaminase activity using ^{3}H-dCMP as substrate. After incubation, percentage of radioactivity in dUMP fraction was determined.

5. DISCUSSION

Data, presented here, prove that dCK activity can be increased by CdA treatment of human tonsillar lymphocytes. ^{3}H-dCyd phosphorylating capacity of the crude extracts showed a substrate dependent increase either measured by ^{3}H-dCyd or by ^{3}H-CdA, as substrates. On the other hand, there was no change either in TK2 or in dCMP deaminase activity in crude extracts of treated lymphocytes. These data shows that activation of dCK is responsible for the increase of ^{3}H-dCyd phosphorylating capacity. Further investigations are in progress, what kind of DNA synthesis inhibitors can activate dCK, beside CdA, and what is the molecular mechanism of dCK activation.

ACKNOWLEDGMENTS

These work was made in the frame of BIOMED 2 (BMH4-CT96-0479)-OMFB E0019 and also supported by National Research Grant, Hungary (OTKA T 022608).

REFERENCES

1. Xu, Y.Z., Huang, P., Plunkett, W. (1995) J. Biol. Chem. 270, 631–637
2. Arnér, E.S.J, Flygar, M., Bohman,C., Wallström,B. and Eriksson, S. (1988) Exp. Cell Res. 178, 335–342
3. Spasokukotskaja, T., Sasvári-Székely, M., Taljanidisz, J. and Staub, M. (1992) FEBS Letters 297, 151–154
4. Björklund, S., Skog, S., Tribukait, B. and Thelander, L. (1990) Biochemistry 29 (23), 5452–5458
5. Arner, E.S., Eriksson, S. (1995) Pharmacol. Ther. 67, 155 - 186
6. Bohman, C. and Eriksson, S. (1988) Biochemistry, 27 (12), 4258–65
7. Szyfter, K., M. Sasvári-Székely, T. Spasokukotskaja, F. Antoni and M. Staub

8. Owens, J.K., Shewash, D.S., Ullman, B. and Mitchell, B.S. (1992) Cancer Res. 52 (9), 2389–93
9. Hengstschlager, M., Denk, C. and Wawra, E. (1993) FEBS Lett. 321, 237–240
10. Sasvári-Székely, M., Zs. Piróth, Z. Kazimierczuk and M. Staub (1994) Biochem. Biophys. Res. Commun. 203, 1378–1384
11. Xu, Y..Z., Plunkett, W. (1992) Biochemical Pharmacol. 44, 1819–27
12. Ives, D.H., Wang, S.M. (1978) Methods in Enzym. pp. 337–345 eds. P.A. Hoffee and M.E.Jones
13. Spasokukotskaja, T., Arnér, E.S.J., Brosjö,O., Gunvén,P., Juliusson,G., Liliemark, J. and Eriksson, S. (1995) Eur.J.Cancer 31A, 202–208

102

SIMILAR CHANGES WERE INDUCED BY CLADRIBINE AND BY GEMCITABINE, IN THE DEOXYPYRIMIDINE SALVAGE, DURING SHORT TERM TREATMENTS

Zs. Csapó,[1] G. Keszler,[1] M. Sasvári-Székely,[1] K. Smid,[2] P. Noordhuis,[2] G. J. Peters,[2] and M. Staub[1]

[1]Semmelweis Medical University
Department Med. Chem., Molecular Biology, and Pathobiochemistry
Budapest, Hungary
[2]Free University Hospital
Department Med. Oncology
Amsterdam, The Netherlands

1. SUMMARY

Short term treatments (1–2 hrs) of human tonsillar lymphocytes by Cladribine (2-Chloro-deoxyadenosine, CdA) have suggested a new target for CdA, the inhibition of dCMP deaminase (Sasvári et al. 1994; BBRC 203, 1378). Further investigations have shown, that the dCMP-deaminase activity could be inhibited by 2-Cl-dAMP in cell free extracts of lymphocytes. The pool size of dUMP (measured by an antibody against dUMP) was also decreased in WiDr colon cancer cells by CdA. The new antimetabolite against solid tumours, Gemcitabine (2',2'-difluoro-deoxycytidine, dFdC), had similar effects on the salvage of thymidine (dThd) and deoxycytidine (dCyd) as CdA. The Ki values for 3H-dThd and 3H-dCyd incorporation into DNA were 0.16 uM and 1.0 uM dFdC, respectively. The labeling of the TTP pool increased 6–7 times, while of dCTP pool only 1.5–1.7 times, suggesting a decrease of the size of corresponding pools. Similarly to CdA, the labeling as well as the concentration of dUMP was also decreased by dFdC. Both analogues are able to increase the deoxycytidine kinase activity, necessary for their phosphorylation and therapeutic action in cells. The target(s) for the two different drugs seems to be common.

Purine and Pyrimidine Metabolism in Man IX,
edited by Griesmacher *et al.* Plenum Press, New York, 1998.

2. INTRODUCTION

Gemcitabine (2',2'-difluoro-2'-deoxycytidine) and Cladribine (2-chloro-deoxyadenosine) are new antileukemic and anticancer drugs. These compounds have a common feature that they are deoxynucleoside analogues, but gemcitabine is a pyrimidine and Cladribine is a purine analogue. Similarly to the other deoxynucleoside analogues, both gemcitabine and cladribine are activated intracellularly by deoxycytidine kinase, forming deoxymononucleotide analogues, which in turn could influence the metabolism of thymidylate through dCMP deaminase (for a review see 1). Deoxycytidine kinase is expressed in ymphocytes (1,2), however, its activity was shown to be high enough in solid tumors to activation of gemcitabine (3). On the other hand, resistance to these drugs is often caused by dCK deficiency. Molecular mechanism of these drugs include perturbance of the nucleotide pool (4), inhibition of ribonucleotide reductase (5), and direct effects on DNA synthesis. Recently, the inhibition of dCMP deaminase was shown in our laboratory by cladribine (6).

3. METHODS

Cell labeling and the determination of radioactivity incorporated into the nucleotide, liponucleotide, and into the DNA fractions of the cells was performed as described earlier (6). Briefly: Cells were pulse labeled either with [5–^{3}H]-deoxycytidine ([^{3}H]dCyd, 1500 Gbq/mmol), or with [5–^{3}H]-thymidine ([^{3}H]dThd, 1480 GBq/mmol) in a concentration of 0.1 μM for 20 min. at 37°C. After labeling, cells were washed twice and precipitated with ice-cold 60% methanol. Labeling of the deoxynucleotide pools ("pool"), while DNA synthesis was characterized by incorporation of the label into the acid-insoluble fractions ("DNA"). Deoxycytidine kinase (dCK) and thymidine kinase (dTK) enzyme activity were measured under optimal conditions as described (7) using 5-[^{3}H]-deoxycytidine or 5-[^{3}H]deoxythymidine, respectively, in 10 μM final concentration and approx. 20–50 μg protein in a 50 μl final volume. After 10 minutes of incubation at 37°C, radioactivity of phosphorylated products bound to DEAE cellulose paper was measured. Determination of intracellular dUMP level was measured using antibody against dUMP (7). Measurement of growth inhibition: WiDr colon carcinoma cells were plated in 96-well flat-bottom plates. After 24 hours, when the cells reach the logarithmic growth phase, drugs were added for another 72 hours. Cell number was enumerated using the sulforodamide B (SRB) assay.

4. RESULTS

In order to investigate the effect of gemcitabine on the intracellular deoxynucleotide metabolism, human lymphocytes were treated with different concentrations of the drug, and the radioactivity of the labeled deoxynucleoside precursors were measured in the "pools" and in the "DNA" (Fig. 1).

Gemcitabine caused a 6–7 times increase on the labeling of dTTP pool, while the dCTP pool was increased only 1.5–1.7 times (Fig. 1 "pool"). Apparent K_i values regarding incorporation of ^{3}H-dThd and ^{3}H-dCyd into DNA were also determined. ^{3}H-dThd incorporation was more efficiently inhibited (Ki = 0.16 uM) than incorporation of ^{3}H-dCyd (K_i = 1.0 uM) by gemcitabine (dFdC). In vitro activity of deoxycytidine kinase (dCK) was also measured in extracts of lymphocytes treated by different drugs.

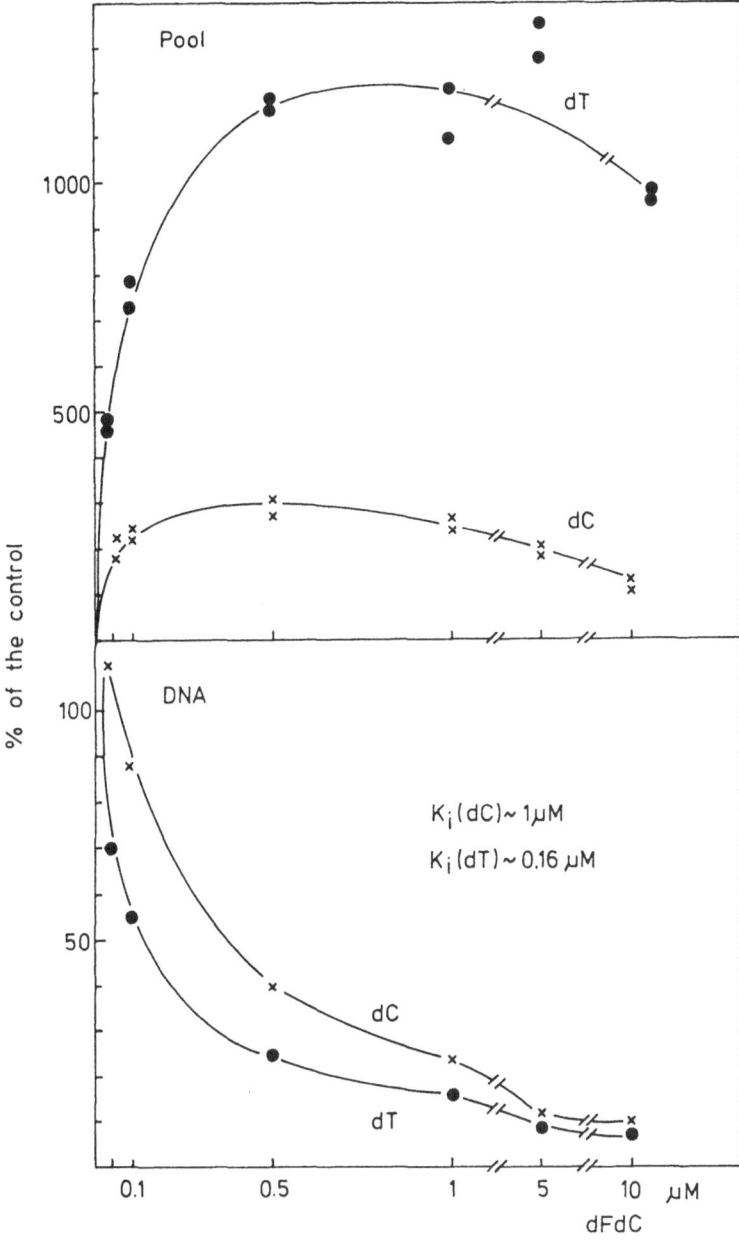

Figure 1. Effect of gemcitabine on ³H-deoxycytidine and ³H-thymidine metabolism in human tonsillar lymphocytes. After labeling the cells for an hour with ³H-dThd and with ³H-dCyd in the presence of different concentrations of gemcitabine (2′,2′-difluoro-2′-deoxycytidine, dFdC), radioactivity incorporated into the methanol soluble pools ("pool"), as well as into the DNA fraction of cells ("DNA") were measured. Apparent K_i values are also shown.

As can be seen on Fig. 2, 2 hours of treatment with gemcitabine increase the activity of dCK in human lymphocytes, while the level of thymidine kinase (dTK) did not change significantly.

Similarly to CdA (7), both the labeling and the concentration of dUMP were decreased by gemcitabine, as measured by thin layer chromatography of labeled products

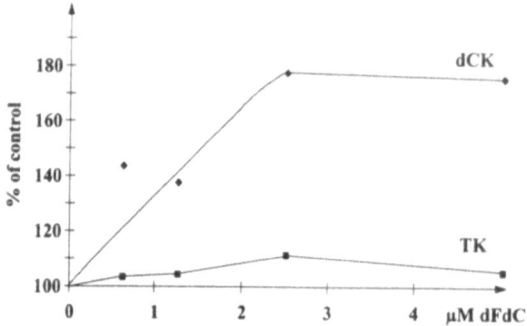

Figure 2. Effect of gemcitabine on the in vitro activity of deoxycytidine kinase (dCK) and deoxythymidine kinase (dTK). Human tonsillar lymphocytes were treated with gemcitabine (dFdC) for 2 hours, and the activities of dCK and dTK were measured in crude protein extracts of the cells. Enzyme activities, measured under optimal conditions at saturating substrate concentration of deoxycytidine, are expressed as the percentage of the untreated cells.

in human lymphocytes (data not shown), and by a specific antibody (8) in WiDr cells (Fig. 3), showing the inhibition of dCMP deaminase by cladribine, and a less extent by gemcitabine. However, intracellular concentration of 5-F-dUMP, as measured after 5-FU treatment, was also decreased on the same way by cladribine and by gemcitabine (Fig. 3).

5. DISCUSSION

Numerous deoxynucleoside analogues are used in anticancer and antiviral therapy. Among them, a comparative study of cladribine and gemcitabine is especially interesting, because both the drugs are activated by deoxycytidine kinase, but cladribine is a purine, while gemcitabine is a pyrimidine analogue. Similar disturbance of the pyrimidine metabolism was found, either with cladribine (7) or with gemcitabine (Fig. 1) Recent studies in our laboratories have shown that treatment of human tonsillar lymphocytes by cladribine increased the activity of deoxycytidine kinase (dCK), as measured in crude protein extracts of the cells (9). A similar, but less pronounced increase of dCK activity was found by gemcitabine, too, under the same conditions (Fig. 2.). However, the effect of deoxypurine analogues on the activation of dCK seemed to be more pronounced, than that of the deoxypyrimidines (see also T. Spasokoukotskaja et al. in the same issue). The decrease of dUMP caused by cladribine (Fig. 3.), which reduces the competition of dUMP with FdUMP to inhibit thymidylate synthase, which supports our previous findings regarding a possible inhibition of dCMP deaminase by 2-Cl-dAMP (6). On the other hand, the common effect of cladribine and gemcitabine on inhibition of FdUMP accumulation might be related to the inhibitory effect of the appropriate deoxynucleotide analogues on ribonucleotide reductase, resulting a decreased conversion of F-UDP to F-dUDP and subsequently to F-dUMP.

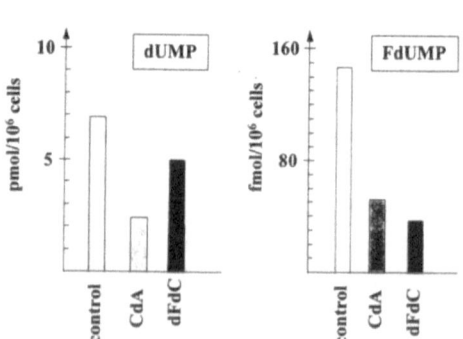

Figure 3. Effect of CdA and gemcitabine on dUMP and FdUMP pools of WiDr cells. WiDr colon cancer cells in logarithmic growth phase were exposed to dFdC or CdA for 2 hours in the presence of 5-FU or its absence for 2 hours. Than cells were extracted and dUMP was measured using specific antibody, and FdUMP was measured with a thymidilate synthase ligand assay.

ACKNOWLEDGMENTS

These work was made in the frame of BIOMED 2 (BMH4-CT96-0479)-OMFB E0019 and also supported by National Research Grant, Hungary (OTKA T 022608).

REFERENCES

1. Arner , ES., Eriksson, S. (1995) Pharmacol. Ther. 67, 155–86.
2. Carson, D.A., Wasson, D.B., Kaye, J., Ullman, B., Martin, D.J., Robins, R.K. and Montgomery, J.A. (1980) Proc. Natl. Acad. Sci. USA 77, 6865–6869.
3. Ruiz van Haperen, V.W.T., Veerman, G., Eriksson, S., Boven, E., Stegmann, A.P.A., Hermsen, M., Vermorken, J.B., Pinedo, H.M., Peters, G.J. (1994) Cancer Res. 54, 4138–4143.
4. Xu, Y..Z., Plunkett, W. (1992) Biochemical Pharmacol. 44, 1819–27
5. Heinemann, V., Xu, Y.Z., Chubb, S., Sen, A., Hertel, L.W., Grindey, G.B. and Plunkett, W. (1990) Molec. Pharmacol. 38, 567–572.
6. Sasvári-Székely, M., Zs. Piróth, V. Kazimierczuk and M. Staub Biochem. Biophys. Res. Commun. (1994), 203,1378–1384.
7. Szyfter K., Sasvári-Székely M., Spasokukotskaja T., Antoni F. and Staub M. (1985) Acta Biochim. Biophys. Acad. Sci. Hung. 20, 173–182. ·
8. Peters, G.J., E. Laurensse, H.W.M. Steinbusch, J.deVente, K. Smid, C.L. Van der wilt, H.P. Pinedo (1993) Anicancer res., 13, 835–840.
9. Staub. M., Spasokoukotskaja, T., Sasvári-Székely, M. and Keszler, G. (1966) Drug Development Res. 37/3 p.178.

103

EFFECT OF CLADRIBINE, FLUDARABINE, AND 5-AZA-DEOXYCYTIDINE ON S-ADENOSYLMETHIONINE (SAM) AND NUCLEOTIDES POOLS IN STIMULATED HUMAN LYMPHOCYTES

K. Fabianowska-Majewska,[1] K. Ruckemann,[2] J. A. Duley,[2] and H. A. Simmonds[2]

[1]Department of General Chemistry
Medical University of Lodz
Lindleya 6, 90-131 Lodz, Poland
[2]Purine Research Laboratory UMDS
Guy's Hospital
London SE1 9RT, United Kingdom

1. INTRODUCTION

Cladribine (2-chloro-2'-deoxyadenosine, 2CdA) and Fludarabine (9-β-D-arabino-syl-2-fluoro-adenine) are 2-halogenated adenosine analogues which have found clinical application for the treatment of lymphocytic and lymphoid leukemias.[1] Cladribine and Fludarabine result in apoptosis of cells[2] and it has been established that the mechanisms of action of both drugs lead *via* their phosphorylated derivates to inhibition of DNA synthesis. Recently, we have reported the block of deoxyadenosine (dAdo) metabolism by Cladribine in lysates of CNS lymphoma cells[3] and in lysates of human normal lymphocytes.[4] A complete inhibition of phosphorylation of dAdo, and 60% of inhibition of adenosine deaminase (ADA) activity were observed in *in vitro* studies.

Moreover, in erythrocyte lysates of patients, we observed a large decrease in the activity of adenosine deaminase and SAH-hydrolase, after one week treatment with Cladribine.[5] Currently, in studies with cultured murine leukemic cells (L-1210), we noted that Cladribine, in contrast to Fludarabine, causes depletion of SAH-hydrolase activity and a large decrease of DNA transmethylation reactions.[6]

The findings suggested that the cytotoxicity of Cladribine results from, apart from others, the inhibitory effect of the drugs on enzymes involved in the metabolism of dAdo.

Purine and Pyrimidine Metabolism in Man IX,
edited by Griesmacher *et al.* Plenum Press, New York, 1998.

It was proposed that the mechanism of Cladribine toxicity, by blocking dAdo metabolism, leads to inactivation of SAH-hydrolase as a consequence of dAdo accumulation.

To investigate this hypothesis we decided to compare the effect of Cladribine with that of Fludarabine and 5-aza-deoxycytidine (5-aza-dCyt), a potent inhibitor of DNA methyltransfe-rase, on S-adenosylmethionine (SAM) and nucleotides pools in stimulated human lymphocytes.

2. MATERIALS AND METHODS

2.1. Chemicals

Histopaque, PHA, Fludarabine and 5-aza-dCyt were purchased from Sigma. RPMI 1610 medium, fetal calf serum, gentamicin and 24 well plates were obtained from Life Technologies, Inc. Cladribine was a kindly donated by Ortho-Biotech (USA).

2.2. Separation of Lymphocytes and Their Stimulation

Heparized blood was obtained immediately before experiments from healthy laboratory personnel. The lymphocytes were isolated by Histopaque-1077 density gradient centrifugation, and cells were washed twice with Hanks' solution. The cells were then counted and resuspen-ded at 1×10^6 in RPMI 1640 medium containing physiological concentrations of phosphate, supplemented with antibiotic (gentamicin), plus 10% fetal calf serum and L-glutamine. The cells were incubated in microwells for 0, 1, 2 or 3 days (designated D1-D4 on the figures), in the presence of PHA and in the absence or presence of inhibitors (5 µM), i.e. Cladribine, Fluarabine and 5-aza-dCyt.

2.3. Assay of Ribonucleotides, Deoxyadenosine (dAdo), S-Adenosylhomocysteine (SAH), and S-Adenosylmethionine (SAM)

The cells, obtained after brief centrifugation, were twice washed in PBS solution, and disrupted with 200 µl of 10% trichloroacetic acid. The precipitate was removed by centrifugation, the supernatant was extracted with water saturated diethyl ether to pH of 5 and analyzed by HPLC. Ribonucleotides were quantified by anion-exchange HPLC (Bofill

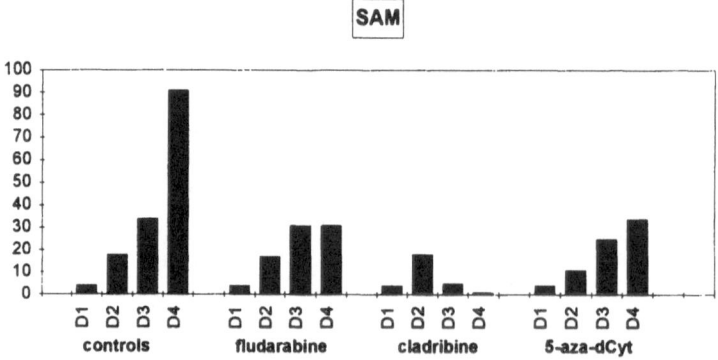

Figure 1. Effects of Fludarabine, Cladribine and 5-aza-dCyt on the SAM pool in stimulated human lymphocytes. Concentration of SAM is shown as pmol/10^6 cells.

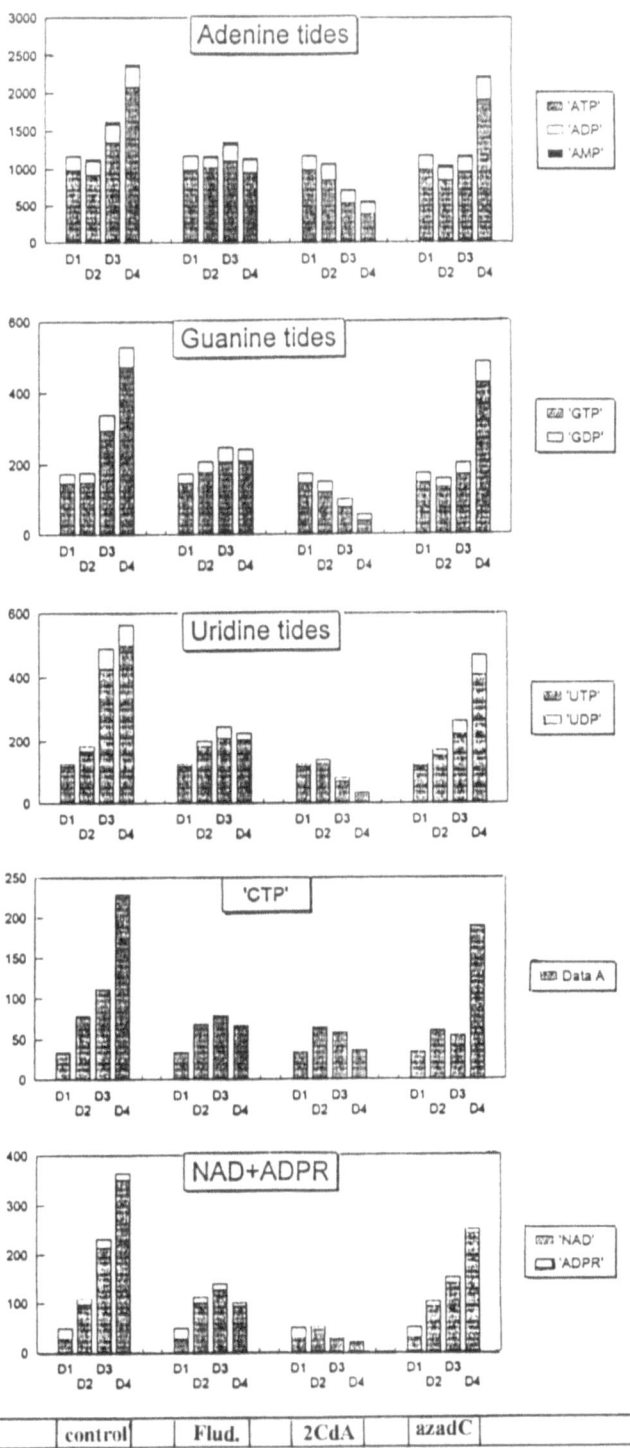

Figure 2. Effects of Fludarabine, Cladribine and 5-aza-dCyt on nucleotides pools in stimulated human lympho-cytes. Inhibitors are designated: Fludarabine: Flud., Cladribine: 2CdA, and 5-aza-dCyt: azadC. Concentration of nucleotides is shown as pmol/10^6 cells.

et al.), using Phenomenex Hypersil 5 µm NH_2–2 column and linear phosphate buffer gradient elution system: buffer A: 5 mM KH_2PO_4 (pH 2.65) and buffer B: 0.5 M KH_2PO_4, 1.0 M KCl (pH 3.5) and a flow rate of 0.5 ml/min. Adenosine compounds were quantified by reverse phase HPLC,[7,8] using Ultrasphere ODS 5 µm column, an isocratic elution with phosphate buffer, pH 3.0 (50 mM NaH_2PO_4 and 10 mM heptanesulfonic acid) and a flow rate of 0.9 ml/min. Retention times of dAdo, SAH and SAM were estimated to be: 11.2, 16.2 and 12.4 min., respectively.

3. RESULTS AND DISCUSSION

dAdo was not detectable in unstimulated lymphocytes and also no accumulation of dAdo was observed in the presence of any inhibitors during growth of stimulated lymphocytes.

Unstimulated lymphocytes contained negligible levels of SAM and SAH (detection limit 10 pmol/10^6 cells), but progressive accumulation of SAM was found in stimulated lymphocytes during the 72 hour growth of cells. The inhibitory effect of Fludarabine, Cladribine and 5-aza-dCyt on the SAM pool at 24 hr intervals in stimulated lymphocytes is documented in Figure 1. Among the inhibitors used, Cladribine caused the most dramatic inhibition of SAM production; at 3 days in the presence of Cladribine, the concentration of SAM decreased to 10% of control, while Fludarabine and 5-aza-dCyt had negligible effects. No accumulation of SAH was noted in either control or drug treated cells.

The accumulation of SAM in stimulated lymphocytes was paralled with increments in the pools of ATP, GTP, UTP, CTP and NAD (Figure 2).

Cladribine caused progressive depletion of all nucleotides. The effect of 5-aza-dCyt, a potent inhibitor of DNA methyltransferase activity, on nucleotides pools was insignificant. Fludarabine did not cause change of nucleotides pools as great as 2CdA.

The results do not accord with the inhibition of methylation by these drugs at the level of SAH-hydrolase and raise the question of what is the exact mechanism of inhibition of SAM production as well as nucleotide depletion, by Cladribine. A possible mechanism is the inhibition of SAM production *de novo,* caused by adenine nucleotide depletion rather then the inhibition of methylation by SAH.

ACKNOWLEDGMENTS

One of us (K.F-M) gratefully acknowledges the support of Fellowship Award from the European Science Foundation (grant No SVF/96/46T).

We would like to thank the Special Trustees of Guy's Hospital for support.

REFERENCES

1. Saven A., Piro L.P., 2-Chlorodeoxyadenosine: a newer purine analog active in the treatment of indolent lymphoid malignancies. Ann.Intern.Med., 1994; 120: 784–791.
2. Robertson L.E, Chubb S., Meyn R.E., Story M., Ford R., Hittelman W.N., Plunkett W. Induction of apoptotic cell death in chronic lymphocytic leukemia by 2-chloro-2'-deoxyadenosine and 9-β-D-arabinosyl-2-fluoroadenine. Blood 1993; 81: 143–150.
3. Fabianowska-Majewska K., Tybor K., Duley J., Simmonds A. The influence of 2-chloro-2'-deoxyadenosine on metabolism of deoxyadenosine in human primary CNS lymphoma. Biochem.Pharmacol., 1995; 9: 1379–1383.

4. Fabianowska-Majewska K., Wyczechowska D., Comparison of the inhibitory effects of 2-chloro-2'-deoxyadenosine and 9-β-D-arabinosyl-2-fluoro-adenine on metabolism of deoxy-adenosine in human lymphocytes and erythrocytes. Nucleosides, Nucleotides, 1997 (in press).
5. Warzocha K., Fabianowska-Majewska K., Blonski J., Krykowski E., Robak T. 2-Chlorodeo-xyadenosine inhibits activity of adenosine deaminase and S-adenosylhomocysteine hydrolase in patients with chronic lymphocytic leukemia. Eur.J.Cancer 1997; 33: 170–173.
6. Wyczechowska D., Fabianowska-Majewska K., Does 2-chloro-deoxyadenosine contribute to alteration of DNA methyltransferase activity? Adv.Exp.Med.Biol. (in press).
7. Bofill M., Fairbanks L.D., Ruckemann K., Lipman M., Simmonds H.A., T- lymphocytes from AIDS patients are unable to synthesize ribonucleotides *de novo* in response to mitogenic stimulation. J. Biol. Chem., 1995; 270: 29690–29679.
8. Wise C., Fullerton F., Analytical procedure for determination of S-adenosylmethionine, S-adenosylhomocysteine, and S-adenosyllethionine in some isocratic HPLC run, with a procedure for preparation and analysis of the analog S-adenosylhomocysteine sulfoxide. J.Liquid Chromat., 1995; 18: 2005–2017.
9. She Q-B., Nagao I., Hayakawa T., Tsuge H., A simple HPLC method for determination of S-adenosylmethionine and S-adenosylhomocysteine in rat tissues: the effect of vitamin B_6 deficiency on these concentrations in rat liver. Biochem.Biophys.Res.Commun., 1994; 205: 1748–1754.

104

INHIBITION OF INOSINE MONOPHOSPHATE DEHYDROGENASE ACTIVITY BY THE PLASMA OF HEART TRANSPLANT RECIPIENTS RECEIVING MYCOPHENOLATE MOFETIL

Andrea Griesmacher,[1] Günter Weigel,[2] Gernot Seebacher,[2] R. Mallinger,[2]
Günther Laufer,[2] and Mathias M. Müller[1]

[1]Ludwig Boltzmann Institute for Cardiosurgical Research at the Institute of
    Laboratory Diagnostics of the Kaiser-Franz-Josef-Hospital
Kundratstraße 3, A-1100 Vienna, Austria
[2]Department of Cardiothoracic Surgery
University Hospital Vienna
Währinger Gürtel 18-20, A-1090 Vienna, Austria

INTRODUCTION

Human IMP dehydrogenase (IMPDH; EC 1.1.1.205) has become a potent targeting enzyme for immunosuppressive drugs since it is the rate-limiting enzyme in de novo purine nucleotide synthesis in leuckocytes. Human IMPDH exists as two isoforms derived from different genes, designated type I and type II (1,2). Type I is constitutively expressed in normal leukocytes, while type II is up-regulated in neoplastic and replicating cells (3, 4). Mycophenolic acid (MPA) inhibits both isoforms of IMPDH.

Mycophenolate Mofetil (MMF; Hoffmann-La Roche), which is nowadays adminis-tered to transplant recipients as adjunctive therapy to cyclosporine A and corticosteroids (5), is in vivo hydrolyzed to MPA. MPA is further glucuronated to a single metabolite MPAG, which is then excreted in urine (6).

Traditionally therapeutic monitoring of immunosuppressive drugs involves the measurement of drug concentrations in blood. An alternative approach is the measurement of the drugs' biological effects to assess the state of immunosuppression. We report here the relationship between MPA and MPAG concentrations and the degree of inhibition of IMPDH deriving from human lymphoblasts (exhibiting type II isoform). For this purpose IMPDH activity was measured with a non-radioactive method in human lymphoblasts, which were incubated on the one hand with plasma of transplanted patients receiving MMF and on the other hand with plasma of healthy volunteers containing various amounts of MPA or MPAG. In addition, the influence of albumin on the MPA and MPAG mediated

Purine and Pyrimidine Metabolism in Man IX,
edited by Griesmacher *et al.* Plenum Press, New York, 1998.

effects were investigated, since the pharmacological activity of MPA is suggested to be a function of unbound drug concentration (7).

PATIENTS AND METHODS

Patients

10 patients, receiving three month after orthotopic heart transplantation (HTX) 2 × 1 g MMF per day in addition to cyclosporine A and corticosteroids, entered this study. Blood was collected two and one weeks before and 7, 14, 28, 48 and 72 days after the beginning of the MMF therapy. Blood samples were drawn in the morning after an overnight fasting period under standardized conditions. Na-ethylenediaminetetraacetic acid was used as anticoagulant. After centrifugation plasma samples were stored at −30°C till the experiments were performed.

Cultivation of Lymphoblasts

MOLT-3 cells (=human lymphoblasts) were cultivated with RPMI 1640 (Sigma), 100 mL/L fetal bovine serum (Sigma), 100 kIU/L penicillin (Sigma), 100 mg/L streptomycin (Sigma) and 2 µmol/L L-glutamine (Sigma). Cell cultivation was carried out in a humidified incubator set at 37°C and 5% CO_2. For the experiments cells were washed twice and diluted with deionized and distilled water to a final concentration of 2,500,000 ± 9,000 cells/10 µl.

Determination of IMPDH Activity

Cells were cracked by freezing and thawing twice. 30 µl cell lysate was preincubated with 30 µl of patients' plasma and water or plasma of healthy volunteers containing varying amounts of MPA or MPAG at 37°C for 30 min before the determination of enzyme activity.

IMPDH activity was measured by means of high performance liquid chromatography (HPLC) using a modified procedure described by Montero et al. (8). The incubation buffer consisted of 80 mmol/L NaH_2PO_4, 200 mmol/L KCl, 500 µMol/L NAD^+, 1000 µMol/L IMP (pH 7.5). The reaction was started by the addition of 60 µL cell lysate and carried out for 2.5 h at 37°C. The incubation was stopped by addition of $HClO_4$.

Inosine 5'-monophosphate (IMP), xanthosine 5'-monophosphate (XMP) and guanosine 5'-monophosphate (GMP) were separated injecting 100 µL of the neutralized supernatant onto a CNU-010 column (Chemcon) using a K_2HPO_4 gradient. Buffer A consisted of 0.015 mol/L K_2HPO_4 (pH 3.45), buffer B of 0.5 mol/L K_2HPO_4 (pH 3.45). A linear gradient rising from 0% B to 50% B in 8 min was used with a total run time of 28 min and an equilibrium delay of 5 min.

Under these test conditions only formation of XMP was detected and was therefore alone considered in the calculation of IMPDH activity (given as µmol/10^6 cells/h). The amount of formed XMP was determined by the ratio of its peak area in relation to XMP standards measured under the same conditions.

Determination of MPA and MPAG

After precipitation with acetonitrile (Merck) the supernatants were evaporated under nitrogen. At dryness samples were dissolved in the mobile phase, which consists of equal amounts of acetonitrile and 50 mmol/l o-phosphoric acid. MPA was determined by means of HPLC using a reversed phase column (Supelcosil LC-18 DB; Supelco Inc.). For the determination of MPAG samples were incubated with β-glucuronidase from helix pomatia (Sigma) in order to deglucuronidate MPAG to MPA as described by (9). MPAG was calculated according to the following formula:

$$MPAG = MPA_{(after\ deglucuronidation)} - MPA_{(before\ deglucuronidation)}$$

Statistical Analysis

Correlations were calculated for all patients under MPA therapy irrespective of the duration of MPA therapy but normalized for patients' albumin and total protein plasma levels, since the non-protein-bound MPA fraction is reported to be mainly immunosuppressive. All experiments were performed in duplicate on at least three separate occasions. Differences between the groups were tested for statistical significance using analysis of variance. All calculations were carried out using the statistical software package SAS/STAT (10). Values are expressed as mean ± standard error of mean (SEM).

RESULTS

The patients exhibited approx. 30-fold higher plasma levels of MPAG than of MPA (Table 1). Our results indicate that inhibition of IMPDH activity occurs also throughout the dosing intervals, since the trough plasma levels led to significant inhibition of enzyme activity by approximately 54.1% compared to controls (Table 1). One exception was observed: the patient R.H. showed only a slight inhibition of IMPDH (−22.9%), although having very high MPA and MPAG trough levels.

Nevertheless, acceptable correlations between IMPDH activity and both MPA and MPAG could be calculated in the overall statistical analysis (Table 2). The best correlations were found between MPA or MPAG levels corrected for patients' total protein or albumin plasma levels, respectively.

Incubation experiments with MOLT-3 lymphoblasts and defined amounts of MPA or MPAG demonstrated that IMPDH activity was not only inhibited by MPA but also by MPAG (Figure 1). Already 2.5 μmol/L MPA decreased lymphoblastic IMPDH activity by 80% and 100 μmol/L MPAG by 72%. In the presence of plasma the inhibitory effects of MPA as well as of MPAG were generally diminished.

CONCLUSION

Our results clearly show that the periodical administration of MMF leads to an inhibition of type II IMPDH activity throughout the complete dosing interval, which should alter the balance of the purine nucleotides within replicating white blood cells (3,4), resulting in impaired synthesis of DNA.

With respect to the optimization of immunosuppression and to the establishing of therapeutic drug ranges, the pharmacodynamic monitoring of the biological effects of MPA and MPAG (=inhibition of IMPDH activity) should provide a meaningful supplement to traditional therapeutic drug monitoring, which only consists of the measurement of drug concentrations. This is supported by our observation that one patient (R.H., Table 1)—although

Table 1. Mean values of albumin, total protein, MPA, MPAG and
inhibition of IMPDH activity measured in lymphoblasts

	n	Albumin (g/l)	SEM	Protein (g/l)	SEM	MPA (μM)	SEM	MPAG (μM)	SEM	IMPDH (%)*	SEM
All patients	63	33.7	1.4	64.5	2.2	5.3	1.2	155.1	15.1	45.9°	3.6
Patient R. H.	7	35.6	0.9	65.9	0.8	10.1	1.1	380.1	33.6	77.1°°	1.4

*IMPDH activity: 100%= 11.5 ± 3.0 nmol/10^6 Cell/h or 395 ± 58 nmol/mg protein/h.
°Statistical significant compared to 100% controls (p<0.0001).
°°Statistical significant compared to 100% controls (p<0.001).

Table 2. Correlation: Inhibition of IMPDH activity versus MPA and
IMPDH activity versus MPAG

	MPA	MPA/albumin*	MPA/total protein**
IMPDH			
r	−0.57907	−0.63292	−0.61152
p	0.0006	0.0001	0.0003
	MPAG	MPAG/albumin*	MPAG/total protein**
IMPDH			
r	−0.61781	−0.6376	−0.64189
p	0.0048	0.0033	0.0003

*Normalized for patients' albumin plasma levels.
**Normalied for patients' protein plasma levels.

having very high MPA and MPAG trough levels—showed only a slight inhibition of IMPDH activity.

To our knowledge, no detailed studies of the possible pharmacological effects of MPAG have been published. The few studies, dealing with the effects of MPAG on IMPDH activity, are contradictory (6,11). For this reason lymphoblastic IMPDH activity

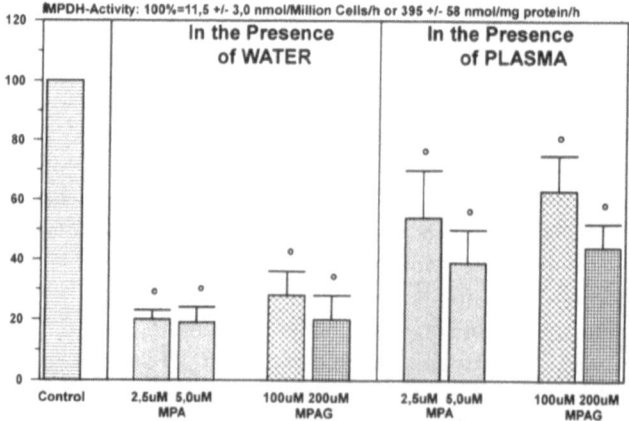

Figure 1. MPA and MPAG inhibition of IMP-DH activity measured in lymphoblasts (type II isoform).

was measured in the presence of defined amounts of MPA or MPAG, which were based upon the blood trough-levels measured in our patients. The data of these experiments clearly show that not only MPA but also MPAG is responsible for the inhibition of IMPDH activity. These inhibitory effects are partially neutralized by albumin or plasma (data not shown). Taking into account that in patients measured MPAG levels were 30 times higher than MPA levels, MPAG levels should be more taken into consideration during the therapy with MMF, since MPAG could not only contribute to the therapeutic but also to the unwanted side effects.

REFERENCES

1. F.R. Collart, and E. Huberman, Cloning and sequence analysis of the human and Chinese hamster inosine 5'-monophosphate dehydrogenase cDNAs. J. Biol. Chem. 63:15769 (1988).
2. Y. Natsumeda, S. Ohno, H. Kawasaki, Y. Konno, G. Weber, and K. Suzuki, Two distinct cDNAf for human IMP dehydrogenase. J. Biol. Chem. 265:5292 (1990).
3. M. Nagai, Y. Natsumeda, Y. Konno, R. Hoffman, S. Irino, and G. Weber, Selective up-regulation of type II inosine 5'-monophosphate dehydrogenase messenger RNA expression in human leukemias. Cancer Res. 51:3886 (1991).
4. M. Nagai, Y. Natsumeda, and Weber G, Proliferation-linked regulation of type II IMP dehydrogenase gene in human normal lymphocytes and HL-60 leukemic cells. Cancer Res. 52:258 (1992).
5. European Mycophenolate Mofetil Cooperative Study Group, Placebo-controlled study of mycophenolate mofetil combined with cyclosporin and corticosteroids for prevention of acute rejection. Lancet 345:1321 (1995).
6. W.A. Lee, L. Gu, A.R. Miksztal, N. Chu, K. Leung, and P.H. Nelson, Bioavailability of mycophenolic acid through amino acid derivatization. Pharm. Res. 7:161 (1990).
7. I. Nowak, and L.M. Shaw, Mycophenolic acid binding to human serum albumin: characterization and relation to pharmacodynamics. Clin. Chem. 41:1011 (1995).
8. C. Montero, J.A. Duley, L.D. Fairbanks, M.B. McBride, V. Micheli, A.J. Cant, and G. Morgan, Demonstration of induction of erythrocyte inosine monophosphate dehydrogenase activity in Ribaverin-treated patients using high performance liquid chromatography linked method. Clin. Chim. Acta 238:169 (1995).
9. N. Sugioka, H. Odani, T. Ohata, H. Kishimoto, T. Yasumura, and K. Takada, Determination of a new immunosuppressant, mycophenolate mofetil, and its active metabolite, mycophenolic acid, in rat and human body fluids by high-performance liquid chromatography. J. Chromatogr. B 654:249 (1994).
10. SAS Institute. SAS/STAT user's guide, version 6, 4th ed. Cary, NC: SAS Institute Inc., (1989).
11. L.J. Langman, D.F. LeGatt, and R.W. Yatscoff, Pharmacodynamic assessment of mycophenolic acid-induced immunosuppression by measuring IMP dehydrogenase activity. Clin. Chem. 41:295 (1995).

105

CHANGES OF POLYMORPHONUCLEAR LEUKOCYTE (PMN) FUNCTIONS UNDER DIETARY NUCLEIC ACID DEFICIENCY AND EFFECTS OF ADMINISTRATION OF NUCLEOTIDE AND NUCLEOSIDES MIXTURE SOLUTION IN MICE

Akihiro Hirai, Makoto Usami, Seiji Haji, Hiroshi Kasahara,
George Kotani, Atsunori Iso, Yuko Kitamura, Hiroyoshi Sen,
Masahiro Yamamoto, and Yoichi Saitoh

First Department of Surgery
Kobe University School of Medicine, 7-5-2
Kusunoki-cho, Chuo-ku, Kobe, 650, Japan

INTRODUCTION

Recent investigations of enteral nutritients or artificial feedings with nucleotides and nucleosides suggested that they potentiate the bio-defense activity in immunologically compromised patients in the intensive care units.[1] Decreased T lymphocyte functions with dietary nuclear acid deficiency and reversal by ribonucleic acid (RNA) supplement are reported.[2-4] In surgical stress, macrophages and/or polymorphonuclear leukocytes (PMNs) are considered as key factors in the pathogenesis of systemic inflammatory response leading to septic shock by producing humoral mediators under endotoxin stimulation. We reported that dietary nucleic acid deficiency inhibited lipopolysaccharide (LPS) uptake and humoral mediators secretion by peritoneal macrophages in rats.[5] However, it is not known whether dietary nucleic acid influences on PMN functions. Blocking of Mac-1 (CD11b/CD18) and/or lymphocyte function-associated antigen-1 (LFA-1, CD11a/CD18) which are the adhesion proteins involved in the integrin superfamily, reduce the number of PMNs accumulating in the tissue and attenuate the oxidant stress of infiltrated PMNs.[6,11] Phagocytes, such as PMNs and macrophages, undergo an oxidative burst in response to phagocytic or membrane stimuli, with generation and release of a variety of reactive oxygen metabolites.

Purine and Pyrimidine Metabolism in Man IX,
edited by Griesmacher *et al.* Plenum Press, New York, 1998.

A nucleotide and nucleosides mixture solution (OG-VI) has been developed to supply purine and pyrimidine nucleosides in the organ/ogoshi. The effect of OG-VI on humoral mediator production is reported in infection models.[7] Therefore, the aim of this study is to evaluate 1) the influences of dietary nucleic acid deficiency and 2) the effects of intraperitoneal administration of OG-VI7) on PMNs functions in mice injected with LPS as an endotoxin shock model.

MATERIALS AND METHODS

Six-weeks-old female BALB/C mice (approximately 20 g) were maintained on nucleic acid free diet (NFD : AIN-B. Oriental Corp.) or standard chow (CE-2, Crea) as a control (C group) for 7 days. Then nucleic acid free diet colony were randomized into two groups, OG group injected OG-VI (3.5 ml/mouse) intraperitoneally on 7th days or NF group with saline injection. OG-VI was composed of 5′-GMP 30 mM, inosine 30 mM, cytidine 30 mM, uridine 22.5 mM, thymidine 7.4 mM (Otsuka Pharmaceutical Co., Tokushima, Japan). Blood samples were obtained by cervical dislocation. Approximately a 500 µl aliquot of heparinized whole blood (heparin: 10 u/ml) was prepared per mouse, and immediately placed on ice. The cells were maintained at 0 to 4°C in all subsequent steps except as specified. After over 90% purity and viability were assessed by trypan blue staining, the cells were analyzed by using flow cytometer (FACScan, Becton Dickinson) within 1 hour and the mean fluorescence intensity (MFI) was measured. Each sample consisted of 5,000 PMNs. The PMNs were selectively gated for analysis by forward right-angle light-scatter properties.

100 microl whole blood were placed in polypropylene tubes in 1 ml of phosphate buffered saline (PBS : Gibco) containing fluorescein isothiocyanate LPS (FITC-LPS : E. coli 0111, B4, Sigma) at a concentration of 10 µg/ml, and placed in a shaking water bath at 37°C for 60 minutes. Then the tubes were placed back on ice, leukocytes were isolated by adding of lysing solution (Becton Dickinson), and the cells were washed twice after centrifugation (1500 rpm, at 4°C for 5 minutes) and resuspended through 30–40 µm mesh in 1 ml of PBS, then MFI per cell was measured. Analysis of Mac-1 and of LFA-1 (CD11a/CD18) expression on PMNs was performed by incubated in 1 ml of PBS containing 0.25 µg of FITC-labeled anti-Mac-1 antibody, or PE-labeled anti-LFA-1 antibody (PharMingen), and placed in shaking water bath at 37°C for 60 minites,[6,8] then MFI per cell was measured. Production of oxidative burst on PMNs was quantified by the addition of 2′,7′-dichlorofluorescein diacetate (DCFH-DA, Eastman Kodak Co.) to the whole blood.[9,10] Aliquots of 25 mM DCFH-DA (250 µM) were stored in absolute ethanol at −20°C until diluted 100 fold in sterile PBS for use in the assay. Phorbol myristate acetate (PMA; Sigma) was dissolved in anhydrous ethanol (25 µg/ml) and stored in the dark at −20°C. A 100 µl heparinized whole blood was added to 2 ml of PBS containing of DCFH-DA, and placed in a shaking water bath at 37°C for 20 minites. A 10 µl aliquot of PMA (+25 mM EDTA) was added and placed in a shaking water bath for 30 minutes. Then isolated leukocytes were washed to determine MFI per cell.

Tumor necrosis factor-alpha (TNF-alpha) level after intraperitoneal administration of 25 mg/kg of LPS (E. coli. 0111: B4, Difco) was measured by ELISA (Genzyme). And 72-hours survival rates after LPS injection were observed to compare among three groups.

All data were expressed as mean Statistical analysis of data was performed by the two tailed Student's t-test, one-way analysis of variance (ANOVA) among means. Generalized Wilcoxon test was used to determine significant differences for survival rates. A p-value less than 0.05 was considered significant.

Table 1. PMN functions in mice (mean fluorescence intensity)

	C	NF	OG
LPS uptake	52.0 ± 7.2	39.5 ± 6.3*	37.2 ± 3.3*
Mac-1	237.5 ± 57	183.1 ± 31*	197.3 ± 32
LFA-1	389.0 ± 90	243.8 ± 57*	358.9 ± 118
Osidative burst	525.1 ± 99	368.1 ± 101	657.9 ± 201#

n = 5, mean ± SD, *; P<0.05 vs. C. #; P<0.05 vs. NF

RESULTS

PMNs functions were demonstrated in Table 1. LPS uptake was significantly decreased in the NF and the OG groups (p<0.05). Expressions of Mac-1 and LFA-1 were lower in the NF group with statistically significance (p<0.05 vs C), but there was no significant difference between the NF and the OG group. Although oxidative burst was inhibited in the NF group compared with that in the C group, there was no significant difference between two groups. OG-VI enhanced oxidative burst in comparison with the NF group with a statistically significance (p<0.05).

Blood TNF-alpha level increased to 1000 ng/ml at 90 minutes after LPS administration in the C group (Figure 1). TNF-alpha level was lower in the NF group without statistical significance. OG-VI suppressed the increase in TNF-alpha level at 45 and 90 minutes (p<0.05).

Over 75% of mice died 12 hours after LPS injection in the C group (Figure 2). The 72 hours survival rates of the C, NF, OG group were 6.7% (1/15), 26.7% (4/15), 53.3% (8/15), respectively. There were significant differences among three groups (p<0.05).

DISCUSSION

In vitro effects of nucleotide free diet in mouse LPS shock model is in accordance with the data obtained in rats.[5] The observation of improved survival rate and the decrease

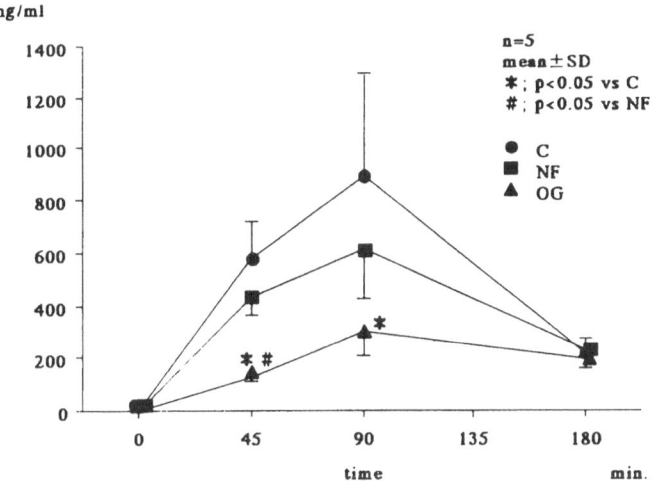

Figure 1. Changes in blood TNF-alpha levels after LPS (25 mg/kg) injection.

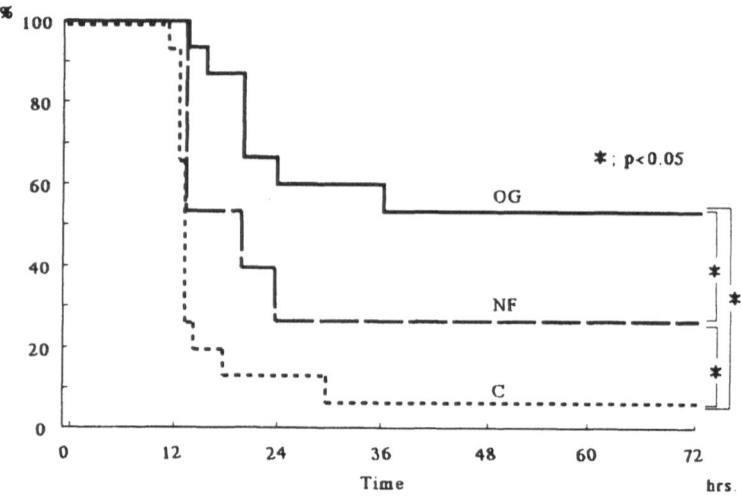

Figure 2. Survival curves after LPS (25 mg/kg) injection.

in blood TNF-alpha level after LPS injection can be explained from the data showing decreased LPS uptake by PMNs and macrophages.[5]

However, these data can not explain the changes of PMN functions, because results in PMN functions were obtained before LPS stimulation. New aspect introduced from this study is the observation that adhesion molecules expression and oxidative burst in PMNs are influenced by dietary nucleotide deficiency. Adhesion molecules have the important roles in cell-to-cell interaction in immune functions, microcirculation changes, and as the sub-signals for T cell activation.[11] Improvement of survival in this LPS shock mice could be caused by functional depression of PMNs involving a series of steps, including rolling along the vessel wall, adherence to endothelial cells and active migration through the vessel wall, in addition to inhibition of macrophage functions.[5] Also, the mechanism of decreased T cell functions in immune responses[13] and allograft models[3] might be explained from our results. Further study is required to explain the detailed mechanism of nucleotide free diet on immune response including surgical stress that is encountered many clinical conditions requiring nutritional supports. PMN functions under in vivo The reason why LPS injection (25 mg/kg) showed the extreme results is considered that higher LPS dose influences various intense effects on immune functions and changes PMN functions difficult to analyze consequently.

In vivo results of OG-VI in mice with survival and TNF-alpha production did not recover the changes observed in the NF group. One shoot intraperitoneal administration of OG-VI could not compensate the effect already formed during 7 days diet change. The separate pharmacological effect should be considered to explain these results. The comparative result of decreased TNF-alpha production by OG-VI in normal diet rats is reported by Yokoyama.[13] However, PMN functions after OG-VI administration showed the recovery of LFA-1 expression and oxidative burst to the control level, and their results are independent with decreased TNF-alpha production.

OG-VI solution has been developed nutritional supplement, however, our data suggested that the role of OG-VI was not only nutritional supports but also a potent chemical agent against endotoxemia. We found the biochemical importance of a nucleotide and nucleosides mixture solution, while it is considered as purine and pyrimidine source for salvage synthesis under nutritionally insufficient diets with total parenteral nutrition. But definitive effects and mechanism of OG-VI on endotoxemia remained to be clarified.

In conclusion, dietary nucleic acid deficiency inhibits PMN functions, especially adhesion molecule expression and oxidative burst in mice. OG-VI stimulated the decreased PMN functions, improved survival more than dietary nucleic acid deficiency.

REFERENCES

1. F.B. Cerra, S. Lehmann, N. Konstantinides, et al, Improvement in immune function in ICU patients by enteral nutrition supplemented with arginine, RNA, and menhaden oil is independent of nitrogen balance, Nutr 7: 193 (1991)

2. J.D. Carver, W.I. Cox, and L.A. Barness, Dietary nucleotide effects upon murine natural killer cell activity and macrophage activation, J Parent Ent Nutr. 14: 18 (1990)

3. C.T. Van Buren, A.D. Kulkarni, W.C. Fanslow, et al, Dietary nucleotides, a requirement for helper/induce T lymphocytes, Transplantation 40: 694 (1985)

4. C.T. Van Buren, F.B. Rudolph, A. Kulkarni, et al, Reversal of immunosuppression induced by a protein-free die, Crit Care Med. 18: S114 (1990)

5. S. Haji, M. Usami, G. Kotani, et al, The effect of dietary nucleic acid deficiency and the administration of a nucleotide and nucleosides mixture solution on endotoxin shock in rats, Purine and Pyrimidine Metabolism in Man VIII: 787 (1995)

6. H. Jaeschke, A. Farhood, A.P. Bautista, et al, Functional inactivation of neutrophils with a Mac-1 (CD11b/CD18) monoclonal antibody protects against ischemia-reperfusion injury in rat liver, Hepatology 17: 915 (1993)

7. S. Ogoshi, M. Iwasa, S. Kitagawa, et al, Effects of total parenteral nutrition with nucleoside and nucleotide mixture on D-galactosamine-induced liver injury in rats, J. Paren. Ent. Nutr. 12: 53 (1988)

8. A. Wollner, S. Wollner, J.B. Smith, et al, Acting via A2 receptors adenosine inhibits the upregulation of Mac-1 (CD11b/CD18) expression on FMLP-stimulated neutrophils, Am. J. Respir. Cell Mol. 9: 179 (1993)

9. D.A. Bass, J.W.Parce, L.R. Dechatelet, et al, Flow cytometric studies of oxidative product formation by neutrophils: A grade response to membrane, J. Immunol. 130: 1910 (1983)

10. L.S. Trinkle, S.R. Wellhausen, K.R. Mcleish, A simultaneous flow cytometric measurement of neutrophil phagocytosis and oxidative burst in whole blood, Diag Clin Immunol. 5: 62 (1987)

11. L.W. Argerbright, L.G. Letter, R. Rothlein, Monoclonal antibodies to the leukocyte membrane CD18 glycoprotein complex and to intercellular adhesion molecule-1 inhibit leukocyte-endothelial adhesion in rabbits, J Leuko Biol 49: 253 (1991)

12. H. Jyonouchi, Nucleotide actions on humoral immune responses, J Nutr 124: 138S (1994)

13. H. Yokoyama, S. Kano, S. Okamoto, et al, Modification of tumor necrosis factor (TNF) production and survival rate by a nucleoside mixture in lipopolysaccharide-injected rats, Purine and Pyrimidine Metabolism in Man VIII: 783 (1995)

106

DIFFERENTIATION AND REDUCTION OF INTRACELLULAR GTP LEVELS IN HL-60 AND U937 CELLS UPON TREATMENT WITH IMP DEHYDROGENASE INHIBITORS

Kunihiro Inai, Hiroshi Tsutani, Takahiro Yamauchi, Toru Nakamura, and Takanori Ueda

First Department of Internal Medicine
Fukui Medical University
School of Medicine
23-3 Shimoaizuki, Matsuoka, Fukui 910-11, Japan

1. INTRODUCTION

Inosine-5'-monophosphate dehydrogenase (IMP dehydrogenase, EC 1.1.1.205) is a rate-limiting enzyme of *de novo* guanylate biosynthesis and catalyzes the formation of XMP from IMP (1). IMP dehydrogenase have two isoforms. Type I transcript is the major form found in normal leukocytes (2), whereas type II is predominant in malignant cells (3), thus type II IMP dehydrogenase is a target for antineoplastic agents. In addition, a variety of human leukemic cells, including K562 erythroid leukemia cells (4), and HL-60 myeloid leukemia cells (5) as well as human breast cancer cells (6) which are treated with IMP dehydrogenase inhibitors, decreases in type II IMP dehydrogenase and the level of guanine nucleotide, especially GTP, are shown to be associated with cell maturation (7,8).

Recently, we clarify the ability of mizoribine, an IMP dehydrogenase inhibitor, to induce differentiation in HL-60 and U937 cells (9). However, decreases of intracellular GTP levels by mizoribine are transient, which is different from those of other IMP dehydrogenase inhibitors as mycophenolic acid and tiazofurin. In the present study, we investigated the detailed association between decreased intracellular guanine nucleotide levels and cell differentiation. We find that IMP dehydrogenase inhibitors can induce cell differentiation through a transient decline of intracellular GTP levels following inhibition of IMP dehydrogenase.

Purine and Pyrimidine Metabolism in Man IX,
edited by Griesmacher *et al.* Plenum Press, New York, 1998.

2. MATERIALS AND METHODS

2.1. Chemicals

Mizoribine (Asahi Chemical Industry Co., Tokyo, Japan) was dissolved in phosphate-buffered saline (PBS) at a concentration of 5 mM and stored at −80°C until use. Mycophenolic acid (Sigma Chemical Co., St. Louis, MO, USA) was dissolved in 150 mM $NaHCO_3$ and stored at −80°C. Guanosine (Nacalai Tesque Inc., Kyoto, Japan) was dissolved in 1N NaOH and neutralized with 1N HCl. Deoxyguanosine (Nacalai) was dissolved in PBS.

2.2. Cells and Culture

Both the HL-60 and the U937 cells were cultured in RPMI 1640 medium (Nissui Pharmaceutical Co., Tokyo) containing 10% heat-inactivated fetal calf serum (GIBCO, Grand Island, NY, USA). For the various experiments, exponentially growing cells were inoculated at a density of 5 to 10×10^4 cells/mL and incubated in the presence of mizoribine or mycophenolic acid. In some experiments, 50 μM of guanosine or 10 μM of deoxyguanosine was added to the culture medium as indicated.

2.3. Differentiation Markers

Cell differentiation was evaluated using three kinds of differentiation markers as described before (9). In brief, morphologic differentiation was evaluated using cytocentrifuged smears stained using the May-Gruenwald-Giemsa technique. Expression of cell surface antigens was studied using an indirect immunofluorescence technique with anti-CD11b (DAKO, Glostrup, Denmark) and anti-CD14 (Coulter, Hialeah, FL, USA) antibodies. To examine the ability of individual cells to generate superoxide, a slide test using nitroblue tetrazolium (NBT) was performed.

2.4. Analysis of Intracellular GTP Levels

Intracellular GTP and ATP levels were determined by high-performance liquid chromatography (HPLC) as previously described (10). The intracellular GTP and ATP levels were quantified from detected peak areas using external standards.

3. RESULTS

3.1. Cell Growth and Differentiation

Addition of mizoribine or mycophenolic acid, resulted in cessation of cell growth dose-dependently. Differentiation specific morphologic antigenic, and enzymatic markers were significantly increased in HL-60 and U937 cells treated with 50 μM of mizoribine, or 2 μM of mycophenolic acid. Although the cells did not show any expressions of differentiation markers in the presence of 50 μM of guanosine, they induced maturation in the existence of 10 μM of deoxyguanosine.

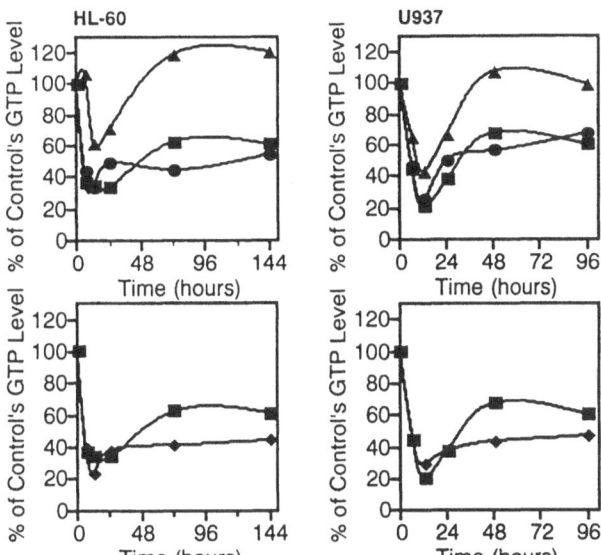

Figure 1. Intracellular guanosine 5'-triphosphate levels in HL-60 and U937 cells treated with either 50 μM mizoribine (Square), 20 μM mizoribine (Round), 5 μM mizoribine (Triangle), or 2 μM of mycophenolic acid (Diamond). The cells were exposed to the inducers for the indicated times and analyzed as described in the Materials and Methods. Values are the mean of three independent experiments.

3.2. Intracellular GTP Levels

Upon treatment of the cells for 12–24 h with more than 20 μM of mizoribine or 2 μM of mycophenolic acid, intracellular GTP levels decreased to approximately 20–30% of control levels, whereas the GTP levels gradually recovered to about 50–60% of control values and then maintained in the presence of mizoribine (Figure 1). In contrast, the GTP pools remained at a lowest level in the presence of mycophenolic acid (Figure 1). Intracellular ATP levels were constant during the incubation period (data not shown).

3.3. Short-Term Treatment with IMP Dehydrogenase Inhibitors

To elucidate the minimal treatment period needed to induce differentiation, the cells were incubated in the presence of the inducers for 24 h, and then incubation was continued in the absence of the inducers or in the presence of 50 μM of guanosine for a total of 96 hours from the beginning of treatment. 24 hr-treatment with the inducers already evoked both HL-60 and U937 cells to express differentiation markers at the end of culture period of 96 hours. When these cells were treated less than 16 h, differentiation markers did not increased (Figure 2).

4. DISCUSSION

In previous studies, several IMP dehydrogenase inhibitors have been reported to induce maturation by decrease of intracellular GTP levels in a time-dependent manner (11,12), suggesting that a longer duration of low GTP levels more effectively induced differentiation. In the present study, treatment with mycophenolic acid rapidly decreased in intracellular GTP levels to 30% of control values and then maintained at a lowest level during culture period. However, mizoribine-induced decreases of in intracellular GTP levels to 30% of control values were only detected for less than 24 h, same degree of dif-

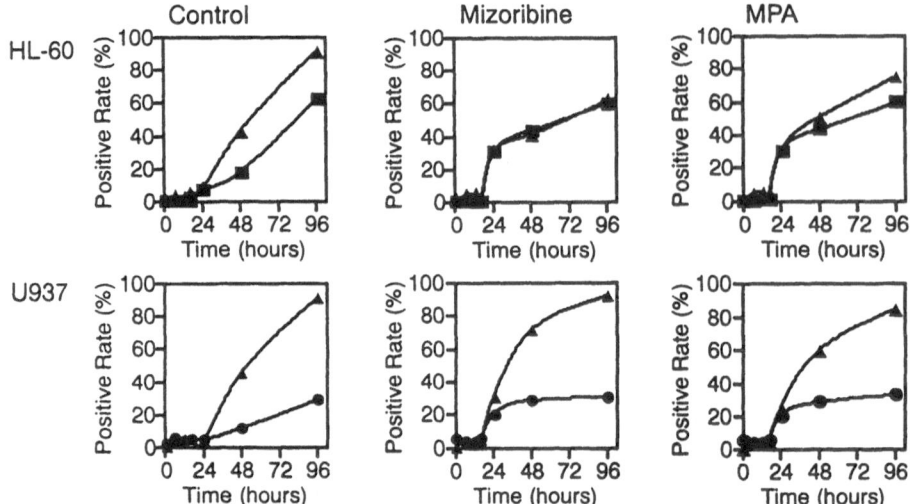

Figure 2. Kinetics of differentiation induction in HL-60 and U937 cells. Cells were cultured in the presence of 50 μM of mizoribine or 2 μM of mycophenolic acid for the indicated times (control), and culture was then continued in the absence of mizoribine or mycophenolic acid for a total of 96 h. The induction of differentiation was determined by assessing expression of CD11b (Square) and CD14 (Round), as well as NBT (Triangle) positivity. Values represent means from three independent experiments. MPA: mycophenolic acid.

ferentiation as mycophenolic acid did was observed by mizoribine. Exposure to the inducers for as short a time period as 24 h resulted in induction, whereas treatment for less than 16 h did not induce maturation. Moreover, addition of exogenous guanosine 24 h after initiation of the treatment with the IMP dehydrogenase inhibitors did not prevent differentiation. These findings suggest that differentiation induction by IMP dehydrogenase inhibitors does not occur in a time-dependent manner, but rather occurs when GTP levels decrease to a specific level at least 24 h following inhibition of IMP dehydrogenase.

It is still unclear how decreases in intracellular GTP levels by IMP dehydrogenase inhibitors induce differentiation. In the present study, treatment with mycophenolic acid induced differentiation in the absence or presence of deoxyguanosine that was synthesized to deoxyGTP (dGTP) (13). 10 μM of exogenous deoxyguanosine was enough to keep on intracellular dGTP levels under treatment with IMP dehydrogenase inhibitors (13). These findings suggest that differentiation induction by IMP dehydrogenase inhibitors does not need to inhibit DNA synthesis through lower intracellular dGTP levels. Several studies have suggested a relationship between intracellular GTP levels and GTP-dependent signal transduction (14,15). Low GTP levels may affect intracellular signal transduction and induce differentiation. This study demonstrates that low GTP levels for at least 24 h to approximately 30% of control values may turn on a switch to induce maturation through a signal transduction mechanism.

ACKNOWLEDGMENTS

The authors thank Miss Yukie Nishikiori for her valuable technical assistance and help with the preparation of this manuscript.

REFERENCES

1. Jackson RC, et al. IMP dehydrogenase, an enzyme linked with proliferation and malignancy. Nature 1975;256:331–333.
2. Konno Y, et. al. Expression of human IMP dehydrogenase type I and II in *Escherichia coli* and distribution in human normal lymphocytes and leukemic cell lines. J Biol Chem 1991;266:506–509.
3. Collart FR, et al. Cloning and sequence analysis of the human and Chinese hamster inosine-5'-monophosphate dehydrogenase cDNAs. J Biol Chem 1988;263:15769–15772.
4. Olah E, et al. Induction of erythroid differentiation and modulation of gene expression by tiazofurin in K-562 leukemia cells. Proc Natl Acad Sci USA 1988;85:6533–6537.
5. Lucus DC, et al. Purine metabolism in myeloid precursor cells during maturation. J Clin Invest 1983;72:1889–1900.
6. Sidi Y, et al. Growth inhibition and induction of phenotypic alterations in MCF-7 breast cancer cells by an IMP dehydrogenase inhibitor. Br J Cancer 1988;58:61–63.
7. Nagai M, et al. Proliferation-linked regulation of type II IMP dehydrogenase gene in human normal lymphocytes and HL-60 cells. Cancer Res 1992;52: 258–261.
8. Nagai M, et al. Selective up-regulation of type II inosine 5'-monophosphate dehydrogenase messenger RNA expression in human leukemias. Cancer Res 1991;51:3886–3890.
9. Inai, K, et al. Differentiation induction in non-lymphocytic leukemia cells upon treatment with mizoribine. Int J Hematol 1997 (*in press*).
10. Yamauchi T, et al. A new sensitive method for determination of intracellular 1-β-D-arabinofuranosylcytosine 5'-triphosphate content in human materials *in vivo*. Cancer Res 1996;56:1800–1804.
11. Kiguchi K, et al. Induction of cell differentiation in melanoma cells by inhibitors of IMP dehydrogenase: Altered patterns of IMP dehydrogenase expression and activity. Cell Growth Differ 1990;1:259–270.
12. Kiguchi K, et al. Cell differentiation and altered IMP dehydrogenase expression induced in human T-lymphocytoid leukemia cells by mycophenolic acid and tiazofurin. Exp Cell Res 1990;187:47–53.
13. Eugui EM, et al. Immunosuppressive activity of mycophenolate mofetil. Ann NY Acad Sci 1993;685:309–329.
14. Sokoloski JA, et al. Alterations in glycoprotein synthesis and guanosine triphosphate levels associated with the differentiation of HL-60 leukemia cells produced by inhibitors of inosine 5'-phosphate dehydrogenase. Cancer Res 1986;46:2314–2319.
15. Kharbanda SM, et al. Effects of tiazofurin on guanine nucleotide binding regulatory proteins in HL-60 cells. Blood 1990;75:583–588.

107

ISOENZYMES OF 5'-NUCLEOTIDASE IN HUMAN LYMPHOCYTES

E. Marinello, A. Tabucchi, F. Carlucci, P. Galieni, and F. Rosi

Institute of Biochemistry and Enzymology
Department of Haematology
University of Siena, Italy

1. INTRODUCTION

Ecto-5'-nucleotidase (ecto-5'-NT, CD73) is a 69 KDa purine catabolic-pathway enzyme attached via a GPI-linkage to the external plasma membrane of various cell types, including subsets of human lymphocytes.[1] In addition to purine catabolism, ecto-5'-NT serves as a maturation marker.[2,3]

Other forms of 5'-nucleotidase exist in the cytoplasm of different species and tissues and can be distinguished by substrate and certain properties. They control intracellular levels of nucleoside 5'-monophosphates; the first form has a preference for AMP, the second for IMP as substrate. These isoenzymes are indicated in the literature[4] as e-Ns (soluble ecto-5'-nucleotidase) and c-N-II respectively. Ecto-5'-nucleotidase and c-N-II have different catalytic, structural and genomic properties. The nucleotide sequence of human c-N-II cDNA and the aminoacid sequence are known and are different from the human placental ecto-form.[5-7] The relationships between ecto-5'-nucleotidase and e-Ns are unclear: only a few similarities have been noted.

The behaviour of 5'-nucleotidase isoenzymes was studied in lymphocytes of patients affected by B-cell lymphocytic leukemia (B-CLL). We were specifically concerned with ecto-5'-NT, which is considered to be the most representative of the three isoforms due to its high activity, and various functions.

2. METHODS

Peripheral blood lymphocytes (PBLs) were isolated by Lymphoprep and suspended, at a concentration of 140,000 cells/µl, in 0.29 M sucrose containing 10 mM Tris-HCl (pH 7.4). B and T populations were isolated from PBLs by an immunomagnetic procedure.[8] The cells were counted with a Delcon cell counter. Determination of ecto-5'-nucleotidase,

Purine and Pyrimidine Metabolism in Man IX,
edited by Griesmacher *et al.* Plenum Press, New York, 1998.

e-Ns and c-N-II activities were carried out by a radiochemical method coupled with HPLC. Preparation of the assay mixture, all details on storage and injection of samples and the HPLC instrument adopted, are reported in a previous paper.[9]

2.1. Immunofluorescence Detection of CD73

For the identification of ecto-5'-NT positive cells (CD73[+] antigen) by FACS, according to Kummer et al.,[10] we used the monoclonal antibody 1E9.28.1 (anti-5'-nucleotidase antibody) as primary antibody and PE (phycoerythrin) conjugated-anti-mouse IgG3 as secondary antibody (single-color immunofluorescence). For the simultaneous expression of CD73 and other lymphocyte differentiation antigens (CD19 and CD20 for B cells and CD3 for T cells), we used two-color immunofluorescence; the anti-CD19, anti-CD20 and anti-CD3 antibodies were labelled with fluorescein isothiocyanate-avidin (FITC-A).

2.2. PCR Analysis

The reverse transcription reaction for the synthesis of first-strand cDNA of ecto-5'-nucleotidase was performed using samples of total RNA from a healthy subject and a leukemia patient as template. cDNA was synthesized using the specific oligonucleotide R (5'-CATCTGGAACCCATCTCC-3') complementary to nt 1513–1530 of human placental 5'-NT cDNA as a primer. About 2 µg of total RNA was mixed with the oligonucleotide R (1 µM final concentration) and incubated at 65°C for 15 min to denature the secondary structure of the template. The RNA was incubated for 60 min at 42°C in a reaction mixture containing 1X RT buffer (50 mM Tris-HCl pH 8.3, 10 mM MgCl$_2$, 60 mM KCl and 1 mM dithiothreitol), 4 mM sodium pyrophosphate, 5 mM dNTPs mix, 1 unit of RNAse inhibitor and 0.5 units of reverse transcriptase.

The cDNA was amplified by PCR to obtain a 595 pb fragment using the same primer R and primer F (5'-CAGCATTCCTGAAGATCC-3') identical to nt 936–953 of human placental 5'-NT cDNA. PCR was carried out in a reaction mixture consisting of RT product (about 100 ng of cDNA), 1X PCR buffer (10 mM Tris-HCl pH 8.3, 1.5 mM MgCl$_2$, 50 mM KCl), 2 mM dNTPs mix, 1 µM of each primer and 1 unit of Taq polymerase for 30 cycles, denaturing at 94°C for 1 min, annealing at 54°C for 2 min and extending at 72°C for 3 min.

3. RESULTS

Severe reduction in ecto- and cytosolic activities was observed. A significant decrease in ecto-5'-nucleotidase, e-Ns and c-N-II was found in B and T populations (Table 1).

It was necessary to ascertain whether this low activity in leukemia patients was due to fewer normally active molecules, production of an altered enzyme by these cells, or genomic alterations.

To check the first hypothesis, we performed immunofluorescence analysis using a specific monoclonal antibody; a good linear correlation was found between the percentage of 5'-NT-positive cells determined by FACS and 5'-NT activity determined by radiochemical assay in healthy subjects, confirming that CD73 identifies with ecto-5'-NT. The percentage of 5'-nucleotidase positive cells (CD73[+]) was reduced in leukemia patients. (Table 2). To ascertain the possibility of an alteration in the expression of the ecto-5'-nucleotidase gene, we analyzed specific mRNA by the RT-PCR.

Table 1. Activity of 5'-NT isoforms in B and T cells
from healthy subjects and leukemia patients

	PBLs		B cells		T cells	
	Healthy subjects	Leukemia patients	Healthy subjects	Leukemia patients	Healthy subjects	Leukemia patients
ecto-5'-NT	21.1 ± 1.8	$*3.8 \pm 0.8$	94.8 ± 17.8	$*1.8 \pm 0.7$	9.9 ± 1.2	$*1.7 \pm 0.6$
	n = 30	n = 14	n = 11	n = 6	n = 7	n = 4
e-Ns	5.3 ± 0.7	$*0.8 \pm 0.2$	15.0 ± 5.8	$*0.7 \pm 0.2$	1.2 ± 0.2	$*0.3 \pm 0.1$
	n = 22	n = 10	n = 8	n = 7	n = 8	n = 5
c-N-II	8.7 ± 0.1	$*2.9 \pm 0.7$	26.7 ± 10.1	$*2.0 \pm 0.6$	2.2 ± 0.6	$*0.3 \pm 0.1$
	n = 21	n = 10	n = 8	n = 7	n = 6	n = 5

The activities are expressed in $nmol/h/10^6$ cells and reported as mean $\pm$ S.E.
n = number of cases.
$* = p < 0.05$.

Table 2. Expression of CD73 on PBLs determined by immunofluorescence

		B cells		T cells
	PBLs	$CD19^+$	$CD20^+$	$CD3^+$
Healthy (n = 2)	18.90	86.65	81.99	14.86
Leukemia (n = 2)	0.15*	0.14*	0.14*	0.33*
% decrease in leukemia patients	99.20	99.84	99.83	97.78

The values are mean percentages of cells positive to CD73 $\pm$ S.D.
n = number of cases.
$* = p < 0.05$.

4. DISCUSSION

The reduced expression of CD73 in lymphocytes from leukemia patients indicates fewer active enzyme molecules on the cell surface.

By RT-PCR, we obtained a fragment of the expected size from lymphocytes of healthy subjects and leukemia patients. This suggests that there are no detectable deletions in the amplified region. We hypothesize that these changes are due to slight alterations in the gene-sequence, or to production of altered protein. The formation of this protein that could responsable for the low activity, may occur at post-translational level, that is, during protein assembly, transport across the Golgi apparatus, or insertion in the cell membrane. Complete sequencing of the cDNA is necessary and will be carried out to answer these questions.

If the low expression of CD73 is due to alteration of the ecto-5'-NT gene, molecular analysis may be used as an early marker of B-CLL familiarity.

REFERENCES

1. Thompson LF, Ruedi JM, Glass A, Moldenhauer G, Moller P, Low MG, Klemens MR, Massaia M, Lucas AH. Production and characterization of monoclonal antibodies to the glycosylphosphatidylinositol-anchored lymphocyte differentiation antigen ecto-5'-nucleotidase (CD73). Tissue Antigens, 35: 9–19, 1990.
2. Edwards NL, Gelfand EW, Burk L, Dosch DM, Fox JH. Distribution of 5'-nucleotidase in human lymphoid tissues. Proc Natl Acad Sci, 76: 3473–3476, 1979.

3. Thompson LF, Ruedi JM, O'Connor RD, Bastian JF. Ecto-5'-nucleotidase expression during human B cell development. An explanation for the heterogeneity in B lymphocyte ecto-5'-nucleotidase activity in patients with hypogammaglobulinemia. J Immunol, 137: 2496–2500, 1986.

4. Zimmermann H. 5'-Nucleotidase: molecular structure and functional aspects. Biochem J, 285: 345–365, 1992.

5. Misumi Y, Ogata S, Ohkubo K, Ikehara Y. Primary structure of human placental 5'-nucleotidase and identification of the glycolipid anchor in the mature form. Eur J Biochem, 191: 563–569, 1990.

6. Hansen KR, Resta R, Webb CF, Thompson LF. Isolation and characterization of the promoter of the human 5'-nucleotidase (CD73)-encoding gene. Gene, 167: 307–312, 1995.

7. Oka J, Matsumoto A, Hosokawa Y, Inoue S. Molecular cloning of human cytosolic purine 5'-nucleotidase. Biochem Biophys Res Comm, 205(1): 917–922, 1994.

8. Brinchman JE, Vartdal F, Gaudernack G, Markussen G, Funderud S, Ugelstad J, Thorsby E. Direct immunomagnetic quantification of lymphocyte subsets in blood. Clin Exp Immunol, 71: 182–186, 1988.

9. Agostinho AB, Rosi F, Tabucchi A, Carlucci F, Pizzichini M, Leoncini R, Vannoni D, Pagani R. Isoforme della 5'-nucleotidasi e tumori (comportamento nelle leucemie linfoproliferative). Riv. It. Biol. Med., 15: 120–125, 1995.

10. Kummer U, Mysliwietz J, Gutensohn W, Buschette s, Jahn H, Neuser D, Munker R. Development and properties of a monoclonal antibody specific for human ecto-5'-nucleotidase. Immunobiol, 166: 203–211, 1984.

108

BIOCHEMICAL BASIS FOR THE IMPAIRED IMMUNE RESPONSE IN CHRONIC RENAL FAILURE?

Katarzyna Ruckemann,[3] Arian Laurence,[1] Lynette D. Fairbanks,[1]
David Richards,[2] Ryszard T. Smolenski,[3] Catherine Hawrylowicz,[2]
Margarita Bofill,[4] and H. Anne Simmonds[1]

[1]Purine Research Laboratory
[2]Allergy and Respiratory Disorders Laboratory
Guy's Hospital, London
[3]Department of Biochemistry
University School of Medicine
Gdansk, Poland
[4]Academic Department of Clinical Immunology
Royal Free Hospital, London

INTRODUCTION

Uraemia has been termed "nature's immunosuppressive device" [1]. Both cellular and humoral immunity are now known to be affected, the first being more severely disrupted [2]. Infections still represent an important source of mortality and morbidity among patients with chronic renal failure (CRF) [2].

Proven connection between inherited severe combined immunodeficiency (SCID) and adenosine deaminase deficiency first drew attention to the importance of intact pathways of purine and pyrimidine metabolism to the normal immune response.

In order to gain a deeper insight into the metabolic mechanisms involved studies were performed first in mitogen stimulated T-lymphocytes from healthy donors to confirm the dependence of the immune response on purine and pyrimidine *de novo* synthesis [4]. In this study we investigated whether any aberrations in the nucleotide responses to mitogen stimulation could be identified in T-lymphocytes from CRF patients which could contribute to our understanding of the immunodeficiency observed.

Purine and Pyrimidine Metabolism in Man IX,
edited by Griesmacher *et al.* Plenum Press, New York, 1998.

METHODS

T Cell Isolation and Stimulation

Heparinised blood was obtained from CRF patients (n=16) prior to haemodialysis and from healthy volunteers (n=16). Patients chosen had no previous renal transplant history and were not suffering from haematological proliferative disorders, HIV, viral hepatitis or acute infections. Patients receiving drugs with known immunosuppressive action and with CRF secondary to autoimmunological diseases were also excluded. Plasma creatinines were in the range of 393–1175 μmol/l.

Mononuclear cells were obtained by Ficoll-Hypaque gradient centrifugation and T-lymphocytes isolated using negative selection. Monocytes, B lymphocytes and NK cells were depleted by adhesion or monoclonal antibodies and magnetic beads. The purity of the cell preparation was >96% CD3+. Viability in trypan-blue exclusion test was higher than 98%.

T-cells were stimulated and incubated as described [4] for 0, 24, 48, 72 hours (D1, D2, D3 and D4).

Radioactive Tracer Studies

At 24 hours intervals cells were resuspended in bicarbonate free medium plus [^{14}C] radiolabelled bicarbonate and incubated for 2 hours at 37°C (6μCi, final concentration 1.08 mM) to evaluate the flux through the purine and pyrimidine *de novo* pathways.

Extraction and Assay of Ribonucleotides

Cells were extracted and analysed by HPLC with in-line Photodiode-array and radiodetection as described [4]. All results are expressed per 10^6 cells based on the initial count. Protein in the pellet was estimated by the method of Lowry.

Statistical analyses were performed using Student's t-test. Values of p<0.05 were considered statistically significant.

RESULTS

1. No significant differences in the baseline (D1) ribonucleotide concentrations for either purines or pyrimidines were observed between the two groups (Fig. 1). (UDPG is included with the uridine nucleotides).
2. A 2–7 fold expansion in all ribonucleotide pools was observed consistently in control lymphocytes by D4 following stimulation (Fig. 1). Triphosphates comprised more than 85% of the total.
3. A 60% impairment of expansion of all ribonucleotide pools was evident in the CRF group by D4 with adenine nucleotides showing no increase at all by D4 compared with D1 (Fig. 1). Again concentrations of triphosphates were more than 85% of the total.
4. [^{14}C] bicarbonate incorporation into the different nucleotide pools was in proportion to the total nucleotide concentration, the pattern being similar in both groups (Fig. 2). Note that resting cells did not incorporate bicarbonate indicating negligible activity of purine and pyrimidine *de novo* synthetic pathways.
5. The increase in the protein concentration in the control group was 62.5% while in the CRF group protein concentration fell on day 2 and 3 but increased again to reach baseline values by 72 hours (Fig. 2).

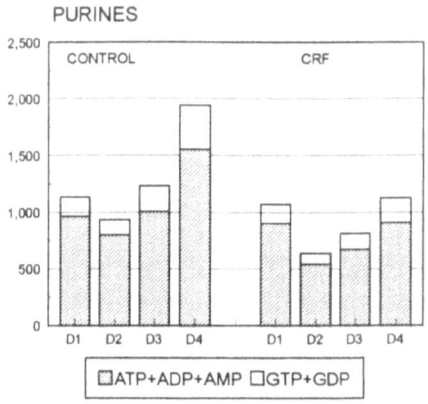
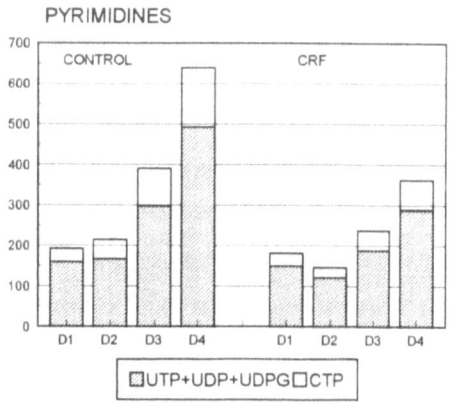

Figure 1. Purine and pyrimidine nucleotide pool changes in stimulated T-lymphocytes 0 (D1), 24 (D2), 48 (D3) and 72 hours (D4) after stimulation. All results in pmols/10^6 cells.

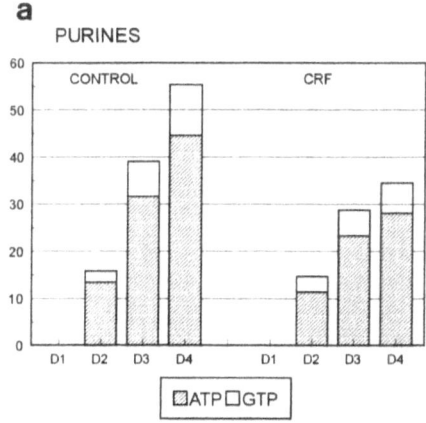
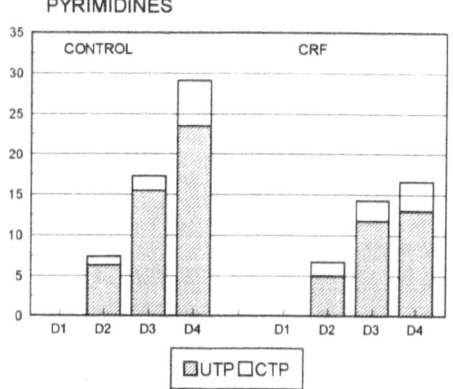

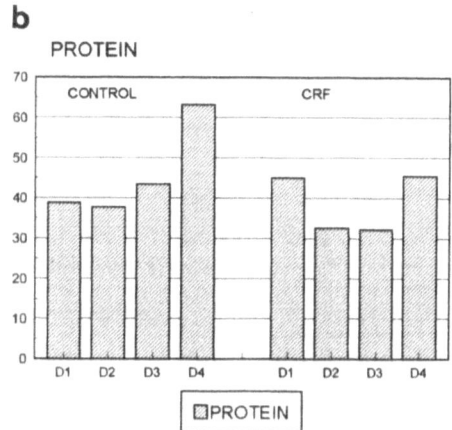

Figure 2. A: [^{14}C] bicarbonate incorporation. Results in radiocounter units/10^6 cells. B: Protein concentration in μg/10^6 cells.

DISCUSSION

In this study we observed that unstimulated CRF lymphocytes do not differ from healthy controls in terms of ribonucleotide pools. However differences were evident within the first 24 hours following stimulation with phytohaemagglutinin, with the response to stimulation by 72 hours being 60% of the control.

The possibility that these discrepancies were due to a defect in either purine or pyrimidine *de novo* synthesis in the renal failure group was excluded by the comparable incorporation of bicarbonate into ATP, GTP, CTP and UTP by the T-cells of both groups. Thus, impaired purine or pyrimidine *de novo* synthesis does not appear to be the underlying mechanism for the immunodeficiency in uraemia.

What then is the explanation? The above results coupled with the lower protein concentrations in CRF suggest that approximately 40% of T-cells died within the first 24 hours after stimulation. However, the remainder responded normally to stimulation thereafter, paralleling the controls over the next 48 hours. The similar pattern of bicarbonate incorporation in both groups of stimulated cells would substantiate this hypothesis. Likewise, the cell count data supported higher cell loss in the 24 hours after stimulation in the CRF group and preliminary flow cytometry studies confirmed that subsequently at least some of the CRF lymphocytes progressed normally through the cell cycle.

The possible mechanism of cell death remains speculative. Several explanations have already been proposed, including cell hyperactivation in vivo [2], decreased IL-2 bioavilability [2], diminished TCR/CD3 expression [2], disregulation of Fas and bcl-2 expression [3] and membrane instability leading to disruption and necrosis [6]. Apoptosis has been reported as a mechanism of T-lymphocyte death in CRF [3] but our preliminary study did not support this finding as by morphology this cells did not show chromatin condensation or nuclei fragmentation. Necrosis is the more likely possibility emerging from our studies.

The normal nucleotide pool expansion observed in the remaining CRF cells implies the existence of at least two groups of T-lymphocytes—one being more susceptible to death after stimulation, the other responding in a manner similar to the controls.

Consequently investigation of responses to stimuli in different T-lymphocyte sub-populations in CRF should provide a deeper insight into the mechanisms which cause the death of some cells but protect others. Such knowledge would contribute to a better understanding of the T-cell defect in CRF and thus development of novel approaches to immunointervention.

ACKNOWLEDGMENTS

We are very grateful to Ms Gosia Furlong from the University of London and to the staff of the Renal Unit and Haematology Department of Guy's Hospital for their support.

These studies were supported by NATO grant CRG. 921365 and the Polish Scholarship financed jointly by the Foreign & Commonwealth Office, Corporation of London, the M B Grabowski Fund and the University of London.

REFERENCES

1. H. S. Lawrence: Uremia: Nature's immunosuppressive device, Ann. Intern. Med., 62:166, 1965

2. L. Chatenoud , P. Jungers , B. Descamps-Latscha, Immunological considerations of the uremic and dialysed patient, Kidney Int. 45, suppl.44: S92-S96, 1994.

3. Y. Matsumoto, T.Shinzato, I.Amano, I. Takai, Y.Kimura, H. Morita, K.Nakane, Y.Yoshikai, K. Maeda, Relationship between Susceptibility to Apoptosis and Fas Expression in Peripheral Blood T Cells from Uremic Patients: A Possible Mechanism for Lymphopenia in Chronic Renal Failure, Biochem and Biophys Research Com, 215 (1): 98:105, 1995.

4. L. D. Fairbanks, M. Bofill, K. Ruckemann, H. A. Simmonds, Importance of Ribonucleotide Availability to Proliferating T-lymphocytes from Healthy Humans, J. Bio. Chem., 270 (50): 29682–29689, 1995.

5. I. Nicoletti, G. MIgliorati, M. C. Pagliacci, F. Grignani, C. Riccardi, A rapid and simple method for measuring thymocyte apoptosis by propidium iodide staining and flow cytometry, J. Immun. Meth., 139: 271–279, 1991.

6. Iu. I. Grinshtein, I. V. Osetrov, V. P. Tereshchenko, The morphofunctional instability of the lymphocyte membrane in patients with chronic kidney failure, Urol. Nefrol. (Mosk), 6: 46 - 48, 1993

109

MYCOPHENOLIC ACID INFLUENCES THE EXPRESSION OF ICAM-1 ON HUMAN ENDOTHELIAL CELLS

P. Bertalanffy, P. Dubsky, G. Seebacher, A. Griesmacher, G. Weigel, and E. Wolner

Department of Cardiothoracic Surgery
University of Vienna

1. INTRODUCTION

Mycophenolic acid (MPA) is a potent immunosuppressive agent, which selectively decreases the intracellular guanine nucleotide pools by inhibiting the inosine-5'-monophosphate dehydrogenase (IMP-DH) thereby suppressing the proliferation of rapidly proliferating cells e.g. alloantigen stimulated lymphocytes (1,2).

In the endothelial cells the effect of IMP-DH inhibition must be seen in a different light due to the ability of using predominately the salvage pathway of purine nucleotide synthesis (3). Nevertheless, alterations of adhesion molecules (AM) surface expression whether due to changes in intracellular guanine nucleotide pools or due to other effects of MPA were subject to our investigations (4).

Endothelial AM initiate the inflammatory response of the graft as T-cells bind to them and transmigrate the endothelium (e.g. the interaction of ICAM-1 and the leukocyte function associated antigen 1 (LFA-1)). Thus, the blockade of AM has emerged as important immunological target site for new clinical stategies in immunosuppressive therapy (5).

We aimed at investigating the effects of MPA on the expression of ICAM-1 on human umbilical vein endothelial cells (HUVEC). Experiments were performed on both unstimulated endothelium and endothelium stimulated by Th1-Cytokines (IFN-γ) or Th2-Cytokines (IL-4, IL-10).

2. MATERIALS AND METHODS

2.1. Cell Isolation and Culture

Endothelial cells were isolated following collagenase digestion (incubation for 5 minutes at 37°C, using 0.1% collagenase solution) from human umbilical cord veins, by

Purine and Pyrimidine Metabolism in Man IX,
edited by Griesmacher *et al.* Plenum Press, New York, 1998.

565

perfusion with M199 medium (Sigma, Germany) containing 20% fetal bovine serum (FBS) (ph 7.4) (Gibco,USA). Cells were collected from the perfusate by centrifugation at 200 g for 5 minutes and seeded into 6-well culture plates, precoated with human fibronectin (Upstate Biotechnology, USA).

Cells were cultured until confluency in M199 medium (pH 7.4) containing 20% PBS, 100.000 U/l penicillin (Gibco,USA), 100.000 µg/l streptomycin (Gibco,USA), 100.000U/l low molecular weight heparin (Sigma,Germany) and 30 mg/l bovine hypothalamic growth factor (Upstate Biotechnology, USA). Only cells from the primary monolayers were used for the experiments described below.

2.2. Incubation Experiments

Human umbilical vein endothelial cells (HUVEC) were preincubated with 25 µM MPA for 24 hours and thereafter stimulated with 250 U/ml IFN-γ, or 10 ng/ml IL-4, or 10 ng/ml IL-10 for another 24 hours. Control experiments were performed with medium, cytokine or MPA alone.

Furthermore HUVEC were incubated with a combination of MPA (25 µM) with either 100 µmol/l guanine or guanosine for 48 hours or pretreated with MPA (25 µM) for 24 hours and subsequently 100 µmol/l guanine or guanosine were added for further 24 hours.

2.3. Measurement of AM Surface Expression

HUVEC were evaluated for expression of ICAM-1 using flow-cytometry. At the end of the incubation the cells were washed and removed from the plates by adding 1.25 mmol/l EDTA followed by gentle scraping on ice. Thereafter the cells were centrifuged at 200 g for 5 minutes and resuspended in ice cold DPBS without Ca^{2+} and Mg^{2+} ($DPBS^{--}$) containing 5% FBS and 0.1% sodium acide and incubated with a saturating concentration of a fluoresceinisothiocyanate (FITC)-conjugated monoclonal antibody (Mab) at 4°C for 10 minutes. Anti-ICAM-1 (IgG1, FITC) was purchased from R&D Systems (UK). After washing with $DPBS^{--}$ containing 5% FBS and 0.1% sodium acide cells were centrifuged and resuspended. Cell viability was detected by co-staining with propidium iodide (Sigma, USA). Purity of the cells was checked by staining with anti-CD-31 Mab, non-specific fluorescence was assessed by staining with an irrelevant isotype matched Mab. Samples were analyzed with an EPICS Profile 2 flow cytometer (Coulter Corp., USA). Data for 5000 viable (propidium iodide negative) cells were collected.

2.4. Statistical Analysis

For statistical analysis paired Student's t-tests were used. All data are given as mean ± SEM. p-values < 0.05 were considered to be significant.

3. RESULTS AND DISCUSSION

Figure 1 shows the ICAM-1 surface expression was not altered after incubation with the Th2-cytokines IL-4 and IL-10 (data not shown), whereas IFN-γ was shown to induce ICAM-1. The increase of ICAM-1 surface expression after stimulation with IFN-γ was not changed significantly when preincubated with MPA.

In Figures 2, 3 MPA alone induced a highly significant increase in ICAM-1 expression concerning the mean fluorescence as well as the percent of ICAM-1 positive cells. This effect of MPA on ICAM-1 expression could be reduced by coincubating HUVEC with MPA and guanine or guanosine and also be partially reversed by the subsequent treatment with MPA and guanine or guanosine.

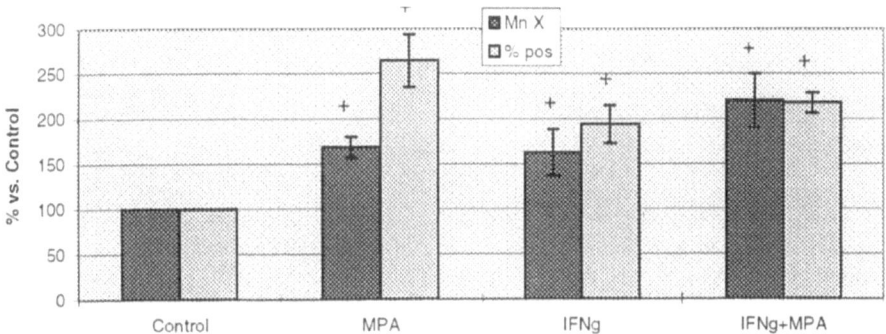

Figure 1. MnX: Mean fluorescence; % pos: Percentage of Mab positive cells; +: Statistically significant vs. Control p < 0.05.

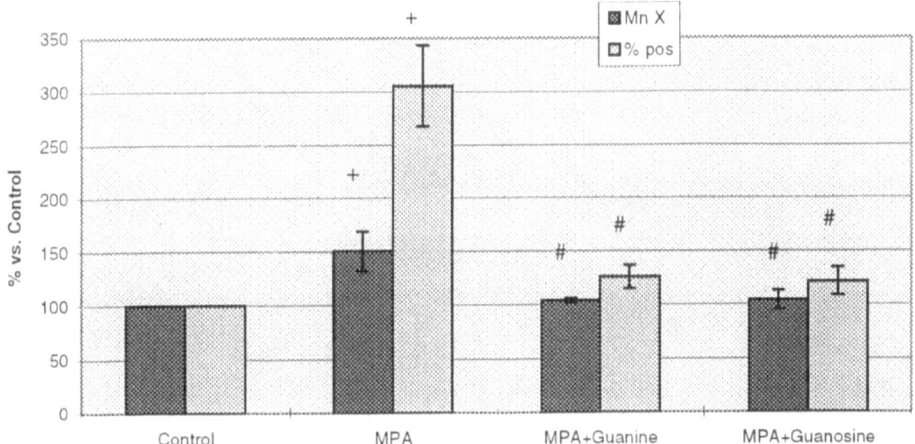

Figure 2. MnX: Mean fluorescence; % pos: Percentage of Mab positive cells; +: Statistically significant vs. Control p < 0.05; #: Statistically significant vs. MPA p < 0.05.

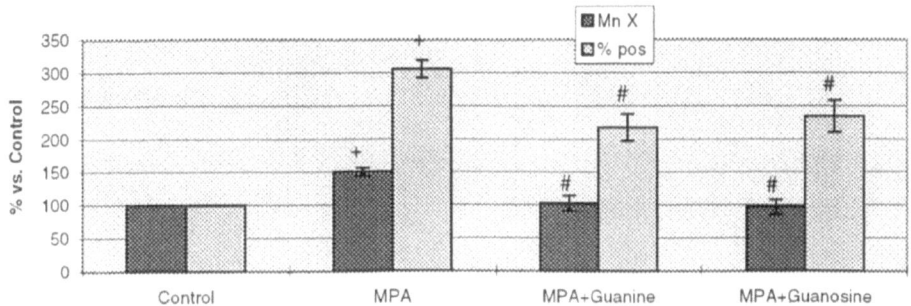

Figure 3. MnX: Mean fluorescence; % pos: Percentage of Mab positive cells; +: Statistically signifikant vs. Control p < 0.05; #: Statistically significant vs. MPA p < 0.05.

This data might imply that the effect observed under MPA is due to alterations in guanine nucleotide metabolism of the endothelial cells. Experiments performed in our laboratory revealed that MPA leads to measurable GTP depletion (−70%) and UTP increase (+52%) in endothelial cells. Since both are responsible for the synthesis of glycoproteins (3) these changes could account for the effect of MPA on endothelial ICAM-1. In summary we have provided evidence that MPA on its own might critically affect leukoendothelial interactions.

REFERENCES

1. T.J. Franklin, J.M. Cook, The inhibition of nucleic acid synthesis by mycophenolic acid. Biochem. J. 113: 2515 (1969).
2. E.M. Eugui, S.J. Almquist, C.D. Muller, A.C. Allison, Lymphocyte selective cytostatic and immunosuppressive effects of mycophenolic acid in vitro: role of deoxyguanosine nucleotide depletion. Scand. J. Immunol. 33: 175 (1991).
3. A. Griesmacher, M.M. Mueller, Purine metabolism in endothelial cells. Z. Med. Lab. Diagn. 31(3): 139 (1990).
4. J.A. Sokoloski, and A.C. Sartorelli,. Effects of the inhibitors of IMP dehydrogenase tiazofurin and mycophenolic acid on glycoprotein biosynthesis. Mol. Pharmacol. 28: 567 (1986).
5. M. Isobe, H. Yagita, K. Okumura, A. Ihara, Specific Acceptance of Allograft After Treatment with Antibodies to ICAM-1 and LFA-1. Science. 255: 1125 (1992).

110

REACTION OF 2′, 3′-DIDEOXYNUCLEOTIDES TRIPHOSPHATES WITH RECOMBINANT HUMAN NUCLEOSIDE DIPHOSPHATE KINASE

Dominique Deville-Bonne, Benoit Schneider, Julie Bourdais,[*] and Michel Véron

Unité de Régulation Enzymatique des Activités Cellulaires
CNRS, UMR 321, Institut Pasteur
25 rue du Dr Roux
75724 Paris Cedex 15, France

1. INTRODUCTION

Nucleoside diphosphate (NDP) kinase catalyzes the formation of triphospho-nucleotides from diphospho-nucleotides.[1] The enzyme displays little specificity for the nucleobase and accepts both oxy and deoxy ribonucleotides as substrates. In the cell, ATP is considered as the major phosphate donor. Due to its lack of specificity, NDP kinase is also believed to phosphorylate antiviral nucleoside analogs and in particular AZT, ddC or ddI now used in AIDS therapies. Since these drugs are delivered as nucleosides and the viral transcription is inhibited by the tri-phosphoderivative, it is important to better understand their phosphorylation by cellular kinases.[2] In human, two NDP kinases, NDPK-A and NDPK-B, have been isolated[3] which are the product of two different genes respectively named *nm23-H1* and *nm23-H2*.[4,5] The X-ray structure of free human NDPK-B[6] or complexed to GDP[7] is very similar to that of other eukaryotic NDP kinases as *Dictyostelium*[8,9] and *Drosophila*.[10] In all cases, NDP kinases are homo-hexamers with a single binding site per subunit for nucleotide di or triphosphates.

The NDP kinase reaction can be described by a ping-pong bi-bi mechanism according the following two half-reactions:

$$N_1TP + E \Longleftrightarrow N_1DP + E–P \qquad \text{(reaction A)}$$

$$E–P + N_2DP \Longleftrightarrow N_2TP + E \qquad \text{(reaction B)}$$

* Present address: Instituto de Biotecnologia, UNAM, Cuernavaca, 62170 Morelos, Mexico.

Recently the steady state kinetics parameters of the recombinant human NDPK-B were characterized for several antiviral nucleotides analogs.[11] AZT and ddA derivatives are poor substrates either as phosphate donor or acceptor with a k_{cat}/K_M less than 0.5% of that of natural analogs. We recently showed that the fluorescence of the unique tryptophan in the *Dictyostelium* enzyme is quenched upon the phosphorylation of the catalytic histidine.[12]

We show here that human NDPK-B also presents a similar quenching of fluorescence upon phosphorylation of the enzyme. We have used this signal to monitor the time course of human NDPK-B phosphorylation by dideoxynucleotides triphosphates.

2. MATERIALS AND METHODS

Human NDPK-B was overexpressed in *E. coli* (XL1-Blue) as described[7] and purified according to.[11] The activity was measured at 20°C using either a coupled assay[13] or a radioactive assay.[11] Enzyme concentration (expressed in subunit concentration) was determined using an absorbance coefficient of $\Delta A^{280nm} = 1.248$ for a 1 mg/mL solution and a M_W of 17 000.

All fluorescence measurements were performed at 20°C in T buffer (Tris-HCl 50 mM pH=7.5, containing 5 mM $MgCl_2$, 75 mM KCl and 2 mM DTT), on a PTI spectrofluorometer Quantamaster™. The excitation wavelength was 295 nm to minimize the contribution of tyrosyl residues to the total fluorescence as well as light absorption by ATP. The excitation and emission bandwidths were both set to 2 nm. Correction for fluctuations of the excitation light was achieved using the reference detector accessory provided by the manufacturer. The spectra were corrected by substracting background intensities of a blank solution made of buffer with the added nucleotide. When ddGTP is used, the excitation wavelength was 304 nm and the emission bandwidth was enlarged to 4 nm. The reaction of NDP kinase with nucleotides was initiated by hand mixing of the two reactants. Equation (1) was fitted to the time course of the fluorescence change:

$$F(t) = F_0 + (F_1 - F_0) \cdot \exp(-k_{obs} \cdot t) \tag{1}$$

F(t) is the observed signal, $(F_1 - F_0)$ is the maximum amplitude of the variation, F_1 and F_0 being respectively the initial and final fluorescence intensity. The pseudo first-order rate constants (k_{obs}) were determined as a function of substrate concentration (S), with [S] >> [E] and submitted to linear fitting using the software Kaleidagraph.

3. RESULTS AND DISCUSSION

3.1. Steady State Fluorescence of Human NDPK-B

The intrinsic fluorescence of human NDPK-B, at an excitation wavelength of 295 nm, is primarily due to its three Trp residues in position 78, 133 and 149. Trp 133 in the human enzyme, homologous to the single Trp 137 in *Dictyostelium* NDPK, is in close proximity to the catalytic histidine (His 118), with the side chain of His 51, equivalent to His 55 in the *Dictyostelium* enzyme, bridging the two. Trp 78 is buried and remote from the active site by 10–12 Å. Trp 149 from the C-terminal sequence is also far from His 118 (18 Å). It is exposed in an extended conformation pointing towards the neighbouring subunit.[6,7]

Figure 1 shows that the fluorescence of the native enzyme is quenched by 10% in the presence of ATP without any spectrum distorsion. ADP or AMPPCP, a non-hydrolys-

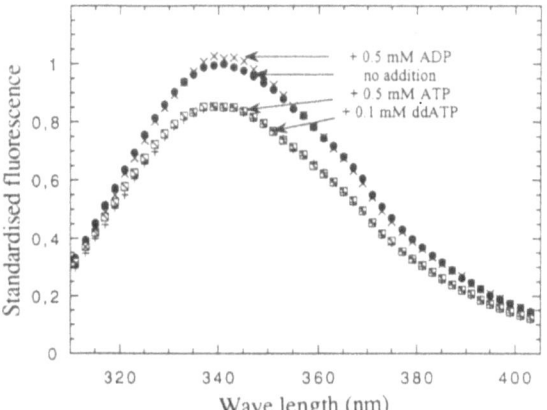

Figure 1. Fluorescence emission spectrum of 2 μM human NDPK-B at 20°C (λ_{exc} = 295 nm, bandwidths = 2 nm) in T buffer and different conditions: no addition (●–●), with 0.5 mM ADP (×–×), with 0.5 mM ATP (+–+) and with 0.1 mM ddATP (□–□).

able ATP analog, have no effect. We conclude that the observed quenching of fluorescence is due to the phosphorylation of the catalytic histidine, similarly to what was observed with the *Dictyostelium* enzyme.[12] Dideoxyadenosine-triphophate (ddATP) as well as GTP and ddGTP also induce a quenching of the fluorescence by 10–15%, while the diphospho derivative (ddADP) does not (not shown). In conclusion, the fluorescence quenching provides an useful signal to monitor the time course of the reaction independent of the nature of the base.

At the molecular level, the environment of Trp 133 of the human enzyme is similar to its counterpart in *Dictyostelium* enzyme with its N hydrogen bounded to His 51, homologous to His 55 in *Dictyostelium* enzyme. It is likely that the fluorescence quenching is mainly due to Trp 133. The smaller amplitude of the quenching as compared to that of the *Dictyostelium* enzyme, likely results from the presence of the Trp 78 and Trp 149. The fluorescence of these residues, remote from the active site, is probably insensitive to phosphorylation of His 118.

3.2. Pre-Steady State Kinetics of the Reaction of NDPK-B with Dideoxynucleotides

When 1.5 μM NDPK-B is reacted with ATP by hand mixing, the change in the fluorescence is instantaneous (not shown), indicating that the reaction is completed within the mixing-time, a result in agreement with the very high turn over of the enzyme $k_{cat} > 500$ s^{-1}. In contrast, when the enzyme is mixed with ddATP, a slow decrease of protein fluorescence is observed in the minutes time-scale (Fig. 2). Curves corresponding to ddATP varying from 20–150 μM can be fitted by monoexponential indicating that the reaction is pseudo first-order with a time constant depending of the concentration of the nucleotide.

As shown in Fig. 3, this rate constant varies linearly with the concentration of ddATP. Similar results were obtained with ddGTP with slightly higher values for the rate constants.

These results can be described by a two steps mechanism as proposed in the following minimal model of a fast reaction followed by a slow isomerisation.

$$E + ddXTP \underset{k_{-1}}{\overset{k_1}{\rightleftharpoons}} E{:}ddXTP \underset{k_{-2}}{\overset{k_2}{\rightleftharpoons}} E{-}P + ddXDP$$

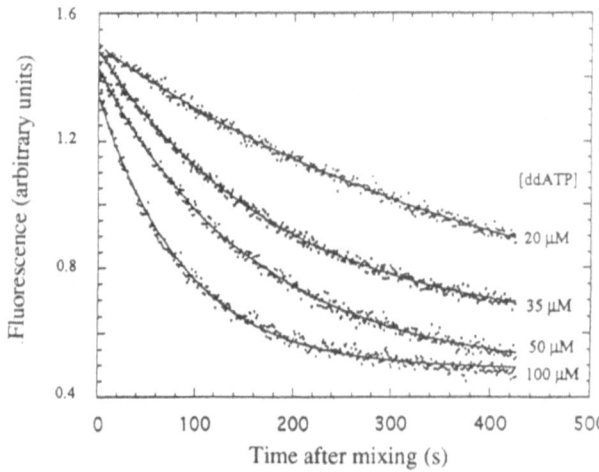

Figure 2. Presteady state kinetics of reaction of NDP kinase with ddATP. The enzyme (2 μM in subunits) in T buffer at 20°C was hand mixed with ddATP at various concentrations: 20, 35, 50 and 100 μM. The lines represent the graph of monoexponential functions with time constants of respectively 0.00229, 0.00534, 0.00634 and 0.0114 s^{-1}.

The first step leads to the non covalent enzyme-substrat complex and the second step corresponds to the phosphorylation of the enzyme.

The association and dissociation rate constants of a nucleotide with a macromolecule are usually diffusion limited with values as high as 10^6 to 10^8 $M^{-1}.s^{-1}$. ddGTP and ddATP interact with the enzyme slowly, indicating that our experiments may correspond to the phosphorylation step. This hypothesis should be confirmed by stopped flow experiments allowing to measure faster kinetics.

Within the scheme of proposed model, the slope of the curves in Fig. 3 represent a measure of the ratio k_2/K_D where k_2 is the rate of phosphate transfer and K_D the true dissociation constant of the nucleotide for the enzyme.[14,15] The ratio k_2/K_D is similar to the ratio k_{cat}/K_M determined by enzymatic assays. The difference in rate constants oberved for ddGTP and ddATP (Fig. 3) is relatively small but we find it significant. Indeed, ddGTP is three times more efficient than ddATP, indicating that NDP kinase is able to discriminate among dideoxynucleotides triphosphates, in contrast to natural ones.

The present results show that fluorescence quenching is a valuable tool in measuring the kinetics of interaction of NDP kinase with its substrates for slow reactions. They suggest further investigation of the antiviral drugs to assess their ability to be phosphorylated efficiently within the cell.

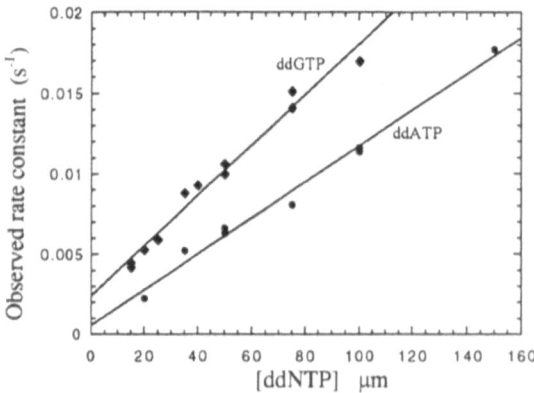

Figure 3. Concentration dependence of the pseudo first-order rate constant k_{obs} as a function of different dideoxynucleotides triphosphate: ddGTP (♦) and ddATP (●) (same experiment as in Figure 2). The lines represent the linear fit obtained using Kaleidagraph.

ACKNOWLEDGMENTS

We thank G. Lebras for his help in providing some purified NDP kinase and J. Janin for stimulating discussions.

REFERENCES

1. Parks, R.E.J. & Agarwal, R.P., Nucleoside diphosphokinases, *The Enzymes* 8, 307–334 (1973).
2. Balzarini, J., Herdewijn, P. & De Clercq, E., Differential patterns of intracellular metabolism of 2',3'-dide-hydro-2', 3'-dideoxythymidine (D4T) and 3'-azido-2',3'-dideoxythymidine (AZT), two potent anti-HIV compounds, *J. Biol. Chem.* 264, 6127–6133 (1989).
3. Gilles, A.M., Presecan, E., Vonica, A. & Lascu, I., Nucleoside diphosphate kinase from human erythrocytes. Structural characterization of the two polipeptide chains responsible for heterogeneity of the hexameric enzyme, *J. Biol. Chem.* 266, 8784–8789 (1991).
4. Steeg, P.S., Bevilacqua, G., Kopper, L., Thorgeirsson, U. P., Talmadge, J. E., Liotta, L. A. and Sobel, M. E., Evidence for a novel gene associated with low tumor metastasic potential, *J. Natn. Cancer Inst.* 80, 200–204 (1988).
5. Stahl, J.A., Leone, A., Rosengard, A. M., Porter, L., King, C. R. and Steeg, P. S., Identification of a second human nm23 gene, nm23-H2, *Cancer Res.* 51, 445–449 (1991).
6. Webb, P.A., Perisic, O., Mendola, C.E., Backer, J.M. & Williams, R.L., The crystal Structure of a Human Nucleoside Diphosphate Kinase, NM23-H2, *J. Mol. Biol.* 251, 574–587 (1995).
7. Moréra, S., Lacombe, M.-L., Xu, Y., LeBras, G. & Janin, J., X-Ray structure of nm23 Human Nucleoside Diphophate Kinase B complexed with GDP at 2 Å resolution, *Structure* 3, 1307–1314 (1995).
8. Dumas, C., Lascu, I., Moréra, S., Glaser, P., Fourme, R., Wallet, V., Lacombe, M.-L., Véron, M. and Janin, J., X-ray structure of Nucleoside diphosphate kinase, *EMBO J.* 11, 3203–3208 (1992).
9. Cherfils, J., Moréra, S., Lascu, I., Véron, M. & Janin, J., X-ray structure of nucleoside diphosphate kinase complexed with dTDP and Mg^{2+} at 2Å resolution, *Biochemistry* 33, 9062–9069 (1994).
10. Chiadmi, M., Moréra, S., Lascu, I., Dumas, C., LeBras, G., Véron, M. and Janin, J., The Awd Nucleoside Diphosphate Kinase from *Drosophila*, structure, *Structure* 1, 283–293 (1993).
11. Bourdais, J., Biondi, R., Lascu, I., Sarfati, S., Guerreiro, C., Lascu, I., Janin, J. and Véron, M., Cellular phosphorylation of anti-HIV nucleosides: role of nucleoside diphosphate kinase, *J. Biol Chem.* 271, 7887–7890 (1996).
12. Deville-Bonne, D., Sellam, O., Mérola, F., Lascu, I., Desmadril, M. and Véron, M., Phosphorylation of Nucleoside Diphosphate Kinase at the active site studied by steady-state and time-resolved fluorescence, *Biochemistry* 35, 14643–14650 (1996).
13. Lascu, I., Chaffotte, A., Limbourg-Bouchon, B. & Véron, M., A Pro/Ser substitution in nucleoside diphosphate kinase of of *Drosophila melanogaster* (mutation Killer of prune) affects stability but not catalytic efficiency of the enzyme, *J. Biol. Chem.* 267, 12775–12781 (1992).
14. Bernasconi, C.F., Relaxation times in two-step systems, *in Relaxation Kinetics, Academic Press, New York* , 20–39 (1976).
15. Johnson, K., Transient-state kinetic analysis of enzyme reaction pathways, *In The Enzymes, Ed. Sigman DS, Academic Press, New-York* 20, 2–62 (1992).

111

ANTITUMOR ACTIVITY OF YUNNANMYCIN

A Novel Pyrimidine Nucleoside Antibiotic

Yong-su Zhen,[*] Chang-qing Qi, Ji-lan Hu, Yu-chuan Xue, and Wen-jun Chen

Institute of Medicinal Biotechnology
Chinese Academy of Medical Sciences and Peking Union Medical College
Beijing 100050, China

1. INTRODUCTION

Nucleoside salvage in nucleotide biosynthesis is a subject of great concern in cancer chemotherapy. Antimetabolites that currently used for cancer treatment inhibit *de nono* pathways of nucleotide synthesis fail to block nucleoside salvage pathways. Nucleoside transport is the first step in the process of nucleoside salvage. It has been demonstrated that dipyridamole, an inhibitor of nucleoside transport, potentiated the effects of antimetabolite.[1–4] A number of compounds of natural origin including green tea polyphenols, salvianolic acid A, antibiotics C3368-A and C3368-B were found to be active nucleoside transport inhibitors and acting as biochemical modulators.[5–7] In recent years nucleoside transporter has been set as the molecular target for new drug research in our laboratory. Because of the high diversity of climate conditions and vegetations in Yunnan Province, China, one of our projects was to investigate novel nucleoside transport inhibitors or nucleoside analogs produced by actinomycetes which were isolated from this area. Yunnanmycin was detected after screening of 5,000 fermentation liquor samples from actinomycetes.

2. MATERIALS AND METHODS

2.1. Properties

Yunnanmycin (YNM) was obtained from the fermentation liquor of a streptomyces strain isolated from a soil sample which was collected in Guan-Ping Nature Conservatory Zone of Yunnan Province, China. The structure of YNM has been elucidated. As shown in

* Correspondence: Yong-su Zhen, phone: 86 10 6301 0985; fax: 86 10 6301 7302; e-mail: ys.zhen@bj.col.co.cn.

Purine and Pyrimidine Metabolism in Man IX,
edited by Griesmacher *et al.* Plenum Press, New York, 1998.

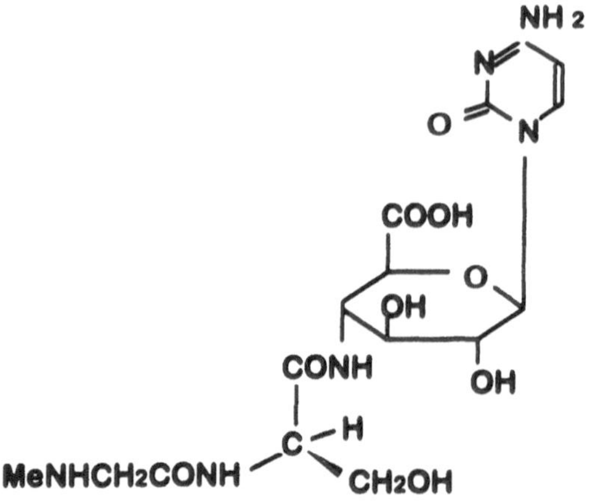

Figure 1. Chemical structure of yunnanmycin.

Figure 1, YNM is a novel antibiotic and the molecule consists of a pyrimidine nucleoside analog and a sarcosylserine dipeptide side chain.

2.2. Precursor Incorporation Assay

Cultured KB carcinoma cells in logarithmically growing stage were harvested and seeded into 24-well plates at 10^5 cells/well. After culture for 24 hr, the medium was removed and serum-free Hank's solution was added. ^{3}H-labeled thymidine, uridine or leucine was added at 1 μCi/well and incubated for 60 min at 37°C. The followed up steps of the procedure were performed as previously described.[8]

2.3. Clonogenic Assay

Logarithmically growing cells were seeded in 24-well plates at 100 cells/well. After 24 hr, various concentrations of YNM were added to the well. Triplicate wells were used for each concentration. After further 6-day incubation, colonies were counted under an inverted microscope. The percentage of colony inhibition was calculated accordingly.

2.4. Inhibition of Tumor Growth in Mice

Kunming mice weighing 18–20 g were obtained from the Institute of Laboratory Animal Science, Chinese Academy of Medical Sciences. Hepatoma H22 or sarcoma 180 cells were inoculated subcutaneously in the axillar region at 10^6 cells per mouse. YNM was administered i.v. or p.o. started 24 hr after tumor inoculation. Mice were sacrificed on the 11th day and therapeutic effects were evaluated by tumor weight.

3. RESULTS

3.1. Effects of YNM on Macromolecular Biosyntheses

As shown in the radio-labeled precursor incorporation assay (Figure 2), YNM moderately inhibited DNA and protein biosyntheses in cancer cell.

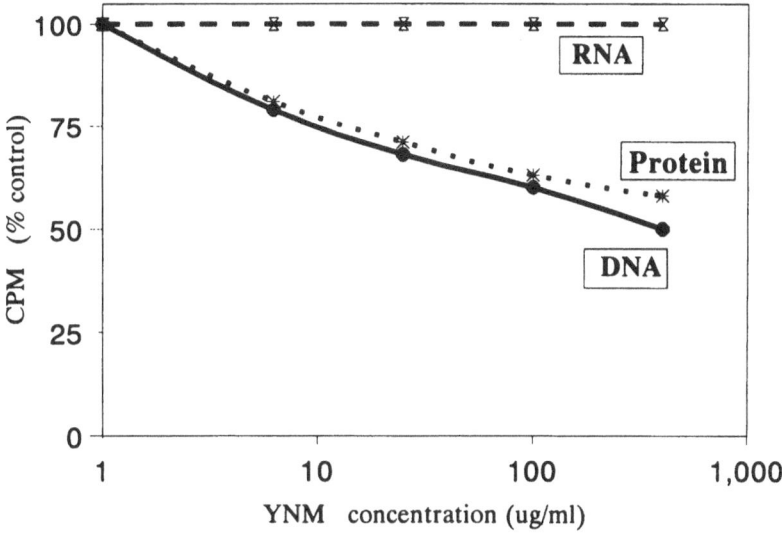

Figure 2. Effect of yunnanmycin on macromolecular biosyntheses in cancer cells. Radio-labeled precursor incubation time, 1 hr.

At a concentration of 400 µg/ml, YNM inhibited [³H]thymidine and [³H]leucine incorporation by 50% approximately; however, YNM showed no suppression on [³H]uridine incorporation at tested concentrations.

3.2. Cytotoxicity of YNM

By clonogenic assay, YNM showed moderate cytotoxicity to cultured human cancer cell lines including epidermoid carcinoma KB cells, hepatoma BEL-7402 cells and lung carcinoma A549 cells. The IC50 values ranged from 3 to 12 µg/ml.

3.3. Antitumor Effects on Hepatoma H22 in Mice

As shown in Figure 3, YNM exerted marked suppression on the growth of mouse hepatoma H22 at tolerable doses that caused no death of treated animals and no body weight loss of over 10%. YNM administered p.o. or i.v. was effective; the tumor inhibition rate reached 83% ($P < 0.01$) or 77% ($P < 0.01$), respectively.

3.4. Antitumor Effects of YNM on Sarcoma 180 in Mice

As shown in Figure 4, YNM exerted remarkable suppression on the growth of mouse sarcoma 180 at tolerable doses. As administered by p.o. route or i.v. route, the tumor inhibition rate by YNM reached 72% ($P < 0.01$) or 66% ($P < 0.01$), respectively.

4. DISCUSSION

The present study has demonstrated that YNM is highly effective against solid tumors in mice. According to the chemical structure, YNM is closely related to gougerotin. In YNM molecule, there exists a carboxyl group in the sugar moiety; however, in gougerotin there is

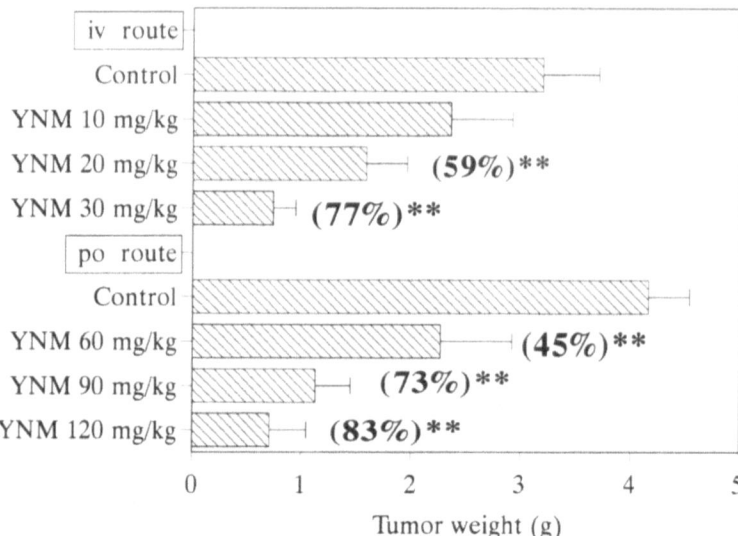

Figure 3. Effect of yunnanmycin on the growth of hepatoma H22 in mice. (n = 10). Tumor, sc; YNM, iv or po, × 2; ** $P < 0.01$.

a formamyl group instead. As described in the original report,[9] gougerotin was an antibiotic with weak antibacterial activity and no antitumor activity was documented. Gougerotin was highly active of inhibiting protein synthesis in cell-free system.[10] The inhibitory activity of gougerotin on protein synthesis in intact cell system was rather weak, this may be related to it poor permeation into the cell. Virus-infected cells in which the permeation increased were more sensitive to gougerotin.[11,12] It was reported that asteromycin, identical with gougerotin, showed cytotoxicity to HeLa cells and was active against sarcoma 180 ascites tumor when

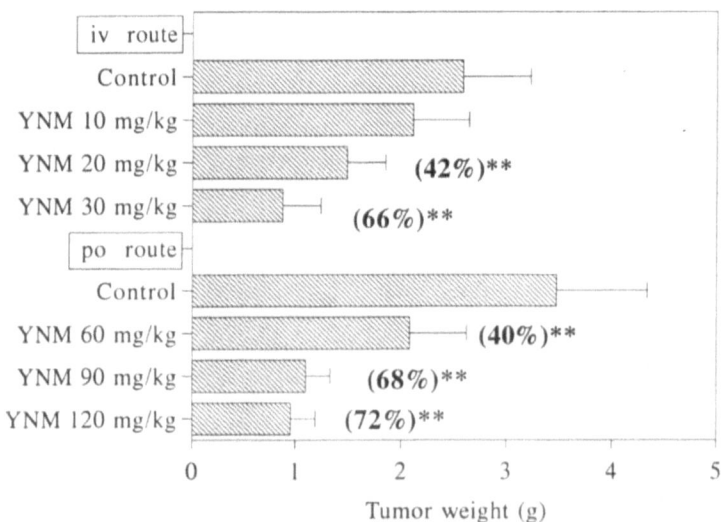

Figure 4. Effect of yunnanmycin on the growth of sarcoma 180 in mice. (n = 10). Tumor, sc; YNM, iv or po, × 2; ** $P < 0.01$.

administered intraperitoneally.[13] No efficacy of gougerotin against solid tumors was mentioned in the early reports. Recently, our study demonstrated that yungumycin, verified to be identical with gougerotin, while it was toxic by i.v. or i.p. route administration, showed remarkable effectiveness against mouse solid tumors including colon carcinoma 26 and sarcoma 180 by p.o. administration.[14] In comparative studies, YNM was less toxic than gougerotin in acute toxicity examination. YNM showed high antitumor efficacy both by p.o. route and by i.v. route. However, gougerotin (yungumycin) showed highly effective only by the p.o. route. Because of its toxicity, i.p. or i.v. administration of gougerotin was less effective. YNM was compared with 5-FU at equitoxic doses, one third of the LD^{50} respectively; the inhibition rates of hepatoma H22 and sarcoma 180 by YNM were 83% and 81%, and by 5-Fu were 74% and 66%, respectively. Less suppression on bone marrow hemopoiesis was found in YNM-treated mice than that in 5-FU-treated mice.

The present studies indicate that yunnanmycin, a novel pyrimidine nucleoside antibiotic, is of high efficacy against experimental tumors, particularly by p.o. administration, and is potentially useful in cancer chemotherapy.

ACKNOWLEDGMENTS

This work was supported by grants (No. 9390012-03 and No. 39522004) from National Natural Sciences Foundation of China.

REFERENCES

1. Y.S. Zhen, M.S. Lui and G. Weber, Effects of acivicin and dipyridamole on hepatoma 3924A cells, *Cancer Res.* 43:1616 (1983).
2. J.A. Nelson and S. Drake, Potentiation of methotrexate toxicity by dipyridamole, *Cancer Res.* 44:2493 (1984).
3. J.L. Grem and P.H. Fischer, Augmentation of 5-fluorouracil cytotoxicity in human colon cancer cells by dipyridamole, *Cancer Res.* 45:2967 (1985).
4. T,C,K, Chan and S.B. Howell, Mechanism of synergy between N-phosphonacetyl-L-aspartate and dipyridamole in a human ovarian carcinoma cell line, *Cancer Res.* 45:3598 (1985).
5. Y.S. Zhen, S.S. Cao, Y.C.Xue and S.T. Wu, Green tea extract inhibits nucleoside transport and potentiates the antitumor effect of antimetabolite, *Chin. Med. Sci. J.* 6:1 (1991).
6. Y.S. Zhen, S. Jian, Y.C. Xue, C.Q. Qi and J.L. Hu, Novel nucleoside transport inhibitors of natural origin, *Adv. Exp. Med. Biol.* 370:779 (1994).
7. Y.S. Zhen, S. Jian and Y.C. Xue, Salvianolic acid A inhibits nucleoside transport and potentiates the activity of antitumor drugs. *Proc. Am. Assoc. Cancer Res.* 37:287 (1996).
8. Y.S. Zhen, T. Taniki and G. Weber, Azidothymidine and dipyridamole as biochemical response modifiers: Synergism with methotrexate and 5-fluorouracil in human colon and pancreatic carcinoma cells, *Oncology Res.* 4:73 (1992).
9. T. Kanzaki, E. Higashide, H. Yamamoto, M. Shibata, K. Nakazawa, H. Iwasaki, T. Takewaka and A. Miyake, Gougerotin, a new antibacterial antibiotic, *J. Antibiotics, Ser. A,* 15:93 (1962).
10. M. Yukioka, Gougerotin, In: J.W. Corcoran (ed), *Antibiotics*, New York, Spring-Verlag, Vol. 3:448 (1975).
11. A. Contreas, D. Vazquez and L. Carrasca, Inhibition by selected antibiotics of protein synthesis in cells growing in tissue culture, *J. Antibiotics,* 31:598 (1978).
12. J.C. Lacal, d. Vazquez, J.M. Fernandez-Sousa and L. Carrasca, Antibiotics that specifically block translation in virus-infected cells, *J. Antibiotics,* 33:441 (1980).
13. T. Ikeuchi, F. Kitame, M. Kikuchi and N. Ishida, An antimycoplasma antibiotic asteromycin: its identity with gougerotin, *J. Antibiotics,* 25:548 (1972).
14. Y.C. Xue, S.Y. Wu, Y. Li, S.H. Zhang and Y.S. Zhen, Antitumor activity of yungumycin, *Acta Pharm. Sin.* 31:171 (1996).

112

MECHANISMS OF SYNERGISM BETWEEN GEMCITABINE AND CISPLATIN

C. J. A. van Moorsel, G. Veerman, J. B. Vermorken, D. A. Voorn, J. R. Kroep, A. M. Bergman, H. M. Pinedo, and G. J. Peters

Department of Medical Oncology
University Hospital Vrije Universiteit
P. O. Box 7057
1007 MB Amsterdam, The Netherlands

1. INTRODUCTION

2',2'-difluorodeoxycytidine (Gemcitabine, dFdC) is an antineoplastic agent with clinical activity against several cancer types.[1] cis-Diamminedichloroplatinum (cisplatin, CDDP) is a drug with long established anticancer activity, which acts by Platinum (Pt)-DNA adduct formation.[2,3] Because of the low toxicity profile of dFdC and the differences in mechanism of cytotoxicity, preclinical studies were performed that demonstrated synergism between dFdC and CDDP in several cancer cell lines and in vivo,[4–8] which is likely to be related to increased formation of Pt-DNA adducts.[8] Pre-treatment with dFdC gave the best results both in vitro and in vivo.[6,7,8] Several potential mechanisms underlying the synergism were studied in vitro, based on these results several schedules were studied in patients.

2. MATERIALS AND METHODS

2.1. Chemicals

dFdC and [5-³H]-dFdC (16.7 Ci/mmol) were a kind gift of Eli Lilly Inc. (Indianapolis, IN, USA). CDDP was obtained from Bristol-Myers Squibb (Woerden, the Netherlands). All other chemicals were of analytical grade and commercially available.

2.2. Cell Culture

The human ovarian cancer cell lines A2780, AG6000 and ADDP and the non-small cell lung cancer cell lines Lewis Lung (LL) and H322 were maintained in exponential growth in Dulbecco's Modification of Eagle's Medium (Gibco Laboratories, Grand Island

Purine and Pyrimidine Metabolism in Man IX,
edited by Griesmacher *et al.* Plenum Press, New York, 1998.

NY, USA) or in RPMI medium supplemented with 5 or 10% heat-inactivated fetal calf serum (Gibco) at 37°C and 5% CO_2.

2.3. Patient Treatment

In a Phase I study patients were treated with either dFdC 4 and 24 hr before CDDP or CDDP 4 and 24 hr before dFdC (dFdC 30 min infusion 800 mg/m^2, CDDP 1 hr infusion 50 mg/m^2). Blood samples were taken at different time points up to 48 hr to determine dFdCTP accumulation in WBC which were isolated on a Ficoll-Paque gradient (Pharmacia, Sweden).

2.4. Accumulation of dFdCTP

Determination of dFdCTP accumulation and the effect of CDDP was performed as described previously.[5]

2.5. [5–³H]-dFdC incorporation into DNA and RNA

Incorporation of ³H-dFdC into DNA and RNA of cells treated with 0.1 μM dFdC with/without 20 μM CDDP for 24 hr was performed essentially as described previously.[9] Incorporation into RNA was determined by adding DNAse I and incorporation into DNA was determined by adding RNAse A/T1.

3. RESULTS

3.1. dFdCTP Accumulation

3.1.1. dFdCTP Accumulation in Cell Lines. In order to determine a possible role of dFdCTP in the interaction between dFdC and CDDP we measured the accumulation of dFdCTP after 24 hr exposure to 1 μM dFdC alone or in combination with 200 μM CDDP (Fig. 1). Co-incubation with CDDP resulted in a decrease of dFdCTP compared to dFdC alone in all cell lines.

3.1.2. dFdCTP Accumulation in WBC. To determine a possible schedule dependency of dFdC and CDDP treatment, patients received both compounds with two different intervals and dFdCTP accumulation was determined in WBC (Fig. 2). No significant difference between both schedules in dFdCTP accumulation was found with the 4 hr interval, however, dFdCTP accumulation was higher in patients treated with CDDP 24 hr before dFdC.

3.2. dFdC Incorporation into DNA and RNA in Cell Lines

To determine the possible effect of CDDP on dFdC incorporation into DNA and RNA, incorporation of [5–³H]-dFdC into DNA and RNA was studied. No significant differences in dFdC incorporation into DNA were found in all cell lines (Fig. 3). However, the incorporation of dFdC into RNA was higher after co-incubation with CDDP than after incubation with dFdC alone in the ovarian cancer cell lines A2780 and its CDDP and dFdC resistant variant ADDP, (2- and 3-fold, respectively).

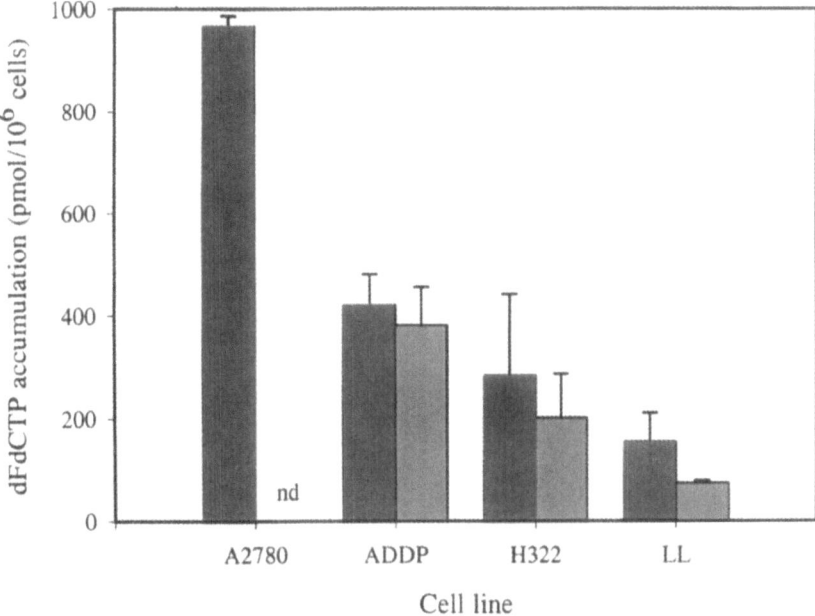

Figure 1. Effect of CDDP on the accumulation of dFdCTP. A2780, ADDP, H322 and LL cells were exposed for 24 hr to either 1 μM dFdC alone (closed bars) or in combination with 200 μM CDDP (hatched bars). Values are means ± SEM of three to five experiments. nd = not detectable; in A2780 because of cell death after combination of both compounds at these concentrations.

4. DISCUSSION

The present studies were performed to elucidate the mechanism of the synergistic interaction between dFdC and CDDP in vitro and to determine how these data translate into patients.

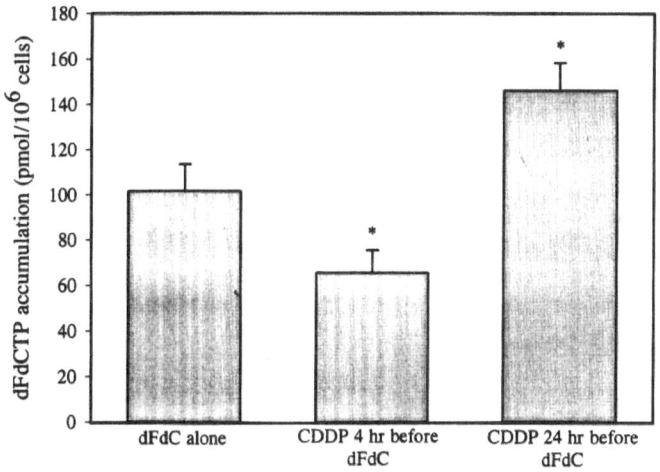

Figure 2. dFdCTP accumulation in WBC of patients treated with dFdC and CDDP 2 hr after start of dFdC infusion. Values are means ± SEM of at least 10 patients. *p-value < 0.05.

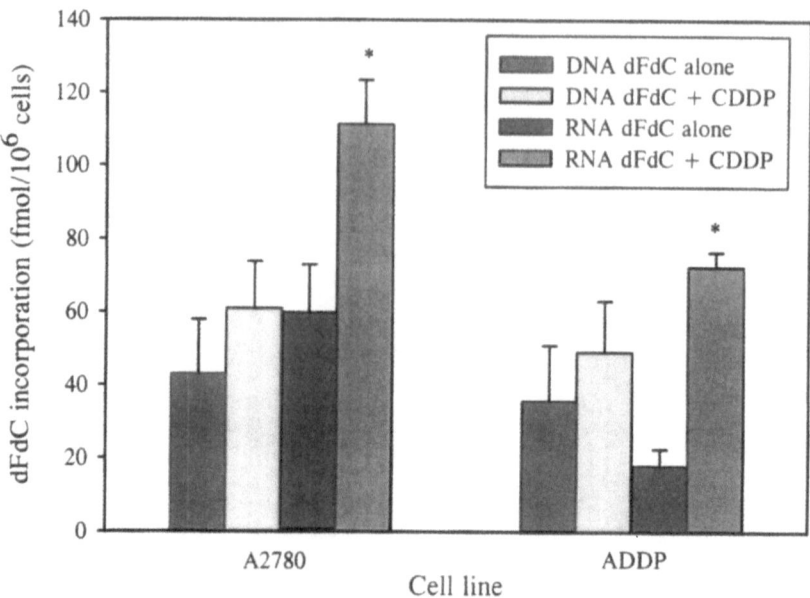

Figure 3. Effect of CDDP on the incorporation of ^{3}H-dFdC into DNA and RNA in A2780 and ADDP cells. Cells were exposed to either 0.1 μM dFdC alone or in combination with 20 μM CDDP for 24 hr. Values are means ± SEM of three to five experiments. *p-value < 0.05.

Two possible mechanisms for the synergism related to the mechanism of action of dFdC were investigated in vitro. CDDP decreased the dFdCTP accumulation in all cell lines, which is in agreement with a previous study in C26-10 and 22B cells.[5] This phenomenon might be the result of a rise in CTP and UTP pools, caused by both CDDP and dFdC. Both CTP and UTP can inhibit the activity of dCK,[10] therefore a rise in these nucleotides might decrease the accumulation of dFdCTP. In the human ovarian cancer cell line A2780 and its CDDP and dFdC resistant variant ADDP, CDDP is favoring the incorporation of dFdC into RNA, possibly due to the inhibition of ribonucleotide reductase (RR) by both CDDP and dFdC,[11] which leads to a disturbance in the ribonucleotide pools of the cells. Another factor possibly influencing the dFdC incorporation, is the inhibition of RNA synthesis. The addition of CDDP decreases RNA synthesis 2-fold compared to dFdC alone in both cell lines (results not shown).

There seems to be a pharmacokinetic interaction between CDDP and dFdC resulting in an increased dFdCTP accumulation in WBC which is more pronounced when CDDP preceeded dFdC by 24 hr than by 4 hr. CTP, UTP and GTP pools were higher in patients treated with CDDP 24 hr before dFdC compared to dFdC 24 hr before CDDP. The reverse effect was found for the 4 hr interval treatments (results not shown) indicating a possible relation with the difference in dFdCTP accumulation found. However, the underlying mechanism is still unclear.

In conclusion, the mechanism underlying the synergistic interaction between dFdC and CDDP appears not to be related to dFdCTP accumulation in vitro. dFdC incorporation into RNA, however, is higher in two out of four cell lines after exposure to the combination of dFdC and CDDP, compared to dFdC alone. In patients a schedule dependent dFdCTP accumulation was observed, indicating that treatment schedule and duration of the interval is important in patients similar to what was found in vitro and in vivo.

ACKNOWLEDGMENTS

This research was supported by a grant from the Dutch Cancer Society (VU-94-753) and Eli Lilly Inc.

REFERENCES

1. Van Moorsel CJA, Peters GJ and Pinedo HM: Gemcitabine: future prospects of single agent and combination studies. The Oncologist 2:127–134, 1997.
2. Terheggen PMAB, Emondt JY, Floot BGJ, Dijkman R, Schrier PI, Den Engelse L and Begg AC: Correlation between cell killing by cis-diamminedichloroplatinum(II) in six mammalian cell lines and binding of a cis-diamminedichloroplatinum(II)-DNA antiserum. Cancer Res 50:3556–3561, 1990.
3. Parker RJ, Gill I, Tarone R, Vionnet J, Grunberg S, Muggia F and Reed E: Platinum DNA-damage in leukocyte DNA of patients receiving cisplatin and carboplatin chemotherapy, measured by atomic absorption spectrometry. Carcinogenesis 12:1253–1258, 1991.
4. Peters GJ, Bergman AM, Ruiz van Haperen VWT, Veerman G, Kuiper CM and Braakhuis BJM: Interaction between gemcitabine and cisplatin in vitro and in vivo. Semin Oncol 22(suppl 11):72–79, 1995.
5. Bergman AM, Ruiz van Haperen VWT, Veerman G, Kuiper CM and Peters GJ: Synergistic interaction between cisplatin and gemcitabine in vitro. Clin Cancer Res 2:521–530, 1996.
6. Van Moorsel CJA, Veerman G, Kuiper CM, Bergman AM, Van der Vijgh WJF and Peters GJ: Synergism between gemcitabine and cisplatin in ovarian and non-small cell lung cancer cell lines. Proc 9th NCI-EORTC Ann Oncol 7(suppl 1):65 [abstract 219], 1997.
7. Braakhuis BJM, Ruiz van Haperen VWT, Welters MJP and Peters GJ: Schedule-dependent therapeutic efficacy of the combination of gemcitabine and cisplatin in head and neck cancer xenografts. Eur J Cancer 31A:2335–2340, 1995.
8. Van Moorsel CJA, Veerman G, Bergman AM, Guechev A, Vermorken JB, Postmus PE and Peters GJ: Combination chemotherapy studies with gemcitabine. Semin Oncol, in press.
9. Langeveld CH, Jongenelen CAM., Theeuwes JWM, Baak JPA, Heimans JJ, Stoof JC and Peters GJ: The antiproliferative effect of 8-chloro-adenosine, and active metabolite of 8-chloro-cyclic adenosine monophosphate, and disturbances in nucleic acid synthesis and cell cycle kinetics. Biochem Pharmacol 53:141–148, 1997.
10. Ruiz van Haperen VWT, Veerman G, Vermorken JB, Pinedo HM and Peters GJ:. Regulation of deoxycytidine kinase from solid tumor cell lines by CTP and UTP. Biochem Pharmacol 51:911–918, 1996.
11. Chiu CSM, Chan, AK, and Wright, JA: Inhibition of mammalian ribonucleotide reductase by cis-diamminedichloroplatinum(II). Biochem. Cell Biol 70:1332–1338, 1992.

113

CROSS-RESISTANCE OF THE GEMCITABINE RESISTANT HUMAN OVARIAN CANCER CELL LINE AG6000 TO STANDARD AND INVESTIGATIONAL DRUGS

Andries M. Bergman, Herbert M. Pinedo, Veronique W. T. Ruiz van Haperen, Gijsbert Veerman, Catharina M. Kuiper, and Godefridus J. Peters

Department Medical Oncology
Free University Hospital
P. O. Box 7057
1007 MB Amsterdam, The Netherlands

1. INTRODUCTION

2′,2′-Difluorodeoxycytidine (gemcitabine, dFdC) is a deoxycytidine (dCyd) analog with proven activity in solid tumors, including ovarian cancer, both *in vitro* and *in vivo* (1–3). In the cell dFdC is phosphorylated by deoxycytidine kinase (dCK) to its monophosphate and subsequently to its triphosphate dFdCTP, which can be incorporated into DNA and RNA (4,5). In the DNA, exonuclease activity is unable to excise dFdCMP (5). Resistance to cytostatic agents commonly occurs during cancer treatment and is often associated with cross-resistance to other related and unrelated drugs. AG6000 is a variant of the human ovarian cancer cell line A2780, which was made resistant to dFdC. This resistance was associated with a total absence of dCK activity (8). In the initial characterization of this cell line we not only observed cross-resistance to related compounds such as other deoxynucleoside analogs, but also to unrelated compounds used in the treatment of ovarian cancer (8). Therefore we extended this study to a panel of other drugs used in and of potential interest for treatment of ovarian cancer. Several parameters known to be important in the metabolism of the different drugs were determined in A2780 and AG6000.

2. METHODS

Growth inhibition tests: cells were plated in 96-wells plates and exposed for 72 hr to the drugs. Cell growth was evaluated by the sulforhodamine B (SRB) assay and growth inhibition was expressed as the 50% growth inhibiting concentration (IC50).

Purine and Pyrimidine Metabolism in Man IX,
edited by Griesmacher *et al.* Plenum Press, New York, 1998.

　　　Enzyme assays: were performed as described previously; the dCK activity assay was performed with 230 μM ^{3}H-deoxycytidine (10), the dGK activity assay was performed with 330 μM ^{3}H-deoxyguanosine (10,11).

　　　DNA damage: was evaluated at iso-toxic drug concentrations as percentage double-strand DNA (ds-DNA) after 24 hr using a Fluorometic Analysis of DNA Unwinding (FADU) assay (12).

3. RESULTS

　　　Growth inhibition tests: AG6000 was not only resistant to dFdC, but as expected also to other compounds which are substrates for dCK (Fig. 1.) Cross-resistance was found in AG6000 for the TS inhibitors 5FU and ZD1694, but not for the anti-folate methotrexate. Trimetrexate, another antifolate, however, showed some cross-resistance, as well as the topo-isomerase II inhibitor Etoposide (VP-16) in order to explain this wide range of resistance we determined several enzymes related to activity of several of these compounds; dCK activity was very low in AG6000 cells, but thymidylate synthase (TS) activity, which is the target enzyme of 5-fluorouracil (5FU) and ZD1694, is 2 fold increased compared to A2780 (Tab. 1) (8,13). No difference in deoxyguanosine kinase (dGK), which activates the deoxyguanosine analog 2′,2′-Difluorodeoxyguanosine (dFdG), activity was found between A2780 and AG6000.

　　　Since dFdC and VP-16 are known to introduce DNA damage, we determined whether differences in DNA damage would be related to the cross-resistance. DNA damage was concentration dependent. An decreased percentage ds-DNA was found in A2780 compared to AG6000 at either the IC50 for A2780 or for AG6000. At 1,000 nM dFdC 56% ds-DNA was found in A2780 and 97% ds-DNA in AG6000. VP-16 introduced the highest amounts of DNA damage at its IC50, which was 35% ds-DNA in A2780 and 55% ds-DNA in AG6000.

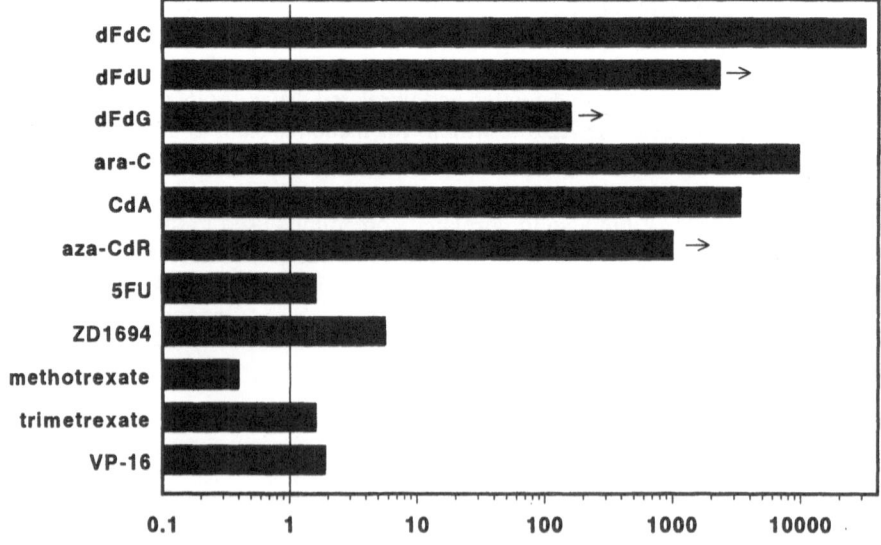

Figure 1. Resistance factors (=IC50$_{resistant}$/IC50$_{wild\ type}$) of the dFdC resistant human ovarian cancer cell line AG6000 to a panel of drugs used in, and of potential interest for the treatment of ovarian cancer.

Table 1. Enzyme activities of the human
ovarian cell lines A2780 and its dFdC
resistant variant AG6000

Enzyme	A2780	AG6000
dCK [*1]	1.42 ± 0.63	0.16 ± 0.03
TS [*2]	0.67 ± 0.21	1.17 ± 0.26
dGK	1.39 ± 0.17	1.17 ± 0.27

Activity values are means ± SEM of at least 3 separate experiments in nmol/hr/10^6 cells.
[*1] V.W.T. Ruiz van Haperen *et al.* (8)
[*2] V.W.T. Ruiz van Haperen *et al.* (16)

4. DISCUSSION

In this paper we demonstrate a cross-resistance in AG6000 to a wide variety of anti-cancer agents. Exept for the dihydrofolate reductase inhibitor methotrexate. However, for the lipophilic anti-folate trimetrexate cross-resistance was observed, as well as for the folate-based TS inhibitor ZD1694. Altough TS activity is increased 2 fold, this does not appear sufficient to explain the marked cross-resistance. Some as yet unclarified distur-bance in folate metabolism may be responsible for this. The absence of dCK activity clari-fies the resistance to dFdC and probably the cross-resistance to ara-C and dFdG.

The large difference in DNA damage between A2780 and AG6000, correlates with the absence of dCK activity and the 50,000 fold resistance to dFdC in AG6000. The cross-resistance to VP-16 might be clarified by differences in DNA damage between A2780 and AG6000, which was possibly due to differences in the activity of DNA excision/repair mechanisms.

In conclusion: AG6000 is a cell line highly resistant to dFdC and cross-resistant to a wide variety of cytostatic agents, which could partially be clarified by differences in enzyme activities and amount of DNA damage between A2780 and AG6000.

ACKNOWLEDGMENTS

This research was supported by grants from the Dutch Cancer Society.

REFERENCES

1. L.W. Hertel, G.B. Boder,, J.S. Kroin, S.M. Rinzeel, G.A. Poore, G.C. Todd, G.B. Grindey, Evaluation of the antitumor activity of a (2′,2′-difluoro-2′-deoxycytidine), *Cancer Res.*, 50: 4417–4422 (1990).
2. E. Boven, H. Schipper, C.A.M. Erkelens, S.A. Hatty, H.M. Pinedo, The influence of the schedule and the dose of gemcitabine on the anti-tumour efficiacy in experimental human cancer, *Br. J. Cancer*, 68: 52–56 (1993).
3. B. Lund, O.P. Hansen, K. Theilade, M. Hansen, J.P. Neijt, Phase II study of gemcitabine (2′,2′-difluorode-oxycytidine) in previously treated ovarian cancer patients, *J. Natl. Cancer Inst.*, 86: 1530–1533 (1994).
4. V.W.T. Ruiz van Haperen, G. Veerman, J.B. Vermorken, G.J. Peters, 2′,2′-Difluoro-deoxycytidine (gemci-tabine) incorporation into RNA and DNA from tumour cell lines, *Biochem. Pharmacol.*, 46: 762–766 (1993).
5. P. Huang, S. Chubb, L.W. Hertel, G.B. Grindey, W. Plunkett, Action of 2′,2′-difluorodeoxycytidine on DNA synthesis, *Cancer Res.*, 51: 6110–6117 (1991).

6. J.A. Endicott and V. Ling, The biochemistry of P-glycoprotein-mediated multidrug resistance, *Annu. Rev. Biochem.*, 58: 137–171 (1989).
7. C.E. Grant, G. Valdimarsson, D.R. Hipfner, K.C. Almquist, S.P.C. Cole, R.G. Deeley, Overexpression of multidrug resistance-associated protein (MRP) increases resistance to natural product drugs, *Cancer Res.*, 54: 357–361 (1994).
8. V.W.T. Ruiz van Haperen, G. Veerman, S. Eriksson, E. Boven, A.P.A. Stegman, M. Hermsen, J.B. Vermorken, H.M. Pinedo, G.J. Peters, Development and molecular characterization of a 2′,2′-difluorodeoxy-cytidine-resistant variant of the human ovarian carcinoma cell line A2780, *Cancer Res.*, 54: 4138–4143 (1994).
9. Y.P. Keepers, P.E. Pizao, G.J. Peters, J. van Ark-Otte, B. Winograd, H.M. Pinedo, Comparison of the sulforhodamine B protein and tetrazolium (MTT) assays for in vitro chemosensitivity testing, *Eur. J. Cancer*, 27: 897–900 (1991).
10. V.W.T. Ruiz van Haperen, G. Veerman, B.J.M. Braakhuis, J.B. Vermorken, E. Boven, A. Leyva, G.J. Peters, Deoxycytidine kinase and deoxycytidine deaminase activities in human xenografts, *Eur. J. Cancer*, 29A: 2132–2137 (1993).
11. G.J. Peters, A. Oosterhof, J.H. Veerkamp, Metabolism of purine nucleosides in human and ovine lymphocytes and rat thymocytes and their influence on mitogenic stimulation, *Biochem. Bioph. Acta*, 755: 127–136 (1983).
12. A.M. Bergman, V.W.T Ruiz van Haperen, G. Veerman, C.M. Kuiper, G.J. Peters, Synergistic interaction between cisplatin and gemcitabine in vitro, *Clin. Cancer Res.*, 2: 521–530 (1996).
13. V.W.T. Ruiz van Haperen, G. Veerman, K. Smid, H.M. Pinedo, G.J. Peters, 2′,2′-Difluorodeoxycytidine (Gemcitabine) and 2′,2′-difluorodeoxyuridine inhibit thymidylate synthase activity in solid tumour cell lines, submitted.

114

INCREASED SENSITIVITY TO GEMCITABINE OF P-gP AND MRP OVEREXPRESSING HUMAN NON-SMALL CELL LUNG CANCER CELL LINES

Andries M. Bergman, Herbert M. Pinedo, Gijsbert Veerman,
Catharina M. Kuiper, and Godefridus J. Peters

Department Medical Oncology
Free University Hospital
P.O. Box 7057
1007 MB Amsterdam
The Netherlands

1. INTRODUCTION

Gemcitabine ($2',2'$-difluorodeoxycytidine, dFdC) is a deoxycytidine analog with proven activity in ovarian and Non-Small Cell Lung Cancer (NSCLC) both *in vivo* and *in vitro* (1–3). Deoxycytidine (dCyd) and dFdC are phosphorylated by deoxycytitidine kinase (dCK) to their monophosphates, which are subsequently phosphorylated to dCTP and dFdCTP, respectively. DFdCTP can be incorporated into both DNA and RNA (4,5). Other enzymes involved in deoxynucleoside metabolism are the mitochondrial enzyme thymidine kinase 2 (TK2) which phosphorylates the natural nucleosides thymidine (TdR), dCyd and deoxyuridine. This in contrast to the cytosolic enzyme thymidine kinase 1, which only phosphorylates thymidine but not dCyd (6). Since dCTP is the major feedback inhibitor of dCK and competes with dFdCTP for DNA polymerase, an increase in dCTP pools will decrease dFdC sensitivity (7,8).

Cross-resistance to some structurally and functionally unrelated natural-derived drugs (e.g., daunomycin, etoposide, vincristine) is called Multidrug Resistance (MDR). MDR is caused by overexpression of the plasma membrane drug efflux pumps P-gP or MRP (9,10). We tested the sensitivities to dFdC of the human NSCLC cell line SW1573 and its daunomycin resistant variants 2R120 (expressing MRP) and 2R160 (expressing P-gP). The sensitivities of SW1573, 2R120 and 2R160 were related to dCyd phosphorylating activities, the dFdCTP accumulation, the incorporation of dFdC into DNA and RNA and the amount of double strand DNA (dsDNA) breaks as a result of dFdC incorporation.

Purine and Pyrimidine Metabolism in Man IX,
edited by Griesmacher *et al.* Plenum Press, New York, 1998.

2. METHODS

Growth inhibition was determined by exposure of cells in 96-wells plates to dFdC for 72 hr, using the sulforhodamine B (SRB) assay (11,12).

The dCK activity was assayed with ^{3}H-dCyd as a substrate at 230 μM dCyd with 1 mM TdR to block TK2 activity. Substrate and product were separated on a TLC plate, as described previously (13).

For dFdCTP accumulation, cells were exposed to 0.1 or 1.0 μM dFdC for 24 hr. Nucleotides were extracted and analyzed by HPLC as described (14).

For dFdC incorporation into DNA, cells were plated in 96-wells filter bottom plates and exposed to 0.1 or 1.0 μM ^{3}H-dFdC for 24 hr. nucleic acids were precipitated with 8% TCA and RNA was degraded by RNAse and DNA was reprecipitated. The filters were removed and radioactivity was estimated in a liquid scintillation counter (15).

To determine DNA damage, cells were exposed to 0.1 or 1.0 μM dFdC for 24 hr and a Fluorometic Analysis of DNA Unwinding (FADU) was performed to determine dsDNA breaks (16).

3. RESULTS

SW1573 was the least sensitive cell line, the doxorubicin resistant variants 2R120 and 2R160 were 9 and 28 fold more sensitive to dFdC, respectively (Tab. 1). As an initial explanation for this difference we determined the dCK activity which was lowest in SW1573 and 6.6 and 4.0 fold higher in 2R120 and 2R160, respectively (Tab. 1). Despite this difference in dCK activity no significant difference in dFdCTP accumulation at both dFdC concentrations was found. In contrast, 2R120 and 2R160 incorporated 1.3 and 1.7 fold more ^{3}H-dFdC into DNA than SW1573, respectively. At 1.0 μM dFdC caused 1.4 fold more DNA damage in 2R120 than in SW1573. Surprisingly, in 2R160 more DNA damage was found at 0.1 than at 1.0 μM dFdC.

4. DISCUSSION

In this paper we describe that the daunomycin resistant NSCLC cell lines 2R120 and 2R160 are more sensitive to dFdC than the parental cell line SW1573. A partial explanation for this phenomenon seems to be the low dCK activity in SW1573 compared to

Table 1. IC50 values for dFdC and dCK activities in the human Non-Small Cell Lung Carcinoma cell lines SW1573 and its daunomycin resistant variants 2R120 and 2R160

Cell line	IC50 (nM)	dCK activity (nmol/hr/106 cells)
SW1573	17.00 ± 1.15	0.32 ± 0.05
2R120	1.92 ± 0.68	2.10 ± 0.18
2R160	0.61 ± 0.21	1.29 ± 0.40

Values are means ± SEM of 3 experiments. In the growth inhibition tests the cells were exposed for 72 hr.

2R120 and 2R160. Despite the differences in dCK activities no difference in accumulation of dFdCTP was found between the cell lines. However, SW1573 incorporated less ^{3}H-dFdC into DNA than its daunomycin resistant variants, but the concentration dependency of ^{3}H-dFdC incorporation into DNA (not shown) and correlation with sensitivity, is in agreement with previous studies in ovarian, colon and leukemia cell lines (4,5). It seems that the difference in ^{3}H-dFdC incorporation between these cell lines is related to the significant difference in DNA damage as found between SW1573 and 2R120 after exposure to dFdC.

In conclusion: MRP and P-gP overexpressing daunomycin resistant human NSCLC cell lines are more sensitive to gemcitabine than the wild type cell line. Also Jensen *et al.* (17) reported that several Small Cell Lung Cancer cell lines with various resistance phenotypes, including P-gP, were more sensitive to dFdC and ara-C. This interesting result is of major interest for application of dFdC to treat tumors with P-gP or MRP overexpression.

ACKNOWLEDGMENTS

This research was supported by grants from the Dutch Cancer Society.

REFERENCES

1. R.P Abratt, W. Rezwoda, G. Falkson, L. Goedhals, D. Hacking, Efficacy and safety profile of gemcitabine in non-small cell lung cancer. Phase II study, *J. Clin. Oncol.*, 12: 1535–1540 (1994).
2. B. Lund, O.P. Hansen, K. Theilade, M. Hansen, J.P. Neijt, Phase II study of gemcitabine (2',2'-difluorodeoxycytidine) in previously treated ovarian cancer patients, *J. Natl. Cancer Inst.*, 86: 1530–1533 (1994).
3. C.J.A. van Moorsel, G. Veerman, A.M. Bergman, A. Guechev, J.B. Vermorken, P.E. Postumus, G.J. Peters, Combination chemotherapy studies with gemcitabine, *Sem. Oncol.*, in press.
4. V.W.T. Ruiz van Haperen, G. Veerman, J.B. Vermorken, G.J. Peters, 2',2'-Difluoro-deoxycytidine (gemcitabine) incorporation into RNA and DNA from tumour cell lines, *Biochem. Pharmacol.*, 46: 762–766 (1993).
5. P. Huang, S. Chubb, L.W. Hertel, G.B. Grindey, W. Plunkett, Action of 2',2'-difluorodeoxycytidine on DNA synthesis, *Cancer Res.*, 51: 6110–6117 (1991).
6. S. Eriksson, B. Kierszuk, B. Munch-Petersen, B. Oberg, N.G. Johansson, Comparison off the substrate specifities of human thymidine kinase 1 and 2 and deoxycytidine kinase toward antiviral and cytostatic nucleoside analogs, *Biochem. Biophys. Res. Comm.*, 176: 586–592 (1991).
7. V. Heinemann, Y-Z Xu, S. Chubb, A. Sen, L.W. Hertel, G.B. Grindey, W. Plukett, Inhibition of ribonucleotide reduction in CCRF-CEM cells by 2',2'-difluorodeoxycytidine, *Mol. Pharmacol.*, 38: 567–572 (1990).
8. V.W.T. Ruiz van Haperen, G.J. Peters, New targets for pyrimidine antimetabolites for the treatment of solid tumours. II: Deoxycytidine kinase, *Pharmacy World Sci.*, 16: 104–112 (1994).
9. J.A. Endicott and V. Ling, The biochemistry of P-glycoprotein-mediated multidrug resistance, *Annu. Rev. Biochem.*, 58: 137–171 (1989).
10. C.E. Grant, G. Valdimarsson, D.R. Hipfner, K.C. Almquist, S.P.C. Cole, R.G. Deeley, Overexpression of multidrug resistance-associated protein (MRP) increases resistance to natural product drugs, *Cancer Res.*, 54: 357–361 (1994).
11. B. Van Triest, H.M. Pinedo, F. Telleman, C.L. Van der Wilt, G. Jansen, G.J. Peters, Cross-resistancce to anti-folates in multi drug resisttant cell lines with P-glycoprotein or multi drug resistance related protein expression, *Biochem. Pharm.*, in press.
12. Y.P. Keepers, P.E. Pizao, G.J. Peters, J. van Ark-Otte, B. Winograd, H.M. Pinedo, Comparison of the sulforhodamine B protein and tetrazolium (MTT) assays for in vitro chemosensitivity testing, *Eur. J. Cancer*, 27: 897–900 (1991).
13. V.W.T. Ruiz van Haperen, G. Veerman, B.J.M. Braakhuis, J.B. Vermorken, E. Boven, A. Leyva, G.J. Peters, Deoxycytidine kinase and deoxycytidine deaminase activities in human xenografts, *Eur. J. Cancer*, 29A: 2132–2137 (1993).

14. V.W.T. Ruiz van Haperen, G. Veerman, E. Boven, P. Noordhuis, J.B. Vermorken, G.J. Peters, Schedule dependence of sensitivity to 2',2'-difluorodeoxycytidine (gemcitabine) in relation to accumulation and retention of its triphosphate in solid tumour cell lines and solid tumours, *Biochem. Pharmacol*, 48: 1327–1339 (1994).

15. C.H. Langeveld, C.A.M. Jongenelen, J.W.M. Theeuwes, J.P.A. Baak, J.J. Heimans, J.C. Stoof, G.J. Peters, The antiproliferative effect of 8-chloro-adenosine, an metabolite of 8-chloro-cyclic adenosine monophosphate, and disturbances in nucleic acid synthesis and cell cycle kinetics, *Biochem. Pharm.*, 53:1441–148 (1997).

16. A.M. Bergman, V.W.T Ruiz van Haperen, G. Veerman, C.M. Kuiper, G.J. Peters, Synergistic interaction between cisplatin and gemcitabine in vitro, *Clin. Cancer Res.*, 2: 521–530 (1996).

17. P.B. Jensen, B. Holm, M. Sorensen, I.J. Christensen, M. Sehested, In vitro cross-resistance and collateral sensitivity in seven resistant small-cell lung cancer cell lines: preclinical identification of suitable drug partners to taxotere, taxol, topotecan and gemcitabine, *Brit. J. Cancer*, 75: 869–877 (1997).

115

DOES 2-CHLORODEOXYADENOSINE CONTRIBUTE TO ALTERATION OF DNA METHYLTRANSFERASE ACTIVITY?

D. Wyczechowska and K. Fabianowska-Majewska

Department of General Chemistry
Medical University of Lodz
Lindleya 6, 90-131 Lodz, Poland

1. INTRODUCTION

2-Chlorodeoxyadenosine (2CdA) is a new and effective drug for indolent lymphoid malignan-cies. However, the mechanisms which link 2CdA action and tumour cell death (by apoptosis) are not fully clarified. 2CdA is rapidly taken up by the target cells and phosphorylated by cytosolic deoxycytidine kinase (dCK). Its triphosphate is a potent inhibitor of human ribonucleotide reductase and a good substrate for human DNA polymerases.[1] However, phosphorylation of 2CdA is clearly not the only event responsible for its cytotoxic effect. The 2CdA phosphorylation in hairy cell leukemia (HCL) was not higher than in chronic lymphocytic leukemia (CLL), despite a better response to 2CdA therapy. Using different cell lines, it was shown that a wide range of cell sensitivity to 2CdA could not be explained by different levels of 2CdA nucleotide.[2,3] 2 CdA was also shown to inhibit the growth of myeloid progenitor cell in which the levels of dCK are low.[4] Finally, we have recently shown that *in vitro* the inhibitory effect of 2CdA results from complete suppression of deoxyadenosine phosphorylation by dCK in both human normal lymphocytes and in lymphoma cells obtained from patients with a central nervous system involvement.[5] Also an inhibition of adenosine deaminase activity in lysate of both types of cells was observed.[5] Moreover, we observed a large decrease in the activity of adenosine deaminase and S-adenosylhomocysteine (SAH) hydrolase, in the erythrocyte lysates of patients, after one week treatment with 2CdA.[6] We assumed that inhibitory effect of 2CdA on deoxyadenosine metabolism could lead to inactivation of SAH-hydrolase with perturbation of methylation reactions.

The present studies were designed to investigate this hypothesis by estimation of 2CdA effect on activity of two enzymes: SAH-hydrolase and DNA methyltransferase (C-5 MT-ase) in cultured murine leukemic cells (L-1210). The effect of 2CdA was compared with that of 9-β-D-arabinosyl-2-fluoroadenine (fludarabine, another effective drug used

Purine and Pyrimidine Metabolism in Man IX,
edited by Griesmacher *et al.* Plenum Press, New York, 1998.

for indolent malignancies) and of 5-aza-deoxycytidine (5-aza-dCyt), known as a potent hinhibitor of C-5 MT-ase.

2. MATERIALS AND METHODS

2.1. Chemicals

Basal reagents, cold substrates, fludarabine, 5-aza-dCyt, proteinase K, (poly [dI-dC] • poly [dI-dC]), were products of Sigma Chemical Co.; labelled substrates: [^{14}C] adenosine (Ado) and [^{3}H] S-adenosylmethionine (SAM) were products of Amersham Life Science. 2CdA was a kind gift from Dr.P. Grieb (Dept. of Neurophysiology, PAN, Warsaw, Poland).

2.2. Enzyme Assay

SAH-hydrolase activity was determined in the SAH synthesis direction by radio-chemical paper chromatography analysis, at 24 hour intervals, in extracts of L-1210 cells, cultured in the absence or presence of the above mentioned inhibitors. Concentration of inhibitors in cultured medium was 3 µM. Incubation mixture for the analysis of SAH-hydrolase activity was prepared according to the method used previously,[7] in which the concentration of Ado was approximately 0.4 mM.

C-5 MT-ase activity was analysed according to Issa's method.[8] The amount of methyl group incorporated from [^{3}H] SAM to cytidine of both poly [dI-dC] • poly [dI-dC], used as acceptor of the methyl groups, and cytidine of DNA of L-1210 cells was determined. After reaction, DNA of both types were extracted (phenol-chloroform), precipitated by ethanol and eluted by alkali. Then solution with DNA was spotted on Whatman GF/C filters, which were counted in scintillation counter.

3. RESULTS AND DISCUSSION

The inhibitory effect of 2CdA, fludarabine and 5-aza-dCyt on the activity of SAH-hydrolase at 24 hr intervals in extracts of L-1210 cells is documented in Figure 1. The results show that the activity of SAH-hydrolase increases over 2-fold within 48 hr, when cells are incubated without inhibitors. The presence of 2CdA in cultured medium, at 3 µM

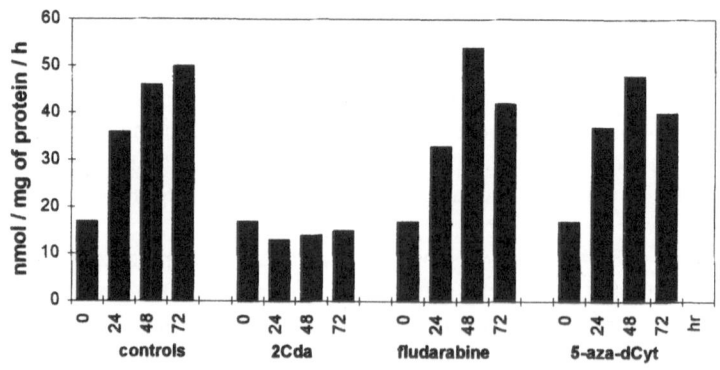

Figure 1. SAH-hydrolase activity in extracts of L-1210 cells during 72 hour growth of culture without/with inhibitors. Each value represent the average of at least four experiments.

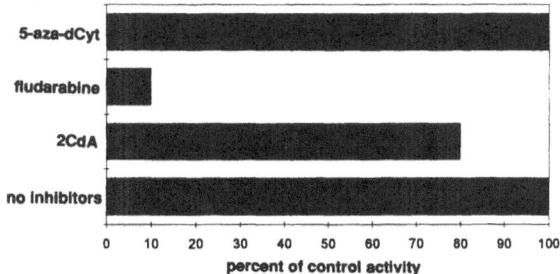

Figure 2. Direct influence of 2CdA, fludarabine and 5-aza-dCyt on the activity of SAH-
hydrolase in extracts of L-1210 cells. Each
value represent the average of four experiments.

concentration, causes 50% inhibition of SAH-hydrolase activity, compared to control cells
incubation. The alteration of the enzyme activity is insignificant in the presence of other
inhibitors, i.e. fludarabine and 5-aza-dCyt.

Direct *in vitro* effect of the tested inhibitors on the activity of SAH-hydrolase in
extracts of L-1210 cells were also estimated. The results of this experiment are presented
in Figure 2. They show that 2CdA has only a slight influence on the change of activity of
SAH-hydrolase (not more than 20% inhibition), and this result confirms our previous findings for human normal lymphocytes and CNS lymphoma cells.[5] 5-aza-dCyt has no effect,
whereas fludarabine is a very potent inactivator of SAH-hydrolase (90% inhibition).

The activity of C-5 MT-ase in extracts of L1210 cells at 24 hr intervals during 72
hour growth of cells in the presence or absence of tested inhibitors is summarized in Figure 3 (poly [dI-dC] • poly [dI-dC] as acceptor of [³H] methyl group of SAM) and in Figure
4 (incorporation of methyl group to DNA of L-1210 cells).

The activity of C-5 MT-ase increased after 24 hr incubation without inhibitors,
achieving approximately 100% increase. After next two days, a decrease of the enzyme
activity was observed, but the activity remained higher than on the first day of incubation.
5-aza-dCyt, as specific inhibitor of C-5 MT-ase, and 2CdA inhibited incorporation of radiolabelled methyl group in the 24th hr of incubation at 98% and 60%, respectively, comparing
with activity on the first day. In the case of 5-aza-dCyt, this low level of the enzyme activity
remained for the next two growing days. In 2CdA case, for the next two days, a consequent
decrease of inhibition was noted. The effect of fludarabine on C-5 MT-ase activity was
insignificant.

In the case of incorporation of radiolabelled methyl group from SAM to DNA of
L-1210 cells, the effect of the tested inhibitors was quite similar to experiments with poly

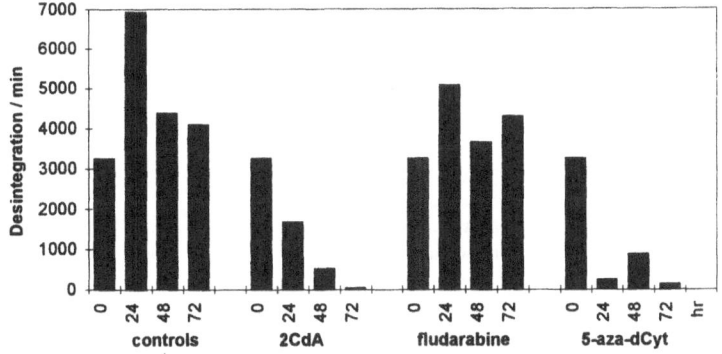

Figure 3. Effect of 2CdA, fludarabine and 5-aza-dCyt on activity of C-5 MT-ase, when poly [dI-dC] • poly [dI-dC]
is acceptor of [³H] methyl groups. Each value represent the average of four experiments.

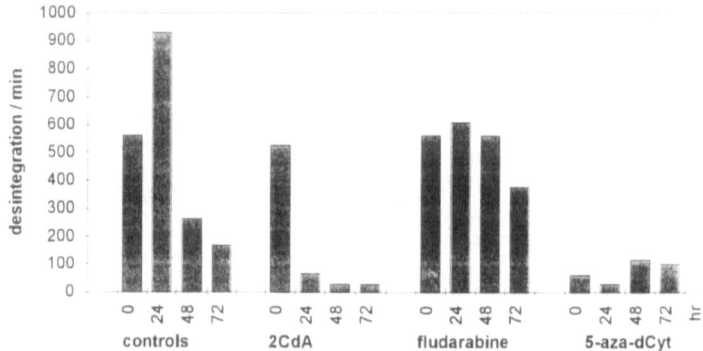

Figure 4. Effect of 2CdA, fludarabine and 5-aza-dCyt on incorporation of [³H] methyl group from SAM to DNA of L-1210 cells.

[dI-dC] • poly [dI-dC] with one exception: in the case of 2CdA, decrease of methylation of L-1210 cells DNA (Fig. 4) was deeper.

The present studies clearly demonstrate that, although 2CdA is a very slight direct inactivator of SAH-hydrolase (Fig. 2), 2CdA unlike fludarabine contributes to depletion of SAH-hydrolase activity (Fig. 1) during the 72 hr growth of L1210 cells, and to deep decrease of DNA transmethylation reaction (Figs. 3 and 4). It may be a consequence of perturbation of "active methyl" cycle enzymes, mainly due to dramatic downfall of ATP and NAD concentrations.

ACKNOWLEDGMENT

This work was supported by the Medical University of Lodz, grant No. 502-11-332.

REFERENCES

1. Plunkett W., Saunders PP. Metabolism and action of purine nucleoside analogs. Pharmacol. Ther., 1991; 49: 239–268.
2. Spasokoukotskaja T., Arner EJS., Brosjo O., Gunven P., Juliusson G., Liliemark J., Eriksson S. Expression of deoxycytidine kinase and phosphorylation of 2-chlorodeoxyadenosine in human normal and tumour cells and tissues. Eur.J.Cancer 1995; 31A: 202–208.
3. Petzer AL., Bilgeri R., Zilian U., Haun M., Geisen FH., Pragnell I., Braunsteiner H., Konwalinka G. Inhibitory effect of 2-chlorodeoxyadenosine on granulocytic, erythroid, and T-lymphocytic colony growth. Blood 1992; 78: 2583–2587.
4. Robak T., Korycka A. The comparison of 2-chlorodeoxyadenosine (2CdA) in combination with interferon α (IFNα) or interferon γ (IFNγ) on granulocyte-macrophage progenitor cell (CFU-GM) and clonogenic blasts (CFU-L) in in vitro cultures. Leuk.Lymph., 1996; 21: 161–168.
5. Fabianowska-Majewska K., Tybor K., Duley J., Simmonds A. The influence of 2-chloro-2'-deoxyadenosine on metabolism of deoxyadenosine in human primary CNS lymphoma. Biochem.Pharmacol., 1995; 9: 1379–1383.
6. Warzocha K., Fabianowska-Majewska K., Blonski J., Krykowski E., Robak T. 2-Chlorodeoxyadenosine inhibits activity of adenosine deaminase and S-adenosylohomocysteine hydrolase in patients with chronic lymphocytic leukemia. Eur.J.Cancer 1997; 33: 170–173.
7. Smolenski RT., Fabianowska-Majewska K., Montero C., Duley JA., Fairbanks LD., Marlewski M., Simmonds HA. A novel route of ATP analysis. Biochem. Pharmacol., 1997; 43: 2053–2057.
8. Issa J-PJ., Vertino PM., Wu J., Sazawal S., Celano P., Nelkin BD., Hamilton SR., Baylin SB. Increased cytosine DNA- methyltransferase activity during colon cancer progression. J.Nat.Cancer Inst. 1993; 85: 1235–1240.

116

IRON BINDING CAPACITY OF DIDOX (3,4 DIHYDROXYBENZOHYDROXAMIC ACID) AND AMIDOX (3,4 DIHYDROXYBENZAMIDOXIME) TWO INHIBITORS OF THE ENZYME RIBONUCLEOTIDE REDUCTASE

Monika Fritzer-Szekeres,[1] Ladislav Novotny,[2] Anna Vachalkova,[2] Rainer Göbl,[1] Howard L. Elford,[3] and Thomas Szekeres[1]

[1]Clinical Institute of Medical and Chemical Laboratory Diagnostics
University of Vienna
Waehringerguertel 18-20
1090 Vienna, Austria
[2]Cancer Research Institute
Slovak Academy of Sciences
Spitalska 21, SK-81232 Bratislava, Slovak Republic
[3]Molecules for Health Inc.
Richmond, Virginia

1. SUMMARY

Ribonucleotide reductase is the rate limiting enzyme of deoxynucleoside triphosphate synthesis and is considered to be an excellent target of cancer chemotherapy. Didox and amidox are newly synthesized compounds, which inhibit this enzyme and have in vitro and in vivo antitumor activity. We have now investigated the capability of didox and amidox to interfere with the iron metabolism. We show by photometric and polarographic methods, that didox and amidox are capable of forming an iron complex. However, their cytotoxic action cannot be circumvented by addition of Fe-ammoniumcitrate, indicating the iron complexing capacity not to be responsible for the mechanism of action of these compounds. When L1210 leukemia cells were incubated with the didox-iron or amidox-iron complex itself, only slight changes of the 50% growth inhibitory capacity of the complex in comparison with didox or amidox alone could be shown. We conclude, that didox and amidox are capable of forming an iron complex, but in contrast to other agents, the anticancer activity cannot be contributed to this effect alone. Further studies will have to elucidate the molecular mechanism of action of these new and promising anticancer agents.

Purine and Pyrimidine Metabolism in Man IX,
edited by Griesmacher *et al.* Plenum Press, New York, 1998.

2. INTRODUCTION

Ribonucleotide reductase (EC 1.17.4.1) is the rate limiting enzyme of de novo deoxyribonucleotide synthesis and was shown to be increased in activity linked with proliferation and malignant transformation (1–3). It was therefore considered as a good target for cancer chemotherapy (4–6). A number of compounds inhibiting this enzyme were synthesized. Only hydroxyurea, a relatively weak inhibitor of ribonucleotide reductase, is clinically used for the treatment of hematological malignancies.

Investigators have synthesized several polyhydroxy-substituted benzohydroxamates as inhibitors of ribonucleotide reductase (7–9). Among these compounds were didox (3,4,-dihydroxybenzohydroxamic acid) and amidox (3,4 dihydroxybenzamidoxime) (Figure 1). Both compounds inhibited the growth of various tumor cell lines more effectively than hydroxyurea. Didox was effective as an anticancer agent in a number of animal tumor systems including L1210, P388 murine leukamias, lewis lung carcinoma and several human tumor xenografts (8,9). On the basis of these data, didox was also investigated in phase I and II clinical trials (10,11).

We demonstrated previously, that trimidox (3,4,5 trihydroxybenzamidoxime) one of these polyhydroxy-substituted benzohydroxamic acid derivates is able to form a stable complex with iron (12). This effect might play a role in the mechanism of enzyme inhibition. The aim of the present study was to investigate whether didox and amidox are also capable of forming complexes with iron salts. We also elucidated the role of didox-iron and amidox-iron intercalation regarding the biochemical and biological properties of these compounds.

3. MATERIALS AND METHODS

3.1. Chemicals and Supplies

Didox and amidox were synthesized as described (8). All other chemicals and reagents were commercially available and of highest purity.

3.2. Cell Culture

K562 and L1210 leukemia cell lines were purchased from ATCC (American Type Culture Collection, Rockville, MD, USA). Cells were grown in RPMI 1640 medium sup-

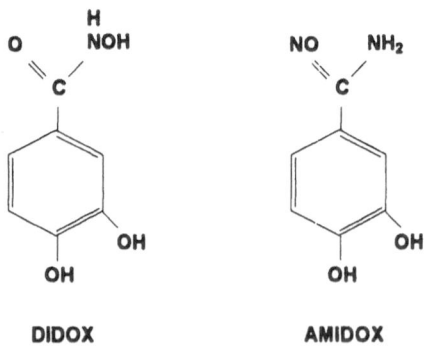

Figure 1. Structure of Didox and Amidox.

plemented with 10% heat inactivated fetal calf serum (FCS) (GIBCO, Grand Island Biological Co., Grand Island, NY, USA), supplemented with 1% penicillin-streptomycin in a humidified atmosphere containing 5% CO_2. Cell counts were determined using the microcellcounter CC-108 (Sysmex, Kobe, Japan). Cells growing in logarithmic phase of growth were used for all the studies described below.

3.3. Growth Inhibition Assay

Cells were seeded in 25 cm^2 flasks at a cell density of 0.1 × 10^9/l and incubated with various drug concentrations (2–128 µmol/l of didox or amidox). Cells were counted and IC_{50} concentrations (drug concentration at 50% inhibition of cell growth) were determined 4 days later.

3.4. Polarographic Measurements

Polarographic measurements were performed using a PA 4 polarographic analyzer, which was interfaced to a two-line XY 4106 plotter (Laboratorni Pristroje, Praha, The Czech Republic). A dropping mercury electrode with a drop time of 3s and a flow rate of 2.27 mg/s at mercury column height h_{Hg} of 81 cm was used as the indicating electrode. A saturated calomel electrode (SCE) was used as the reference electrode. A platinum electrode was the auxiliary electrode. All polarographic measurements were carried out at room temperature in a stream of gaseous argon to exclude atmospheric oxygen from the polarographic cell. The pH was 7.016.

$FeSO_4 \cdot 7H_2O$ was used as a source for Fe^{2+} and Fe-ammoniumcitrate was the source for Fe^{3+}.

4. RESULTS

4.1. Formation of an Didox-Iron and Amidox-Iron Complex

4.1.1. Photometric Determination. When didox or amidox (50–400 µmol/l) were incubated with various concentrations of Fe-Ammoniumcitrate (0–200 µmol/l) in an aqueous solution, we could observe the formation of a greybrown colour complex. When 200 µmol/l didox was incubated with 100 µmol/l Fe-Ammoniumcitrate in an aqueous solution for 2 hours at 37°C the photometric determination showed maximal absorption at 530 nm. When 200 µmol/l amidox was incubated with 100 µmol/l Fe-Ammoniumcitrate in an aqueous solution for 2 hours at 37°C the photometric determination showed maximal absorption at 550 nm, respectively. The absorbance of the greybrown complex increased with increasing didox or amidox concentrations, however when Fe-Ammoniumcitrate was preincubated with 200 µmol/l desferroxamin for 1 hour prior to incubation with didox or amidox, the formation of the complex could be inhibited.

4.1.2. Polarographic Determination. Didox could not be reduced under the conditions used (pH range from 3.4 to 10.0 in buffer solutions). We therefore could not perform polarographic measurements with didox.

4.1.2.1. Polarographic Reduction of Amidox in the Presence of Fe^{2+}. The equimolar ratio of concentrations amidox:$FeSO_4$ were 1: 0.5; 1: 0.8; 1:1. We could observe a one step reduction of Fe^{2+} in Britton Robinson buffer at pH 7.00, at the half wave potential of $E_{1/2}$ = -1.340 V (versus standard calomel electrode). Under the same conditions the reduction of amidox had two steps ($E_{1/2}$: -1.550 and -1.770 V, respectively) (Figure 2). The reduction of amidox was influenced by the presence of Fe^{2+}-ions as demonstrated by the change in altitude of the polarographic wave of Fe^{2+}, which decreased in a concentration dependent manner with increasing Fe^{2+} concentrations. These changes were accompanied by the formation of a dark greybrown colour (equimolar ratio of concentrations amidox:$FeSO_4$ = 1:1).

4.1.2.2. Polarographic Reduction of Amidox in the Presence of Fe^{3+}-Ions. The equimolar ratios of concentrations amidox:Fe^{3+} were 1:0.3; 1:0.5; 1:0.8. The reduction of Fe^{3+} at pH 7.00 in Britton Robinson Buffer is a one step reduction with a half wave potential of $E_{1/2}$: -0.490 V (versus standard calomel electrode). The reduction of amidox could be significantly influenced by the presence of Fe^{3+} ions. The first polarographic wave of amidox increased markedly even at low Fe^{3+} concentrations and its $E_{1/2}$ shifted to more positive values. The wave generated by Fe^{3+} alone decreased with increasing amidox concentrations whereas an increase of the wave caused by amidox could be observed (not shown). With increasing iron concentration the development of a dark greybrown colour could be observed.

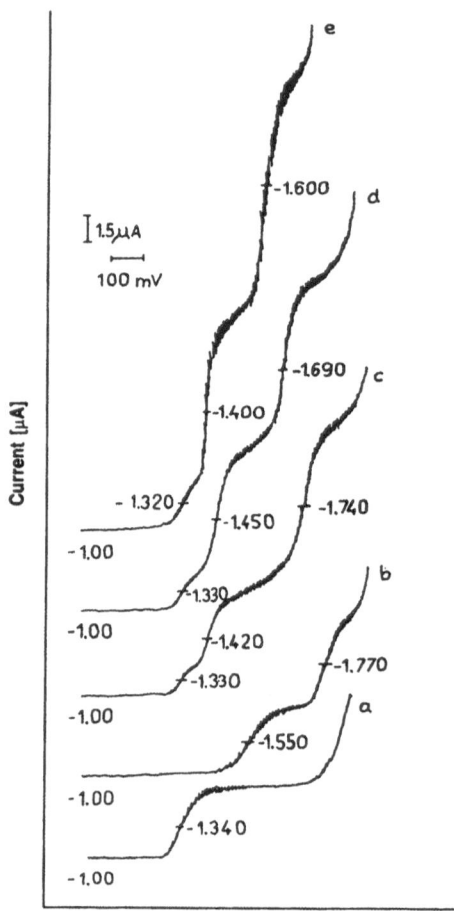

Figure 2. Polarographic reduction of Amidox in Britton Robinson buffer in the presence of $FeSO_4 \cdot 7H_2O$, pH 7.00. a) 1 mmol/l $FeSO_4$, b) 1 mmol/l amidox, c) mixture of 1 mmol/l amidox with 0.5 mmol/l $FeSO_4$, d) mixture of 1 mmol/l amidox with 0.8 mmol/l $FeSO_4$, e) mixture of 1 mmol/l amidox with 1 mmol/l $FeSO_4$. Scan rate 5 mV/s.

Table 1. Effect of Fe^{3+} on the IC_{50} s of didox and amidox
on leukemia L1210 and K562 cell lines

Cell line	IC_{50} (μM)			
	Didox	Didox+Fe^{3+}	Amidox	Amidox+Fe^{3+}
L1210	25	30	6	10
K562	44	55	32	50

Cells were seeded at a density of 0.1×10^9 cells/l and incubated for 4 days; then viable cell count was determined.

4.2. Effect of Iron on in Vitro Growth Inhibitory Capacity of Didox and Amidox

When K562 erythroleukemic cells were incubated with various concentrations of didox or amidox, addition of iron (40 μmol/l Fe-Ammoniumcitrate) did slightly increase the IC_{50}s (50% inhibitory capacity) of the compounds. Similar results were observed for L1210 mouse leukemia cells (Table 1).

4.3. Cytotoxic Effects of the Iron-Didox and Iron-Amidox Complex in Leukemia Cell Lines

Various concentrations of didox or amidox were incubated with 40 μmol/l Fe-Ammoniumcitrate 30 minutes prior to incubation of the cells to allow the formation of the iron-didox or iron-amidox complex. Then the L1210 leukemia cells were incubated with didox or amidox or with the respective iron complex. On day 4 cell numbers were counted. In L1210 cells didox showed an IC_{50} of 25 μM and the iron-didox complex an IC_{50} of 12,5 μM, respectively. Amidox showed an IC_{50} of 6 and the iron-amidox complex an IC_{50} of 25μM, respectively (Table 2).

5. DISCUSSION

Ribonucleotide reductase is the rate limiting enzyme of de novo deoxynucleoside triphosphate (dNTP) synthesis. Its activity is significantly increased in malignant cells and therefore the enzyme is considered to be an excellent target of cancer chemotherapy (1–5).

Ribonucleotide reductase consists of two distinct subunits which are required for enzymatic activity. One subunit was identified as the non-heme iron (M2) and the other as the effector binding subunit (M1). The non-heme iron subunit needs a free tyrosyl radical to be active; iron is a cofactor for generation and stabilization of this free tyrosyl radical. Inhi-

Table 2. Effect of didox and amidox or their respective
iron-complexes on leukemia L1210 cells

Cell line	IC_{50} (μM)			
	Didox	Didox-Fe complex	Amidox	Amidox-Fe complex
L1210	25	12.5	6	25

Cells were seeded at a density of 0.1×10^9 cells/l and incubated for 4 days; then viable cell count was determined.

bition of the enzyme can therefore be caused by direct scavenging of the tyrosyl radical or iron deprivation (5). Gallium for instance was recently shown to inhibit ribonucleotide reductase by preventing iron incorporation and scavenging the tyrosyl radical (13).

We recently demonstrated that trimidox upregulates the transferrin receptor on HL-60 human promyelocytic leukemia cells and is able to form a stable complex with iron (12,14), the aim of the present study was to investigate, whether didox and amidox are also forming complexes with iron salts. We could now clearly prove by photometric and polarographic determinations, that didox as well as amidox are capable of forming complexes with iron. Then, we examined whether the cytotoxic effect of didox or amidox can be circumvented by addition of access Fe-ammonium salt to the culture medium. However, addition of iron-ammonium salt did only slightly increase the IC_{50} concentrations of didox or amidox. When cells were incubated with the iron-didox or iron-amidox complex itself, only slight differences of the IC_{50}s were observed, when compared with the 50% growth inhibitory concentration of the pure compound. The IC_{50} of didox dropped from 25 to 12,5 µM and the 50% growth inhibitory concentration of amidox increased from 6 to 25 µM in L1210 cells.

Although didox and amidox are capable of forming stable complexes with iron, the mechanism of action of both compounds ought to be due to other chemical properties as well. It might be caused by the free radical scavenging capacity of didox or amidox, which was demonstrated (15) or to other direct enzyme inhibitory effects. Further ongoing studies will have to elucidate the exact mechanisms of action of these promising anticancer agents.

ACKNOWLEDGMENTS

This work was supported by the "Anton Dreher Gedächtnisschenkung für Medizinische Forschung", a grant of the "Jubläumsfonds der österr. Nationalbank, Grant No.: 5267" and by a grant of the Slovak Grant Agency Nr. 1331/94.

REFERENCES

1. G. WEBER, New Engl. J. Med. **296** 486–93 (1977).
2. E. TAKEDA and G. WEBER, Life Sci. **28** 1007–14 (1981).
3. H.L. ELFORD, M. FREESE, M. PASSAMANI, H.P. MORRIS, J. Biol. Chem. **245** 5228- 33 (1970).
4. J.G. CORY, G.L. CARTER, Advan. Enzyme Regul. **24** 385–401 (1985).
5. J.G. CORY, Advan. Enzyme Regul. **27** 437–55 (1988).
6. H.L. ELFORD, B. VAN'T RIET, G.L. WAMPLER, A.L. LIN, R.M. ELFORD, Advan. Enzyme Regul. **19** 151–68 (1981).
7. H.L. ELFORD, G.L. WAMPLER, B. VAN'T RIET, Cancer Res. **39** 844–51 (1979).
8. B. VAN'T RIET, L.B. KIER, H.L. ELFORD, J. Pharm. Sci. **69** 856–7 (1980).
9. H.L. ELFORD, B. VAN'T RIET, G.L. WAMPLER, U.S. patent 4623659 United States Patent Office. Washington, D.C. (1986).
10. D. VEALE, J. CARMICHAEL, B.M.J. CANTWELL, H.L. ELFORD, R. BLACKIE, D.J. KERR, S.B. KAYE, A.L. HARRIS, Br. J. Cancer. **58** 70–72 (1988).
11. R.D. RUBENS, S.B. KAYE, M. SOUKOP, C.J. WILLIAMS, M.H. BRAMPTON, A.L. HARRIS, Br. J. Cancer. **64**, 1187–1188 (1991).
12. T. SZEKERES, E. VIELNASCHER, L. NOVOTNY, A. VACHALKOVA, M. FRITZER, G. FINDENIG, R. GOEBL, H.L. ELFORD, H. GOLDENBERG, Eur. J. Clin. Chem. Clin. Biochem. **33** 785–789 (1995).
13. J. NARASIMHAN, W.E. ANTHOLINE, C.R. CHITAMBAR, Biochem. Pharmacol. **44** 2403–2408 (1992).
14. T. SZEKERES, M. FRITZER, H. STROBL, K. GHAREHBAGHI, G. FINDENIG, H.L. ELFORD, C. LHOTKA, H.J. SCHÖN and H.N. JARAYAM, Blood. **84** 4316–21 (1994).
15. H.L. ELFORD, B. VAN'T RIET, E.A. MIKHAIL, J.L. ZWEIER, Proc. Am. Ass. Cancer Res. **36** 296 (abstract)(1995).

117

IN VITRO AND IN VIVO INHIBITORY ACTIVITY OF THE DIFFERENTIATION-INDUCING AGENT 9-(2-PHOSPHONYLMETHOXYETHYL)ADENINE (PMEA) AGAINST RAT CHORIOCARCINOMA

S. Hatse,[1] L. Naesens,[1] B. Degrève,[1] M. Vandeputte,[1] M. Waer,[2]
E. De Clercq,[1] and J. Balzarini[1]

[1]Rega Institute for Medical Research
Katholieke Universiteit Leuven, Belgium
[2]Laboratory of Nephrology
University Hospitals
3000 Leuven, Belgium

1. ABSTRACT

The acyclic nucleoside phosphonate 9-(2-phosphonylmethoxyethyl)adenine (PMEA) has previously been shown to be a strong inducer of differentiation in several tumor cell lines. We have now investigated the in vitro differentiation-inducing and the in vivo antitumor properties of PMEA in a rat choriocarcinoma tumor cell model. PMEA at 2 to 50 µM induced choriocarcinoma RCHO cell differentiation in vitro in a concentration-dependent manner, as monitored by morphological changes, induction of alkaline phosphatase and production and secretion of progesterone. Likewise, a clear dose-response relationship was established for the in vivo antitumor activity of PMEA in choriocarcinoma-bearing rats. (R)-PMPA, a structural analogue of PMEA which is much less effective than PMEA in inducing differentiation in vitro did not demonstrate any in vivo antitumor activity. This observation points to the specificity of the differentiation-inducing potential of PMEA.

2. INTRODUCTION

Acyclic nucleoside phosphonates represent a unique class of compounds, endowed with potent activity against herpesviruses, hepatitis B virus and retroviruses, including HIV.[1–5] Recently, 9-(2-phosphonylmethoxyethyl)adenine (PMEA), the prototype congener of this class of compounds, was also found to exert a remarkable differentiation-inducing

Purine and Pyrimidine Metabolism in Man IX,
edited by Griesmacher *et al.* Plenum Press, New York, 1998.

605

effect in human erythroleukemia K562 cells, human promyelocytic HL-60 cells and rat choriocarcinoma RCHO cells.[6]

The RCHO cell line consists of proliferating cytotrophoblast cells and differentiated, hormonally active giant cells.[7,8] The morphological changes accompanying differentiation of cytotrophoblasts into giant cells can easily be observed microscopically.[9] Moreover, ecretion of progesterone[10] and induction of alkaline phosphatase activity[11] are appropriate and easy measurable differentiation markers. When grafted under the kidney capsule of syngeneic WKA/H rats, RCHO cells give rise to extremely aggressive, hemorrhagic umors. Hence, this rat choriocarcinoma model may be particularly useful to investigate both the *in vitro* and an *in vivo* differentiation-inducing properties of test compounds.

We used the choriocarcinoma RCHO cell line to study the *in vitro* differentiation-inducing potential of PMEA, its closely related structural analogue (*R*)-9-(2-phosphonyl-methoxy-propyl)adenine [(*R*)-PMPA] and methotrexate. In addition, the *in vivo* antitumor activity of these compounds was evaluated in choriocarcinoma-bearing rats.

3. MATERIALS AND METHODS

3.1. Compounds

The synthesis and anti-retroviral activity of the acyclic nucleoside phosphonates used in this study (i.e., 9-(2-phosphonylmethoxyethyl)adenine [PMEA] and (*R*)-9-(2-phos-phonylmethoxypropyl)adenine [(*R*)-PMPA], have been described previously.[12–14]

3.2. Cell Culture

Rat choriocarcinoma RCHO cells were routinely cultured in tissue culture flasks coated with 0.1% gelatin (Sigma Chemical, St. Louis, MO) in RPMI-1640 medium, supplemented with 10% fetal calf serum (GIBCO, Paisley, UK), 1 mM sodium pyruvate (GIBCO), 2 mM L-glutamine (GIBCO), 2 ng/ml epidermal growth factor (GIBCO) and 5×10^{-5} M ß-mercaptoethanol (UCB, Belgium).[15] Cultures were maintained at 37°C in a humidified, CO_2-controlled atmosphere. *In vitro* passages were done every 2–3 days by digestion with dispase grade II (BOEHRINGER, Mannheim, Germany).

3.3. *In Vitro* Differentiation Experiments

RCHO cells were seeded at 2.5×10^3 cells/well onto non-coated 96-well microtiter plates (Falcon, Becton Dickinson) in RPMI 1640-based growth medium. Test compounds were added at the appropriate concentrations and the cells were incubated for 96 hours at 37°C in a humidified, CO_2-controlled atmosphere.

3.3.1. Alkaline Phosphatase Assay. Drug-treated cells were lysed by repeated freezing and thawing. Then, to each well 100 µl substrate was added [2 mg/ml disodium p-nitrophenyl phosphate (PNPP) (Sigma Chemical, St. Louis, MO) in 50 mM Tris-HCl buffer pH 9.5 containing 0.1% Tween-20]. After incubation at 37°C for 90 min, the optical density was measured at 405 nm and at a reference wavelength of 620 nm.

3.3.2. Detection of Progesterone in Culture Supernatant. Progesterone levels in the culture supernatant were determined by a radioimmunoassay using a monoclonal antibody (Laboratory for Experimental Medicine and Endocrinology, Leuven, Belgium).

3.4. *In Vivo* Experiments

0.75 to 1 × 10^4 RCHO cells were grafted under the kidney capsule of syngeneic WKA/H rats on day 0. Test compounds were dissolved in phosphate-buffered saline at the appropriate concentrations and injected intraperitoneally for 11 sequential days, starting on the day before tumor cell inoculation (day −1). Control rats were injected with PBS without test compound.

4. RESULTS AND DISCUSSION

4.1. Effects of PMEA, Methotrexate, and (*R*)-PMPA on the Proliferation and Differentiation of Rat Choriocarcinoma Cells *in Vitro*

Differentiation of cytotrophoblasts into giant cells implies pronounced morphological changes which do not allow an accurate comparison of cell numbers of non-differentiated *versus* strongly differentiated RCHO cell cultures by standard procedures (automated cell counting, MTT staining). Therefore, total cell derived protein concentration was used to estimate the inhibitory effect of the test compounds on cell proliferation. After exposure of RCHO cell cultures to PMEA at 20 and 50 μM and to MTX at 0.004 and 0.02 μM, total protein concentrations were markedly decreased (Table 1). In contrast, exposure of RCHO cells to (*R*)-PMPA at concentrations up to 250 μM did not result in a significant decrease in the amount of cell material (Table 1). Alkaline phosphatase activity and extracellular concentration of progesterone, synthezised and secreted by the highly differentiated giant cells, were used as biochemical parameters to estimate the differentiation stage of drug-treated RCHO choriocarcinoma cell cultures. Table 1 shows the gradual increase in alkaline phosphatase activity from 100% to 491% for PMEA concentrations varying from 0 to 50 μM. For the hormonal activity of the PMEA-treated RCHO cell cultures, a similar tendency was found. Compared to the background level of 0.39 μg extracellular progesterone/100 ml culture medium of untreated control cells, progesterone concentrations of 0.47, 0.92 and 2.61 μg/100 ml were noted for RCHO cell cultures exposed to PMEA at 10, 20 and 50 μM, respectively (Table 1). This concentration-dependent effect of PMEA is

Table 1. Total protein concentration, alkaline phosphatase activity and extracellular progesterone concentration of RCHO cell cultures exposed to differentiation-inducing agents[a]

Compound	Concentration (μM)	Total protein amount (% of control)	Alkaline phosphatase (% of control)	Progesterone secretion (μg/100ml)
Control	–	100	100	0.39
PMEA	2	101	131	0.39
	5	97	182	0.42
	10	92	269	0.47
	20	84	360	0.92
	50	60	491	2.61
MTX	0.004	73	238	1.31
	0.02	43	271	4.57
(*R*)-PMPA	50	104	120	0.29
	250	95	153	0.39

[a]Data represent the mean of 3 independent experiments. Values are normalized to reflect an equal amount of cell material.

consistent with the gradually increasing degrees of giant cell formation observed in the PMEA-treated RCHO cell cultures. Also, the *in vitro* differentiation-inducing potential of PMEA was found to be equivalent to that of methotrexate (MTX) (Table 1), which is currently the drug of choice for the treatment of choriocarcinoma in humans.[16] In contrast, (*R*)-PMPA, a closely related analogue of PMEA, did not induce marked differentiation at a concentration as high as 250 μM (Table 1). This observation suggests that the induction of RCHO cell differentiation by PMEA is a specific phenomenon.

4.2. Effects of PMEA, MTX, and (*R*)-PMPA on Choriocarcinoma Tumor Growth *in Vivo*

RCHO choriocarcinoma-inoculated rats were treated by daily intraperitoneal injections of the test compounds, starting on the day before tumor cell grafting. At day 10 after RCHO inoculation, tumor outgrowth in the injected kidney was evaluated and scored from [−] (no macroscopically visible tumor tissue) to [5+] (massive tumor growth, associated with a marked enlargement of the kidney). The results in Fig. 1 clearly demonstrate a dose-dependent inhibitory effect of PMEA on choriocarcinoma tumor development. Neither macroscopic nor microscopic signs of tumor growth could be found in the kidney of animals treated with PMEA at 250 mg/kg/day. In most cases, a tiny focus of tumor tissue was found in the injected kidney of the rats after treatment with a daily PMEA dose of 100 mg/kg. When administered at 50 mg/kg/day, PMEA markedly retarded choriocarcinoma tumor growth. At 25 mg/kg/day, PMEA afforded a minor reduction in tumor size, compared to untreated tumor-inoculated control rats. Also, methotrexate proved to be a potent inhibitor of choriocarcinoma : a daily dose of 0.8 mg/kg/day sufficed to completely inhibit

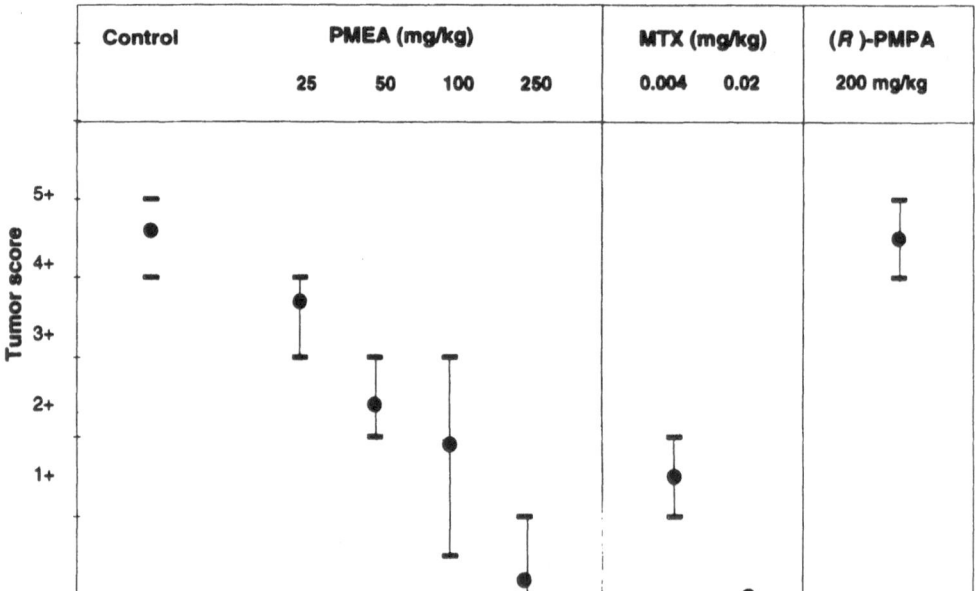

Figure 1. Antitumor activities of PMEA, MTX and (*R*)-PMPA in RCHO choriocarcinoma-bearing rats. Tumor growth was scored after 11 days of treatment. Results shown represent the mean score of 2 to 10 animals, performed in 2 to 5 independent experiments. The range of highest and lowest values between which the individual scores varies is also indicated.

tumor development (Fig. 1). In contrast, (R)-PMPA did not show any significant antitumor activity at a dose as high as 200 mg/kg/day (Fig. 1). This observation is in full agreement with the inability of (R)-PMPA to induce marked differentiation of RCHO cells *in vitro*.

From these findings, we can conclude that PMEA is a potent inducer of differentiation of rat choriocarcinoma cells both *in vitro* and *in vivo*. The strong antitumor activity exhibited by PMEA in the rat choriocarcinoma RCHO model justifies further studies on the therapeutic potential of PMEA in particular, and acyclic nucleoside phosphonates in general, for the treatment of neoplasms that are sensitive to differentiation induction.

ACKNOWLEDGMENTS

This work was supported by the Belgian A.S.L.K. Cancer Fund. We are grateful to Mrs. Marie-Jeanne Leemput (Laboratory for Experimental Medicine and Endocrinology, Leuven, Belgium), Mr. Constant Segers and Mr. Willy Zeegers for excellent technical assistance.

REFERENCES

1. De Clercq, E., Holy, A., Rosenberg, I., Sakuma, T., Balzarini, J., and Maudgal, P.C. Nature 323: 464–467 (1986).
2. De Clercq, E., Sakuma, T., Baba, M., Pauwels, R., Balzarini, J., Rosenberg, I. and Holy, A. Antiviral Res. 8: 261–272 (1987).
3. Heijtink, R.A., De Wilde, G.A., Kruining, J., Berk, L., Balzarini, J., De Clercq, E., Holy, A., and Schalm, S.W. Antiviral Res. 21: 141–153 (1993).
4. Balzarini, J., Naesens, L., Herdewijn, P., Rosenberg, I., Holy, A., Pauwels, R., Baba, M., Johns, D.G. and De Clercq, E. Proc. Natl. Acad. Sci. USA, 86: 332–336 (1989).
5. Balzarini, J., Naesens, L., and De Clercq, E. Int. J. Cancer 46: 337–340 (1990).
6. Balzarini, J., Verstuyf, A., Hatse, S., Goebels, J., Sobis, H., Vandeputte, M., and De Clercq, E. Int. J. Cancer 61: 130–137 (1995).
7. Verstuyf, A., Sobis, H., Goebels, J., Fonteyn, E., Cassiman, J.J., and Vandeputte, M. Int. J. Cancer 45: 752–756 (1990).
8. Verstuyf, A., Sobis, H., and Vandeputte, M. Int. J. Cancer 44: 879–884 (1989).
9. Friedman, S.J., and Skehan, P. Cancer Res. 39: 1960–1967 (1979).
10. Botte, V., Tramontana, S., and Chieffi, G. J. Endocrinol. 40: 189–194 (1968).
11. Campbell, W.J., Larsen, D., Deb, S., Kwok, S.C.M., and Soares, M.J. Placenta 12: 227–237 (1991).
12. Holy, A., and Rosenberg, I. Collect. Czech. Chem. Commun. 52: 2801–2809 (1987).
13. Balzarini, J., Hao, Z., Herdewijn, P., Johns, D.G., and De Clercq, E. Proc. Natl. Acad. Sci. USA 88: 1499–1503 (1991).
14. Balzarini, J., Holy, A., Jindrich, J., Naesens, L., Snoeck, R., Schols, D. and De Clercq, E. Antimicrob. Agents Chemother. 37: 332–338 (1993).
15. Verstuyf, A., Goebels, J., Sobis, H., and Vandeputte, M. Tumor Biol. 14: 46–54 (1993).
16. Pommier, Y., and Lokiec, F. Chimiothérapie anticancereuse 31: 3083–3089 (1981).

118

INCREASED CELL KILLING AND MUTAGENECITY BY DNA ALKYLATING AGENTS IN CELLS WITH DECREASED TTP POOLS

Masaru Kubota, Yoshihiro Wakazono, and Kenshi Furusho

Department of Pediatrics
Kyoto University
Kyoto 606, Japan

1. INTRODUCTION

Numerous studies have demonstrated that an imbalance of deoxyribonucleoside triphosphates (dNTPs) pools is mutagenic and can alter the lethality and the mutagenicity of DNA-damaging agents [1]. In order to examine the effect(s) of the changes in dNTP pools on DNA fidelity, various drugs, including methotrexate (MTX), fluorodeoxyuridine (FdUrd) and fluorouracil (FUra), have been used. The attempts to utilize these drugs as inhibitors of dNTP synthesis, however, have been hampered because they have several sites of action. For the purpose of investigating whether the reduced TTP pool affected the lethality and the mutagenesis induced by DNA-alkylating agents, we have employed two strategies. First, a thymidylate synthase (TS) inhibitor, ICI D1694 [2], was used. ICI D1694 is a specific and potent TS inhibitor, which exerts its action mainly through metabolizing to its polyglutamate derivatives. Although the monoglutamate of ICI D1694 has similar inhibitory activity against TS and dihydrofolate reductase, the polyglutamate forms are considered to be much more specific for TS [3]. Therefore, treatment of cells with ICI D1694 is expected specifically to decrease intracellular TTP levels. Second is to introduce a mutational loss of thymidine kinase (TK), which is an important salvage enzyme for maintaining TTP pools. In a human promyelocytic leukemia cell line, HL-60, TK deficient mutant had only one fifth of TTP pools of the wild type cells [4]. In the present study, we try to determine the effect of the diminished TTP pool on the killing or mutagenicity induced by DNA-alkylating agents using these strategies.

Purine and Pyrimidine Metabolism in Man IX,
edited by Griesmacher *et al.* Plenum Press, New York, 1998.

2. MATERIALS AND METHODS

2.1. Cell Line

A human promyelocytic leukemia cell line, HL-60, was grown in RPMI-1640 medium supplemented with 10% heat-inactivated fetal bovine serum, L-glutamine (2 mM), penicillin (100 U/ml) and streptomycin (100 μg/ml) (regular medium). For the selection of a mutant deficient in TK, wild type cells were mutagenized by incubation with ethylmethane sulfonate (EMS) (100–200 μg/ml) for 5 hr. Surviving cells were expanded and grown for several months with increasing concentrations of trifluorothymidine. The drug-resistant population were cloned by limiting dilution (0.5 cells/well) in 96-well tissue culture trays. TK activity as measured by a radiochemical method with a use of [^{14}C]-thymidine as a substrate was less than 1% of that of the wild type cells.

2.2. Cytotoxicity Assay

For the cytotoxicity assay, exponentially growing cells were diluted to 2×10^5/ml in regular medium. Cells were incubated with various concentrations of N-methyl-N'-nitro-nitrosoguanidine (MNNG) or EMS for 12 hours, after which they were washed three times with regular medium before being reseeded at the same cell density. After 72 hours, cell viability was enumerated by the trypan blue dye exclusion test. Percent cell survival was calculated as the ratio of the viable cell number in the drug-treated cultures to that in control cultures. Treatment with ICI D1694, kindly provided by Dr.B.M.Vose (ZENECA Pharmaceuticals), was initiated 2 hours prior to the addition of alkylating agents and continued until the end of exposure of DNA-alkylating agents.

2.3. Mutation Frequency

To measure mutation frequency (Mf) at the hypoxanthine phosphoribosyl transferase (HPRT) locus, cells were exposed to various concentrations of alkylating agents for 12 hours. After washing and resuspension in fresh medium, the cells were grown for 7–10 days to allow phenotypic expression. Cells were then cloned in the presence or absence of

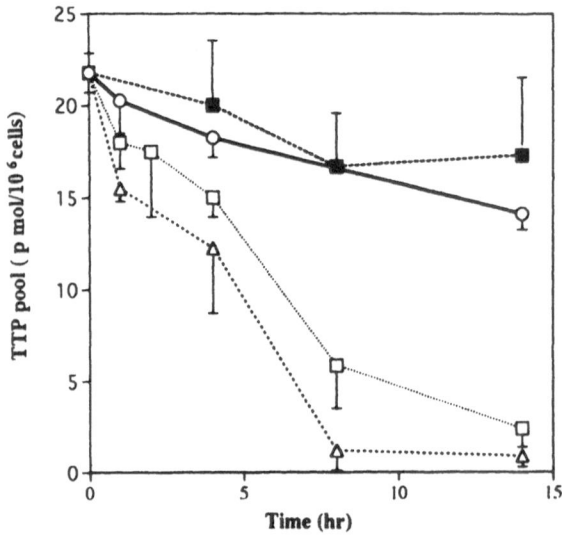

Figure 1. Changes of intracellular TTP pools in HL-60 cells during treatment with various concentrations of ICI D1694; (O) 0.01 μM, (□) 0.02 μM, (△) 0.04 μM, (■) 0.02 μM plus 20 μM thymidine. Results are expressed as pmol/10^6 cells and are the mean ± SD of three experiments.

5 µg/ml 6-thioguanine (6-TG) in 96-well, flat-bottomed microtiter plates. Each well without 6-TG (control plate) received one cell and each well with 6-TG (selection plate) received 4×10^3 or 2×10^4 cells. On day 15, each well was observed with an inverted microscope to determine the colony formation. Cloning efficiencies (CEs) were calculated from the number of colony-negative wells, assuming that cells had Poisson distribution. Mf was expressed as the ratio of CE with 6-TG to CE without 6-TG.

2.4. Measurement of TTP Pool

Intracellular nucleotides were extracted overnight at -20°C with 60 % methanol. The supernatants were lyophilized and dissolved in distilled water immediately before analysis. TTP pools were evaluated by the method of Garrett and Santi with high pressure liquid chromatography [5].

3. RESULTS

3.1. Effect of ICI D1694 on TTP Pool

Treatment of HL-60 cells with ICI D1694 decreased TTP pools in a dose and time dependent fashion (Fig. 1). Simultaneous addition of 20 µM thymidine with 0.02 µM ICI D1694 returned TTP levels to levels almost comparable to those of untreated control cells. Three other dNTP pools did not change significantly during treatment of ICI D1694 (data not shown).

3.2. Effect of ICI D1694 on the Cytotoxicity and the Mutagenesis Induced by DNA-Alkylating Agents

Based on the experiments described above, cells were pretreated with 0.02 µM ICI D 1694 for 2 hours followed by treatment with MNNG or EMS for 12 hours without washings. ICI D1694 pretreatment augmented several-fold the sensitivity to the cytotoxic effect (Fig. 2) and Mfs at the HPRT locus (Fig. 3) induced by these DNA-alkylating agents. In fact, ID_{50} values as determined from Fig. 2 indicate that ICI D 1694-treated cells became approximately 15- and 3.6-fold more sensitive to MNNG and EMS, respectively, than did

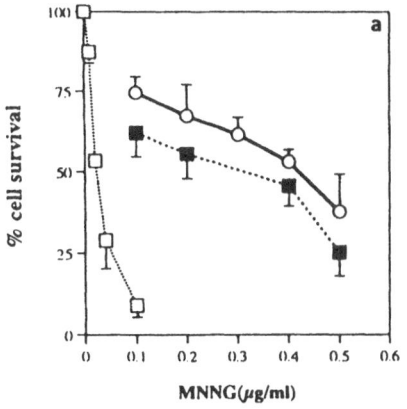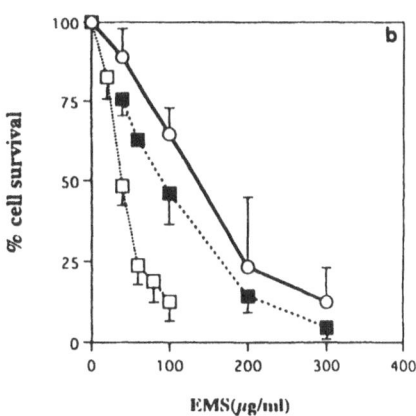

Figure 2. Dose response of cytotoxic effects of MNNG (a) or EMS (b) treated with ICI D1694; (O) 0 µM, (□) 0.02 µM, (■) 0.02 µM plus 20 µM thymidine, in HL-60 cells. Results are shown as the mean ± SD of four experiments.

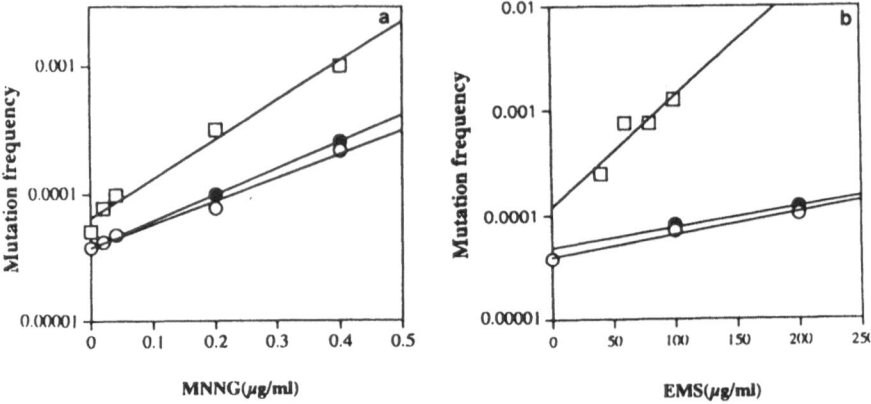

Figure 3. Mf at the HPRT locus induced by MNNG (a) or EMS (b) treated with ICI D1694; (O) 0 μM, (□) 0.02 μM, (■) 0.02 μM plus 20 μM thymidine, in HL-60 cells. Data represent the mean of three experiments.

untreated cells. Mfs were also found to increase 3.1-fold by 0.4 μg/ml, MNNG and 12-fold by 100 μg/ml, EMS. When HL-60 cells were treated with 0.02 μM ICI D 1694 and 20 μM thymidine, TTP levels returned to the levels of untreated cells (Fig. 1). Concomitantly, these cells had a similar cell survival and Mf at the HPRT locus following treatment with MNNG or EMS to those of control cells without any pretreatment (Figs. 2 and 3).

3.3. TTP and Other dNTP Pools in the Wild Type and the TK Deficient Mutant

TK⁻ mutants had approximately one-fifth of TTP levels of the parent HL-60 (WT; 23.5 pmol/10^6 cells, TK⁻; 4.9 pmol/10^6 cells). On the other hand, the wild type and TK deficient cells had comparable levels of dATP, dGTP and dCTP (data not shown).

3.4. Comparison of the Cytotoxicity and Mutation at the HPRT Locus Induced by DNA-Alkylating Agents in the Wild Type and TK⁻ HL 60

Fig. 4 demonstrates the dose response curves of the cytotoxicity induced by MNNG or EMS. HL-60 TK⁻ cells turned out to be more sensitive to these agents than were wild type cells, showing the ID_{50} values approximately 15- and 7-fold less against MNNG and EMS, respectively. Spontaneous Mfs at the HPRT locus in the wild type and TK⁻ cells were comparable (data not shown). Although pretreatment of these cells with either MNNG or EMS significantly increased Mfs in a dose dependent manner, the effect was, however, much more striking in TK⁻ cells (Fig. 5).

4. DISCUSSION

A balanced supply of dNTPs is essential not only for the fidelity of DNA replication but also for the repair of DNA damage. If the synthesis of dNTPs were perturbed, the genotoxicity of various DNA-damaging agents would be expected to increase. Among four dNTPs, the effects of perturbation of TTP have been most extensively studied [6]. Two major methods are available to reduce intracellular TTP levels. First strategy is to use inhibitors for the enzyme(s) involved in TTP synthesis, especially TS. For this purpose,

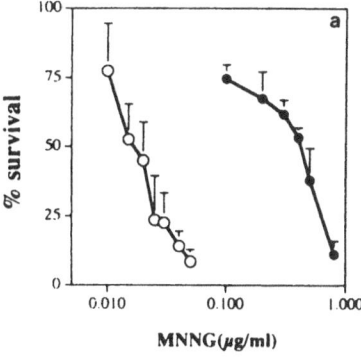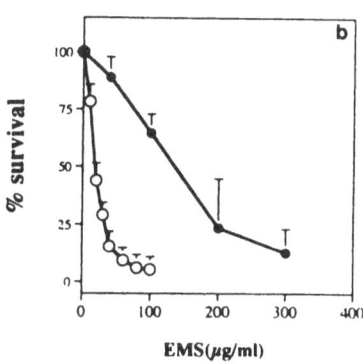

Figure 4. Dose response of cytotoxic effects of MNNG (a) or EMS (b) in HL-60 wild type (●), and TK⁻ cells (O). Results are shown as the mean ± SD of 4 experiments.

MTX, FdUrd and FUra have been widely used [6], but these drugs are also known to have multiple sites of action. On the other hand, ICI D1694, a structural analogue of 5,10-CH_2FH_4, is a specific and potent TS inhibitor. In this report, we decreased TTP levels by using ICI D1694 and showed an increased sensitivity to the cytotoxic and mutagenic effects of DNA-alkylating agents. The specificity of ICI D1694 in our experiments has been supported by the following observations. First, treatment with ICI D1694 did not have any significant influence on three other dNTP pools. Second, a simultaneous addition of thymidine (20 μM) to the culture canceled the effect of ICI D 1694 in parallel with a replenishment of TTP levels. These results clearly demonstrate that TTP pool imbalance plays an important role in the sensitivity to DNA-alkylating agents. Moreover, 0.01 μM ICI D1694, which diminished TTP levels to a lesser extent, did not have any significant effects on the sensitivity to alkylating agents (data not shown). This fact may indicate that there is a threshold for TTP level to exhibit a defect in repair capability after DNA damage by alkylating agents.

Second is the use of mutant strains which lacks the enzyme(s) responsible for the synthesis of TTP. For example, Ayusawa et al. have demonstrated an increase of killing effect by bleomycin in mouse FM3A cell mutants with altered ribonucleotide reductase and TS [7]. McKenna et al. described that TK deficiency in P388 mouse lymphoma cells led to increased ultraviolet-, EMS- and MNNG- induced cell killing and mutagenesis [8]. Rat glioma cell lines deficient in TK was also hypersensitive to gamma irradiation [9]. Although

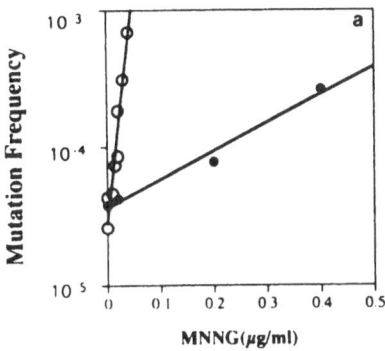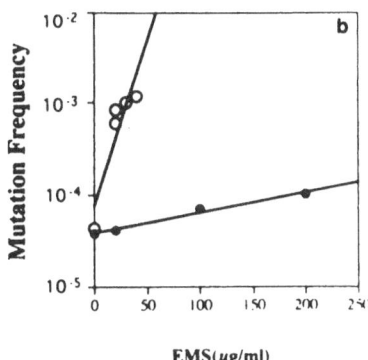

Figure 5. Mf at the HPRT locus induced by MNNG (a) or EMS (b) in HL-60 wild type (●) and TK⁻ cells (O). Data represent the mean of three experiments.

both investigators postulated the role of TTP pool perturbation as a consequence of TK loss, they did not measure TTP levels. We established HL-60 TK deficient strains which had about 20% of TTP levels of parent cells [4] and found that this mutant had an increased lethality and Mfs against DNA-alkylating agents [10].

The reason why the diminished TTP pool was associated with increased cytotoxicity and mutagenicity by DNA alkylating agents is not clear at present. MNNG and EMS can cause a variety of alkylated-DNA lesions, among which O^6-alkylguanine is considered to be most mutagenic. This adduct mispairs with thymine during DNA synthesis, resulting in G:C to A:T transition [11]. G:C to A:T transition is quite likely to occur in cells with high TTP/dCTP ratio, i.e. increased TTP or decreased dCTP levels. In opposition, Kohalmi et al. have reported that the elevated dCTP/TTP ratio decreased the frequency of G:C to A:T transition in yeast [12]. Therefore, our present data seem to be inconsistent with these previous investigations. To further clarify this issue, we are going to undertake a molecular analysis of HPRT mutants induced by DNA-alkylating agents in either cells treated with ICI D1694 or TK$^-$ cells.

ACKNOWLEDGMENTS

We thank Elsevier Science-NL (Amsterdam, The Netherlands) for·kind permission of reprinting Figs. 4 and 5 from Mutation Res., 362, 119–126, 1996; Thymidine kinase deficient cells with decreased TTP pools are hypersensitive to DNA alkylating agents, by Wakazono Y. et al.

REFERENCES

1. Kunz BA, Mutagenesis and deoxyribonucleotide pool imbalance. Mutation Res 200: 133–147, 1988.
2. Ward WH, Kimbell R and Jackmann AL, Kinetic characteristics of ICI D 1694: a quinazoline antifolate which inhibits thymidylate synthase. Biochem Pharmac 43: 2029- 2031, 1992.
3. Jackman AL, Marsham PR, Moran RG, Kimbell R, O'Connor BM, Hughes LR and Calvert AH, Thymidilate synthase inhibitors: the in vivo activity of a series of hetero- cyclic benzoyle ring modified 2-desamino-2-methyl-N10-substituted-5,8-dideazo-folates. Adv Enzyme Regul 31: 13–27, 1991.
4. Wakazono Y, Kubota M, Sano H, Matsubara K, Hirota H, Kuwakado K and Adachi S, Different effect of thymidine kinase loss on TTP pools: comparison among human leukemia cell lines. Mutation Res 304: 295–300, 1994.
5. Garrett C and Santi DV, A rapid and sensitive high pressure liquid chromatography assay for deoxyribonucleoside triphosphates in cell extract. Analyt Biochem 99: 268–273, 1979.
6. Kunz BA, Kohalmi SE, Kunkel TA, Mathews CK, McIntosh EM and Reidy JA, International commission for protection against environmental mutagens and carcinogens. Deoxyribonucleotide triphosphate levels: a critical factor in the maintenance of genetic instability. Mutation Res 318: 1–64, 1994.
7. Ayusawa D, Iwata K and Seno T, Unusual Sensitivity to bleomycin and joint resistance to 9-β-D-Arabino-furanosyladenine and 1-β-D-Arabinofuranosylcytosine of mouse FM3A cell mutants with altered ribonucleotide reductase and thymidylate synthase. Cancer Res 43: 814–818, 1983.
8. Mckenna PG, Yasseen AA and McKelvey VJ, Evidence for indirect involvement of thymidine kinase in excision repair processes in mouse cell lines. Somat Cell Mol Genet 11: 239–246,1985.
9. Al-Nabulsi I, Takamiya Y, Voloshin Y, Dritschilo A, Martuza RL and Jorgensen TJ, Expression of thymidine kinase is essential to low dose radiation resistance of rat glioma cells. Cancer Res 54: 5614–5617, 1994.
10. Wakazono Y, Kubota M, Furusho K, Liu L and Gerson SL, Thymidine kinase deficient cells with decreased TTP pools are hyper -sensitive to DNA alkylating agents. Mutation Res. 362: 119–126, 1996.
11. Loechler EL, Green CL and Essigman JM, in vivo mutagens and by O^6-methylguanine built into a unique site in a viral genome. Proc Natl Acad Sci USA 81: 6271–6275, 1984.
12. Kohalmi SE, Roch H and Kunz BA, Elevated intracellular dCTP levels reduce the induction of GC→AT transitions in yeast by ethyl methanesulfonate or N-methyl-N'-nitro-N-nitrosoguanidine but increase alkylation-induced GC→CG transversions. Mutagenesis 8: 457–465, 1993.

119

N[4]-HYDROXY-5-HALOGENO-2'-DEOXYCYTIDINES AND THEIR 5'-MONO-PHOSPHATES AS INHIBITORS OF THYMIDYLATE SYNTHASE AND *IN VITRO* ANTILEUKEMIC AGENTS

K. Felczak,[1] M. Bretner,[1] J. M. Dzik,[2] B. Gołos,[2] Z. Zieliński,[2] W. Rode,[2] and T. Kulikowski[1]

[1]Institute of Biochemistry and Biophysics
Polish Academy of Sciences
5a Pawińskiego St., 02-106 Warszawa, Poland
[2]Nencki Institute of Experimental Biology
Polish Academy of Sciences
3 Pasteur St., 02-093 Warszawa, Poland

1. INTRODUCTION

A rare example of dUMP analogue that is C(4)-substituted and, nevertheless, a strong inhibitor of thymidylate synthase, a target in chemotherapy, is N[4]-hydroxy-2'-deoxycytidine-5'-mono-phosphate (N[4]-OH-dCMP). It inactivates the enzyme *via* time-dependent formation of a ternary covalently bound complex with thymidylate synthase and N[5,10]-methylenetetrahydro-folate,[1,2] with 5-fluoro substitution (in N[4]-OH-FdCMP) potentiating this process.[3] In order to test antitumour activity of N[4]-hydroxy-FdCyd (N[4]-OH-dCyd), as well as its 5-chloro (N[4]-OH-CldCyd) and 5-bromo (N[4]-OH-BrdCyd) congeners, they were studied as murine leukemia L5178Y cell growth inhibitors. Inhibitory activities with purified L1210 cell thymidylate synthase of N[4]-hydroxy-5-chloro-dCMP (N[4]-OH-CldCyd) and N[4]-hydroxy-5-bromo-dCMP (N[4]-OH-BrdCMP) were also studied.

2. MATERIALS AND METHODS

2.1. General Procedure for the Synthesis of N[4]-Hydroxy-5-Halogeno-2'-Deoxycytidines 5a–c

Corresponding 5-halogeno-2'-deoxyuridine 1a-c (2.43 mmol) was dissolved in 30 ml Ac$_2$O, DMAP added and stirred overnight at room temperature. Clear solution was coeva-

Purine and Pyrimidine Metabolism in Man IX,
edited by Griesmacher *et al.* Plenum Press, New York, 1998.

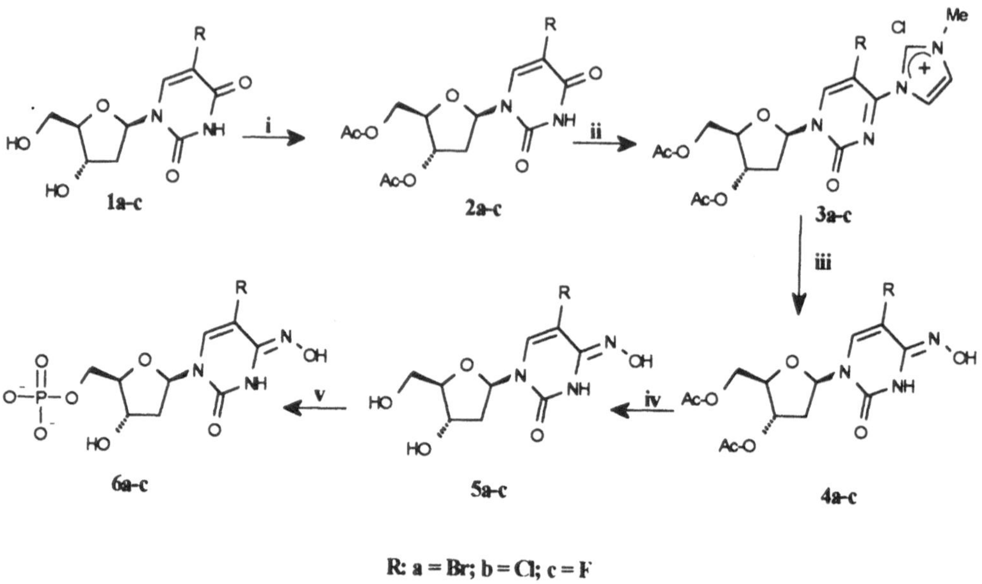

R: a = Br; b = Cl; c = F

Scheme 1. Synthesis of N^4-hydroxy-5-halogeno-2'-deoxyuridines and their 5'-monophosphates i: Ac₂O/Py; ii: MeIm, POCl₃, Py; iii: NH₂OH; iv: NH₃/MeOH; v: p-nitrophenyl phosphate/wheat-shoot phosphotransferase, 37°C.

porated with the mixture of toluene and ethanol to give crystalline compounds 2a–c in nearly quantitative yield. These were dissolved in 15 ml of dry acetonitrile and added to the N-methylphosphoimidazolide (prepared from 30 mmol of N-methylimidazole and 9 mmol POCl₃). The mixture was stirred for 2h at room temperature and methanolic solution of hydroxylamine added. Stirring was continued for 2h at room temperature and evaporated to dryness in vacuo. The residue was dissolved in 20 ml water and extracted with ethylacetate. The extract was washed with 30 ml water, dried over MgSO₄, evaporated to dryness in vacuo to give ca 1g of crude product(s), which was dissolved in methanol and deposited on Dowex 50W(H⁺) column and eluted with a gradient of TEA in MeOH (0–1M). The fractions containing compounds 4a-c were deblocked with methanolic sodium methanolate to give free N^4-hydroxy nucleosides 5a-c in ca 80% yield. UV, λ_{max}nm (ϵ). 5a: (pH 0) 225 (9000), 297 (11350); (pH 12) 244.5 (10300), 282 (6700)(shoulder). 5b: (pH 0) 226 (11950), 294 (15300); (pH 12) 250 (14200), 284 (9400). 5c: (pH 0) 222 (10600), 286 (14600); (pH 12) 253 (13500).

2.2. General Procedure for Synthesis of Nucleoside 5'-Phosphates

To a solution of 0.02 mmol of appropriate nucleoside analogue (5a-c) in 0.6 ml of 0.1 M acetate buffer pH 4, was added 112 mg (0.3 mmol) of p-nitrophenylphosphate, and the pH was brought to 4 by addition of concentrated acetic acid. To this was added 0.6 ml of a crude extract of wheat shoot nucleoside phosphotransferase. The mixture was incubated at 37°C for 40 h, concentrated to half-volume and extracted 3 times with ether. The aqueous layer was subjected to chromatography on Whatman paper 3MM and developed with the solvent iPrOH-H₂O-NH₄OH (7:2:1). The 5'-phosphate band was eluted with water, and the eluate brought to dryness to yield appropriate nucleoside 5'-phosphates 6a–c.

2.3. General Procedure for Enzymatic Hydrolysis of Nucloside 5'-Phosphates

To a solution of 40 μl 0.1 M Tris/HCl buffer pH 8.8 + 20 μl 0.1 M $MgCl_2$ was added 0.05 μmol of nucleoside 5'-phosphate, followed by 5 μl of a 10 mg/ml stock solution of *Crotalus adamanteus* (E.C. 3.1.3.5.) snake venom phosphodiesterase. Following 1h incubation at 37°C, nucleoside 5'-phosphate was quantitively converted to the parent nucleosides 5a–c. An additional control, 2'(3')-GMP was unaffected, pointing to the absence of non-specific phosphatases.

2.4. Cell Lines

Mouse leukemia L5178Y cells were grown as reported earlier.[4] FdUrd-resistant line was developed by growing in the presence of the drug in the cell medium. FdUrd concentration was increased (stepwise and after cells became adapted to the previous concentration) up to 860 nM.

2.5. *In Vitro* Cell Growth Inhibition

The influence of each analogue on exponentially growing cell viability, and [¹⁴C]leucine and [³H]thymidine incorporation was followed, and IC_{50} values determined as previously described.[4]

2.6. Thymidylate Synthase

Highly purified preparations of thymidylate synthases from parental and FdUrd-resistant L1210 cells are described in more detail elsewhere.[3,5] The [5-³H]dUMP tritium release activity assay was performed as previously described.[3] The dUMP analogues were added to the reaction mixtures as neutral aqueous solutions.

2.7. Kinetic Studies

To identify the type of inhibition involved, the effects of the N⁴-OH-dCMP analogues on the dependence of reaction rate on dUMP concentration, in the form of Liveweaver-Burk plots, were analyzed as previously reported.[3] Quantitative analyses of thymidylate synthase inhibition, leading to time-dependent inactivation of the enzyme, were performed as earlier described.[4]

3. RESULTS AND DISCUSSION

Novel, convenient "one-pot" procedure based on activation of position 4 of corresponding O'-acetylated 5-halogeno-2'-deoxyurydines 1a-c with the use of N-methylimidazolide was applied to the synthesis of N⁴-hydroxy-5-halogenocytidines 5a–c. This procedure gave better yields (~80%) than previosly reported methods involving hydroxylamination of 2'-deoxycytidines[6] and 4-thio2'-deoxyuridines,[7] and eliminated the formation of undesired intermediate—4,6-dihydroxylamino-5,6-dihydroderivative. Among N⁴-hydroxy-5-halogeno-2'-deoxycytidines only N⁴-OH-FdCyd caused strong growth in-

Table 1. Inhibition of cell growth by N^4-hydroxy-5-halogeno-2'-deoxynucleosides

	Drug	Growth assay	[^{14}C]Leu incorporation	[^{3}H]Thd incorporation
			IC_{50}[a] (μM)	
L5178Y parental cells	FdUrd[b]	0.0020	0.0024	0.0020
	N^4-OH-FdCyd	0.020[c]	0.018[c]	0.026[c]
	N^4-OH-CldCyd	1.15[d]	1.24[d]	1.48[d]
L5178Y FdUrd-resistant cells	FdUrd	0.61[c]	0.67[d]	0.64[c]
	N^4-OH-FdCyd	0.051[c]	0.053[d]	–
	N^4-OH-CldCyd	1.66[c]	3.01[c]	–

[a]IC_{50} is drug concentration required for 50% reduction in cell number, [^{14}C]Leu or [^{3}H]Thd incorporation; [b]Ref. 4; [c-e]Results are means of two experiments which did not differ by more than 5%[c], 10%[d], 20%[e].

hibition with both parental and FdUrd-resistant L5178Y cell lines, while its 5-chloro congener was 10^2-fold less potent.

Both N^4-OH-BrdCMP and N^4-OH-CldCMP were found to be moderately potent competitive, slow-binding (K_i values with both 10^3-fold higher than with previously studied N^4-OH-FdCMP) inhibitors of two thymidylate synthases differing in sensitivity to time-dependent inactivation by FdUMP.

Thus potency of cell growth inhibition by the N^4-OH-dCyd analogues was correlated with thymidylate synthase sensitivity for the corresponding nucleotides. 5-chloro and 5-bromo (in contrast to 5-fluoro) substitutions at C(5) of N^4-hydroxy-dCMP weakened inhibition of the enzyme. In conclusion it can be assumed that N^4-OHFdCyd may be regarded as the potential cytotoxic agent against FdUrd resistant leukemias.

Table 2. Parameters for inactivation by FdUMP, N^4-OH-FdCMP, N^4-OH-CldCMP and N^4-OH-BrdCMP of thymidylate synthase from L1210 cells parental (L1210P) and FdUrd-resistant (L1210R) cells

	Enzyme source	K_i' (μM)	K_i'' (μM)	k_2' (min^{-1})	k_2'' (min^{-1})
FdUMP[a]					
	L1210P	0.0018	0.020	0.17	0.12
	L1210R	0.0122	0.014	0.25	0.06
N^4-OH-dCMP[a]					
	L1210P	0.063	0.226	0.17	0.02
	L1210R	0.184	1.46	0.20	0.09
N^4-OH-FdCMP[a]					
	L1210P	0.073	0.056	0.24	0.07
	L1210R	0.093	0.071	0.24	0.06
N^4-OH-CldCMP					
	L1210P	20 ± 2 (3)[b]	16 ± 2 (3)	0.17 ± 0.01 (3)	0.06 ± 0.00 (3)
	L1210R	31 ± 9 (3)	15 ± 6 (3)	0.19 ± 0.03 (3)	0.06 ± 0.02 (3)
N^4-OH-BrdCMP					
	L1210P	33 ± 8 (3)	26 ± 3 (3)	0.23 ± 0.05 (3)	0.07 ± 0.01 (3)
	L1210R	36 ± 5 (3)	11 ± 1 (3)	0.22 ± 0.03 (5)	0.05 ± 0.01 (5)

[a]From ref. 3, [b]Results are presented as means ± SEM, followed by the number of separate experiments in parentheses.

ACKNOWLEDGMENT

This work was supported by KBN grants 4P05F03011 p01 and 4P05F03011 p02.

REFERENCES

1. Lorenson, M.Y., Maley, G.F. & Maley, F. The purification and properties of thymidylate synthetase from chick embryo extracts. *J. Biol. Chem.* 242: 3332 (1967).
2. Goldstein, S., Pogolotti, A.L., Jr., Garvey, E.P. & Santi, D.V. Interaction of N⁴-hydroxy-2'-deoxycytidilic acid with thymidylate synthetase. *J. Med. Chem.* 27: 1259 (1984).
3. Rode W., Zieliński, Z., Dzik J.M., Kulikowski T., Bretner M., Kierdaszuk B., Cieśla, J. and Shugar D. Mechanism of inhibition of mammalian tumor and other thymidylate synthases by N⁴-hydroxy-dCMP, N⁴-hydroxy-5-fluoro-dCMP, and related analogues. *Biochemistry* 29: 10835 (1990).
4. Dzik, J.M., Bretner, M., Kulikowski, T., Gołos, B., Jarmuła, A., Poznański, J., Rode, W., Shugar, D. Synthesis and interactions with thymidylate synthase of 2,4-dithio-analogues of dUMP and 5-fluoro-dUMP. *Biochim. Biophys Acta* 1293: 1 (1996).
5. Zieliński Z., Dzik J.M., Rode W., Kulikowski T., Bretner M., Kierdaszuk B. and Shugar D. Interaction of N⁴-hydroxy-dCMP and N⁴-hydroxy-5-FdCMP with L1210 Thymidylate synthase differing in sensitivity towards 5-FdUMP inhibition, in "Chemistry and Biology of Pteridines 1989. Pteridines and Folic Acid Derivatives", Curtius, H.-Ch., Ghisla, S. and Blau, N., eds., Walter de Gruyter, New York, pp 817–820, 1990.
6. Brown, D.M., Shell, P. The reaction of hydroxylamine with cytosine and related compounds. *J. Mol. Biol.* 3: 709 (1961).
7. Fox, J.J., Van Praag, d., Wempen, I., Doerr, I.L., Cheong, L., Knoll, J.E., Eidinoff, M.L., Bendich, A., and Brown, G.B. Thiation of nucleosides. II. Synthesis of 5-methyl-2'-deoxycytidine and related pyrimidine nucleosides. *J. Am. Chem. Soc.* 81: 178 (1959).

120

SUBSTRATE/INHIBITOR SPECIFICITIES OF HUMAN DEOXYCYTIDINE KINASE (dCK) AND THYMIDINE KINASES (TK1 AND TK2)

Borys Kierdaszuk,[1][*] Krzysztof Krawiec,[1] Zygmunt Kazimierczuk,[1]
Ulla Jacobsson,[2] Nils G. Johansson,[3] Birgitte Munch-Petersen,[4]
Staffan Eriksson,[5] and David Shugar[1,6]

[1]University of Warsaw
Department of Biophysics
93 Żwirki i Wigury St.
PL-02089 Warsaw, Poland
[2]Department of Organic Chemistry
Royal Institute of Technology
S-10044 Stockholm, Sweden
[3]Medivir AB
S-14144 Huddinge, Sweden
[4]Roskilde University
Department of Life Sciences and Chemistry
DK-4000 Roskilde, Denmark
[5]Biomedical Centre
Department of Veterinary Medical Chemistry
S-75123 Uppsala, Sweden
[6]Polish Academy of Sciences
Institute of Biochemistry and Biophysics
PL-02106 Warsaw

1. ABSTRACT

Substrate/inhibitor specificities of nucleoside analogues with modified sugar moieties towards highly purified deoxycytidine kinase (dCK) and thymidine kinases (TK1 and TK2) from human leukemic spleen have been examined. Substrate activities of cytosine nucleosides vs dCK were as follows: 2′-fluoro-dC > 2′-O-methyl-C > araC > 2′-fluoro-2′-deoxy-araC > 3′-O-methyl-dC = 3′-fluoro-2′,3′-ddC > cytosine β-L-riboside >

[*] Corresponding author: fax: +48-(0)22-220248; e-mail: borys@asp.biogeo.uw.edu.pl.

Purine and Pyrimidine Metabolism in Man IX,
edited by Griesmacher *et al.* Plenum Press, New York, 1998.

2′,3′-ddC > C = 1-(4-hydroxy-1,2-butadienyl)-cytosine (cytalene) = 2′-azido-dC. Modified purine nucleosides were only feeble substrates: ara-A > 2′-fluoro-2′, 3′-dideoxy-ara-A = 2′-O-methyl-A. With TK1 and TK2, similar sugar-modified analogues of dU and dT were feeble substrates. Surprisingly α-dT was a relatively good substrate, as well some β-L-ribonucleo-sides. Several 5′-substituted analogues of dC were good non-substrate inhibitors of dCK and, to a lesser extent, of TK2. The overall data are relevant to the role of these enzymes in "activation" (by phosphorylation) of nucleoside analogues with antiviral and antitumor activities.

2. INTRODUCTION

Deoxycytidine kinase (dCK, NTP:deoxycytidine-5′-phosphotransferase, EC 2:7.1.74) catalyzes the phosphorylation of dC to dCMP in the presence of a nucleoside-5′-triphosphate phosphate donor. The cytosolic enzyme has been isolated and purified to apparent homogeneity from human leukemic spleen[1] and leukemic human T-lymphoblasts.[2] The enzyme also phosphorylates purine deoxyribonucleosides and some nucleoside analogues.[3–5]

Thymidine kinases (ATP:thymidine-5′-phosphotransferases, EC 2:7.1.21) catalyze the phosphorylation of dT to dTMP. In eucaryotic cells, two thymidine kinases (TK) are known: TK1 - in cytoplasm and TK2 - predominantly localized in mitochondria. TK1 has been purified to homogeneity from HeLa cells[6] and human leukemic spleen.[7] TK2, most highly purified from human leukemic spleen[7] exhibits substrate specificity partially overlapping with dCK.

All three enzymes exhibit very broad substrate and inhibitor specificities.[3,4,7] The present investigation is an extension of the foregoing, to obtain further insight into structural requirements of the enzymes for the sugar moiety of substrates and inhibitors, and to provide a guide to the role of these enzymes in phosphorylation of nucleoside antitumor and antiviral agents.

3. MATERIALS AND METHODS

3.1. Materials

[5-³H]-2′-deoxycytidine (19.3 Ci/mmol), [6-³H]-2′-deoxythymidine (29 Ci/mmol) and [γ-³²P]-ATP (~3000 Ci/mmol) were from Amersham, UK. [2,8–³H]-2′-deoxyadenosine (46 Ci/mmol) was from Moravek Biochemicals, Inc. (Brea, California). Unlabeled nucleosides (dC, dT, dA), ATP and BSA (fraction V, for enzyme stabilisation) were from Sigma. Nucleoside analogues were checked for purity by various procedures, including NMR.

3.2. Enzymes and Assays

All three enzymes were highly purified to apparent homogenity from human leukemic spleen: dCK 6000-fold to a specific activity of 260 nmols dCMP formed per min per mg, TK1 20000-fold to a specific activity of 9.5 μmols dTMP formed per min per mg, TK2 approximately 20000-fold to a specific activity of 0.5 μmols dTMP formed per min per mg.

Activity of dCK was routinely followed by a radiochemical procedures described by Ives and Wang[8] with modifications reported by Kierdaszuk et al.[9] Activity of thymidine kinases was measured as described by Munch-Petersen.[10] The phosphate transfer assay was performed as previously.[4,11] The dC kinase activity of dCK and dT kinase activities of TK1 and TK2 were set to 1.0 as indicated in Table 1. The sensitivity of this transfer assay was in the same range as for the usual assay with radiolabelled nucleoside substrates.

Table 1. Relative phosphorylation of 100 μM nucleoside analogs by pure human dCK and TK1 and TK2 using 100 μM [γ-^{32}P]ATP as phosphate donor[a]

Nucleoside analogs		Relative activity		
Base	Pentose ring[b]	dCK	TK1	TK2
Cytosine	2′-deoxy	1.0	0	0.9
	3′-deoxy	0.13	0	0
	2′-O-methyl	1.9	n.d.	n.d.
	3′-O-methyl	0	n.d.	n.d.
	3′-O-methyl-2′-deoxy	0.6	0	0.06
	3′-O-acetyl-2′-deoxy	0.18	n.d.	n.d.
	3′-O-ethyl-2′-deoxy	0.08	0	0
	2′-fluoro-2′-deoxy	3.0	0	0.3
	2′-fluoro-2′-deoxyarabinose	1.0	n.d.	n.d.
	3′-fluoro-2′,3′-dideoxy	0.6	0	0
	2′-azido-2′-deoxy	0.2	n.d.	n.d.
	2′,3′-dideoxy	0.3	0	0
	2′,3′-dideoxy-2′,3′-didehydro	0.04	n.d.	0
	2′,3′-isopropylidene	0.09	n.d.	n.d.
	ribose	0.2	0	0
	β-L-ribose	0.4	n.d.	0.07
	arabinose	1.2	n.d.	0.05
	cytalene	0.2	n.d.	0
Adenine	2′-deoxy	3.5	0	0
	2′-O-methyl	0.05	n.d.	n.d.
	3′-O-acetyl-2′-deoxy	0.02	n.d.	n.d.
	2′-fluoro-2′,3′-dideoxyarabinose	0.07	0	0
	arabinose	0.5	0	0
Guanine	2′-deoxy	2.5	0	0
	arabinose	0.06	0	0
	β-L-ribose	0.1	n.d	n.d.
Uracil	2′-deoxy	0.06	1.0	1.0
	2′-chloro-2′-deoxy	0	n.d.	0
	3′-O-methyl-2′-deoxy	0	0	0.03
	3′-O-ethyl-2′-deoxy	0	0	0.01
	ribose	0	0	0.04
	2′,3′-dideoxy	0	0.1	0.02
	arabinose	0	0	0.2
Thymine	2′-deoxy	0.02	1.0	1.0
	2′,3′-dideoxy	0	0.4	0.04
	3′-azido-2′,3′-dideoxy	0	0.4	0.05
	3′-fluoro-2′,3′-dideoxy	0	0.3	0
	ribose	0	0.02	0.03
	arabinose	0	0	0.6
	α-ribose	0	0.36	0.35
	β-L-ribose	n.d.	0.05	0.13

[a]dC activity of dCK and dT activity of TKs were taken as 1.0.
[b]This refers to ribose, except where otherwise indicated.
n.d., not determined.

4. RESULTS AND DISCUSSION

Phosphorylation of nucleoside analogs by dCK and by TK1 and TK2 are key processes in metabolic activation, very often the rate-limiting step in the metabolic flux of nucleoside drugs to their active forms at the triphosphate level. Here we compare the activities of all three enzymes with nucleoside analogs modified on the sugar ring using a $[\gamma\text{-}^{32}P]$ATP phosphate transfer assay with use of unlabelled phosphate acceptors. The relative activities of the enzymes with various nucleosides, grouped according to the base moiety, are shown in Table 1.

The enzymes are relatively specific for the pentose moiety of nucleoside substrates, although one cytosine acyclonucleoside, 1-(4-hydroxy-1,2-butadienyl)-cytosine (cytalene) exhibited significant substrate activity with dCK, which is also able to phosphorylate cytosine nucleoside analogs with 2′- and 3′-modifications, while modification of the sugar moiety of purine 2′-deoxynucleosides led to drastically decreased activity *vs* dCK, and to undetectable activity with TK1 and TK2. Substrate activities of cytosine nucleosides *vs* dCK were as follows: 2′-fluoro-dC > 2′-O-methyl-C > araC > 2′-fluoro-2′-deoxy-araC > 3′-O-methyl-d = 3′-fluoro-2′,3′-ddC > cytosine L-riboside > 2′,3′-ddC > Cyd = cytalene = 2′-azido-dC (Table 1). Modified purine nucleosides exhibited lower substrate activities: ara-A > 2′-fluoro-2′, 3′-dideoxy-ara-A = 2′-O-methyl-A.

Similar sugar-modified analogues of dU and dT were feeble substrates *vs* TK1 and TK2. But, surprisingly, α-dT was a relatively good substrate. Several dU analogues, non-substrates of TK1 and only feeble substrates of TK2, proved to be inhibitors which distinguished between these two enzymes, e.g. K_i values for 2′-chloro-dU *vs* TK1 and TK2 were >5 and 0.025 mM, for 3′-O-methyl-dU, 0.3 and 0.015 mM, and for 3′-O-ethyl-dU, 1.1 and 0.01 mM. Some analogues of deoxyadenosine with a modified sugar ring (3′- deoxy-, 2′,3′-dideoxy-, 2′-tosyl-2′,3′-dideoxy-, 3′-O-methyl-2′-deoxy-, 2′,3′-di-O-acetyl, 3′-fluoro-2′,3′-dideoxy-, 3′-azido-2′,3′-dideoxy-, 2′-azido-2′-deoxyxyloribose) as well as of adenosine (2′,3′-O-p-anisylidene-, 2′,3′-isopropylidene-, 2′-O-methyl-2′-xyloribose) and acycloadenosine, 9-(3,4-dihydroxybutyl)adenine (DHPA) and 9-(4-hydroxy-1,2-butadienyl)adenine (adenalene) are inactive (data not shown). Similar modifications of guanosine also led to inactive compounds.

It should be noted that β-L-ribose nucleosides exhibit substrate activity, usually more so than their β-D-counterparts, recently extensively investigated for β-L-2′-deoxyribonucleosides *vs* dCK.[12]

β-D-2′-Deoxyribonucleosides are more efficient substrates or inhibitors than the corresponding β-D-arabinonucleosides (Table 1), including those with antitumor activity (ara-C, ara-U, ara-T, ara-A), and ribonucleosides (see also[3,4,7]). Amongst β-D-ribonucleosides of cytosine, thymine, uracil, adenine and guanine,[3,4] 5-aza-C[3] and C (Table 1, see also [4]) were the only β-D-ribonucleosides with substrate activity *vs* dCK.

For 2′-, 3′-, or 5′-C-hydroxymethyl analogues of dC, dA and dG, considered as antiviral compounds,[13] only cytidine analogues had significant substrate activity with dCK, while adenosine and guanosine analogues are not phosphorylated by these three enzymes (data not shown). Remarkably, 3′-C-hydroxymethyl-2′,3′-ddC exhibited higher activity than dC, and initial analysis does not exclude possible phosphorylation of the 3′-C-hydroxymethyl group.

Methylation of sugar hydroxyls led to very low (if any) substrate activity of purine nucleosides, while 2′-O-methyl-C exhibited very high substrate activity with dCK (Table 1). From the table it should be noted that blocking of the 3′-OH of cytosine and uracil nucleosides leads to liquidation or marked reduction of substrate activity. This may be related to interaction of this hydroxyl with the enzyme (perhaps by hydrogen bonding), as in the crystal structure of the complex of dT with HSV-1 TK, where the 3'-OH is hydrogen-bonded to Tyr-101.[14,15] Furthermore, with the HSV enzyme, the 5'-OH interacts with Arg-163, and should also be sensitive to chemical modifications. We have previously shown that 5'-O-methyl- and 5'-O-ethyl analogues of 2',5'-ddC are nonsubstrate inhibitors of dCK, like 5'-amino-ddC,[16] which exhibited a pH-dependent inhibition pattern, better at pH 8.6 ($\approx$50 % of neutral form) than at pH 7.2 ($\leq$10 % of neutral form), suggesting that the neutral form has a higher affinity vs dCK.

2'-Deoxyinosine (dIno) exhibited 2- and 3- times lower phosphorylation (Table 1) than dGuo and dAdo, respectively. Their sugar-modified analogues, like 2',3'-ddI, 3'-fluoro-2',3'-ddI, 3'-azido-2',3'-ddI, I and ara-I, have no detectable substrate activity vs dCK (data not shown).

ACKNOWLEDGMENTS

Supported by the Polish State Committee for Scientific Research (KBN, 6P04A03812, UM-418, UM-430), the Swedish Natural Science Research Council and, in part (B.K., K.K., D.S.), by an International Research Scholar's award from the Howard Hughes Medical Institute (HHMI 75195-543401).

REFERENCES

1. Bohman C. & Eriksson S., Biochemistry 27, 4265–4273 (1988).
2. Kim M.-Y, Ikeda S. & Ives D. H., Biochem. Biophys. Res. Commun. 156, 92–97 (1988).
3. Krenitsky T.A., Tuttle J.V., Koszalka G.W., Chen I.S., Beacham L.M., Rideout J.L. & Elion G.B., J. Biol. Chem., 251, 4055–4061 (1976).
4. Eriksson S., Kierdaszuk B., Munch-Petersen B., Öberg B. & Johansson N.G., Biochem. Biophys. Res. Commun. 176, 586–592 (1991).
5. Kierdaszuk B., Bohman C., Ullman B. Eriksson S., Biochem. Pharmacol., 43, 197–206 (1992).
6. Sherley J. L. & Kelly T.J., J. Biol. Chem. 263, 375–382 (1988).
7. Munch-Petersen B., Cloos L., Tyrsted G. & Eriksson S. J. Biol. Chem. 266, 9032–9038 (1991).
8. Ives D. H. & Wang S.-M., Methods Enzymol. 51, 337–345 (1978).
9. Kierdaszuk B. & Eriksson S. Biochemistry 29, 4109–4114 (1990).
10. Munch-Petersen B., Mol. Cell. Biochem. 64, 173–185 (1984).
11. Krawiec K., Kierdaszuk B., Munch-Petersen B., Eriksson S., & Shugar D., Biochem. Biophys. Res. Commun. 216, 42–48 (1995).
12. Verri A., Focher F., Priori G., Imbach J.L., Capobianco M., Gabresi A., Spadari S. Mol. Pharmacol., 51, 132–138 (1997).
13. LePage G.A., Banks P.A., Noujaim M.J. & Buzzell G.R. Cancer. Chemother. 5, 127–131 (1980).
14. Brown D.G., Visse R., Sandhu G., Davies A., Rizkallah P.J., Melitz C., Summers W.C. & Sanderson M.R. Nature Struct. Biol., 2, 876–880 (1995).
15. Wild K., Bohner T., Aurby A., Folkers G., Schultz G.E. FEBS Lett., 368, 289–292 (1995).
16. Krawiec K., Kierdaszuk, B. and Shugar, D. Nucleosid. Nucleotid. 14, 495–499 (1995).

121

INVESTIGATION OF ISOENZYMES OF ADENYLOSUCCINASE IN HUMAN BLOOD CELLS

R. Pagani, R. Guerranti, G. Caldarelli, L. Brogi, G. Landriscina, and
E. Marinello

Institute of Biochemistry and Enzymology
University of Siena

1. OBJECTIVES

Adenylosuccinate lyase (ASL: E.C. 4.3.2.2.) is an important enzyme involved in AMP biosynthesis. The existence of different isoenzymes is not yet clearly demonstrated. The first evidence of specific tissue isoenzymes was reported in preliminary research carried out on mammalian[1] and human tissue: after starvation, rat liver enzyme decrease, while the activity of rat muscle enzyme does not change. Studies on human tissue from a subset of autistic patients,[2] showed that ASL deficiency has severe symptoms in liver, kidney and skin cells, than in skeletal muscle cells.

More direct evidence of ASL isoenzymes was obtained by isoelectrofocusing. Yeast ASL showed only one form while partially purified rat muscle enzyme showed three different bands of enzyme activity.

An increase in ASL activity was reported in human breast and prostate cancer.[3] On the other hand, we found a decrease in this enzyme activity in lymphocytes of B-cell chronic lymphocytic leukemia patients.

These differences in tumor and leukemia blood cells may be due to different isoenzymes.

To clarify the problem of the existence of different isoforms of ASL, we studied its activity and distribution in blood cells. Pure cell fractions are required to demonstrate the existence of isoforms but separation by density gradient centrifugation is not suitable for all fractions because lymphocytes are still contaminated by large number of platelets. To eliminate this contamination we developed a new immunochemical procedure using an antibody specific for a surface platelets receptor. We also developed a procedure for electrophoretic separation of ASL and its visualization which will be useful for detecting isoenzymes in cell extracts.

Purine and Pyrimidine Metabolism in Man IX,
edited by Griesmacher *et al.* Plenum Press, New York, 1998.

2. DESIGN AND METHODS

2.1. Cell Separation

Platelets were eliminated by centrifugating 30 ml blood of healthy patients in EDTA at 500 × g for 10 minutes at 15°C and discarding the supernatant (procedure A). The pellet (12 ml) was resuspended in 10 ml of PBS and centrifuged in 30 ml of polymorphprep[4] at 500 × g for 40 min at 15 °C and then we separated erythrocytes, granulocytes, lymphocytes and the remaining platelets. All the fractions were washed in PBS and centrifuged 3 times at 2000 × g for 10 min, resuspended in the same buffer and counted using a CELL DYN 3000 (Abbott Lab.).

The lymphocyte fraction, still containing platelets, was treated with monoclonal antibody (mouse IgM) raised against the thrombospondin (CD 36) receptor on the surface of platelets. Dynabeads M-450, coated with a secondary rat monoclonal antibody for mouse IgM, were added to the mixture and the complex containing platelets, was blocked on the surface of the tube using a magnet.[5] The purified lymphocytes were recovered.

2.2. Preparation of Cell Extracts

The different fractions were frozen, thawed and sonicated in ice for 20 sec three times, at 40 sec intervals. They were then centrifuged at 50,000 × g for 60 min at 4°C to remove the membrane. Red cell lysis was obtained by dilution (1:4) with 5 mM hypotonic buffer KH_2PO_4 (pH 7.4). Supernatants were recovered. Protein concentration ranged from 0.5 to 1.5 mg/ml.

2.3. Assay of Activity and Protein

Enzyme activity was measured according to Casey[6] by continuous spectrophotometric assay, in a Shimadzu UV 160 double beam spectrophotometer, following the decrease of absorbance at 287 nm due to the transformation of adenylosuccinic acid to adenylic acid. The assay mixtures contained 0.05 M Tris-HCl (pH 7.4, 37°C) and 100 µM adenylosuccinate; the protein content was 200 µg. Protein concentration was determined with Bio-Rad Protein Reagent, using crystalline bovine serum albumin as standard.

2.4. Agarose Gel Electophoresis

We used standard yeast ASL from Sigma (0.2 units/mg protein).

The electrophoretic separation was carried out in 1% agarose gel (agarose M - low electroendosmosis from Pharmacia) which enable the direct assay of the enzyme activity in gel after separation. The 1.5 mm gel prepared in running buffer, was cooled and placed in a humidity chamber for one hour before use. The run was carried out in Tris-Tricine lactate buffer (pH 8.6) for 45 minutes at 10°C with a constant voltage of 300 V in a Multiphor II electrophoresis unit. We ran 3 µl of yeast ASL using "sample foil application" (Pharmacia) and in parallel a marker serum protein mixing with an equal volume of Bromophenol Blu solution. The protein concentration of the samples ranged from 50 to 100 µg. The fractionated protein was stained with a Coomassie Blue.

ASL activity was visualized by a two step reaction producing a fluorescent substance (NADPH) that can be detected with a transilluminator apparatus. In the first step, fumarate (one of the direct products of the ASL reaction) is converted to malate by fuma-

Table 1. Platelets contaminations of PBL fraction
before and after removal with antibody

	Platelets		Lymphocytes	
	$\times 10^6$	% rec	$\times 10^6$	% rec
Blood (30 ml)	3114	100	38	100
Lymphocyte fraction before treatment with antibody	25	0.8	12	32
Lymphocyte fraction after treatment with antibody	0	0	6	16

rase; in the second step malic enzyme trasforms malate and NADP to pyruvate and NADPH.

We overlapped the agar gel containing the incubation mixture (20 mM AMPS, 10 mM Tris-Cl pH 7.4, 2 µl fumarase and malic enzyme from Sigma, 20 µM NADP⁺) into agarose gel for 1 hour at 37°C. The reaction was slow but gave positive results.

3. RESULTS

Platelets were about twice as numerous as lymphocytes before the immunochemical procedure which completely eliminated them. The final recovery of purified lymphocytes was 16% which is sufficient for the assay and for electrophoresis of the enzyme.

In Table 2 we report the ASL activity of each cell fraction. The enzyme was found in all cells even also in erythrocytes and platelets, which do not synthesize DNA.

4. CONCLUSIONS

All blood cells have ASL activity. Platelets ASL activity expressed as nmol/h/mg protein was the highest of all confirming the importance of purification of the other fractions.

No direct evidence of ASL isoenzymes in the human blood cells was found. We obtained the following results:

1. we obtained purified lymphocytes by a three step procedure: a) elimination of most platelets from whole blood (procedure A); b) separation of the cell fraction with polymorphprep; c) immunochemical purification.
2. we developed an electrophoretic separation of the protein and visualized the enzyme in gel, using yeast ASL as reference;

Table 2. Activity of adenylosuccinate lyase in human
blood cells. The activity of enzyme in blood isn't
detectable due to high protein concentration

	pmoles h $\times 10^6$ cell	nmoles h $\times$ mg prot
Platelets	310	222
Lymphocytes	5340	79
Granulocytes	4821	67
Erythrocytes	3800	92

3. we also demonstrated that at pH 8,6 the enzyme was always detectable directly in the gel. The present results we show that yeast ASL has a single band of fluorescence.

Our future aim is to apply this procedure to all human blood cells. The different electric mobility of the enzyme in each cell fraction would be a direct demonstration of the presence of isoenzymes and would also explain the different activity found in platelets, lymphocytes, granulocytes and erythrocytes.

REFERENCES

1. Mack D.O., Smith L.D., (1991), Biochemistry International 23 (5), 855–860
2. Stone R.L., Aimi J., Barshop B.A., Jaeken J., Van den Berghe G., Zalkin H., Dixon J. (1992), Nature Genetics (1), 59–63
3. Reed V.L., Mack D.O., Smith L.D. (1987) Clin. Biochem. 20, 349–351
4. Boyum A. (1968), Scand. J. Clin. Invest. 21, Suppl. 97
5. Aakhus AM., Staven P., Hovig T., Pederson TM., Solum N.O. (1990), British Journal of Haematology 74, 320–329
6. Casey P.J., Abeles R.H., Lowenstein J.M. (1986) J. Biol. Chem. 261, 13637–13642

122

DIHYDROPYRIMIDINE DEHYDROGENASE IN LIVERS FROM MOUSE AND RAT, AND IN HUMAN LIVER, COLON TUMORS, AND MUCOSA IN RELATION TO ANABOLISM OF 5-FLUOROURACIL

G. J. Peters, C. J. van Groeningen, and H. M. Pinedo

Department of Oncology
University Hospital Vrije Universiteit
P. O. Box 7057
1007 MB Amsterdam
The Netherlands

1. INTRODUCTION

Dihydropyrimidine dehydrogenase (DPD) is the first enzyme in the catabolic pathway of pyrimidine bases, uracil and thymine,[1] and of the anticancer agent 5-fluorouracil (5FU) to 5-fluoro-dihydrouracil (5FDHU). In most tissues, including peripheral blood lymphocytes the degradation pathway is limited to DPD.[1] In liver and kidney, which have a high DPD activity FDHU can be degraded subsequently to fluoro-ureidopropionate (FUPA) and fluoro-β-alanine (FBAL).[1] In rat hepatomas with different growth rates DPD activity was lower in highly proliferating tissues;[2] thus tumors usually have a low DPD activity compared to liver. DPD activity may be related to the response to 5FU.[4] DPD is also considered to be the major enzyme regulating 5FU bioavailablity after i.v. or oral administration, i.e. 80–90% of the 5FU dose is degraded and excreted as a breakdown product.[4] In order to increase the availablity of 5FU, inhibition of DPD is an attractive target.[6] Since mice and rats are frequently used model system for the evaluation of new anticancer agents and new 5FU formulations,[7,8] we determined the DPD activity in livers of these animals in comparison to human liver, and determined potential inhibition of 5FU degradation by normal substrates. These data were related to the activity of 5FU anabolic enzymes, and to the activity of these enzymes in colon tumors and colon tumor cell lines.

Purine and Pyrimidine Metabolism in Man IX,
edited by Griesmacher *et al.* Plenum Press, New York, 1998.

2. METHODS

5FU metabolism was studied with $[6-^{14}C]$-5FU. The origins of the tumors and cell lines used for this study are reported previously.[8,9] Livers from Balb-c mice and Wistar rats were always obtained in the morning to prevent alterations due to circadian variation.[11] Human livers were obtained from patients undergoing surgery for non-cancer related diseases. Livers were homogenized in Tris-EDTA (50/1 mM) buffer pH 7.4 using a Potter-Elvehjem apparatus and a 10,000 g supernatant was prepared.

DPD assay was performed with 1 mM NADPH as the co-substrate at 270 µM 5FU. The reaction time varied from 10–60 min and was terminated by heating for 3 min at 95°C. The products were separated from 5FU using thin-layer chromatography (TLC) on Silicagel 60 sheets, as described[11] with 1 M NH_4Ac and Ethanol (1/5; v/v). 5FU could be visualized by its u.v. absorbance but for FUPA and FBAL we used ureidopropionate and β-alanine as carriers which could be visualized using ninhydrin staining. FDHU could be visualized with acidic di-methylamino-benzaldehyd.[1]

The 5FU anabolic assays were performed as described previously.[8] 5FU was separated from its products using polyethyleneimine (PEI) cellulose TLC using water as the eluent or HPLC.

3. RESULTS

In livers total DPD activity (degradation to FDHU, FUPA and FBLA) varied between 10 and 30 nmol/hr per mg protein (Table 1); co-incubation with 0.6 mM uracil or thymine resulted in a significant 80% enzyme inhibition (Fig. 1). Since thymidine has also been reported to inhibit DPD[12] we tested DPD inhibition by thymidine without and with preincubation. Preincubation clearly increased DPD inhibition, indicating that degradation seems to be essential for effective DPD inhibition.

In intact rat hepatocytes degradation of 5FU exceeded that of anabolism 4-fold, but catabolism could be inhibited effectively by 0.5 mM thymine. Formation of FdUMP and FUTP were about 2–3-fold increased, while RNA incorporation was not significantly affected. Since cisplatin is an anticancer agent frequently administered in combination with 5FU, we tested the effect of 20 µM cisplatin; total degradation decreased to 63 ± 7%.

In human colon tumors activity of DPD was much lower than in the liver and varied from 0.06 to 2.73 nmol/hr/mg protein, and was in the same range as adjacent mucosa. In human colon cancer cell lines a similar patters was observed, varying from not detectable to 75 pmol/hr/10^6 cells.

Table 1. Comparison of the activities of anabolic and catabolic enzyme in murine, rat, and human liver

Species	DPD	UP	TP
Mouse	28.6 ± 18.6 (3)	37.5 ± 10.9 (3)	581 ± 148 (3)
Rat	27.5 ± 9.3 (6)	111.5 ± 22.2 (4)	552 ± 173 (4)
Human	14.5 ± 1.5 (3)	54.4 ± 10.3 (3)	1015 ± 225 (3)

Values (nmol/hr/mg protein) are means ± SD of 3-6 separate experiments. In human and murine liver less FBAL was formed in the DPD assay than in rat liver. UP and TP were assayed with $[6-^{14}C]$-5FU as substrates and ribose-1-P and deoxyribose-1-P as co-substrates, respectively.

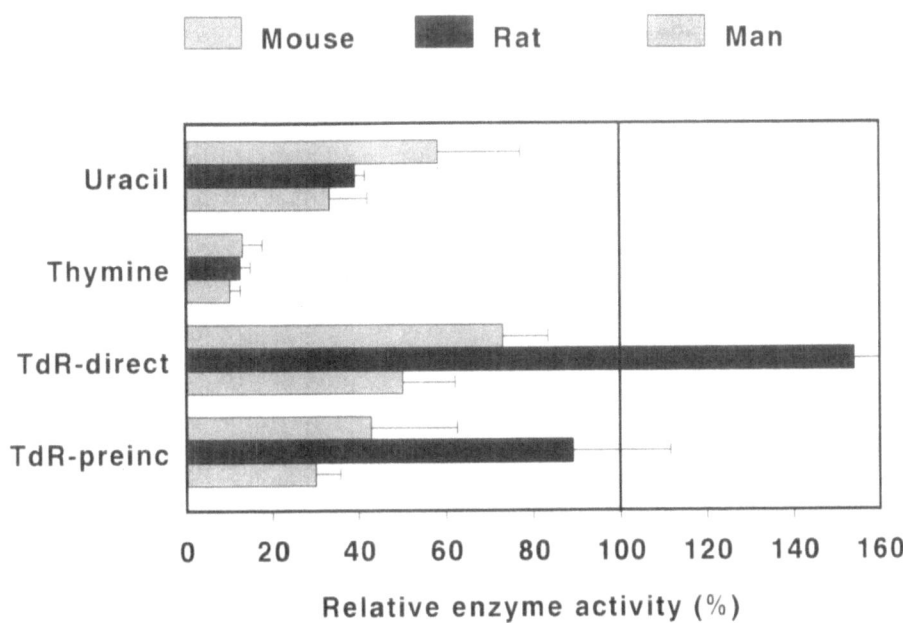

Figure 1. Inhibition of DPD from rat, murine and human livers by uracil, thymine and thymidine (added directly or 10 min before the assay). Values are % (± SE) from the DPD activity without addition. Control activity from each experiment was set at 100%. Final concentrations of each compound were 0.6 mM.

The anabolic pathway of 5FU to fluorouridine in liver was comparable to higher than the catabolic pathway (Table 1). In human colon tumors, however, the activities of two anabolic enzymes were much higher than that of DPD; uridine phosphorylase varied from 100–300 and that of "fluorouracil"-phosphoribosyltransferase from 1–5 nmol/hr/mg protein.[9] In mucosa these values were 1.5, 12–50 and 0.2–1 nmol/hr mg protein.[9] In two murine colon tumors DPD activity was similar to that in human colon tumors; Colon 26A, 0.40 ± 0.12; Colon 38 0.72 ± 0.02, and murine mucosa 4.34 ± 1.3 nmol/hr/mg protein.

4. DISCUSSION

This study shows that DPD activities in rat and murine liver are in the same range as human liver and are affected similarly by uracil and thymine as human liver. DPD activity is much lower in mucosal tissues compared to liver. In colon tumors DPD activity varies considerably but is usually lower than in adjacent mucosa. The difference is, however, not very high. These data on tissue distribution of DPD are in agreement with data reported previously.[1] However, in the tumors DPD activity is considerably lower than that of all anabolic enzymes, even compared to the 'FU"phosphoribosyl transferase, the anabolic enzyme with the lowest activity compared to the other anabolic enzymes. Considering the relevance of this equilibrium in colon tumors, it can be concluded that both based on overall enzyme activity and specific activity, the equilibrium is towards anabolite formation. Only when 5FU is administered at high doses, the overal degradation pathway can be predominant due to saturation of the anabolic enzymes, with a very long retention of the breakdown product (10–20 hr).[4]

These data indicate an important role of DPD in 5FU metabolism. The very high activity of DPD in the liver, results in a rapid and significant degradation of 5FU. During the first pass a major part of 5FU will be degraded resulting in 90% overall degradation. Thus inhibition or deficiency of DPD will result in a much higher 5FU bioavailability. Indeed DPD deficiency has been associated with increased and even lethal toxicity due to increased 5FU plasma and tissue levels.[13] Inhibition of DPD with *e.g.* uracil (in the UFT protocol) will result in increased 5FU bioavailability as has also been observed in cell culture,[14] but also in patients.[15] Recently several new DPD inhibitors have been introduced into the clinic with the aim to increase the 5FU bioavailability and thus the antitumor activity.[15] The present study demonstrates that due to the similarities of humans, rat and murine DPD, these two rodent species are suitable for the study of DPD inhibitors.

REFERENCES

1. Naguib FNM, El Kouni MH, Cha S (1985). Enzymes of uracil catabolism in normal and neoplastic human tissues. Cancer Res 45: 5405–5412.
2. Weber G (1983). Biochemical strategy of cancer cells and the design of chemotherapy: GHA Clowes memorial lecture. Cancer Res: 518–524.
3. Etienne MC, Chéradame S, Fischel JL, Formento P, Dassonville O, Renée N, Schneider M, Thyss A, Demard F, Milano G (1995). Response to Fluorouracil Therapy in Cancer Patients: The Role of Tumoral Dihydropyrimidine Dehydrogenase Activity. J Clin Oncol 13: 1663–1670.
4. Heggie GD, Sommadossi J-P, Cross DS, Huster WJ, Diasio RB (1987). Clinical Pharmacokinetics of 5-Fluorouracil and its metabolites in Plasma, Urine, and Bile. Cancer Res 47: 2203–2206.
5. Peters GJ, Van der Wilt CL, Van Groeningen CJ (1994). Predictive value of thymidylate synthase and dihydropyrimidine dehydrogenase. Eur J Cancer 30A: 1408–1411.
6. Spector T, Porter DJT, Nelson DJ, et al (1994). 5-Ethynyluracil (776C85), a modulator of the therapeutic activity of 5-fluorouracil. Drugs Future 19: 565–571.
7. Cao S, Frank C, Shirasaka T, Rustum YM (1995). 5-Fluorouracil Prodrug: Role of Anabolic and Catabolic Pathway Modulation in Therapy of Colorectal Cancer. Clin Cancer Res 1: 839–845.
8. Peters GJ, Laurensse E, Leyva A, Lankelma J, Pinedo HM (1986). Sensitivity of human, murine and rat cells to 5-fluorouracil and 5'deoxy-5-fluorouridine in relation to drug metabolizing enzymes. Cancer Res 46: 20–28.
9. Peters GJ, Van Groeningen CJ, Laurensse EJ, Pinedo HM (1991). A Comparison of 5-Fluorouracil Metabolism in Human Colorectal Cancer and Colon Mucosa. Cancer 68: 1903–1909.
10. Harris BE, Song R, He Y-J, Soong S-J, Diasio RB (1988). Circadian Rhythm of Rat Liver Dihydropyrimidine Dehydrogenase. Possible Relevance to Fluoropyrimidine Chemotherapy. Biochem Pharmacol 37: 4759–4762.
11. Ikenaka K, Shirasaka T, Kitano S, Fujii S (1979). Effect of uracil on metabolism of 5-fluorouracil *in vitro*. Gann 70: 353–359.
12. Tuchman M, Ramnaraine MLR, O'Dea RF (1985). Effects of Uridine and Thymidine on the Degradation of 5-Fluorouracil, uracil, and Thymine by Rat Liver Dihydropyrimidine Dehydrogenase. Cancer Res 45: 5553–5556.
13. Diasio RB, Beavers TL, Carpenter JT (1988). Familial deficiency in dihydropyrimidine dehydrogenase: Biochemical basis for familial pyrimidinenia and severe 5-fluorouracil-induced toxicity. J Clin Invest 81: 47–51.
14. Spoelstra EC, Pinedo HM, Dekker H, Peters GJ, Lankelma J (1991). Measurement of in vitro cellular pharmacokinetics of 5-fluorouracil in human and rat cancer cell lines and rat hepatocytes using a flow-through system. Cancer Chemother Pharmacol 27: 320–325.
15. Peters GJ, Ackland SP (1996). New antimetabolites in preclinical and clinical development. Expert Opinion Invest Drugs 5: 637–679.

123

INDUCTION OF *IN VIVO* RESISTANCE AGAINST GEMCITABINE (dFdC, 2',2'-DIFLUORO-DEOXYCYTIDINE)

V. W. T. Ruiz van Haperen,[1] G. Veerman,[2] C. J. A. van Moorsel,[2] and G. J. Peters[2]

[1]Department Clinical Investigation
M.D. Anderson Cancer Center
Houston, Texas
[2]Department Oncology
University Hospital Vrije Universiteit
Amsterdam, The Netherlands

1. INTRODUCTION

Gemcitabine (2',2'-difluorodeoxycytidine, dFdC) is a cytidine analog with major activity against several solid tumors (1,2). Until now acquired resistance has only been associated with deoxycytidine kinase (dCK) deficiency, after continuous exposure to dFdC *in vitro* (3). For 1-β-D-arabinofuranosylcytosine (ara-C) the main *in vitro* resistance mechanism also is dCK deficiency (4,5). In patients however, the mechanism for ara-C resistance is less clear (5). Since dFdC is now widely used in the treatment of non-small cell lung and pancreatic cancer, it is very likely that resistance to gemcitabine will develop. Therefore it is important to determine mechanisms of resistance in an *in vivo* tumor model. For this purpose we used a tumor with moderate *in vivo* sensitivity to dFdC (6), Colon 26-A. We induced resistance by repeated dFdC treatment. To facilitate mechanism studies, cell lines were derived form the parental and resistant tumors. The questions we sought to answer were: 1. Is *in vivo* resistance to dFdC associated with dCK deficiency? And 2. If not, what mechanism is responsible for this resistance?

2. METHODS

1. Induction of resistance: Balb/C mice bearing Colon 26-A tumors were treated at the maximal tolerated dose of 120 mg/kg dFdC every three days for four times (6). One day after the last treatment the most resistant tumor was transplanted and

Purine and Pyrimidine Metabolism in Man IX,
edited by Griesmacher *et al.* Plenum Press, New York, 1998.

treatment repeated. After several passages mice were treated 17 times without transplantation of the tumor, resulting in tumors resistant against dFdC (Colon 26-G) with a stable phenotype.

2. Cell lines, C26-A and C26-G, were derived from the Colon 26-A and Colon 26-G tumors, respectively, by dissociating the tumors. The cells were cultured in Dulbecco's MEM, supplemented with 5% fetal bovine serum, at 5% CO_2, 37°C.

3. Growth inhibition (IC50) for dFdC in the C26-A and C26-G cell lines was estimated using the sulforhodamine B (SRB) assay (7). The sensitivity pattern of other compounds (2',2'-difluorodeoxyguanosine (dFdG); 2-chlorodeoxyadenosine (CdA); ara-C; 2',2'-difluorodeoxyuridine (dFdU)) was assessed to find a possible mechanism for resistance.

4. dFdCTP accumulation *in vivo*: mice were treated with one dose of 120 mg/kg dFdC and tumors were taken out after 2, 6, 8 or 24 hours. Tumors were pulverized, nucleotides extracted and dFdCTP analyzed by anion exchange HPLC (8). dFdCTP accumulation *in vitro*: cells were exposed to 10 μM dFdC for 4 and 24 hours. 24 hour retention was also included. After the incubation cells were harvested, nucleotides extracted and analyzed by anion exchange HPLC (8).

5. Enzyme activities were determined for deoxycytidine kinase, deoxycytidine deaminase and thymidine kinase (3,9). In all assays radioactive substrate (^{14}C-TdR or ^{3}H-CdR) was added to a partially purified cell extract or pulverized tumor material. DCK assays were performed ± dTTP in the reaction mixture to correct for CdR phosphorylation by TK2 (11). TK assays were done ± dCTP to estimate the contribution of TK2 in the total TK activity (10). Enzyme activity was expressed as product formed per hour per mg protein or per million cells.

3. RESULTS

Resistance to dFdC could only be induced after prolonged repeated treatment of Colon 26-A bearing mice with 120 mg/kg dFdC, q3d for several months (Fig. 1).

Values are means ± SEM of at least 6 tumors.

Determination of enzyme activities (Table 1) showed that the resistant Colon 26-G tumor was not deficient for dCK, although a 1.7-fold decrease in activity was observed.

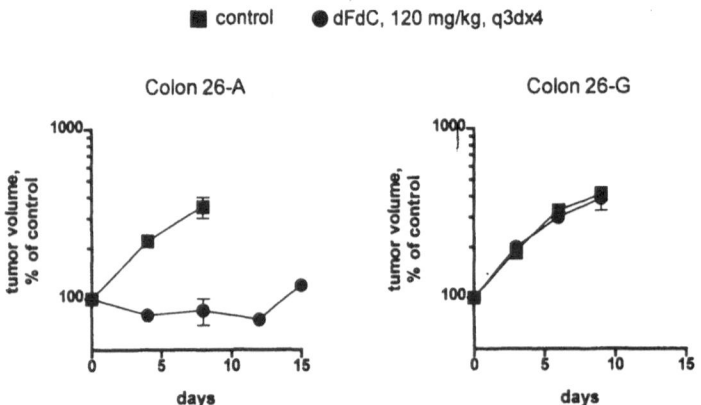

Figure 1. Tumor volumes of parental (Colon 26-A) and resistant (Colon 26-G) tumors. (■) control; (●) dFdC, 120 mg/kg, q3dx4.

Table 1. Enzyme activities in tumors and cell lines

	in vivo		in vitro	
	Colon 26-A	Colon 26-G	C26-A	C26-G
dCK	4.9 ± 0.7	2.9 ± 0.4	0.27 ± 0.04	0.31 ± 0.04
TK	7.9 ± 0.7	1.0 ± 0.2	0.71 ± 0.08	0.18 ± 0.08
TK + dCTP	0.8 ± 0.1	0.4 ± 0.1	0.21 ± 0.03	0.07 ± 0.02
dCDA	2.0 ± 0.2	2.3 ± 0.7	0.50 ± 0.50	0.20 ± 0.10

Enzyme activity was expressed as nmol/h/mg protein (in vivo) or nmol/h/10^6cells (in vitro).

DCDA activity was not changed, but a significant decrease in total TK activity was observed. Since dCTP could inhibit 90% of total TK in Colon 26-A and only partly in Colon 26-G, the reduction in TK activity was mainly due to a decrease in TK2.

To determine whether resistance to dFdC was associated with a decrease in dFdCTP accumulation we treated tumor bearing mice with dFdC and studied accumulation patterns. The resistant tumor Colon 26-G accumulated equal concentrations of dFdCTP as the parental tumor Colon 26-A: both 15 nmol dFdCTP / mg wet weight, 6 h after dFdC injection. The cell lines derived from the parental and resistant tumors showed a similar sensitivity pattern i.e. C26-G was resistant to dFdC as compared to C26-A (Fig. 2).

Additional experiments showed a minor cross-resistance to ara-C, CdA, dFdU and most prominently dFdG (resistance factor of 160). Determination of enzyme activities in cell lines revealed the same pattern as in the tumors (Table 1). And as in the tumors, C26-G cells accumulated higher dFdCTP levels than C26-A cells (6000 and 3000 pmol/10^6cells, respectively).

4. DISCUSSION

The resistance induced in the murine colon carcinoma model Colon 26-A was not due to a deficiency in dCK. The 2-fold decrease in dCK activity as found in the Colon 26-G tumor would account for a part of the resistance, if the dFdCTP pools in this tumor would have been lower as well. Since this is not the case, dCK deficiency is unlikely to be the cause for resistance. An increase in dCDA activity is a known mechanism for ara-

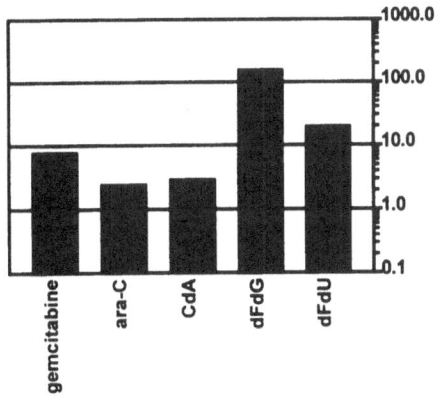

Figure 2. Resistance factors of resistant C26-G versus parental C26-A cells.

C resistance (5). In the resistant Colon 26-G tumor and C26-G cell line dCDA activities were not different from the parental tumor and cell line, precluding dCDA as a mechanism for resistance. Accumulation of dFdCTP was not affected, so the resistance mechanism is not to be found at this level. The only significant change found was in TK activity levels. It is not clear how this would explain dFdC resistance. Mitochondrial TK2 is able to phosphorylate dCyd and also dFdC, albeit less efficiently than dCyd (10,11). Mitochondria have been shown to be important in signaling for apoptosis (12). In sensitive cells phosphorylation of dFdC by TK2 would lead to a disturbance in the mitochondrial homeostasis, releasing a signal for apoptosis. In the resistant tumor and cell line, TK2 is reduced significantly, rendering the cells resistant to the effects of dFdC. Future work will test this hypothesis.

REFERENCES

1. Hertel LW, Boder GB, Kroin KS, Rinzel SM, Poore GA, Todd GC, Grindey GB. Cancer Res. 50: 4417–4422, 1990.
2. Lund B, Kristjanssen PAG, Hansen HH. Cancer Treatm. Rev. 19: 45–55, 1993.
3. Ruiz van Haperen VWT, Veerman G, Eriksson S, Boven E, Stegmann APA, Hermsen M, Vermorken JB, Pinedo HM, Peters GJ. Cancer Res. 54: 4138–4143, 1994.
4. Ruiz van Haperen VWT, Peters GJ. Pharm., World & Sci. 16: 104–112, 1994.
5. Grant S. Pharmacol. Ther. 49: 29–44, 1990.
6. Veerman G, Ruiz van Haperen VWT, Vermorken JB, Pinedo HM, Peters GJ. Cancer Chemother. Pharmacol. 38: 335–342, 1996.
7. Keepers YPAM, Pizao PE, Peters GJ, Van Ark-Otte J, Winograd B, Pinedo HM. Eur. J. Cancer 27: 897–900, 1991.
8. Ruiz van Haperen VWT, Veerman G, Boven E, Noordhuis P, Vermorken JB, Peters GJ. Biochem. Pharmacol. 48: 1327–1339, 1994.
9. Ruiz van Haperen VWT, Veerman G, Braakhuis BJM, Vermorken JB, Boven E, Leyva A, Peters GJ. Eur. J. Cancer 29A: 2132–2137, 1993.
10. Eriksson S, Kierdaszuk B, Munch-Petersen B, Öberg B, Johansson NG. Biochem. Biophys. Res. Comm. 176:586–592, 1991.
11. Wang L, Munch-Petersen B, Herrström-Sjöberg A, Bergman T, Jörnvall H, Hellman U, Eriksson S. Manuscript in preparation.
12. Zamzami N, Susin SA, Marchetti P, Hirsch T, Gómez-Monterrey, Castedo M, Kroemer G. J. Exp. Med. 183: 1533–1544, 1996.

124

ACTIVATION OF DEOXYCYTIDINE KINASE BY VARIOUS NUCLEOSIDE ANALOGUES

Tatjana Spasokoukotskaja,[1] Maria Sasvári-Székely,[1] Lehel Hullán,[2] Freidoun Albertioni,[3] Staffan Eriksson,[4] and Maria Staub[1]

[1]Semmelweis University of Medicine
Department of Medical Chemistry, Molecular Biology, and
    Pathobiochemistry
Budapest, Hungary
[2]National Institute of Oncology
Budapest, Hungary
[3]Karolinska Hospital
Department of Clinical Pharmacology
Stockholm, Sweden
[4]Department of Veterinary Medical Chemistry
Biomedical Centre
Uppsala, Sweden

1. SUMMARY

The effect of different nucleoside analogues on deoxycytidine kinase (dCK) and thymidine kinase (TK) was compared in normal human lymphocytes and various leukemic cell lines. G-phase enriched tonsilar lymphocyte subpopulation treated by CdA showed more profound stimulation of dCK activity than S-phase cells. No substantial changes in TK activity were detected. CdA treatment increased the activity of dCK 4-fold in peripheral blood mononuclear cells (PBMC) and 2-fold in promyelocytic cell line HL60, too. However, no significant stimulation was detected either in CCRF-CEM or in K562 cell lines. 2-Cl-2'deoxy-2'F-adenine arabinoside (CAFdA), 2F-adenine arabinoside (F-araA) and cytosine arabinoside (AraC) had the same effect as CdA, although higher concentrations were needed for maximal activation. In contrast, treatment by dCyd caused slight inhibition of dCK. The possibility of interference of nucleoside analogues with the mechanisms of posttranslational modification of dCK was proposed.

Purine and Pyrimidine Metabolism in Man IX,
edited by Griesmacher *et al.* Plenum Press, New York, 1998.

2. INTRODUCTION

Deoxycytidine kinase (dCK, EC 2.7.1.74) is a key enzyme in salvage of deoxyribonucleosides. Besides dCyd the enzyme catalyzes the 5' phosphorylation of purine deoxynucleosides and of various nucleoside analogues, such as arabinosylcytosine (AraC), 2-fluoro-arabinosyladenine (FaraA), 2-chloro-2'-deoxyadenosine (CdA) and the recently developed 2-chloro-2'-fluoro-2'-deoxy-arabinosyladenine (CAFdA).[1-3] dCK is constitutively expressed in lymphoid tissues, where it supplies cells with precursors for replicative and repair DNA synthesis, as well as for synthesis of liponucleotides.[4] Conflicting results have been published regarding the cell cycle regulation of dCK. No cell cycle variations in dCK mRNA levels were found, while 2–3-fold differences in dCK activity were indicated in different phases of the cell cycle.[5,6] Several studies have demonstrated that dCK is feed-back inhibited by dCTP and that dCTP level may play an important role in regulation of dCK activity in cells.[6]

Previously we have shown that treatment of freshly prepared tonsil lymphocytes by CdA leads to substantial increase in dCK activity in cell free extracts, which was not accompanied by increase in dCK mRNA or protein level. Thus the possibility of post-translational modification of dCK was presumed.[7] The effect of CdA and several other nucleoside analogues on cells with different origin being in different phases of cell cycle were investigated.

3. MATERIALS

CdA and Cl-Ado were synthesized by Dr. Zygmunt Kazimierczuk (Warsaw, Poland); CAFdA was a gift from Dr. Howard Cottam (University of California, USA); F-araA, AraC and dCyd were products of Sigma; 5-^{3}H-dCyd and (methyl-^{3}H)-dThd were from Amersham.

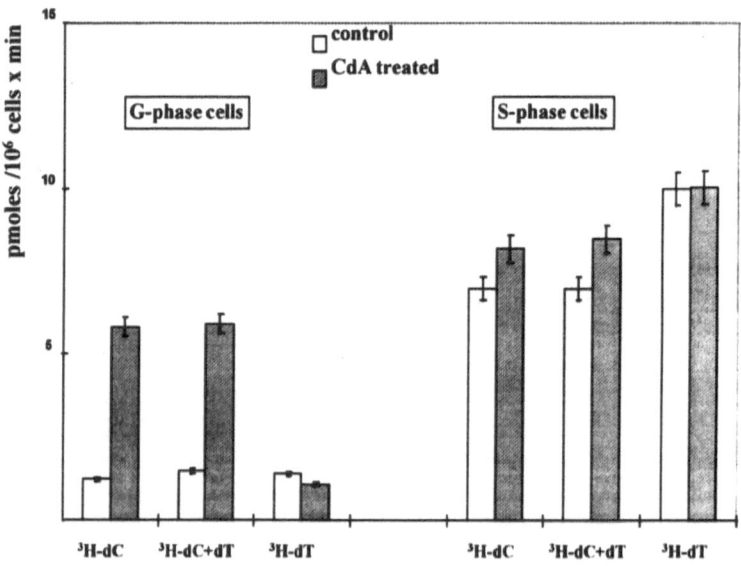

Figure 1. The Effect of CdA on dCK and TK in Tonsilar Lymphocytes. Lymphocytes were separated on S-phase and G-phase, then were incubated in Eagle's MEM without analogue (C) or in the presence of 1μM CdA. Enzyme activities were determined in crude extracts from 0.25×10^6 cells in 50 μl reaction mixture.

4. METHODS

Tonsilar lymphocytes were separated on an albumin gradient for blast (S-phase en-riched) and small (G-phase enriched) cell fractions according to their density. Mononuclear cells from normal peripheral blood (PBMC) were isolated by Ficoll-Hypaque (Pharmacia) density centrifugation. Human T-lymphoblastic (CCRF-CEM), chronic myelogenous (K562) and promyelocytic (HL60) leukemic cell lines were exponentially grown in RPMI 1640 medium supplemented with fetal calf serum (10%). Cells used in experiments were washed twice, resuspended in serum-free medium, incubated at 37°C for 2 hours with or without nucleoside analogues, washed in PBS, extracted as described in ref. 8. The cell-free extracts were used for determination of enzyme activities. dCK and TK were assayed with ^{3}H-dCyd or ^{3}H-dThd as a substrates (both 10 μM) in reaction mixture containing 50 mM TRIS-HCl pH 7.6, 5 mM MgCl$_2$, 5 mM ATP, 2 mM DTT, 10 mM NaF and cell extract. The enzyme assays were carried out at 37°C, and after appropriate time intervals to yield a linear reaction rates aliquots were spotted on DEAE cellulose filters, which were then washed, eluted and counted.

5. RESULTS AND DISCUSSION

5.1. Effect of CdA on Deoxycytidine Kinase and Thymidine Kinase in Tonsilar Lymphocytes

Human tonsils are natural sources of B lymphocytes mainly in a resting state, how-ever, 10–20% of tonsilar lymphocytes are activated *in vivo* and display relatively high rate of DNA synthesis. Centrifugation of lymphocytes across 25% bovine serum albumin re-sults in separation of cells on low density fraction enriched in S-phase cells and on high density fraction of G-phase cells. dCK and TK activities in these fractions differ several times (Fig. 1). After treatment with 1μM CdA dCK activity in cell extracts increases more than 4-fold in G-phase, but only 1.2-fold in S-phase subpopulation. Apart from dCK, TK2 is also known to phosphorylate dCyd.[1] To exclude the possibility that TK2 is accounted for increase in dCyd phosphorylation we measured dCK in the presence of 1 mM dThd, too. However, no substantial changes in activities of either TK1, or TK2 were detected.

CldATP has been previously shown to inhibit ribonucleotide reductase and to de-crease the dNTP pools.[2] Tonsilar lymphocytes contain 1–3 pmoles dCTP/10^{6}cells,[9] thus it can be calculated, that in condition used in Fig. 1 the dCTP originating from cells should not exceed 15 nM, while *in vitro* inhibition of dCK by dCTP can be detected only at μM concentrations (data not presented). Thus, dCTP pool depletion after CdA treatment can-not be responsible for higher dCK activity in cell extracts. No changes in dCyd deami-nase, dCMP deaminase, thymidylate synthase and dCMP catabolism could be detected (data not presented). It can be concluded that CdA or CdA nucleotides cause modification on structure of deoxycytidine kinase molecule, more clear-cut in G-phase and resulted in increase of dCK activity.

5.2. CdA Increases dCK Activity in PBMC and in HL60, But Not in CCRF-CEM and K562 Cell Lines

Normal human peripheral blood was chosen as a source for resting Go lymphocytes, while exponentially growing CCRF-CEM cells represented S-phase cells. dCK activity

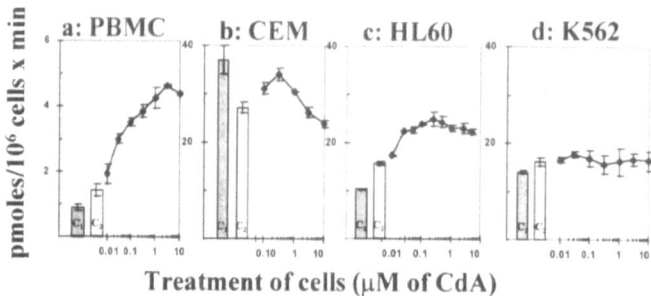

Figure 2. Changes in dCK Activity in PBMC and in CCRF-CEM, HL60, K562 cell lines after treatment by CdA. Cells were extracted immediately (C1), or after incubation in RPMI with 0 (C2) - 10 μM CdA. Crude extracts were used for activity measurements.

calculated per 10^6 cell was much lower in blood lymphocytes, than in CEM. Incubation of PBMC in serum-free RPMI without CdA led to some activation of dCK. Treatment of PBMC by CdA resulted in further dose dependent increase of the dCK activity up to 3 μM CdA (Fig. 2a). In contrast, the dCK activity in CCRF-CEM cells slightly diminished upon incubation and did not achieve the initial values at any CdA concentration used. Although CdA is intensively phosphorylated up to CldATP in both CEM and PBMC,[10] neither CldATP accumulation, nor subsequent dNTP pool changes[2] were able to potentiate the deoxycytidine kinase activity in CEM cells (Fig. 2b). Similar discrepancies in action of CdA were observed in HL60, a promyelocytic cell line, compared to K562 (Fig. 2c and d). More than 2 fold activation of dCK was detected in HL60 cells, while less than 35% differences were obtained for K562 after 2 hours CdA treatment. We proposed that mechanisms leading to dCK activation can be modified in several leukemic cell lines compare to natural lymphocytes.

5.3. Activation of dCK in PBMC by CAFdA, F-araA, and AraC

Like CldATP, 5'-triphosphates of both F-araA and CAFdA are potent inhibitors of ribonucleotide reductase,[2,3,6] while AraC treatment of cells does not affect the dNTP pool sizes.[6] Treatment of PBMC with various concentrations of purine analogues resulted in more than 3 times activation of dCK (Fig. 3). Differences in activating concentrations of analogues were similar to differences in the affinity of dCK for these nucleosides (dates not shown). AraC treatment also caused increase in dCK activity, though extent of stimulation was lower with AraC compare to purine analogues. No stimulation was achieved with dCyd, the natural substrate of dCK, moreover it had slight inhibitory effect.

Stimulation of AraCTP accumulation by F-araA was previously explained by the decrease in dNTP levels leading to activation of dCK.[11] Moreover, direct stimulation of dCK by F- in extracts from K562 cells was also shown, while AraCTP supposed to be a negative regulator.[11] Our results show, that all three arabinosyl nucleosides, as well as CdA, are effective stimulators of dCK, although the mechanisms can be different.

6. CONCLUSIONS

In general evidences are presented that active metabolites, probably triphosphates, of purine deoxyribonucleoside analogue CdA, as well as arabinosylnucleoside analogues

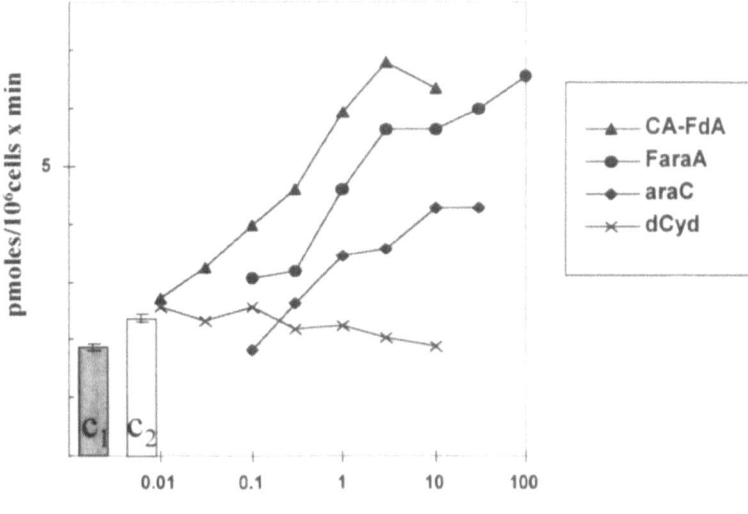

Treatment of PBMC (μM of analogue)

Figure 3. dCK activities in extracts of peripheral blood mononuclear cells treated by dCyd or by various arabinosyl nucleoside analogues. PBMC were extracted immediately (C1) or after incubation with various concentrations of CAFdA, AraC, F-araA and dCyd.

CAFdA, F-araA and AraC are able to stimulate the dCK *ex vivo*. This stimulation can be measured *in vitro*, in spite of several hundred times dilution of the cell-derived nucleotides pools, that makes improbable the interference of feed-back, allosterical or other types of reversible regulation with activity measurements. We supposed that action of several analogues results in a covalent modification of dCK, but neither the nature, nor processes leading to this modification are known yet. The role of dNTP pool changes, of DNA synthesis inhibition or signal transduction processes, as well as metabolic significance of such modification have to be clarified.

REFERENCES

1. Arnér, E.S. and Eriksson, S. (1995) Pharmacol. Ther. 67, 155–186
2. Parker, W.B., Shaddix, S.C., Chang, C-H., White, E.L., Rose, L.M., Brockman, R.W., Shortnacy, A.T., Montgomery, J.A., Secrist III, J.A. and Bennett, L.L., Jr (1991) Cancer Res. 51, 2386–2394
3. Xie, K. Ch. and Plunkett, W. (1996) Cancer Res. 56, 3030–3037
4. Spasokoukotskaja T., Spyrou G., Staub M. (1988) Biochem. Biophys. Res. Commun. 155, 923–927
5. Hengstschlager, M., Mudrak, I., Wintersberger, E. and Wawra, E. (1994) Cell Growth & Differentiation 5, 1389–1394
6. Gandhi, V. and Plunkett, W. (1992) Cancer Chemother. Pharmacol. 31, 11–17
7. Staub, M., Spasokukotskaja, T., Sasvári, M. and Keszler, G. (1996) Drug Development Res. 37/3, 178/11 Special Issue, Purines '96
8. Arnér, E.S.J., Spasokoukotskaja, T. and Eriksson, S. (1992) Biochem. Biophys. Res. Commun. 188, 712–718
9. Spasokukotskaja, T., Sasvári-Székely, M., Taljanidisz, J. and Staub, M. (1992) FEBS Lett. 297, 151–154
10. Reihelova, V., Albertioni, F. and Liliemark, J. (1996) J. Chromatogr. B, 682, 115–123
11. Gandhi, V. and Plankett, W. (1988) Cancer Res. 48, 329–334

125

CLINICAL PHARMACOLOGY OF INTERMEDIATE AND LOW-DOSE CYTOSINE ARABINOSIDE (ara-C) THERAPY IN PATIENTS WITH HEMATOLOGIC MALIGNANCIES

Takanori Ueda,[1] Sugako Matsuyama,[1] Takahiro Yamauchi,[1] Shinji Kishi,[1] Toshihiro Fukushima,[1] Hiroshi Tsutani,[1] Toru Nakamura,[1] Nobuyuki Gotoh,[2] and Mikio Masada[2]

[1]First Department of Internal Medicine
[2]Department of Pharmacy
Fukui Medical School
Matsuoka, Fukui, 910-11, Japan

INTRODUCTION

1-β-D-arabinofuranosylcytosine (ara-C) is one of the most effective drugs for the treatment of hematologic malignancies, such as acute leukemia and malignant lymphoma. ara-C is one of the deoxycytidine analogs. In its chemical structure OH residue is introduced at the 2′ position of the sugar. One of the shortcomings of the drug is its rapid metabolism to inactive metabolite, uracil arabinoside (ara-U) in human bodies.[1,2] On the other hand, when ara-C is incorporated into the cells, it is converted to ara-CTP, its active form, by deoxycytidine kinase[3] (Fig. 1). Thus, the efficacy of ara-C as an antileukemic agents are mainly determined by two factors, deamination to ara-U and activation to ara-CTP.

A special feature of ara-C is the wide variation of the administration dose. The drug has been used in high dose (2~3 g/m^2), intermediate dose (0.5~1 g/m^2), regular dose (50~200 mg/m^2), low dose (10 mg/m^2) and very low dose (3 mg/m^2). However, the precise plasma and intracellular pharmacokinetics including ara-U and ara-CTP has not yet been studied. Here, we reported the pharmacokinetic behavior in intermediate dose therapy by iv infusion and also in low dose therapy by s.c. injection to clarify its pharmacokinetic characteristics including intracellular ara-CTP.

Purine and Pyrimidine Metabolism in Man IX,
edited by Griesmacher *et al.* Plenum Press, New York, 1998.

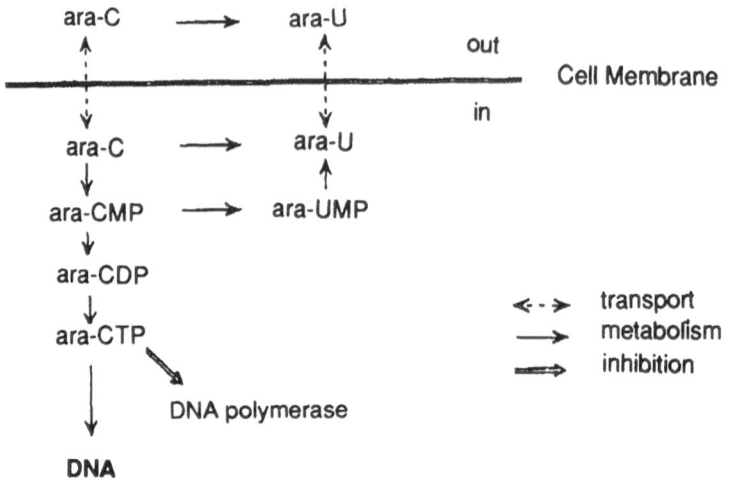

Figure 1. Metabolism of ara-C.

DESIGN AND METHODS

Eleven and nine patients were treated with intermediate dose ara-C and low dose ara-C therapy, respectively. ara-C was administered as 1 hr iv infusion for intermediate dose (1.0 g/m^2). Plasma levels of ara-C and ara-U were measured using HPLC as described previously.[4] Briefly, for determination of ara-C levels, Partisil-10 SCX (Whatman, USA) was used. The sample was injected onto the column and eluted with 10 mM $NH_4H_2PO_4$ at a rate of 1.5 ml/min at ambient temperature. The quantity was detected by absorbance at 280 nm. The limit of detection was 0.05 µg/ml. ara-U concentration was determined by the method of Linssen et al.[5] A limit of detection was 0.1 µg/ml.

For low dose ara-C therapy, 10 mg/m^2 was injected s.c. Concentrations of both ara-C and ara-U were determined by radioimmunoassay method of Sato et al.[6] Antibodies for ara-C and ara-U were kindly provided by Nippon Kayaku, Ltd., Japan.

The data of plasma ara-C and ara-U concentrations were analyzed by a computer program of nonlinear least squares regression analysis, named automated pharmacokinetic analysis system (APAS)[7], which is the modification of MULTI.[8] The AUC from time zero to infinity (AUC inf) was computed following the trapezoidal rule.

ara-CTP levels were measured by new combination method of HPLC and RIA in patients with hematologic malignancies.[9] Briefly, ara-CTP fraction was collected by HPLC, lyophilized and dephosphorylated to ara-C by incubation with alkaline phosphatase.

The obtained ara-C fraction was measured by RIA with anti-ara-C serum (kindly given by Asahi Chemical Industry, Ltd., Japan).

RESULTS AND DISCUSSION

In patients treated with intermediate dose, the plasma ara-C levels increased up to 14.9 µg/ml (mean value) at the end of infusion, then declined and reached the level below the limit of detection (0.05 µg/ml). The curves corresponded to a two-compartment open model. The main pharmacokinetic parameters of ara-C are as follows: t 1/2 β, 18.3 min; AUC, 15.3 mg·hr/liter; total clearance, 139 liter/m^2/hr. The plasma ara-U levels elevated slowly up to 36.9 µg/ml at 20 min after the end of the infusion, then was persistent in plasma and was 4.2 µg/ml even at 11 hr after the end of infusion. The t 1/2 β, AUC and total clearance were 277 min, 222 mg·hr/liter and 4.7 liter/m²/hr, respectively (Figs. 2 and 3).

In patients treated with low-dose ara-C therapy, the plasma ara-C curves also corresponded to the two compartment open model with t 1/2 β of 2.3 hr, AUC inf of 102 ng/ml·hr and total clearance with 132 liter/m²/hr. Regarding ara-U, the curves also corresponded to the two-compartment open model with t 1/2 β of 7.5 hr, AUC inf of 2750 ng/ml² hr and total clearance of 4.06 liter/m²/hr (Fig. 2).

These finding showed that the plasma ara-C and ara-U concentration-versus-time curves were essentially the same between intermediate dose and low dose, suggesting that the deamination process proceeded without saturation even in intermediate dose. There was a wide interpatient variability in ara-C but not in ara-U levels. In addition, ara-U is persistent much longer in plasma than ara-C. These findings suggest that the variability and rapid decline of plasma ara-C levels was mainly due to the variation in the deamination process but not due to difference in the excretion or transport.

Intracelluar ara-CTP levels were detected even in low dose by the new method which is about 60 times more sensitive compared with already existing methods.[10] Pre-

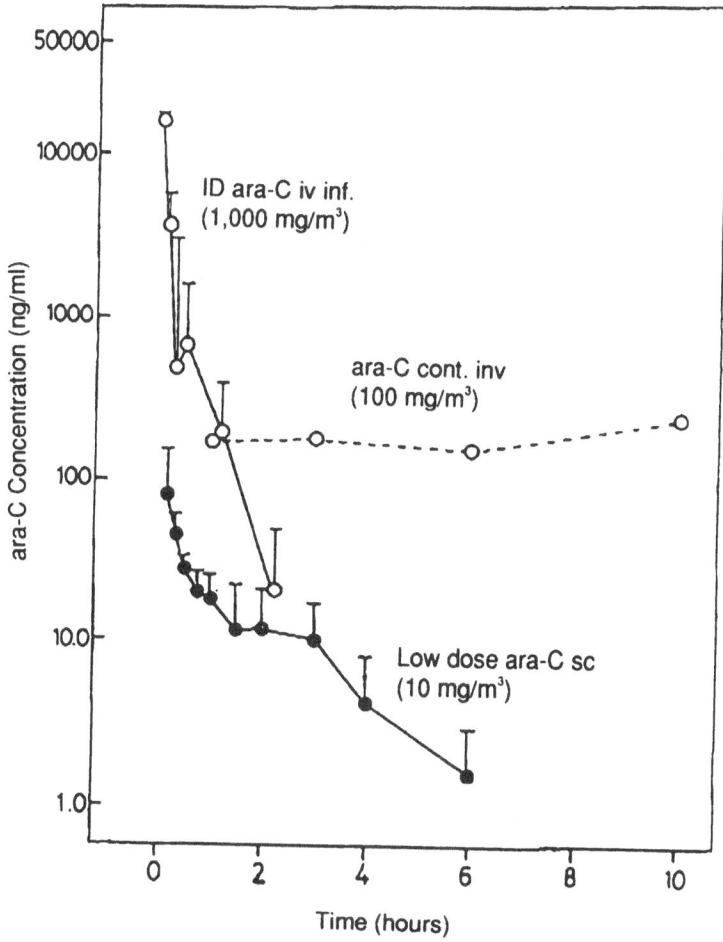

Figure 2. Pharmacokinetics of ara-C therapy.

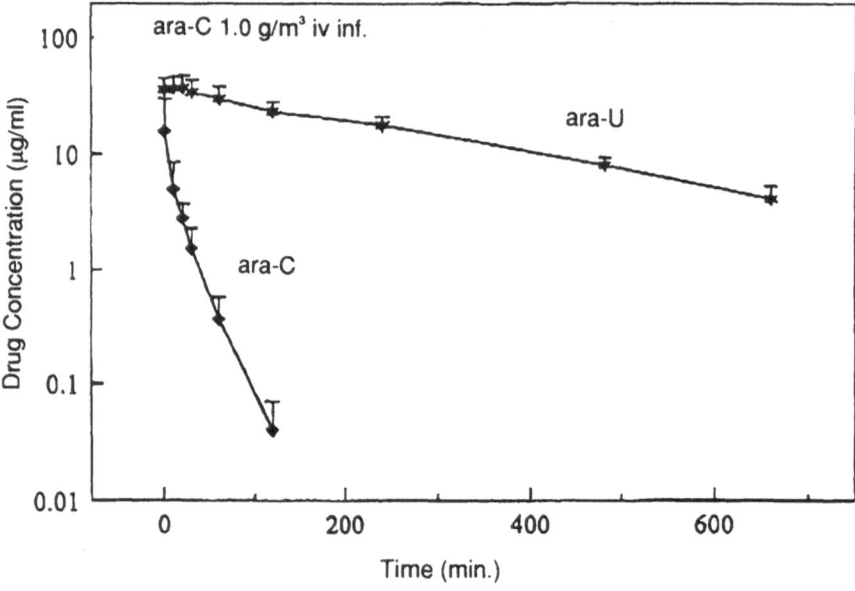

Figure 3. Pharmacokinetics of intermediate dose ara-C therapy (Matsuyama, S. et al., in preparation).

liminary results showed the wide interpatients variability in ara-CTP levels (Table 1), which were not necessarily correlated to plasma ara-C levels (Fig. 4).

CONCLUSIONS

1. The deamination process to ara-U (inactive metabolite) is crucial for ara-C pharmacokinetics in both intermediate and low dose therapy.
2. Intracellular ara-CTP levels were not necessarily correlated with plasma ara-C levels.
3. To estimate the antineoplastic effect of ara-C, direct determination of intracellular ara-CTP level (active metabolite) is essential.

Table 1. Intracellular are-CTP content in patients treated with ara-C or its derivatives

Age/sex/disease	WBC (/μl) blasts (%)	Drug delivery (mg/sqm/day)	Sampling time	ara-CTP (pmol/5×10⁶ cells)
73/F	50200	BHAC (100)	4 h	47
M4*	61**	I.V. inf 2 h		
59/F	24700	BHAC (125)	4 h	3.6
M2	94	I.V. inf 2 h		
60/M	40000	ara-C (1000)	2 h	260
L2	90	I.V. inf 2 h		
80/M	73900	ara-C (6×2)	1 h	3.1
CML	51	S.C.		
61/F	8600	ara-C (14)	9 days	2.8
MDS-LT**	55	cont. I.V.		

BHAC: N4-behenoyl-1-β-D-arabinofuranosyl cytosine; *: FAB classification of acute leukemia;
: Myelodysplastic syndrome (leukemic transformation); *: after the administration.

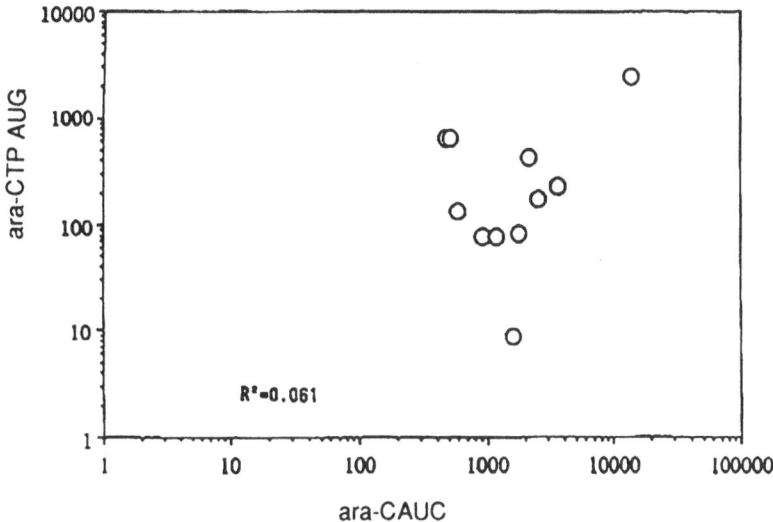

Figure 4. Relation between AUC of ara-CTP and that of ara-C (Yamauchi, T. et al. in preparation).

REFERENCES

1. Camiener GW. Studies of the enzymatic deamination of ara-cytidine : V. Inhibition in vitro and in vivo by tetrahydrouridine and other reduced pyrimidine nucleosides. Biochem Pharmacol 17: 1981, 1968.
2. Nakamura T, et al. Deamination of cytosine arabinoside in human leukemia cells and deaminase activity in serum. In: S Seno, F Takaku, S Irino(eds.), Topics in Hematology, Amsterdam: Excerpta Medica, p.1044, 1977.
3. Momparler R. Effects of cytosine arabinoside 5'-triphosphate on mammalian DNA polymerase. Biochem Biophys Res Commun 34: 465, 1969.
4. Ueda T, et al. Clinical pharmacology of intermediate dose ara-C therapy in patients with acute leukemia. Jpn J Clin Pharmacol Ther 20: 119, 1989. (in Japanese)
5. Linssen P, et al. Determination of 1- β-D-arabinofuranosylcytosine and 1- β -D-arabinofuranosyluracil in human plasma by high performance liquid chromatography. J Chromatogr 223: 371, 1981.
6. Sato T, et al. Sensitive radioimmunassay for cytarabine and uracil arabinoside in plasma. Cancer Treat Rep 68: 1357, 1984.
7. Yamaoka K, et al. In Automated pharmacokinetic analysis system. A Primer on Pharmacokinatics Using a Microcomputer: Nankoudo, Tokyo p.159, 1983. (in Japanese)
8. Yamaoka K, et al. A pharmacokinetic analysis program (MULTI) for microcomputer. J Pharmacobiodyn 4: 879, 1981.
9. Yamauchi T, et al. A new sensitive method for determination of intracellular 1- β-D-arabinofuranosylcytos-ine 5'- triphosphate content in human materials in vivo. Cancer Res 56: 1800, 1996.
10. Chiba P, et al. Determination of pyrimidine deoxynucleoside triphosphates in leukemia cell extracts containing 1- β-D-arabinofuranosylcytosine triphosphate. Eur J Clin Chem Clin Biochem 29: 385, 1991.

126

THE ROLE OF THYMIDINE KINASE ACTIVITY IN MURINE COLON TUMOURS TREATED WITH 5-FLUOROURACIL

C. L. van der Wilt, K. Smid, G. Veerman, and G. J. Peters

Academic Hospital Vrije Universiteit
Department of Medical Oncology
P. O. Box 7057
1007 MB Amsterdam, The Netherlands

1. INTRODUCTION

5-Fluorouracil (5FU) is widely used in the treatment of colon cancer. To exert its antitumour effect 5FU has to be converted into toxic metabolites: FUTP, FdUTP and FdUMP. FUTP and FdUTP are incorporated into RNA and DNA, respectively. FdUMP provides a potent inhibitor of thymidylate synthase (TS) and thereby inhibits *de novo* formation of dTMP and DNA synthesis. The latter is considered the main mechanism of antitumour activity.[1,2]

Several murine colon tumour models have been used to evaluate the activity of 5FU and combination therapies containing 5FU.[3,4] However the antitumour activity of 5FU mediated through inhibition of TS might be influenced by high plasma thymidine (TdR) levels in mice.[5] These levels are about 10-fold higher than in man.[5,6] The salvage pathway for thymidine nucleotides, that is the conversion of TdR into dTMP by thymidine kinase could bypass the inhibition of TS and thereby reduce the activity of 5FU.[6]

We evaluated the activity of TK in several murine colon tumours: Colon 38 and three Colon 26 subtypes Colon 26A, Colon 26G and Colon 26–10. The origin and characteristics of these tumours as they are maintained in our laboratory have been described by Van Laar et al.,[4] with exception of Colon 26G.[7] The latter is a variant of Colon 26A, which is resistant to the cytidine analog gemcitabine. The sensitivity of the tumours for 5FU was established *in vivo*. We also determined TK activity during 5FU therapy by measuring the activity of TK in Colon 26A and Colon 38 tumours obtained at 1 or 4 days after injection with 5FU. In this way the effect of TK activity on the antitumour activity of 5FU could be studied.

Purine and Pyrimidine Metabolism in Man IX,
edited by Griesmacher *et al.* Plenum Press, New York, 1998.

2. MATERIALS AND METHODS

Female Balb/c mice (Harlan/Cpb, Zeist, the Netherlands) were used for subcutaneous transplantation of tumour pieces of the Colon 26 variants. Colon 38 was maintained in C57Bl mice (Harlan/Cpb). At a tumour size of approximately 100 mm³ 5FU treatment was started. 5FU (ABIC, Netanya, Israel) was administered as a bolus injection of 100 mg/kg i.p., weekly during 3 weeks. Tumour volume was measured with callipers twice weekly. Antitumour activity was evaluated by calulation of the T/C (relative tumour size of the treated mice divided by the relative tumour size of the control mice) and growth delay factor (GDF), which indicated the gain of doubling time by treatment.[4]

The total activity of TK (both cytosolic TK1 an mitochondrial TK2) was measured in control tumours (Colon 26A, Colon 26G, Colon 26–10 and Colon 38) and in tumours obtained 1 and 4 days after treatment with 5FU (Colon 26A and Colon 38). Frozen tumors were homogenized in a TRIS/EDTA buffer. TK activity was measured with ¹⁴C-TdR (specific activity 62.8 mCi/mmol, NEN Dupont, Dreiech, Germany) as a substrate. We used 21.9 µM ¹⁴C-TdR in the assay as described by Ruiz van Haperen et al[8].

3. RESULTS

The chosen tumour panel shows a wide variation in response to 5FU therapy (Table 1). The antitumour activity of 5FU was most pronounced in mice bearing Colon 26–10. Complete remissions were observed with this tumour type. Mice bearing Colon 38 were moderately sensitive to 5FU treatment, but in Colon 26G bearing mice no effect of the drug was observed. Colon 26A was also rather resistant to 5FU.

Total TK activity in control tumours (Table 2) also varied a lot for this tumour panel, from 7834 to 1000 pmol/h/mg protein. After treatment with 5FU we observed about 5-fold decrease in TK activity (Fig. 1) in both Colon 26A and Colon 38. In Colon 38 tumours the TK activity remained low till at least 4 days after treatment. In Colon 26A TK activity had recovered to control level 4 days after 5FU administration.

4. DISCUSSION

This study shows that TK activity in two murine colon tumours decreased after treatment with 5FU. Therefore it seems unlikely that high thymidine levels in murine plasma or initially high TK activity in murine tumours will directly influence the sensitivity of

Table 1. Antitumour activity of 5FU*

Tumour	T/C	GDF
Colon 38	0.14 ± 0.05	4.4
Colon 26A	0.43 ± 0.19	0.7
Colon 26G	0.97	0.03
Colon 26-10	0.095 ± 0.05	ND (>10)

*Data (except those of Colon 26G) are derived from a retrospective study on large number of antitumour activity experiments by Van Laar et al[4] T/C values are means ± SD, ND = not detectable, because most of the tumours went into complete remission.

Table 2. Total TK activity in control tumours

Tumour	pmol/h/mg protein (n)
Colon 38	4728 ± 2042 (5)
Colon 26A	7577 ± 1746 (5)
Colon 26G	1000 ± 202 (4)
Colon 26-10	4348; 7858

Values are means ± SD, followed by the number of measurements (n).

these tumours to 5FU. Even administration of high dose thymidine did not reverse the 5FU antitumour activity.[9]

In the subtypes of Colon 26 we observed a large difference in sensitivity to 5FU, but resistance to 5FU was not related to high TK activity. The only tumour which did not respond at all to 5FU treatment Colon 26G had the lowest control level of TK activity. The decrease of TK activity shortly after 5FU administration coincided with a strong inhibition of TS at this time point as shown in a previous study.[10] So despite high plasma thymidine levels only a small amount of dTMP may be formed via the salvage pathway. Whether this should be enough for a tumour cell to survive remains debatable, since both TK and TS are not completely inhibited. However the antitumour activity experiments clearly showed that tumour growth was inhibited.

One factor that appeared to be important for the sensitivity to 5FU was not the extent of TK inhibition, but the duration of this inhibition, as illustrated by the differences between Colon 26A and Colon 38. A similar trend was observed for the inhibition of TS in these murine tumours concerning extent and duration.[10] TS and TK activities in murine tumours after 5FU treatment were also evaluated in CD8F1 mammary tumours with a similar effect.[3] Nord and Martin[3] observed that TK activity in Colon 26 was decreased by 50%, while we measured about 75% decrease in Colon 26A. For Colon 38 the inhibition was comparable (90% to 85%, respectively). Our absolute values of TK activity in both tumours were much higher (2.5 to 5-fold). This might be due to differences in sensitivity of the TK assays used, or to differences in subtype of the Colon 26 and Colon 38 since they were maintained in different institutions. In a previous study we also observed that this may considerably influence characteristics of a tumor.[4]

One of the mechanism by which TK activity was reduced after exposure to 5FU could be a decrease in the number of dividing cells. TK expression and activity is strongly cell cycle dependent and used as a proliferation marker.[11] When drug treatment induces

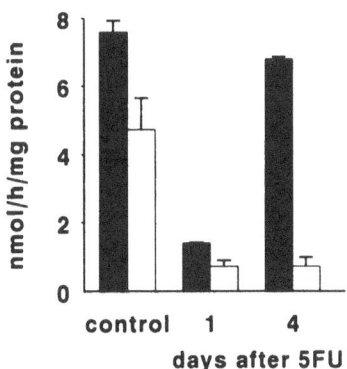

Figure 1. Residual TK activity in murine colon tumours 1 and 4 days after treatment with 100 mg/kg 5FU. Values are means ± SEM (n=4). Colon 26A (solid) and colon 38 (white).

growth inhibition TK activity, usually measured as incorporation of ^{3}H-TdR into DNA is reduced. However, two studies performed *in vitro* describe an increase of TK activity after 24 h exposure to 5FU[12] or two folate based inhibitors of TS.[13] The feedback inhibition of dTTP on TK activity would be release by the depletion of dTTP after inhibition of TS by 5FU. Whether this is a relevant mechanism *in vivo* in solid tumours remains unclear. Both *in vivo* studies confirm the conclusion that in some tumours high dose of 5FU can inhibit salvage and *de novo* synthesis of thymidylate providing an increased block of DNA synthesis and an increased therapeutic advantage.

REFERENCES

1. Pinedo HM, Peters GJ. Fluorouracil, biochemistry and pharmacology. J Clin Oncol 1988, 6:1653–1664
2. Peters GJ, Van der Wilt CL, Van Groeningen CJ, Meyer S, Pinedo HM. Thymidiylate synthase inhibition after administration of fluorouracil with or without leucovorin in colon cancer patients: Implications for treatment with fluorouracil. J Clin Oncol 1994, 12:2035–2042
3. Nord LD, Martin DS. Loss of murine tumor thymidine kinase activity *in vivo* following 5-fluorouracil (FUra) treatment by incorporation of FUra into RNA. Biochem Pharmacol 1991, 12:2369–2375
4. Van Laar JAM, Rustum YR, Van der Wilt CL, Smid K, Pinedo HM, Peters GJ. Tumor size and origin determine antitumor activity of ciplatin or 5-fluorouracil and its modulation by leucovorin in murine colon carcinomas. Cancer Chemother Pharmacol 1996, 39:79–89
5. Houghton JA, Williams LG, Loftin SK, Cheshire PJ, Morton CL, Houghton PJ, Dayan A, Jolivet J. Factors that influence the therapeutic activity of 5-fluorouracil [6RS]leucovorin combinations in colon adenocarcinoma xenografts. Cancer Chemother Pharmacol 1992–30:423–432
6. Traut TW. Physiological concentrations of purines and pyrimidines. Mol Cell Biochem 1994 140:1–22
7. Ruiz van Haperen VWT, Veerman G, Van Moorsel CJA, Peters GJ. Induction of *in vivo* resistance against gemcitabine (gem, 2',2'-difluorodeoxycytidine). This volume
8. Ruiz van Haperen VWT, Veerman G, Erikson S, Boven E, Stegmann APA, Hermsen M, Vermorken JB, Pinedo HM, Peters GJ. Development and molecular characterization of a-2',2'-difluorodeoxycytidine-resistant variant of the human ovarian carcinoma cell line A2780. Cancer Res 1994 54:43–138–4143
9. Van der Wilt CL, Pinedo HM, Smid K, Cloos J, Noordhuis P, Peters GJ. Effect of folinic acid on fluorouracil activity and expression of thymidylate synthase. Sem Oncol 1992 19(suppl 2):16–25
10. Van der Wilt CL, Pinedo HM, Smid K, Peters GJ. Elevation of thymidylate synthase levels following 5-fluorouracil treatment is prevented by the addition of leucovorin in murine colon tumors. Cancer Res 1992 52:4922–4928
11. Kralovanski J, Prajda N, Kerpel-Fronius S, Bagrij T, Kiss E, Peters GJ. Biochemical consequences of 5-fluorouracil gastrointestinal toxicity in rats; effect of high dose uridine. Cancer Chemother. Pharmacol. 1993, 32:243–248
12. Darnowski J, Goulette F.A. Increased azidodeoxythymidine metabolism in the presence of fluorouracil reflects increased thymidine kinase activity. Proc Amer Ass Cancer Res 1993 34:302
13. Pressacco J, Mitrovski B, Erlichman C, Hedley D. Effects of thymidylate synthase inhibition on thymidine kinase activity and nucleoside transporter expression. Cancer Res 1995–55:1505–1508

127

ROLE OF DEOXYCYTIDINE KINASE (dCK), THYMIDINE KINASE 2 (TK2), AND DEOXYCYTIDINE DEAMINASE (dCDA) IN THE ANTITUMOR ACTIVITY OF GEMCITABINE (dFdC)

J. R. Kroep,[1] C. J. A. van Moorsel,[1] G. Veerman,[1] D. A. Voorn,[1] R. M. Schultz,[2] J. F. Worzalla,[2] L. R. Tanzer,[2] R. L. Merriman,[3] H. M. Pinedo,[1] and G. J. Peters[1]

[1]Department of Med. Oncology
University Hospital VU
Amsterdam, The Netherlands
[2]Eli Lilly
Indianapolis, Indiana
[3]Warner-Lambert Park-Davis
Ann Arbor, Michigan

1. INTRODUCTION

Deoxycytidine kinase (dCK) and deaminase (dCDA) are as activating and inactivating enzymes, respectively, in the metabolism of several chemotherapeutically important deoxynucleoside analogues [1]. 2'2'-Difluorodeoxycytidine (dFdC; gemcitabine) has considerable antitumor activity against solid tumors, such as against the chemoresistant non-small cell lung cancer (NSCLC) and pancreatic cancer [2]. dCK catalyses the rate-limiting phosphorylation of CdR and its analogues to their corresponding monophosphates [1,3]. To avoid an overestimation of the dCK activity by thymidine kinase 2 (TK2), which can also efficiently phosphorylate CdR [3], dCK activity was measured in the presence of thymidine (TdR) to inhibit TK2 [4]. dCDA inactivates cytidine (CR), CdR and its analogues to their deaminated products [5,6]. Previously we could not establish a relationship between antitumor activity and the dCK and dCDA activities [6], while in a cell line study more precise measurements of dCK showed a relation between sensitivity to dFdC and efficiency of dCK [7]. We now reevaluated the role of dCK, TK2 and dCDA in the antitumor effect of dFdC against different solid tumors.

Purine and Pyrimidine Metabolism in Man IX,
edited by Griesmacher *et al.* Plenum Press, New York, 1998.

2. MATERIALS AND METHODS

2.1. Materials

CdR and deoxyuridine (UdR) were purchased from Sigma Chemical Co. (ST Louis, Missouri, U.S.A). Deoxy-[5-^{3}H]-cytidine (^{3}H-CdR, 21.5 Ci/mmol) was obtained from Amersham International and [2-^{14}C]-Thymidine (^{14}C-TdR, 59.3 Ci/mol) from Dupont de Nemours NEN. All other chemicals were of analytic grade and are commercially available.

2.2. Tumors

Human tumor xenografts grown at The Lilly Research laboratories, Indianapolis IN, USA [8] were made available for enzyme activity characterization. Lewis Lung was grown at Amsterdam. Sensitivity to dFdC was determined as described previously [8] and evaluated by calculation of the T/C values, the ratio between tumor volume of treated versus control.

2.3. dCK, TK, and dCDA Assay

Frozen tumor tissues were pulverized using a microdismembrator [9]. The powder was reconstituted in dCK buffer (0.3 mol/L Tris-HCL, 50 μmol/L β-mercaptoethanol, pH 8.0) at 1 g tumor/3 ml buffer. dCK, total TK and TK1 activity were assayed using ^{3}H-CdR and ^{14}C-TdR and radiolabeled monophosphate product was separated on TLC from the substrate; TK2 was determined by substracting TK1 from the total TK. dCDA activity was measured with HPLC as described previously [6]. Enzyme activities were expressed as pmol product formed per hour per mg protein (pmol/h/mg protein).

3. RESULTS

Sensitivity of the human carcinoma xenografts and the Lewis lung mouse tumor varied from resistant (BxPC3 pancreas; T/C 0.91) to moderately sensitive (H460 lung; T/C 0.55) to sensitive (CALU-6 lung; T/C 0.25, Lewis Lung; T/C 0.23). The resistant BxPC3 appeared to have the highest deoxycytidine phosphorylating activity (4 times CALU-6). However this was due to a 90% contribution of TK2 at low, near physiological CdR concentration; the corrected dCK activity assayed in the presense of TdR was lowest in this tumor (as Table 1). dCK activity itself correlated with sensitivity (Fig. 1), in contrary to TK2 efficiency. Moreover a significant relation between dCDA activity and response to dFdC was found. The dCDA activity was 45 times higher in the resistant pancreas carcinoma compared to the mean of the dCDA activity in the other non-resistant tumors.

4. DISCUSSION

The results show that dCK activity itself, but not the total deoxycytidine phosphorylating activity (dCK + TK2), can be correlated with the sensitivity for dFdC in these solid tumors. Conflicting results have been published regarding the relation between the deoxycytidine phosphorylating activity and efficiency of inhibition of DNA synthesis by deoxycytidine analogs [6]. Previously we demonstrated in 28 solid tumors of four different histological origins that there was no correlation between total CdR phosphorylation,

Table 1. Enzyme activities and sensitivities to dFdC treatment
in NSCLC- and pancreas tumors in vivo

Tumor	T/C	dCK 10μM CdR	10μM CdR + TdR	dCDA	dCK/dCDA	TK2
BxPC3 (pancreas)	0.91	2355 ± 348	225 ± 57	144390 ± 7661	0.0016	4399
NCI-H460 (NSCLC)	0.55	553 ± 65	402 ± 49	4105 ± 1462	0.098	3002
CALU-6 (NSCLC)	0.25	668 ± 144	528 ± 69	4707 ± 836	0.11	3934
Lewis Lung (NSCLC)	0.23	377 ± 54	321 ± 19	n.d.	n.d.	1783

T/C: The ratio between tumor volume of treated versus control.
n.d: not done
Activity values are means ± SEM of 3-4 experiments in pmol/h/mg protein.

dCDA or the ratio of total CdR phosphorylation to dCDA activity and sensitivity [6].
Even at the lowest dCK levels a good antitumor activity of dFdC was observed, demonstrating that the total dCK activity in solid tumors is high enough to convert CdR analogues to the active metabolites. In the current panel of solid tumors including a resistant pancreas carcinoma we found significant higher dCK activity and lower dCDA activities towards the more sensitive tumors. It remains however unclear what is the minimal activity of dCK required for sufficient activation of dFdC in solid tumors. This paper demonstrates that exact measurement of specific enzyme activities is required in order to establish the relationship between dCK activity and antitumor activity. This can be established by the current approach (inhibition of interfering enzymes), by using substrates which are specific for dCK and not for TK 2, such as 2-chlorodeoxyadenosine (CdA) or by using a specific immunoblotting technique [10].

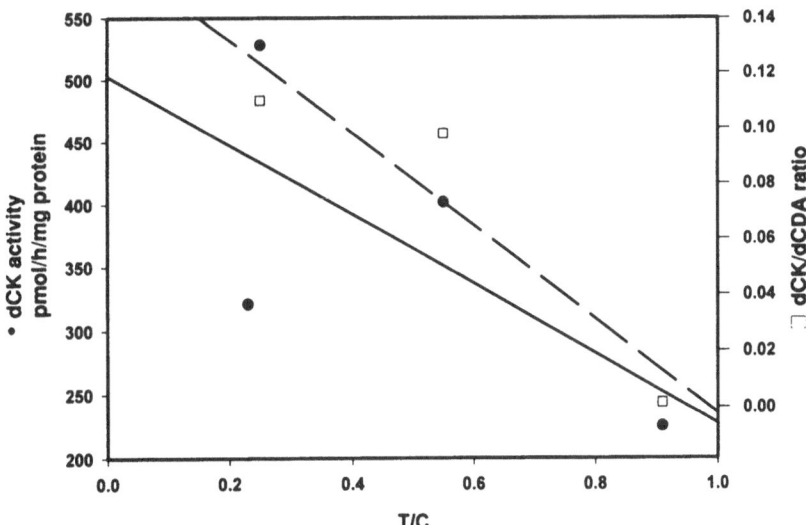

Figure 1. Correlation between dCK activity (bullets), the ratio dCK/dCDA (squares) and tumor sensitivity to dFdC (r = 0.69 and r = 0.93, respectively). Values are means of 3–4 separate measurements.

ACKNOWLEDGMENTS

This research was supported by a grant from the Dutch Cancer Society IKA VU-94-753 and by Eli Lilly, Inc.

REFERENCES

1. Ruiz van Haperen VWT, Peters GJ: New targets for pyrimidine antimetabolites in the treatment of solid tumors. 2. Deoxycytidine kinase. Pharmacy World & Science (1994) 16:104–112.
2. van Moorsel CJA, Peters GJ, Pinedo HM: gemcitabine: future prospects of single-agent and combination studies. The Oncologist (1997) 2:127–134.
3. Arner ESJ, Eriksson S: Mammalian deoxyribonucleoside kinases. Pharmac Ther (1995) 67:155–186.
4. Munch-Petersen B, Cloos L, Tyrsted G, Eriksson S: Diverging substrate specificity of pure human thymidine kinases 1 and 2 against antiviral dideoxynucleosides. J Biol Chem (1991) 266:9032–9038.
5. HoDHW: Distribution of kinase and deaminase of 1-ß-D-arabinofuranosylcytosine in tissues of man and mouse. Cancer Res (1973) 33:2816–2820.
6. Ruiz van Haperen VWT, Veerman G, Braakhuis BJM, Vermorken JB, Boven E, Leyva A, Peters GJ: Deoxycytidine kinase and deoxycytidine deaminase activities in human tumour xenografts. Eur J Cancer (1993) 29A:2132–2137.
7. Ruiz van Haperen VWT, Veerman G, Vermorken JB, Pinedo HM, Peters GJ: Regulation of deoxycytidine and 2',2'-difluorodeoxycytidine (gemcitabine); effects of cytidine 5'-triphosphate and uridine 5'-triphosphate in relation to chemosensitivity for 2',2'-difluorodeoxycytidine. Biochem Pharmacol (1996) 51:911–918.
8. Merriman RL, Hertel LW, Schultz RM, Houghton, PJ, Houghton JA, Rutherford PG, Tanzer LR, Boder GB, Grindey GB: Comparison of the antitumor activity of gemcitabine and ara-C in a panel of human breast, colon, lung and pancreatic xenograft models. Invest New Drugs (1996)14:243–247.
9. Peters GJ, Laurensse EJ, Leyva A, Pinedo HM: Tissue homogenisation using a micro-dismembrator for the measurement of enzyme activities. Clin Chim Acta (1986)158:193–198.
10. Spasokoukotskaja T, Arner ESJ, Brosjo O, Gunven P, Juliusson G, Liliemark J and Eriksson S: Expression of deoxycytidine kinase and phosphorylation of 2-chlorodeoxy-adenosine in human normal and tumor cells and tissues. Eur J Cancer (1995) 31A:202–8.

128

DIFFERENTIAL EFFECTS OF GEMCITABINE ON NUCLEOTIDE POOLS OF 19 SOLID TUMOR CELL LINES

C. J. A. van Moorsel,[1] G. Veerman,[1] V. W. T. Ruiz van Haperen,[2]
A. M. Bergman,[1] P. B. Jensen,[3] M. Sehested,[3] and G. J. Peters[1]

[1]Department Medical Oncology
University Hospital Vrije Universiteit
P. O. Box 7057
1007 MB Amsterdam, The Netherlands
[2]MD Anderson
Houston, Texas
[3]Finsen Center, Rigshospitalet
Lab. Eksperimental Medical Oncology
Copenhagen, Denmark

1. INTRODUCTION

2,2-Difluorodeoxycytidine (dFdC, Gemcitabine, Gemzar) is a deoxycytidine ana-logue, which is very active against murine experimental tumor models, human tumor xenografts and in clinical Phase II studies.[1,2] After entering the cell gemcitabine is phos-phorylated to its active metabolite gemcitabine triphosphate (dFdCTP) by deoxycytidine kinase (dCK),[3,4] incorporated into the DNA, and then, after adding one more nucleotide, polymerisation stops.[5] Gemcitabine not only acts on DNA but has several selfpotentiating interactions within the cell, such as incorporation into RNA[6] and inhibition of ribonu-cleotide reductase, thereby depleting the cell of deoxyribonucleotides and favouring its own incorporation into DNA.[7]

Since previous studies showed a marked change in normal ribonucleotides after exposing human ovarian and colon carcinoma cell lines to dFdC,[8] a study was performed to gain better insight in the metabolism of dFdC and its effect on normal ribonucleotide metabolism in solid tumor cell lines. The sensitivity to dFdC in a panel of 19 solid tumor cell lines (3 ovarian-, 9 lung-, 1 head and neck- (HNC) and 6 colon cancer cell lines) with different histological phenotypes was related to the accumulation of dFdCTP and changes in normal ribonucleotides.

Purine and Pyrimidine Metabolism in Man IX,
edited by Griesmacher *et al.* Plenum Press, New York, 1998.

2. MATERIALS AND METHODS

2.1. Chemicals

dFdC and dFdCTP were kindly provided by Eli Lilly & Co., Indianapolis IN, USA. ATP, ADP, CTP, GTP and UTP were purchased from Sigma Chemical Co., St. Louis MO, USA. All other chemicals were of analytical grade and commercially available.

2.2. Cell Culture

Cells were maintained in exponential growth in Dulbecco's Modification of Eagle's Medium (Gibco Laboratories, Grand Island NY, USA) or in RPMI medium supplemented with 5% or 10% heat-inactivated fetal calf serum (Gibco) at 37°C and 5% CO_2. The following cell lines were used in this study: the human ovarian cancer cell lines A2780, ADDP and AG6000, the lung cancer cell lines LewisLung, H322, SW1573, SW1573/2R120, SW1573/2R160, H69, H69/DAU, NYH and NYH/VM, the colon cancer cell lines C26-10, WiDr, HT29, SW620, C26-A and C26-G and the human HNC cell line 22B.

2.3. Accumulation of dFdCTP and Ribonucleotides

For determination of the accumulation of dFdCTP, $2–4.10^5$ cells were seeded in 6-wells plates, in triplicate. After two days, when cells were growing in log-phase dFdC was added to a final concentration of 1 µM (10 µM for C26-A and C26-G cells) for 24 h. As a control non-exposed cells were cultured for the same period. Cells were harvested and extracted for the nucleotides, as described previously.[8]

The applied HPLC method enabled the measurement of not only dFdCTP, but also that of ADP, ATP, UTP, CTP and GTP in one run and thus the assurance of the different nucleotides not interfering with each other. The HPLC methodology was reported previously.[8]

3. RESULTS

3.1. dFdCTP Accumulation

dFdCTP accumulation was cell line dependent (Fig. 1). Colon tumor cells accumulated the highest dFdCTP concentrations (median of all cell lines, C26-A and C26-G excluded (for which 10 µM instead of 1 µM was used): 925 pmol/10^6 cells), followed by the lung cancer cell lines (median: 816 pmol/10^6 cells) and the H&N cancer cell line (614 pmol/10^6 cells).

The ovarian cancer cell lines accumulated the lowest amount of dFdCTP (450 pmol/10^6 cells). The amount of dFdCTP accumulation was correlated with the sensitivity for dFdC after 24 h exposure (Fig. 2), the cell line AG6000 with induced dFdC resistance, accumulated the least dFdCTP whereas more sensitive cell lines accumulated higher amounts of dFdCTP.

3.2. Effects on Normal Ribonucleotide Pools

The normal ribonucleotide pools were affected considerably by dFdC treatment. In Table 1 the changes in the pools of ATP, CTP, UTP and GTP of cells treated with dFdC are summarized. In some cell lines dFdC resulted in a depletion of CTP pools (2R120, WiDr

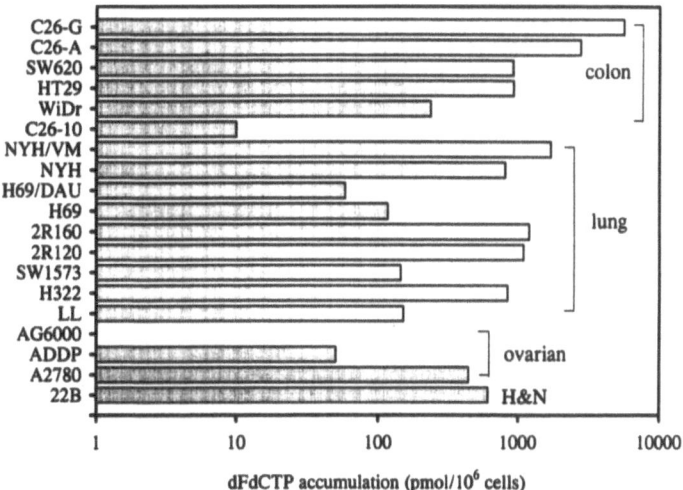

Figure 1. Accumulation of dFdCTP after exposure to 1 μM dFdC in 19 solid tumor cell lines (10 μM dFdC for C26-A and C26-G cells). Values shown represent the mean ± SEM of 3–5 separate experiments.

and C26-G), but in most cell lines CTP pools increased. An increase in UTP pools was shown in 15 out of 19 cell lines. The most predominant effect was observed in NYH lung cancer cells (3.5-fold increase). dFdC also caused perturbations of the GTP and ATP pools, which increased in all cell lines. The highest increases were evident in the colon carcinoma cell line SW620.

4. DISCUSSION

In this study we demonstrate that dFdCTP can be accumulated in a large panel of tumor cell lines and generally correlates with sensitivity to dFdC in vitro. This is in

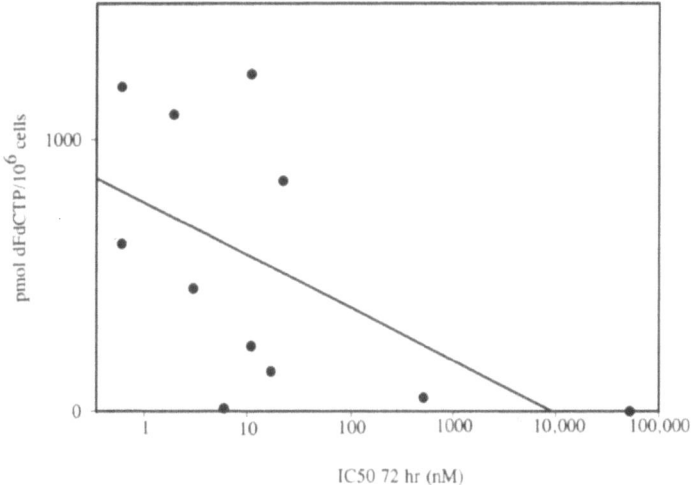

Figure 2. Relation between dFdCTP accumulation and in vitro sensitivity to 72 h exposure of dFdC in 11 evaluable tumor cell lines. Values shown represent the mean of 3–5 separate experiments. Correlation coefficient r = 0.56.

Table 1. Changes in ribonucleotide pools after treatment for 24 hr with 1 μM dFdC in 19 different tumor cell lines (treatment with 10 μM dFdC in C26-A and C26-G cells)

Cell line	Changes compared to control			
	CTP	UTP	GTP	ATP
A2780	1.6	1.4	1.8	1.6
ADDP	1.5	1.5	1.5	2
AG6000	=	=	=	=
LL	3	3	2	1.8
H322	=	1.5	1.3	1.7
SW1573	1.2	1.7	1.7	2
SW1573/2R120	0.5	=	=	1.2
SW1573/2R160	1.2	=	=	1.1
H69	2.5	2	2	2
H69/DAU	2	=	1.9	2
NYH	3	3.5	3	3
NYH/VM	1.25	1.8	1.5	2
22B	=	2	2	2
C26-10	2.7	2	2	1.8
WiDr	0.3	1.5	1.6	1.6
HT29	=	2	1.5	2.1
SW620	1.5	3	3	4
C26-A	=	1.5	1.2	1.2
C26-G	0.5	2	1.5	1.2

Changes are expressed as increase or decrease factors of the mean values of accumulation compared to the control values.

accordance with the results of a previous study performed in our laboratory with a small panel of cell lines.[8]

In contrast to the observations of Gandhi & Plunkett[10] in K562 cells, but in accordance to our previous study, we observed substantial changes in ribonucleotide pools. The decrease in CTP, as observed in three out of 19 cell lines, might be explained by an inhibition of CTP-synthetase, since this would lead to an increase in UTP pools, the substrate for CTP synthetase. Heinemann & Plunkett[11] reported inhibition of CTP-synthetase in CCRF-CEM cells. The inhibition of this important enzyme, rate limiting in the synthesis of cytidine nucleotides, may have consequences for the activity of dCK, the activator of dFdC. Furthermore, White & Cappizzi[12] and Shewach et al.[13] showed that in leukemic cells, UTP is a more efficient phosphate donor of dCK than ATP. However, in solid tumor cell lines ATP was found to be a better phosphate donor for dCK than UTP.[14] Increases of ATP as well as UTP levels could therefore be favorable for dCK activity.

The increase in GTP pools might be indirectly related to the inhibition of CTP-synthetase. When this enzyme is inhibited, the L-glutamine normally used for amination of UTP, will accumulate. This may enhance activity of GMP-synthetase, which catalyzes the conversion of xanthosine-5'-phosphate into GMP.[15] Another factor in the increase in UTP and GTP pools in all cell lines and, in some cases the increase of CTP pools, may be the inhibition of RNA synthesis.[8] In both C26–10 cells and A2780 cells RNA synthesis was inhibited 10–30% at a 4 h exposure to 1 μM dFdC. This impaired RNA synthesis could also explain the further increase of ribonucleotide concentrations after removal of dFdC from the cells. More research is warranted to clarify this phenomenon.

In conclusion, in this study we demonstrate that dFdCTP accumulation in solid tumor cell lines is clearly related with sensitivity. The antitumor effect of dFdC may also be related to the marked changes in ribonucleotide pools, leading to and due to both inhibition and stimulation of several important enzymes involved in RNA and DNA synthesis. The changes, possibly leading to selfpotentiation, suggest that the mechanism of action of dFdC cannot only be attributed to the damage to DNA. The disturbance of normal pyrimidine and purine metabolism in solid tumor cells may be of equal importance.

ACKNOWLEDGMENTS

This research was supported by a grant from the Dutch Cancer Society (IKA-VU-94-753).

REFERENCES

1. Braakhuis BJM, Ruiz van Haperen VWT, Boven E, Veerman G and Peters GJ: Schedule-dependent antitumor effect of gemcitabine in in vivo model systems. Sem Oncol 22 (suppl 11):42–45, 1995.
2. Van Moorsel CJA, Peters GJ and Pinedo HM: Gemcitabine: future prospects of single agent and combination studies. The Oncologist 2:127–134, 1997.
3. Heinemann V, Hertel LW, Grindey GB and Plunkett W: Comparison of the cellular pharmacokinetics and toxicity of 2',2'-difluorodeoxycytidine and 1-B-D-arabinofuranoxylcytosine. Cancer Res 48:4024–4031, 1988.
4. Ruiz van Haperen VWT and Peters GJ: New targets for pyrimidine antimetabolites in the treatment of solid tumours II: Deoxycytidine kinase. Pharm World & Science 26:104–112, 1994.
5. Huang P, Chubb S, Hertel LW, Grindey GB and Plunkett W: Action of 2',2'-difluoro-deoxycytidine on DNA synthesis. Cancer Res 51:6110–6117, 1991.
6. Ruiz van Haperen VWT, Veerman G, Vermorken JB and Peters GJ: 2',2'-Difluoro-deoxycytidine (Gemcitabine) incorporation into RNA and DNA of tumour cell lines. Biochem Pharmacol 46:762–766, 1993.
7. Plunkett W, Huang P, Xu YZ, Heinemann V, Grunewad R and Gandhi V: Gemcitabine: metabolism, mechanisms of action, and self-potentiation. Sem Oncol 22 (suppl. 11):42–46, 1995.
8. Ruiz van Haperen VWT, Veerman G, Boven E, Noordhuis P, Vermorken JB and Peters GJ: Schedule dependence of sensitivity to 2',2'-difluorodeoxycytidine (gemcitabine) in relation to accumulation and retention of its triphosphate in solid tumour cell lines and solid tumours. Biochemical Pharmacology 48: 1327–1339, 1994.
9. Bergman AM, Ruiz van Haperen VWT, Veerman G, Kuiper CM and Peters GJ: Synergistic interaction between cisplatin and gemcitabine in vitro. Clin Cancer Res 2:521–530, 1996.
10. Gandhi V and Plunkett W: Modulatory activity of 2',2'-difluorodeoxycytidine on the phosphorylation ans cytotoxicity of arabinosyl nucleosides. Cancer Res 50: 3675–3680, 1990.
11. Heinemann V and Plunkett W: Inhibitory action of 2',2'-difluorodeoxycytidine (dFdC) on cytidine 5'-triphosphate synthetase. Ann Oncol 3 (suppl. 1): 187 (abstr. 510), 1992.
12. White JC and Capizzi RL: A critical role for uridine nucleotides in the regulation of deoxycytidine kinase and the concentration dependence of 1-ß-D-arabinofuranosylcytosine phosphorylation in human leukemia cells. Cancer Res 51: 2559–2565, 1991.
13. Shewach DS, Reynolds KK and Hertel LW: Nucleotide specificity of human deoxycytidine kinase. Mol Pharmacol 42: 518–524, 1992.
14. Ruiz van Haperen VWT, Veerman G, Vermorken JB, Pinedo HM and Peters GJ: Regulatory effects of CTP and UTP on deoxycytidine kinase activity in solid tumor cell lines. Biochem Pharmacol 51:911–918, 1996).
15. Weber G: Biochemical strategy of cancer cells and the design of chemotherapy: GHA Clowes Memorial Lecture. Cancer Res 43: 3466–3492, 1983.

129

INCREASED ACTIVITY OF *CYTIDINE TRIPHOSPHATE SYNTHETASE* IN PEDIATRIC ACUTE LYMPHOBLASTIC LEUKEMIA

A. C. Verschuur, A. H. van Gennip, E. J. Muller, P. A. Voûte, and
A. B. P. van Kuilenburg

Academic Medical Centre, University of Amsterdam
Department of Clinical Chemistry and Department of Pediatrics
P. O. Box 22700
1100 DE Amsterdam, The Netherlands

1. INTRODUCTION

Children suffering from acute lymphoblastic leukemia (ALL) possess an increased concentration of cytidine triphosphate (CTP) in their lymphoblasts compared to resting lymphocytes.[1] This might be due to either an enhanced flux through the pyrimidine "de novo" and/or uridine salvage pathway or to an increased flux through the cytidine salvage pathway. By studying ribonucleotide fluxes in a MOLT-3 lymphoblastic cell line it has been shown that the increased CTP concentration is the result of an enhanced activity of CTP synthetase (CTPS).[2] We now analyzed the "in vitro" CTPS activity in lymphoblasts of children with ALL. If increased, CTPS might be inhibited by drugs like cyclopentenylcytosine (CPEC).

2. PATIENTS AND METHODS

2.1. Cell Isolation

The CTPS activity was measured in lymphoblasts of 16 pediatric patients with ALL at diagnosis and compared with proliferating and quiescent lymphocytes. Leukemic cells were obtained by bonemarrow aspiration of the posterior iliac crest or by venapuncture in case of a white blood cell count exceeding 100 10^6 mononuclear cells/ml. The material was anticoagulated by EDTA. After collection, the cell suspension was diluted 1:1 with a "supplemented phosphate-buffered saline" (supplemented PBS) solution (9.2 mM Na_2HPO_4, 1.3 mM NaH_2PO_4, 140 mM NaCl, 5 mM glucose, 0.2% bovine serum albumin, 13 mM trisodiumcitrate and 5 mM EDTA, pH 7.4) and centrifuged (200 g, 10 min) after which the platelet-enriched plasma was discarded. The remaining cell pellet was again diluted 1:1 with

Purine and Pyrimidine Metabolism in Man IX,
edited by Griesmacher *et al.* Plenum Press, New York, 1998.

supplemented PBS and was carefully layered on top of a 1.077 gr/L Percoll solution with a 2:1 volume ratio. After centrifugation for 20 minutes at 800 g the interphase was collected and washed with supplemented PBS, after which the pellet was resuspended in 5 ml ice-cold erythrocyte-lysisbuffer (8.29 gr/L NH_4Cl, 1.0 gr/L $KHCO_3$, 37.2 mg/L EDTA, pH 7.4); after 5 minutes of lysis 9 ml of supplemented PBS was added. The cells were washed twice and the final cell pellet was resuspended in a buffer containing PBS (9.2 mM Na_2HPO_4, 1.3 mM NaH_2PO_4, 140 mM NaCl, pH 7.4) and 5 mM glucose. Cells were counted with a Coulter Counter ZF cellcounter. Vitality was assessed by the trypan blue exclusion test and cell purity was assessed morphologically by Jenner Giemsa staining and light microscopy.

Lymhocytes were isolated in the same way, but after collection of the mononuclear cells, lymphocytes were purified by magnetic cell sorting (Miltenyi, Germany), using CD11b microbeads or the pan-T isolation microbeads, and washed with PBS supplemented with 5 mM glucose. Proliferation of lymphocytes was achieved using round-bottom culture plates coated with anti-CD3 monoclonal antibody (CLB 16A9, 1:1000, 4 hr, 37°C), after which lymphocytes were cultured for 72 hours in RPMI 1640 containing 10% fetal bovine serum, 100 IU/ml penicilline, 100 µg/ml streptomycine, 200 µg/ml gentamycine, 0.125 µg/ml amphotericin B, 2 mM glutamine, 100 µM β-mercaptoethanol and 50 U/ml Il-2.

2.2. CTP Synthetase Assay

The enzyme activity was measured by the method described by Van Kuilenburg[3] et al. Briefly, a cell homogenate was obtained by sonication, after which the homogenate was centrifuged and the supernatant was used for the assay. The protein content of the supernatant was determined with a modified Lowry method.

The assay mixture contained 1 mM UTP, 4 mM ATP, 10 mM glutamine, 1 mM GTP, 20 mM $MgCl_2$, 17 U/ml pyruvate kinase, 15 mM PEP, 10 mM dithiotreitol, 2.5 mM phenylmethylsulphonyl fluoride, 1 mM EGTA. After a 10 minute preincubation, the assay-mixture and cell homogenate were coïncubated for 2 hours, after which the nucleoside triphosphates were extracted by 0.55 N perchloric acid and neutralized with 0.35 M K_2CO_3. The nucleoside triphosphates were separated by anion-exchange high performance liquid chromatography (HPLC), under isocratic elution conditions using a 0.594 M NaH_2PO_4 buffer, pH 4.55. The retention times were 2.90, 4.15, 5.05 and 6.85 minutes for UTP, CTP, ATP and GTP respectively. Response factors were calculated using an external standard solution containing authentic nucleotides UTP, CTP, ATP and GTP. Absorption was measured at a wavelength of 280 nm.

The HPLC system consisted of a Gilson 231XL and 402 sampling device, a Perkin Elmer Binary LC 250 pump, a Whatman Partisphere SAX 4.6 × 125 mm column (5 µm particles) and Whatman 10 × 2.5 mm AX guard column, Waters UV detector, a Nelson 900 series Interface and Nelson PC Integrator Software version 5.1.5.

3. RESULTS

There were 11 cases of precursor B-lymphocytic leukemia (common ALL) and 5 cases of T-lymphocytic leukemia. Ages ranged from 14 months to 15 years (mean 5.6 years). White blood cell counts ranged from $2.4 × 10^9/L$ to $610 × 10^9/L$.

The purified leukemic cells of the patients showed a purity of at least 80%. The control lymphocytes were > 95% pure and > 90% viable.

The mean activity of CTP synthetase proved to be significantly higher in lympho-blasts compared to quiescent lymphocytes (6.8 versus 1.8 nmol CTP/mg protein/hr,

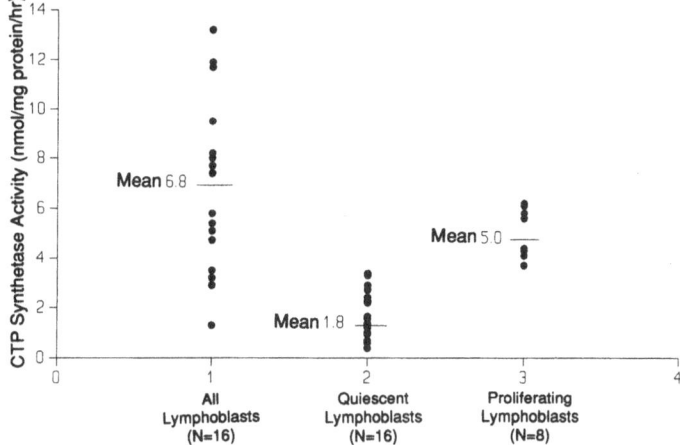

Figure 1. The specific CTPS activity is shown for ALL lymphoblasts (column 1), quiescent lymphocytes (column 2) and proliferating lymphocytes (column 3). The difference in CTPS activity between column 1 and 2 is significant (p = 0.002).

p=0.002) (Fig. 1). The activity in lymphoblasts seemed slightly higher compared to proliferating lymphocytes although not yet significant (6.8 versus 5.0 nmol CTP/mg protein/hr, p=0.17) (Fig. 1).

No correlation could be observed between the activity of CTP synthetase and the white blood cell count, nor with the percentage of lymphoblasts of the peripheral blood. A comparable mean enzyme activity was observed in T-ALL and B-ALL.

Preliminary results of incubation experiments with lymphoblasts of a patient with CPEC (kindly provided by the National Cancer Institute, Bethesda, Maryland) showed that CPEC is metabolized to its active triphosphate form, which inhibited CTPS activity and led to a CTP depletion (Fig. 2).

At a CPEC concentration of 1.25 µM a CTP depletion was observed with an increased UTP concentration, reflecting inhibition of CTPS. At higher CPEC concentration (>5 µM) both CTP and UTP concentrations decreased, which could be attributed to competition of CPEC with uridine and cytidine at the uridine-cytidine kinase level, in addition to CTPS inhibition. Similar patterns were observed in two other patients.

Figure 2. Incubation experiments of ALL lymphoblasts using increasing CPEC concentrations showing the effects on intracellular CTP (■), UTP (▲), CPEC-TP (♦), and ATP (□) concentrations, demonstrated as the relative percentages of ribonucleotide concentrations compared to the control sample without CPEC.

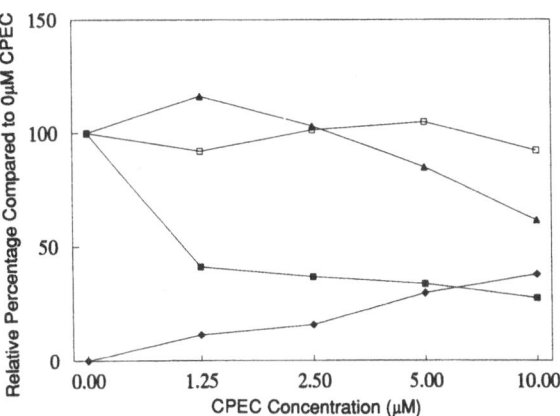

CPEC incubation experiments with proliferating T-lymphocytes also showed a profound CTP depletion, which was paralelled by a decreased proliferation rate. Therefore, CTP synthetase plays an important role not only in malignant lymphoblasts but probably also in proliferating T-lymphocytes.

4. DISCUSSION

A high CTP synthetase activity in lymphoblasts of pediatric patients is in line with studies of other types of malignancies (hepatoma, renal cell carcinoma, colon carcinoma, lymphoma[4-6]). Adults suffering from ALL also showed an increased enzyme activity.[6] However, the mean activity of CTPS observed in lymphoblasts of children with ALL was two-fold higher compared to that observed in lymphoblasts of adults with ALL.[6] This phenomenon might be explained by the profound effect of protein concentration on the specific activity of CTPS.[3]

Proliferating lymphocytes also showed a high CTPS activity, which might reflect the higher needs for nucleotides as their metabolism is accelerated to facilitate cell proliferation. Surprisingly, no correlation between the enzyme activity and the white blood cell count was observed in pediatric patients with ALL. This could suggest that the high enzyme activity is independent of the leukemic cells' proliferation rate, but it should be noted that lymphoblasts do not proliferate anymore "in vitro".

CTPS inhibition was accomplished in our experiments since CTP depletion occurred in our patients' lymphoblasts treated with CPEC, concurrent with an increased UTP concentration. CPEC also seemed to cause CTP depletion in proliferating T-lymphocytes, suggestive of CTPS inhibition. However, a direct effect of CPEC on cytidine salvage in proliferating lymphocytes can not be excluded since the majority of CTP seems to be produced by salvage of cytidine in proliferating T-lymphocytes,[2] and CPEC might compete with cytidine as a substrate for uridine/cytidine kinase.

CPEC is not only a potential cytotoxic drug, but is also capable of enhancing the cytotoxic effect of conventional drugs like arabinofuranosyl cytosine (AraC),[7] which is currently being used for the treatment of pediatric ALL. AraC must be phosphorylated to AraCTP in order to become cytotoxic, and a high dCTP concentration, that results from high CTP levels, will inhibit the first phosphorylation step of AraC (catalyzed by deoxycytidine kinase). Furthermore dCTP and AraCTP compete with each other in their affinity for DNA polymerase. Lowering the CTP concentrations by inhibition of CTPS, might therefore lead to lower dCTP levels and enhance the cytotoxic effect of AraC.

5. CONCLUSION

Our results provide the first evidence of an increased CTPS activity in pediatric ALL. Therefore, inhibiting CTPS by a drug like CPEC might be promising, and may cause a direct cytotoxic effect on leukemic cells, as well as an AraC-cytotoxicity modulating effect.

ACKNOWLEDGMENTS

The authors wish to express their gratitude for the generous financial support provided by the Dutch Cancer Society and the Foundation for Paediatric Cancer Research.

REFERENCES

1. D. De Korte, W.A. Haverkort, D. Roos, H. Behrendt, A.H. Van Gennip. Imbalance in the ribonucleotide pools of lymphoid cells from acute lymphoblastic leukemia patients. Leukemia Res. 10:389–96 (1986).
2. A.A. Van den Berg, H. Van Lenthe, S. Busch D. De Korte, D. Roos, A.B.P. Van Kuilenburg, A.H. Van Gennip. Evidence for transformation-related increase in CTP synthetase activity in situ in human lymphoblastic leukemia. Eur. J. Biochemistry 216:161–7 (1993).
3. A.B.P. Van Kuilenburg, L. Elzinga, A.C. Verschuur, A.A. Van den Berg, R.J. Slingerland, A.H. Van Gennip. Determination of CTP synthetase activity in crude cell homogenates by a fast and sensitive non-radiochemical assay using anion- exchange HPLC. J. Chromatogr 1997, in press.
4. J.C.Williams, H. Kizaki, G. Weber. Increased CTP synthetase activity in cancer cells. Nature 271:71–2 (1978).
5. H. Kizaki, J.C. Williams, H.P. Morris, G. Weber. Increased Cytidine 5'-triphosphate synthetase activity in rat and human tumors. Cancer. Res. 40:3921–7 (1980).
6. P.H Ellims, T.E. Gan, G. Medley. Cytidine triphosphate synthetase activity in lymphoproliferative disorders. Cancer Res 43:1432–5 (1983).
7. J.O. Lillemark, W. Plunkett. Regulation of 1-β-D- Arabinofuranosylcytosine 5-triphosphate accumulation in human leukemia cells by deoxycytidine 5-triphospha te. Cancer Res 46:1079–83 (1986).

130

EXPRESSION OF HUMAN CYTOSOLIC THYMIDINE KINASE IN *ESCHERICHIA COLI*

A Bacterial Whole Cell System for Screening of Potential Antiproliferative Nucleoside Analogs

Jianghai Wang,[1] Jan Neuhard,[2] and Staffan Eriksson[1]

[1]Department of Veterinary Medical Chemistry
The Swedish University of Agricultural Sciences
The Biomedical Center, Uppsala, Sweden
[2]Center for Enzyme Research
Institute of Molecular Biology
University of Copenhagen, Copenhagen, Denmark

INTRODUCTION

In mammalian proliferating cells cytosolic thymidine kinase (TK1) phosphorylate thymidine and deoxyuridine to the corresponding monophosphate, thus providing precursors for DNA synthesis through a salvage route. In addition, TK1 is capable of phosphorylating many nucleoside analogs. The expression of TK1 is highly cell cycle dependent, which leads to large variations among different cells in their capacity to phosphorylate deoxynucleosides (Arnér and Eriksson, 1995).

The deoxynucleoside salvage in procaryotes shows large variations (Munch-Petersen, 1983). In case of *Escherichia coli*, the bacteria have been shown to contain two functional nucleoside-transport systems, but express only one type of deoxynucleoside kinase, i.e. thymidine kinase. This bacterial enzyme shows considerable similarity in its amino acid sequence to human TK1, while several important differences still exist between these two enzymes (Black and Hruby, 1991).

The most popular systems for screening pharmacologically active nucleoside analogs are *in vitro* cell culture sytems and direct assays with purified target enzymes (Mitsuya *et al.*, 1990). Although they are reasonable and comparatively successful, the difference between species and cell types and the variability and cost involved in cell culture research cause considerable problems. The purification of target enzymes is also costly and complicated and *in vitro* assays are always subject to criticisms as to their *in vivo* relevance.

Purine and Pyrimidine Metabolism in Man IX,
edited by Griesmacher *et al.* Plenum Press, New York, 1998.

In this report we describe a bacterial system based on genetically modified *E. coli* strains which may be applied to screen the toxicity of nucleoside analogs by monitoring the selective inhibition of bacterial growth caused by the compounds. The main deoxypyrimidine catabolic enzymes were mutated in *E. coli* and the cDNA for human TK1 were introduced. Growth inhibition caused by nucleoside analogs was compared with the expression of human TK1. The results show that for several cytostatic and antiviral analogs there is an overall close correletion between their anti-proliferative effects observed with the bacterial system and with human T-lymphoblast CEM cells. Thus, *E. coli* expressing human salvage enzymes can be a rapid and inexpensive model system which may complement other screening methods in drug discovery projects.

MATERIALS AND METHODS

The bacterial strains used were all derivatives of *E. coli* K12. New strains for this study were constructed by P1 mediated transduction.

Plasmid pTrc99-A (Pharmacia) was used as cloning and expression vector. The coding sequence of human TK1 cDNA was amplyfied by PCR and cloned into pTrc99-A, yielding pTrcHUMTK1. Transcription of the cloned cDNA is from the IPTG inducible *trc* promoter in the vector. pTrcHUMTK1 was introduced into SØ5286, yielding SØ5288. SØ5286 is unable to catabolize thymidine and deoxyuridine due to mutational inactivation of thymidine phosphorylase (*deoA*) and uridine phosphorylase (*udp*, Neuhard and Kelln, 1996). In addition SØ5286 carries the *tdk-1* mutation inactiviating the endogenous *E. coli* thymidine kinase (Hiraga *et al.*, 1967). For the control strain, plasmid pTrc99-A was transformed into SØ5286, yielding SØ5292.

The incorporation of thymidine into the bacterial DNA by the recombinant *E. coli* was measured in the presence of IPTG. The culture was incubated with labelled thymidine for 30 minutes and the alkaline resistant acid precipitable radioactivity was determined (Karlström, 1970). To detect the TK level in the bacteria, the cells were collected by centrifuging the culture at 13 000 × g for 30 minutes and resuspended in a buffer containing lysozyme. After incubation at room temperature for 30 minutes and centrifugation at 13 000xg for 30 minutes, a supernant was achieved for enzyme activity analysis.

For screening of growth inhibition of the recombinant *E. coli* by nucleoside analogs, a fresh overnight culture was diluted into a fresh minimal media supplemented with required antibiotics and 1 mM IPTG to achieve an initial A_{600} of 0.01. This culture was then divided into tubes, 1 ml in each, and 20 μl analog solution was added to each tube. The tubes were shaked at 37°C for 3–4 hours until the control tube, which contained no analog, reached an A_{600} of about 0.6. The optical absorbance at 600 nm of all tubes was then measured. The relative growth of the bacteria in each sample was calculated by comparing its A_{600} to that of the control tubes, which was set as 100%.

The method for growth rate experiments of human T-lymphoblast cells with analogs was described by Törnevik *et al.* (1995) and the cells were cultured for 72 hours.

RESULTS AND DISCUSSION

To demonstrate that human TK1 was successfully expressed in the recombinant *E. coli* and its activity enables thymidine or analogs to be phosphorylated and incorporated into bacterial DNA, an incorporation experiment was performed and TK levels in the

Table 1. Effect of some thymidine analogs on the growth
of recombinant *E. coli* expressing human TK1

Analog	Concentration (μM)	Relative growth of SØ5288 (%)	Relative growth of SØ5292 (%)
5-FdU	1	15	40
5-Cl-dU	50	69	103
5-BrdU	20	30	85
5-IdU	20	45	77
AZT	20	21	89
3'-FdU	20	87	92
3'-F-5-F-dU	20	84	86
3'-F-5-propyl-dU	20	85	84
3'-F-5-I-dU	20	93	84
3'-F-5-Cl-dU	20	85	89
3'-Azido-5-I-dU	20	86	85
ddT	50	98	98
AraT	10	100	101
FIAU	50	90	104
FMAU	50	72	104
FLT	50	76	104

crude extract of the recombinant bacteria were measured. In the presence of 1 mM IPTG, the strain expressing human TK1 incorporated thymidine into its DNA more than 20 times higher than the control strain (540 and 25 pmole/ml/cell mass, respectively). When the TK level was compared between these two strains, it was found that SØ5292, the control strain, had no TK activity detectable, while SØ5288, the expressing strain, had a specific activity for TK of 22.7 nmole/min/mg. The TK level of *E. coli* origin was also measured and it was 0.5% of that in SØ5288 (94 pmole/min/mg). These results show that human TK1 is overexpressed in the recombinant *E. coli* and the incorporation of thymidine is dependent on the activity of the human enzyme.

The effects of a number of deoxythymidine analogs have been tested with SØ5288 and SØ5292. Results from the screening showed that most of the 5-halogenated dUrd analogs, as well as 5-fluoro-deoxyuridine (5-FdU), 3'-azido-2',3'-dideoxythymidine (AZT), 1-(2-deoxy-2-fluoro-β-D-arabinofuranosyl)-5-methyluracil (FMAU), and 1-(2-deoxy-2-fluoro-β-D-arabinofuranosyl)-5-iodouracil (FIAU) showed selective inhibition of bacterial growth (Table 1). The concentration curves were determined for several active analogs, and their EC_{50} values are shown in Table 2. Except for AZT, these analogs exhibited a very similar toxicity to both the recombinant *E. coli* and the CEM cells.

Table 2. EC_{50} value for some nucleoside analogs
with recombinant *E. coli* expressing human TK1
and human T-lymphoblast CEM cells

Analog	EC_{50} for *E. coli* (μM)	EC_{50} for CEM cells
5-FdU	8 nM	10 nM
5-IdU	20 μM	8 μM
AZT	500 nM	150 μM
3'Br-dU	5 μM	10 μM
FLT	60 μM	7 μM

AZT is the most used anti-HIV compound at present and it is known to cause side effects mainly by inhibiting bone marrow derived stem cells, but the inhibition of SØ5288 by AZT occured at considerably lower concentrations than those observed with many different human cell culture lines, including CEM cells (Mitsuya *et al.*, 1990; Törnevik *et al.*, 1995). AZT is toxic to many members of *Enterobacteriaceae*, including *E. coli* (Elwell *et al.*, 1987), and its antibacterial effect could be involved in its gastro intestinal side effect . In CEM cells the rate limiting step in the anabolism of AZT is the reaction catalyzed by thymidylate kinase and presumably this is also the case in bacterial cells. We have observed that the sensitivity of *E. coli* with its own thymidine kinase is very close to that of TK-deficient *E. coli* expressing human TK1 (data not shown), althought the TK levels are very different.

There are several important differences in the overall biosynthesis and catabolism of nucleosides and nucleotides between *E. coli* and mammalian cells. Nevertheless, an overall good correlation can be found between the antiproliferative effects of several important pyrimidine analogs on the modified *E. coil* system and the CEM cells and this should encourage similar attempts to construct bacterial test systems that may enhance future drug discovery projects.

ACKNOWLEDGMENT

We thank C. Ljungcrantz for her technical assistance in obtaining the data for CEM cells. Funds are from Swedish Medical Research Council and EU Biomedical Project BMH4-CT96-0479.

REFERENCES

E. S. J. Amér and S. Eriksson. Pharmac. Ther. **67**:155–186 (1995).
M. E. Black and D. E. Hruby. Mol. Microbiol. **5**:373–379 (1991).
L. P. Elwell, *et al.* Antimicrob. Agents Chemother. **31**:274–280 (1987).
S. Hiraga, K. Igarashi, and T. Yura. Biochim. Biophys. Acta **145**:41–51 (1967).
H.O. Karlström. J. Biochem. **17**:68–71 (1970).
H. Mitsuya, R. Yarchoan, and S. Broder. Science **249**:1533–1544 (1990).
A. Munch-Petersen (ed.). Metabolism of nucleosides and nucleobases in microorganisms. Academic Press, London (1983).
J. Neuhard and R.A. Kelln. p. 580–599. *In* F.C Neidhardt, *et al.* (eds.), *Escherichia coli* and *Salmonella:* Cellular and Molecular Biology, 2nd ed. American Society for Microbiology, Washington, D.C (1996).
Y. Törnevik, *et al.* Biochem. Phar. **49**:829–837 (1995).

131

STRUCTURE-ACTIVITY RELATIONSHIP IN FTORAFUR (TEGAFUR) AND RELATED 5-FU PRODRUGS

R. Zhuk

Latvian Institute of Organic Synthesis
Aizkraukles 21, Riga, LV-1006, Latvia

1. INTRODUCTION

5-Fluorouracil (5-FU) developed by Ch. Heidelberger and R. Duschinsky in 1957 is still one of the most widely used anticancer drug. Ftorafur (FT, Tegafur) introduced by S. Hiller, R. Zhuk and M. Lidak in 1964[1] is recognized as a prodrug of 5-FU with better bioavailability and lower gastrointestinal and bone marrow toxicity. This observation stimulated considerably the interest in 5-FU derivatives. A large number of compounds including 1-substituted-5-fluorouracils have been synthesized and their antitumour activity evaluated. The physico-chemical properties of these compounds essential for the antitumour activity, such as electron density distribution and lipophilicity have been less studied. The metabolism of 1-substituted-5-fluorouracils is still not clearly understood. It has been postulated that these compounds can act as effective prodrugs of 5-FU, releasing 5-FU by various metabolic pathways or exhibit antitumour activity unrelated to 5-FU release. However, there is no evidence for any 1-substituted 5-FU derivative uncapable of releasing 5-FU and being still active against tumours. Because of the lack of systematic studies on quantitative structure-activity relationship in this series of compounds design of new effective drugs remains mainly empirical.

This paper deals with three series of compounds, namely, (5-substituted-2-tetrahydrofuryl)-, 1-phthalidyl- and 1-alkoxyalkyl-5-fluorouracils. The synthesis of these compounds has been reported by our group as well as by other investigators.[2-8] We have measured the hydrolitic stability, partition coefficients (lg P) in n-octanol-phosphate buffer (pH 7.4) and binding constants with bovine serum albumin (lg K) for 32 5-FU derivatives. An attempt has been undertaken to correlate these properties with antitumour activity in leukemia L-1210 bearing mice as well as with inhibition of 2–14C-UdR incorporation into DNA of Ehrlich ascites carcinoma cells *in vivo*.

Purine and Pyrimidine Metabolism in Man IX,
edited by Griesmacher *et al.* Plenum Press, New York, 1998.

2. MATERIALS AND METHODS

Compounds I–IV were obtained from 5-FU by silyl procedure.[2-7] For compounds I a–g 2-chloro-5-substituted tetrahydrofurans were used as alkylating agents . If mixtures of cis- and trans-isomers had been formed in the reaction, they were resolved by Dry-Column Chromatography (DCC) on Silica Woelm and cis-isomers were used for physico-chemical and biological evaluation.

Partition coefficients have been determined by the "shake-flask" method as well as by the reversed phase HPLC on Zorbax ODS. Binding with albumin was assessed by the equilibrium dialysis and by the fluorescent probe with 1,8-ANS (1-anilino-8-naphtalene sulfonate). The results obtained by both the methods agreed.[9-10]

Life span increase in leukemia L-1210 bearing BDF_1 mice (ILS) on i/p drug administration and inhibition of 2-^{14}C-UdR incorporation into DNA of Ehrlich ascites carcinoma cells in vivo upon i/p drug administration were determined as described previously.[11]

3. RESULTS AND DISCUSSION

3.1. 1-(5-Substituted 2-Tetrahydrofuryl)-5-Fluorouracils

a $R = CH_2OH$
b $R = H$
c $R = COOC_2H_5$
d $R = CF_3$
e $R = C(CF_3)_3$
f $R = CH_2Cl$
g $R = CH_2F$

I

Tetrahydrofuryl derivatives I (a–g) are rather stable in neutral and alkaline media and less stable in acidic solutions. The most unstable in the series is I b (Ftorafur), which is characterized by $\tau_{1/2}$ = 587 h in 0.01 N HCl at 37°C. Therefore, hydrolysis hardly influences the biological activity in vivo in this series of compounds. Lipophilicity of compounds I a–g is varied in the range $-1.30 < \lg P < 0.57$ and the binding constants with albumin are in the range $3.45 < \lg K < 5.00$. The most active antitumour compound I b (Ftorafur) is characterized by $\lg P = -0.31$ and $\lg K = 3.74$. However, no correlation was found between physico-chemical properties and antitumour activity in this series of compounds.

3.2. 1-Phthalidyl-5-Fluorouracils

Contrary to the tetrahydrofuryl derivatives I a–g 1-phthalidyl-5-FU II a–d are stable in neutral and acidic pH, but decompose to 5-FU and 2-carboxybenzaldehyde in alkaline solutions (pH 8.0–11.5). A linear correlation between hydrolysis rate constants and pH was obseved. Compounds II a–d differ considerably in their rate of hydrolysis. For compounds II a–c $\tau_{1/2}$ values are 3.3; 2.8 and 3.0 min, respectively, at pH 10.5 and t=40°C. However, compound II d is much more resistant to hydrolysis with $\tau_{1/2}$=7.3 h under the same conditions.

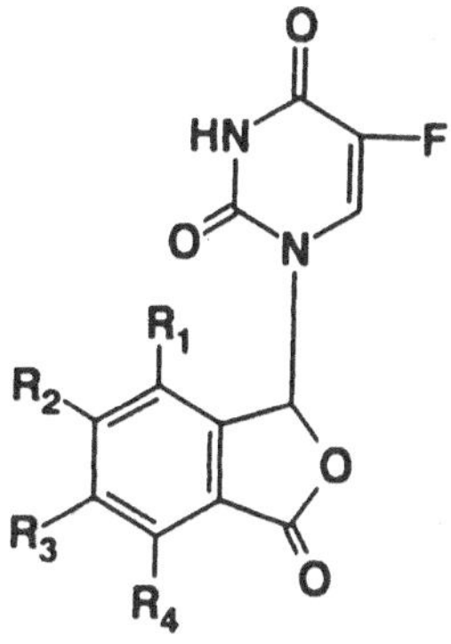

$$a \quad R_1 = R_2 = R_3 = R_4 = H$$
$$b \quad R_1 = R_2 = H; \ R_3 = R_4 = OCH_3$$
$$c \quad R_1 = R_2 = R_3 = OCH_3, \ R_4 = H$$
$$d \quad R_1 = R_3 = OCH_3, \ R_2 = OH, \ R_4 = H$$

II

Compounds **II** a-d differ in their antitumour activity as well. Compounds **II** a–c increase life span of L-1210 bearing mice following i/p drug administration in equimolar doses by 91, 70 and 82%, correspondingly. Compound **II** d is completely inactive. It allows us to assume that 5-FU released by the hydrolysis may be important for antitumour activity of these compounds along with the metabolism by liver hydrolases described previously.[12]

3.3. 1-Alkoxyalkyl-5-Fluorouracils

1-Alkoxyalkyl-5-fluorouracils (III, IV) may be considered as acyclic analogs of ftorafur:

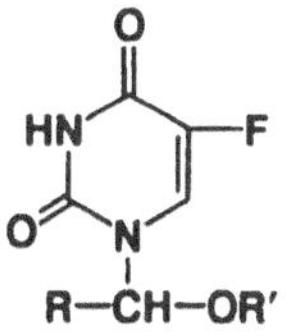

III 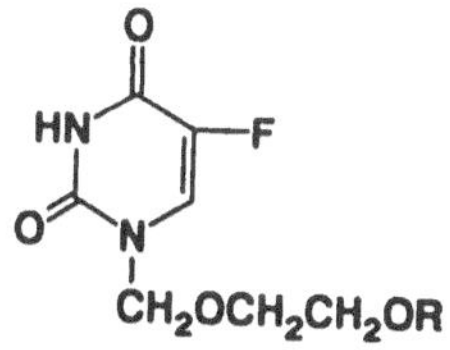   **IV**

a $R = H$; $R' = CH_3$	a $R = COCH_3$
b $R = H$; $R' = C_2H_5$	b $R = H$
c $R = H$; $R' = C_3H_7$	c $R = CH_3$
d $R = H$; $R' = C_4H_9$	d $R = C_2H_5$
e $R = CH_3$; $R' = C_2H_5$	
f $R = CH_3$; $R' = C_3H_7$	
g $R = CH_3$; $R' = C_4H_9$	

Compounds **III** and **IV** are stable to hydrolysis either in acidic or in alkaline media. They are metabolized less efficiently than Ftorafur,[13-14] nevertheless the release of 5-FU is sufficient to cause a considerable antitumour effect, which correlates with the lipophilicity of the compounds. The parabolic relationship has been found between lg ILS, % and lg P:

$$\lg (ILS, \%) = -0.70 \ (\lg P)^2 - 0.23 \lg P + 1.73$$
$$n = 6 \qquad r = 0.962 \qquad s = 0.123$$

Although the number of the data is insufficient to obtain significance in the regression curve we might be able to get a rough estimation. The same tendency was observed for inhibition of 2-^{14}C-2'-UdR incorporation in Ehrlich ascites carcinoma cells *in vivo* following i/p drug administration. Most active compounds have the partition coefficients in the range −0.40 < lg P < 0.40. The binding constants with albumin (lgK) are considered to be related to lipophilicity, as they are mainly determined by hydrophobic interactions between the drug and the protein. The linear correlation between lgK and lgP was observed in the homologic series of compounds **III** a–d. Branching in the radical (compounds **III** e–g) lowers lipophilicity, as compared with the normal alkyl of the same length, but increases binding with albumin. Therefore for the whole series of compounds **III** and **IV** the correlation was rather poor.

4. CONCLUSIONS

Only those 1-substituted 5-FU derivatives which release 5-FU in the organism either chemically or enzymatically exhibit anticancer activity. The optimal range for lipophylicity is −0.4 < lg P < 0.4 and that for the binding with albumin is 3.0 < lg R < 4.0. Compounds with different physico-chemical characteristics as well as with different pathways and rates of metabolic transformation may vary in their clinical behaviour and, therefore, have distinctive modes of application.

REFERENCES

1. Hiller, S.A.; Zhuk, R.A.; Lidak, M.Yu. *Dokl.Akad.Nauk* SSSR, 1967, 176(2), 332 (in Russ.).
2. Karpeiskii, M.Ya.; Mikhailov, S.N.; Tsieminya, A.S.; Zidermane, A.A.; Kravchenko, I.M.; Lidak, M.Yu.; Zhuk, R.A. *Khim.Geterotsikl.Soed.*, 1980, 11, 1541 (in Russ.). Engl. transl. *Chem.Heterocycl. Compd.*, 1981, 5, 1176.
3. Zhuk, R.A.; Ludzisha, A.S.; Shpaer, E.G.; Lidak, M.Yu.; Zidermane , A.A.; Meirena, D.V. *Khim.Geterotsikl.Soed.*, 1985, 10, 1406 (in Russ.). Engl. transl . *Chem.Heterocycl.Compd.*, 1985, 10, 1159.
4. Yagupolskii, L.M.; Vechirko, E.P.; Kondratenko, N.V.; Liepinsh, E.E., Shpaer, E.G.; Zhuk, R.A. *Zh.Org.Khim.*, 1981, 17(1), 186 (in Russ.).
5. Kaulinya, L.T.; Liepinsh, E.E.; Lidak, M.Yu., Zhuk, R.A. Khim.Geterotsikl.Soed., 1982, 1, 101 (in Russ.). Engl. transl. *Chem.Heterocycl.Compd.*, 1982, 1, 85.
6. Kaulinya, L.T.; Yagupolskii, L.M.; Kondratenko, N.V., Vechirko, E.P.; Berzinya, A.E.; Silinya, V.N.; Liepinsh, E.E.; Lidak, M.Yu.; Zhuk, R.A. *Khim.Geterotsikl.Soed.*, 1982, 2, 256 (in Russ.). Engl. transl. *Chem.Heterocycl.Compd.*, 1982, 2, 202.
7. Kemme, A.A.; Bleidelis, Ya.Ya.; Lidak, M.Yu.; Zhuk, R.A. *Zh.Org.Khim.*, 1982, 19, 1537 (in Russ.).
8. Kametani, T.; Kigasawa, K.; Hiiragi, M.; Wakisawa, K.; Nakazato, N.; Ishikawa, K.; Fukawa, K.; Irino, O.; Nishimura, N,; Okada, T. *J.Med.Chem.*, 1982, 25, 1219.
9. Zhuk, R.A.; Kaulinya, L.T.; Ludzisha, A.S., Liepinsh, E.E.; Lidak, M.Yu. *Nucleic Acids, Symp.Ser.*, 1981, 9, 123.
10. Ludzisha, A.S.; Shpaer, E.G.; Zhuk, R.A.; Eninsh, N.I.; Chuyeva, N.I.; Eiduss, Ya.A. *Nucleic Acids, Symp.Ser.*, 1984, 14, 227.
11. Zhuk, R.; Lidak, M.; Zidermane, A.; Gilev, A. In *Ftorafur*, Ed. By R.A. Zhuk, Riga, Zinatne, 1985, 16.
12. Hirata, M.; Tonda, K. *Chem.Biol.Interactions*, 1984, 52, 141.
13. Germane, S.; Kozhukhov, A. In *Ftorafur*, Ed. by R.A.Zhuk, Riga, Zinatne, 1985, 57.
14. Fuji, S.; Fukushima, M.; Shimamoto, Y.; Ohshimo, H., Imaoka, T.; Shirasaka, T. *Jpn.J.Cancer.Res.*, 1989, 80, 173.

132

MEASUREMENT OF MUTATION FREQUENCY AT THE HPRT LOCUS IN PERIPHERAL LYMPHOCYTES

Is This a Good Method to Evaluate a Cancer Risk in Pediatric Patients?

Ying-Wei Lin, Masaru Kubota, Yuichi Akiyama, Machiko Sawada, and Kenshi Furusho

Department of Pediatirics
Faculty of Medicine
Kyoto University/54 Kawahara-cho
Shogoin, Sakyo-ku, Kyoto 606, Japan

1. ABSTRACT

Validity of measurement of somatic cell mutation frequency (Mf) at the hprt locus for evaluating cancer risk of the given individual was determined in pediatic patients. Peripheral lymphocytes (PL) from patients with various diseases, including acute lymphoblastic leukemia (ALL) and Hodgkin's disease (HD), DNA repair deficient syndromes or short stature receiving growth hormone (GH), were isolated through Ficoll-Hypaque sedimentation with informed consent. Mf at the hprt locus of PL was determined by limiting dilution assay using 6-thioguanine (6-TG). Results were as follows. (1) ALL patients after chemotherapy had higher Mf than that of age-matched controls. (2) Patients with HD tended to have higher Mf after chemotherapy. (3) Among DNA-repair deficient syndromes, diseases which are susceptible to cancer (Xeroderma pigmentosum, Ataxia telangiectasia) have high Mf, but those without any cancer disposition (Cockayne syndrome, Rothmund-Thomson syndrome) have normal Mf. (4) GH-receiving patients have normal Mf, regardless of total doses of GH. Measurement of Mf at HPRT locus may be useful for evaluating cancer risk of pediatric patients.

2. INTRODUCTION

A genetic instability is now considered to be a cause of cancer. There are several methods to evaluate the genetic instability, such as microsatelite instability, somatic cell

Purine and Pyrimidine Metabolism in Man IX,
edited by Griesmacher *et al.* Plenum Press, New York, 1998.

Mfs at any genetic loci, and mutational assay of DNA repair genes. Mfs have been studied at the hprt locus,[1] human leukocyte antigen A locus,[2] T cell antigen receptor locus,[3] and glycophorin A locus.[4] We have examined Mf at the hprt locus (hprt-Mf) in pediatric patients with ALL[5,6] or HD receiving chemotherapy and/or radiation therapy, DNA repair deficient diseases,[7] and pituitary dwarfism receiving GH therapy.[8] Several kinds of DNA repair deficient diseases and GH users have been reported to be cancer prone. We have discussed the usefulness of hprt-Mf to evaluate a cancer risk in pediatric patients.

3. MATERIALS AND METHODS

3.1. Study Population

The 32 children with ALL, 3 children with HD, 10 children with DNA repair deficient diseases including six Cockayne syndrome (CS), one Xeroderma pigmentosum (XP), one Rothmund-Thomsom syndrome (RT), one Ataxia telangiectasia (AT), and one Fanconi anemia (FA), 84 children with pituitary dwarfism receiving GH therapy, and 50 healthy children were enrolled in the present study. The details of each group were summarized in Table 1.

3.2. Blood Sampling

Blood samples were obtained with informed consent. Patients with ALL and HD were in complete remission at the time of evaluation and had been free from any treatment for at least 1 month. Chemotherapy for ALL consisted of prednisolone (PRD), daunorubicin, l-asparaginase, and methotrexate. HD patients were treated with PRD, vinblastine, vincristine, dacarbazine, doxorubicin, and bleomycin. Patients with DNA repair deficient diseases have not received any DNA-damaging agents. During the course of GH therapy, none of the children with pituitary dwarfism received any other kind of therapy. None of the patients had choromosomal abnormalities. They had received GH 2 to 6 times per week at doses of 0.5 IU/kg per week for at least 3 months.

Table 1. Details of the patients (Lin, Y. W., et al.)

Category	Disease	Number	Mean age ± SD, (range)
Malignancy	ALL	32	10.3 ± 5.0, (2.6–21.2)
	HD	3	11.7 ± 2.9, (7.7–14.0)
DNA repair	CS	6	9.5 ± 5.3, (4–16)
deficient	XP	1	19.2
	RT	1	12.2
	FA	1	15.8
	AT	1	4.1
GH users	pituitary dwarfism	84	13.4 ± 2.8, (6.9–20.6)
Controls		50	9.5 ± 5.3, (0.4–22.1)

CS: Cockayen syndrome, XP: Xeroderma pigmentosum, RT: Rothmund-Thomson syndrome, FA: Fanconi anemia, AT: Ataxia telangiectasia.

3.3. Measurement of hprt-Mf in Peripheral Lymphocytes

The hprt-Mf was measured by the method of Albertini et al.[1] with several modifications. Peripheral blood mononuclear cell (PBMC) fraction was isolated through Ficoll-Hypaque sedimentation, and the cells were washed 3 times prior to use. Separated PBMCs were seeded in 96-well round-bottom microtiter dishes at a density of 1 cell/well (non-selection) or 2×10^4 cells/well (6-TG selection). These cells were cultured in RPMI 1640 meidum supplemented with 20% fetal calf serum, 1% human serum, 50 IU/ml of recombinant human interleukin-2, 5.0 μg/ml of phytohemagglutinin-P, and 1×10^4 cells/well of irradiated (100 Gy) Raji cells as feeder cells. Non-selection plates contained 2×10^4 cells/well of irradiated (50 Gy) autologous PBMC, and selection plates contained 30 μM of 6-TG. After 2 weeks of culture, colony formation was observed under an inverted microscope. Plating efficiencies (PEs) were calculated from the number of negative wells, assuming that cells had a Poisson distribution. The hprt-Mf was expressed by the equation; Mf = [−ln{(number of colony negative wells / total wells in 6-TG selection plates) / 2×10^4}] / [−ln{(number of colony negative wells / total wells in non-selection plates) / 1}].

4. RESULTS

The hprt-Mf in both ALL and HD patients as a cohort was higher than that of age-matched normal controls (Fig. 1). All six CS patients, one RT patient, and one FA patient had normal hprt-Mf. One XP patient and one AT patient had significantly increased hprt-Mf (Fig. 2). The hprt-Mf in GH users as a cohort was not different from those of normal controls (Fig. 3). No significant increase was observed during GH therapy (Fig. 4). A serial evaluation of hprt-Mf during GH therapy of the same individual did not show any changes (Fig. 5).

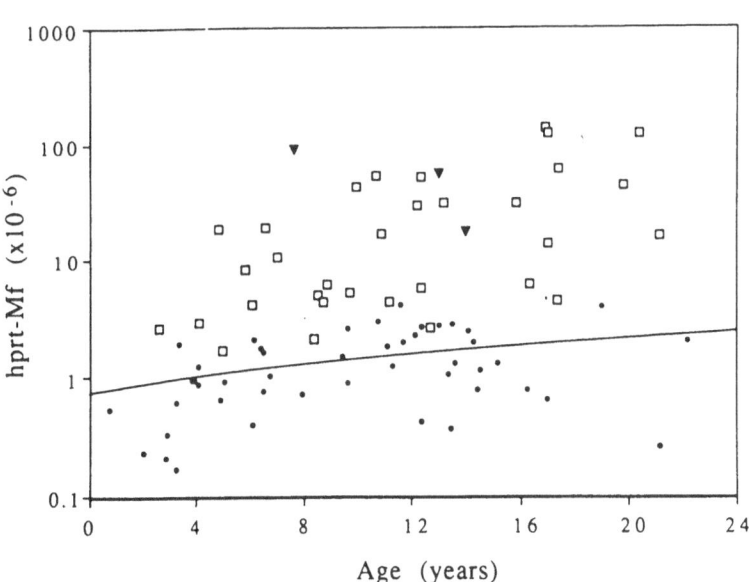

Figure 1. hprt-Mf in the patients with ALL and HD. The solid line represents the correlation between ages and Mf in healthy controls. (Closed circles) Controls; (Open squares) ALL.; (Closed inverted triangles) HD.

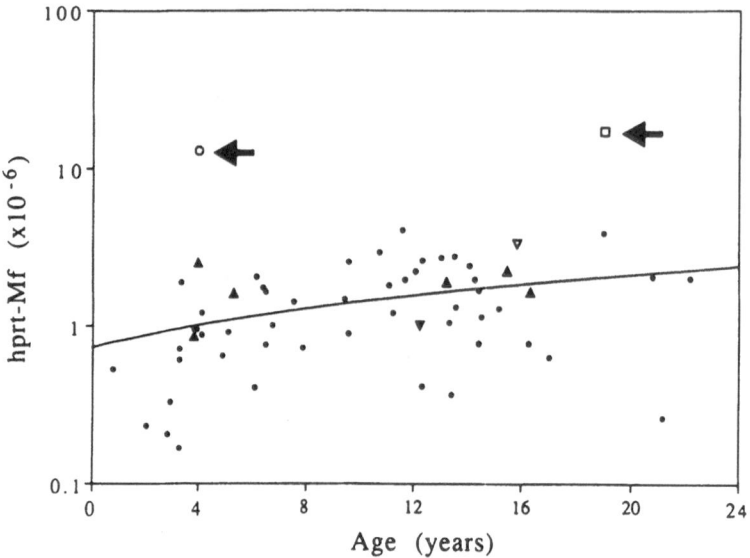

Figure 2. hprt-Mf in the patients with DNA repair deficient diseases. The solid line represents the correlation between ages and Mf in healthy controls. (Closed circles) Controls; (Open square) XP; (Closed inverted triangle) RT; (Open inverted triangle) FA; (Open circle) AT; (Closed triangles) CS.

5. DISCUSSION

Childhood ALL and HD have been reported to have an increased frequency of secondary cancer after chemotherapy and/or radiotherapy.[9,10] The patients with ALL after anti-cancer treatment had significantly higher hprt-Mf than did healthy controls. HD chil-

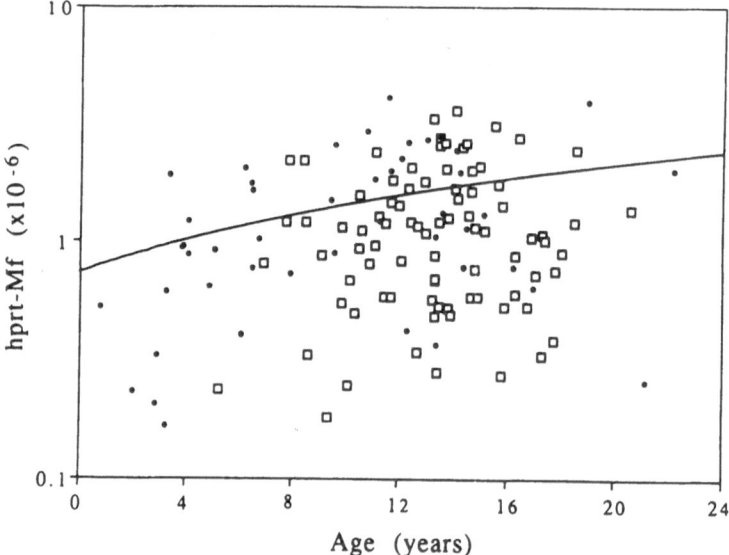

Figure 3. hprt-Mf in GH users. The solid line represents the correlation between ages and Mf in healthy controls. (Closed circles) Controls; (Open squares) GH users.

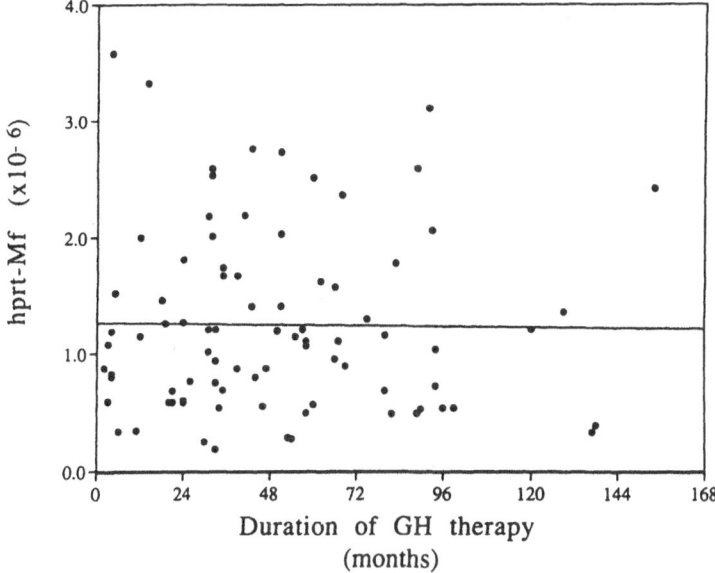

Figure 4. hprt-Mf and duration of GH therapy. The line indicates simple regression line.

dren during chemotherapy had also increased Mf. CS and RT have no increased incidence of any cancer. Hprt-Mf in patients with CS and RT had not been elevated. XP, FA, and AT are known to be cancer prone disease. Although the patients with XP and AT showed high hprt-Mf, the patient with FA showed normal hprt-Mf. Papadopoulo, et al. (1990)[11] reported that FA patient had marked tendency for deletions at the hprt locus in spite of normal Mf. Thus, our FA patient may have normal Mf because of the large deletion at the hprt-locus.

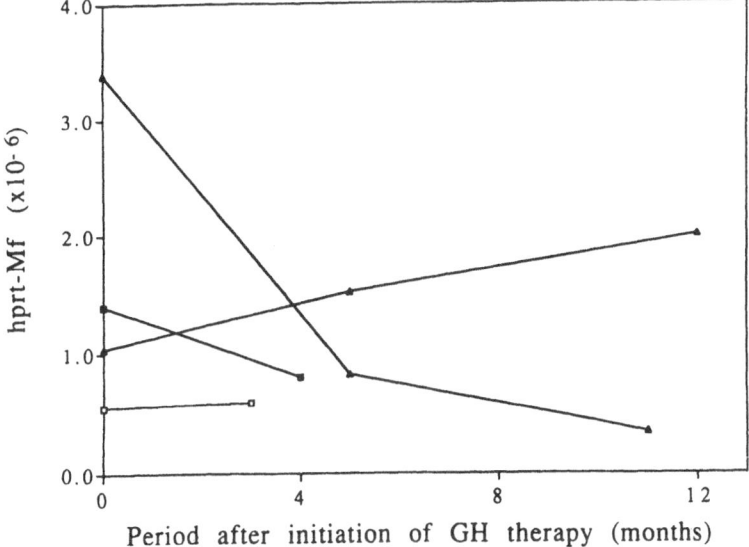

Figure 5. Changes of hprt-Mf during GH therapy. Each symbol denotes the same individual.

The patients with pituitary dwarfism who are receiving GH therapy have been reported to be at high risk for cancer. But, after starting to use recombinant GH for the therapy of dwarfism, the incidence of any cancer decreased among the GH users. The patients who were enrolled in this study had received recombinant GH and had normal hprt-Mf. There have been no occurence of any cancer among these children.

In summary, the measurement of hprt-Mf may be a useful method for evaluation of a cancer risk in pediatric patients. Further molecular investigation of the mutants is underway in our laboratory to get more insight into this issue.

ACKOWLEDGMENTS

Figures are reprinted from Mutation Research, 337, Y.W. Lin, et al., Somatic cell mutation frequency at the HPRT, T-cell antigen receptor and glycophorin A loci in Cockayne syndrome, 49–55, 1995, and Mutation Research, 362, Y. W. Lin, et al., Normal mutation frequencies of somatic cells in patients receiving growth hormone therapy, 97–103, 1996, with kind permission of Elsevier Science - NL, Sara Burgerhartstraat 25, 1055 KV Amsterdam, The Netherlands.

REFERENCES

1. Albertini, R.J., et al., (1982) T-cell cloning to detect the mutant 6-thioguanine-resistant lymphocytes present in human preipheral blood. Proc Natl Acad Sci USA, 79, 6617–6621.
2. Janatipour, M., et al., (1988) Mutations in human lymphocytes studied by HLA selection system. Mutat Res, 198, 221–226
3. Kyoizumi, S., et al., (1990) Spontaneous loss and alteration of antigen receptor expression in mature CD4+ T cells. J Exp Med, 171, 1981–1999.
4. Langlois, R.G., et al., (1986) Measurement of the frequency of human erythrocytes with gene expression loss phenotypes at the glycophorin A locus. Hum Genet, 74, 353–362.
5. Hirota, H., et al., (1993) Analysis of hprt gene mutation following anti-cancer treatment in pediatric patients with acute leukemia. Mutat Res, 319, 113–120.
6. Hirota, H., et al., (1994) Somatic mutations at T-cell antigen receptor and glycophorin A loci in pediatric leukemia patients following chemotherapy: comparison with HPRT locus mutation. Mutat Res, 315, 95–103.
7. Lin, Y.W., et al., (1995) Somatic cell mutation frequency at the HPRT, T-cell antigen receptor and glycophorin A loci in Cockayne syndrome. Mutat Res, 337, 49–55.
8. Lin, Y.W., et al., (1996) Normal mutation frequencies of somatic cells in patients receiving growth hormone therapy. Mutat Res, 362, 97–103.
9. Pui, C.H., et al., (1989) Secondary acute myeloid leukemia in children treated for acute lymphoid leukemia. N Engl J Med, 321, 136–142.
10. Bhatia, S., et al., (1996) Breast cancer and other second neoplasms after childhood Hodgkin's diasease. N Engl J Med, 334, 745–751.
11. Papadopoulo, D.C., et al., (1990) Hypomutability in Fanconi anemia cells is associated with increased deletion frequency at the HPRT locus. Proc Natl Acad Sci USA, 87, 8383–8387.

133

THIOPURINE TREATMENT IN CHILDHOOD LEUKEMIA

Metabolic Aspects and Sensitivity

R. A. De Abreu, J. P. M. Bokkerink, C. W. Keuzenkamp-Jansen, E. H. Stet, and J. F. M. Trijbels

Center for Pediatric Oncology S.E. Netherlands
Department of Pediatrics
University Hospital Nijmegen, St. Radboud
P. O. Box 9101
6500 HB Nijmegen, The Netherlands

INTRODUCTION

The thiopurine antimetabolites 6-mercaptopurine (6MP) and 6-thioguanine (TG) are important chemotherapeutic drugs in the treatment of acute lymphobastic leukemia. Both drugs are initially converted by the purine enzyme hypoxanthine guanine phosphoribosyl transferase (HGPRT) into 6-thio-inosine monophoshate (TIMP) and 6-thio-guanosine monophosphate (TGMP), respectively. Further conversion of TIMP to TGMP via 6-thio-xanthine monophosphate (TXMP) is catalyzed by two enzymes, IMP-dehydrogenase (IMPDH) and GMP-synthetase, respectively. IMPDH is the rate-limiting enzyme in these two enzymatic steps (1). Subsequently, TGMP is converted by consecutive steps to the 6-thio-guanosine triphosphates: 6-thio-GTP (TGTP) and 6-thio-deoxy-GTP (TdGTP), which are incorporated into RNA and DNA, respectively.

Several reports have presented data from *in vitro* studies, showing that cytotoxicity of 6MP and TG is mainly caused by incorporation of TdGTP into DNA. The incorporation, as a deoxy-TG residue, induces DNA damage such as single strand breaks, DNA-protein crosslinks, interstrand cross-links and sister chromatid exchanges (2–5).

6MP seems to provoke cytotoxicity in a more complex manner than TG. Thiometabolites from 6MP can also be methylated to methyl-6-thiopurines. The 6-thiopurine methylation is catalysed by the thiopurine methyltransferase (TPMT) and the conversion is S-adenosyl-L-methionine (SAM) dependent (6,7). Methyl-thio-IMP (meTIMP), the predominant anabolic metabolite formed by methylation of TIMP, is a strong inhibitor of PRPP amidotransferase the second enzyme of *de novo* purine synthesis (6–9).

Purine and Pyrimidine Metabolism in Man IX,
edited by Griesmacher *et al.* Plenum Press, New York, 1998.

In recent studies we revealed a new pathway leading to 6MP cytotoxicity *in vitro* (10,11). Conversion of methionine to SAM is ATP dependent. We could demonstrate that depleted ATP concentrations as well as consumption of SAM during incubation with 6MP caused an increase of the intracellar methionine pools which was inversily correlated with a decrease of ATP and of SAM. Depleted SAM pools may lead to a low capacity of intracellular methylation processes, e.g. DNA methylation, and ultimately this may also cause cell death. Indeed, we could proof that treatment with 6MP as well as methylmercaptopurine (meMPR) caused hypomethylation of DNA (12).

TREATMENT WITH ORAL 6MP

Generally, studies revealing more insight into intracellular therapeutic efficacy of 6MP *in vivo* are performed in red blood cells (RBC) from patients who were receiving low daily oral doses of 6MP (50–100 mg/m^2) during maintenance treatment (13–22). Several authors presented reports from these studies, suggesting that during oral therapy the formation of TGN is mainly responsible for treatment versus outcome and for drug toxicity (13–20). This is supported by a significant positive correlation between high TGN levels in RBC and bone marrow toxicity (18).

In contrast, despite the convincing data from *in vitro* studies that thiopurine methylation is involved in 6MP cytotoxicity, conflicting ideas still exist with respect to its clinical relevance. Levels of TPMT activity in human tissue are controlled by a common genetic polymorphism, approximately 88.6% of humans have a high, 11.1% have an intermediate and 0.3% have very low activity. Lennard (14,15,19) observed a negative correlation between TGN concentrations and TPMT activity in RBC from children with ALL on long-term oral 6MP therapy (75 mg/m^2). Children below the group median had higher TPMT activities and a higher relapse rate. Fifty out of 105 long-term survivors of ALL—no longer on treatment—had been treated with 'gentle' low-dose protocols and this subgroup contained an excess of children with lower TPMT activities. From these findings she concluded that TPMT activity may be a substantial regulator of the cytotoxicity of 6MP and in turn could be important in influencing the outcome of therapy for childhood ALL. In this view, methylation can be considered as a detoxification pathway for 6MP.

On the other hand, several other features indicate that the methylation of 6MP may also be positively involved in 6MP cytotoxicity *in vivo* and that the production of TGN alone is not sufficient to explain the cytotoxic action of 6MP. Weinshilboum (23) reported that the concentration of 6MP required to induce 50% inhibition of DNA synthesis in phytohemagglutinin stimulated peripheral blood lymphocytes was higher in subjects with a genetically low TPMT activity. Another study showed that meTIMP and TGN concentrations as well as levels of TPMT activity in RBC, were below the population mean in six out of seven patients who relapsed (21). Preliminary results of a recent study show identical effects on bone marrow toxicity in patients who were treated with either 6MP or TG, while the TGN concentrations found in RBC from the patients treated with TG were five times higher than those found in RBC from the patients treated with 6MP (22). So, it seems that in addition to the TG nucleotide formation, the methylation of intracellular 6MP metabolites may also contribute to bone marrow toxicity in vivo. This raises the question whether, during oral maintenance treatment, low levels of TG nucleotides alone or in combination with low meTIMP levels are responsible for the worse prognosis.

It should be taken into account, however, that during maintenance treatment the patients are in remission, which means that during this period no lymphoblasts are available.

Until now all correlations are only based on observations of 6TG nucleotide concentrations and TPMT measured in RBC from patient receiving oral 6MP during maintenance treatment, so these correlations were not drawn from studies of the leukemic blasts of these patients. This means that the question still remains how much of the drug really reaches the leukemic blasts and how effective the drug is on these target cells during oral treatment. It is possible that too little of the drug reaches the leukemic blasts after low-dose oral 6MP treatment and that the intracellular meTIMP concentrations are too low to inhibit purine *de novo* synthesis, even in the patients having a high TPMT activity. Low intracellular drug concentrations may e.g. be due to poor bioavailibility of oral 6MP, with plasma concentrations of approximately 20–140 nM (24).

TREATMENT INTRAVENOUS HIGH-DOSE 6MP VERSUS ORAL 6MP AFTER HIGH-DOSE METHOTREXATE INFUSION

During intravenous (i.v.) infusion, 6MP bioavailability is expected to be at a much higher level. In a study in 9 children with ALL receiving an i.v. infusion during 12 to 24 hrs at a dose rate of 50 mg/m^2/hr, mean steady-state plasma concentrations of 6MP were measured of ± 6.2 µM, with less than two-fold difference between the highest and the lowest plasma concentrations. Under these conditions it may be possible that enough me-TIMP is formed to inhibit PDNS in leukemic blasts *in vivo*.

It is very important to know in what way methylation of 6MP contributes to clinical relevance, since this may have clinical implications for choosing between treatment with 6MP or with TG. The latter drug is directly converted into TG nucleotides. If methylation of 6MP only results in detoxification of the drug, it would be better to administer with TG instead of 6MP, especially in patients with a high TPMT activity. However, if formation of meTIMP would add to the cytotoxic action of 6MP *in vivo*, high TPMT activity in lymphoblasts may even enhance the therapeutic efficacy of 6MP. Moreover, cytotoxicity as a consequence of inhibition of purine *de novo* synthesis would not be so important when 6TG is used instead of 6MP, since 6TG nucleotides are twelve-fold less inhibitors of purine *de novo* synthesis than metIMP (25).

In a recent study, we studied the metabolic conversions of 6MP *in vivo*, investigating 6MP metabolism in a randomized trial—during consolidation phase of treatment—comparing daily low-dose oral 6MP (25mg/m^2) with high-dose 24 hr intravenous 6MP (1300 mg/m^2/24 hr) after 24 hr high-dose methotrexate infusion (26).

We designed HPLC methods to measure all known extracellular and intracellular metabolites of 6MP separately. So, in contrast to most other studies, where total (methyl-)thionucleotide concentration was measured after hydrolysis to the base, in our study we were able to quantitate the intracellular thio- and methylthionucleotides in their original forms as mono-, di- and triphosphate nucleotides (27).

In the oral 6MP group of our randomized study, we found lower mean plasma levels of 6MP than in the oral group we described previously, where the patients received a daily dose of 50 mg/m^2 (24). The intravenous group of our randomized trial contained much higher plasma levels of 6MP and its metabolites. We measured steady state 6MP plasma levels of 14.1 µM (sem 1.0). Studying this group we discovered a metabolite of 6MP, 6-methylmercapto-8-hydroxypurine (meM8OHP), whose presence in human plasma has not been reported before (28). The detection of meM8OHP has lead to new ideas about the catabolism of 6MP *in vivo*. It appeared that uric acid can be considered as the main catabolite of thiopurine nucleotide degradation and meM8OHP of methyl-thiopurine degradation, with steady state plasma levels of 4.2 μM (sem 0.2) and 3.2 μM (sem 0.2), respectively. Other catabolites of 6MP: Thio-xanthine, methylmercaptopurine, its riboside and methylthiouric acid reached lower steady state plasma levels (< 1 μM) (26).

TGN were the main metabolites of 6MP in RBC (total: 90–130 pmoles/10^9 RBC) of the oral group, with thio-GTP accounting for 50–70% of the TGN pool. Methylthio-inosinenucleotides (meTIN) were undetectable in RBC of the oral group. So, it seems that during the oral treatment meTIMP are too low to inhibit PDNS and methylation of the thiopurine metabolites leads to detoxification of the drug.

In contrast to the oral group, the intravenous group had significantly higher meTIN levels in RBC (mean 52 nmoles/10^9 RBC, range 6–122) than TGN levels (mean 402 pmol/10^9 RBC, range 162–741). MeTIMP and TGTP accounted for 60–70% of the total meTIN and TGN pool, respectively. So, it appears that meTIMP and probably meTIDP are poor substrates for purine kinases, whereas TGMP and TGDP are better substrates for these enzymes. MeTIN levels are further increased after termination of the 6MP infusion, whereas TGN remained constant. Also, meTIN levels in RBC were significantly higher during the four successive courses, whereas TGN levels remained comparable with previous courses (except for course four where TGN are significantly higher than in course one). We achieved similar data studying the intracellular conversion in the peripheral mononuclear cells of the intravenous group. No significant correlation was observed between 6TGN and meTIN levels.

Extrapolating the findings of our *in vitro* study (7–11) to those of the intravenous group, it appears that the meTIMP concentrations of the intravenous group are high enough to completely inhibit PDNS in still circulating leukemic blasts. If this is true, then methylation will contribute to cytotoxicity during intravenous administration of 6MP. Again, it should be stressed that this still has to be proven.

The results of the clinical outcome of both arms of the trial are still under evaluation by the SNWLK committee, so the future will tell whether treatment with intravenous 6MP may have a greater benefit than oral treatment.

TPMT POLYMORPHISM

The gene for TPMT is located on human chromosome 6 and was found to be approximately 34 kb in length and consisted of 10 exons and 9 introns (29).

TPMT-deficiency is an autosomal recessive trait, two mutant alleles have been isolated recently from TPMT deficient and heterogenous patients. The one mutant (TMPT*2) cDNA revealed a single point mutation: G238 → C leading to an amino acid substitution at codon 80 (Ala80 → Pro), this mutation led to an 100-fold reduction in TPMT catalytic activity (30). A second mutant allele (TPMT*3) revealed two point mutations: a G → A transition at nucleotide 460, leading to an amino acid substitution at codon 154 (Ala154 → Thr) and an A → G transition at nucleotide 719, leading to an amino acid substitution at codon 240 (Tyr240 → Cys) (29). The presence of both mutations leads to complete loss of activity, while either of TPMT*3 point mutation alone leads to a partial reduction of activity: G460 A, 9-fold and A719 → G, 1.4-fold (31). The authors indicated that this allele is the most prevalent mutant allele associated with TPMT-deficiency in Caucasians.

ROLE OF KEY-ENZYMES

Higher activities have been reported for TPMT in RBC during maintenance of childhood ALL compared to those at diagnosis (32) and those of healthy controls (15). After cessation of maintenance treatment the elevated levels of TPMT activity in RBC decreased to normal values (15,32). Results from another study showed that TPMT activity in RBC is elevated at the end of intensive induction and consolidation treatment, i.e. just before maintenance, and that TPMT activity is decreasing during maintenance (33).

Other key-enzymes of 6MP metabolism may also predict the effectiveness of 6MP treatment. Two studies reported that IMPDH activity is higher in malignant lymphoblasts and myeloblasts compared to normal lymphocytes (34,35), and that its activity is increased in cultured human fibroblasts during incubation with 6MP (1), suggesting an induction of the enzyme during 6MP treatment. Also, increased RBC HGPRT activity has been described in patients during maintenance treatment in comparison to age-matched normal controls (36), suggesting a high conversion of 6MP into its active metabolites during maintenance treatment. Other authors reported decreased HGPRT activity and PRPP levels (a cosubstrate for HGPRT) in lymphoblasts of patients at relapse (37). A Dutch study found correlations between high 5NT activity and *in vitro* TG resistance (38), which makes it worthwhile to study this drug in more detail during 6MP treatment *in vivo*.

In conclusion, the data from literature indicate correlations between some enzyme activities and 6MP efficacy and that changes in enzyme activity may be induced by the drug. However, most studies are performed on a limited number of patients and little is known about the real effects of 6MP on the target cells *in vivo*, i.e. patient's lymphoblasts.

REFERENCES

1. Leyva A, Holmes EW Jr, Kelley WN. Effect of 6-mercaptopurine on inosinic acid dehydrogenase in cultured human fibroblasts. Biochem. Pharmacol.1976: 527–532.
2. Christie N, Drake S, Meyn R, Nelson J. 6-Thioguanine-induced DNA damage as a determinant of cytoxicity in cultured chinese hamster ovary cells. Cancer Res. 1984; 44: 3665–3671.
3. Covey J, D'Incalci M, Kohn W. Production of DNA-protein crosslinks (DPC) by 6-thioguanine (TG) and 2'-deoxy-6-thioguanosine (TGdR) in L1210 cells in vitro. Proc. Am. Assoc. Cancer Res 1986; 27: 17.
4. Pan B, Nelson J. Characterization of the DNA damage in 6-thioguanine-treated cells. Biochem. Pharmacol. 1990; 40: 1063–1069.
5. Bodell W. Molecular dosimetry of sisterchromatid exchange induction in 9L cells treated with 6-thioguanine. Mutagen. 1991; 6: 175–177.
6. Bennett Jr L, Allan P. Formation and significance of 6-methylmercaptopurine ribonucleotide as a metabolite of 6-mercaptopurine. Cancer Res. 1971; 31:152–158.
7. Weinshilboum R, Sladek S, Mercaptopurine pharmacokinetics: monogenic inheritance of erythrocyte thiopurine methyltransferase activity. Am. J. Hum. Genet.(1980; 32: 651–662.
8. Vogt M, Stet E, De Abreu R, Bökkerink J, Lambooy L, Trijbels J. The importance of methylthio-IMP for methylmercaptopurine ribonucleoside (MeMPR) cytotoxicity in Molt F4 human malignant T-lymphoblasts. Biochem. Biophys. Acta 1993; 1181: 189–194.
9. Stet E, De Abreu R, Bökkerink J, Vogels-Mentink T, Lambooy L, Trijbels J, Trueworthy R. Reversal of of 6-mercaptopurine and methylmercaptopurine cytotoxicity by amino imidazole carboxamide riboside. Biochem. Pharmacol. 1993; 46: 547–550.
10. Stet E, De Abreu R, Bökkerink J, Lambooy L, Vogels-Mentink T, Keizer-Garritsen J, Trijbels F. Reversal of methylmercaptopurine ribonucleoside cytotoxicity by purine ribonucleosides and adenine. Biochem. Pharmacol. 1995; 49: 49–56.
11. Graaf-Hess A, van Raay-Selten B, Trijbels F. Decrease in s-adenosylmethionine synthesis by 6-mercaptopurine and methylmercaptopurine ribonucleoside in Molt F4 human malignant lymphoblasts. Biochem. J. 1994; 304: 163–168.
12. De Abreu R, Lambooy L, Stet E, Vogels-Mentink T, van den Heuvel L, Thiopurine induced disturbance of DNA methylation in human malignant cells. Adv. Enzyme Regul. 1995; 35: 251–263
13. Lennard L, Van Loon J, Lilleyman J,Weinshilboum R. Thiopurine pharmacogenetics in leukemia: correlation of erythrocyte thiopurine methyltransferase activity and 6-thioguanine nucleotide concentrations. Clin. Pharmacol.Ther. 1987; 41: 18–25.

14. Lennard L, Lilleyman J. Variable mercaptopurine metabolism and treatment outcome in childhood lymphoblastic leukemia. J. Clin. Oncol. 1989; 7: 1816–1823.

15. Lennard L, Lilleyman J, Van Loon J, Weinshilboum R. Genetic variation in response to 6-mercaptopurine for childhood acute lymphoblastic leukemia. Lancet 1990; 336: 225–229.

16. Lennard L, Gibson B, Nicole T, Lilleyman J. Congenital thiopurine methyltransferase deficiency and 6-mercap-topurine toxicity during treatment for acute lymphoblastic leukemia. Arch. Dis. Child. 1993; 69: 577–579.

17. Lennard L, Lilleyman JS. Individualizing therapy with 6-mercaptopurine and 6-thioguanine related to the thiopurine methyltransferase genetic polimorphism. Ther. Drug. Monit. 1996; 18: 328–334.

18. Lennard L, Rees C, Lilleyman J, Maddocks J. Childhood Leukemia: a relationship between intracellular 6-mercaptopurine metabolites and neutropenia. Br. J. Clin. Pharmacol. 1983; 16: 359–363.

19. Lilleyman J, Lennard L. Mercaptopurine metabolism and risk of relapse in childhood lymphoblastic leukemia. Lancet 1994; 343; 1188–1190.

20. Evans W, Horner M, Chu Y, Kalwinsky D, Roberts W. Altered mercaptopurine metabolism, toxic effects and dosage requirement in a thiopurine methyltransferase-deficient child with acute lymphoblastic leukemia. J. Pediatr. 1991; 119: 985–989.

21. Bostrom B, Erdmann G. Association of relapse with 6-mercaptopurine (6MP) cellular pharmacokinetics in children with acute lymphoblastic leukemia. Proc. Am. Soc. Clin. Oncol 1992; 11: 278. (Abstract)

22. Janka-Schaub G, Erb N, Harms D. Randomized comparison of 6-mercaptopurine (6MP) vs 6-thioguanine (6TG) in maintenance treatment of childhood acute lymphoblastic leukemia (ALL): differing metabolism and hematologic toxicity. Med. Ped. Oncol. 1994; 23: 197. (Abstract)

23. Van Loon JA, Weinshilboum RM. Human lymphocyte thiopurine methyltransferase pharmacokinetics: effects of phenotype on 6-mercaptopurine-induced inhibition of mitogen stimulation. J. Pharmacol. Exp. Ther.1987; 242: 21–26.

24. Schouten TJ, De Abreu RA, De Bruyn CHMM, Van de Kleijn E, Oosterbaan MJM, Schretlen EDAM, De Vaan GAM. 6-Mercaptopurine: Pharmacokinetics in animals and preliminary results in children. Adv. Exp. Med. Biol. 165B (1984): 367–370.

25. Allan P. and Bennet L. Biochem. Pharmacol.1971; 20: 847–852.

26. Keuzenkamp-Jansen CW. The metabolic enigma of 6-mercaptopurine. Intravenous administration of high-dose 6-mercaptopurine to children with acute lymphoblastic leukemia and non-Hodgkin lymphoma. *Thesis*, Nijmegen, The Netherlands (1996).

27. Keuzenkamp-Jansen CW, De Abreu RA, Bökkerink JPM, Trijbels JMF. Determination of extracellular and intracellular thiopurines and methylthiopurines with HPLC. J. Chromatogr. Biomed. Appl. 1995; 672:53–61.

28. Keuzenkamp-Jansen C.W., van Baal JM, De Abreu RA, de Jong JGN, Zuiderent R, Trijbels JMF. Detection and Identification of 6-methylmercato-8-hydroxy-purine: a major metabolite of 6-mercaptopurine in plasma during intraveneous administration. Clin.Chem. 1996; 42: 380–386.

29. Szmulanski C, Otterness D, Her C, Lee D, Brandriff B, Kelsell D, Spurr N, Lennard L,Wieben E, Weinshilboum R. Thiopurine methyl transferase Pharmacogenetics: human gene cloning and characterization of a common polymorphism. DNA. Cell. Biol. 1996; 15: 17–30.

30. Krynetski EY, Scheultz JD, Gaplin AJ, Pui CH, Relling MV, Evans WE. A single point mutation leading to loss of catalytic activity in human thiopurine S-methyltransferase. Proc. Natl. Acad. Sci. USA 1995; 92: 949–953.

31. Tai HL, Krynetski EY, Yates CR, Loennechen T, Fessing MY, Krynetskaia NF, Evans WE. Thiopurine S-methyltransferase deficiency: two nucleotide transitions define the most prevalent mutant allele associated with loss of catalytic activity in Caucasians. Am. J. Hum. Genet. 1996; 58: 694–702.

32. McLeod H, Relling M, Liu Q, Pui C, Evans W. Polymorphic thiopurine methyltransferase in erythrocytes is indicative of activity in leukemic blasts from children with acute lymphoblastic leukemia. Blood 1995; 85: 1897–1902.

33. Capdeville R, Mousson B, Bax G, Betrand Y, Philippe N. Interactions between 6-mercaptopurine therapy and thiopurine methyltransferase (TPMT) activity. Eur. J. Clin. Pharmacol. 1994; 46: 385–386.

34. Becher H, Löhr G. Inosine 5'phosphate dehydrogenase activity in normal and leukemic blood cells. Klin. Wochenschr. 1979; 57: 1109–1115.

35. Price G, Hoffbrand V, Taheri M, Evans J. Inosine monophosphatte dehydrogenase activity in acute leukemia. Leukemia Res. 1987; 11: 525–528.

36. Lennard L, Hale J, Lilleyman J. Red blood cell hypoxanthine phosphoribosyltransferase activity measured using 6-mercaptopurine as a substrate: a population study in children with acute lymphoblastic leukemia. Br. J. Pharmacol. 1993; 36: 277–284.

37. Zimm S, Reaman G, Murphy R, Poplack D. Biochemical parameters of mercaptopurine activity in patients with acute lymphoblastic leukemia. Cancer Res. 1986; 46: 1495–1498.

38. Pieters R, Huismans DR, Loonen AH, Peters GJ, Hälen K, Van der Does-Van den Berg A, Van Wering ER, Veerman AJP et al. Relation of 5'-nucleotidase and phosphatase activities with immunophenotype, drug resistance and clinical prognosis in childhood leukemia. Leukemia Res. 1992; 873–880.

134

RELATIONSHIP BETWEEN CLADRIBINE (CdA) PLASMA, INTRACELLULAR CdA-5'-TRIPHOSPHATE (CdATP) CONCENTRATION, DEOXYCYTIDINE KINASE (dCK), AND CHEMOTHERAPEUTIC ACTIVITY IN CHRONIC LYMPHOCYTIC LEUKEMIA (CLL)

Freidoun Albertioni, Synnöve Lindemalm, Staffan Eriksson, Gunnar Juliusson, and Jan Liliemark

Department of Clinical Pharmacology
Karolinska Hospital, Stockholm
Veterinary Medical Chemistry
The Biomedical Center, Uppsala
Department of Hematology
University Hospital, Linköping
Department of Clinical Oncology Karolinska Hospital
Stockholm, Sweden

1. SUMMARY

Seventeen patients with CLL were treated with oral 2-chloro-2'-deoxyadenosine (cladribine, CdA, 10 mg/m^2) on 3 consecutive days and the pharmacokinetic parameters of CdA in patient plasma and its intracellular nucleotides (CdAMP, CdATP) in circulating leukemic cells were studied after the last dose intake and up to 72 h thereafter. The median terminal half life ($t_{1/2}$) of CdA in plasma was 21.1 h and the area under the curve (AUC) was median 1.2 μMh. The median $t_{1/2}$ was 14.6 h for CdAMP and 9.7 h for CdATP. The AUC of CdATP in leukemic cells is lower than the AUC of CdAMP (median ratio 0.60). There was no correlation between cellular CdATP and plasma CdA concentrations or dCK activity. The clinical response was related to higher C_{max} values for plasma CdA ($p=0.05$) and higher products of dCK activity and CdA C_{max} of plasma ($p=0.02$). The activity of dCK alone was not related to the clinical outcome in this patient group. The results suggest that further steps in the mechanism of action of CdA

Purine and Pyrimidine Metabolism in Man IX,
edited by Griesmacher *et al.* Plenum Press, New York, 1998.

beyond its bioactivation may be more important, e.g. the extent of DNA fragmentation or the ability of the leukemic cell to go into apoptosis, than the concentration of CdA nucleotides alone.

2. INTRODUCTION

CdA is one of the most interesting nucleoside developed during the last 2 decades. It differs from the essential naturally occurring nucleoside, deoxyadenosine (dAdo), in the substitution of hydrogen to a chlorine at the 2-position of the purine ring. This structural alteration made CdA resistant to attack by adenosine deaminase (ADA). The clinical activity of CdA in lymphoproliferative disorders has been intensively documented durig the last 10 years. It has outstanding activity in hairy cell leukemia (HCL), with an 85% complete remission rate and the activity in chronic lymphocytic leukemia (CLL), non-Hodgkin's lymphoma (NHL) and chronic progressive multiple sclerosis is impressive (1,2).

CdA is a prodrug and intracellularly it is phosphorylated to CdA-5'-monophosphate (CdAMP) by the cytosolic enzyme deoxycytidine kinase (dCK) (3) and by the mitochondrial enzyme deoxyguanosine kinase (dGK) (4). CdA-5'-diphosphate (CdADP) and CdA-5'-triphosphate (CdATP) are formed through phosphorylation of CdAMP by nucleoside monophosphate kinase and nucleoside diphosphate kinase. Since CdA is resistant to deamination by ADA, CdA mono-, di- and triphosphates accumulate in cells rich in dCK and poor in cytosolic 5'-nucleotidase (5'-NT) activity.

It has been postulated that resistance to CdA may be mainly due to an decreased activity of the phosphorylating enzyme dCK. Reduction in dCK activity was a major determinant of CdA acquired resistance to CdA in W1L2 human B lymphoblastoid and L1210 murine leukemia cell lines (5) and in the lymphoid cell line, CEM/CCRF (6). This has also been demonstrated in patients with CLL where a relationship between the response to CdA chemotherapy in CLL and the pretreatment levels of dCK was found (7).

No study has so far investigated the pharmacokinetics of specific CdA nucleotides in patients cells during the CdA therapy. We report here the pharmacokinetics of CdA nucleotides in patients cells during CdA treatment using a newly developed ion-pair HPLC method and correlate that to dCK levels.

3. MATERIALS AND PATIENTS

3.1. Chemicals and Reagents

CdATP, synthesized by Sierra Bioresearch (Tucson, AZ, USA), was provided by Dr. William Plunkett. CdA and 2-chloroadenine (CAde) were synthesized by Dr. Zygmunt Kazimierczuk at the Foundation for the Development of Diagnostics and Therapy, Warsaw, Poland.

3.2. Patients and Treatment

Seventeen patients, nine male and 8 female, with B-cell chronic lymphocytic leukemia, participated in the study after giving their informed consent prior to the study. The study was approved by the local Ethics committee at Huddinge hospital and the Swedish Medical Products Agency. Response to treatment was evaluated according to guidelines for clinical protocols for CLL as recommended by the National Cancer Institute (8).

3.3. Leukemic Cell Isolation

Mononuclear cells were isolated by Ficoll-Hypaque density centrifugation. The number of cells in the samples and the median cell volume of the samples analyzed were determined by a Coulter Multisizer (Coulter Electronics, Luton, UK).

3.4. Determination of CdA and CdA Nucleotides

A previously published reversed-phase high-performance liquid chromatographic (HPLC) method was used to quantitate CdA in plasma (9).

In order to investigate the pharmacokinetics of CdA nucleotides in leukemic cells from patients and to study the correlation between plasma and cellular pharmacokinetics to clinical response, we developed a specific isocratic ion-pair HPLC method for the specific quantitation of mono-, di-, and triphosphates of CdA in leukemic cells from patients (10). Briefly, nucleotides were extracted from cells with a solution of perchloric acid (0.4 M, PCA). In order to avoid degradation of CdATP, the addition of the ion-pairing agent triethylammonium phosphate (0.08 M, TEAP) to the extracting solution is essential as a threefold difference in the half-lives of CdATP in the PCA solutions with and without TEAP was observed ($t_{1/2}$ in PCA-TEAP was 4.8 h vs 1.6 h for the half-life in PCA alone). Isocratic separation was achieved on an Ultrasphere ODS column with a mixture of TEAP buffer (0.08 M) and methanol (11%) as the eluent.

3.5. Pharmacokinetic Calculation

Pharmacokinetics of CdA and CdA nucleotides were calculated using SIPHAR, program version 4.0 (Siphar Societe Simed, Creteil, France).

3.6. Measurement of Deoxycytidine Kinase

The activity of dCK in extracts of leukemic cells were measured according to described procedure using tritiated CdA as substrate (7).

4. RESULTS AND DISCUSSION

Seventeen patients with CLL were treated with CdA (10 mg/m^2) orally on three consecutive days and the pharmacokinetic parameters of CdA in plasma and CdA nucleotides, CdAMP and CdATP in circulating leukemia cells were studied after the last dose intake and up to 72 h thereafter. The pharmacokinetic parameters of CdA in plasma and CdATP are presented in Table 1.

The CdA nucleotides accumulated rapidly in the leukemic cells and the C_{max} of CdAMP and CdATP after orally administered CdA was more than 50 and 35 times higher than that of plasma CdA, respectively. In most patients, CdAMP could be detected in leukemic cells for up to 48 h after the last CdA dose and the AUC was generally higher than that of CdATP. However, in 4 of 17 patients the reverse relation was seen (Figure 1). The median ratio between the AUC for CdATP and that for CdAMP was 0.60. The concentration of the 5'-diphosphate was much lower, in most patients undetectable. The median initial $t_{1/2}$ was 0.9 and 0.8 h, respectively, for CdAMP and CdATP, whereas the terminal phase had a median $t_{1/2}$ of 14.6 and 9.7 h.

Table 1. Summary of pharmacokinetic parameters of CdA in plasma and CdATP in leukemic cells after administration of single an oral dose of CdA in 17 patients with CLL

Compound	Parameters	Median	95% CI
CdA	$AUC_{0-\infty}$, µmol/L × h	1.19	(0.82–2.08)
	T_{max}, h	1.0	(1.0–1.9)
	C_{max}, µmol/L	0.107	(0.086–0.140)
	Elimination $t_{1/2}$, h	21.1	(18.2–44.8)
CdATP	$AUC_{0-\infty}$, µmol/L × h	36.0	(24.1–77.9)
	T_{max}, h	2.0	(1.9–4.1)
	C_{max}, µmol/L	3.5	(2.3–4.8)
	Elimination $t_{1/2}$, h	9.7	(5.3–15.7)

There was no statistical significant correlation between plasma CdA and cellular CdATP kinetics at C_{max} or AUC. We found, however, a weak but significant correlation between the C_{max} of CdA in plasma and the C_{max} of CdAMP (r=0.56, p=0.02). No correlation was observed between cellular CdATP and the plasma CdA concentrations or dCK activity. The activity of dCK alone was not related to the clinical outcome in this patient group.

The results in cell lines (11) and in patients suggest events beyond its bioactivation may be more important for cytotoxicity, e.g. a reduced activity of the DNA polymerase, DNA fragmentation and induction of apoptosis. Recently it has been shown that CdA increased the amount of p53 in human thymocytes suggesting that induction of apoptosis is a major determinant of CdA cytotoxicity (12). Furthermore, compartmentalization of nucleotides has been suggested in bacteriophage T4, E. coli, and in eukaryotic cells (13). There are dNTP pools that are not readily utilized for DNA synthesis and thus the dNTPs levels in whole cells may not be representative of the concentrations available at the site of action. Moreover, dCTP derived from the salvage pathway was predominantly utilized for DNA repair, whereas dCTP for DNA replication was supplied from the de novo pathway (14). As CdA nucleotides are formed by the salvage enzyme, dCK it can be assumed

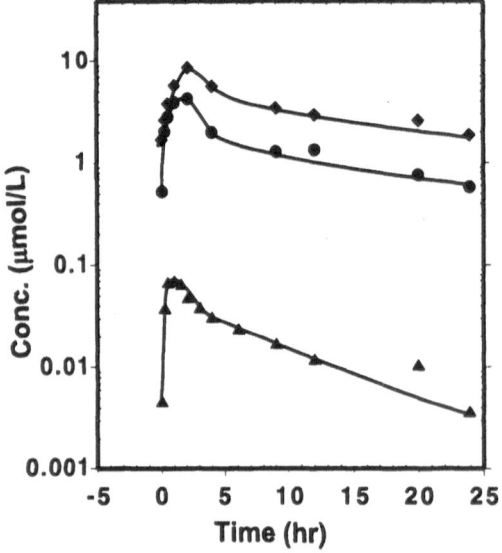

Figure 1. CdA (▲) in plasma and concentration of the CdA nucleotides CdAMP (●) and CdATP (■) in leukemic cells in a patient treated with oral CdA and showing a reversed relation of the CdATP to CdAMP concentration.

that inhibition of DNA repair is a more important mechanism of action than inhibition of DNA replication. This is supported by the finding that CdA is as active in resting as dividing cells.

REFERENCES

1. Beutler, E., Sipe, J. C., Romine, J. S., Koziol, J. A., Mcmillan, R., and Zyroff, J. Proceedings of the National Academy of Sciences of the United States of America. *93:* 1716–1720, 1996.
2. Piro, L. D., Ellison, D. J., and Saven, A, Leuk Lymph. *1:* 121–5, 1994.
3. Carson, D. A., Wasson, D. B., Kaye, J., Ullman, B., Martin, D. J., Robins, R. K., and Montgomery, J. A. Proceedings of the National Academy of Sciences of the United States of America. *77:* 6865–9, 1980.
4. Wang, L., Karlsson, A., Arner, E. S., and Eriksson, S. J Biol Chem. *268:* 22847–52, 1993.
5. Talbot, D., Orr, R., Serafimowski, P., and Karrap, K. Proc. Am. Assoc. Cancer Res. *34:* 305, 1993.
6. Albertioni, F., Reichelova, B., Pettersson, B., Eriksson, S., and Liliemark, J. Proc Annu Meet Am Assoc Cancer Res. *37:* 316, 1996.
7. Arner, E. S., Spasokoukotskaja, T., Juliusson, G., Liliemark, J., and Eriksson, S. Br J Haematol. *87:* 715–8, 1994.
8. Cheson, B. D., Bennett, J. M., Rai, K. R., Grever, M. R., Kay, N. E., Schiffer, C. A., Oken, M. M., Keating, M. J., Boldt, D. H., Kempin, S. J., and et, a. l. Am J Hematol. *29:* 152–63, 1988.
9. Albertioni, F., Pettersson, B., Reichelova, V., Juliusson, G., and Liliemark, J. Ther Drug Monit. *16:* 413–8, 1994.
10. Reichelova, V., Albertioni, F., and Liliemark, J. J Chromatogr B: Biomed Appl. *682:* 115–23, 1996.
11. Avery, T. L., Regh, J. E., Lumm, W. C., Harwood, F. C., Santana, V. M., and Baker, R. L. Cancer Res. *49:* 4972–8, 1989.
12. Szondy, Z. Biochem J. *311:* 585–8, 1995.
13. Mathews, C. K. Progress in Nucleic Acid Research & Molecular Biology. *44:* 167–203, 1993.
14. Xu, Y. Z., Huang, P., and Plunkett, W. J Biol Chem. *270:* 631–7, 1995.

135

DETERMINANTS OF THE THERAPEUTIC EFFICACY OF THYMIDYLATE SYNTHASE INHIBITORS

G. J. Peters, C. M. Kuiper, B. van Triest, H. Backus, C. L. van der Wilt, J. A. M. van Laar, G. Jansen, C. J. van Groeningen, and H. M. Pinedo

Department of Oncology
University Hospital Vrije Universiteit
P. O. Box 7057
1007 MB Amsterdam
The Netherlands

1. THYMIDYLATE SYNTHESIS

Thymidylate synthase (TS) catalyses a critical step in the pathway of DNA synthesis.[1,2] Using the co-substrate 5,10-methylene tetrahydrofolate (CH_2-THF) as a methyl donor, thymidylate synthase converts dUMP by methylation to dTMP (Fig. 1). This process is the only *de novo* source of dTMP which is subsequently metabolized to dTTP, exclusively for incorporation into DNA during synthesis and repair. The other source for dTMP is thymidine kinase, which uses thymidine as its precursor.[2] TS inhibition results in a thymineless state, which is lethal to most actively dividing cells. In addition, the circulating thymidine concentrations in humans may not be adequate to prevent the lethal thymineless state. TS is therefore an ideal target for anticancer therapy.[1,2]

2. THYMIDYLATE SYNTHASE INHIBITORS

Several types of TS inhibitors have been developed in the last decades; the pyrimidine analog 5-fluorouracil (5FU) inhibits TS via its metabolite FdUMP by formation of a ternary complex of TS, CH_2-THF and FdUMP[2]. FdUMP is a competitive TS inhibitor with a Ki values of about 3 nM both for purified enzyme and for TS assayed in extracts from 14 cultured tumor cell lines the Ki was 1.6 ± 0.7 nM with Km values for dUMP of 2.6 ± 1.6 µM. Low TS levels and pronounced inhibition predict good therapeutic efficacy of 5FU in colon tumors.[3]

Purine and Pyrimidine Metabolism in Man IX,
edited by Griesmacher *et al.* Plenum Press, New York, 1998.

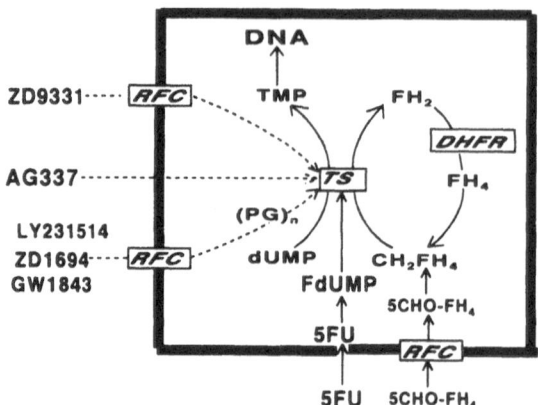

Figure 1. Schematic representation of 5FU and (anti)-folate uptake and cellular metabolism. ZD9331, LY231514 (multitargeted antifolate, MTA), ZD1694, GW1843 and 5CHO-FH$_4$ (leucovorin) are taken up by the reduced folate carrier (RFC); AG337 is lipophilic and enters by passive diffusion. LY231514, ZD1694 and GW1843 can be polyglutamylated (PG)n; ZD9331 and AG337 not. TS, thymidylate synthase; DHFR, dihydrofolate reductase; CH$_2$FH$_4$, 5,10-methylene tetrahydrofolate; FH$_4$, tetrahydrofolate; FH$_2$; dihydrofolate. Solid lines indicate reactions, broken lines inhibition.

In the last decade numerous anti-folates have been developed with the aim to achieve specific inhibition of TS by competition with the natural folate co-substrate. These compounds (Tomudex, LY231514, GW1843U89, Thymitaq) are targeted to the folate binding site of TS (Ki's 0.09–60 nM) (Fig. 1). Polyglutamates of ZD1694 and LY231514 are 10–100-fold better TS inhibitors.[4,5] Tomudex (Raltitrexed; ZD1694) is now active against colon cancer while AG337 and LY231514 have a similar activity and are active against head and neck cancer and non-small cell lung cancer, respectively.

Although these compounds have been developed as TS inhibitors, TS activity is not the only determinant for their activity. In cells with increased TS, resistance to these compounds has indeed been observed. However, in a panel of 13 unselected colon cancer cell lines, we have observed that although TS activity is a determinant for the growth-inhibitory effects of 5FU[6] this does not determine the activity of the antifolates (Fig. 2). Several folate specific properties are more important factors in their activity; such as a limited uptake by the reduced folate carrier and decreased polyglutamylation, leading to reduced cellular retention and decreased TS inhibition. These will limit the activity of ZD1694 and LY231514, but not of the non-polyglutamylable AG337. In addition we have observed that intracellular folate pools are important determinants for the growth-inhibition by several antifolates.[7]

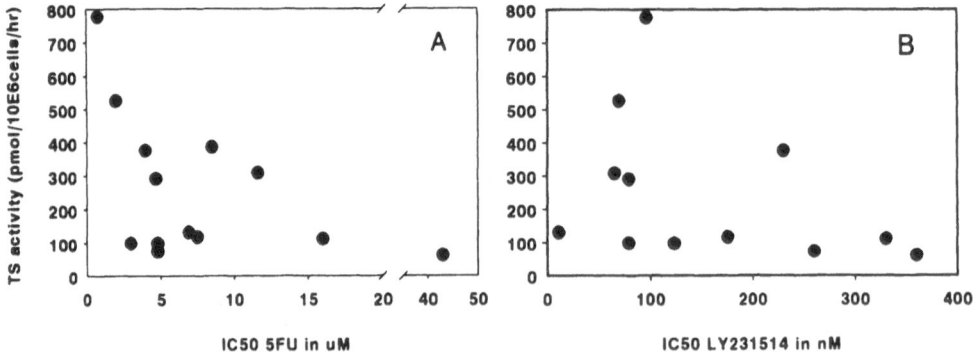

Figure 2. Relation between TS catalytic activity and growth inhibition by 5FU (A) or LY231514 (B). A significant relationship was found for the relation between 5FU growth inhibition and TS activity, but not for LY231514 (MTA).

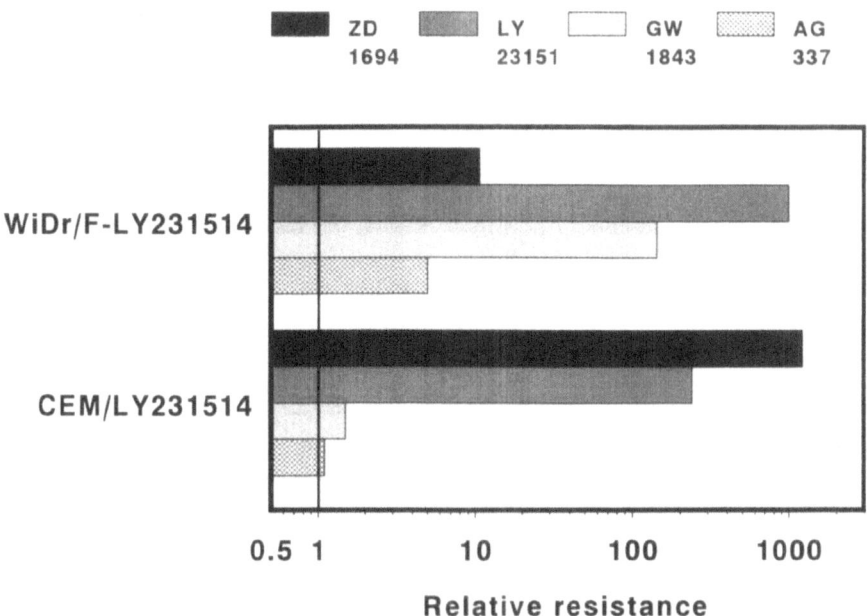

Figure 3. Cross-resistance patterns for CEM/LY231514 and WiDr-F/LY231514. Cells were made resistance by weekly exposure of cells to 5 µM drug for 24 hr. Cross-resistance patterns were determined by 72 hr exposure to the drugs. Cross-resistance is defined as the ratio between IC50 values for the resistant variant compared to the parent cells.

3. INDUCED RESISTANCE TO THYMIDYLATE SYNTHASE INHIBITORS

As for most other anticancer agents resistance to TS inhibitors can be divided in intrinsic and acquired resistance.[8] We exposed CEM-CCRF leukemia cells and WiDr colon cancer cells weekly for 24 hr to the anti-folates at 5 µM final concentration. Resistance in the leukemia cells developed rather rapidly, but in the colon cancer cells it lasted considerably longer. The pattern of resistance/sensitivity for the various antifolates was very different for the various sublines derived from the CEM leukemia cell line and between the CEM leukemia and the WiDr colon cancer cell lines made resistant to LY231514 (Fig. 3). The cross-resistance pattern in CEM/ZD1694 and CEM/LY231514 (made resistant to ZD16694 and LY231514, respectively) followed a similar pattern: cross-resistant to each other, some decreased sensitivity to GW1843, but no resistance to the lipophilic AG337 at all. In contrast the WiDr/F-LY231514 showed cross-resistance to ZD1694, to GW1843 and to AG337. The CEM/LY231514 showed a reduced activity of FPGS and accumulation of MTX-polyglutamates, but not of MTX transport. There were no indications as yet that TS activity was changed in these cell lines.

These data indicate that induction of resistance to the anti-folate TS inhibitors not necessarily results in an induction of TS. Resistance is more likely to be related to decreased folate uptake and polyglutamylation. This does however not exclude increased TS as a mechanism of resistance. H630 colon cancer cells made resistant to 5FU with more than 100-fold increased TS levels, were also resistant to ZD1694, LY231514, GW1843 and AG337. Both the experiments with unselected colon cancer cell lines and that with induced resistance demonstrate that primary resistance mechanisms may be at the level of folate transport and metabolism. In addition LY231514 has more targets [5].

4. THYMIDYLATE SYNTHASE INHIBITION AND APOPTOSIS

Tumor cell kill mediated by apoptosis has been characterized extensively in leukemic cells[9]. Morphologically this process is characterized by oligonucleosomal DNA fragmentation and the formation of apoptotic bodies. From a molecular point of view apoptosis is considered to be triggered by DNA damage; TS inhibitors can deplete dTTP and increase of dUTP (and FdUTP after 5FU exposure) followed by misincorporation into DNA. dNTP imbalance induces various enzymes, e.g. endonucleases, and changes in (onco) apoptosis proteins which finally can lead to cell death. DNA damage can result in upregulation of wild-type p53 and p21 inducing cell cycle arrest, leading to either DNA repair or apoptosis. In leukemic cells an excess of bcl-2 (as homodimers) can protect cells against apoptosis. A heterodimer with bax can be formed, while homodimers of bax will favor cell death. Normal induction of apoptosis is deregulated in many tumor cells due to p53 mutations, thereby resulting in drug resistance.

5FU and ZD1694 caused DNA damage in four colon carcinoma cell lines (about 25% DNA strand breaks at IC_{50} concentrations in Lovo, HT29, WiDr, WiDr/F). 5FU and ZD1694 at concentrations inducing 80–100% growth inhibition upregulated p53 and p21 after 24 hr in Lovo cells which have wt-p53. However, in cells with mt-p53, WiDr, WiDr/F and HT29, p53 and p21 were not induced. Only in Lovo and WiDr/F (1–10%) cells apoptotic bodies were observed.[10] In none of the colon cancer cell lines DNA laddering was found in contrast to the leukemic HL60 cells. Untreated cells showed no or a 5×

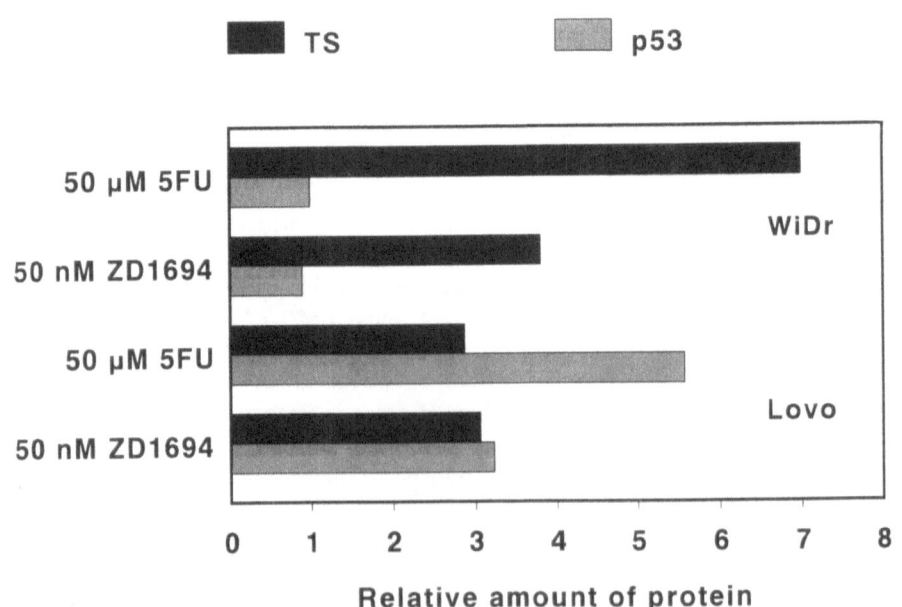

Figure 4. Response of cells to exposure with 5FU. Cells were exposed to 50 μM 5FU or 50 nM ZD1694 causing 100% growth inhibition. Expression of TS and p53 were determined after 24 hr using Western Blotting with a polyclonal and monoclonal antibody.

lower (Lovo) expression of bcl-2 than HL60 cells, while bax levels were in the same range. Drug exposure increased the bcl-2/bax ratio in apoptotic HL60, but decreased it in Lovo cells. The ratio bcl-2/bax did not predict apoptosis. 5FU and ZD1694 upregulated TS in all cell lines. The induction of TS seemed higher in cells with mt-p53 (Fig. 4). This might implicate that not only TS binds to its own mRNA and that of mRNA of p53[11] but that (wt)p53 may also bind to TS mRNA.

These data demonstrate that in colon cancer cells induction of apoptosis is not easily achieved at concentrations which would cause 100% growth inhibition. In addition the changes in bcl-2 and bax do not follow the pattern as expected from the proposed mechanism of action of bcl-2 and bax. The role of mut-p53 in this aspect is unclear; although a moderate increased of p53 protein in mutant cells is expected, a connection with TS induction was not.

5. CONCLUSIONS

There is a clear relation between the clinical efficacy of 5FU and the levels of TS and the extent of TS inhibition. Although compounds such as ZD1694 have been developed as *specific, direct* TS inhibitors, their activity is, surprisingly, not associated with TS levels. Folate-dependent parameters seem to determine the efficacy of these drugs. Therefore it seems advisable that in prognostic studies not only TS levels are evaluated but also the expression of the RFC and of FPGS. Another factor, possibly of major importance is the immediate response of cells to cytotoxic stress caused by TS inhibitors, such as the increase in TS and p53 protein levels. One would expect that the high increase in both TS and p53 would not be in favor for a therapeutic effect. Indeed high TS and p53 expression were both poor therapeutic prognostic factors. Combined determination of folate-dependent parameters and several cell death proteins possibly gives the best predictive value.

REFERENCES

1. Peters GJ & Van der Wilt CL (1995). Thymidylate synthase as a target in cancer chemotherapy. Biochem. Soc. Trans. 23: 884–888.
2. Peters GJ, Ackland SP (1996). New antimetabolites in preclinical and clinical development. Expert Opinion Invest. Drugs 5; 637–679.
3. Peters GJ, Van der Wilt CL, Van Groeningen CJ, Meijer S, Smid K, Pinedo HM (1994). Thymidylate synthase inhibition after administration of 5-fluorouracil with or without leucovorin; implications for treatment with 5-fluorouracil. J Clin Oncol 12: 2035–2042.
4. Jackman AL, Taylor GA, Gibson W, et al (1991). ICI D1694, a quinazoline antifolate thymidylate synthase inhibitor that is a potent inhibitor of L1210 tumor cell growth *in vitro* and *in vivo*: A new agent for clinical study. Cancer Res. 51: 5579–81.
5. Shih C, Chen VJ, Gossett LS, Gates S, MacKellar W, Habeck LL, Shackelford KA, Mendelsohn L, Soose D, Patel V, Andis SL, Bewley JR, Rayl EA, Moroson BA, Beardsley GP, Kohler W, Ratnam M Schultz RM (1996). LY231514, a pyrrolo[2,3-d]pyrimidine based antifolate that inhibits multiple folate requiring enzymes. Cancer Res. 57; 1116–1123.
6. Van Triest B, Pinedo HM, Van Hensbergen Y, Telleman F, Smid K, Van der Wilt CL, Jansen G, Peters GJ (1997). Polyglutamylation as predicter for sensitivity to new thymidylate synthase inhibitors. Proc. AACR 38; 476 (Abstract 3172)
7. Jansen G, Mauritz RM, Assaraf YG, Sprecher H, Drori S, Kathman I, Westerhof GR, Priest DG, Bunni M, Pinedo HM, Schornagel JH, Peters GJ (1997). Regulation of carrier-mediated transport of folates and antifolates in methotrexate-senstitive and resistant leukemia cells. Adv. Enzyme Reg 37; in press
8. Peters GJ, Jansen G (1996). Resistance to antimetabolites. Chapter 28 in: Schilsky RL, Milano GA, Ratain MJ. Principles of antineoplastic drug development and pharmacology. Marcel Dekker, Inc, New York, USA. pp. 543–585.

9. Stewart BW (1994). Mechanisms of apoptosis: Integration of genetic, biochemical, and cellular indicators. J Nat Cancer Inst 86; 1286–1296.

10. Kuiper CM, Van Triest B, Pinedo HM, Peters GJ (1997). Downstream events of thymidylate synthase (TS) inhibition: induction of apoptosis? Proc. AACR 38; 91 (Abstract 608).

11. Chu E, Allegra CJ (1996). The vole of thymidylate synthase as an RNA binding protein. Bioessays 18; 191–198.

136

PYRIMIDINE ANTAGONISTS AND ANTIFOLATES AS ANTIMALARIAL DRUGS

Richard I. Christopherson, Kristen K. Seymour, and Anthony E. T. Yeo

Department of Biochemistry
University of Sydney
Sydney NSW 2006, Australia

1. INTRODUCTION

The malarial parasite, *Plasmodium falciparum*, synthesises pyrimidines and folate *de novo* and is unable to synthesise pyrimidines via the alternative salvage pathway from preformed uridine or cytidine (Scheibel and Sherman, 1988). By contrast, purines are not synthesised *de novo* and must be salvaged from preformed purines such as adenosine which is readily available within the human erythrocyte (Szabados *et al.*, 1996). Drug resistance has become a major problem for the treatment of malaria and many potential antagonists of nucleotide and folate biosynthesis have been tested for selective toxicity against the parasite. Because the parasite only has the *de novo* pyrimidine pathway and humans have both the *de novo* and salvage pathways, inhibitors of the *de novo* pyrimidine pathway could be effective antimalarial drugs. We have synthesised several potent inhibitors of the enzyme dihydroorotase (CA-asp → DHO, eq 1; Christopherson *et al.*, 1989), 6-*L*-thiodihydroorotate (TDHO) blocks *de novo* pyrimidine biosynthesis in parasites, inducing accumulation of CA-asp and depletion of UTP and CTP (Seymour *et al.*, 1994). Atovaquone blocks the respiratory chain of malarial mitochondria at Complex III, leading to inhibition of dihydroorotate dehydrogenase (DHO → Oro, eq 1) and consequent accumulation of DHO and CA-asp (Seymour *et al.* 1994). Pyrazofurin as the nucleoside 5'-monophosphate derivative, inhibits OMP decarboxylase (OMP → UMP, eq 1) with accumulation of orotidine and Oro in cells (Seymour *et al.*, 1994). 5-Fluoroorotate (FOro) is a potent inhibitor of the growth of *P. falciparum* and has an ID_{50} value of 42 nM while 5-fluorouracil (FUra) is far less toxic with an ID_{50} of 5.2 μM (Queen *et al.*, 1990). FOro is converted to FUTP which may inhibit the malarial carbamyl phosphate synthetase (Seymour *et al.*, 1994).

$$HCO_3^- \longrightarrow CAP \longrightarrow CA\text{-}asp \longrightarrow DHO \longrightarrow Oro \longrightarrow$$
$$OMP \longrightarrow UMP \longrightarrow UDP \longrightarrow UTP \longrightarrow CTP \tag{1}$$

Purine and Pyrimidine Metabolism in Man IX,
edited by Griesmacher *et al.* Plenum Press, New York, 1998.

In P. falciparum, tetrahydrofolate (FH_4) is synthesised via the reaction sequence:

$$GTP \longrightarrow H_2neopterinTP \longrightarrow H_2neopterin \longrightarrow H_2pterinCH_2OH \longrightarrow$$

$$H_2pterinCH_2OPP \longrightarrow H_2pteroate \longrightarrow FH_2 \longrightarrow FH_4 \qquad (2)$$

Only the last reaction ($FH_2 \rightarrow FH_4$), catalysed by dihydrofolate reductase is found in humans. Dapsone is an antimalarial drug which inhibits the parasitic dihydropteroate synthase ($H_2pterinCH_2OPP \rightarrow H_2pteroate$; McCulloch and Maren, 1974) and the folate analogues, cycloguanil and WR99210, selectively inhibit the malarial dihydrofolate reductase (Ferone *et al.*, 1969; Rieckmann, 1973). These antifolates have potent antimalarial activity which may be attributed to an induced deficiency of FH_4, which as the one-carbon derivative $N^{5,10}$ methylene FH_4 is required for the synthesis of dTMP.

Antagonists of pyrimidine and folate biosynthesis induce deficiencies or imbalances in the pools of 2′-deoxynucleoside-5′-triphosphates (dNTPs) within the parasite which would result in arrest of DNA synthesis or genetic miscoding, respectively, and toxicity against the parasite. We have developed assay procedures for measurement of the levels of intermediates of the *de novo* pyrimidine pathway (Seymour *et al.*, 1994) and of dNTPs (Crisp *et al.*, 1996) in *P. falciparum* growing in erythrocytic culture. The results obtained provide new insight into how pyrimidine antagonists and antifolates kill the malarial parasite.

2. EXPERIMENTAL PROCEDURES

2.1. Materials

TDHO was synthesised as described by Christopherson *et al.* (1987). Atovaquone, cycloguanil, dapsone, proguanil, and PS-15 were obtained from Jacobus Pharmaceutical (Princeton, NJ, USA). Pyrazofurin, FOro, FUra, Oro and folinic acid (N^5-formyl tetrahydrofolate) were from the Sigma Chemical Co. (St. Louis, MO, USA). For growth of *P. falciparum*, human O^+ erythrocytes and serum were obtained from the Australian Red Cross Society Blood Transfusion Service (Sydney, Australia), RPMI 1640 medium (bicarbonate-free and containing 25 mM K.Hepes buffer) was from Flow Laboratories (Sydney, Australia), and the gas mixture 90% N_2/5% O_2/5% CO_2 was from Commonwealth Industrial Gases (Sydney, Australia).

2.2. Effects of Drugs on Parasites

The multi-drug resistant K1 strain of *P. falciparum* was maintained in low p-aminobenzoate, low folate RPMI medium containing 32 mM $NaHCO_3$ in 10% v/v human serum (Trager and Jensen, 1976) and cultures were synchronised with 5% w/v sorbitol (Lambros and Vanderberg, 1979). To determine the effects of drugs on levels of nucleoside 5′-triphosphates (NTPs) and dNTPs, drug was added to the culture approximately 24 h into the 48-h asexual life cycle where parasites are moving from trophozoites to schizonts and the *de novo* pyrimidine pathway is most active (Seymour *et al.*, 1994). Parasites were exposed to drug for 6 h and metabolites were extracted from 50-ml samples of the culture with saponin lysis of erythrocytes as described by Seymour *et al.* (1994). Acid-soluble metabolites were separated by gradient anion exchange HPLC on a Partisil 10-SAX column and were quantified using a UV 2000 ultraviolet detector (Spectra-Physics Analytical, San Jose,

CA, USA) set at 260 nm (Seymour *et al.*, 1994). For determination of levels of dNTPs, neutralised extracts were oxidised with periodate to remove NTPs prior to analysis for the less abundant dNTPs (Crisp *et al.*, 1996). Nucleotide concentrations are expressed as amol per parasitised erythrocyte (amol/pe) and as a fraction of control parasites.

3. RESULTS AND DISCUSSION

The effects upon levels of NTPs and dNTPs in *P. falciparum* exposed to a pyrimidine antagonist or an antifolate were determined in the absence and presence of a potential "antidote". For the pyrimidine antagonists, TDHO induced decreases in UTP, CTP and dTTP, but dCTP did not decrease (Table 1). Oro added as an antidote for TDHO toxicity would be converted Oro → OMP → UMP thus bypassing the blockade at dihydroorotase. Parasites exposed to TDHO plus Oro showed recovery of UTP and CTP to above control levels, dTTP increased to 23.3-fold control levels while dCTP was not affected (Table 1). The moderate increases in UTP and CTP observed for TDHO + Oro indicate some stimulation of pyrimidine biosynthesis via the salvage pathway while the mechanisms for the major increase in dTTP and maintenance of dCTP levels are unknown. It should be noted that the imbalance between dTTP and dATP levels is more severe for the culture treated with TDHO + Oro, where levels are 23.3- and 0.258-fold, respectively, of the control levels. Such an imbalance could lead to genetic miscoding and the death of the parasite.

The effects of atovaquone, in the absence and presence of Oro, on levels of NTPs and dNTPs are similar to those for TDHO (Table 1). However, atovaquone induces a 1.95-fold elevation of dATP while TDHO depresses dATP to 0.546-fold. Atovaquone + Oro also induces a severe imbalance in dTTP and dATP to 17.8- and 0.240-fold, respectively. For both TDHO and atovaquone, the co-administration of Oro, a potential "antidote", could enhance parasite killing due to enhanced genetic miscoding. FOro is highly toxic to *P. falciparum* as discussed above and induces elevation of CTP, and imbalance in levels of dTTP and dATP to 0.20- and 2.71-fold, respectively of the control (Table 1). The depletion of dTTP in this case may be attributed to inhibition of thymidylate synthase by FdUMP synthesised:

$$FOro \longrightarrow FOMP \longrightarrow FUMP \longrightarrow FUDP \longrightarrow$$
$$FdUDP \longrightarrow FdUTP \longrightarrow FdUMP \tag{3}$$

Table 1. The effects of pyrimidine antagonists on nucleotides in *P. falciparum*

Nucleotide	Control (amol/pe)	Atovaquone (2.5 µM)	Atovaquone (2.5µM) + Oro (500 µM)	TDHO (250 µM)	TDHO (25 µM) + Oro (250 µM)	FOro (25 µM)	FUra (25 µM)
GTP	2.99 ± 0.84	1.59	0.520	1.20	0.199	1.11	1.05
ATP	11.6 ± 2.9	1.26	0.563	0.804	0.247	1.13	1.06
UTP	3.51 ± 1.03	0.356	5.11	0.277	4.45	1.16	1.03
CTP	1.21 ± 0.54	0.695	2.13	0.481	1.62	1.94	0.952
dATP	0.218 ± 0.092	1.95	0.240	0.546	0.258	2.71	0.730
dTTP	0.239 ± 0.080	0.285	17.8	0.228	23.3	0.20	0.607
dCTP	0.421 ± 0.151	0.966	1.15	1.18	0.705	0.95	1.22

Cells were exposed to these drugs for 6 h. Levels of nucleotides in the control cultures (n=5) are expressed as amol/ parasitised erythrocyte (pe), dGTP was not detectable using 50-ml culture samples. Levels of nucleotides in drug-treated cultures are expressed as fractions of the control. Further details appear in Experimental Procedures.

Table 2. The effects of antifolates on nucleotides in *P. falciparum*

Nucleotide	Control (amol/pe)	Dapsone (250 µM)	Proguanil (100 µM)	PS-15 (10 µM)	Cycloguanil (2.5 µM)	WR99210 (1 µM)
GTP	5.49	0.920	0.668	0.234	1.48	1.22
ATP	20.2	0.984	0.729	0.279	1.62	1.29
UTP	5.65	1.22	0.624	0.248	1.96	1.86
CTP	2.24	1.00	0.828	0.372	1.87	2.74
dATP	0.302	1.12	0.841	0.222	2.14	2.46
dTTP	0.377	0.787	0.715	0.267	0.664	0.0438
dCTP	0.543	1.06	0.786	0.459	1.63	1.00

Levels of nucleotides are expressed as fractions of control values, experimental procedures were as for Table 1.

FUra is far less toxic than FOro to *P. falciparum* and FUra induces minor decreases in the levels of dTTP and dATP (Table 1). It is interesting to note that both TDHO and FOro have a negatively-charged carboxylate group but readily enter the parasitised erythrocyte and *P. falciparum*. The parasitised erythrocyte may have increased permeability to these negatively-charged pyrimidine analogues when compared with normal human cells which could provide another level of selectivity against the parasite. An important generalisation can be made from the data of Table 1, the level of dCTP is maintained in *P. falciparum* under conditions of pyrimidine deficiency or excess. The mean plus or minus the standard error for the fractional levels of dCTP in the 6 drug-treated cultures of Table 1 is 1.03 ± 0.19. Such conservation of dCTP levels suggests compartmentation of dNTPs within the parasite and/or some mechanism of metabolic regulation of levels of dCTP.

For the antifolates, dapsone induced a minor decrease to 0.787-fold in dTTP (Table 2) consistent with inhibition of dihydropteroate synthase, a consequent deficiency of $N^{5,10}$-methylene FH_4 and inhibition of the conversion of dUMP → dTMP. Cycloguanil induced elevations of UTP, CTP, dATP and dCTP, with a decrease in dTTP to 0.664-fold. Again, this deficiency in dTTP may be attributed to a deficiency of FH_4 and consequent inhibition of thymidylate synthase. WR99210 induces similar but more potent effects with dTTP reduced to 0.044-fold and dATP elevated to 2.46-fold (Table 2). Addition of folinic acid (N^5-formyl FH_4) with either antifolate did not restore dTTP levels in the parasites, suggesting that cycloguanil and WR99210 may have an additional or different mechanism of toxicity (Yeo *et al.*, 1997). Cycloguanil and WR99210 may induce genetic miscoding in malaria due to imbalances in the levels of dTTP and dATP. Proguanil and PS-15 are biguanide precursors of cycloguanil and WR99210, respectively. These prodrugs are converted into the antifolates in the liver and therefore the effects on malaria in erythrocytic culture are non-specific with a general reduction in the levels of NTPs and dNTPs (Table 2).

ACKNOWLEDGMENTS

KKS was suported in part by a grant from UNDP/World Bank/WHO Special Programme for Research and Training in Tropical Diseases (TDR). We thank Professor K.H. Rieckmann for the malarial parasites used in this research.

REFERENCES

Christopherson, R.I., Schmalzl, K.J., and Sharma, S.C. ,1987, Enzyme inhibitors. *Complete patent specification:* Australia 77692/87, Japan 220095/87, USA 091, 761, South Africa 7/6552, Europe 87307744.0.

Christopherson, R.I., Schmalzl, K.J., Szabados, E., Goodridge, R.G., Harsanyi, M.C., Sant, M.E., Algar, E.M., Anderson, J.E., Armstrong, A., Sharma, S.C., Bubb, W.A. and Lyons, S.D. ,1989, Mercaptan and dicarboxylate inhibitors of hamster dihydroorotase. *Biochemistry,* 28: 463.

Crisp, L.B., Smith, S.M., Mathers, M.A.G., Young, G.A.R., Lyons, S.D. and Christopherson, R.I., 1996, Effects of cytosine arabinoside on human leukaemia cells. *Int. J. Biochem. Cell Biol.*, 28: 1061.

Ferone, R., Burchall, J.J. and Hitchings, G.H., 1969, *Plasmodium berghei* dihydrofolate reductase. Isolation, properties and inhibition by antifolates. *Mol. Pharmacol.,* 5: 49.

McCulloch, J.L. and Maren, T.H., 1974, Dihydropteroate synthase from *Plasmodium berghei*: isolation, properties and inhibition by dapsone and sulfadiazine. *Mol. Pharmacol.,* 10: 140.

Queen, S.A., Vanderjagt, D.L. and Reyes, P., 1990, *In vitro* susceptibilities of *Plasmodium falciparum* to compounds which inhibit nucleotide metabolism. *Antimicrob. Agents Chemo.* 34: 1393.

Rieckmann, K.H., 1973, The *in vitro* activity of experimental antimalarial compounds against strains of *Plasmodium falciparum* with varying degrees of sensitivity to pyrimethamine and chloroquine. Chemotherapy of malaria and resistance to antimalarials. *World Health Organisation Technical Report Series* 529:58.

Scheibel, LW. and Sherman,, I.W., 1988, In *Malaria: Principles and Practiice of Malariology* (Wernsdorfer, W.H. and McGregor, I. eds.). Vol 1, p 234. Churchill Livingstone. Melbourne.

Seymour, K.K., Lyons, S.D., Phillips, L., Rieckmann, K.H., and Christopherson, R.I., 1994, Cytotoxic effects of inhibitors of *de novo* pyrimidine biosynthesis upon *Plasmodium falciparum. Biochemistry,* 33: 5268.

Szabados, E., Duggleby, R.G. and Christopherson, R.I., 1996, Metabolism of adenosine and deoxyadenosine by human erythrocytes and CCRF-CEM leukemia cells. *Int. J. Biochem. Mol. Biol.,* 28: 1405.

Trager, W. and Jensen, J.B., 1976, Human malaria parasites in continuous culture. *Science,* 193: 673.

Yeo, A.E.T., Seymour, K.K., Rieckmann, K.H. and Christopherson, R.I., 1997, Effects of folic and folinic acid on the activities of cycloguanil and WR99210 against *Plasmodium falciparum* in in erythrocytic culture.*Ann. Int. Med. Parasitol.,* 91: 17.

137

ADENINE NUCLEOSIDE PHOSPHORYLASES IN TREMATODE *FASCIOLA HEPATICA*, THE MAMMALIAN PARASITE

Halina Trembacz and Maria M. Jeżewska

Institute of Biochemistry and Biophysics
Polish Academy of Sciences
5A Pawińskiego St.
02-106 Warszawa, Poland

1. INTRODUCTION

The trematode *Fasciola hepatica* is one of the most common parasite of cattle and sheep, pathogenic also to man. In the adult form of *F. hepatica*, which lives in the bile ducts of vertebrate host, we have previously found[1] two adenosine phosphorylases (AdoPho). Ado phosphorylase found in many parasitic organisms, Mycoplasmatales, protozoa and trematode *Schistosoma mansoni*, does not occur in the vertebrate hosts of parasites. The metabolic role of AdoPho is not clearly established, and a great variety in the properties of this enzyme from various sources have been reported. In the living organisms, the three adenine nucleosides are the potential AdoPho substrates: adenosine (Ado), 2'-deoxyadenosine (dAdo) and 5'-methylthioadenosine (MTA), arising from adenylates (and RNA), from DNA, and during polyamine biosynthesis, respectively. Indeed, AdoPho from the trypanosomes catalyses the reversible phosphorolysis of all these substrates to adenine (Ade).[2,3] In contrast, in vertebrate hosts of parasites Ado and dAdo are deaminated by adenosine deaminase (ADA), whereas only MTA is phosphorolyzed to Ade by a specific MTA phosphorylase. It seems that these differences in the metabolic pathways between the hosts and parasites could be exploited in the chemotherapy against the parasitic diseases. The present report concerns the properties of the adenosine phosphorylases previously found[1] in adult form of *F. hepatica*.

2. MATERIALS AND METHODS

2.1. Chemicals

The sources of chemicals were as described.[4,5] Xanthine oxidase, phosphate free, was purchased from Sigma.

Purine and Pyrimidine Metabolism in Man IX,
edited by Griesmacher *et al.* Plenum Press, New York, 1998.

2.2. Enzyme Preparation

The adult trematodes, *F. hepatica*, were isolated from the bile ducts of cows, and the enzyme preparation was made as previously described.[1] Recalling, the gradient elution from the Cellulose DE-52 column of active fraction A gave a single peak (A1), that of the fraction B—two peaks (B1 and B2) which exhibited AdoPho activity. The fractions corresponding to each peak were pooled, desalted on the Sephadex G-25 column and concentrated in an Amicon Ultrafiltration Cell. The obtained protein solutions were stored at −20°C and used in the experiments without further purification which caused a quick decrease of the AdoPho activity.

2.3. Enzymatic Assays

The activities of AdoPho, PNPase, APRTase, HGPRTase and ADA were determined with the use of (8-^{14}C)-labelled purine substrates as described previously.[4,5] The hydrolysis, phosphorolysis and deamination of various nucleosides were followed by scanning of UV spectra (λ range 230–330 nm) at the chosen time intervals or by monitoring the changes in the UV absorbance at suitable wavelength. The enzymic activities were also identified and measured using a coupled enzyme assay with commercial PNPase and xanthine oxidase as the ancillary enzymes; the UV absorbance increases given by final products, 2,8-dihydroxyadenine or uric acid, were followed at 305 or 293 nm, respectively. The reaction mixtures (1 ml) were incubated at +37°C in the stoppered cuvettes (10 mm path length) in the thermostatically controlled cell compartment of the Kontron (Vien, Austria) Uvikon 860 double-beam spectrophotometer equipped with the kinetic program (Software Pack 3).

2.4. Kinetics

The apparent Michaelis constants of Ado phosphorylases were determined using the initial velocity method and the linear regression analysis of the Hanes-Woolf plot in the case of radiochemical assays (the conversion of substrate to product was below 7%). In the case of spectrophotometric measurements, the kinetic program of the Software 3 pack was used (reaction time 1–3 min). The optimum pH for AdoPho was determined to be 6.0–6.5.

3. RESULTS AND DISCUSSION

3.1. Enzymes of Purine Compound Metabolism in *F. hepatica*

Neither in crude extracts nor in the enzyme preparations, a nucleosidase hydrolytic activity towards Ado, dAdo, MTA and Ino was found. The activities (nmol · min^{-1} per mg protein of crude extract) of HPRTase, APRTase, ADA and PNPase were: 0.29, 3.4, 0.09 and 7.9, respectively, and the activity of AdoPho was as much as 24.5. Curiously, we have found[5] that AdoPho is many times less active than PNPase in larval forms of several trematode species, growing in the hepatopancreas of their gastropod intermediate hosts. This difference between the trematode larval and adult forms could reflect the variations in the purine compounds available to the parasite during its life cycle. AdoPho has been postulated[6] to take part in the assimilation of the nutritive materials by *S. mansoni*.

3.2. Properties of *F. hepatica* Ado Phosphorylases

The three peaks appearing (see 2.2 section) on the elution diagrams (Fig. 1a,b) contained: (A1) 33–39, (B1) 22–27 and (B2) 36–39 per cent of the total AdoPho activity. With the peaks A1 and B2 different Ado phosphorylases were associated as it was evidenced by their different molecular weight determined previously[1] to be 76 000 and 109 000, respectively. In turn, the identity of AdoPho in the peaks A1 and B1 was suggested by the same position of these peaks on the elution diagrams.

AdoPho in both these peaks cleaved Ado and dAdo, was inactive towards MTA, and had the same K_m for Ade (Table 1); however, the relative rates of the nucleoside phosphorolysis and K_m for Rib-1-P varied, which could point to the presence of different molecular forms of Ado phosphorylase. In contrast, Ado phosphorylase in the peak B2 cleaved all three natural adenine nucleosides (Table 1), with the highest affinity to MTA. As compared to Ado/dAdo phosphorylase, this Ado/dAdo/MTA phosphorylase had the K_m for Ado and dAdo about two times higher, and K_m for P_i 2.5 times lower. All three Ado phosphorylases from *F. hepatica* cleaved cladribine (2-Cl-2'-dAdo), an antileukemic drug, at the highest rate.

In comparison with Ado/dAdo/MTAPho of *F. hepatica*, the host liver MTA phosphorylase cleaves MTA about 30 times more effectively, exhibits much lower activity towards Ado, and is inactive towards dAdo.[8,9] The difference between the host and parasite enzymes was even greater in the case of *F. hepatica* Ado/dAdo phosphorylase which did not phosphorolyze MTA, and had a greater affinity to Ado and dAdo. This last enzyme seemed to be a better target for the chemotherapy, as it could be possible to find an nucleoside analogue not interfering with the action of host MTAPho.

3.3. Course of the Reversible Reaction Catalysed by Ado/dAdoPho

The apparent K_m's for all four substrates were determined at the two concentrations of respective cosubstrate (see a legend to Table 1). It turned out that K_m for each substrate was independent of the concentration of the cosubstrate (Fig. 2a,b). Also the addition of a reaction product (P_i or Rib-1-P) to the reaction medium had no effect on K_m, diminishing only the reaction rate. This evidenced that the binding of one substrate to the enzyme did not influence the binding of the other substrate, in contrast to the other nucleoside phosphorylases.

Table 1. Kinetic data for Ado/dAdo (peaks A1 and B1) and Ado/dAdo/MTA (peak B2) phosphorylases

Substrate	Apparent K_m (µM)			V_{rel} (%)*		
	A1	B1	B2	A1	B1	B2
Ade	51	53	74			
Ado	61		141	100	100	100
2'-dAdo	47		96	117	164	115
MTA	n.a.	n.a.	42	0	0	27
2-Cl-2'-dAdo	–	–		169	348	327
Rib-1-P (Ade)	470	675	620			
P_i (Ado)	1500		600			

Apparent K_m was determined: for Ade (at concentration range of 10–180 µM) with Rib-1-P at fixed concentrations (1.0 or 2.0 mM); for Ado (35–540 µM) with P_i (10 or 20 mM); for Rib-1-P (0.15–1.0 mM) with Ade (360 or 720 µM); for P_i (1–200 mM) with Ado (360 or 720 µM).
*Calculated from the initial velocities (nmol/min) of phosphorolysis of nucleosides at the initial 100 µM conc.
n.a. - non active towards this substrate.

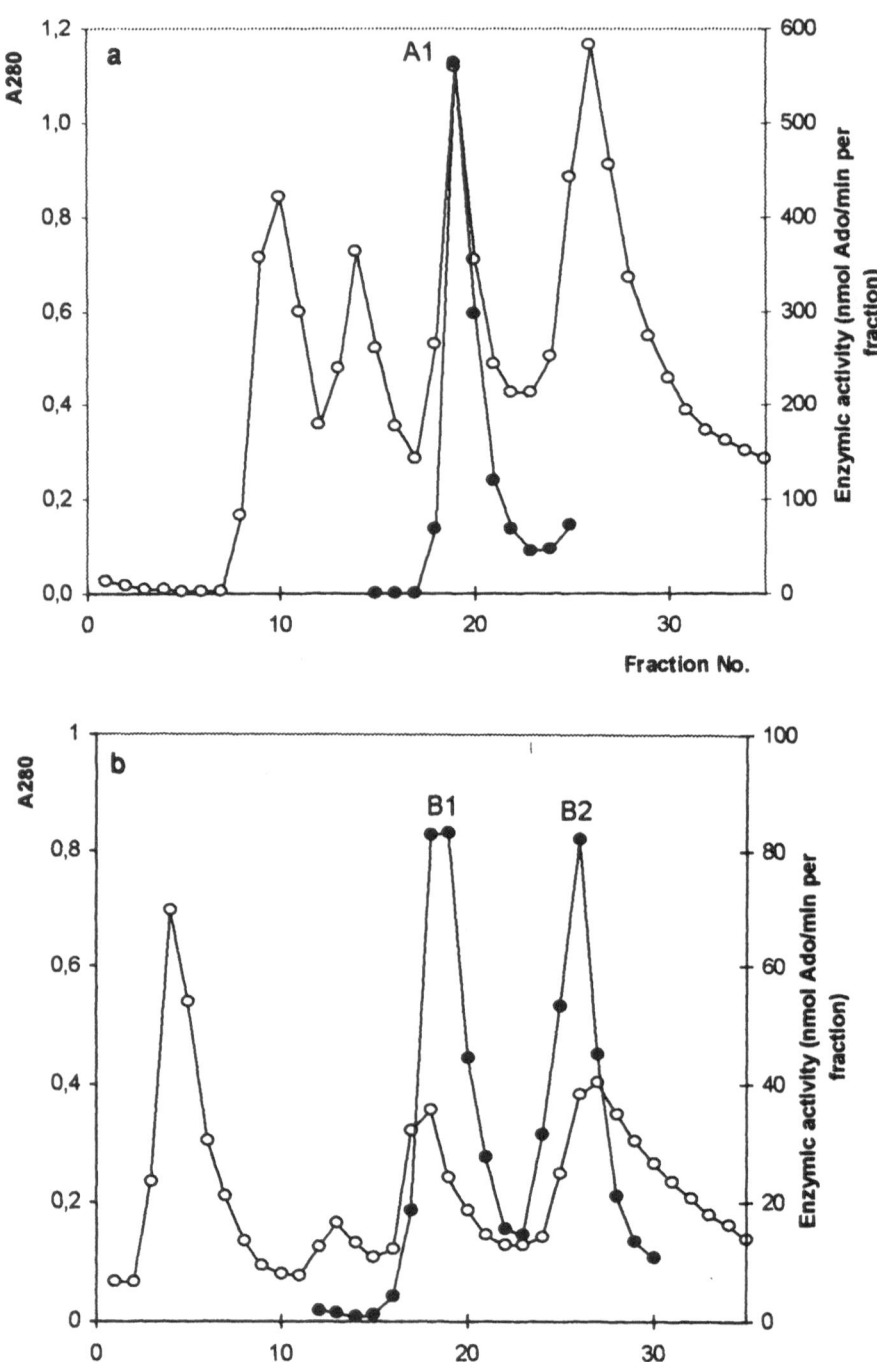

Figure 1a and b. Separation of *F. hepatica* Ado phosphorylases. Elution diagrams from Cellulose DE-52 column of proteins precipitated within the ranges: a) 0.45–0.60 and b) 0.60–0.80 ammonium sulphate saturation. Gradient elution with 50–450 mM NaCl in 50 mM HEPPS-KOH buffer, pH 8.0. –o–o– protein content, –•–•– AdoPho activity.

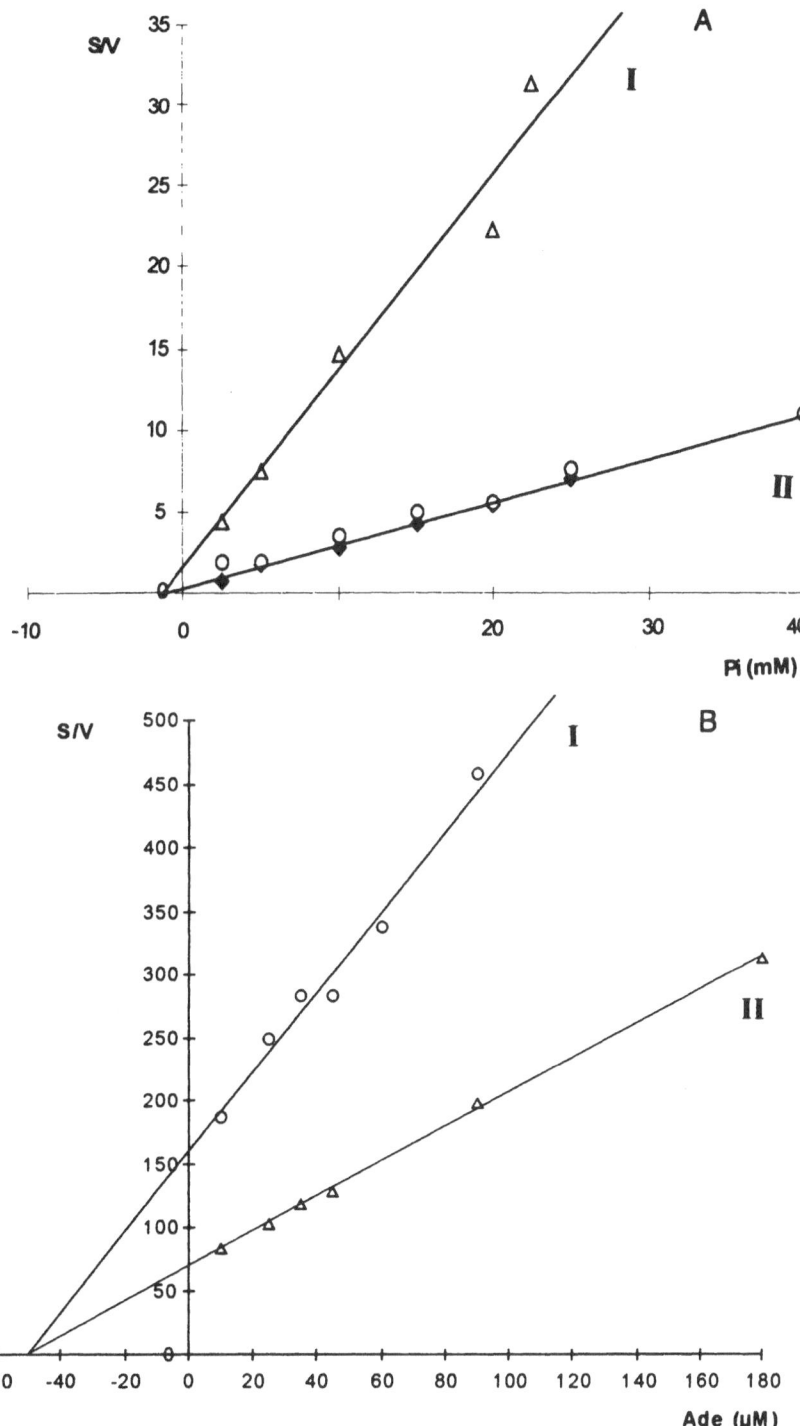

Figure 2. A) Ado phosphorolysis by Ado/dAdoPho. Various conc. of P_i, and fixed conc. of Ado: 360 μM and 720 μM (curve I and II, respectively). B) Ado synthesis by Ado/dAdoPho. Various conc. of Ade, and fixed conc. of Rib-1-P: 1 mM and 2 mM (curve I and II, respectively).

Table 2. Adenosine analogues with ribose moiety modified
as substrates and inhibitors of ADA and Ado/dAdo Pho

Ade nucleoside (100 μM)	Deamination V rel.(%)	Phosphorolysis V rel. (%)	Inhibition of Ado synthesis* (%)
Ado	100	100	–
2'-dAdo	142	116	–
3'-dAdo	63	0	0
2',3'-didAdo	55[a]	0	12
MTA	35[a]	0	–
5'-dAdo	76[a]	0	18
5'-Cl-5'-dAdo	14[a]	0	10[b]
Ade arabinoside	44[a]	0	8

*360 μM (8-^{14}C)-Ade, 2 mM Rib-1-P and 100 μM adenosine analogue.
[a]After an initial burst the reaction slowed down or ceased.
[b]Ado phosphorolysis and deamination were also inhibited by 10 and 75%, respectively.

The equilibrium of the reaction was shifted towards nucleoside synthesis: 14% of 360 μM Ado was cleaved to Ade at 2.5 mM P_i, and 90% of 360 μM Ade was converted to Ado in the presence of 1 mM Rib-1-P (reaction time 3 hr). Although Ado and dAdo were cleaved at the comparable rates, the dAdo synthesis from Ade and 2'-deoxy-Rib-1-P was about 10 times slower than the Ado synthesis, and Xyl-1-P did not serve as substrate for nucleoside synthesis.

3.4. Ado/dAdo Phosphorylase and Adenine Nucleoside Analogues

F. hepatica Ado/dAdoPho appeared to be strictly specific towards ribose and 2'-deoxyribose moiety of adenine nucleoside (Table 2), whereas the *T. cruzi* phosphorolyzed 3'-dAdo, 2',3'-didAdo and MTA. However, several analogues with modified ribose moiety exerted a weak inhibitory effect on the synthesis of Ado by the *F. hepatica* enzyme. The spectrophotometric measurements (see 2.2. section) revealed the presence of adenosine deaminase activity associated together with Ado/dAdo phosphorylase in the peak A1. Only Ado, dAdo and 3'-dAdo were deaminated by this ADA (Table 2). With other nucleosides deamination after an initial burst slowed or entirely ceased. 5'-Cl-5'-dAdo was deaminated at the slowest rate, but this analogue inhibited Ado phosphorolysis and its deamination by 10 and 75%, respectively.

Several nucleoside analogues with the purine ring modified (Table 3) were tested as the potential inhibitors of Ado/dAdo phosphorylase from *F. hepatica*. None of them served

Table 3. Effects of nucleoside analogues with modified
purine ring on the Ado/2'-dAdo phosphorylase activity

Nucleoside analogue	Analogue conc. (μM)	Inhibition Ado[a] → Ade (%)	Analogue conc. (μM)	Inhibition Ade[b] → Ado (%)
Formycin B			(200–300)*	7–14
Formycin A	(20–100)	4–15	(200–300)*	22–25
N⁶-methylformycin A	(1.25)	5	(1.75)*	14
Coformycin			(100)	31
Tubercidin			(100)	10

[a]100 μM Ado;   [b]360 μM Ade;   *enzyme preincubated with analogue for 10 min.

as a substrate for this enzyme. Formycin B, a potent inhibitor of PNPase, inhibited weakly trematode adenosine phosphorylase, similarly as that enzyme from snail *H. pomatia*.[4] Formycin A was a more potent inhibitor of the enzyme from both these sources; this Ado analogue inhibits efficiently Ado phosphorylase from *S. mansoni*.[6] Coformycin which strongly inhibited *F. hepatica* adenosine deaminase, exerted also a significant inhibitory effect on Ado/dAdo phosphorylase. Tubercidin inhibited weakly *F. hepatica* Ado phosphorylase as well as the *S. mansoni* enzyme.[6]

The nucleoside analogues modified in the purine ring seems to be more efficacious inhibitors of *F. hepatica* Ado/dAdo phosphorylase than the analogues modified in the ribose moiety.

REFERENCES

H. Trembacz, and M.M. Jeżewska, Specific enzyme synthesizing adenosine from adenine and ribose-1-phosphate in invertebrates, *Adv. Exp. Med. Biol.*, 370:509 (1995).

R. Miller, C.L.K. Sabourini, and T. Krenitsky, *Trypanosome cruzi* adenine phosphorylase, purification and substrate specificity, *Biochem. Pharmacol.*, 36:553 (1987).

L.Y. Ghoda, T.M. Savarese, C.H. Northup, R.E. Parks, Jr., J. Garofalo, L. Katz, B.B. Ellenbogen, and C.J. Bacchi, Substrate specificity of 5'-deoxy-5'-methylthioadenosine phosphorylase from *Trypanosoma brucei brucei* and mammalian cells, *Mol. Biochem. Parasitol.*,27:109 (1988).

H. Trembacz, and M.M. Jeżewska, Specific adenosine phosphorylase from hepatopancreas of gastropod *Helix pomatia*", *Comp.Biochem. Physiol.*, 104B:481 (1993).

H. Trembacz, and M.M. Jeżewska, Adenosine phosphorylase and other enzymes of purine salvage in Pulmonata snails and their Trematoda parasites, *Comp. Biochem. Physiol.*, 107B:135 (1994).

R.P. Miech, A.W. Senft, and D.G. Senft, Pathways of nucleotide metabolism in *Schistosoma mansoni* - VI Adenosine phosphorylase, *Biochem. Pharmacol.*, 24:407 (1975).

W.E Gutteridge, and M.J. Davies, Enzymes of purine salvage in *Trypanosome cruzi*, *FEBS Letters*, 127:211 (1981).

D. Torchen, and R.L. Miller, Purification and characterization of 5'-deoxy-5'-methylthioadenosine (MTA) phosphorylase from human liver, *Biochem. Pharmacol.*, 41:2023 (1991)

K. Fabianowska-Majewska, J. Duley, L. Fairbanks. A. Simmonds, and T. Wasiak, Substrate specificity of methylthioadenosine phosphorylase from human liver, *Acta Biochim. Polon.*, 41:391 (1994).

138

INHIBITION BY 3'-AZIDO-3'-DEOXYTHYMIDINE (AZT) OF *TRYPANOSOMA CRUZI* GROWTH IN MAMMALIAN CELLS AND A POSSIBLE MECHANISM OF ACTION

Junko Nakajima-Shimada and Takashi Aoki

Department of Parasitology
Juntendo University
School of Medicine
Hongo, Bunkyo-ku, Tokyo 113, Japan

INTRODUCTION

In mammalian hosts, the protozoan parasite *Trypanosoma cruzi*, which causes Chagas' disease in Latin America, occurs in two different forms, trypomastigotes and amastigotes. Trypomastigotes circulate in the bloodstream, invading host cells and transforming into amastigotes. Amastigotes grow by binary fission, and eventually transform back to trypomastigotes that destroy the host cells and again appear in the circulation. We have recently established an in vitro culture system of HeLa cells infected with *T. cruzi*, that enables us to determine quantitatively the time course of parasite growth inside the host cells (N.-Shimada et al., 1994 & 1996). This in vitro system is now available as a primary screening for anti-*T. cruzi* compounds. We expect that effective compounds may decrease the infection rate of host cells and/or the number of amastigotes per infected HeLa. Here we report the inhibitory action of an anti-human immunodeficiency virus (anti-HIV) agent, AZT, on the growth of *T. cruzi* in mammalian cells in vitro.

MATERIALS AND METHODS

The Tulahuene strain of *T. cruzi* and the human cancer cell line HeLa, used as in vitro host cells, were maintained and passaged as described previously (N.-Shimada et al., 1994). HeLa cells were infected with *T. cruzi* trypomastigotes essentially as described (N.-Shimada et al., 1996). Determinations of the host-cell infection rate and the average number of amastigotes per infected cell were as described previously.

Purine and Pyrimidine Metabolism in Man IX,
edited by Griesmacher *et al.* Plenum Press, New York, 1998.

Partial purification of the *T. cruzi* reverse transcriptase (RT) was carried out at 0 to 4°C. Culture forms (epimastigotes, approximately 2×10^9) of *T. cruzi* were homogenized in 50 mM Tris-HCl buffer (500 μl), pH 7.5, containing various protease inhibitors, namely, 1 mM PMSF, 50 μM each of leupeptin, antipain, o-phenanthroline, and NEM, 20 μg/ml of pepstatin, and 1 mg/ml of bovine serum albumin. To the high speed supernatant obtained, the same buffer solution (1 ml) was added and the crude enzyme was adsorbed to and eluted from a column (0.7 × 2.5 cm) of DEAE-Sepharose CL-6B. The enzyme elution was done by a step-wise procedure with 50 mM Tris-HCl buffer, pH 7.5, containing 0.1, 0.2, 0.4, 0.6, 0.8, and 1.0 M NaCl (1 ml each).

A non-radioactive method developed by Boehringer Mannheim (Cat. No. 1468 120) was used for the parasite RT assay following the manufacturer's instructions. Briefly, samples containing RT were added to the poly(A) and oligo(dT) complex, used as RNA templates and primers for RT, respectively, in the presence of a dNTP mixture containing digoxigenin (DIG)- and biotin-labeled dUTP. Then, the RT reaction proceeded to incorporate these labeled dUTPs into the newly synthesized DNA strand. The resulting hybrid of RNA and DNA strands was connected to a streptavidin-coated microtiter-plate, and then coupled with an anti-DIG-antibody conjugated to peroxidase. Finally, peroxidase activity was measured using the substrate ABTS. RT activity was determined as absorbance at 405 nm using a microtiter plate reader.

RESULTS AND DISCUSSION

In the present study, we tested the efficacies of pyrimidine analogs, AZT, 3'-deoxythymidine, 5'-deoxythymidine, 3'-azido-2',3'-dideoxyuridine, 3'-deoxyuridine, 2'-azido-2'-deoxyuridine, 2',3'-dideoxyuridine (ddU), and 2',3'-dideoxycytidine (ddC), on the rate of HeLa cell infection by *T. cruzi* and on the amastigote number per infected cell, using allopurinol as a positive control. The test compounds possess normal pyrimidine bases and modification(s) in their ribose moiety.

Figure 1 shows that, in a control culture, the rate of HeLa cell infection by *T. cruzi* increases in a time-dependent manner up to day 7. On day 4 after infection, 0.1, 1.0, 10, or 50 μM AZT lowered the infection rate to 94.8, 82.5, 72.2, or 61.9% of control, respectively. Similarly, the anti-HIV agent decreased the number of amastigotes per infected

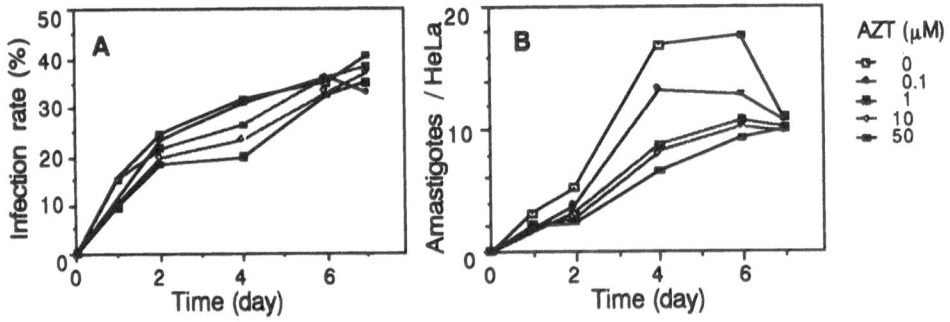

Figure 1. Effects of AZT on the infection rate of HeLa cells and on *T. cruzi* amastigote growth. HeLa cells were preincubated for 2 days and then infected with *T. cruzi* trypomastigotes. See the "Materials and Methods" for details of the conditions. (A) Infection rates of host cells; (B) average numbers of amastigotes per infected cell. The values represent the means of four separate experiments.

HeLa cell to 78.1, 49.5, 46.9, or 37.5% of control, respectively. The IC_{50} value for AZT was in the range of 0.5 to 1 µM, comparable to the value of 3 µM for allopurinol (Aoki et al., 1995). The inhibition of parasite growth by AZT is not dose-dependent, particularly at the higher concentrations of 10 and 50 µM (Figure 1). In this context, it has been reported that the intracellular half-life of an AZT metabolite, AZT-triphosphate, is approximately 1.3 hrs, whereas that of the metabolite of 2′,3′-dideoxyinosine (ddI) is 12 to 24 hrs (Yarchoan et al., 1990). The AZT concentration might be very low at the time of evaluation on day 4 after its addition, because AZT was added to the culture medium only once in the present study.

Other pyrimidine analogs tested in this study affected neither the infection rate nor parasite growth at any concentration tested from 1 to 50 µM. Since the anti-HIV agent AZT inhibited parasite growth, we also examined the effects of other related compounds. 2′,3′-Dideoxyadenosine (ddA) and ddI, purine analogs, demonstrated dose-dependent growth inhibition to a lesser extent than AZT. The pyrimidine analogs ddC and ddU had no effect. Among the pyrimidine analogs tested, only AZT, a thymidine derivative, was a potent inhibitor. On the contrary, a uridine derivative with the same modified ribose (3′-azido-2′,3′-dideoxyuridine) was not inhibitory. These data suggest that AZT interferes with DNA synthesis rather than RNA synthesis in *T. cruzi* amastigotes. This is in contrast to the action of allopurinol that interferes with RNA synthesis (Marr and Berens, 1983).

Since the reverse transcriptase (RT) of HIV is a target for AZT, we attempted to detect enzyme activity in culture forms (epimastigotes) of *T. cruzi*, although it is unusual for eukaryotic cells to possess RT activity. It has been reported that the RT genes in the *T. cruzi* genome occur as a family of spliced leader-associated retrotransposons (Villanueva et al., 1991). Our preliminary experiments suggest that a RT gene is transcribed in all the three developmental stages of the parasite, epimastigotes, trypomastigotes, and amastigotes. Because of the difficulty we experienced in achieving a reproducible assay for the parasite RT, a partially purified *T. cruzi* enzyme was prepared. The supernatant fraction contained a total activity of about 1000 units (the retroviral RT equivalency of activity is expressed as one unit), while a DEAE-Sepharose fraction eluted with 0.1 M NaCl contained about 3,300 units of activity, suggesting that the DEAE fraction was free from an endogenous inhibitor. Using the partially purified enzyme, we compared the effects of AZT on HIV- and *T. cruzi*-RT activities. The results of a typical experiment are shown in Table 1. The compound inhibited retroviral RT activity, but did not inhibit *T. cruzi* RT activity. These results suggest that AZT might target some other enzymes, presumably including a DNA polymerase in *T. cruzi*. The trypanosomal DNA polymerase is unique in showing a susceptibility to aphidicholine different from those of other species (Solari et al., 1983).

Table 1. Effect of AZT on the HIV- and *T. cruzi*- reverse transcriptase (RT) activities

Source	AZT (mM)	Reverse transcriptase activity	
		Abs. at 405 nm	% of control
HIV	0	1.05	100
	0.15	0.50	48
	1.0	0.38	36
T. cruzi	0	0.28	100
	0.15	0.29	104
	1.0	0.31	111

ACKNOWLEDGMENTS

This work was supported by a Grant-in-Aid for Scientific Research from the Ministry of Education, Science, and Culture of Japan (No. 07557211) and by a grant from the Ministry of Health and Welfare.

REFERENCES

Aoki, T., Nakajima-Shimada, J., and Hirota, Y. (1995): Quantitative determination of *Trypanosoma cruzi* growth inside host cells in vitro and effect of allopurinol. Adv. Exp. Med. Biol., 370, 499–502.

Marr, J.J., and Berens, R.L. (1983): Pyrazolopyrimidine metabolism in the pathogenic Trypanosomatidae. Mol. Biochem. Parasitol., 7, 339–356.

Nakajima-Shimada, J., Hirota, Y., Kaneda, Y., and Aoki, T. (1994): Quantitative determination of growth of amastigotes and trypomastigotes in an in vitro cultivation system of HeLa cells infected with *Trypanosoma cruzi*. J. Protozool. Res., 4, 10–17.

Nakajima-Shimada, J., Hirota, Y., and Aoki, T. (1996): Inhibition of *Trypanosoma cruzi* growth in mammalian cells by purine and pyrimidine analogs. Antimicrob. Agents Chemother., 40, 2455–2458.

Solari, A., Tharaud, D., Repetto, Y., Aldunate, J., Moewllo, A., and Litvak, S. (1983): In vitro and in vivo studies of *Trypanosoma cruzi* DNA polymerase. Biochem. Int., 7, 147–157.

Villanueva, M.S., Williams, S.P., Beard, C.B., Richards, F.F., and Aksoy, S. (1991): A new member of a family of site-specific retrotranspozons is present in the spliced leader RNA genes of *Trypanosoma cruzi*. Mol. Cell. Biol., 11, 6139–6148.

Yarchoan, R., Pluda, J.M., Perno, C.F., Mitsuya, H., Thomas, R.V., Wyvill, K.M., and Broder, S. (1990): Initial clinical experience with dideoxynucleosides as single agents and in combination therapy. Ann. N. Y. Acad. Sci., 616, 328–343.

139

METABOLISM OF EICAR (5-ETHYNYL-1-β-D-RIBOFURANOSYLIMIDAZOLE-4-CARBOXAMIDE), A POTENT INHIBITOR OF INOSINATE DEHYDROGENASE

J. Balzarini,[1] L. Stet,[1*] A. Matsuda,[2] L. Wiebe,[3] E. Knauss,[3] and E. De Clercq[1]

[1]Rega Institute for Medical Research
Katholieke Universiteit Leuven
B-3000 Leuven, Belgium
[2]Faculty of Pharmaceutical Sciences
Hokkaido University
Sapporo-060, Japan
[3]Faculty of Pharmacy and Pharmaceutical Sciences
University of Alberta
Edmonton, Alberta, Canada T6G 2N8

1. ABSTRACT

The cytostatic agent 5-ethynyl-1-β-D-ribofuranosylimidazole-4-carboxamide (EICAR) causes a rapid and marked inhibition of inosinate (IMP) dehydrogenase activity in intact tumor cells. [³H]EICAR is metabolised in L1210 cells to its 5'-mono-, 5'-di- and 5'-triphosphate in a concentration-dependent manner. The metabolites accumulate proportionally with the initial extracellular EICAR concentrations (ranging from 0.25 to 200 μM). The nicotinamide adenine dinucleotide (NAD) analogue of EICAR, designated EAD, also accumulates within the cells and becomes the major metabolite after 48 hr incubation with 5 μM [³H]EICAR. EAD has a markedly longer intracellular half-life than EICAR 5'-mono-, 5'-di- and 5'-triphosphate. An additional EICAR metabolite elutes on an anion exchange Partisphere SAX HPLC chromatogram between EICAR 5'-di- and 5'-triphosphate. Its intracellular levels are ~10-fold lower than those of EAD and the nature of this metabolite has still to be identified. The differential role of EAD and EICAR 5'-monophosphate in the inhibition of IMP dehydrogenase is currently under investigation.

---

* Present address: Organon, 5340 BH Oss, The Netherlands.

Purine and Pyrimidine Metabolism in Man IX,
edited by Griesmacher *et al.* Plenum Press, New York, 1998.

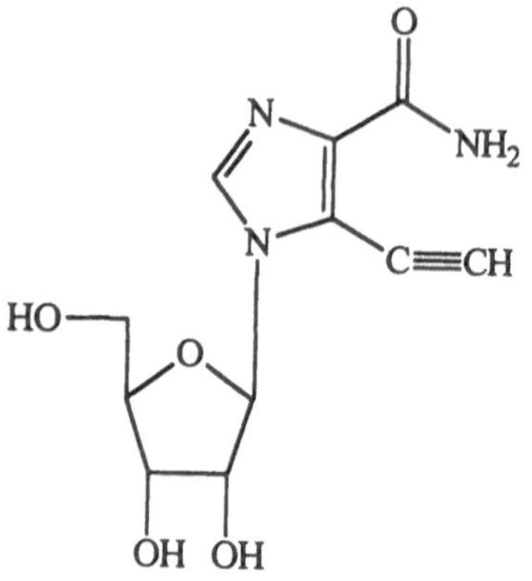

Figure 1. Structural formula of EICAR.

2. INTRODUCTION

EICAR (5-ethynyl-1-β-D-ribofuranosylimidazole-4-carboxamide) (Fig. 1). belongs to a series of novel compounds that have been identified as broad-spectrum antiviral agents. It shows an antiviral potency that is about 10- to 100-fold greater than that of ribavirin.[1] EICAR has a similar activity spectrum as ribavirin. It displays a pronounced inhibitory activity against several members of the poxviruses, togaviruses, arenaviruses, reoviruses, orthomyxoviruses and paramyxoviruses. Interestingly, EICAR has also proved cytostatic for rapidly growing cells (i.e murine leukemia L1210 and human lymphocyte CEM) with a 50% inhibitory concentration between 0.5 and 2 μM.[2,3] Moreover, in addition to its antiviral and cytostatic activity, EICAR also exhibits a pronounced differentiation-inducing activity in human erythroleukemia K562 cells[3] and antileukemic activity in mice.[4] We have previously demonstrated that EICAR causes a rapid and marked inhibition of inosinate (IMP) dehydrogenase (IMP-D) activity in intact L1210 and CEM cells, resulting in a selective accumulation of IMP and depletion of GTP and dGTP levels.[2] We also found that EICAR 5'-monophosphate (EICAR-MP) is a potent inhibitor of purified L1210 cell IMP-D (K_i/K_m: 0.06). The inhibition of IMP dehydrogenase by EICAR-MP was noted at a similar drug concentration as for mycophenolic acid, a well-known potent IMP-D inhibitor. Based on its inhibitory properties against IMP-D, EICAR was combined with 2',3'-dideoxyinosine (DDI) in human immunodeficiency virus type 1 (HIV-1)-infected CEM cells, and found to potentiate the antiretroviral activity of DDI in cell culture.[5] Murine L1210 cells that were selected for resistance to the cytostatic effect of EICAR proved adenosine kinase-deficient, and studies with adenosine kinase-deficient cell lines revealed that adenosine kinase, as well as at least one other alternative pathway, is responsible for the conversion of EICAR to its 5'-monophosphate.

3. MATERIALS AND METHODS

3.1. Compounds

The synthesis of radiolabeled [^{3}HEICAR will be published elsewhere. The radiospecificity of [^{3}HEICAR was 1.56 Ci/mmole or 76 mCi/mmol.

3.2. Metabolism of [³H]EICAR in Murine Leukemia L1210 Cells

The metabolism of radiolabeled EICAR was monitored as follows: L1210 cells were seeded at $2–4 \times 10^5$ cells per ml and incubated with different concentrations of [³H]EICAR (varying from 0.25 to 200 µM). The total amount of radiolabel per cell culture was kept constant (2 µCi/5 ml). At different time intervals (i.e. 0, 6, 24, 30 and 48 hrs), cells were centrifuged at 4°C, thoroughly washed three times with ice-cold medium (without serum), and precipitated with 60% ice-cold methanol. After centrifugation at 10,000 rpm, the supernatants were filtered and quantitation of [³H]EICAR and its metabolites was accomplished by HPLC analysis using a Partisil-SAX-10 radial compression column.

3.3. Intracellular Retention of [³HEICAR and Its Metabolites in L1210 Cells after Removal of the Drug from the Culture Medium

L1210 cells were seeded at $2–4 \times 10^5$ cells per ml and incubated with 5 µM [³H]EICAR (2 µCi per 5-ml culture) for 24 hrs. Then, the extracellular drug was removed by centrifugation of the cells and washing the cell cultures three times with warm culture medium. At 0, 2, 4, 6 and 24 hrs after removal of [³H]EICAR, cell extracts were prepared and [³H]EICAR and its metabolites were determined by HPLC.

4. RESULTS AND DISCUSSION

Radiolabeled [³HEICAR was converted to its 5'-mono-, 5'-di- and 5'-triphosphate in a dose-dependent manner at an initial EICAR concentration ranging between 0.25 and 200 µM (Fig. 2). The 5'-mono- and 5'-triphosphates of EICAR were formed to a marked extent, whereas EICAR-DP was present at 10- to 30-fold lower levels. Interestingly, the nicotinamide adenine dinucleotide (NAD) analogue of EICAR, designated EAD, was formed at intracellular levels comparable to those of EICAR-TP, together with an unknown metabolite that eluted between EICAR-DP and EICAR-TP. It is speculated that the unknown metabolite represents the phosphorylated EAD analogue EADP, but the identity of this metabolite has not yet been confirmed with chemically synthesized compound.

When the metabolism of EICAR was investigated in function of time, intracellular 5'-mono-, 5'-di- and 5'-triphosphate metabolites of EICAR reached the highest levels within 6 hrs post incubation, and progressively decreased upon longer incubation times (Fig. 3). After 48 hrs, only 20% of the amount of the metabolites compared to those recorded after the 6-hr drug incubation period could be recovered.

However, in contrast to the EICAR phosphates, EAD tended to accumulate upon longer incubation times. EAD levels had increased by 8-fold after 48 hrs as compared with the EAD levels recorded after 6 hrs. Interestingly, EAD was the predominant EICAR metabolite after 48 hrs of incubation. The unknown metabolite (HPLC retention time: 19–20 min) levels were virtually kept constant during the whole incubation period.

Wash-out experiments revealed that the intracellular 5'-mono-, 5'-di- and 5'-triphosphorylated EICAR metabolites disappeared quickly. After 4 hrs post-removal of the radiolabeled compound from the extracellular medium, no EICAR-MP or EICAR-DP could be detected. Only small amounts of EICAR-TP were still present after 4 to 6 hrs. In contrast, EAD persisted markedly longer and was still present for more than 50% of the initial intracellular levels 6 hrs after the drug had been removed (data not shown).

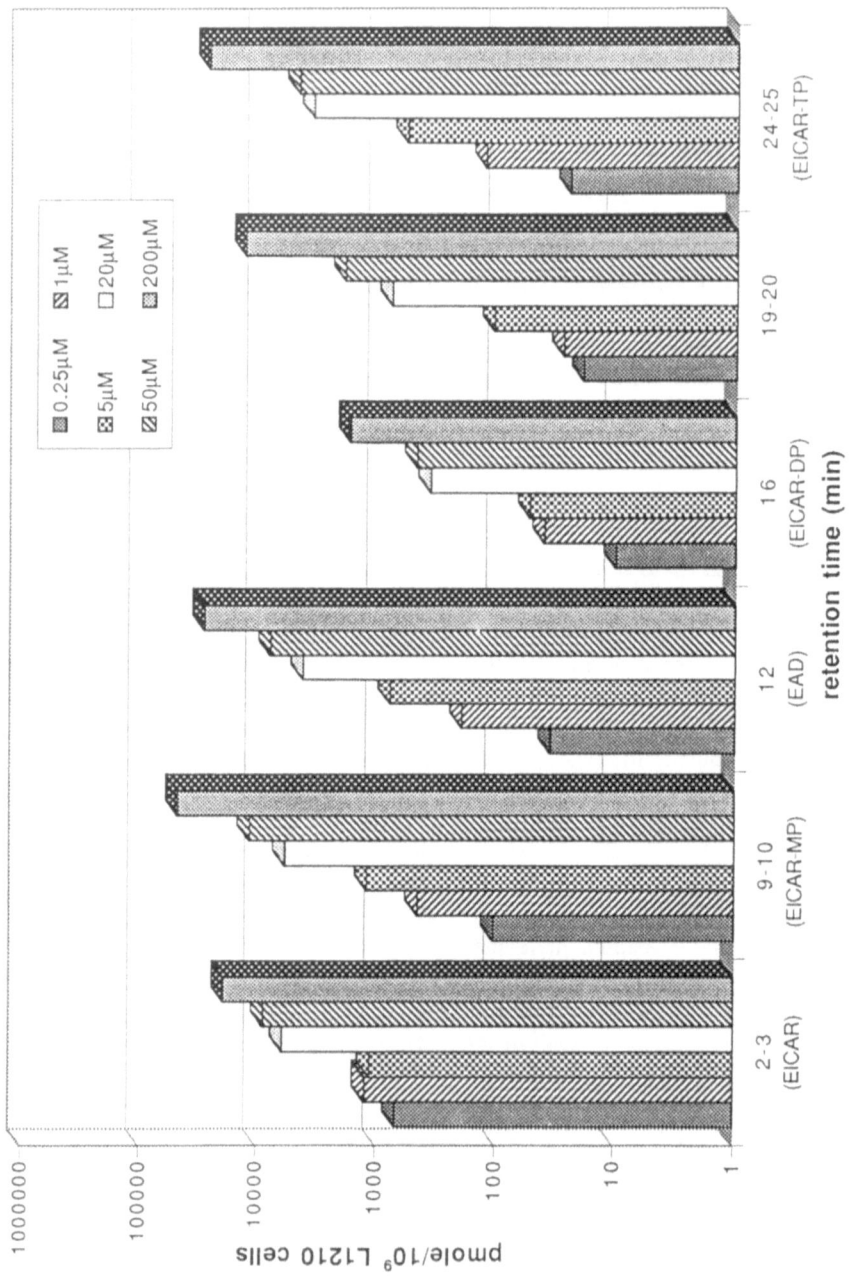

Figure 2.

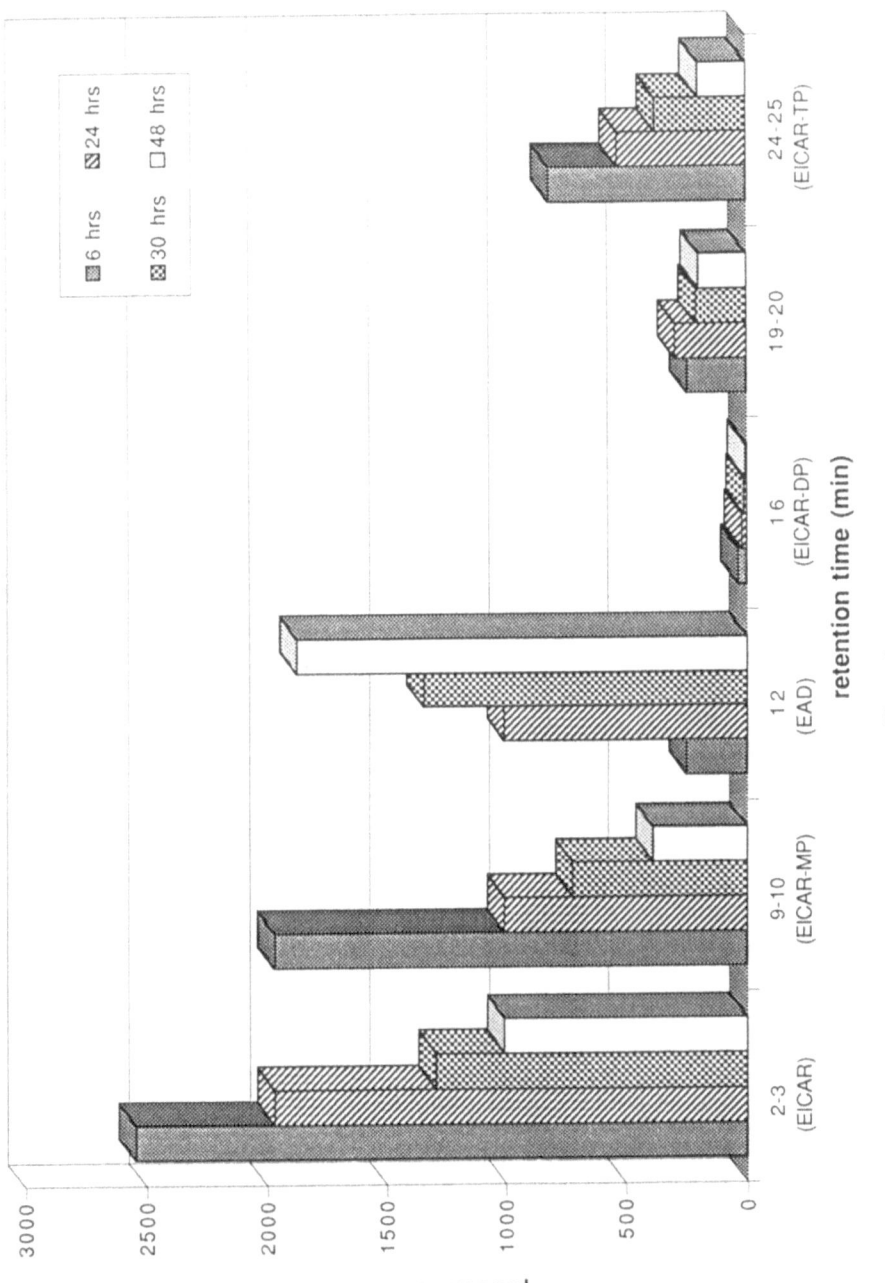

Figure 3.

At least three EICAR metabolites can contribute to its eventual cytostatic activity in cell culture. EICAR-MP has previously proven to be a potent inhibitor of IMP dehydrogenase.[2] However, also EAD, the NAD analogue of EICAR, is produced in EICAR-exposed L1210 cells. It has been reported that the NAD analogue of thiazofurin (designated TAD) exerts potent inhibitory activity against IMP dehydrogenase.[6] Since EAD accumulates in the L1210 cells and has a considerably longer half-life than EICAR-MP, it may well contribute to the eventual inhibition of tumor cell proliferation. Studies are now ongoing to reveal the role of EAD in the inhibition of IMP dehydrogenase. Finally, the 5'-triphosphate of EICAR may achieve a cytostatic activity through interference with the RNA polymerization process. It is presently unknown whether EICAR may be incorporated into RNA. The individual contribution of the different EICAR metabolites to the eventual cytostatic activity of EICAR is currently under investigation.

In conclusion, exposure of L1210 cells to EICAR resulted in the formation of the EICAR 5'-mono-, 5'-di- and 5'-triphopshate derivatives. In addition, the nicotinamide adenine dinucleotide derivative of EICAR (EAD) was also formed together with a yet to be identified metabolite (presumably the phosphorylated EAD derivative). EAD persisted for a much longer time within the cells than any of the other EICAR metabolites.

ACKNOWLEDGMENTS

This work was supported by the Belgian Geconcerteerde Onderzoeksacties (GOA 95/5) and the Belgian ASLK Cancer Fund. We are grateful to Mrs. Ria Van Berwaer for excellent technical help and Mrs. Christiane Callebaut for dedicated editorial assistance.

REFERENCES

1. De Clercq, E., Cools, M., Balzarini, J., Snoeck, R., Andrei, G., Hosoya, M., Shigeta, S., Ueda, T., Minakawa, N., and Matsuda, A. Antiviral activities of 5-ethynyl-1-β-D-ribofuranosylimidazole-4-carboxamide and related compounds. Antimicrob. Agents Chemother. 35: 679–684 (1991).
2. Balzarini, J., Karlsson, A., Wang, L., Bohman, C., Horská, K., Votruba, I., Fridland, A., Van Aerschot, A., Herdewijn, P., and De Clercq, E. EICAR (5-ethynyl-1-β-D-ribofuranosylimidazole-4-carboxamide). A novel potent inhibitor of inosinate dehydrogenase activity and guanylate biosynthesis. J. Biol. Chem. 268: 24591–24598 (1993).
3. Hatse, S., De Clercq, E., and Balzarini, J. Evidence for distinction of the differentiation-inducing activities and cytostatic properties of 9-(2-phosphonylmethoxyethyl)adenine and a variety of differentiation-inducing agents in human erythroleukemia K562 cells. Mol. Pharmacol. 50: 1231–1242 (1996).
4. Matsuda, A., Minakawa, N., Sasaki, T., and Ueda, T. The design, synthesis and antileukemic activity of 5-alkynyl-1-β-D-ribofuranosylimidazole-carboxamides. Chem. Pharm. Bull. 36: 2730–2733 (1988).
5. Balzarini, J., Lee, C.-K., Schols, D., and De Clercq, E. 1-β-D-Ribofuranosyl-1,2,4-triazole-3-carboxamide (ribavirin) and 5-ethynyl-1-β-D-ribofuranosylimidazole-4-carboxamide (EICAR) markedly potentiate the inhibitory effect of 2',3'-dideoxyinosine on human immunodeficiency virus in peripheral blood lymphocytes. Biochem. Biophys. Res. Commun. 178: 563–569 (1991).
6. Jayaram, H.N., Pillwein, K., Nichols, C.R., Hoffman, R., and Weber, G. Selective sensitivity to tiazofurin of human leukemic cells. Biochem. Pharmacol. 35: 2029–2032 (1986).

140

A HPLC METHOD FOR THE MONITORING OF HUMAN RED CELL 6-THIOGUANINE AND METHYL 6-MERCAPTOPURINE IN A SINGLE RUN

Thierry Dervieux[1,2] and Roselyne Boulieu[1,2]

[1]Service Pharmaceutique
Hôpital Neuro-Cardiologique
Lyon, France
[2]Laboratoire de Pharmacie clinique
Institut des Sciences Pharmaceutiques et Biologiques
Université Lyon 1, Lyon, France

1. INTRODUCTION

Azathioprine is an immunosuppressive drug coadministered with ciclosporine and corticoides to prevent rejection after transplantation.[1] Azathioprine acts as an antimetabolite of purine after conversion into 6-mercaptopurine leading to 6-thioguanine nucleotides (6-TGN) via hypoxanthine guanine phosphorybosyl transferase, and methyl 6-mercaptopurine nucleotides after methylation of 6-thioinosine monophosphate into methyl 6-thioinosine monophosphate (Me6TIMP) via thiopurine methyl transferase (TPMT).[2]

Administration of azathioprine in transplant patients leads rapidly to severe myelosuppression when the homozygote allele for TPMT deficiency is expressed.[3] Then high 6-TGN concentrations in red blood cells could be correlated with low erythrocyte TPMT activity. However, the relation between high 6-TGN levels in red blood cells and myelosuppression induced by azathioprine remains controversed.[4] More recently, Me6TIMP was shown to strongly inhibits purine *de novo* synthesis in a concentration dependant manner leading to cytotoxicity in Molt F4 cells suggesting that methyl 6-mercaptopurine could explain a part of the pharmacological activity of azathioprine.[5]

Among the high performance liquid chromatographic methods[6,7] described for the analysis of 6-thiopurine and methylated 6-thiopurine nucleotides after hydrolysis of nucleotides in bases only one reports a method for the simultaneous analysis of 6-thioguanine and methyl 6-mercaptopurine nucleotides in a single run. The bases formed after acid hydrolysis of nucleotides were extracted by a laborious double step extraction using toxic phenyl mercuric acid (PMA) and toluene. However, Me6MP methylated on the thiol function cannot be extracted by PMA adduct formation like 6-thiopurines. So, the free base analysed by this

Purine and Pyrimidine Metabolism in Man IX,
edited by Griesmacher *et al.* Plenum Press, New York, 1998.

method is not methyl 6-mercaptopurine *per se* but a compound synthesised from Me6MP during the hydrolysis step. Analytical recoveries for Me6MPN were less than 40%.

In order to investigate more closely the role of thiopurine nucleotides in the pharmacodynamic effect of the drug, we report a reversed phase HPLC method using a simple and rapid sample treatment procedure for the simultaneous determination of 6TGN and Me6MPN in red blood cells.

2. EXPERIMENTAL

2.1. Reagents and Stock Solutions

6-mercaptoguanosine (6-TGR), 6-thioguanine (6-TG), methyl 6-mercaptopurine ribonucleoside (Me6MPR), methyl 6-mercaptopurine (Me6MP), dithiotreitol (DTT) and sodium bicarbonate were obtained from Sigma. Methanol, potassium dihydrogenphosphate and perchloric acid were obtained from Merck. Stock solutions of the compounds of interest were prepared in 0.1 M hydrochloric acid and stored at −80°C.

2.2. Apparatus and Chromatographic Conditions

The liquid chromatograph consisted of a model 510 pump connected with a photodiode array detector model 960 and a Wisp 715 solvent delivery model (all from Waters).

6-TG, 6-TGR, Me6MP, Me6MPR and Me6MP derivative were analysed using a reversed phase high performance liquid chromatographic method. The separation was performed on a Purosphere RP-18e column, 5 μm particle size (Merck) with a linear gradient elution mode using 0.02 M KH_2PO_4 pH 3.5 and 0.02 M KH_2PO_4 pH 3.5-methanol (40 : 60, v/v). The concentration of methanol varied from 0 to 20.0% over a period of 12 min. The flow rate was 1.2 ml/min. Detection of 6-TG and 6-TGR were performed at 341 nm, Me6MP and Me6MPR at 291 nm and methyl 6-mercaptopurine derivative was detected at 304 nm.

2.3. Sample Collection and Treatment

Blood samples of 5 ml were collected in heparinized tubes containing 1 mg of DTT and centrifuged without delay at low temperature (4°C). Plasma was discarded, the leukocytes and the upper layer of the erythrocytes were removed. The erythrocytes were counted and drug concentrations were normalized to 10^8 cells.

500 μl of the remaining erythrocytes were transferred in a tube containing 5 mg of DTT and rapidly deproteinized by 50 μl of 70% perchloric acid. The deproteinized samples were centrifuged at 3000 g for 15 min at 4°C. The supernatants were decanted. In these conditions, the pH of the supernatant was less than 0.1.

In order to investigate the effect of the pH of acid hydrolysis on analytical recoveries, the pH of the supernatant was adjusted between 0 and 2.0 with 1 M sodium bicarbonate, pH 13, prior to hydrolysis. Then, the supernatant was heated for 45 min at 100°C to hydrolyse thiopurine nucleotides into their bases.

2.4. Analytical Recoveries

Analytical recoveries were determined by adding known concentrations of compounds to erythrocytes and comparison of peak heights with those obtained by direct injection of aqueous standards. For the determination of the recoveries of methyl 6-mercaptopurine derivative, 500 µl of aqueous standard of Me6MP was heated for 45 min at 100 °C with 50 µl perchloric acid to allow conversion of Me6MP into its derivative.

3. RESULTS

3.1. Deproteinization Step

Previously, we have reported that the recoveries of 6-thiopurines like 6-TG were greatly influenced by the addition of dithiotreitol during the sample treatment procedure.[8] DTT protects thiol function from oxidation during the deproteinization step but does not contribute to an improvement of recoveries for Me6MP, the thiol group being protected from oxidation by a methyl function.

3.2. Hydrolysis Step

During acid hydrolysis, methyl 6-mercaptopurine is converted into a compound as previously reported,[6] which contributes to a significant decrease in analytical recoveries. However, between pH 1.0 and 2.0, analytical recoveries of methyl 6-mercaptopurine are stable and higher than 70%. 6-thioguanine is stable towards hydrolysis. Moreover, the efficacy of acid hydrolysis of thiopurine nucleotides into their free bases was tested. Acid hydrolysis of 6-TGR and Me6MPR into their free bases is complete in 45 min at pH less than 1.5.

Consequently, two procedures could be developed for the determination of methyl 6-mercaptopurine nucleotides in red blood cells, either a procedure for the analysis of methyl 6-mercaptopurine *per se*, or one for the analysis of methyl 6-mercaptopurine derivative.

However, the analysis of Me6MP *per se* appears technically difficult due to the instability of this compound at pH value required for the hydrolysis of thiopurine nucleotides into their free bases.

So, a procedure for the analysis of Me6MP derivative was chosen. Optimal analytical recoveries were obtained when the supernatant is adjusted at pH less than 0.1 with mean values of 73.1% and 84.0% for 6-TG and Me6MP derivative respectively (Figure 1).

3.3. Linearity and Precision

The linearity was tested for each compound and was excellent up to 50 nmol/ml packed cells for 6-TGN and 120 nmol/ml packed cells for Me6MP derivative with a correlation coefficient greater than 0.998. The intra-day and inter-day relative standard deviations were less than 6.0% for 6-TGN at a concentration of 0.3 nmol/ml packed cells and less than 7.0% for Me6MP derivative at a concentration of 3.0 nmol/ml packed cells. The quantification limit was 0.1 nmol/ml and 0.3 nmol/ml for 6TGN and Me6MP derivative respectively.

3.4. Selectivity

No interference with the related endogenous compounds, uric acid, hypoxanthine, xanthine, guanine, guanosine or with commonly coadministered drugs in transplant recipi-

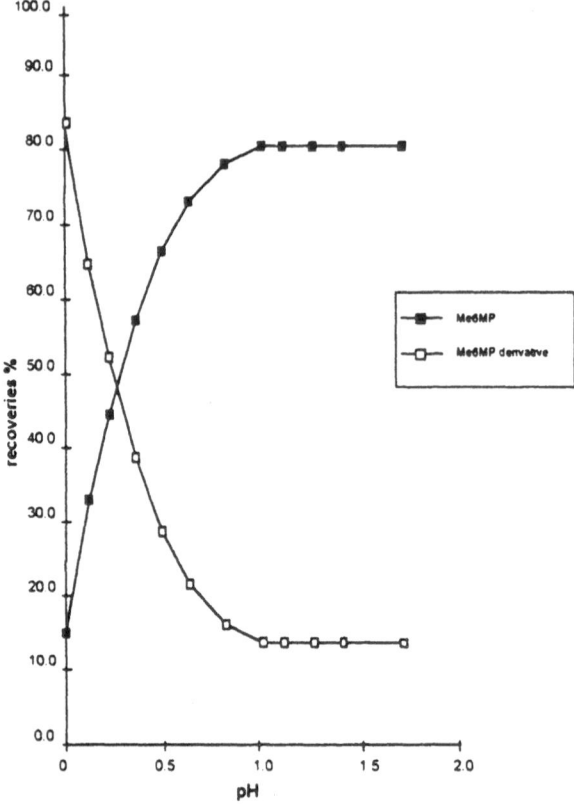

Figure 1. Influence of the pH of the acid supernatant on analytical recoveries.

ents such as ciclosporine, azathioprine, ganciclovir, acyclovir, prednisone, dexamethazone was found. Futhermore, 6-mercaptopurine, 6-thioxanthine and methyl 6-thioguanine were separated from the compounds of interest.

4. ANALYSIS OF PATIENT SAMPLE

The method described was used for the determination of 6-thioguanine and methyl 6-mercaptopurine nucleotide contents in erythrocytes from heart/lung transplant patients under azathioprine therapy.

Chromatograms of a blank erythrocyte and erythrocyte sample from a patient who received azathioprine (2.0 mg/kg per os daily) are presented in Figure 2. The concentrations of thiopurine nucleotides found were 48 pmol/10^8 cells for 6-TGN and 1042 pmol/10^8 cells for Me6MPN.

5. CONCLUSION

The HPLC method described is sensitive and selective for the analysis of 6-thioguanine and methyl 6-mercaptopurine nucleotides in a single run. The sample treat-

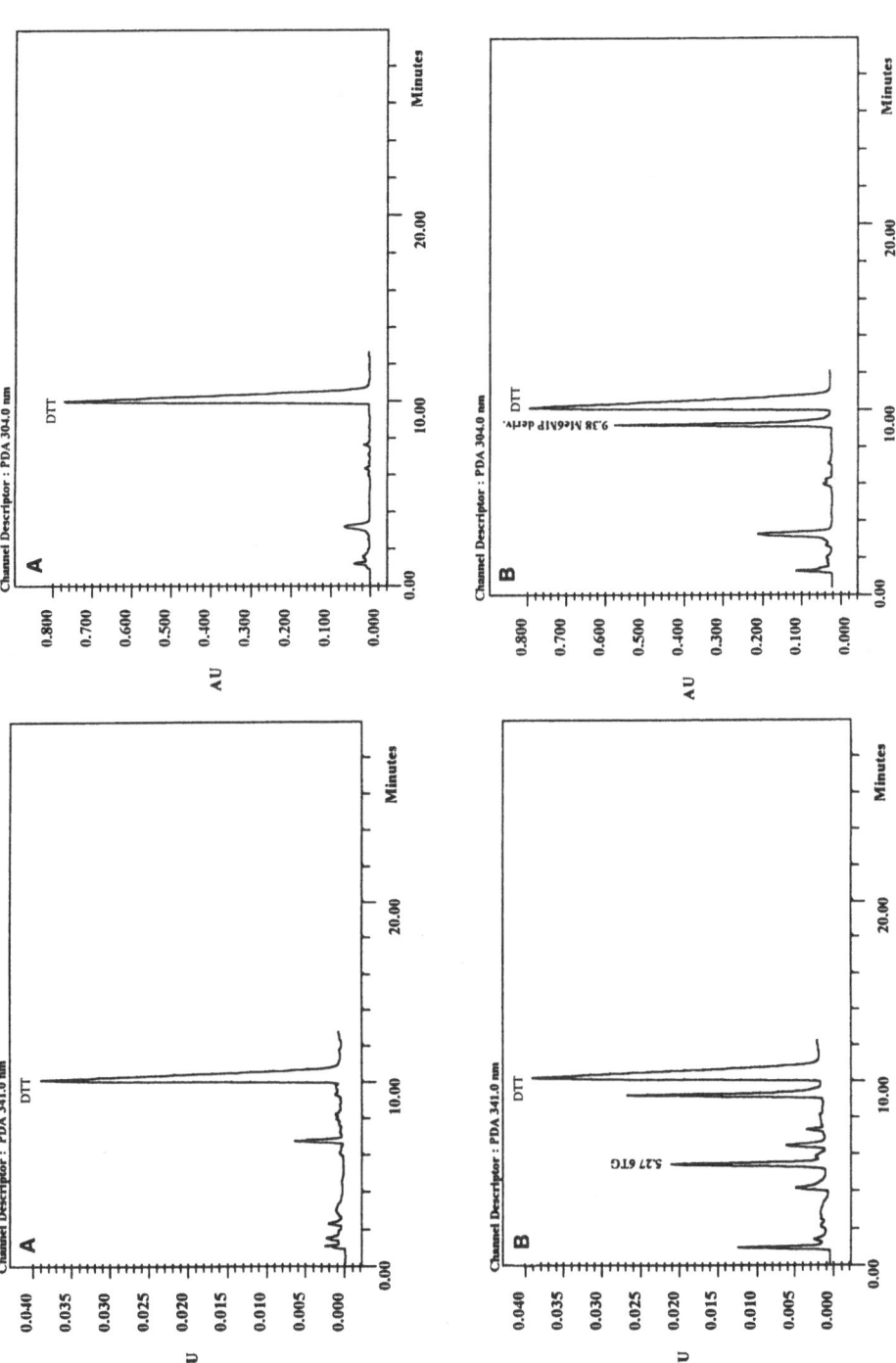

Figure 2. Chromatograms of a blank erythrocyte (A) and erythrocyte sample (B) at 304 and 341 nm from a patient receiving 2.0 mg/kg per os daily of azathioprine. Peaks: 6-TGN, 48 pmol/10^8 cells; Me6MPN, 1042 pmol/10^8 cells.

ment procedure described is simple and rapid with recoveries higher than 73%. Moreover, the use of Purosphere column contributes to improve the peak symmetry of Me6MP derivative. This method should be useful for the monitoring of intracellular thiopurine nucleotides in organ transplant patients receiving azathioprine therapy.

REFERENCES

1. R.L. Simmonds, D.M. Canafax, S.D. Fryd, N.L. Asher, W.D. Payne, W. Sutherland and J.S. Narajan, 1986, New immunosuppressive drug combination for mismatched related and cadaveric renal transplant. *Transplant. Proc.*, **18**:76–81.
2. L. Lennard, 1992, The clinical pharmacology of 6-mercaptopurine, *Eur. J. Clin. Pharmacol.*, **43** : 329–339.
3. A. Anstey, L. Lennard, S.C. Mayou, J.D. Kirby, 1992, Pancytopenia related to azathioprine an enzyme deficiency caused by a common genetic polymorphism : a review, *J. Royal Soc. Med.*, **85**:752–756.
4. R. Boulieu, A. Lenoir, J.F. Mornex, M. Bertocchi, 1997, Intracellular thiopurine nucleotides and azathioprine myelotoxicity in organ transplant patients, *Br. J. Clin. Pharmacol.*, **43**:116–118.
5. M. Vogt, E. Stet, R. De Abreu, J. Bökkerink, L. Lambooy, J. Trijbels, 1993, The importance of methylthio-IMP for methylmercaptopurine ribonucleoside (Me-MPr) cytotoxicity in Molt F4 human malignant T lymphoblasts, *Biochem. Biophys. Acta*, **1181**: 189–194.
6. L. Lennard, H. Singleton, 1992, High-performance liquid chromatographic assay of the methyl and nucleotide metabolites of 6-mercaptopurine quantification of red blood cell 6 thioguanine nucleotide, 6-thioinosinic acid and methylmercaptopurine metabolites in a single sample. *J. Chromatogr.*, **58**:383–390.
7. G. Erdman, L. France, B. Bostrom, M. Canafax, 1990, A reversed phase high performance liquid chromatographic approach in determining total red blood cell concentrations of 6-thioguanine, 6-mercaptopurine, methylthioguanine and methylmercaptopurine nucleotides in a patient receiving thiopurine therapy, *Biomed. Chromatogr.*, **4**:47–51.
8. R. Boulieu, A. Lenoir, M.A. Bory, 1993, High-performance liquid chromatographic determination of thiopurine metabolites of azathioprine in biological fluids, *J. Chromatogr.*, **615**:352–356.

141

EXPRESSION AND PROPERTIES OF RECOMBINANT *P. FALCIPARUM* HYPOXANTHINE-GUANINE PHOSPHORIBOSYLTRANSFERASE

D. T. Keough,[1,2] A.-L. Ng,[1] B. T. Emmerson,[2] and J. de Jersey[1]

[1]Department of Biochemistry and
[2]Department of Medicine
University of Queensland
Brisbane, Australia 4072

INTRODUCTION

Plasmodium falciparum is the most deadly of the *Plasmodium* species causing human malaria. Due to the development of drug resistant strains of *P. falciparum*, there is a need for the development of new antimalarial drugs.

Hypoxanthine-guanine phosphoribosyltransferase (HPRT; E.C. 2.4.2.8) plays a major role in purine salvage. It catalyses the formation of the mononucleotides GMP and IMP from the purine bases guanine and hypoxanthine. These mononucleotides can then be converted to nucleoside triphosphates for nucleic acid synthesis.

Unlike humans, all parasitic protozoans known to date are unable to synthesise purines *de novo*. Hence, they rely totally on the availability of preformed purine bases and/or nucleosides from their host organism. Purine bases are then converted to mononucleotides by the salvage pathway within the parasite. For this reason, it appears that HPRT has a function that is crucial for the replication of these parasites and is therefore a potential target for new therapeutic approaches. Parasite HPRT is a particularly attractive target because the human enzyme can be inhibited by up to 90% with the only effect on the host being an elevated urate level which is readily amenable to treatment by allopurinol.

Human HPRT and HPRTs from several parasites (*Leishmania donovani*[1], *Trypanosoma brucei*[2], *Trypanosoma cruzi*[3], *Tritrichomonas foetus*[4], *Schistosoma mansoni*[5], *Eimeria tenella*[6], *Giardia lamblia*[7] and *Toxoplasma gondii*[8]) have been purified and characterised. Much is already known about some of these HPRTs: extensive kinetic studies have been performed with the human[9] and *S. mansoni*[10] enzymes and the crystal structures of human, *E. coli*, *T. foetus* and *T. gondii* HPRTs have been solved.[8,11–13] However, detailed

Purine and Pyrimidine Metabolism in Man IX,
edited by Griesmacher *et al.* Plenum Press, New York, 1998.

characterisation of *P. falciparum* HPRT has been impeded by the difficulty in obtaining sufficient amounts of the enzyme from either natural or recombinant bacterial sources and by its inherent instability.

Here, we report the first account of high level expression of recombinant *P. falciparum* HPRT, its purification to homogeneity as assessed by SDS-PAGE and mass spectrometry and preliminary characterisation.

METHODS

Cloning of *P. falciparum* HPRT cDNA

P. falciparum HPRT cDNA cloned into the *Eco*R I site of pUC was obtained from Dr Ross Coppel, Walter and Eliza Hall Institute. An *Nde* I restriction site was then created at the start codon by PCR amplification of the 5' fragment of HPRT cDNA (containing the *Dde* I site and the start codon) using primers containing a two base pair change to create the *Nde* I site. The amplified fragment was then cut with *Dde* I and *Nde* I, ligated to the 3' HPRT cDNA fragment (containing *Dde* I and *Eco* RI sites) and inserted into pT7-7 to produce pT7-PFAL. This was performed to ensure that the expressed protein did not contain any extra amino acids.

Bacterial Cultures and Expression of Recombinant *P. falciparum* HPRT

Escherichia coli Sϕ606 [*ara*, Δ *pro-gpt-lac*, *thi*, *hpt*, *F⁻*] cells were transformed with pGP1–2 and pT7-PFAL and grown in enriched Sϕ606 media. Expression was induced and cells harvested as described by Free *et al.*[14]

Purification

Lysozyme (4 mg lysozyme /mL of cell paste) was added to the cell paste and incubated on ice for 2 hours. The cells were then freeze-thawed in liquid nitrogen and centrifuged at 35000 g at 4°C for 15 minutes. *P. falciparum* HPRT was purified by affinity chromatography of the lysate.

HPRT Activity Assays

HPRT activity was determined by a continuous spectrophotometric assay measuring the conversion of guanine, hypoxanthine or xanthine to GMP, IMP or XMP at 257.5 nm, 245 nm and 255 nm respectively.[15]

RESULTS AND DISCUSSION

An *Nde* I restriction site was created immediately before the start codon and *P. falciparum* HPRT cDNA cloned into the pT-7 expression vector using two restriction enzyme sites, *Eco*R I and *Nde* I. The cDNA was then sequenced and found to contain the same sequence as that reported by Vasanthakumar *et al.*[16] (having threonine at position 101 instead of methionine in that position as reported by King and Melton[17]).

In order to perform structural and kinetic studies on *P. falciparum* HPRT, large amounts of purified enzyme will be required. Queen *et al.*[18] found that it was difficult to obtain the milligram quantities required from parasites grown in culture. Hence, the cDNA was cloned and attempts made to express recombinant *P. falciparum* HPRT in *E. coli* and *S. typhimurium* using the ptac-85 and λP_L expression systems respectively.[16,19] Only limited success was achieved in these attempts since the specific activity of the lysates was very low. This suggests that the enzyme was not being overexpressed, that most of the expressed enzyme was inactive or incorrectly folded, or that the expressed enzyme was very unstable.

In contrast, high levels of expression have been achieved in our laboratory using the pT7-7 expression system in the HPRT⁻ strain of *E. coli*, Sφ606. This same expression system has been used to obtain high levels of recombinant human and *E. coli* HPRT activity. The specific activity (using guanine as a substrate) of the crude lysate of pT7-PFAL transformed cells was approximately 483 nmol min^{-1} mg^{-1}, compared with 5.5 and 1.6 nmol min^{-1} mg^{-1} for the expression systems of Vasanthakumar *et al.*[16] and Shahabuddin and Scaife[19] respectively. Using a specific activity of 8 µmol min^{-1} mg^{-1} for highly purified enzyme (Table 1), approximately 10 mg of *P. falciparum* HPRT was present in the lysate from a 2 L culture. This should be a sufficient level of expression for kinetic and structural characterisation of the enzyme and for site directed mutagenesis.

Recombinant *P. falciparum* HPRT was purified to homogeneity by affinity chromatography, as assessed by a single band migrating in the expected position on SDS-PAGE (Figure 1A). When analysed by electrospray mass spectrometry, the molecular mass of the enzyme was 26 239.8 ± 3.8 Da, in good agreement with the calculated mass of 26 232 Da (Figure 1B). This result showed that recombinant *P. falciparum* HPRT had the expected sequence and that the N-terminal methionine residue had been cleaved.

The specific activity of the purified enzyme (assayed with guanine) was approximately 8 µmol min^{-1} mg^{-1}. Although this is significantly lower than that of the human enzyme, it is similar to the values obtained for some other parasite HPRTs (Table 1). The lower specific activity observed for purified *P. falciparum* HPRT may be due to the fact that the pH and magnesium concentrations in the assay did not reflect physiological conditions within the parasite.

Reyes *et al.*[20] showed that HPRT is abundantly expressed within the parasite. *P. falciparum* HPRT had a specific activity of 88 nmol min^{-1} mg^{-1} of protein in the crude lysate of *P. falciparum*, whereas in human red blood cell lysate, the specific activity of HPRT was 1.4 nmol min^{-1} mg^{-1} of protein. King and Melton[17] found that 3 µg of total *P. falciparum* RNA contained approximately the same amount of HPRT mRNA as 30 µg of total mouse RNA. This strongly suggests that the salvage reaction catalysed by HPRT is important to the parasite.

Table 1. Specific activities of purified HPRTs from various organisms

Source	Specific activity (µmol min^{-1} mg^{-1} protein)		
	Guanine	Hypoxanthine	Xanthine
Human[a]	40	23	not detectable
P. falciparum[a,b]	8.0	3.1	4.0
S. mansoni[c]	8.1	5.8	not detectable
T. foetus[d]	2.2	2.3	1.4

[a]100 mM Tris-HCl, pH 8.5, 0.11 M MgCl$_2$, 1 mM PRib-PP and 60 µM guanine or hypoxanthine or 200 µM xanthine at 25°C.
[b]Specific activity values for hypoxanthine and xanthine were calculated based on the ratios of activity between guanine, hypoxanthine and xanthine in the crude lysate.

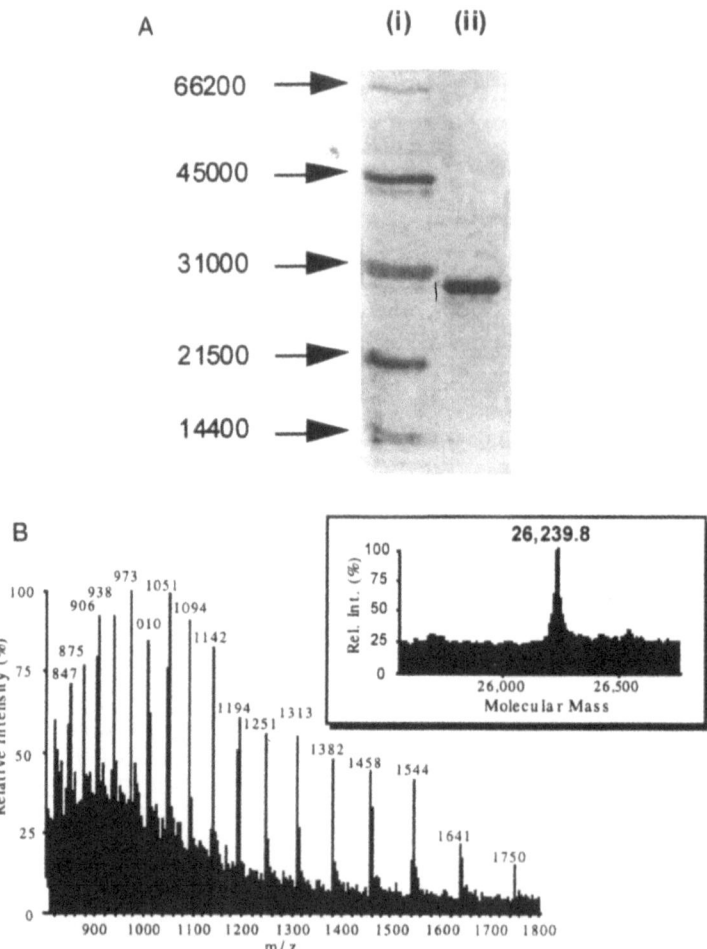

Figure 1. (A) SDS-PAGE of purified *P. falciparum* HPRT assessed by silver staining. (i) Low molecular weight standards; (ii) *P. falciparum* HPRT after affinity chromatography. (B) Electrospray mass spectrometry of purified recombinant *P. falciparum* HPRT.

Like the naturally occurring enzyme, purified recombinant *P. falciparum* HPRT was also found to utilise the three purine bases, guanine, hypoxanthine and xanthine as substrates (Table 1). This specificity differs from that of human and some parasite HPRTs (*C. fasciculata*[21], *T. brucei*[2], *T. cruzi*[3], *L. donovani*[22], *S. mansoni*[5] and *P. lophurae*[23]) which are only able to use hypoxanthine and guanine as substrates. Other parasite HPRTs known to use xanthine have been isolated from *T. foetus*, *E. tenella*, *G. lamblia* and *T. gondii*[4,6–8]. It may be possible to exploit this difference in substrate specificity for the design of inhibitors specific to *P. falciparum* HPRT and not the human enzyme so that the side effects of antimalarial drugs derived from such inhibitors could be minimized.

In this paper, we report the expression of recombinant *P. falciparum* HPRT in *E. coli* in milligram quantities and the purification of small amounts of homogeneous enzyme. It should now be possible to isolate the milligram quantities of *P. falciparum* HPRT required for kinetic and structural characterisation of the enzyme.

ACKNOWLEDGMENTS

The authors would like to thank Dr Ross Coppel from the Walter and Eliza Hall Institute, Melbourne, Australia for the gift of the plasmid pUC containing cDNA coding for *P. falciparum* HPRT. They are also grateful to Alun Jones for mass spectrometry measurements.

REFERENCES

1. Allen, T., Henschel, E. V., Coons, T., Cross, L., Conley, J. and Ullman, B. (1989). *Mol. Biochem. Parasitol.* **33**:273–282.
2. Allen, T. E. and Ullman, B. (1993). *Nucl. Acids Res.* **21**:5431–5438.
3. Allen, T. E. and Ullman, B. (1994). *Mol. Biochem. Parasitol.* **65**:233–245.
4. Beck, J. T. and Wang, C. C. (1993). *Mol. Biochem. Parasitol.* **60**:187–194.
5. Yuan, L., Craig, S. P., McKerrow, J. H. and Wang, C. C. (1990). *J. Biol. Chem.* **265**:13528–13532.
6. Wang, C. C. and Simashkevich, P. M. (1981). *Proc. Natl. Acad. Sci. USA.* **78**:6618–6622.
7. Aldritt, S. M. and Wang, C. C. (1986). *J. Biol. Chem.* **261**:8528–8533.
8. Schumacher, M. A., Carter, D., Roos, D. S., Ullman, B. and Brennan, R. G. (1996). *Nature Structural Biology.* **3**:881–887.
9. Xu, Y., Eads, J., Sacchettini, J. C., Grubmeyer, C. (1997). *Biochemistry* **36**:3700–3712.
10. Yuan, L., Craig, S. P., McKerrow, J. H. and Wang, C. C. (1992). *Biochemistry* **31**:806–810.
11. Eads, J. C., Scapin, G., Xu, Y., Grubmeyer, C. and Sacchettini, J. C. (1994). *Cell* **78**:325–334.
12. Vos, S., de Jersey, J. and Martin, J. L. (1997). *Biochemistry* **36**:4125–4134.
13. Somoza, J. R., Chin, M. S., Focia, P. J., Wang, C. C. and Fletterick, R. J. (1996). *Biochemistry* **35**:7032–7040.
14. Free, M. L., Gordon, R. B., Keough, D. T., Beacham, I. R., Emmerson, B. T. and de Jersey, J. (1990). *Biochim. Biophys. Acta.* **1087**:205–211.
15. Keough, D. T., McConachie, L. A., Gordon, R. B., de Jersey, J. and Emmerson, B. T. (1987). *Clin. Chim. Acta.* **163**:301–308.
16. Vasanthakumar, G., Davis, R. L., Sullivan, M. A. and Donahue, J. P. (1990). *Gene* **91**:63–69.
17. King, A. and Melton, D. W. (1987). *Nucl. Acids Res.* **15**:10469–10481.
18. Queen, S. A., Jagt, D. V. and Reyes, P. (1988). *Mol. Biochem. Parasitol.* **30**:123–134.
19. Shahabuddin, M. and Scaife, J. (1990). *Mol. Biochem. Parasitol.* **41**:281–288.
20. Reyes, P., Rathod, P. K., Sanchez, D. J., Mrema, J. E. K., Rieckmann, K. H. and Heidrich, H.-G. (1982). *Mol. Biochem. Parasitol.* **5**:275–290.
21. Jiang, Y., Allen, T. E., Carter, D., Ray, D. S. and Ullman, B. (1996). *Experimental Parasitol.* **82**:73–75.
22. Allen, T. E., Hwang, H., Jardin, A., Olafson, R. and Ullman, B. (1995). *Mol. Biochem. Parsitol.* **73**:133–143.
23. Schimandle, C. M., Mole, L. A. and Sherman, I. W. (1987). *Mol. Biochem. Parasitol.* **23**:39–45.

142

DETERMINATION OF THIOPURINE METHYLTRANSFERASE ACTIVITY IN ERYTHROCYTES USING 6-THIOGUANINE AS THE SUBSTRATE

T. Kröplin, N. Weyer, and H. Iven

Institute of Pharmacology
Medical University of Lübeck
Ratzeburger Allee 160
D-23538 Lübeck, Federal Republic of Germany

The antimetabolites 6-Thioguanine (6-TG), 6-Mercaptopurine (6-MP) and the pro-drug Azathioprine (Aza) are purine analogs and interfere with nucleic acid metabolism. They have antiproliferative and immunsuppressive properties and are therefore part of cytostatic protocols in the treatment of malignant haematologic diseases. In addition 6-MP and Aza are also used for the treatment of autoimmune diseases and after organ transplantation.

After uptake of 6-MP into the cell, three competitive metabolic pathways are known so far (Fig.1):

- Oxidation to 6-thiouric acid via xanthine oxidase
- Metabolism to thiopurine nucleotides by the hypoxanthine-guanine phosphoribo-syltransferase (HGPRT)
- S-methylation catalysed by the enzyme thiopurine methyl-transferase (TPMT)

The latter enzyme displays a genetic polymorphism which leads to a wide interindividual variation in enzyme activity. Population studies showed, that 89% have a high TPMT-activity, 11% have an intermediate TPMT-activity, however, 0.3% have a very low enzyme activity (1–3). This deficiency is associated with a severe Aza induced bone-marrow toxicity and cases with a lethal outcome have occurred. For this reason, all patients should be screened for TPMT-activity before starting Aza therapy to exclude those with TPMT-deficiency from Aza treatment.

Purine and Pyrimidine Metabolism in Man IX,
edited by Griesmacher *et al.* Plenum Press, New York, 1998.

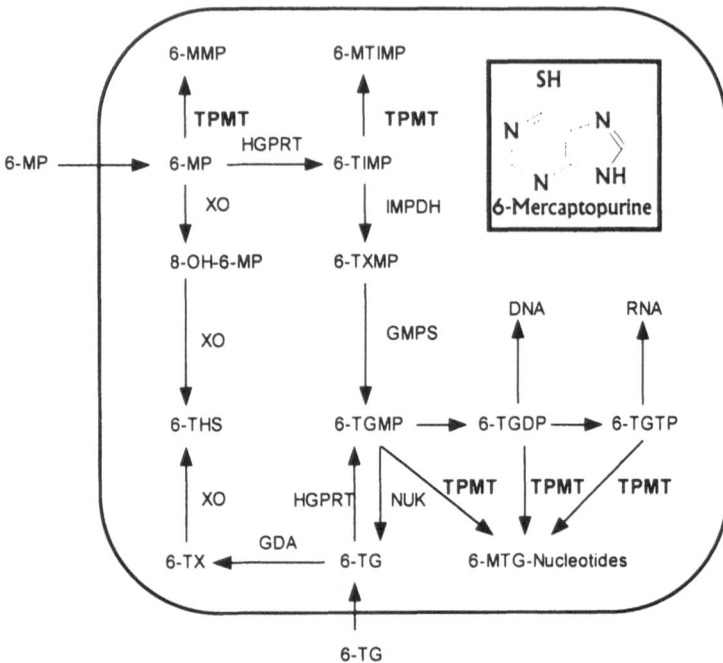

Figure 1. Proposed metabolism of 6-Mercaptopurine in nucleated cells TPMT: Thiopurine-Methyltransferase; XO: Xanthineoxidase; HGPRT: Hypoxanthine-Guanine Phosphoribosyltransferase; IMPDH: Inosinemonophosphate-ehydrogenase; GMPS: Guanosinemonophosphate-Synthetase; NUK: Nucleotidase; GDA: Guanine-desaminase; 6-MP: 6-Mercaptopurine; 8-OH-6-MP: 8-Hydroxy-6-Mercaptopurine; 6-TG (MP): 6-Thioguanine (monophosphate); 6-TX (MP): 6-Thioxanthine (monophosphate); 6-THS: Thiouric acid; 6-MMP: 6-Methyl-Mercaptopurine; 6-(M)TIMP: 6-(Methyl-) monophosphate; 6-MTG; 6-Methyl Thioguanine.

The established HPLC-assays for TPMT activity with 6-MP as substrate are cumbersome, e.g. with an acid hydrolysis step for 2 hours (4).Others use radioactive S-adenosyl-methionine (SAM) as methyl donor (5), however, the handling of radioactive substances is problematic for many laboratories.

We developed a new, non-radioactive TPMT assay, using 6-TG as the substrate and taking advantage of the high fluorescence of the reaction product 6-methyl-thioguanine for very sensitive quantitation.

MATERIALS AND METHODS

Blood Samples and Preparations

1 ml venous blood is collected in EDTA-tubes. After centrifugation, the plasma and the "buffy coat" are removed. Red blood cells (RBC) are washed once with an 0.9% sodium chloride solution. The packed RBCs are diluted 1:5 with 0.02 M phosphate buffer, pH 7.4, and then kept frozen at −30°C until the day of analysis. Hemoglobin content of the hemolysate is determined by the cyanmethemoglobin method (Sigma Diagnostics). One unit of enzyme activity represents the formation of 1 nmol of 6-methyl-thioguanine (6-MTG) per hour of incubation at 37°C.

Incubation Mixture and Incubations

50 µl of 600 µM/l 6-Thioguanine (Sigma, St. Louis)
+50 µl of 21 µM/l S-Adenosyl-Methionine (Sigma, St. Louis)
+100 µl 0.1 M KH_2PO_4, pH 7.4

The reaction is started by adding 50µl of RBC-lysate. The incubation is performed for 1 hour at 37°C in a thermo mixer (Eppendorf 5436) or shaking water bath.

Extraction

The reaction is stopped by adding 1 ml of chloroform/2-propanol (90/10; v/v).After shaking for 10min and centrifugation, 800 µl of the organic phase is evaporated to dryness and the residue resuspended in 200 µl of the mobile phase. Injection volumes are 30 µl.

Chromatographic Conditions

Column: RP-18 LichroCart 125-4 with guard column (E. Merck, Darmstadt)
Detection: Fluorescence, Ex 315 nm, Em 390 nm
Flow-rate: 1.2 ml/min
Retention time: ~ 5 min (Fig. 2)
Mobile phase: −0.05 M potassium dihydrogenphosphate (KH_2PO_4), acetonitrile, tetrahydrofuran (91/5/4; v/v/v), adjusted to pH 6.2

RESULTS

Whole blood samples kept at room temperature or at 4°C for up to 6 days showed no significant loss of TPMT activity. When the RBC-lysate was stored at room temperature, the TPMT activity already decreased after 24 hours. Stored at −30°C, the TPMT activity in RBC-lysate was stable over a period of several months.

Using the assay conditions given, TPMT-activity was found linear up to an incubation period of 90 min. We fixed the reaction time at 60 min. The enzyme activity was also linear with respect to the hemoglobin content of the RBC lysate up to 14 g Hb/dl.

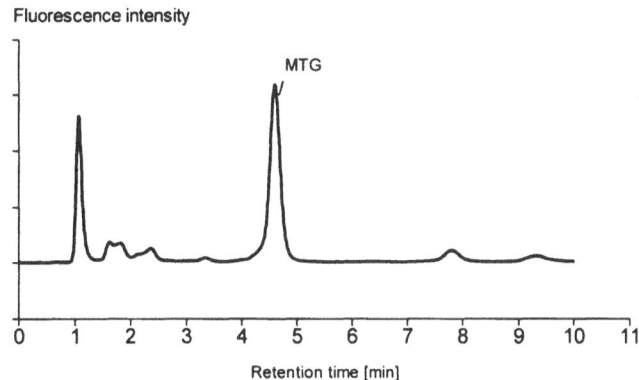

Figure 2. Typical chromatogram of 6-Methyl-Thioguanine under the conditions described above.

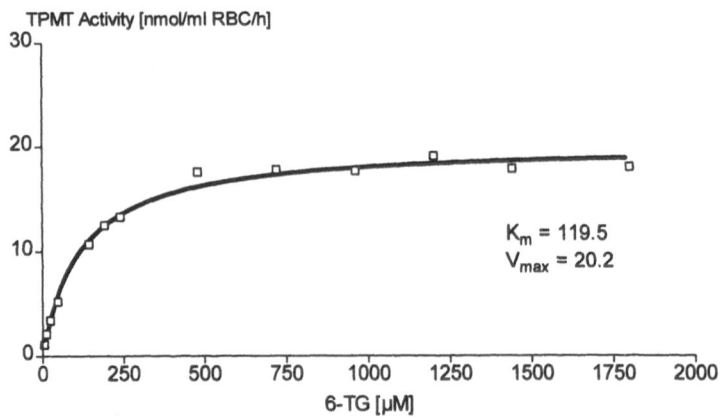

Figure 3. Substrate curve of the TPMT reaction.

In addition, there was no significant difference in enzyme activity observed over the pH range from 7.1 to 7.9 in the incubation mixture. The recovery of 6-MTG in the extraction procedure was calculated 50%.

The lower limit of detection in the HPLC- system is about 40 pg 6-MTG absolute. The amount of 6-MTG injected after incubation and extraction amounts to 1.2 ng absolute when TPMT activity is 20 Units/gHb.

The K_m and V_{max} for 6-MTG were calculated from quadruplicate experiments using the Lineweaver-Burk method.The calculated K_m values ranged from 118 to 155. We used 15 concentrations of 6-TG from 0 to 2000 µmol/l at a constant concentration of 21 µmol/l of SAM (Fig. 3). For the assay we fixed the substrate concentration at 600 µmol/l 6-TG. This is the 5-fold K_m value and close to the V_{max} of the reaction.

Calibration curves were made by spiking substrate blank incubates with 5 concentrations (6.25 to 75 ng / 250 µl) of 6-MTG. They were treated in parallel with the samples. The curves were linear with a $r^2 > 0.98$. The intra assay C.V. is 5%, n=20. In every TPMT incubation a reference sample for quality control was included. The inter assay C.V. of this control sample was 6.7%, n=36.

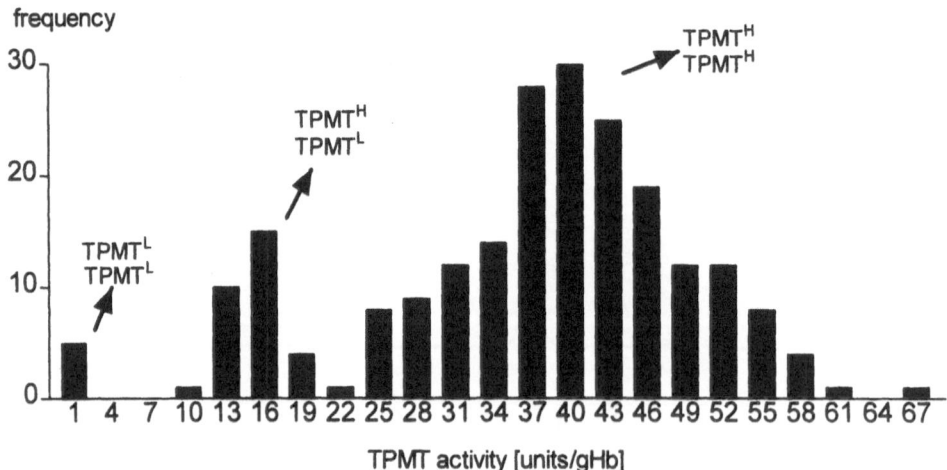

Figure 4. Frequency distribution of erythrocyte thiopurine methyltransferase in 218 blood donors.

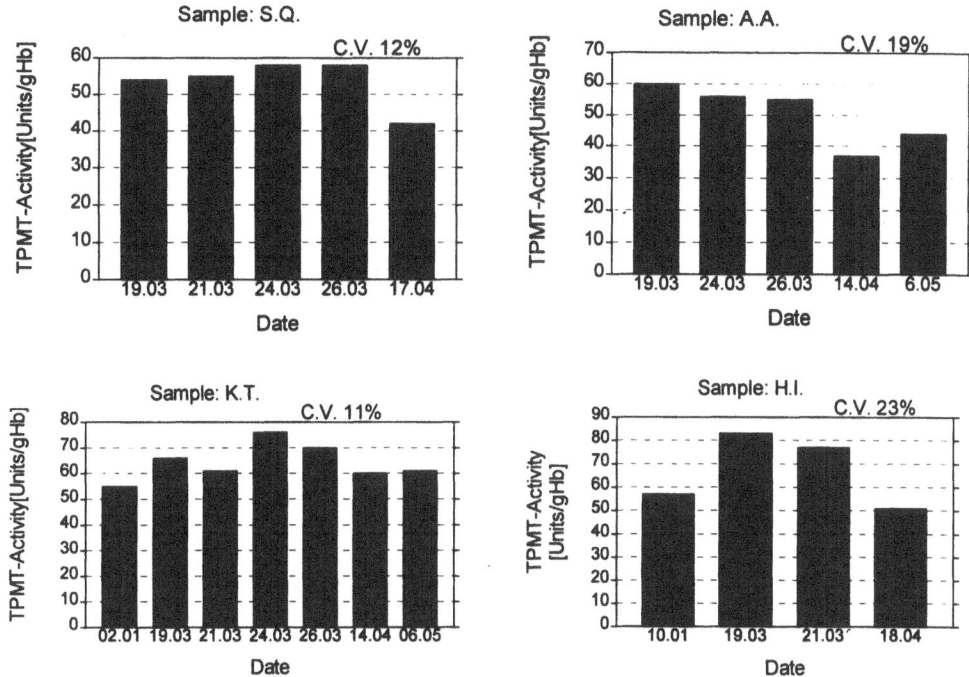

Figure 5. Variability of TPMT -activity in healthy persons.

TPMT-activity of 218 Lübeck blood donors showed the expected bimodal distribution (Fig. 4). However, five of our blood donors had a very low TPMT-activity and would be at high risk of myelosuppression when treated with 6-MP or AZA.

The TPMT activity seems not to be constant in time. Figure 5 shows the enzyme-activities of 4 healthy members of our department, who were without any medication. With respect to the observed variability of the TPMT activity in the same subject, the consequences of intermediate or high activity for the individual therapy remains unclear. But there is no doubt, that the screening of patients scheduled for Aza treatment is important to exclude the TPMT deficient from thiopurine therapy.

REFERENCES

1. Weinshilboum RM, Sladek SL. Mercaptopurine pharmacogenetics: monogenetic inheritance of erythrocyte thiopurine methyltransferase activity. Am J Hum Genet 1980; 32: 651–62.
2. Tinel M, Berson A, Pessayre D, Letteron P, Cattoni MP, Horsmans Y, Larrey D. Pharmacogenetics of human erythrocyte thiopurine methyltransferase activity in a French population. Br J Clin Pharmacol 1991; 32: 729–34.
3. McLeod HL, Lin J-S, Pui C-H, Scott EP, Evans WE. Thiopurine methyltransferase activity in American white and black subjects. Clin Pharmacol Ther 1994; 55: 15–20.
4. Lynne Lennard, Helen J. Singleton. High-performance liquid chromatographic assay of human red blood cell thiopurine methyltransferase activity. J of Chromatography B 1994; 661: 25–33.
5. Weinshilboum RM, Raymond FA, Pazmino PA. Human erythrocyte thiopurine methyltransferase: radiochemical microassay and biochemical properties. Clin Chim Acta 1978; 85: 323–33.

143

ANTIVIRAL ACYCLIC NUCLEOSIDE PHOSPHONATE ANALOGUES AS INHIBITORS OF PURINE NUCLEOSIDE PHOSPHORYLASE

E. Kulikowska,[1] A. Bzowska,[1] A. Holy,[2] L. Magnowska,[1] and D. Shugar[1,3]

[1]Department of Biophysics
Institute of Experimental Physics
University of Warsaw, Poland
[2]Institute of Organic Chemistry and Biochemistry
Academy of Sciences of the Czech Republic
Praha, Czech Republic
[3]Insitute of Biochemistry and Biophysics
Polish Academy of Sciences
Warsaw, Poland

1. INTRODUCTION

Purine nucleoside phosphorylase (PNP, E.C.2.4.2.1) catalyses reversible phosphorolysis of purine ribo- and deoxyribonucleosides. Mammalian phosphorylases are specific for 6-ketopurines while some bacterial PNPs (e.g. from *E. coli* and *S. typhimurium*) have very broad specificty.[1–4] Genetic deficiency of PNP leads to a loss of cellular immunity.[5] The potential clinical applications of PNP inhibitors include treatment of T-cell leukemias, suppression of the host-*vs*-graft response in organ transplantation and potentation of the chemotherapeutic action of PNP-cleavable nucleosides.[6]

A new class of PNP inhibitors comprises phosphonylalkoxyalkyl analogues of nucleotides[7] which are broad-spectrum antiviral agents with potent and selective *in vitro* and *in vivo* activity *vs* DNA viruses and retroviruses.[8] It has been suggested that cellular PNP may play a role in the cytotoxic action of these compounds.[7]

2. MATERIALS AND METHODS

The acyclic phosphonate analogues were prepared by earlier described procedures.[9] All other compounds and enzymes (human erythrocyte and calf spleen PNPs, xanthine

Purine and Pyrimidine Metabolism in Man IX,
edited by Griesmacher *et al.* Plenum Press, New York, 1998.

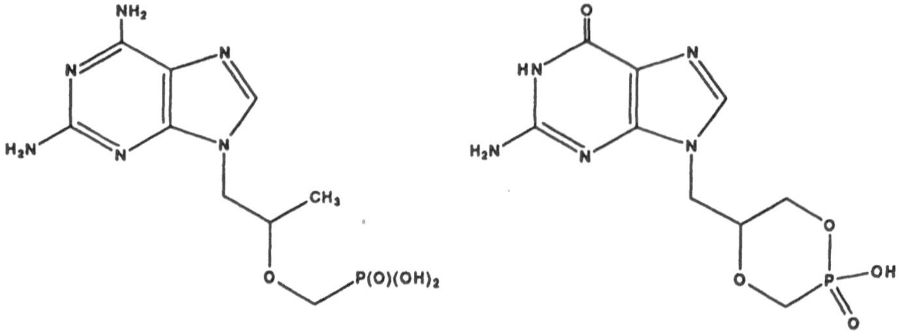

Scheme 1. Structures of acyclic phoshonate analogues. Left: 9-(2-phosphonylmethoxypropyl)-2, 6-diaminopurine, PMPDAP, (S) enantiomer (9). Right: cyclic 9-(3-hydroxy-2-phosphonylmethoxypropyl)-guanine, cyclic HPMPG, (R) and (S) enantiomers, compounds (3) and (4), respectively.

oxidase) were commercial products with the exception of PNP from *E.coli* (a gift of Dr. G. Koszalka, Wellcome Research Laboratories). Initial screening of inhibitors was carried out in the presence of 50 mM or 1 mM phosphate buffer, pH 7, at 25°C with 7-methylguanosine (m^7G) as substrate. Inhibition constants were obtained in the presence of 1 mM phosphate in 50 mM Hepes buffer pH 7.

Phosphorolysis and inhibition of phosphorolysis of inosine and m^7G were followed by the coupled xanthine oxidase procedure[10] and direct spectrophotometric method[11], respectively. With human erythrocyte PNP initial velocity method was employed, while for the calf phosphorylase continuous monitoring assay was used.[11] With the initial velocity the kinetic parameters $(V_{max}/K_m)^{app}$ and V_{max}^{app} were determined by linear regression analysis from slopes and intercepts of Lineweaver-Burk plots of $1/v_o$ *vs* $1/c_o$.[12] With continuous monitoring, 10–20 experimental points, taken from the curve representing the total course of phosphorolysis, were fitted by a weighted linear least-squares procedure to the integrated form of the Michaelis-Menten equation[12,13]. Inhibition constants were then determined using the kinetic parameters for substrate phosphorolysis as standards. K_i^{app} values were calculated from $(V_{max}/K_m)_{app}$, from the equation for competitive inhibition:

$$K_i^{app} = I / [(V_{max}/K_m)_o^{app}/(V_{max}/K_m)^{app} - 1]$$

where $(V_{max}/K_m)^{app}$ and $(V_{max}/K_m)_o^{app}$ are parameters for phosphorolysis in the presence and absence of inhibitor, respectively, and I is inhibitor concentration.[12]

To demonstrate competition between inosine and (S)-PMPDAP (9) data were analysed by both Dixon plot of $1/v_o$ *vs* I, and replots *vs* I of slopes and slopes/intercepts of the Lineweaver-Burk plot, and inhibiton constants were determined from all three plots.[12]

3. RESULTS AND DISCUSSION

A study has been made of inhibition of phosphorylases from human erythrocytes, calf spleen and *E. coli*, by a series of acyclic phosphonate analogues: 9-(3-hydroxy-2-phosphonylmethoxy)propyl-, (HPMP), cyclic-HPMP, 9-(2-phosphonylmethoxy)propyl-, (PMP), 9-(2-phosphonylmethoxy)ethyl-, (PME) and 9-(2-phospho-nylethoxy)ethyl-, (PEE), derivatives of guanine (G), 2,6-diaminopurine, (DAP), 2-aminopurine (AP), adenine (A) and of some analogues.

Table 1. Initial screening of inhibitory activities of a series of purine 9-phosphonylalkoxyalkyl derivatives. Preliminary experiments with m^7G (100 µM) as substrate and PNP from human erythrocytes, calf spleen and *E. coli*, in 50 and 1 mM phosphate buffer pH 7.0, 25°C. v_o and v_i - initial velocities of the reactions without and with inhibitor, respectively. Inhibitor concentration ~100 µM, if not otherwise indicated

Compound		Human erythrocytes PNP (A) c_{Pi}[mM]		Calf spleen PNP (B) c_{Pi}[mM]		*E. coli* PNP (C) c_{Pi}[mM]	
Phosphonyl-alkoxyalkyl moiety	Base	50	1	50	1	50	1
1 (R)-HPMP	G	79	8	55	4	74	>100
2 (S)-HPMP	G	100	56	51	18	84	10
3 cyclic (R)-HPMP	G	71	7	64	20	78	>100
4 cyclic (S)-HPMP	G	85	38	76	14	84	>100
5[a] PEE	G	91	32	53	26	>100	84
6 (S)-PMP	G	98	102	66	54	88	99
7[b] PME	G (N7)			58	60	85	93
8[a] PME	s⁶G			90	62	92	73
9 (S)-PMP	DAP			82	13	82	91
10 PEE	DAP	82	41	82	18	95	81
11 (S)-HPMP	DAP	100	43	88	10	94	70
12 PME	DAP			100	85	95	61
13 PME	AP			60	100	92	93
14[c] cyclic (S)-HPMP	A	94	81	60	70	>100	>100
15[a] cyclic (S)-HPMP	3-deaza-A			>100	82	>100	>100
16[a] cyclic (S)-HPMP	1-deaza-A			58	71	>100	>100

a,b,c - Inhibitor concentrations ~120 µM, ~290 µM and ~140 µM respectively.

Table 1 depicts results of initial screening for inhibitory activity of the compounds in 50 mM and 1 mM phosphate buffer with m^7G as substrate. It will be seen from Table 1 that the inhibition patterns for phosphorylysis of m^7G, catalyzed by both mammalian enzymes, are similar, and that the activities of the analogues are higher at the lower phosphate concentration. Interestingly, all compounds are virtually inactive *vs E. coli* PNP (Table 1). This is in striking contrast to formycins A and B, and especially N^6-methylformycin A, which are good inhibitors of the *E. coli,* but not of the mammalian, enzymes.[14]

Table 2 shows K_i^{app} values for some G and DAP derivatives with both mammalian enzymes and m^7G as a substrate. Surprisingly, (S)-PMPDAP (9, see Scheme 1) proved to be the most potent inhibitor of human PNP (K_i 0.8 µM), although 2,6-diaminopurine riboside is an inhibitor of *E. coli,* but not mammalian, PNP.[15]

Figures 1A,B,C and D depict Dixon and Lineweaver-Burk plots, as well as replots of slopes and slopes/intercepts of the latter, for inhibition of inosine phosphorolysis by (S)-PMPDAP (9) catalyzed by human PNP. The plots demonstrate competitive inhibition (with K_{iapp} values 1.6, 1.3 and 1.2 µM from Dixon plot, slope *vs* I, and slope/intercepts *vs* I, respectively, as compared to a value of 1.4 µM from Table 2, calculated from equation for competitive inhibition), notwithstanding that the base does not possess 6-keto substituent and N(1) proton known to be necessary for binding by mammalian PNPs.[4] Inhibition patterns for all seven analogues with the calf enzyme and m^7G as substrate (Table 2B) were also consistent with a simple competitive model.

The increase in inhibitory potency of the analogues on lowering the phosphate concentration (Table 1A,B) suggests that they may be bisubstrate analogue inhibitors which are bound by PNPs *via* both the base and the phosphate-like moiety.[16]

Table 2. Apparent inhibition constants (K_i^{app} calculated from V_{max}/K_m from equation for a competitive inhibition, see Methods) of some purine 9-phosphonylalkoxyalkyl derivatives with human erythrocytes and calf spleen PNP and with m^7G as a substrate in 1mM phosphate and 50 mM Hepes buffers, pH 7, 25°C

| | | | K_i^{app} [µM] | |
| | Compound | | A[a] Human | B[a] Calf |
	Phosphonylalkoxyalkyl moiety	Aglycone	erythrocyte PNP	spleen PNP
1	(R)-HPMP	G		4
2	(S)-HPMP	G		~8
3	cyclic (R)-HPMP	G	~2	4
4	cyclic (S)-HPMP	G	11	~7
9	(S)-PMP	DAP	0.81.4[b]	~6
10	PEE	DAP		17
11	(S)-HPMP	DAP	1514[b]	24

[a] Error ≤ 25% (A) and ≤ 30% (B) unless otherwise indicated (~).
[b] With inosine as substrate.

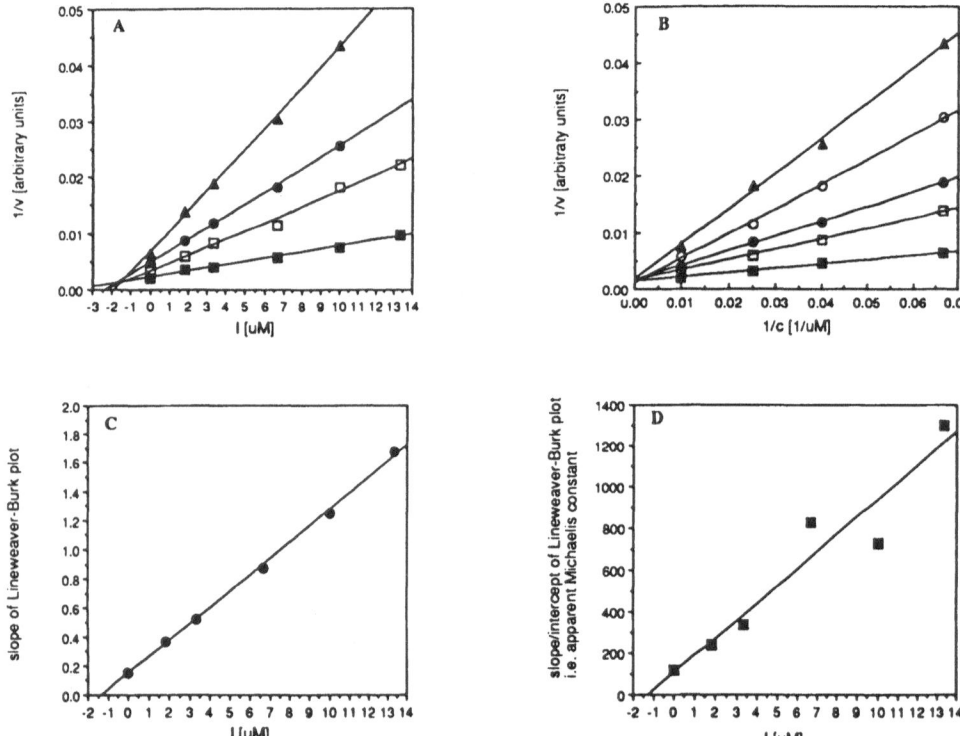

Figure 1. Inhibition of inosine phosphorolysis with human PNP by (S)-PMPDAP (9). A. Dixon plot of $1/v_o$ *vs.* inhibitor concentration, I inosine concentrations: (■) 15 µm, (□) 25 µm, (●) 40 µm, and (▲) 100 µm. B. Lineweaver-Burk plot of $1/v_o$ *vs.* $1/c_o$ inhibitor concentrations: (■) 0 µm, (□) 1.78 µm, (●) 3.34 µm, (○) 6.68 µm, and (▲) 10.02 µm. C. Slopes of the Lineweaver-Burk plots *vs.* inhibitor concentration, i.e. $(K_m/V_{max})^{app}$ *vs.* I. D. Slopes/intercepts of the Lineweaver-Burk plots *vs.* inhibitor concentration, i.e. $(K_m)^{app}$ *vs.* I.

Although cyclic (S)-HPMPA (14) is only slightly active (Table 1B), the HPMP moiety (both enantiomers, cyclic and linear) appears to have good affinity for both mammalian enzymes (especially when linked to G, cf. compounds (1), (2), (3) and (4), Table 2, Scheme 1). It is of interest, in this context, that the racemate of HPMPG was reported to be a rather feeble inhibitor of the enzymes from human cancer tissues, from L1210 mouse leukemic cells and from diverse organs of the rabbit.[7]

In the case of human enzyme enantiomers differ markedly in their inhibitory activity (compare (1) and (2) from Table 1A, and (3) and (4) from Table 2A). For calf phosphorylase smaller differences between enantiomers were observed (compare (1) and (2), (3) and (4) from Table 2B).

4. CONCLUSIONS

i. Several HPMP and PMP derivatives of G and DAP are specific inhibitors of mammalian, but not E. coli, PNPs, in contrast to formycins A and B, specific inhibitors of the E. coli enzyme.

ii. Inhibition patterns for human erythrocytic and calf spleen enzymes are similar.

iii. With the calf spleen enzyme, kinetics of inhibition, with μM K_i values, are consistent with simple competition vs nucleoside substrate. With inosine as substrate, (S)-PMPDAP was a competitive inhibitor of the human enzyme, notwithstanding that it does not have a proton at N(1).

iv. The compounds appear to be bisubstrate analogue inhibitors.

v. In some cases enantiomers differ several-fold in their inhibitory activity.

ACKNOWLEDGMENTS

Supported by the Polish Committee for Scientific Research (KBN grant 6 P04A 043 12), by an International Research Scholar's award from the Howard Hughes Medical Institute (HHMI 75195-543401), and by the Grant Agency of the Czech Republic (grant 203/96/K001).

REFERENCES

1. J.D. Stoeckler (1984) in Developments in Cancer Chemotherapy (R.I. Glazer, ed.) 35–60, CRC Press, Boca Raton, FL.
2. J.A. Montgomery (1993) Med. Res. Rev. 13, 209–228, John Wiley and Sons.
3. K.J. Jensen and P. Nygaard (1975) Eur. J. Biochem. 51, 253–265.
4. A. Bzowska, E. Kulikowska and D. Shugar (1990) Z. Naturforsch. 45c, 59–70
5. Giblett, E.R., Amman, A.J., Wara, D.W., Sandman, R. and Diamond, L.K. (1975) Lancet 1, 1010–1013.
6. Stoeckler, J.D., Ealick, S.E., Bugg, C.E. and Parks, R.E. Jr (1986) Fed.Proc. 45, 2773–2778.
7. Sediva, K., Ananiev, A.V., Votruba, I., Holy, A., Rosenberg, I. (1991) Int.J.Purine Pyrimidine Res. 2, 35–39.
8. Naesens, L., Snoeck, R., Andrei, G., Balzarini, J., Neyts, J. and De Clercq, E. (1997) Antiviral Chem.Chemother. 8, 1–23.
9. (a)Holy, A. and Rosenberg, I. (1987) Collect.Czech.Chem.Commun. 52, 2775–2791, (b) Holy, A. and Masojidkova, M. (1987) Collect.Czech.Chem.Commun., 52, 1196–1212, (c) Holy, A. and Rosenberg, I. (1987) Collect.Czech.Chem.Commun. 52, 2801–2809.
10. Kalckar, H.M. (1947) J.Biol.Chem. 167, 477–486.

11. Kulikowska, E., Bzowska, A., Wierzchowski, J., Shugar, D. (1986) Biochim.Biophys.Acta 874, 355–363.
12. Segal, I.H. (1975) Enzyme Kinetics, John Wiley and Sons, New York.
13. Wierzchowski, J., Lassota, P. and Shugar, D. (1984) Biochim.Biophys.Acta 786, 170–178.
14. Bzowska, A., Kulikowska, E. and Shugar, D. (1992) Biochim.Biophys.Acta 1120, 239–247.
15. Wielgus-Kutrowska, B., Kulikowska, E., Wierzchowski, J., Bzowska, A. and Shugar, D. (1997) Eur.J.Biochem 243, 408–414.
16. Krenitsky, T.A., Tuttle, J.W., Miller, W.H., Moorman, A.R., Orr, G.F. and Beauchamp, L. (1990) J.Biol.Chem 265, 3066–3069.

144

METABOLISM AND ANTI-HIV ACTIVITY OF PHOSPHORAMIDATE DERIVATIVES OF D4T-MP WITH VARIATIONS IN THE AMINO ACID MOIETY

L. Naesens,[1] D. Cahard,[2] A. Salgado,[2] L. Bidois,[2] E. De Clercq,[1]
C. McGuigan,[2] and J. Balzarini[1]

[1]Rega Institute for Medical Research
Katholieke Universiteit Leuven
B-3000 Leuven, Belgium
[2]Welsh School of Pharmacy
University of Wales
Cardiff, CF1 3XF, United Kingdom

1. ABSTRACT

The metabolism of different phosphoramidate prodrugs of d4T-MP, in which the phosphate group is linked to a phenyl group and the alkyl ester of an amino acid was studied in crude CEM cell extracts. Significant (80–100%) conversion to the amino acyl d4T-MP metabolite was obtained with derivatives containing L-alanine or methyl-L-aspartic acid. A lower degree of conversion was seen with derivatives containing L-phenylalanine, L-methionine, methyl-L-glutamic acid or L-leucine. Derivatives containing D-alanine, β-alanine, glycine, L-valine or L-lactate showed no conversion to the amino acyl d4T-MP metabolite. Overall, there was a close correlation between the anti-HIV activity of these prodrugs and their conversion rate to the amino acyl d4T-MP metabolite. Our data suggest that the enzymes involved in the formation of the amino acyl d4T-MP metabolite have a rather stringent specificity for L-alanine as the amino acid moiety. In addition, these enzymes were found to be markedly species-dependent, their activities being highest in mouse serum, followed by guinea pig serum, but only minimal in human serum. Mouse serum therefore appears to be the medium of choice to isolate and identify the enzymes that are involved in the metabolism of these phosphoramidate prodrugs.

2. INTRODUCTION

The intracellular activation of the anti-HIV compound 2',3'-dideoxy-2',3'-didehydrothymidine (d4T) is markedly impeded by the low efficiency of the first phosphorylation

Purine and Pyrimidine Metabolism in Man IX,
edited by Griesmacher *et al.* Plenum Press, New York, 1998.

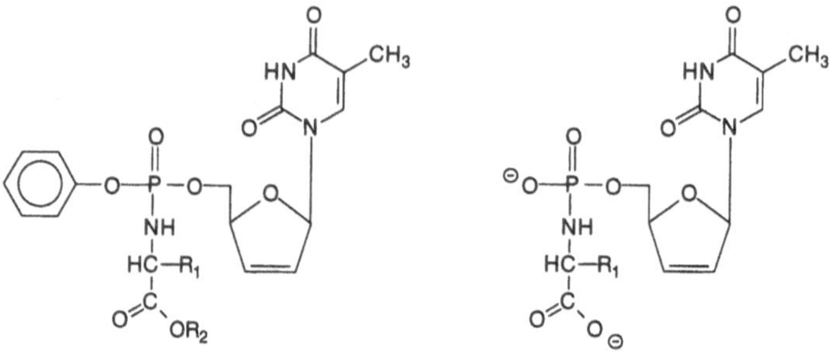

Figure 1. Basic structure of the d4T-MP phosphoramidate prodrugs (left) and their amino acyl d4T-MP metabolite (right). R_1 is an amino acid-specific substituent while R_2 is an alkyl group (methyl, ethyl or benzyl) (cfr. Table 1 for individual compounds).

step, catalyzed by thymidine kinase. This can be circumvented by the design of membrane-soluble prodrugs of d4T-MP.[1] In our approach, one acid function of the phosphate group of d4T-MP is esterified to a phenyl group, while the other is attached via a phosphoramidate bridge to an amino acid, of which the free carboxyl group is linked to an alkyl group (methyl, ethyl or benzyl) (Figure 1). We have previously reported the anti-HIV activity and metabolism of the prototype compound So-324, that contains L-alanine as the amino acid moiety.[2,3] The optimal antiviral activity of So-324 in thymidine kinase-deficient CEM cells was found to be based on the significant metabolism of So-324 to alaninyl d4T-MP, that shows marked intracellular accumulation, thus acting as a depot form for d4T-MP. In the study presented here, we determined whether the formation of the amino acyl d4T-MP metabolite is a prerequisite for the phosphoramidate prodrugs to be antivirally active.

3. MATERIALS AND METHODS

3.1. Compounds

The d4T-MP phosphoramidate derivatives were synthesized according to the methodology described elsewhere.[3]

3.2. Anti-HIV Activity

Activity against HIV-1 or HIV-2 was determined in human T-lymphoid CEM cells (CEM/0) or in a thymidine-kinase deficient cell line (CEM/TK⁻) derived thereof. Antiviral efficacy was expressed as EC_{50} or compound concentration that results in 50% inhibition of virus-induced cytopathicity, as determined by microscopic inspection of the infected cell cultures.

3.3. Metabolism of the Test Compounds in CEM Cell Extract or in Animal Serum

Crude CEM cell extracts were prepared by spinning down exponentially growing CEM/0 cells, followed by sonication and clarification (25,000 g; 15 min; 4°C). To study the metabolism of the phosphoramidate prodrugs in CEM cell extract or serum (i.e., human, mouse, guinea pig, or heat-inactivated fetal calf serum), the compounds were incubated at

37°C, at a concentration of 200 µM, in crude cell extract or serum containing 10 mM $MgCl_2$. At the end of the incubation period (0, 15, 20 or 30 min; or 2 or 6 hr), the reaction mixture was extracted with ice-cold methanol, centrifuged and analyzed by HPLC. Samples were separated on a strong-anion exchange column (Partisphere® SAX from Whatman), to determine the metabolites amino acyl d4T-MP [retention time (RT): 17 min] and d4T-MP [RT: 10 min]. In parallel, the samples were analyzed on a C8 reverse phase column (LiChrospher®60 RP-Select B from Merck), to determine the parent compound [RT: 17–20 min], d4T [RT: 9 min], and any other intermediate metabolites.

4. RESULTS AND DISCUSSION

4.1. Thymidine Kinase Independence of Antiviral Activity of the Test Compounds

Significant activity against HIV-1 and HIV-2 was observed for the d4T-MP phosphoramidate prodrugs containing either L-alanine, L-phenylalanine, L-leucine, L-methionine, or methyl-L-aspartic acid (EC_{50} 1–2 µM) (Table 1). These compounds fully retained their anti-HIV activity in thymidine kinase-deficient CEM/TK⁻ cells in contrast to the parent compound d4T, that proved to be highly dependent on thymidine kinase (EC_{50}: 0.8 and 25 µM in CEM/0 and CEM/TK⁻ cells, respectively). These data prove that the phosphoramidate prodrugs are successful in directly delivering d4T-MP inside the cells, thus bypassing the activation by thymidine kinase. The antiviral efficacy of the prodrugs proved to be markedly dependent on the nature of the amino acid moiety. The highest antiviral potency (EC_{50} < 0.1 µM) was noted for derivatives containing L-alanine, while "unnatural" substitutions such as D-alanine, β-alanine, or L-lactate resulted in a significant reduction or even total loss of antiviral activity (EC_{50}: 3 µM, 250 µM and 40 µM, respectively). The alkyl group linked to the amino acid's carboxyl group (R_2) only had a minor impact on the antiviral efficacy (Table 1: compare compounds So-324, So-730 and Cf-951).

Table 1. Anti-HIV activity of d4T-MP phosphoramidate derivatives in wild-type and thymidine kinase-deficient CEM cells

			EC_{50} (µM)		
Compound	Amino acid	alkyl (R_2)	HIV-1 in CEM/0	HIV-2 in CEM/0	HIV-2 in CEM/TK
So-324	L-alanine	methyl	0.08	0.08	0.08
So-730	L-alanine	benzyl	0.02	0.02	0.06
Cf-951	L-alanine	ethyl	0.10	0.07	0.07
Cf-863	L-phenylalanine	methyl	0.80	1.4	0.33
Cf-932	D-alanine	methyl	3.0	2.0	2.5
Cf-950	β-alanine	methyl	250	250	250
Cf-866	glycine	methyl	6.0	6.0	7.0
Cf-978	L-lactate	methyl	40	50	250
Cf-864	L-leucine	methyl	1.1	2.2	0.4
Cf-865	L-valine	methyl	13	13	4.0
Cf-867	L-methionine	methyl	0.6	0.8	0.34
Cf-1078	methyl-L-aspartic acid	methyl	0.6	0.7	0.3
Cf-1081	methyl-L-glutamic acid	methyl	8.0	5.3	1.6
d4T	–	–	0.8	0.8	25

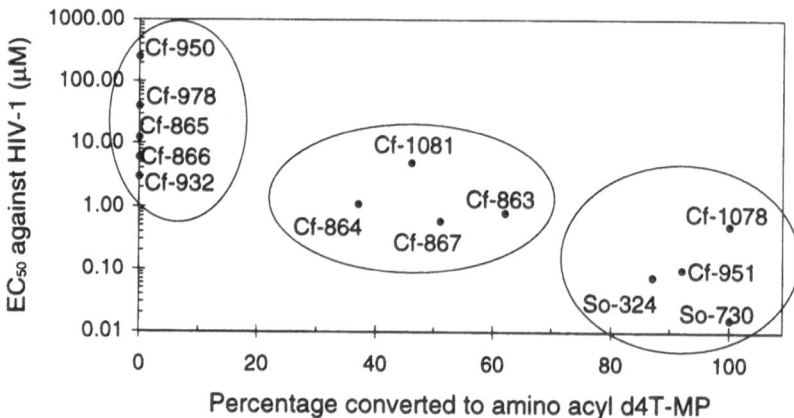

Figure 2. D4T-MP phosphoramidate prodrugs: correlation between anti-HIV activity (determined in CEM/0 cells) and conversion rate to the amino acyl d4T-MP derivative upon 2 hours incubation in crude CEM cell extract. See Table 1 for structure of the compounds.

4.2. Metabolism in Crude CEM Cell Extract

Several d4T-MP phosphoramidate prodrugs were extensively metabolized to the amino acyl d4T-MP metabolite upon incubation with crude CEM cell extract. The conversion rate was markedly dependent on the amino acid moiety (Figure 2). Three subgroups could be distinguished: non-converters (no conversion to the amino acyl d4T-MP metabolite after 2 hours incubation): derivatives containing β-alanine, L-lactate, L-valine, glycine or D-alanine, low-converters (conversion rate between 40 and 60%): derivatives containing methyl-L-glutamic acid, L-leucine, L-phenylalanine or L-methionine, and high-converters (80–100% conversion) derivatives containing L-alanine or methyl-L-aspartic acid. As can be seen from Figure 2, there was a nice correlation beween the anti-HIV activity of the d4T prodrugs and their ability to convert to the amino acyl d4T-MP metabolite. This strongly supports the crucial role of the amino acyl d4T-MP metabolite in the metabolic activation of the phosphoramidate prodrugs. The finding that this correlation was stronger for a 2 hr than for a 20 min incubation period suggests that some prodrugs are converted rather slowly (data not shown). The complete metabolic activation pathway of the d4T phosphoramidate prodrugs remains to be resolved. HPLC analysis of the CEM cell extract-treated d4T prodrugs revealed the presence of unknown metabolites, which, based on their HPLC characteristics, were tentatively identified as "intermediate" metabolites in which either the alkyl ester group (at the amino acid moiety) or the phenyl ester group at the phosphate moiety had been split off (Figure 1). Further characterization of these metabolites, and the enzymes involved, is underway. Other groups have studied slightly different phosphoramidate prodrugs of nucleoside analogues, and suggested an activation pathway via carboxyesterases, phosphodiesterases, or ribonucleoside phosphoamidase.[4,5]

4.3. Metabolism in Animal Sera

The rate of metabolism of the d4T-MP phosphoramidate prodrugs upon incubation in animal sera proved to be markedly species-dependent. Considerable metabolism was seen in mouse serum. For instance, the L-alanine derivatives So-730, So-324 and Cf-951 were completely converted to the L-alaninyl d4T-MP after only 15 min incubation in mouse serum.

No amino acyl d4T-MP was formed from compounds containing β-alanine, L-valine or L-lactate. While the distribution between non-, medium- and high-converters was similar for mouse serum and CEM cell extract, the conversion rate was markedly faster in mouse serum than in CEM cell extract. In contrast, human serum was found to be virtually inert; all prodrugs tested were stable during as long as 6 hours incubation in human serum. Serum from guinea pigs or cats showed a metabolism rate that was intermediate between mouse and human. This species-dependent metabolism was also observed in antiviral efficacy studies in different cell types,[3] and complicates the choice of an animal model to study the pharmacokinetics or *in vivo* antiviral efficacy of these phosphoramidate compounds. Also, it is clear that metabolism studies in isolated serum cannot entirely predict the situation in the intact animal, where metabolism involves whole tissues such as the liver.

5. CONCLUSION

Phosphoramidate derivatives of d4T-MP represent a useful approach to bypass the inefficient thymidine kinase-dependent phosphorylation of d4T. This was clearly demonstrated in our antiviral efficacy studies in thymidine kinase-deficient CEM cells. Whether these d4T prodrugs will be useful in the clinical setting, will mainly depend on their pharmacological properties. In the present study, we observed a clear correlation between antiviral activity of the prodrugs, and their ability to convert to the amino acyl d4T-MP metabolite. This finding points to the crucial role of this metabolite in the activation pathway, as demonstrated previously in our studies on the metabolism of the L-alanine derivative (So-324) in CEM cells. The favorable antiviral and biochemical properties of So-324 and the closely related derivatives So-730 and Cf-951, make these prodrugs interesting derivatives for further evaluation of their therapeutic potential in animal models for HIV infection. However, the complex metabolic pathway of the phosphoramidate derivatives still needs further investigation, the main issues being: characterization of the enzymes involved, metabolism in different tissues (such as the liver), and metabolism of phosphoramidate derivatives of dideoxynucleoside analogues other than d4T, such as d4A, ddA, AZT, 3TC.[6-10]

REFERENCES

1. McGuigan, C., Cahard, D., Sheeka, H. M., De Clercq, E. and Balzarini, J. 1996. J. Med. Chem., 39, 1748–1753.
2. Balzarini, J., Karlsson, A., Aquaro, S., Perno, C.-F., Cahard, D., Naesens, L., De Clercq, E. and McGuigan, C. 1996. Proc. Natl. Acad. Sci. USA, 93, 7295–7299.
3. Balzarini,, J., Egberink, H., Hartmann, K., Cahard, D., Vahlenkamp, T., Thormar, H., De Clercq, E. and McGuigan, C. 1996. Mol. Pharmacol., 50, 1207–1213.
4. Valette, G., Pompon, A., Girardet, J.-L., Cappellaci, L., Franchetti, P., Grifantini, M., La Colla, P., Loi, A. G., Périgaud, C., Gosselin, G. and Imbach, J.-L. 1996. J. Med. Chem., 39, 1981–1990.
5. Winter, H. Maeda, Y., Mitsuya, H. and Zemlicka, J. 1996. J. Med. Chem., 39, 3300–3306.
6. McGuigan, C., Pathirana, N., Mahmood, N., Devine, K. G. and Hay, A. J. 1992. Antiviral Res., 17, 311–321.
7. McGuigan, C., Wedgwood, O. M., De Clercq, E. and Balzarini, J. 1996. Bioorg. Med. Chem. Letters, 6, 2359–2362.
8. Balzarini, J., Kruining, J., Wedgwood, O., Pannecouque, C., Aquaro, S., Perno, C.-F., Naesens, L., Witvrouw, M., Heijtink, R., De Clercq, E. and McGuigan, C. 1997. FEBS Letters, in press.
9. McGuigan, C., Cahard, D., Sheeka, H. M., De Clercq, E. and Balzarini, J. 1996. Bioorg. Med. Chem. Letters, 6, 1183–1186.
10. Balzarini, J., Wedgwood, O., Kruining, J., Pelemans, H., Heijtink, R., De Clercq, E. and McGuigan, C. 1996. Biochem. Biophys. Res. Commun., 225, 363–369.

145

DETERMINATION OF PURINE ENZYME ACTIVITIES IN HUMAN ERYTHROCYTES BY CAPILLARY ELECTROPHORESIS

Tomáš Adam,[1,2] Juraj Ševčík,[2] Lynnette D. Fairbanks,[3] and Petr Barták[2]

[1]Laboratory for Inherited Metabolic Disorders
Medical Hospital
I.P. Pavlova 6
775 20 Olomouc, Czech Republic
[2]Laboratory of Bioanalytical Research
Palacký University
Třída Svobody 8
771 26 Olomouc, Czech Republic
[3]Purine Research Laboratory
Guy's and St. Thomas's Medical and Dental School
London Bridge
London SE1 9RT, United Kingdom

1. INTRODUCTION

Assay of purine enzymes in cells is important in determining diagnosis and prognosis in purine enzyme deficiencies. Earlier methods for determination of purine enzyme activities have relied on the use of radiolabelled substrates in conjunction with thin layer chromatographic techniques.[1] More recently HPLC measurements of enzyme activities in lysed erythrocytes were introduced, which eliminated the hazards due to isotopes, but also reduces costs and shortened the time taken to obtain a diagnosis.[2]

Capillary electrophoresis (CE) is a current rapidly developing technique in bioanalytical sciences.[3,4] Several papers dealing with diagnosing inborn errors of purine and pyrimidine metabolism published so far[5-9] have shown the possibilities of CE in this field. CE, with its ability to separate compounds of very similar structure in short analyses from small samples, offers excellent capabilities for enzyme studies.[10-11]

Purine and Pyrimidine Metabolism in Man IX,
edited by Griesmacher *et al.* Plenum Press, New York, 1998.

2. EXPERIMENTAL

2.1. Chemicals

All chemicals were of analytical reagent grade. Purine bases, nucleosides and nucleotides were obtained from Sigma Chemicals (St. Louis, MO), boric acid, phosphoric acid and sodium hydroxide were purchased from Merck (Darmstadt, Germany). Deionised water (Milli-Q, 18 MΩ .cm) was used for preparation of all solutions.

2.2. Apparatus and Conditions

Experiments were performed on Beckman PACE 5510 capillary electrophoretic system equipped with diode array detector (mobility studies, APRT and HPRT assays) and on modular system SpectraPhoresis 100 equipped with fast scanning Spectra FOCUS detector (PNP assay). All electrophoretic separations were carried out in uncoated fused-silica capillaries (75 μm I.D. × 375 μm O.D., Polymicro Technologies Inc., Phoenix, Arizona). At the beginning of each working day, the capillary was washed with water, 0.1 M KOH, water and separation buffer for 5 min. All samples were filtered through a 0.45 μm membrane filter (Millipore, Milford, MA) before loading. HPLC method was used as described elsewhere.[12]

2.2.1. Mobility Studies. Capillary: effective length of 30 cm (total length 37 cm). BGE: 40 mM boric acid - 10 mM phosphoric acid adjusted to pH 6.0–10.0 with 50% NaOH. Samples (solution of individual standard compounds at concentration 100 μmol/l) were injected hydrodynamicaly for 4 s. Applied voltage: 30 kV (electric field of 810 V/cm). Capillary was thermostated at 25°C. UV detection 200–300 nm was used. After each run the capillary was washed with BGE for 0.5 min.

2.2.2. PNP Assay. Capillary: effective length of 45 cm (total length 72 cm). Background electrolyte: 100 mmol/L boric acid, adjusted to pH 9.5 with 50% NaOH. Sample loaded for 0.5 s by vacuum injection. Applied voltage: 30 kV (electric field of 415 V/cm). Capillary was at ambient temperature (25°C). UV detection at 210 nm was used for quantification. After each run the capillary was washed with BGE for 3 min.

2.2.3. APRT and HPRT Assays. Capillary: effective length of 30 cm (total length 37 cm). BGE: 40 mM boric acid - 10 mM phosphoric acid adjusted to pH 9.0 with 50% NaOH. Sample was injected by pressure for 4 s. Applied voltage: 30 kV (electric field of 810 V/cm). Capillary was thermostated at 25°C. UV detection at 254 nm was used for quantification. After each run the capillary was washed with BGE for 0.5 min.

2.3. Erythrocytes and Incubation

Procedure used here ascribed briefly is described in detail elsewhere.[12] Erythrocytes were obtained from patients in which enzyme activity was measured as a part of diagnostic screening for genetic purine disorders. Erythrocytes were separated from fresh heparinised blood, washed twice with isotonic saline and lysed by freezing-thawing in added water (1:6). Centrifuged lysate (stroma removed) was diluted and added to preincubated appropriate substrate in buffer (HPRT and APRT assays include phosphoribosylpyrophosphate in the incubation mixture). The mixture was incubated for exactly 15 minutes at 37°C and the reaction was terminated by adding 40% trichloracetic acid while vortex mixing. Precipitated protein was removed by centrifugation and the supernatant was extracted with water-saturated diethylether to a pH above 5.0. Samples were immediately quantified by HPLC and aliquots were freeze-dried. The freeze dried samples were reconstituted in water and used for CE determination. Samples after enzymatic digestion were directly loaded into the capillary without preconcentration or purification after filtering through a 0.45 μm membrane filter.

3. RESULTS AND DISCUSSION

In order to optimise separation conditions, dependence of mobilities on pH of BGE were tested (see Fig. 1). All compounds of interest were separated at pH above 9.0. This high pH of BGE causes high electroosmotic flow in the capillary and shortens the analysis time. Electropherograms of samples from PNP, APRT and HPRT assays are shown in Fig. 2. Results from CE methods compared favourably with standard HPLC method. Analytical parameters of presented methods are shown in Tab. 1.

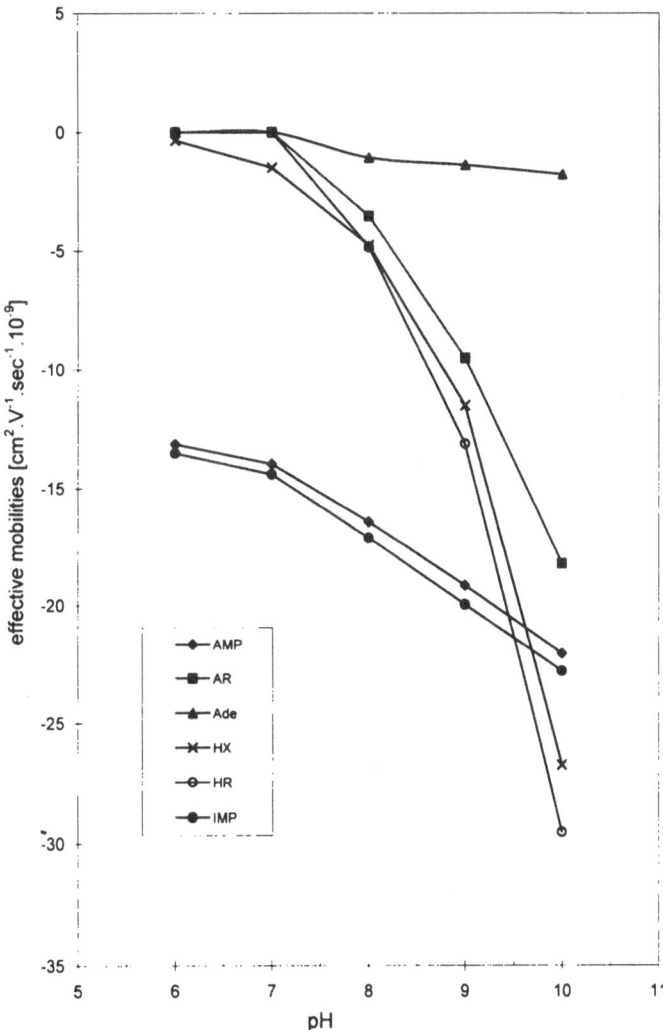

Figure 1. Dependence of mobilities on pH of BGE. Conditions : see 2. Experimental. Ade - adenine, AR - adenosine, HR - inosine, HX - hypoxanthine, AMP - adenosine monophosphate, IMP - inosine monophosphate.

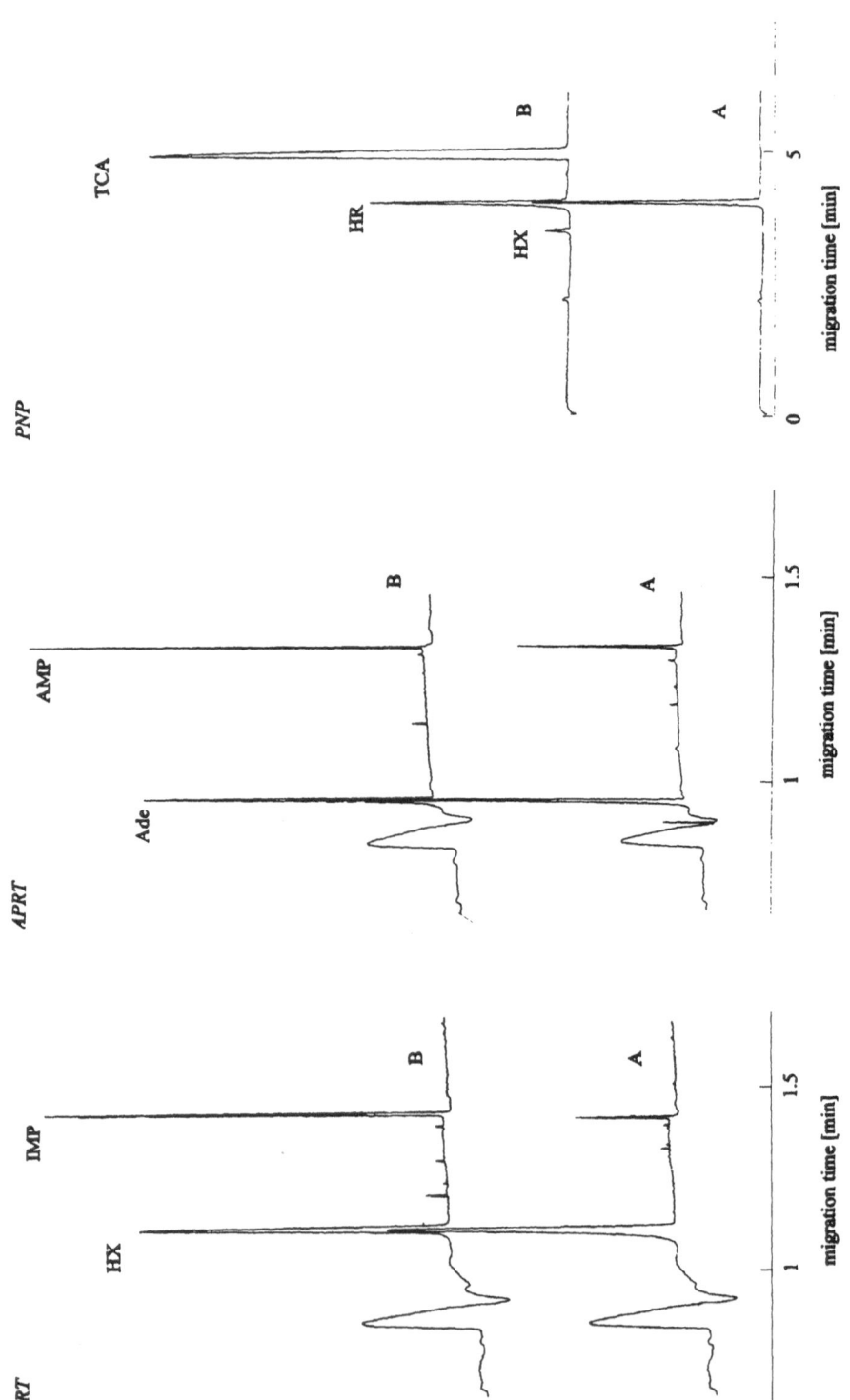

Figure 2. Original electropherograms pre- (A) and postincubation (B) sample from PNP, HPRT and APRT assays. TCA - trichloracetic acid. Conditions: see 2. Experimental.

Table 1. Analytical parameters of CE methods for purine
enzyme measurement and comparison with HPLC

Assay	LOD	Linearity equation	R1	Correlation	R2
PNP	5	y = 83.47x − 0.02	0.999	y = 0. 826x + 14.973	0.989
HPTR	1	y = 148.73x − 812.92	0.999	y = 1.298x − 37.354	0.996
APRT	1	y = 189.84x − 651.43	0.998	y = 1.051x − 3.667	0.998

LOD - limit of detection [μmol/l] at signal to noise (S/N) = 3. Linearity equation - linearity equation for
product up to 500 μmol/l. Correlation : Correlation between HPLC (x) and CE (y).

4. CONCLUSION

The data presented indicates that PNP, APRT and HPRT activities in human erythrocytes can be measured by CE. The methods are as reliable, faster and less expensive than
routinely performed HPLC method. CE is potentially applicable for the measurement of
other purine and pyrimidine enzyme activities.

ACKNOWLEDGMENT

The authors would like to thank Dr. A. H. Simmonds (London) for help in performing this study. This study was supported by grant No. 3439-3 of the Ministry of Health of
the Czech Republic and grant No. VS 96021 MŠ MT of the Czech Republic.

REFERENCES

1. Cartier P, Hamet M. Dosage de l'activite adenosine desaminasique dans les erythrocytes et les lymphocytes
humains. Clin Chim Acta 1976;71:429–433.
2. Fairbanks LD, Goday A, Morris GS, Simmonds HA, Gibson T. Rapid determination of purine enzyme activity in intact and lysed cells using high performance liquid chromatography with and without radiolabeled substrates. J Chromatogr 1983;276:427–432.
3. Landers JP. Clinical capillary electrophoresis. Clin Chem 1995;41:495–509.
4. Jellum E. Capillary electrophoresis for medical diagnosis.J Cap Elec 1994;1:97–105.
5. Grune T, Perrett D. Rapid simultaneous measurement of nucleotides, nucleosides and bases in tissues by
capillary electrophoresis. In: Sahota A, Taylor MW, eds. Purine and Pyrimidine Metabolism in Man VIII.
New York: Plenum Press,1995;805–810.
6. Grune T, Ross GA, Schmidt H, Siems W, Perrett D. Optimized separation of purine bases and nucleosides
in human cord plasma by capillary zone electrophoresis. J Chromatogr 1993;636:105–111.
7. Gross M, Gathof BS, Kolle P, Gresser U. Capillary electrophoresis for screening of adenylosucinate lyase
deficiency. Electrophoresis 1995;16,1927–1929.
8. Ševčík J, Adam T, Mazáčová H. A fast and simple screening method for detection of 2,8-dihydroxyadenine
urolithoiasis by capillary zone electrophoresis. Clin Chim Acta 1996;245:85–92.
9. Ševčík J, Adam T, Sázel V. A fast and simple screening method for detection of orotic aciduria by capillary
zone electrophoresis. Clin Chim Acta 1997;259:73–81.
10. Salerno C, Crifo C. Microassay of adenylosuccinase by capillary electrophoresis. Anal Biochem
1995;226:377–379.
11. Bao JM, Regnier FE. Ultramicro enzyme assays in a capillary electrophoretic system. J Chromatogr
1992;608:217–224.
12. Simmonds HA, Duley JA, Davies PM. Analysis of purines and pyrimidines in blood, urine, and other
physiological fluids. In: Hommes FA, ed. Techniques in Diagnostic Human Biochemical Genetics: A Laboratory Manual. New York: Wiley-Liss, 1991;397–423.

146

CAPILLARY ELECTROPHORETIC ANALYSIS OF HYPOXANTHINE AND XANTHINE FOR THE DIAGNOSIS OF XANTHINURIA

Claude Bory,[1] Christiane Chantin,[1] and Roselyne Boulieu[2]

[1]Laboratoire d'Etudes des Maladies Métaboliques
Hôpital Debrousse
69322 Lyon Cedex 05, France
[2]Laboratoire de Pharmacie Clinique
Institut des Sciences Pharmaceutiques et Biologiques
69373 Lyon Cedex 08, France

1. INTRODUCTION

Hypoxanthine and xanthine determination is necessary in purine metabolism disorders such as xanthine oxidase deficiency. Until now these two compounds have been generally measured by high performance liquid chromatography (HPLC).[1] Because of its high efficiency and rapidity,[2] capillary electrophoresis (CE) was chosen to develop a method for the determination of hypoxanthine and xanthine in urine to diagnose xanthinuria.

2. MATERIALS AND METHODS

2.1. Subjects

2.1.1. Patients. Thirteen unrelated patients, five neonates and height adults were included in this study.

2.1.2. Control Subjects. Ten healthy neonates and ten healthy adults were included for comparison.

2.2. Electrophoretic Conditions

CE was performed with a automated P/ACE 2000 system (Beckman Instruments, Gagny, France) controlled by Beckman System Gold software. Separations were achieved

Purine and Pyrimidine Metabolism in Man IX,
edited by Griesmacher *et al.* Plenum Press, New York, 1998.

within a fused-silica tubing (37 cm × 50 μm I.D., 30 cm from sample-end electrode to the detector) and using 0.04 M $(NH_4)_2HPO_4$ pH 8.1 as the running buffer.

Injections were made hydrodynamically by pressure at the anodic end for 5 s. The applied voltage was 20 kV giving a current of 80 μA. The temperature was kept at 25°C. Detection was performed at 254 nm. The daily initial preparation of the capillary involved a 10-min rinse with 1 M NaOH followed by a 10-min rinse with running buffer. Before each run, the capillary was rinsed with 0.05 M NaOH for 2 min and equilibrated with running buffer for 5 min.

2.3. Urine Samples

Urine was collected during 24 h and stored at −20°C. The urine samples were diluted fivefold before analysis.

2.4. Peak Identification

The identification of hypoxanthine and xanthine was confirmed by the enzymatic peak shift technique using xanthine oxidase as enzyme.

3. RESULTS AND DISCUSSION

Fig. 1a shows the electropherogram of a standard solution including hypoxanthine, xanthine, uric acid and Fig. 1b shows the representative electropherogram of a urine sam-

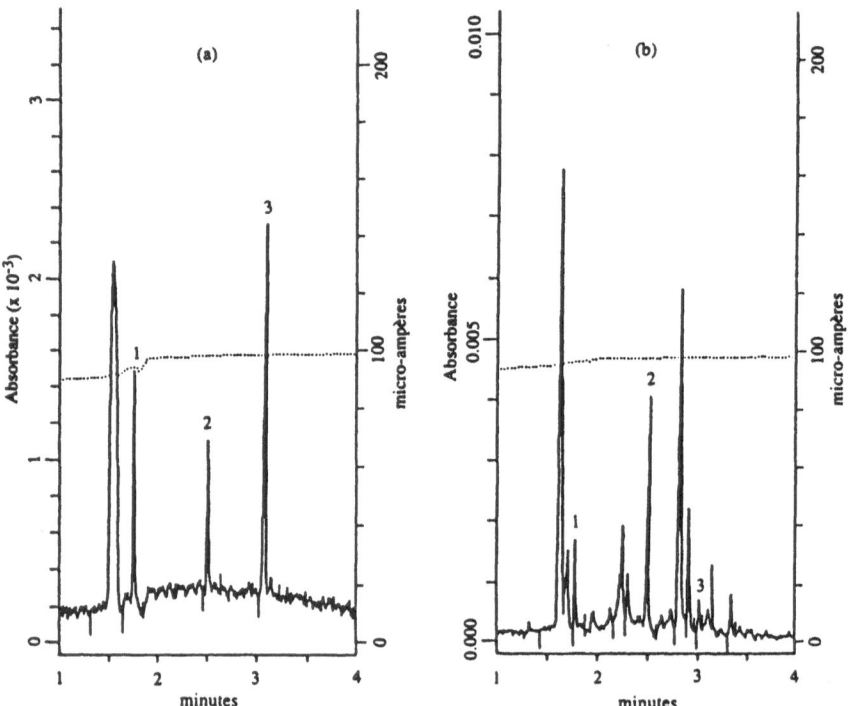

Figure 1. Electropherograms of a standard solution (a), and of urine sample (b), dilutede 5 times, from a patient with xanthinuria. The analysis were carried out as outlined under Method. Peaks: 1=hypoxanthine [(a) 20 μmol l^{-1}, (b) 113 μmol l^{-1} (136 μmol 24 h^{-1})]; 2=xanthine [(a) 20 μmol l^{-1}, (b) 450 μmol l^{-1} (540 μmol 24 h^{-1})]; 3=uric acid [(a) 70 μmol l^{-1}, (b) traces].

ple from a patient with xanthinuria. From these figures it can be seen that the compounds of interest can be separated with high resolution in only 3.5 min. The concentration of hypoxanthine and especially the concentration of xanthine are increased in the urine sample of patient with xanthinuria. The increase is associated with a very low uric acid concentration. Concentrations of the compounds were calculated from the respective prepared standard curves. The calibration curves were linear up to 200 μmol l^{-1}. For quantification the correction of the peak area by the migration time is not necessary, the migration time being constant from run to run.

The run-to-run and the day-to-day reproducibilities were tested at 20 μmol l^{-1} (n = 10). The relative standard deviations (R.S.D.) for the migration times were 0.4% and 1.7% for hypoxanthine, 0.6% and 2.1% for xanthine. The R.S.D.$_s$ for the concentrations were less than 3% for the two compounds. The minimum detectable concentration based on a 3:1 signal to noise ratio was about 2 μmol l^{-1}.

There was no interference with the compounds of interest by related endogenous compounds such as uric acid, guanine, guanosine or by drugs such as allopurinol, oxypurinol, 6-mercaptopurine, caffeine and theophylline.

With regard to selectivity, sensibility and reproducibility the results obtained with CE are similar than those previously reported with high pressure liquid chromatography. With regard to analysis times, using HPLC, the separation of the compounds of interest lasts 6 minutes but the elution of all the compounds present in normal urines needs at least 30 minutes. Using CE, the separation lasts 4 minutes but the analysis includes a preconditioning of the capillary for 7 minutes. Therefore the analysis time in CE is three times shorter than that in HPLC.

In conclusion, CE and LC represent suitable and useful tools for the diagnosis of xanthinuria but CE is more rapid than LC and easier to set up and to use.

REFERENCES

1. F.T.A. Chen, C.M. Liu, Y.Z. Hsich and J.C. Sternberg, 1991, Capillary electrophoresis-a new analytical tool, *Clin. Chem.* **37**:14–19.
2. R. Boulieu, C. Bory and P. Baltassat, 1982, High-performance liquid chromatographic determination of hypoxanthine and xanthine in biological fluids, *J. chromatogr.* **233**:131–140.

147

ECTO-NUCLEOTIDASES IN ISOLATED INTACT RAT VAGI, NODOSE GANGLIA, AND SUPERIOR CERVICAL GANGLIA

G. P. Connolly,[1] C. Demaine,[2] and J. A. Duley[1]

[1]Purine Research Laboratory
UMDS Guy's Hospital
[2]Physiology Group
King's College, London

1. INTRODUCTION

The effects of the adenosine and ATP on the nervous system have been extensively studied, but the effects of pyrimidine bases and their 5'-nucleotides, e.g., UTP, have received little attention. Sympathetic ganglia may possess distinct extracellular receptors for purines and pyrimidines based on the fact that UTP is far more potent than alpha, beta-methylene-ATP (α,β-Me-ATP)(a metabolically stable ATP analogue) or ATP itself (Connolly *et al.*, 1993) in depolarising intact rat superior cervical ganglia (SCG). It is however, known that the potency of an agonist on a tissue is dependent on whether it is susceptible to metabolism by the tissue and the metabolism of purines and pyrimidines by intact rat SCG has yet to be reported. Here we report on studies to examine if and how intact rat SCG (sympathetic ganglia), nodose ganglia (parasympathetic ganglia) and its nerve trunk, the vagus, metabolise ATP and UTP.

Further evidence for distinct receptors for ATP and UTP on rat SCG is based upon the ability of purine receptor antagonists such as suramin to selectively antagonise depolarisations of intact SCG evoked by α,β-Me-ATP and ATP but not those evoked by UTP (Connolly *et al.*, 1993; Connolly, 1995). However, prolonged contact with suramin at 10 mM has been shown to inhibit the degradation of ATP by strips of guinea-pig urinary bladder, (Hourani & Chown, 1989) and more recently suramin has been reported to inhibit ecto-ATPase activity of other cell types. Therefor the potentiation of UTP-evoked depolarisations of rat SCG by suramin and/or their lack of antagonism by other purine receptor antagonists could be explained by an inhibition UTP metabolism by ecto-ATPase.

Thus, we also report on further studies examining the actions of suramin on UTP-evoked depolarisation of the rat SCG, and on the effect of suramin and ARL 67156 (6-N,N-diethyl-D-b,g-dibromomethylene-ATP), a novel ecto-ATPase inhibitor (Crack *et*

Purine and Pyrimidine Metabolism in Man IX,
edited by Griesmacher *et al.* Plenum Press, New York, 1998.

al., 1995) on the metabolism of ATP and UTP by rat SCG, and on ATP metabolism by nodose ganglia and vagi.

2. MATERIALS AND METHODS

2.1. D.C Recording of Electrophysiological Activity

Ganglia from male Wistar rats (200–300 g) killed with a lethal dose of urethane were prepared for recording of d.c. potentials as described before (Connolly & Stone, 1993; Connolly *et al.*, 1993). Single SCG were submerged between the greased barriers of a three compartment recording bath, with the central chamber containing the ganglion body. SCG were superfused (2 ml min^{-1}, at 25 ± 1°C) with a physiological salt solution (PSS) at pH 7.4, containing (mM) NaCl, 125; NaHCO$_3$, 25; KCl, 1; KH$_2$PO$_4$, 1; MgSO$_4$, 1; glucose, 10; CaCl$_2$, 0.1 and pre-oxygenated (5% CO$_2$/95% O$_2$). The potential difference between the earthed central chamber and the internal carotid nerve was recorded using Ag/AgCl electrodes, amplified and recorded on a pen recorder. UTP (1000 µM; 2 min applications) was applied at 45 min intervals or multiples of this time.

2.2. Metabolism Studies

Isolated tissues were placed in an agitated chamber containing 1400 ml PSS (as above) at 25 ± 1°C, equilibrated with 5% CO$_2$/95% O$_2$. Samples (20 ml) of PSS for assay by HPLC (Simmonds, *et al.*, 1990) were collected 2, 20 and 40 min after ATP or UTP (both 100 µM) addition. To measure loss and 'physical' breakdown, nucleotides were incubated in PSS for up to 40 min. 'Leakage' of enzyme from tissues was measured by pre-incubating tissues in PSS for 30 min before their removal and addition of nucleotide. Metabolism by ganglia and vagi was determined by pre incubating tissues for 30 min in PSS, suramin or ARL 67156, before incubation with nucleotides.

2.3. Statistical Analysis

Experimental data are presented as mean ± s.e. mean and analysis was performed on raw non-transformed data. Differences between UTP-evoked depolarisations of rat SCG or metabolite concentrations with time and in the absence or presence of suramin or ARL 67156 were analysed for statistically significant changes using paired and unpaired Student's t-tests (2 tailed) as appropriate. Where probability values of less than 5% ($P<0.05$) were obtained they were considered significant.

3. RESULTS

3.1. D.C Recording of Electrophysiological Activity

The effect of applying UTP repeatedly to SCG at 45 min intervals or longer is shown in Figure 1. Depolarisations evoked by UTP were not significantly different if applied at greater than 45 min intervals (Fig. 1A), but significantly declined if applied at 45 min intervals (Fig. 1B). Suramin significantly potentiated UTP-evoked responses evoked upon a second application of UTP (applied after 90 min) to 124 ± 9% their control

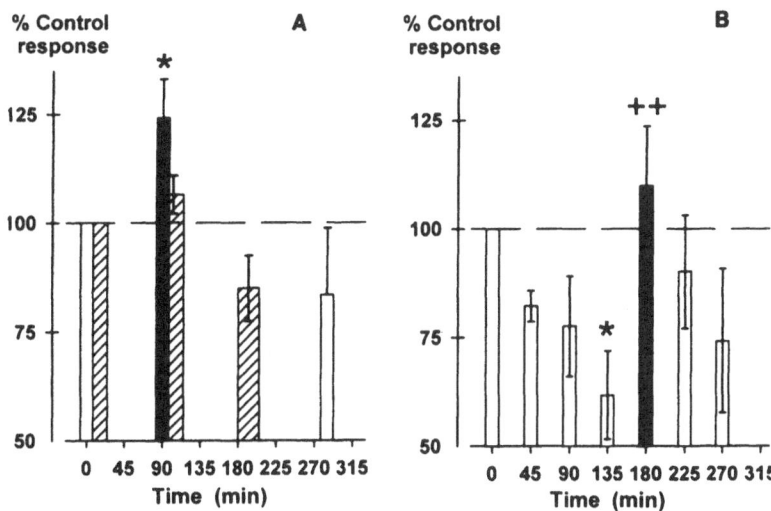

Figure 1. Effect of suramin (300 μM) on UTP(1000 μM)-evoked depolarisation of isolated rat SCG. (A) & (B) Repeated applications of UTP in the absence (hatched (time controls) and open bars) and in the presence of suramin (dark bar). N.B. Depolarising responses are expressed as a % of the initial 'control' response. Suramin was applied 20 min before, during and 20 min after UTP. In (B) application of UTP every 45 min caused desensitisation of subsequent responses which by their 4th application were significantly different (P < 0.05, *) to initial responses. Suramin significantly potentiated UTP-evoked depolarisations (P < 0.01, ++) when compared to its previous response (c.f. 4th & 5th responses) restoring depolarisations to initial levels (i.e., 1st and 5th responses not significantly different). In this and subsequent figures, vertical bars represent the mean and crossed bars the standard error of the mean response, the number of tissues used was between four and six; for abbreviations see text.

value whereas time control responses were not significantly different (Fig. 1A). In contrast when suramin was tested upon UTP-evoked responses after repeated applications of UTP at 45 min intervals, the response to UTP was dramatically and significantly potentiated to 198 ± 53% their control values (Fig. 1B).

3.2. Metabolism Studies

3.2.1. Metabolism of ATP by Rat SCG. A small amount of spontaneous 'physical' breakdown of ATP to ADP and AMP was detected in the absence of any tissue (Fig. 2). Similar levels of ADP and AMP to those produced by 'physical' breakdown of ATP were observed in the 'leakage' experiment, but higher levels were produced in the presence of SCG (Fig. 2). Suramin and ARL 67156 significantly reduced the formation of ADP or AMP by SCG, to levels seen during 'physical' breakdown or 'leakage' (Fig. 3). The total amount of adenine nucleotides remained constant with time for all treatments (~5 ± 5% total) except SCG in the presence of PSS which declined by 18.5 ± 11%.

3.2.2. Metabolism of ATP by Rat Nodose Ganglia. The amount of ADP and AMP produced in 'leakage' experiments with nodose ganglia was similar to levels observed in 'leakage' experiments with SCG (c.f. Fig. 4 and Fig. 2) or upon 'physical' breakdown of ATP (cf. Fig. 4 and Fig. 1). Nodose ganglia in PSS produced ADP and AMP, however, both suramin and ARL 67156 did not alter their production by nodose ganglia (Fig. 4). The total amount of adenine nucleotides was constant with time for all treatments, e.g., for nodose ganglia in PSS a change of 3 ± 6.9% was observed.

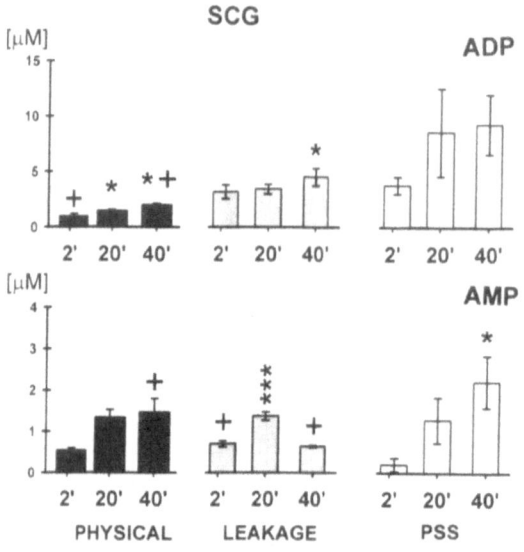

Figure 2. Production of ADP and AMP from ATP due to spontaneous breakdown (first column), pre incubation with isolated rat SCG (second column) and SCG in PSS (third column). In this and subsequent figures, P<0.05, 0.01 & 0.001 are indicated by +, ++ and +++ for values compared to those produced in PSS, and by *, ** & *** for values produced after 20 and 40 min compared to those after 2 min.

3.2.3. Metabolism of ATP by Rat Vagi. The amount of ADP and AMP produced in 'leakage' experiments was comparable to that seen using SCG or nodose ganglia (cf. Figs. 5, 4 & 3). Vagi metabolised ATP to ADP and AMP, and to a similar extent to that seen using SCG (cf. Figs. 5 & 3). The total amount of adenine nucleotides was constant with time for all treatments, e.g., for vagi in PSS a change of 6.2 ± 2.9% was observed. Both suramin and ARL 67156 reduced the formation of ADP and AMP (Fig. 5).

3.2.4. Metabolism of UTP by Rat SCG. A high level of 'contaminating' UDP was present in our samples of UTP accounting for around 10% of UTP content. UDP was not significantly altered by incubation in PSS (i.e., to determine 'physical' breakdown), cf. 10.2 ± 0.9 and 10.4 ± 1.8 μM or by pre incubation with SCG (i.e., 'leakage' experiment), c.f. 10.1 ± 1.0 and 9.4 ± 0.4 μM, after 2 and 40 min respectively, and UMP and uridine were not detected in these samples. SCG metabolised to UTP to UDP, UMP and uridine

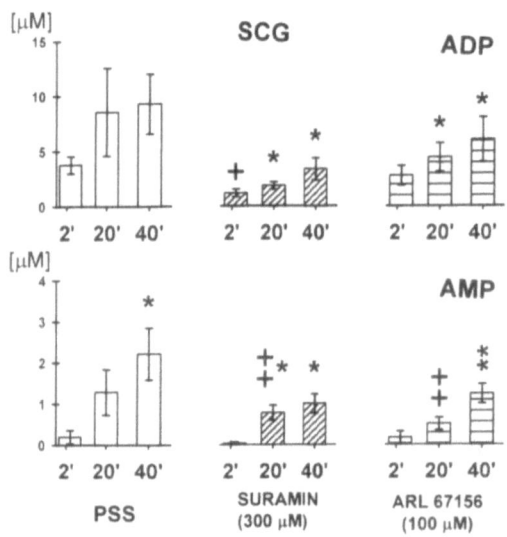

Figure 3. Production of ADP and AMP from ATP by isolated rat superior cervical ganglia (SCG) in PSS (first column), by SCG in PSS and suramin (second column) and by SCG in PSS and ARL 67156 (third column). N.B. Data in first column also shown in Fig. 2.

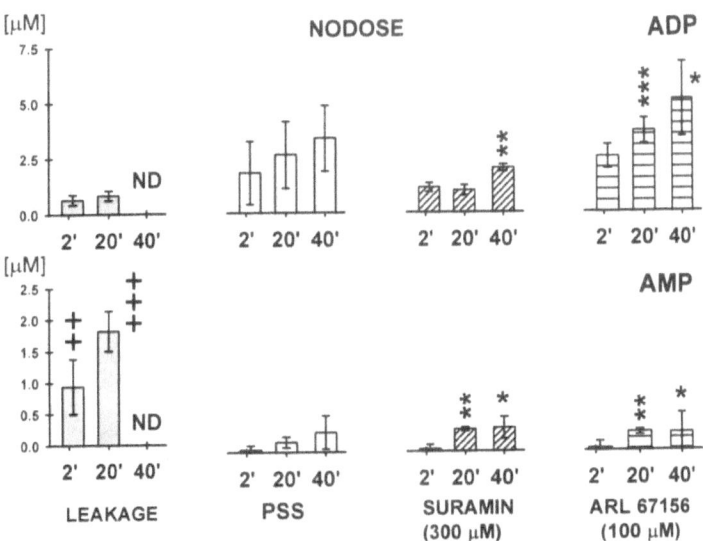

Figure 4. Production of ADP and AMP from ATP by pre incubation with isolated rat nodose ganglia (first column), by nodose ganglia in PSS (second column), by nodose ganglia in PSS and suramin (third column) and by nodose ganglia in PSS and ARL 67156 (forth column).

(Fig. 6). After 2 min suramin did not significantly alter the production of UDP, UMP or UR, but by 40 min significantly reduced the total nucleotide production (P<0.01), and more specifically the production of UDP and UMP (Fig. 6). After 2 min ARL 67156 educed the formation of both UDP and UMP and by 40 min reduced UMP formation (Fig. 6). Over 40 min in PSS, suramin or ARL 67156 the total amount of uridine and its nucleotides remained constant (~5 ± 5%).

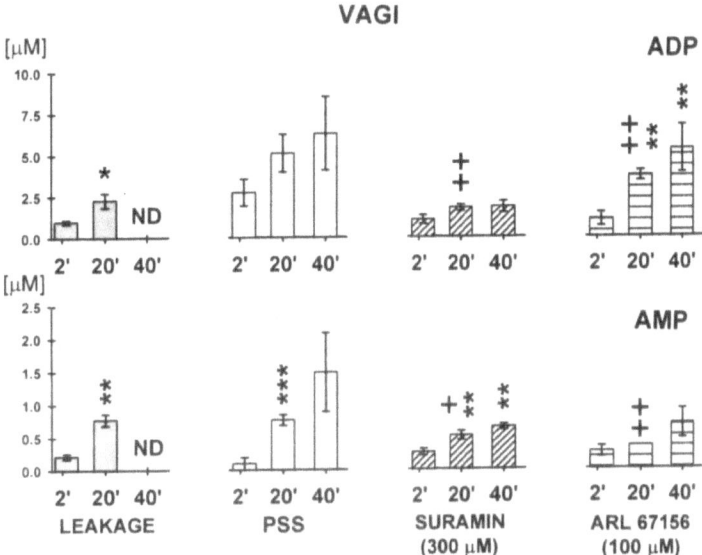

Figure 5. Production of ADP and AMP from ATP by pre incubation with isolated rat vagi (first column), vagi in PSS (second column), vagi in PSS and suramin (third column) and by vagi in PSS and ARL 67156 (forth column).

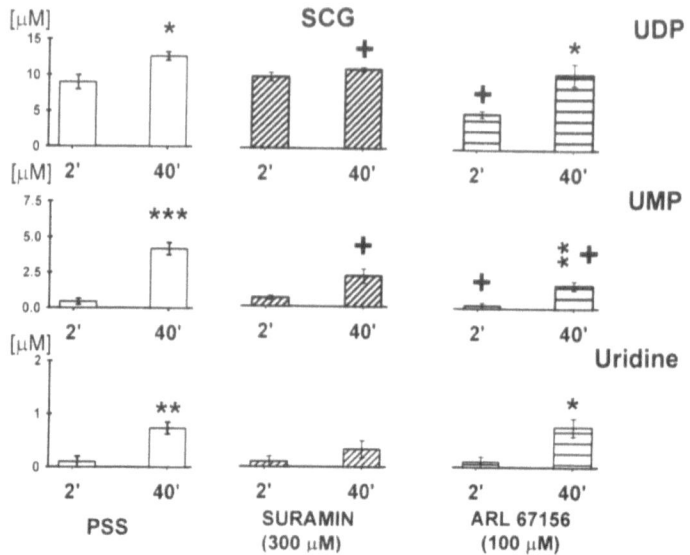

Figure 6. Production of UDP, UMP and uridine from UTP due by rat superior cervical ganglia (SCG) in PSS (first column) and SCG in PSS and suramin (second column) and by SCG in PSS and ARL 67156 (third column).

4. DISCUSSION

4.1. Roles of 'Leaked Enzymes' and Ecto-Enzymes in Metabolism

The role of 'leaked' enzymes from tissues appears to be small given that the level of nucleotide production was small and similar to the production of nucleotides produced during 'spontaneous physical breakdown'. Any increase in nucleotides observed during 'leakage' experiments was unlikely to be due spontaneous released compounds because no nucleotides were detected after pre-incubation with tissues in PSS for 30 min. Further studies underway using cultured neurones should help in discovering if the source of the 'leakage' is due to the release of enzymes by normal or damaged cells. Prior purification by HPLC of ATP and UTP should help alleviate some of the problems caused by contamination by their respective diphosphates.

4.2. Roles of Suramin in Potentiation of UTP-Evoked Depolarisations of Isolated Rat SCG

Here we confirm that suramin potentates UTP-evoked depolarisations of the isolated rat SCG at a concentration that is near it maximal effective dose for depolarising SCG. Interestingly the degree of enhancement caused by suramin was greater after desensitisation of UTP-evoked responses. Although suramin inhibited the breakdown of UTP to UDP and UMP, the small but significant decrease in the level of UDP and UMP in the presence of suramin (see Fig. 6) make it unlikely that this was the cause of the dramatic potentiation caused by suramin. We suggest that suramin may modulate the effectiveness of UTP by reducing the degree of receptor desensitisation. A mechanism that could account for the potentiating effect of suramin is if in addition to being an antagonist of purine receptors it is also an allosteric modulator of nucleotide receptors. Some support for an allosteric effect of suramin is provided by the observation that suramin reversed the muscle paralysis of the rat diaphragm caused by structurally unrelated non-depolarising skeletal muscle relaxants (Henning *et al.*, 1992). The inability of suramin to antagonise UTP-evoked depolarisations of isolated rat SCG is therefore most easily explained by the presence of receptors distinct to those for ATP that are suramin-insensitive, namely pyrimidinoceptors.

4.3. Evidence for Ectonucleotidases on Isolated Rat SCG

In contrast to rat vagi and nodose ganglia, isolated rat SCG metabolised ATP and UTP not only to their di- and monophosphates but also to their nucleosides, adenosine and uridine, suggesting the presence of 5'-nucleotidase (5'NT) activity. Support for the presence of purine 5'NT on intact rat SCG was found using cytochemical studies by Nacimiento & Kreutzber (1990), who reported its confinement to ganglion capillaries and complete absence of 5'NT on the neuropil. The absence or lack of detectable 5'NT activity on the rat nodose ganglion and vagus is thus consistent with their low density of vascular tissue. The presence of an uptake system for adenosine and adenosine deaminase activity on intact rat SCG (Connolly & Stone, 1993) is consistent with the presence of enzymes capable of metabolising ATP to adenosine. Given that rate of production of ADP + AMP and UDP + UMP over 38 minutes (difference between the 40 and 2 min samples) were 30 ± 4 μM and 30 ± 5 μM respectively, we suggest that the same ecto-nucleotidases metabolises ATP and UTP. The similar potency of degradative enzymes on the rat SCG for ATP and UTP, is consistent with a lack of selectivity of ecto-ATPase for ATP or UTP present on other tissues, including those of the nervous system (Stefanovic *et al.*, 1976). The ability of ecto-nucleotidases to metabolise both ATP and UTP also suggests that the notably higher potency of UTP compared to ATP in depolarising isolated SCG is not due to due to selective catabolism of ATP and not UTP. In fact, our results suggest that the potency of UTP may have been under estimated due its potential metabolism in situ.

4.4. Evidence for Ectonucleotidases on Isolated Rat Nodose Ganglia and Vagus

Our studies have also shown that isolated rat nodose ganglia and it corresponding nerve trunk, i.e. the vagus both possess enzymes capable of metabolising ATP to ADP and AMP. These tissue may not or only weakly metabolise ATP to adenosine given we were unable to detect any change in the total concentration of adenine nucleotides with time. Recently it was reported that the absence of extracellular Ca^{2+} and Mg^{2+} augmented the depolarising response of the rat vagus evoked by ATP (Trezise *et al.*, 1994), an effect that might be mediated by inhibition of ectonucleotidases. Our results support and provide functional evidence for the presence of ecto-nucleotidases on the rat vagus that are capable of metabolising ATP to ADP and AMP (Fig. 5). The ability ARL 67156 and of suramin, two reported inhibitors of ecto-ATPases, to inhibit the metabolism of ATP by the rat vagus also advocates the presence of ectonucleotidases on this tissue. In addition, rather than antagonising the response produced by 1 mM ATP on rat vagi, suramin was reported to potentiated it, and also increased the maximum depolarising effect of a metabolically stable analogue of ATP, i.e., α,β-MeATP (Trezise *et al.*, 1994). These observations along with others showing suramin potentiates α,β-MeATP-evoked responses, corroborate our proposal that suramin may act as an allosteric modulator of nucleotide receptors. Interestingly, the ability of UTP to depolarise rat vagi has probably been under estimated, because it would be expected to be metabolised by the same enzymes that metabolise ATP. Unpublished results from our laboratory substantiate this latter possibility.

4.5. Evidence for Selective Actions of Suramin and ARL 67156 on Isolated Rat SCG and Vagi

ARL 67156 and suramin are inhibitors of the metabolism of ATP to ADP and AMP by isolated rat SCG and rat vagi, but did not alter ATP metabolism by rat nodose ganglia. This was an unexpected observation may indicate that rat nodose ganglia express ARL 67156 and suramin-insensitive ectonucleotidases. Further studies will be required to test this hypothesis. The ability of suramin and ARL 67156 to inhibit the metabolism of both ATP and UTP suggests that these drugs may not only inhibit ecto-ATPase but may also inhibitors of ectonucleotidases.

5. CONCLUSION

We suggest that the existence of ecto-nucleotidases on sympathetic ganglia (SCG) and parasympathetic ganglia (nodose) and their nerve trunks, capable of metabolising purines and pyrimidines to their nucleosides is entirely consistent with the presence of extracellular receptors for these compounds and their role as regulators of the peripheral nervous system.

ACKNOWLEDGMENTS

We thank King's College Fund and the Wellcome Trust for financial support. We also wish to thank Dr. Paul Leff (Astra-Charnwood) for the gift of ARL 67156 and Roy R. Gurprashad for technical assistance during part of this project. G.P.C. is currently funded by Action Research and Purine Metabolic Patients Association (PUMPA).

REFERENCES

Connolly, G. P. (1995). Differentiation by pyridoxal 5-phosphate, PPADS and IsoPPADS between depolarisations mediated by UTP and those evoked by α,β-methylene-ATP on rat sympathetic ganglia. Br. J. Pharmacol., 114, 727–731.
Connolly, G. P. & Stone, T.W. (1993). On the site of action and inactivation of adenosine by the rat superior cervical ganglion. J. Auton. Pharmacol., 13, 237–247.
Connolly, G. P., Harrison, P.J., & Stone. T.W. (1993).Action of purine and pyrimidine nucleotides on the rat superior cervical ganglion. Br. J. Pharmacol. 110., 1297–1304.
Crack, B.E., Pollard, C.E., Beukers, et al., (1995). Pharmacological and biochemical analysis of FPL 67156, a novel, selective inhibitor of ecto-ATPase. Br. J. Pharmacol., 114, 475–481.
Henning, R.H., Nelemans, A., Scaf., A.H.J., et al., (1992). Suramin reverses non-depolarising neuromuscular blockade in rat diaphragm. Eur. J. Pharmacol., 216, 73–79.
Hourani, S.M.O. & Chown, J.A. (1989). The effects of some possible inhibitors of ectonucleotidases on the breakdown and pharmacological effects of ATP in the guinea-pig urinary bladder. Gen. Pharmacol., 20, 413–416.
Nacimento, W. & Kreutzberg, G.W. (1990). Cytochemistry of 5'-nucleotidase in the superior cervical ganglion of the rat: Efffects of pre- and postganglionic axotomy. Expt. Neurology, 109, 362–373.
Simmonds, H.A., et al., (1990). Techniques In Diagnostic Human Biochemical Genetic. Laboratory manual. pp 397–424. Ed. Fritz A. Holmes. Wiley-Liss.
Stefanovic, V., Ledig, M. & Mandel. P. (1976). Divalent cations-activated ecto-nucleoside triphosphatase activity of nervous system cells in tissue culture. J. Neurochem., 27, 799–805.
Trezise, D. J., Kennedy, I. & Humphrey. P.P.A. (1993). Characterisation of purinoceptors mediating depolarisation of rat isolated vagus nerve. Br. J. Pharmacol., 110, 1055–1060.
Trezise, D. J. Bell., N.J., Kennedy, I. et al., (1994). Effect of divalent cations on the potency of ATP and related agonists in the rat isolated vagus nerve: Implications for P₂ purinoceptor classification. Br. J. Pharmacol., 113, 463–470.

148

HYPOXANTHINE AND XANTHINE AS MARKERS IN EARLY DIAGNOSIS OF FOETAL DISEASES

F. Gizzi, M. Papponetti, G. D. Palka, I. Ruffini, C. Di Ilio, M. Odorisio, and G. Spoto[*]

Dipartimento di Scienze Biomediche
Università of Chieti "Gabriele D'Annunzio"
66013 Italy

1. INTRODUCTION

Human chorionic gonadotrophin (hCG), alpha-feto protein (αFP), unconjugated 3-oestriol (uE3) are the markers currently used in prenatal serum screening programs for neural tube defects and chromosomal disorders (1). However the use of triple markers, when screening for foetal Down syndrome, still fails to detect 40% of Down syndrome in women less than 35 years of age. Other screening methods are currently under investigation (2). In Down syndrome, abnormalities of purine metabolism have been detected (3–5); gene mutation of PDEa and PDEb have been identified in cases of retinitis pigmentosa and of macular degeneration (6–8).

In the present study the preliminary results of the use of hypoxanthine (Hy) and xanthine (X) concentration and of phosphodiesterase (PDE) activities as screening analytes are reported.

2. MATERIALS AND METHODS

2.1. Substances

Purine and pyrimidine compounds were purchased from Wellcome, Boehringer-Mannheim or Sigma; all other reagents were of analytical grade from Merck.

---

[*] Mailing address: Giuseppe Spoto, Dipartimento di Scienze Biomediche, Università 'G.D'Annunzio', Chieti, via dei vestini 31-66013 Chieti, Italy. E-Mail: Spoto@isb.unich.it.

Purine and Pyrimidine Metabolism in Man IX,
edited by Griesmacher *et al.* Plenum Press, New York, 1998.

777

2.2. Plasma and Amniotic Fluids

After drawing blood from pregnant women, plasma was separated by centrifugation within 30 minutes (9).

Plasma and amniotic fluid were prepared, according to Gerrits et al. methods for purine bases (10).

2.3. Pellet for PDE Measurement

Method and analysis for PDE was from Spoto et al. (11).

2.4. Apparatus

HPLC was a Beckman model 110A with two pumps, autosampler Promis, connected with a variable wavelenght spectrophotometer Kratos mod. 783 Spectroflow measuring at 254 nm.

2.5. HPLC Conditions

For elution of purine bases, the column was a 5μm Spherisorb S5 ODS2 (250 × 4 mm) with pre-column; running conditions are described in ref. 12.

Peak identities were confirmed by coelution with standards and by disappearance of the peaks after incubation with the appropriate enzymes.

3. RESULTS AND DISCUSSION

Results obtained for hypoxanthine and xanthine are summarized in Tables 1–3. Table 1 shows that patients with a positive triple test had higher plasma Hy and Hy/X ratio without significant differences between true and false positive. These differences were not observed in amniotic fluid (Table 2).

Patients with very high or very low values of hCG, αFP and uE3 had very high serum hypoxanthine concentrations and, in particular, serum Hy/X ratios (Table 3).

Although phosphodiesterase activities were measured in only few cases, in two of four true positive cases very high PDE activities were measured in cells from maternal blood (One case was associated with Down syndrome and another case with microcephaly).

For the early diagnosis of foetal chromosomic disease, presently the most useful method is the determination of foetal karyotype, after sampling of amniotic fluid, in the

Table 1. Hypoxanthine, xanthine concentrations and their ratios in maternal serum at 15–18 weeks of gestation in true positive (TP), false positive (FP) and true negative (TN) cases classified according to the results of the Triple-Test

	Hy (μM)	X (μM)	Hx/X (μM)
Plasma *TP* (n = 7)	4.263 ± 1.39	1.222 ± 0.95	4.083 ± 2.54
Plasma *FP* (n = 15)	3.986 ± 1.43	0.925 ± 0.55	4.556 ± 2.70
Plasma *VN* (n = 13)	1.787 ± 0.55	1.572 ± 1.16	1.170 ± 0.47
Controls (n = 10)	2.066 ± 0.43	1.753 ± 0.92	1.227 ± 0.95

Table 2. Hypoxanthine, xanthine concentrations in amniotic fluid in true positive (TP) and true negative (TN) pregnants classified according to the results of amniocentesis

	Hy (µM)	X (µM)
Amniotic fluid *TN* (n = 29)	1.551 ± 1.20	1.523 ± 1.53
Amniotic fluid *TP* (n = 5)	1.333 ± 0.77	1.805 ± 1.48
Amniotic fluid (*)	0.6 ± 0.5	1.0 ± 0.4

(*):Mateos FA, Puig JG (1993) "Molecular Genetics, Biochemistry and Clinical Aspects of inherited disorders of Purine and Pyrimidine Metabolism"U.Gresser Ed.(Springer-Verlag,Berlin Heidelberg) 27–32.

Table 3. Hypoxanthine, xanthine concentrations and their ratios in maternal serum at 15–18 weeks of gestation in true positive (TP), false positive (FP) and true negative (TN) cases classified according to the multiple of median index (MoM) results of the Triple-Test

	Hy (µM)	X (µM)	Hx/X
Plasma with very low or very high MoM values (n = 10)*	4.366 ± 2.04	0.953 ± 0.65	6.063 ± 3.80
Plasma with positive "Triple-Test" (n = 22)	4.099 ± 1.39	1.046 ± 0.74	3.918 ± 1.89
Controls (n = 13)	1.787 ± 0.55	1.572 ± 1.16	1.170 ± 0.47

*MoM hCG ≥ 2.5, ≤ 0.6; MoM αFP ≥ 2.5, ≤ 0.6; MoM uE3≤ 0.6

second trimester, or after biopsy of chorionic villi, in the first trimester of pregnancy. These procedures have some disadvantages: they are expensive and are performed only in over 35 years or, in some cases, over 38, in current prenatal screenings. Moreover, there is a very high risk for the foetus: in fact, aroused aborts are 1:200 for amniocentesis, and 1:50 for biopsia.

Useful markers for easy sampling and indicative of pregnancy status should reduce the number of invasive surgeries, bounding them to situations with documented risk.

4. CONCLUSIONS

These preliminary results suggest that Hy to X ratio and PDE activities should be further investigated for use in the first and second trimester as screening tests for fetal Down syndrome and neural tube defects.

REFERENCES

1. Mancini G., Perona M., Carbonara A., 1992: *Marcatori sierici materni nelle gravidanze con feti affetti da aberrazioni cromosomiche.* Suppl. Analysis- *7.92*, 63–68
2. Wald N.J. e coll.,1988: *Maternal serum screening for Down's syndrome.* Br. Med. J., *297*, 883–887.
3. Peeters MA,MegarbaneA.,Cattaneo F.,Rethore MO, Lejeune J., 1993: *Differences in purine metabolism in patients with Down's syndrome.* J.Intellect Disabil Res 37:491–505.
4. Puukka R.,Puukka M.,Perkkila L.,Kouvalainen K.,1986: *Levels of some purine metabolizing enzymes in linfocites from patients with Down's syndrome.*Biomed Med Metab Biol 36:45–50.
5. Spoto G. et al., 1996: *Hypoxanthine and xanthine levels in human aqueous humor from cataractous eyes.* Life Sci., *59* :387–90.
6. Wells J. et al., 1993: *Mutation in the human retinal degeneration slow (RDS) gene can cause either retinitis pigmentosa or macular distrophy.* Nat. Genet., *3*, 213–218.

7. Nicols B.E. et al., 1993: *Butterfly shaped pigment distrophy of the fovea caused by a point mutation in co-don 167 of the RDS gene.* Nat. Gen., *3*, 202–207.
8. Natik I. Priev et al., 1995: *Gene structure and amino-acid sequence of the human cone photoreceptor cGMP-PDEα1 subunit (PDEA2) and its chromosomal localization to 10q24.* Genomics, *28*, 429–435.
9. Boulieu R e coll., 1983: *Hypoxanthine and xanthine levels: the problem of hypoxanthine level evolution as a function of time.* Anal. Biochem. *129*, 398–404.
10. Gerrits G.P.J.M. et al., 1988: *Reference values for nucleosides and nucleobases in cerebrospinal fluid of children.* Clin. Chem. *34* 1439–1457.
11. Spoto G., Whitehead E., Ferraro A., Di Terlizzi P.M., Turano C.and F. Riva.1991.*A reverse-phase HPLC method for cAMP phosphodiesterase activity.* Anal. Biochem. 196: 207–210.
12. Spoto G., Berardi S., Ajerba G.and V. De Laurentis.1995. *A reverse-phase HPLC method for cyclic nucleo-tide phosphodiesterases activity and classification.* Adv.Exp.Med.and Biol.370:815–819.

149

EFFECTS OF DIPYRIDAMOLE AND ADENINE/RIBOSE ON ATP CONCENTRATION AND ADENOSINE PRODUCTION IN CARDIAC MYOCYTES

Kameljit K. Kalsi,[1] Ryszard T. Smolenski,[1] Ronald D. Pritchard,[1] Asghar Khaghani,[1] Anne-Marie L. Seymour,[2] and Magdi H. Yacoub[1]

[1]Heart Science Centre
National Heart and Lung Institute
Harefield Hospital
Harefield, United Kingdom
[2]Department of Biological Sciences
University of Hull
Hull, United Kingdom

INTRODUCTION

Heart failure remains one of the major clinical problems in humans. One important hypothesis to explain this is that the hypertrophied heart is in an energy deprived state resulting from an imbalance between energy production and utilization.[1,2] Thus, impaired energy metabolism in the hypertrophied and failing hearts could be one of the underlying mechanisms of deteriorated function and increased susceptibility to ischemia.[3,4] An imbalance between the provision of substrates and the increased demand this in turn may lead to a disturbance of nucleotide. We evaluated the concentrations of high energy phosphate metabolites in the myocardium.

METHOD

Nineteen patients undergoing heart or heart/lung transplantation were studied. Fourteen patients were diagnosed with dilated cardiomyopathy (DCM). Five patients had hypertrophic cardiomyopathy (HCM). All heart transplant recipients were in end stage heart failure (N.Y.H.A. class IV) at the time of transplantation. Control specimens were obtained from three unused donor hearts and one from a patient undergoing retransplantation after heart-lung transplant due to chronic lung rejection.

Purine and Pyrimidine Metabolism in Man IX,
edited by Griesmacher *et al.* Plenum Press, New York, 1998.

Approximately 1 g of tissue was removed from the left ventricle prior to cross clamping of the aorta and freeze clamped within 15 seconds after excision of tissue using aluminium tongs precooled in liquid nitrogen. The tissue was stored in liquid nitrogen until further analyses. To measure total creatine and total nucleotide concentrations, the frozen tissue was weighed and ground to a fine powder. The powder was added to 5 volumes of ice cold 6% perchloric acid, mixed vigorously and left on ice for 10 minutes. Homogenates were subsequently centrifuged (3500 g for 10 minutes at 4°C) and a known volume of supernatant collected for neutralization with a solution of 6 M KOH and 2 M K_2CO_3. The neutralized sample was centrifuged for 10 minutes at 3500 g and 20 µl of the resulting supernatant was analyzed for PCr, Cr, ATP, ADP, AMP, hypoxanthine, inosine, adenosine, GTP, GDP, GMP, NAD and ADPR.

For determination of creatine metabolites an anion exchange HPLC method described previously was used.[5] The chromatographic analysis of nucleotides and their catabolites was performed using a reversed-phase method.[6]

RESULTS

Table 1 shows the concentrations of phosphocreatine, creatine, total adenylates, NAD and ADPR in the two heart failure groups and control specimens. The creatine concentration was 70% lower in the DCM group and 60% lower in the HCM group compared to control donor hearts. There was less variation in the levels of ATP, ADP, AMP and purine catabolites between the groups. Similarly NAD and ADPR remained comparable in all of the groups.

Table 2 gives the concentration of the guanine nucleotides. There was a trend towards higher GTP concentrations in the DCM and HCM groups.

Table 1. Phosphocreatine, creatine, total adenylates
and NAD µmol/g wet weight of tissue

	Control hearts (n = 4)	DCM (n = 14)	HCM (n = 5)
Phosphocreatine	0.23 ± 0.16	0.76 ± 0.11	0.99 ± 0.39
Creatine	10.50 ± 3.53	3.24 ± 0.48*	4.10 ± 0.99*
ATP	1.52 ± 0.59	2.46 ± 0.26	2.62 ± 0.72
ADP	0.70 ± 0.04	0.76 ± 0.10	0.92 ± 0.14
AMP	0.95 ± 0.70	0.28 ± 0.05	0.67 ± 0.39
Hypoxanthine	0.19 ± 0.14	0.03 ± 0.01	0.07 ± 0.04
Inosine	0.73 ± 0.56	0.07 ± 0.04*	0.34 ± 0.31
Adenosine	0.01 ± 0.01	0.02 ± 0.01	0.04 ± 0.01*
NAD	0.32 ± 0.03	0.32 ± 0.03	0.37 ± 0.06
ADPR	0.06 ± 0.01	0.04 ± 0.01	0.08 ± 0.03

*$p < 0.05$ vs. concentration in the control hearts.

Table 2. Guanine nucleotides and nucleosides
µmol/g wet weight of tissue

	Control hearts (n = 4)	DCM (n = 14)	HCM (n = 5)
GTP	0.16 ± 0.04	0.22 ± 0.02	0.20 ± 0.04
GDP	0.06 ± 0.01	0.08 ± 0.01	0.06 ± 0.01
GMP	0.04 ± 0.01	0.02 ± 0.01	0.04 ± 0.02

Total creatine pool (phosphocreatine + creatine), total adenine (TAN = ATP + ADP + AMP) and guanine (TGN = GTP + GDP + GMP) nucleotide pools and total NAD pool (NAD + ADPR) are presented in Fig. 1.

Total creatine (Fig. 1A) was markedly reduced in the two heart failure groups as compared to the donor tissue, TAN (Fig. 1B) and TGN (Fig. 1C) were moderately elevated in failing hearts but differences were not significant. Similarly total NAD (Fig. 1D) was not significantly different among the groups.

DISCUSSION

This study identified a specific and common pattern of changes in the myocardial energy metabolism in hearts with dilated cardiomyopathy (DCM) and in hearts with

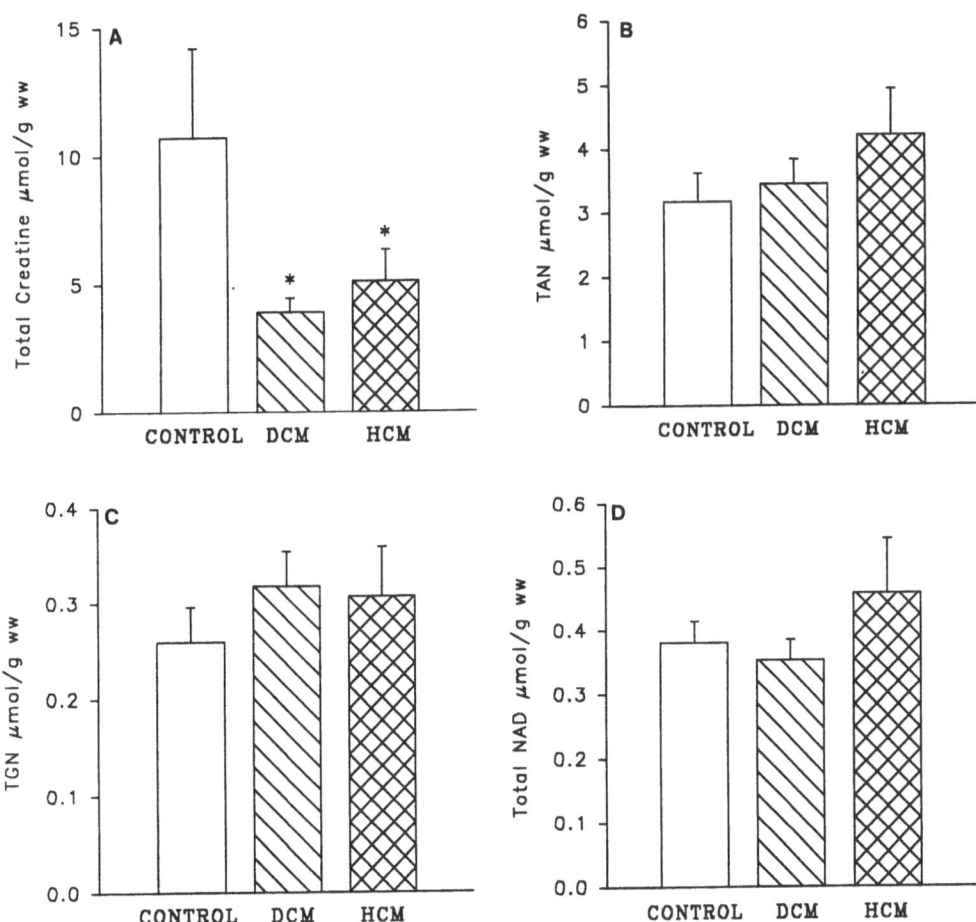

Figure 1. Total Creatine (A), total adenine nucleotides (TAN) (B), total guanine nucleotides (TGN) (C), and total NAD (NAD + ADPR) (D), in control hearts (n=4), DCM (n=14), and HCM (n=5). Values are means ± SEM. *p<0.05 vs the concentration in the control hearts.

hypertrophic cardiomyopathy (HCM). Decreased levels of creatine metabolites observed in these patients with heart failure was the most prominent difference observed. This observation is consistent with *in vivo* NMR studies showing that low PCr/ATP ratio was found in patients with heart failure[7-10] which could be explained by a depletion of creatine pool but also, alternatively by decreased phosphocreatine/creatine ratio. Our data are important in providing direct evidence that this low phosphocreatine/ATP ratio is caused by a decrease in the creatine pool.

Adenine nucleotide pool seems to be adequately maintained in end stage heart failure both in HCM and DCM. NAD pool was also not different among the groups. There was an enhancement in the guanine nucleotide pool which was elevated in the two patient groups. This may reflect enhanced RNA and protein synthesis, consistent with previous studies in response to physiological changes in the heart.[11]

In summary, the results highlight a decrease in the creatine pool in the failing hearts in the course of both, dilated and hypertrophic cardiomyopathies. Furthermore, energy resources and efficiency of energy transfer between mitochondrial and cytosolic compartments could be affected in the hearts of these patients at end stage heart failure. This, in turn, underlines possible areas for pharmacological intervention but its effects on mechanical function in the failing hearts require further detailed studies.

ACKNOWLEDGMENTS

We wish to thank British Heart Foundation and the Harefield Heart Transplant Trust for their support.

REFERENCES

1. A.M. Katz, Pathiophysiology of congestive heart failure, *Journal of Applied Cardiology* 5:427(1990).
2. J. Zhang, H. Merkle, K. Hendrich, M. Garwood, A.H. From, K. Ugurbil, and R.J. Bache, Bioenergetic abnormalities associated with severe left ventricular hypertrophy, *J. Clin. Invest.* 92:993(1993).
3. A.M. Katz, Metabolism of the failing heart, *Cardioscience* 4:199(1993).
4. P.E. Pool, J.F. Spann, Jr., R.A. Buccino, E.H. Sonnenblick, and E. Braunwald, Myocardial high energy phosphate stores in cardiac hypertrophy and heart failure, *Circ. Res.* 21:365(1967).
5. E. Harmsen, P.P. de Tombe, and J.W. de Jong, Simultaneous determination of myocardial adenine nucleotides and creatine phosphate by high-performance liquid chromatography, *J. Chromatogr.* 230:131(1982).
6. R.T. Smolenski, D.R. Lachno, S.J.M. Ledingham, and M.H. Yacoub, Determination of sixteen nucleosides, nucleotides and bases using high performance liquid chromatography and its application to the study of purine metabolism in hearts for transplantation. *Journal of chromatography* 527:414(1990).
7. M.A. Conway, J. Allis, R. Ouwerkerk, T. Niioka, B. Rajagopalan, and G.K. Radda, Detection of low phosphocreatine to ATP ratio in failing hypertrophied human myocardium by 31P magnetic resonance spectroscopy, *Lancet* 338:973(1991).
8. F. Brunotte, B. Peiffert, J.M. Escanye, G. Pinelli, J. Zamorano, P.M. Walker, J. Robert, and J.P. Villemot, Nuclear magnetic resonance spectroscopy of excised human hearts, *Br. Heart J.* 68:272(1992).
9. C.J. Hardy, R.G. Weiss, P.A. Bottomley, and G. Gerstenblith, Altered myocardial high-energy phosphate metabolites in patients with dilated cardiomyopathy, *Am. Heart J.* 122:795(1991).
10. S. Neubauer, T. Krahe, R. Schindler, M. Horn, H. Hillenbrand, C. Entzeroth, H. Mader, E.P. Kromer, G.A. Riegger, K. Lackner, and et al, 31P magnetic resonance spectroscopy in dilated cardiomyopathy and coronary artery disease. Altered cardiac high-energy phosphate metabolism in heart failure, *Circulation* 86:1810(1992).
11. J.L. Swain, R.L. Sabina, R.B. Peyton, R.N. Jones, A.S. Wechsler, and E.W. Holmes, Derangements in myocardial purine and pyrimidine nucleotide metabolism in patients with coronary artery disease and left ventricular hypertrophy, *Proc. Natl. Acad. Sci. U. S. A.* 79:655(1982).

150

EVALUATION OF ADENINE CONCENTRATION IN PLASMA OF PATIENTS WITH RENAL FAILURE USING IMPROVED ULTRAFILTRATION TECHNIQUE

Maciej Marlewski,[1] Ryszard T. Smolenski,[1] Arian Laurence,[2] H. Anne Simmonds,[2] and Julian Swierczynski[1]

[1]Department of Biochemistry
Medical University of Gdansk, Poland
[2]Purine Research Laboratories
Guy's Hospital
London, United Kingdom

INTRODUCTION

Changes in nucleotide pattern is one of the most prominent changes observed in the erythrocytes of patients with renal failure,[1,2] but the mechanism of this phenomenon is unclear. To evaluate possible role of increased precursor availability as potential cause of increased erythrocyte ATP and GTP concentrations in renal failure we analysed adenine and adenosine concentration in plasma. Since acid extraction before HPLC analysis may lead to a breakdown of unstable metabolites with release of adenine we applied ultrafiltration technique.

MATERIALS AND METHODS

Blood samples were collected into heparinized tubes from seven patients with renal failure, while nine healthy subjects served as controls. Plasma was separated by centrifugation immediately after collection and passed through Milipore ultra-free ultrafiltration membrane with M=10000 cut-off fitted into centrifuge tube insert. Tubes were centrifuged for 15 min. Ultrafiltrate was directly analysed using reversed-phase HPLC method described in detail previously.[3] Identity of adenine and adenosine quantified in plasma was verified using diode-array detector and comparison of UV spectra.

Purine and Pyrimidine Metabolism in Man IX,
edited by Griesmacher *et al.* Plenum Press, New York, 1998.

RESULTS AND DISCUSSION

Concentration of adenine in plasma ultrafiltrates is presented in Figure 1A. As may be seen, concentration of adenine was 17 times higher in patients with renal failure than in control subjects. Results of determination of adenosine concentration is presented in Figure 1B. As may be seen, there were no differences in concentration between renal failure and control subjects.

It is difficult to explain such a prominent increase in plasma adenine concentration in renal failure only by its insufficient excretion. Activation of formation of this base or its reduced utilisation seems to be an obligatory component. Adenine is predominantly produced in a human body as a by-product of polyamine synthesis in the 5'-deoxy-5'-

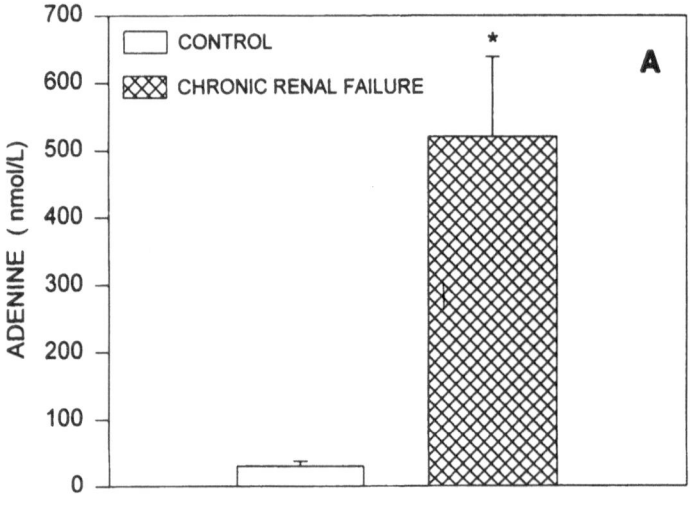

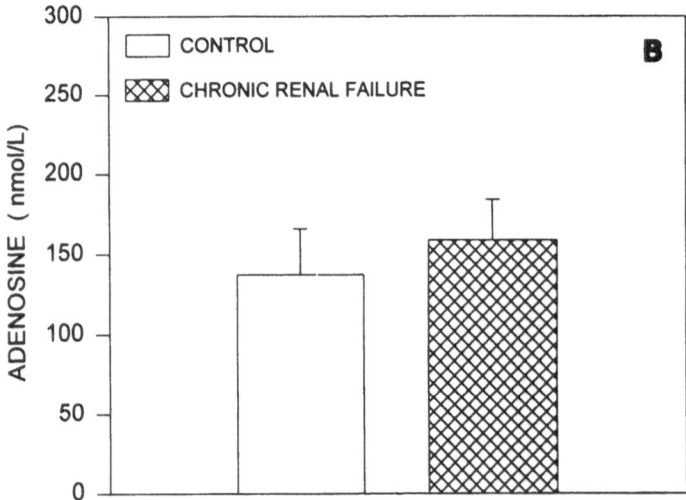

Figure 1. Concentration of adenine A or adenosine B in plasma ultrafiltrates collected from patients with renal failure (n=7) or from control subjects (n=9). Values are mean ± SEM. *p < 0.001.

methylthioadenosine phosphorylase reaction. High concentration of polyamines in blood of patients with renal failure seems to support a possibility that adenine production is enhanced in renal failure.[4,5]

Since adenine is a good substrate for nucleotide synthesis in the erythrocytes, it is possible that elevated adenine concentration contribute to increase in nucleotide pool in the erythrocytes of patients with renal failure. However, clarification whether this is a predominant mechanism or whether other processes are involved require further detailed studies.

ACKNOWLEDGMENTS

This study was supported by the Polish State Committee for Scientific Research (St-42 and W-922) and the Collaborative Research Grant from NATO (CRG 921365).

REFERENCES

1. G.A. Hurt and A. Chanutin, Organic phosphate compounds of erythrocytes from individuals with uremia, *J. Lab. Clin. Med.* 64:675(1964).
2. M.A. Lichtman and D.R. Miller, Erythrocyte glycolysis, 2,3-diphosphoglycerate and adenosine triphosphate concentration in uremic subjects: Relationship to extracellular phosphate concentration, *J. Lab. Clin. Med.* 76:267(1970).
3. R.T. Smolenski, D.R. Lachno, S.J.M. Ledingham, and M.H. Yacoub, Determination of sixteen nucleotides, nucleosides and bases using high-performance liquid chromatography and its application to the study of purine metabolism in hearts for transplantation, *J. Chromatogr.* 527:414(1990).
4. J. Gumprecht, M. Snit, W. Grzeszczak, E. Zukowska-Szczekowska, M. Kaminski, D. Moczulski, and J. Starzyk, Serum putrescine, spermidine and spermine concentrations in uraemic patients, *Ann. Acad. Med. Gedan.* 25:53(1995).
5. A. Saito, T. Takagi, T.G. Chung, and K. Ohta, Serum levels of polyamines in patients with chronic renal failure, *Kidney Int.* 24:234(1983).

151

A SYNDROME OF SEIZURES AND PERVASIVE DEVELOPMENTAL DISORDER ASSOCIATED WITH EXCESSIVE CELLULAR NUCLEOTIDASE ACTIVITY

Theodore Page,[1] Alice Yu,[2] John Fontenessi,[2] and William Nyhan[2]

[1]Department of Neurosciences and
[2]Department of Pediatrics
University of California, San Diego
La Jolla, California 92093

INTRODUCTION

Pervasive Developmental Disorder (PDD) is defined broadly as impaired reciprocal social interaction, communication, and imaginative activity combined with a markedly restricted repertoire of activities and interests.[1] PDD includes such disorders as childhood schizophrenia, infantile autism, and Asperger's syndrome. PDD is often associated with neurological symptoms such as seizures and ataxia. A number of single-gene defects and chromosomal abnormalities are associated with PDD; these include Fragile X, Down's syndrome, histidenemia, phenylketonuria, neurolipidosis, and tuberous sclerosis.

We report here a unique type PDD which is associated with seizures and other neurological symptoms, abnormal speech and behavior, and increased susceptibility to infection. All of these patients show excessive 5' nucleotidase activity in their fibroblast lysates. Metabolic therapy with oral uridine brings about dramatic improvement in every case.

CLINICAL PRESENTATION

The clinical presentation of four patients is summarized in Table 1, and was fairly consistent. All patients were markedly delayed in their developmental milestones, especially language. All had seizures, ataxia, an awkward gait, and mildly impaired fine motor control. All four displayed an unusual behavioral phenotype which was characterized by extreme hyperactivity, distractability, a strange "delirious" quality to their affect, and abnormal social interaction. All four patients experienced frequent ear and sinus infections but no consistent reason for immunodeficiency (such as reduced antibody titre or abnormal T-cell

Purine and Pyrimidine Metabolism in Man IX,
edited by Griesmacher *et al.* Plenum Press, New York, 1998.

Table 1. Clinical presentation of four patients

	Symptom	Patient 1	Patient 2	Patient 3	Patient 4
General	sex	F	F	M	M
	age first studied (years)	3	4	2	8
	growth retardation	+	−	+	+
Behavior	hyperactive	+	+	+	+
	inability to focus	+	+	+	+
	extreme distractability	+	+	+	+
	occasionally aggressive	+	−	+	+
	impulsive	+	+	+	+
	"delirious" affect	+	+	+	+
	compulsiveness	+	+	+	+
	abnormal social interaction	+	+	+	+
Speech	speech delay	+	+	+	+
	slurred speech	+	+	+	+
	tremulous speech	+	+	+	+
	short, telegraphic sentences	+	+	+	+
Neurological	seizures	+	+	+	+
	abnormal EEG	+	+	+	+
	ataxia	+	+	+	+
	impaired fine motor control	+	+	+	+
	awkward gait	+	+	+	+
Immunological	frequent infections	+	+	+	+
	abnormal immunoglobulins	+	+	−	−
	abnormal T-cell response	+	+	−	−
Other	developmentally delayed	+	+	+	+
	sparse hair/ hair loss	+	+	−	+
	skin rash	+	+	+	−
	urate excretion mg/mg creatinine	0.76	0.50	0.41	0.45

response) could be found. All patients excreted reduced quantities of uric acid when compared with age-matched controls[2], but other metabolic tests, such as plasma and urinary amino acids and organic acids were found to be within normal limits.

MATERIALS AND METHODS

Incorporation of purine and pyrimidine precursors into nucleotides in intact cultured fibroblasts was done as previously described[3]. Concentrations of intracellular nucleotides were measured by harvesting 4×10^6 cultured skin fibroblasts in the log phase of growth by trypsinization, extraction of the nucleotides by 100 ul of 0.5 M perchloric acid, neutralization with potassium phosphate buffer, and analysis by HPLC. Individual enzymes were assayed in fibroblast lysates by previously described methods.[4]

RESULTS AND DISCUSSION

Studies of the incorporation of purine and pyrimidine precursors into nucleotides as well as individual enzyme assays were performed to determine the basis of the observed

Table 2. Incorporation of precursors into nucleotides

Precursor	Pt 1	Pt 2	Pt 3	Pt 4	Controls (n)
Adenine	9472	9792	8993	9025	9170 (6)
Hypoxanthine	3246	3686	3071	3387	3043 (6)
Guanine	3429	2970	2893	2913	3242 (6)
Formate	5046	6681	6269	5890	7842 (3)
Uridine	3982	5840	4876	6137	8655 (3)

Incorporation of formate into nucleotides is in units of pmol/ 100 nmol UV/24 hr; all others are in units of pmol/100 nmol UV/2 hr

reduced uric acid excretion. The results of the incorporation studies are shown in Table 2. Incorporation of adenine, guanine, and hypoxanthine into nucleotides appears to be normal. Incorporation of formate into nucleotides, a measure of *de novo* purine synthesis, was found to be slightly reduced. Incorporation of uridine into pyrimidine nucleotides was also reduced. In all of the incorporation studies, no abnormal distribution of labeled nucleotides which would suggest a deficiency of any of the enzymes of nucleotide interconversion was seen.

Individual enzymes of pyrimidine metabolism which might affect the incorporation of uridine were assayed in fibroblast lysates. As seen in Table 3, all of these enzyme activities were normal with the exception of 5' nucleotidase, which was consistently 6- to 8-fold greater in the patients compared to age-matched controls. This increased nucleotidase activity was seen with purine substrates, such as AMP, as well.

To determine whether the increased nucleotidase acitivity caused any abnormality in steady-state nucleotide concentrations, nucleotide concentrations were measured in fibroblasts. All nucleotide levels appeared to be within normal limits (Table 4).

With consideration to the possibility that pyrimidine nucleotide deficiency in some cell type might be responsible for these symptoms, pyrimidine replacement therapy was initiated with uridine at a dose of 1000 mg/kg/day.

Table 3. Enzyme activities in fibroblast lysates

Enzyme (substrate)	Pt 1	Pt 2	Pt 3	Pt 4	Controls (n)
5' Nucleotidase (UMP)	7.48	8.21	7.61	8.67	1.14 (9)
5' Nucleotidase (AMP)	7.94	9.54	8.87	9.41	1.54 (6)
UMP Synthetase (Orotate)	2.55	1.56	1.95	2.27	1.99 (5)
Uridine Kinase (uridine)	3.41	3.09	4.06	3.53	3.85 (3)
Nucleoside Phosphorylase (Uridine)	0.59	0.39	0.65	0.44	0.48 (5)

Enzyme activities are in units of nmol/min/mg protein.

Table 4. Nucleotide concentrations in cultured fibroblasts

Nucleotide	Pt 1	Pt 2	Pt 3	Pt 4	Controls (5)
ATP	11.1	12.2	10.7	11.5	13.7
ADP	3.4	2.5	3.7	3.1	3.4
GTP	3.8	4.7	4.3	4.3	4.9
GDP	0.5	0.3	0.2	0.3	0.5
UTP	3.2	3.7	4.1	4.5	4.1
CTP	0.8	0.8	1.1	0.8	1.0

Nucleotide concentrations are in units of nmol/10^6 cells.

At this dose seizure activity was eliminated or greatly reduced, and other neurological symptoms improved as well. Speech became more normal and age-appropriate. Improvements in behavior and social interaction were also noted, and infections became less frequent. As the patients seemed to derive considerable benefit from uridine, a double-blind crossover study was undertaken to test the efficacy of uridine. The switch to placebo was accompanied by regression to the pretreatment state in virtually all areas. Performance on standardized tests of cognitive function also deteriorated while on placebo.

The metabolic basis of these symptoms is presently unknown. There is no evidence of pyrimidine nucleotide deficiency in the cells of these patients, and the disorder bears little resemblance to orotic aciduria. The rapid and impressive response to uridine and the rapid regression when uridine is discontinued argues against a neurodevelopmental defect. Rather, the increased nucleotidase activity appears to produce a continual deficiency of some essential metabolite, and this deficiency is corrected by uridine.

ACKNOWLEDGMENT

Funded by a grant from the Epilepsy Foundation of America.

REFERENCES

1. Young JG, Brasic JR, Leven L, Genetic causes of autism and the pervasive developmental disorders, in *Applications of Basic Neuroscience to Child Psychiatry*, Deutsch SI, Weizman A, Weizman R eds. New York, Plenum, 1990, pp 183–216.
2. Kaufman JM, Greene ML, Seegmiller JE, Urine uric acid to creatinine ratio - a screening test for inherited disorders of purine metabolism, *J Pediatr* 73 (1968) 583–592.
3. Page T, Bakay B, Nissinen E, Nyhan WL, Hypoxanthine-guanine phosphoribosyl-transferase variants: correlation of clinical phenotype with enzyme activity, *J Inher Metab Dis* 4 (1981) 203–206.
4. Page T, Nyhan WL, Yu AL, Yu J, A syndrome of megaloblastic anemia, immunodeficiency, and excessive nucoetide degradation, in *Purine and Pyrimidine Metabolism in Man VII* Harkness RA, Elion GB, Zoellner N eds, New York, Plenum Press, 1991, pp 345–348.

152

DE NOVO PURINE SYNTHESIS IS INCREASED IN THE FIBROBLASTS OF PURINE AUTISM PATIENTS

Theodore Page[1] and Mary Coleman[2]

[1]University of California, San Diego
La Jolla, California 92093
[2]Department of Medicine
Georgetown University School of Medicine
Washington, DC 20008

INTRODUCTION

A number of cases have been reported in the medical literature over the past thirty years describing patients with autistic symptoms combined with prominent hyperurico-suria.[1-3] A number of these patients exhibited other neurological symptoms as well, including seizures, ataxia, and spasticity. In some of these reports, the autistic and/or neurological symptoms seem to have been ameliorated by antihyperuricosuric metabolic therapy. Studies undertaken to determine the percentage of the autistic population which overproduces uric acid arrived at a figure of 22–29%, as compared to < 3% of the normal population.[3,4]

Two well-studied defects of purine metabolism which are associated with uric acid overproduction are hypoxanthine phosphoribosyltransferase (HPRT) deficiency and phos-phoribosyl pyrophosphate synthetase (PRPPS) superactivity. In both of these disorders it is possible to demonstrate increased *de novo* purine synthesis in the cultured fibroblasts of patients.[5] The present study was undertaken to determine if increased *de novo* purine synthesis could be demonstrated in the cultured fibroblasts of the purine autism patients as well, and if so, to compare the rate of purine synthesis with that of these other hyperurico-suric metabolic disorders.

AUTISTIC SUBJECTS

The subjects for this study were chosen both on the basis of classic autistic symptoms and uric acid production greater that two standard deviations above the normal mean.[3] Uric acid was determined by the uricase method on a 24-hr urine sample.

Purine and Pyrimidine Metabolism in Man IX,
edited by Griesmacher *et al.* Plenum Press, New York, 1998.

Table 1. Clinical presentation of purine autism patients

Symptom	Patient					
	1	2	3	4	5	6
Early onset of symptoms	+	+	+	+	+	+
Autistic aloneness	+	+	+	+	+	+
Communication deficit	+	+	+	+	+	+
Stereotypic behaviors	+	+	+	+	+	+
Apparent mental retardation	+	+	+	−	+	−
Abnormal sensory response	+	+	+	+	+	+
Self-injurious behavior	−	+	+	−	+	+
Seizures	+	−	−	+	+	+
Savant skills	+	−	−	−	+	−
Dietary sensitivities	+	−	+	+	−	−

The clinical presentation of six patients is summarized in Table 1. All the patients exhibited the classic autistic symptoms, including lack of interest in social contact, severely restricted communication, and lack of imaginative activity. These symptoms were apparent by the third year in every case. Stereotypic behavior was seen in all patients and consisted of twirling, repetitive motions, toe-walking, and hand-flapping. Though accurate assessment of intellectual capacity was complicated by impaired communicative ability, 4 of 6 patients were judged to be mildly retarded, though two of these retarded individuals were thought to have savant skills. All showed some sort of abnormal sensory response, such as increased sensitivity to audititory stimuli. Self-injurious behavior was seen in 4 of 6 patients and included eye-poking, head-banging, or hair-pulling. Clinical seizures and abnormal EEGs were seen in 4 of 6 patients and were of varied types. As is often noted with purine autism patients,[1-4] dietary sensitivities were noted; ingestion of certain foods seemed to exacerbate both behavioral and neurological symptoms.

MATERIALS AND METHODS

Fibroblasts from punch biopsies were grown in Coon's F12 media with 10% FBS in 100 mm plates. Cells were harvested in the log phase of growth between passages 6 and 10 and replated at a density of 10^6 cells per plate and switched to Earl's MEM. After 72 hrs, the cells were harvested and replated at 10^6 cells per plate, and 7 ml of MEM containing 10 uCi ^{14}C sodium formate was added. Cells were incubated for an additional 24-hr and harvested by trypsinization. The cell pellet was extracted with perchloric acid, neutralized with potassium phosphate and analyzed by HPLC as previously described.[6]

RESULTS AND DISCUSSION

The measurement of *de novo* purine synthesis in the cultured fibroblasts of the purine autism patients is shown in Table 2. For comparison, *de novo* purine synthesis was measured by the same method in the cultured fibroblasts of normal controls, a normouricosuric autistic patient, a Lesch-Nyhan patient, an HPRT-deficient hyperuricosuric patient, and a patient with PRPPS superactivity. Owing to the difficulty of finding age-matched normal control cell lines, urate excretion was not measured in the individuals from whom these

Table 2. Urate excretion and *de novo* purine synthesis

Patient	Age	Urate excretion mg urate/kg/day	*De novo* purine synthesis pmol/100 nmol/24 hr[4]
Normal 1	12	(8.3 ± 2.0)	8412
Normal 2	4	(8.3 ± 2.0)	8186
Normal 3	6	(8.3 ± 2.0)	7323
Normal 4	3	(8.3 ± 2.0)	10786
Normal 5	7	(8.3 ± 2.0)	9174
			8776 ± 1302
Autistic 1	5	26.5	30126
Autistic 2	7	19.8	36533
Autistic 3	12	17.9	31874
Autistic 4	4	16.2	29093
Autistic 5	5	35.4	37258
Autistic 6	9	18.1	28650
Autistic 7	6	5.9	8348
PRPPS superactivity	3	43	48273
HPRT-deficiency gout	4	37	52015
Lesch-Nyhan	6	48	56375

lines were established but is assumed to be normal; instead, the literature normal value is presented in the Table.[3]

Most of the purine autism patients excreted 2–3 times the normal amount of urate, and *de novo* synthesis in their fibroblasts was found to be elevated 3- to 4-fold. Analysis of the radiolabeled nucleotides produced from radiolabeled formate in the cultured fibroblast of these patients showed a distribution of both the type (adenosine, inosine, and guanosine) of nucleotides and degree of phosphoribosylation (base, nucleoside, mono-, di-, and triphosphate) which was comparable to normal controls (data not shown). This suggests that there is no gross deficiency in any of the enzymes of purine nucleotide interconversion which might provide a basis for increased purine synthesis.

The patients with other defects of purine metabolism showed a greater increase in both *de novo* purine synthesis in cultured fibroblasts and production of uric acid. The symptoms of urate overproduction, including urate crystaluria, renal colic, and gouty arthritis are often seen in these disorders,[6] whereas none of these symptoms was noted in any of the purine autism patients. The serum urate of purine autism patients is consistently within the normal range;[3] without antihyperuricemic therapy, the serum urate concentration often exceeds the limit of solubility of urate (6.4 mg/100 ml) in HPRT deficiency and PRPPS superactivity.[7] Apparently, excretion of urate by the kidneys is able to maintain a normal plasma urate concentration in the case of the smaller elevation seen in the purine autism patients.

The enzyme defect responsible for increased *de novo* purine synthesis in purine autism is thus far unknown. Symptomatically, purine autism bears little resemblance to any of the known genetic disorders which cause urate overproduction. HPRT deficiency and PRPPS superactivity are associated with a variety of neurological and behavioral abnormalities, but these are distinctly different from classic infantile autism.[5] Glycogen storage disease type I (glucose-6-phosphatase deficiency) is also associated with purine overproduction but has not been associated with any unusual behavioral phenotype.[8] Identification of the enzyme defect in purine autism would be valuable not only for diagnostic purposes but perhaps also in the design of a rational metabolic therapy. A number of patients seem to benefit from low purine diets and/or allopurinol[1-3,9] and this suggests that the disorder may be amenable to metabolic therapy.

ACKNOWLEDGMENTS

Funded by grants from the the Purine Research Society, the Stallone Fund for Autism Research, Cure Autism Now, and the Barbara, Jean and Norton L. Smith Trust.

REFERENCES

1. C Hooft, C Van Nevel, AF DeSchaepdryver: Hyperuricosuric encephalopathy without hyperuricemia. Arch Dis Childhood 43 (1968) 734–737.
2. H Kaihara: Two autistic retarded children with different inborn errors of metabolism. Bul Tokyo Rehab Center 1974 1–17.
3. M Coleman, MA Landgrebe, AR Landgrebe: Purine autism. Hyperuricosuria in autistic children: does this identify a subgroup of autism? in M Coleman (ed) The Autistic Syndromes. New York, Elsevier, 1976, 183–214.
4. C Gilberg, M Coleman: *The Biology of the Autistic Syndromes*, London, MacKeith Press, 1992, pp 211–226.
5. T Page, WL Nyhan: "Abnormalities of Purine and Pyrimidine Metabolism" in *Neurologic Manifestations of Pediatric Diseases*, ed B Berg, Boston: Butterworth, 1992, pp 177–191.
6. T Page, B Bakay, E Nissinen, WL Nyhan: Hypoxanthine-guanine phosphoribosyl transferase variants: correlation of clinical phenotype with enzyme activity, J Inher Metab Dis 4 (1981) 203–206.
7. JB Wyngaarden, WN Kelley: Gout. in JB Stanbury, JB Wyngaarden, DS Fredrickson, JL Goldstein, MS Brown (eds) The Metabolic Basis of Inherited Disease, Fifth Ed., New York: McGraw-Hill (1983) 1043–1114.
8. RR Howell, JC Williams: The glycogen storage diseases. in JB Stanbury, JB Wyngaarden, DS Fredrickson, JL Goldstein, MS Brown (eds) The Metabolic Basis of Inherited Disease, Fifth Ed., New York: McGraw-Hill (1983) 141–166.
9. M Coleman, M Landgrebe, A Landgrebe: Progressive seizures with hyperuricosuria reversed by allopurinol. Arch Neurol 31 (1974) 238–242.

153

DETERMINATION AND SEPARATION OF ALLANTOIN, URIC ACID, HYPOXANTHINE, AND XANTHINE BY CAPILLARY ZONE ELECTROPHORESIS

M. Pizzichini, L. Arezzini, C. Billarelli, F. Carlucci, and L. Terzuoli

Institute Biochemistry and Enzimology
University of Siena, Italy

1. OBJECTIVES

Hypoxanthine, xanthine, uric acid and allantoin are the main products of purine nucleotide catabolism and they are formed through the sequence: nucleotide → hypoxanthine → xanthine → uric acid → allantoin. Under normal conditions, they represent the balance between synthesis and breakdown of nucleotides. Their levels change, for example, under oxidizing conditions and may be useful for quantifying oxidant generation in human extracts. Uric acid is oxidized to allantoin by variuos reactive oxygen species and is thought to act as an antioxidant in human bodily fluids. Allantoin concentrations may therefore indicate free radical damage *in vivo*.

Here we report the separation of allantoin, uric acid, xanthine and hypoxanthine by capillary zone electrophoresis (CE). One of the greatest advantages of CE is that it requestes only small amounts of sample, which is useful where the amount of biological material is limited. CE is excellent for the determination of allantoin. Preliminary observations indicated that the peak falls in a zone free of interference by other signals. Allantoin is widely determined by the colorimetric method of Young and Conway (1), according to which allantoin is converted to glyoxylic acid by sequential hydrolysis under alkaline and acidic conditions, and the glyoxylic acid reacts with pheylhydrazine to form a hydrazone which is measured colorimetrically. This procedure is tedious and time critical. Many methods for the determination of allantoin based on HPLC have been reported (2–4). Some employ long columns and resolution is often unsatisfactory, requiring derivatization of allantoin or one of its degradation products (5–7). All these problems are overcome by CE.

Purine and Pyrimidine Metabolism in Man IX,
edited by Griesmacher *et al.* Plenum Press, New York, 1998.

2. DESIGN AND METHODS

Allantoin, uric acid, hypoxanthine and xanthine were purchased from Sigma. Disodium tetraborate and NaOH were obtained from Merck.

Standard solutions of allantoin, uric acid, hypoxanthine and xanthine were freshly prepared in double-distilled water before analysis. Plasma samples were thawed at room temperature and 500 µl plasma or serum was extracted with 100 µl 4 N perchloric acid. The extracts were centrifuged for 10 min at 12,000 × g. The supernatant was neutralized with KOH 5 N (*ca.* 35 µl), cooled for 1 hour and centrifuged (12,000 × g for 10 min). The final supernatant was stored at −20°C until analysis.

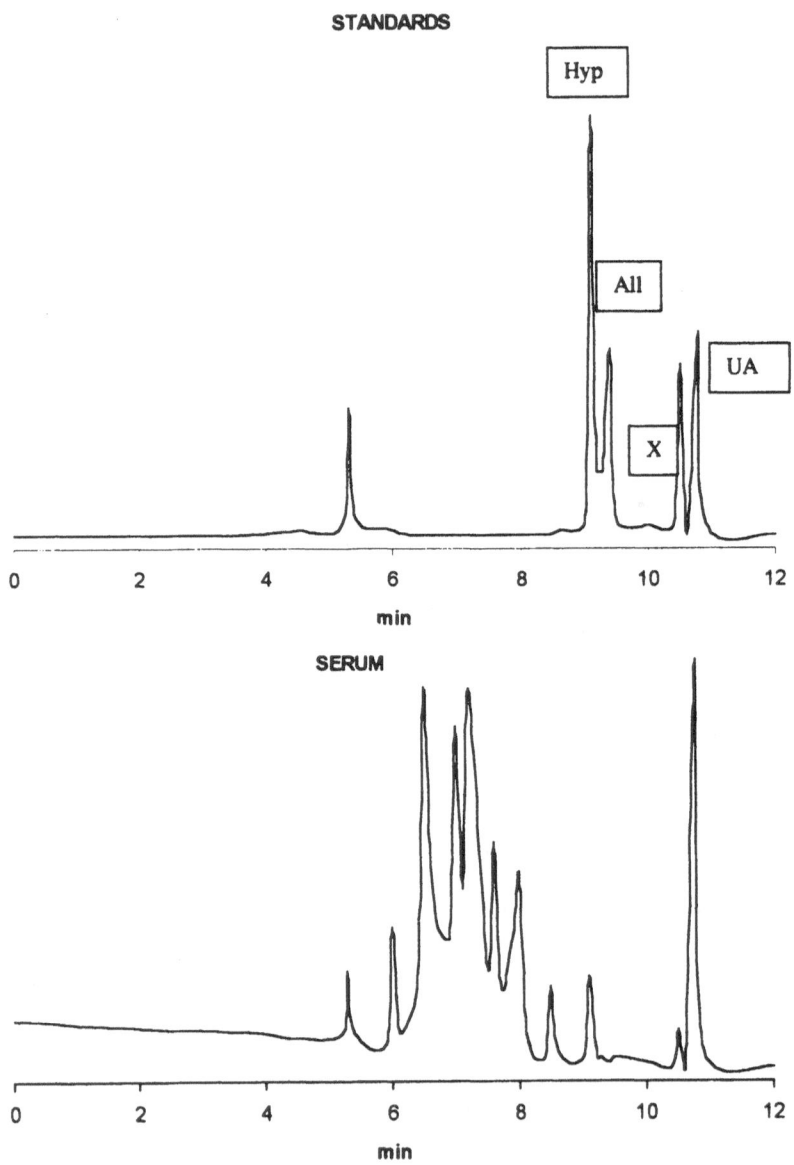

Figure 1. The eletrophoregram of allantoin, uric acid, xanthine and hypoxanthine and a serum.

For all electrophoretic separations a Waters Quanta 4000 instrument was used. Analyses were performed in an uncoated silica capillary (44 cm × 70 μm i.d.) with the window located at a distance of 67 cm. UV detection at 214 nm and hydrostatic injection (9 sec.) were used. For separation, a 20 mM borate buffer, pH 9.1, was employed, with the current stable around 30–35 μA. The capillary was cleaned by flushing sequentially with 100 mM NaOH, duoble-distilled water, and separation buffer for 12 min each. Between the analyses, the capillary was flushed for 1 min with the separation buffer prior to injection of the next sample.

3. RESULTS

The method permits the resolution of four compounds in 12 min. The migrations time were: 9.62 for hypoxanthine, 10.08 for allantoin, 11.68 for xanthine and 11.90 for uric acid The electropherogram of standard solutions and a perchloric acid extract of serum is shown in Figure 1.

The relationship between the concentration and peak are was a linear over a wide range of concentrations (20–100 μM). The sensitivity of the method is sufficient to measure hypoxanthine, xanthine, uric acid and allantoin in biological samples.

The reproducibility of the retention times and peak areas investigated for a standard mixture under optimized conditions (20 mM borate buffer, pH 9.1) is reported in Figure 2.

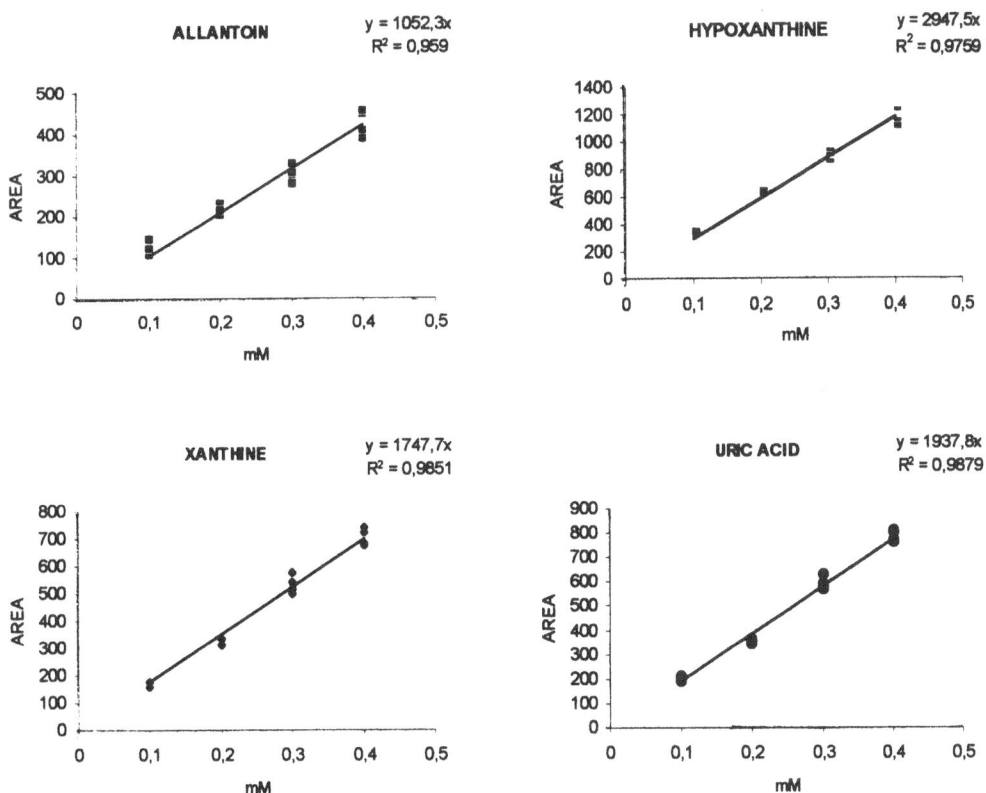

Figure 2. Correlation coefficients and regression equations of allantoin, hypoxanthine, xanthine and uric acid.

4. CONCLUSION

The CE method enables rapid analysis of purine derivatives using small-volume samples and short running times. It may also be used for assaying allantoin, uric acid, xanthine and hypoxanthine in biological extracts.

REFERENCES

1. E.G. Young and C.W. Conway. On the estimation of allantoin by the Rimini-Schryver reaction. (1942) J. Biol. Chem. *142*:839–853.
2. W. Tiemeyer and D. Giesecke. Quantitative determination of allantoin and oxypurines in biological fluids by reversed-phase high-performance liquid chromatography. (1992) Anal. Biochem. *123*:11–13.
3. J.A. Resines, M.J. Arin and M.T. Diez. Determination of creatinine and purine derivatives in ruminants' urine by reversed-phase high-performance liquid chromatography. (1992) J. Chrom. *607*:199–202.
4. J. Balcells, J.A. Guada, J.M. Peirò and D.S. Parker. Simultaneous determination of allantoin and oxypurines in biological fluids by high-performance liquid chromatography. (1992) J.Chrom. *575*: 153–157.
5. J. Lagendijk, J.B. Ubbinnk and W.J. Hayward Vermaak. The determination of allantoin, a possible indicator of oxidant status, in human plasma. (1995) J.Chrom.Sc. *33*:186–193.
6. K. Hirota, M. Kawase, S. Ohmori and T. Kishie. High-performance liquid chromatographic method for the determination of plasma allantoin. (1983) J.Chrom. *277*:165–172.
7. X.B. Chen, D.J. Kile and E.R. Orskov. Measurament of allantoin in urine and plasma by high-performance liquid chromatography with pre-column derivatization. (1993) J.Chrom. *617*:241–247.

154

PHARMACOKINETICS OF MYCOPHENOLATE MOFETIL IN HEART TRANSPLANT RECIPIENTS

G. Seebacher,[1] R. Mallinger,[1] G. Laufer,[1] M. Grimm,[1] A. Griesmacher,[2] G. Weigel,[1] E. Wolner,[1] and M. M. Müller[2]

[1]Department of Cardiothoracic Surgery
University of Vienna
[2]Ludwig Boltzmann Institute for Cardiothoracic Research at the Institute of
    Laboratory Diagnostics
Kaiser-Franz-Josef-Hospital
Vienna, Austria

INTRODUCTION

Mycophenolate Mofetil (MMF) is an ester prodrug of the immunosuppressant mycophenolic acid (MPA), which selectively inhibits the inosine monophosphate dehydrogenase (IMPDH). IMPDH converts inosine monophosphate (IMP) into xanthosine monophosphate (XMP) which will further be metabolized to ATP and GTP.

Lymphocytes are highly dependent on this de novo pathway of purine synthesis while other human cells can use the salvage pathway for purine formation. This means that MPA selectively blocks the proliferation of lymphocytes and is therefore used as an immunosuppressive substance (1). GTP is required for the formation of GDP-fucose and GDP-mannose, substances that are further metabolized to oligosaccharides. These oligosaccharides, functionally ligands, are altered in their expression on lymphocytes and endothelial cells. Therefore, lymphocyte-endothelium interactions are reduced leading to immunologic impairment (1,2).

Concerning the pharmacokinetics the following is known about mycophenolic acid so far: MPA is metabolized into a single glucuronidated metabolite, MPAG, which is mostly excreted in urine, whereby a small part can also be detected in the bile. This part is partly excreted in the feces and partly re-uptaken via an enterohepatic circulation. Some of the substance is also deglucuronidated so that MPA and MPAG show a second peak after the interval of 6 to 8 hours for MPA and 8 to 10 hours for MPAG (3).

Purine and Pyrimidine Metabolism in Man IX,
edited by Griesmacher *et al.* Plenum Press, New York, 1998.

Recently, we were able to demonstrate that there is a highly significant correlation of IMPDH activity and MPA plasma levels (unpublished data). Therefore, MPA should be the appropriate compound to analyze for assessment of the immunosuppressive activity of MMF.

PATIENTS AND METHODS

Seven patients, free from infection and transplant rejection, were enrolled in this study three months after orthotopic heart transplantation.

Patients received 1 g MMF twice a day orally (4,5); blood samples were taken at different time points and anticoagulated with EDTA. Samples were centrifuged immediately after venipuncture at 2000 RPM for 5 minutes at 4°C. Plasma was withdrawn and 25 µmol/l indomethacin was added as internal standard. Acetonitrile was used for deproteinization of the samples. After centrifugation and evaporation under nitrogen stream samples were analyzed by means of HPLC.

RESULTS AND CONCLUSION

Plasma time course profiles of MPA and MPAG were broadly similar after oral administration, although the early MPAG profile was a delayed version of that for MPA (Fig. 1).

This is consistent with the fact that MPA is the metabolic precursor of MPAG. Individual time course profiles showed a second peak of MPA plasma levels around 6 hours after intake of MMF as it was also measured for MPAG around 10 hours after administration (Fig. 2).

This observation might be attributable to the enterohepatic circulation which contributes to the bioavailability of MPA exceeding 6 hours and indicates that twice daily administration of MMF would be appropriate clinically.

Nevertheless, the second rise of plasma levels was not detectable in the overall statistical analysis due to the high variability of individual plasma levels.

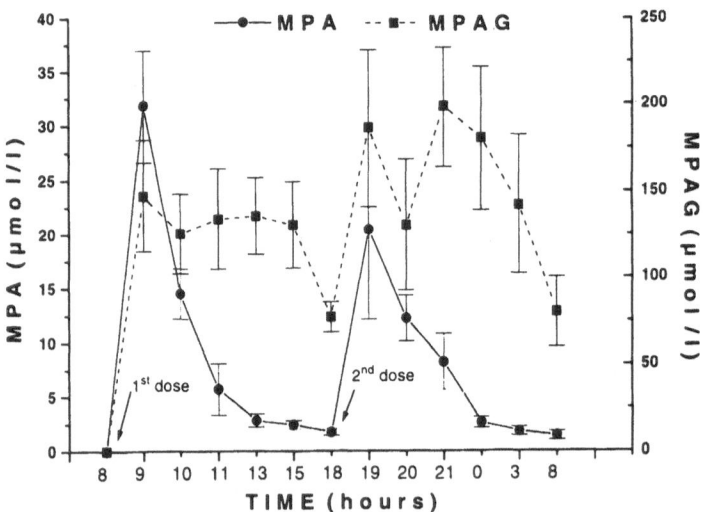

Figure 1. Plasma time course profile of MPA and MPAG.

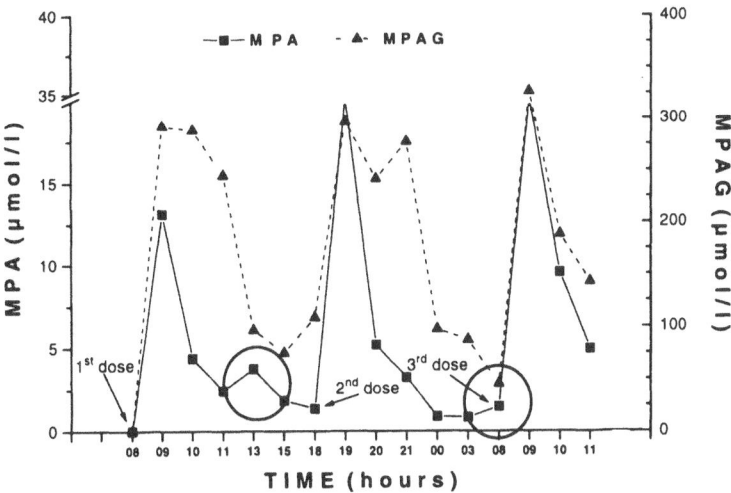

Figure 2. Individual time course profile of MPA and MPAG.

Based upon these results (e.g. correlation of IMP-activity, enterohepatic circulation) we have started a routine MPA monitoring of transplant recipients at our department which revealed to be important during the following situations:

1. During the first weeks after onset of MMF therapy the patients' response on the new drug can easily be evaluated and adverse events can be prevented.
2. Control of the patients' compliance especially during periods of cyclosporine A and steroid reduction.
3. Possibility of control in case of impaired renal function or in case of change of dosage.

REFERENCES

1. J. T. Ransom, Mechanism of Action of Mycophenolate Mofetil. Therapeutic Drug Monitoring 17:681–684 (1995).
2. A. C. Allison, W. J. Kowalski, C. J. Muller, R. V. Waters, and E. M. Eugui, Mycophenolic Acid and Brequinar, Inhibitors of Purine and Pyrimidine Synthesis, Block the Glycosylation of Adhesion Molecules. Transplant Proceedings 25, Suppl. 2: 67–70 (1993).
3. L. M. Shaw, H. W. Sollinger, P. Halloran, R. E. Morris, R. W. Yatscoff, J. Ransom, I. Tsina, P. Keown, D. W. Holt, R. Lieberman, A. Jaklitsch, and J. Potter, Mycophenolate Mofetil: A Report of the Consensus Panel. Therapeutic Drug Monitoring 17:690–699 (1995).
4. European Mycophenolate Mofetil Cooperative Study Group, Placebo-controlled study of mycophenolate mofetil combined with cyclosporin and corticosteroids for prevention of acute rejection. Lancet 345:1321–1325 (1995).
5. H. W. Sollinger for the U. S. Renal Transplant Mycophenolate Mofetil Study Group, Mycophenolate Mofetil for the prevention of acute rejection in primary cadaveric renal allograft recipients. Transplantation 60: 225–232 (1995).

155

IMMUNOHISTOCHEMICAL ANALYSIS OF P185 AND ASL ACTIVITY IN PRE-NEOPLASTIC LESIONS AND INTESTINAL MUCOSA OF HUMAN SUBJECTS

L. Terzuoli,[1] B. Frosi,[1] B. Porcelli,[1] F. Carlucci,[1] C. Minacci,[2] R. Vernillo,[3] L. Baldi,[1] A. Tabucchi,[1] and E. Marinello[1]

[1]Institute of Biochemistry and Enzimology
[2]Institute of Anatomo-pathology
[3]Institute of General Surgery and Surgical Specialities
University of Siena, Italy

1. INTRODUCTION

Colorectal cancer is the second neoplasia in terms of incidence and mortality (after lung-cancer in men and breast cancer in women), with 20,000 new cases in Italy every year and 10,000 deaths. The starting point of the multiphasic process leading to colorectal cancer (somatic and inherited genetic alterations which involve activation of oncogenes and inactivation of oncosuppressor genes) seems to be adenomatous polyposis. Since mortality due to colorectal cancer is high in our country, new perspectives for early diagnosis and chemotherapy, based on molecular and biochemical patterns at different stages of the pathological process, are needed.

The HER2 gene encodes a 185–190 kDa glycoprotein which has tyrosine-specific kinase activity and is homologous to epidermal growth factor receptor.[1–3] Overexpression of the HER2 product (p185) induces cell transformation in different experimental models, occurs in a wide variety of adenocarcinoma, and is associated with aggressive breast and ovarian tumors. This gene product was recently considered as a potential target for immunotherapy. The present study was undertaken to delineate the expression pattern of the HER2 gene in preneoplastic lesions.

Adenylosuccinate lyase (ASL, EC 4.3.2.2) is also increased in certain malignancies, like colorectal, breast and prostatic cancer.[4–5] This enzyme is a tetrameric protein, that catalyzes two distinct steps of de novo purine nucleotide biosynthesis. It converts succinylaminoimidazolo-carboxyamide ribotide (SAICAR) into AICAR, and also splits adenylosuccinate (S-AMP) into AMP and fumaric acid. In order to evaluate the potential of ASL as a tumor marker, its activity was compared with the expression of p185.

Purine and Pyrimidine Metabolism in Man IX,
edited by Griesmacher *et al.* Plenum Press, New York, 1998.

2. MATERIALS AND METHODS

The patients were under care at the Institute of General Surgery and Surgical Specialities of Siena University; the histopathological diagnosis was made by the Institute of
Anatomo- pathology of the same University.

For the study of p185 expression, we analyzed 16 subjects with pre-neoplastic
lesions with the following histotypes: 11 with tubular adenoma, 4 with tubulo-villous
adenoma and 1 with polypus. Paraffin-embedded tissue sections were processed for the
immunohistochemical analysis of p185, dewaxed by xylol incubation and dehydrated with
alcohol. Endogenous peroxidase activity was destroyed with hydrogen peroxide solution,
followed by washing in phosphate buffered saline solution. Tissue sections were immu-
nostained by the avidin-biotin peroxidase method.[6] Murine 300G9 antibody, specific for
the extracellular domain of human HER2 oncogene product,[7] was used at a concentration
of 10 μg/ml. Immunoenzymatic staining, performed according to the manufacturer's
recommendations (LSAB DAKOPATTS, Italy), employed 3-aminoethylcarbazol as chro-
mogenic substrate and Mayer's hematoxylin as nuclear counterstain followed by mounting
in buffered glycerol. The intensity of immunoperoxidase staining was graded as follows:
+, weakly positive; ++, moderately positive; +++, intensely positive.

Analysis of ASL activity was performed in 23 specimens intestinal mucosa adjacent to the neoplasia of colorectal cancer patients (A group) and in 14 of the 16 pre-
neoplastic colorectal lesions (B group), in which analysis of p185 was carried out. All
patients had undergone endoscopic biopsy by diathermal loop. The specimens were frozen in liquid nitrogen and stored at $-70°C$. Subsequently they were homogenized in four
volumes of 300 mM sucrose (pH 7.8), containing 10 mM Tris and 5 mM DTT. After centrifugation at $50,000 \times g$ for 60 minutes, supernatants were dialyzed with Slide A-Lyzer
10K Dialysis Cassettes by Pierce (Rockford, IL, U.S.A.) for 2 hours against the same
buffer. Protein content was determined by the Comassie brilliant blue binding procedure
of Bradford,[8] with Bio-Rad protein reagent, using crystalline bovine serum albumin as
standard.

The enzyme activity was evaluated following the disappearance of substrate and
the formation of product, in a mixture containing 200 mM Tris (pH 7.4) and 0.5 mM
S-AMP. The dialyzed extract was added at a concentration of 200 μg of protein /400 μl
of mixture. The reaction was stopped adding $HClO_4$ (0.21 N final concentration) and neutralized with KOH (0.22 N final concentration). S-AMP and AMP were separated by
high performance liquid chromatography (Beckman, Mod 110B pumps and Mod 166 UV
detector set at 254 nm), with an anionic Partisil 5 Sax (4.6 × 100 mm) column. Gradient
elution was performed with low ionic-strength buffer (A: 5 mM ammonium monobasic
phosphate pH 2.8) and high ionic-strength buffer (B: 500 mM ammonium monobasic
phosphate pH 4.8) The gradient was: 3.5 min 0 → 80% B; 1 min 80 → 100% B; 8.5 min
100% B; 2 min 100 → 0% B; equilibration of the column at 0% B for 5 min, for a total
analysis time of 20 min. The flow rate was 1.5 ml/min. The enzyme activity was
expressed as nmol/h/mg of protein and nmol/h/g of tissue, for S-AMP disappearance and
AMP formation.

3. RESULTS

Immunohistochemicals analysis of the expression of p185 in the three histotypes
showed a high immunohistochemical reactivity (+++) in 12 subjects, with strong homoge-

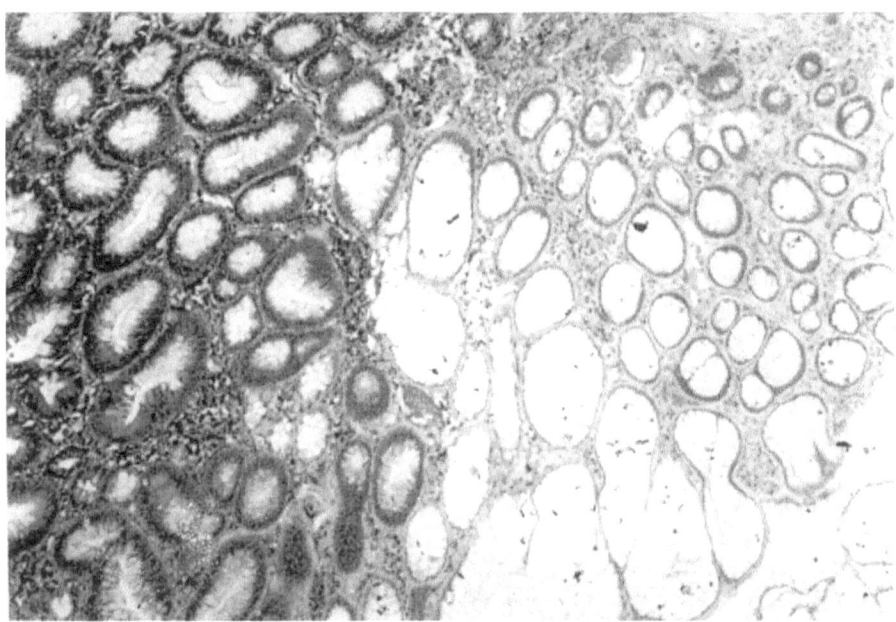

Figure 1. Expression of p185 in preneoplastic lesions. High immunohistochemical reactivity is expressed by cells of a colon adenoma.

neous staining of the more dysplastic glandular structures with respect to normal areas (Fig. 1). Three subjects showed diffuse, moderately positive staining, prevalently basal (++). One subject had a weak non homogeneous staining (+).

Table 1 showed ASL activity in the 14 biopsy fragments and ASL activity in the mucosa adjacent to neoplasias of colorectal cancer patients.

High enzyme activity was observed in the preneoplastic lesion (B), whether expressed as nmoles of substrate consumed or as nmoles of reaction product formed per gram of tissue. When expressed in terms of protein content, the enzyme activity showed only slight, non significant differences.

Table 1. Activity of ASL in the mucosa adjacent to neoplasia of colorectal cancer patients (A) and in pre-neoplastic colorectal lesions (B)[a]

		nmoles/h/g tissue		nmoles/h/mg protein	
		AMP	AMP-S	AMP	AMP-S
A	n=23	2074.37	2596.80	93.34	118.93
	s.e.	416.08	471.42	13.44	18.94
B	n=14	4974.19*	5864.13*	123.42	131.91
	s.e.	900.33	1140.14	14.20	18.60

n = number of cases. *$P \leq 0.05$
[a]Values are reported as nmoles of AMP-S disappearing and AMP formed /h/g of tissue and nmoles of AMP-S disappearing and AMP formed /h/mg of protein.

4. DISCUSSION

Among the antigens implicated in neoplastic cell growth, those encoded by activated oncogenes expressed on the cell membrane and restricted to specific tissues may be of clinical relevance. In our study we attempted to obtain more information on the biology and the potential value of p185 in clinical oncology. Our results show a high level of expression of this antigen in pre-neoplastic colon lesions, indicating that oncogene products such as p185-HER2 are candidates for markers of this disease.

The association of high ASL activity and preneoplastic lesions is not surprising in alterations that implicate the loss of replicative control. A variety of alterations in enzyme activities of purine metabolism have been observed in tumor cells, associated with neoplastic transformation and/or progression.[9-10]

The entity of the alterations in ASL activity is different if we express the activity in nmoles/h/mg protein or nmoles/h/g of tissue. This dissimilar behaviour is due to the different protein content of adenomatous mucosa and area adjacent to cancer (data not shown), being 100% higher in the former.

The activity expressed in terms of tissue weight represents the absolute activity of the enzyme, that in terms of protein content, only the behaviour of the protein with respect to the protein pool of the cell. Moreover in certain neoplasias some proteins show a relative increment as a consequence of higher resistance to degradation.[10]

The pattern of ASL activity and p185 expression are associated with pre-neoplastic lesions recognized as the starting-point of colorectal cancer, in which we have already demonstrated an increase in ASL activity.[11] In order to evaluate the possibility of using these determinations as markers for the early diagnosis of colorectal cancer and the possibility of a correlation between the degree of p185 expression and ASL activity, we need to extend this study to a larger population.

ACKNOWLEDGMENT

We thank Prof. Natali, Institute Regina Elena, Roma, for the gift of Mab 300G9.

REFERENCES

1. King C.R., Kraus M., Aaronson S.A.:Amplification of a novel v-erbB related gene in human mammary carcinoma. Science (Wash. DC),229: 974–976, 1985
2. Semba K., Kamata N., Toyoshima K., Yamamoto T. A v-erbB related protooncogene, c-erbB-2, is distinct from the c-erb-1 epidermal growth factor receptor gene and is amplified in human salivary gland adenocarcinoma. Proc.Natl. Acad. Sci. Usa, 82: 6497–6501,1985.
3. Yokota J., Toyoshima K., Yamamoto T., Terada M., Battifora H., Cline M.J. Amplification of c-erbB-2 oncogene in human adenocarcinomas in vivo. Lancet 2: 765–766,1986.
4. Jackson R.C., Morris H.P., Weber G. Enzymes of the purine ribonucleotide cycle in rat hepatomas and kidney tumors. Cancer Res. 37: 3057- 3065,1977.
5. Weber G. Biochemical strategiy of cancer cells and the design of chemotherapy: G.H.A. Clowes Memorial Lecture. Cancer Res. 43: 3466–3492, 1983.
6. Hsu S.M., Raine L., Fanger H.J. Histochem. Cytochem 29,577–580, 1981.
7. Digiesi G., Giacomini P., Mariani M., Nicotra M.R., Segatto O. Production and characterization of murine Mabs to the extracellular domain of human neu oncogene product GP185HER2. Hybridoma 11:519–527,1992.
8. Bradford M.M. A rapid and sensitive method for the quantitation of microgram quantities of protein utilizing the principle of protein-dye binding. Anal Biochem 72: 248- 254 , 1976.

9. Reed V.L., Mack D.O., Smith L.D. Adenylosuccinate lyase as an indicator of breast and prostate malignancies: a preliminary report. Clin. Bichem. 20: 349–351, 1987.

10. Jackson R.C., Morris H.P., Weber G. Neoplastic transformation-linked alterations in adenylosuccinate synthetase activity. Biochem. Biophys. Res. Comm. 66: 528–532, 1975.

11. Lorenzini L., De Martino A., Testi W., Sorbellini F., Dispensa E., Tabucchi A., Carlucci F., Rosi F. Behavior of enzymes involved in purine nucleotide metabolism in tumors. Purine and Pyrimidine Metabolism in Man VIII 213–218, Plenum Press Newe York, 1995.

156

NO CIRCADIAN VARIATION OF DIHYDROPYRIMIDINE DEHYDROGENASE, URIDINE PHOSPHORYLASE, β-ALANINE, AND 5-FLUOROURACIL DURING CONTINUOUS INFUSION OF 5-FLUOROURACIL

A. B. P. Van Kuilenburg,[1] R. L. Poorter,[1] G. J. Peters,[2] A. H. Van Gennip,[1] H. Van Lenthe,[1] A. E. M. Stroomer,[1] K. Smid,[2] P. Noordhuis,[2] P. J. M. Bakker,[1] and C. H. N. Veenhof[1]

[1]Academic Medical Center
University of Amsterdam
Department Clinical Chemistry and Division Emma Children's Hospital
P. O. Box 22700
1100 DE Amsterdam, The Netherlands
[2]Free University Hospital
Department of Oncology
P. O. Box 7057
1007 MB Amsterdam, The Netherlands

1. INTRODUCTION

Dihydropyrimidine dehydrogenase (DPD, EC 1.3.1.2) is the initial and rate-limiting enzyme in the catabolism of the pyrimidine bases and it catalyses the reduction of thymine and uracil to 5,6-dihydrothymine and 5,6-dihydrouracil, respectively. In mammals, the degradation of uracil by DPD is the only pathway leading to the biosynthesis of β-alanine. Furthermore, DPD is also responsible for the breakdown of the widely used antineoplastic agent 5-fluorouracil (5FU), thereby limiting the efficacy of the therapy. 5FU is one of the few drugs that shows some antitumour activity against various otherwise untreatable tumours including carcinomas of the gastrointestinal tract, breast, ovary and skin. Furthermore, 5FU is one of the few drugs for which a limited clinical effect has been shown when applied as a single agent during the treatment of advanced colorectal cancer. In order to exert its cytotoxic effect against cancer, 5FU must first be anabolised to the nucleotide level. 5FU can be converted into FUMP by a sequential, two-step reaction consisting of the initial addition of a ribose by uridine phosphorylase (UP) to yield 5-fluorouridine (5FUrd), followed by phos-

Purine and Pyrimidine Metabolism in Man IX,
edited by Griesmacher *et al.* Plenum Press, New York, 1998.

811

phorylation to FUMP by uridine kinase. The direct conversion of 5FU to FUMP is catalyzed by orotate phosphoribosyl transferase (OPRT) which transfers the ribose-phosphate moiety from phosphoribosyl pyrophosphate (PRPP) to 5FU. Although the cytotoxic effects of 5FU are probably directly mediated by the anabolic pathways, the catabolic route plays a significant role since more than 80% of the administered 5FU is catabolised by DPD. It has been reported that the bioavailability, efficacy as well as host-toxicity of 5FU follows a circadian rhythm in rodents[1] and cancer patients[2]. In the present study, we investigated whether a circadian variation could be observed of the activity of DPD, UP and the plasma concentration of β-alanine and 5FU in patients treated with continuous infusion of 5FU.

2. MATERIALS AND METHODS

2.1. Patients

Five patients with metastatic colorectal cancer were treated with continuous infusion of 5FU; four patients received a dose of 300 mg/m²/day and one patient received a dose of 450 mg/m²/day. Before treatment a venous access device (Port-a-Cath®, Pharmacia Deltec Inc, St. Paul, USA) was implanted and the chemotherapy was delivered by an ambulatory portable infusion pump (Pharmacia CADD-Plus®, Pharmacia Deltec Inc, St. Paul, USA). Treatment was scheduled for 14 days every 4 weeks. At day 7 and 14, halfway and at the end of a chemotherapy course, patients were hospitalized for blood sampling. Blood samples were taken every 4 hours during one day

2.2. Isolation of Blood Cells

Peripheral blood mononuclear cells were isolated from EDTA-anticoagulated blood by centrifugation over lymphopaque (spec. gravity 1.086 g/ml, 350 mOsm). Cells from the interface were collected and treated with ice-cold NH_4Cl to lyse the contaminating erythrocytes.

2.3. Determination of the DPD Activity and UP Activity

The activity of DPD was determined in a reaction mixture containing 35 mM potassium phosphate (pH 7.4), 2.5 mM $MgCl_2$, 1 mM dithiothreitol, 250 μM NADPH and 25 μM $[2-^{14}C]$thymine. Separation of radiolabeled thymine from radiolabeled dihydrothymine was performed isocratically (50 mM NaH_2PO_4 (pH 4.5) at a flow rate of 2 ml/min) by reversed-phase HPLC on a Supelcosil LC-18-S column (250 × 4.6 mm, 5 μm particle size) with online detection of the radioactivity.[3]

The activity of UP was determined in a reaction mixture containing 19 mM Tris-HCl (pH 7.4), 0.4 mM EDTA, 3.8 mM $MgCl_2$, 9.2 mM ribose-1-phosphate and 68 μM $[6-^{14}C]$5FU. Separation of radiolabeled 5FU from radiolebeled 5Furd was performed by thin layer chromatography on PEI cellulose with subsequent quantification of the spots with liquid scintillation counting.

2.4. Determination of 5FU and β-Alanine in Plasma

The plasma concentrations of 5FU were determined with a GC-MS procedure, as described before.[4] The concentration of β-alanine in plasma was determined by HPLC separation with postcolumn o-phtaldialdehyde detection.

2.5. Statistical Analysis

The correlation analysis was performed by determination of the Pearsons correlation coefficient and linear regression. The data were analysed by the cosinor method whereby the data were fitted to a cosine by the least-squares method. Four parameters were calculated: the mesor (rhythm-adjusted mean), the amplitude (maximum or minimum value from the mean), the acrophase (time of maximum or minimum) and ψ (the differences between the time of maximum of UP, β-alanine and 5FU with the time of maximum of DPD).Analyses were performed with the Statistical Package for the Social Sciences (SPSS Inc. Chicago, USA).

3. RESULTS

Surprisingly, analysis of the individual data of the patients treated with 5FU did not reveal a significant circadian rhythm for DPD, UP, β-alanine or 5FU in almost all cases ($p > 0.05$). Due to the large inter-patient variability of the obtained values for the activity of DPD, UP and the levels of β-alanine and 5FU, the data of each patient were normalized in order to establish an overall pattern for all patients. Therefore, the data were expressed as a percentage of the 24-h mean for that patient. A large inter-patient variability was observed for the time of maximum of DPD, UP and 5FU (Table 1). In contrast, a profound variation with time was observed for β-alanine and maximum concentrations of β-alanine were observed at 20.00 h (Table 1). For the normalized data the maximum activity of DPD exceeded the minimum activity by a factor of 1.13, thus indicating that almost no variation with time was observed for DPD. Furthermore, the ψ values from Table 1 also showed that their is no apparent time correlation between the peak of the DPD activity and the peak of the UP activity or the peaks of the 5FU and β-alanine concentration.

Fig. 1A shows that there is a negative correlation between the mean DPD activity and the mean 5FU levels in plasma. Similarly, a negative correlation was observed between the UP activity and the 5FU levels in plasma (Fig. 1D). A positive correlation between the mean DPD activity and the mean UP activity was observed (Fig. 1C), whereas no correlation was found between the mean DPD activity and the β-alanine concentration (Fig. 1B).

4. DISCUSSION

5FU has been shown to be an effective antimetabolite with antitumour activity in a number of different tumors including carcinomas of the gasterointestinal tract, breast, ovary and skin. In order to exert its cytotoxic effect 5FU must first be anabolised to the nucleotide level. In rodents a circadian rhythm has been demonstrated for OPRT, UP, thymidine kinase and DPD which are crucial enzymes in the anabolism and catabolism of 5FU.[5-8] In our study, we could not detect a clear circadian variation for the activity of DPD and UP nor for the concentrations 5FU and β-alanine in patients treated with continuous infusion with 5FU. Even if the circadian rhythm would have been significant in all patients the maximum activity of DPD exceeded the minimum activity of DPD only with a factor of 1.13. Our results are not in line with those of Harris and coworkers who provided evidence for a clear circadian variation of DPD and 5FU.[2] In their patients the maximum/minimum ratio for the activity of DPD and the plasma levels of 5FU were 1.9 and

Table 1. Parameters describing the analysis of the data by the cosinor method

	Mesor[1] (%)	Amplitude[2] (%)	Maximum[3] (%)	Time[4] of maximum (h)	Minimum[5] (%)	Time[4] of minimum (h)	Maximum/ minimum	ψ[6] (h)
DPD	99.6 ± 1.2	13.3 ± 4.4	112.9 ± 4.0	11.7 ± 7.5	86.4 ± 4.9	23.7 ± 7.5	1.13 ± 0.04	
UP	99.7 ± 1.5	15.4 ± 5.5	115.3 ± 4.6	15.5 ± 7.4	84.5 ± 6.3	3.5 ± 7.4	1.16 ± 0.05	−1.0 ± 6.4
5FU	99.7 ± 5.3	41.4 ± 25.7	140.8 ± 25.0	12.9 ± 9.3	58.6 ± 25.7	0.9 ± 9.3	1.41 ± 0.25	−0.1 ± 6.9
β-alanine	102.7 ± 2.1	51.6 ± 12.9	155.8 ± 15.4	19.4 ± 1.1	49.6 ± 12.9	7.4 ± 1.1	1.52 ± 0.13	2.8 ± 6.1

[1] Rhythm-adjusted mean
[2] Maximum or minimum value from the mean
[3] Calculated by adding the amplitude to the mean
[4] Calculated from the acrophase parameter from the cosinor analysis
[5] Calculated by substracting the amplitude from the mean
[6] Difference between time of maximum with time of DPD maximum

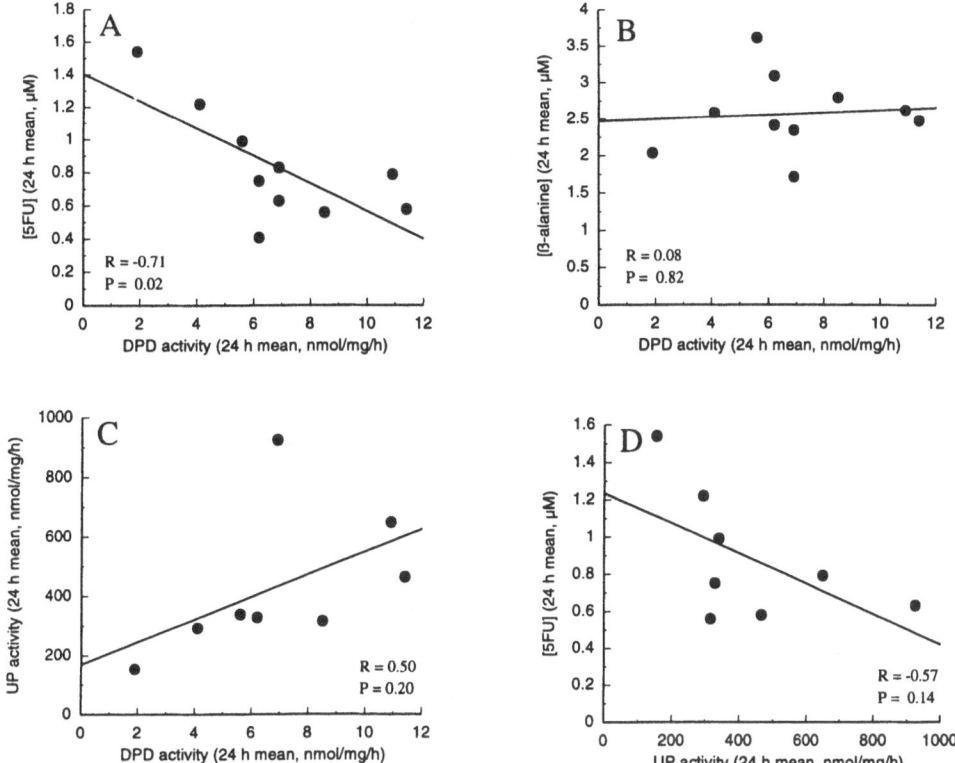

Figure 1. The correlation between the mean activities of DPD, UP and the concentrations of 5FU and β-alanine.

4.8, respectively. Surprisingly, in our patients a profound variation with time was observed for β-alanine, which could not be fitted according to a cosine, and maximum concentrations of β-alanine were observed at 20.00 h.

Although the large variation in ψ values demonstrated that there was no correlation between the time of maximum of the activity of DPD and the time of maximum of 5FU levels in plasma, a negative correlation was observed for the mean plasma levels of 5FU and the DPD activity. This observation is in line with the correlation that has been observed between the pretreatment activity of DPD in peripheral blood mononuclear cells and the systemic clearance of 5FU in cancer patients.[9,10] The negative correlation between the mean activity of UP and the plasma levels of 5FU might suggest that both the catabolic pathway (DPD) and the anabolic pathway (UP) are involved in regulating the levels of 5FU.

REFERENCES

1. Peters, G.J., Van Dijk, J., Nadal, J.C., Van Groeningen, C.J., Lankelma, J. and Pinedo, H.M. (1987) Diurnal variation in the therapeutic efficacy of 5-fluorouracil against murine colon cancer. *In Vivo* 1, 113–118.
2. Harris, B.E., Song, R. and Diasio, R.B. (1990) Relationship between Dihydropyrimidine Dehydrogenase Activity and Plasma 5-Fluorouracil Levels with Evidence for Circadian Variation of Enzyme Activity and Plasma Drug Levels in Cancer Patients Receiving 5-Fluorouracil by Protracted Continuous Infusion. *Cancer Res.* **50**, 197–201.

3. Van Kuilenburg, A.B.P., Van Lenthe, H., and Van Gennip, A.H. (1996) Identification and Tissue- Specific Expression of a NADH-Dependent Activity of Dihydropyrimidine Dehydrogenase in Man. *Anticancer Res.* **16**, 389–394.

4. Peters, G.J., Lankelma, J., Kok, R.M., Noordhuis, P., Van Groeningen, C.J., Van der Wilt, C.L., Meyer, S. and Pinedo, H.M. (1993) Prolonged retention of high concentrations of 5-fluorouracil in human and murine tumors as compared with plasma. *Cancer Chemother. Pharmacol.* **31**, 269–276.

5. Harris, B.E., Song, R., He, Y.-J., Soong, S.-J. and Diasio, R.B. (1988) Circadian rhythm of rat liver dihydropyrimidine dehydrogenase. Biochem. Pharmacol. 37, 4759–4762.

6. El Kouni, M.H., Naguib, F.N.M., Park, K.S., Cha, S., Darnowski, J.W. and Soong, S.-J. (1990) Circadian rhythm of hepatic uridine phosphorylase activity and plasma concentrations of uridine in mice. *Biochem. Pharmacol.* **40**, 2479–2485.

7. Naguib, F.N.M., Soong, S.-J. and El Kouni, M.H. (1993) Circadian rhythm of orotate phosphoribosyltransferase, pyrimidine nucleoside phosphorylases and dihydrouracil dehydrogenase in mouse liver. Possible relevance to chemotherapy with 5-fluoropyrimidines. *Biochem. Pharmacol.* **45**, 667–673.

8. Zhang, R., Lu, Z., Liu, T., Soong, S.-J. and Diasio, R.B. (1993) Circadian rhythm of rat spleen cytoplasmic thymidine kinase. *Biochem. Pharmacol.* **45**, 1115–1119.

9. Fleming, R.A., Milano, G., Thyss, A., Etienne, M.-C., Reneé, N., Schneider, M. and Demard, F. (1992) Correaltion between Dihydropyrimidine Dehydrogenase Activity in Peripheral Mononuclear Cells and Systemic Clearance of Fluorouracil in Cancer Patients. *Cancer Res.* **52**, 2899–2902.

10. Etienne, M.C., Lagrange, J.L., Dassonville, O., Fleming, R., Thyss, A., Reneé, N., Schneider, M., Demard, F. and Milano, G. (1994) Population Study of Dihydropyrimidine Dehydrogenase in Cancer Patients. *J. Clin. Oncology* **12**, 2248–2253.

157

SUBCELLULAR LOCALIZATION OF DIHYDROPYRIMIDINE DEHYDROGENASE

André B. P. Van Kuilenburg, Henk Van Lenthe, Ronald J. A. Wanders, and Albert H. Van Gennip

Academic Medical Center
University of Amsterdam
Department Clinical Chemistry and Division Emma Children's Hospital
P. O. Box 22700
1100 DE Amsterdam
The Netherlands

1. INTRODUCTION

Dihydropyrimidine dehydrogenase (DPD, EC 1.3.1.2) is the initial and rate-limiting step in the catabolism of the pyrimidine bases and it catalyses the reduction of uracil and thymine to 5,6-dihydrouracil and 5,6-dihydrothymine, respectively. In children, a deficiency of DPD is often accompanied by a neurological disorder, but a considerable variation in the clinical presentation among these patients has been reported.[1] DPD is also responsible for the breakdown of the widely used antineoplastic agent 5-fluorouracil, thereby limiting the efficacy of the therapy.

The activity of DPD can be detected in a variety of tissues with the highest activity being found in liver[2] and peripheral blood monocytes.[3] The enzyme has been purified and extensively characterized from liver tissues of various mammals but conflicting data exist on its subcellular localization. Although the activity of DPD can be detected in a soluble fraction obtained during the purification process, one can not conclude from this observation that its localization is exclusively cytosolic. Moreover, in most of these studies frozen tissue has been used which implies that all subcellular structures have been disrupted. So far, the only studies which were entirely dedicated to the subcellular localization provided evidence that the majority of the DPD activity was present in subcellular organelles.[4,5] For these reasons, we feel that the subcellular localization of DPD has not been properly investigated and therefore a thorough study is warranted. In this paper, we have investigated the subcellular localization of DPD in rat liver using differential and equilibrium density centrifugation.

Purine and Pyrimidine Metabolism in Man IX,
edited by Griesmacher *et al.* Plenum Press, New York, 1998.

2. MATERIALS AND METHODS

2.1. Differential Centrifugation and Density Gradient Fractionation of Rat Liver Homogenates

Rat liver (6.9 g) was finely minced and homogenized in a medium containing 250 mM sucrose, 2.5 mM EDTA and 5 mM Mops-NaOH (pH 7.4) with a Teflon-glass homogenizer. The resulting homogenate was subjected to differential centrifugation to prepare a heavy mitochondrial, light mitochondrial, microsomal and cytosolic subcellular fraction.[6] Dithiothreitol was added to a final concentration of 1 mM to aliquots of the subcellular fractions used for the determination of the activity of DPD.

A light-mitochondrial fraction was prepared from rat liver (13.1 g) of a male Wistar rat fed a standard laboratory diet supplemented with 1% (w/w) di-(2-ethylhexyl)-phthalate for 7 days. The light-mitochondrial fraction was subjected to equilibrium density gradient centrifugation on a preformed linear Nycodenz gradient (10 → 50% (w/v)) essentially as described by Van Roermund et al.[7] After centrifugation the gradient was unloaded and dithiothreitol was added to a final concentration of 1 mM to aliquots of the gradient fractions used for the determination of the activity of DPD. All gradient fractions were subsequently stored at –80°C until further analysis.

2.2. Determination of Enzyme Activities

The activity of glutamate dehydrogenase (EC 1.4.1.3), lactate dehydrogenase (EC 1.1.1.27) and esterase (EC 3.1.1.2) were measured essentially as described by Wanders and coworkers.[6,8]

The activity of catalase (EC 1.11.1.6) was measured using a newly developed method based on the peroxidative action of catalase in the presence of excess ethanol.[9] The assay medium contained the following components: 100 mM potassium phosphate (pH 7.4), 1 mM NAD^+, 10 mM pyrazole, 1.7 M ethanol, 0.05% (w/v) Triton X-100 and 1 U/ml aldehyde dehydrogenase. The reaction was started by addition of H_2O_2 to a final concentration of 5 mM. The increase in absorbance at 340 nm was measured in time at 10 s time intervals using a COBAS-BIO centrifugal analyzer (Hoffmann-LaRoche, Basle, Switzerland).

The activity of dihydropyrimidine dehydrogenase was determined essentially as described before[10] with minor modifications: the reaction mixture contained 35 mM potassium phosphate (pH 7.4), 40 μM [2-^{14}C] thymine, 250 μM NADPH, 2.5 mM $MgCl_2$, 2 mM dithiothreitol and 0.1% (w/v) Triton X-100. After an appropriate time of incubation (1.5–3 h) the reaction catalyzed by dihydropyrimidine dehydrogenase was terminated with perchloric acid and the samples were stored at 4°C for 18h allowing the NaOH solution to trap the $^{14}CO_2$. The separation of radiolabeled thymine and the reaction products 5,6-dihydrothymine and ß-ureidoisobutyric acid was accomplished by reversed-phase HPLC as described before.[10]

3. RESULTS

To investigate the subcellular localization of DPD we prepared a heavy-mitochondrial, light mitochondrial, microsomal and a soluble fraction of a rat liver homogenate by differential centrifugation. In each of the fractions the activities of marker enzymes were

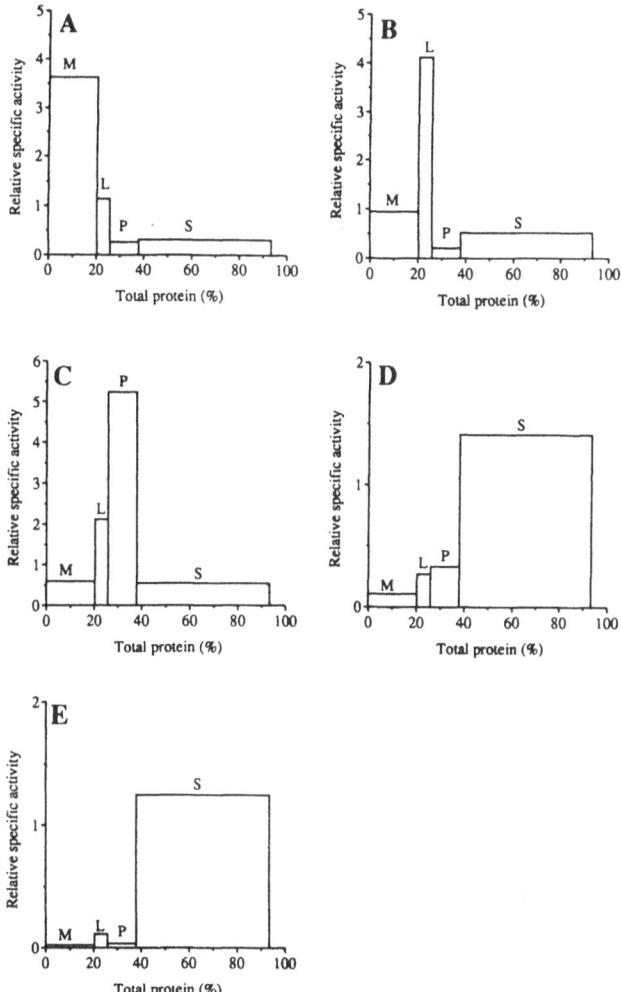

Figure 1. Differential centrifugation of a rat liver homogenate. A rat liver homogenate was fractionated into a heavy-mitochondrial (M), light-mitochondrial (L), microsomal (P) and soluble (S) fraction. The activities of the marker enzymes glutamate dehydrogenase (panel A), catalase (panel B), esterase (panel C), lactate dehydrogenase (panel D) and that of DPD (panel E) were measured in each of the different fractions.

measured for mitochondria (glutamate dehydrogenase), peroxisomes (catalase), microsomes (esterase) and cytosol (lactate dehydrogenase). The specific activities of the marker enzymes as well as that of DPD are shown in Figure 1.

These results show that the activity profile of DPD closely resembles that of lactate dehydrogenase, but differs markedly from those observed for the other enzymes. Therefore, these results indicate that the vast majority of DPD is located in the cytosol.

To exclude any particulate-bound DPD a light mitochondrial fraction was subsequently subjected to equilibrium density gradient centrifugation. In this way, a good separation between the various subcellular organelles was obtained which was reflected by the distinct activity patterns of the marker enzymes (Fig. 2).

A unimodal pattern of activity was observed for the marker enzymes glutamate dehydrogenase (Fig. 2A) and esterase (Fig. 2C). In contrast, a bimodal pattern of activity

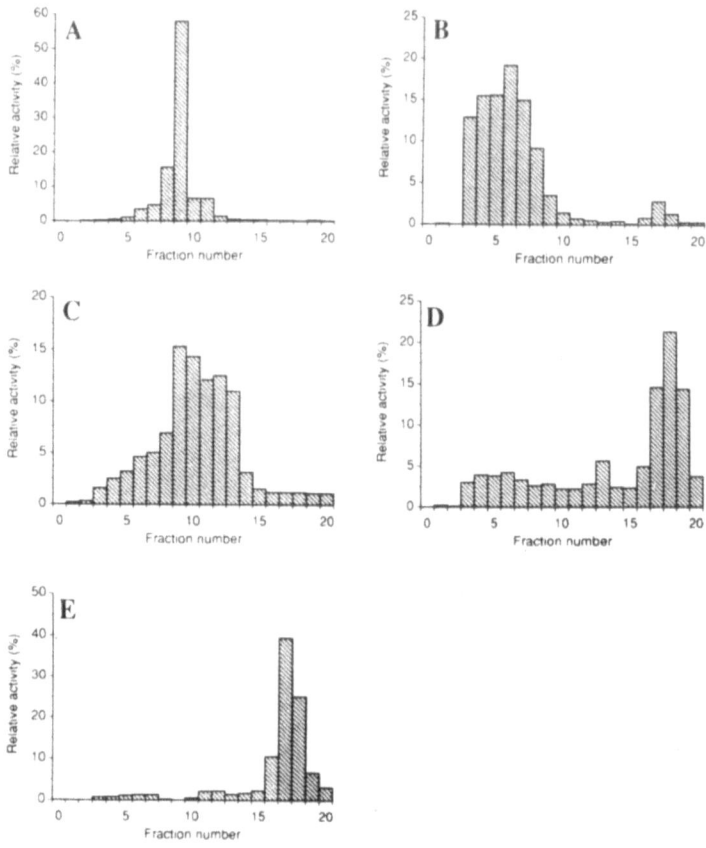

Figure 2. Subcellular localization of DPD. A light-mitochondrial fraction obtained by differential centrifugation was subjected to equilibrium density centrifugation on Nycodenz. The activities of the marker enzymes glutamate dehydrogenase (panel A), catalase (panel B), esterase (panel C), lactate dehydrogenase (panel D) and that of DPD (panel E) were measured in each of the fractions and expressed as a percentage of the total activity in the whole gradient.

was observed for catalase (Fig. 2B). Although the majority of the activity was located in the high density region of the gradient, a small but distinct peak was observed in the upper part of the gradient. The catalase detected in the low density region of the gradient most likely represents the soluble enzyme released from peroxisomes broken during the homogenization process.[7] A broad distribution profile was observed for lactate dehydrogenase (Fig. 2D) with some activity recovered at the high density region of the gradient. This small amount of activity of lactate dehydrogenase in the region where the peroxisomes are located has been observed by others as well,[11,12] suggesting that peroxisomes might contain lactate dehydrogenase.[12] Fig. 2E shows that an unimodal pattern of activity was observed for DPD with the highest activity being found in the low density part of the gradient where the cytosolic enzymes are located.

4. DISCUSSION

In this study we investigated the localization of DPD in rat liver in subcellular fractions obtained by differential centrifugation and equilibrium density gradient centrifugation.

The distribution profile of the activity of DPD and the marker enzymes in the various subcellular fractions indicated that DPD from rat liver was almost exclusively located in the cytosol since no significant activity of DPD could be detected in any subcellular organelle. In this respect, the results obtained by Smith and Yamada suggesting that the NADPH-dependent activity of DPD was distributed mainly between mitochondria, plasma membranes and endoplasmic reticulum membranes are enigmatic.[4,5] However, this discrepancy might be attributed to the spectrophotometric assay used to determine the activity of DPD. They attempted to measure the activity of DPD by following the decrease in absorbance at 290 nm and 295 nm which is far beyond the absorbance maximum of uracil (259 nm) and thymine (265 nm), respectively. Therefore, such assay conditions might easily evoke unreliable results. Another conceivable possibility might be that the particulate-bound DPD has been caused by ionic interaction between the DPD protein and mitochondrial and other membranes. It is now well established that several enzymes such as the dehydrogenases of the pentose phosphate pathway adhere to the outside of peroxisomal membranes and, therefore, they have been initially classified as peroxisomal proteins.[11] Indeed, Smith and Yamada also observed that by slightly increasing the ionic strength of their homogenization buffer a considerable increase in cytosolic DPD activity was obtained.[4] Nevertheless, our results obtained with a highly sensitive and accurate radiochemical assay of DPD demonstrate that DPD is exclusively located in the cytosol.

REFERENCES

1. Van Gennip, A.H., Abeling, N.G.G.M., Stroomer, A.E.M., Van Lenthe, H. and Bakker, H.D. (1994) Clinical and Biochemical Findings in Six Patients with Pyrimidine Degradation Defects. *J. Inher. Metab. Dis.* **17**, 130–132.
2. Naguib, F.N.M., el Kouni, M.H. and Cha, S. (1985) Enzymes of Uracil Catabolism in Normal and Neoplastic Human Tissues. *Cancer Res.* **45**, 5405–5412.
3. Van Kuilenburg, A.B.P., Blom, M.J., Van Lenthe, H., Mul, E. and Van Gennip, A.H. (1997) The activity of dihydropyrimidine dehydrogenase in human blood cells. *J. Inher. Metab. Dis.* (in press).
4. Smith, A.E. and Yamada, E.W. (1971) Dihydrouracil Dehydrogenase of Rat Liver. Separation of Hydrogenase and Dehydrogenase Activities. *J. Biol. Chem.* **246**, 3610–3617.
5. Hallock, R.O. and Yamada, E.W. (1976) Pyrimidine reducing enzymes of rat liver. *Can. J .Biochem.* **54**, 178–184.
6. Wanders, R.J.A., Romeyn, G.J., Schutgens, R.B.H. and Tager, J.M. (1989) L-pipecolate oxidase: A distinct peroxisomal enzyme in man. *Biochem. Biophys. Res. Commun.* **164**, 550–555.
7. Roermund, C.W.T., Van den Berg, M. and Wanders, R.J.A. (1995) Localization of peroxisomal 3- oxoacyl-CoA thiolase in particles of varied density in rat liver: implications for peroxisome biogenesis. *Biochim. Biophys. Acta* **1245**, 348–358.
8. Wanders, R.J.A., Van Roermund, C.W.T., De Vries, C.T., Van den Bosch, H., Schrakamp, G., Tager, J.M., Schram, A.W. and Schutgens, R.B.H. (1986) Peroxisomal ß-oxidation of palmitoyl-CoA in human liver homogenates and its deficiency in the cerebro-hepato-renal (Zellweger) syndrome. *Clin. Chim. Acta* **159**, 1–10.
9. Chance, B., Sies, H. and Boveris, A. (1979) Hydroperoxide metabolism in mammalian organs. *Physiol. Rev.* **59**, 527–604.
10. Van Kuilenburg, A.B.P., Van Lenthe, H., and Van Gennip, A.H. (1996) Identification and Tissue- Specific Expression of a NADH-Dependent Activity of Dihydropyrimidine Dehydrogenase in Man. *Anticancer Res.* **16**, 389–394.
11. Antonenkov, V.D. (1989) Dehydrogenases of the pentose phosphate pathway in rat liver peroxisomes. *Eur. J. Biochem.* **183**, 75–82.
12. Vandor, S.L. and Tolbert, N.E. (1970) Glyoxylate metabolism by isolated rat liver peroxisomes. *Biochim. Biophys. Acta* **215**, 449–455.

158

THE ACTIVITY OF DIHYDROPYRIMIDINE DEHYDROGENASE IN HUMAN BLOOD CELLS

André B. P. Van Kuilenburg,[1] Henk Van Lenthe,[1] Maarten J. Blom,[1] Erik P. J. Mul,[2] and Albert H. Van Gennip[1]

[1]Academic Medical Center
University of Amsterdam
Department Clinical Chemistry and Division Emma Children's Hospital
P.O. Box 22700
1100 DE Amsterdam, The Netherlands
[2]Central Laboratory of The Netherlands Red Cross Blood Transfusion Service
Plesmanlaan 125
1066 CX Amsterdam, The Netherlands

1. INTRODUCTION

Dihydropyrimidine dehydrogenase (DPD, EC 1.3.1.2) is the initial and rate-limiting enzyme in the catabolism of the pyrimidine bases and it catalyses the reduction of thymine and uracil to 5,6-dihydrothymine and 5,6-dihydrouracil, respectively. In children, the deficiency of DPD is often accompanied by a neurological disorder but a considerable variation in the clinical presentation among these patients has been reported.[1] In these patients, a large accumulation of uracil and thymine has been detected in urine, blood and in cerebrospinal fluid.[2] DPD is also responsible for the breakdown of the widely used antineoplastic agent 5-fluorouracil (5FU), thereby limiting the efficacy of the therapy. A correlation has been observed between the pretreatment activity of DPD in peripheral blood mononuclear cells (PBM cells) and the systemic clearance of 5FU in cancer patients.[3,4] However, a large interpatient and intrapatient variability in the activity of DPD was observed.[3] PBM cells are also frequently used to identify patients suffering from a complete or partial deficiency of DPD. Although the activity of DPD can be detected in a variety of human tissues, with the highest activity being found in liver and lymphocytes,[5] it is not known yet whether the activity of DPD is also present in other blood cell types as well. In this study, we determined the activity of DPD in monocytes, lymphocytes, platelets, granulocytes and erythrocytes obtained from healthy volunteers.

Purine and Pyrimidine Metabolism in Man IX,
edited by Griesmacher *et al.* Plenum Press, New York, 1998.

2. MATERIALS AND METHODS

2.1. Patients

Thirteen patients with cancer entered the study. Median age was 65 years (range 41–70). Nine patients were treated with an intermittent continuous infusion of fluoropyrimidines and four patients were treated with other types of chemotherapy and/or radiotherapy. Blood samples were taken at day 7 half way the chemotherapy course. Seventeen healthy volunteers entered the study. In all cases, blood samples were collected between 8.00 and 10.00 a.m. to minimize the influence of a possible circadian rhythm of the activity of DPD.

2.2. Isolation of Blood Cells

Peripheral blood mononuclear cells were isolated from EDTA-anticoagulated blood by centrifugation over lymphopaque (spec. gravity 1.086 g/ml, 350 mOsm). Cells from the interface were collected and treated with ice-cold NH_4Cl to lyse the contaminating erythrocytes. Platelets, monocytes and lymphocytes were purified from buffy coats from healthy volunteers by centrifugation over Percoll followed by elutriation centrifugation as described before.[6,7] Granulocytes were purified from the pellets of the centrifugation step over Percoll and the contaminating erythrocytes were lysed with ice-cold isotonic NH_4Cl.[6,7] Erythrocytes were prepared from the same pellet with subsequent separation from the granulocytes by centrifugation.

2.3. Determination of the DPD Activity

The activity of DPD was determined in a reaction mixture containing 35 mM potassium phosphate (pH 7.4), 2.5 mM $MgCl_2$, 1 mM dithiothreitol, 250 µM NADPH and 25 µM [2-^{14}C]thymine. Separation of radiolabeled thymine from radiolabeled dihydrothymine was performed isocratically (50 mM NaH_2PO_4 (pH 4.5) at a flow rate of 2 ml/min) by reversed-phase HPLC on a Supelcosil LC-18-S column (250 × 4.6 mm, 5 µm particle size) with online detection of the radioactivity.[8]

2.4. Statistical Analysis

The difference between the mean activities of DPD between various blood cells were analyzed with the two sample *t* test. In case of unequal variances, as indicated by Levene's test for equality of variances, the log transformed data were used. The correlation's between the activity of DPD and the percentage of monocytes or lymphocytes was studied by determination of the Pearsons correlation coefficient and linear regression. The level of significance was set *a priori* at P ≤ 0.05. Analyses were performed with the Statistical Package for the Social Sciences (SPSS Inc. Chicago, USA).

3. RESULTS

To investigate the activity of DPD in human blood cells we purified the various blood cell types from healthy volunteers with density centrifugation and elutriation centrifugation. The highest activity was found in monocytes (13.7 ± 5.5 nmol/mg/h) followed by that of lymphocytes (5.6 ± 1.6 nmol/mg/h), granulocytes (2.2 ± 0.5 nmol/mg/h) and platelets (1.5 ± 0.9 nmol/mg/h). We could detect only a very low activity in red blood cells

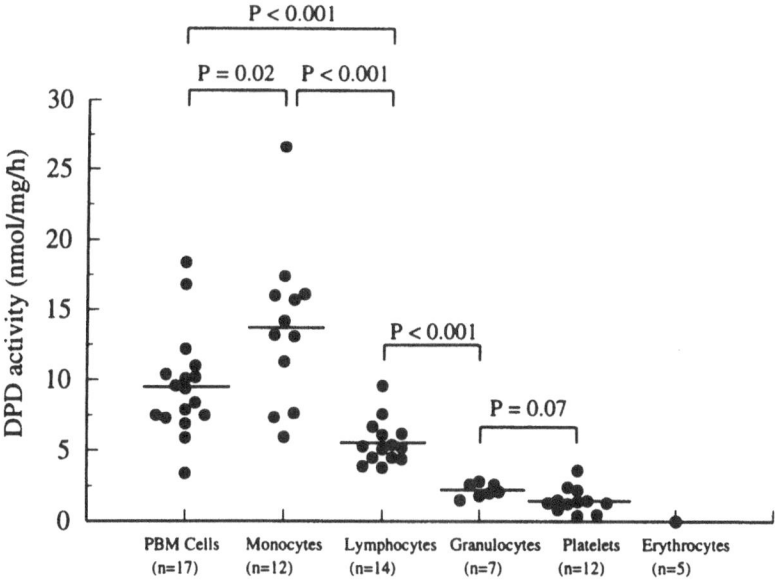

Figure 1. The activity of DPD. The mean activities of DPD in the blood cells are indicated by solid lines.

(0.0044 nmol/mg/h) which might have been caused by the presence of some residual granulocytes (< 0.2%). Furthermore, the activity of DPD in PBM cells proved to be intermediate compared to the DPD activity observed in monocytes and lymphocytes. This observation is in line with the fact that a differential count showed that the PBM cells contained on average 23 ± 12% monocytes (range 5–52%).

Since our results demonstrate that the highest activity of DPD is present in monocytes and the fact that PBM cells purified with density centrifugation over Percoll contain a substantial amount of monocytes we investigated whether a correlation exists between the DPD activity measured in PBM cells and the percentage of monocytes and lymphocytes in PBM cell suspensions. Figure 2 shows that a positive correlation exists between the DPD activity of PBM cells obtained from healthy volunteers and cancer patients and the percentage of monocytes (Fig. 2A). Consequently, a negative correlation was observed between the DPD activity in PBM cells and the percentage of lymphocytes (Fig. 2B).

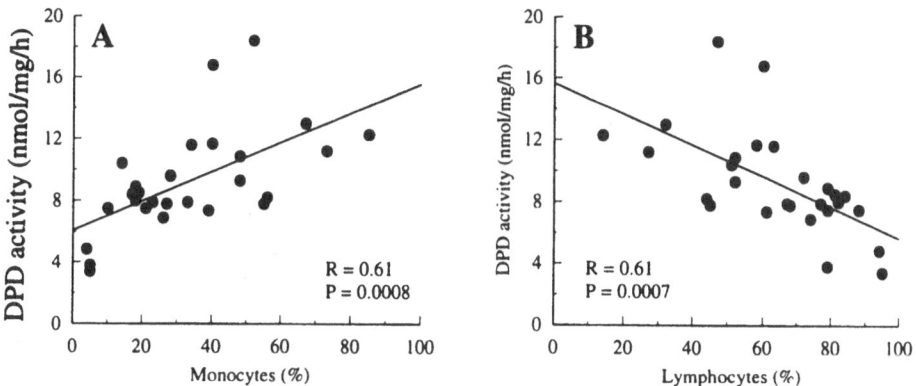

Figure 2. Correlation between the activity of DPD and the percentage of monocytes (A) and lymphocytes (B).

4. DISCUSSION

In this study we showed that a profound activity of DPD can be detected in all blood cell types except red blood cells. The presence of DPD activity in platelets is in line with the observation of Pero and coworkers[9] who showed that platelets were able to catabolise thymine. Until now, the highest specific activity of DPD has been found in lymphocytes.[5] However, our results clearly demonstrate that the specific activity of DPD in monocytes even exceeds that of lymphocytes. Surprisingly, the PBM cells obtained from healthy volunteers and cancer patients contained substantial amounts of monocytes. The fact that the activity of DPD is not restricted to lymphocytes but is also present in monocytes and platelets might therefore at least partly explain the large variation in DPD activity measured in PBM cells isolated from cancer patients. So far, only a weak correlation has been observed between the activity of DPD in PBM cells and the clearance of 5FU. Therefore, these authors stated that the pretreatment activity of DPD in PBM cells is not a sufficient reliable predictor of the clearance of 5FU and that the pretreatment activity of DPD can not be used to optimize the dose of 5FU to be administered to each individual cancer patient. However, we feel that this statement should be addressed again by using a homogenous cell population for the determination of the activity of DPD. Although the majority of the DPD activity is located in liver other tissues also contribute to the catabolism of pyrimidine bases since the systemic clearance of 5FU significantly exceeds the blood liver flow.[10] From our results it can be estimated that in peripheral blood the majority of the DPD activity is located in platelets. Whether the diagnosis of patients and carriers of a DPD deficiency can be equally well established in platelets as in mononuclear cells is currently under investigation.

REFERENCES

1. Van Gennip, A.H., Abeling, N.G.G.M., Stroomer, A.E.M., Van Lenthe, H. and Bakker, H.D. (1994) Clinical and Biochemical Findings in Six Patients with Pyrimidine Degradation Defects. *J. Inher. Metab. Dis.* **17**, 130–132.
2. Bakkeren, J.A.J.M., De Abreu. R.A., Sengers. R.C.A., Gabreëls. F.J.M., Maas. J.M. and Renier WO (1984) Elevated urine, blood and cerebrospinal fluid levels of uracil and thymine in a child with dihydropyrimidine dehydrogenase deficiency. *Clin. Chim. Acta.* **140**, 247–256.
3. Fleming, R.A., Milano, G., Thyss, A., Etienne, M.-C., Reneé, N., Schneider, M. and Demard, F. (1992) Correaltion between Dihydropyrimidine Dehydrogenase Activity in Peripheral Mononuclear Cells and Systemic Clearance of Fluorouracil in Cancer Patients. *Cancer Res.* **52**, 2899–2902.
4. Etienne, M.C., Lagrange, J.L., Dassonville, O., Fleming, R., Thyss, A., Reneé, N., Schneider, M., Demard, F. and Milano, G. (1994) Population Study of Dihydropyrimidine Dehydrogenase in Cancer Patients. *J. Clin. Oncology* **12**, 2248–2253.
5. Naguib, F.N.M., el Kouni, M.H. and Cha, S. (1985) Enzymes of Uracil Catabolism in Normal and Neoplastic Human Tissues. *Cancer Res.* **45**, 5405–5412.
6. De Korte, D., Haverkort, W.A., Van Gennip, A.H. and Roos, D. (1985) Nucleotide Profiles of Normal Human Blood Cells Determined by High-Performance Liquid Chromatography. *Anal. Biochem.* **147**, 197–209.
7. De Boer, M. and Roos, M. (1986) Metabolic comparison between basophils and other leukocytes from human blood. *J. Immunol.* **136**, 3447–3454.
8. Van Kuilenburg, A.B.P., Van Lenthe, H., and Van Gennip, A.H. (1996) Identification and Tissue- Specific Expression of a NADH-Dependent Activity of Dihydropyrimidine Dehydrogenase in Man. *Anticancer Res.* **16**, 389–394.
9. Pero, R.W., Johnson, D. and Olsson, A. (1984) Catabolism of Exogenously Supplied Thymidine to Thymine and Dihydrothymine by Platelets in Human Peripheral Blood. *Cancer Res.* **44**, 4955–4961.
10. Diasio, R.B. and Harris, B.E. (1989) Clinical Pharmacology of 5-fluorouracil. *Clin. Pharmacokinet.* **16**, 215–237.

159

ENDOTHELIAL PURINE CONTENT

An Alternative Model for Testing Antibiotic Solutions for Intravenous Use

H. Vorbach,[1] G. Weigel,[2] B. Robibaro,[3] M. Reiter,[1] M. Hlousek,[1]
C. Armbruster,[3] A. Griesmacher,[2] A. Georgopoulos,[1] and W. Graninger[1]

[1]Department of Infectious Diseases
[2]Department of Cardiothoracic Surgery
[3]Department of Pulmonary Medicine
University Hospital of Vienna, Austria

1. INTRODUCTION

The fluoroquinolones are a commonly well tolerated class of antibiotics with a low incidence of systemic side effects (1–3). Hooper et al. reported local reactions at the site of infusion after intravenously administered ciprofloxacin and fleroxacin (4). Local reactions have been more common during infusions in the small veins of the dorsum of the hand compared to administration via an anticubital vein (5,6). For this reason, we presumed that fluoroquinolone solutions might be directly toxic for endothelial cells, therby resulting in subsequent phlebitis.

The tolerance of intravenously applied antibiotics has usually been tested in animal models (7). To provide an alternative test system, we have set up an in vitro culture system using endothelial cells. We compared the effects of different fluoroquinolones available for intravenous application and their solvents on these cultures. Cell viability was analyzed by measuring intracellular ATP levels (8). TXA_2, a potent vasoconstrictory and pro aggregatory, and PGI_2 a strong vasodilatatory and anti aggregatory compound, served as parameters for cell function (9).

2. MATERIALS AND METHODS

2.1. Cell Culture

Endothelial cells were isolated and cultured according to a modified standard procedure (10). Cells were collected from the perfusate by centrifugation ($300 \times$ g, $4°C$) and

Purine and Pyrimidine Metabolism in Man IX,
edited by Griesmacher *et al.* Plenum Press, New York, 1998.

seeded into culture T-75 flasks precoated with human fibronectin (Upstate Biotechnology Incorporated, Lake Placid, USA).

The confluent primary monolayers were washed and trypsinized. The cell suspensions were transferred into each well of a 6 well culture plate and cultivated for four days. Only cells from these first subcultures were used for the experiments. The cells were identified as endothelial cells by the typical cobblestone, contact inhibited morphology (11) and by von Willebrand Factor (F VIII : vWF) production (12).

2.2. Antibiotics

The commercially available i.v. preparations of ciprofloxacin (Bayer, Wuppertal, Germany), fleroxacin (Hoffman-La Roche, Basel, Switzerland) and ofloxacin (Hoechst, Frankfurt, Germany) were used.

2.3. Composition of the Antibiotic Diluents

Solvents for the three antibiotics tested were prepared according to the instructions of the manufacturers. These solvents alone were also tested for their toxicity on HUVEC.

Solvent of ciprofloxacin: 11.06 mg lactic acid and 900 mg NaCl in 100 ml distilled water (pH = 3.5). Solvent of fleroxacin: Glucose, lactic acid, NaOH jn 100 ml distilled water (pH = 4.4) (No data on composition available). Solvent of ofloxacin: 900 mg NaCl in 100 ml distilled water, pH was adjusted to 4.5 by adding HCl.

2.4. Incubation Experiments

For the experiments the culture medium was removed and the cell layers were gently washed with Dulbecco's phosphate buffered saline (Gibco, Paisly, Scotland). Thereafter, the cells were incubated with medium 199 (pH 7.4) containing 10% bovine calf serum. Antibiotics were added in different dilutions yielding final concentrations ranging from 20 µg/ml to 200 µg/ml. For the high dose experiments, pure solvents as well as the commercially available fluoroquinolones were incubated for 20 minutes at final concentrations of 2 mg/ml (ciprofloxacin, ofloxacin) and 4 mg/ml (fleroxacin). The diluted 50:50 (v/v) solvents and fluoroquinolones, reaching a concentration of 1 mg/ml (ciprofloxacin, ofloxacin) and 2 mg/ml (fleroxacin), were dissolved in medium 199 containing 10% bovine calf serum.

2.5. Determination of PGI_2 and TXA_2

Direct radio immunoassays (RIA; BIOTECX) were carried out to determine the concentrations of 6-keto-prostaglandin F $_{1\alpha}$ (6-keto-PGF $_{1\alpha}$) and TXB_2, the stable degradation products of PGI_2 and TXA_2, respectively (5, 19).

2.6. Determination of ATP

The determination of the intracellular ATP-content was carried out in a cell lysate prepared by adding 300 µl of a 0.5 mol/l $HClO_4$ to each well after removal of the culture supernatant. The lysate was neutralized with 1 mol/l K_2CO_3, centrifuged and the supernatant analyzed for ATP by means of bioluminescence using an ATP-Monitoring Reagent and an Autolumat 953 (Berthold, Germany) as described earlier (15).

Table 1. Effects of concentrated and diluted preparations of fluoroquinolones
on ATP levels, PGI$_2$ and TXA$_2$ production of HUVEC after 20 minutes

	pH	ATP	PGI2	TXA2
Control	7.4	100	16.43 ± 1.08	4.83 ± 0.62
Ciprofloxacin 2 mg/ml	4.2	24.1 ± 1.9*	0.55 ± 0.09*	0.26 ± 0.06*
Solvent	4.0	21.9 ± 2.3*	0.40 ± 0.26*	0.23 ± 0.07*
Ciprofloxacin 1mg/ml	6.6	98.3 ± 2.9	11.43 ± 0.95*	2.72 ± 0.51*
Fleroxacin 4 mg/ml	4.2	17.9 ± 0.6*	0.46 ± 0.10*	0.30 ± 0.09*
Solvent	4.1	15.1 ± 1.5*	0.39 ± 0.08*	0.26 ± 0.11*
Fleroxacin 2 mg/ml	6.3	98.8 ± 2.0	11.73 ± 2.38*	2.08 ± 0.62*
Ofloxacin 2 mg/ml	4.5	96.7 ± 2.5	0.54 ± 0.10*	0.14 ± 0.03*
Solvent	4.5	97.3 ± 4.4	0.53 ± 0.13*	0.33 ± 0.10*
Ofloxacin 1 mg/ml	6.9	99.3 ± 3.2	10.12 ± 1.49*	2.47 ± 0.45*

Percentual changes of ATP are given as mean ± SD. The mean ATP level of controls (=100%) was 12.46 ± 0.84 nmol/10^6 cells. PGI$_2$ and TXA$_2$ levels are given in pmol/10^6 cells (mean ± SD). *p < 0.001, statistically significant differences compared to controls.

2.7. Statistical Analysis

Data from twelve different representative experiments are expressed as mean ± SD. For evaluating statistical significances the Mann-Whitney U test was used.

3. RESULTS

3.1. Effects of Ciprofloxacin on HUVEC

Both 2000 µg/ml ciprofloxacin and the solvent alone led to the same degree of cell damages (Table 1). Although application of diluted ciprofloxacin (50:50, v/v) did not affect the cell viability within 20 minutes at a final concentration of 1000 µg/ml (Table 1), functional impairment was observed (reduced eicosanoid release)(Table 1). Incubation of cells with 20 µg/ml ciprofloxacin for 24h, 48h and 72h had no statistically significant adverse effect on cell viability (Table 2). Experiments with 100 µg/ml ciprofloxacin resulted in an increase of PGI$_2$ synthesis (Table 1). A rapid decrease of the intracellular high energy phosphates was observed at concentration of 200 µg/ml after 48 hours (Table 2).

Table 2. Effects of fluoroquinolones on intracellular ATP levels
in human umbilical vein endothelial cells

	24 hours	48 hours	72 hours
Ciprofloxacin 20 µg/ml	99.2 ± 5.2	98.5 ± 5.9	100.1 ± 5.3
Ciprofloxacin 100 µg/ml	98.5 ± 6.7	96.3 ± 6.8	84 ± 4.1*
Ciprofloxacin 200 µg/ml	102 ± 4.1	24.13 ± 2.1*	6.7 ± 1.9*
Fleroxacin 20 µg/ml	100.4 ± 5.6	101.6 ± 5.3	100.3 ± 6.1
Fleroxacin 100 µg/ml	114.4 ± 6.8*	118 ± 6.2*	106.6 ± 4.4
Fleroxacin 200 µg/ml	130.3 ± 7.9*	133.7 ± 7.6*	113.6 ± 6.5*
Ofloxacin 20 µg/ml	99.8 ± 6.3	99.6 ± 5.5	98.8 ± 5.1
Ofloxacin 100 µg/ml	98.8 ± 7.6	97.4 ± 6.3	104.2 ± 4.2
Ofloxacin 200 µg/ml	96.2 ± 5.4	95.4 ± 4.3	105.4 ± 5.9

Percentual changes of ATP are given as mean ± SD. The mean ATP level of controls (=100%) was 12.46 ± 0.84 nmol/10^6 cells. *p < 0.001, statistically significant differences compared to controls.

3.2. Effects of fleroxacin on HUVEC

Concentrations of 4000 µg/ml of fleroxacin, as well as the solvent on its own induced massive decreases in intracellular ATP and diminished PGI_2 and TXA_2 liberation
(Table 1). A 50:50 (v/v) dilution of this substance at a final concentration of 2000 µg/ml
did not alter cell viability compared to controls, but reduced the PGI_2 and TXA_2 production. Fleroxacin at a concentration of 20 µg/ml did neither affect intracellular ATP levels
(Table 2) nor the TXA_2/PGI_2 ratio (Table 1). Concentrations of 100 and 200 µg/ml led to
statistically significant releases of PGI_2 within 20 min (Table 1) and increased intracellular
ATP values up to 72 hours (Table 2).

3.3. Effects of Ofloxacin on HUVEC

Ofloxacin at a dose of 2000 µg/ml did not influence the cellular energy state within
20 min. However, a concentration dependent effect for PGI_2 and TXA_2 could be observed
at this dose range (Table 1). Ofloxacin in concentrations ranging from 20 to 200 µg/ml had
no significant effects on HUVEC over time periods up to 72 h (Table 2).

4. DISCUSSION

The tolerance of intravenously applied antibiotics has usually been tested in animal
models (7). On the other hand there is a high incidence of injection site reaction (5,6)
showing the problem in finding a responsive animal model for prognostic venous irritation
studies. In vitro studies with cultured human cells have proved to be a promising method
for predicting toxicity and for clarifying mechanisms of toxicity (13). For investigating
cellular impairment during incubation with quinolones intracellular purine content was
measured. Since the trypan blue exclusion test seem to be inappropriate for describing
cellular disturbances and the measurement of LDH release is a late phenomenon in cell
dysfunction/death, this essay was chosen.

Since the tested antibiotics are available at concentrations of 2000 µg/ml (ciprofloxacin, ofloxacin) and 4000 µg/ml (fleroxacin), this high dose range was used in the in vitro
experiments of this study (2000–4000 µg/ml) to mimic as close as possible clinical concentrations at the site of infusion. Similar concentrations may be reached by retrograde
intravenous pressure infusion techniques. This local high dose injection of antibiotics has
been recommended in ischaemic ulcers and diabetic feet to achieve higher tissue concentrations as the extent of amputation is often influenced by infection (16,17). We have
shown that ciprofloxacin and fleroxacin but not ofloxacin at these concentrations lead to
considerable endothelial cell damage. These findings are in line with the clinical observations of Hooper et al. that local reactions at the site of infusion are common with intravenous application of ciprofloxacin and fleroxacin (2,4). Ofloxacin has recently become
commercially available as double concentrated solution (4000 µg/ml). This product meets
the recommendation of Physicians' Desk Reference and is in line with our observation
that this product was relatively best tolerated by endothelial cells (15).

Table 1 shows that a detrimental effect after 20 min occurs only using undiluted
solutions, whereas a dilution of 50:50 (v/v) renders the solutions more compatible to
HUVEC (no cell death, but abrogation of TXA_2 and PGI_2 production) and simultaneously
buffers the pH-values to a more physiological range of 6.33–7.08. These data are in line
with the observation of Thorstenson et al that intravenous administration of ciprofloxacin

was associated with erythema or burning at the infusion site in 20–48% of patients when the infusion was administered through small vein in the hand or forearm but in only 4–12% when the infusion was given via an antecubital vein (5,6). Alternatively, the occurrence of phlebitis could be minimized by diluting the manufacturers' preparations at least to a ratio of 50:50 (v/v). Since such precautions may be easily overlooked in clinical practice, it seems most desirable that the solvents of the manufactured solutions would be optimized. By using human venous endothelial cells for testing antibiotic solutions for intravenous compatibility, we provide a valuable alternate model being helpful for reducing local side effects. Our data clearly show that commercially available preparations of ciprofloxacin and fleroxacin should be diluted or at least applied slowly into large veins.

REFERENCES

1. Christ W, Lehnert T, Ulbrich B 1988. Specific toxicologic aspects of the fluoroquinolones. Rev Infect Dis 10 Suppl 1: 141–146.
2. Hooper DC, Wolfson JS 1991. Fluorquinolone antimicrobial agents. N Engl J Med 324: 384–394.
3. Forsgren A, Bredberg A, Riesbeck K 1989. New quinolones: in vitro effects as a potential source of clinical toxicity. Rev Infect Dis 11 Suppl. 5: 1382–1389.
4. Hooper DC, Wolfson JS. 1993. Quinolone Antimicrobial Agents, 2 nd. ed. American Society for Microbiology, Washington, DC.
5. Thorsteinsson SB, Bergan T, Johannesson G, Thorsteinsson HS, Rohwedder R 1987 Tolerance of ciprofloxacin at injection site, systemic safety and effect on electroencephalogram. Chemotherapy 33: 448–51.
6. Thorsteinsson SB, Rahm V, Bergan T 1989 Tolerance of intravenous ciprofloxacin. Scand J Infect Dis Suppl 60: 116–119.
7. Forgue ST, Shyu WC, Gleason CR, Pittman KA, Barbhaiya RH 1987 Pharmacokinetics of the novel cephalosporin cefepime (BMY - 28142) in rats and monkeys. Antimicrobial Agents and Chemotherapy 31: 799–804.
8. Griesmacher A, Weigel G, Windischbauer A, Mueller MM 1991 Purine metabolism in human endothelial cells. Int J Purine and Pyrimidine Res 2: 23–127.
9. Weigel G, Griesmacher A, Mueller MM 1991 Regulation of eicosanoid release in human umbilical vein endothelial cells. Thromb Res 62: 685–695.
10. Jaffe EA, Nachman RL, Becker CG, Minck CR 1973 Culture of human endothelial cells derived from umbilical veins. J Clin Invest 52: 2745–2756.
11. Haudenschild CC, Cotran RS, Gimbrone MA, Folkman J 1975 Fine structure of vascular endothelium in culture. J of Ultrastruct Res 50: 22–34.
12. Jaffe EA, Hoyer LW, Nachman RL 1973. Synthesis of antihaemophilic factor antigen by human endothelial cells. J Clin Invest 52: 2757–2763.
13. Griesmacher A, Weigel G, David M, Horvath G, Mueller MM 1992 Functional implications of cAMP and Ca^{2+} on prostaglandin I_2 and thromboxane A_2 synthesis by human endothelial cells. Arterioscler Thromb 12: 512–518.
14. Griesmacher A, Weigel G, David M, Schimke I, Mueller MM 1992 Influence of oxygen radicals generating agents on eicosanoid metabolism of human endothelial cells. Thromb Res 65: 721–731.
15. Weigel, G., A. Griesmacher, Toma, C., Leukauf, C., Schreiner, W., Wolner, E. 1993 The protective potency of two commonly used cardioplegic solutions on cultured endothelial cells exposed to free-oxygen radicals injury. Free Radical Biology & Medicine, 17: 577–585.
16. Burgmann H, Georgopoulos A, Graninger W, Koppensteiner R, Maca T, Minar E, Schneider B, Stümpflen A, Ehringer H 1996 Tissue concentration of clindamycin and gentamicin near ischaemic ulcers with transvenous injection in Bier's arterial arrest. Lancet 368: 781–83.
17. Acevedo A, Schoop W, Schnell A, Toledo T 1990 Antibiotic treatment for diabetic foot. Advantages of intravenous regional route as alternative for systemic route. Rev Med Chi 118: 881–888.
18. Physicians' Desk Reference. 1994 Floxin® for intravenous infusion 48: 1352–1357.

160

INFLUENCE OF GLYCOPEPTIDE ANTIBIOTICS ON PURINE METABOLISM OF ENDOTHELIAL CELLS

B. Robibaro,[1] H. Vorbach,[2] G. Weigel,[3] A. Weihs,[2] M. Hlousek,[2] E. Presterl,[2] A. Georgopoulos,[2] A. Griesmacher,[3] and W. Graninger[2]

[1]Department Pulmonary Medicine
[2]Department Infectious Diseases
[3]Department Cardiothoracic Surgery
University Hospital of Vienna, Austria

1. INTRODUCTION

The glycopeptides are generally well tolerated and a low incidence of systemic side effects has been described (1). The adverse effects of vancomycin includes local phlebitis at the site of infusion (2,3). For this reason, we presumed that these solutions might be directly toxic for endothelial cells (EC), resulting in subsequent phlebitis.

The tolerance of intravenously applied antibiotics has been tested in animal models (4). To provide an alternative test system, we have set up an in vitro culture system using human umbilical venous endothelial cells (HUVEC). We tested the effects of vancomycin and teicoplanin available for intravenous application on these cultures. Cell viability was analysed by measuring intracellular ATP levels (5).

2. MATERIALS AND METHODS

2.1. Cell Culture

Endothelial cells were prepared using human umbilical veins. Cells were isolated and cultured according to a modified standard procedure (6). Briefly, fresh human umbilical veins were filled with 0.1% collagenase solution and incubated at 37°C. Thereafter, the veins were perfused with medium 199 (Sigma, St. Louis, USA) containing 20% bovine calf serum (HyClone, Utah, USA). Cells were collected from the perfusate by centrifugation (300 × g, 4°C) and seeded into culture T-75 flasks precoated with human fibronectin (Upstate Biotechnology Incorporated, Lake Placid, USA).

Purine and Pyrimidine Metabolism in Man IX,
edited by Griesmacher *et al.* Plenum Press, New York, 1998.

Cells were cultured in medium 199 containing 20% bovine calf serum, 50,000 U/l penicillin-streptomycin (Gibco, Paisley, Scotland), 50 mg low molecular weight heparin (Sigma, St. Louis, USA) and 15 mg/l H- Neurext® (endothelial cell growth supplement), (Upstate Biotechnology Incorporated, Lake Placid, USA). The confluent primary monolayers (approx. 8,000,000 cells/flask) were washed and trypsinized. The cell suspensions were transferred into each well of a 12 well culture plate and cultivated for four days. Only cells from these first subcultures were used for the experiments. The cells were identified as endothelial cells by the typical cobblestone, contact inhibited morphology (7) and by von Willebrand Factor (F VIII: vWF) production (8).

2.2. Antibiotics

The commercially available preparations of vancomycinhydrochlorid (Eli Lilly, Indianapolis, USA) and teicoplanin were dissolved in aqua ad injectionem and diluted further with 0.9% NaCl for the experiments (Merrell Dow Pharma GmbH, Rüsselsheim, Germany).

2.3. Incubation with Vancomycin and Teicoplanin

For the experiments the culture medium was removed and the cell layers were gently washed with Dulbecco's phosphate buffered saline (Gibco, Paisly, Scotland). Thereafter, vancomycinhydrochlorid and teicoplanin solutions were added to the endothelial cells and incubated for 20, 60 and 120 minutes at final concentrations of 2, 5 and 10 mg/ml. All incubations were carried out in a humidified incubator at 37°C and 5% CO_2.

2.4. Measurement of ATP

The determination of the intracellular ATP-content was carried out in a cell lysate prepared by adding 300 μl of a 0.5 mol/l $HClO_4$ to each well after removal of the culture supernatant. The lysate was neutralised with 1 mol/l K_2CO_3, centrifuged and the supernatant analysed for ATP by means of bioluminescence using an ATP-Monitoring Reagent and an Autolumat 953 (Berthold, Germany) as described earlier (9).

2.5. Statistical Analysis

Data from twelve different representative experiments are expressed as mean ± SD. The statistical significance was determined by means of Mann Whitney U-Test. $P < 0.001$ (*) was considered to be significant.

3. RESULTS

3.1. Effects of Vancomycin on HUVEC

Incubation of cells with 5 and 10 mg/ml vancomycin resulted in a rapid decrease of the intracellular high energy phosphates to 5.8 ± 0.73 nmol/million cells and 0.95 ± 0.53 nmol/million cells, respectively, after 120 minutes (Fig. 1). Experiments with vancomycin at concentrations of 2 mg/ml (Fig. 1) resulted in no significant decline of intracellular high energy phosphates (11.2 ± 0.75 nmol/million cells) after 120 minutes.

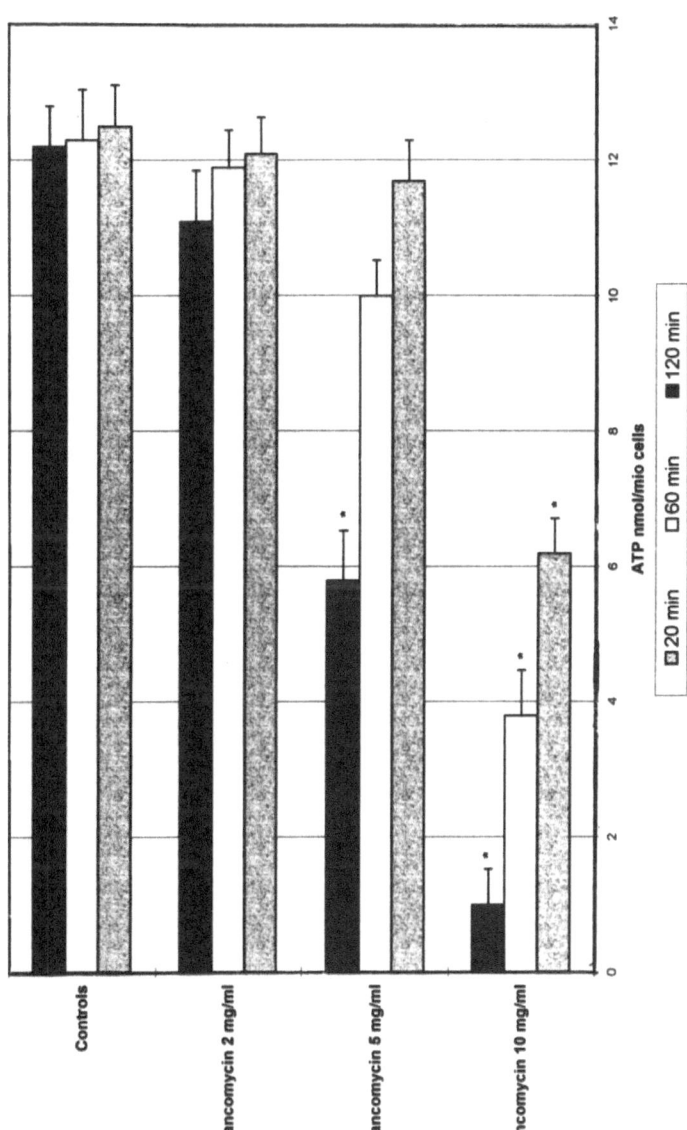

Figure 1. Effects of vancomycin on intracellular ATP levels. Columns represent percentual changes of ATP levels after 20, 60 and 120 minutes (mean ± SD). *p < 0.001, statistically significant differences compared to controls.

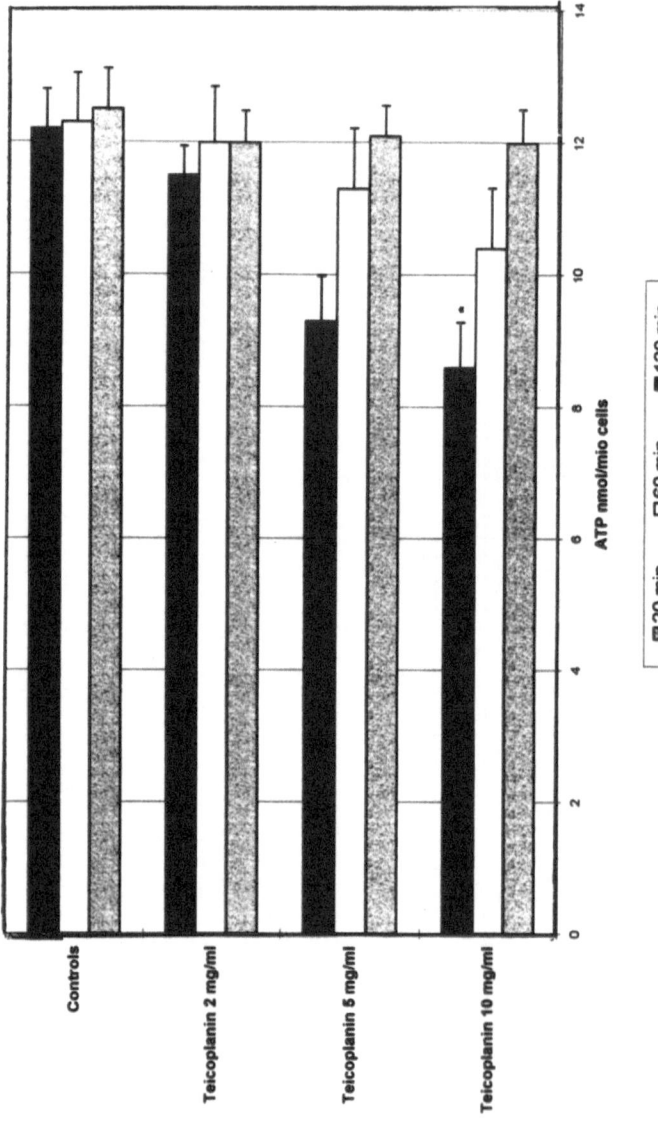

Figure 2. Effects of teicoplanin on intracellular ATP levels. Columns represent percentual changes of ATP levels after 20, 60 and 120 minutes (mean ± SD). *p < 0.001, statistically significant differences compared to controls.

3.2. Effects of Teicoplanin on HUVEC

Teicoplanin at a concentration of 5 and 10 mg/ml did reduce intracellular ATP levels to 9.3 ± 0.69 nmol/million cells and 8.6 ± 0.68 nmol/million cells respectively (Fig. 2). 2 mg/ml once again led to no ATP decline in comparison to controls.

4. SUMMARY

We provide evidence that the commercially available preparations of glycopeptides for intravenous application are well tolerated by endothelial cells when applied in concentrations less than 5 mg/ml.

Since the antibiotics tested are administered at maximal concentrations of 10 mg/ml, the dose range used in our in vitro experiments (5 and 10 mg/ml) mimics possible clinical concentrations at the site of infusion. Similar concentrations may be reached by retrograde intravenous pressure infusion techniques (10–12). We have demonstrated that these high concentrations lead to considerable endothelial cell damage. These findings may explain the common side effect associated with intravenously applied glycopeptides namely pain and phlebitis at the site of infusion (2,13).

Figure 1 shows that a detrimental effect measurable after 20 min occurs only using vancomycin solutions at concentrations of 10 mg/ml, whereas already a dilution to 5 mg/ml renders the solutions more compatible to HUVEC. These data are in line with the observation that slow intravenous application of glycopeptides into large veins can largely prevent the occurrence of local phlebitis. Alternatively, the occurrence of phlebitis should be avoidable by diluting the manufacturers' preparations at least to 2–5 mg/ml and not 10 mg/ml as recommended by the manufacturer of vancomycin. The same aspects need to be considered for use of glycopeptides for retrograde high pressure infusion.

The tolerance of intravenously applied antibiotics has previously been tested in animal models (4). Our model of human venous endothelial cells for testing antibiotic solutions for intravenous compatibility provides a valuable alternate model.

In conclusion our data show that the commercial preparation of teicoplanin is more compatible for HUVEC than those of vancomycin.

REFERENCES

1. Wood, M.J. (1996). The comparative efficacy and safety of teicoplanin and vancomycin. *Journal of Antimicrobial Chemotherapy* **37**, 209–222.
2. Phillips, G. & Golledge, C. (1992). Vancomycin and teicoplanin: something old, something new. *The Medical Journal of Australia* **156**, 53–57.
3. Downs, N. J., Neihart R. E., Dolezal, J. M. & Hodges, G. R. (1989). Mild nephrotoxicity associated with vancomycin use. *Archivs of Internal Medicine* **149** (8), 1777–1781.
4. Forgue, S. T., Shyu, W. C., Gleason, C. R., Pittman, K. A. & Barbhaiya R. H. (1987). Pharmacokinetics of the novel cephalosporin cefepime (BMY - 28142) in rats and monkeys. *Antimicrobial Agents and Chemotherapy* **31** (5), 799–804.
5. Griesmacher, A., Weigel, G., Windischbauer, A. & Mueller M. M. (1991). Purine metabolism in human endothelial cells. *International Journal of Purine and Pyrimidine Research* **2**, 123–127.
6. Jaffe, E. A., Nachman, R. L., Becker, C. G. & Minck, C. R. (1973). Culture of human endothelial cells derived from umbilical veins. *Journal of Clinical Investigation* **52**, 2745–2756.
7. Haudenschild, C. C., Cotran, R. S., Gimbrone, M. A. & Folkman, J. (1975). Fine structure of vascular endothelium in culture. *Journal of Ultrastructural Research* **50**, 22–34.

8. Jaffe, E. A., Hoyer, L. W. & Nachman, R. L. (1973). Synthesis of antihaemophilic factor antigen by human endothelial cells. *Journal of Clinical Investigation* **52**, 2757- 2763.
9. Schopf, G., Rumpold, H. & Mueller, M. M. (1986). Alterations of purine salvage pathways during differentiation of rat heart myoblasts towards myocytes. *Biochimistry Biophysics Acta* **884**, 319–327.
10. Acevedo, A., Schoop, W., Schnell, A. & Toledo, L. (1990). Antibiotic treatment for diabetic foot. Advantages of intravenous regional route as alternative for systemic route. *Rev. Med.Chil.* **118**, 881–888.
11. Partsch, H. (1990). Bier´s occlusion. An unjustly neglected therapeutic possibility in resistant tissue defects of the extremities. *Acta Med. Austriaca* **17**, 35–39.
12. Schoop, W. & Acevedo, A. (1993). Antibiotic concentrations after intravenous and retrograde intravenous injections. *Wiener Medizinische Wochenschrift* **143**, 199–200.
13. Farber, B. F. & Moellering, R. C. Jr. (1983). Retrospective study of the toxicity of preparations of vancomycin from 1974 to 1981. *Antimicrobial Agents and Chemotherapy* **23** (1), 138–141.

161

EFFECT OF TOFU (BEAN CURD) INGESTION ON URIC ACID METABOLISM IN HEALTHY AND GOUTY SUBJECTS

Jun-ichi Yamakita, Tetsuya Yamamoto, Yuji Moriwaki, Sumio Takahashi, Zenta Tsutsumi, and Kazuya Higashino

Third Department of Internal Medicine
Hyogo College of Medicine
Mukogawa-cho 1-1
Nishinomiya, Hyogo 663, Japan

SUMMARY

The effect of Tofu (bean curd) ingestion on uric acid metabolism was examined in 8 healthy and 10 gout subjects. Ingestion of Tofu increased plasma concentration of uric acid, together with increases in uric acid clearance and urinary excretion of uric acid. However, the increase in plasma concentration of uric acid was fairy small. Interestingly, no significant rise in the plasma, urinary and clearance of uric acid was observed in gout patients with uric acid clearance > 6.0 mL/min (lower normal limit). The results suggest that Tofu is a preferable source of protein, especially in gout patients with uric acid clearance > 6.0 mL/min.

INTRODUCTION

It has been demonstrated that ingestion of large amount of protein increases uric acid excretion and thereby, it is expected that protein diet decreases serum uric acid concentration. However, in general protein diet also contains large quantities of purines. Therefore, it may rather increase serum concentration of uric acid. Although Tofu (bean curd) is rich in protein, much of purines is lost in the process of its manufacturing, and it is classified as low or no purine containing diet according to the food table, being seemingly preferable to patients with gout or hyperuricemia. These facts prompted us to study the effect of Tofu ingestion on the uric acid metabolism in healthy subjects and gout patients.

Purine and Pyrimidine Metabolism in Man IX,
edited by Griesmacher *et al.* Plenum Press, New York, 1998.

Table 1. Nutrients of Tofu (bean curd)
per 100 g

Energy	69 Cal
Protein	6.3 g
Fat	5.6 g
Carbohydrate	2.5 g
Purine body	13 mg

MATERIALS AND METHODS

Eight healthy male and 10 male gout patients aged 30 to 50 were included in the study. Gout patients were divided into two groups according to uric acid clearance (below 6 mL/min; underexcretor of uric acid 6 patients, others 4). After an overnight fasting, urine was completely voided and 1-h urine was collected, then 4 g/kg of tofu was ingested. Then, 1-hour urine was collected three times succesively. Blood samples were drawn at the mid-points of the intervals. Uric acid concentrations in urine and plasma were measured by the uricase method, and the clearance and 1-h urinary excretion of uric acid were calculated.

RESULTS

Table 1 indicated the nutrients of Tofu. Tofu contained 6.3 g of protein and 13 mg of purines per 100 grams, which was unexpectedly rich.

Effect of Tofu Ingestion on Uric Acid Clearance

Uric acid clearance was significantly increased from 6.86 ml/min to 7.86 ml/min, 8.65 ml/min and 8.45 ml/min ($p < 0.05$, $p < 0.05$, $p < 0.05$, respectively) after Tofu ingestion (Table 2). There was no significant differences of the increase in uric acid clearance between healthy and gout patients. However, uric acid clearance was not increased in gout patients with uric acid clearance > 6.0 mL/min, compared with healthy and gout patients with decreased uric acid clearance (Fig. 1a).

Effect of Tofu Ingestion on the Urinary Excretion of Uric Acid

Urinary excretion of uric acid was significantly increased from 34.24 mg/h to 39.88 mg/h, 44.30 mg/h and 42.63 mg/h ($p < 0.05$, $p < 0.05$, $p < 0.05$, respectively) after Tofu

Table 2. Effect of Tofu ingestion on uric acid clearance of the subjects. Significant increases in uric acid clearance were observed after 1, 2, and 3 hours of Tofu ingestion

	1	2	3	4
Control	8.08 ± 0.50	8.81 ± 0.84	9.98 ± 0.76*	9.95 ± 0.67*
Gout	5.89 ± 0.71	7.11 ± 0.74	7.58 ± 0.52**	7.21 ± 0.54**
Total	6.86 ± 0.52	7.87 ± 0.57*	8.65 ± 0.52**	8.43 ± 0.53**

*Denotes $p < 0.05$ compared with the value of period 1.
**Denotes $p < 0.01$ compared witht he value of period 1.

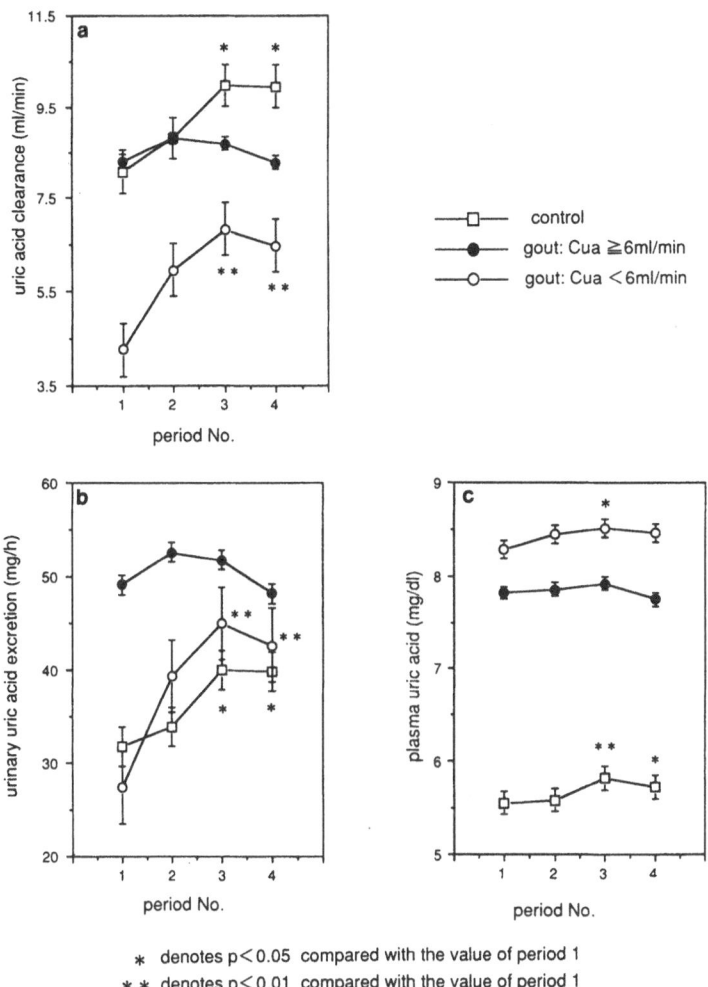

Figure 1. Changes in uric acid clearance (a), urinary excretion of uric acid (b) and plasma concentration of uric acid (c) No significant rise in uric acid clearance, urinary excretion of uric acid and plasma concentration of uric acid was observed in gout patients with uric acid clearance > 6.0 mL/min.

ingestion (Table 3). There was no significant differences of the increase in urinary uric acid excretion between healthy and gout patients. However, urinary uric acid excretion was not increased in gout patients with uric acid clearance > 6.0 mL/min, compared with healthy and gout patients with decreased uric acid clearance (Fig. 1b).

Effect of Tofu Ingestion on the Plasma Concentrations of Uric Acid

Plasma uric acid concentration was significantly increased from 6.97 mg/dL to 7.05 mg/dL, 7.19 mg/dL and 7.09 mg/dL ($p < 0.05$, $p < 0.05$, $p < 0.05$, respectively) after Tofu ingestion (Table 4). There was no significant differences of the increase in plasma uric acid concentration between healthy and gout patients. However, plasma uric acid concentration was not increased in gout patients with uric acid clearance > 6.0 mL/min, compared with healthy and gout patients with decreased uric acid clearance (Fig. 1c).

Table 3. Effect of Tofu ingestion on urinary excretion of uric acid of the subjects Significant increases in the urinary uric acid excretion were observed after 1, 2, and 3 hours of Tofu ingestion

	1	2	3	4
Control	31.84 ± 2.91	33.81 ± 1.97	40.06 ± 1.51*	39.92 ± 2.72*
Gout	36.16 ± 4.29	44.68 ± 4.94	47.68 ± 2.73**	44.86 ± 2.80**
Total	49.18 ± 6.01	52.60 ± 5.36*	51.79 ± 4.09**	48.17 ± 5.64**

*Denotes $p < 0.05$ compared with the value of period 1.
**Denotes $p < 0.01$ compared with the value of period 1.

Table 4. Changes in plasma uric acid of the subjects

	1	2	3	4
Control	5.56 ± 0.30	5.59 ± 0.27	5.83 ± 0.30	5.73 ± 0.28
Gout	8.10 ± 0.33	8.21 ± 0.33	8.27 ± 0.32	8.12 ± 0.33
Total	6.98 ± 0.38	7.05 ± 0.38	7.19 ± 0.36	7.09 ± 0.37

* Denotes $p < 0.05$ compared with the value of period 1.
** Denotes $p < 0.01$ compared with the value of period 1.

DISCUSSION

The present study demonstrated that Tofu, a low purine and high protein diet, increases plasma concentrations of uric acid, despite a significant increases in uric acid clearance. The results indicated that an increase in uric acid clearance by protein in Tofu could not prevent an increase in the plasma concentration of uric acid by purines in Tofu. However, the increases in plasma concentration of uric acid was fairly small.

Interestingly, no significant increases in plasma concentration, clearance and urinary excretion of uric acid were observed in gout patients with normal uric acid clearance. These results may be ascribable to more increased urinary excretion of uric acid in gout patients with normal uric acid clearance than those with low uric acid clearance and normal subjects. Tofu seems one of preferable protein sources at least in these patients with no detrimental effect of plasma uric acid concentration.

REFERENCE

1. Matzkies F, Berg G, Madl H: The uricosuric action of protein in man. Adv. Exp. Med. Biol. 122A:227–231,1980.

162

PURINES AND PYRIMIDINES DETERMINATION IN URINE USING HIGH-PERFORMANCE LIQUID CHROMATOGRAPHY

G. Minniti, U. Caruso, R. Cerone, and E. de Toni

Laboratory for the Study of Stand. Verif. Metabolic Disease
G.Gaslini institute-University
Department of Pediatrics I
Genova, Italy

SUMMARY

Single purine and pirimidine bases are involved in two fundamental metabolic pathways that lead to formation of the building stones of DNA and RNA. Purine and pyrimidine nucleotides are also critically important metabolites in many cellular functions. The main breakdown of purines and pyrimidines produces uric acid and B-minoacids, respectively. Therefore, the study of purine and pirimidine compounds in body fluid has high clinical relevance. We report, in this work, our experience in purines and pyrimidines determination in urine from children presenting with a clinical picture suggesting an inborn these pathways.

INTRODUCTION

Purine and pyrimidines bases are etherocyclic compounds containing nitrogen and one or more either aminic or iminic group.

Single purine and pyrimidine free bases are involved in two fundamental metabolic pathqays that, through the monophosphates of the bases guanine, adenine, cytosine, tymine and uracil, lead to the formation of nucleotides, building stones for nucleic acids, DNA and RNA.

Nucleotides are the phosphates esters of ribo- and deoxyribo-nucleosides, formed by a purine or pyrimidine bases and ribose or deoxyribose, respectively. The chemical identity of the whole molecule so formed cames from the nature of the purine or the pyrimidine bases.

Purine and pyrimidines nucleotides are critically important metabolites that participate in many cellular function like high-energy source and intermediates, physiological

Purine and Pyrimidine Metabolism in Man IX,
edited by Griesmacher *et al.* Plenum Press, New York, 1998.
843

mediators of hormone action, components of coenzymes, carriers of activated intermediates, allosteric effectors (1).

Purine free bases are formed de novo in mammalian cells by aminoacids and simple molecules like formate and CO_2. Their metabolism can be divided in three patways: 1) purine nucleotide synthesis, that, starting from 5-phosphoribosyl-1-pyrophosphate (PRPP) and aminoacids, produces inosine 5'-monophosphate (IMP), precursor of the other nucleotides; 2) interconversion pathway, to mantain an appropriate balance of purine nucleotides; 3) breakdown way, leading to the production of uric acid.

Similarly pyrimidines compounds are synthesized from glutamine, aspartate and CO_2, through carbamoyl phosphate, and degradated to the B-aminoacids, B-alanine and B-amino isobutiric acid, an interconversion pathway allows the appropriate balance of deoxyribopyrimidine nucleosides (2).

CLINICAL CORRELATIONS

The clinical relevance of the disorders affecting these pathways is obviously high and the study of purine and pyrimidine compounds (bases, nucleosides and nucleotides) in biological fluids is highly meaningful.

To date fifteen primary defects have been recognized, located in biosynthesis, interconversion, degradation pathways (Table 1) (3).

Aim of this work is to present our experience in determination of purine and pyrimidine compounds in urine from children presenting with a clinical picture suggesting an inborn error involving their metabolism.

METHOD

Purine and pyrimidine compounds, because their chemical properties, cannot be easily separated by an extraction procedure from the other components of urine, that must be analyzed without any preliminary clean-up step.

Table 1.

Defects of purine metabolism	Defect of pyrimidine metabolism
Degradation defects • adenosine deaminase deficiency • purine nucleoside phosporilase deficiency • xanthine oxidase deficiency	Degradation defects • pyrimidine-5'-nucleotidase deficiency • dihdropyrimidine deydrogenase deficiency • dihydropyrimidinase
Biosynthesis defects • increased PRPP synthetase • adenylosuccinase deficiency	Biosynthesis defects • orotic aciduria type I • orotic aciduria type II
Interconversion defects • adenosine monophosphate deaminase deficiency • S-adenosylhomocysteine hydrolase deficiency • hypoxanthine-guanine phosphoribosyltransferase deficiency • adenine phosphoribosyltransferase deficiency • inosine triphosphatase deficiency	

Several methods have been proposed and used for purine and pyrimidine compounds determination in body fluid, including mono- and bi-dimensional TLC and ion-exchange chromatography (4). More recently the reverse-phase HPLC was appied to the separation and quantitation of purine and pyrimidine compounds: nanomoles of these substances can easily separated and quantified in a short time.

The metod used in our laboratory is essntialy based on that described by Symmonds et al. (5) with slght adaptations (6).

Urine is collected for 24 hrs after some days of low purine diet in a clean bottle containing 10 ml of toluene as preservative. Material can be immediately used as such or stored at −20°C until analysis.

Before analysis urine is diluited 1:30 with the first HPLC eluent (see below), after sonication for 30 mins at 51°C, to dissolve adequately all bases, while urine has been stored deep frozen.

HPLC analysis is performed using an Hewlett Packard 1090L liquid chromatograph equipped with a ternary-gradient pump system, Rehodyne valve injector and an Hewlett Packard 1040M Diode Array Detector (DAD).

The column is a Spherisorb-ODS 2, 25 cm long. × 0.4 cm i.d., particle size 5 μm (Hewlett Packard 7992402-584).

50 μl of diluited urine are injected.

Eluition is performed at flow rate of 0.8 ml/min using two eluents:

A. ammonium actate 40 mM in water, pH 5.00;
B. methanol, acetonitrile, tetrahydrofuran, 80/10/10, v/v/v

Elution is performed in gradient mode, starting with 100% of A and reaching 20% of B in 20 mins. After equilibration with 10 mins of pure A eluent, the column is ready for a new injection.

For UV signals are monitored and recorded: 210, 254, 280 and 310 nm with a λ reference of 550 nm. UV spectra are aquired between 210 and 350 nm (band whidth 16 nm) when signal is higher than 1 mAU.

The instrument is managed by a workstation (HP 3000 series, Pascal program) that allows chromatograms, spectra storage, spectra retriving and comparition, library search, peak integration and quantitative report.

Metabolites are identified by maching of retention times and UV spectra with that from authentical reference substances.

RESULTS AND DISCUSSION

Eighteen metabolites have been separated and identified in our system, ancluding some compounds other than purines and pyrimidines (mainly drugs and their metabolites) with similar structure. Table 2 reports, for each compound, retention time and λ max.

Purine and pyrimidine compounds are almost undetectable in control urine <, the main component is represented by uric acid. Uric acid is not really a purine compound; but, being himself the final product of purine catabolism, the measurement of its excretion in urine is a significant diagnostic parameter. The reference values change slightly with age and sex: 1.5 ± 0.3 mmol/24 hrs in children, 3.0 ± 0.5 mmol/24 hrs in adults male and 2.7 ± 0.5 mmol/24 hrs in adults female.

Hyperuricuria is present in increased activity of the first enzyme of the synthetic pathway (PRPP synthetase) and in hypoxanthine-guanine phosphoribosyl transferase defi-

Table 2. Retention times (rt) and λ max for identified compounds

Compound	RT (min)	λ max
Orotidine	2.53	265
Oroticacid	3.12	280
Uracil	4.15	280
Di hydrouracil	5.01	254
Uric acid	5.30	210
Uridine	7.05	290
Dihydroxyadenine	7.32	260
Hypoxanthine	7.37	310
Xanthine	7.61	254
Timine	8.32	260
Oxypurinol	8.74	254
Allopurinol	9.25	254
Inosine	10.42	254
Guanosine	11.45	254
Adenine	13.24	254
Deoxyinosine	14.43	254
Adenosine	14.88	280
Deoxyguanosine	14.92	254

ciency, in both cases due to an overcharge of PRPP; hypouricuria is characteristic of the defects of IMP breakdown, namely purine nucleoside phosphorilase and xanthine oxidase deficiencies.

Using this method we performed, in the last five years, 561 analyses, studying 390 patients. In three patients hypouricuria and abnormal xanthine and hypoxanthine excretion have been found, meaning a defect in xanthine oxidase (Fig. 1). In two the defct was associated to sulphite oxidase deficiency (7), consequence of a defect of molibdenum cofactor, essential component of both enzymes (8).

In one case the urinary purine profile, mainly characterized by a massive hyperuricuria, was consistent with the clinical suspect of hypoxantine-guanine phosphoribosyltransferase deficiency (Lesch-Nyhan syndrome). In one male child the high orotic acid excretion, in addition to suggestive plasma aminoacid alterations led to a diagnosis of ornitine carbamyltransferase (OCT) deficiency. Indeed, a mitochondrial overproduction of carbamoyl-phosphate, due to a reduced OCT activity, can result in high orotic acid and/or orotidine excretion in urine (9).

The measurement of urine orotic acid and orotidine in the diagnostic work-up in patients with a suspect of urea cycle disorders or lysinuric protein intolerance (as well as in the biochemical follow-up of already diagnosed patients) represents almost the 70% of the whole work in analysis of purine and pyrimidine compounds.

Dynamic studies are performed in presence of suggestive clinical signs and/or plasma aminoacids alterations, when basal urine is not informative. Orotic (and/or orotidine) excretion can, indeed, be stimulated, in patients with OCT (even partial) deficiency either by a protein load or by allopurinol administration: allopurinol produces osipurinol and, by the action of orotate phosphoribosyl-pyrophosphate transferase, oxipurinol ribonucleotide that inibits orotidine monophosphate decarboxylase activity (9).

Allopurinol loading test was clearly positive in five patients, two males and three females; in one male patient OCT reduced activity was demonstrated in intestinal biopsy (10). Enzyme assay is still in progress in the others.

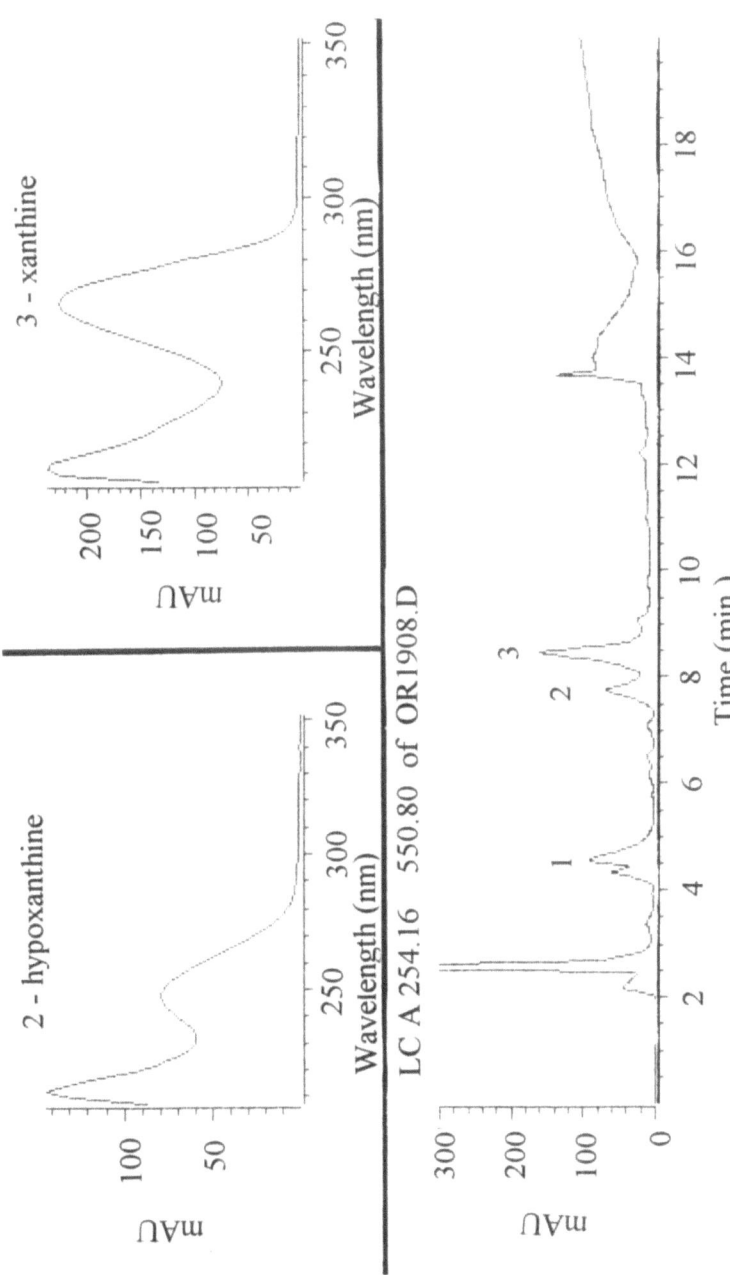

Figure 1. Chromatogram from urine of a child with xanthine oxidase deficiency, showing low urix acid (1) and high hypoxanthine (2) and xanthine (3). The UV spectra are reported at the top.

The use of DAD allows simultaneous monitoring of up four wavelenght and UV spectra recording: The UV spectra are particurarly helpful for the identification of diagnostically significant metabolites. Indeed, could be difficult to interpret the chromatograms on the bases of the retention times alone, due to interferences of several unidentified substances present also in normal urine (must be remembered that urine is injected without any clean-up step). The opportunity of four chromatograms (each one at different wavelenght) leads to a consistent increase of sensitivity and specificity. Indeed, even doing only one injection, for each compound is possible to choose, for quantitation, the chromatogram corresponding to the maximum or the more specific absorbance.

This method suitable to be used in HPLC/MS systems (both used aluents are volatile!), giving substantial structural informations: thermospray and electrospray interfaces allow a measure of peak purity and the exact molecular mass, while particle-beam interface lead to a whole electron-impact spectrum, suitable for lybrary search and definitive compound identification.

ACKNOWLEDGMENTS

We thank Dr. M. Duran (Utrecht) who helped us in builting-up this method in our laboratory and shared with us his large experience.

REFERENCES

1. JG Cory, in TM Devlin (Ed.): Textbook of Biochemistry With Clinical Correlations Wiley-Liss, Neww York, 1992, Ch. 13, p. 529.
2. G. van den Berghe, in J. Fernandes, JM Saudubray, K Tada (Eds): Inborn Metabolic Diseases Springer-Verlag, Berlin, 1990,Ch. Ix, p. 455
3. HA Simmonds, in JB Holton (Ed): The inherited metabolic diseases Churchill Livingstone, Edinburgh, 1994, Ch. 8 p. 297.
4. PK de Bree, SK Wadman, M Duran, H Fabery de Jonge. Clinica Chimica Acta 156 (1986)279.
5. HA Simmonds, JA Duley, PM Davies in: F. A. Hommes (Ed) Tecniques in Diagnostic Human Biochemical Genetics Wiley-Liss, New York, 1991, Ch. 25, p. 397.
6. M Duran. Personal Communication. 1991
7. V Briscioli, S Manoukian, et al. (Abstract) VI International Congress Inborn Errors of Metabolism. Milano, May 1994, W15.1
8. SK Wadman, M Duran,et al. J Inher Metab: Dis: 6 Suppl 1 (1983), 78.
9. D Parker Suttle, DMO Becroft,et al in: CR Scriver, AL Beaudet, WS Sly, D Valle (Eds): The metabolic Basis of Inherited Disease McGraw-Hill, New York, 1989, Ch. 43 p. 1095
10. R Cerone,G.Minniti, et al.: 33rd Annual Symposium SSIEM Toledo, 1995.

163

DESIGN, SYNTHESIS, AND STRUCTURE-ACTIVITY RELATIONSHIPS OF THE FIRST HIGHLY POTENT, SELECTIVE, AND BIOAVAILABLE ADENOSINE 5′-MONOPHOSPHATE DEAMINASE INHIBITORS

Srinivas Rao Kasibhatla, Brett C. Bookser, James R. Appleman, Gary Probst, Wei Xiao, James M. Fujitaki, and Mark D. Erion

Metabasis Therapeutics Inc.
San Diego, California 92121

INTRODUCTION

In the preceding article[1] we described our strategy for the discovery of a series of cell penetrable AMP deaminase (AMPDA) inhibitors for postulated use in the treatment of ischemia, especially of the myocardium, and any other conditions expected to benefit from site- and event-specific enhancement of extracellular adenosine. The initial set of compounds, prepared as a preliminary test of our design concept and represented by **1**, exhibited only modest potency and selectivity. We thus viewed **1** as a lead structure which would require further optimization to maximize the potency, and accordingly initiated a study to probe the structure-activity profile of this new series. Herein, we report our SAR study which led to the discovery of a series of highly potent and highly selective AMPDA inhibitors.

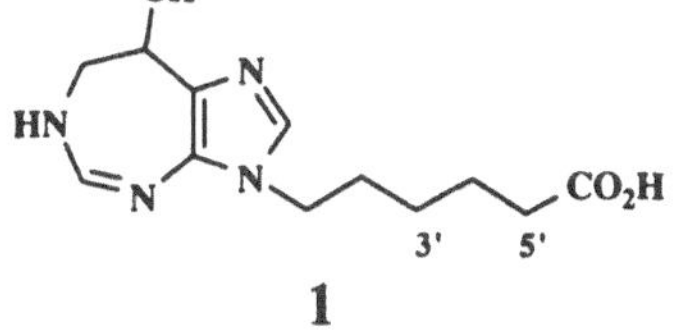

1

Purine and Pyrimidine Metabolism in Man IX,
edited by Griesmacher *et al.* Plenum Press, New York, 1998.

849

SYNTHESIS

The target compounds (**2**, **4** and **5**) described in this article were prepared by the same basic strategy reported in our previous article.[1] The majority of electrophiles were prepared as shown below. Palladium catalyzed vinylation followed by hydroboration-oxidation and then alcohol to bromide conversion provided the electrophile.

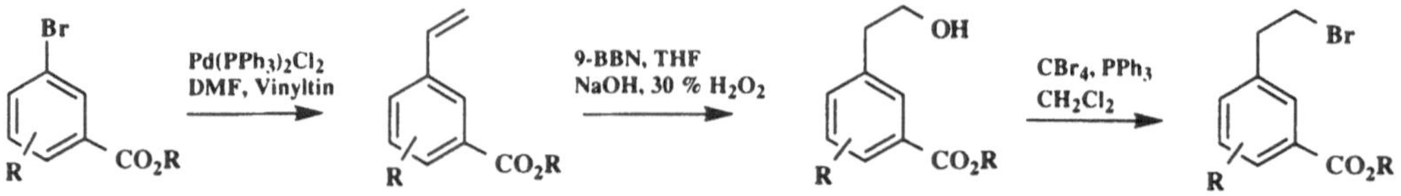

RESULTS AND DISCUSSION

It was our hypothesis that conformational restriction of the side-chain of **1** should increase AMPDA binding affinity. Through an extensive SAR study[2] to confine the flexibility in the side chain of compound **1** (AMPDA K_i = 4,000 nM; ADA K_i = 280,000 nM), we discovered that incorporation of a phenyl ring into the sidechain as represented in **2a**, substantially increased both potency (AMPDA K_i = 500 nM) and maintained selectivity (ADA K_i = $\gg$ 7500 nM). We next examined the effect of substitutions on the phenyl ring (Table 1). Methyl or ethyl groups at C-4' position increased potency > 5-fold (**2b**, 97 nM and **2c**, 79 nM). However, larger alkyl substituents (**2d**, 190 nM) or a halogen (**2e**, 180 nM) at this position showed a modest improvement. Both small substituents like ethyl (**2g**, 60 nM) and large substituents like phenyl (**2h**, 68 nM) and bromide (**2f**, 60 nM) at C-5' provided significant enhancement in binding. This suggests that the active site contains a relatively large hydrophobic pocket. Alkyl or halogen substitution at C-6' position was well tolerated but much inferior (**2i**, 570 nM; **2j**, 160 nM).

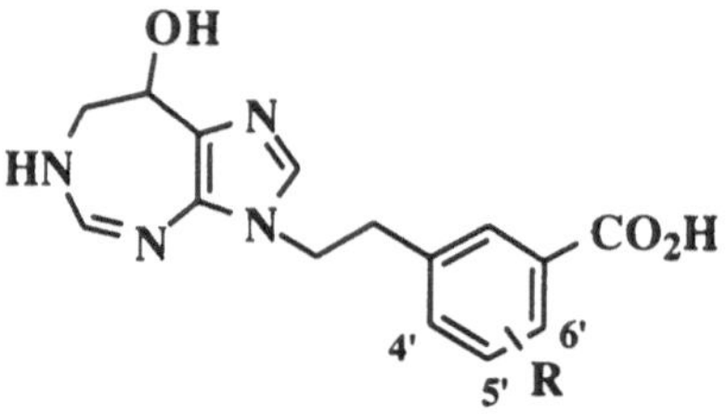

2

Table 1. SAR of *m*-carboxyphenethyl coformycin aglycon analogs

No.	R	AMPDA K_i (nM)	ADA K_i (nM)
2a	H	500	>7,500
2b	4'-Me	97	>100,000
2c	4'-Et	79	>7,500
2d	4'-^{n}Pr	190	>7,500
2e	4'-F	180	>7,500
2f	5'-Br	81	>7,500
2g	5'-Et	60	>7,500
2h	5'-Ph	68	>7,500
2i	6'-Me	570	>7,500
2j	6'-F	160	>7,500

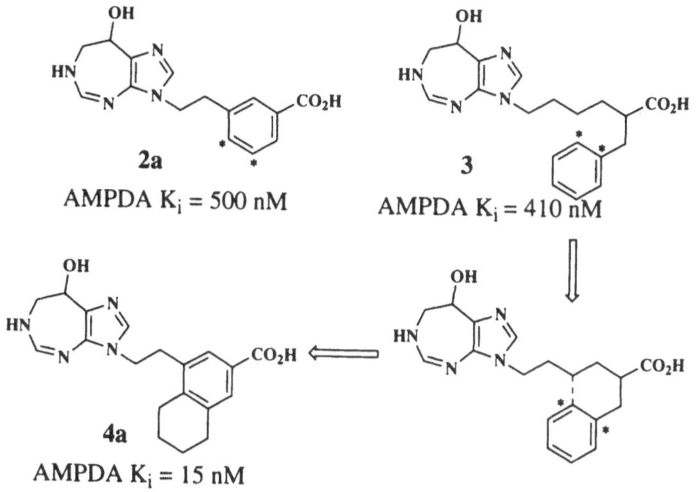

Figure 1. Design of inhibitor with favorable entropy and hydrophobicity.

We then designed a new series by examination of the two new leads, **2a** and **3**[1] (Figure 1). Our strategy was to improve further the potency and abolish the additional chiral center (see structure **3**) by restricting conformation and including increased hydrophobicity. As expected, this union resulted in a more favored analog, **4a** (AMPDA K_i = 15 nM), which binds 30-fold more tightly to AMPDA than does either of the parents. This new lead was further optimized (Table 2). Bromo substitution at C-1' position in this tetrahydronapthyl series led to the identification of our most potent inhibitor in this program, **4g** (3 nM). This represents a remarkably potent inhibitor considering that the K_m of AMP is ca. 1 mM ($K_m/K_i > 10^5$). This large improvement over AMP suggests that the compound is capable of binding in a manner that resembles the transition state of deamination.

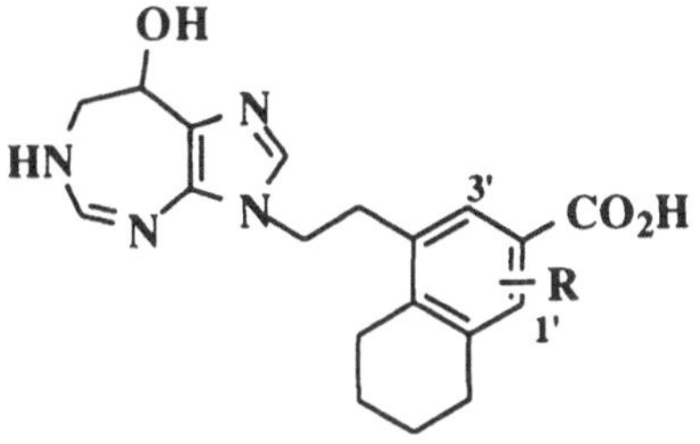

4

Table 2. SAR of 3-carboxytetrahydronapthylethyl coformycin aglycon analogs

No.	R	AMPDA K_i (nM)	ADA K_i (nM)
4a	H	15	>7,500
4b	1'-OCH$_3$	30	>7,500
4c	1'-OBn	10	>7,500
4d	1'-O^{i}Pr	90	>7,500
4e	3'-OCH$_3$	40	>7,500
4f	1',3'-Cl$_2$	50	>7,500
4g	1'-Br	3	>7,500

While these carboxylic acid inhibitors have good potency at AMPDA, they exhibited low oral bioavailability when administered to rats. In an effort to enhance the oral bioavailability, a variety of esters **5** of compound **4a** were prepared. A trend towards increasing oral bioavailability with increasing hydrophobicity of the ester was observed, with the benzyl ester derivative **5d** the most orally bioavailable of the series (Table 3).

5

Table 3. Prodrug esters **5** of **4a** for increased oral activity

No.	R	Oral bioavailability (% F) in rat
4a		5
5a	Et	10
5b	^{i}Pr	7
5c	^{n}Bu	13
5d	Bn	28
5e	$CH_2OC(O)^t$Bu	20
5f	$CH(Ph)_2$	5

SUMMARY

Structure-activity studies have been performed to optimize the potency of this novel series of AMPDA inhibitors. Conformational rigidification of the N-3 sidechain resulted in substantial effect on the potency. Addition of the hydrophobic groups provided further benefit. The most potent compound identified, **4g** (K_i = 3 nM), bears little structural resemblance to AMP and exhibits a remarkable improvement (10^3 and 10^5) in binding affinity relative to the original lead and AMP, respectively. The application of prodrug strategy achieved a large improvement (benzyl ester **5d**) in oral bioavailability, resulting in compounds that should be useful in evaluating the role of AMPDA in normo- and pathophysiological states.

REFERENCES

1. Bookser, B. C., Kasibhatla, S. R., Appleman, J. R., Erion, M. D. Preceding article this volume.
2. Kasibhatla, S. R., Bookser, B. C., Appleman, J. R., Probst, G., Xiao, W., Fujitaki, J. M., Erion, M. D. unpublished results.

164

DESIGN AND SYNTHESIS OF THE FIRST POTENT, SELECTIVE, AND CELL PENETRATING ADENOSINE 5′-MONOPHOSPHATE DEAMINASE INHIBITORS

Brett C. Bookser, Srinivas Rao Kasibhatla, James R. Appleman, and
Mark D. Erion

Metabasis Therapeutics Inc.
San Diego, California 92121

INTRODUCTION

Adenosine (Ado) is known to be an endogenous cardio- and neuroprotective agent, but is in general unsuitable for use as a drug due to systemic side effects, short half-life in the circulatory system, and lack of oral bioavailability.[1,2] Adenosine 5′-monophosphate deaminase (AMPDA), an enzyme involved indirectly in regulation of adenosine levels, catalyzes the hydrolytic deamination of AMP to IMP and ammonia ($K_m \sim 1$ mM). Flux through the enzyme is reported to be significant primarily during periods of net ATP breakdown when the concentration of the substrate AMP is increased many-fold.[3] This nucleotide is subject to further metabolism via two competing routes: (i) AMPDA-catalyzed deamination to IMP followed by conversion to inosine and downstream products, and (ii) dephosphorylation by 5′-AMP-specific endonucleotidases to adenosine. Inhibition of AMPDA should make more AMP available for dephosphorylation, and consequently should result in a site- and event-specific therapeutic benefit by increasing the local concentration of the endogenous cellular protectant adenosine exclusively under conditions of cellular stress. Flux is believed to be limited through AMPDA under normoxic conditions so that this enzyme should be more "silent" than targets like adenosine kinase. We disclose herein the first appropriate compounds to evaluate these hypotheses.

AMP binding sites represent a particular challenge to medicinal chemists because the phosphate moiety of AMP is often required for high binding affinity but at the same time prevents cell penetration of the nucleotide which forces the necessity of prodrug strategies.[4] Efforts to design potent and specific AMPDA inhibitors began with compounds 1 and 2. These two natural products are potent inhibitors of AMPDA: coformycin 5′-monophosphate 1 with $K_i = 0.055$ nM for rabbit muscle enzyme[5] and coformycin 2 with $K_i = 1000$ nM for recombinant human erythrocytic enzyme (hE-AMPDA).[6] The former is

Purine and Pyrimidine Metabolism in Man IX,
edited by Griesmacher *et al.* Plenum Press, New York, 1998.

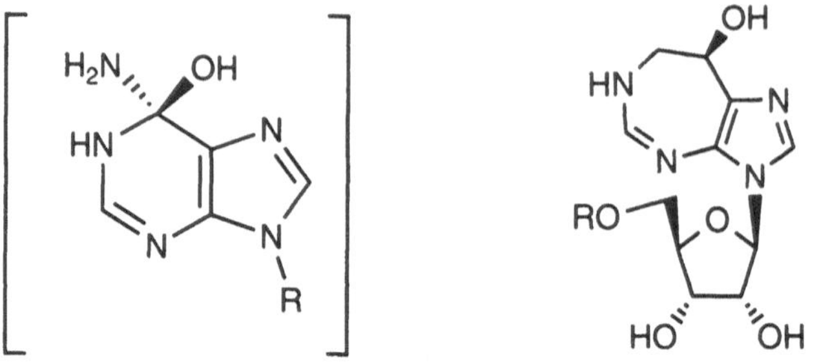

Figure 1. Tetrahedral intermediate (R = ribose 5' monophosphate for AMPDA).

completely selective for inhibition of AMPDA over adenosine deaminase (ADA) but is inappropriate for pharmacological investigation because of its potential for poor cell penetration and rapid dephosphorylation in plasma. The latter is also inappropriate because it is an extremely potent and tight-binding inhibitor of ADA, deficiencies in which are known to result in severe combined immunodeficiency.[7] Compounds **1** and **2** are believed to be such potent inhibitors by virtue of their ability to mimic the tetrahedral intermediate in the deamination reaction (Figure 1).[8] This unique heterocycle when attached to ribose or ribose monophosphate also provides for total selectivity toward the inhibition of these deaminases when compared to other Ado or AMP binding enzymes.[9]

RESULTS

We sought to replicate the potency and selectivity observed with compound **1** in new compounds having physicochemical properties appropriate for membrane penetration by replacement of the ribose monophosphate with a simpler group containing some anionic substituent. This was realized by carboxyalkyl substitution, resulting in the discovery of a series of potent and selective inhibitors more appropriate for pharmacological investigation. The compounds were prepared according to the method outlined in the following sequence:

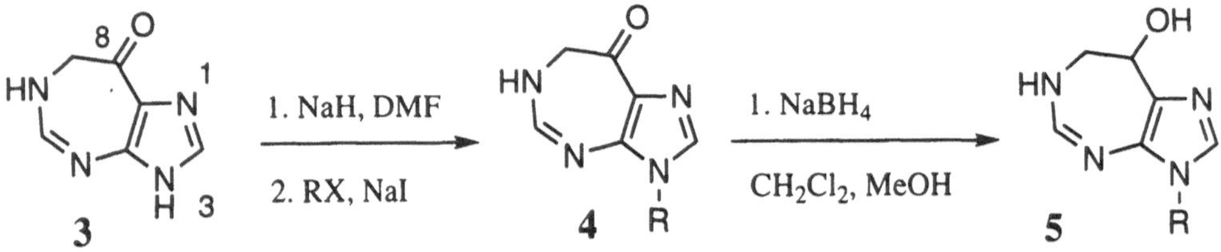

The heterocycle **3**[10] was deprotonated with sodium hydride and then alkylated regioselectively at N-3 with an electrophile RX where X is a leaving group such as a primary alkylbromide to provide **4** in yields ranging from 20–40%. Reduction of **4** provided the coformycin aglycon analog **5** as a 50:50 C-8 *R/S* mixture. If the R-group sidechain contained an ester, it could be hydrolyzed with aqueous 1 N NaOH in dioxane (1:1) and the resulting carboxylic acid isolated by Dowex-1 acetate ion exchange resin treatment.

The structure activity relationship for inhibition of recombinant hE-AMPDA and calf intestinal ADA is dependent on the R-group substituent (see Table 1). Simple *n*-alkyl chains represented by **5a**–**5c** show increasing potency with increasing chain lengths at AMPDA and ADA. However, potency and selectivity at AMPDA is achieved with carboxyalkyl

Table 1. Structure activity relationship of N-3 substituted coformycin aglycon analogs

no.	R	AMPDA K_i (μM)	ADA K_i (μM)
5a	~~~~CH$_3$	84	17
5b	~~~~~CH$_3$	20	0.48
5c	~~~~~~CH$_3$	10	0.35
5d	~~~~CO$_2$H	> 125	> 7.5
5e	~~~~~CO$_2$H	4.2	280
5f	~~~~~~CO$_2$H	10	250
5g	~~~~~tetrazole	6	> 7.5
5h	~~~~~~P(=O)(OH)(OMe)	140	> 500
5i	~~~~~PO$_3$H$_2$	35	> 500
5j	~~O~~CO$_2$H	85	> 500
5k	~~~~~C(CH$_3$)$_2$CO$_2$H	8.2	> 7.5
5l	~~~~~CH(Ph)CO$_2$H	2.1	> 7.5
5m	~~~~~CH(CH$_2$Ph)CO$_2$H	0.41	> 200
5n	~~~~~CH(CH$_2$(3-Br-Ph))CO$_2$H	0.79	> 7.5

substituents when the spacer group is at least 5 methylene units long (see **5d–5f**). The tetra-
zole **5g** exhibits similar potency to **5e** but the phosphonates **5h** and **5i** are surprisingly weak.
An ether oxygen in the chain does not boost potency (see **5j**). Finally, substitution on the
methylene α to the carboxy leads to as much as a 10-fold increase in potency at AMPDA
when the group is large and hydrophobic (see **5k–5n**).[11]

Compound **5n** was evaluated for its ability to penetrate cells *in vitro*. This compound
was able to equilibrate rapidly across cell membranes ($t_{1/2} \leq 5$ min) for a variety of cell
types. No evidence for concentrative or active transport was observed with this or like
compounds.

Ethyl esters of compounds **5l** and **5n** were evaluated for oral bioavailability in the
rat by comparing P.O. to I.V. urinary excretion of intact parent compound as detected by
HPLC after a 48 h collection period. They were found to be 44 and 35% orally bioavailable respectively. A total of 78% of each compound was recovered in the urine after I.V.
administration of the compounds which is suggestive of metabolic stability. Furthermore,
these compounds were stable in *in vitro* assays with hepatic microsomes. Finally, the good
oral bioavailability seems to indicate good stability in the gastrointestinal tract, thus low
pHs encountered do not significantly decompose the molecules prior to oral absorption.

SUMMARY

A major milestone in purine metabolism research has been achieved with the discovery of these potent and selective AMPDA inhibitors. These inhibitors of AMPDA are
based on carboxypentyl substitution on N-3 of the coformycin aglycon. They are simpler
than coformycin ribose 5'-monophosphate, more stable, selective against other AMP binding enzymes as well as ADA and have good cell penetration and good oral bioavailability.
These compounds and their more potent analogs[12] are the first compounds with suitable
characteristics to allow a definitive analysis of the role of AMPDA in cellular metabolism
and AMPDA as a therapeutic target.

ACKNOWLEDGMENTS

The following individuals made important scientific contributions to this work: Dr.
Christie Anderson, Ms. Maureen Cottrell, Dr. James M. Fujitaki, Mr. Colin Ingraham and
Mr. J. Patrick McCurley.

REFERENCES

1. Williams, M. in *Adenosine and Adenosine Receptors*, M. Williams, Ed. (Humana Press, Clifton NJ, 1990),
   pp.1–15.
2. The only approved clinical uses of adenosine are treatment of supraventricular tachycardia (Adenocard)
   and as a pharmacological stress agent in assessment of cardiovascular function (Adenoscan).
3. Skladanowski, A. C. in *Myocardial Energy Metabolism*, J. W. de Jong, Ed. (Martinus Nijhoff, Dordrecht,
   1988), pp. 53–65.
4. (a) Srivastva, D. N.; Farquhar, D. 1984. Bioorg. Chem., *12*, 118; (b) Starrett, J. E., Jr.; Tortolani, D. R.;
   Hitchcock, M. J. M.; Martin, J. C.; Mansuri, M. M. 1992. Antiviral Res., *19*, 267.
5. Frieden, C.; Kurz, L. C.; Gilbert, H. R. 1980. Biochemistry, *19*, 5303.
6. Takabayashi, K.; Appleman, J. R. unpublished results.

 7. Kredich, N. M.; Hershfield, M. S. in *The Metabolic Basis of Inherited Disease*, C. R. Scrivner *et. al.*, Eds. (McGraw-Hill, New York, ed. 6, 1989), pp. 1045–1075.
 8. Marrone, T. J.; Straatsma, T. P.; Briggs, J. M.; Wilson, D. K.; Quiocho, F. A.; McCammon, J. A. 1996. J. Med. Chem., *39*, 277.
 9. Bzowska, A.; Lassota, P.; Shugar, D. 1985. Z. Naturforsch., *40c*, 710.
10. Chan, E.; Putt, S. R.; Showalter, H. D. H.; Baker, D. C. 1982. J. Org. Chem., *47*, 3457.
11. Compounds **5l** to **5n** have two chiral centers and are an equally distributed mixture of four diastereomers.
12. Kasibhatla, S. R.; Bookser, B. C.; Appleman, J. R.; Probst, G.; Xiao, W.; Fujitaki, J. M.; Erion, M. D. This volume.

AUTHOR INDEX

SUBJECT INDEX